Environmental Materials and Waste

Resource Recovery and Pollution Prevention

Environmental Materials and Waste

Resource Recovery and Pollution Prevention

Edited by

M.N.V. Prasad

Kaimin Shih

ELSEVIER

AMSTERDAM • BOSTON • HEIDELBERG • LONDON
NEW YORK • OXFORD • PARIS • SAN DIEGO
SAN FRANCISCO • SINGAPORE • SYDNEY • TOKYO
Academic Press is an imprint of Elsevier

Academic Press is an imprint of Elsevier
125 London Wall, London EC2Y 5AS, UK
525 B Street, Suite 1800, San Diego, CA 92101-4495, USA
50 Hampshire Street, 5th Floor, Cambridge, MA 02139, USA
The Boulevard, Langford Lane, Kidlington, Oxford OX5 1GB, UK

British Library Cataloguing-in-Publication Data
A catalogue record for this book is available from the British Library

Library of Congress Cataloging-in-Publication Data
A catalog record for this book is available from the Library of Congress

ISBN: 978-0-12-803837-6

For information on all Academic Press publications
visit our website at https://www.elsevier.com/

Publisher: Candice G. Janco
Acquisition Editor: Laura S Kelleher
Editorial Project Manager: Emily Thomson
Production Project Manager: Mohanapriyan Rajendran
Designer: Victoria Pearson

Typeset by TNQ Books and Journals

Contents

List of Contributors

T. Astrup
Technical University of Denmark, Lyngby, Denmark

S. Bashir
University of Agriculture Faisalabad, Faisalabad, Punjab, Pakistan

P. Beckett
Laurentian University, Sudbury, ON, Canada

I. Bibi
University of Agriculture Faisalabad, Faisalabad, Punjab, Pakistan; Southern Cross University, East Lismore, NSW, Australia

E. Bocharnikova
Institute of Physical-Chemical and Biological Problems in Soil Science, Pushchino, Russia

N.S. Bolan
University of Newcastle, Callaghan, NSW, Australia

S.K. Butti
CSIR-Indian Institute of Chemical Technology (CSIR-IICT), Hyderabad, Telangana, India

E.N.V.M. Carrilho
Universidade Federal de São Carlos, Araras, SP, Brazil

G. Choppala
Southern Cross University, East Lismore, NSW, Australia

S. Chowdhury
Hannam University, Daejeon, Republic of Korea

A. Demir
Mersin University, Mersin, Turkey

R.A. de Toledo
Department of Civil and Environmental Engineering, University of Macau, Avenida da Universidade, Taipa, Macau, SAR, China

N.S. Dionisiou
Aristotle University of Thessaloniki, Thessaloniki, Greece

E. Evangelou
Hellenic Agricultural Organization, Institute of Soil Mapping and Classification, Larissa, Greece

P.J.C. Favas
University of Trás-os-Montes e Alto Douro, Vila Real, Portugal; MARE – Marine and Environmental Sciences Centre, University of Coimbra

Y. Fei
Guangdong Institute of Eco-Environment and Soil Sciences, Guangzhou, People's Republic of China; Chinese Academy of Sciences, Guiyang, People's Republic of China

N. Geekiyanage
Kyoto University, Kyoto, Japan; Rajarata University, Anuradhapura, Sri Lanka

E.R. Graber
Agricultural Research Organization, The Volcani Center, Bet Dagan, Israel

D.M. Gurumurthy
Institute of Urban Environment, Chinese Academy of Sciences, Xiamen, China

J. Harris
Cranfield University, Cranfield, Bedford, United Kingdom

I. Herath
National Institute of Fundamental Studies, Kandy, Sri Lanka

J. Icenhower

Y. Jayawardhana
National Institute of Fundamental Studies, Kandy, Sri Lanka

M.Y. Kamogawa
Universidade de São Paulo - ESALQ, Piracicaba, SP, Brazil

R. Karunanithi
University of Newcastle, Callaghan, NSW, Australia

N. Khan
University of South Australia, Mawson Lakes, SA, Australia

G.-H. Kim
Hannam University, Daejeon, Republic of Korea

J. Koelmel
University of Hyderabad, Hyderabad, Telangana, India; USIEF Fulbright-Nehru Scholar; Present address: University of Florida, Gainesville, FL, USA

N. Koleli
Mersin University, Mersin, Turkey

G. Koptsik
M.V. Lomonosov Moscow State University, Moscow, Russia

S. Koptsik
M.V. Lomonosov Moscow State University, Moscow, Russia

S. Krishnamurthy
ICAR-Indian Institute of Soil and Water Conservation, Chandigarh, India

A. Kumar
Agricultural Research Organization, The Volcani Center, Bet Dagan, Israel

A. Kumar
Ural Federal University Named After the First President of Russia B.N. Yeltsin, Ekaterinburg, Russia

P. Kumarathilaka
National Institute of Fundamental Studies, Kandy, Sri Lanka

A. Kunhikrishnan
National Academy of Agricultural Science, Wanju-gun, Jeollabuk-do, Republic of Korea

G. Labuto
Universidade Federal de São Paulo, Diadema, SP, Brazil

C.-Z. Liao
University of Hong Kong, Hong Kong, Hong Kong SAR, China

J. Ling
Department of Civil and Environmental Engineering, University of Macau, Avenida da Universidade, Taipa, Macau, SAR, China; School of Environmental Science and Engineering, Guangdong University of Technology, Guangzhou, China

C. Liu
Guangdong Institute of Eco-Environment and Soil Sciences, Guangzhou, People's Republic of China; Chinese Academy of Sciences, Guiyang, People's Republic of China

S.-L. Lo
National Taiwan University, Taipei, Taiwan, Republic of China (ROC)

P. Longhurst
Cranfield University, Cranfield, Bedford, United Kingdom

S.K. Maiti
Indian School of Mines, Dhanbad, Jharkhand, India

S. Mandal
University of South Australia, Mawson Lakes, SA, Australia

V. Matichenkov
Institute of Basic Biological Problems, Pushchino, Russia

T. Matsi
Aristotle University of Thessaloniki, Thessaloniki, Greece

S. Venkata Mohan
CSIR-Indian Institute of Chemical Technology (CSIR-IICT), Hyderabad, Telangana, India

A. Muntoni
University of Cagliari, Cagliari, Italy

B. Murtaza
COMSATS Institute of Information and Technology, Vehari, Pakistan

G. Murtaza
University of Agriculture Faisalabad, Faisalabad, Punjab, Pakistan

R. Naidu
University of Newcastle, Callaghan, NSW, Australia; Cooperative Research Centre for Contamination Assessment and Remediation of the Environment, Callaghan, NSW, Australia

M.F. Nawaz
University of Agriculture Faisalabad, Faisalabad, Punjab, Pakistan

T. Naz
University of Agriculture Faisalabad, Faisalabad, Punjab, Pakistan

N.K. Niazi
University of Agriculture Faisalabad, Faisalabad, Punjab, Pakistan; Southern Cross University, East Lismore, NSW, Australia

Y.S. Ok
Kangwon National University, Chuncheon, Korea

R.K. Patel
National Institute of Technology, Rourkela, Odisha, India

A. Polettini
University of Rome "La Sapienza," Rome, Italy

R. Pomi
University of Rome "La Sapienza," Rome, Italy

M.N.V. Prasad
Ural Federal University Named After the First President of Russia B.N. Yeltsin, Ekaterinburg, Russia; University of Hyderabad, Hyderabad, Telangana, India

D. Rakshit
University of Calcutta, Calcutta, West Bengal, India

M.K. Sahu
National Institute of Technology, Rourkela, Odisha, India

C. Saint
University of South Australia, Adelaide, SA, Australia

S.K. Sarkar
University of Calcutta, Calcutta, West Bengal, India

I.J. Schreiter
Technische Universität Darmstadt, Darmstadt, Hessen, Germany

C. Schüth
Technische Universität Darmstadt, Darmstadt, Hessen, Germany

B. Seshadri
University of Newcastle, Callaghan, NSW, Australia

M. Shahid
COMSATS Institute of Information and Technology, Vehari, Pakistan

M.B. Shakoor
University of Agriculture Faisalabad, Faisalabad, Punjab, Pakistan

K. Shih
University of Hong Kong, Hong Kong, Hong Kong SAR, China

H. Shim
Department of Civil and Environmental Engineering, University of Macau, Avenida da Universidade, Taipa, Macau, SAR, China

G. Spiers
M.V. Lomonosov Moscow State University, Moscow, Russia; Laurentian University, Sudbury, ON, Canada

A. Surapaneni
South East Water, Melbourne, VIC, Australia

A. Szogi
United States Department of Agriculture, Florence, SC, United States

Tripti
Ural Federal University Named After the First President of Russia B.N. Yeltsin, Ekaterinburg, Russia

C.D. Tsadilas
Hellenic Agricultural Organization, Institute of Soil Mapping and Classification, Larissa, Greece

L. Tsechansky
Agricultural Research Organization, The Volcani Center, Bet Dagan, Israel

V.K. Tyagi
The Energy and Resources Institute (TERI) University, New Delhi, India

T. Van Gerven
Katholieke Universiteit Leuven, Leuven, Belgium

A. Van Zomeren
Energy Research Centre of The Netherlands (ECN), Petten, The Netherlands

G. Velvizhi
CSIR-Indian Institute of Chemical Technology (CSIR-IICT), Hyderabad, Telangana, India

P. Venkatachalam
Periyar University, Salem, Tamil Nadu, India

M. Vithanage
National Institute of Fundamental Studies, Kandy, Sri Lanka

H. Wang
Zhejiang A & F University, Lin'an, Hangzhou, Zhejiang Province, China; Guangdong Dazhong Agriculture Science Co. Ltd., Dongguan, China

A. Wefer-Roehl
Technische Universität Darmstadt, Darmstadt, Hessen, Germany

J.C. White
The Connecticut Agricultural Experiment Station, New Haven, CT, United States

H. Wijesekara
University of Newcastle, Callaghan, NSW, Australia

H. Yan
University of Hong Kong, Hong Kong, Hong Kong SAR, China

F. Zhao
Institute of Urban Environment, Chinese Academy of Sciences, Xiamen, China

Foreword

Environmental sustainability is an essential aspect of sustainable development. In 2015, the United Nations adopted 17 sustainable development goals (see *Transforming Our World: The 2030 Agenda for Sustainable Development*, United Nations, 2015). Within these goals appear several phrases that relate to this book on environmental materials and waste: "sustainable management of water," "sustainable industrialization," "make cities sustainable," "ensure sustainable consumption and production patterns," "sustainable use of terrestrial ecosystems," and "halt and reverse land degradation." The contributors to this book are addressing important environmental sustainability issues that relate to the sustainability goals.

Environmental considerations in the management of waste are growing in importance because of population growth and increasing production of a wide variety of products. This book addresses many topics that are essential to include in efforts to advance environmental sustainability.

Land and water have been affected by waste for many years, and there are portions of the earth that are in need of reclamation. The problems addressed include the management of wastes that are being generated daily as well as the reclamation of land that has been contaminated by past practices. Resource recovery includes the beneficial materials from waste so they can be used again to make new products. Historically landfills have been used to deposit waste; however, it is becoming more important to recycle, recover materials, and reuse them to reduce the amount of land that is being used as landfills. If by-products can be incorporated into soil to improve soil quality or used for mine site restoration, this is much better than adding to a landfill. The chapters in this book address a number of complex issues encompassing recovery of resources from biowaste for pollution prevention, biochar production from biowaste and municipal solid waste for environmental protection and sustainable agriculture production, use of biowastes for mine site rehabilitation, biodiesel production from wastewater, energy and manure recovery from sludge, and resource recovery from mining industry waste. These aspects have interest and application in many parts of the world.

Real solutions exist for many of the waste management issues that we have today, and it is beneficial to provide information on successful approaches that have application in many locations. Dissemination of these advances can address significant environmental challenges related to environmental sustainability.

Larry E. Erickson
Professor of Chemical Engineering and Director
Center for Hazardous Substance Research
Kansas State University
Manhattan, KS, USA
lerick@ksu.edu

Preface

Concerns about the environment are growing in the wake of climate change. In this regard, "environmental sustainability" is globally important in the contemporary world. A wide variety of natural resources is being heavily exploited to meet the demands of a worldwide growing population and economy. Despite technological developments, pressure on natural capital of resources is constantly increasing. Thus, the efficient recovery and processing of resources for recycling is becoming significant worldwide. Awareness and opportunities to obtain valuable resources from process and bleed streams are of paramount importance in recent times. This would create revenue and achieve environmental sustainability.

The sustainable recovery of environmental material including wastewater offers tremendous new benefits. The various chapters presented in this book center on how studies on environmental materials and waste can facilitate the recovery of resources to prevent environmental pollution. The collection of chapters brings together recent findings and state-of-the-art developments in resource recovery and pollution prevention.

Therefore, the various chapters reinforce the importance of multiple *R*'s, such as *R*ecycle, *R*euse, *R*emediation, *R*eclamation, *R*enovation, *R*eplenish, *R*esilience, *R*everence for nature, and *R*ainwater harvest, which are being increasingly valued in many parts of the world.

Sound policies and affordable technologies are required for resource recovery from waste materials to prevent environmental pollution. The contemporary world is overloaded with a wide array of waste. Handling waste is a big issue across the world, and thus there is a strong need to address such a challenge in waste management. For example, the new innovation wave of biovalorization of biowaste for biomaterials and biofuels is rapidly expanding.

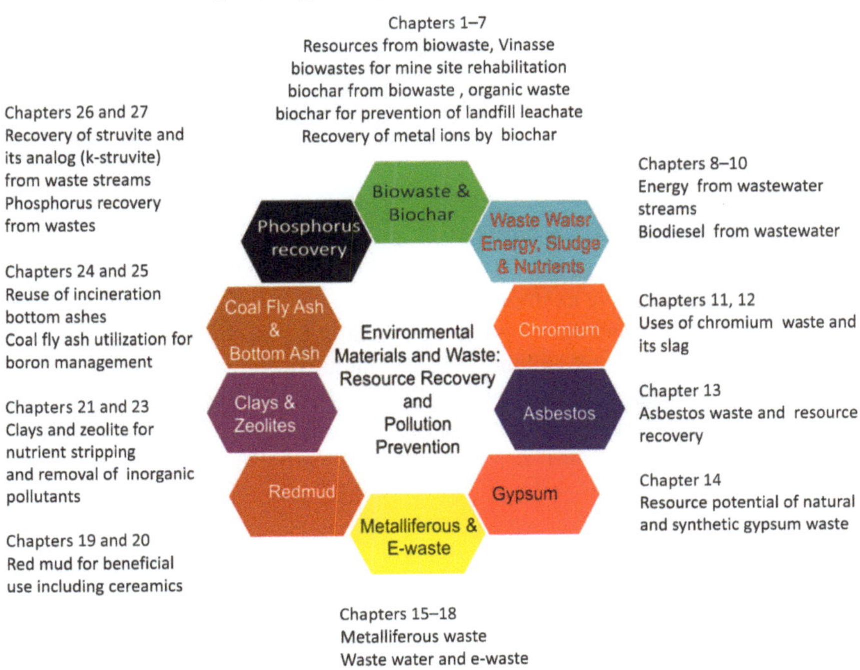

Chapters 1 through 7 deal with the recovery of resources from biowaste, vinasse as a resource, the use of biowastes for mine site rehabilitation, the application of biochar produced from biowaste materials for environmental protection and sustainable agriculture, the use of biochar produced from organic wastes for pollutant cleanup under site-specific conditions, municipal solid waste–produced biochar applied to prevent landfill leachate pollution, and the recovery of metal ions by biochar derived from biowastes.

Chapters 8 through 10 cover the use of wastewater to produce bioenergy, biodiesel production using oleaginous yeast and microalgae, the use of sludge as manure.

Chromite waste minimization and use, including resource recovery of chromium slag, are covered in Chapters 11 and 12.

Asbestos industry waste management and the resource recovery potential of natural and synthetic gypsum waste are dealt with in Chapters 13 and 14, respectively.

Chapters 15 through 18 discuss fluxes of metalliferous waste in India, with possible strategies for the recovery and prevention of pollution, resource recovery from wastewater and electronic waste based on extracellular electron transfer; acid mine drainage from abandoned mines of north Portugal: hydrochemistry, resource recovery and prevention of pollution; and landscape restoration in nickel-processing affected areas.

The beneficial use of red mud material for the ceramic industry is covered in Chapters 19 and 20.

Clay minerals, zeolite for nutrient stripping from wastewater streams, and surfactant-modified zeolites for inorganic pollutant (anion) removal from natural water and wastewater are covered in Chapters 21 through 23.

Chapters 24 and 25 are devoted to the reuse of treated incineration bottom ash and coal fly ash for boron management in soils and wastewaters.

Chapters 26 and 27 are devoted to the crystallization of struvite and its analog (K-struvite) for nutrient recovery from waste streams.

This book is the collective effort of 89 contributors from 20 countries/regions: Australia, Belgium, Brazil, Canada, China, Denmark, Germany, Greece, Hong Kong, India, Israel, Italy, Macau, Pakistan, People's Republic of China, Portugal, Republic of Korea, Russia, Sri Lanka, The Netherlands, Turkey, United Kingdom and United States.

The book provides comprehensive and new information about resource recovery from environmental and waste materials to reduce environmental pollution.

Moreover, as a pioneering attempt to link resource recovery with in-depth environmental materials studies, we think the comprehensive examples in this book will inspire more new strategies and technologies to achieve sustainable and environmentally friendly developments.

Other unique features of this book include: (1) substantial color figures, flow diagrams, tables, and key literature; (2) a main focus on the strategies of using waste materials; (3) highlights of feasible options in establishing large-scale installations for resource recovery and energy production operations.

Acknowledgments

We would like to thank all the authors of this volume for their cogent and comprehensive contributions.

We are thankful to Candice Janco (ELS-WAL) Senior Acquisitions Editor, Environmental Science, Laura Kelleher (ELS-WAL), Rowena Prasad (ELS-SDG) for excellent co-ordination of this fascinating project.

Editors are also thankful to Ms. Emily Joy Grace Thomson for suggestions and quick correspondence through hundreds of emails; thanks are also due to Mr. Rajendran, Mohanapriyan (Monu) and his team for excellent help in many ways that resulted in record time of publication of the volume.

The editors are thankful to Larry E. Erickson, Professor of Chemical Engineering and Director, Center for Hazardous Substance Research, Kansas State University (USA) for writing the foreword at a short notice.

We wish to thank numerous colleagues from overseas, our students and research associates for sharing knowledge, ideas and assistance which helped in developing and shaping this book.

CHAPTER

RECOVERY OF RESOURCES FROM BIOWASTE FOR POLLUTION PREVENTION

1

M.N.V. Prasad

University of Hyderabad, Hyderabad, Telangana, India

CHAPTER OUTLINE

1.1 INTRODUCTION

Applying the scientific knowledge of recycling to solve soil problems in the environment is the Milky Way: waste can be defined as right or useful material in the wrong place. Biowaste should not be treated as a problem; rather, it should be considered as an opportunity. Instead of costly disposal, new markets could be found for previously discarded biomaterial by improving its quality through conventional processing or biotechnology. For example, the meal obtained after extracting oil from novel crops can be processed into animal feed (Bothast and Schlicher, 2005)

As such, there is no waste until it is wasted. Thus, all waste is a reusable resource (Fig. 1.1). Some selected sayings about recycling of waste are that (1) one used aluminum beverage can saves enough energy to run a 100-W bulb for 20 h, a computer for 3 h, or a TV for 2 h; (2) 1 ton of aluminum saves 37 barrels of oil; (3) 1 ton of glass saves 9 gallons of fuel (oil), (4) 1 ton of newspaper saves 15 mature trees; and (5) plastics require 100 to 400 years to break down in a landfill.

The field of biowaste disposal and use toward the production of diverse biofuels (solid, liquid, and gaseous) has gained considerable momentum globally following the biorefinery approach (Table 1.1). In the process of biowaste management, bioremediation contractors must derive a profit from this activity. The primary driver is regulatory compliance rather than manufacturing a profit. It is an

Environmental Materials and Waste. http://dx.doi.org/10.1016/B978-0-12-803837-6.00001-9

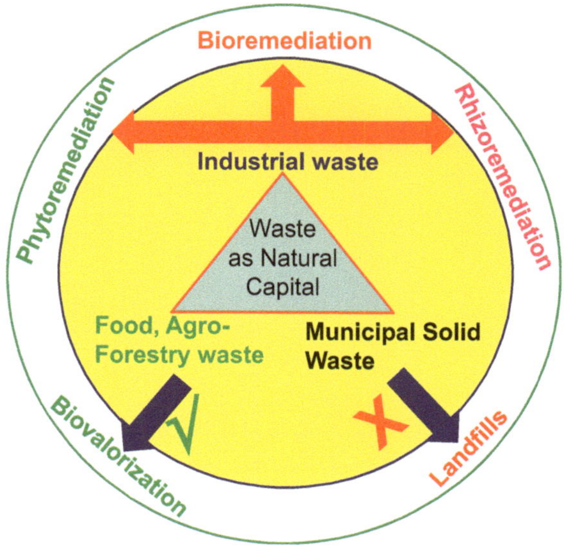

FIGURE 1.1

Waste categories: industrial waste, biowaste (food and agro-forestry waste), and domestic waste (municipal solid waste (MSW)). The most common disposal is composting and landfills. Biovalorization and bioremediation are emerging tools.

attractive technology in the context of bioeconomy. Nature's cure using biodiversity (bioremediation) is a sustainable solution for environmental decontamination. Starting with a few in the early 1990s, several articles have been published on various aspects of using biological resources for environmental cleanup (Figs. 1.2 and 1.3). Often policy and decision makers, academics, and civic governors think that bioremediation is a temporary solution for transferring pollutants and contaminants from one place to another.

Environmental decontamination through bioremediation is an integral part of the bioeconomy and sustainable development. There has been growing interest in using biodiversity as a raw material for environmental decontamination. On the other hand, the volume and diversity of contaminated substrates (water, soil, and air) are increasing owing to anthropogenic and technogenic sources. Toxic metals are the most prevalent inorganic pollutants/contaminants and are widely used for a wide variety of needs, from building materials to information technology. Therefore, the bioavailable fraction of metals is increasing in waste.

1.2 BIOWASTE MANAGEMENT

With the growing economy, extensive industrialization and extraction of natural resources have resulted in environmental contamination and pollution. Large amounts of toxic waste have been dispersed in thousands of contaminated sites spread across the globe. These adverse and intrinsic effects of polluted/contaminated sites are shown in Fig. 1.4. These pollutants belong to two main classes: inorganic and organic. The challenge is to develop innovative and cost-effective solutions to

Table 1.1 Wide Range of Products From Biowaste After Biovalorization

Product/Process From Biowaste*	References
Adsorbent for methylene blue dye, utilization of oxalic acid-modified spent mushroom substrate for removal of methylene blue from aqueous solution	Bretanha et al. (2014) and Yan et al. (2015)
Anaerobic digestion	Romero-Guiza et al. (2014)
Antioxidants	Kumar and Suresh (2014) and Wei et al. (2014)
Bioactive	Sinha et al. (2014)
Biofilms	Franzoso et al. (2015)
Biogas	Dach et al. (2014)
Biogas production chains	Uusitalo et al. (2014a,b)
Biomethane	San Miguel and Corona (2014)
Bio-oil	Bordoloi et al. (2015)
Biopolymers	Franzoso et al. (2015)
Biorefinery	Fava et al. (2015)
Biosorbent	Kim et al. (2015)
Biosurfactants	Vargas et al. (2014)
Biopolymer production for biosorption of heavy metals	Kennedy and Roig (2011)
Carboxymethyl cellulose	Haleem et al. (2014)
Chitosanase	Sinha et al. (2014)
Electricity generation	Messagie et al. (2014)
Extracellular protease and antioxidant materials	Kumar and Suresh (2014)
Hazardous organic compounds in biogas plant	Dach et al. (2014)
Hydrochar	Zhu et al. (2014)
Hydrogen and polyhydroxybutyrate	Patel et al. (2015)
Hydroxyapatite from fish scale	Sukaimi et al. (2015)
Hydroxyapatite nanowires	Bramhe et al. (2014)
Lactic acid and methane	Dreschke et al. (2015)
Poly(3-hydroxybutyrate)	Angelini et al. (2014)
Polyhydroxyalkanoate	Kumar et al. (2014)
Polyhydroxybutyrate granules	Martino et al. (2014)
Production of nanoporous carbon for removal of anionic dye	Güzel et al. (2014)
Pyrolysis characteristics and products	
Solid biofuel	Zhang et al. (2014)
Uranium(vi) from aqueous solution by *Citrus limetta* peels	Gondhalekar and Shukla (2014)
Synthesis of Zeolite Socony Mobil–type zeolotes from biowaste gasification ashes	Lin et al. (2003)

**not exhaustive*

decontaminate polluted environments. Bioremediation is emerging as an invaluable tool for environmental cleanup. Globally, agro-forestry, domestic, industrial, and food-processing waste is increasing. All of this waste is no longer wasteful and is being used as an invaluable resource for the production of a wide variety of materials including energy. Every ton of biowaste is equivalent to 4.5 tons of CO_2 emissions, which contribute to environmental pollution. One estimate shows that in the

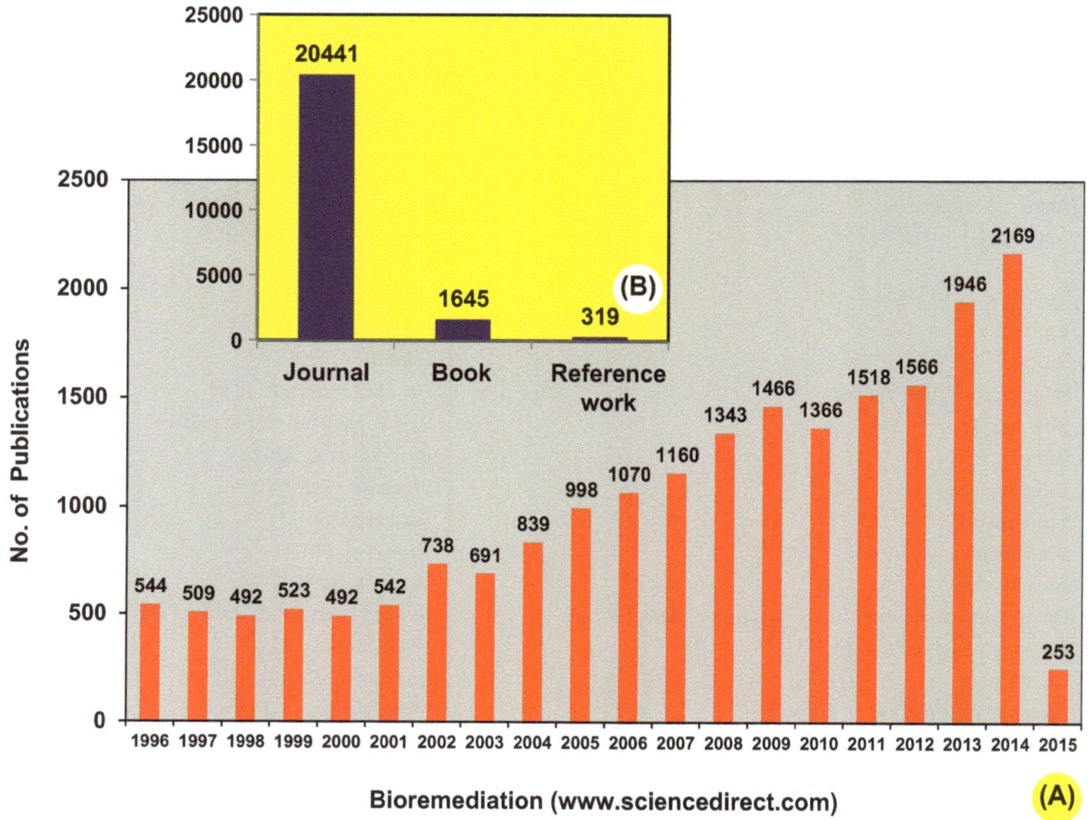

FIGURE 1.2

(A) Articles published in the field of bioremediation steeply increasing *(data based on* http://www.sciencedirect.com*)*. (B) Journals, books, and references in the field of bioremediation.

United States about 45 billion kg of fresh fruit, vegetables, milk, and grain products is lost as waste in storage, transport, or use. The US Environmental Protection Agency (EPA) estimates that about US $1 billion is required to dispose of the cost. In the United Kingdom, about 20 million tons of food is wasted each year (Lin et al., 2013).

Over a period of time, by natural attenuation (monitored natural attenuation, or bioremediation), any waste would become stabilized. However, it takes a long time with no disturbances. The normal practice is to dump waste at designated sites as "landfills," which may be single lined or double lined. Because of the scarcity of land and the explosion of knowledge of green chemistry, there is a search for new innovations. Furthermore, there are negative health effects from landfills because of the emission of the toxic gas ammonia.

People who inhale 50,000 parts per billion (ppb) of ammonia for less than 1 day experience minor and temporary irritation. People who inhale 100,000 ppb of ammonia in the air for more than 6 weeks experience eye, nose, and throat irritation. In addition, odors originating from certain compounds and

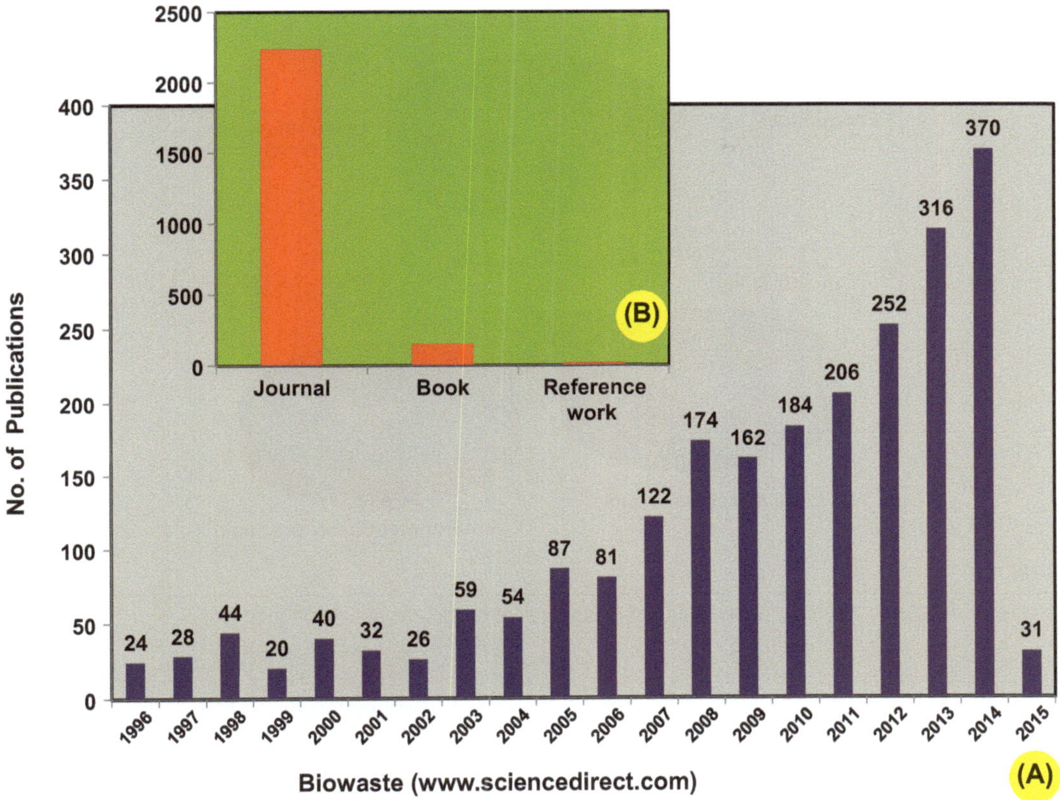

FIGURE 1.3

(A) Articles published on "biowaste" steeply increasing *(data based on* http://www.sciencedirect.com*)*.
(B) Journals, books, and references in the field of bioremediation.

wastes in landfills discourage their use. Human exposure to landfill odors can cause headaches and nausea.

Collection of methane gas from landfills is known but only at a few sites. Four main goals of management of biowaste are: (1) protecting the environment, (2) promoting partnerships, (3) strengthening the marketplace, and (4) sustaining reuse.

A landfill site is an area of land that has been specifically engineered to allow for the deposition of waste onto and into it. MSW landfill sites are a large source of human-related methane emissions; in some countries such as the United States, they can account for up to 25% of these emissions. At the same time, methane emissions from landfills represent a lost opportunity to capture and use a significant energy resource.

Landfills may include internal waste disposal sites (where a producer of waste carries out its own waste disposal at the place of production) as well as sites used by many producers. Many landfills are also used for other waste management purposes, such as temporary storage, consolidation, and transfer, or processing of waste material (sorting, treatment, or recycling). A landfill also may refer to

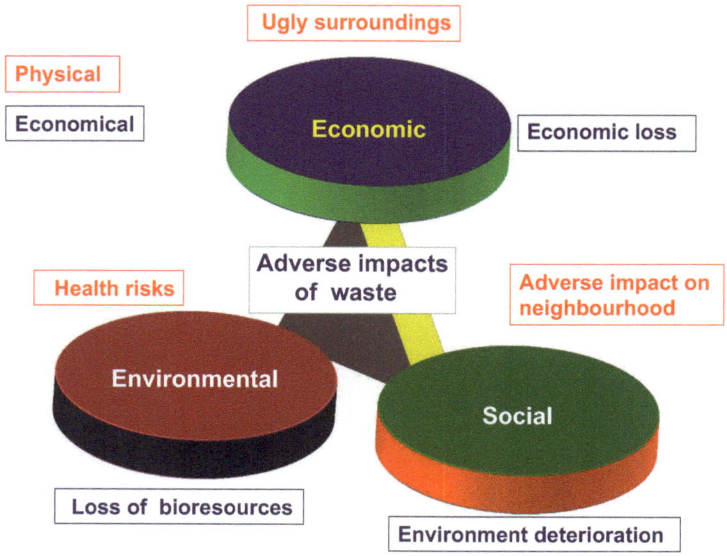

FIGURE 1.4

Adverse and intrinsic effects of biowaste (*red text*, physical loss; *blue text*, economic loss).

ground that has been filled in with soil and rocks instead of waste materials, so that it can be used for a specific purpose, such as for building houses.

1.3 BIOREMEDIATION OF WASTE DISPOSED LAND OR EXISTING LANDFILLS

Excavate soil and remove it to a disposal site away from ready pathways for human or sensitive ecosystem contact through one of the possible options. Aeration of soils at the contaminated site (with the attendant risk of creating air pollution) bioremediation, involving microbial digestion of certain organic chemicals. Extraction of groundwater or soil vapor with an active electromechanical system. Containment of the soil contaminants, recycling reduces the ill effects of landfills and makes the environment cleaner and safer. Modernization and policy must emphasize their ability to both the environment and contribute to the economy (Figs. 1.5 and 1.6).

1.4 SOLID WASTE MANAGEMENT: AN INDIAN PERSPECTIVE

India is perhaps the world's second most populous country; it has about 18% of the global population on 2.4% of the land area. Rapid industrialization and economic growth in the past few decades have imposed tremendous pressure on its natural resources as well as an increase in the generation of a wide variety of wastes. The indiscriminate disposal of waste has resulted in ugly scenes in many places and in squandered resources (Fig. 1.7). On October 2, 2014, the Indian Prime Minister initiated a "Swachh

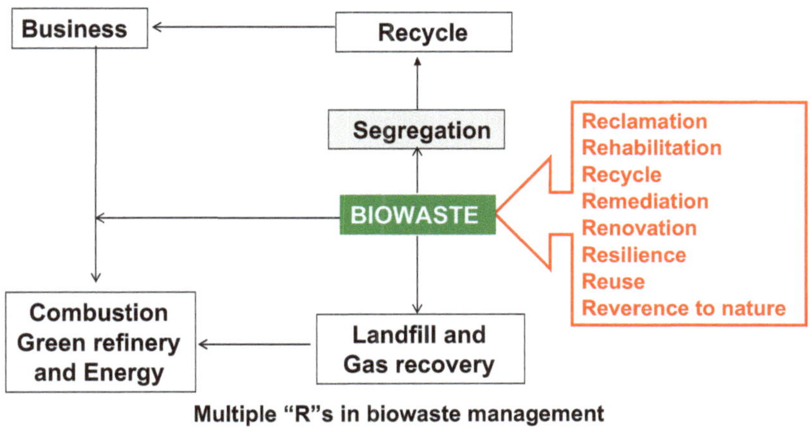

FIGURE 1.5

Multiple R's in handling biowaste to generate resources.

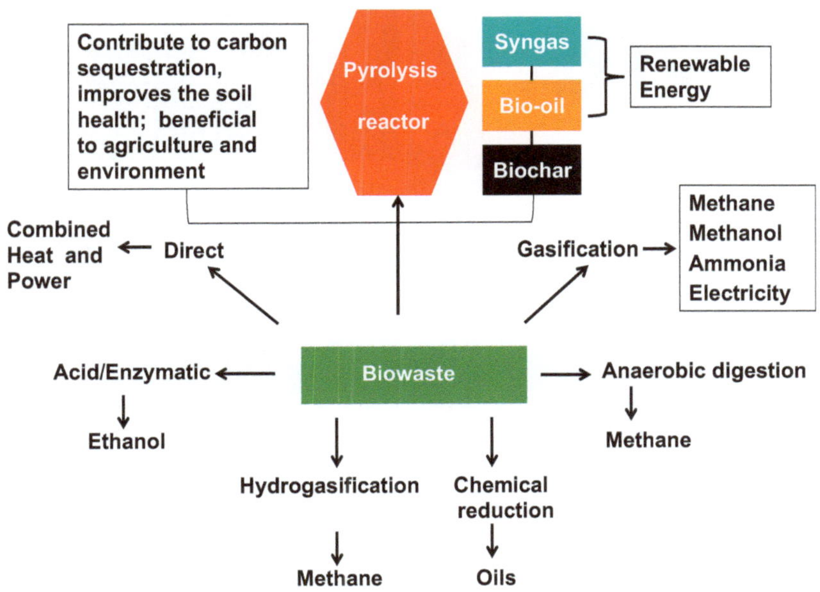

FIGURE 1.6

Energy sources from biowaste including production of biochar, which is beneficial for agriculture and the environment.

Bharat Mission" ("Clean India program"). Various ministries and civic bodies are actively taking part in this mission to better manage wastes of different kinds (domestic and industrial), because the failure of current methods have detrimental effects on human health. Therefore, the multiple "R's" are of considerable importance for environmental sustainability and waste management (Fig. 1.5).

India is forging ahead to attain a high gross domestic product growth rate to achieve the status of an industrialized and developed world by 2020. To achieve this, the heavy consumption of natural resources is inevitable. The other side of the coin of this task is enormous waste generation. The waste management summit held in Chennai, India in 2012 estimated that the poised industrial growth would generate 100 million tons of nonhazardous solid waste, 6 to 7 million tons of hazardous waste annually. There is an urgent need to have a robust national program of waste collection, segregation, and recovery of resources by phasing out the landfill system (Figs. 1.8 and 1.9). Power plants using the

FIGURE 1.7

(A) The Greater Hyderabad Municipal Corporation collects domestic solid waste and transports by trucks to designated landfill site (Jawaharnagar MSW disposal site). (B–C) Physical appearance of MSW disposed landfill site, a health hazard. (D) Leachate collection at the base of the MSW dump site (*white arrow*). Gradual leaching causing ground water polluting.

FIGURE 1.8

Color-coded bins for segregation of domestic waste, a mandatory regulation in many European countries. For example, (A) In Poland (i) paper, (ii) metal (iii) plastic, (iv and v) glass. (B) In Hungary, (i) plastic, (ii) glass, (iii) paper, (iv) metal.

concept of "Build–Own–Operate–Transfer" employing MSW are established in Hyderabad are in progress in other cities in India (Fig. 1.10).

1.4.1 USE OF BIOWASTE FROM BIOREMEDIATION: THE CASE OF AQUATIC WEEDS

Eichhornia crassipes (water hyacinth), *Pistia stratiotes* (water lettuce), and other high–biomass producing aquatic weeds are considered for the bioremediation of polluted and contaminated water. (Fig. 1.11). These water weeds spread rapidly and clog water channels, inconveniencing irrigation and watershed management. Therefore, an approach based on applying phytoremediation and using phytoremediated phytomass for bioenergy would be a welcome solution. Estimates show that 1 hectare of water hyacinth grown on sewage can purify waste water and produce 0.8 ton of dry

FIGURE 1.9

(A–C) Color-coded bags are used to dispose different categories of domestic waste in Europe: (A) Household degradable waste; (B) paper waste for recycling; (C) plastic waste; and (D) kitchen waste for compost.

matter per day; this can be converted to 200 m^3 gas, which is enough to generate 250 kW power (Fig. 1.11).

Water bodies are receptacles for numerous types of industrial waste, sewage effluent discharge, urban and agricultural runoff, and so forth. Aquatic plants are known since time immemorial for their ability to treat contaminants from the surrounding environment; they are a widely recognized and applied phenomenon in constructed wetlands. Floating water weeds such as duckweeds, water hyacinth, and water lettuce are ecosystem service providers and have been studied extensively for their ability to remove metal from water and their application in constructed wetlands for waste water treatment (Vymazal et al., 1998; Hadad et al., 2009; Zhang et al., 2009). They have a crucial role in phytosanitation and phytofiltration: for example, heavy metal removal by *E. crassipes* (Dhote and Dixit, 2009), *P. stratiotes* (Maine et al., 2001; Suñe et al., 2007), and *Salvinia herzogii* (Maine et al., 2004; Miretzki et al., 2004; Odjegba and Fasidi, 2007).

FIGURE 1.10

(A) Premises of MSW-based power plant in the vicinity of Hyderabad, Telangana State of India. (B, C) Solar drying system of MSW. (D) Briquettes (pellets) made of dried MSW for use in power plant as fuel.

Courtesy: Solid Waste to Electricity Company (SELCO) International Limited, Hyderabad.

Different methods have been used to remove metallic elements from water (Fu and Wang, 2011). However, with most of these suggested methods, it can be expensive to treat waste waters contaminated with low concentrations of metals from a non–point source. Different aquatic plants exhibit variable efficiencies for different metals in low-level metal-contaminated water. Aquatic plants are key players in phytofiltration and therefore have been proposed as a promising environmentally friendly technology for removing heavy metal concentration from contaminated water. In phytofiltration, high metal–accumulating plants function as biofilters, and thereby effectively sequester metals from polluted or contaminated waters. Countries with vast biodiversity that cannot afford sophisticated water treatment plants can derive substantial benefit by adopting phytoremediation approaches. Ultimately the success of phytofiltration depends on the plant growth rate and the ability to uptake a high concentration of metals in the phytomass. Many aquatic plants, usually those found in polluted water bodies, have been suggested for waste water treatment; ie, they have the ability to accumulate an unusually high concentration of heavy metals without experiencing phytotoxicity. However, most

FIGURE 1.11

High—biomass producing aquatic macrophytes: (A) *Eichhornia crassipes* and (B) *Pistia stratiotes*. (C) Mechanical harvesting of the aquatic weed biomass. (D) Use of the aquatic weed biomass for biogas production.

Prasad (2015).

species identified so far are not suitable for on-site phytofiltration because of their small root and shoot biomass and slow growth rate.

Biomass gasification is basically the conversion of wood chips and wood waste into a combustible gas mixture called a producer gas (low-BTU gas). This process involves the partial combustion of phytomass. Because phytomass contains carbon, hydrogen, and oxygen, complete combustion would produce carbon dioxide and water vapor. Partial combustion produces carbon monoxide and hydrogen, both of which are combustible gases. The gas thus generated in a "gasifier" could be used to generate captive power (a few kilowatts to several hundred kilowatts). It can also be used for direct thermal and shaft power applications. Fuel wood direct combustion has several disadvantages owing to its limited applications, and it has low efficiency. Direct thermal applications for power generation require steam boilers with a steam engine or turbine with additional necessary equipment. Such a conversion technology is not only capital intensive but of low conversion efficiency. Conversion of the same phytomass (wood chips or wood waste) to combustible gases in a gasifier to generate power has innumerable advantages. Phytomass gasification (using gasifiers of various capacities) generates electricity, and the gas can be used for direct thermal energy and shaft power (Kathi, 2016; Bini et al., 1999).

1.5 FOOD WASTE TO FEED FISH

Faced with escalating commercial fish feed prices, local aquaculture farms have begun trying to recycle food waste for use as fish feed. The aquaculture industry requires large quantity of fish meal. The demand for farmed fish for human consumption is increasing worldwide. As the demand rises, the demand for fish meal to feed the farmed fish also rises. The food waste is being processed as fish feed at a cheaper cost in Hong Kong and China (Fig. 1.12)

FIGURE 1.12

(A) Plastic Containers for collecting food waste (B) Plastic container with food waste (C) Food waste processing machinery (D) Food waste conversion to fish feed.

Food waste recovery centre, South China Reborn Resources (Zhongshan) Co. Ltd, Hong Kong.

1.6 BIOLOGICAL RECULTIVATION OF PETROLEUM INDUSTRY–RAVAGED LAND

The indiscriminate disposal of waste appears to be a legacy of developing nations. The solution to this problem requires dedication and a huge investment to recover resources and safeguard the environment.

Among many types of industrial waste, polycyclic aromatic hydrocarbons (PAHs) are generated in large quantities at exploration sites and refineries. PAHs are recognized worldwide as priority pollutants because of their toxicity and carcinogenicity. Along with halogenated hydrocarbons, PAHs represent important challenges for bioremediation. Because they are hydrophobic, PAHs accumulate and sequester in organic matter in soil. Thus, their desorption from soil limits the efficiency of biodegradation in situ. The rhizosphere, in which soil microflora may metabolize these compounds, may also act as a sink for

FIGURE 1.13

(A) Petroleum exploration in the vicinity of Baku, Azerbaijan. (B, C) Greenhouse with stocks of suitable plants for biological recultivation of petroleum hydrocarbon–contaminated sites. (D) Biological recultivation of petroleum hydrocarbon–contaminated sites. Baku, Azerbaijan.

heavy metals and high-molecular-weight PAHs. The success rate of any bioremediation method depends on the adaptation of the microbe and/or plant to the local environment. The US EPA listed 16 PAHs as priority pollutants. Those with a molecular weight of up to 202 g/mol can be degraded through growth-linked aerobic reactions, including high-molecular-weight PAHs such as pyrene and fluoranthene, whereas the rest of the PAHs, such as benzo[a]pyrene, are only susceptible to co-metabolic removal. Plant–microbe combinations using indigenous organisms are preferred (Fig. 1.13).

Large areas of land contaminated with PAHs are remediated following the concept of biological recultivation, using selected plants often inoculated with microbial combinations. Selected plant stocks are raised in greenhouses near the contaminated sites and planted for PAH-contaminated site remediation (Fig. 1.13).

1.7 CONCLUSION

Biorefineries for the valorization of biowaste to produce products and bioenergy are gaining considerable momentum in developed counties. The age-old concept of landfills is discouraged because of a variety of problems: (1) from a global perspective, land resources are under immense pressure; (2) landfills produce greenhouse gases; and (3) valuable resources are being buried (Table 1.2).

For several centuries, waste management has been dumping, burning, composting, and converting waste into a useful thing (Deka et al., 2011; Van Wyk, 2011). However, the concept of biowaste valorization with emphasis on the production of higher-value materials has attained tremendous importance. The old concept of incineration and composting is no longer valid in all situations. In this context, the circular economic model is appropriate to the problem of waste management because it integrates waste into a resource-based economy applicable to all categories of waste (Fig. 1.14).

Key value drivers in achieving these goals are:

1. Environmental: lower footprint, less waste
2. Economic: cost of feedstock, carbon, or waste taxation
3. Strategic: license to operate, branding, renewable

Thus waste is no longer to be considered wasted; indeed, it is a valuable resource for the production of chemicals, materials, and energy. Fig. 1.15 depicts challenges and opportunities for the management and recovery of resource from biowaste.

Table 1.2 Food Waste: Some Facts as per the United Nations Environment Programme (2009)

World food production for human consumption is about 3.8 billion tons	1.3 Billion tons is wasted owing to mismanagement
Consumers in rich countries waste 222 million tons	Food production of sub-Saharan Africa is about 230 million tons
About 1.2 billion tons of food is wasted	During 2009–10, the world cereal crop was about 2.3 billion tons

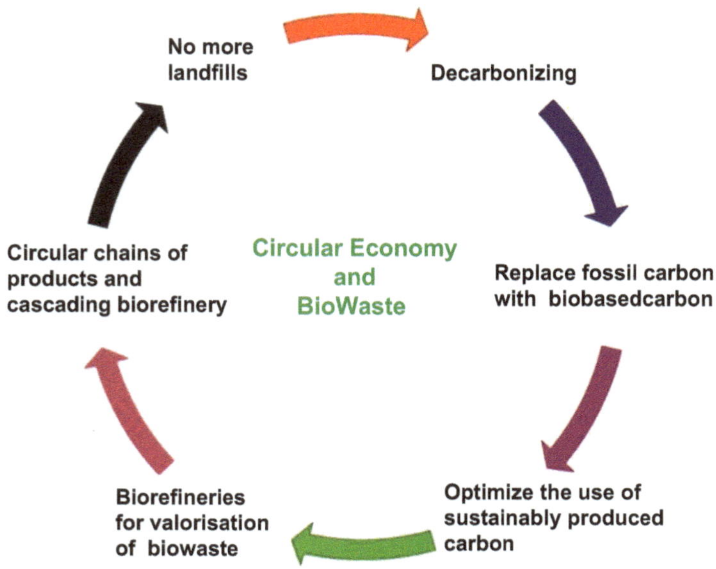

FIGURE 1.14

Circular economy, carbon sequestration, and biorefinery for cogeneration of products.

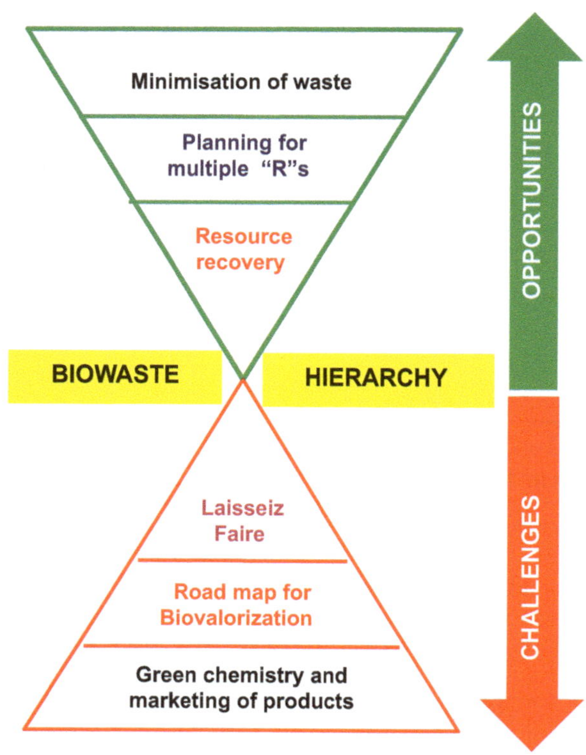

FIGURE 1.15

Challenges and opportunities for the management and resource recovery from biowaste.

ACKNOWLEDGMENTS

The author is grateful to the authorities of EcoPark, Baku, Azerbaijan to visit the Biological recultivation of petroleum hydrocarbon contaminated sites; South China Reborn Resources (Zhongshan) Co. Ltd, Hong Kong Food waste recovery centre, and SELCO International Limited, Hyderabad.

REFERENCES

Angelini, S., Cerruti, P., Immirzi, B., Santagata, G., Scarinzi, G., Malinconico, M., 2014. From biowaste to bioresource: effect of a lignocellulosic filler on the properties of poly (3-hydroxybutyrate). International Journal of Biological Macromolecules 71, 163–173.

Bini, L.M., Thomaz, S.M., Murphy, K.J., Camargo, A.F.M., 1999. Aquatic macrophyte distribution in relation to water and sediment conditions in the Itaipu reservoir, Brazil. Hydrobiologia 415, 147–154.

Bordoloi, N., Narzari, R., Chutia, R.S., Bhaskar, T., Kataki, R., 2014. Pyrolysis of *Mesua ferrea* and *Pongamia glabra* seed cover: characterization of bio-oil and its sub-fractions. Bioresource Technology 178, 83–89.

Bothast, R.J., Schlicher, M.A., 2005. Biotechnological processes for conversion of corn into ethanol. Appl Microbiol Biotechnol 67, 19–25.

Bramhe, S., Kim, T.N., Balakrishnan, A., Chu, M.C., 2014. Conversion from biowaste *Venerupis* clam shells to hydroxyapatite nanowires. Materials Letters 135, 195–198.

Bretanha, M.S., Rochefort, M.C., Dotto, G.L., Lima, E.C., Dias, S.L.P., Pavan, F.A., 2014. *Punica granatum* husk (PGH), a powdered biowaste material for the adsorption of methylene blue dye from aqueous solution. Desalination and Water Treatment 57 (7), 3194–3204.

Dach, J., Boniecki, P., Przybył, J., Janczak, D., Lewicki, A., Czekała, W., Witaszek, K., Rodríguez- Carmona, P.C., Cieślik, M., 2014. Energetic efficiency analysis of the agricultural biogas plant in 250 kWe experimental installation. Energy 69, 34–38.

Deka, H., Deka, S., Baruah, C.K., Das, J., Hoque, S., Sarma, H., Sarma, N.S., 2011. Vermicomposting potentiality of *Perionyx excavatus* for recycling of waste biomass of java citronella – an aromatic oil yielding plant. Bioresource Technology 102 (24), 11212–11217.

Dhote, S., Dixit, S., 2009. Water quality improvement through macrophytes - a review. Environmental Monitoring and Assessment 152, 149–153.

Dreschke, G., Probst, M., Walter, A., Pümpel, T., Walde, J., Insam, H., 2015. Lactic acid and methane: improved exploitation of biowaste potential. Bioresource Technology 176, 47–55.

Fava, F., Totaro, G., Diels, L., Reis, M., Duarte, J., Carioca, O., Poggi-Varaldo, H.M., Ferreira, B.S., 2015. Biowaste biorefinery in Europe: opportunities and research & development needs. New Biotechnology 32 (1), 100–108.

Franzoso, F., Tabasso, S., Antonioli, D., Montoneri, E., Persico, P., Laus, M., Mendichi, R., Negre, M., 2015. Films made from poly(vinyl alcohol-co-ethylene) and soluble biopolymers isolated from municipal biowaste. Journal of Applied Polymer Science 132 (4), 41359.

Fu, F., Wang, Q., 2011. Removal of heavy metal ions from wastewaters: a review. Journal Environmental Management 92 (2011), 407–418.

Gondhalekar, S.C., Shukla, S.R., 2014. Equilibrium and kinetics study of uranium(VI) from aqueous solution by *Citrus limetta* peels. Journal of Radioanalytical and Nuclear Chemistry 302 (1), 451–457.

Güzel, F., Sayğili, H., Sayğili, G.A., Koyuncu, F., 2014. Elimination of anionic dye by using nanoporous carbon prepared from an industrial biowaste. Journal of Molecular Liquids 194, 130–140.

Haleem, N., Arshad, M., Shahid, M., Tahir, M.A., 2014. Synthesis of carboxymethyl cellulose from waste of cotton ginning industry. Carbohydrate Polymers 113, 249–255.

Hadad, H.R., Maine, M.A., Pinciroli Mufarrege, M.M., 2009. Nickel and phosphorous sorption efficiencies, tissue accumulation kinetics and morphological effects on Eichhornia crassipes. Ecotoxicology 18 (5), 504–513.

Kennedy, J.F., Roig, M.G., 2011. Biosorption of heavy metals from acid mine drainage onto biopolymers (chitin and α (1,3) β-D-glucan) from industrial biowaste exhausted brewer's yeasts (*Saccharomyces cerevisiae* L.). Biotechnology and Bioprocess Engineering 16 (6), 1262–1272.

Kim, N., Park, M., Park, D., 2015. A new efficient forest biowaste as biosorbent for removal of cationic heavy metals. Bioresource Technology 175, 629–632.

Kumar, P., Singh, M., Mehariya, S., Patel, S.K.S., Lee, J.-K., Kalia, V.C., 2014. Ecobiotechnological approach for exploiting the abilities of bacillus to produce co-polymer of polyhydroxyalkanoate. Indian Journal of Microbiology 54 (2), 151–157 in press.

Kumar, P.K.A., Suresh, P.V., 2014. Biodegradation of shrimp biowaste by marine *Exiguobacterium* sp. CFR26M and concomitant production of extracellular protease and antioxidant materials: production and process optimization by response surface methodology. Marine Biotechnology 16 (2), 202–218.

Kathi, S., 2016. Bioenergy from phytoremediated phytomass of aquatic plants via gasification. In: Prasad, M.N.V. (Ed.), Bioremediaiton and Bioeconomy. Elsevier, USA, pp. 111–128.

Lin, C.S.K., Pfaltzgraff, L.A., Herrero-Davila, L., Mubofu, E.B., Abderrahim, S., Clark, J.H., Koutinas, A.K., Kopsahelis, N., Stamatelatou, K., Dickson, F., Thankappan, S., Mohamed, Z., Brocklesby, R., Luque, R., 2013. Food waste as a valuable resource for the production of chemicals, materials and fuels. Current situation and global perspective. Energy and Environment Science 6, 426–464.

Lin, K.S., Wang, H.P., Chang, Ni.-B., Jou, C.J.G., Hsiao, M.C., 2003. Synthesis of ZSM-type zeolites from biowaste gasification ashes. Energy Sources 25 (6).

Maine, M.A., Duarte, M., Suné, N., 2001. Cadmium uptake by floating macrophytes. Water Res. 35, 2629–2634.

Maine, M.A., Suné, N., Lagger, S.C., 2004. Chromium bioaccumulation: comparison of the capacity of two floating aquatic macrophytes. Water Research 38, 1494–1501.

Martino, L., Cruz, M.V., Scoma, A., Freitas, F., Bertin, L., Scandola, M., Reis, M.A.M., 2014. Recovery of amorphous polyhydroxybutyrate granules from *Cupriavidus necator* cells grown on used cooking oil. International Journal of Biological Macromolecules 71, 117–123.

Messagie, M., Mertens, J., Oliveira, L., Rangaraju, S., Sanfelix, J., Coosemans, T., Van Mierlo, J., Macharis, C., 2014. The hourly life cycle carbon footprint of electricity generation in Belgium, bringing a temporal resolution in life cycle assessment. Applied Energy 134, 469–476.

Miretzky, P., Saralegui, A., Fernandez, A., 2004. Aquatic macrophytes potential for the simul-taneous removal of heavy metals (Buenos Aires, Argentina). Chemosphere 57, 997–1005.

Odjegba, V.J., Fasidi, I.O., 2007. Phytoremediation of heavy metals by Eichhornia crassipes Environmentalist, 27, pp. 349–355.

Patel, S.K.S., Kumar, P., Singh, M., Lee, J.-K., Kalia, V.C., 2015. Integrative approach to produce hydrogen and polyhydroxybutyrate from biowaste using defined bacterial cultures. Bioresource Technology 176, 136–141.

Prasad, M.N.V., 2015. Phytoremediation crops and biofuels. In: Lichtfouse, E. (Ed.), Sustainable Agriculture Reviews 17. © Springer International Publishing, Switzerland, pp. 159–261. http://dx.doi.org/10.1007/978-3-319-16742-8_7.

Romero-Güiza, Peces, M.S., Astals, M., Benavent, S., Valls, J., Mata-Alvarez, J., 2014. Implementation of a prototypal optical sorter as core of the new pre-treatment configuration of a mechanical-biological treatment plant treating OFMSW through anaerobic digestion. Applied Energy 135, 63–70.

San Miguel, G., Corona, B., 2014. Hybridizing concentrated solar power (CSP) with biogas and biomethane as an alternative to natural gas: analysis of environmental performance using LCA(2014). Renewable Energy 66, 580–587.

Sinha, S., Chand, S., Tripathi, P., 2014. Microbial degradation of chitin waste for production of chitosanase and food related bioactive compounds. Applied Biochemistry and Microbiology 50, 125–133.

Sukaimi, J., Hamzah, S., Mohd Ghazali, M.S., 2015. Green synthesis and characterization of hydroxyapatite from fish scale biowaste. Applied Mechanics and Materials 695, 235–238.

Suñe, N., Sánchez, G., Caffaratti, S., Maine, M., 2007. Cadmium and chromium removal kinetics from solution by two aquatic macrophytes. Environ. Pollut. 145, 467–473.

UNEP, 2009. The Environmental Crisis: The Environments Role in Averting Future Food Crisis.

Uusitalo, V., Havukainen, J., Manninen, K., Höhn, J., Lehtonen, E., Rasi, S., Soukka, R., Horttanainen, M., 2014a. Carbon footprint of selected biomass to biogas production chains and GHG reduction potential in transportation use. Renewable Energy 66, 90–98.

Uusitalo, V., Havukainen, J., Kapustina, V., Soukka, R., Horttanainen, M., 2014b. Greenhouse gas emissions of biomethane for transport: uncertainties and allocation methods. Energy Fuels 28, 1901–1910.

Vargas, A.K.N., Bianco Prevot, A., Montoneri, E., Le Roux, G.C., Savarino, P., Cavalli, R., Guardani, R., Tabasso, S., 2014. Use of biowaste-derived biosurfactants in production of emulsions for industrial use. Industrial and Engineering Chemistry Research 53, 8621–8629.

Van Wyk, J.P.H., 2011. Biowaste as a Resource for Bioproduct Development Environmental Earth Sciences, pp. 875–883.

Vymazal, J., Brix, H., Cooper, P.F., Green, M.B., Haberl, R., 1998. Constructed wetlands for wastewater treatment in Europe. Backhuys Publishers, Leiden, The Netherlands, 366 pp. Meppe. Gorteria 27: 133–135.

Wei, C.-C., Yu, C.-W., Yen, P.-L., Lin, H.Y., Chang, S.T., Hsu, F.L., Liao, V.H.C., 2014. Antioxidant activity, delayed aging, and reduced amyloid-β toxicity of methanol extracts of tea seed pomace from *Camellia tenuifolia*. Journal of Agricultural and Food Chemistry 62, 10701–10707.

Yan, T., Wang, P., Wang, L., 2015. Utilization of oxalic acid-modified spent mushroom substrate for removal of methylene blue from aqueous solution. Desalination and Water Treatment 55, 1007.

Zhang, B.Y., Zheng, J.S., Sharp, R.G., 2010. Phytoremediation in Engineered Wetlands. Mechanisms and Applications Procedia Environmental Sciences 2, 1315–1325.

Zhang, G., Ma, D., Peng, C., Liu, X., Xu, G., 2014. Process characteristics of hydrothermal treatment of antibiotic residue for solid biofuel. Chemical Engineering Journal 252, 230–238.

Zhu, X., Liu, Y., Luo, G., Qian, F., Zhang, S., Chen, J., 2014. Facile fabrication of magnetic carbon composites from hydrochar via simultaneous activation and magnetization for triclosan adsorption. Environmental Science and Technology 48, 5840–5848.

CHAPTER

2 DESTINATION OF VINASSE, A RESIDUE FROM ALCOHOL INDUSTRY: RESOURCE RECOVERY AND PREVENTION OF POLLUTION

E.N.V.M. Carrilho[1], G. Labuto[2], M.Y. Kamogawa[3]

Universidade Federal de São Carlos, Araras, SP, Brazil[1]; Universidade Federal de São Paulo, Diadema, SP, Brazil[2]; Universidade de São Paulo - ESALQ, Piracicaba, SP, Brazil[3]

CHAPTER OUTLINE

2.1 INTRODUCTION

The worldwide need for production to meet the needs of human consumption has driven the search for alternative methods of energy generation that satisfy the demands of the industrial sector. Among these alternatives, renewable fuels have gained prominence, including environmental appeals trumpeted by governments and productive sectors. As a result, ethanol demand has increased, mainly owing to the demand for energy sources that reduce the emission of greenhouse gases. The main biomass employed in the fermentation process is corn and sugarcane.

According to the Organization for Economic Cooperation and the Development Food and Agriculture Organization of the United Nations, the global trend of production growth and use of ethanol

Environmental Materials and Waste. http://dx.doi.org/10.1016/B978-0-12-803837-6.00002-0

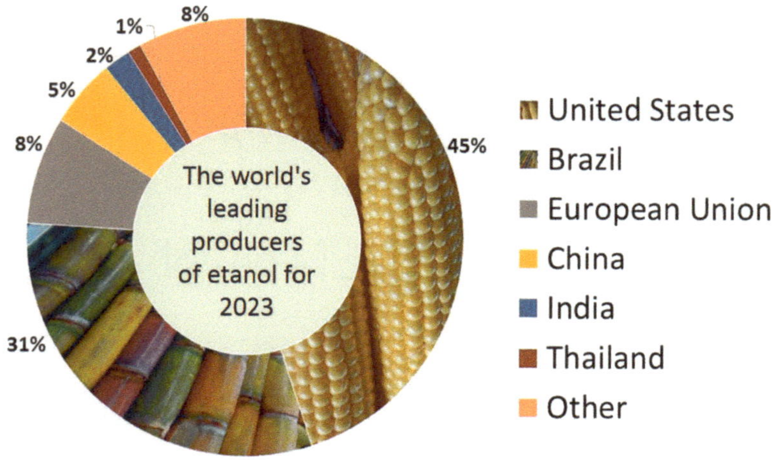

FIGURE 2.1

Projection of ethanol production among the largest producers scheduled for 2023.

Adapted from http://ethanolproducer.com/articles/11264/oecd-fao-outlook-report-predicts-growth-in-ethanol-production.

indicates a 158 billion—liter production forecast for 2023. Fig. 2.1 shows the projection of the contribution among the largest producers of ethanol scheduled for 2023.

Considering that each liter of ethanol produced generates around 9-14 L of vinasse, a byproduct of biomass distillation, it has been forecast that about 6 trillion liters of this material will need to be managed in 2023 (Silva et al., 2007, http://www.oecd.org/tad/agricultural-trade/38893266.pdf). Fig. 2.2 shows the vinasse production projection, based on the minimum ethanol produced by the world's largest ethanol industries.

The destination or disposal of vinasse has been one of the biggest challenges involved in the production of ethanol, because, as a result of its composition, it has a high oxygen demand for degradation. Vinasse composition is mostly 93% water and 7% solids, and it has high levels of salt and around 40% of organic matter, mainly composed of organic acids, which results in this material having a low pH (Freire and Cortez, 2000). As a result, the regular disposal of vinasse in soils and water bodies has caused environmental problems because this practice alters the chemical and physicochemical properties of these environmental entities.

Consequently, alternative uses for vinasse have been evaluated and developed, among which are the use of fertirrigation, compost production, and soil recovery as an addition in animal feed; and biodegradation by anaerobic and aerobic fermentation and as a means of cultivation for the production of unicellular protein (Larrahondo et al., 2000). Several technological possibilities have been the subject of research and development. During the 1980s, several promising paths were developed, including biodigestion, a biotechnological treatment of vinasse that produces biogas, recycling of vinasse in fermentation, and concentration by evaporation (Corazza and Salles-Filho, 2014).

The ethanol factory produces as its main waste vinasse and sugarcane bagasse, which currently have applications in fertirrigation and as biomass for thermal power generation, respectively. Fig. 2.3 shows a flowchart depicting the processing of sugar, first- and second-generation ethanol production and possible vinasse destinations.

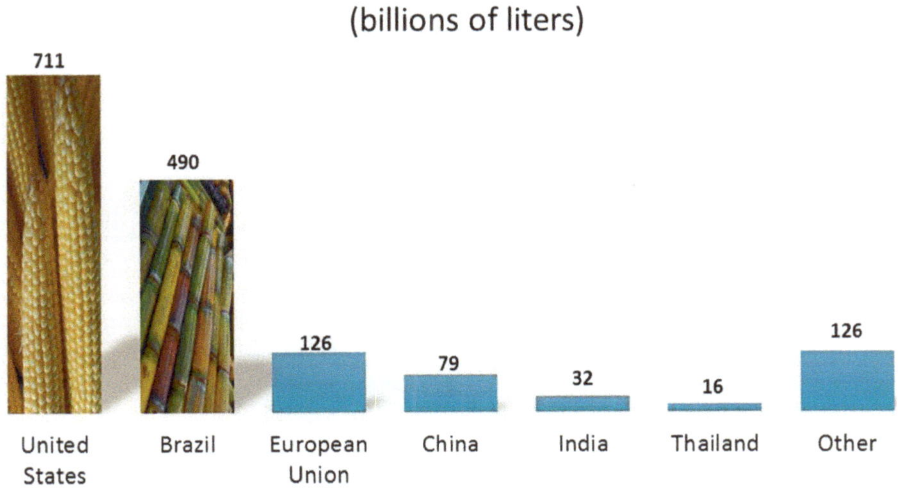

FIGURE 2.2

Projection of vinasse production for 2023 among the world's largest ethanol producers.

This chapter examines available data on the subject as a contribution to update the information on vinasse, its characteristics and chemical composition, environmental impacts, and alternative uses.

2.2 CHARACTERIZATION OF VINASSE

The composition of vinasse is mostly water (93%) and solids (7%). It has a sharp smell and dark and light brown color. Its acidic pH (pH 3.5–5) and highly corrosive nature make this effluent approximately 100 times more polluting than domestic sewage (Freire and Cortez, 2000; Ribeiro et al., 2007). All of these features make vinasse an environmentally worrisome effluent.

Vinasse is a rich source of salts and organic contents, ranging from 24,000 to 80,000 mg L^{-1} of minerals and 4000 to 64,000 mg L^{-1} of organic matter. It has high contents of potassium, calcium, magnesium, sulfate, total phosphorus, and nitrogen, and significant concentrations of cumulative heavy metals such as copper (2.2–37.8 mg kg^{-1}), lead (0.5–8.8 mg kg^{-1}), zinc (2.7–47.7 mg kg^{-1}), and other metals that are detected, such as As, Cd, and Hg. All of these can contaminate soil, groundwater, sediments, and superficial water (Chen and Cho, 1993; Fuess and Garcia, 2014).

Therefore, vinasse can be used as a fertilizer owing to its nutrient content, mainly calcium and potassium, and its high organic material content and dissolved solids (Prasad et al., 2008), which could be toxic or contaminating under certain conditions. Thus, vinasse can pollute soil and groundwater if disposed of in the environment. The content of dissolved organic matter in vinasse is high, varying from 10,973 to 14,801 mg L^{-1} of dissolved organic carbon (Christofoletti et al., 2013). This high organic content may cause problems in groundwater through an increase in nutrient supply (España-Gamboa et al., 2011).

Vinasses with a high concentration of soluble solids are obtained when sugarcane, sugar beet, sweet sorghum, grape, and agaves are used. The characteristics of vinasses depend mainly on the raw

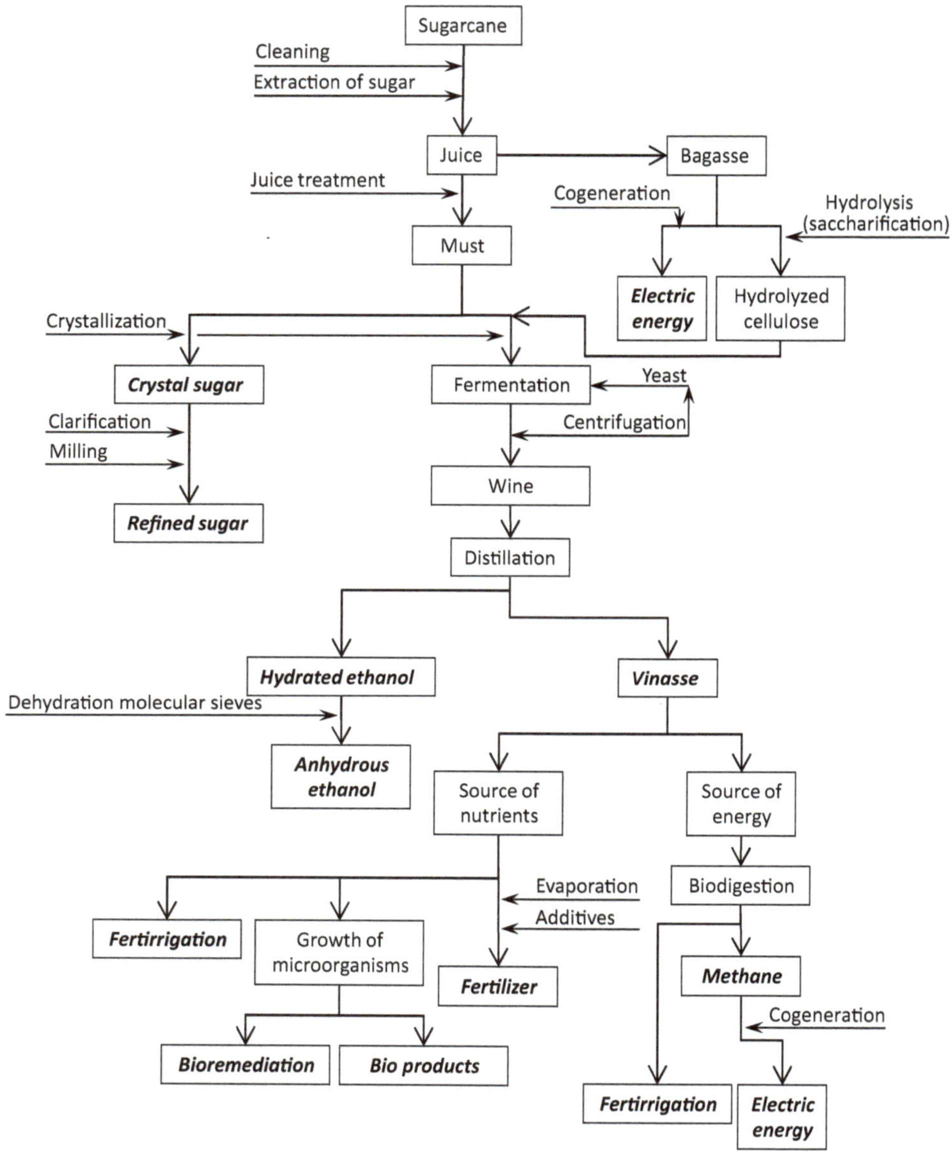

FIGURE 2.3

Simplified flowchart of first- and second-generation ethanol production and possible vinasse destination.

material used for bioethanol production. In the case of sugarcane, this includes molasses, the substrate used in the fermentation process, the type and efficiency of fermentation and distillation, and the variety and maturation of the sugarcane. From these, 9–14 L of vinasse can be obtained per liter of alcohol produced (España-Gamboa et al., 2011). This residue has high organic matter with elevated biochemical oxygen demand (BOD) and chemical oxygen demand (COD), which vary between 6–25 and 15–65 g L^{-1} O_2, respectively (Freire and Cortez, 2000; Ribeiro et al., 2007). These waste waters also contain phytotoxic, antibacterial, and recalcitrant compounds such as phenols, polyphenols, and heavy metals, which have negative effects on microorganisms and plants at disposal sites.

Highly colored compounds are also found in vinasse. This characteristic leads to reduced sunlight penetration in rivers and lakes, thereby reducing photosynthetic activity and dissolved oxygen concentrations and causing hazardous conditions for aquatic life (Prasad et al., 2008).

With regard to the concentration of solids in vinasse (Irisarri, 2006), this can vary according to the degree of dilution. Therefore, for diluted, semiconcentrated, concentrated, and solid vinasses, their amounts of total solids are 8%–10%, 20%–30%, 55%–60%, and 99%–99.9%, respectively. The composition of concentrated vinasse at 60 °Brix is 60%–65% dry matter, 16%–20% ashes, 4%–8% crude protein, 35%–42% carbohydrates, 5% sugars, and 4% potassium (Pérez and Garrido, 2006). In general, diluted vinasse contains 377.0 g kg^{-1} dry matter of organic carbon and 380.1 g kg^{-1} dry matter of total carbon, whereas concentrated vinasse is composed of 359.8 and 363.9 g kg^{-1} dry matter of organic and total carbon, respectively (Parnaudeaua et al., 2008). Concentrated vinasse is used as an alternative to overcome some of the problems encountered when diluted vinasse has to be handled. The use of concentrated vinasse as a fertilizer avoids contamination of water bodies and facilitates its transport over longer distances, resulting in a lower-cost alternative. In addition, water from diluted vinasse is an important resource which can be used for different applications in the sugarcane industry, as well as to irrigate crops.

The increase in sugarcane production and processing to obtain important products such as sugar and ethanol has the negative aspect of increasing industrial residues as well. Several physicochemical characteristics of vinasse have been studied (España-Gamboa et al., 2011; Pérez and Garrido, 2006; Robertiello, 1982; Scull et al., 2012). Vinasse characteristics such as density, viscosity, electric conductivity, boiling point, specific heat capacity, and heat of combustion have average values of 1.031 kg m^{-3}, 1.38 cP, 16.4 dS m^{-1}, 100.25°C, 0.934 cal g^{-1}°C^{-1}, and 3.39 cal g^{-1}, respectively (Pérez and Garrido, 2006). The low pH (average of 4.3), electric conductivity, and chemical elements present in sugarcane vinasse may cause changes in the chemical and physicochemical properties of soils and watercourses, with frequent discharges over a long time, and may also cause adverse effects on agricultural soils and biota in general. In view of this, studies and green methods need to be developed aimed at sugarcane vinasse recycling and proper disposal.

2.2.1 INORGANIC CONTENT OF VINASSE

Among the macronutrients, potassium is the element found in highest concentration in vinasse, followed by sulfur, magnesium, nitrogen, calcium, and phosphorus. With regard to the micronutrients, the highest concentration is iron, and then manganese, zinc, and copper at lower concentrations (CETESB, 2015). The potassium ion, a monovalent cation, has the characteristic of promoting the dispersion of clay particles, in the same way as sodium, causing a reduction in the pore space of the soil and, in its turn, reducing permeability (Corazza and Salles-Filho, 2014).

In general, hydrolyzed bagasse and vinasse contain interesting components, such as Ca, Mg, Na, K, and P. Ca is present, probably as a result of chemical processes that use $Ca(OH)_2$ or CaO as a reagent. The similar use of magnesium hydroxide and sodium hydroxide also explains the high contents of Mg and Na in the samples. Vinasse is valuable for application in agriculture because of the K and P contents. Fe, Ca, and K were found in all of the residues analyzed; the highest concentrations were in the filter cake and vinasse samples. Vinasse contains a considerable amount of S as a consequence of the chemical treatment, which explains its use as a fertilizer (Crespi et al., 2011).

Vinasse has been characterized by its qualitative and quantitative characteristics (Crespi et al., 2011; España-Gamboa et al., 2011; Prada et al., 1998; Robertiello, 1982; Scull et al., 2012). Among them, the mineral content is found in high amounts in this residue. Table 2.1 shows the qualitative and quantitative characteristics of sugarcane vinasse, molasses, juice, and molasses–juice mix. Vinasse resulting from the distillation of fermented molasses has an interesting elemental composition and contains all components of the must, which are entrained by water steam flow, as well as amounts of residual sugar and volatile components. In general, the constituents are soluble inorganic compounds, mainly K^+, Ca^{2+}, and SO_4^{2-}, dead yeast cells, organic compounds from the metabolic processes of contaminant yeast and microorganisms, alcohol and residual sugar, insoluble organic compounds, and volatile organic compounds (Ocampo and Cruz, 2005).

In Brazil, specific legislation controls the application of sugarcane residue as a fertilizer in agriculture, mainly regarding soil contamination by metals (CETESB, 2015). According to the Companhia Ambiental do Estado de São Paulo (CETESB), the Environmental Sanitation Technology Company in São Paulo, Brazil, for it to be applied to soil, vinasse must be characterized regarding the following parameters: pH, total nitrogen ($mg\,N\,L^{-1}$), nitrogen as nitrite ($mg\,N\,L^{-1}$), ammoniac nitrogen ($mg\,N\,L^{-1}$), Kjeldahl N, sodium ($mg\,Na\,L^{-1}$), calcium ($mg\,Ca\,L^{-1}$), potassium ($mg\,K\,L^{-1}$), sulfate ($mg\,SO_4^{2-}\,L^{-1}$), total phosphorus ($mg\,P\,L^{-1}$), and chlorine ($mg\,Cl^-\,L^{-1}$). Other parameters may be requested by CETESB. This characterization must be a result of at least two samplings at the location where vinasse is produced between the last crop harvested and the vinasse application plan. The methodologies of analysis for these parameters are described in CETESB Regulation P4. 231 of 2015.

2.2.2 ORGANIC CONTENT OF VINASSE

Organic components make up the major part of the must and are formed mainly by components present in the molasses that are not able to ferment as remaining sugars (glucose and fructose), products formed in the fermentation and that are not extracted in the distillation (glycerol and organic acids), and the remains of yeast that have not been separated in the industrial process.

Table 2.2 shows the main organic compounds present in vinasse. Identification and quantification of these substances were carried out employing solvent extraction, derivatization, and solid-phase microextraction (Larrahondo et al., 2000; Morales, 2000; D'Mello, 1995). Fifteen organic compounds were identified; those with the highest concentration were glycerol (2.7%), aconitic acid (1.8%), sorbitol (1.4%), lactic acid (1.3%), quinic acid (0.7%), beta-fructofuranose (0.5%), and alpha-glucopyranose (0.3%) (Larrahondo et al., 2000; Morales, 2000). In addition, by employing solid-phase microextraction, volatile compounds were identified as 2,3-butanodiol, furfuryl alcohol, benzaldehyde, and 3-metoxiacetofenona. Polysaccharides were also quantified as 3.38% $w\,w^{-1}$. It was also found that 2.3% of a nondialyzable fraction of vinasse was made of a mix of polymeric material and pigments of high molecular weight (over 12,000 Da).

A large range of organic compounds has also been found in vinasse: alcohols, aldehydes, ketone, esters, acids, and sugars. The presence of these compounds in vinasse comes from different origins (for example, molasses, the fermentation process of the alcohol) or is produced by thermal degradation in

Table 2.1 Qualitative and Quantitative Characteristics of Vinasse From Sugarcane, Molasses, Juice, and Molasses–Juice Mix

Parameter	Molasses	Juice	Molasses and Juice Mix
pH	4.2–5.0	3.7–4.6	4.4–4.6
Temperature (°C)	80–100	80–100	80–100
BOD (mg O_2 L^{-1})	25,000	6000–16,500	19,800
COD (mg O_2 L^{-1})	65,000	15,000–33,000	45,000
Total solids (mg L^{-1})	81,500	23,700	52,700
Volatile solids (mg L^{-1})	60,000	20,000	40,000
Nitrogen (mg N L^{-1})	450–1610	150–700	480–710
Phosphorus (mg P_2O_5 L^{-1})	100–290	10–210	9–200
Potassium (mg K_2O L^{-1})	3740–7830	1200–2100	3340–4600
Calcium (mg CaO L^{-1})	450–5180	130–1540	1330–4570
Magnesium (mg MgO L^{-1})	420–1520	200–490	580–700
Sulfate (mg SO_4^{2-} L^{-1})	6400	600–760	3700–3730
Carbon (mg C L^{-1})	11,200–22,900	5700–13,400	8700–12,100
Cupper (mg Cu L^{-1})	0.27–1.71	4	Na
Cadmium (mg Cd L^{-1})	0.04–1.36	Na	Na
Lead (mg Pb L^{-1})	0.02–0.48	Na	Na
Iron (mg Fe L^{-1})	12.8–157.5	16	Na
Proportion C/N	16–16.27	19.7–21.07	16.4–16.43
Organic matter (mg L^{-1})	63,400	19,500	3800

	Sugarcane Vinasse (mg L^{-1})
Hardness	2493
Zinc	1.66
Nickel	0.03
Molybdenum	0.008
Mercury	0.0019
Barium	0.41
Chromium	0.04
Sodium	52.2
Manganese (mg kg^{-1})	167

BOD, *biochemical oxygen demand;* COD, *chemical oxygen demand;* Na, *not available.*
Adapted from Crespi, M.S., Martins, Q.V., Almeida, S., Barud, H.S., Kobelnik, M., Ribeiro, C.A., 2011. Characterization and thermal behavior of residues from industrial sugarcane processing. Journal of Thermic Analytical Calorimetry 106:753–757; España-Gamboa, E., Mijangos-Cortes, J., Barahona-Perez, L., Dominguez-Maldonado, J., Hernandez-Zarate, G., Alzate-Gaviria L., 2011. Vinasses: characterization and treatments. Waste Management & Resource 29:1235–1250; Pérez, I., Garrido, N., 2006. Tratamiento de residuos: Aprovechamento integral de vinazas de destilaría, una revisión actual. Instituto Cubano de Investigaciones de los Derivados de la Caña de Azúcar. ICIDCA. Available in: http://www.inese.es/html/files/pdf/amb/iq/435/09ARTICULOAB.pdf; Prada et al., 1998; Robertiello, A., 1982. Upgrading of agricultural and agroindustrial wastes: the treatment of distillery effluents (vinasses) in Italy. Agriculture Wastes 4, 387–395.

Table 2.2 Organic Composition of Vinasse (64.8 °Brix)

Compound	Concentration
Alcohols	**$\% w\ w^{-1}$**
2,3-Butanediol	0.01
2-Methyl-1,3-butanediol	0.20
Glycerol	2.70
3-Methoxy-4-hydroxyphenyl glycerol	5.4
Sorbitol	1.40
Organic acids	**$\% w\ w^{-1}$**
Pyroglutamic acid	16.5
Itaconic acid	9.3
Fumaric acid	9.3
p-Hydroxybenzoic acid	4.8
Lactic acid	1.30
Succinic acid	0.07
Malic acid	0.23
Aspartic acid	0.05
Citric acid	0.80
2,4-Dihydroxypentanoic acid	0.71
Quinic acid	0.70
Aconitic acid	1.80
Sugars	**$\% w\ w^{-1}$**
Glucose + fructose	1.30
β-Fructofuranose	0.50
α-Glucopyranose	0.30
Trehalose	0.30
Saccharose	0.2
Amino acids	**$g\ kg^{-1}$**
Aspartic acid	76.9–88.5
Threonine	56.6–73.1
Serine	46.4–8.1
Glutamic acid	121.1–132.3
Proline	49.8–60.2
Glycine	40.8–36.1
Alanine	25–37.4
Cysteine	18.1
Valine	56.6–71.3

Table 2.2 Organic Composition of Vinasse (64.8 °Brix)—cont'd

Compound	Concentration
Methionine	nd−45.3
Isoleucine	57.6−67.9
Leucine	31−45.3
Tyrosine	49.9−62.3
Phenylalanine	47.3−50.9
Histidine	47.2−50.9
Lysine	9.28−9.96
Arginine	7.47−10.57

nd, *not determined.*
Adapted from Larrahondo, J.E., Morales, A.A., Victoria, M.H., Jaramillo, A., 2000. Compuestos orgánicos en vinaza. Carta Trimestral − CENICANA 22, 5−7; Morales, A.A., 2000. Extractión e identificación de compuestos orgánicos em vinaza. Universidad del Valle, Tesis Magister en Química Orgánica, Cali, 113 p.; D'Mello, J.P.F., 1995. Leguminous leaf meals in non − ruminant nutrition. In: D'Mello, J.P.F., Devendra, C. (Eds.), Tropical Legumes in Animal Nutrition. CAB International, p. 247.

the distillation process of the must obtained in the fermentation step. Among the identified compounds, those with a higher concentration were glycerol, lactic acid, and sorbitol. Because of their market value, it is important to make progress in technical-economic evaluations to study the route of extraction and the use of these and other vinasse organic compounds. There are also volatile compounds such as benzaldehyde, benzoic acid, ethyl palmitate, 2-acetyl pyrrole, alcohol 2-phenylethyl, furfuryl acid, acetic acid, formic acid, 1−3-butanediol, ketone, and ethanol.

Table 2.2 lists the profile of amino acids in vinasse. To evaluate the use of vinasse in animal feed, it is important to assess its amino acid content. Both samples had high values of essential amino acids (leucine, isoleucine, valine, tyrosine, histidine, phenylalanine, lysine, and threonine) and were comparable or superior to those of the US Food and Agriculture Organization's pattern for monogastric species (D'Mello, 1995). An aspect to highlight, and of great importance for these species, is the presence of methionine, because unconventional feeds either do not contain this or contain it only in small amounts, although sometimes an analysis is not conducted. The results allow this residue to be classified as being of good biological value, so it can be used in animal feed (Barros, 2010).

The most important residues that are created during processing are bagasse, hydrolyzed bagasse, filter cake, vinasse, and ash. Thermal gravimetric analysis, X-ray diffractometry (EDX), and colorimetric and Fourier transform infrared spectroscopy (FTIR) have been conducted and revealed that these residues have peculiar properties (Crespi et al., 2011). EDX and FTIR showed the presence of Si in the residues. Elemental composition was assessed by atomic absorption and atomic emission spectrometry, and the contents of Zn, Ni, Fe, Mn, Cu, Cr, Ca, Mg, Na, K, and P were determined. The identification (qualitative determination) of these elements was carried out by XRD. Functional organic groups were identified through FTIR.

To determine total flavonoids from sugarcane (*Saccharum officinarum*) leaves and vinasse, a modification of the UV-visible spectrometric procedure described by the French Pharmacopoeia for the determination of these compounds was proposed (Ferreira and Colombo, 2011). The modified method was validated and the precision and accuracy revealed a low relative standard deviation (<5.0%) and high recovery percentages (99.79% and 98.34%). The results were compared with those obtained by high-performance liquid chromatography (HPLC)-UV and demonstrated fair compatibility between the modified UV-visible and HPLC-UV methods.

2.3 APPLICATIONS OF VINASSE

2.3.1 THE USE OF VINASSE AS A RESOURCE

Its high amounts of elements make vinasse an attractive source of nutrients which can be applied in irrigation or used as a microbiological culture medium. The key focus of interest is the high concentration of potassium, calcium, magnesium, sulfur, nitrogen, and organic carbon. Potassium comes from sugarcane obtained during the sugar production. Calcium, magnesium, and sulfur are added to the sugarcane must treatment (pH adjustment), and nitrogen and organic materials are derived from sugarcane and the fermentation step.

Currently, the main destination of vinasse is its direct application to the soil, aimed at providing nutrients and water to crops such as sugarcane (Carvalho et al., 2013; Gómez and Rodríguez, 2000), corn (Basso et al., 2013), and soybean (Mascarenhas et al., 1994). However, because of limitations such as transport costs and environmental laws, alternative treatments have been applied, such as evaporation to concentrate the vinasse and the reuse of water (Salomon and Lora, 2009). The incorporation of additives or other ethanol waste, aimed at raising the nutrient content and/or improving the physical and chemical properties of the material (eg, reducing viscosity and odor) (Gurgel et al., 2015) has also been frequently used.

The high content of nutrients, organic matter, and water make vinasse an excellent reaction medium for the cultivation of microorganisms. This characteristic has been exploited for yeast protein production and fungi and bacteria cultivation, which produce chemical exudates (eg, organic acids, surfactants, or amino acids), or self-degradation in anaerobic reactors producing electricity by methane combustion. Applications of vinasse are discussed here as a nutrient source for plants and microorganisms, as well as for other purposes.

2.3.2 NUTRIENT SOURCE FOR PLANTS

Currently, the main destination for sugarcane vinasse produced in Brazil is fertirrigation, mainly owing to its cost-effectiveness. This use is justified because of the high water volume, nutrients, and organic matter present in vinasse.

Brazilian law regulates the application of sugarcane vinasse in accordance with the upper limit of potassium concentration, which may not exceed 5% of the cation exchange capacity (CEC) of the soil, at pH 7.0. Because of the pollution potential and the strong odor produced by the decomposition of vinasse, the application should be avoided in populated or watershed areas (CETESB, 2015).

According to the CETESB, the equation used to estimate the volume of vinasse, which can be applied to the soil, is described as:

$$m^3 \text{ vinasse } ha^{-1} = [(0.05 \times CEC - Ks) \times 3744 + 185]/Kvi$$

where 0.05 = 5% of CEC; CEC = cation exchange capacity, expressed in $cmol_c\ L^{-1}$, determined by the sum of calcium, magnesium, potassium (exchange resin), hydrogen, and aluminum (SMP buffer) concentrations; Ks = concentration of potassium in the soil, expressed in $cmol_c\ L^{-1}$, the depth of 0–0.80 m; 3744 = constant to transform the results of fertility analysis, expressed in $cmol_c\ L^{-1}$ or $mEq/100\ cm^3$ for kilograms of potassium in 1 hectare (soil volume, 8000 m^3); 185 = weight in kilograms of K_2O extracted by the crop per hectare per cut; Kvi = potassium concentrations in the vinasse, expressed in kilograms of K_2O/m^3 determined experimentally from the annual average.

The application volume varies; however, owing to the characteristics of vinasse in Brazilian soils, this volume seldom exceeds 300 m^3ha^{-1}.

An evaluation of 54 Brazilian distilleries showed that 82.5% of the vinasse produced was distributed by canals and 17.5% was delivered by trucks. The main form of application, around 82%, was by sprinkling using pneumatic pumps employing irrigation reels, or by hydraulic cannons fed by tanks in trucks or a canal system. Application using trucks, infiltration canals, irrigation systems with a central pivot, and drip are other forms of soil irrigation (Nunes et al., 2004). Fig. 2.4 shows these forms of sugarcane vinasse application to the soil and vinasse reservoirs around sugarcane crops and distilleries areas.

The nutrient concentration in vinasse is variable and is influenced mainly by parameters such as the origin of the materials (ie, sugars, starch, and cellulosic products), the sampling process, and the fermentation and distillation processes. The composition of vinasses from different raw materials, cane (juice and molasses), grape (wine), agave (tequila), sweet sorghum, and beet molasses, has high contents of nutrients and organic compounds (España-Gamboa et al., 2011). All of the raw materials have different concentrations, but potassium, sulfur, and organic carbon have the highest concentrations of all.

The main crop irrigated with vinasse in Brazil is sugarcane, because it surrounds the ethanol factory. However, fertilization of corn (Basso et al., 2013), soybeans (Mascarenhas et al., 1994), sunflowers (Ramos et al., 2008, 2009), castor beans, and peanuts (Ramos et al., 2008) with vinasse has also been reported. In general, the application of vinasse to soil has to take into account supplementation with potassium and nitrogen; it can fully replace mineral fertilizer (España-Gamboa et al., 2011). However, vinasse primarily contributes to the supply of water and works as a soil conditioner owing to its organic content. Overall, the application of vinasse produces chemical, physical, and microbiological changes in soil, mainly owing to the influence of organic matter, which improves the formation of soil aggregates, consequently improving the aeration, percolation, and retention of nutrients.

Areas with more than 50 years of vinasse application were compared with sites without application; fertilization with this residue showed significant improvements in the soil's chemical properties, favoring the formation of more polymerized alkaline-soluble humic substances (Canellas et al., 2003).

The influence of vinasse application was investigated in a long-term experiment in areas cultivated with sugarcane. These areas were treated with vinasse at different times, including 1, 2, 3, 4, 12, and 20 years, and were compared with a reference area that had no vinasse irrigation. For areas treated with

FIGURE 2.4

Vinasse storage around sugarcane crops and distilleries (A), (B), and (C), and forms of vinasse application to the soil (D), (E), and (F).

vinasse, each year corresponds to an application of 150 $m^3 h^{-1}$. The annual vinasse application increases the content of organic carbon and potassium, positively influencing the density, CEC, and total soil porosity, and the sugarcane crop yield had a direct relationship with the increase in potassium (Zolin et al., 2011).

Studies showed that vinasse application delays the emergence of seedlings and hampers the initial growth of sunflowers, with a reduction of around 24%–46% compared with soil without vinasse. However, this outcome does not affect the cultivars' development. The probable cause of the delay in emergence was the great volume of water in the soil irrigated with vinasse (Ramos et al., 2008, 2009).

The large volume of water is also reported as causing an increase in sugarcane's aging time. When available in abundance, water increases silicon absorption, reflecting a higher ash content and more difficult processing; and potassium uptake proportionally increases the aconitic acid concentration, requiring higher calcium hydroxide concentrations in the juice neutralization step, which can lead to flocculating yeast processes (Gutierrez et al., 1988).

Owing to the increase in vinasse transport costs and the need for water sources, the industry has sought an alternative to the treatment of vinasse at the factory, such as vinasse concentration strategies, a mixture of vinasse with other wastes from the alcohol industry, such as bagasse, filter cake, and fly ash (Gurgel et al., 2015; Salomon and Lora, 2009).

Several processes can be used to concentrate vinasse: flocculation followed by sedimentation and filtration, bioconcentration, nanofiltration, reverse osmosis, and evaporation by thermal processes (Gomes et al., 2011). Among these, the thermal process is most used, especially because the distillation and cogeneration plants can be integrated into the process of sugar and ethanol production. This process generates a vinasse 2–10 times more concentrated than the original residue. The main advantages of concentrating vinasse are the reduction in transport and application costs, lower water collection, minimized leaching and runoff of the vinasse, as well as a less problematic odor.

Economic costs expended on the evaporation of vinasse are minimized by the benefits provided by this process, such as: (1) a lower possibility of contamination of shallow groundwater; (2) ease of transport and application to areas with difficult access; and (3) lower water consumption. Vinasse concentration plants currently in operation employ a more efficient fermentation process that reduces the volume of vinasse produced, lowering it to 6 L of vinasse per 1 L of ethanol.

In Brazil, the application of concentrated vinasse has to follow the same recommendations for nonconcentrated vinasse, respecting the 5% limit of potassium in the CEC. It is usually applied with tank truck assistance with direct application to the crop area in average volumes not exceeding 30 $m^3 ha^{-1}$.

An alternative use of concentrated vinasse is the preparation of organic fertilizer combined with other wastes from the ethanol industry, such as ash/boiler dust and filter cake at an average of 6 and 35 $kg\, t^{-1}$ of sugarcane, respectively. The product formed from this mixture is ground and can be bagged, with the advantage of easy storage and transportation (Gurgel et al., 2015).

The mixture of vinasse with other wastes aims to improve the nutritional characteristics of the fertilizer by increasing the nitrogen and phosphorus content, which is often in low concentrations in the vinasse. Sewage sludge, gypsum, and mineral fertilizers, such as urea, monomethyl phosphate, and dimethyl phosphate are the most common combinations. An example of these applications is the development of an organic fertilizer named Biofom, produced by mixing concentrated vinasse, filter cake, boiler ash, and soot from chimneys and supplemented with mineral fertilizers. It was reported that there is no difference in corn production when employing Biofom or conventional mineral fertilizer (Gurgel et al., 2015).

Because of its acidity, vinasse is also used to increase the solubility of basalt rock powder. The material resulting from this treatment results in an increase in Ca^{2+}, Mg^{2+}, and K^{+} concentration, which give this material characteristics that make it potentially applicable to agricultural fertilization. This could provide an important opportunity to add value to wastes resulting from the ethanol and mining industries (Lopes et al., 2014).

Responsibly applied, taking into account environmental and socioeconomic aspects, the use of vinasse as a source of nutrients for crops is feasible and currently stands as one of the best alternatives for a cost-effective destination for this residue.

2.3.3 NUTRIENT SOURCE FOR MICROORGANISMS

An interesting employment of vinasse is its use as solution for microbiologic culture in bioprocesses or bioremediation. Among the vinasse characteristics that allow this application are the high volume available, its low cost, and its favorable chemical composition, which contains macronutrients and micronutrients, including a variety of organic acids such as malonic, formic, tartaric, malic, succinic, glutamic, acetic, lactic, and glutamic acid, various amino acids, and polymers of high molecular weight, which are responsible for the dark coloring called melanoidins.

The use of vinasse as a medium for growing microorganisms is shown in Table 2.3. Its applications may be related to the use of this effluent as a nutrient for the bioaugmentation of beneficial soil microorganisms, which, among other effects, may act as biological pesticides to control pathogenic organisms. Vinasse also grows microorganisms in biotechnological processes for the production of chemical compounds of industrial interest (Crivelaro et al., 2010) and diesel (Mariano et al., 2009).

An example of the use of vinasse as a solution for microorganism growth is the cultivation of *Bacillus subtilis*, which has a biological function in the control of nematodes. A culture medium prepared with 25% vinasse achieved a performance similar to conventional bases in aqueous suspension, promoted the growth of sugarcane, and reduced the reproduction of the nematode galls. However, the performance of the microorganism grown on vinasse provides no benefits in sugarcane development in soil contaminated by nematodes (Cardozo and Araújo, 2011).

In relation to biotechnological production, the production of citric acid from the sugarcane vinasse and bagasse was evaluated using solid-state cultivation incubated with *Aspergillus niger*. The bagasse was used as solid support for the microorganism growth and the vinasse provided the necessary humidity and nutrients for this process. The *A. niger* produces citric acid during their development, which can be used for industrial or chemical applications (Bastos et al., 2015).

Other interesting applications of vinasse as a nutrient source for microorganisms are the production of surfactants by *Bacillus pumilus*, using vinasse and waste frying oil (Oliveira and Garcia-Cruz, 2013); the production of laccase, peroxidase, and manganese-peroxidase by three strains of *Pleurotus* and *Trichoderma reesei* cultivated in media with pretreated bagasse (Aguiar et al., 2010); and microalgae cultivation (*Scenedesmus* sp), with potential for biodiesel production (Ramirez et al., 2014).

In Brazil, the burning of sugarcane to remove remaining straw after harvesting used to be a common practice. However, this management method is in the process of being banned and the straw left on the field has brought about some problems. It reduces the infiltration of fertilizers, lime, and pesticides and increases costs in an effort to incorporate the straw into the soil. Because of the high

Table 2.3 Use of Vinasse as a Nutrient Source for Microbial Growth

Microorganism	Application	Remarks	References
Soil microorganism	Oil and diesel biodegradation in soil	Use of 0.11–0.33 mL vinasse per gram dry soil; increase in soil microbial population with vinasse application, but not demonstrated adequate biodegradation	Crivelaro et al. (2010), Mariano et al. (2009)
Microalgae *Scenedesmus* sp	Use of vinasse as nutrient source for microalgae cultivation	Was possible to cultivate microalgae at concentrations of up to 40% of vinasse in culture medium	Ramirez et al. (2014)
Pleurotus and *Trichoderma reesei*	Use of vinasse and sugarcane bagasse for production of enzymes by lignocellulolytic fungi	High lignolytic activity; *T. reesei* produced laccase, peroxidase, and manganese-peroxidase	Aguiar et al. (2010)
Bacillus pumilus	Use of vinasse and waste frying oil to produce biosurfactant	Maximum production of crude biosurfactant was 27.7 and 5.7 g L^{-1} for vinasse and waste frying oil, respectively; alternative substrates could be used to produce efficient biosurfactant	Oliveira and Garcia-Cruz (2013)
Aspergillus niger	Use of vinasse and sugarcane bagasse as support for solid-state cultivation and citric acid production	Results indicate that citric acid production and glucose consumption depend on oxygen availability, which can be modulated by selection of bed height and air flow in packed-bed bioreactors	Bastos et al. (2015)

cellulose and lignin content, the sugarcane straw degradation time under the soil is long, and vinasse has been applied to it to stimulate microbial activity and consequently accelerate the degradation of organic matter. In this case, the vinasse acts as a stimulator of microorganism growth, which can collaborate with the improvement of soil quality (Santos et al., 2008). Nevertheless, studies have shown that the effect of vinasse on the degradation of sugarcane straw incorporated into the soil has no significant difference from the control test, treatment with urea, or mixtures of urea and vinasse (Gava et al., 2003; Oliveira et al., 1999). Currently, the straw is collected and pressed into bales, which are transported for use in energy production in the ethanol industry. The remaining straw on the field is incorporated into the soil before new planting takes place (Lombardi et al., 2012).

The chemical composition of vinasse also allows it to be exploited for methane production in anaerobic digesters. Different reactor configurations have been tested; however, in general, the combination of methane production in generation plants with cogeneration power plants has been advantageous, increasing the potential for electric/thermic generation; the effluent from digestion may also be used in fertirrigation (Salomon and Lora, 2009).

2.3.4 OTHER APPLICATIONS

Beyond those applications already mentioned in the search for new commercial destinations, vinasse has been evaluated in animal feed studies owing to its mineral composition, but mainly because of the presence of amino acids in its composition (Scull et al., 2012). Liquid vinasse has been used in the diet of rabbits for growth performance, at doses of up to 100 g kg^{-1}, and differences were observed in final or daily weights, and in the mortality as well as in the carcass yield. On the other hand, the inclusion of vinasse in feed affected the duodenum crypt and the ilium villus. As a result, the maximum concentration of vinasse recommended for rabbit diets is 87.8 g kg^{-1} (Oliveira et al., 2013).

Chickens fed a diet containing vinasse or a conventional formulation did not show significant differences between treatments at the end of 42 days. Chickens fed a vinasse diet showed an increase of around 17% in the weight of the comestible viscera; however, no changes were observed in the neck, gizzard, and heart. The researchers concluded that vinasse may be used as a supplement for chickens (Hidalgo et al., 2009).

High doses of vinasse are described as the cause of diarrhea in young pigs. However, at a dose of 2.5% vinasse in the feed, an increment of around 7 kg in weight was observed over a period of 26 days. This result shows the advantage of using vinasse in the feed diet for a more productive performance (Gonzalez et al., 2012).

2.4 ENVIRONMENTAL CONCERNS REGARDING VINASSE MANAGEMENT

Environmental concerns about vinasse have been the object of research and social reaction regarding regulating its disposal and applications, owing to the volume of vinasse produced by the fermentative process and its physicochemical characteristics. Therefore, regulatory frameworks worldwide have evolved over the years. The most detailed are those from the United States and Brazil, respectively, the first and second major producers of ethanol and consequently the highest generators of this byproduct.

In the United States, the Resource Conservation and Recovery Act regulated by the Environmental Protection Agency (EPA) deals with the disposal of hazardous wastes, in particular in Parts 260 to 273. Although many countries permit the use of vinasse in fertirrigation, the Renewable Fuel Standard Program (RFS2) Regulatory Impact Analysis (EPA EPA-420-R-10-006, 2010) states that this practice is not currently allowed in the United States. Nevertheless there are no specific restrictions on the use of vinasse as a fertilizer in the US Code of Federal Regulations. On the other hand, environmental legislation forbids the inappropriate disposal of vinasse into rivers, lakes, the ocean, and soil, owing to the high nutrient loading and the potential for polluting waterways (EPA EPA-420-R-10-006, 2010).

Even though in Brazil the disposal of vinasse in watercourses has been forbidden since 1978 (Portaria MINTER no. 323, from November 29, 1978), no Federal Act has been applied regarding the disposal of this residue. On the other hand, a Law Project has been filed which establishes rules for vinasse storage, transport, and soil application as a byproduct of the sugarcane industry (PL 5182/2013). Despite this, the National Water Resources Policy (Act No. 9.433/1997) and the National Solid Waste Policy (Act No. 12.305/2010) require the Brazilian states to establish their own laws and regulatory frameworks based on the national policies.

In this way, São Paulo state, the largest Brazilian producer of ethanol and therefore of vinasse, has a technical normative which defines the criteria and procedures for the storage, transportation, and application of vinasse in agricultural soil as well as the control of groundwater sampling monitoring (CETESB P 4231/2006). This normative reports all federal and state acts to be observed regarding storage, transportation, and application of vinasse.

Despite the regulatory frameworks and the potential of vinasse as a contaminant, the disposal of this residue in watercourses and soils has been the most common practice over the past 30 years. In the early 1980s, different destinations for vinasse were applied worldwide. In Brazil, vinasse used to be discharged into an adjacent waterway or land area and used for fertirrigation; in Australia, this residue was released in marine outfall, carried out to a conventional sewage or lagoon treatment, and or used in anaerobic digestion. On the other hand, in Japan, vinasse was incinerated to an ash to be used as fertilizer, and in the United States it was evaporated to be employed as animal and aquaculture feed (Willington and Marten, 1982).

Most recently, vinasse has been disposed of in sacrifice zones, which consist of ponds or swales (Marchi et al., 2006), soaked with untreated vinasse, and thus became unusable for any other purpose (Corazza, 2006). Besides this procedure, other actions such as transportation and even storage for cooling are potential causes of soil and water contamination.

It has been possible to identify the infiltration of vinasse from conducting channels using porous cup probes and groundwater reservoirs by monitoring their pH, electrical conductivity, BOD, and complete nitrogen series: N-Kjeldahl, ammoniacal-N, NO_3^-, and NO_2^- (Ludovice, 1997). Fig. 2.5 illustrates the main forms of the introduction of vinasse into water bodies.

Besides contamination resulting from storage, vinasse has been used in fertirrigation with direct application to the soil acting as a fertilizer owing to its high concentration of essential macronutrients and micronutrients. This procedure has as its main advantages economic agricultural inputs and clean water for irrigation (BNDES and CGEE, 2008; Christofoletti et al., 2013; Fuess and Garcia, 2014; Macedo, 2005; Pimentel et al., 2007; Sheehan and Greenfield, 1980; Smeets et al., 2008; Willington and Marten, 1982).

Nevertheless, these practices may have adverse effects both for the environment and for agriculture. The first impacts of the disposal of vinasse on soil are those of the microbiota that comprise the vinasse, leading to development of fungi and bacteria that work in the mineralization and immobilization of nitrogen and its nitrification, denitrification, and biological fixation, as well as microorganisms that participate in biogeochemical cycles of other chemical elements (Giachini and Ferraz, 2009).

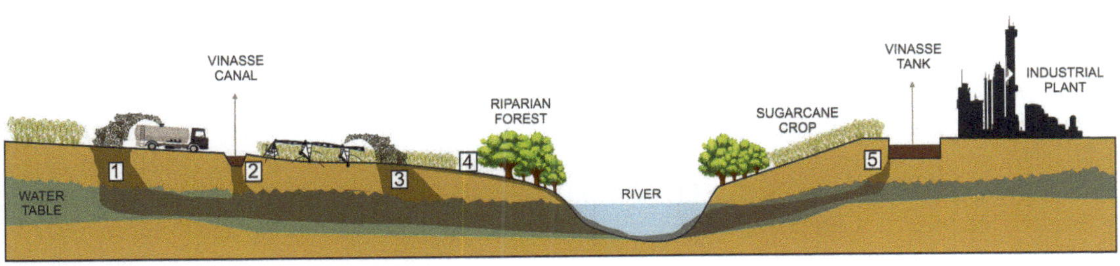

FIGURE 2.5

Main forms of vinasse introduction in water bodies: (1) and (3) Percolation from fertirrigation; (2) Percolation from a vinasse canal spill; (4) Runoff over soil; (5) Percolation from a vinasse tank.

In the first 10 days after vinasse application, the pH experiences a considerable reduction and then rises abruptly, reaching values greater than 7. This is a result of the action of microorganisms, favoring the development of bacteria responsible for the decomposition of the vinasse, which returns the soil to its original pH value before the application of vinasse (Rossetto, 1987; Silva and Ribeiro, 1998).

When the organic matter of vinasse is incorporated into the soil, colonization by fungi takes place, converting it into humus and neutralizing soil acidity (Silva et al., 2007). In this way, the process of microbiological decomposition of vinasse affects the physical and chemical characteristics of the soil.

With respect to the chemical contamination of soils, there are reports of increased concentrations of macronutrients such as K and P and metals such as Na, Fe, Zn, Mn, and C (Barros, 2010). The organic matter present in the vinasse gives the soil a lot of negative charge, which hinders the leaching of cations and increases the CEC (Mutton et al., 2014). In addition, the continued use of vinasse for fertirrigation can promote the saturation of cations in the soil (mainly K^+), resulting in, problems for groundwater (owing to the leaching of components), which may be contaminated, even with heavy metals.

Depending on the soil characteristics, the vinasse can affect its resistivity. The resistivity measurement of clay soils contaminated by vinasse has low resistivity values (between 10 and 90 Ω m), probably because of the influence of salts present in this effluent associated with water retention in the soil. These values are compared with those caused by contamination from manure, which is also conductive (Cruz et al., 2008).

Moreover, direct and continuous application of vinasse in watercourses and soil can lead to serious environmental problems (Fuess and Garcia, 2014; Mohana et al., 2009; Pant and Adholeya, 2007). The contamination of groundwater will depend on factors such as the soil composition and permeability/compression, which influence the velocity of the contaminant's transit into the aquifer; the shallowness of the water table; the relief of the region; rainfall and/or irrigation; soil pH; the amount of soil organic matter; and the composition of vinasse (Bedient et al., 2000; Saracino and Phipps, 2002).

Fertirrigation areas in Brazil, monitored for 15 years, denote the maintenance of soil fertility and show an increase of NO_3^- in groundwater (Cruz et al., 1992). In addition, pH is reported as being the main factor of fish toxicity when vinasse is disposed of in watercourses, including the development of liver disease in tilapia (Botelho et al., 2012; Ferreira et al., 2011; Marinho et al., 2014). In Pernambuco, Brazil, the contamination of groundwater and of the Ipojuca River by a vinasse storage lagoon was indicated by changes in electrical conductivity, BOD, COD, concentrations of total dissolved solids, Cl^-, K^+, Ca^{2+}, and Mg^{2+} (Rolim et al., 2013).

Some contaminants introduced into the environment by vinasse are of concern to human and animal health. As an example, NO_3^- ions are one of the main components of vinasse (around 150–1600 mg L^{-1}) and are soluble in water and not promptly adsorbed in the soil, which contributes to their transport to groundwater. The EPA's Safe Drinking Water Interim Health Standard established 10 mg L^{-1} of NO_3^- for nitrogen, which equals 45 mg L^{-1} of NO_3^- (Saracino and Phipps, 2002).

The contamination by NO_3^- of water for human consumption can lead to serious health problems, and the intake of NO_3^- by children causes a drop in oxygen levels in the blood, leading to a potentially fatal condition known as methemoglobinemia or "blue baby" disorder (Spalding and Exner, 1993). The authors also found the contamination of surface and groundwater by phosphate and nitrate, which has impacts on human and animal health and promoting interference in the development of plants and environmental eutrophication (Meurer et al., 2000; Stevenson, 1986).

The impacts of practicing fertirrigation may affect other components of the environment. For instance, a study of the emission of greenhouse gases, especially N_2O, CH_4, and CO_2 from stored vinasse and from its application to soil, created an increase of 15% in greenhouse gas emissions in the agricultural phase of the ethanol production process. From the emitted gases, 99.8% was CH_4, and when applied to the soil, vinasse potentiated N_2O emissions and increased CO_2 emissions in 46 and 31 kg/hectare in areas of burned and raw sugarcane, respectively (Oliveira et al., 2015).

The disposal of vinasse can also cause population explosions of insects. An example is the *Stomoxys calcitrans* outbreaks, a hematophagous fly that attacks animals and humans and can transmit diseases in sugarcane-producing regions of Brazil (Correa et al., 2013). The influence of different sugarcane sub-products, bagasse, straw, straw with vinasse, and filter cake, on increasing this bug population was smaller on straw with vinasse than that observed for flies on the filter cake. Furthermore, the large area covered by filter cake would be potentially capable of producing around 24 million *S. calcitrans*/month in only 12.23% of the cultivated area where vinasse was applied monthly.

Table 2.4 shows the main reported environmental impacts promoted by using fertirrigation vinasse in soil.

Table 2.4 Environmental Impacts of Vinasse Use in Fertirrigation

- Commitment of soil structure and water bodies owing to high acidity and concentration of sulfates with consequent reduction in crop yields
- Soil salinization
- Increased salt concentrations in groundwater
- Increase in soil instability
- Overload of organic substances in soil, which may lead to clogging of pores of the soil and reduction in microbial activity caused by to low concentrations of dissolved oxygen, generating unpleasant odors as result of consumption of organic matter by anaerobic digestion
- Hyper-soil fertilization that can lead to the destabilization of soil structure and eutrophication of water bodies
- Contamination by abundant ions in the composition of vinasse such as Cl^- and NO_3^- and heavy metals such as copper, lead, and zinc
- Permanent acidification of water bodies
- Interference in process of photosynthesis by aquatic plants and microalgae owing to high color and turbidity of vinasse
- Inhibition of seed germination with consequent reduction in crop yields
- Increase in soil resistivity
- Moderate increase in population of *S. calcitrans* (fly that attacks animals)
- Gas emissions that cause the greenhouse effect

Adapted from Fuess, L.T., Garcia, M.L., 2014. Implications of stillage land disposal: a critical review on the impacts of fertirrigation. Journal of Environmental Management 145, 210–229; Mohana, S., Acharya, B.K., Madamwar, D., 2009. Distillery spent wash: treatment technologies and potential applications. Journal of Hazardous Materials 163, 12–25; Pant, D., Adholeya, A., 2007. Biological approaches for treatment of distillery wastewater: a review. Bioresource Technology 98, 2321–2334.

ACKNOWLEDGMENTS

The authors are thankful to undergraduate Matheus Gomes Nirschi, from Universidade de São Paulo - ESALQ, for providing the pictures taken at Buriti distillery in 2015, to be used in Fig. 2.4. Thanks are also due to Dr. Décio Luis Semensatto Junior for his drawing in Fig. 2.5.

REFERENCES

Aguiar, M.M., Ferreira, L.F.R., Monteiro, R.T.R., 2010. Use of vinasse and sugarcane bagasse for the production of enzymes by lignocellulolytic fungi. Brazilian Archives of Biology and Technology 53, 1245–1254.

Barros, P.V., 2010. Evaluación de un subproducto de destilería de alcohol (vinaza) como aditivo en la alimentación de pollos de engorde (Graduate thesis). Instituto de Ciencia Animal, Mayabeque, Cuba, p. 10.

Basso, C.J., Lamego, A.L.S.F.P., Somavilla, L., Brigo, T.J., 2013. Vinasse as a source of potassium: black oat/corn silage/short-season corn succession response and soil chemical alterations in the Northwest region of the state of Rio Grande do Sul. Ciência Rural 43, 596–602.

Bastos, R.G., Morais, D.V., Volpi, M.P.C., 2015. Influence of solid moisture and bed height on cultivation of *Aspergillus niger* from sugarcane bagasse with vinasse. Brazilian Journal of Chemical Engineering 32, 377–384.

Bedient, P., Rifai, H., Newell, C., 2000. Ground Water Contamination Transport and Remediation, second ed. Hardback, 604 p.

BNDES e CGEE, 2008. Sugarcane-Based Bioethanol: Energy for Sustainable Development accessed in October 2015, Available in: http://www.cgee.org.br/publicacoes/conferencia_bioetanol.php.

Botelho, R.G., Tornisielo, V.L., Olinda, R.A., Maranho, L.A., Machado-Neto, L., 2012. Acute toxicity of sugarcane vinasse to aquatic organisms before and after pH adjustment. Toxicological & Environmental Chemistry 94, 2035–2045.

Canellas, L.P., Velloso, A.C.X., Marciano, C.R., Ramalho, J.F.G.P., Rumjanek, V.M., Rezende, C.E., Santos, G.A., 2003. Propriedades químicas de um cambissolo cultivado com cana-de-açúcar, com preservação do palhiço e adição de vinhaça por longo tempo. Revista Brasileira de Ciência Solo 27, 935–944.

Cardozo, R.B., Araújo, F.F., 2011. Multiplicação de *Bacillus subtilis* em vinhaça e viabilidade no controle da meloidoginose, em cana-de-açúcar. Revista Brasileira de Engenharia Agrícola e Ambiental 15, 1283–1288.

Carvalho, J.M., Andreotti, M., Buzetti, S., Carvalho, M.P., 2013. Produtividade de cana soca sem queima em função do uso de gesso e vinhaça. Pesquisa Agropecuária Tropical 43, 1–9.

CETESB, 2015. Stillage – Criteria and Procedures for Agricultural Soil Application. Norma Técnica P4.231, third ed. 15 p., accessed in October 2014. http://www.cetesb.sp.gov.br/wp-content/uploads/sites/11/2014/12/DD-045-2015-C.pdf.

Chen, J.C.P., Cho, C.C., 1993. Cane Sugar Handbook: A Manual for Cane Sugar Manufactures and Their Chemists, twelfth ed. John Wiley & Sons, United States of America, New York. 436 pp.

Christofoletti, C.A., Escher, J.P., Correia, J.E., Marinho, J.F.U., Fontanetti, C.S., 2013. Sugarcane vinasse: environmental implications of its use. Waste Management 33, 2752–2761.

Corazza, R.I., 2006. Impactos ambientais da vinhaça: controvérsias científicas e lock-in na fertirrigação. In: XLIV Congresso da Sociedade Brasileira de Economia e Sociologia Rural. Sessão em debate. Fortaleza, Ceará.

Corazza, R.I., Salles-Filho, S., 2014. Vinasse. Treatment in Brazil from 1970s to 1990s: a lock-in case study into the ethanol agroindustry. Análise Econômica 62, 177–198.

Corrêa, E.C., Ribas, A.C.A., Campos, J., Barros, A.T.M., 2013. Abundância de *Stomoxys calcitrans* (Diptera: *Muscidae*) em diferentes subprodutos canavieiros. Pesquisa Veterinária Brasileira 33, 1303–1308.

Crespi, M.S., Martins, Q.V., Almeida, S., Barud, H.S., Kobelnik, M., Ribeiro, C.A., 2011. Characterization and thermal behavior of residues from industrial sugarcane processing. Journal of Thermic Analytical Calorimetry 106, 753–757.

Crivelaro, S.H.R., Mariano, A.P., Furlan, L.T., Gonçalves, R.A., Seabra, P.N., Angelis, D.F., 2010. Evaluation of the use of vinasse as a biostimulation agent for the biodegradation of oily sludge in soil. Brazilian Archives of Biology and Technology 53, 1217–1224.

Cruz, J.I., Portugal, R.S., Lucendo, M.C.H., Elis, V.R., Fachin, S.J.S., Ustra, A.T., Borges, W.R., 2008. Detecção de contaminação de solo por vinhaça através de análise de dados de eletrorresistividade. Revista Brasileira de Geofísica 26, 481–492.

Cruz, R.L., Righetto, A.M., Nogueira, M.A., 1992. Experimental Investigation of soil and ground water impacts caused by vinasse disposal. Water Science & Technology 24, 77–85.

D'Mello, J.P.F., 1995. Leguminous leaf meals in non – ruminant nutrition. In: D'Mello, J.P.F., Devendra, C. (Eds.), Tropical Legumes in Animal Nutrition. CAB International, p. 247.

España-Gamboa, E., Mijangos-Cortes, J., Barahona-Perez, L., Dominguez-Maldonado, J., Hernandez-Zarate, G., Alzate-Gaviria, L., 2011. Vinasses: characterization and treatments. Waste Management & Research 29, 1235–1250.

Ferreira, C.H.A., Colombo, R., 2011. Validation of an analytical method and spectrometric determination of flavonoids from sugarcane leaves and vinasse and comparison with an HPLC-UV method. Química Nova 34, 1651–1655.

Ferreira, L.F.R., Aguiar, M.M., Messias, T.;G., Pompeu, G.B., Lopez, A.M.Q., Silva, D.P., Monteiro, R.T., 2011. Evaluation of sugar-cane vinasse treated with *Pleurotus sajor-caju* utilizing aquatic organisms as toxicological indicators. Ecotoxicology and Environmental Safety 74, 132–137.

Freire, W.J., Cortez, L.A.B., 2000. Contaminação de águas subterrâneas por pesticidas: avaliação preliminar. In: Vinhaça de cana-de-açúcar (Ed.), Agropecuária, Guaíba-Brazil, 203 p.

Fuess, L.T., Garcia, M.L., 2014. Implications of stillage land disposal: a critical review on the impacts of fertirrigation. Journal of Environmental Management 145, 210–229.

Gava, G.J.C., Trivelin, P.C.O., Vitti, A.C., Oliveira, M.W., 2003. Recovery of nitrogen (^{15}N) from urea and cane trash by sugar cane ratoon (*Saccharum* spp.). Revista Brasileira de Ciência do Solo 27, 621–630.

Giachini, C.F., Ferraz, M.V., 2009. Benefícios da utilização de vinhaça em terras de plantio de cana-de-açúcar-revisão de literatura. Revista Científica Eletrônica de Agronomia 15, 1–5.

Gomes, M.T.M.S., Eça, K.S., Viotto, L.A., 2011. Vinasse concentration by microfiltration followed by nanofiltration with membrane. Pesquisa Agropecuária Brasileira 46, 633–638.

Gómez, J., Rodríguez, O., 2000. Effects of vinasse on sugarcane (*Saccharum officinarum*) productivity. Revista Faculdade de Agronomia (LUZ) 17, 318–326.

Gurgel, M.N.A., Correa, S.T.R., Dourado-Neto, D., Paula-Júnior, D.R., 2015. Technology for sugarcane agro-industry waste reuse as granulated organomineral fertilizer. Engenheria Agrícola 35, 63–75.

Gutierrez, L.E., Ferrari, S.E., Gerald, L.T.S., Orelli Jr., A.A., 1988. Efeito da aplicação da vinhaça como fertilizante sobre os teores de ácido Trans-aconítico de caldo de cana-de-açúcar (*Saccharum* spp). Anais da Escola Superior de Agricultura Luiz de Queiroz 45, 453–462.

González, Y.M., Borlet, M.G., Rodríguez, Z., Salabarría, R.B., Barbarú, A.D., Mederos, M., Noviembre, 2012. La vinaza, como aditivo, en la alimentación de cerdos en la fase crecimiento -destete bajo condiciones de producción. In: I Taller Internacional de Extensión, Innovación y Transferencias de Tecnologías para La Producción Animal. Mayabeque, Cuba.

Hidalgo, K., Rodríguez, B., Valdivié, M., Febles, M., 2009. Utilización de la vinaza de destilería como aditivo para polos en ceba. Revista Cubana de Ciencia Agrícola 43, 281–284.

Irisarri, D., 2006. Usos industriales y agrícolas de la vinaza de caña de azúcar. Revista Tecnicaña 17, 20–25.

Larrahondo, J.E., Morales, A.A., Victoria, M.H., Jaramillo, A., 2000. Compuestos orgánicos en vinaza. Carta Trimestral – CENICANA 22, 5–7.

Lombardi, G.M.R., Giroto, V.S., Lombardi, N.M.R., Peres, M.M., Anjos, S.D.S., Alves, C.E.S., Abílio, R.S., 2012. Uso da palha de cana de açúcar como fonte de bioenergia versus a sua contribuição nutricional quando mantida no solo. In: 7° Congresso internacional de Bioenergia São Paulo – SP.

Lopes, O.M.M., Carrilho, E.N.V.M., Lopes-Assad, M.L.R.C., 2014. Effect of rock powder and vinasse on two types of soils. Revista Brasileira de Ciência do Solo 38, 1547–1557.

Ludovice, M.T.F., 1997. Estudo do efeito poluente da vinhaça infiltrada em canal condutor de terra sobre o lençol freático (Masters dissertation). School of Civil Engineering – Unicamp, 117 p.

Macedo, I.C. (coord.), 2005. Sugar Cane's Energy – Twelve Studies on Brazilian Sugar Cane Agribusiness and its Sustainability. Única, São Paulo.

Marchi, G., Guilherme, L.R.G., Chang, A.C., Curi, N., Guerreiro, M.C., 2006. Changes in isoelectric point as affected by anion adsorption on two Brazilian oxisols. Communications in Soil Science and Plant Analysis 37, 1357–1366.

Mariano, A.P., Crivelaro, S.H.R., Angelis, D.F., Bonotto, D.M., 2009. The use of vinasse as an amendment to ex-situ bioremediation of soil and groundwater contaminated with diesel oil. Brazilian Archives of Biology and Technology 52, 1043–1055.

Marinho, J.F., Correia, J.E., Marcato, A.C., Pedro-Escher, J., Fontanetti, C.S., 2014. Sugar cane vinasse in water bodies: impact assessed by liver histopathology in tilapia. Ecotoxicology and Environmental Safety 110, 239–245.

Mascarenhas, H.A.A., Tanaka, R.T., Pereira, J.C.V.N.A., Gallo, P.B., Bataglia, O.C., 1994. Efeito de adubos potássicos na Produção de soja. Scientia Agricola 51, 82–89.

Meurer, E.J., Bissani, C.A., Selbach, P.A., 2000. Poluentes do solo e do ambiente. In: Meurer, E.J. (Ed.), Fundamentos de química do solo. Genesis 1, Porto Alegre, pp. 151–168.

Mohana, S., Acharya, B.K., Madamwar, D., 2009. Distillery spent wash: treatment technologies and potential applications. Journal of Hazardous Materials 163, 12–25.

Morales, A.A., 2000. Extractión e identificación de compuestos orgánicos em vinaza. Universidad del Valle, Tesis Magister en Química Orgánica, Cali, 113 p.

Mutton, M.A., Rossetto, R., Mutton, M.J.R., 2014. Agricultural use of stillage. In: Cortez, L.A.B. (Ed.), Sugarcane Bioethanol — R&D for Productivity and Sustainability. Edgard Blücher, São Paulo, pp. 423–440. Available in: http://dx.doi.org/10.5151/BlucherOA-Sugarcane-SUGARCANEBIOETHANOL_40.

Nunes Jr., D., Pinto, R.S., de, A., Trento Filho, E., Elias, A.I., 2004. Indicadores Agrícolas Do Setor Sucroalcooleiro Safra 2003/2004. Grupo IDEA, Ribeirão Preto, 111 p.

Ocampo, A.G., Cruz, C.A.R., 2005. Posibilidades de Uso de la Vinaza en la Agricultura de Acuerdo con su Modo de Acción en los Suelos. Revista Tecnicaña 17, 3–13.

Oliveira, J.G., Garcia-Cruz, C.H., 2013. Properties of a Biosurfactant produced by *Bacillus pumilus* using vinasse and waste frying oil as alternative carbon sources. Brazilian Archives of Biology and Technology 56, 155–160.

Oliveira, B.G., Carvalho, J.L.N., Cerri, C.E.P., Cerri, C.C., Feigl, B.J., 2015. Greenhouse gas emissions from sugarcane vinasse transportation by open channel: a case study in Brazil. Journal of Cleaner Production 94, 102–107.

Oliveira, M.C., Silva, D.M., Faleiros, C.A., Carvalho, R., Alves, M.F., Dias, D.M.B., Martins, P.C., Bonifácio, N.P., Souza Júnior, M.A.P., 2013. Effect of including liquid vinasse in the diet of rabbits on growth performance. Revista Brasileira Zootecnia 42, 259–263.

Oliveira, M.W., Trivelin, P.C.O., Gava, G.J.C., Penatti, C.P., 1999. Sugarcane trash degradation. Scientia Agricola 56 (4). Available in: http://www.scielo.br/scielo.php?=sci_arttext&pid=S0103–90161999000400006.

Pant, D., Adholeya, A., 2007. Biological approaches for treatment of distillery wastewater: a review. Bioresource Technology 98, 2321–2334.

Parnaudeaua, V., Condom, R., Oliver, R., Cazevieilleb, P., Recousa, S., 2008. Vinasse organic matter quality and mineralization potential, as influenced by raw material, fermentation and concentration processes. Bioresource Technology 99, 1553–1562.

Pérez, I., Garrido, N., 2006. Tratamiento de residuos: Aprovechamento integral de vinazas de destilaría, una revisión actual. Instituto Cubano de Investigaciones de los Derivados de la Caña de Azúcar. ICIDCA. Available in: http://www.inese.es/html/files/pdf/amb/iq/435/09ARTICULOAB.pdf.

Pimentel, D., Patzek, T.W., Cecil, G., 2007. Ethanol production: energy, economic, and environmental losses. Reviews of Environmental Contamination and Toxicology 189, 25–41.

Prada, S.M., Guekezian, M., Suarez-Iha, M.E.V., 1998. Metodologia analítica para a determinação de sulfato em vinhoto. Química Nova 21, 249–252.

Prasad, K.R., Kumar, R.R., Srivastava, S.N., 2008. Design of optimum response surface experiments for electro-coagulation of distillery spent wash. Water, Air and Soil Pollution 191, 5–13.

Ramirez, N.N.V., Farenzena, M., Trierweiler, J.O., 2014. Growth of microalgae *Scenedesmus* sp in ethanol vinasse. Brazilian Archives of Biology and Technology 57, 630–635.

Ramos, N.P., Novo, M.C.S.S., Ungaro, M.R.G., Lago, A.A., Marin, G.C., 2008. Oilseed crops initial development as influenced by sugarcane vinasse soil application. Bragantia 67, 685–692.

Ramos, N.P., Novo, M.C.S.S., Lago, A.A., Ungaro, M.R.G., 2009. Sunflower – emergence and initial plant growth under sugarcane residues. Ciência Rural 39, 45–51.

Ribeiro, M.L., Lourencetti, C., Pereira, S.Y., Marchi, M.R.R., 2007. Contaminação de águas subterrâneas por pesticidas: Avaliação preliminar. Química Nova 30, 688–694.

Robertiello, A., 1982. Upgrading of agricultural and agroindustrial wastes: the treatment of distillery effluents (vinasses) in Italy. Agriculture Wastes 4, 387–395.

Rolim, M.M., Lyra, M.R.C.C., Duarte, A.S., Medeiros, P.R.F., Silva, Ê.F.F., Pedrosa, E.M.R., 2013. Influência de uma lagoa de distribuição de vinhaça na qualidade da água freática. Revista Ambiente & Água – An Interdisciplinary Journal of Applied Science 8, 155–171.

Rossetto, A.J., 1987. Utilização agronômica dos subprodutos e resíduos da indústria açucareira e alcooleira. In: Paranhos, S.B. (Ed.), Cana-de-açúcar: cultivo e utilização, 2. Fundação Cargill, Campinas, pp. 435–504.

Salomon, K.R., Lora, E.E.S., 2009. Estimate of the electric energy generating potential for different sources of biogas in Brazil. Biomass and Bioenergy 33, 1101–1107.

Santos, M., Diánez, F., de Cara, M., Tello, J.C., 2008. Possibilities of the use of vinasses in the control of fungi phytopathogens. Bioresource Technology 99, 9040–9043.

Saracino, A., Phipps, H., 2002. Groundwater Contaminates and Contaminant Sources Sacramento Califórnia accessed October 2015, Available in: http://groundwater.ucdavis.edu/files/136257.pdf.

Scull, I., Savón, L., Gutiérrez, O., Valiño, E., Orta, I., Mora, P.O., Ortal, H., Ramos, Y., Molineda, A., Coto, G., Noda, A., 2012. Physic-chemical composition of concentrated vinasse for their assessment in animal diets. Cuban Journal of Agricultural Science 46, 385–390.

Sheehan, G.J., Greenfield, P.F., 1980. Utilization, treatment and disposal of distillery wastewater. Water Research 14, 257–277.

Silva, A.J.N., Ribeiro, M.R., 1998. Caracterização de um Latossolo Amarelo sob cultivo contínuo de cana-de-açúcar no Estado de Alagoas: propriedades químicas. Revista Brasileira de Ciência do Solo 22, 291–299.

Silva, M.A.S., Griebeler, N.P., Borges, L.C., 2007. Uso de vinhaça e impactos nas propriedades do solo e lençol freático. Revista Brasileira de Engenharia Agrícola e Ambiental 11, 108–114.

Smeets, E., Junginger, M., Faaij, A., Walter, A., Dolzan, P., Turkenburg, W., 2008. The sustainability of Brazilian ethanol: an assessment of the possibilities of certified production. Biomass Bioenergy 32, 781–813.

Spalding, R.F., Exner, M.E., 1993. Occurrence of nitrate in groundwater – a review. Journal of Environmental Quality 22, 392–402.

Stevenson, F.J., 1986. Cycles of sil-carbon, nitrogen, phosphorus, sulfur, micronutrientes. In: Tan, K.H. (Ed.), Principles of Soil Chemistry, second ed. Marcell Dekker, New York. 362 p.

Willington, I.P., Marten, G.G., 1982. Options for handling stillage waste from sugar-based fuel ethanol production. Resources and Conservation 8, 111–129.

Zolin, C.A., Paulino, J., Bertonha, A., Freitas, P.S.L., Folegatti, M.V., 2011. Estudo exploratório do uso da vinhaça ao longo do tempo. I. Características do solo. Revista Brasileira de Engenharia Agrícola e Ambiental 15, 22–28.

CHAPTER

3

BIOSOLIDS ENHANCE MINE SITE REHABILITATION AND REVEGETATION*

H. Wijesekara[1], N.S. Bolan[1], P. Kumarathilaka[2], N. Geekiyanage[3,4], A. Kunhikrishnan[5], B. Seshadri[1], C. Saint[6], A. Surapaneni[7], M. Vithanage[2]

University of Newcastle, Callaghan, NSW, Australia[1]; National Institute of Fundamental Studies, Kandy, Sri Lanka[2]; Kyoto University, Kyoto, Japan[3]; Rajarata University, Anuradhapura, Sri Lanka[4]; National Academy of Agricultural Science, Wanju-gun, Jeollabuk-do, Republic of Korea[5]; University of South Australia, Adelaide, SA, Australia[6]; South East Water, Melbourne, VIC, Australia[7]

CHAPTER OUTLINE

*This chapter is a condensed version of the following review paper: H. Wijesekara, N.S. Bolan, M. Vithanage, Y. Xu, S. Mandal, S.L. Brown, G.M. Hettiarachchi, G.M. Pierzynski, L. Huang, Y.S. Ok, M.B. Kirkham, C. Saint, A. Surapaneni. UTILIZATION OF BIOWASTE FOR MINE SPOIL REHABILITATION. Advances in Agronomy. (Accepted full manuscript for volume 138)

Environmental Materials and Waste. http://dx.doi.org/10.1016/B978-0-12-803837-6.00003-2

3.1 INTRODUCTION

Modern wastewater and sewage treatment facilities generate biosolids in large quantities. Globally, around 10×10^7 tons year^{-1} of biosolids is generated (Thangarajan et al., 2013). In 2050, the world's population is projected to be 9.6 billion and will potentially generate 17.5×10^7 tons year^{-1} of biosolids at the rate of 50 g persons^{-1} day^{-1}. Furthermore, 66% of the world's population (7.4 billion) is forecasted to be urban by 2050 (UN, 2015). This urban population will generate 13.5×10^7 tons year^{-1} of biosolids. To manage this large quantity, the most appropriate strategy would be reuse and recycling. Biosolids are reused in different ways. A significant quantity of biosolids is used as a fertilizer or soil conditioner. For example, around 59% of generated biosolids is used in agricultural lands in Australia (ANZBP, 2013). Furthermore, 6% and 4% are used in landscaping (composting) and land rehabilitation, respectively. Public opposition has been identified as a major challenge to be overcome for the use of biosolids in agriculture by many countries; therefore, biosolids have been stockpiled at sewage treatment plants indefinitely. These stockpiles cause environmental issues such as the direct emissions of greenhouse gases (GHGs): methane (CH_4), nitrous oxide (N_2O), and carbon dioxide (CO_2), which contribute significantly to global warming (Majumder et al., 2014). Globally, the overall contribution of the waste industry to the emission of GHGs is as small as <3%, with total emissions of approximately 1446 $MtCO_{2\text{-eq}}$ in 2010. However, around 54% and 43% of this emission is contributed by wastewater handling (eg, stockpiling of biosolids) and land disposal of solid waste, respectively (Blanco et al., 2014). Therefore, biosolids must be used in an eco-friendly manner to avoid environmental pollution and ensure their sustainable reuse. Among sustainable uses, biosolids-associated mine site rehabilitation is an important strategy and has a significant role according to the perspective of environmental remediation.

The excavation of economically important resources from terrestrial landmasses, known as "mining," is an integral part of the current development trajectory of the world. Mining advances global economic prosperity by generating a large quantity of valuable natural resources. For example, Australia is the world's leading producer of a variety of minerals. The income from sales and services in the mining industry has been reported to be $217.8 billion between 2012 and 2014 for the Australian economy (ABS, 2014; AM, 2015). However, mining severely disturbs the land. Several thousands of historic and currently operating mines are scattered throughout the earth, with further estimates of around 0.4×10^6 km^2 land area disturbed by mining activities (Hooke and Martin-Duque, 2012). Unfortunately, most of these areas have never been properly reclaimed, thereby causing extensive damage to the environment (Hooke and Martin-Duque, 2012). Mining causes disturbance to the land in many ways. Generation and land disposal of waste materials such as mine tailings, subsoils, oxidized wastes, fireclay, and mudstone are the main causes of land disturbance. These waste materials contain potentially hazardous substances such as heavy metals in elevated concentrations, causing groundwater and soil contamination (Evangelou and Zhang, 1995). Production of acids associated with these waste materials can increase heavy metal pollution through enhanced metal solubility at an acidic pH (Bolan et al., 2003; Lindsay et al., 2015). The excavation of land or other cut-in engineering works causes shortfalls and overburdens topsoil and subsoil. As a result of poor soil characteristics such as low-level organic matter and poor soil texture and structure, these shortfalls and overburdens are identified as drastically disturbed unproductive landmasses (Diamond, 1999; Ghose, 2001; Johnson, 2003; Sopper, 1992). Ultimately, infertile soils and harsh conditions associated with these disturbed lands can adversely affect the establishment of soil microbial life, earthworms, other soil fauna, and plant growth (Boyer et al., 2011; Castillejo and Castello, 2010; Larney and Angers, 2012). Therefore, disturbed lands need to be rehabilitated to avoid potential environmental consequences and to restore the lost ecological wealth.

Depending on the level of land disturbance, environmental pollution, and cost, different methods are used to rehabilitate mine sites. Additions of organic or inorganic amendments (ie, biosolids, compost, fly ash, biochar), nano-enhanced materials (ie, iron oxides and iron sulfide nanoparticles), uncontaminated soils, heavy metal–tolerant genotypes, and forest floor and peat materials are current methods in mine site rehabilitation (Beasse, 2012; Liu and Lal, 2012; O'Reilly, 1997). Application of liner material over mine wastes and the isolation of mine wastes from the environment by storage in impoundments under water or behind dams are also used (Kossoff et al., 2014; Lamb et al., 2013). Among these methods, mine site rehabilitation by incorporating biowaste, in particular the application of biosolids, has been identified as a sustainable strategy, and therefore has been rewarded with greater attention.

This chapter provides a wider outlook at the compositions of different types of biosolids, their current regulations for use, and benefits of mine site rehabilitation. Special attention has been given to linking biosolids with afforestation of reclaimed mine sites. In addition, environmental challenges and pollution prevention strategies related to biosolids land application are discussed. Finally, future research needs and strategies are identified in terms of sustainable biosolids use in mine site rehabilitation.

3.2 GENERATION AND COMPOSITION OF BIOSOLIDS

3.2.1 GENERATION OF BIOSOLIDS

Sewage and wastewater treatment facilities generate biosolids. Biosolids are often referred to as treated sewage sludge. Sewage sludge is defined as any solid, semisolid, or liquid residue generated during the municipal wastewater and sewage treatment process. Biosolids can be defined as stabilized organic solids derived from sewage treatment processes (mostly resulting from the biological treatment of wastewater) which can be managed safely to be used beneficially for their nutrient, soil conditioning, energy, or other values (Shammas and Wang, 2008). An increased number of high-capacity wastewater treatment facilities generate large quantities of biosolids (Fig. 3.1). As a brief overview about the biosolids generation process, wastewater or sewage treatment plants receive wastewater and sewage from domestic, industrial, and agricultural sources. First, large objects, grit, or scum (naturally floating materials) are removed by screening at a grit chamber and the rest is forwarded to a sedimentary tank where primary sludge is produced. Primary sludge contains mainly fecal solids and further treatments are performed to meet the criteria for biosolids production. For this, different wastewater and sewage treatment plants use different techniques such as an activated sludge method, thickening (ie, flotation, gravity, centrifugation), digestion (ie, anaerobic or aerobic digestion), dewatering (ie, vacuum filters, centrifuges, belt press), alkaline treatment (ie, at high pH using lime), conditioning (ie, using coagulants and separation of water), drying or heat techniques (ie, at high temperature), and beta- or gamma-ray irradiation processes (ie, pathogen control) (Kajitvichyanukul et al., 2008; Wang et al., 2008). Some of these methods have their own advantages and disadvantages. For instance, biological stabilization leads to CO_2 emission via the decomposition of organic matter in

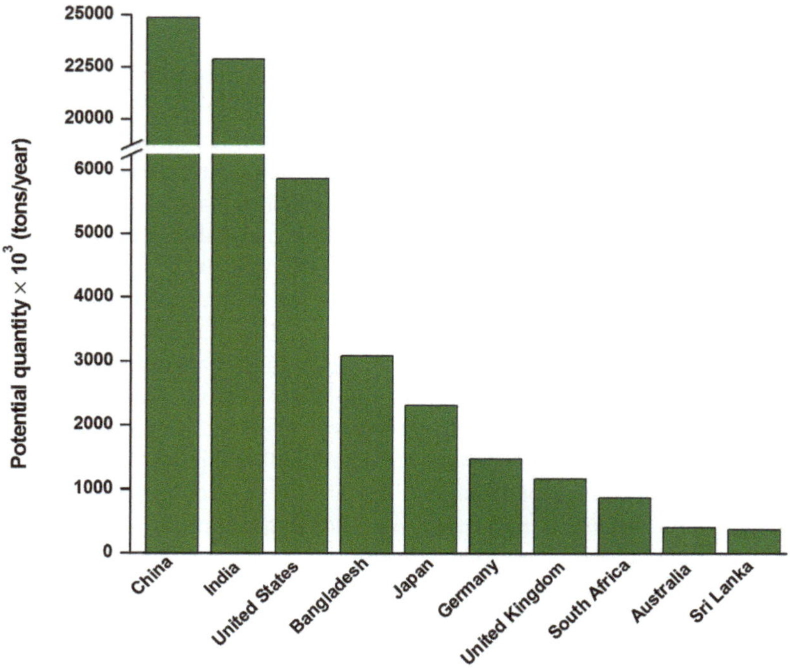

FIGURE 3.1

Potential quantity of biosolids produced in selected countries; values were calculated based on biosolids generation assuming 50 g person^{-1} day^{-1}.

sewage sludge and reduces odors (Wang et al., 2008). Most of these treatment techniques control pathogens (ie, some types of bacteria, viruses, protozoa, parasitic worms) that cause diseases and reduce the potential to attract flies, mosquitoes, or other disease-carrying vectors (Wang et al., 2008). In addition, in many countries it is mandatory to store biosolids for several months before their beneficial use to ensure elimination or inactivation of pathogens. For instance, in some sewage treatment plants in the United States, anaerobically digested sludge is stored for a minimum of 18 months for further stabilization before use (Tian et al., 2009).

Biosolids are defined in different groups by different countries. Table 3.2 shows selected criteria for grouping biosolids in the United States. Three types of biosolids have been defined by the US Environmental Protection Agency's (EPA's) sludge management program in terms of pathogen requirements and site restrictions (USEPA, 1994): "exceptional quality" (EQ) biosolids, Class "A" biosolids, and Class "B" biosolids. For the EQ biosolids, no restrictions are associated with the application of these biosolids to land, which can also be sold even as compost or soil amendments. Strict limits must be met on pollutants and pathogens to become eligible in this biosolids category. Class A biosolids (no detectable pathogens and used like commercial fertilizer) and Class B biosolids (a reduced level of pathogens) are considered as safe; however, certain requirements such as limiting public access to site application of biosolids, livestock grazing limitations, and crop harvesting schedule restrictions are attached with Class B biosolids application. Australia and New Zealand national guidelines consider recycling or disposal of biosolids and are grouped into seven categories (NWQMS, 2004). At the state level in Australia, the New South Wales (NSW) EPA has grouped biosolids into three groups: unrestricted use products, restricted use 1 products, and restricted use 2 products (Ang and Sparkes, 1997). The NSW EPA has strict regulations including application rates, contaminant limits for use of biosolids in agricultural land, and mine site rehabilitation. The European Union has implemented directives to regulate biosolids since 1986 (Evans, 2012): landfill, sludge, and waste incineration.

3.2.2 COMPOSITION OF BIOSOLIDS

Biosolids can be identified as a complex heterogeneous matrix. The composition and properties of biosolids vary with many factors such as the process of generation, age of biosolids, environmental conditions such as temperature and humidity (Wang et al., 2008). For example, alkaline material such as lime, kiln dust, portland cement, and fly ash are used by many biosolids producers to reduce pathogen content and immobilize heavy metals (Kajitvichyanukul et al., 2008; Pichtel and Hayes, 1990; Wang et al., 2008). However, the alkaline stabilization of biosolids with lime is reported to cause the loss of nitrogen (N) because of the volatilization of ammonia (NH_3), thereby reducing the N fertilizer value from biosolids (Wang et al., 2008). Nevertheless, the alkaline stabilization enhances carbon (C) stabilization (Chowdhury et al., 2015; Palumbo et al., 2004).

Total solids (TS) and volatile solids (VS) are some important parameters describing the organic matter and solid contents of biosolids. TS include suspended solids and dissolved solids. Liquid, dewatered, dried, or compost biosolids contain 2–12%, 12–30%, and 50% TS, respectively (Wang et al., 2008). VS indicates the availability of readily decomposable organic matter in biosolids (Wang et al., 2008). The VS fraction has a strong correlation with odor emissions in biosolids, and hence VS is a critical determinant when applying biosolids to land (Wang et al., 2008). Biosolids is composed of a high amount of organic matter ranging from 50% to 70% (Li, 2012). It has been reported that fatty acids constitute the predominant polar fraction, thereby representing 51% of the organic compounds in biosolids, whereas steroids and aliphatic compounds contribute to 13% and 14%, respectively (Torri and Alberti, 2012). Furthermore, biosolids' organic matter significantly differs in chemistry with soil organic matter, because biosolids are composed of more alkyl carbon than soil (Smernik et al., 2003). A study based on ^{13}C-nuclear magnetic resonance spectroscopy revealed the increased alkyl C and decreased aromatic C content in soil humic acids with the application of biosolids (Chiu and Tian, 2011). Another, comparable study revealed the addition of high doses of biosolids resulting in fatty acids, amino acids, and/or paraffinic structures in soils (Antilén et al., 2014). Also, biosolids are rich in carbon-based compounds, mainly microbial biopolymers (Torrecillas et al., 2013). For instance, polysaccharide and proteins can be found in biosolids as their pure forms or in association with other compounds such as glycoproteins and lipopolysaccharides. Lipids, nucleic acids, and humic substances are other major organic constituents in biosolids (García Becerra et al., 2010).

Biosolids are a significant source of inorganic nutrients such as N and phosphorus (P). The ammonium-N $(NH_4^+ - N)$ content in biosolids depends on sewage treatment, biosolids stabilization, and storage process, because NH_4^+ is readily volatilized into NH_3 gas (Wang et al., 2008). Biosolids contain around 3% N with low nitrate-N (NO_3-N) (ie, <0.05%) content, but most of N exists as organic-N in complex molecules such as proteins, nucleic acids, amines, and other cellular materials (Kajitvichyanukul et al., 2008; Wang et al., 2008). To become an N source for crops, these complex molecules must be converted into inorganic ammonium and NO_3-N through degradation (García Becerra et al., 2010). In addition, biosolids contain P, indicating its additional nutritional value, and thereby exhibiting the importance of P recovery and conservation (Evans, 2012).

The metal fraction of biosolids is well understood. Heavy metals such as zinc (Zn), copper (Cu), nickel (Ni), and lead (Pb) have been reported to be found in significant concentrations in biosolids. Table 3.1 shows the trace element concentrations of biosolids and sewage sludge from different sewage treatment facilities. Heavy metals in biosolids are thought to be sorbed by both organic matter (McBride, 1995) and inorganic constituents (Hettiarachchi et al., 2003). The role of sulfide minerals in binding and precipitating heavy metals (ie, Zn and Cu) in freshly produced and stockpiled biosolids has been reported elsewhere (Donner et al., 2011). A metal speciation approach involving synchrotron techniques concluded the importance of a heavy metal—iron oxides association, but not as the dominant mechanism governing metal binding in biosolids as previously suggested. Zn sorption with iron oxides and Cu sorption with organic matter has been reported, which highlights the need for future studies on metal association with the mineralizable biosolids fraction (Donner et al., 2012). In the Australian context, Victorian water authorities produce a variety of biosolid types and have reported different physicochemical parameters. Depending on the process of generation and the stockpiled duration or age of biosolids, diverse characteristics of individual elements (ie, C, N, P, sulfur [S], silicon [Si], and aluminum [A]), organic matter, clay, silt, and sand have been reported (Albuquerque et al., 2014). Furthermore, the sludge-drying process by clay-lined drying pans contributes to the clay mineral matter content and obviously increases in Al and Si fractions (Albuquerque et al., 2014). The same study revealed that biosolids more than 3 years old had less organic C than those less than 3 years old, which suggests C loss owing to mineralization with time and dilution of biosolids with clay impurity that influences the sludge-drying process (Albuquerque et al., 2014). In a study in India, a higher percentage of heavy metals (ie, iron [Fe], Zn, and cadmium [Cd]) in residual fraction and a lower percentage of exchangeable and carbonate-bound fraction were reported in biosolids (Bai et al., 2012). That particular study revealed that the concentrations of heavy metals in the biosolids were lower than those established by the European legislations.

3.3 LAND APPLICATION OF BIOSOLIDS AND POLLUTION PREVENTION

3.3.1 HEAVY METALS

The occurrence of heavy metals in biosolids can be a major cause of soil and water pollution. The repeated application of biosolids to Genesee silt loam soils has increased the lability of chromium (Cr), Pb, cobalt (Co), Zn, Cu, Ni, and arsenic (As) by 20%, 11%, 9%, 9%, 8%, 6% and 4%,

Table 3.1 Selected References for Trace Element Concentrations in Municipal Sewage Sludge and Biosolids

Sample	Concentration (mg kg^{-1} or %)														References
	Al	As	Cd	Co	Cr	Cu	Hg	Mn	Fe	Mo	Ni	Pb	Se	Zn	
Athens sewage sludge		4.97	11.2		75.1	54.7		1248			53.4	2.47	1.24	294	Jackson and Miller (2000)
Urban compost			0.48		71	119		214			15	324		328	de Abreu et al. (1996)
Urban compost			0.45		65	89		350			4304	85.3		354	de Abreu et al. (1996)
Sewage sludge			11.4		645	870		497			479	226		1788	de Abreu et al. (1996)
Denver sewage sludge		8.1	26	7.1	280	816	7.8	220		84.9		950	4.57	1672	Capar et al. (1978)
City sewage sludge		14.3	104	9.6	1441	1346	8.6	194		14.3		1832	3.1	2132	Capar et al. (1978)
Austin sewage sludge (Autinite)		9.4	3.3	4.10	106	300	1.5	430			36.7	86.9	2.57	563	Raven and Loeppert (1997)
Milwaukee sewage sludge (Milorganite)			7.2	4.07	2940	–	1.1	142			31.2	130	1.04	450	Raven and Loeppert (1997)
Anaerobically digested, dried by centrifugation, air-dried in wind rows, and finally stockpiled	5660								32,200						Bolan et al. (2013) Donner et al. (2012)
Anaerobically digested, dried by centrifugation, air-dried in wind rows, and finally stockpiled	6.3%					438		246	2.8%					134	
Digested sewage in an Imhoff tank, dewatered on sand bed, and aged in storage area	1.9%					1121		298	2.1%					1157	Donner et al. (2012)
Anaerobic digested and lagoon-dried sewage	1.4%					913		340	1.7%					1068	Donner et al. (2012)
Activated sludge-derived biosolids		5.6	2.3	2.1	29	463	1.7	245	9792		23	59	5.0	482	Galvín et al. (2009)
Biosolids derived through biodisks as biological process		18	1.4	6.5	70	649	1.2	687	16,717		47	172	12	1117	Galvín et al. (2009)

respectively (Islam et al., 2013). This study revealed that the extractable fractions of Pb, As, Zn, and Cu concentrations were significantly higher at 0–15 cm soil depth (Islam et al., 2013). Consequently, the accumulated heavy metals may mobilize from the soils to groundwater and surface water bodies. Organic matter in biosolids, hydrous oxides (ie, iron [Fe] and manganese [Mn] oxides in clays), and mineral phases (ie, phosphates and aluminum compounds) immobilize heavy metals (Hettiarachchi et al., 2006). Nevertheless, the rapid decomposition of organic matter releases organically bound heavy metals and increases the bioavailability or toxicity (Obrador et al., 2001). The increased bioavailability of heavy metals causes their excessive uptake by plants or leaching down to the soil profile. The plant uptake of heavy metals is correlated with extractable forms of metals rather than the total metal contents in soils. Interestingly, some plant species can protect the food chain by providing an effective barrier against the uptake of most heavy metals (Lu et al., 2012). The influence of biosolids on the availability of heavy metals and their effects on seed germination has been also reported (Walter et al., 2006). The results revealed that pelletization of sludges increased the availability of Ni, whereas dewatering sludge led to the greater availability of Cr and Mn. Furthermore, biosolids affected seed germination and most serious adverse effects were caused by dewatered sludge.

The protection of soil systems would be better guaranteed if the input heavy metal concentration were evaluated systematically in biosolids before land application. Alkaline stabilization of biosolids is identified as a better strategy to immobilize heavy metals. Fly ash–stabilized sludge remarkably reduced the availability of heavy metals by chemical modification into less available forms (Su and Wong, 2004). The study revealed that loamy acid soil amended with fly ash–stabilized sludge was able to reduce the availability of Cu, Zn, Ni, and Cd in the sludge. With increasing fly ash amendment rates, the percentages of Cu, Zn, and Ni in the residual fraction increased. Furthermore, concentrations of Zn and Cu in shoot tissues of corn were reduced remarkably after the application of ash amendments (Su and Wong, 2004).

3.3.2 NUTRIENTS

The land application of biosolids has the potential to contaminate surface and groundwater bodies with excess nutrients such as N and P (Haney et al., 2015). Application of biosolids followed by irrigation resulted in the excessive leaching of nutrients in agricultural fields (Price et al., 2015). Biosolids are rich in nutrients; therefore, excess nutrients can be released into surface and groundwater bodies, contributing to eutrophication (He et al., 2015). The potential for nutrient leaching resulting from the application of biosolids depends on several factors: topography, rainfall intensity, soil texture, soil pH, cation exchange capacity (CEC), redox conditions, organic matter content, cropping systems, and time and methods of biosolids application (Nkoa, 2014; Zinati et al., 2004). Nitrogen leaching has gained significant attention in studies, possibly owing to the high concentration of N in biosolids (Oladeji et al., 2013; Yager and McMahon, 2012). On the contrary, high amounts of N leaching can be attributed to the low efficiency of animal N use. It has been reported that only 20–30% of N consumed by dairy cows is converted into meat and milk; the rest is excreted as feces and urine (Nkoa, 2014). When applying biosolids to lands NH_4^+ can be adsorbed by negatively charged soil particles or uptake by plant root cells or oxidized to nitrate (NO_3^-) by nitrifying microorganisms. It is not possible to adsorb NO_3^- by negatively charged soil particles since the repulsion of ions (Nkoa, 2014). Therefore, NO_3^- is highly mobile along the soil profile, causing increased surface and groundwater contamination. A biosolids-applied farmland has been identified as the main source in which to observe fairly high concentrations of nitrate in shallow groundwater monitoring wells in Colorado, United States (Yager and McMahon, 2012). Groundwater nitrate concentrations were reported in high concentrations in biosolids-amended strip-mined lands compared with reference fields, but they were below the regulatory limit of 10 mg L^{-1} (Oladeji et al., 2013).

P accumulates as different fractions in soils, such as water-soluble, exchangeable, carbonate, Fe-Mn oxides, organic, and residual (Zinati et al., 2004). P concentrations in these fractions have different immobilization behaviors under different environmental conditions. The temperature and moisture content of a biosolids-amended soil may determine the forms and availability of phosphorus (O'Connor et al., 2004). Although P is immobile in most soils, coarse-textured soils enhance the P transport through soil profile. Most P in biosolids is available as inorganic P (Elliott et al., 2002). Several studies have reported P release in soils amended with biosolids. A kinetics study revealed that biosolids-amended soil released eight times more P compared with bare soil (Islas-Espinoza et al., 2014). The degree of P saturation (DPC) is a major indicator to determine potential P loss from a particular soil (Alleoni et al., 2014). A long-term field experiment was conducted to investigate the DPC in Oxisol amended with biosolids; it revealed critical levels of DPC in the top soil layer (Alleoni et al., 2014). Proper management practices included cultivation against slopes, maintaining sufficient gaps between biosolids applications and sowing, consideration of climatic conditions before applying biosolids to the land, and systematic soil testing. In addition, decision making should be taken into account to minimize nutrient leaching from biosolids.

3.3.3 PATHOGENS

Land application of biosolids may be responsible for spreading human pathogens. Biosolids primarily originate from human feces, hospitals, abattoirs, and wastewater treatment plants, and therefore consist of a wide range of pathogens (Lasobars et al., 1999; Sidhu and Toze, 2009). These pathogens include viruses (eg, hepatitis A, rotavirus), bacteria (eg, *Escherichia coli*, *Salmonella*), parasites (eg, *Cryptosporidium*, *Giardia*), and helminths (eg, *Ascaris*) (Brewster et al., 2003; Kato et al., 2003; Monpoeho et al., 2004; Pepper et al., 2012; Zaleski et al., 2005). Human exposure to pathogens can be through direct exposure to biosolids or indirectly via biosolids-based soil amendments. Sewage treatment plant workers and the general public using biosolids-amendment products may be directly exposed to pathogens through accidental ingestion owing to the contamination of hands and clothes. Food crops grown in biosolids-amended soil and water contaminated with biosolids are indirect pathways to spreading pathogens to human beings who reside in remote areas. In other words, pathogens in biosolids may create a transmission route to spread pathogenic diseases between urban and rural areas (Sidhu and Toze, 2009). Some vector animals such as birds, mammals, and insects might propagate infections to humans in remote areas by transporting pathogens over long distances. Human pathogens can survive for a long time, possibly owing to the resistance to inactivation. Different physicochemical and biological parameters such as temperature, moisture content, oxygen, pH, sunlight, soil type, texture, and predation may influence the inactivation of pathogens in biosolids (Sidhu et al., 2001). Nevertheless, the influence of those factors acts differently from one pathogen to another. A number of studies have shown the numbers and survival patterns of pathogens in biosolids, whereas few studies have reported the detection limits of the method used (Mocé-Llivina et al., 2003; Payment et al., 2001; Pourcher et al., 2005).

Viruses are infectious agents and are more concentrated in biosolids during the wastewater treatment process. Different types of enteric viruses, such as norovirus, hepatitis A virus, rotavirus, and enterovirus, are commonly reported in biosolids (Bofill-Mas et al., 2000). Noroviruses have gained worldwide significance owing to the most common cause of acute gastroenteritis (Lodder et al., 1999). Hepatitis A virus has worldwide distribution and is more prominent in developing countries. Rotaviruses are responsible for causing gastroenteritis in young children and in immune-compromised individuals worldwide. Human enteroviruses are commonly detected in all types of biosolids (Sidhu and Toze, 2009).

Bacterial pathogens are a major cause of gastroenteritis all over the world. In general, enteric viruses, parasites, and protozoa are considered to be obligatory parasites; hence, they are unable to multiply in biosolids (Gibbs et al., 1997). Nevertheless, bacteria are able to multiply under favorable conditions in biosolids. For instance, a study with soil amendments and biosolids storage trials determined the density of *E. coli* and *Salmonella* (Gibbs et al., 1997). This study revealed the indicator organisms at undetectable concentrations in a hot, dry summer period whereas their repopulation occurred during the rainfall at the beginning of winter in both trials. *Salmonella* sp. is the most prominent bacteria in raw and treated biosolids and one of the major causes of diarrhea, fever, and abdominal cramps, which can last for 4 to 7 days (Bicudo and Goyal, 2003). *Salmonella* sp. can survive up to 3 months in stored conditions and can regrow under favorable conditions (Bicudo and Goyal, 2003). In the case of *E. coli* O157:H7, they survive in pastures for more than 11 years and more than 6 weeks under winter conditions (Bicudo and Goyal, 2003). In favorable conditions, *E. coli* O157: H7 is able to regrow in biosolids. The symptoms of *E. coli* infection are similar to *Salmonella*, but in approximately 10–20% of patients it causes serious diseases such as hemorrhagic colitis or hemolytic uremic syndrome (Bicudo and Goyal, 2003). *Shigella* sp. is a prominent cause of gastric illness globally and is commonly found in sewage worldwide. Another prominent bacterial pathogen spread by biosolids is *Helicobacter pylori*, which is responsible for gastric ulcers and is linked to gastric cancer (Peng et al., 2002).

Among protozoans of concern, genera *Cryptosporidium* and *Giardia* are highly resistant to environmental stress. They are a significant cause of gastroenteritis, possibly owing to low infectious doses (Caccio et al., 2003). These protozoans need a host for reproduction and are impossible to inactivate as a result of their oocysts, which gain a thick spore coat that protects them from many disinfectants (Bicudo and Goyal, 2003). *Cryptosporidium* oocysts are able to survive in soil for up to 3 months (Kato et al., 2003). *Cyclospora cayetanensis* is an oocyst-forming intracellular protozoan reported to be a significant cause of gastroenteritis. Human pathogenic microsporidia such as *Encephalitozoon intestinalis* and *Enterocytozoon bieneusi* are responsible for diarrheal illness worldwide (Sidhu and Toze, 2009).

Helminth infections are leading concerns, especially in developing countries, possibly as a result of the lack of sanitation facilities. As a result of high settling velocities, most helminth eggs concentrate rapidly in biosolids (Nelson, 2003). *Ascaris lumbricoides* is a major parasite worldwide, and almost a quarter of the world's population is infected by *Ascaris* (Johnson et al., 1998). Available information about the density and fate of helminths in biosolids is reported mainly in *Ascaris* (Johnson et al., 1998), which may be due to the higher prevalence of *Ascaris* than other helminths. *Ascaris* eggs can tolerate environmental conditions (ie, storage lagoons in temperate locations) and hence remain infective for long time, possibly several years. *Ascaris* eggs have been found after 29 weeks in stored biosolids (Johnson et al., 1998). Moreover, hookworms such as *Ancylostoma duodenale* also infect a quarter of human population at any one time (Johnson et al., 1998). Overall, improper treatment practices of sewage, utilization, and management practices of biosolids can produce an exposure pathway causing pathogenic diseases in humans. Consequently, biosolids handling processes need to be designed to ensure safety from numbers of pathogens and to safeguard public health.

3.3.4 ODOR EMISSIONS

The land application of biosolids releases odorous volatile organic compounds (VOCs) (eg, terpenes, alcohols, ketones, furans, sulfur-containing compounds, and amines) and ammonia (Komilis et al., 2004; Maulini-Duran et al., 2013). These unpleasant odors are released during biodegradation and cause discomfort. Exposure to higher concentrations of these odors leads to toxicological effects such as sensory irritation and psychohygenic effects to humans (Lleo et al., 2013). The initial substrate chemical composition, pH, moisture content, redox potential, temperature, microbial activity, and physical and chemical properties of VOCs in biosolids affect the extent of odor generation when applied to land (Rosenfeld et al., 2001). In one study, terpenes and ketones were identified as the most abundant VOCs in overall emissions (Rosenfeld et al., 2001). Odor emissions from two anaerobically digested biosolids from Washington, the United States, consisted of ammonia, dimethyl disulfide, carbon disulfide, trimethyl amine, acetone, and methyl ethyl ketone, whereas dry biosolids produced volatile compounds including methyl ethyl disulfide, methane thiol, acetic acid, propionic acid, and butyric acid (Rosenfeld et al., 2001). Identification and semiquantitative determination of volatile and semivolatile organic compounds were assessed by Kotowska et al. (2012), and approximately 170 VOCs were detected, including aliphatic and aromatic hydrocarbons, alcohols, esters, carbonyls, as well as sulfur-, nitrogen-, and chlorine-containing compounds. Ethyl ether, *n*-hexane, *p*-xylene, *o*-xylene, mesitylene, *m*-ethylbenzene, limonene, *n*-decane, *n*-undecane, and *n*-dodecane were detected as more prominent substances (Kotowska et al., 2012).

3.3.5 GREENHOUSE GAS EMISSIONS

Biosolids contribute to GHG emissions in different stages, mainly through stockpiles and their land application. The direct emission of GHG generated from biosolid stockpiles in Melbourne, Australia was reported (Majumder et al., 2014). According to this study, the youngest biosolids (<1 years) released higher amounts of CH_4 and N_2O emissions of 60.2 kg of CO_2-e per Mg of biosolids per year. In comparison, stockpiles aged between 1 and 3 years emitted higher overall GHGs (about 29 kg CO_2-e Mg^{-1} $year^{-1}$) compared with the oldest stockpiles (about 10 kg CO_2-e Mg^{-1} $year^{-1}$). Moreover, the youngest stockpiles released two-thirds of the GHG as N_2O, whereas equal amount of N_2O and CO_2 were emitted from the 1- to 3-year-old stockpiles and CO_2 emissions were dominant in old stockpiles. This particular study also assessed GHG flux from a large biosolids stockpiles and two shallow stockpiles, one of which was planted with willow (*Salix reichardtii*) trees. Results revealed that all stockpiles emitted large amounts of GHGs ranging from 38 to 65 kg CO_2-e Mg^{-1} $year^{-1}$. Furthermore, GHG emissions were dominated by N_2O and CO_2, whereas CH_4 emissions were negligible and accounted for less than 2% (Majumder et al., 2014). Consequently, this particular study concluded that the biosolids stockpiles were significant sources for GHG emissions, but planting willow trees did not reduce GHG emissions. Charcoal amending onto biosolids may reduce the emissions of GHGs. Aguilar-Chavez et al. (2012) monitored the effect of charcoal application on GHG emissions from biosolids-amended arable land cultivated with wheat. Results revealed lower cumulative GHG emissions over 45 days as 2%, 34%, and 39% in the presence of 1.5%, 3.0%, and 4.5% charcoal, respectively, than for the control soils.

The emission of N_2O could have resulted from biochemical (ie, nitrification and denitrification) and chemical (ie, chemodenitrification) pathways in soil; these processes are controlled by the availability of O_2 and soil pH (Nkoa, 2014). Nitrifying bacteria such as *Nitrosomas*, *Nitrosococcus*, *Nitrobacter*, and *Nitrococcus* are able to convert nitrite (NO_2^-) to N_2O. Denitrifiers such as *Paracoccus*, *Pseudomonas*, and *Thiobacillus* are able to release N_2O when reducing nitrate (NO_3^-) to dinitrogen (N_2) during denitrification. The global warming potential of N_2O is significantly high (approximately 300 times higher than that of CO_2) and therefore it is considered to be one of the major GHGs and ozone-depleting substances released from the earth (Borjesson and Berglund, 2007).

Methane is produced as a result of microbial respiration under anaerobic conditions when carbon is available as the only electron acceptor (Maulini-Duran et al., 2013). When energetically favorable electron acceptors such as oxygen, iron, nitrogen, sulfur, and manganese are exhausted, carbon is used as an electron acceptor (Brown et al., 2008). In the case of NH_3 emission, ammonium, urea, and other organic nitrogen compounds in biosolids are possible sources. Temperature, pH, and aeration may govern the release of NH_3 into the environment (Nkoa, 2014). Once NH_3 gas is emitted, it reacts with atmospheric gases such as H_2SO_4, HCl, and HNO_3 and produces soluble aerosol salts such as NH_4Cl, NH_4NO_3, and NH_4SO_4 (McMurry et al., 1983). The atmospheric deposition of NH_3 through acid rain results in the acidification of terrestrial and aquatic ecosystems. Eutrophication of freshwater bodies could occur from atmospheric NH_3 deposition (Chowdhury et al., 2014).

Some actions such as selecting remote sites, minimizing the length of time for storage of biosolids, and avoiding land application when wind conditions favor the transport of odors should be taken into account to minimize odor and GHG emissions from biosolids applications.

3.3.6 EMERGING CONTAMINANTS

Over the past couple of decades, significant attention has been given to selected groups of persistent organic pollutants in biosolids, including chlorinated dioxins/furans, polychlorinated biphenyls, and polycyclic hydrocarbons (Clarke and Smith, 2011). Most of these compounds do not affect human health when biosolids are recycled to farmland, possibly because of effective source control (Hundal et al., 2008; Wilson et al., 1997). Nevertheless, a number of emerging organic contaminants were identified in biosolids based on environmental persistence, human toxicity, and evidence of bioaccumulation in humans and the environment. For instance, perfluorinated chemicals, polychlorinated naphthalenes, triclocarban, benzothiazoles, synthetic musks, steroids, and reactive nanomaterials were recognized for priority attention because they can enter into living organisms via biosolids-amended soil (Brunetti et al., 2015). Thereby, precautionary actions such as quality control during biosolids generation and continuous monitoring (long-term) after biosolids land application need to be taken to ensure long-term sustainability and environmental security. The production of quality-controlled biosolids and their stabilization methods during land application could be used to avoid issues with respect to nutrients, pathogens, emerging contaminants, and vectors. Fig. 3.2 shows possible pathways of chemical and biological contaminants released by biosolids into the environment.

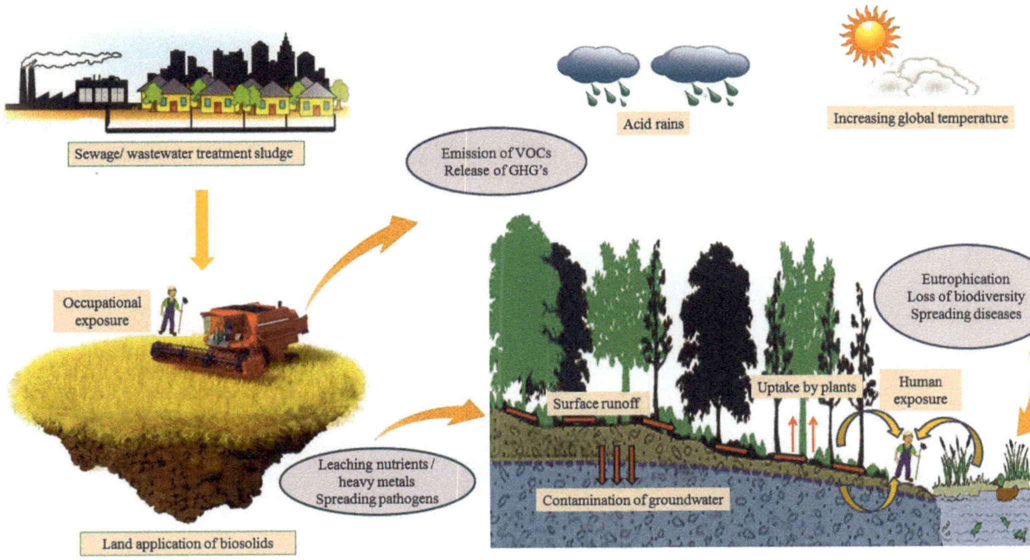

FIGURE 3.2

Possible pathways of chemical and biological contaminants released by biosolids in the environment.

3.4 REGULATIONS OF BIOSOLIDS USE

A number of countries and organizations around the world have developed regulations for the application of biosolids to agricultural lands or disturbed lands for rehabilitation. Mainly, three reasons can be identified for the development of these regulations regarding biosolids. The elevated production and fertilization or soil conditioning value of biosolids has increased their land application; therefore, regulations are required for their sustainable management (Tjell, 1986; USEPA, 1994). Potential risks (ie, spread of pathogens) associated with the land application of biosolids has led to the formulation of regulations to ensure minimal harm to the environment and to minimize public health risks (NWQMS, 2004).

Among various regulations regarding biosolids in different countries, regulations in the United States draw significant attention and their guidelines are frequently used as model regulations by some countries. For example, the pathogen and vector attraction reduction criteria of the US EPA's Code of Federal Regulations Part 503 rule for land application of sewage sludge are used by the NSW EPA in Australia (Ang and Sparkes, 1997). However, before formulating the regulations for biosolids use in the United States, the background was not environmentally responsive. For example, tons of partially treated wastewater from many wastewater treatment plants were discharged into the rivers, lakes, and bays in the United States before the 1950s (Lu et al., 2012). On the contrary, sludge disposal into the ocean from wastewater treatment plants continued until the Ocean Dumping Ban Act was ratified in 1988. The National Environmental Policy Act of 1969 regulated wastewater sludge use to protect the environment (USEPA, 1979). The Federal Water Pollution Control Act was enacted in 1972, further contributing deed restrictions against environmental pollution from wastewater treatment plants in the United States. Under provisions of the Resource Conservation and

Table 3.2 Ceiling Concentrations for 10 Heavy Metals (Defined as Pollutants) That Define Different Options on Biosolids Under US Environmental Protection Agency's Sludge Management Program (USEPA, 1994)

Pollutant	Ceiling Concentration Limits for All Biosolids Applied to Land (mg kg^{-1})[a]	Pollutant Concentration Limits for EQ and PC Biosolids (mg kg^{-1})[a]	Cumulative Pollutant Loading Rate Limits for CPLR Biosolids (kg ha^{-1})	Annual Pollutant Loading Rate Limits for APLR Biosolids (kg ha^{-1} $year^{-1}$)
As	75	41	41	2
Cd	85	39	39	1.9
Cr	3000	1200	3000	150
Cu	4300	1500	1500	75
Pb	840	300	300	15
Hg	57	17	17	0.85
Mo[b]	75	–	–	–
Ni	420	420	420	21
Se	100	36	100	5
Zn	7500	2800	2800	140

EQ, *exceptional-quality biosolids;* PC, *pollutant concentration biosolids*; CPLR, *cumulative pollutant loading rate biosolids;* APLR, *annual pollutant loading rate biosolids.*
[a]*Dry-weight basis.*
[b]*Pending EPA consideration.*

Recovery Act of 1976 and the Clean Water Act of 1977, the US EPA was established. The US EPA is the federal regulative authority responsible for regulating sludge management in the United States. Currently, the United States regulates land application of sludge at the federal, state (ie, the Pennsylvania Department of Environmental Resources guidelines), and local levels (PDER, 1997). The US EPA focuses on the use of sludge based on three end-use groups: as a soil amendment, a source of heat, and a source of other useful products including reclamation of disturbed lands from mining (USEPA, 1979). In 1983, the US EPA identified four related options for the safe disposal of sludge: agricultural use, land reclamation (ie, strip-mined lands, mine tailings, and other disturbed or marginal lands), forest use, and dedicated land use (USEPA, 1983). Furthermore, benefits (ie, reduction in commercial fertilizer use, retardation of soil erosion), environmental consequences (ie, risk of phytotoxicity from excess metals and salts), and remediation strategies (ie, sludge application rates, phytoremediation) have been identified by the US EPA's 1983 regulations (USEPA, 1983). Based on comprehensive requirements for the reduction of selected metals and pathogens in biosolids for land reclamation/or application, the US EPA's Part 503 regulation was developed (Table 3.2) (USEPA, 1994).

3.5 EFFECTS OF ADDITION OF BIOSOLIDS IN MINE SITE REHABILITATION

To date, extensive work has been carried out on the use of biosolids for the rehabilitation of mine sites. During the process of mine site rehabilitation using biosolids, direct (eg, decreasing bulk density,

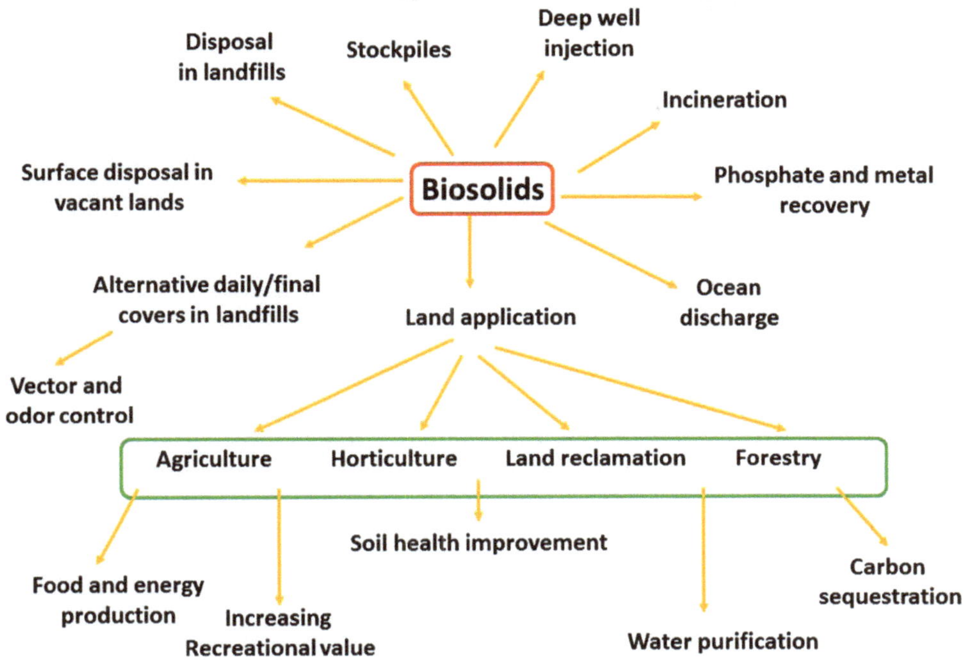

FIGURE 3.3

Various approaches to the utilization of biosolids.

avoiding soil erosion) and indirect (ie, improving chemical, physical, and biological properties) effects of receiving degraded soils could be achieved to restore a degraded environment. Fig. 3.3 illustrates various approaches to using biosolids, including mine spoil rehabilitation.

3.5.1 PHYSICAL CHARACTERISTICS

The addition of biosolids enhances most soil physical properties in degraded lands from mining. The high organic matter of biosolids is the main cause for improvement in physical properties in mine spoil soils. Improvements in physical properties including decreased bulk density and temperature, increased porosity and aggregation, increased hydraulic conductivity and water holding capacity, increased infiltration, maintained soil texture, and reduced erosion and sedimentation have been reported in mine spoil rehabilitation with biosolids (Brofas et al., 2000; Gardner et al., 2010; Sopper, 1992). The application of biosolids causes a reduction in bulk density in degraded soils by increasing pore space by enhancing macropores, mesopores, and micropores and developing soil texture, thereby increasing the field capacity (Jones et al., 2010, 2012). A remarkable increase in porosity and aggregate stability has been reported in acidic tailing rehabilitation with sewage sludge in Spain (Zanuzzi et al., 2009). A long-term study revealed that the application of biosolids caused short-term improvements but no lasting improvement in the physical properties of degraded soils (Bendfeldt et al., 2001).

3.5.2 CHEMICAL CHARACTERISTICS

The application of biosolids raises chemical properties such as pH, electrical conductivity (EC), CEC, nutrient contents, and organic matter in soils. Mostly, the low pH of degraded mine soils (ie, acidic tailing-rich soils) is raised with biosolids or sewage sludge application (Basta et al., 2001; Bendfeldt et al., 2001; Sopper, 1992). The high pH of alkaline mine spoils experiences decreased pH with biosolids application (Brofas et al., 2000; Jones et al., 2011). However, degraded land characteristics such as the oxidation of minerals over time can decrease in pH (Brofas et al., 2000; Sopper, 1992). As a result of an increase in soluble salts, the EC increases (Gardner et al., 2010; Jones et al., 2011; Mingorance et al., 2014). However, a reduction in EC after biosolids application has also been reported (Arocena et al., 2012). This may be the result of the leaching of soluble salts and the immobilization of metal ions. The presence of clay, mineral particles, and organic colloids in biosolids increases the CEC in the receiving degraded sites (Bendfeldt et al., 2001; Brofas et al., 2000; Gardner et al., 2010). The application of biosolids to mine spoils increases key plant nutrients such as N, P, Ca, and S, thereby increasing fertility (Bendfeldt et al., 2001; Brofas et al., 2000; Castillejo and Castello, 2010; Larney and Angers, 2012). To increase the supply of plant nutrients, biosolids combined with chemical fertilizers are also used in mine site rehabilitation (Li et al., 2013). Biosolids increase organic matter in spoil soils, accelerating mine site rehabilitation (Brofas et al., 2000; Castillejo and Castello, 2010; Larney and Angers, 2012). An increase in the immobilized or reduced concentrations of extractable metal ions in plant species has been reported from biosolids application with woody debris, paper sludge, and iron by-products at some superfund sites (Brown et al., 2003, 2004).

3.5.3 BIOLOGICAL CHARACTERISTICS

Biosolids application increases microbial biomass carbon and microbial enzymatic activities in receiving mine spoils (Gardner et al., 2010; Mingorance et al., 2014; Mora et al., 2005). Biosolids creates an energy-rich soil environment favorable to soil microorganisms by increasing soil organic matter, which is the main energy source of microbes. An increased amount of soil microorganisms such as denitrifiers, sulfate reducers, iron reducers, total aerobic microorganisms, and total anaerobic heterotrophs was observed in a mine site rehabilitated with biosolids in Canada (Gardner et al., 2010). Microbial enzyme concentrations of dehydrogenase, β-glucosidase, alkyl phosphatase, protease, and arylsulfatase increased after the addition of stabilized sewage sludge in a mine site (Mingorance et al., 2014). Furthermore, populations of fungi, bacteria, actinomycetes, and total microorganisms along with soil enzymatic activity have been reported with the application of sewage sludge combined with chemical fertilizers (Li et al., 2013). The addition of fresh sewage sludge was reported to result in a higher population density of earthworms in open-cast coal mine site rehabilitation (Emmerling and Paulsch, 2001). The mortality of earthworms increased owing to biosolids application in a mine site rehabilitation in New Zealand, which further suggests the need for ecological tests before implementing remediation strategies (Waterhouse et al., 2014). In a comprehensive case study, microbial functions (ie, CO_2 respiration and ammonia oxidation), plant diversity and metal uptake, earthworm (*Eisenia fetida*) toxicity, and deer mice (*Peromyscus maniculatus*) total body burden by selected heavy metals were reported with biosolids application to overburden and tailings (Brown et al., 2014). This study indicated no potential for ecosystem transfer of contaminants and showed high rates of recolonization of the restored sites by a range of small mammals owing to biosolids-enhanced rehabilitation.

Considering all of these physical, chemical, and biological changes with the use of biosolids, beneficial effects such as food and energy production by enhancing vegetation, nutrient cycling and carbon sequestration, water purification and pest/or disease control, and enhancement of recreational value are also achieved (Larney and Angers, 2012; Sopper, 1992). Compared with the beneficial effects, above- and below-ground carbon sequestration is identified as a major opportunity. Table 3.3 shows biosolids used in various vegetation sites in terms of carbon sequestration.

Large-scale mine sites are highly altered from their historical levels of biotic and abiotic integration, which make them unsuitable for the natural succession of historical species or direct planting of tree species. The concept of novel ecosystems, which is a new mixture of physical and biological components, including both native and nonnative species, is proposed as a feasible option for the rehabilitation of mine sites. As discussed in the previous section, after improving the physical, chemical, and biological properties of soil, such mine sites suitable for establishing novel ecosystems (Doley and Audet, 2013; Hobbs et al., 2009). The role of biosolids is important to reduce the time taken for restoration by improving soil biological and physicochemical properties. The application of biosolids provides levels of stability and functionality of an ecosystem (Doley and Audet, 2013; Li et al., 2013). In this section, we propose the land application of biosolids for soil rehabilitation and subsequent establishment of novel ecosystems as a feasible option for mine site rehabilitation and increasing carbon sequestration.

The early establishment of photosynthetic organisms in a bare land, such as lichens proceeded by bryophytes and annual grasses in a mine site, gradually enhances the thickness of the fine soil. As the natural succession proceeds, such lands will be suitable for tree establishment. For instance, in Estonia, long-term monitoring of the natural colonization of a pine stand on a reclaimed mine site showed that nearly 25 years is required to develop 6.2 cm thickness of fine soil with 4.38% of organic carbon (Laarmann et al., 2015). Moreover, 7-year-old stands of black alder, silver birch, and Scots pine with various planting densities sequestered 2563, 161, and 1899 kg ha^{-1} of aboveground biomass, respectively (Kuznetsova et al., 2011). That study further suggested black alder as a promising tree species for the reclamation of oil shale postmining areas in Estonia. Similarly, ecological assessments focusing on changes in the floristic compositions, impact from wildfires on the persistence of restored species, and factors affecting the plant growth in abandoned mine sites are available for major mining areas across the world, especially in Australia (Herath et al., 2009; Martínez-Ruiz et al., 2007; Ohsowski et al., 2012; Pallavicini et al., 2013; Vickers et al., 2012; Wiegleb and Felinks, 2001). However, integrated studies focusing on both land application of biosolids and restoration of the ecosystem as a whole are rare. Pallavicini et al. (2013) studied factors affecting herbaceous richness and biomass accumulation in reclaimed coal mines and found that pH, soil erosion severity, and soil nutrient availability are the key determinants. In tropical environments in which tree growth rates are faster, Tripathi et al. (2014) studied a mine spoil in India and showed that total carbon accumulation in total plant biomass, mine soil, and soil microbial biomass is 44.5, 22.9, and 1.8 Mg ha^{-1}, respectively. Therefore, afforestation followed by the land application of biosolids is a promising alternative because afforestation increases the net ecosystem carbon sequestration in reclaimed mine sites and enhance opportunities for other potential uses such as recreation. Therefore, we recommend the integration of land application of biosolids and establishment of novel ecosystems in mine sites as a feasible and more economical strategy for rehabilitation.

Table 3.3 Selected References for Biosolids Application to Various Vegetation Sites in Terms of Carbon Sequestration

Country or Region	Application Rate (Mg ha^{-1})	Tested Plants	Carbon Sequestration (Mg C^{-1} ha^{-1} $year^{-1}$)	Remarks	References
United States	455–1654	Corn, wheat, sorghum, soybean, grass	1.73	• 8–23 years in rotation application of biosolids controlled the SOC (−0.07–0.17) • World's largest reclamation project with biosolids	Tian et al. (2009)
Australia	25 or 50	Indian mustard, sunflower	3.11–7.54 for mustard; 2.48–4.79 for sunflower	• Effect of biosolids application rate on carbon sequestration performed • Effect of plant species to SOC sequestration studied	Bolan et al. (2013)
United States	10 and 18		Biomass yields 35–40%	Degradation rate model applications identify 2 SOC phases in soil	Zhai et al. (2014)
United States	2.5, 5, 10, 21, and 30	Semiarid grassland	Soil total C% increased from 1.63 to 3.58 for short term biosolids application rate at 30 Mg $year^{-1}$	Long-term (13–14 years) and short-term (2–3 years) effects of biosolids application to soils 0–8 cm deep	Ippolito et al. (2010)
Spain	50		Mean increase 1.7 g kg^{-1}, peak increases 3.8 g kg^{-1}	• Sampled from ploughed layer (0–30 cm) • Short-term SOC pool in 60 contrasting agricultural soils	Soriano-Disla et al. (2010)
	4.2 kg m^{-2}		Increased crop residue C from 11.8% to 32.5%	• 13-year study • Crop residue C analyzed by ^{13}C technique	Tian et al. (2015)
Europe	1.0		SOC accumulation rate 0.49% $year^{-1}$	Maximum yearly carbon mitigation potential 2.65 Tg $year^{-1}$	Smith et al. (2000)
England	33		0.18	Organic C in biosolids, when applied to soil, is less decomposable than that in farm manures	Powlson et al. (2012)

3.6 CONCLUSIONS, CHALLENGES, AND FUTURE RESEARCH NEEDS

A large quantity of biosolids is generated as a consequence of increased human population and the subsequent expansion of wastewater treatment industries. Because of the presence of high amounts of organic matter and nutrients, biosolids are used extensively as a soil amendment or fertilizer for agricultural, land reclamation, and revegetation purposes. Biosolids enhance physical, chemical, and biological properties of degraded lands. Advantages associated with the use of biosolids in the remediation of mine sites are food and energy production through enhancing vegetation or afforestation, nutrients cycling and carbon sequestration, water purification and pest/or disease control, and enhancement of recreational and scientific discovery. However, a number of challenges are linked with biosolids land application, in particular mine site rehabilitation: groundwater and soil contamination from enhanced leaching of inorganic (ie, N and P) and organic nutrients, heavy metals, pathogens, and emerging contaminants (ie, pharmaceuticals and reactive nanomaterials). GHGs and odorous gaseous emission from biosolids stockpiles and during their handling (ie, land application) have been identified as contributing to air pollution and global warming. Therefore, a number of countries have developed regulations for the production and land application of biosolids. Most regulations focus on minimizing pathogens and environmental contamination from heavy metals and major inorganic nutrients. For example, chemical stabilization (ie, alkaline treatment) and pretreatment (ie, composting) are used to minimize the contaminants (ie, immobilization of heavy metals, disinfection of pathogen) during the production of biosolids. These options also ensure the quality of biosolids (ie, composting of biosolids with biodegradable wastes) and enhance their efficacy.

This chapter recognizes the challenges and future research needs that have to be addressed carefully to minimize negative environmental effects associated with biosolids:

- To minimize GHG emission and loss of carbon value from biosolids, minimizing decomposition of biosolids is an important factor to be considered. Options include minimizing the duration of biosolids stockpiles, applying efficient stabilizing agents, and/or using methods to convert easily degradable organic fractions into a more recalcitrant fraction, thereby minimizing microbial decomposition.
- To convert degraded mine sites as "future carbon storage sites," unique characteristics (ie, extremely low pH and diminished microorganisms in soils) associated with degraded lands and incorporation of biosolids in terms of increasing residual carbon and carbon sequestration from afforestation need to be studied further. Also, a scientific approach is needed to estimate the carbon sequestration resulting from biosolids use.
- To encourage biosolids use, issues such as the availability of insufficient quantities for remote areas or sites, the high cost of transportation, and the loss of nutrients or organic matter at field conditions need to be resolved.
- To recover the nutrients and their conservation from biosolids (ie, P), future research should be carried out.

ACKNOWLEDGMENTS

This research was partly supported by Australian Research Council Discovery Projects (DP140100323). This chapter is one of the outcomes of the research project on "Carbon sequestration from land application of biosolids." We thank South East Water (Dr. Aravind Surapaneni), Western Water (William Rajendran), Gippsland Water (Mark Heffernan), City West Water (Sean Hanrahan), Yarra Valley Water (Andrew Schunke), and Transpacific Industries (Chris Hetherington) for supporting this research project.

REFERENCES

ABS, 2014. Mining Operations. Australian Bureau of Statistics, Australia. Document viewed on November 15, 2015. http://www.abs.gov.au/ausstats/abs@.nsf/mf/8415.0.

Aguilar-Chavez, A., Diaz-Rojas, M., Cardenas-Aquino, M.D.R., Dendooven, L., Luna-Guido, M., 2012. Greenhouse gas emissions from a wastewater sludge-amended soil cultivated with wheat (*triticum* spp. L.) as affected by different application rates of charcoal. Soil Biology & Biochemistry 52, 90–95.

Albuquerque, N., Meehan, B., Surapaneni, A., Hughes, J., Bolan, N., 2014. Overlooked characteristics of biosolids products for land application. In: AWA Biosolids and Source Management National Conference, Melbourne, 25–26 June 2014.

Alleoni, L., Fernandes, A., de Campos, M., 2014. Degree of phosphorus saturation of an oxisol amended with biosolids in a long-term field experiment. Environmental Science and Pollution Research 21, 5511–5520.

Ang, C., Sparkes, J., 1997. Environmental Guidelines: Use and Disposal of Biosolids Products. Environment Protection Authority, New South Wales, Sydney, Australia, ISBN 0 7310 3792 8. Document viewed on January 10, 2015. http://www.epa.nsw.gov.au/resources/water/BiosolidsGuidelinesNSW.pdf.

Antilén, M., Silva, K., Acevedo, S., Amiama, F., Faúndez, M., Knicker, H., Pizarro, C., 2014. Characterization of humic acids extracted from biosolid amended soils. Journal of Soil Science and Plant Nutrition 14, 1005–1020.

AM, 2015. Australian Mining. Document viewed on November 12, 2015. http://www.thisisourstory.com.au/.

ANZBP, 2013. The Australia & New Zealand Biosolids Partnership. Document viewed on November 12, 2015. http://www.biosolids.com.au/bs-australia.php.

Arocena, J.M., Van Mourik, J.M., Cano, A.F., 2012. Granular soil structure indicates reclamation of degraded to productive soils: a case study in southeast Spain. Canadian Journal of Soil Science 92, 243–251.

Bai, S., Srikantaswamy, S., Krishnanandan, V., Naik, O., 2012. Speciation of heavy metals in biosolids of wastewater treatment plants at Mysore, Karnataka, India. Environmental Monitoring and Assessment 184, 239–249.

Basta, N.T., Gradwohl, R., Snethen, K.L., Schroder, J.L., 2001. Chemical immobilization of lead, zinc, and cadmium in smelter-contaminated soils using biosolids and rock phosphate. Journal of Environmental Quality 30, 1222–1230.

Beasse, M., 2012. Microbial Communities in Organic Substrates Used for Oil Sands Reclamation and Their Link to Boreal Seedling Growth. University of Alberta (M.Sc thesis).

Bendfeldt, E.S., Burger, J.A., Daniels, W.L., 2001. Quality of amended mine soils after sixteen years. Soil Science Society of America Journal 65, 1736–1744.

Bicudo, J., Goyal, S., 2003. Pathogens and manure management systems: a review. Environmental Technology 24, 115–130.

Blanco, G., Gerlagh, R., Suh, S., Barrett, J., de Coninck, H.C., Morejon, C.F.D., Mathur, R., Nakicenovic, N., Ahenkora, A.O., Pan, J., Pathak, H., Rice, J., Richels, R., Smith, S.J., Stern, D.I., Toth, F.L., Zhou, P., 2014. Drivers, trends and mitigation. In: Edenhofer, O., Pichs-Madruga, R., Sokona, Y., Farahani, E., Kadner, S., Seyboth, K., Adler, A., Baum, I., Brunner, S., Eickemeier, P., Kriemann, B., Savolainen, J., Schlömer, S., von Stechow, C., Zwickel, T., Minx, J.C. (Eds.), Climate Change 2014: Mitigation of Climate Change. Contribution of Working Group III to the Fifth Assessment Report of the Intergovernmental Panel on Climate Change. Cambridge University Press, Cambridge, United Kingdom and New York, NY, USA.

Bofill-Mas, S., Pina, S., Girones, R., 2000. Documenting the epidemiologic patterns of polyomaviruses in human populations by studying their presence in urban sewage. Applied and Environmental Microbiology 66, 238–245.

Bolan, N.S., Adriano, D.C., Curtin, D., 2003. Soil acidification and liming interactions with nutrient and heavy metal transformation and bioavailability. In: Spark, D.L. (Ed.), Advances in Agronomy, vol. 78. Academic Press, pp. 215–272.

Bolan, N.S., Kunhikrishnan, A., Naidu, R., 2013. Carbon storage in a heavy clay soil landfill site after biosolid application. Science of the Total Environment 465, 216–225.

Borjesson, P., Berglund, M., 2007. Environmental systems analysis of biogas systems—Part II: the environmental impact of replacing various reference systems. Biomass and Bioenergy 31, 326–344.

Boyer, S., Wratten, S., Pizey, M., Weber, P., 2011. Impact of soil stockpiling and mining rehabilitation on earthworm communities. Pedobiologia 54 (Suppl.), S99–S102.

Brewster, J., Oleszkiewicz, J., Bujoczek, G., Reimers, R., Abu-Orf, M., Bowman, D., Fogarty, E., 2003. Inactivation of *Ascaris suum* eggs in digested and dewatered biosolids with lime and fly ash at bench scale and full scale. Journal of Environmental Engineering and Science 2, 395–400.

Brofas, G., Michopoulos, P., Alifragis, D., 2000. Sewage sludge as an amendment for calcareous bauxite mine spoils reclamation. Journal of Environmental Quality 29, 811–816.

Brown, S., Chaney, R.L., Hallfrisch, J.G., Xue, Q., 2003. Effect of biosolids processing on lead bioavailability in an urban soil. Journal of Environmental Quality 32, 100–108.

Brown, S., Chaney, R., Hallfrisch, J., Ryan, J.A., Berti, W.R., 2004. In situ soil treatments to reduce the phyto- and bioavailability of lead, zinc, and cadmium. Journal of Environmental Quality 33, 522–531.

Brown, S., Kruger, C., Subler, S., 2008. Greenhouse gas balance for composting operations. Journal of Environmental Quality 37, 1396–1410.

Brown, S., Mahoney, M., Sprenger, M., 2014. A comparison of the efficacy and ecosystem impact of residual-based and topsoil-based amendments for restoring historic mine tailings in the Tri-State mining district. Science of the Total Environment 485-486, 624–632.

Brunetti, G., Donner, E., Laera, G., Sekine, R., Scheckel, K.G., Khaksar, M., Vasilev, K., De Mastro, G., Lombi, E., 2015. Fate of zinc and silver engineered nanoparticles in sewerage networks. Warter Research 77, 72–84.

Caccio, S.M., De Giacomo, M., Aulicino, F.A., Pozio, E., 2003. *Giardia* cysts in wastewater treatment plants in Italy. Applied and Environmental Microbiology 69, 3393–3398.

Capar, S.G., Tanner, J.T., Friedman, M.H., Boyer, K.W., 1978. Multielement analysis of animal feed, animal wastes, and sewage sludge. Environmental Science & Technology 12, 785–790.

Castillejo, J.M., Castello, R., 2010. Influence of the application rate of an organic amendment (Municipal Solid Waste [MSW] Compost) on gypsum quarry rehabilitation in semiarid environments. Arid Land Research and Management 24, 344–364.

Chiu, C.Y., Tian, G., 2011. Chemical structure of humic acids in biosolids-amended soils as revealed by NMR spectroscopy. Applied Soil Ecology 49, 76–80.

Chowdhury, M.A., de Neergaard, A., Jensen, L.S., 2014. Potential of aeration flow rate and bio-char addition to reduce greenhouse gas and ammonia emissions during manure composting. Chemosphere 97, 16–25.

Chowdhury, S., Bolan, N.S., Seshadri, B., Kunhikrishnan, A., Wijesekara, H., Xu, Y., Yang, J., Kim, G.H., Sparks, D., Rumpel, C., 2015. Co-composting solid biowastes with alkaline materials to enhance carbon stabilization and revegetation potential. Environmental Science and Pollution Research. http://dx.doi.org/10.1007/s11356-015-5411-9.

Clarke, B.O., Smith, S.R., 2011. Review of 'emerging' organic contaminants in biosolids and assessment of international research priorities for the agricultural use of biosolids. Environment International 37, 226–247.

de Abreu, M.F., Berton, R.S., de Andrade, J.C., 1996. Comparison of methods to evaluate heavy metals in organic wastes. Communications in soil Science and Plant Analysis 27, 1125–1135.

Diamond, B., 1999. Recent developments in mountaintop removal mining: West Virginia rivers are not the coal industry's private Dump. Environmental Law 6, 891–912.

Donner, E., Howard, D.L., de Jonge, M.D., Paterson, D., Cheah, M.H., Naidu, R., Lombi, E., 2011. X-ray absorption and micro X-ray Fluorescence spectroscopy investigation of copper and zinc speciation in biosolids. Environmental Science & Technology 45, 7249–7257.

Donner, E., Ryan, C.G., Howard, D.L., Zarcinas, B., Scheckel, K.G., McGrath, S.P., de Jonge, M.D., Paterson, D., Naidu, R., Lombi, E., 2012. A multi-technique investigation of copper and zinc distribution, speciation and potential bioavailability in biosolids. Environmental Pollution 166, 57–64.

Doley, D., Audet, P., 2013. Adopting novel ecosystems as suitable rehabilitation alternatives for former mine sites. Ecological Processes 2, 1–11.

Elliott, H.A., O'Connor, G.A., Brinton, S., 2002. Phosphorus leaching from biosolids-amended sandy soils. Journal of Environmental Quality 31, 681–689.

Emmerling, C., Paulsch, D., 2001. Improvement of earthworm (*Lumbricidae*) community and activity in mine soils from open-cast coal mining by the application of different organic waste materials. Pedobiologia 45, 396–407.

Evangelou, V.P., Zhang, Y.L., 1995. A review: pyrite oxidation mechanisms and acid mine drainage prevention. Critical Reviews in Environmental Science and Technology 25, 141–199.

Evans, T.D., 2012. Biosolids in Europe. In: 26th WEF Residuals & Biosolids Conference, Raleigh, NC, USA.

Galvín, R.M., Cardenete López, J.M., Rodríguez Mellado, J.M., 2009. Chemical characterization of biosolids from three Spanish WWTPs: transfer of organics and metallic pollution from urban wastewater to biosolids. CLEAN – Soil, Air, Water 37, 52–59.

García Becerra, F.Y., Acosta, E.J., Allen, D.G., 2010. Alkaline extraction of wastewater activated sludge biosolids. Bioresource Technology 101, 6972–6980.

Gardner, W.C., Broersma, K., Naeth, A., Chanasyk, D., Jobson, A., 2010. Influence of biosolids and fertilizer amendments on physical, chemical and microbiological properties of copper mine tailings. Canadian Journal of Soil Science 90, 571–583.

Gibbs, R., Hu, C., Ho, G., Unkovich, I., 1997. Regrowth of faecal coliforms and *salmonellae* in stored biosolids and soil amended with biosolids. Water Science and Technology 35, 269–275.

Ghose, M., 2001. Management of topsoil for geo-environmental reclamation of coal mining areas. Environmental Geology 40, 1405–1410.

Haney, R.L., Jin, V.L., Johnson, M.-V.V., White, M.J., Arnold, J.G., 2015. On-site assessment of extractable soil nutrients after long-term biosolid applications to perennial forage. Communications in Soil Science and Plant Analysis 46, 873–887.

He, L., Zhu, T., Cao, T., Li, W., Zhang, M., Zhang, X., Ni, L., Xie, P., 2015. Characteristics of early eutrophication encoded in submerged vegetation beyond water quality: a case study in Lake Erhai, China. Environmental Earth Sciences 74, 3701–3708.

Herath, D.N., Lamont, B.B., Enright, N.J., Miller, B.P., 2009. Impact of fire on plant-species persistence in post-mine restored and natural shrubland communities in southwestern Australia. Biological Conservation 142, 2175–2180.

Hettiarachchi, G.M., Pierzynski, G.M., Oehme, F.W., Sonmez, O., Ryan, J.A., 2003. Treatment of contaminated soil with phosphorus and manganese oxide reduces lead absorption by Sprague-Dawley rats. Journal of Environmental Quality 32, 1335–1345.

Hettiarachchi, G.M., Scheckel, K.G., Ryan, J.A., Sutton, S.R., Newville, M., 2006. μ-XANES and μ-XRF investigations of metal binding mechanisms in biosolids. Journal of Environmental Quality 35, 342–351.

Hobbs, R.J., Higgs, E., Harris, J.A., 2009. Novel ecosystems: implications for conservation and restoration. Trends in Ecology & Evolution 24, 599–605.

Hooke, R.L., Martin-Duque, J.F., 2012. Land transformation by humans: a review, 22. GSA Today, pp. 4–10.

Hundal, L.S., Cox, A., Granato, T.C., Abedin, Z., 2008. Levels of dioxins in soil and corn tissues after 30 years of biosolids application. Journal of Environmental Quality 37, 1497–1500.

Ippolito, J.A., Barbarick, K.A., Paschke, M.W., Brobst, R.B., 2010. Infrequent composted biosolids applications affect semi-arid grassland soils and vegetation. Journal of Environmental Management 91, 1123–1130.

Islam, K.R., Ahsan, S., Barik, K., Aksakal, E.L., 2013. Biosolid impact on heavy metal accumulation and lability in soiln under alternate-year no-till corn–soybean rotation. Water Air and Soil Pollution 224, 1–10.

Islas-Espinoza, M., Solís-Mejía, L., Esteller, M.V., 2014. Phosphorus release kinetics in a soil amended with biosolids and vermicompost. Environmental Earth Sciences 71, 1441–1451.

Jackson, B.P., Miller, W.P., 2000. Soil solution chemistry of a fly ash-, poultry litter-, and sewage sludge-amended soil. Journal of Environmental Quality 29, 430–436.

Johnson, P., Dixon, R., Ross, A., 1998. An in-vitro test for assessing the viability of *Ascaris suum* eggs exposed to various sewage treatment processes. International Journal of Parasitology 28, 627–633.

Johnson, D.B., 2003. Chemical and microbiological characteristics of mineral spoils and drainage waters at abandoned coal and metal mines. Water Air and Soil Pollution 3, 47–66.

Jones, B.E.H., Haynes, R.J., Phillips, I.R., 2010. Effect of amendment of bauxite processing sand with organic materials on its chemical, physical and microbial properties. Journal of Environmental Management 91, 2281–2288.

Jones, B.E.H., Haynes, R.J., Phillips, I.R., 2011. Influence of organic waste and residue mud additions on chemical, physical and microbial properties of bauxite residue sand. Environmental Science and Pollution Research 18, 199–211.

Jones, B.E.H., Haynes, R.J., Phillips, I.R., 2012. Addition of an organic amendment and/or residue mud to bauxite residue sand in order to improve its properties as a growth medium. Journal of Environmental Management 95, 29–38.

Kajitvichyanukul, P., Ananpattarachai, J., Amuda, O.S., Alade, A.O., Hung, Y.T., Wang, L.K., 2008. Landfilling engineering and management. In: Wang, L.K., Shammas, N.K., Hung, Y.T. (Eds.), Biosolids Engineering and Management. Humana Press, New Jersey, pp. 415–442.

Kato, S., Fogarty, E., Bowman, D., 2003. Effect of aerobic and anaerobic digestion on the viability of *Cryptosporidium parvum* oocysts and *Ascaris suum* eggs. International Journal of Environmental Health Research 13, 169–179.

Komilis, D.P., Ham, R.K., Park, J.K., 2004. Emission of volatile organic compounds during composting of municipal solid wastes. Warter Research 38, 1707–1714.

Kossoff, D., Dubbin, W.E., Alfredsson, M., Edwards, S.J., Macklin, M.G., Hudson-Edwards, K.A., 2014. Mine tailings dams: characteristics, failure, environmental impacts, and remediation. Applied Geochemistry 51, 229–245.

Kotowska, U., Żalikowski, M., Isidorov, V., 2012. HS-SPME/GC–MS analysis of volatile and semi-volatile organic compounds emitted from municipal sewage sludge. Environmental Monitoring and Assessment 184, 2893–2907.

Kuznetsova, T., Lukjanova, A., Mandre, M., Lõhmus, K., 2011. Aboveground biomass and nutrient accumulation dynamics in young black alder, silver birch and Scots pine plantations on reclaimed oil shale mining areas in Estonia. Forest Ecology and Management 262, 56–64.

Laarmann, D., Korjus, H., Sims, A., Kangur, A., Kiviste, A., Stanturf, J.A., 2015. Evaluation of afforestation development and natural colonization on a reclaimed mine site. Restoration Ecology 23, 301–309.

Lamb, D.T., Venkatraman, K., Bolan, N., Ashwath, N., Choppala, G., Naidu, R., 2013. Phytocapping: an alternative technology for the sustainable management of landfill sites. Critical Reviews in Environmental Science and Technology 44, 561–637.

Larney, F.J., Angers, D.A., 2012. The role of organic amendments in soil reclamation: a review. Canadian Journal of Soil Science 92, 19–38.

Lasobars, J., Dellunde, J., Jofre, J., Lucena, F., 1999. Occurrence and level of phages proposed as surrogate indicators of enteric viruses in different types of sludges. Journal of Applied Microbiology 86, 723–729.

Li, J., 2012. Effects of Biosolids on Carbon Sequestration and Nitrogen Cycling. Virginia Polytechnic Institute and State University (PhD thesis).

Li, S., Di, X., Wu, D., Zhang, J., 2013. Effects of sewage sludge and nitrogen fertilizer on herbage growth and soil fertility improvement in restoration of the abandoned open-cast mining areas in Shanxi, China. Environmental Earth Sciences 70, 3323–3333.

Lindsay, M.B.J., Moncur, M.C., Bain, J.G., Jambor, J.L., Ptacek, C.J., Blowes, D.W., 2015. Geochemical and mineralogical aspects of sulfide mine tailings. Applied Geochemistry 57, 157–177.

Liu, R., Lal, R., 2012. Nanoenhanced materials for reclamation of mine lands and other degraded soils: a review. Journal of Nanotechnology 2012, 1–17.

Lleo, T., Albacete, E., Barrena, R., Font, X., Artola, A., Sanchez, A., 2013. Home and vermicomposting as sustainable options for biowaste management. Journal of Cleaner Production 47, 70–76.

Lodder, W., Vinje, J., van De Heide, R., de Roda Husman, A., Leenen, E., Koopmans, M., 1999. Molecular detection of norwalk-like caliciviruses in sewage. Applied and Environmental Microbiology 65, 5624–5627.

Lu, Q., He, Z.L., Stoffella, P.J., 2012. Land application of biosolids in the USA: a review. Applied and Environmental Soil Science 2012.

Majumder, R., Livesley, S.J., Gregory, D., Arndt, S.K., 2014. Biosolid stockpiles are a significant point source for greenhouse gas emissions. Journal of Environmental Management 143, 34–43.

Martínez-Ruiza, C., Fernández-Santos, B., Putwain, P.D., Fernández-Gómez, M.J., 2007. Natural and man-induced revegetation on mining wastes: changes in the floristic composition during early succession. Ecological Engineering 30, 286–294.

Maulini-Duran, C., Artola, A., Font, X., Sanchez, A., 2013. A systematic study of the gaseous emissions from biosolids composting: raw sludge versus anaerobically digested sludge. Bioresource Technology 147, 43–51.

McBride, M.B., 1995. Toxic metal accumulation from agricultural use of sludge: are USEPA regulations protective? Journal of Environmental Quality 24, 5–18.

McMurry, P.H., Takano, H., Anderson, G.R., 1983. Study of the ammonia (gas)-sulfuric acid (aerosol) reaction rate. Environmental Science & Technology 17, 347–352.

Mingorance, M.D., Oliva, S.R., Valdes, B., Gata, F.J.P., Leidi, E.O., Guzman, I., Pena, A., 2014. Stabilized municipal sewage sludge addition to improve properties of an acid mine soil for plant growth. Journal of Soils and Sediments 14, 703–712.

Mocé-Llivina, L., Muniesa, M., Pimenta-Vale, H., Lucena, F., Jofre, J., 2003. Survival of bacterial indicator species and bacteriophages after thermal treatment of sludge and sewage. Applied and Environmental Microbiology 69, 1452–1456.

Monpoeho, S., Maul, A., Bonnin, C., Patria, L., Ranarijaona, S., Billaudel, S., Ferre, V., 2004. Clearance of human-pathogenic viruses from sludge: study of four stabilization processes by real-time reverse transcription-pcr and cell culture. Applied and Environmental Microbiology 70, 5434–5440.

Mora, A.P.D., Ortega-calvo, J.J., Cabrera, F., Madejon, E., 2005. Changes in enzyme activities and microbial biomass after "in situ" remediation of a heavy metal-contaminated soil. Applied Soil Ecology 28, 125–137.

Nelson, K., 2003. Concentrations and inactivation of *Ascaris* eggs and pathogen indicator organisms in wastewater stabilization pond sludge. Water Science and Technology 48, 89–95.

Nkoa, R., 2014. Agricultural benefits and environmental risks of soil fertilization with anaerobic digestates: a review. Agronomy for Sustainable Development 34, 473–492.

NWQMS, 2004. National Water Quality Management Strategy, Guidelines for Sewerage Systems Biosolids Management. Natural Resource Management Ministerial Council, Australian Water Association, Artarmon, NSW, ISBN 0-9581875-3-3.

Obrador, A., Rico, M., Alvarez, J., Novillo, J., 2001. Influence of thermal treatment on sequential extraction and leaching behaviour of trace metals in a contaminated sewage sludge. Bioresource Technology 76, 259–264.

O'Connor, G.A., Sarkar, D., Brinton, S.R., Elliott, H.A., Martin, F.G., 2004. Phytoavailability of biosolids phosphorus. Soil Science 175, 154–161.

Ohsowski, B.M., Klironomos, J.N., Dunfield, K.E., Hart, M.M., 2012. The potential of soil amendments for restoring severely disturbed grasslands. Applied Soil Ecology 60, 77–83.

Oladeji, O.O., Tian, G., Cox, A.E., Granato, T.C., O'Connor, C., Abedin, Z., Pietz, R.I., 2013. Effect of long-term application of biosolids for mine land reclamation on groundwater chemistry: nutrients and other selected qualities. Journal of Environmental Quality 42, 94–102.

O'Reilly, J.L., 1997. An Incubation Study to Assess the Effect of Waste Sludge Additions on Some Chemical Characteristics of Mine Spoils (M.Sc thesis Massey University, New Zealand).

Pallavicini, Y., Alday, J.G., Martínez-Ruiz, C., 2013. Factors affecting herbaceous richness and biomass accumulation patterns of reclaimed coal mines. Land Degradation & Development 26, 211–217.

Palumbo, A.V., Mccarthy, J.F., Amonette, J.E., Fisher, L.S., Wullschleger, S.D., Daniels, W.L., 2004. Prospects for enhancing carbon sequestration and reclamation of degraded lands with fossil-fuel combustion by-products. Advances in Environmental Research 8, 425–438.

Payment, P., Plante, R., Cejka, P., 2001. Removal of indicator bacteria, human enteric viruses, giardia cysts, and cryptosporidium oocysts at a large wastewater primary treatment facility. Canadian Journal of Microbiology 47, 188–193.

PDER, 1977. Utilization of Municipal Wastewater and Sludge for Land Reclamation and Biomass Production. The Pennsylvania Department of Environmental Resources, Washington, DC.

Peng, X., Luo, W., Zhang, J., Wang, S., Lin, S., 2002. Rapid detection of shigella species in environmental sewage by an immunocapture pcr with universal primers. Applied and Environmental Microbiology 68, 2580–2583.

Pepper, I.L., Zerzghi, H.G., Bengson, S.A., Lker, B.C., Banerjee, M.J., Brooks, J.P., 2012. Bacterial populations within copper mine tailings: long-term effects of amendment with Class A biosolids. Journal of Applied Microbiology 113, 569–577.

Pichtel, J.R., Hayes, J.M., 1990. Influence of fly ash on soil microbial activity and populations. Journal of Environmental Quality 19, 593–597.

Pourcher, A.M., Morand, P., Picard-Bonnaud, F., Billaudel, S., Monpoeho, S., Federighi, M., Ferré, V., Moguedet, G., 2005. Decrease of enteric micro-organisms from rural sewage sludge during their composting in straw mixture. Journal of Applied Microbiology 99, 528–539.

Powlson, D.S., Bhogal, A., Chambers, B.J., Coleman, K., Macdonald, A.J., Goulding, K.W.T., Whitmore, A.P., 2012. The potential to increase soil carbon stocks through reduced tillage or organic material additions in England and Wales: a case study. Agriculture, Ecosystems & Environment 146, 23–33.

Price, G.W., Astatkie, T., Gillis, J.D., Liu, K., 2015. Long-term influences on nitrogen dynamics and pH in an acidic sandy soil after single and multi-year applications of alkaline treated biosolids. Agriculture, Ecosystems & Environment 208, 1–11.

Raven, K.P., Loeppert, R.H., 1997. Trace element composition of fertilizers and soil amendments. Journal of Environmental Quality 26, 551–557.

Rosenfeld, P.E., Henry, C.L., Dills, R.L., Harrison, R.B., 2001. Comparison of odor emissions from three different biosolids applied to forest soil. Water Air and Soil Pollution 127, 173–191.

Shammas, N.K., Wang, L.K., 2008. Transport and pumping of sewage sludge and biosolids. In: Wang, L.K., Shammas, N.K., Hung, Y.T. (Eds.), Biosolids Engineering and Management. Humana Press, New Jersy, USA, pp. 1–64.

Sidhu, J., Gibbs, R., Ho, G., Unkovich, I., 2001. The role of indigenous microorganisms in suppression of salmonella regrowth in composted biosolids. Water Research 35, 913–920.

Sidhu, J.P.S., Toze, S.G., 2009. Human pathogens and their indicators in biosolids: a literature review. Environment International 35, 187–201.

Smernik, R.J., Oliver, I.W., Merrington, G., 2003. Characterization of sewage sludge organic matter using solid-state Carbon-13 nuclear magnetic resonance spectroscopy. Journal of Environmental Quality 32, 1516–1522.

Smith, P., Powlson, D.S., Smith, J.U., Falloon, P., Coleman, K., 2000. Meeting Europe's climate change commitments: quantitative estimates of the potential for carbon mitigation by agriculture. Global Change Biology 6, 525–539.

Sopper, W.E., 1992. Reclamation of mine land using municipal sludge. In: Lal, R., Stewart, B.A. (Eds.), Soil Restoration. Springer, New York, pp. 351–431.

Soriano-Disla, J., Navarro-Pedreño, J., Gómez, I., 2010. Contribution of a sewage sludge application to the short-term carbon sequestration across a wide range of agricultural soils. Environmental Earth Sciences 61, 1613–1619.

Su, D., Wong, J., 2004. Chemical speciation and phytoavailability of Zn, Cu, Ni and Cd in soil amended with fly ash-stabilized sewage sludge. Environment International 29, 895–900.

Thangarajan, R., Bolan, N.S., Tian, G., Naidu, R., Kunhikrishnan, A., 2013. Role of organic amendment application on greenhouse gas emission from soil. Science of the Total Environment 465, 72–96.

Tian, G., Granato, T.C., Cox, A.E., Pietz, R.I., Carlson, C.R., Abedin, Z., 2009. Soil carbon sequestration resulting from long-term application of biosolids for land reclamation. Journal of Environmental Quality 38, 61–74.

Tian, G., Chiu, C.-Y., Franzluebbers, A.J., Oladeji, O.O., Granato, T.C., Cox, A.E., 2015. Biosolids amendment dramatically increases sequestration of crop residue-carbon in agricultural soils in western Illinois. Applied Soil Ecology 85, 86–93.

Tjell, J.C., 1986. Trace metal regulations for sludge utilization in agriculture; a critical review. In: Hermite, P.L. (Ed.), Processing and Use of Organic Sludge and Liquid Agricultural Wastes. Springer, Netherlands, pp. 348–361.

Torrecillas, C., Martínez-Sabater, E., Gálvez-Sola, L., Agulló, E., Pérez-Espinosa, A., Morales, J., Mayoral, A.M., Moral, R., 2013. Study of the organic fraction in biosolids. Communications in Soil Science and Plant Analysis 44, 492–501.

Torri, S., Alberti, C., 2012. Characterization of organic compounds from biosolids of Buenos Aires city. Journal of Soil Science and Plant Nutrition 12, 143–152.

Tripathi, N., Singh, R.S., Nathanail, C.P., 2014. Mine spoil acts as a sink of carbon dioxide in Indian dry tropical environment. Science of the Total Environment 468-469, 1162–1171.

UN, 2015. World Urbanization Prospects: The 2014 Revision, Highlights (ST/ESA/SER.A/352). Department of Economic and Social Affairs, Population Division (2014).

USEPA, 1979. Process Design Manual for Sludge Treatment and Disposal. United States Environmental Protection Agency. EPA 625/1-79-011. Document viewed on December 21, 2014. http://nepis.epa.gov/Exe/ZyPURL.cgi?Dockey=P10037H6.TXT.

USEPA, 1983. Process Design Manual for Land Application of Municipal Sludge. United States Environmental Protection Agency, EPA-625/1-83-016. Document viewed on December 23, 2014. http://nepis.epa.gov/Exe/ZyPDF.cgi/300044ZS.PDF?Dockey=300044ZS.PDF.

USEPA, 1994. A Plain English Guide to the EPA Part 503 Biosolids Rule. United States Environmental Protection Agency. EPA/832/R-93/003. Document viewed on December 23, 2014. http://nepis.epa.gov/Exe/ZyPURL.cgi?Dockey=200046QX.TXT.

Vickers, H., Gillespie, M., Gravina, A., 2012. Assessing the development of rehabilitated grasslands on post-mined landforms in North West Queensland, Australia. Agriculture, Ecosystems & Environment 163, 72–84.

Walter, I., Martinez, F., Cala, V., 2006. Heavy metal speciation and phytotoxic effects of three representative sewage sludges for agricultural uses. Environmental Pollution 139, 507–514.

Wang, L.K., Shammas, N.K., Evanylo, G., 2008. Biosolids engineering and management. In: Wang, L.K., Shammas, N.K., Hung, Y.T. (Eds.), Engineering and Management of Agricultural Land Application. Humana Press, New Jersy, pp. 343–414.

Waterhouse, B.R., Boyer, S., Adair, K.L., Wratten, S.D., 2014. Using municipal biosolids in ecological restoration: what is good for plants and soil may not be good for endemic earthworms. Ecological Engineering 70, 414–421.

Wiegleb, G., Felinks, B., 2001. Primary succession in post-mining landscapes of Lower Lusatia: chance or necessity. Ecological Engineering 17, 199–217.

Wilson, S.C., Alcock, R.E., Sewart, A.P., Jones, K.C., 1997. Persistence of organic contaminants in sewage sludge-amended soil: a field experiment. Journal of Environmental Quality 26, 1467–1477.

Yager, T.J., McMahon, P.B., 2012. Preliminary Assessment of Sources of Nitrogen in Groundwater at a Biosolids-application Area Near Deer Trail. US Geological Survey.

Zaleski, K.J., Josephson, K.L., Gerba, C.P., Pepper, I.L., 2005. Potential regrowth and recolonization of salmonellae and indicators in biosolids and biosolid-amended soil. Applied and Environmental Microbiology 71, 3701–3708.

Zanuzzi, A., Arocena, J.M., Van Mourik, J.M., Cano, A.F., 2009. Amendments with organic and industrial wastes stimulate soil formation in mine tailings as revealed by micromorphology. Geoderma 154, 69–75.

Zinati, G.M., Li, Y., Bryan, H.H., Mylavarapu, R.S., Codallo, M., 2004. Distribution and fractionation of phosphorus, cadmium, nickel, and lead in calcareous soils amended with composts. Journal of Environmental Science and Health. Part. B, Pesticides, Food Contaminants, and Agricultural Wastes 39, 209–223.

Zhai, W., Moschandreas, D.J., Tian, G., Venkatesan, D., Noll, K.E., 2014. Degradation rate model formulation to estimate soil carbon sequestration from repeated biosolids application. Soil Science Society of America Journal 78, 238–247.

CHAPTER

4

APPLICATION OF BIOCHAR PRODUCED FROM BIOWASTE MATERIALS FOR ENVIRONMENTAL PROTECTION AND SUSTAINABLE AGRICULTURE PRODUCTION

S. Mandal[1], A. Kunhikrishnan[2], N.S. Bolan[3], H. Wijesekara[3], R. Naidu[3,4]

University of South Australia, Mawson Lakes, SA, Australia[1]; National Academy of Agricultural Science, Wanju-gun, Jeollabuk-do, Republic of Korea[2]; University of Newcastle, Callaghan, NSW, Australia[3]; Cooperative Research Centre for Contamination Assessment and Remediation of the Environment, Callaghan NSW, Australia[4]

CHAPTER OUTLINE

Environmental Materials and Waste. http://dx.doi.org/10.1016/B978-0-12-803837-6.00004-4

4.1 INTRODUCTION

Biochar is a carbon-rich compound which can be produced by the thermal decomposition of organic biomass in the absence of oxygen at a relatively low temperature (<700°C) (Clough et al., 2013). Biochar has created a lot of interests because of its unique properties of sustainable agricultural production and environmental protection. Biochar is known to reduce nitrogen (N) loss from soils in terms of nitrous oxide (N_2O) emission and ammonia (NH_3) volatilization, improve nutrient retention capacity and structural and chemical properties of soil, and increase plant growth and productivity (Singh et al., 2010; Clough et al., 2013; Clough and Condron, 2010; Lehmann, 2007).

Biowaste materials such as biosolids, municipal waste, and paper mill sludge are effective raw materials for biochar production. These materials are rich in nutrient contents, and the biochars produced using these waste products are highly effective for agricultural production and environmental remediation. This chapter will first discuss the different types of biowaste materials and their applications, then the synthesis and characterization of biochars, and finally the environmental and agricultural applications of biochars.

4.2 SOURCES OF BIOWASTE

Large quantities of organic biowastes, including municipal solid waste (MSW), biosolids, animal and poultry manure, and paper mill sludge, are currently being produced (Torri et al., 2014). Municipalities generate two major sources of organic wastes that can be used as soil amendments. These organically based by-products include MSW and biosolids (often referred to as sewage sludge). Only some of the MSW is eventually recycled. The global MSW production rate was 1.3 billion tons in 1994 and rose by 31% to 1.7 billion tons in 2008 (Foo and Hameed, 2009).

4.2.1 MUNICIPAL SOLID WASTE

MSW is mainly derived from the disposal of general waste streams that include green waste, food waste, and miscellaneous products (ie, leather, textile, metal scraps), which can be separated as noncompostable materials (Hargreaves et al., 2008; Menikpura and Basnayake, 2009). Although most MSW ends up in unsightly landfill sites, a significant quantity has been used to produce different value-added products such as compost, feedstuffs, and biogas (García et al., 2005, Vithanage et al., 2014). MSW compost (MSWC) is increasingly being used in agriculture by a number of countries as a soil conditioner and a fertilizer (Hargreaves et al., 2008). MSWC has a high content of organic matter, which is a good soil conditioner. The high content of N and phosphorus (P) in MSWC increases its nutritional value (Table 4.1) (Hargreaves et al., 2008; Garcı;a-Gil et al., 2000). However, elevated levels of heavy metals and a high salt concentration in MSWC negatively affect soil health (Hargreaves et al., 2008; García-Gil et al., 2000).

4.2.2 BIOSOLIDS

An increased number of high-capacity wastewater treatment facilities generate large quantities of biosolids. Biosolids typically contain 40–70% organic matter, with organic carbon (C) content ranging from 20% to 50% (Torri et al., 2014). Biosolids contain around 3.2%, 2.3% and 0.3% of N, P, and potassium (K), respectively, which add nutritional value (Kajitvichyanukul et al., 2008). Some heavy metal(loid)s in biosolids are present in high concentrations. For instance, zinc (Zn), copper (Cu), nickel (Ni), and lead (Pb) have been observed in significant concentrations in biosolids (Table 4.1). Although there are several end uses of biosolids in agriculture, eg, landscaping, forestry, and land rehabilitation, large quantities of biosolids end up in landfills and stockpiles because of public opposition to their use on land. Potential health risks associated with the pathogens and contaminants in biosolids, plus attraction to disease-causing vectors, are some major factors for local public opposition (Wang et al., 2008).

4.2.3 ANIMAL AND POULTRY MANURE

With the continuous decline in the availability of land area for crop production, the increase in food demand for human consumption is likely to be met mainly through intensive poultry and livestock production (Thangarajan et al., 2013). Manure is considered to be a by-product that contains almost all of the essential elements necessary for plants. The nutritional value of livestock and poultry manure for crop and pasture growth has been well-described in previous studies (Bolan et al., 2010). Metal contamination in manure occurs in many ways. Feed additives that are rich in metals serve as a direct source of contamination (Bolan et al., 2004). However, soil ingestion by livestock (ie, soil ingestion is an important source of Cd contamination by grazing sheep and cattle) and metal contamination during manure collection (ie, usage of pen made of crashed coal ash results in higher concentration of Cu and Zn) and serves as an indirect source (Loganathan et al., 1999; Bolan et al., 2004; Sweeten et al., 1995). A number of feed additives such as arsenic (As), cobalt (Co), Cu, iron (Fe), manganese (Mn), and selenium (Se) are used in intensive livestock management to prevent or reduce diseases, improve weight gain, and improve egg production in poultry (Bolan et al., 2004; Jackson and Bertsch, 2001).

4.2.4 PAPER MILL SLUDGE

The paper mill and pulp industry produces enormous quantities of paper and pulp products each year. It is the sixth largest polluting industry after the oil, cement, leather, textile, and steel industries, and many environmental contaminants are associated with the discharge of paper and pulp mill sludge (Ali and Sreekrishnan, 2001). About 60–94% of organic content is available in paper mill sludge, which has the potential for use as a soil amendment in disturbed lands (Marko and Polonca, 2012). Sludge rich in organic matter is generated in high content in the paper and pulp industries. Although paper and pulp mill sludge is rich in organic matter, it contains less N and P than biosolids and compost (Park et al., 2011). Hence, paper mill sludge often needs additional nutrient input to be used in mine spoil rehabilitation (Park et al., 2011). Paper and pulp mill sludge is managed through its use in landfills and as landfill capping materials, in land spreading, composting, land reclamation, and in employment in brick, light aggregate, and cement production (Marko and Polonca, 2012).

4.2.5 OTHER ORGANIC WASTES

A range of other organic waste products, such as digestates and yard and wood wastes, are generated in significant quantities (Allen et al., 2007; Nada et al., 2012). These organic wastes are derived from the biological treatment or anaerobic digestion of organic wastes such as biosolids, MSW, animal manure, or some other industrial wastes (ie, sugar and ethanol production, wastewater from food and agro-allied industries). These digestates are often rich in organic matter as well as nutrient content, including N and P (Möller and Müller, 2012). Yard wastes include plant leaves, shrub or tree trimmings, and lawn clippings. Municipalities collect yard waste and produce compost or wood chips (ie, to produce fuel, mulch, or compost). Wood wastes are mainly derived from wood-processing facilities (WPF) that include tree chips and bark chips. Wood wastes serve as a significant and beneficial source for an organic matter amendment or mulching material. They are being increasingly used as fuel in industrial boilers and therefore are not readily available for other purposes (Fig. 4.1) (Venner et al., 2011).

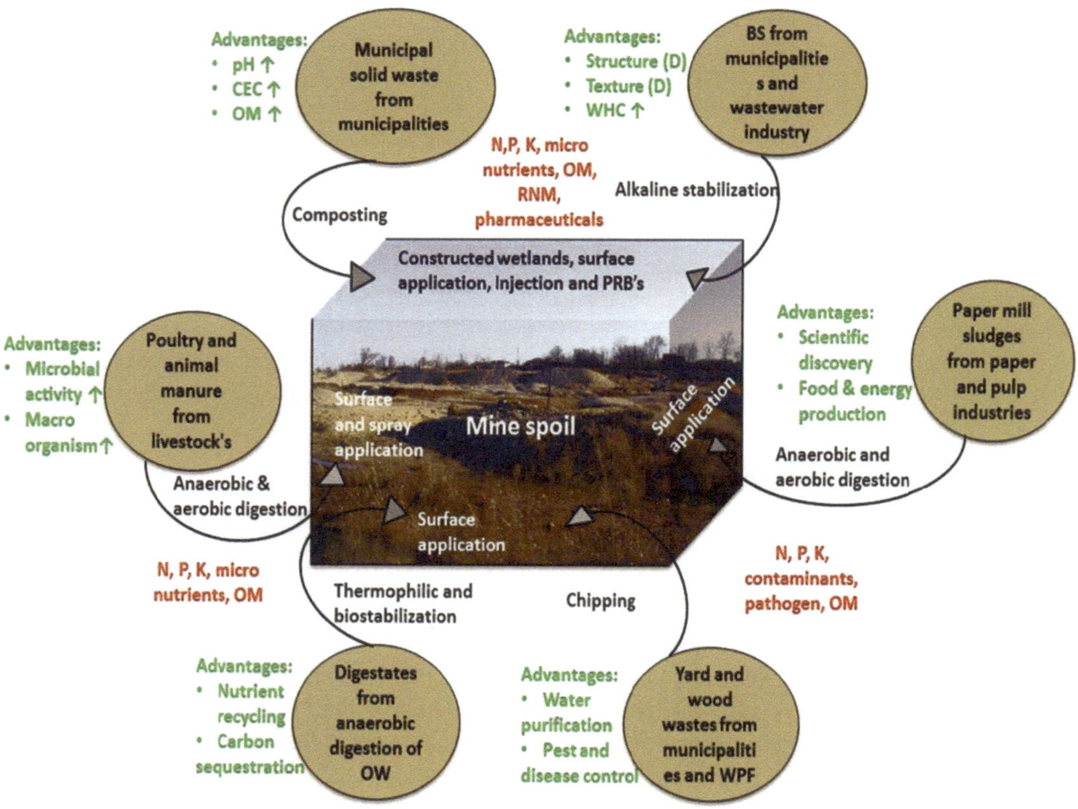

FIGURE 4.1

Various approaches to the use of different biowastes for mine spoil rehabilitation. *BS*, biosolids; *D*, develop; *CEC*, cation exchange capacity; *WHC*, water-holding capacity; *OM*, organic matter; *RNM*, reactive nanomaterials; *PRB*, permeable reactive barriers; *OW*, organic waste; *WPF*, wood-processing facilities.

4.3 SYNTHESIS AND CHARACTERIZATION OF BIOCHAR

4.3.1 PRODUCTION METHODS

Pyrolysis is the thermochemical decomposition of organic matter into noncondensable gases, condensable liquids, and a solid residual coproduct, biochar or charcoal in an inert environment (ie, in the absence of oxygen). The end product of pyrolysis can be controlled by optimizing pyrolysis parameters such as temperature and residence time. Biochar is known to reduce N loss from the soil in terms of nitrous oxide (N_2O) emission and ammonia (NH_3) volatilization, improve nutrient retention capacity and the cation exchange capacity (CEC) of soils, and improve soil physical, chemical and biological properties and thus improve the soil health and plant growth (Knowles et al., 2011; Lehmann, 2007; Mandal et al., 2015).

The types of pyrolysis used to produce biochar and some of the properties of these pyrolysis techniques are described in Table 4.1.

Table 4.1 Pyrolysis Production Techniques and Operating Parameters

Pyrolysis Systems	Main Products	By-Products	Temperature (°C)	Heating Rate (°C/s)
Slow pyrolysis	Biochar, biogas	Bio-oil	300–550	0.01–2
Fast pyrolysis	Gas	Liquid and very low biochar	>600	$>10^5$
Torrefaction	Torrified biomass or biochar	Gas	200–300	0.83
Gasification	Gas	Liquid and char	>800	–
Hydrothermal carbonization	Biocoal/biochar and water	Gas	180–220	–

Slow pyrolysis is a low-temperature pyrolysis in which the biomass is less finely crushed and subjected to longer residence times and low heating rates to obtain high quantities of biochar at the expense of significant bio-oil formation (Klass, 1998). In fast pyrolysis, feedstock is finely ground to allow fast heat transfer. This finely ground feedstock is subjected to a moderate temperature with shorter residence time to produce high-quality ethylene-rich gases (syngas).

Torrefaction is a modified pyrolysis technique in which biomass is slowly heated to within a specified temperature range and retained there for a particular time interval in an inert or limited oxygen environment (Özçimen and Ersoy-Meriçboyu, 2010). Gasification rearranges the molecular structure of solid or liquid biomass to produce gaseous fuel as the primary product and liquid/solid as secondary products. Unlike pyrolysis, gasification requires a gasifying medium such as steam, air, or oxygen to rearrange the molecular structure; it also adds hydrogen to the feedstock and strips hydrocarbon from the feedstock (Lv et al., 2004). It is an exothermic process in which biomass is subjected to elevated temperatures in aqueous media (water) in a confined system (Bridgwater and Grassi, 1991). This process transforms all C from the biomass to coal (100% C efficiency) with no gaseous products.

4.3.2 BIOCHAR CHARACTERIZATION

The physical and chemical properties of biochar greatly depend on the feedstock properties and production parameters such as the pyrolysis temperature, residence time, and reaction rate. The magnitude of interactions between biochar and the soil environment can be explained by the surface area of biochar. In addition to the feedstock properties, the surface area of biochar greatly depends on pyrolysis conditions including the temperature, gas flow, and presence of air. The porosity of biochar is defined as the amount of total pore volume relative to the total bulk biochar volume. Pore sizes of less than 2 nm, 2–50 nm, and greater than 50 nm are termed micro-, meso-, and macropores, respectively, and are often related to their respective adsorption capacity. For example, micropores (<2 nm) have high adsorptive capacities for small molecules such as gases and common solvents (Lehmann and Joseph, 2009). Thus, the addition of biochar with high micropores to soil may reduce N loss as NH_3 volatilization. Similar to surface area (SA) the pore structure greatly depends on the pyrolysis temperature. Scanning electron microscopy (SEM) is a potential technique for studying the physical structure of biochar. SEM images are useful for obtaining accurate details about the pore structure of biochars (Özçimen and Ersoy-Meriçboyu, 2010).

The biochar pH can be measured using a standard laboratory pH meter. An increase in pyrolysis temperature has shown to either increase (Cao et al., 2009) or decrease (Uchimiya et al., 2010) total acidity. The addition of biochar had shown to increase pH, thereby enhancing the immobilization of heavy metals (Cd(II), Cu(II), Ni(II), and Pb(II)) (Uchimiya et al., 2010). Metal(loid) retention by biochar can also be explained by its precipitation as hydroxide, carbonate, or phosphate compounds. Elemental or proximate analysis is the quantification of organic content (C) in the biochar (Litaor et al., 2005). Traditionally, thermogravimetric method was used; elemental or CHN analysis that quantifies C, H, and N content in the biochar is currently being used (Inyang et al., 2012).

Aromaticity is defined as the fraction of C in biochar that forms aromatic bands. Biochars with dense aromatic structures are more resistant to oxidation and microbial degradation (Spokas et al., 2009). Low-temperature biochars are likely to have lower aromaticity than are high-temperature biochars (Keiluweit et al., 2010). The ash composition of a biochar sample can be quantified by digesting or leaching the sample and then measuring the concentration of ions in the resulting solution by atomic absorption spectroscopy or inductively coupled plasma atomic emission spectroscopy (Singh et al., 2010). Several techniques including diffuse reflectance infrared Fourier transform spectroscopy, X-ray photoelectron spectroscopy, energy-dispersive X-ray spectroscopy, and near-edge X-ray absorption fine structure spectroscopy have been used to examine the surface and functional chemistry of biochar (Lehmann et al., 2006).

4.4 ENVIRONMENTAL APPLICATIONS

4.4.1 HEAVY METAL(LOID)S

The contamination of agricultural and industrially affected urban lands with heavy metal(loid)s, nutrients, and pesticides is increasing rapidly. Among the organic materials used to remediate heavy metal(loid)s' contaminated sites, the use of biochar has received considerable interest. In the case of heavy metal(loid)s, biochar decreases bioavailability through adsorption, precipitation, and redox reactions (Fig. 4.2).

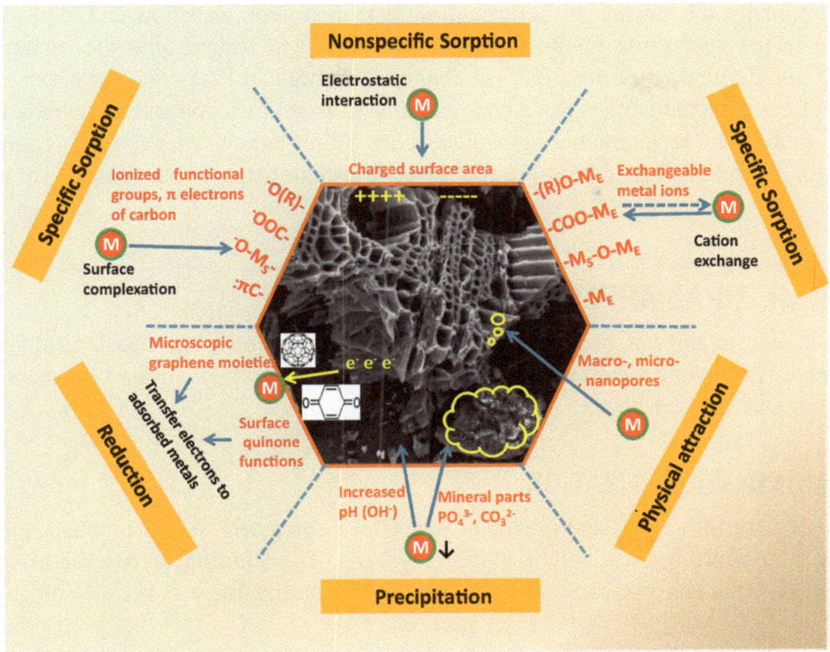

FIGURE 4.2

Schematic diagram of biochar sorption for metals.

Biochar has a high specific surface area, and the reactive functional groups in biochar may complex and sorb metal(loid)s in contaminated soils. The sorption capacity of biochars depends on its feedstock properties and the pyrolysis temperature. Oxygen-containing functional groups present on biochar surface stabilize heavy metal(loid)s such as Pb(II) and Cu(II) in acidic and low-organic C soils (Uchimiya et al., 2011). The sorption of mercury (Hg) on a biochar surface is an irreversible reaction and sorbs via complexation with phenolic, hydroxyl, and carboxylic functional groups on low-temperature biochars (Dong et al., 2013). High Fe(III) oxide content in biochar may increase the anionic contaminants' sorption via precipitation. However, the adsorption capacity of anionic contaminants by biochars is not strong because a significant amount of adsorbed metal(loid)s are released during desorption experiments (Agrafioti et al., 2014). Biochar can influence soil biogeochemistry and reduce chromate (Cr(VI)) contamination by a reduction reaction through oxygen-containing functional groups (Choppala et al., 2012; Kappler et al., 2014).

4.4.2 ORGANIC CONTAMINANTS

Organic contaminants include herbicides, pesticides, and plant and animal tissues, and are usually expected to cause adverse impacts on the environment. Trace levels of organic contaminant residues present in the soil, water, air, and sometimes food may result in harmful effects for human and environmental health (Kookana et al., 1998). Biochar application can reduce the toxicity of these organic contaminants; the mechanism is through surface adsorption, partition, and sequestration (Tang et al., 2013). In the adsorption method, organic pollutants get sorbed onto the surface of biochar because of a surface interaction reaction. Biochars with a high surface area and good pore size distribution are effective in reducing organic contaminants from soils (Kwon and Pignatello, 2005). Other mechanisms such as π–π interaction as well as pore filling have been proven to contribute to the increased sorption and decreased desorption of organic contaminants in the presence of biochar (Qiu et al., 2014).

4.5 AGRICULTURAL APPLICATIONS

Biochar application to soil has been shown to decrease the loss of N through leaching and gaseous emission, thereby increasing the efficiency of N use by plants (Fig. 4.3). Similarly, biochar application enhances the uptake of P by plants.

4.5.1 AMMONIA VOLATILIZATION

The physical (eg, surface area) and chemical (eg, CEC) characteristics of biochars influence their effectiveness in controlling NH_3 volatilization (Fig. 4.4). The addition of biochar to soil normally increases soil pH, but it can also capture NH_3 and reduce volatilization loss because of its high surface

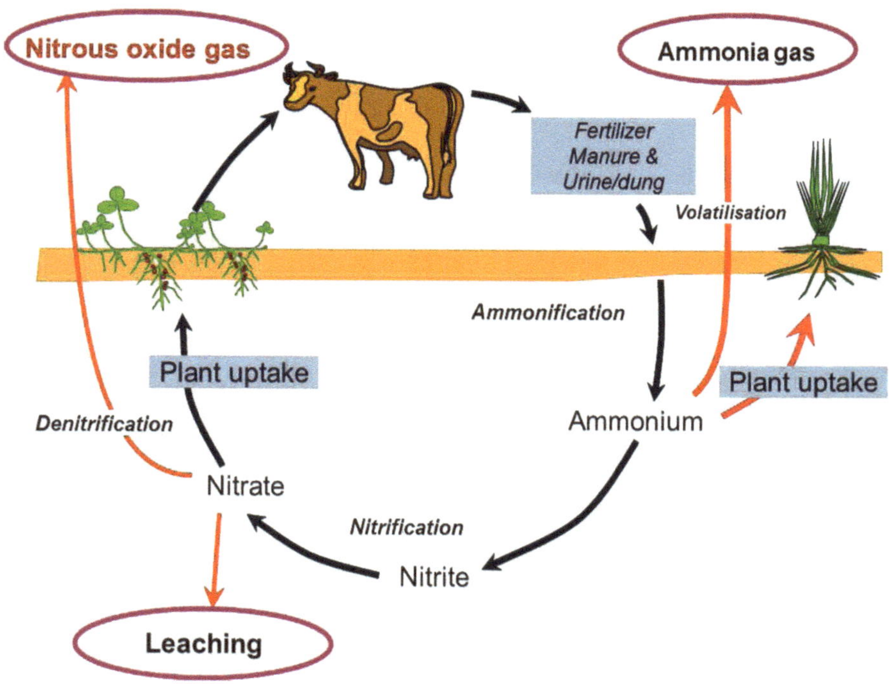

FIGURE 4.3

Dynamics of nitrogen transformation in soil (Bolan et al., 2004).

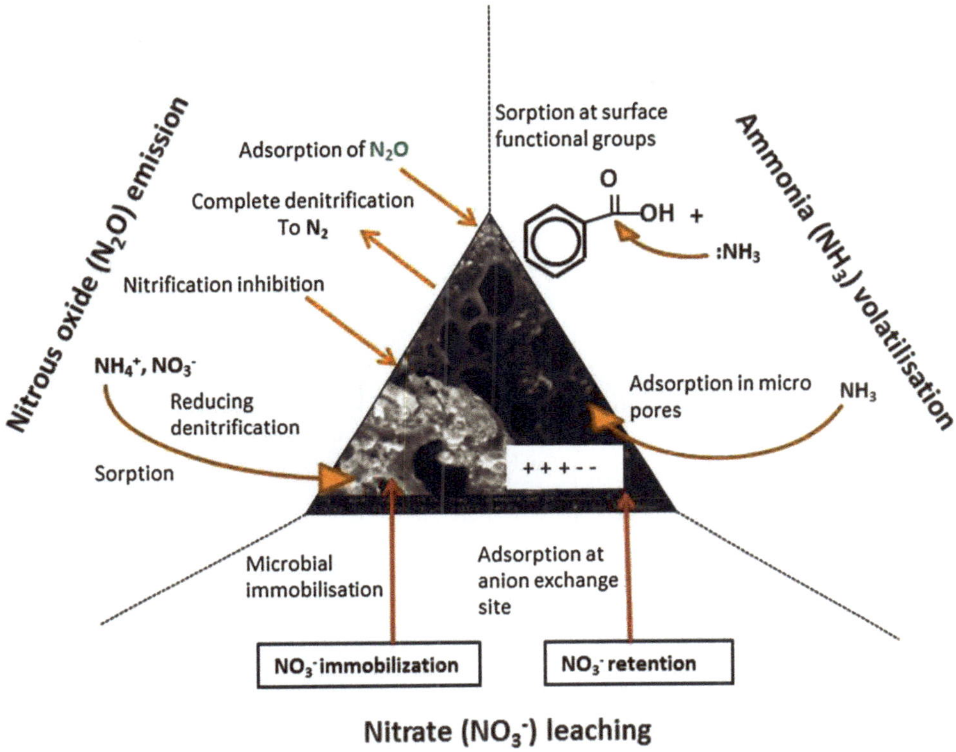

FIGURE 4.4

Schematic diagram of mechanisms for biochar-induced mitigation of ammonia volatilization, nitrous oxide emission, and nitrate leaching.

area and functional groups (Mandal et al., 2015). Biochar with a high specific surface area contains several surface functional groups that can capture NH_3. Because ammonia is an alkaline gas, the acidic surface groups on biochar with a low pH can protonate NH_3 gas to form NH_4^+ ions, thereby promoting their adsorption onto cation exchange sites (Bandosz, 2006) (Eq. [4.1]). The NH_3 sorbed by biochar can subsequently become available for plants (Taghizadeh-Toosi et al., 2011) (Eq. [4.1]):

$$NH_3 + H^+ \rightarrow NH_4^+ \qquad [4.1]$$

4.5.2 NITROUS OXIDE EMISSION

The addition of biochar has often been shown to decrease total N_2O emission from soil treated with N sources such as manures, urea, and composts (Bruun et al., 2012; Singh et al., 2010; Spokas et al., 2009; Van Zwieten et al., 2009) (Table 4.2). Denitrification is the biological process leading to increased N_2O emission from soil. Possible explanations for reduced N_2O loss from biochar-treated soils include reduced denitrification and/or complete denitrification (Bruun et al.,

Table 4.2 Effect of Biochar Addition on Ammonia (NH_3) Volatilization and Nitrous Oxide (N_2O) Emission From Soils

Biochar	N Source (Amount Added)	Soil Type	% NH_3 or N_2O Reduction	References
Monterey pine biochar	Ruminant urine patches	Templeton silt loam soil	N_2O, >50	Taghizadeh-Toosi et al. (2011)
Monterey pine wood chips	Urine (509 kg N/ha)	Silt loam	NH_3−45	Taghizadeh-Toosi et al. (2011)
Wood biochar, poultry manure biochar	–	Alfisol Vertisol	N_2O Alfisol, 14–73 Vertisol, 23–52	Singh et al. (2010)
Wheat straw biochar	No urea Urea (300 kg N/ha)	Hydroagric Stagnic Anthrosol	N_2O No urea, 40–51 Urea, 21–28	Zhang et al. (2014)
Peanut hull biochar, maize biochar, wood chip biochar, beech charcoal biochar	–	Sandy loam	N_2O, about 60	Kammann et al. (2012)
Wood chip biochar	–	Waukegan silt loam	N_2O, 20, 40, and 60 w/w biochar → 57, 57, and 74% reduction, respectively	Spokas et al. (2009)

2012; Spokas et al., 2012). A decrease in denitrification is likely to occur as a result of sorption of mineral N compounds (NH_4^+, NO_3^-) to biochar surface and lattice, thus reducing the substrate for denitrification (Taghizadeh-Toosi et al., 2012). Complete denitrification leading to N_2 emission owing to biochar addition was explained by the presence of labile C in biochar, elevated soil pH, and enhanced microbial activities (Anderson et al., 2011). Lehmann et al. (2006) also hypothesized that biochar can reduce N_2O emissions by inducing microbial immobilization of the mineral N in the soil.

4.5.3 NITRATE LEACHING

Nitrate (NO_3^-) leaching is another major loss of N from soils. The properties of biochar, such as the surface area, CEC, EC, and pH, facilitate the sorption of ions from the soil solution by a combination of electrostatic and capillary forces on their surface, thereby reducing nutrient leaching from soils. For example, it has been shown that biochar produced at high temperatures reduced NO_3^- leaching more significantly compared with low-temperature biochar, which may be attributed to greater sorption by former samples (Clough et al., 2013; Knowles et al., 2011). Similarly, Yao et al. (2012) found that the application of biochar reduced NO_3^- leaching by 34% through adsorption. Therefore, soil amended with biochar can adsorb NO_3^- through its anion exchange sites, thereby reducing its loss and increased NO_3^- retention in the soil.

4.5.4 PHOSPHORUS LEACHING

Excessive application of P fertilizers has caused P leaching from agricultural fields to aquatic systems (Karunanithi et al., 2014; Loganathan et al., 2014). Biochar has been proven to alter P availability in soils by reducing P leaching through sorption/adsorption characteristics (Fig. 4.5). In a column study, biochar produced from Brazilian pepperwood at 600°C reduced the total amount of phosphate in the leachates by about 20.6% (Yao et al., 2012). Doydora et al. (2011) observed that application of peanut hull biochar increased the amount of phosphate in soil solution by 39%. Possible mechanisms suggested for the influence of biochar on P availability are changes in soil pH and the subsequent influence on the interaction of P with other cations, or enhanced retention through anion exchange (Atkinson et al., 2010). In the natural environment, P is strongly adsorbed onto the surface of Fe(III)-(hydr)oxides in soils (Jaisi et al., 2010). The addition of biochars reduced the amount and rate of P sorbed onto ferrihydrite (the most effective Fe-oxide for P) (Cui et al., 2011). For example, Chen et al. (2011) demonstrated that biochars with magnetized Fe^{3+}/Fe^{2+} enhanced phosphate sorption compared with nonmagnetic char.

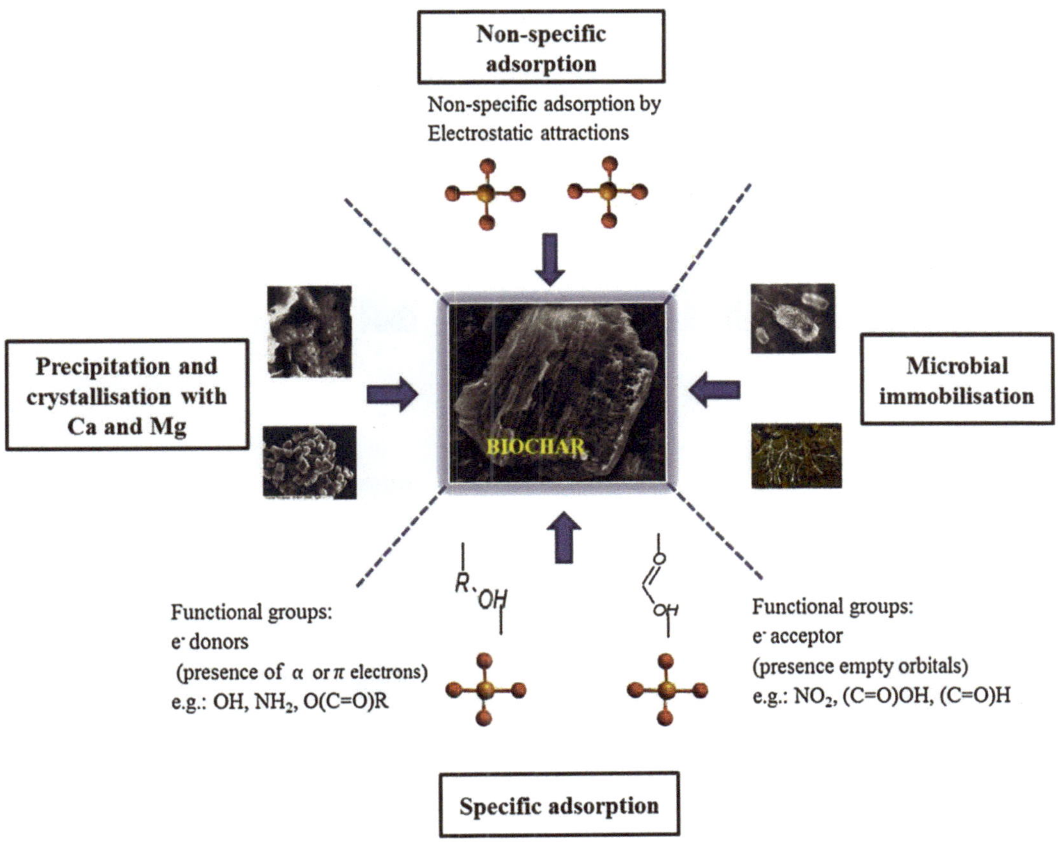

FIGURE 4.5

Schematic diagram of mechanisms of biochar-induced immobilization of phosphorus.

4.6 CASE STUDIES

In a laboratory column study, Kelly et al. (2014) assessed biochar made from pine wood as a soil amendment by mixing soil material from two mine sites (Bonner mine and Joe John mine) collected near Silverton, Colorado, United States. The mineralization event resulted from sulfur-rich hydrothermal fluids that produced mineral assemblages, which generally include abundant pyrite. The Bonner mine and Joe John mine had an initial pH of 2.99 and 4.41, respectively, and acidity content of 1.11 and 0.26 meq/g soil, respectively (Fig. 4.6).

Pine wood biochar was applied at four application rates (0%, 10%, 20%, and 30% V:V). Columns were leached seven times over 65 days and the leachate pH and concentration of toxic elements and base cations were measured at each leaching. The soil pH increased in both tailings with the addition of biochar , but significant differences were detected only in the Bonner material at the highest biochar addition level, increasing from 3.33 in the pure tailing to 3.63 with 30% biochar ($P = 0.0044$). In the

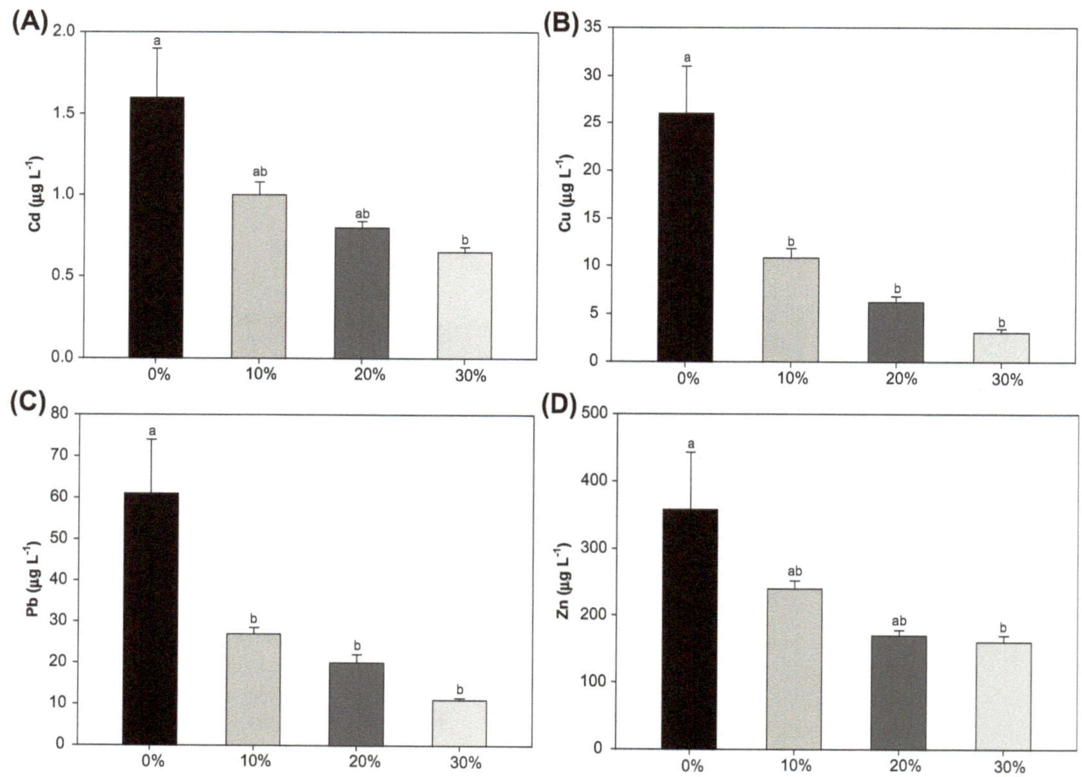

FIGURE 4.6

Concentration of dissolved metals in leachate from the Joe John material as affected by biochar treatments ($n = 3$). Means followed by different letters are considered different owing to biochar application according to Tukey's honestly significantly difference (at $\alpha = 0.05$). Error bars show standard error of the mean (Kelly et al., 2014).

Joe John tailing, a significant increase in pH was noted with just 10% biochar added; it ranged from 4.07 in the pure material to 4.77 with 30% biochar ($P < 0.0001$).

In the leachate solution, biochar increased base cations from both materials and reduced the concentrations of Cd, Cu, Pb, and Zn in the leachate solution from Joe John material. The pine wood biochar was more effective at increasing the soil pH and increasing the cation content in the tailing with a relatively higher initial pH and was less effective in the tailing with a lower pH. This study shows that biochar may be an effective strategy for the reclamation of soils if the application rate is great enough to provide sufficient sorption capacity and sufficient to raise the pH to mitigate potential toxicity.

In a study by Singh et al. (2010), biochars were produced at two different temperatures (400°C and 550°C) from poultry manure mixed with rice hull (PM) and chipped stem-wood of *Eucalyptus saligna* (W). To develop a highly porous structure, toward the end of the pyrolysis process the high temperature biochars (PM550 and W550) were activated using steam.

Soil—biochar and soil—sand (control) mixtures were set up in a specially designed column with an outlet for leachate collection (NH_4^+-N, NO_3^--N analysis) and an air-tight screw cap for gas sampling (N_2O analysis). Columns were subjected to three wetting—drying (W-D) cycles to achieve a range of water-filled pore space. Leachate and gas samples were collected during every W-D cycle.

During the first two W-D cycles, PM400 had a higher N_2O emission than control treatment, which was attributed to the presence of higher labile intrinsic N content in PM400 compared with the other biochars. Nitrous oxide emission from other treatments (PM550, W400, and W550) was inconsistent during the first two W-D cycles. The possible explanation for N_2O emission during these cycles was attributed to the denitrification activity triggered owing to the high moisture content along with the

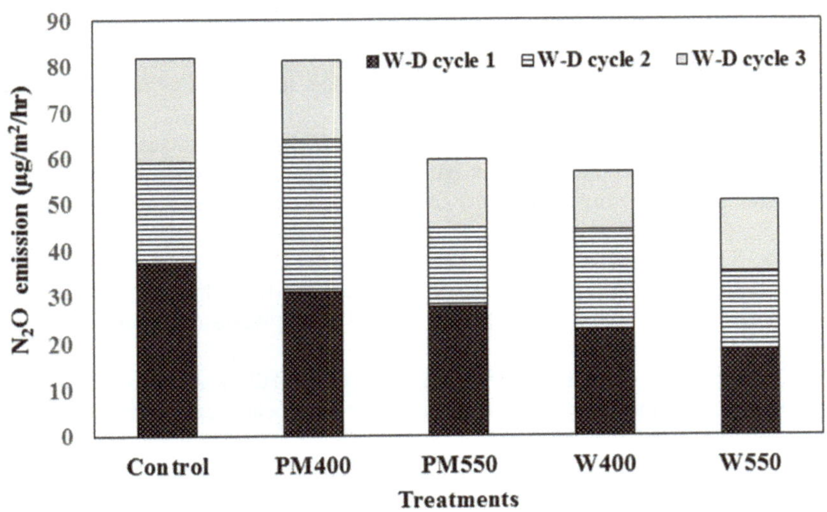

FIGURE 4.7

Mean nitrous oxide emissions (µg/m^2/h) from control soils (Alfisol and Vertisol) and biochar treatments during the three wetting-drying (W-D) cycles. PM400 and PM550—poultry manure and rice hull mixture biochar were produced at 400°C and 550°C, respectively; W400 and W550- chipped stem wood of *Eucalyptus saligna* biochar were produced at 400°C and 500°C, respectively (Singh et al., 2010).

application of nutrient solution and glucose (Ahmad et al., 2014). Nitrous oxide emission declined after the initial peak in all of the treatments and was significantly lower in biochar treatments than that in control treatment, which was attributed to the biochar-induced increase in soil pH and the occurrence of complete denitrification (N_2O to N_2) (Van Zwieten et al., 2009). The limited anion exchange capacity of biochars was one reason for the low retention of NO_3^--N on the biochars. However, compared with the control, significantly lower NO_3^--N in the leachate from the biochar treatments was observed during the later stages of leaching, which was attributed to NH_4^+-N fixation in the interlayer, thereby decreasing the substrate for nitrification. The authors concluded that the biochars decreased N loss as N_2O by 73% compared with the control treatment (Fig. 4.7). They also demonstrated that oxidation of biochar surfaces over the study period potentially increased the sorption capacity of the biochars.

4.7 CONCLUSION

Irrespective of their raw materials and characterization, biochars have multiple roles in improving soil properties, reducing heavy metal toxicity, and increasing plant growth and development. Biochar produced from biowaste materials is an effective soil amendment in terms of reducing greenhouse gas emissions (N_2O) and increasing agricultural productivity. However, appropriate production and application strategies should be followed to achieve agricultural productivity and reduce environmental contaminants.

REFERENCES

Agrafioti, E., Kalderis, D., Diamadopoulos, E., 2014. Arsenic and chromium removal from water using biochars derived from rice husk, organic solid wastes and sewage sludge. Journal of Environmental Management 133, 309–314.

Ahmad, M., Rajapaksha, A.U., Lim, J.E., Zhang, M., Bolan, N., Mohan, D., Vithanage, M., Lee, S.S., Ok, Y.S., 2014. Biochar as a sorbent for contaminant management in soil and water: a review. Chemosphere 99, 19–33.

Ali, M., Sreekrishnan, T.R., 2001. Aquatic toxicity from pulp and paper mill effluents: a review. Advances in Environmental Research 5, 175–196.

Allen, H.L., Brown, S.L., Chaney, R., Daniels, W.L., Henry, C.L., Neuman, D.R., Rubin, E., Ryan, J.A., Toffey, W., 2007. The Use of Soil Amendments for Remediation, Revitalization, and Reuse, EPA Office of Superfund Remediation and Technology Innovation (OSRTI). Solid Waste and Emergency Response (5203P). US, EPA, Cincinnati, OH.

Anderson, C.R., Condron, L.M., Clough, T.J., Fiers, M., Stewart, A., Hill, R.A., Sherlock, R.R., 2011. Biochar induced soil microbial community change: implications for biogeochemical cycling of carbon, nitrogen and phosphorus. Pedobiologia 54, 309–320.

Atkinson, C.J., Fitzgerald, J.D., Hipps, N.A., 2010. Potential mechanisms for achieving agricultural benefits from biochar application to temperate soils: a review. Plant and soil 337, 1–18.

Bandosz, T.J., 2006. Chapter 5 Desulfurization on activated carbons. Interface Science and Technology 7.

Bolan, N., Adriano, D., Mahimairaja, S., 2004. Distribution and bioavailability of trace elements in livestock and poultry manure by-products. Critical Reviews in Environmental Science and Technology 34, 291–338.

Bolan, N.S., Szogi, A.A., Chuasavathi, T., Seshadri, B., Rothrock, M.J., Panneerselvam, P., 2010. Uses and management of poultry litter. World's Poultry Science Journal 66, 673–698.

Bridgwater, A.V., Grassi, G., 1991. Biomass Pyrolysis Liquids: Upgrading and Utilisation. Springer.

Bruun, E.W., Ambus, P., Egsgaard, H., Hauggaard-Nielsen, H., 2012. Effects of slow and fast pyrolysis biochar on soil C and N turnover dynamics. Soil Biology and Biochemistry 46, 73–79.

Cao, X., Ma, L., Gao, B., Harris, W., 2009. Dairy-manure derived biochar effectively sorbs lead and atrazine. Environmental Science & Technology 43, 3285–3291.

Chen, B., Chen, Z., Lv, S., 2011. A novel magnetic biochar efficiently sorbs organic pollutants and phosphate. Bioresource Technology 102, 716–723.

Choppala, G.K., Bolan, N.S., Megharaj, M., Chen, Z., Naidu, R., 2012. The influence of biochar and black carbon on reduction and bioavailability of chromate in soils. Journal of Environmental Quality 41, 1175–1184.

Clough, T.J., Condron, L.M., 2010. Biochar and the nitrogen cycle: introduction. Journal of Environmental Quality 39, 1218–1223.

Clough, T.J., Condron, L.M., Kammann, C., Müller, C., 2013. A review of biochar and soil nitrogen dynamics. Agronomy (Basel) 3, 275–293.

Cui, H.-J., Wang, M.K., Fu, M.-L., Ci, E., 2011. Enhancing phosphorus availability in phosphorus-fertilized zones by reducing phosphate adsorbed on ferrihydrite using rice straw-derived biochar. Journal of Soils and Sediments 11, 1135–1141.

Dong, X., Ma, L.Q., Zhu, Y., Li, Y., Gu, B., 2013. Mechanistic investigation of mercury sorption by Brazilian pepper biochars of different pyrolytic temperatures based on X-ray photoelectron spectroscopy and flow calorimetry. Environmental Science and Technology 47, 12156–12164.

Doydora, S.A., Cabrera, M.L., Das, K.C., Gaskin, J.W., Sonon, L.S., Miller, W.P., 2011. Release of nitrogen and phosphorus from poultry litter amended with acidified biochar. International Journal of Environmental Research and Public Health 8, 1491–1502.

Foo, K., Hameed, B., 2009. An overview of landfill leachate treatment via activated carbon adsorption process. Journal of Hazardous Materials 171, 54–60.

García-Gil, J.C., Plaza, C., Soler-Rovira, P., Polo, A., 2000. Long-term effects of municipal solid waste compost application on soil enzyme activities and microbial biomass. Soil Biology and Biochemistry 32, 1907–1913.

García, A.J., Esteban, M.B., Márquez, M.C., Ramos, P., 2005. Biodegradable municipal solid waste: characterization and potential use as animal feedstuffs. Waste Management 25, 780–787.

Hargreaves, J.C., Adl, M.S., Warman, P.R., 2008. A review of the use of composted municipal solid waste in agriculture. Agriculture, Ecosystems and Environment 123, 1–14.

Inyang, M., Gao, B., Yao, Y., Xue, Y., Zimmerman, A.R., Pullammanappallil, P., Cao, X., 2012. Removal of heavy metals from aqueous solution by biochars derived from anaerobically digested biomass. Bioresource Technology 110, 50–56.

Jackson, B.P., Bertsch, P.M., 2001. Determination of arsenic speciation in poultry wastes by IC-ICP-MS. Environmental Science and Technology 35, 4868–4873.

Jaisi, D.P., Blake, R.E., Kukkadapu, R.K., 2010. Fractionation of oxygen isotopes in phosphate during its interactions with iron oxides. Geochimica et Cosmochimica Acta 74, 1309–1319.

Kajitvichyanukul, P., Ananpattarachai, J., Amuda, O.S., Alade, A.O., Hung, Y.T., Wang, L.K., 2008. Landfilling engineering and management. In: Wang, L.K., Shammas, N.K., Hung, Y.T. (Eds.), Biosolids Engineering and Management. Humana Press, New Jersy.

Kammann, C., Ratering, S., Eckhard, C., Müller, C., 2012. Biochar and hydrochar effects on greenhouse gas (carbon dioxide, nitrous oxide, and methane) fluxes from soils. Journal of environmental quality 41, 1052–1066.

Kappler, A., Wuestner, M.L., Ruecker, A., Harter, J., Halama, M., Behrens, S., 2014. Biochar as an electron shuttle between bacteria and Fe (III) minerals. Environmental Science and Technology Letters 1, 339–344.

Karunanithi, R., Szogi, A., Bolan, N., Naidu, R., Loganathan, P., Hunt, P.G., Vanotti, M.B., Saint, C.P., Ok, Y.S.A., Krishnamoorth, S., 2014. Phosphorus recovery and reuse from waste streams. Advances in Agronomy 131, 1–78.

Keiluweit, M., Nico, P.S., Johnson, M.G., Kleber, M., 2010. Dynamic molecular structure of plant biomass-derived black carbon (biochar). Environmental Science and Technology 44, 1247–1253.

Kelly, C.N., Peltz, C.D., Stanton, M., Rutherford, D.W., Rostad, C.E., 2014. Biochar application to hardrock mine tailings: soil quality, microbial activity, and toxic element sorption. Applied Geochemistry 43, 35–48.

Klass, D.L., 1998. Biomass for Renewable Energy, Fuels, and Chemicals. Academic press.

Knowles, O.A., Robinson, B.H., Contangelo, A., Clucas, L., 2011. Biochar for the mitigation of nitrate leaching from soil amended with biosolids. Science of the Total Environment 409, 3206–3210.

Kookana, R.S., Baskaran, S., Naidu, R., 1998. Pesticide fate and behaviour in Australian soils in relation to contamination and management of soil and water: a review. Australian Journal of Soil Research 36, 715–764.

Kwon, S., Pignatello, J.J., 2005. Effect of natural organic substances on the surface and adsorptive properties of environmental black carbon (char): pseudo pore blockage by model lipid components and its implications for N_2-probed surface properties of natural sorbents. Environmental Science and Technology 39, 7932–7939.

Lehmann, J., 2007. Bio-energy in the black. Frontiers in Ecology and the Environment 5, 381–387.

Lehmann, J., John, G., Rondon, M., 2006. Bio-char sequestration in terrestrial ecosystems – a review. Mitigation and Adaptation Strategies for Global Change 11, 403–427.

Lehmann, J., Joseph, S., 2009. Biochar for Environmental Management: Science and Technology. Earthscan.

Litaor, M., Seastedt, T., Walker, M., Carbone, M., Townsend, A., 2005. The biogeochemistry of phosphorus across an alpine topographic/snow gradient. Geoderma 124, 49–61.

Loganathan, P., Louie, K., Lee, J., Hedley, M.J., Roberts, A.H.C., Longhurst, R.D., 1999. A model to predict kidney and liver cadmium concentrations in grazing animals. New Zealand Journal of Agricultural Research 42, 423–432.

Loganathan, P., Vigneswaran, S., Kandasamy, J., Bolan, N.S., 2014. Removal and recovery of phosphate from water using sorption. Critical Reviews in Environmental Science and Technology 44, 847–907.

Lv, P., Xiong, Z., Chang, J., Wu, C., Chen, Y., Zhu, J., 2004. An experimental study on biomass air–steam gasification in a fluidized bed. Bioresource Technology 95, 95–101.

Mandal, S., Thangarajan, R., Bolan, N.S., Sarkar, B., Khan, N., Ok, Y.S., Naidu, R., 2015. Biochar-induced concomitant decrease in ammonia volatilization and increase in nitrogen use efficiency by wheat. Chemosphere 142, 120–127.

Marko, L., Polonca, T., 2012. In: Show, K.-Y. (Ed.), Recent Advances in Paper Mill Sludge Management, Industrial Waste. InTech.

Menikpura, S.N.M., Basnayake, B.F.A., 2009. New applications of "Hess Law" and comparisons with models for determining calorific values of municipal solid wastes in the Sri Lankan context. Renewable Energy 34, 1587–1594.

Möller, K., Müller, T., 2012. Effects of anaerobic digestion on digestate nutrient availability and crop growth: a review. Engineering in Life Sciences 12, 242–257.

Nada, W., Blumenstein, O., Claassens, S., Rensburg, L., 2012. Effect of wood compost on extreme soil characteristics in the Lusatian lignite region. Open Journal of Soil Science 2, 347–352.

Özçimen, D., Ersoy-Meriçboyu, A., 2010. Characterization of biochar and bio-oil samples obtained from carbonization of various biomass materials. Renewable Energy 35, 1319–1324.

Park, J.H., Lamb, D., Paneerselvam, P., Choppala, G., Bolan, N., Chung, J.-W., 2011. Role of organic amendments on enhanced bioremediation of heavy metal(loid) contaminated soils. Journal of Hazardous Materials 185, 549–574.

Qiu, L., Shao, Z., Xiang, P., Wang, D., Zhou, Z., Wang, F., Wang, W., Wang, J., 2014. Study on novel functional materials carboxymethyl cellulose lithium (CMC-Li) improve high-performance lithium-ion battery. Carbohydrate Polymers 110, 121–127.

Singh, B.P., Hatton, B.J., Singh, B., Cowie, A.L., Kathuria, A., 2010. Influence of biochars on nitrous oxide emission and nitrogen leaching from two contrasting soils. Journal of Environmental Quality 39, 1224–1235.

Spokas, K.A., Cantrell, K.B., Novak, J.M., Archer, D.W., Ippolito, J.A., Collins, H.P., Boateng, A.A., Lima, I.M., Lamb, M.C., Mcaloon, A.J., Lentz, R.D., Nichols, K.A., 2012. Biochar: a synthesis of its agronomic impact beyond carbon sequestration. Journal of Environmental Quality 41, 973–989.

Spokas, K.A., Koskinen, W.C., Baker, J.M., Reicosky, D.C., 2009. Impacts of woodchip biochar additions on greenhouse gas production and sorption/degradation of two herbicides in a Minnesota soil. Chemosphere 77, 574–581.

Sweeten, J.M., Kantor, T., King, S., Amosson, S., Ruah, W., 1995. Manure Analysis Summary—Coal Ash Surfacing vs. Control Treatments for Beef Cattle Feed Yards, Texas Agricultural Extension Service (Result Demonstration Report, Amarillo, Texas).

Taghizadeh-Toosi, A., Clough, T.J., Condron, L.M., Sherlock, R.R., Anderson, C.R., Craigie, R.A., 2011. Biochar incorporation into pasture soil suppresses in situ nitrous oxide emissions from ruminant urine patches. Journal of Environmental Quality 40, 468–476.

Taghizadeh-Toosi, A., Clough, T.J., Sherlock, R.R., Condron, L.M., 2012. Biochar adsorbed ammonia is bioavailable. Plant and Soil 350, 57–69.

Tang, J., Zhu, W., Kookana, R., Katayama, A., 2013. Characteristics of biochar and its application in remediation of contaminated soil. Journal of Bioscience and Bioengineering 116, 653–659.

Thangarajan, R., Bolan, N.S., Tian, G., Naidu, R., Kunhikrishnan, A., 2013. Role of organic amendment application on greenhouse gas emission from soil. Science of The Total Environment 465, 72–96.

Torri, S.I., Correa, R.S., Renella, G., 2014. Soil carbon sequestration resulting from biosolids application. Applied and Environmental Soil Science 38 (1), 61–74.

Uchimiya, M., Chang, S., Klasson, K.T., 2011. Screening biochars for heavy metal retention in soil: role of oxygen functional groups. Journal of hazardous materials 190, 432–441.

Uchimiya, M., Lima, I.M., Klasson, K.T., Wartelle, L.H., 2010. Contaminant immobilization and nutrient release by biochar soil amendment: roles of natural organic matter. Chemosphere 80, 935–940.

Van Zwieten, L., Singh, B., Joseph, S., Kimber, S., Cowie, A., Chan, K.Y., 2009. Biochar and Emissions of Non-CO_2 Greenhouse Gases from Soil. Biochar for Environmental Management: Science and Technology. Earthscan, London, pp. 227–249.

Venner, K., Preston, C., Prescott, C., 2011. Characteristics of wood wastes in British Columbia and their potential suitability as soil amendments and seedling growth media. Canadian Journal of Soil Science 91, 95–106.

Vithanage, M., Wijesekara, S.S.R.M.D.H.R., Siriwardana, A.R., Mayakaduwa, S.S., Ok, Y.S., 2014. Management of municipal solid waste landfill leachate: a global environmental issue. In: Malik, A., Grohmann, E., Akhtar, R. (Eds.), Environmental Deterioration and Human Health. Springer, Netherlands.

Wang, L.K., Shammas, N.K., Evanylo, G., 2008. Biosolids engineering and management. In: Wang, L.K., Shammas, N.K., Hung, Y.T. (Eds.), Engineering and Management of Agricultural Land Application. Humana Press, New Jersy, USA.

Yao, Y., Gao, B., Zhang, M., Inyang, M., Zimmerman, A.R., 2012. Effect of biochar amendment on sorption and leaching of nitrate, ammonium, and phosphate in a sandy soil. Chemosphere 89, 1467–1471.

Zhang, X., Zhang, S., Yang, H., Feng, Y., Chen, Y., Wang, X., Chen, H., 2014. Nitrogen enriched biochar modified by high temperature CO_2-ammonia treatment: characterization and adsorption of CO_2. Chemical Engineering Journal 257, 20–27.

CHAPTER

PRODUCTION AND UTILIZATION OF BIOCHAR FROM ORGANIC WASTES FOR POLLUTANT CONTROL ON CONTAMINATED SITES

5

A. Kumar[1], I.J. Schreiter[2], A. Wefer-Roehl[2], L. Tsechansky[1], C. Schüth[2], E.R. Graber[1]

Agricultural Research Organization, The Volcani Center, Bet Dagan, Israel[1]; Technische Universität Darmstadt, Darmstadt, Hessen, Germany[2]

CHAPTER OUTLINE

5.1 INTRODUCTION

Since the industrial revolution, anthropogenic activities such as mining and smelting of metal ores, the use of pesticides, and the extensive use of industrial solvents and petroleum hydrocarbons (PHCs), combined with improper storage facilities, poor waste disposal practices, waste water irrigation, sewage sludge applications, and accidental releases have led to widespread

Environmental Materials and Waste. http://dx.doi.org/10.1016/B978-0-12-803837-6.00005-6

contamination of soils and groundwater with both inorganic and organic pollutants. Inorganic pollutants include heavy metals and metalloids such as cadmium (Cd), lead (Pb), mercury (Hg), chromium (Cr), nickel (Ni), copper (Cu), zinc (Zn), and arsenic (As). Organic pollutants include hydrocarbons such as polycyclic aromatic hydrocarbons (PAHs), polychlorinated biphenyls (PCBs), halogenated hydrocarbon compounds (HHCs) (ie, tetrachloroethylene, trichloroethylene, carbon tetrachloride), and PHCs (ie, benzene, toluene). These compounds are a serious global concern for rural and urban development as well as the agricultural sector, because they may accumulate in the soil and have a deleterious effect on environmental health (Khan et al., 2008; Beesley et al., 2010).

Soil remediation of both inorganic and organic pollutants may rely on in situ or ex situ technologies. In situ technologies include aerobic and anaerobic bioremediation and phytoremediation (Gomes et al., 2013; Ma et al., 2011), natural attenuation and enhanced natural attenuation (Wang et al., 2015), reactive barrier technologies, physical methods as capping with adsorbent (Tang et al., 2013), and chemical application, especially advanced oxidation methods (Liang et al., 2014). Ex situ technologies include land farming, soil slurry bioreactors, thermal desorption, and chemical treatments such as soil washing (Gomes et al., 2013; Tomei et al., 2013). The combination of natural attenuation and physical capping technologies is one of the most cost-effective and environmentally friendly remediation methods and provides the opportunity to use waste and plant biomass to induce immobilization, sorption, stabilization, and degradation of these contaminants (Gomes et al., 2013; Wang et al., 2015; Tang et al., 2013).

A relatively new technology tested over the past decade for its suitability for in situ remediation of contaminated soils is the amendment of contaminated soils with biochar. Biochar is the solid carbonaceous product of incomplete, anaerobic combustion of carbon (C)-rich biomass by a process known as pyrolysis. Biochar is especially attractive for in situ application because it is an excellent adsorbent for many inorganic and organic chemicals, causing their immobilization. In addition, it has a number of agronomic benefits when added to soil (Ghani et al., 2013; Gwenzi et al., 2015; Lehmann and Joseph, 2015; Waqas et al., 2014). Biochar may improve soil properties by retaining higher soil moisture and nutrient levels (Uchimiya et al., 2011a), by increasing microbial activity and diversity (Anderson et al., 2011; Kolton et al., 2011; Elad et al., 2012), and by improving soil productivity and plant health (Butnan et al., 2015; Jaiswal et al., 2015; Meller Harel et al., 2012; Graber et al., 2010). Unlike other organic soil amendments, biochar typically is not readily decomposed, and thus persists in the soil for extended periods of time. It has therefore also the potential for C-sequestration because it transfers C to a slower cycling pool compared with the biosphere pool (Kuhlbusch and Crutzen, 1995). For these reasons, and because pyrolysis is a technology that is suitable for treating various organic wastes, biochars gained considerable scientific and commercial interest over the past decade (Lehmann and Joseph, 2015).

5.2 BIOCHAR PRODUCTION AND PROPERTIES

Organic material can be thermally decomposed by pyrolysis, typically at temperatures between 350 and 800°C, to an array of solid, liquid (bio-oil), and gaseous products. The resulting C-rich, porous, and low density solid product is known as biochar (Fig. 5.1) (Ghani et al., 2013; Duku et al., 2011). A variety of feedstocks can be used, eg, forestry residues, agricultural by-products, coconut coir, wood waste, crop residue, poultry waste, dairy and cattle manure, the organic fraction of municipal solid waste, and industrial sewage. Generally, feedstocks with high lignin concentration have the highest biochar yield (Duku et al., 2011).

Also, the pyrolysis process itself has a number of variables, eg, the heating rate, final temperature (highest treatment temperature), and total processing time. These affect the quality and quantity of produced biochar directly and indirectly (Butnan et al., 2015; Kleber et al., 2015). In general, an increase in pyrolysis temperature decreases the biochar yield, which is typically between 10% and 30% of the dry biomass, but increases its specific surface area (SSA), porosity, and aromaticity, and the ash and fixed C content (Uchimiya et al., 2011a). Pyrolysis temperature also affects the concentration of oxygenated functional groups at the surfaces of biochars (Uchimiya et al., 2011a), such that biochars produced at lower temperature have more oxygenated functional groups than do higher temperature chars (Das and Sarmah, 2015). Low pyrolysis temperatures also result in a larger fraction of volatile carbon in the biochar, and increasing pyrolysis temperatures result in more biochars with increased stability in the soil (Uchimiya et al., 2011a; Sun et al., 2014). Predicted half-lives for biochar in soils with oxygen (O):C ratios greater than 0.6 are less than 100 years, whereas for O:C ratios lower than 0.2, half-lives of more than 1000 years were predicted (Spokas, 2010).

However, because of the number of variables influencing biochar properties during pyrolysis, biochar does not have a distinct structure or chemical composition, but rather a range of properties.

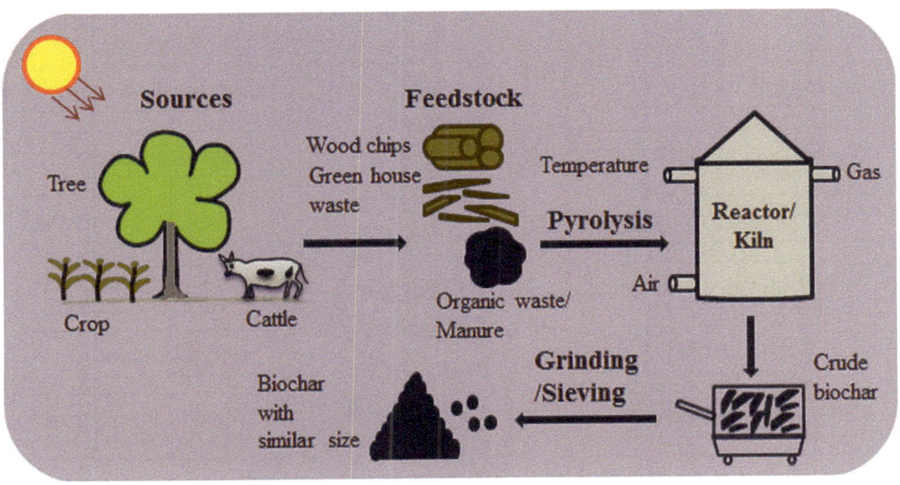

FIGURE 5.1

Schematic representation of biochar production from different feedstocks.

By using an appropriate combination of feedstock and pyrolysis conditions, it is therefore possible to influence the properties of biochar and thus tailor biochars for specific applications (Zhao et al., 2013).

5.3 BIOCHAR APPLICATION AT CONTAMINATED SITES

Several potential applications for biochar have been proposed, including energy production, C sequestration, improvement of soil fertility, and as a supplement in animal feed. The potential of biochars for the removal of various pollutants from aqueous solution has also been studied in detail (Anyika et al., 2015). Fig. 5.2 schematically shows different uses of biochar.

In addition, biochars have been suggested as potential amendment for soils at brownfields and other contaminated sites. Such sites are typically characterized by elevated concentrations of a variety of pollutants and by poorly developed soils with little or no vegetation cover (Bellamy et al., 2005). To reduce remediation costs, the employment of in situ addition of amendments to contaminated soils for contaminant immobilization and promoting plant growth and ecological restoration is receiving much attention (Vangronsveld et al., 2009). The former is usually based on C-rich amendments such as activated carbon (AC), used to reduce the bioavailability of the inorganic and organic pollutants in sediments (Cho et al., 2009). The latter, referred to as phytoremediation or phytomanagement (Bardos et al., 2011; Suthersan, 2001), is based on different approaches such as phytocontainment and phytostabilization, and prevents access to contaminants or reduces their mobility, or phytoextraction, in which soil contaminants, especially heavy metals, are incorporated into harvestable plant biomass. Organic contaminants can also be mineralized by phytodegradation directly in the root zone, or by phytostimulation of root zone microorganisms. However, brownfield soil quality is often poor and not suitable for cultivation, and might even be phytotoxic, inhibiting plant growth. In such cases, a series of

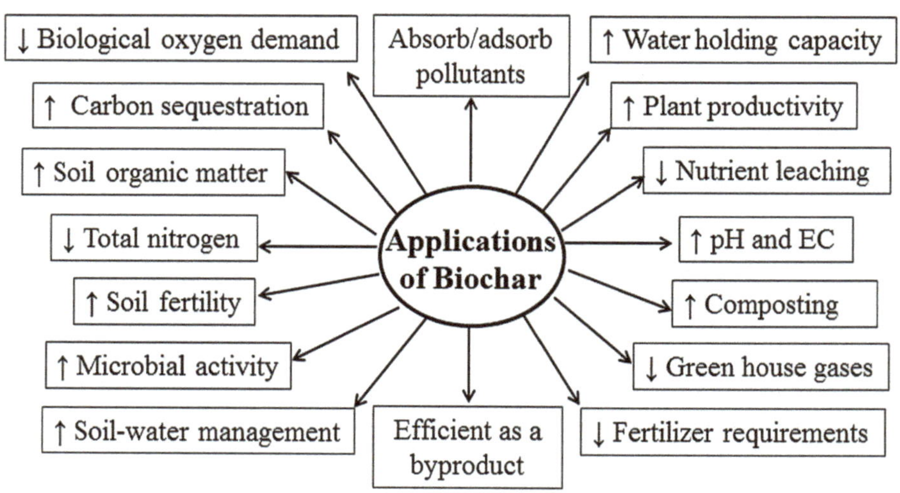

FIGURE 5.2

Schematic overview of primary biochar application: Induction (*up arrow*) and reduction (*down arrow*) mechanisms. *EC*, electrical conductivity.

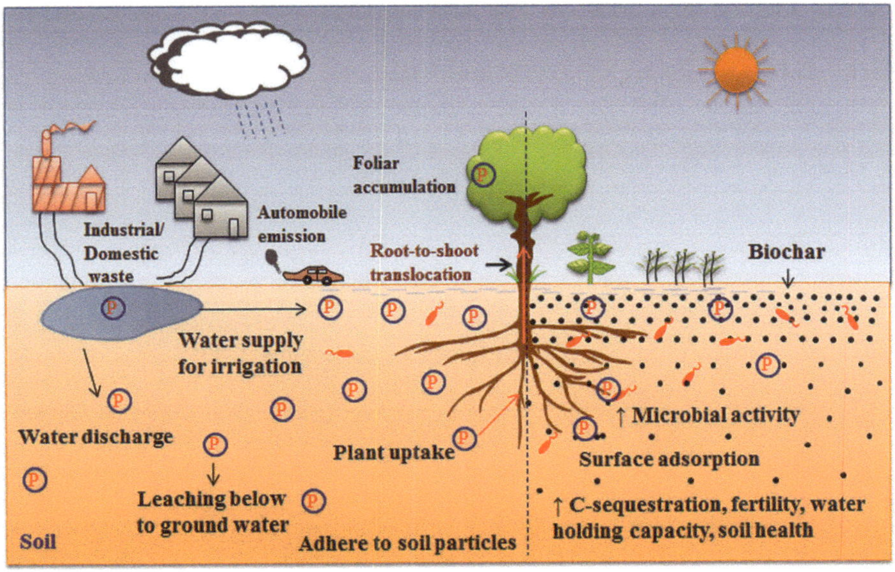

FIGURE 5.3

Functions of biochar to mitigate pollutants and improve soil health: interaction between biochar, pollutants, and plant. Emissions of toxic pollutants from domestic sewage, industrial waste, and automobile exhaust. Role of biochar in removing pollutants and environmental cleanup. *Dashed line* represents the difference between the function of biochar; the left side is without biochar and the right side is with biochar amendments. Biochar amendment to soil reduces pollutant bioavailability and its uptake by plants, adsorption/absorption of pollutants, and their degradation by increased microbial colonies; its activity leads to sustainable agriculture. *Circled P* represents pollutants.

"soil-forming" interventions are needed (Bardos et al., 2011), mainly based on adding inorganic or organic fertilizers to the soil.

The addition of biochar to contaminated brownfield soils can potentially contribute to site restoration by combining both contaminant immobilization and improvement of soil fertility. This is because of the ability of biochar to enhance soil physical, chemical, and biological functioning, reduce soil toxicity, promote phytocontainment and phytostabilization processes, augment abiotic stabilization through adsorption, and stimulate biotic and abiotic degradation of contaminants. This approach has therefore received much attention (Lehmann and Joseph, 2015; Uchimiya et al., 2010b; Beesley et al., 2011). In this chapter, we focus on interactions between pollutants and biochar, examining the key mechanisms involved in immobilization and remediation of various pollutants, shown schematically in Fig. 5.3. Some case studies are highlighted in Table 5.1. We also address the role of biochar in soil improvement.

5.3.1 BIOCHAR IN SOIL IMPROVEMENT

The addition of biochar to soil offers several opportunities for soil improvement and greenhouse gas mitigation by providing mineral nutrients and organic matter, and sequestering C. Biochar helps

Table 5.1 Biochar Production and Its Application

Raw Material/ Feedstock	PT	Chemical Composition	Characteristics	Products Application/ Process Control	Country	References
Tree wastes (*Conocarpus* species)	400°C	C:H:N:O (%) 76.18:2.53:0.42:18.67	Significant increase in pH, MC, and EC; decrease in SD	Reduced shoot heavy metal (HM) level in maize plants in response to increasing application rates	Saudi Arabia	Al-Wabel et al. (2015)
Pine sawdust (PSD), sewage sludge (SS), paunch grass (PG), broiler litter (BL), dewatered pond sludge (DWPS), and dissolved air floatation sludge (DAFS)	680°C	C:H:N:O (%) PSD-90.9:1.3:0.11:6.1 SS-77.9:2.1:0.5:19.3 PG-64.7:2.28:1.8:12.4 BL-86.7:1.89:1.3:10.7 DWPS-78.6:2.33:0.3:18.5 DAFS-66.8:1.9:0.7:30.53	Scanning electron micrograph showed that PSD and all other biochars are smooth, glassy, elongated, and lacking visible porosity	PSD biochar has potential for contaminant remediation and C-sequestration owing to its high surface area, C content, and aromaticity. All biochars are also applicable for land application	New Zealand, Australia	Srinivasan et al. (2015)
Pine wood chip (PB) and maize stubble (MB)	450°C	C:N:P:K:OM:Zn:Cu:As ($g\ kg^{-1}$) PB-837:3.6:148:1708:982:42:134:1.7 MB-686:7.9:2981:22331:794:99:41:0	Reduces contaminant bioavailability in soil	Positive effects on maize growth and negative effect on PAHs uptake and potentially toxic elements and useful for field-scale remediation efforts	United Kingdom	Brennan et al. (2014)
Hickory wood (HW), bagasse (BG), and bamboo (BB)	300, 450, and 600°C	Unit = %; C:H:N:O HW-45.51:6.17:0.15:47.83 BG-45.82:6.25:0.36:47.18 BB-46.52:6.11:0.20:46.89	Feedstock and PT affected production rate and physicochemical and biological properties	Biochars made at lower PT had higher production rates and higher PT increased C content and thermal stability and can be useful as soil amendments	China	Sun et al. (2014)

Sewage sludge	300 and 500°C	C:H:N (%) 300°C-39.7:4.1: 12.1500°C-9.8: 0.4:2.1 HMs (mg kg^{-1}): Cd-0.78; Cu-176.51; Ni-22.86; Pb-91.23; Cr-23.86	Increase PT reduces yield and increases SA	BC is more effective in adsorbing cations than anions and adsorbed 70% Cr(III) and 30% As(V) from aqueous solution.	Greece	Agrafioti et al. (2013)
Wheat straw (WS) and olive tree pruning (OTP)	370°C (WS), 450°C (OTP)	C:N:P:K:Ca:Mg: Ash (g kg^{-1}) WS-8:628:0.8: 8.0:5.4:1.4:251 OTP-10:484: 0.8:5.4:86:2.0: 391	WS showed higher EC and CEC than OTP and have germination indices 90% and 60%, respectively	Beneficial effect of biochar on wheat yield. Applicable for biochar-based soil management strategies and better crop production with mitigating global warming	Spain	Alburquerque et al. (2013)
Pig manure	600°C	C:H:N:O:Ash: VM (g kg^{-1}) 452:51:45:219: 222:697	Sawdust and compost addition influenced physical and chemical properties	Positive effect of sawdust and compost addition on biochar yield and renewable energy	Ireland	Troy et al. (2013)
Fraxinus excelsior (FE), *Fagussylvatica* (FS), and *Quercusrobur* (QR)	450°C		Chip size: 17% (0–2 mm), 19% (2–5 mm), 32% (5–7.5 mm), 32% (7.5–10 mm) and, dry BD of 0.20 ± 0.01 g cm^{-3}	Biochar showed significant induction on soil quality ie, P, K, EC, SM, OM, microbial activity, mycorrhizal root colonization and, crop performance	United Kingdom, Sweden	Quilliam et al. (2012)
Eight species of green tide filamentous algae	307, 414, 450, 512°C (±5°C)	C:H:N:Ash (%) 20.5–32.1: 1.9–4.0: 2.5–4.1: 10.5–33.8	Low in C, SA, and CEC, but high in pH, ash, N, and P, K, Ca, and mg.	Supplied nutrient to soils and thereby to crop product and will be useful for application on acidic soils	Australia	Bird et al. (2011)

PT, *pyrolysis temperature;* MC, *moisture content;* EC, *electrical conductivity;* SA, *surface area;* CEC, *cation exchange capacity;* OM, *organic matter;* SD, *soil density.*

improve soil organic matter, soil water-holding capacity, crop resistance to disease, and soil fertility (Elad et al., 2012; Chan et al., 2007; Graber and Elad, 2013). Biochar itself is chemically and biologically recalcitrant, and thus helps maintain soil health by decreasing its biological oxygen demand, nitrogen (N) and C requirements, greenhouse gas emissions, nutrient leaching from soil, and fertilizer requirements.

Biochar added to hard-setting soils was found to reduce the overall tensile strength of the soil and hence improve vegetation growth (Lehmann and Joseph, 2015). Biochar addition to contaminated soil helped to decrease the tensile strength and facilitated seed germination and root elongation deep into the soil (Lehmann et al., 2011). This is important because soils at contaminated sites are frequently of poor quality and have high tensile strengths. Biochar can further improve soil physical properties such as bulk density, soil moisture content, and soil aggregation (Al-Wabel et al., 2015). The large inner surface area and porous structure of biochar may have a role in improving soil water retention (Ghani et al., 2013). Another key aspect of soil improvement caused by adding biochar involves its influence on nutrient transformation and the nutrient cycle. Biochar produced from nutrient-rich biomass can act as a prime source of highly available nutrients, resulting in better soil nutrient condition and hence improving soil health (Al-Wabel et al., 2015). Whereas N and other volatile compounds are evaporated during pyrolysis, other essential nutrients become concentrated and remain in the product (Lehmann and Joseph, 2015).

5.3.2 INTERACTION OF CONTAMINANTS WITH BIOCHAR

Many biochars are excellent adsorbents for various types of pollutants, and adsorption behavior for a specific contaminant can be correlated to pollutant properties and biochar characteristics. Adsorption is typically a function of the biochar SSA, which is especially important for the adsorption of organic contaminants (Graber et al., 2011; Smernik, 2009), and the content of oxygenated surface acidic groups, which is important for the adsorption of metallic contaminants (Beesley and Marmiroli, 2011; Uchimiya et al., 2011b, 2012a). Surface acidic groups, which are oxygen-containing carboxyl, lactonic, and phenolic ligands, are substantial in biochars produced at relatively lower temperatures (Uchimiya et al., 2010b; Amonette and Joseph, 2009; Tsechansky and Graber, 2014). When biochar ages in the soil, its oxygenated functional group content increases significantly (Cheng et al., 2008); thus, its sorption capacity for metals is expected to increase with time in the soil. Alkaline biochars also induce decreases in dissolved metals concentrations as a result of precipitation of insoluble metal hydroxide, carbonate, and phosphate compounds (Kołodyńska et al., 2012; Xu and Zhao, 2013).

5.3.2.1 Inorganic Pollutants

Among the inorganic contaminants, heavy metals and metalloids are of greatest concern because of their high toxicity at low concentrations. Heavy metals in soil can be either bound to soil solid phases or freely available for uptake by organisms. In contrast to many organic contaminants, inorganic contaminants are generally sensitive to changes in pH and redox conditions in the soil environment. Therefore, not only do physical and chemical properties of the added biochar have an important role in the dynamic system of contaminant sorption, but so do the soil conditions and environment. Biochar has been shown to be effective in immobilizing, adsorbing, and sequestering a number of heavy metals including Pb, Cd, Cr, Hg, Ni, Cu, Zn, and metalloid As (Beesley et al., 2015). Application of biochar could change soil physiochemical properties by increasing the alkalinity or pH, surface area, and cation exchange capacity (CEC) of soil. These changes tend to result in immobilization of most heavy metals, resulting in a decrease of their bioavailability. Immobilization of heavy metals by biochar generally encompasses surface chemistry and surface adsorption, precipitation reactions, and complexation reactions (Beesley et al., 2015; Cao et al., 2009).

The thermal alterations of feedstock that accompany the pyrolysis process create various metal-binding functional groups on the surface of biochar. The amphoteric nature of biochar provides both acidic and basic functional groups including carboxylic, phenolic, pyridine, other heterocyclic N, and basic O and N donor groups. These functional groups all can take part in binding or adsorption of heavy metals (Lehmann and Joseph, 2015; Uchimiya et al., 2011a). Metal adsorption to biochar may occur in multiple stages and two levels, starting with fast but weaker interaction, followed by slower but stronger interactions (Limousin et al., 2007).

Metal solubility is influenced by the pH of the soil environment and is generally lower at higher pH. On the other hand, As, a metalloid most commonly present as an anionic oxide form, behaves differently from other metals with respect to pH, with its solubility and mobility rising as the pH increases (Uchimiya and Bannon, 2013). Interestingly, however, biochar amendments of a multielement polluted soil showed potential for immobilizing both water-soluble Cu and As in soil, which further resulted in their reduced toxicity to the plant (Beesley et al., 2010). Hartley et al. (2009) also observed no increase in As mobility to plants in three different brownfield soils amended with hardwood biochar (20% v/v) and planted with *Miscanthus*. The increase in alkalinity or pH of the soil resulting from amendment with biochar is generally caused by the available ash content and dissolution of hydroxides, carbonates, and basic cations, which facilitate the precipitation of metal ions by hydrolysis (Butnan et al., 2015; Lucchini et al., 2014). Biochar-induced reduction in bioavailability and mobility of heavy metals such as Cd, Pb, Cr, Cu, and Zn concentrations have been observed in various studies (Park et al., 2011; Houben et al., 2013). For example, adding 10% biochar to a contaminated soil reduced bioavailable metal concentrations by 71%, 87%, and 92% for Cd, Zn, and Pb, respectively, and promoted phytoremediation by rapeseed (Houben et al., 2013).

Regarding metal capture, in situ oxidation of aromatic C at biochar surface creates carboxyl groups which also result in higher CEC values (Denyes et al., 2012). CEC is a measure of negatively charged sites on the surface, which help in holding positively charged ions and nutrients such as sodium (Na), potassium (K), calcium (Ca), magnesium (Mg), and Zn (Liang et al., 2006). The CEC of biochar increases as the pH increases (Silber et al., 2010), and the effectiveness of a biochar in attracting cationic metals generally depends on its CEC (Uchimiya et al., 2011a). As discussed previously, biochar produced at lower pyrolysis temperatures have greater numbers of negatively charged functional groups. Over time, the number of negatively charged functional groups increases owing to oxidation of the surfaces (Uchimiya et al., 2011a).

A study by Hmid et al. (2014) investigated the effects of biochar produced from olive mill wastes on a Zn smelter-contaminated soil. The sampled topsoil was amended with different ratios of biochar (0%, 5%, 10%, and 15% w/w), moistened until saturation, equilibrated for 30 and 90 days in a greenhouse, and analyzed for calcium nitrate—extractable Zn, Pb, and Cd. The authors found that with an increasing rate of biochar application and longer equilibration time, the amount of extractable metals decreased significantly. This immobilization effect was attributed to several mechanisms. The pH of the soil increased as a result of the amendment of biochar, which led to decreased metal mobility in the soil. The number of negative surface charges also increased owing to amendment with biochar, which also favored the binding of the metal cations.

The importance of biochar surface functional groups on metal retention was demonstrated in several studies. Uchimiya et al. (2011a) attributed the increased retention of Ni, Cu, Pb, and Cd in an acidic sandy loam soil with low organic carbon content mainly to properties of the surface functional groups of the added biochars (pH, point of zero charge (PZC), volatile matter (VM), and O content). The biochars were all produced from cottonseed hulls at varying pyrolysis temperatures (200−800°C). Results revealed that biochar prepared at the lowest temperature had the greatest metal retention, which was controlled mainly by VM, ratios of O/C and N/C, pH, and PZC, all properties that are linked to oxygen-containing surface functional groups. These findings are supported by results published in Uchimiya et al. (2011b), who investigated the retention of heavy metals (Cu, Ni, Cd, and Pb) in a loamy sand and in a clay-rich soil. By evaluating the experimental sorption data with a statistical approach based on principal-component analysis and positive matrix factorization, they concluded that O-containing functional groups are the main factor ruling heavy metal stabilization. Concurrent with the uptake of the heavy metals, release of Na, Ca, K, Mg, phosphorus (P) and sulfur (S) was observed, pointing to ion exchange at the oxygenated sites as a major mechanism.

Uchimiya et al. (2012a) showed the significance of oxygen-containing functional groups by comparing the retention of Pb, Cu, and Zn on surface-oxidized versus untreated biochar. Biochar treated with different acids had a greater content of functional groups and consequently greater adsorption of Pb, Cu, and Zn from a contaminated soil. However, the Pb oxyanion was mobilized by the addition of the oxidized biochars, as a result of repulsive electrostatic forces. In an earlier study, Uchimiya et al. (2010a) examined the immobilization of Cu, Cd, Ni, and Pb in soil and water by different broiler-litter manure-derived biochars. They reported a significant pH drop after 24 h equilibration with metal ions in water as well as in experiments with soil. This indicated that protons were released from the biochar while the metals were taken up, which points to a cation exchange process as an important mechanism. Cation exchange and the importance of pyrolysis temperature on the retention of Pb was studied in batch equilibration experiments (Uchimiya et al., 2012b). Two biochars produced from broiler-litter at 350 and 650°C were added to a Pb-contaminated small arms range soil. It was reported that the biochar produced at lower temperature had a greater adsorption capacity for Pb and simultaneously higher nutrient release (P, K, Ca, and others). This clearly points to cation exchange as a main immobilization mechanism and suggests that biochar added to a metal-contaminated soil may help detoxify it via metal immobilization and also enhance the release of nutrients to growing plants.

In addition to ion exchange, precipitation can be important in immobilizing soil contaminants. One example from Cao et al. (2011) showed that dairy manure biochar was more effective than a commercial AC in immobilizing Pb in contaminated soil. The biochar was rich in P and the major mechanism was found to be transformation of the labile Pb contaminants into the more stable Pb mineral hydroxylpyromorphite [$Pb_5(PO_4)_3OH$]. The properties of biochar that make it suitable for heavy metal immobilization and soil remediation may sometimes render it unsuitable for other specific applications in which the objective is to increase bioavailability of elements such as Cu, Zn, and manganese. These are nutrients and are essential for plants, microbes, and humans, but in excess they are toxic (Beesley et al., 2015). Biochar may lock up such nutrients and actually cause deficiencies in agricultural systems.

5.3.2.2 Organic Pollutants

It is well accepted that soil organic matter is the most important parameter governing the sorption/desorption and thus the fate and transport of organic pollutants in soils and sediments. Many models, both linear and nonlinear, have been proposed to describe sorption of a single solute to natural solids. Studies have shown that adsorption on mineral surfaces, condensation in micropores, and partitioning into both hard and soft components of soil organic matter may contribute to the overall pollutant uptake. The affinity of organic pollutants to organic matter, and more specifically to its organic C fraction, is also used in many engineering applications: for example, water decontamination by filtering through AC, an engineered highly porous carbonaceous material with a large surface area.

Biochar differs from AC in certain respects: Depending on the feedstock, it may have high mineral ash content, and its surface area and porosity is not maximized by chemical or physical activation processes. Biochar typically has lower C content and possesses a higher degree of surface heterogeneity, with the presence of a variety of carboxyl, hydroxyl, and phenolic surface functional groups. This varied functionality means that biochars can be suitable sorbents for organic pollutants with a variety of chemical characteristics: for example, hydrophobic compounds such as PAHs and PCBs, as well as more hydrophilic compounds such as pesticides and HHCs.

Polycyclic Aromatic Hydrocarbons, Polychlorinated Biphenyls, and Halogenated Hydrocarbon Compounds

PAHs, PCBs, and HHCs are the most commonly studied groups of organic pollutants frequently found in brownfields. Studies indicate that bioavailable fractions of these pollutants are a better proxy than total concentrations for understanding their environmental effects (Lehmann and Joseph, 2015). Related studies include the direct residue analysis of exposed organisms, such as plants and invertebrates, to the contaminated soils, or studies investigating the chemical activity or accessibility of these pollutants in contaminated soil (Khan et al., 2008; Lehmann and Joseph, 2015; Anyika et al., 2015).

Biochar amendments often reduce the bioavailable concentrations of organic pollutants in soils, and with this, their toxic effects (Brennan et al., 2014; Cao et al., 2011; Zielińska and Oleszczuk, 2015). This was shown for PAHs, in which biochar amendment consistently reduced both total and bioavailable PAH concentrations in soils because of their strong sorption on the biochar (Waqas et al., 2014; Yang et al., 2006). Zhang and colleagues investigated the influence of biochar produced from *Pinius radiate* at 350 and 700°C on the sorption of phenanthrene in three different soils (Zhang et al., 2010). The addition of 0.5% and 1% (weight) of the biochar produced at low temperature resulted in 2-fold to 51-fold higher phenanthrene sorption compared with the control samples (soils only). Addition of the same amount of the high-temperature biochar resulted in an increase in phenanthrene sorption by a factor of up to 700. Hale and colleagues reported sorption coefficients normalized to 1 ng L^{-1} for the sorption of the PAH pyrene in soil mixed with biochar produced from corn stover residues at 600°C. By incorporating 5% of biochar, log K_d values (L kg^{-1}) increased from 3.53 ± 0.12 to 4.95 ± 0.35 (Hale et al., 2011).

The synergistic effect of adding biochar together with green-waste compost to contaminated soil was also tested (Beesley et al., 2010). Biochar together with the compost was less effective in reducing the bioavailable pollutant concentrations compared with biochar added alone. It was concluded that biochar is the better sorbent for both inorganic and organic pollutants than green-waste compost, although because of competitive effects, sorption on the compost also occurred (Beesley et al., 2010). For effective immobilization, the biochar–water distribution coefficient, K_d (L kg^{-1}), has to be high compared with the soil–water distribution coefficient, K_d (Anyika et al., 2015).

Pyrolysis temperature was also reported to have a significant influence on the adsorption efficiency of biochar. Biochars produced from pine needles at 100, 300, 400, and 700°C were tested for their effect on sorption of pyrene, naphthalene, and phenanthrene in a paddy soil (Chen and Yuan, 2011). Soil samples amended with different amounts of the four biochars all showed enhanced PAH sorption. This effect was positively correlated with biochar pyrolysis temperature, in which biochar produced under high pyrolytic temperature showed higher sorption.

Several experiments were conducted with hexachlorobenzene in agricultural soil with biochars produced from wheat straw with a maximum treatment temperature of 500°C to assess sorption and dissipation behavior of the pollutant (Song et al., 2012, 2013). Biochar sorbed 42 times more hexachlorobenzene than soil alone (K_d (L g^{-1}) values calculated at 1 μg L^{-1} of 2.01 and 86.16 for soil and biochar, respectively) (Song et al., 2012). Immobilization efficiency of biochar can also change over time, because aging of the biochar might significantly alter the physiochemical properties of biochar. The dissipation and volatilization of hexachlorobenzene in spiked soil in the presence of 0%, 0.1%, 0.5%, 1%, and 2% biochar after 24 weeks' incubation revealed that the dissipation of hexachlorobenzene was rapid in the first 2 weeks and slowed significantly after 4 weeks (Song et al., 2012). The hexachlorobenzene residual concentration in soil increased with increasing biochar content and volatilization decreased with increasing biochar concentration (Song et al., 2012). Incubation and sorption experiments were further conducted with same soil and biochar, at similar amendment levels, for three other persistent organic pollutants, the semivolatile compounds pentachlorobenzene, tetrachlorobenzene and trichlorobenzene (Song et al., 2013). Batch isotherm experiments showed that soil sorption increased significantly with the addition of biochar and increasing amendment concentration. Pentachlorobenzene log K_d values at 1 μg mL^{-1} changed from 2.58 (soil only) to 3.08 for 2% wheat straw biochar. The isotherms became progressively more linear as biochar concentrations increased, which suggests that partitioning dominated the sorption process at high amendment concentrations. Furthermore, the impacts of biochar on immobilization of chlorobenzenes increased with decreases in the chlorination of the compound.

The addition of a hardwood-derived biochar reduced the total and bioavailable PAH concentrations in a historically contaminated soil (Beesley et al., 2010). In 60-day pot experiments, at a soil to biochar ratio of 2:1, total extractable as well as estimated bioavailable PAH concentrations were determined by acetone/hexane and cyclodextrine extraction, respectively. Results suggested that biochar was effective at immobilizing PAHs in the contaminated soils because it reduced the total extractable as well as bioavailable PAH content significantly. The total extractable fraction was reduced by more than 50% for the heavy four- and five-ring PAHs and by about 40% for the lighter two- and three-ringed PAHs (Beesley et al., 2010). In another study, the effect of biochar on the bioavailability of PAHs was monitored over 56 days in a calcareous contaminated soil in the presence or absence of the earthworm, *Eisenia fetida* (Gomez-Eyles et al., 2011). Biochar reduced total (449–306 mg kg^{-1}) and bioavailable (276–182 mg kg^{-1}) PAHs in the soil. In the presence of *E. fetida*, biochar significantly reduced bioavailable PAHs (heavier five-ring) concentrations 50% relative to untreated soil after 56 days (Gomez-Eyles et al., 2011).

Pesticides and Herbicides

The impact of biochar on sorption and bioavailability of pest control products such as herbicides, insecticides, and fungicides in agricultural soils is of specific interest because biochar is often intended for use in improving agricultural soil fertility (Graber and Kookana, 2015). Biochars in soil may affect pesticide efficacy by reducing bioavailability. At the same time, biochars might prevent pesticide leaching and reduce their potential to biodegrade (Spokas et al., 2009; Wang et al., 2010).

Yang et al. (2006) examined the sorption of the urea herbicide diuron in soil as influenced by the addition of wheat straw biochar obtained simply by burning dried straw under natural conditions. The soil was classified as a silty loam and amended with 0.05% to 1% by weight biochar for sorption and bioavailability experiments. Soil containing 1% biochar sorbed seven to 80 times more of the herbicide than pure soil. At only 0.5% biochar, diuron was less bioavailable to soil microorganisms, and hence its degradation was reduced (Yang et al., 2006). Additional batch equilibrium experiments with a sandy loam soil amended with two different biochars (amendment rates of 0.1% to 5%) produced from red gum wood chips at 450°C (BC450) and 850°C (BC850) showed a significant effect of biochar on diuron sorption (Yu et al., 2006). Sorption coefficients at an equilibrium concentration of 1 mg L^{-1} were seven to 80 and five to 125 times higher for 450 and 850°C biochars, respectively. Sorption also increased as the amount of biochar added increased. Desorption of diuron from samples amended with the higher-temperature biochar (BC850) was more strikingly hysteretic than desorption from samples amended with the lower-temperature biochar (BC450). This was attributed to greater micropore content in higher-temperature biochar (Yu et al., 2006).

The fungicide pyremethanil was tested using the same red gum wood chip biochars and soil (sandy loam) as previously described (Yu et al., 2006). Sorption isotherm and desorption experiments conducted with BC450 (0–1%) and BC850 (0–5%) biochar and precontaminated samples showed that sorption capacity increased as the biochar concentration increased. Freundlich coefficients at 1 mg L^{-1} increased from 4.36 (L kg^{-1}) for soil by up to 263.76 and 273.25 (L kg^{-1}) for BC450 and BC850, respectively, as a function of biochar dose. Desorption showed a reverse trend in which increasing the biochar content resulted in decreased amounts of contaminant desorbed. Also, BC850 was found to immobilize pyremethanil much more than the low-temperature biochar. At the highest amendment rate, 13.6% of the pyremethanil was desorbed from BC450, but only 1.49% from BC850. In equilibrium batch and one-step desorption experiments, the effect of rice hull biochar amendment at 0.5%, 1%, and 2% on the behavior of the acidic herbicide fomesafen in five different soils was evaluated (Khorram et al., 2015). Sorption on all five soil types was significantly enhanced with increasing biochar application rate. Calculations revealed that biochar contributed more than 60%, 80%, and 90% of the overall sorption for 0.5%, 1%, and 2% biochar, respectively. Wang and colleagues investigated the influence of *Pinus radiate*–derived biochar on the sorption efficiency of New Zealand soils. They reported that the addition of biochar resulted in an increased soil sorption of the polar herbicide terbuthylazine (Wang et al., 2010).

Bioassays with *Setaria viridis* tested the influence of two biochars (differing in SSA) on the phytoavailability of two pesticides (S-metolachlor and sulfentrazone). Results showed that biochar amendment to soil reduces the efficacy of soil-applied herbicides owing to sorption (Graber et al., 2012). The study also revealed that the biochar with a high SSA (242 m^2 g^{-1}) absorbed the herbicides by one or more orders of magnitude stronger than the biochar with a low SSA (3.6 m^2 g^{-1}). Similarly, a soil mixture with 1% corn straw biochar (SSA 3.0 m^2 g^{-1}) showed a strong nonlinear and significantly greater sorption of 1,3-dichloropropene, a soil fumigant, than sorption by the soil alone

(Graber et al., 2011). Biochar amendment resulted in a substantial reduction in fumigant vaporization along with a significant reduction in the aqueous phase of the soil, where it works against soil nematodes. The application of biochar with high SSA for agronomic purposes can considerably reduce the availability of soil-applied pesticides, depending on the sorption strength of the particular biochar (Graber et al., 2011, 2012). As a result, biochars with a low SSA would best serve pest control requirements (Graber et al., 2011).

The simultaneous immobilization of the herbicide atrazine and lead in soil amended with 2.5% and 5% of dairy manure–derived biochar produced at 450°C resulted in interesting findings (Cao et al., 2011). Total atrazine (determined by extraction with acetone/hexane, 1:1) as well as total lead concentration (determined with HNO_3/H_2O_2 digestion) in the soil did not change over a 210-day incubation period. However, the bioavailable fraction to soil organisms (determined by $CaCl_2$ extraction) and mobility (assessed by toxicity characteristic leaching procedure) changed significantly. The concentration of both contaminants in the two extracts decreased with increasing biochar/soil ratio and incubation time. This fact and a strong correlation of extractable atrazine to the organic C content in the biochar suggested that mostly adsorption to the organic C is the ruling factor in immobilization of atrazine. Furthermore, the sorption of atrazine as well as a second herbicide, acetochlor, showed a significant increase after incorporation of biochar produced from mixed sawdust (500°C) (Spokas et al., 2009). Sandy loam soil, spiked with two active pharmaceutical ingredients, carbamazepine and propranolol, showed significant changes in sorption capacity after amendment with biochar produced from wheat chaff, oil mallee eucalyptus, and blue mallee eucalyptus. In all cases, sorption increased notably with biochar additions (Williams et al., 2015).

Experiments investigating the retention of the antibiotic sulfamethoxazole in a silt loam pasture soil with the addition of biochars from green waste (350, 450, and 550°C), *P. radiate* sawdust (700°C), and corncob (600°C) revealed that only sawdust and corn straw biochar had significant effects on sorption behavior (Srinivasan and Sarmah, 2015). The addition of 0.5% sawdust biochar resulted in a 10-fold increase in sorption; the increase in sorption was linearly associated with increasing biochar dose. The effect of corncob biochar on sorption was smaller than that of sawdust biochar; nevertheless, K_d (calculated at a value of 0.5 mg L^{-1}) increased by two- to fourfold as a function of the biochar/soil ratio. In contrast, there were no significant changes in K_d with any of the green-waste biochars. Bushnaf et al. (2011) investigated the effect of biochar on the dissipation of 13 volatile PHC in a sandy soil. Tests were conducted with a wood chip–derived biochar produced at 800°C and an amendment rate of 2%. K_d values increased from 1.1-fold (for cyclohexane) up to 36-fold (for toluene) of the control samples without biochar (Bushnaf et al., 2011).

5.3.3 BIOCHAR-BASED SOIL REMEDIATION FOR SITE-SPECIFIC CONTAMINANTS: CASE STUDIES

Compared with laboratory and pot experiments, relatively few reports are available demonstrating the effects of biochar amendments for soil remediation on a field scale. Cui et al. (2011, 2012), conducted field experiments to understand the effect of biochar on Cd mobilization and its uptake by summer rice and winter wheat in a contaminated paddy soil. Soil surface contaminated with Cd (21.84 mg kg^{-1}) was amended with wheat straw biochar (550°C) at 0, 10, 20, 40 t ha^{-1} and thoroughly mixed by manual plowing in 2009 and, in the following year, by mechanized plowing (Cui et al., 2011). Over the course of the 2-year field experiment (2009–2010), the added biochar resulted in increased soil pH and organic carbon content along with significantly decreased Cd mobility by 32–52.5% (2009) and 5.5–43.4% (2010). The biochar-induced decrease in Cd mobility caused a reduction in uptake by the rice plants. In biochar-treated soils, the Cd concentration in the rice grain was reduced by 16.8–45% and 42.7–61.9%, and in the total plant by 28.1–54.2% and 14.4–45.9% in 2010 and 2011, respectively. Following a conventional cultivation system, wheat planting followed the harvesting of the rice (Cui et al., 2012). For the wheat crop, the biochar reduced soil Cd mobility by 10.1–40.2% and 10–57%, total Cd uptake by wheat by 16.8–37.3% and 6.5–28.3%, as well as wheat grain Cd concentrations by 24.8–44.2% and 14–39.2% in 2010 and 2011, respectively (Cui et al., 2012).

Another field experiment conduced in a heavy metal–contaminated rice paddy in China showed promising results. Over 3 years (2010–2012), Bian et al. (2014) monitored the effect of wheat straw–derived biochar on Cd and Pb mobility. Biochar was applied at a rate of 0–40 t ha^{-1} and mixed into the soil at a depth of 15 cm. After biochar amendment, rice was cultivated and harvested every year. For all 3 years, a significant decrease in extractable soil Cd and Pb was found. Also, the uptake of the metals by rice plants was significantly reduced. Analysis of fresh and contaminated biochar samples suggested that cation exchange and precipitation of Pb- and Cd-containing mineral phases may have been the main immobilizing processes. The contaminant metals were mainly associated with Al, Fe, and P phases in the biochar. Also, aging of the biochar in the soil increased the amount of surface functional groups and therefore the net negative charge. This also could have led to increased metal binding. The yearly wetting and drying cycle of the contaminated field was suggested to increase biochar aging and contribute to increased ability of the biochar to immobilize metal over the course of the 3-year experiment. This study gives a good overview of the complexity of mechanisms involved in biochar–pollutant interactions in soil, which also may change over time as the biochar ages in the soil.

As far as we know, few studies are available investigating the influence of biochar on the immobilization of organic contaminants under field conditions. A plot trial showed that contaminant and nutrient behavior in PHC and PAH-contaminated soils treated with a wood-derived biochar and a biochar–compost substrate (BCS) (Vogel et al., 2015). The BCS amendment resulted in about a 90% reduction in total PHC concentrations in the soil over the 1.5-year experiment compared with the no-amendment control treatment. In contrast, no significant reduction in total PAH concentrations was observed in the PAH-contaminated plots. This might be a result of the less biodegradable nature of the four- to six-ring PAHs (Worzyk et al., 2014). However, a cyclodextrin extraction revealed that the PAH bioavailability was significantly reduced in both biochar- and BCS-amended soils (Vogel et al., 2015). This was also confirmed by reduced soil–plant transfer factors determined in lysimeter trials (Worzyk et al., 2014).

Persistent organic pollutants such as dieldrin in soils have also been shown to be amenable to immobilization by biochar (Saito et al., 2011). This pesticide has been prohibited since 1975 but remains present in many soils because of its strong stability in soil. Three types of biochar added to dieldrin-contaminated soils successfully reduced the pesticide concentration in cucumber fruits by 9- to 10-fold compared with those in untreated soil. The best-performing biochar, produced from wood chips at 450°C, had a high pH and total C (8.0 and 746 g kg^{-1}, respectively) compared with the other two biochars, one produced from rice husk (6.6 and 404 g kg^{-1}, respectively) and one from bamboo (7.2 and 719 g kg^{-1}, respectively). Moreover, the effect of the wood chip biochar was enhanced by increasing its pyrolysis temperature (1000°C) and crushing it to a finer particle size (Saito et al., 2011). Nevertheless, the immobilization effect of the wood chip biochar was weaker than that of AC.

Furthermore, in a field experiment the AC (mixed only to a depth of 15 cm) efficiently remediated dieldrin from contaminated soils to a depth of 80 cm and reduced its concentration in cucumber fruit. The effect of C material continued to a certain degree for at least 4 years after the application (Saito et al., 2011). These results point to the important role of the biochar or AC surface area in the remediation of organic micropollutants.

More work has been done involving the use of AC in remediating organic pollutants in the environment. AC amendment to polluted soil or sediment is an emerging treatment technique that reduces freely dissolved pore water concentrations. A pilot study showed that PCB exposure from moderately contaminated river sediments may be managed effectively through AC amendment in sediments (Beckingham and Ghosh, 2011). The uptake of persistent organic pollutants such as PCBs was strongly influenced by applying AC to sediments in the field, which reduces bio-uptake of PCBs from the contaminated river by benthic organisms. AC amendments reduced the bioaccumulation of PCBs by 69−99% in freshwater oligochaete worms compared with preamendment conditions. PCBs in sediment pore water were reduced by greater than 93% at all treatment sites for up to 3 years of monitoring (Beckingham and Ghosh, 2011).

In situ field remediation of a marine site contaminated with PAHs and PCB using AC showed strong sorption of contaminants to the added carbonaceous material (Cornelissen et al., 2011, 2012). AC amendment to the site also affected the benthic macrofauna composition and biodiversity. Innovative monitoring methods confirmed that capping sediments with AC could effectively reduce the sediment-to-water fluxes and pore water concentrations of PAHs (Cornelissen et al., 2011) and polychlorinated dibenzodioxin and dibenzofuran (Cornelissen et al., 2012). Field aging of AC for 12−32 months effectively reduced pore water concentrations of both native (30−95%) and spiked (10−90%) hydrophobic organic compounds compared with the unamended sediment or soil (Oen et al., 2012). In addition, the adsorbing power of AC was maintained several years after the amendment (Oen et al., 2012).

Activated carbon (500 kg) was plowed into a 34.4 m^2 plot of marine sediment up to 0.3 m deep, a depth that includes most of the biologically active zone, to assess its impact on the stabilization of PCBs (Cho et al., 2007, 2009). After 7 months, the AC amendment had no impact on sediment resuspension or PCB release into the water column over the treatment plot. On the other hand, there was a significant reduction in PCB bioaccumulation in the AC-amended plot compared with the control plot. The AC treatment showed long-term effectiveness after 18 months, as denoted by a great reduction (90%) in aqueous PCB concentrations. Furthermore, it was shown that the overall effectiveness can be improved by increasing the dose of the AC and reducing its particle size (Cho et al., 2009).

5.3.4 INFLUENCE ON MICROBIAL ACTIVITY/BIODEGRADATION OF ORGANIC CONTAMINANTS

In addition to immobilizing pollutants, adding biochar to soil can have a significant impact on soil microbial activity and community structure. This change in microbial activity can influence nutrient mineralization and transformation along with the degradation and transformation of toxic pollutants, organic and inorganic (Meynet et al., 2014; Gul et al., 2015). Anyika et al. (2015) discussed four major ways in which biochar may influence soil microbiology. With its extensive SSA, biochar provides an important shelter for microbes in soil to reproduce and hide from predators. This pore space also leads to a significant amount of water-holding capacity, as well as the retention of O and nutrients, which can aid microbial growth. Another important factor is the soil pH, which can be significantly altered by adding biochar, and has an important influence on microbial activity (Gul et al., 2015). In addition, biochar itself can be a food source for microbes, including carbon and other nutrients (Ameloot et al., 2013). These aspects influence the overall microbial activity in soil and therefore can have a significant impact on the biodegradation of organic contaminants. Although the addition of biochar to soil frequently stimulates soil microbial activity, both sorption of organic contaminants and desorption hysteresis decrease compound bioavailability because aqueous phase concentrations of the contaminants are significantly reduced (Gul et al., 2015; Zhang et al., 2005). This usually results in hindered biodegradation.

Experiments were conducted with a benzonitrile degrading species called *Nicardia* isolated from an agricultural soil, and untreated and HCl-treated biochar (Zhang et al., 2005). At high initial contaminant concentrations in a silt loam soil, the addition of biochar initially stimulated benzonitrile degradation via the addition of nutrients that were otherwise limiting, which was especially true for P. In slurry samples amended with untreated biochar, initial degradation occurred faster and microbial cell density was higher than in soil alone or in HCl-treated biochar samples. However, after the benzonitrile concentration decreased by 88%, further degradation significantly slowed, indicating that at low available benzonitrile concentrations, the availability of the contaminant and therefore sorption became the limiting factor (Zhang et al., 2005). This corresponded with reports that sorption of benzonitrile to wheat-derived biochar substantially reduced biodegradation (Zhang et al., 2004, 2006).

A similar effect was observed by Jones et al. (2011) for the ^{14}C-labeled herbicide simazine; they reported significantly lower simazine mineralization in soils with added wood-derived biochar. Over a 21-day incubation period, they observed a significant suppression of simazine degradation. Biodegradation was most effective in the first week and was overall greater in soil-only treatments compared with biochar-only treatments ($10 \pm 4\%$ vs. $0.34 \pm 0.04\%$ of total ^{14}C simazine degraded, respectively). Experiments with low initial fertility and high initial fertility soils showed that biodegradation of the herbicide simazine was significantly higher in the high-fertility soil and was suppressed significantly by adding biochar. These results support the finding that a rapid and strong binding of simazine by biochar reduces its microbial availability and therefore its biotic breakdown (Jones et al., 2011). Wood chip biochar prepared at 800°C showed significant effects on the behavior of 12 volatile PHCs (mainly benzene, toluene, ethyl benzene, and xylene [BTEX]) in a sandy soil (Bushnaf et al., 2011). In sterile and live batches as well as in column experiments, it was found that biochar amendment has no negative impact on the overall microbial degradation of BTEX. Increased sorption of BTEX in biochar-amended soil did not lead to less CO_2 evolution, which was used as an indicator for microbial activity. This was attributed to the case that sorption of one compound may result in enhanced degradation of other outcompeted contaminants, because they are preferentially found in the solution phase where degradation takes place. In this case, strong sorption of monoaromatic compounds (eg, toluene) on the biochar resulted in enhanced degradation of straight chain (eg, n-octane), cyclic, and branched alkanes (eg, methylcyclohexane) (Bushnaf et al., 2011).

In another study, evolution of the microbial community and biodegradation of volatile PHCs in an oil spill–contaminated soil amended with 2% rice straw biochar (pyrolyzed at 500°C) was monitored in a 180-day incubation experiment (Qin et al., 2013). A rapid decrease in total PHCs concentration was reported in the soil (without biochar) during the first 60 days; thereafter, no further degradation occurred. Degradation of hydrocarbons in the biochar-amended soil was not significantly different from that in the control soil in the early stages, but degradation continued after 60 days, and final removal efficiency was 78% in the biochar-treated soil compared with 61% in the control soil. Adding biochar to the soil after the initial degradation period was completed improved total degradation further to 85% removal. This temporal effect was not observed for saturated or aromatic compounds but was strong for polar fractions, which suggests that metabolites produced by earlier degradation could be adsorbed when the biochar was added at a later stage. This led to reduced soil toxicity and further degradation (Qin et al., 2013). Yet over the whole course of the experiment, no significant effect of biochar was observed on the microbial community composition. In a similar experiment, the kinetic rate constants for the biodegradation of total PHC from crude oil in soil increased significantly after the addition of biochar produced from plantain peels at 350°C. They also showed that degradation correlated positively with biochar dosage (2% < 3% < 4%) and was more efficient at smaller grain sizes (Agarry et al., 2015).

Incubation experiments with a field-contaminated soil were conducted with four biochars produced from dairy manure and rice hulls at 350 and 500°C (Liu et al., 2015). The effect of adding 4% and 0% biochar on the dissipation of 16 US Environmental Protection Agency priority PAHs was analyzed in sterilized and nonsterilized microcosms. For all four biochars, amendment showed a greater removal of two- to four-ring PAHs compared with five- and six-ringed compounds, in which rice hull 350°C showed the greatest removal. Average removal efficiencies of two- to four-ring PAHs in unsterilized soil increased about 35% and 37% for dairy manure and rice hull biochar compared with the control samples, respectively. This shows that natural degradation was stimulated by the addition of biochar. This was attributed to a similar effect as that described by Qin et al. (2013), in which polar metabolites were adsorbed, resulting in reduced soil toxicity and enhanced biodegradation of other compounds. Molecular biology surveillance showed further that biochar stimulated the activity of PAH-metabolizing microorganisms, which led to an effective enhancement in degradation.

A further step in bioremediation with the aid of biochar was taken by Chen et al. (2012). They applied an immobilized-microorganism technique in which biochar served as a carrier for PAH-degrading microorganisms. Two-ring PAHs were easily degraded in all samples, but three- to six-ring PAHs concentrations hardly changed in unamended soil. In contrast, a significant degradation of all PAHs was detected for samples treated with biochar-immobilized bacteria over a 90-day incubation period. They suggested that biochar can preconcentrate PAHs owing to sorption, shortening the distance between the compounds and microorganisms, and therefore enhance the degradation (Chen et al., 2012).

5.4 BIOCHAR, POLLUTANTS, AND PLANT INTERACTIONS

Pollutants generally cause phytotoxic effects which limit plant growth and reduce plant productivity. Therefore, suitable soil amendments which may detoxify the soil and promote plant productivity, such as biochar, are worth considering and testing. One example of this was the significant reduction in bioavailability and mobility of various heavy metals upon the addition of 10% biochar (Houben et al., 2013). This led to a great decrease in metal uptake by *Brassica napus* (rapeseed), which resulted in a significant reduction in toxicity symptoms such as leaf chlorosis, desiccation, and growth retardation, and moreover promoted the growth of the plants and their uptake of K and P (Houben et al., 2013).

The rapeseed did not contain excessive metals, and therefore could potentially be used as feedstock for additional biochar. This cyclic process can improve the nutrient cycle, pH, CEC, and water-holding capacity while simultaneously sequestering C in the soil. It also can reduce the irrigation requirement and the need of other soil amendments for soil remediation (Houben et al., 2013).

Different types of biochars may affect plant physiology differently along with various metabolic conditions of the plant at all levels of development. The effect of sewage sludge biochar (10%) upon biomass yield and the bioaccumulation of PAHs into lettuce plants grown in PAH-contaminated soil showed that biochar significantly increased lettuce biomass by reducing the bioaccumulation of PAHs by 56–67% (Khan et al., 2013). Furthermore, the effect of the same biochar (2% and 5%) was estimated using turnip (*B. rapa*) plants. In this study, biochar amendment of contaminated soil significantly not only reduced the bioavailable PAHs in soil and bio-accumulated PAHs in *B. rapa*, but also significantly reduced the bioaccumulation of As, Cd, Cu, Pb, and Zn concentrations in *B. rapa* (Khan et al., 2015).

Some studies have shown that plant growth is generally a function of biochar dose (Jaiswal et al., 2015), perhaps owing to nutrient immobilization by adsorbing nitrogen (Taghizadeh-Toosi et al., 2012) or the excess of phytotoxic compounds and elements (Jaiswal et al., 2015). It is possible that negative effects of biochar can be treated by adding mineral nutrients and organic matter, including by co-composting (Alburquerque et al., 2013; Steiner et al., 2010). Song et al. (2014) investigated the effect of biochar from sewage sludge as a soil amendment on plant cultivation with special consideration paid to heavy metal accumulation in *Allium sativum* plants (Song et al., 2014). Results revealed that biochars produced at four different temperatures were rich in nutrients and improved *A. sativum* cultivation, final dry matter, and plant yield. Furthermore, adding biochar produced from sewage sludge at 550°C resulted in an increase in cherry tomato yield by 64% (Hossain et al., 2010) and was effective in reducing heavy metal (As, Pb, and Cu) bioaccumulation in *Cucumis sativa* (Waqas et al., 2014) when cultivated in polluted soil. Maize stalk and pine wood–derived biochars added to soil contaminated with PAHs and heavy metals demonstrated positive effects on *Zea mays* plant growth characteristics, such as chlorophyll content and shoot to root biomass, along with a reduction in contaminant uptake to shoots (Brennan et al., 2014). Denyes et al. (2012) showed that adding increasing amounts of biochar (from 0% to 11%) to a soil contaminated with two different PCBs (136 ± 15.3 and $3.1 \pm 0.75\ \mu g\ g^{-1}$) resulted in a biochar concentration–dependent decrease in PCB uptake in the *Cucurbita pepo* root tissues along with increased plant biomass and increased earthworm survival rate (Denyes et al., 2012).

5.5 SUMMARY AND FUTURE PERSPECTIVES

Biochar application may improve soil health through its impacts on pH, EC, CEC, microbial activity, and soil C content. Soil C management strategies have strong benefits for soil health as well as the potential for carbon sequestration and mitigation of climate change. Moreover, biochar can be beneficial as a soil conditioner and for reducing the bioavailability and mobility of inorganic and organic pollutants in soil. Yet, responses may vary because of the heterogeneous nature of different biochars, as a function of feedstock and production temperature. Optimizing feedstock and pyrolysis conditions is necessary to produce the desired products for site-specific conditions and applications. This is especially important for mixed-contaminant sites where several immobilization mechanisms might be important. Studies have been conducted to understand the mechanisms of immobilization and remediation of toxic pollutants under pot or laboratory conditions over short time periods. Unfortunately, little information is available regarding the long-term sorption—desorption behavior of biochar, especially under field conditions. Yet, the establishment of long-term field trials is a requirement to understand and manage undesirable effects, such as leaching of certain metals or the reduced biodegradation of organic pollutants. Also, the effect of aging of biochars under field conditions over time on plant—contaminant—soil—biochar interactions is not yet understood. These aspects are highly important to apply biochar safely on a commercial large-scale basis. Therefore, it is expected that in coming years there will be a growing number of studies investigating interactions between biochar and contaminated soils under field conditions to understand the mechanisms of biochar-induced remediation of a wide range of pollutants.

ACKNOWLEDGMENT

A.K. gratefully acknowledges The Volcani Center, Agriculture Research Organization, Israel for a Postdoctoral fellowship, as well as a grant from the German—Israel Foundation (GIF), Research Grant Agreement Number G-1263-307.8/2014. I.J.S. also gratefully acknowledges support for doctoral studies from the same GIF grant.

REFERENCES

Amonette, J., Joseph, S., 2009. Characteristics of biochar: microchemical properties. In: Lehmann, J., Joseph, S. (Eds.), Biochar for Environmental Management: Science and Technology. Routledge, pp. 33—52.

Anderson, C.R., Condron, L.M., Clough, T.J., Fiers, M., Stewart, A., Hill, R.A., Sherlock, R.R., 2011. Biochar induced soil microbial community change: implications for biogeochemical cycling of carbon, nitrogen and phosphorus. Pedobiologia 54, 309—320.

Anyika, C., Abdul Majid, Z., Ibrahim, Z., Zakaria, M., Yahya, A., 2015. The impact of biochars on sorption and biodegradation of polycyclic aromatic hydrocarbons in soils—a review. Environmental Science and Pollution Research 22, 3314—3341.

Al-Wabel, M.I., Usman, A.R.A., El-Naggar, A.H., Aly, A.A., Ibrahim, H.M., Elmaghraby, S., Al-Omran, A., 2015. *Conocarpus* biochar as a soil amendment for reducing heavy metal availability and uptake by maize plants. Saudi Journal of Biological Sciences 22, 503—511.

Agrafioti, E., Bouras, G., Kalderis, D., Diamadopoulos, E., 2013. Biochar production by sewage sludge pyrolysis. Journal of Analytical and Applied Pyrolysis 101, 72—78.

Alburquerque, J., Salazar, P., Barrón, V., Torrent, J., del Campillo, M.D., Gallardo, A., Villar, R., 2013. Enhanced wheat yield by biochar addition under different mineral fertilization levels. Agronomy for Sustainable Development 33, 475—484.

Ameloot, N., Graber, E.R., Verheijen, F.G.A., De Neve, S., 2013. Interactions between biochar stability and soil organisms: review and research needs. European Journal of Soil Science 64, 379—390.

Agarry, S.E.O., Moses, K., Solomon, B.O., 2015. Kinetic modelling and half life study of adsorptive bioremediation of soil artificially contaminated with Bonny light crude oil. Journal of Ecological Engineering 16, 13.

Beesley, L., Moreno-Jiménez, E., Gomez-Eyles, J.L., 2010. Effects of biochar and greenwaste compost amendments on mobility, bioavailability and toxicity of inorganic and organic contaminants in a multi-element polluted soil. Environmental Pollution 158, 2282—2287.

Butnan, S., Deenik, J.L., Toomsan, B., Antal, M.J., Vityakon, P., 2015. Biochar characteristics and application rates affecting corn growth and properties of soils contrasting in texture and mineralogy. Geoderma 237–238, 105–116.

Bellamy, P.H., Loveland, P.J., Bradley, R.I., Lark, R.M., Kirk, G.J.D., 2005. Carbon losses from all soils across England and Wales 1978–2003. Nature 437, 245–248.

Bardos, R.P., Bakker, L.M., Slenders, H.A., Nathanail, C.P., 2011. Sustainability and remediation. In: Swartjes, F.A. (Ed.), Dealing with Contaminated Sites. Springer, Netherlands, pp. 889–948.

Beesley, L., Moreno-Jiménez, E., Gomez-Eyles, J.L., Harris, E., Robinson, B., Sizmur, T., 2011. A review of biochars' potential role in the remediation, revegetation and restoration of contaminated soils. Environmental Pollution 159, 3269–3282.

Brennan, A., Moreno Jiménez, E., Alburquerque, J.A., Knapp, C.W., Switzer, C., 2014. Effects of biochar and activated carbon amendment on maize growth and the uptake and measured availability of polycyclic aromatic hydrocarbons (PAHs) and potentially toxic elements (PTEs). Environmental Pollution 193, 79–87.

Bird, M.I., Wurster, C.M., de Paula Silva, P.H., Bass, A.M., de Nys, R., 2011. Algal biochar – production and properties. Bioresource Technology 102, 1886–1891.

Beesley, L., Marmiroli, M., 2011. The immobilisation and retention of soluble arsenic, cadmium and zinc by biochar. Environmental Pollution 159, 474–480.

Beesley, L., Moreno-Fimenez, E., Fellet, G., Melo, L., Sizmur, T., 2015. Biochar and heavy metals. In: Lehmann, J., Joseph, S. (Eds.), Biochar for Environmental Management: Science, Technology and Implementation. Routledge, pp. 563–594.

Bushnaf, K.M., Puricelli, S., Saponaro, S., Werner, D., 2011. Effect of biochar on the fate of volatile petroleum hydrocarbons in an aerobic sandy soil. Journal of Contaminant Hydrology 126, 208–215.

Bian, R., Joseph, S., Cui, L., Pan, G., Li, L., Liu, X., Zhang, A., Rutlidge, H., Wong, S., Chia, C., Marjo, C., Gong, B., Munroe, P., Donne, S., 2014. A three-year experiment confirms continuous immobilization of cadmium and lead in contaminated paddy field with biochar amendment. Journal of Hazardous Materials 272, 121–128.

Beckingham, B., Ghosh, U., 2011. Field-scale reduction of PCB bioavailability with activated carbon amendment to river sediments. Environmental Science & Technology 45, 10567–10574.

Cho, Y.-M., Ghosh, U., Kennedy, A.J., Grossman, A., Ray, G., Tomaszewski, J.E., Smithenry, D.W., Bridges, T.S., Luthy, R.G., 2009. Field application of activated carbon amendment for in-situ stabilization of polychlorinated biphenyls in marine sediment. Environmental Science & Technology 43, 3815–3823.

Chan, K.Y., Van Zwieten, L., Meszaros, I., Downie, A., Joseph, S., 2007. Agronomic values of greenwaste biochar as a soil amendment. Soil Research 45, 629–634.

Cheng, C.-H., Lehmann, J., Thies, J.E., Burton, S.D., 2008. Stability of black carbon in soils across a climatic gradient. Journal of Geophysical Research: Biogeosciences 113 n/a-n/a.

Cao, X., Ma, L., Gao, B., Harris, W., 2009. Dairy-manure derived biochar effectively sorbs lead and atrazine. Environmental Science & Technology 43, 3285–3291.

Cao, X., Ma, L., Liang, Y., Gao, B., Harris, W., 2011. Simultaneous immobilization of lead and atrazine in contaminated soils using dairy-manure biochar. Environmental Science & Technology 45, 4884–4889.

Chen, B., Yuan, M., 2011. Enhanced sorption of polycyclic aromatic hydrocarbons by soil amended with biochar. Journal of Soils and Sediments 11, 62–71.

Cui, L., Li, L., Zhang, A., Pan, G., 2011. Biochar amendment greatly reduces rice Cd uptake in a contaminated paddy soil: a two-year field experiment. BioResources 6 (3), 2605–2618.

Cui, L., Pan, G., Li, L., Yan, J., Zhang, A., Bian, R., Chang, A., 2012. The reduction of wheat Cd uptake in contaminated soil via biochar amendment: a two-year field experiment. BioResources 7, 5666–5676.

Cornelissen, G., Elmquist Kruså, M., Breedveld, G.D., Eek, E., Oen, A.M.P., Arp, H.P.H., Raymond, C., Samuelsson, G., Hedman, J.E., Stokland, Ø., Gunnarsson, J.S., 2011. Remediation of contaminated marine sediment using thin-layer capping with activated carbon—a field experiment in Trondheim harbor, Norway. Environmental Science & Technology 45, 6110–6116.

Cornelissen, G., Amstaetter, K., Hauge, A., Schaanning, M., Beylich, B., Gunnarsson, J.S., Breedveld, G.D., Oen, A.M.P., Eek, E., 2012. Large-scale field study on thin-layer capping of marine PCDD/F-contaminated sediments in Grenlandfjords, Norway: physicochemical effects. Environmental Science & Technology 46, 12030–12037.

Cho, Y.-M., Smithenry, D.W., Ghosh, U., Kennedy, A.J., Millward, R.N., Bridges, T.S., Luthy, R.G., 2007. Field methods for amending marine sediment with activated carbon and assessing treatment effectiveness. Marine Environmental Research 64, 541–555.

Chen, B., Yuan, M., Qian, L., 2012. Enhanced bioremediation of PAH-contaminated soil by immobilized bacteria with plant residue and biochar as carriers. Journal of Soils and Sediments 12, 1350–1359.

Das, O., Sarmah, A.K., 2015. The love—hate relationship of pyrolysis biochar and water: a perspective. Science of the Total Environment 512–513, 682–685.

Duku, M.H., Gu, S., Hagan, E.B., 2011. Biochar production potential in Ghana—a review. Renewable and Sustainable Energy Reviews 15, 3539–3551.

Denyes, M.J., Langlois, V.S., Rutter, A., Zeeb, B.A., 2012. The use of biochar to reduce soil PCB bioavailability to *Cucurbita pepo* and *Eisenia fetida*. Science of the Total Environment 437, 76–82.

Elad, Y., Cytryn, E., Meller Harel, Y., Lew, B., Graber, E.R., 2012. The biochar effect: plant resistance to biotic stresses. Phytopathologia Mediterranea 50, 335–349.

Gomes, H.I., Dias-Ferreira, C., Ribeiro, A.B., 2013. Overview of in situ and ex situ remediation technologies for PCB-contaminated soils and sediments and obstacles for full-scale application. Science of the Total Environment 445–446, 237–260.

Ghani, W.A.W.A.K., Mohd, A., da Silva, G., Bachmann, R.T., Taufiq-Yap, Y.H., Rashid, U., Al-Muhtaseb, A.A.H., 2013. Biochar production from waste rubber-wood-sawdust and its potential use in C sequestration: chemical and physical characterization. Industrial Crops and Products 44, 18–24.

Gwenzi, W., Chaukura, N., Mukome, F.N.D., Machado, S., Nyamasoka, B., 2015. Biochar production and applications in sub-Saharan Africa: opportunities, constraints, risks and uncertainties. Journal of Environmental Management 150, 250–261.

Graber, E., Meller Harel, Y., Kolton, M., Cytryn, E., Silber, A., Rav David, D., Tsechansky, L., Borenshtein, M., Elad, Y., 2010. Biochar impact on development and productivity of pepper and tomato grown in fertigated soilless media. Plant and Soil 337, 481–496.

Graber, E.R., Elad, Y., 2013. Biochar impact on plant resistance to disease. In: Biochar and Soil Biota. CRC Press, pp. 41–68.

Graber, E.R., Tsechansky, L., Khanukov, J., Oka, Y., 2011. Sorption, volatilization, and efficacy of the fumigant 1,3-dichloropropene in a biochar-amended soil. Soil Science Society of America Journal 75, 1365–1373.

Gomez-Eyles, J.L., Sizmur, T., Collins, C.D., Hodson, M.E., 2011. Effects of biochar and the earthworm *Eisenia fetida* on the bioavailability of polycyclic aromatic hydrocarbons and potentially toxic elements. Environmental Pollution 159, 616–622.

Graber, E.R., Kookana, R.S., 2015. Biochar and retention/efficacy of pesticides. In: Lehmann, J., Joseph, S. (Eds.), Biochar for Environmental Management: Science, Technology and Implementation. Routledge, New York, pp. 655–678.

Graber, E.R., Tsechansky, L., Gerstl, Z., Lew, B., 2012. High surface area biochar negatively impacts herbicide efficacy. Plant and Soil 353, 95–106.

Gul, S., Whalen, J.K., Thomas, B.W., Sachdeva, V., Deng, H., 2015. Physico-chemical properties and microbial responses in biochar-amended soils: mechanisms and future directions. Agriculture, Ecosystems & Environment 206, 46–59.

Hartley, W., Dickinson, N.M., Riby, P., Lepp, N.W., 2009. Arsenic mobility in brownfield soils amended with green waste compost or biochar and planted with *Miscanthus*. Environmental Pollution 157, 2654–2662.

Houben, D., Evrard, L., Sonnet, P., 2013. Beneficial effects of biochar application to contaminated soils on the bioavailability of Cd, Pb and Zn and the biomass production of rapeseed (*Brassica napus* L.). Biomass and Bioenergy 57, 196–204.

Hmid, A., Mondelli, D., Fiore, S., Fanizzi, F.P., Al Chami, Z., Dumontet, S., 2014. Production and characterization of biochar from three-phase olive mill waste through slow pyrolysis. Biomass and Bioenergy 71, 330–339.

Hale, S., Hanley, K., Lehmann, J., Zimmerman, A., Cornelissen, G., 2011. Effects of chemical, biological, and physical aging as well as soil addition on the sorption of pyrene to activated carbon and biochar. Environmental Science & Technology 45, 10445–10453.

Hossain, M.K., Strezov, V., Yin Chan, K., Nelson, P.F., 2010. Agronomic properties of wastewater sludge biochar and bioavailability of metals in production of cherry tomato (*Lycopersicon esculentum*). Chemosphere 78, 1167–1171.

Jaiswal, A., Frenkel, O., Elad, Y., Lew, B., Graber, E., 2015. Non-monotonic influence of biochar dose on bean seedling growth and susceptibility to *Rhizoctonia solani*: the "Shifted R_{max}-Effect". Plant and Soil 395, 125–140.

Jones, D.L., Edwards-Jones, G., Murphy, D.V., 2011. Biochar mediated alterations in herbicide breakdown and leaching in soil. Soil Biology and Biochemistry 43, 804–813.

Khan, S., Aijun, L., Zhang, S., Hu, Q., Zhu, Y.-G., 2008. Accumulation of polycyclic aromatic hydrocarbons and heavy metals in lettuce grown in the soils contaminated with long-term wastewater irrigation. Journal of Hazardous Materials 152, 506–515.

Kolton, M., Harel, Y.M., Pasternak, Z., Graber, E.R., Elad, Y., Cytryn, E., 2011. Impact of biochar application to soil on the root-associated bacterial community structure of fully developed greenhouse pepper plants. Applied and Environmental Microbiology 77, 4924–4930.

Kuhlbusch, T.A.J., Crutzen, P.J., 1995. Toward a global estimate of black carbon in residues of vegetation fires representing a sink of atmospheric CO_2 and a source of O_2. Global Biogeochemical Cycles 9, 491–501.

Kleber, M., Hockaday, W., Nico, P.S., 2015. Characterstics of biochar: physical and structural properties. In: Lehmann, J., Joseph, S. (Eds.), Biochar for Environmental Management. Routledge, Taylor and Francis Group, pp. 89–110.

Kołodyńska, D., Wnetrzak, R., Leahy, J.J., Hayes, M.H.B., Kwapiński, W., Hubicki, Z., 2012. Kinetic and adsorptive characterization of biochar in metal ions removal. Chemical Engineering Journal 197, 295–305.

Khorram, M.S., Wang, Y., Jin, X., Fang, H., Yu, Y., 2015. Reduced mobility of fomesafen through enhanced adsorption in biochar-amended soil. Environmental Toxicology and Chemistry 34, 1258–1266.

Khan, S., Wang, N., Reid, B.J., Freddo, A., Cai, C., 2013. Reduced bioaccumulation of PAHs by *Lactuca satuva* L. grown in contaminated soil amended with sewage sludge and sewage sludge derived biochar. Environmental Pollution 175, 64–68.

Khan, S., Waqas, M., Ding, F., Shamshad, I., Arp, H.P.H., Li, G., 2015. The influence of various biochars on the bioaccessibility and bioaccumulation of PAHs and potentially toxic elements to turnips (*Brassica rapa* L.). Journal of Hazardous Materials 300, 243–253.

Lehmann, J., Joseph, S., 2015. Biochar for Environmental Management: Science, Technology and Implementation, second ed. Routledge.

Liang, X., Han, J., Xu, Y., Sun, Y., Wang, L., Tan, X., 2014. In situ field-scale remediation of Cd polluted paddy soil using sepiolite and palygorskite. Geoderma 235–236, 9–18.

Lehmann, J., Rillig, M.C., Thies, J., Masiello, C.A., Hockaday, W.C., Crowley, D., 2011. Biochar effects on soil biota – a review. Soil Biology and Biochemistry 43, 1812–1836.

Limousin, G., Gaudet, J.P., Charlet, L., Szenknect, S., Barthès, V., Krimissa, M., 2007. Sorption isotherms: a review on physical bases, modeling and measurement. Applied Geochemistry 22, 249–275.

Lucchini, P., Quilliam, R.S., DeLuca, T.H., Vamerali, T., Jones, D.L., 2014. Does biochar application alter heavy metal dynamics in agricultural soil? Agriculture, Ecosystems & Environment 184, 149–157.

Liang, B., Lehmann, J., Solomon, D., Kinyangi, J., Grossman, J., O'Neill, B., Skjemstad, J.O., Thies, J., Luizão, F.J., Petersen, J., Neves, E.G., 2006. Black carbon increases cation exchange capacity in soils. Soil Science Society of America Journal 70, 1719–1730.

Liu, L., Chen, P., Sun, M., Shen, G., Shang, G., 2015. Effect of biochar amendment on PAH dissipation and indigenous degradation bacteria in contaminated soil. Journal of Soils and Sediments 15, 313–322.

Ma, Y., Prasad, M.N.V., Rajkumar, M., Freitas, H., 2011. Plant growth promoting rhizobacteria and endophytes accelerate phytoremediation of metalliferous soils. Biotechnology Advances 29, 248–258.

Meller Harel, Y., Elad, Y., Rav-David, D., Borenstein, M., Shulchani, R., Lew, B., Graber, E., 2012. Biochar mediates systemic response of strawberry to foliar fungal pathogens. Plant and Soil 357, 245–257.

Meynet, P., Moliterni, E., Davenport, R.J., Sloan, W.T., Camacho, J.V., Werner, D., 2014. Predicting the effects of biochar on volatile petroleum hydrocarbon biodegradation and emanation from soil: a bacterial community finger-print analysis inferred modelling approach. Soil Biology and Biochemistry 68, 20–30.

Oen, A.M.P., Beckingham, B., Ghosh, U., Kruså, M.E., Luthy, R.G., Hartnik, T., Henriksen, T., Cornelissen, G., 2012. Sorption of organic compounds to fresh and field-aged activated carbons in soils and sediments. Environmental Science & Technology 46, 810–817.

Park, J., Choppala, G., Bolan, N., Chung, J., Chuasavathi, T., 2011. Biochar reduces the bioavailability and phytotoxicity of heavy metals. Plant and Soil 348, 439–451.

Quilliam, R.S., Marsden, K.A., Gertler, C., Rousk, J., DeLuca, T.H., Jones, D.L., 2012. Nutrient dynamics, microbial growth and weed emergence in biochar amended soil are influenced by time since application and reapplication rate. Agriculture, Ecosystems & Environment 158, 192–199.

Qin, G., Gong, D., Fan, M.-Y., 2013. Bioremediation of petroleum-contaminated soil by biostimulation amended with biochar. International Biodeterioration & Biodegradation 85, 150–155.

Silber, A., Levkovitch, I., Graber, E.R., 2010. pH-dependent mineral release and surface properties of cornstraw biochar: agronomic implications. Environmental Science & Technology 44, 9318–9323.

Sun, Y., Gao, B., Yao, Y., Fang, J., Zhang, M., Zhou, Y., Chen, H., Yang, L., 2014. Effects of feedstock type, production method, and pyrolysis temperature on biochar and hydrochar properties. Chemical Engineering Journal 240, 574–578.

Spokas, K.A., 2010. Review of the stability of biochar in soils: predictability of O:C molar ratios. Carbon Management 1, 289–303.

Suthersan, S.S., 2001. Natural and Enhanced Remediation Systems. Lewis Publisher, New York.

Srinivasan, P., Sarmah, A.K., Smernik, R., Das, O., Farid, M., Gao, W., 2015. A feasibility study of agricultural and sewage biomass as biochar, bioenergy and biocomposite feedstock: production, characterization and potential applications. Science of the Total Environment 512–513, 495–505.

Smernik, R., 2009. Biochar and sorption of organic compounds. In: Lehmann, J., Joseph, S. (Eds.), Biochar for Environmental Management: Science and Technology. Earthscan, pp. 289–300.

Song, Y., Wang, F., Bian, Y., Kengara, F.O., Jia, M., Xie, Z., Jiang, X., 2012. Bioavailability assessment of hexachlorobenzene in soil as affected by wheat straw biochar. Journal of Hazardous Materials 217–218, 391–397.

Song, Y., Wang, F., Kengara, F.O., Yang, X., Gu, C., Jiang, X., 2013. Immobilization of chlorobenzenes in soil using wheat straw biochar. Journal of Agricultural and Food Chemistry 61, 4210–4217.

Spokas, K.A., Koskinen, W.C., Baker, J.M., Reicosky, D.C., 2009. Impacts of woodchip biochar additions on greenhouse gas production and sorption/degradation of two herbicides in a Minnesota soil. Chemosphere 77, 574–581.

Srinivasan, P., Sarmah, A.K., 2015. Characterisation of agricultural waste-derived biochars and their sorption potential for sulfamethoxazole in pasture soil: a spectroscopic investigation. Science of the Total Environment 502, 471–480.

Saito, T., Otani, T., Seike, N., Murano, H., Okazaki, M., 2011. Suppressive effect of soil application of carbonaceous adsorbents on dieldrin uptake by cucumber fruits. Soil Science and Plant Nutrition 57, 157–166.

Steiner, C., Das, K.C., Melear, N., Lakly, D., 2010. Reducing nitrogen loss during poultry litter composting using biochar. Journal of Environmental Quality 39, 1236–1242.

Song, X.D., Xue, X.Y., Chen, D.Z., He, P.J., Dai, X.H., 2014. Application of biochar from sewage sludge to plant cultivation: influence of pyrolysis temperature and biochar-to-soil ratio on yield and heavy metal accumulation. Chemosphere 109, 213–220.

Tang, J., Zhu, W., Kookana, R., Katayama, A., 2013. Characteristics of biochar and its application in remediation of contaminated soil. Journal of Bioscience and Bioengineering 116, 653–659.

Tomei, M.C., Mosca Angelucci, D., Annesini, M.C., Daugulis, A.J., 2013. Ex situ remediation of polluted soils by absorptive polymers, and a comparison of slurry and two-phase partitioning bioreactors for ultimate contaminant degradation. Journal of Hazardous Materials 262, 31–37.

Troy, S.M., Nolan, T., Leahy, J.J., Lawlor, P.G., Healy, M.G., Kwapinski, W., 2013. Effect of sawdust addition and composting of feedstock on renewable energy and biochar production from pyrolysis of anaerobically digested pig manure. Biomass and Bioenergy 49, 1–9.

Tsechansky, L., Graber, E.R., 2014. Methodological limitations to determining acidic groups at biochar surfaces via the Boehm titration. Carbon 66, 730–733.

Taghizadeh-Toosi, A., Clough, T., Sherlock, R., Condron, L., 2012. Biochar adsorbed ammonia is bioavailable. Plant and Soil 350, 57–69.

Uchimiya, M., Lima, I.M., Thomas Klasson, K., Chang, S., Wartelle, L.H., Rodgers, J.E., 2010a. Immobilization of heavy metal ions (CuII, CdII, NiII, and PbII) by broiler litter-derived biochars in water and soil. Journal of Agricultural and Food Chemistry 58, 5538–5544.

Uchimiya, M., Wartelle, L.H., Lima, I.M., Klasson, K.T., 2010b. Sorption of deisopropylatrazine on broiler litter biochars. Journal of Agricultural and Food Chemistry 58, 12350–12356.

Uchimiya, M., Wartelle, L.H., Klasson, K.T., Fortier, C.A., Lima, I.M., 2011a. Influence of pyrolysis temperature on biochar property and function as a heavy metal sorbent in soil. Journal of Agricultural and Food Chemistry 59, 2501–2510.

Uchimiya, M., Chang, S., Klasson, K.T., 2011b. Screening biochars for heavy metal retention in soil: role of oxygen functional groups. Journal of Hazardous Materials 190, 432–441.

Uchimiya, M., Bannon, D.I., Wartelle, L.H., 2012a. Retention of heavy metals by carboxyl functional groups of biochars in small arms range soil. Journal of Agricultural and Food Chemistry 60, 1798–1809.

Uchimiya, M., Bannon, D.I., Wartelle, L.H., Lima, I.M., Klasson, K.T., 2012b. Lead retention by broiler litter biochars in small arms range soil: impact of pyrolysis temperature. Journal of Agricultural and Food Chemistry 60, 5035–5044.

Uchimiya, M., Bannon, D.I., 2013. Solubility of lead and copper in biochar-amended small arms range soils: influence of soil organic carbon and pH. Journal of Agricultural and Food Chemistry 61, 7679–7688.

Vangronsveld, J., Herzig, R., Weyens, N., Boulet, J., Adriaensen, K., Ruttens, A., Thewys, T., Vassilev, A., Meers, E., Nehnevajova, E., van der Lelie, D., Mench, M., 2009. Phytoremediation of contaminated soils and groundwater: lessons from the field. Environmental Science and Pollution Research 16, 765–794.

Vogel, I., Rößler, K., Worzyk, F., Wagner, R., Schatten, R., Friede, K., Krüger, C., Terytze, K., 2015. Five years biochar research at Freie Universität – lessons learned for application in horticulture and contaminated site management. In: Glaser, B., Kammann, C., Loewen, A. (Eds.), Joint Biochar Symposium 2015. ANS e.V, Braunschweig, Geisenheim, Germany, p. 194.

Wang, L., Tsang, D.C.W., Poon, C.-S., 2015. Green remediation and recycling of contaminated sediment by waste-incorporated stabilization/solidification. Chemosphere 122, 257–264.

Waqas, M., Khan, S., Qing, H., Reid, B.J., Chao, C., 2014. The effects of sewage sludge and sewage sludge biochar on PAHs and potentially toxic element bioaccumulation in *Cucumis sativa* L. Chemosphere 105, 53–61.

Wang, H., Lin, K., Hou, Z., Richardson, B., Gan, J., 2010. Sorption of the herbicide terbuthylazine in two New Zealand forest soils amended with biosolids and biochars. Journal of Soils and Sediments 10, 283–289.

Williams, M., Martin, S., Kookana, R., 2015. Sorption and plant uptake of pharmaceuticals from an artificially contaminated soil amended with biochars. Plant and Soil 1–12.

Worzyk, F., Schatten, R., Krüger, C., Terytze, K., Vogel, I., 2014. Auswirkung von Biokohle-Substraten und Biokohle auf Bodenparameter und Pflanzenwachstum MKW- und PAK- kontaminierter Böden. Altlasten Spektrum 2014, 13.

Xu, R.-K., Zhao, A.-Z., 2013. Effect of biochars on adsorption of Cu(II), Pb(II) and Cd(II) by three variable charge soils from southern China. Environmental Science and Pollution Research 20, 8491–8501.

Yang, Y., Sheng, G., Huang, M., 2006. Bioavailability of diuron in soil containing wheat-straw-derived char. Science of the Total Environment 354, 170–178.

Yu, X.-Y., Ying, G.-G., Kookana, R.S., 2006. Sorption and desorption behaviors of diuron in soils amended with charcoal. Journal of Agricultural and Food Chemistry 54, 8545–8550.

Zhao, L., Cao, X., Mašek, O., Zimmerman, A., 2013. Heterogeneity of biochar properties as a function of feedstock sources and production temperatures. Journal of Hazardous Materials 256–257, 1–9.

Zielińska, A., Oleszczuk, P., 2015. The conversion of sewage sludge into biochar reduces polycyclic aromatic hydrocarbon content and ecotoxicity but increases trace metal content. Biomass and Bioenergy 75, 235–244.

Zhang, H., Lin, K., Wang, H., Gan, J., 2010. Effect of *Pinus radiata* derived biochars on soil sorption and desorption of phenanthrene. Environmental Pollution 158, 2821–2825.

Zhang, P., Sheng, G., Feng, Y., Miller, D.M., 2005. Role of wheat-residue-derived char in the biodegradation of benzonitrile in soil: nutritional stimulation versus adsorptive inhibition. Environmental Science & Technology 39, 5442–5448.

Zhang, P., Sheng, G., Wolf, D.C., Feng, Y., 2004. Reduced biodegradation of benzonitrile in soil containing wheat-residue-derived ash. Journal of Environmental Quality 33, 868–872.

Zhang, P., Sheng, G., Feng, Y., Miller, D., 2006. Predominance of char sorption over substrate concentration and soil pH in influencing biodegradation of benzonitrile. Biodegradation 17, 1–8.

CHAPTER

6

MUNICIPAL SOLID WASTE BIOCHAR FOR PREVENTION OF POLLUTION FROM LANDFILL LEACHATE

Y. Jayawardhana, P. Kumarathilaka, I. Herath, M. Vithanage
National Institute of Fundamental Studies, Kandy, Sri Lanka

CHAPTER OUTLINE

Environmental Materials and Waste. http://dx.doi.org/10.1016/B978-0-12-803837-6.00006-8

6.1 INTRODUCTION

Municipal solid wastes (MSW) mainly include household or domestic waste collected through community sanitation services. However, in some cases it also contains commercial and industrial waste, excluding hazardous industrial waste. This is true especially in the developing countries. MSW are generated from three main sources: (1) domestic solid waste (from households and public areas, including waste collected from residential buildings, litter bins, streets, marine areas, and country parks), (2) commercial solid waste (from shops, restaurants, hotels, offices, and markets in private housing estates), and (3) industrial solid waste (waste from industries, but not including construction and demolition waste, chemical waste, or other special types of waste) (Chandler et al., 1997; SUTD, 2012).

6.1.1 GENERATION AND DISPOSAL OF MUNICIPAL SOLID WASTE

The generation of refuse, garbage, and sludge from anthropogenic activities is of considerable importance in the solid waste sector (Leton and Omotosho, 2004). As a result of changes in the consumption patterns of certain societies, especially in urban societies, more heterogeneous consumption patterns have arisen, causing a substantial increase in the generation of solid waste (Babayemi and Dauda, 2009; Buenrostro and Bocco, 2003). It has been estimated that about 1.3 billion tons of MSW are generated globally each year, which is increasing rapidly as urbanization, mass consumption, and throw-away lifestyles become more prevalent (Hoornweg et al., 2012). The volume of MSW generated globally is projected to double by 2025, particularly in developing countries (Hoornweg, 1999). Most food waste is disposed in landfills (Matteson and Jenkins, 2007). A report on the assessment of California's compost- and mulch-producing infrastructure concluded that only about 15% of food waste which is diverted to landfills is converted into bioproducts (Matteson and Jenkins, 2007).

Regardless, MSW ends up in landfills which may be sanitary, controlled, engineered, or open dump sites in most cases. Open dumps are the worst-case scenario among the different types of landfills and is a common practice in developing countries (Diaz et al., 2005). An open dump site is a land disposal site where solid wastes are disposed of in a manner which does not protect the environment, with no landfill liner or cover, is susceptible to open burning, and is exposed to the elements to spread disease vectors (Joseph et al., 2002). Open dumps are unplanned heaps of uncovered wastes, often burned (many open dump operators set fire to the MSW at the dump site to reduce the waste and increase the disposal area at the site) and surrounded by pools of stagnated polluted water (Ali et al., 2014), with rat and fly infestations and with domestic animals roaming freely, together with families of scavenging humans picking through the wastes (Diaz et al., 2005; Hazra and Goel, 2009). Three main groups of landfills are classified as young (less than 5 years old), intermediate (5–10 years old), and old or stabilized (more than 10 years old).

6.1.2 OPEN DUMP SITES AND ENVIRONMENTAL POLLUTION

MSW landfills may contain a variety of pollutants such as heavy metals, nutrients, dissolved solids and organic carbon, various organic pollutants including xenobiotic compounds, pesticides, and hydrocarbons (Kjeldsen et al., 2002). Similarly, the decomposition of the MSW produces various gases such as volatile organic compounds (VOCs) and particularly benzene, toluene, ethyl-benzene, and xylene isomers (collectively called BTEX) through biochemical oxidation of CO_2, CH_4, and their derivatives (Kjeldsen et al., 2002; Christensen et al., 2001). These pollutants end up in the leachate that forms as water percolates through the waste mass. Generation of polluted water containing a wide range of contaminants is a common occurrence in most of the landfills of developing countries (Mahmud et al., 2012; Manahan, 2000).

Leachate is the fluid percolating through the landfills and is generated from liquids present in the waste and from outside water, including rainwater, percolating though the waste. The quantity and concentration of the generated leachate depend on several factors, the most important of which are (1) the amount and characteristics of the discharged waste, (2) climatic conditions, (3) the cell size and the phasing of the disposal area, (4) operational techniques applied at the landfill, and (5) the nature of final top cover applied. Moreover, its quality and quantity depend on several factors including the geography, climatic conditions, the hydrogeology of the area, and the age of waste pile (Armstrong and Rowe, 1999). Potential toxicity analyses including the monitoring of test organisms (rotifers, crustaceans, protozoans, luminescent bacteria, etc.) reveal the toxic impacts of landfill leachates (Bernard et al., 1996). Therefore, landfill leachate characterization is one of the most important aspects in the development of compatible remediation strategy and remediation technique (El-Fadel et al., 2002). Even at trace concentrations, the toxics in the leachate may pose a risk to groundwater quality and human health if allowed to leach into the subsurface and into the groundwater table.

However, other impacts may arise from gases generated and leachate formation if not controlled. Such impacts include fires and explosions, vegetation damage, unpleasant odors, landfill settlement, groundwater pollution, air pollution, and global warming (US-EPA, 1984). Hence, the open dump sites pollute the atmosphere through various gases and volatile compounds produced, whereas water and soils are contaminated via heavy metals and organic pollutants (Visvanathan and Trankler, 2003). The existence of several VOCs in landfill leachate has been reported, such as the aromatic hydrocarbons BTEX together with halogenated hydrocarbons (trihalomethane, trichloroethelene, and tetrachloroethylene) (Baun et al., 2004).

6.1.3 STRATEGIES FOR PREVENTION OF POLLUTION FROM LANDFILLS

Landfill leachates and gases produced from landfills are the most obvious sources for groundwater contamination and for CH_4 release phenomena in the world. Landfill gas (LFG) emission contributes to total greenhouse gas emissions and thus contributes to global warming (US-EPA, 1984; Forster et al., 2007).

6.1.3.1 Prevention of Air Pollution

Many studies have reported different kinds of mechanisms and technologies for the prevention of gas emissions from open dump sites. Effective design of landfill covers implies the transformation of CH_4 and other trace gas VOCs (Scheutz et al., 2009). Consequently, the composition of gases produced within landfills differ from the atmospheric release through cover material, owing to various kinds of oxidation and adsorption phenomena (Garg and Achari, 2010). In addition, landfill covers incorporate measures to mitigate infiltration, thus avoiding a bathtub effect within the landfill. Thus proper landfill covers enable management of the leachate in a landfill operation. A complete type of landfill cover contain several types of layers including a surface protection layer, a drainage layer, a composite barrier, and a gas collection layer (Koerner and Daniel, 1997). There are several types of cover techniques including the conventional (barrier cover) and biocovers (Bogner et al., 2011).

Biocovers are suited for old and massive landfill sites that are not eligible for harvesting; such covers retain gases using a collection system reducing the emission of trace fluxes of gases (eg, CH_4, VOC) through the surface (Sadasivam and Reddy, 2014). Soil cover systems represent conventional technology and are elongated with soil capping to optimize LFG emission by methanotrophic microbial oxidation while controlling the generation of leachate by hindering percolation (Sadasivam and Reddy, 2014). However, it has been shown that natural attenuation by soil cover does not remove halogenated carbon compounds which remain and are emitted from the landfill under the simulated oxidative condition of CH_4 (Scheutz et al., 2004). Currently, technological development enables the replacement of soil cover by biocover, biowindow, and biofilter systems that are expected to mitigate LFG emissions (Huber-Humer et al., 2008). Biocovers minimize CH_4 emission by effective degradation. The cover contains a layer for gas collection with fluxes of methane and basal gases to achieve optimal oxidation and fixation (Scheutz et al., 2009). Exponential levels of amended covers were invented by researchers to control LFG emissions. Therefore, biocovers consist of rich organic materials (ie, compost, sewage sludge, peat) and are a promising engineered technology among other technologies mentioned previously, because they minimize leachate generation while optimizing LFG emission (Scheutz et al., 2009; Koerner and Daniel, 1997).

6.1.4 PREVENTION OF WATER AND SOIL POLLUTION

Landfills are the most common source of soil and water pollution; this is true concerning both saturated and unsaturated zones of water (Ahmed and Sulaiman, 2001). Comprehensive studies involving electrical resistivity and groundwater monitoring techniques have evaluated and confirmed emerging concerns about landfill leachate plumes (Wijesekara et al., 2014; Reyes-López et al., 2008).

6.1.4.1 Characteristics of Leachates

Leachate characteristics depend heavily on several factors; waste composition and the age of waste, degree of decomposition, decomposition phase, waste-filling procedure, moisture content of waste, rate of water movement, and temperature (Armstrong and Rowe, 1999). Environmental toxicity and the health impacts of generated wastewaters are strictly governed by their composition compared with the prescribed permissible levels for the environment (Bhalla et al., 2013). Leachate can be categorized as a specific type of wastewater because of its relative variability of composition (Ehrig and Robinson, 2011). That variability results from numerous site-specific parameters including geography, climate, the geology of the area, the landfill design, and waste composition, but also the control measures and management practices applied. Understanding the natural attenuation of leachate pollutants in aquifers is mandatory for evaluating environmental risks associated with leachates entering the groundwater, for coherent interpretation of groundwater samples from monitoring wells, and for determining appropriate remedial action, including monitoring natural attenuations (Christensen et al., 2001). The first phase of degradation, which normally is short, is characterized by the aerobic degradation of organic matter. When the oxygen is depleted, degradation continues anaerobically. The anaerobic degradation process consists of two major fermentation phases: the acidogenic phase, generating a "young" biodegradable leachate, and the methanogenic phase, generating "old," stabilized leachates (Welander, 1998). Table 6.1 summarizes the typical characteristics of the leachate according to the age of the landfill (Renou et al., 2008).

6.1.4.2 Treatment of Landfill Leachate

One of the most important issues related to the siting, planning, design, operation, and long-term management of an MSW landfill is the management of leachates. A sustainable requirement for leachate management is to prevent its release by controlling the generation of liquid emissions and minimizing the percolation of wastes into the environment. Rates of leachate production may be low when sites are selected, designed, and constructed so as to minimize the production of leachate as a major objective of the process. A well-maintained, low-permeability cap over the landfill, which will be descriptively discussed later under the section on MSW—biochar (BC) use as a landfill cover, is an important requirement for leachate minimization (Manahan, 2000). However, most landfills of the South and Southeast Asian region are implemented as nonengineered, nonplanned waste dump sites. Estimated values and figures show that 90% of landfills come under this category within this region (Tränkler et al., 2005). Numerous kinds of toxicological and epidemiological studies have revealed the potential risks of groundwater contamination from leachates. Therefore, removal of organic material based on the chemical oxygen demand (COD), the biochemical oxygen demand (BOD), and the ammonium content of the leachate is the usual prerequisite before discharging the leachates into natural waters (Renou et al., 2008). Thus, proper leachate treatment is a foremost concern, and an economically viable and environmentally friendly option for leachate treatment has to be sought to resolve emerging environmental impacts resulting from leachate contamination.

6.1.4.3 Parameter Considerations for Leachate Treatment

The specifics of leachate treatment options depend heavily on the characteristics of the leachate. A suitable action on the leachate can be considered according to the type of treatment options available and their respective efficiencies. For instance, the stripping process reduces the ammoniacal-N concentration by stripping ammonia gas (Cheung et al., 1997). Furthermore, the evaporation process is designed to segregate solidified contaminants from the leachate system, and it consists mainly of two processes: direct evaporation of the raw leachate and evaporation after reverse osmosis concentration (Di Palma et al., 2002). Reverse osmosis concentration or the membrane process can be classified as a physical approach, which is capable of treating permeance streams from highly contaminated plumes. This process can be reclassified according to the aperture size of the membrane separator as microfiltration (0.05—2.0 μm), ultrafiltration (0.005—0.2 μm), nanofiltration (0.001—0.01 μm), and reverse osmosis (<0.002 μm). The membrane aperture range optimizes the filtration process (Christensen, 2011). Precipitation and flocculation processes can be used to remove larger particulates from the leachates and is especially suitable for humic substances. Usually flocculation and precipitation agents are used to stimulate the precipitation process (Renou et al., 2008). Chemical oxidation processes are targeted for the chemical derivation of toxic substances into less harmful substances by adding an oxidizing agent (Renou et al., 2008). In the case of constructed treatment wetlands, polluted water is allowed to flow through man-made wetlands that simulate natural wetland remediation processes (Bulc et al., 1997).

Table 6.1 General Characteristics of Different Experimental Results for Landfill Leachate

Pollutants	Parameter	Concentration	Country and Place	References
Organic matter	pH	8.1	Nigeria, Lagos	Aderemi et al. (2011)
		7.1	Canada, Toronto	Armstrong and Rowe (1999)
		7.3	South Korea, Kyungjoo	Im et al. (2001)
	EC	124	Nigeria, Lagos	Aderemi et al. (2011)
		245	India, Gazipur	Mor et al. (2006)
		15.74	Iran, Isfahan	Asadi et al. (2011)
	Chemical oxygen demand (COD)	160	Nigeria, Lagos	Aderemi et al. (2011)
		24,000	South Korea, Kyungjoo	Im et al. (2001)
		320	Denmark	Kjeldsen and Christophersen (2001)
		12,971	Canada, Toronto	Armstrong and Rowe (1999)
		5436[a]	China, Guangdong	Zhou et al. (2014)
		27,200	India, Gazipur	Mor et al. (2006)
	Biological oxygen demand (BOD_5)	44	Denmark	Kjeldsen and Christophersen (2001)
		7777	Canada, Toronto	Armstrong and Rowe (1999)
		19,000	India, Gazipur	Mor et al. (2006)
		10,800	South Korea, Kyungjoo	Im et al. (2001)
	BOD_5/COD ratio	0.11	Denmark	Kjeldsen and Christophersen (2001)
		0.6	Canada, Toronto	Armstrong and Rowe (1999)
		0.3	Bangladesh, Matuail	Mahmud et al. (2012)
	Total organic carbon (TOC)	30–29,000[c]	–	Christensen et al. (2001)
Inorganic macrocomponent	Sodium	7863	Nigeria, Lagos	Aderemi et al. (2011)
		1343[c]	Germany, northern part	Ehrig (1983)
		545	India, Gazipur	Mor et al. (2006)
	Potassium	1085[c]	Germany, northern part	Ehrig (1983)
		545	India, Gazipur	Mor et al. (2006)
	Calcium	690[c]	Germany, northern part	Ehrig (1983)
		588	Canada, Toronto	Armstrong and Rowe (1999)
	Magnesium	425[c]	Germany, northern part	Ehrig (1983)
		215[c]	Canada, Toronto	Lee et al. (2010)

Table 6.1 General Characteristics of Different Experimental Results for Landfill Leachate—cont'd

Pollutants	Parameter	Concentration	Country and Place	References
Inorganic trace element	Iron	6.5	Nigeria, Lagos	Aderemi et al. (2011)
		76	South Korea, Kyungjoo	Im et al. (2001)
		70.62	India, Gazipur	Mor et al. (2006)
	Cadmium	8.8	Nigeria, Lagos	Aderemi et al. (2011)
		0.035[a]	China, Guangdong	Zhou et al. (2014)
		0.06	India, Gazipur	Mor et al. (2006)
	Arsenic	0.04[a]	China, Guangdong	Zhou et al. (2014)
		0.23	South Korea, Kyungjoo	Im et al. (2001)
	Copper	2.72[a]	China, Guangdong	Zhou et al. (2014)
		0.78	South Korea, Kyungjoo	Im et al. (2001)
	Lead	10.2	Nigeria, Lagos	Aderemi et al. (2011)
		1.54	India, Gazipur	Mor et al. (2006)
	Mercury	79.4[b]	China, Guiyang	Li et al. (2005)
		1016[b]	China, Laohang	Tang et al. (2003)

EC, *electrical conductivity (mS cm^{-1}).*
All other values are in mg L^{-1} except pH and EC.
[a]*Dry season.*
[b]*ng L^{-1} concentration.*
[c]*Average value.*

If nitrification and denitrification are being considered, reduced nitrogenous species (ammoniacal nitrogen compounds) are oxidized to nitrates in the phase of nitrification. Denitrification is the process of reduction of nitrate-N to N_2 gas by involving microorganisms. Concerning the biological processes, two methods may be discussed in this category: conventional and the sequential batch reactor (SBR)-activated sludge treatment (Renou et al., 2008; Christensen, 2011; Laitinen et al., 2006). The SBR-activated sludge process typically differs from conventional methods of sludge processing. In SBR technology, aeration, biological degradation, and settling take place in a single tank that governs the functions of equalization, aeration, and sedimentation according to a time sequence. Adsorption is the phenomenon of phase transformation that retains a liquid phase's inorganic and organic contaminants onto a solid phase (adsorbents) using their surface characteristics. Pertaining to this, the use of BC as an adsorbent potentially implies its crucial role in leachate treatment (Ehrig and Robinson, 2011; Renou et al., 2008).

6.1.4.4 Major Challenges in Leachate Treatment

The selection of leachate treatment processes depend on the characteristics of the leachate to be treated. However, major drawbacks can be observed only during the process of onsite field application. Some drawbacks arise in connection with variations in flow, changes in leachate composition over time, and changes in waste composition.

The biggest challenge is in removing many different pollutants at the same time. Therefore, some studies are conducted over a wide range of issues, including managing major groups of contaminants, total suspended solid polycyclic hydrocarbons (PAH), pathogens, and heavy metals (Chang et al., 2011). However, not many studies have shown a high potential for removing many pollutants simultaneously. Chang et al. (2011) revealed that an adsorption method studied incorporates high efficiency along such a leachate stream, and a study by Kurniawan and Lo (2009) explained a method in which a raw leachate stream along granular activated carbon (GAC) reduces the BOD and COD values to a desirable level. However, incorporating activated carbon (AC) into leachate treatment in the developing world involves a high cost. At the moment, BC can be an alternative for AC. The dual behavior of ash and char components in BC explains the high effectiveness of BC in the simultaneous management of inorganic, organic, and microbe contaminants not only as a sorbent but also as a medium (Ahmad et al., 2014). A management focus on cost-effectiveness and environmental friendliness inherent this method and its sustainable approach would be a great asset to reduce the global pressure of waste accumulation.

6.2 A GREENER TRANSFORMATION OF MUNICIPAL SOLID WASTE TO MUNICIPAL SOLID WASTE—BIOCHAR

6.2.1 OVERVIEW

Open dumps and landfills are the most common methods for MSW disposal (Sharholy et al., 2008) because of their easy operation and economic benefits (low cost). However, open dumps and landfills adversely affect the natural environment because it takes a long time to stabilize such disposed waste materials (Li et al., 2015b). Restoration and management of such landfills may incur a huge cost to the government and relevant authorities. Incineration and biodegradation are other applicable methods for treating MSW. Although incineration leads to a reduction of more than 90% of the waste load, it also leads to the emission of toxic chemicals into the atmosphere and leaves behind of a metal-rich ash which ultimately has to be disposed under environmental pressure. Biodegradation can be considered an environmentally friendly treatment relative to incineration. However, this method is not efficient when the MSW does not contain adequate amounts of easily biodegradable material, and it takes a long time to stabilize.

6.2.2 A SUSTAINABILITY PLATFORM FOR MUNICIPAL SOLID WASTE TRANSFORMATION

There are different applications and strategies for managing different types of waste biomass with added value. The biotransformation of waste including MSW is currently developing into a bio-based economy. The sustainable approach of MSW used for biotransformation (eg, char, gas, bio-oil) contributes vital solutions to the environmental crisis (Kaur et al., 2014). Different types of advantages are elaborated by Roy (1988) regarding the consumption of waste biomass for biotransformation via the production of BC (Fig. 6.1). A volume reduction of raw waste accumulation of up to 90% and various types of valuable products produced, including solids, liquids and gases, are some advantages outlined.

FIGURE 6.1

MSW-BC production and application process.

Currently, a growing number of small cities are increasingly moving toward sustainability through several processes of solid waste management in which pyrolysis and composting are the most common practices used. Pyrolysis plants seem to be a viable alternative, with their controlled conditions for the production of char, oil, and gases (Chen et al., 2014). With a low environmental impact, fuel and chemical products can be obtained from waste biomass. A closed sustainable system of energy with a variable material flow and variable output according to the process cycle ensures the desirable cost-effectiveness.

The word "char" means any carbonaceous solid material processed under thermal decomposition without or under starved O_2 conditions. Lehmann and Joseph (2012) described BC as an application for soil improvement that contains charred organic matter. Attention has been drawn to BC because of its usefulness as a platform for climate change mitigation along with CO_2 and greenhouse gas control (Verma et al., 2014). Time limits for the sequestration of carbon in BC extend to 1000 years; therefore, it can be considered a tool to inhibit global warming. Environmental benefits of the MSW-BC production process can be distinguished along several streams within open dumps (Fig. 6.2). Use of the organic fraction from MSW for transformation into BC has double benefits: namely, the reduction of emissions from landfills and reduction of the waste mass (Oonk and Boom, 1995).

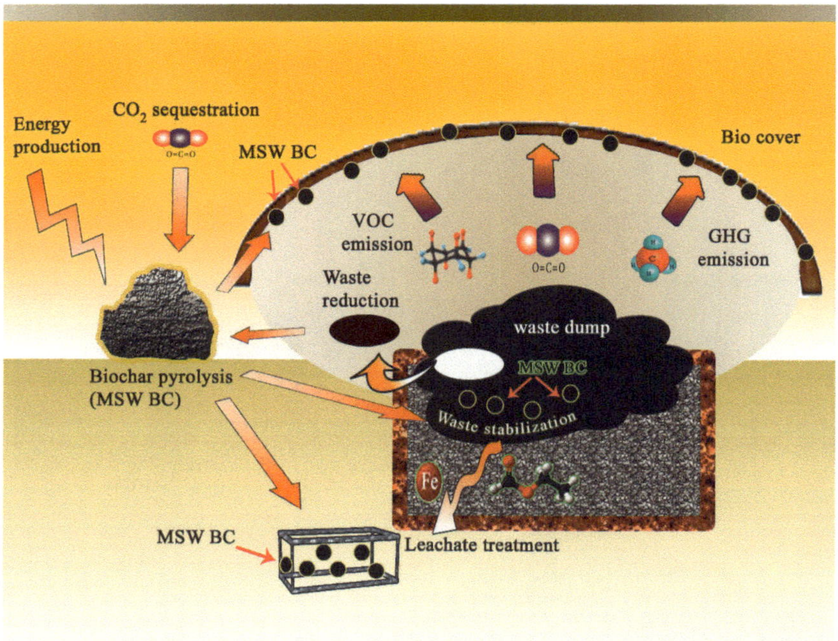

FIGURE 6.2

Potential environmental benefits of the MSW-BC stream.

6.2.3 MUNICIPAL SOLID WASTE–BIOCHAR PRODUCTION

The thermal decomposition and thermochemical conversion of biomass are two main processes of BC production; gasification focuses mainly on the conversion of biomass to hydrocarbons and pyrolysis is driven by high temperature, high-pressure liquefaction, ultrapyrolysis, and so on (Goyal et al., 2008).

The most abundantly used terminologies for MSW-BC production processes are given in Fig. 6.3. Pyrolysis represents a developing scenario owing to its diversification along a new evolutionary pathway. Exploration of energy stream pyrolysis marks a milestone because of its ability to direct conversion of biomass into gases, solids, and liquids with preserved high-energy content. In addition, this process has the capacity for specific applicability over a wide range of applications, including environmental remediation (Goyal et al., 2008). From the MSW-BC perspective, pyrolysis is subdivided into several processes such as fast pyrolysis, moderate pyrolysis, slow pyrolysis, flash pyrolysis, and flash carbonization (Fig. 6.3). Fast pyrolysis can be distinguished from other processes by its shortened residence time (<2 s) and with respect to its yield of about 75% bio-oil products (Wang et al., 2015; Velghe et al., 2011). Intermediate and slow pyrolysis are characterized by a much longer residence time than the fast process and spread up to a few days from a few minutes. Also, about 25–30% BC production can be observed as a result of this process (Ahmad et al., 2014; Velghe et al., 2011; Buah et al., 2007; He et al., 2010; Kabir et al., 2015; Phan et al., 2008; Williams and Besler, 1992). Flash carbonization can be accounted for as a different subpathway of pyrolysis, in which the

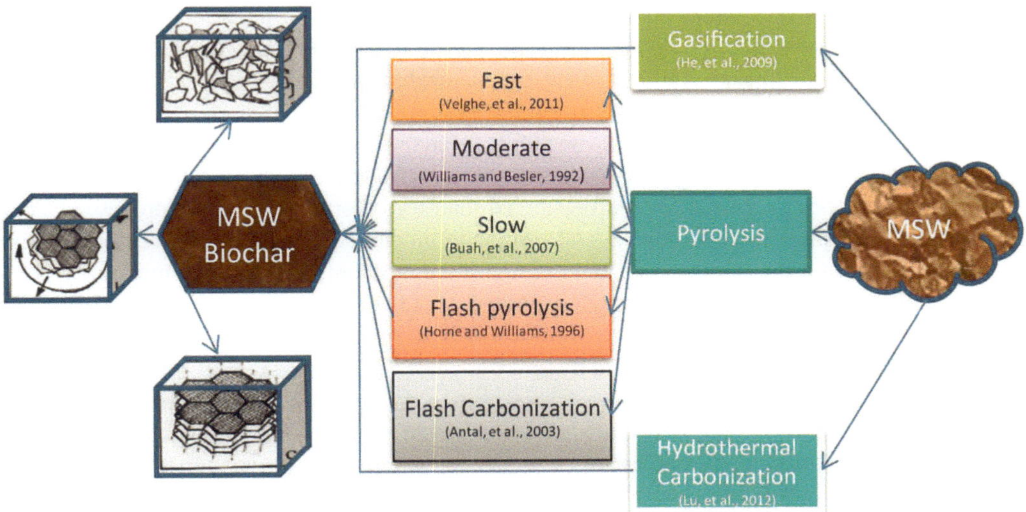

FIGURE 6.3

Thermal pathways for MSW-BC production.

production of charcoal or carbonized charcoal is driven under low-energy conditions. Controlled ignition with the application of elevated pressure shortens the process equilibration time (Antal et al., 2003). Flash pyrolysis is a fast-heating process in which a heating rate of up to 10^4 K s^{-1} is used and the ideal temperature is beyond 650°C. This process is mainly employed to produce a liquid product that is similar to the product of fast pyrolysis (Horne and Williams, 1996). Gasification differs from conventional pyrolysis in that it leads to the generation of "syn" gas under steam or high temperature with controlled oxygen (Ahmad et al., 2014; He et al., 2009). Hydrothermal carbonization (HTC) is the process of wet thermal conversion via a low-temperature (180−350°C) pathway under ambient pressure and its derivatives are mainly known as hydrochar (Lu et al., 2012; Berge et al., 2011).

A wide range of product characteristics and appearances is obtainable from MSW-BC owing to the drastic variations possible according to the environmental production and reactor conditions (Verma et al., 2014). The selection of the thermal decomposition mode fluctuates according to the requirements of the products and available feedstock properties. Therefore, it is important to select the production technology and the associated environmental evaluation before going into production and use.

6.2.3.1 Technology

Feedstock Properties

The technological aspects of BC preparation involve three major steps: feedstock selection preparation of the feedstock, and selection of the thermal pathway and reactor. The first step, the energy requirement, is a factor to be considered for BC production. Regulation of the feedstock moisture content is considered an important parameter. Food and organic waste has a high moisture content. Therefore, additional vaporization takes place during decomposition (Chen et al., 2014). However, the complex behavior and composition of MSW and the different types of processing in the chemical and physical industries contribute to a more diversified array for the desired output. Lignocellulose and moisture content are the main factors to consider when producing biochar using MSW (Berge et al., 2011; Mitchell et al., 2013). Research have been carried out to study the parent material composition and ash content during pyrolysis of MSW. Proximate and ultimate analysis of different types of MSW (Table 6.2) shows the drastic variation in results from different studies. Therefore, for the MSW-BC derivation process, more feedstock properties should be studied than for other BC derivation processes. Considering the production pathways for tropical and wet environments, in which moist feedstocks are most abundant, relevant associated problems should be taken into account, among other matters.

Comprehensive characteristics of thermal degradation of the organic phase can be observed by the related findings of Keiluweit et al. (2010). Moreover, the thermal pathway for the production of char and their derivation process is characterized by lower H/C and O/C ratios in the product. Consequently, dehydration and depolymerization occurs with respect to the original plant material. However, the tolerability of lignocellulosic substances to heat because of their crystalline matrix is observed by infrared spectrum results. Confirmation of a sequence of thermal degradation of organics, from hemicelluloses to cellulose and then to lignin, was observed by Yang et al. (2007). The resultant degradation product that contains a high amount of lignocellulosic residue is BC. Thermal degradation of a mixture of complex substances is explained by several researchers, and in relation to MSW, it represents the different patterns of degradation. A study by Omari (2015) explain three stages of thermal degradation with the loss of moisture and volatile matter as the first step followed by the degradation of lignocellulosic feedstock, and finally the degradation of plastic substances; whereas overlapping patterns were observed in the differential thermogravimetric curve for the hemicelluloses. Sørum et al. (2001) revealed the same pattern of degradation in a separate cellulose fraction (ie, paper, cardboard, wood) of MSW as well as in a mixture of MSW. Lignin and associated material was degraded at a temperature above 400°C whereas hemicelluloses and cellulose materials degraded at a lower temperature. Moreover, a catalytic interaction contributes to separating the cellulose fraction from inorganic species such as ash, which may cause prior cellulosic decomposition. In a mixture of MSW, char formation occurs at lower temperatures, with possible catalytic exhibitory effects postulated to occur by dehydrochlorination with polyvinyl chloride through acid hydrolysis. Therefore, MSW and its char formation is a complex process compared with other BC feedstock production processes. Ateş et al. (2013), Miskolczi et al. (2013) explained that the copyrolysis of a mixture of MSW products is not the same as the separate pyrolysis of biomass and plastic.

The Thermal Pathway

Different thermal conversion processes and respective MSW char yields are given in Table 6.3. Information about methods related to BC production are described in Section 6.2.3; this subsection deals more with the applicability and selectivity of those methods for sorbent-compatible MSW-BC production. Hydrothermal carbonization is the best method for a wet tropical environment because it avoids additional vaporization, and therefore reduces drying energy (Chen et al., 2014; Kumar and Gupta, 2009; Shackley et al., 2010). Among the different methods, the emission of toxic substances is minor through the pyrolysis of MSW compared with the combustion and gasification processes (Agarwal et al., 2015). It has been reported that the pyrolysis of MSW can control and limit the emission of polychlorinated dibenzo-*p*-dioxins (PCDD), polychlorinated dibenzofurans (PCD), and polychlorinated biphenyl (PCB) (Li et al., 2015b). However, gasification has proven difficult to

Table 6.2 Solid Waste Feedstock Properties (Proximate and Ultimate Analysis)

Waste Type/ Composition	Proximate Analysis				Ultimate Analysis					References
	V (%)	FC (%)	A (%)	M (%)	C (%)	H (%)	O (%)	N (%)	S (%)	
Dry MSW	82	12	6	9^{a}	52	6	30	0.3	0.4	He et al. (2009)
Separate MSW (paper, wood, cloth, organic waste, plastic, rubber)	99–8	0–20	47–0.1	87–0.1	36^{b}	5^{b}	27^{b}	0.2^{b}	0.1^{b}	Li et al. (1999)
MSW (plastic, paper, organic waste, textiles)	91^{c}	–	4.7	5.2	59	21	19	–	–	Ateş et al. (2013)
MSW	62	9.6	28.4	6.3	28.5	3.8	38.7	0.5	0.05	Berge et al. (2011)
Tea waste	62.4	23.4	3.2	10.8	51.7	6.6	37.7	3.6	0.2	Rajapaksha et al. (2014)
MSW	0.6	97.5	0.5	–	44.7	12.9	38.5	1.4	1.1	Jin et al. (2014)
MSW	65.8^{d}	2.2^{d}	32.0^{d}	58.3^{a}	–	–	–	–	–	Agrafioti et al. (2014)

MSW, *municipal solid waste;* V, *volatile matter;* FC, *fixed carbon;* A, *ash;* M, *moisture content.*
[a] *Initial moisture content.*
[b] *Average value.*
[c] *Combustibles.*
[d] *Dry basis.*

Table 6.3 Municipal Solid Waste–Derived Char Product Yield Under Different Thermal Conditions

Thermal Condition	Biomass Type	Temperature/ Heating Rate	Char Yield (%)	References
Pyrolysis				
Fast	MSW	500°C, NS	30	Wang et al. (2015)
		480–550°C[a]	8.1–7.3[b]	Velghe et al. (2011)
		830°C, NS	30	Roy (1988)
Moderate	MSW	600–20°C min^{-1}	35.2	Williams and Besler (1992)
Slow	MSW	550–4°C min^{-1}	18 ± 2.0[b]	Velghe et al. (2011)
		900–10°C min^{-1}	15.86	He et al. (2010)
		900–10°C min^{-1}	14.92[c]	
		400–10°C min^{-1}	49.8	Buah et al. (2007)
		700–10°C min^{-1}	32.3	
	Segregated (wood, cardboard, textile residues)	350–700–10°C min^{-1}	34–19	Phan et al. (2008)
	MGW	500–1°C min^{-1}	48.4	Kabir et al. (2015)
	MSW	500°C, NS	50	Ateş et al. (2013)
Flash carbonization	GW	400°C to a few min	40–45	Antal et al. (2003)
Flash pyrolysis	MSW	NS	12	Barducci and Neri (1997), Johnke (2002)
	WW shavings	400–550°C, 2.5 s	17–24	Horne and Williams (1996)
Gasification	MGW (kitchen garbage; 69%)	109.6°C	7.92	He et al. (2009)
	MSW	NS	10	Barducci and Neri (1997), Johnke (2002)
Hydrothermal carbonization	MSW	180–250, NS	50–80	Lu et al. (2012)
Combustion	MSW	NS	35	Barducci and Neri (1997), Johnke (2002)

MSW, *municipal solid waste;* MGW, *municipal green waste;* GW, *green waste;* WW, *waste wood;* NS, *not specified.*
[a]*Time in a few seconds.*
[b]*Pyrolysis liquid (water rich/oil/wax), solid, and gas yields.*
[c]*Under catalyst conditions.*

implement on a large scale and is not yet cost-competitive with combustion (Brian et al., 2009). Therefore, the following content tends to be closer to pyrolysis technology for MSW-BC production. Typically, slow pyrolysis processes are favorable for higher BC product output compared with other processes. The fast pyrolysis process is much more conducive to bio-oil production than to BC production.

Pyrolysis of complex substances is a more difficult issue compared with the degradation of other unimodal compounds. In the pyrolysis process, a major hindrance has been observed by several researchers in the removal of oxygen from the MSW mass. This step has to take into account the wider requirements of large-scale reactors, in which a large mass has to undergo degradation, also ensuring the desired output and its quality (Schilli, 2004). Selection of a pyrolysis reactor and its design is also a complex issue, but it is vital to gain a quality output. Proper reactor selection is important because of the size and the heterogeneity of MSW feedstock; this creates a more complex scenario regarding the pyrolysis process compared with small-scale reactors. MSW pyrolysis and its technology, product characteristics, and associated pollution as reported are not at a sufficient level, especially for systems design and the assessment of environmental impacts during reactor selection (Chen et al., 2015).

6.2.3.2 Characteristics of Municipal Solid Waste–Biochar

Characteristics of the MSW-BC product depend on the thermal pathway and the MSW feedstock composition. The slow pyrolysis process for MSW-BC derivation can be considered sound compared with other methods, and this process has been reported on by several MSW-BC studies. The fibrous fraction of the MSW contributes more sorbent properties to the MSW-BC product with minimized environmental drawbacks (Bernardo et al., 2012; Sharypov et al., 2002). Apart from that, the heterogeneous mixture present in MSW frequently contributes toxic substances and organic pollutants (ie, phenols, ketones, and PAHs) to the surrounding environment. Characteristics of industrial processed waste plastics, paper, and cardboard contribute a higher ash content to BC than wood-based BC materials (Van Zwieten et al., 2010; Enders et al., 2012). Concerning the slow pyrolysis process, constituents of three different fiber materials (lignin, cellulose, and hemicellulose) contained within the fibrous fraction and their elongated phase-by-phase degradation pattern are distinguished by several thermogravimetric studies (Yang et al., 2007; Zhou et al., 2013). The superimposed degradation patterns were also observed via different MSW-BC studies (Omari, 2015; Sørum et al., 2001; Miskolczi et al., 2013). Therefore, the characteristics of MSW-BC can be compared with those of BC, which has an elemental composition near that of lignin (Venderbosch and Prins, 2010). It has also been mentioned that a catalytic environment is much more favorable for the degradation of MSW substances under low energetic conditions. Properties of BC are also improved through catalysis. The presence of Ni-Mo catalyst in a mixture of MSW-derived MSW-BC shows significant surface properties (BET-17.4 $m^2\ g^{-1}$) compared with for bare controls and other catalysts, and the same catalyst (Ni-Mo) shows a reduction in heavy metal content (Fe concentration from 827 to 779 ppm) for the pyrolysis oil product (Miskolczi et al., 2013). The required surface properties and characteristics of MSW-BC obtained from leachate treatment studies are summarized in Table 6.3. A similar pattern of yield drawdown was observed when the temperature of pyrolysis was increased, which was barely observed from MSW-BC, but incremental surface properties could be accounted for by results of 10% surface area increments within a 100°C fluctuation (Jin et al., 2014). The phenomenon of increments in surface properties was explained by a study by Liu et al. (2011). This study explained that volatile matter is released from the biomaterial with increasing retention temperatures, generating char with more pores.

Various adsorption studies revealed the potential applicability of MSW-BC used as a promising sorbent within the landfill leachate treatment pathway (Jin et al., 2014; Agrafioti et al., 2014; Li et al., 2015; Parshetti et al., 2014) and in methane degradation (Sadasivam and Reddy, 2015) along with LFG mitigation. An advanced comparative cost–benefit analysis of adsorbent GAC with BC was successfully conducted. The break-even price of the relevant study for BC and AC, respectively, was US $246/t and US $1500/t (Ahmad et al., 2012; McCarl et al., 2009). Therefore, the free availability of feedstock for MSW-BC, together with this calculation describes MSW-BC as an economically viable and sustainable alternative to AC and other sorbents.

However, the preparation of MSW-BC may accumulate heavy metals and organic contaminants (ie, phenols, ketones, PAHs) in the BC product (Verheijen et al., 2010), and such contaminants may be released into the environment with their respective applications. Hence, more detailed investigations are needed in that direction.

6.3 MUNICIPAL SOLID WASTE—BIOCHAR FOR LANDFILL COVER

6.3.1 MUNICIPAL SOLID WASTE—BIOCHAR AS AN ADSORBENT AND A SUBSTRATE FOR LANDFILL COVER

The representation of BC as a landfill cover has many different aspects, such as LFG control, leachate and contaminant control, plant growth enhancement, soil ecology improvement, and soil property alteration (Chen et al., 2015). A positive effect of the use of BC in MSW composting and incorporation with soil is that it increases the uptake of CH_4 and nonmethane organic compounds from the soil (Reddy et al., 2014; Wang et al., 2015; Barlaz et al., 2004). Evidence has shown the extent of the contribution of physicochemical and environmental factors (ie, temperature, humidity, degree of compaction, and particle size) in LFG remediation and transformation on the presence of BC (Wong et al., 2015; Yaghoubi, 2011). A study investigated the potential ability of BC to ameliorate landfill cover; results obtained were at an acceptable level for use in waste wood BC (WW-BC) for this purpose (Yaghoubi, 2011). LFG adsorption and degradation depend on the moisture content of the material (Yaghoubi, 2011). This study showed a high level of CH_4 and CO_2 adsorption at 10% and 20% moisture content (Yaghoubi, 2011). The highest adsorption among all treatments was observed at the lowest moisture content. Also, the use of WW-BC as a soil mitigation agent showed high efficiency with LFG mitigation (Sadasivam and Reddy, 2014). In addition, evidence for the reduction of other LFGs and emissions of N_2O from pollution sources was investigated by Yanai et al. (2007) using incubation studies; compatible results were acquired. The proportion of BC used in amendment is important for identifying LFG behavior. Therefore, a study of 10% MSW-BC with bare soil incubated for N_2O determination showed a comparatively significant reduction by MSW-BC than for the parent bare soil. A lower value of 383 $\mu g\ Nm^{-2}$ was observed for MSW-BC and a higher value of 2620 $\mu g\ Nm^{-2}$ was obtained for bare soil (Yanai et al., 2007). Transportation and adsorption of CH_4 with WW-BC were reported by another study, and a content of 2% (w/w) of BC showed a higher gas permeability and adsorption. Also, hard WW-BC showed a porosity increment with the addition of BC (Sadasivam and Reddy, 2014). A study via an LFG transformation experiment revealed that with an increase in BC content and a greater degree of compaction, the gas permeability for BC-amended *clay* material was reduced (Wong et al., 2015) (Fig. 6.4).

Particle size is also an important parameter in LFG adsorption and gas transfer along a layer because a lower diffusion rate caused by a smaller particle size tends to lower performance. A study involving a 10% (w/w) BC-amended soil with two different particle sizes for the BC (0.8 and 0.4 mm) showed that the higher amount of CH_4 adsorbed (75 $mL\ kg^{-1}$) was for the 0.4-mm fraction whereas the lower amount adsorbed (64 $mL\ kg^{-1}$) was for the 0.8-mm fraction (Yaghoubi, 2011). Furthermore,

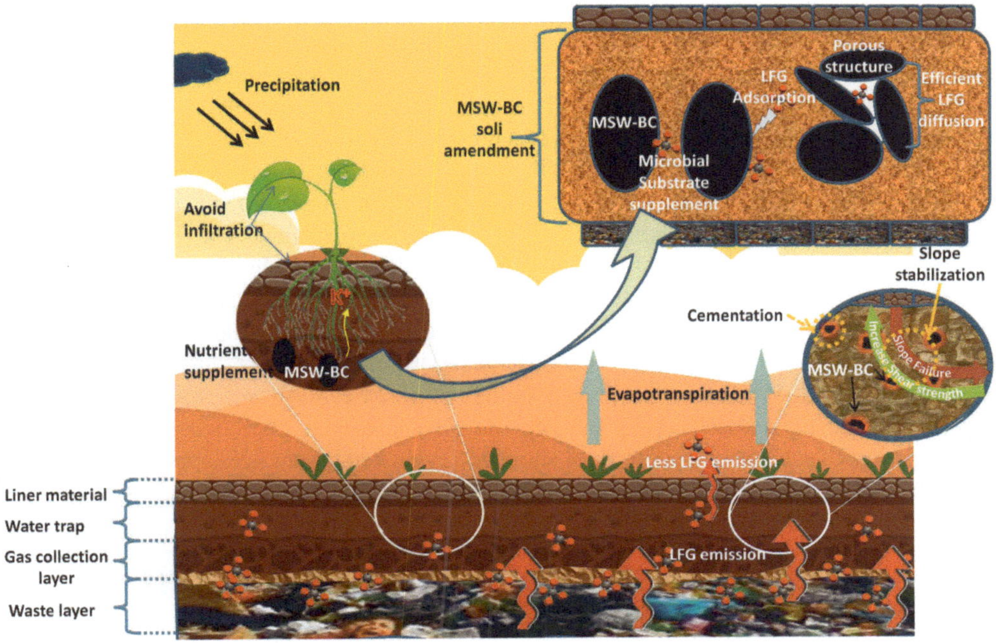

FIGURE 6.4

Multidisciplinary applicability of MSW-BC on landfill biocover.

curve fitting carried out for the adsorption study correlated to 0.99 for CO_2 and CH_4 on a pseudo–second-order model and maximum adsorption for CH_4 showed significant improvement compared with bare soil. The values were 32,400, and 417 mL kg^{-1}, respectively, for bare soil, 0.8 mm BC addition, and 0.4 mm BC addition. The bulk density and the inverse porosity increased to 45% from 25% with BC amendment in the soil. Testing with BC types revealed that the water-holding capacity also increased with the amendments. Diffusion of gas increased with the development of a porous structure as revealed by scanning electron microscopy images (Sadasivam and Reddy, 2015). BC characteristics depend on the nature of the feedstock and the production process. Also, different types of soil inhibit or exhibit the amendment properties of BC with respect to their physicochemical properties. These soil properties regulate the capacity for attenuation of CH_4 with amendment. Meanwhile, the BC amendments significantly alter soil properties (pore size, texture, porosity, consistency, structure, and gradation) and their combined effect on the sorbate is observed (Brady and Weil, 2008). BC has a crucial role in such amendments and the high surface area of BC leads to an increase in the surface area as a combined effect with soil (Lehmann and Joseph, 2012). When all of these requirements and respective facilitations are considered, it can be concluded that MSW-BC can be used as a viable prospective sorbent for LFG mitigation. However, Spokas and Reicosky (2009) postulated that different behavior patterns of 16 different LFGs studied posed difficulties in correlating their identification. Also, some studies emphasize that further correlations between gas permeability properties with the particle size distribution of BC-amended soil should be carried out before field applications (Wong et al., 2015).

Soil properties (physical/chemical) contribute to geotechnical and hydrogeological parameters, which are important for more durable, efficient cover design. Poor attention paid to these properties leads to landfill failure (Chen et al., 2015; Seed et al., 1990). The increasing effect of infiltration with BC amendment was investigated (Yaghoubi, 2011) and this factor negatively affected the stability of slopes because of the development of pore water pressure. In contrast, the same study by Yaghoubi (2011) showed an increment in shear strength with BC amendment which positively contributed against such failure. Furthermore, BC with a clay amendment shows a cementation effect and the elongated processes were cementation, surface deposition, and pore filling. The resultant cover was of higher strength and higher erosion resistance (Haque et al., 2014). According to all of these studies, the properties of BC were of considerable importance in design implementation. The importance of optimizing such properties were highlighted by Chen et al. (2015) in their study. A wide distribution of properties could be observed in the MSW-BC products, which necessitates identifying of the application with the best fit. Therefore, prolonged studies of gases evolved and their behavior with BC amendments should be studied, taking into account such factors as microbial activity. Slope engineering properties of such amended mixtures should be evaluated before field application (Chen et al., 2015; Lamb et al., 2014).

6.4 MUNICIPAL SOLID WASTE–BIOCHAR IN THE TREATMENT OF LANDFILL LEACHATE

6.4.1 MUNICIPAL SOLID WASTE–BIOCHAR AS SORBENT

The pyrolysis of MSW is a pathway which converts waste to energy and produces gas, liquid, and solid hydrocarbons or other valuable products. Consequently, BC produced from MSW can be used for energy production such as to substitute for coal or the production of adsorbent (Mayakaduwa et al., 2015). A few studies have indicated that MSW-derived BC could be successfully used for the remediation of environmental contaminants such as heavy metals and dyes (Agrafioti et al., 2014; Li et al., 2015; Parshetti et al., 2014).

6.4.2 THE POTENTIAL OF MUNICIPAL SOLID WASTE–BIOCHAR FOR THE REMOVAL OF HEAVY METAL AND ORGANIC POLLUTANTS

The application of raw MSW-BC for remediation was successfully performed for arsenic and chromium removal from an aqueous solution (Agrafioti et al., 2014). For most BCs, MSW-BC could be used directly or modified to enhance their qualities with the objective of attaining a high capacity for the removal of heavy metals and organic pollutants. Chemical modification of MSW-BC might rearrange their porous structure and the amount and types of functional groups as well, resulting in a change in adsorption capacity (Uchimiya et al., 2011). Furthermore, owing to the combination of carbonation and activation in a single step, chemical activation could be performed at relatively lower temperatures with an enhanced in-built catalytic oxidation ability (Li et al., 2015a). Therefore, chemically activated MSW-BC may provide an efficient, simple, and low-cost approach to removing environmental contaminants (Liu et al., 2012). Chloride predecessors such as NH_4Cl have been used successfully to modify MSW-BC chemically because of its high efficiency and low cost (Li et al., 2015a). In addition, Jin et al. (2014) activated MSW-BC with a high adsorption capacity by using 2 M potassium hydroxide. Besides chemical activation, physical activation using steam and microwave activation have been demonstrated for the production of MSW-BC (Li et al., 2015a). Table 6.4 summarizes the characteristics of MSW-BC produced via different processes with the objective of obtaining a product capable of removing various contaminants from the environment.

MSW-BC may contain various types of heavy metals. Thus, the use of such types of BC derived from MSW leads to direct contamination of the environment. A study of MSW applications further revealed (Li et al., 2005) that raw MSW contains different kinds of heavy metals including Cd, Cr, Cu, and Pb. In addition, the concentration of all of these heavy metals almost doubles after pyrolysis. Hence, it is obvious that most of the heavy metals are not removed during pyrolysis. Particularly as a result of high volatile content, heavy metals are not sufficiently removed under the low temperature of pyrolysis. Similarly, Jin et al. (2014) showed that metal concentrations for As, Cd, Cr, Cu, Zn, Ni, and Pb increased gradually with increasing pyrolysis temperatures. For example, the concentration of metals such as Cd, Cr, Cu, Zn, Ni, and Pb in MSW-BC pyrolyzed at 500°C was 5, 64, 101, 213, 143, and 10 mg kg^{-1}, respectively. By contrast, the concentration of the same metals in raw MSW was 3, 42, 30, 53, 34, and 0 mg kg^{-1}, respectively. Nevertheless, MSW-BC could be used safely as an excellent sorbent despite the heavy metals contained in it (Jin et al., 2014).

Modeling experimental data may indicate the efficiency of and different mechanisms in the removal of pollutants via MSW-BC. Most studies indicated that heavy metal adsorption onto MSW-BC is perfectly matched with pseudo–second-order kinetic model attributed to chemisorption and the Langmuir isotherm model indicating monolayer adsorption (Table 6.5). Agrafioti et al. (2014) suggested that the possible mechanism for heavy metal adsorption onto MSW-BC could be electrostatic attractions between the negatively charged BC surface and the metal cation. In addition, the large surface area and high pore volume provide faster mass transfer of heavy metals into the MSW-BC pores, thus enhancing more opportunities for metal-binding sites (Ahmad et al., 2014). Consequently, physically and chemically activated MSW-BC has a more precisely gained high adsorption capacity than nonactivated MSW-BC, possibly owing to the larger surface area and higher pore volume. Moreover, three mechanisms were proposed, Van der Waals forces, electrostatic interactions, and hydrogen bonding, to explain the adsorption of Acridine Orange and Rhodamine 6G dyes onto urban food waste–BC (Parshetti et al., 2014). Furthermore, there is little available information about the thermodynamic parameters regarding the adsorption of contaminants onto MSW-BC. A remediation study involving the adsorption of Acridine Orange and Rhodamine 6G dyes onto urban food waste–BC reported that higher-temperature environmental conditions are more favorable for the process, and indicated that the adsorption reaction is endothermic in nature (Parshetti et al., 2014).

Batch desorption studies have been reported for As(V) adsorption onto different kinds of BC derived from different feedstocks such as MSW, rice husk, and sewage sludge (Agrafioti et al., 2014). As far as maximum desorption is concerned, MSW-derived BC desorbed only 4% of the adsorbed amount (about 98%) whereas corresponding percentages for rice husk and sewage sludge-derived BC were 0% and 10%, respectively, but the adsorbed amounts of As(V) were 56% and 80%, respectively.

Hence, MSW-derived BC seems to have a potential to be used as an efficient adsorbent in removing environmental contaminants from landfill leachates. Apart from heavy metals and dyes, adsorption of volatile organic compounds and other organic pollutants such as pesticides and humic acids onto MSW-BC has not yet been reported in the literature. Nevertheless, plenty of research articles are available regarding the adsorption of heavy metals, organic contaminants, and so on using municipal sewage sludge-derived BC (Chen et al., 2014; Zielinska and Oleszczuk, 2015).

Table 6.4 Characteristics of MSW-BC Produced Under Different Pyrolysis Conditions to Remove Environmental Contaminants

Feedstock	Target Contaminant	Pyrolysis Temperature (°C)	pH	Moisture	Mobile Matter (%)	Fixed Matter (%)	Ash (%)	C (%)	H (%)	O (%)	N (%)	Surface Area ($m^2 g^{-1}$)	Pore Volume ($cm^3 g^{-1}$)	References
Organic fraction of MSW	As(V), Cr(III), Cr(VI)	300	–	58.3	65.8	2.2	32.0	–	–	–	–	–	–	Agrafioti et al. (2014)
MSW	Hg(0)	400	–	0.4	26.2	23.3	50.1	80.2	7.4	8.5	2.8	2.0	0.007	Li et al. (2015)
		600	–	1.0	15.1	30.1	53.8	85.0	2.3	6.7	3.6	4.0	0.029	
MSW	As(V)	400	8.0	–	22.3	65.2	6.1	48.6	12.2	31.7	1.3	20.7	0.027	Jin et al. (2014)
		500	8.5	–	26.4	63.8	9.2	59.5	9.1	20.8	1.4	29.1	0.039	
		600	9.0	–	15.1	78.2	6.2	70.1	8.4	13.7	1.3	29.8	0.038	
		500 activated	8.3	–	1.3	94.1	0.7	67.5	6.4	23.5	0.8	49.1	0.357	
Urban food waste	Acridine Orange, Rhodamine	250	–	–	–	–	–	52.2	5.7	38.2	3.4	–	–	Parshetti et al. (2014)
		350	–	–	–	–	–	56.8	4.3	36.4	2.1	–	–	

Table 6.5 Batch Test Result Obtain From Different Studies

Adsorbent	Adsorbate	Conc.	Temp. (K)	Contact Time (h)	Adsorption Capacity	Isotherm	Kinetic Model	References
Organic fraction of MSW (300°C)	As(V)	0–100 μg L^{-1}	RT	3	–	Freundlich	Pseudo–second-order	Agrafioti et al. (2014)
Organic fraction of MSW (300°C)	Cr(III)	0–10 μg L^{-1}	RT	24	–	NF	Pseudo–second-order	Agrafioti et al. (2014)
Organic fraction of MSW (300°C)	Cr(VI)	0–800 μg L^{-1}	RT	96	–	Freundlich	NF	Agrafioti et al. (2014)
MSW (600°C)	Hg (0)	42.0 ± 2 μg m^3	393	7.5	26.8 μg g^{-1}	–	Pseudo–first-order	Li et al. (2015a)
MSW (600°C) + chemical activation	Hg (0)	42.0 ± 2 μg m^3	393	7.5	118.1 μg g^{-1}	–	Pseudo–second-order	Li et al. (2015a)
MSW (600°C) + chemical and microwave activation	Hg (0)	42.0 ± 2 μg m^3	393	7.5	157.7 μg g^{-1}	–	Pseudo–second-order	Li et al. (2015a)
MSW (400°C)	As(V)	5–400 mg L^{-1}	298 ± 1	24	24.20 mg g^{-1}	Langmuir	Pseudo–second-order	Jin et al. (2014)
MSW (500°C)	As(V)	5–400 mg L^{-1}	298 ± 1	24	24.49 mg g^{-1}	Langmuir	Pseudo–second-order	Jin et al. (2014)
MSW (600°C)	As(V)	5–400 mg L^{-1}	298 ± 1	24	18.06 mg g^{-1}	Langmuir	Pseudo–second-order	Jin et al. (2014)
MSW (500°C) activated	As(V)	5–400 mg L^{-1}	298 ± 1	24	30.98 mg g^{-1}	Langmuir	Pseudo–second-order	Jin et al. (2014)
Urban food waste (250°C)	Acridine orange	10–100 mg L^{-1}	298	2	60.25 mg g^{-1}	Langmuir	Pseudo–second-order	Parshetti et al. (2014)
Urban food waste (250°C)	Acridine orange	10–100 mg L^{-1}	303	2	75.18 mg g^{-1}	Langmuir	Pseudo–second-order	Parshetti et al. (2014)
Urban food waste (250°C)	Acridine orange	10–100 mg L^{-1}	313	2	79.36 mg g^{-1}	Langmuir	Pseudo–second-order	Parshetti et al. (2014)
Urban food waste (250°C)	Rhodamine	10–100 mg L^{-1}	298	2	51.05 mg g^{-1}	Langmuir	Pseudo–second-order	Parshetti et al. (2014)
Urban food waste (250°C)	Rhodamine	10–100 mg L^{-1}	303	2	62.34 mg g^{-1}	Langmuir	Pseudo–second-order	Parshetti et al. (2014)
Urban food waste (250°C)	Rhodamine	10–100 mg L^{-1}	313	2	71.42 mg g^{-1}	Langmuir	Pseudo–second-order	Parshetti et al. (2014)

RT, *room temperature;* NF, *not fitted.*

6.5 POTENTIAL TO BE USED AS A MATERIAL FOR PERMEABLE REACTIVE BARRIER

6.5.1 OVERVIEW OF PERMEABLE REACTIVE BARRIER APPLICABILITY

Contaminant plume movement in open dump sites with no liners is considered a challenging issue in groundwater protection. Permeable reactive barriers (PRB) are a novel concept evaluated for the remediation of such contaminants in real environments under in situ remediation (Simon and Meggyes, 2000). Zero valent iron (ZVI) is used as the most common PRB medium for the heterogeneous containment of inorganic metals, organic contaminants, and the removal of radioactive compounds from the plumes (Turner et al., 2005; Tratnyek et al., 2003). However, because of the site specificity and ready availability and the cost involved, alternative materials such as carbon sources (compost and waste materials), limestone, AC, and other relevant materials were tested as reactive media (ITRC, 2011; Gibert et al., 2003). Major mechanisms behind the PRB adsorb or decompose the contaminant or transform it into less toxic substances. Various kinds of techniques are employed, such as adsorption, precipitation, biodegradation, and/or chemical oxidation (Scherer et al., 2000). For the purpose of adsorption, precipitation, and biodegradation, AC listed as a PRB material because of its crucial role in many different pollutants (Thiruvenkatachari et al., 2008). Chiu et al. (2013) investigated the potential of PRB for the effective removal of organic contaminants (ie, methyl tert-butyl ether) using two different types of granular AC. BC can be categorized as a substance with characteristics similar to GAC. BC contains not only a carbonized portion but also an additional non-carbonized fraction, which increases the binding ability of soil contaminants. As a result of phenolic, carboxylic, and other functional groups, BC is a feasible option in media selections for PRB (Uchimiya et al., 2011). On the other hand, MSW-BC can be incorporated with NZVI for a better performance. Chlorinated compound removal in a batch sorption study was evaluated by Liu and Zhang (2010), and the incorporation of $FeCl_3$ with a waste biomass (pinewood sawdust) pyrolysis product was performed for the preparation of porous carbon (PC). Furthermore, the highest adsorption (almost complete) was observed for PCB with Fe-incorporated PC, whereas low-PCB removal was observed with bare PC (Liu and Zhang, 2010). Therefore, waste BC would be favorable and has the potential to be used as a composite material in PRB.

6.5.2 POTENTIAL APPLICABILITY OF MUNICIPAL SOLID WASTE–BIOCHAR INSTEAD OF BIOCHAR

A study by Carl and Craig (2000) concluded that the altered field geotechnical characteristics of amended material would enable more pollutant to flow through the material. The ability of pollutant transferring depends on the saturated hydraulic conductivity (K). A higher K of foreign material rather than the in situ field environment would stimulate plume diversion into foreign material. Different studies of BC amendments with soil showed a wide range of behavior in the bulk density of the final material and K of the final material. With some applications of BC (Herath et al., 2013), for example, corn stover BC with alfisol under 350 and 550°C pyrolysis conditions reported a higher K value with a lower bulk density. However, some sandy soils conditioned by BC deviate from PRB applicability requirements, such as a tendency toward a decline in K. Also, an increase in pyrolysis temperature had a negative effect on PRB applicability. The lowest K value was observed in black locust BC of 500°C pyrolysis with a higher application rate for sandy soil (Uzoma et al., 2011). Also, acacia green waste BC with planosol soil shows an increase in K with a decreasing trend in bulk density within the saturated range (Hardie et al., 2014). However, the optimization of BC amendments and holding material conditions led to better hydrological and physical properties of the media (Barnes et al., 2014), which made them more suitable for PRB application. Likewise, optimization of MSW-BC feedstock and composite holding material optimization would be enable compliance with PRB guidelines.

PRBs are available for remediation of radioactive compounds. The literature have shown the potential of greenwaste BC produced by HTC on removing radioactive compounds (Kumar et al., 2011). Descriptive results have been obtained for U removal phenomena with low temperature carbonization processed BC. This study indicated a high adsorption capacity (0.52 mg U g^{-1}) for U(VI), which was significantly higher than the adsorption values for quartz under the same conditions (Kumar et al., 2011). Agrafioti et al. (2014) reported a batch sorption study for As(V), Cr(III), and Cr(VI) removal from aqueous solution using MSW-derived BC and three other different BCs and a sandy loam soil. Results showed higher irreversible adsorption for the three different BCs whereas the bare soil deviated negatively from adsorption with a simultaneous desorption visualization (Table 6.5). A toxic heavy metal adsorption study with composite PRBs, containing waste organic material and elemental sulfur, proved the ability of PRB to bring a leachate contamination of 10 mg^{-1} Cr(VI) and 0.6 M COD to acceptable levels. Design efficiency for Cr(VI) adsorption onto composite PRB recorded a value of 7.92 mg g^{-1} over 270 days.

A batch sorption study of sulfamethazine adsorption onto tea waste BC under different preparation condition with aqueous media showed that the Langmuir adsorption capacity or q value for different BCs was (50–58%) greater than the respective biomass (Rajapaksha et al., 2014). MSW-BC is potentially be useful as an alternative reactive material for composite PRB. Furthermore, treatment of humic acids using solid waste derivatives is increasingly discussed processing fly ash. The topic of humic acid adsorption on fly ash and its derived unburned carbon was covered by Wang and Zhu (2007), and the adsorption capacity of Langmuir isotherm for the fly ash was 11 mg g^{-1} whereas for unburned carbon it was 72 mg g^{-1} at optimum pH and temperature. Composite matrixes of waste-derived BC, incinerated bottom ash, and MSW compost with other materials have been widely carried out for better performance, identification, and optimization. In a study of waste-derived BC with MSW-incinerated bottom ash (MSWIBA), researchers suggested that the combination of bottom ash and BC contains metal immobilization and chemically enforces inorganic fertility. A study revealed that the combination of MSWIBA and BC had the potential to condition the soil and manage soil contamination (Milla and Huang, 2013). Therefore, by applying a PRB material, MSW-BC combined with MSWIBA can be used for safe disposal of MSWIBA into the environment.

Plumes associated with dyes and derived pharmaceutical products can also be seen in the vicinity of the environment (Slack et al., 2005). The MSW-BC that produced by nonhazardous organic fraction of MSW, can be used as a material to remediate hazardous waste, including organic products, dyes, and pharmaceutical products. A study (Rajapaksha et al., 2014) of sulfamethazine (pharmaceutical) removal was compiled using tea waste BC under the different pyrolytic conditions; the results revealed that modification of steam activation led to higher sorption for high temperature (700°C)-derived BC. According to that batch test, optimization could be used for further column and PRB applicability (Rajapaksha et al., 2014). The MSW-BC derived from cellulosic fraction of MSW employed for azo dye removal under laboratory conditions was investigated by Agarwal et al. (2015). However, those data were for PRB application rather than BC (Agarwal et al., 2015). Different types of studies revealed the potential of BC use for the remediation of contaminant sites such agricultural soils by PAH; the enhancement of sorption capacity of soil for PAH was suggested during mitigation of further groundwater contamination; for poor conditioned BC, however, the contribution to PAH contamination has been investigated by several researchers.

Landfill VOCs are a vital subject regarding the contaminant stream and its elongate; their hazardous toxicity even at a trace level as well can be expressed as carcinogenic, teratogenic, and organogenic, but may also be acutely toxic to organisms. Most commonly PRB are used at the point of plume saturation areas for remediation. The management of diffusion in VOCs through the unsaturated zone and the surrounding closer environment is difficult because of the behavior of VOCs. Horizontal permeable reactive barriers (HPRBs) are technology that has gained less attention in contaminant site remediation. Oxidizing agent–amended HPRB was successfully evaluated by Mahmoodlu et al. (2015) for trichloroethylene, toluene, and ethanol. With HPRB, pH leading to manage oxidation along the chemical reaction and waste level fluctuation inhibit the oxidation by destroying pH regulation. But with the use of BC as an amendment to PRB, the pH of PRB experimental solutions can be optimized significantly owing to the buffering effect of BC samples on the environment (Trakal et al., 2014). Amendment of soil with BC in HPRB can be contribute to degradation of organic materials that are recalcitrant to oxidative degradation from oxidizing agent and are capable of adsorbing while contributing to microbial growth and activity within HPRB (Sadasivam and Reddy, 2014). Often PRB compromises economical viable options by dealing with in situ remediation rather than a pumping and remediation tank system and uses MSW-BC with freely available materials (Fig 6.5).

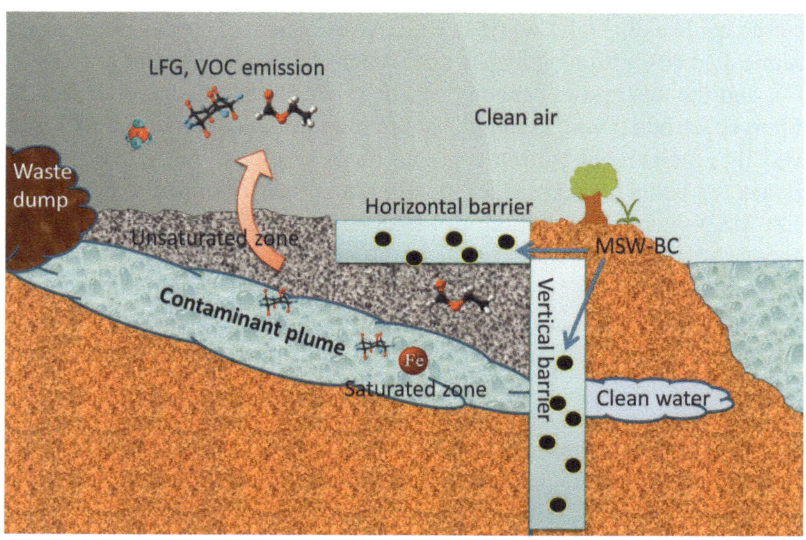

FIGURE 6.5

MSW-BC amended PRB application.

6.6 POTENTIAL OF MUNICIPAL SOLID WASTE-BIOCHAR FOR LANDFILL STABILIZATION

Currently, the use of MSWIBA is discouraged because the higher accumulation of toxicity in it acts as a pollution source. A considerable extent of support for using incinerated bottom ash to stabilize waste landfill was revealed in a laboratory-level investigation that best fit a mix of waste bottom ash. In addition, problems with co-disposal on bottom ash were identified. However, the use of both char and ash can be more effective than incinerated bottom ash, in that cost-effective MSW-BC application can simulate the landfill stabilization application process. The experiment by Li et al. (2014) explored the ratio of bottom ash and MSW of 1:10 and 1:5; owing to the organic (45.5%) higher moisture content, only the 1:10 ratio was compatible with the stabilization process. Stabilization of landfill pulp amendment contributes to improved leachate quality by neutralizing acidity and depleting nitrogen and organic content from the leachate recirculation. Ahmad et al. (2014) and Mitchell et al. (2013) elaborated on the dual composition behavior of BC such as ash and char, and stated that because of the extensive industrial process of lignocellulose bio material, the high ash content of MSW-BC would be have a high potential for application in amending waste stabilization as a substitution for incinerated bottom ash.

6.7 REMARKS

A universal sorbent application would be greatly effective for landfill leachate treatment. In this sense, MSW-BC has been successfully used as a promising sorbent over the past decade. Landfill leachate heavy metal and organic removal studies have proven that it has a superficial effect on contributing to relieving pressure on global waste management while improving leachate quality. The potential applicability of PRB to expanded contaminant plumes including heavy metal, VOCs, and organic and humic substances is emphasized by the results of batch sorption studies as well as closer types of BC-amended PRB results. In addition, the renewable energy production pathway of the MSW stream is widely helpful for fulfilling energy requirements possibly owing to the cost–benefit performance of the pyrolysis process rather than incineration and plasma gasification (Wilson et al., 2013). Therefore, energy production and contaminant management are achieved simultaneously as a result of MSW-BC production. In addition, minimized global warming, carbon sequestration, and reduced solid waste accumulation (ie, weight and volume) are some potential benefits of MSW-BC production. It is obvious that MSW-BC application as landfill capping greatly reduces greenhouse gases, results less global warming. The potential identification of landfill stabilization and leachate quality improvement using a mixture of MSW-BC and MSW would be a greater achievement. That can be helpful for further MSW volume reduction within the landfill site. MSW-BC would be a double benefit in terms of waste reduction and pollution remediation taking place.

In contrast, environmental drawbacks have been reported during MSW-BC production and application. The mixture of MSW contributes to more a diverse array of heavy metal accumulation over MSW-BC (Verheijen et al., 2010) and subsequently MSW-BC application may contaminate the environment. On the other hand, because of the difficulties of small plastic and electronics waste mixture segregation at a large scale, toxic gases (ie, PCDD, PCD, PCB) release extensively into the surrounding environment. In addition, PAHs accumulation on BC, leading to unavoidable circumstances in MSW-BC preparation, and it can be act as a self-polluting source for the environment.

Nevertheless, the application of catalyst to pyrolysis oils shows a significant reduction in contamination with respect to the mixture of MSW in pyrolysis (Miskolczi et al., 2013). Therefore, to apply MSW-BC, it is suggested that MSW-BC be further purified and heavy metals be removed. Another suggestion is that separation of the cellulose fraction of MSW instead of a mixture of MSW, which enhances organic pollutants (ie, phenols, ketones, and PAHs) in BC products, would minimize organic pollutants in the resultant MSW-BC. Furthermore, optimization of the properties of MSW-BC were an emerging concern before their field application because the properties were subsequently effected upon amendment and drastically altered, as observed when applied technologically (Chen et al., 2015). Taking these facts into consideration, it appears that the net effect of MSW-BC application as a green technology for landfill enhances environment sustainability in the long term. However, regular monitoring, proper management practices, and future research are required to ensure the temporal suitability and environmental viability of MSW-BC for landfill leachate treatment.

ACKNOWLEDGMENTS

This study was supported by SATREPS-JICA grant to Sri Lanka on the Development of Pollution Control and Environmental Restoration Technologies of Waste Landfill Sites and partially support from by the NRC grant 15-024.

REFERENCES

Aderemi, A.O., et al., 2011. Assessment of groundwater contamination by leachate near a municipal solid waste landfill. African Journal of Environmental Science and Technology 5 (11), 933–940.

Agarwal, M., Tardio, J., Mohan, S.V., 2015. Pyrolysis biochar from cellulosic municipal solid waste as adsorbent for azo dye removal: equilibrium isotherms and kinetics analysis. International Journal of Environmental Science and Development 6, 67–72.

Agrafioti, E., Kalderis, D., Diamadopoulos, E., 2014. Arsenic and chromium removal from water using biochars derived from rice husk, organic solid wastes and sewage sludge. Journal of Environmental Management 133, 309–314.

Ahmad, M., et al., 2012. Effects of pyrolysis temperature on soybean stover-and peanut shell-derived biochar properties and TCE adsorption in water. Bioresource Technology 118, 536–544.

Ahmad, M., et al., 2014. Biochar as a sorbent for contaminant management in soil and water: a review. Chemosphere 99, 19–33.

Ahmed, A.M., Sulaiman, W.N., 2001. Evaluation of groundwater and soil pollution in a landfill area using electrical resistivity imaging survey. Environmental Management 28 (5), 655–663.

Ali, S.M., et al., 2014. Open dumping of municipal solid waste and its hazardous impacts on soil and vegetation diversity at waste dumping sites of Islamabad city. Journal of King Saud University – Science 26 (1), 59–65.

Antal, M.J., Mochidzuki, K., Paredes, L.S., 2003. Flash carbonization of biomass. Industrial & Engineering Chemistry Research 42 (16), 3690–3699.

Armstrong, M., Rowe, R., 1999. Effect of landfill operations on the quality of municipal solid waste leachate. In: Proc. 3rd Int. Landfill Symp., Cagliari.

Asadi, A., et al., 2011. Effect of MSW leachate on soil consistency under influence of electrochemical forces induced by soil particles. International Journal of Electrochemical Science 6, 2344–2351.

Ateş, F., Miskolczi, N., Borsodi, N., 2013. Comparison of real waste (MSW and MPW) pyrolysis in batch reactor over different catalysts. Part I: product yields, gas and pyrolysis oil properties. Bioresource Technology 133, 443–454.

Babayemi, J., Dauda, K., 2009. Evaluation of solid waste generation, categories and disposal options in developing countries: a case study of Nigeria. Journal of Applied Sciences and Environmental Management 13 (3).

Barducci, P., Neri, G., 1997. An IGCC plant in Italy for power generation from biomass. In: Bioelettrica Internal Report.

Barlaz, M., et al., 2004. Evaluation of a biologically active cover for mitigation of landfill gas emissions. Environmental Science & Technology 38 (18), 4891–4899.

Barnes, R.T., et al., 2014. Biochar-induced changes in soil hydraulic conductivity and dissolved nutrient fluxes constrained by laboratory experiments. PLoS One 9 (9). http://journals.plos.org/plosone/article?id=10.1371/journal.pone.0108340.

Baun, A., et al., 2004. Xenobiotic organic compounds in leachates from ten Danish MSW landfills—chemical analysis and toxicity tests. Water Research 38 (18), 3845–3858.

Berge, N.D., et al., 2011. Hydrothermal carbonization of municipal waste streams. Environmental Science & Technology 45 (13), 5696–5703.

Bernard, C., et al., 1996. Estimation of the hazard of landfills through toxicity testing of leachates—I. Determination of leachate toxicity with a battery of acute tests. Chemosphere 33 (11), 2303–2320.

Bernardo, M., et al., 2012. Physico-chemical properties of chars obtained in the co-pyrolysis of waste mixtures. Journal of Hazardous Materials 219, 196–202.

Bhalla, B., Saini, M., Jha, M., 2013. Effect of age and seasonal variations on leachate characteristics of municipal solid waste landfill. IJRET: International Journal of Research in Engineering and Technology 2 (8). http://ijret.esatjournals.org/Archive?VI=20130208.

Bogner, J.E., Spokas, K.A., Chanton, J.P., 2011. Seasonal greenhouse gas emissions (methane, carbon dioxide, nitrous oxide) from engineered landfills: daily, intermediate, and final California cover soils. Journal of Environmental Quality 40 (3), 1010–1020.

Brady, N.C., Weil, R.R., 2008. An Introduction to the Nature and Properties of Soils, 14th ed. Pearson Prentice Hall.

Brian, G., et al., 2009. Reconsidering Municipal Solid Waste as a Renewable Energy Feedstock. Environmental and Energy Study Institute, Washington.

Buah, W., Cunliffe, A., Williams, P., 2007. Characterization of products from the pyrolysis of municipal solid waste. Process Safety and Environmental Protection 85 (5), 450–457.

Buenrostro, O., Bocco, G., 2003. Solid waste management in municipalities in Mexico: goals and perspectives. resources. Conservation and Recycling 39 (3), 251–263.

Bulc, T., Vrhovsek, D., Kukanja, V., 1997. The use of constructed wetland for landfill leachate treatment. Water Science and Technology 35 (5), 301–306.

Carl, R.E., Craig, H.B., 2000. Evaluation and design of permeable reactive barriers amidst heterogeneity. In: Civil and Environmental Engineering. University of Wisconsin Madison.

Chandler, A.J., et al., 1997. Municipal Solid Waste Incinerator Residues. Elsevier.

Chang, N.B., Wanielista, M.P., Henderson, D., 2011. Temperature effects on functionalized filter media for nutrient removal in stormwater treatment. Environmental Progress & Sustainable Energy 30 (3), 309–317.

Chen, D., et al., 2014. Pyrolysis technologies for municipal solid waste: a review. Waste Management 34 (12), 2466–2486.

Chen, T., et al., 2014. Influence of pyrolysis temperature on characteristics and heavy metal adsorptive performance of biochar derived from municipal sewage sludge. Bioresource Technology 164, 47–54.

Chen, D., et al., 2015. Reprint of: pyrolysis technologies for municipal solid waste: a review. Waste Management 37, 116–136.

Chen, X.-W., et al., 2015. Feasibility of biochar application on a landfill final cover—a review on balancing ecology and shallow slope stability. Environmental Science and Pollution Research 1−15.

Cheung, K., Chu, L., Wong, M., 1997. Ammonia stripping as a pretreatment for landfill leachate. Water, Air, and Soil Pollution 94 (1−2), 209−221.

Chiu, C.-H., Lin, T.-F., Hung, W.-N., 2013. Performance of laboratory adsorbent-based permeable reactive barrier for treating MTBE-contaminated groundwater. Sustainable Environment Research 23 (1), 53−60.

Christensen, T.H., et al., 2001. Biogeochemistry of landfill leachate plumes. Applied Geochemistry 16 (7), 659−718.

Christensen, T.H., 2011. Solid Waste Technology & Management, vol. 2. Wiley Online Library.

Di Palma, L., et al., 2002. Treatment of industrial landfill leachate by means of evaporation and reverse osmosis. Waste Management 22 (8), 951−955.

Diaz, L.F., Savage, G.M., Eggerth, L.L., 2005. Alternatives for the treatment and disposal of healthcare wastes in developing countries. Waste Management 25 (6), 626−637.

Ehrig, H.J., Robinson, H., 2011. Landfilling: leachate treatment. Solid Waste Technology & Management 1 & 2, 858−897.

Ehrig, H.J., 1983. Quality and quantity of sanitary landfill leachate. Waste Management and Research 1, 53−68.

El-Fadel, M., et al., 2002. Temporal variation of leachate quality from pre-sorted and baled municipal solid waste with high organic and moisture content. Waste Management 22 (3), 269−282.

Enders, A., et al., 2012. Characterization of biochars to evaluate recalcitrance and agronomic performance. Bioresource Technology 114, 644−653.

Forster, P., et al., 2007. Changes in atmospheric constituents and in radiative forcing. In: Climate Change 2007. The Physical Science Basis (Chapter 2).

Garg, A., Achari, G., 2010. A comprehensive numerical model simulating gas, heat, and moisture transport in sanitary landfills and methane oxidation in final covers. Environmental Modeling & Assessment 15 (5), 397−410.

Gibert, O., et al., 2003. Evaluation of municipal compost/limestone/iron mixtures as filling material for permeable reactive barriers for in-situ acid mine drainage treatment. Journal of Chemical Technology and Biotechnology 78 (5), 489−496.

Goyal, H., Seal, D., Saxena, R., 2008. Bio-fuels from thermochemical conversion of renewable resources: a review. Renewable and Sustainable Energy Reviews 12 (2), 504−517.

Haque, A., et al., 2014. Biochar sequestration in lime-slag treated synthetic soils: a green approach to ground improvement. Journal of Materials in Civil Engineering 26 (12).

Hardie, M., et al., 2014. Does biochar influence soil physical properties and soil water availability? Plant and Soil 376 (1−2), 347−361.

Hazra, T., Goel, S., 2009. Solid waste management in Kolkata, India: practices and challenges. Waste Management 29 (1), 470−478.

He, M., et al., 2009. Hydrogen-rich gas from catalytic steam gasification of municipal solid waste (MSW): influence of catalyst and temperature on yield and product composition. International Journal of Hydrogen Energy 34 (1), 195−203.

He, M., et al., 2010. Syngas production from pyrolysis of municipal solid waste (MSW) with dolomite as downstream catalysts. Journal of Analytical and Applied Pyrolysis 87 (2), 181−187.

Herath, H., Camps-Arbestain, M., Hedley, M., 2013. Effect of biochar on soil physical properties in two contrasting soils: an Alfisol and an Andisol. Geoderma 209, 188−197.

Hoornweg, D.T.L., et al., 2012. What a Waste: A Global Review of Solid Waste Management. In: Urban Development Series. World Bank.

Hoornweg, D.T.L., 1999. What a Waste: Solid Waste Management in Asia. The World Bank: The International Bank for Reconstruction and Development.

Horne, P.A., Williams, P.T., 1996. Influence of temperature on the products from the flash pyrolysis of biomass. Fuel 75 (9), 1051−1059.

Huber-Humer, M., Gebert, J., Hilger, H., 2008. Biotic systems to mitigate landfill methane emissions. Waste Management & Research 26 (1), 33–46.

Im, J.-H., et al., 2001. Simultaneous organic and nitrogen removal from municipal landfill leachate using an anaerobic-aerobic system. Water Research 35 (10), 2403–2410.

ITRC, 2011. Permeable Reactive Barrier: Technology Update. Interstate Technology & Regulatory Council, PRB: Technology Update Team: PRB-5, Washington, D.C.

Jin, H., et al., 2014. Biochar pyrolytically produced from municipal solid wastes for aqueous As(V) removal: adsorption property and its improvement with KOH activation. Bioresource Technology 169, 622–629.

Joseph, K., Nagendran, R., Palanivelu, K., 2002. "Open Dumps to Sustainable Landfills", CES ENVISION, Monthly newsletter of the ENVIS node under EMCBTA project at CES (AU/CES/NL-02) supported by Ministry of Environment and Forests, Government of India, vol. 1, No. 2.

Johnke, B., 2002. In: Grover, V.I., Grover, V.K., Hogland, W. (Eds.), Current Situation of Waste Incineration and Energy Recovery in Germany, pp. 195–200.

Kabir, M.J., Chowdhury, A.A., Rasul, M.G., 2015. Pyrolysis of municipal green waste: a modelling, simulation and experimental analysis. Energies 8 (8), 7522–7541.

Kaur, S., et al., 2014. Waste biomass: a prospective renewable resource for development of bio-based economy/processes. In: Biotransformation of Waste Biomass into High Value Biochemicals. Springer, pp. 3–28.

Keiluweit, M., et al., 2010. Dynamic molecular structure of plant biomass-derived black carbon (biochar). Environmental Science & Technology 44 (4), 1247–1253.

Kjeldsen, P., Christophersen, M., 2001. Composition of leachate from old landfills in Denmark. Waste Management & Research 19 (3), 249–256.

Kjeldsen, P., et al., 2002. Present and long-term composition of MSW landfill leachate: a review. Critical Reviews in Environmental Science and Technology 32 (4), 297–336.

Koerner, R.M., Daniel, D.E., 1997. Final Covers for Solid Waste Landfills and Abandoned Dumps. Thomas Telford.

Kumar, S., et al., 2011. An assessment of U(VI) removal from groundwater using biochar produced from hydrothermal carbonization. Journal of Environmental Management 92 (10), 2504–2512.

Kumar, S., Gupta, R.B., 2009. Biocrude production from switchgrass using subcritical water. Energy & Fuels 23 (10), 5151–5159.

Kurniawan, T.A., Lo, W.-H., 2009. Removal of refractory compounds from stabilized landfill leachate using an integrated H_2O_2 oxidation and granular activated carbon (GAC) adsorption treatment. Water Research 43 (16), 4079–4091.

Laitinen, N., Luonsi, A., Vilen, J., 2006. Landfill leachate treatment with sequencing batch reactor and membrane bioreactor. Desalination 191 (1), 86–91.

Lamb, D.T., et al., 2014. Phytocapping: an alternative technology for the sustainable management of landfill sites. Critical Reviews in Environmental Science and Technology 44 (6), 561–637.

Lee, A.H., Nikraz, H., Hung, Y.T., 2010. Influence of waste age on landfill leachate quality. International Journal of Environmental Science and Development 1 (4), 347–350.

Lehmann, J., Joseph, S., 2012. Biochar for Environmental Management: Science and Technology. Routledge.

Leton, T.G., Omotosho, O., 2004. Landfill operations in the Niger delta region of Nigeria. Engineering Geology 73 (1–2), 171–177.

Li, A., et al., 1999. Experimental studies on municipal solid waste pyrolysis in a laboratory-scale rotary kiln. Energy 24 (3), 209–218.

Li, Z., et al., 2005. Mercury in landfill leachate and its removal by the on-site sewage treatment plant. Earth and Environment 33, 52–56.

Li, W.-B., et al., 2014. Impact of MSWI bottom ash codisposed with MSW on landfill stabilization with different operational modes. BioMed Research International 2014.

Li, G., et al., 2015a. Elemental mercury removal using biochar pyrolyzed from municipal solid waste. Fuel Processing Technology 133, 43–50.

Li, X.-D., et al., 2015b. Emission characteristics of hazardous components in municipal solid waste incinerator residual ash. Journal of Zhejiang University-Science A 4 (16), 316–325.

Liu, W.-J., et al., 2011. Preparation of high adsorption capacity bio-chars from waste biomass. Bioresource Technology 102 (17), 8247–8252.

Liu, P., et al., 2012. Modification of bio-char derived from fast pyrolysis of biomass and its application in removal of tetracycline from aqueous solution. Bioresource Technology 121, 235–240.

Liu, Z., Zhang, F.-S., 2010. Nano-zerovalent iron contained porous carbons developed from waste biomass for the adsorption and dechlorination of PCBs. Bioresource Technology 101 (7), 2562–2564.

Lu, X., Jordan, B., Berge, N.D., 2012. Thermal conversion of municipal solid waste via hydrothermal carbonization: comparison of carbonization products to products from current waste management techniques. Waste Management 32 (7), 1353–1365.

Mahmoodlu, M.G., et al., 2015. Evaluation of a horizontal permeable reactive barrier for preventing upward diffusion of volatile organic compounds through the unsaturated zone. Journal of Environmental Management 163, 204–213.

Mahmud, K., Hossain, M.D., Shams, S., 2012. Different treatment strategies for highly polluted landfill leachate in developing countries. Waste Management 32 (11), 2096–2105.

Manahan, S.E., 2000. Environmental Science, Technology, and Chemistry, fifth ed. In: Environmental Chemistry

Matteson, G.C., Jenkins, B., 2007. Food and processing residues in California: resource assessment and potential for power generation. Bioresource Technology 98 (16), 3098–3105.

Mayakaduwa, S., Weerasundara, L., Vithanage, M., 2015. Municipal solid waste derived biochar for removal of humic acid in landfill leachate. In: International Conference on Solid Wastes 2015. Hong Kong-SAR, P.R. China.

McCarl, B.A., et al., 2009. Economics of biochar production, utilization and greenhouse gas offsets. Biochar for Environmental Management: Science and Technology 341–358.

Milla, O.V., Huang, W.-J., 2013. Identifying the advantages of using municipal solid waste bottom ash in combination with rice husk and bamboo biochar mixtures as soil modifiers: enhancement of the release of polyphenols from a carbon matrix. Journal of Hazardous, Toxic, and Radioactive Waste 17 (3), 204–210.

Miskolczi, N., Ateş, F., Borsodi, N., 2013. Comparison of real waste (MSW and MPW) pyrolysis in batch reactor over different catalysts. Part II: contaminants, char and pyrolysis oil properties. Bioresource Technology 144, 370–379.

Mitchell, P.J., Dalley, T.S., Helleur, R.J., 2013. Preliminary laboratory production and characterization of biochars from lignocellulosic municipal waste. Journal of Analytical and Applied Pyrolysis 99, 71–78.

Mor, S., et al., 2006. Leachate characterization and assessment of groundwater pollution near municipal solid waste landfill site. Environmental Monitoring and Assessment 118 (1–3), 435–456.

Omari, A.M., 2015. Characterization of municipal solid waste for energy recovery. Multidisciplinary Engineering Science and Technology 2 (1).

Oonk, J., Boom, A., 1995. Landfill Gas Formation, Recovery and Emissions. TNO Inst. of Environmetal and Energy Technology.

Parshetti, G.K., Chowdhury, S., Balasubramanian, R., 2014. Hydrothermal conversion of urban food waste to chars for removal of textile dyes from contaminated waters. Bioresource Technology 161, 310–319.

Phan, A.N., et al., 2008. Characterisation of slow pyrolysis products from segregated wastes for energy production. Journal of Analytical and Applied Pyrolysis 81 (1), 65–71.

Rajapaksha, A.U., et al., 2014. Pyrolysis condition affected sulfamethazine sorption by tea waste biochars. Bioresource Technology 166, 303–308.

Reddy, K.R., et al., 2014. Enhanced microbial methane oxidation in landfill cover soil amended with biochar. Journal of Geotechnical and Geoenvironmental Engineering 140 (9), 04014047.

Renou, S., et al., 2008. Landfill leachate treatment: review and opportunity. Journal of Hazardous Materials 150 (3), 468–493.

Reyes-López, J.A., et al., 2008. Assessment of groundwater contamination by landfill leachate: a case in México. Waste Management 28, S33–S39.

Roy, G., 1988. Municipal solid waste recycle – an economic proposition for a developing nation. Indian Journal of Environmental Protection 8 (1). http://ijep.in/.

Sadasivam, B.Y., Reddy, K.R., 2014. Landfill methane oxidation in soil and bio-based cover systems: a review. Reviews in Environmental Science and Biotechnology 13 (1), 79–107.

Sadasivam, B.Y., Reddy, K.R., 2015. Adsorption and transport of methane in landfill cover soil amended with waste-wood biochars. Journal of Environmental Management 158, 11–23.

Scherer, M.M., et al., 2000. Chemistry and microbiology of permeable reactive barriers for in situ groundwater clean up. Critical Reviews in Microbiology 26 (4), 221–264.

Scheutz, C., et al., 2009. Microbial methane oxidation processes and technologies for mitigation of landfill gas emissions. Waste Management & Research.

Scheutz, C., Mosbæk, H., Kjeldsen, P., 2004. Attenuation of methane and volatile organic compounds in landfill soil covers. Journal of Environmental Quality 33 (1), 61–71.

Schilli, J.W., 2004. Using Gasification to Process Municipal Solid Waste. HDR Innovations. In: http://www.hdrinc.com/Assets/documents/Publications/InnovationsPFS/Winter2004/UsingGasificationtoProcessMSW.pdf.

Seed, R.B., Mitchell, J.K., Seed, H.B., 1990. Kettleman hills waste landfill slope failure. II: Stability analyses. Journal of Geotechnical Engineering 116 (4), 669–690.

Shackley, S., et al., 2010. An Assessment of the Benefits and Issues Associated with the Application of Biochar to Soil. Department for Environment, Food and Rural Affairs, UK Government, London.

Sharholy, M., et al., 2008. Municipal solid waste management in Indian cities—a review. Waste Management 28 (2), 459–467.

Sharypov, V., et al., 2002. Co-pyrolysis of wood biomass and synthetic polymer mixtures. Part I: influence of experimental conditions on the evolution of solids, liquids and gases. Journal of Analytical and Applied Pyrolysis 64 (1), 15–28.

Simon, F.-G., Meggyes, T., 2000. Removal of organic and inorganic pollutants from groundwater using permeable reactive barriers. Land Contamination & Reclamation 8 (2), 103–116.

Slack, R., Gronow, J., Voulvoulis, N., 2005. Household hazardous waste in municipal landfills: contaminants in leachate. Science of the Total Environment 337 (1), 119–137.

Sørum, L., Grønli, M.G., Hustad, J.E., 2001. Pyrolysis characteristics and kinetics of municipal solid wastes. Fuel 80 (9), 1217–1227.

Spokas, K.A., Reicosky, D.C., 2009. Impacts of sixteen different biochars on soil greenhouse gas production. Annals of Environmental Science 3 (1), 4.

SUTD, 2012. D.o. Revenue. In: Sludges Excluded from Municipal Solid Waste. Sales and Use Tax Division, North Carolina.

Tang, Q., et al., 2003. Mercury pollution of water system in landfill site. Urban Environment and Urban Ecology 16, 15–17.

Thiruvenkatachari, R., Vigneswaran, S., Naidu, R., 2008. Permeable reactive barrier for groundwater remediation. Journal of Industrial and Engineering Chemistry 14 (2), 145–156.

Trakal, L., et al., 2014. Copper removal from aqueous solution using biochar: effect of chemical activation. Arabian Journal of Chemistry 7 (1), 43–52.

Tränkler, J., et al., 2005. Influence of tropical seasonal variations on landfill leachate characteristics—results from lysimeter studies. Waste Management 25 (10), 1013–1020.

Tratnyek, P.G., et al., 2003. Permeable Reactive Barriers of Iron and Other Zero-Valent Metals. In: Environmental Science and Pollution Control Series, pp. 371–422.

Turner, M., et al., 2005. Permeable Reactive Barriers: Lessons Learned/New Directions. DTIC Document.

Uchimiya, M., Chang, S., Klasson, K.T., 2011. Screening biochars for heavy metal retention in soil: role of oxygen functional groups. Journal of Hazardous Materials 190 (1–3), 432–441.

US-EPA, 1984. In: A Ground Water Protection Strategy for the Environmental Protection Agency. United States Environmental Protection Agency Office of Drinking Water, p. 11.

Uzoma, K., et al., 2011. Influence of biochar application on sandy soil hydraulic properties and nutrient retention. Journal of Food, Agriculture and Environment 9 (3 & 4), 1137–1143.

Van Zwieten, L., et al., 2010. Effects of biochar from slow pyrolysis of papermill waste on agronomic performance and soil fertility. Plant and Soil 327 (1–2), 235–246.

Velghe, I., et al., 2011. Study of the pyrolysis of municipal solid waste for the production of valuable products. Journal of Analytical and Applied Pyrolysis 92 (2), 366–375.

Venderbosch, R., Prins, W., 2010. Fast pyrolysis technology development. Biofuels, Bioproducts & Biorefining 4 (2), 178.

Verheijen, F., et al., 2010. Biochar Application to Soils. A Critical Scientific Review of Effects on Soil Properties, Processes, and Functions. EUR, p. 24099.

Verma, M., et al., 2014. Thermochemical transformation of agro-biomass into biochar: simultaneous carbon sequestration and soil amendment. In: Biotransformation of Waste Biomass into High Value Biochemicals. Springer, pp. 51–70.

Visvanathan, C., Trankler, J., 2003. Municipal solid waste management in Asia: a comparative analysis. In: Workshop on Sustainable Landfill Management.

Wang, X., et al., 2015. Mitigation of non-methane organic compounds through landfill soil cover and its environmental implications. Journal of Material Cycles and Waste Management 17 (4), 616–625.

Wang, S., Zhu, Z., 2007. Humic acid adsorption on fly ash and its derived unburned carbon. Journal of Colloid and Interface Science 315 (1), 41–46.

Wang, H., Wang, L., Shahbazi, A., 2015. Life cycle assessment of fast pyrolysis of municipal solid waste in North Carolina of USA. Journal of Cleaner Production 87, 511–519.

Welander, U., 1998. Characterization and Treatment of Municipal Landfill Leachates. Lund University.

Wijesekara, S., et al., 2014. Fate and transport of pollutants through a municipal solid waste landfill leachate in Sri Lanka. Environmental Earth Sciences 72 (5), 1707–1719.

Williams, P.T., Besler, S., 1992. The pyrolysis of municipal solid waste. Journal of the Institute of Energy 65, 192–200.

Wilson, B., et al., October 2013. A Comparative Assessment of Commercial Technologies for Conversion of Solid Waste to Energy.

Wong, J.T.F., et al., 2015. Gas permeability of biochar-amended clay: potential alternative landfill final cover material. Environmental Science and Pollution Research 1–6.

Yaghoubi, P., 2011. Development of Biochar-Amended Landfill Cover for Landfill Gas Mitigation. University of Illinois at Chicago.

Yanai, Y., Toyota, K., Okazaki, M., 2007. Effects of charcoal addition on N_2O emissions from soil resulting from rewetting air-dried soil in short-term laboratory experiments. Soil Science and Plant Nutrition 53 (2), 181–188.

Yang, H., et al., 2007. Characteristics of hemicellulose, cellulose and lignin pyrolysis. Fuel 86 (12–13), 1781–1788.

Zhou, H., et al., 2013. The pyrolysis simulation of five biomass species by hemi-cellulose, cellulose and lignin based on thermogravimetric curves. Thermochimica Acta 566, 36–43.

Zhou, D., et al., 2014. Column test-based optimization of the permeable reactive barrier (PRB) technique for remediating groundwater contaminated by landfill leachates. Journal of Contaminant Hydrology 168, 1–16.

Zielinska, A., Oleszczuk, P., 2015. Evaluation of sewage sludge and slow pyrolyzed sewage sludge-derived biochar for adsorption of phenanthrene and pyrene. Bioresource Technology 192, 618–626.

CHAPTER

7

REMOVAL AND RECOVERY OF METALS BY BIOSORBENTS AND BIOCHARS DERIVED FROM BIOWASTES

N.K. Niazi[1,2], B. Murtaza[3], I. Bibi[1,2], M. Shahid[3], J.C. White[4], M.F. Nawaz[1], S. Bashir[1], M.B. Shakoor[1], G. Choppala[2], G. Murtaza[1], H. Wang[5,6]

University of Agriculture Faisalabad, Faisalabad, Punjab, Pakistan[1]; *Southern Cross University, East Lismore, NSW, Australia*[2]; *COMSATS Institute of Information and Technology, Vehari, Pakistan*[3]; *The Connecticut Agricultural Experiment Station, New Haven, CT, United States*[4]; *Zhejiang A & F University, Lin'an, Hangzhou, Zhejiang Province, China*[5]; *Guangdong Dazhong Agriculture Science Co. Ltd., Dongguan, China*[6]

CHAPTER OUTLINE

Environmental Materials and Waste. http://dx.doi.org/10.1016/B978-0-12-803837-6.00007-X

7.1 INTRODUCTION

Rapid industrial development has led to the contamination of valuable surface and groundwater resources with toxic contaminants such as heavy metals, dyes, nutrient ions (nitrogen and phosphate), and precious metals (PMs) (Won et al., 2014; Abdi and Kazemi, 2015; Shakoor et al., 2015). Several conventional techniques (eg, chemical precipitation, chemical coagulation, ion exchange, electrochemical technology, membrane process, ultrafiltration) have been used to achieve effective and rapid removal of these environmental contaminants, particularly metal ions, because of their toxic nature, and high production cost, and the limited or decreasing availability of metal deposits (Das, 2010; Lin et al., 2015; Dodson et al., 2015). However, these traditional techniques have major constraints to their application such as the high operation cost, use of hazardous chemicals, and generation of sludge, which can be a source of secondary contamination. Conversely, sorption using biosorbents and biochars has seen significant advancements and is considered a promising technology for immobilizing (anionic/cationic) PMs and heavy metals, with subsequent recovery from water and wastewaters. This is attributed to the low cost and high efficiency, the negligible production of chemical or biological sludge, regenerability and reusability, and recovery of metals by the desorption process (Vijayaraghavan and Yun, 2008).

A wide variety of biomaterials have been identified and examined to generate low-cost sorbents for the purification of water and wastewater, such as fungi, algae, and biowastes from agricultural and industrial by-products (Vijayaraghavan and Yun, 2008; Park et al., 2010; Imamoglu and Tekir, 2008). In the past few decades, biosorption of heavy metals, PMs, and ionic dyes has been widely and successfully studied for their removal from water and wastewater using agricultural and food industry biowastes: for example, peels of various fruits and vegetables, wheat straw, rice husk, and sugarcane bagasse (Shakoor et al., 2015; Abid et al., 2016). In addition to biosorbents, studies revealed significant advancements in the production of biochar from biowastes which differ in their physicochemical properties from carbonized materials (ie, activated carbons) (Khare et al., 2013).

Heavy metals such as zinc (Zn), nickel (Ni), iron (Fe), chromium (Cr), lead (Pb), manganese (Mn), copper (Cu), and cadmium (Cd) have been used in different industrial manufacturing/catalytic processes and released into the environment through wastewater discharges or from their solid waste resources (Ghaly et al., 2014).

In addition to heavy metals, PMs such as platinum group metals (PGMs) (ruthenium [Ru], rhodium [Rh], palladium [Pd], osmium [Os], iridium [Ir], and platinum [Pt]) are widely used in a variety of "hi-tech" industries owing to their unique physical and chemical properties (details on heavy metals and PMs have been provided in the following sections). The increasing demand of these heavy and PMs is in direct contrast to their reduced availability, particularly of PMs, and this has led to great rises or fluctuations in their prices, as well as efforts to explore new deposits. Hence, there is a need to remove and recover these metals from the water or wastewater and their secondary resources.

Some researchers have tried to use various biomasses to recover PMs, in particular Au, Pt, Pd, and Ru. Table 7.1 shows the various types of biomaterials used to prepare biosorbents and biochars to recover heavy metals and PMs.

In general, the natural biomasses showed low sorption capacities for metals compared with commercial sorbents such as activated carbons and ion-exchange resins. Thus, research interest focused on enhancing the sorption capacity of a biosorbent/biochars using different surface modification methods. This chapter describes PMs and heavy metals and their sources, sources of biowastes, methods of biosorbent and biochar preparation from biowastes, removal and recovery potential of biosorbents and biochars, required specifications of biosorbents/biochars for recovery of metals, and strategies to enhance bioremoval of metals. A schematic summary of this chapter is shown in Fig. 7.1.

Table 7.1 Sources of Precious Metals and Heavy Metals

Precious Metals/ Heavy Metals	Source	Origin	References
Au, Ag, Pt, Rh, Ru	Solid waste	Municipal solid waste (MSW)	Morf et al. (2013)
Au, Ag, Pt, Pd	Solid waste	Incineration bottom ash of MSW	Muchova et al. (2009)
Ag, Au, Pt	Solid waste	Bottom ash fractions of incinerated MSW	Bakker et al. (2007)
Au, Ag, Pt, Ir	Solid waste	Waste electrical and electronic equipment	Cui and Zhang (2008), Park and Fray (2009)
Ru, Ir, Pd	Solid waste	Heterogenous catalytic waste	Paul (2009), Won et al. (2014)
Soluble form of various PMs	Liquid waste	Spent electroplating and other solutions	Cui and Zhang (2008)
Dissolved PMs	Liquid waste	Waste solutions from hydrometallurgy	Won et al. (2014)
Ru, Ir, Rh	Liquid waste	Spent homogenous catalyst	Kwak et al. (2013)
Au, Ag, Pt, Pd	Liquid waste	Hospital wastes, jewelry-processing/ making industry wastewater	Umeda et al. (2011)
Au, Pt	Liquid waste	Municipal sewage	Hartmann and Schuster (2013)
Platinum group metals	Atmospheric air	Automotive catalyst exhaust	Bardi and Stefano (2014)
Pt, Ir	Atmospheric air	Incineration of PM vapors	Gimeno-Fabra et al. (2003)
Au, Ag, Cu	Atmospheric air	Volcanic eruption ashes	Scher et al. (2013)
Zn, Cu	Solid waste	Plastic, paper, ash, kitchen waste from MSW	Long et al. (2011)
Cu, Pb, Zn	Atmospheric air	Motor vehicle emissions, drips of crankcase oil, vehicle tire wear, asphalt road surfaces	Brown and Peake (2006)
Cu, Hg, Pb, Cd, Ni	Liquid waste	Aqueous wastes from industries of metal-plating, mining operations, tanneries, smelting, alloy industries, storage battery manufacture	Kadirvelu et al. (2001)

Table 7.1 Sources of Precious Metals and Heavy Metals—cont'd

Precious Metals/ Heavy Metals	Source	Origin	References
As, Cd, Cu	Agricultural soils	Wastewater irrigation, industrial production, fertilizer and pesticide abuse, fossil fuel combustion	Du et al. (2015)
Cd	Atmospheric air	Tobacco smoking	Järup (2003)
Ni	Atmospheric air	Fumes or dust	Kasprzak et al. (2003)
Ni, Cu, Pb	Oil polluted soil	Oil spillage	Osuji and Onojake (2004)
Cu, Zn	Liquid waste	Sewage sludge	Nicholson et al. (2003)
Cu, Zn	Livestock manure	Provided to animals as dietary supplements	Nicholson et al. (2003)
Hg, As	Agrochemicals	Sprayed as pesticides or fungicides	Nicholson et al. (2003)
Hg	Domestic waste	Household bleach, acid, caustic chemicals, dental amalgam, ink and paper coatings, etc.	Tangahu and Warmadewanthi (2001)
Cu^{2+}	Biochar	Rice husks, olive pomace, orange wastes, compost	Pellera et al. (2012)
Cu^{2+}	Biochar	Peanut, canola, soybean straw biochars	Tong et al. (2011)
Cu^{2+}, Zn^{2+}, Cd^{2+}, Pb^{2+} removal	Biochar	Pig and cow manure	Kołodyńska et al. (2012)
Cu^{2+}, Cd^{2+}	Biochar	Alamo switchgrass biochar	Regmi et al. (2012)
Cd^{2+}, Pb^{2+}	Biochar	Buffalo weed biochars	Yakkala et al. (2013)
Cr^{6+}	Biochar	Sugar beet tailing biochar	Dong et al. (2011)
Hg^{2+}	Biochar	Soybean stalk-based biochar	Kong et al. (2011)
Pb^{2+}	Biochar	Anaerobically digested dairy waste residue	Inyang et al. (2012)
Zn^{2+}	Biochar	Empty fruit branches	Mubarak et al. (2013)
Cu^{2+}	Biochar	Aerobically composted swine manure	Meng et al. (2013)
Pb^{2+}, Cd^{2+}, As^{3+}	Biochar	Bioproduct chars from pine wood, pine bark, oak wood, oak bark	Mohan et al. (2007)
Cr^{3+}	Biochar	Sugarcane pulp residue biochar	Yang et al. (2013)
Ag	Biochar	Brewery waste yeast	Dodson et al. (2015)

Continued

Table 7.1 Sources of Precious Metals and Heavy Metals—cont'd

Precious Metals/ Heavy Metals	Source	Origin	References
Au	Biochar	Rice husk carbon	Dodson et al. (2015)
Ce	Biochar	Grapefruit peel Tangerine peel Corn style *Plantanus orientalis* leaf powder *Pinus brutia* leaf prawn carapace	Dodson et al. (2015)
La	Biochar	Modified rice husk activated carbon, bamboo charcoal	Dodson et al. (2015)
Nd	Biochar	Activated carbon	Dodson et al. (2015)
Ir	Biochar	Polyethyleneimine (PEI)-grafted chitosan beads	
Au(I)	Biochar	Seaweed	Won et al. (2014)
Au(III)	Biochar	Hen eggshell membrane	Won et al. (2014)
Pd(II)	Biochar	PEI-modified *C. glutamicum* biomass	Won et al. (2014)
Ru	Biochar	Decarboxylated PEI-modified *C. glutamicum* biomass	Won et al. (2014)

7.2 PRECIOUS METALS AND HEAVY METALS

PMs such as gold (Au), silver (Ag), Pt, Pd, and Rb are used in various manufacturing fields including jewelry, electronics, dental industries, and the textile industry (Ghaly et al., 2014). The demand for PMs has increased, especially in the emerging economies of the world (McGlone, 2015). However, acquiring these PMs is difficult economically and technologically owing to their restricted supply by specific producing countries or their uneven distribution in production areas (Kawamoto, 2008).

Heavy metals are described as metals which in their standard state have a specific gravity (density) of more than 5 g cm^{-3} and atomic weight of 63.546–200.590 (Aslam, 2011). Heavy metals are responsible for many toxic effects on human health, such as lead poisoning (Pb), immune depression and skin diseases (Zn and Cu), cancer (Cr, Cd), neurological disorders (Mn), and blood disorders (Fe) (Abdalla et al., 2013).

There is growing anxiety about securing a stable supply of these elements; therefore, the development of recycling technologies is important for using heavy metals and PMs resources efficiently, thus reducing their environmental burden through sustainable processes (Umeda et al., 2011).

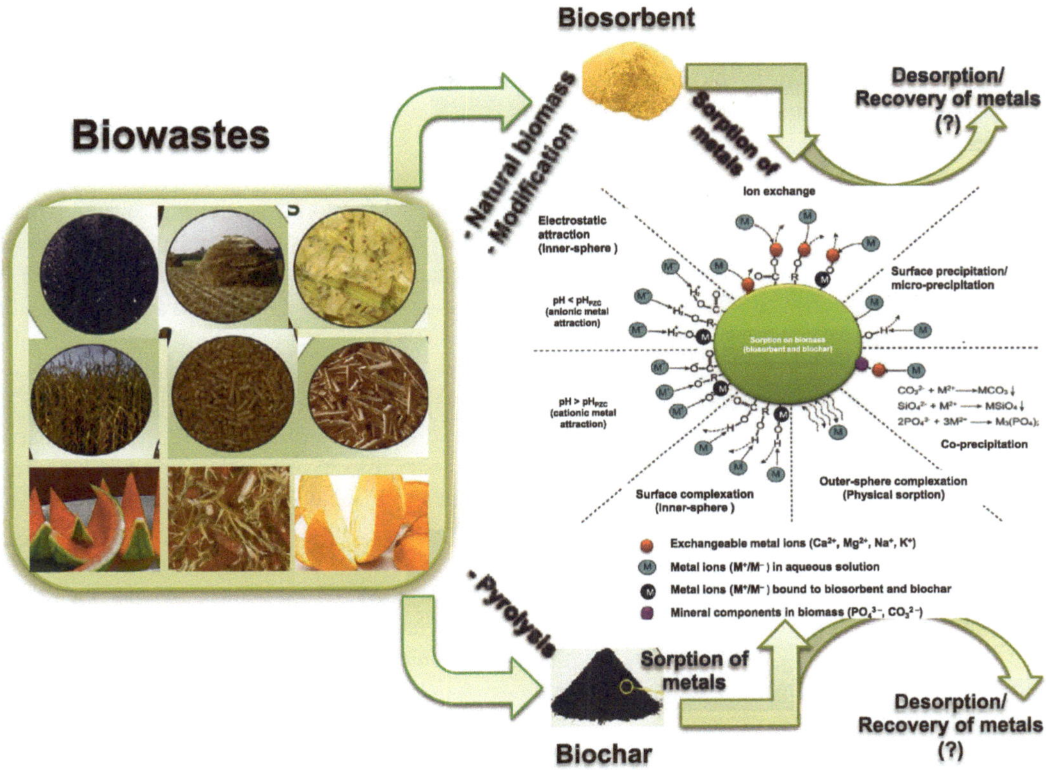

FIGURE 7.1

Schematic diagram of biosorbent and biochar production from biowastes and their role in the metal removal process and mechanisms at the surface of the biomass.

Modified from Tan et al., 2015.

7.3 SOURCES OF HEAVY METALS AND PRECIOUS METALS

7.3.1 SOURCES OF HEAVY METALS

7.3.1.1 Municipal Wastewater and Storm Water

Major sources of municipal wastewater are households, institutions, and commercial buildings which contain large concentrations of nutrients, especially nitrogen (N) and phosphate (P), trace elements such as iron (Fe) and Mn, dissolved salts, particularly NaCl, and in some cases, bicarbonates (HCO_3^-).

Water that falls on roofs or collects on paved areas such as driveways, roads, or footpaths is referred to as storm water. Storm water systems often run from outdoor drains down gutters and flow untreated into natural waterways (creeks, rivers, ground waters, wetlands, and oceans) (Aryal et al., 2010). Contaminants in storm water include pesticides, herbicides, oil, grease, and heavy metals such as Cd, Cr, Cu, Ni, Pb, and Zn. These elements may be present in dissolved form in storm water or bound to colloidal particles (Göbel et al., 2007).

7.3.1.2 Farm Wastewater

Effluent of varying composition from farms, such as those from dairy sheds, poultry, and piggeries, is referred to as farm wastewater and is increasingly used as a source of irrigation water and nutrients (Bolan et al., 2009). Farm wastewater is rich in organic and inorganic components. For instance, Cu and Zn are commonly used as feed additives and growth promoters, and for disease prevention or treatment; their concentration in farm wastewater can reach significant levels (Bolan et al., 2004). Abe et al. (2012) determined the daily output of Zn and Cu in wastewater streams from livestock farms to water systems. The unit output load from piggery and dairy wastewater treatment facilities ranged from 0.13 to 17.8 mg head^{-1} day^{-1} for Zn and from 0.15 to 9.4 mg head^{-1} day^{-1} for Cu, with lower metal concentrations found in dairy wastewater than that of piggery wastewater.

7.3.1.3 Industrial Wastewater

Use of heavy metals in various industries is widespread, and the type and quantity of metal discharges from industries depends on many factors (eg, type of industry, process variables, and contamination abatement practices). For example, Cr is widely used in electroplating; metal finishing; magnetic tapes; pigments; leather tanning operations; wood protection (chromium copper arsenate); brass, electrical and electronic equipment; and catalysis (Espinosa and Tenório, 2001). The solder plates of tin (Sn), Pb, and Ni are widely used as resistant material overplates; Cr and Cd compounds are used in pigment manufacturing units. All of these sources result in the production of a large quantity of wastewater, residues, and sludge that can be categorized as hazardous waste which requires extensive removal of metals in processes (Barakat, 2011).

7.3.2 SOURCES OF PRECIOUS METALS

7.3.2.1 Municipal Solid Waste

Municipal solid waste (MSW) is a pool of various solid wastes by towns and cities from different types of household activities. It may include biodegradable waste, electrical and electronic waste, and composite waste such as clothing, hazardous waste (paints, spray, and chemicals), and medical waste. Information about the quantities of potentially recoverable PMs in MSW was reported by (Morf et al., 2013). Their results indicated that 1 kg of MSW contains 0.41 mg gold (Au), 5.30 mg Ag, 0.06 mg Pt, 0.000094 mg rhodium (Rh), and 0.00051 mg Ru. These are relatively low concentrations but are far higher than their natural ores, particularly in the cases of Au and Pt. Bakker et al. (2007) reported measurable levels of Ag, Au, and Pt in bottom ash fractions of incinerated MSW.

7.3.2.2 Electrical and Electronic Equipment Waste

Generally, electronic waste constitutes a wide range of electronic and electrical products such as large household appliances (ie, refrigerators, air conditioners, washing machines, vacuum cleaners, television sets, tape recorders) and handheld equipment or devices (such as cellular phones, personal stereos, personal computers, printers, and consumer electronics).

Electronics are the main source of different valuable or rare materials: for example, Au, Ag, Pd, and Pt. Electrical and electronic equipment wastes have more PMs than their corresponding metal mines. The removal and recovery of PMs obtained from electrical and electronic equipment waste were investigated by several researchers (Cui and Zhang, 2008). For instance, 1 ton of a scrap of computers has higher Au content than can be extracted from 17 tons of its primary ore deposits (Park and Fray, 2009). Therefore, it is highly recommended to remove and recover PMs from these secondary sources. Although printed circuit boards cover only 6% of the weight of electrical and electronic equipment waste, the value for retrieval of PMs from printed circuit board waste is high. The value of Au and Pd recovered from 1 ton of electrical and electronic equipment waste was estimated to be about $15,000 and $1800, respectively (Wang and Gaustad, 2012).

7.3.2.3 Heterogeneous Catalytic Wastes

Discarded vehicles are another significant source of metal waste. According to estimates, the amount of PMs in scrap or end-of-life vehicles will be approximately 1.86 billion tons by 2030 (Paul, 2009). Therefore, there is strong motivation to process and recover these materials properly. In the United States alone, of the total 568 tons of PGM automotive catalysts, only 10% was recovered, with the remaining discarded as waste. In various chemical industries, PMs containing heterogeneous catalysts (Ru, Ir and Pd) are efficient for oxidation–reduction reactions; thus, used heterogeneous catalysts are a significant solid source of PMs.

7.3.2.4 Aqueous wastes

Bioremoval of PMs through biosorbents and biochar can be applied directly to recover elements from fluid or aqueous phases. The most available aqueous wastes that contain PMs are summarized next.

Used Electroplating and Other Aqueous Solutions

In the electroplating industry, metal objects are coated with a thin layer of different metals. The used or spent aqueous solutions from electroplating are produced as a result of washing and batch dumps containing dissolved forms of various precious as well as base metals themselves. PMs in electronic industries are used as crucial materials in connectors, switches, relay contacts, connecting wires, and strips (Cui and Zhang, 2008). The wastewater or aqueous solutions of these wastes contain various PMs that require recycling after removal from water systems. For example, aqueous waste from the semiconductor electronic industry is a main source of PMs in surface water streams.

Hydrometallurgy Waste Solutions

Various hydrometallurgical methods have been exploited for the recovery of PMs from ores and secondary sources after leaching from solution waste. Hydrometallurgical techniques involve ion exchange, solvent extraction, reduction, and co-precipitation (Das, 2010). As a result of these processes, a huge amount of secondary waste is produced with low concentrations of dissolved PMs. These PMs need to be immobilized from their aqueous phases and recovered for future use in the relevant industries.

Waste Homogeneous Catalysts Solution

In contrast to heterogeneous catalysts, homogeneous catalysts act in the same phase as the reactants. Because they are difficult to isolate from the reaction media, they are frequently detected in the product and waste solution (wastewater) streams. For example, preparation of acetic acid on a large or industrial scale in industries involves the use of PMs as a catalyst (Ru, Ir, or Rh). Thus, runoff from these industries leads to the discharge of PMs in wastewater or their complexation with organic materials (Kwak et al., 2013).

7.4 REMOVAL AND RECOVERY OF HEAVY METALS AND PRECIOUS METALS USING BIOSORBENTS AND BIOCHARS

7.4.1 BIOSORBENTS

Biosorption can be described as any system in which a sorbate (ie, an ion, atom, or molecule) reacts with the dead biomass or biomaterial (referred to as biosorbent), leading to the accumulation of sorbate ions at the biosorbent surface, and thus a decrease in the concentration of sorbate from aqueous solutions. Research on elucidating biosorption mechanisms has gained attention because biomass could be used to immobilize metals from different industries' wastewater (such as from the electroplating and mining industries) or to recover various PM ions present in the processing solutions. Several kinds of biosorbents are derived from various forms of raw biomass, including bacteria (eg, *Bacillus thuringiensis*) (Vijayaraghavan and Yun, 2008), fungi (eg, *Mucorrouxii*), yeasts, and algae (Wang and Chen, 2009). The complex arrangement of natural biomass indicates the involvement of various processes and mechanisms showing how these biosorbents eliminate different contaminants; however, these processes are not yet fully understood. Thus, several chemical or functional groups can attract and sequester contaminants, depending on the choice of biosorbent. These functional groups can be composed of amine, amides carboxyl, carbonyl, hydroxyl, sulfonate, sulfhydryl, phenolic, and phosphate groups (Table 7.3) (Park et al., 2010; Abdi and Kazemi, 2015). Fig. 7.2 shows the mechanisms of metal sorption on a biosorbent surface.

7.4.2 BIOCHAR

Biochar is a stable C-rich solid produced by pyrolysis or as a result of the thermochemical decomposition of organic feedstock material at high temperatures under oxygen-free conditions (Bruno et al., 2009). Biochar has been used to remove a variety of contaminants, such as heavy metal(loid)s from water and wastewater (Inyang et al., 2012; Liu and Zhang, 2009; Xu et al., 2013). It can be produced from different types of biomass residues, including crop plants (eg, rice husk, wheat bran), and as an intermediate product in bioethanol production (biowastes from the food-processing industry) (Cao et al., 2009; Yao et al., 2011) (Fig. 7.2).

Because of the abundance of feedstock materials, both biosorbents and biochars are becoming fundamentally important agents for large-scale environmental applications in water/wastewater treatment processes for immobilization and recovery of precious metals and heavy metals (Gadd, 2009).

7.5 SOURCES AND TYPES OF BIOWASTES FOR BIOSORBENTS AND BIOCHAR PRODUCTION

Biosorption exploits the ability of dead or denatured biomass, such as dried macroalgae and biowastes from agricultural and food industries, to bind metals passively from industrial effluents and wastewater (Gadd, 2009; Volesky, 2007) (Table 7.1).

7.5.1 ALGAL BIOMASS

Dried macroalgae are effective biosorbents because of the high abundance of functional groups that have a strong affinity for dissolved cationic heavy metals and PMs (eg, Zn, Cu, Pt). Many functional

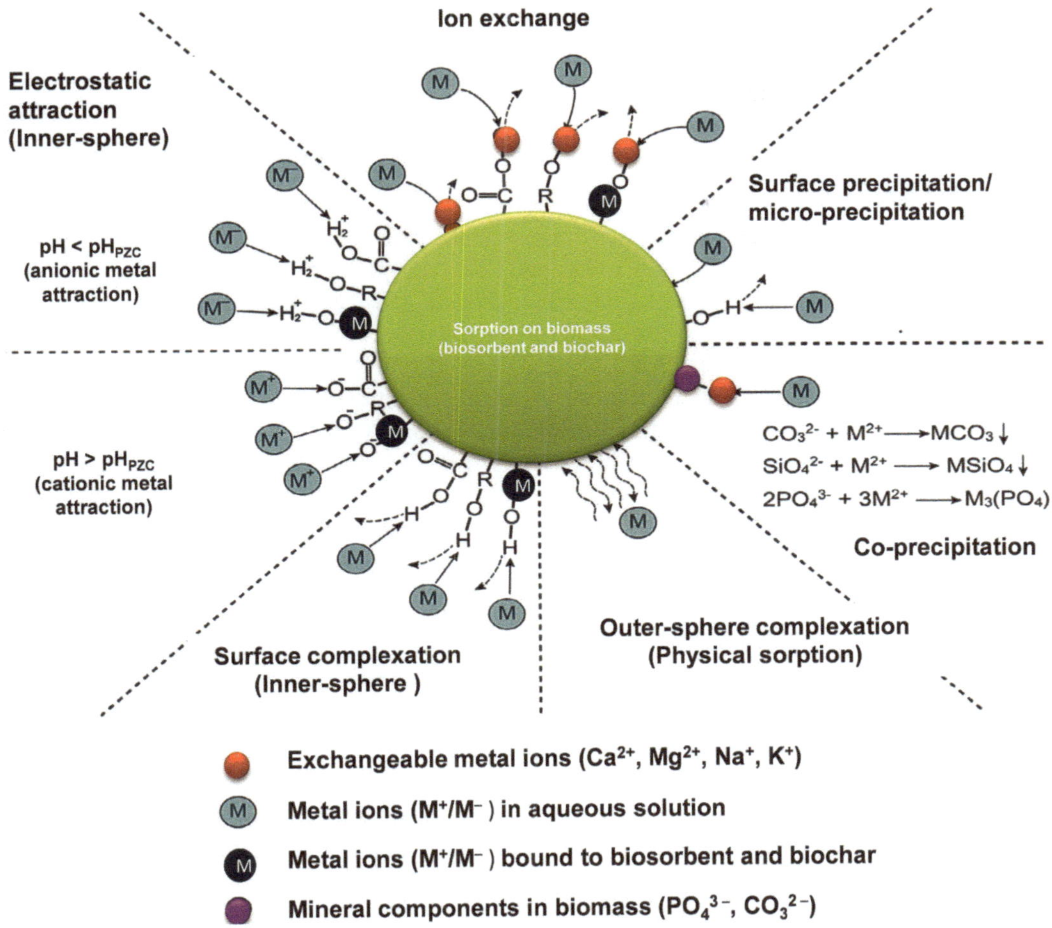

FIGURE 7.2

Metal removal mechanisms at the surface of biosorbents and biochars through sorption process.

Modified from Tan et al., 2015.

groups can be involved in the biosorption of heavy and PMs under aqueous conditions. For example, in brown algae, the carboxylic groups of alginates are typically dominant in the biosorption process, whereas freshwater green algae such as *Oedogonium* have cellulosic cell walls that resemble those of higher plants, and as such immobilize metals from water. These functional groups can passively bind dissolved metals through various processes, including passive electrostatic attraction or ion exchange, or through metal complexation processes (Kidgell et al., 2014a,b).

Several research groups have reported the adsorption of metal ions from aqueous solutions to various types of biomass: eg, agricultural and food industry biowastes, fungal biomass, and algal biomass (Won et al., 2014; Adhikari et al., 2008; Abdul-Talib et al., 2013). Effective use of algal biomass for biosorbent and biochar production to remove and recover heavy metals and PMs from water and wastewater streams could reduce the environmental burden of biowaste and toxic metals, as well as increase the reuse of these metals in their relevant industries through sustainable methods.

7.5.2 AGRICULTURAL AND FOOD INDUSTRY BIOWASTES FOR BIOSORBENTS AND BIOCHAR

Various biosorbents derived from agrowastes, as mentioned earlier, have been used for the removal of different heavy metals (eg, Cu, Zn, Cd, Cr, Pb), and PMs (eg, Pt, Pd, Ru) from water and wastewater (Table 7.1; also see Fig. 7.1 showing a schematic process and mechanisms of metal removal and recovery from water using different biosorbents and biochars).

For example, biosorbents and biochar have been developed from biowastes such as rice straw (Gao et al., 2008), algal biomass (Wan Maznah et al., 2012), plant residue waste (*Helianthus annuus*, *Brassica campestris*, *Typha latifolia*, and *Phragmites communis*) (Yeh et al., 2011; Kour et al., 2012), peanut shell (Witek-Krowiak et al., 2011), Syrian rue (*Peganum harmala*) seeds (Zamani et al., 2013), seaweed (Basha et al., 2008), wood and bark (Mohan et al., 2007), tea waste (Malkoc and Nuhoglu, 2007), maize corn cob, Jatropha oil cake, sugarcane bagasse (Dos Santos et al., 2011), century plant (*Agave salmiana*) bagasse (Velazquez-Jimenez et al., 2013), sawdust (Hashem et al., 2013), charred orange peel (Abid et al., 2016), bagasse fly ash (Rameshraja et al., 2012), almond shells, and impregnated palm shell activated carbon (Owlad et al., 2010).

Applying biochar as a low-cost sorbent is an emerging technology with promising potential to eliminate metal(loid)s from aqueous solutions (Ahmad et al., 2014; Mohan et al., 2014). Biochars modified from agricultural wastes, animal manure, and woody materials have already been examined for their ability to sorb various metal(loid)s such as Cu, As, Cr, Pb, and Cd (Mohan et al., 2014; Agrafioti et al., 2014; Uchimiya et al., 2011). The sorption capacity of biochar can vary significantly with physicochemical properties such as the surface area, pH, and functional groups that result from specific production methods (Kołodyńska et al., 2012). Biochar types such as anaerobically digested biochars, chemically modified biochars, hydrochars, magnetic biochar, Fe-treated biochar, and raw biochar are effective biosorbents of metalloids and metals from wastewater (Kidgell et al., 2014a,b; Roberts et al., 2014). However, Kidgell et al., (2014a,b) tested a multiple biosorbent (*Oedogonium* sp.) approach using Fe-biochar and raw biochar to remediate both metal(loid)s in wastewater from a coal-fired power station. The best treatment was the sequential use of Fe-biochar to eliminate metalloids from the wastewater, followed closely by raw biochar to remove metals.

Most previous research has significantly transformed our understanding of the metal removal efficiency of different biosorbents and biochars and how their composition or pyrolysis temperature could have an influence on the metal sorption capacity of biosorbents/biochars in aqueous environments. However, limited effort has been directed at developing methods for the successful retrieval and recovery of useful heavy metals and PMs from biosorbent and biochar biomass.

7.5.3 PROCESSES FOR BIOCHAR PRODUCTION

As a result of pyrolysis, there are three possible product streams: (1) noncondensable gases; (2) a combustible bio-oil representing condensable liquids; and (3) biochar, a solid residual co-product.

Pyrolysis technology can be differentiated by residence time, pyrolytic temperature, pressure, the size of the adsorbent, and the heating rate and method. Various methods describing the production of biochar are briefly presented next.

7.5.3.1 Slow and Fast Pyrolysis Biochar

In slow pyrolysis, feedstock (biomass) is heated slowly to about 500°C under anoxic conditions for a long period, which results in biochar production containing 30–45% carob (Bruun et al., 2012). Fast pyrolysis requires dry feedstock (biomass) (constituting less than 10 wt% moisture content), a fast heat transfer (approximately 100–1000°C s^{-1}), a rapid increase in temperature by heating small biomass particles (1–2 mm) to 400–500°C and vapor residence times of 1 s (maximum, 5 s) (Lima et al., 2010).

7.5.3.2 Torrefaction Process

Torrefaction, thermochemical biomass treatment process (200–300°C), is achieved by facilitating the decomposition of easily degraded volatile matter (hemicellulose) and repolymerization of cellulose and lignin in oxygen-free environments (Hill et al., 2013). The product is biochar (torrefied biomass) that resembles the original feedstock material in shape and size; however, it is darker, friable, and hydrophobic in nature with a greater calorific content. The hemicellulose, cellulose, and lignin content of biomass are partly decomposed depending on the torrefaction temperature and biomass residence time in the bioreactor.

7.5.3.3 Gasification

Compared with pyrolysis and torrefaction, this process requires higher temperatures (ranging from 600 to 1000°C) and partially oxidizing conditions to transform biomass into biochar. It also produces a clean gas (biogas) that can be used in internal combustion engines or to produce electricity. In gasification processes, the energy in biomass or any other organic matter is converted to combustible gases (carbon monoxide [CO], methane [CH_4], and hydrogen gas [H_2]), with biochar, water, and condensable tar as by-products in small amounts.

7.5.3.4 Hydrochars

The hydrothermal carbonization of C-rich biomass in the presence of water results in the production of a solid material that is referred to as hydrochar (Hu et al., 2010). The advantage of hydrothermal carbonization processes over the previously processes is that it usually takes place at relatively low temperatures (150–350°C, at about 2 MPa pressure) and wet feedstock can be directly used, including wet animal manures, sewage sludge, and algae (Xue et al., 2012). The hydrochar process is also eco-friendly because it does not generate hazardous chemical waste or by-products as does dry pyrolysis.

7.6 CONVENTIONAL TECHNIQUES OF HEAVY METALS AND PRECIOUS METALS REMOVAL AND RECOVERY

Although the recycling of PMs from waste is currently carried out in industry, environmentally friendly and sustainable recycling techniques with high selectivity and efficiency are still desired. PMs are currently being recovered by commercially available methods including pyrometallurgy and hydrometallurgy.

7.6.1 PYROMETALLURGICAL PROCESS

Pyrometallurgical processes have been used to retrieve PMs over the past 2 decades, as well as for removal and recovery of heavy metals. This procedure involves the incineration of wastes, followed by smelting in a blast furnace, drossing, melting, sintering, and reactions in gaseous forms at elevated temperatures. The waste samples are first mechanically crushed into a fine form and then burnt with carbon (coke) in a blast furnace. During this step, plastics are removed and separation and drowning of the slag phase is undertaken, which contains refractory oxides along with some metal (Fe, Cu, and Pt) oxides (Cui and Zhang, 2008). The PGMs are separated in the collector phase and are dissolved in the aqua regia, followed by a purification step to yield a pure form of PMs and heavy metals.

However, the major limitation for pyrometallurgy processes is attributed to a long settling time that is needed for the isolation of slag and metal collection. In addition, high temperatures (1800–2000°C) required for this procedure in the blast furnace and the partial separation of metals (leading to a limited upgrading of the metal value) are considered main drawbacks of the pyrometallurgical process.

7.6.2 HYDROMETALLURGICAL PROCESS

Flexibility in operation, simplicity, clean operating environments, little capital requirement, and the small-scale release of hazardous gases are some major advantages of hydrometallurgy processes. The hydrometallurgical process can be directly and simply applied in the case of liquid sources of PMs and heavy metals (in which metals are present in the dissolved form in water or wastewater). Conversely, leaching of metals from their solid sources, using suitable solvents, is a primary step for extracting these metals from solid wastes (Huisman et al., 2006). Commonly used leaching agents may include aqua regia, hydrogen peroxide, cyanide, halide, thiourea, and thiosulfate, all of which may pose the risk of hazardous waste solution release to the environment. For example, cyanide is a toxic substance that results in serious contamination of surface water and groundwater resources (Cui and Zhang, 2008). Other solvents such as halides (except fluorine and astatine) are used for Au extraction; conversely, chloride has seen industrial-scale application to recover PMs and heavy metals from solid wastes (Shen et al., 2011) owing to ease of recovery from PM–chloro complexes.

State of Precious Metals After Leaching: PMs are in various forms of chloro–metal complexes when leached with the acids such as aqua regia and concentrated HCl in the presence of chlorine gas, to name a few (Cui and Zhang, 2008; Jha et al., 2013). In general, chloro–metal complexes are anionic; typical examples are presented in Table 7.2. Of the different leaching agents, acidic chloride medium is an inexpensive medium in which most PMs can form soluble chloro–complexes. As a result, this approach has been extensively studied with model species. Considering that most PMs are anionic in the liquid phase, it is obvious that biosorbents for recovery of PMs should be able to bind anionic species effectively.

The solvent extraction procedure is mainly used to yield ultra–high purity PMs; it is not commonly employed for heavy metals. This process involves three major steps: (1) a selective extraction step for a particular metal; (2) removal of co-extracted metals (scrubbing step); and (3) removal of the extracted metal from the organic phase (a stripping step).

Table 7.2 State of Precious Metals After Leaching

Metal (Abbreviation)	Complex Form	References
Ruthenium (Ru)	$[RuCl_6]^{3-}$, $[RuCl_5(H_2O)]^{2-}$, $[RuCl_4(H_2O)_2]^-$, $[RuCl_3(H_2O)_3]$, $[RuCl_6]^{2-}$, $[Ru_2OCl_{10}]^{4-}$, $[Ru_2OCl_8(H_2O)_2]^{2-}$	Cui and Zhang (2008), Jha et al. (2013), Won et al. (2014)
Rhodium (Rh)	$[RhCl_6]^{3-}$, $[RhCl_5(H_2O)]^{2-}$, $[RhCl_4(H_2O)_2]^-$, $[RhCl_6]^{2-}$	Cui and Zhang (2008)
Palladium (Pd)	$[PdCl_4]^{2-}$, $[PdCl_6]^{2-}$	Cui and Zhang (2008)
Osmium (Os)	$[OsCl_6]^2$	Cui and Zhang (2008)
Iridium (Ir)	$[IrCl_6]^{3-}$, $[IrCl_5(H_2O)]^{2-}$, $[IrCl_4(H_2O)_2]^-$, $[IrCl_6]^2$	Cui and Zhang (2008)
Platinum (Pt)	$[PtCl_4]^{2-}$, $[PtCl_6]^{2-}$	Cui and Zhang (2008)
Aurum (Au)	$[AuCl_4]^-$	

Ligand exchange and ion-exchange reactions are thought to be the predominant mechanisms by which the selectivity is attributed to a high level of molecular recognition of solid resins. Although research is in progress, effective biosorbents and biochars are produced with enhanced metal removal properties; these biowaste-derived sorbents can be used rather than synthetic resins to control the environmental burden of solid waste in addition to removing and recovering metals.

7.7 SPECIFICATIONS OF BIOSORBENTS AND BIOCHAR FOR METAL RECOVERY

Conventional methods for immobilization of PMs and heavy metals are limited by incomplete metal removal, high energy requirements and capital costs, and generation of secondary hazardous wastes (Mack et al., 2007). These weaknesses haven led to the desire to use new biowastes to produce biochars and biosorbents to develop alternative, feasible, effective, and eco-friendly approaches to remove and recover metals from water and wastewater.

7.7.1 APPLICABILITY OF BIOSORBENTS AND BIOCHAR IN PRECIOUS METAL AND HEAVY METAL REMOVAL AND RECOVERY PROCESS

Although microorganism-based biosorbents (eg, microalgal and fungal biomass) have shown good results for sorption of PMs and heavy metals, these materials have several drawbacks, including low density, small particle size, weak physical strength and low stiffness, and difficult separation from aqueous solutions. Small-size fine powder from microorganism-derived biosorbents could also cause clogging of the column bed in real applications for the water and wastewater remediation process.

The solution to such constraints can be achieved by providing an appropriate size, physical strength, toughness, stability, and high porosity to microbe-based biosorbents, thereby enhancing metal sorption in practical applications (Vijayaraghavan and Yun, 2008). For example, Au was recovered from a simulated Au-bearing aqueous solution containing $AuCl_4$, $CuCl_2$, $FeCl_2$, and $ZnCl_2$ by calcium (Ca)-alginate entrapped cells of *Chlorella vulgaris* L. and *Sargassum platensis* L. in fluid beds (Won et al., 2014).

However, the additional cost of immobilization processes and mass transfer restrictions between the solute (metal ions) and the immobilized sorbent are major shortcomings in their application (Vijayaraghavan and Yun, 2008). Cost-effectiveness is a critically important issue, particularly in the case of treating water and wastewaters contaminated with heavy metals, so that removal and recovery costs during the biosorption process should remain lower than the metal value. However, PMs are extremely expensive and conventional PM removal and recovery methods are costly; thus, the price of biosorbents and biochars for the immobilization and recovery of PMs in water and wastewater may not be a concern. Most important, the superior sorption properties of biosorbents and biochars, including their high sorption capacity, rapid kinetics, and selectivity, over synthetic resins or activated carbons are important advantages.

7.7.2 HIGH SORPTION CAPACITY

High sorption efficiency is an important specification for the removal of PMs and heavy metals (mainly for PMs) on the surface of biosorbents and biochars. The sorption mechanism of PMs and heavy metals with biosorbents and biochars is generally regarded as physicochemical in nature and is derived from the functional groups present on the surface of the solid. The interaction of metal ions with biosorbent and biochar surfaces involves electrostatic interactions, ion exchange and metal ion chelation or complexation, microprecipitation, and reduction coupled with sorption. The carboxyl ($-COOH$), hydroxyl ($-OH$), amine ($-N-H$), sulfhydryl ($-SH$), and phosphoryl functional groups present in the biomass account for metal immobilization in aqueous solutions (Wang and Chen, 2009; Farooq et al., 2011; Ramrakhiani et al., 2011; Mirghaffari et al., 2014) (see Table 7.3 for some major functional groups in biomass).

Table 7.3 Some Major Binding Functional Groups on Surface of Biosorbents and Biochars

Functional Groups	Formula of Functional Groups	Ligand Atom	Occurrence in Selected Biomolecules of Biosorbents and Biochar
Hydroxyl	$-OH$	O	PS, SPS, AA
Carbonyl (ketone)	$>C=O$	O	Peptide bond
Carboxyl	$-C(=O)OH$	O	AA
Sulfhydryl (thiol)	$-SH$	S	AA
Sulfonate	$-S(=O)(=O)OH$	O	SPS
Amine	$-NH_2$	N	Chi, AA
Phosphonate	$-P(=O)(OH)_2$	O	PL

PS, *polysaccharides;* SPS, *sulfated PS;* Chi, *chitosan;* PG, *peptidoglycan;* AA, *amino acids;* TA, *teichoic acid;* PL, *phospholipids.*
Modified from Volesky, B., 2003. Sorption and Biosorption. Published by BV Sorbex, Inc., Montreal, St. Lambert, Quebec, Canada.

7.7.2.1 Ion-Exchange—Electrostatic Interaction

The immobilization of heavy metal cations (such as Zn, Fe, and Cu) in water and wastewater is associated with metal sorption on the biosorbent and biochar surface through electrostatic interactions. The negative charge on the biosorbent and biochar surface, arising mainly owing to $-COO^-$ and $-OH$ functional moieties, leads to the binding of positively charged metals. On the other hand, for PM removal from wastewater, the main forces responsible for the sorption process are the amine sites (−N−H) holding a positive charge. Thus, these functional groups exert an electrostatic attraction with the anionic PM complexes (as mentioned earlier) in chloride-rich acidic conditions, because PMs in secondary sources are found in the form of anionic metal complexes (Mack et al., 2007). To our knowledge, this area of research needs further attention so that specific biowastes can be used to prepare biosorbents and biochars to immobilize and recover anionic PMs or cationic heavy metals with great efficiency (Fig. 7.2).

There are two possible ways to enhance the sorption of PMs and heavy metals: through increasing amine (positive) sites on the biosorbents or biochars, or by minimizing, neutralizing, or diminishing the negatively charged repulsive sites (ie, anionic −COOH groups), or vice versa (Won et al., 2010).

7.7.2.2 Metal Complexation Mechanisms

In the sorption of PMs, complex formation, substitution reactions, and/or strong interactions between acidic−basic functional moieties have a significant role in the binding or complexation of metals on the surface of biosorbents and biochars (Abdi and Kazemi, 2015). To achieve this goal, surface functional groups with a soft basic nature responsible for the complexation of metals should be enhanced on the biosorbent and biochar surface by various chemical pretreatments or by reducing the number of interfering sites or groups, thus increasing the biosorption efficiency. For example, the addition of a sulfur group in glutaraldehyde cross-linked chitin derivates grafted with polyethyleneimine (PEI) and thiourea resulted in the enhanced removal of Pt and Pd from aqueous solutions (Das, 2010). Also, the esterification of biomass minimized the interfering (negatively charged) −COOH functional groups and enhanced the sorptive removal of PMs (Song et al., 2013) (Fig. 7.2).

7.7.2.3 Reduction-Coupled Sorption/Precipitation

The sorption of some PMs or heavy metals in the presence of oxidizing functional groups (eg, −COOH) in the biosorbent or biochar are derived by reduction-coupled sorption reactions. For example, Cr(VI) is reduced to Cr(III) by accepting electrons from −COOH groups in the biomass. Thus, enhancing oxidizing functional groups present in the biosorbent or biochar is desirable. Similarly, glutaraldehyde cross-linked chitosan beads led to the high sorption of Au by increasing the dose of glutaraldehyde, which served as a reducing agent (Park et al., 2013) (Fig. 7.2).

7.8 KINETICS OF METAL REMOVAL BY BIOSORBENT AND BIOCHAR (TIME OF REMOVAL)

The sorption kinetics of PMs and heavy metals depends on the mass transfer of the metal ions from the bulk solution phase to the sites of sorption present at the surface of biosorbents and biochars inside their solid biomass (Van Vinh et al., 2015). The combination of intraparticular diffusion in the sorbents and film diffusion in the aqueous phase is known as mass transfer. Of the two, intraparticular diffusion is considered to be the rate-determining step of the sorption reaction. Therefore, manipulating the biosorbent or biochar so as to enhance the intraparticular diffusion process is an efficient approach to yielding the rapid and fast sorption kinetics of metal ions. To our knowledge, rapid sorption kinetics is one of the fundamentally important specifications needed for effective biosorbents and biochars, so further research should explore this area (Kwak et al., 2013).

7.9 STABILITY IN AQUATIC ENVIRONMENTS

PMs prevail in chloride-rich acidic solutions whereas heavy metals dominate in alkaline aqueous environments existing in wastewater. Thus, biosorbents and biochars should be stable in acidic and alkaline water and wastewater. Previously investigated immobilizing media for biosorbents are based mostly on chitosan or its derivatives, chitin, cellulose, and alginate. However, the efficiency of chitosan in acidic or highly alkaline waters is decreased because it can be swollen and even dissolved in the acidic solution. Conversely, synthetic polymers such as polysulfone and polyurethane have high acid stability and mechanical strength (Zhou et al., 2009). However, they are limited because they do not have surface functional groups responsible for removing the PMs. Won et al. (2013) suggested a new procedure to add binding sites to acid-tolerant polymers, in which cationic polymer-coated polysulfone-based biosorbent fibers recovered Pt(IV) even from highly acidic (1 M HCl) aqueous solutions. Biochars are considered to be more stable than biosorbents in acidic aqueous solution; however, future research is required to expand upon this knowledge in biochar-water systems for the removal and recovery of PMs.

7.10 PROCEDURES FOR INCREASING SORPTION CAPACITY

Although partial development has been achieved to increase the sorption capacity of biomaterials, there is a need to consider a wide variety of surface modification techniques for preparing powerful biosorbents or biochars to immobilize and recover PMs and heavy metals (Fig. 7.3).

7.10.1 PRETREATMENTS

The purpose of biomass pretreatment is to remove impurities from the raw material and/or to activate the sorption sites (Fig. 7.3). Pretreatment of a biomass can be categorized as physical or chemical. Physical pretreatment includes steaming, autoclaving, lyophilization, thermal drying, grinding, and cutting (Park et al., 2010). Generally, physical pretreatments are simple to apply and low-cost; however, these are less efficient than chemical pretreatment (Park et al., 2010). Chemical pretreatment of biomass or biomaterial occurs through mild acids, alkaline solutions, organic solvents, and other chemicals such as $Ca(OH)_2$, Fe salts, or Fe oxide minerals (Vijayaraghavan and Yun, 2008). Chemical pretreatment is considered more effective in terms of the removal of metal from water. For example, the effect of chemical pretreatment on sorption of Cu by *Marrubium globosum* leaves using a number of chemical agents was investigated (Yazıcı et al., 2008). The results demonstrated that the pretreated *M. globosum* leaf biomass showed increased sorption efficiency compared with that of the natural biomass. Notably, pretreatment with laundry detergent, NaOH and $NaHCO_3$, was more efficient than with other pretreatments. This could be because of disruption of the biomass surface structure and the development of additional reactive functional groups which increased metal binding. In the case of

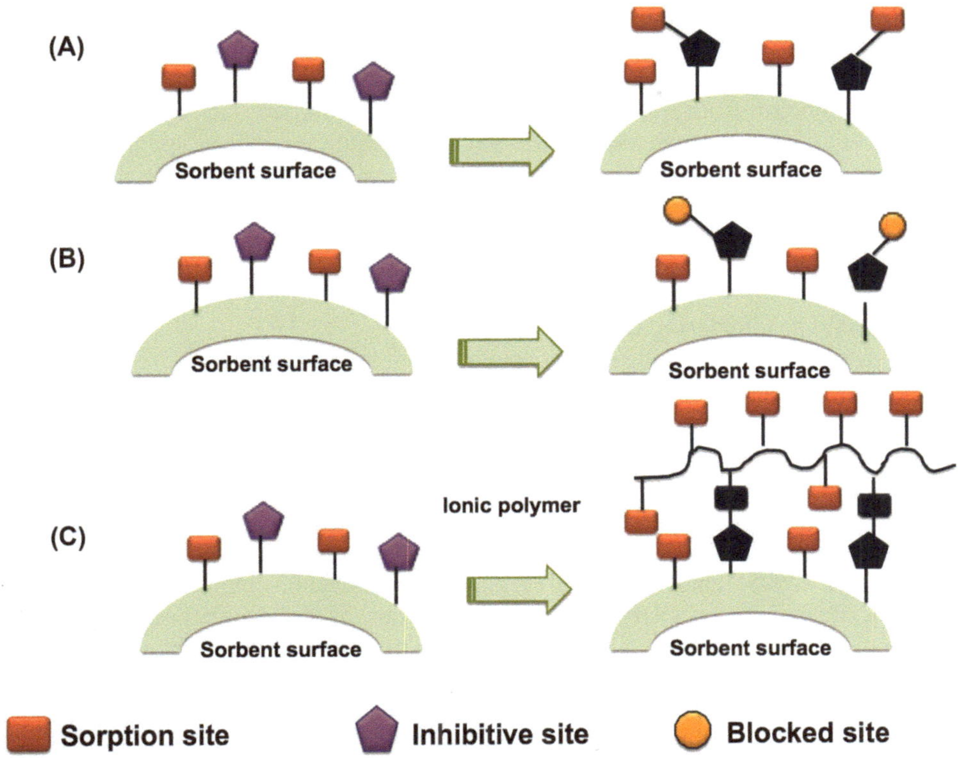

FIGURE 7.3

Various pretreatment methods to increase the sorption of metals by biosorbents (biomass). (A) Increasing the sorption sites. (B) Eliminating the interfering sites. (C) Coating the biosorbent with ionic polymers.

Modified from Won, S.W., Kotte, P., Wei, W., Lim, A., Yun, Y.-S., 2014. Biosorbents for recovery of precious metals. Bioresource Technology 160, 203–212.

pretreatment with acidic solutions (eg, 0.2 M HCl, H_2SO_4), several researchers indicated that acid treatment improved efficiency, particularly for cationic metals, likely owing to removal of some impurities and ions blocking the sorption sites (Lin et al., 2015).

7.10.1.1 Increasing the Number of Sorption/Binding Sites

Particular functional moieties on the biosorbent or biochar surface can be substituted through different chemical pretreatment methods. The introduction of new binding sites to the biosorbent or biochar surface can enhance the sorption capacity. Mainly −N−H, −COOH, phosphonate, sulfonate, and −OH groups have a major role in binding target ions from aqueous solutions (Vijayaraghavan and Yun, 2008) (Fig. 7.3B).

Various chemical surface pretreatment methods are available to increase the sorption sites. For example, positively charged amine (−N−H) groups were added to negatively charged carboxyl (−COOH) groups (inhibition sites) of brown macroalgae (*Stoechospermum marginatum*) by reaction with propylamine (Kousha et al., 2012). It was also reported that the ethylene diamine–treated biomass showed improved biosorption capacity for PMs (Elwakeel et al., 2013). The functional moieties of cationic solutes (ie, carboxyl or phosphonate groups) could also be added successfully to the biosorbent or biochar surface, leading to improvement in the sorption capacity for cationic metals. For example, −OH groups on the biomass were replaced with phosphonate groups via phosphorylation, although the potential for PM sorption has yet to be explored. The biosorption of basic dyes increased when rice straw was modified with phosphoric acid (Gong et al., 2007). Thus, characterizing the influence of various surface modification methods on biosorbents and biochars to enhance PM removal in water and wastewater warrants further investigation.

7.10.1.2 Removing Interfering Sorption Sites

The natural biomass of biosorbents has many surface functional groups, some of which could interfere with the sorption of target metal species. For instance, −N−H groups have a role as sorption sites for anionic metal removal by electrostatic interaction, whereas negatively charged −COOH groups could repel anionic metals. Thus, elimination of interfering binding sites from the biomass surface could result in more efficient biosorbents and biochars (Vijayaraghavan and Yun, 2008). The manipulation of interfering sites has been reported previously as the (1) methylation of −N−H groups, (2) acetylation of −N−H and −OH groups, (3) esterification of −COOH groups, and (4) esterification of phosphonate groups. Gong et al. (2005) removed −COOH groups from peanut hulls by methylation of the −N−H group, esterification of the −COOH group, and acetylation of the −N−H and −OH groups. The methylation of −N−H groups had a minimal influence on the sorption capacity of ionic dyes; however, esterification of the −COOH group greatly affected basic dye sorption. Han and Yun (2007) reported that the maximum uptake of Reactive Red 4 at pH 4 by the esterified biomass of *Corynebacterium glutamicum* was 2.8 times higher than that by the protonated biomass. Therefore, the addition of positively charged −N−H groups and elimination of −COOH groups is a highly attractive strategy for developing more effective biosorbents and biochars to recover negatively charged PMs and positively charged heavy metals in water and wastewater (Fig. 7.3B).

7.10.1.3 Coating With Ionic Polymers

Coating the biosorbents and biochars with ionic polymers could be an efficient way to enhance the recovery of PMs ions (Fig. 7.3C). For example, the combination of an ethanolamine molecule with the biomass can create amine sites on the material surface. In addition, the association of PEI with the biosorbent and biochar biomass can result in the generation of a large number of amine groups. PEI is made of several primary and secondary −N−H groups and can lead to a significant increase in sorption capacity for negatively charged PM ions (Mao et al., 2011). In contrast to methylation pretreatment, in which residual methanol needs to be removed after biosorbent and biochar preparation,, PEI-coated biosorbents and biochars can provide a low-cost and more environmentally friendly solution for metal removal and recovery. Yu et al. (2007) synthesized poly(amic acid)-grafted baker's yeast–derived biomass by reacting with pyromellitic dianhydride and thiourea. The authors reported 15- and 11-fold increases in Cd and Pb removal, respectively. Mao et al. (2013) used a simple method to coat *C. glutamicum* derived biomass electrostatically with anionic polymer polyacrylic acid in aqueous solutions, followed by cross-linking with glutaraldehyde to reinforce the stability of the physically coated polymer. As a result, the modified biomass showed a maximum Cd uptake of 140 mg g^{-1} (a more than threefold enhancement in Cd removal). Thus, physical sorption followed by cross-linking can be a sustainable approach to prepare highly efficiency biosorbents and biochars, although far less research has been directed at such coated biochars. In addition, the application of chemical modification methods coupled with polymer coating could be a potentially beneficial option to increase the sorption capacity of biosorbents and biochars (Song et al., 2013).

7.11 APPROACHES TO ENHANCE SORPTION KINETICS OF BIOSORBENTS AND BIOCHARS

Biosorbents and biochars with rapid sorption kinetics are preferable in scaled-up sorption processes in which a large volume of water or wastewater is treated. The chloro-complexes of PMs are also less reactive than those of other base metal and heavy metal ions. Ligands with soft donor atoms may react more quickly compared with those containing hard donor atoms. Also, the second-row PMs are more reactive than those of the third row (Won et al., 2014).

For the sorption of PMs on biosorbents and biochars under acidic conditions, electrostatic–columbic interactions dominate, with some contribution from complexation reactions in metal removal. In the case of heavy metals, sorption reactions are faster; thus, other factors that are responsible as a result of the sorption process are neglected. Conversely, because PM sorption kinetics are slow, other factors governing the sorption reactions on biosorbent and biochar surfaces are of greater importance. To enhance the sorption kinetics of biosorbents and biochars, one can reduce the particle size, which leads to an enhanced rate of association. For example, thin fiber types of biosorbents and/or activated biochars could have high specific surface areas and small movement path lengths for metal ions. In addition, a thin biosorbent fiber could be used in sorption column beds without a serious pressure drop and column blocking problems. The highly porous structure of biosorbents and biochars results in a fast sorption kinetics for PM removal from water and wastewater (Kwak et al., 2013; Mao et al., 2013; Boddu et al., 2008). Chemical surface modification of the biosorbent and biochar biomass can greatly improve sorption capacity and stability in aqueous environments (Vijayaraghavan and Yun, 2008).

7.12 STRATEGIES FOR STABLE BIOCHARS AND BIOSORBENTS

Stability, particularly of natural biosorbents, is a major consideration in the field of biosorptive removal and recovery of PMs and heavy metals. This is more significant in the case of PMs because they are present in highly acidic chloride-rich aqueous solutions (Mack et al., 2007). The chitosan and/or its derivatives have been used as a matrix of biosorbents. However, as mentioned earlier, chitosan dissociates in highly acidic aqueous media. Thus, to develop stable biosorbent materials as well as stable biochars in actual water or wastewater containing PMs or heavy metals, acid- or alkali-resistant polymers should be used as sorbent matrices. For example, acid-tolerant polymers include polysulfone, polyurethane, cellulose, and calcium alginate (Zhou et al., 2009). Although these polymers lack PM binding functional groups, a study provided a solution for such constraints (Won et al., 2013). The authors prepared biosorbent–polysulfone composite fibers using the biomass functional groups (ie, –COOH groups), and cationic polymer PEI coated the surface of the composite fibers. As a result, the PEI-coated biomass–polysulfone composite fibers demonstrated high stability in acidic solutions and high sorption capacity for Pt(IV) ions. In another study by Kwak et al. (2013), the authors prepared a PEI-coated biomass–chitosan composite fiber that successfully removed and recovered Ru in the water-free waste solution of acetic acid released from Samsung BP Chemicals Co. Ltd., in Ulsan, Korea.

7.13 RECOVERY OF METALS AND REGENERATION OF BIOSORBENTS AND BIOCHARS

PMs and heavy metals are currently recovered from metal-loaded biosorbents and biochars through elution processes using acids, bases, or other types of chemical solutions that are potentially hazardous to the environment. In addition, acidic or basic elution does not provide zero-valent solid-state metal ions; rather, ionic metal-containing solutions are produced (Ishikawa et al., 2002). Some researchers described the procedure for recovering heavy metals by incinerating metal-loaded biosorbents. Won et al. (2010) suggested that incineration could be used to recover metals, particularly for high-price elements (Kwak and Yun, 2010).

Desorption and reuse or regeneration of biosorbents and biochars is a major consideration for cost-effective sorbent materials. To achieve successful desorption, we require efficient and appropriate solutions, depend on the type of biosorbents and biochars, as well as the sorption mechanisms. A good elutant (desorbing solution) is selected based on some key parameters; they need to (1) be nondestructive to the sorbent (biochar or biosorbent biomass); (2) have a low cost; (3) have an environmentally friendly composition; and (4) have high efficiency to desorb metals from the biomass (Syed, 2012). We have briefly summarized typically available and suitable elutants used to desorb and regenerate PM-loaded biosorbents. Yin et al. (2013) investigated the desorption of Au(III) from organophosphonic acid–functionalized consumed buckwheat hulls using thiourea as the desorbing (eluent) solution (ie, acidic in nature). Their results showed that the desorption rate was 90% using a 5% thiourea solution and 0.1 M HCl; the sorbents maintained 88% of the original Au(III) sorption capacity after two cycles of regeneration. Desorption of Au(I) sorbed on eggshell membrane was successfully achieved using NaOH and NaCN; the sorption capacity for Au(I) did not decrease after five continuous sorption–desorption cycles (Won et al., 2011).

Park et al. (2013) reported a new method of two-step desorption for the selective removal and recovery of Pd(II) and Au(III) from glutaraldehyde cross-linked chitosan beads. In the first desorption step using 5 M HCl, only Pd(II) was desorbed, which resulted in the accumulation of pure Pd ions in the desorbing solution. The second desorption step with acidic thiourea solution eluted zero-valent Au that was obtained by the reduction of Au(III) by glutaraldehyde during biosorption, as well as the remaining Pd(II). Those data showed that a pure Pd(II) solution and Au-rich solution (containing 98% Au) resulted from the selective two-step desorption process. The desorption method helps recover PMs in an ionic form, followed by reduction to a metallic form by appropriate reducing agents.

However, if the recovery of PMs is needed in a metallic form directly, a sorption–incineration combined process could be implemented in which ionic metals reduce to metallic forms coupled with the oxidation of organic constituents during incineration (Won et al., 2010). Because PMs are very costly (unlike heavy metals) and biosorbents and biochars are relatively inexpensive compared with synthetic resins, the sorption–incineration method can be another way to remove and recover PMs from biosorbents.

7.14 CONCLUSIONS

Biosorbents and biochars obtained from agricultural and food industry biowastes could be a feasible and sustainable replacement for the costly conventional methods used to remove and recover contaminants (PMs and heavy metal ions, dyes, ammonia, and nitrates from water and wastewater). Ultimately this would help reduce the cost of biowaste disposal and provide a potentially inexpensive and eco-friendly alternative to existing commercial inorganic synthetic resins or activated carbons.

The performance of biosorbents and biochars could be enhanced through different surface pretreatment methods; coating with ionic polymers appears to be the most effective pretreatment method. However, there is a lack of understanding about how chemical surface pretreatment coupled with polymer coating can further increase the metal sorption capacity for maximum recovery. The thin structure of biosorbents and biochars can result in rapid sorption; however, the polymeric matrix must be taken into account while preparing immobilized sorbents for acid and/or solvent stability. It is also important to develop new approaches for biosorbents and biochars with selective sorption processes.

Chemically modified biosorbents and biochars prepared from a wide variety of biowastes (such as rice husks and rice hulls, spent grain, sugarcane bagasse and fly ash, sawdust, wheat bran, corncobs, wheat and soybean straws, corn stalks, weeds, fruit and vegetable wastes, cassava waste fibers, tree barks, azolla (water fern), alfalfa biomass, coir pith carbon, cotton seed hulls, citrus waste, and soybean hulls) have been shown to have excellent sorption capacities for various heavy metals (eg, Cd, Cu, Pb, Zn, Ni). However, the removal and recovery of PMs using biosorbents and particularly via biochars remains only partially resolved. This represents a critically important knowledge gap in this area of research. The available literature shows that the sorption mechanisms and the influence of other factors such as pH conditions in removing and recovering PMs through sorption on the surface of biosorbents/biochars are different from those observed in heavy metals or base metal systems.

The advantage of pretreatment methods for eliminating interfering surface functional groups or adding beneficial functional groups should be a focus of future research. This approach will lead to the development of biowaste-based sorbent materials which should be capable of quantitative recovery of PM and heavy metals from a variety of waste sources.

ACKNOWLEDGMENTS

The authors are grateful to the Grand Challenges Canada—Stars in Global Health (GCC Grant No. S5 0433-01), the International Foundation for Science (IFS Grant No. W/5698-1), and the University of Agriculture Faisalabad, Pakistan, for providing financial support. Thanks also to Faryal Naeem (COMSATS Institute of Information and Technology, Vehari).

REFERENCES

Abdi, O., Kazemi, M., 2015. A review study of biosorption of heavy metals and comparison between different biosorbents. Journal of Materials and Environmental Science 6, 1386–1399.

Abid, M., Niazi, N.K., Bibi, I., Farooqi, A., Ok, Y.S., Kunhikrishnan, A., Ali, F., Ali, S., Igalavithana, A.D., Arshad, M., 2016. Arsenic(V) biosorption by charred orange peel in aqueous environments. International Journal of Phytoremediation 18 (5), 442–449. http://dx.doi.org/10.1080/15226514.2015.1109604.

Aslam, B., Javed, I., Khan, F.H., Rahman, Z.-u., 2011. Uptake of heavy metal residues from sewerage sludge in the milk of goat and cattle during summer season. Pakistan Veterinary Journal 31, 75.

Abdalla, M., Hassabo, A., Elsheikh, N., 2013. Assessment of some heavy metals in waste water and milk of animals grazed around sugar cane plants in Sudan. Livestock Research for Rural Development 25.

Aryal, R., Vigneswaran, S., Kandasamy, J., Naidu, R., 2010. Urban stormwater quality and treatment. Korean Journal of Chemical Engineering 27, 1343–1359.

Abe, K., Waki, M., Suzuki, K., Kasuya, M., Suzuki, R., Itahashi, S., Banzai, K., 2012. Estimation of Zn and Cu unit output loads from animal husbandry facilities. Water Science and Technology 66, 653–658.

Adhikari, C., Parajuli, D., Kawakita, H., Inoue, K., Ohto, K., Harada, H., 2008. Dimethylamine-modified waste paper for the recovery of precious metals. Environmental Science and Technology 42, 5486–5491.

Abdul-Talib, S., Tay, C., Abdullah-Suhaimi, A., Liew, H., 2013. Fungal *Pleurotus ostreatus* biosorbent for cadmium (II) removal in industrial wastewater. Journal of Life Sciences and Technologies 1.

Ahmad, M., Rajapaksha, A.U., Lim, J.E., Zhang, M., Bolan, N., Mohan, D., Vithanage, M., Lee, S.S., Ok, Y.S., 2014. Biochar as a sorbent for contaminant management in soil and water: a review. Chemosphere 99, 19–33.

Agrafioti, E., Kalderis, D., Diamadopoulos, E., 2014. Arsenic and chromium removal from water using biochars derived from rice husk, organic solid wastes and sewage sludge. Journal of Environmental Management 133, 309–314.

Bakker, E., Muchova, L., Rem, P., 2007. Separation of precious metals from MSWI bottom ash. In: Conference Proceeding from 6th International Industrial Mineral Symposium. 1–3 February, Izmir, Turkey, 6.

Bardi, U., Stefano, C., 2014. Precious metals in automotive technology: an unsolvable depletion problem? Minerals 4, 388–398.

Brown, J.N., Peake, B.M., 2006. Sources of heavy metals and polycyclic aromatic hydrocarbons in urban stormwater runoff. Science of the Total Environment 359, 145–155.

Bolan, N.S., Laurenson, S., Luo, J., Sukias, J., 2009. Integrated treatment of farm effluents in New Zealand's dairy operations. Bioresource Technology 100, 5490–5497.

Bolan, N., Adriano, D., Mahimairaja, S., 2004. Distribution and bioavailability of trace elements in livestock and poultry manure by-products. Critical Reviews in Environmental Science and Technology 34, 291–338.

Barakat, M.A., 2011. New trends in removing heavy metals from industrial wastewater. Arabian Journal of Chemistry 4, 361–377.

Bruno, G., Mike, P., Christelle, B., Goodspeed, K., 2009. Biochar is carbon negative. Nature Geoscience 2, 2.

Basha, S., Murthy, Z.V.P., Jha, B., 2008. Biosorption of hexavalent chromium by chemically modified seaweed, *Cystoseira indica*. Chemical Engineering Journal 137, 480–488.

Bruun, E.W., Ambus, P., Egsgaard, H., Hauggaard-Nielsen, H., 2012. Effects of slow and fast pyrolysis biochar on soil C and N turnover dynamics. Soil Biology and Biochemistry 46, 73–79.

Boddu, V.M., Abburi, K., Talbott, J.L., Smith, E.D., Haasch, R., 2008. Removal of arsenic (III) and arsenic (V) from aqueous medium using chitosan-coated biosorbent. Water Research 42, 633–642.

Cui, J., Zhang, L., 2008. Metallurgical recovery of metals from electronic waste: a review. Journal of Hazardous Materials 158, 228–256.

Cao, X., Ma, L., Gao, B., Harris, W., 2009. Dairy-manure derived biochar effectively sorbs lead and atrazine. Environmental Science and Technology 43, 3285–3291.

Das, N., 2010. Recovery of precious metals through biosorption — A review. Hydrometallurgy 103, 180–189.

Dodson, J.R., Parker, H.L., Munoz Garcia, A., Hicken, A., Asemave, K., Farmer, T.J., He, H., Clark, J.H., Hunt, A.J., 2015. Bio-derived materials as a green route for precious & critical metal recovery and re-use. Green Chemistry 17, 1951–1965.

Dong, X., Ma, L.Q., Li, Y., 2011. Characteristics and mechanisms of hexavalent chromium removal by biochar from sugar beet tailing. Journal of Hazardous Materials 190, 909–915.

Dos Santos, V., De Souza, J., Tarley, C., Caetano, J., Dragunski, D., 2011. Copper ions adsorption from aqueous medium using the biosorbent sugarcane bagasse in natura and chemically modified. Water, Air, and Soil Pollution 216, 351–359.

Du, P., Xie, Y., Wang, S., Zhao, H., Zhang, Z., Wu, B., Li, F., 2015. Potential sources of and ecological risks from heavy metals in agricultural soils, Daye City, China. Environmental Science and Pollution Research 22, 3498–3507.

Espinosa, D.C.R., Tenório, J.A.S., 2001. Thermal behavior of chromium electroplating sludge. Waste Management 21, 405–410.

Elwakeel, K.Z., El-Sayed, G.O., Darweesh, R.S., 2013. Fast and selective removal of silver(I) from aqueous media by modified chitosan resins. International Journal of Mineral Processing 120, 26–34.

Farooq, U., Khan, M.A., Athar, M., Kozinski, J.A., 2011. Effect of modification of environmentally friendly biosorbent wheat (*Triticum aestivum*) on the biosorptive removal of cadmium (II) ions from aqueous solution. Chemical Engineering Journal 171, 400–410.

Ghaly, A., Ananthashankar, R., Alhattab, M., Ramakrishnan, V., 2014. Production, characterization and treatment of textile effluents: a critical review. Journal of Chemical Engineering and Process Technology 5.

Gimeno-Fabra, L., Nizard, H., Benien, H., Vlcek, J., Redjaimia, A., 2003. Microstructure control in precious metal coatings for high temperature applications, obtained by metallorganic chemical vapour deposition. Material Science and Engineering Technology 34, 410–414.

Göbel, P., Dierkes, C., Coldewey, W.G., 2007. Storm water runoff concentration matrix for urban areas. Journal of Contaminant Hydrology 91, 26–42.

Gadd, G.M., 2009. Biosorption: critical review of scientific rationale, environmental importance and significance for pollution treatment. Journal of Chemical Technology and Biotechnology 84, 13–28.

Gao, H., Liu, Y., Zeng, G., Xu, W., Li, T., Xia, W., 2008. Characterization of Cr(VI) removal from aqueous solutions by a surplus agricultural waste—Rice straw. Journal of Hazardous Materials 150, 446–452.

Gong, R., Jin, Y., Chen, J., Hu, Y., Sun, J., 2007. Removal of basic dyes from aqueous solution by sorption on phosphoric acid modified rice straw. Dyes and Pigments 73, 332–337.

Gong, R., Sun, Y., Chen, J., Liu, H., Yang, C., 2005. Effect of chemical modification on dye adsorption capacity of peanut hull. Dyes and Pigments 67, 175–181.

Hartmann, G., Schuster, M., 2013. Species selective preconcentration and quantification of gold nanoparticles using cloud point extraction and electrothermal atomic absorption spectrometry. Analytica Chimica Acta 761, 27–33.

Hashem, A., Adam, E., Hussein, H., Sanousy, M., Ayoub, A., 2013. Bioadsorption of Cd (II) from contaminated water on treated sawdust: adsorption mechanism and optimization. Journal of Water Resource and Protection 5, 82–90.

Hill, S.J., Grigsby, W.J., Hall, P.W., 2013. Chemical and cellulose crystallite changes in *Pinus radiata* during torrefaction. Biomass and Bioenergy 56, 92–98.

Hu, B., Wang, K., Wu, L.H., Yu, S.H., Antonietti, M., Titirici, M.M., 2010. Engineering carbon materials from the hydrothermal carbonization process of biomass. Advanced Materials 22, 813–818.

Huisman, J.L., Schouten, G., Schultz, C., 2006. Biologically produced sulphide for purification of process streams, effluent treatment and recovery of metals in the metal and mining industry. Hydrometallurgy 83, 106–113.

Han, M.H., Yun, Y.-S., 2007. Mechanistic understanding and performance enhancement of biosorption of reactive dyestuffs by the waste biomass generated from amino acid fermentation process. Biochemical Engineering Journal 36, 2–7.

Imamoglu, M., Tekir, O., 2008. Removal of copper (II) and lead (II) ions from aqueous solutions by adsorption on activated carbon from a new precursor hazelnut husks. Desalination 228, 108–113.

Inyang, M., Gao, B., Yao, Y., Xue, Y., Zimmerman, A.R., Pullammanappallil, P., Cao, X., 2012. Removal of heavy metals from aqueous solution by biochars derived from anaerobically digested biomass. Bioresource Technology 110, 50–56.

Ishikawa, S.-i., Suyama, K., Arihara, K., Itoh, M., 2002. Uptake and recovery of gold ions from electroplating wastes using eggshell membrane. Bioresource Technology 81, 201–206.

Järup, L., 2003. Hazards of heavy metal contamination. British Medical Bulletin 68, 167–182.

Jha, M.K., Lee, J.-c., Kim, M.-s., Jeong, J., Kim, B.-S., Kumar, V., 2013. Hydrometallurgical recovery/recycling of platinum by the leaching of spent catalysts: a review. Hydrometallurgy 133, 23–32.

Khare, P., Dilshad, U., Rout, P.K., Yadav, V., Jain, S., 2013. Plant refuses driven biochar: application as metal adsorbent from acidic solutions. Arabian Journal of Chemistry.

Kwak, I.S., Won, S.W., Chung, Y.S., Yun, Y.-S., 2013. Ruthenium recovery from acetic acid waste water through sorption with bacterial biosorbent fibers. Bioresource Technology 128, 30–35.

Kadirvelu, K., Thamaraiselvi, K., Namasivayam, C., 2001. Removal of heavy metals from industrial wastewaters by adsorption onto activated carbon prepared from an agricultural solid waste. Bioresource Technology 76, 63–65.

Kasprzak, K.S., Sunderman, F.W., Salnikow, K., 2003. Nickel carcinogenesis. Mutation Research/Fundamental and Molecular Mechanisms of Mutagenesis 533, 67–97.

Kołodyńska, D., Wnetrzak, R., Leahy, J.J., Hayes, M.H.B., Kwapiński, W., Hubicki, Z., 2012. Kinetic and adsorptive characterization of biochar in metal ions removal. Chemical Engineering Journal 197, 295–305.

Kong, H., He, J., Gao, Y., Wu, H., Zhu, X., 2011. Cosorption of phenanthrene and mercury(II) from aqueous solution by soybean stalk-based biochar. Journal of Agricultural and Food Chemistry 59, 12116–12123.

Kawamoto, H., 2008. Japan's policies to be adopted on rare metal resources. Quarterly Review 27, 57–76.

Kidgell, J.T., de Nys, R., Hu, Y., Paul, N.A., Roberts, D.A., 2014a. Bioremediation of a complex industrial effluent by biosorbents derived from freshwater macroalgae. PLoS One 9.

Kour, J., Homagai, P., Pokhrel, M., Ghimire, K., 2012. Adsorptive separation of metal ions with surface modified *Desmostachya bipinnata*. Nepal Journal of Science and Technology 13, 101–106.

Kidgell, J.T., de Nys, R., Paul, N.A., Roberts, D.A., 2014b. The sequential application of macroalgal biosorbents for the bioremediation of a complex industrial effluent. PLoS ONE 9 (7), e101309. http://dx.doi.org/10.1371/journal.pone.0101309.

Kousha, M., Daneshvar, E., Sohrabi, M.S., Jokar, M., Bhatnagar, A., 2012. Adsorption of acid orange II dye by raw and chemically modified brown macroalga *Stoechospermum marginatum*. Chemical Engineering Journal 192, 67–76.

Kwak, I.S., Yun, Y.-S., 2010. Recovery of zero-valent gold from cyanide solution by a combined method of biosorption and incineration. Bioresource Technology 101, 8587–8592.

Lin, S., Wei, W., Wu, X., Zhou, T., Mao, J., Yun, Y.-S., 2015. Selective recovery of Pd(II) from extremely acidic solution using ion-imprinted chitosan fiber: adsorption performance and mechanisms. Journal of Hazardous Materials 299, 10–17.

Long, Y.-Y., Shen, D.-S., Wang, H.-T., Lu, W.-J., Zhao, Y., 2011. Heavy metal source analysis in municipal solid waste (MSW): case study on Cu and Zn. Journal of Hazardous Materials 186, 1082–1087.

Liu, Z., Zhang, F.-S., 2009. Removal of lead from water using biochars prepared from hydrothermal liquefaction of biomass. Journal of Hazardous Materials 167, 933–939.

Lima, I.M., Boateng, A.A., Klasson, K.T., 2010. Physicochemical and adsorptive properties of fast-pyrolysis biochars and their steam activated counterparts. Journal of Chemical Technology & Biotechnology 85, 1515–1521.

Morf, L.S., Gloor, R., Haag, O., Haupt, M., Skutan, S., Lorenzo, F.D., Böni, D., 2013. Precious metals and rare earth elements in municipal solid waste – sources and fate in a Swiss incineration plant. Waste Management 33, 634–644.

Muchova, L., Bakker, E., Rem, P., 2009. Precious metals in municipal solid waste incineration bottom ash. Water Air and Soil Pollution 9, 107–116.

Mubarak, N.M., Alicia, R.F., Abdullah, E.C., Sahu, J.N., Haslija, A.B.A., Tan, J., 2013. Statistical optimization and kinetic studies on removal of Zn2+ using functionalized carbon nanotubes and magnetic biochar. Journal of Environmental Chemical Engineering 1, 486–495.

Meng, J., Wang, L., Liu, X., Wu, J., Brookes, P.C., Xu, J., 2013. Physicochemical properties of biochar produced from aerobically composted swine manure and its potential use as an environmental amendment. Bioresource Technology 142, 641–646.

Mohan, D., Pittman Jr., C.U., Bricka, M., Smith, F., Yancey, B., Mohammad, J., Steele, P.H., Alexandre-Franco, M.F., Gómez-Serrano, V., Gong, H., 2007. Sorption of arsenic, cadmium, and lead by chars produced from fast pyrolysis of wood and bark during bio-oil production. Journal of Colloid and Interface Science 310, 57–73.

McGlone, M., 2015. Precious metals, gold and emerging market. The Journal of Investing 24, 124–131.

Malkoc, E., Nuhoglu, Y., 2007. Potential of tea factory waste for chromium(VI) removal from aqueous solutions: thermodynamic and kinetic studies. Separation and Purification Technology 54, 291–298.

Mohan, D., Kumar, H., Sarswat, A., Alexandre-Franco, M., Pittman, C.U., 2014. Cadmium and lead remediation using magnetic oak wood and oak bark fast pyrolysis bio-chars. Chemical Engineering Journal 236, 513–528.

Mack, C., Wilhelmi, B., Duncan, J.R., Burgess, J.E., 2007. Biosorption of precious metals. Biotechnology Advances 25, 264–271.

Mirghaffari, N., Moeini, E., Farhadian, O., 2014. Biosorption of Cd and Pb ions from aqueous solutions by biomass of the green microalga, *Scenedesmus quadricauda*. Journal of Applied Phycology 27, 311–320.

Mao, J., Kwak, I.-S., Sathishkumar, M., Sneha, K., Yun, Y.-S., 2011. Preparation of PEI-coated bacterial biosorbent in water solution: optimization of manufacturing conditions using response surface methodology. Bioresource Technology 102, 1462–1467.

Mao, J., Won, S.W., Yun, Y.-S., 2013. Development of Poly(acrylic acid)-modified bacterial biomass as a high-performance biosorbent for removal of Cd(II) from aqueous solution. Industrial & Engineering Chemistry Research 52, 6446–6452.

Nicholson, F., Smith, S., Alloway, B., Carlton-Smith, C., Chambers, B., 2003. An inventory of heavy metals inputs to agricultural soils in England and Wales. Science of the Total Environment 311, 205–219.

Osuji, L.C., Onojake, C.M., 2004. Trace heavy metals associated with crude oil: a case study of Ebocha-8 oil-spill-polluted site in Niger Delta, Nigeria. Chemistry & Biodiversity 1, 1708–1715.

Owlad, M., Aroua, M.K., Wan Daud, W.M.A., 2010. Hexavalent chromium adsorption on impregnated palm shell activated carbon with polyethyleneimine. Bioresource Technology 101, 5098–5103.

Park, Y.J., Fray, D.J., 2009. Recovery of high purity precious metals from printed circuit boards. Journal of Hazardous Materials 164, 1152–1158.

Paul, R., 2009. End-of-life management of waste automotive materials and efforts to improve sustainability in North America. In: Fourth International Conference on Sustainable Development and Planning, WIT Transactions on Ecology and the Environment, 120. WIT Press, pp. 853–861.

Pellera, F.-M., Giannis, A., Kalderis, D., Anastasiadou, K., Stegmann, R., Wang, J.-Y., Gidarakos, E., 2012. Adsorption of Cu(II) ions from aqueous solutions on biochars prepared from agricultural by-products. Journal of Environmental Management 96, 35–42.

Park, S.-I., Kwak, I.S., Won, S.W., Yun, Y.-S., 2013. Glutaraldehyde-crosslinked chitosan beads for sorptive separation of Au(III) and Pd(II): opening a way to design reduction-coupled selectivity-tunable sorbents for separation of precious metals. Journal of Hazardous Materials 248–249, 211–218.

Park, D., Yun, Y.-S., Park, J.M., 2010. The past, present, and future trends of biosorption. Biotechnology and Bioprocess Engineering 15, 86–102.

Regmi, P., Garcia Moscoso, J.L., Kumar, S., Cao, X., Mao, J., Schafran, G., 2012. Removal of copper and cadmium from aqueous solution using switchgrass biochar produced via hydrothermal carbonization process. Journal of Environmental Management 109, 61–69.

Rameshraja, D., Srivastava, V.C., Kushwaha, J.P., Mall, I.D., 2012. Quinoline adsorption onto granular activated carbon and bagasse fly ash. Chemical Engineering Journal 181–182, 343–351.

Roberts, D.A., Paul, N.A., Dworjanyn, S.A., Hu, Y., Bird, M.I., de Nys, R., 2014. Gracilaria waste biomass (sampah rumput laut) as a bioresource for selenium biosorption. Journal of Applied Phycology 27, 611–620.

Ramrakhiani, L., Majumder, R., Khowala, S., 2011. Removal of hexavalent chromium by heat inactivated fungal biomass of *Termitomyces clypeatus*: surface characterization and mechanism of biosorption. Chemical Engineering Journal 171, 1060–1068.

Shakoor, M.B., Niazi, N.K., Bibi, I., Rahman, M.M., Naidu, R., Dong, Z., Shahid, M., Arshad, M., 2015. Unraveling health risk and speciation of arsenic from groundwater in rural areas of Punjab, Pakistan. International Journal of Environmental Research and Public Health 12, 12371–12390.

Scher, S., Williams-Jones, A., Williams-Jones, G., 2013. Fumarolic activity, acid–sulfate alteration, and high sulfidation epithermal precious metal mineralization in the crater of Kawah Ijen Volcano, Java, Indonesia. Economic Geology 108, 1099–1118.

Shen, S., Guishen, L., Pan, T., He, J., Guo, Z., 2011. Selective adsorption of Pt ions from chloride solutions obtained by leaching chlorinated spent automotive catalysts on ion exchange resin Diaion WA21J. Journal of Colloid and Interface Science 364, 482–489.

Song, M.-H., Won, S.W., Yun, Y.-S., 2013. Decarboxylated polyethylenimine-modified bacterial biosorbent for Ru biosorption from Ru-bearing acetic acid wastewater. Chemical Engineering Journal 230, 303–307.

Syed, S., 2012. Recovery of gold from secondary sources—a review. Hydrometallurgy 115–116, 30–51.

Tan, X., Liu, Y., Zeng, G., Wang, X., Hu, X., Gu, Y., Yang, Z., 2015. Application of biochar for the removal of pollutants from aqueous solutions. Chemosphere 125, 70–85.

Tangahu, B., Warmadewanthi, I., 2001. Pengelolaan Limbah Rumah Tangga Dengan Memanfaatkan Tanaman Cattail (*Typha Angustifolia*) Dalam Sistem Constructed Wetland. ITS, Surabaya.

Tong, X.-j., Li, J.-y., Yuan, J.-h., Xu, R.-k., 2011. Adsorption of Cu(II) by biochars generated from three crop straws. Chemical Engineering Journal 172, 828–834.

Umeda, H., Sasaki, A., Takahashi, K., Haga, K., Takasaki, Y., Shibayama, A., 2011. Recovery and concentration of precious metals from strong acidic wastewater. Materials Transactions 52, 1462–1470.

Uchimiya, M., Chang, S., Klasson, K.T., 2011. Screening biochars for heavy metal retention in soil: role of oxygen functional groups. Journal of Hazardous Materials 190, 432–441.

Vijayaraghavan, K., Yun, Y.-S., 2008. Bacterial biosorbents and biosorption. Biotechnology Advances 26, 266–291.

Volesky, B., 2007. Biosorption and me. Water Research 41, 4017–4029.

Velazquez-Jimenez, L.H., Pavlick, A., Rangel-Mendez, J.R., 2013. Chemical characterization of raw and treated agave bagasse and its potential as adsorbent of metal cations from water. Industrial Crops and Products 43, 200–206.

Van Vinh, N., Zafar, M., Behera, S.K., Park, H.S., 2015. Arsenic(III) removal from aqueous solution by raw and zinc-loaded pine cone biochar: equilibrium, kinetics, and thermodynamics studies. International Journal of Environmental Science and Technology 12, 1283–1294.

Volesky, B., 2003. Sorption and Biosorption. Published by BV Sorbex, Inc. Montreal, St. Lambert, Quebec, Canada.

Won, S.W., Kotte, P., Wei, W., Lim, A., Yun, Y.-S., 2014. Biosorbents for recovery of precious metals. Bioresource Technology 160, 203–212.

Wang, X., Gaustad, G., 2012. Prioritizing material recovery for end-of-life printed circuit boards. Waste Management 32, 1903–1913.

Wang, J., Chen, C., 2009. Biosorbents for heavy metals removal and their future. Biotechnology Advances 27, 195–226.

Wan Maznah, W., Al-Fawwaz, A., Surif, M., 2012. Biosorption of copper and zinc by immobilised and free algal biomass, and the effects of metal biosorption on the growth and cellular structure of *Chlorella* sp. and *Chlamydomonas* sp. isolated from rivers in Penang, Malaysia. Journal of Environmental Sciences 24, 1386–1393.

Witek-Krowiak, A., Szafran, R., Modelski, S., 2011. Biosorption of heavy metals from aqueous solutions onto peanut shell as a low cost adsorbent. Desalination 265, 126–134.

Won, S.W., Mao, J., Kwak, I.-S., Sathishkumar, M., Yun, Y.-S., 2010. Platinum recovery from ICP wastewater by a combined method of biosorption and incineration. Bioresource Technology 101, 1135–1140.

Won, S.W., Kim, S., Kotte, P., Lim, A., Yun, Y.-S., 2013. Cationic polymer-immobilized polysulfone-based fibers as high performance sorbents for Pt(IV) recovery from acidic solutions. Journal of Hazardous Materials 263 (Part 2), 391–397.

Won, S.W., Park, J., Mao, J., Yun, Y.-S., 2011. Utilization of PEI-modified *Corynebacterium glutamicum* biomass for the recovery of Pd (II) in hydrochloric solution. Bioresource Technology 102, 3888–3893.

Xu, X., Cao, X., Zhao, L., 2013. Comparison of rice husk-and dairy manure-derived biochars for simultaneously removing heavy metals from aqueous solutions: role of mineral components in biochars. Chemosphere 92, 955–961.

Xue, Y., Gao, B., Yao, Y., Inyang, M., Zhang, M., Zimmerman, A.R., Ro, K.S., 2012. Hydrogen peroxide modification enhances the ability of biochar (hydrochar) produced from hydrothermal carbonization of peanut hull to remove aqueous heavy metals: batch and column tests. Chemical Engineering Journal 200–202, 673–680.

Yakkala, K., Yu, M.-R., Roh, H., Yang, J.-K., Chang, Y.-Y., 2013. Buffalo weed (*Ambrosia trifida* L. var. trifida) biochar for cadmium (II) and lead (II) adsorption in single and mixed system. Desalination and Water Treatment 51, 7732–7745.

Yang, Z.-h., Xiong, S., Wang, B., Li, Q., Yang, W.-c., 2013. Cr(III) adsorption by sugarcane pulp residue and biochar. Journal of Central South University 20, 1319–1325.

Yao, Y., Gao, B., Inyang, M., Zimmerman, A.R., Cao, X., Pullammanappallil, P., Yang, L., 2011. Biochar derived from anaerobically digested sugar beet tailings: characterization and phosphate removal potential. Bioresource Technology 102, 6273–6278.

Yeh, T., Lin, C., Chen, C., Pan, C., 2011. Heavy metal biosorption properties of four harvested macrophytes. Journal of Hazardous, Toxic, and Radioactive Waste 15, 108–113.

Yazıcı, H., Kılıç, M., Solak, M., 2008. Biosorption of copper(II) by *Marrubium globosum* subsp. globosum leaves powder: effect of chemical pretreatment. Journal of Hazardous Materials 151, 669–675.

Yu, J., Tong, M., Sun, X., Li, B., 2007. A simple method to prepare poly (amic acid)-modified biomass for enhancement of lead and cadmium adsorption. Biochemical Engineering Journal 33, 126–133.

Yin, P., Xu, M., Qu, R., Chen, H., Liu, X., Zhang, J., Xu, Q., 2013. Uptake of gold (III) from waste gold solution onto biomass-based adsorbents organophosphonic acid functionalized spent buckwheat hulls. Bioresource Technology 128, 36–43.

Zamani, A., Shokri, R., Yaftian, M., Parizanganeh, A., 2013. Adsorption of lead, zinc and cadmium ions from contaminated water onto *Peganum harmala* seeds as biosorbent. International Journal of Environmental Science and Technology 10, 93–102.

Zhou, L.-C., Li, Y.-F., Bai, X., Zhao, G.-H., 2009. Use of microorganisms immobilized on composite polyurethane foam to remove Cu(II) from aqueous solution. Journal of Hazardous Materials 167, 1106–1113.

CHAPTER

8 BIODIESEL PRODUCTION FROM WASTEWATER USING OLEAGINOUS YEAST AND MICROALGAE

J. Ling[1,2], R.A. de Toledo[1], H. Shim[1]

Department of Civil and Environmental Engineering, University of Macau, Avenida da Universidade, Taipa, Macau, SAR, China[1]; School of Environmental Science and Engineering, Guangdong University of Technology, Guangzhou, China[2]

CHAPTER OUTLINE

Environmental Materials and Waste. http://dx.doi.org/10.1016/B978-0-12-803837-6.00008-1

8.1 INTRODUCTION

The world's liquid fuel use is projected to grow from 87 million barrels per day in 2010 to 119 million barrels in 2040, according to the International Energy Outlook 2014 (IEO, 2014). Oil-based fuels are considered to be unsustainable owing to the rapid decline in reserves and emissions of a significant amount of greenhouse gases. Anthropogenic emissions of carbon dioxide are expected to reach 2×10^{10} tons year^{-1}, mainly resulting from the combustion of fossil fuels (Raheem et al., 2015). Under these circumstances, sources of renewable energy have been intensively developed to replace fossil fuels as well as to provide clean energy (Olabi, 2013, 2014). The use of renewable energy such as solar, wind, hydro, biomass, and geothermal is growing worldwide and was responsible for 19% of total global energy consumption (electricity, heating, and transportation) in 2012 (Raupach-Sumiya et al., 2015). That percentage is expected to rise to 30–45% in 2050 owing to fast-growing global investment in renewable energy, to $400–$500 billion annually (Raupach-Sumiya et al., 2015).

Biomass materials often refer to plants or plant-base materials, food waste, and agricultural residues and can be converted directly into liquid fuels, known as biofuels. The two most common types of biofuels currently used are biodiesel and ethanol. Biodiesel is made from the transesterification of triglycerides catalyzed by strong acids or bases and can be produced from a wide range of edible and inedible oleaginous feedstocks. The most common feedstock is soybean oil, which accounts for almost 70% of the total cost of biodiesel and uses arable lands that depend on the volatile climate (Shields et al., 2015).

Oleaginous microorganisms belonging to the genera of algae, bacteria, yeast, and fungi have the capacity to accumulate lipids under specific cultivation conditions; the biodiesels produced have similar chemical composition and energy value compared with oilseed crops and animals (Subramaniam et al., 2010). Oleaginous yeasts are considered to be a promising alternative oil source because of their fast growth rate (hours to several days) and high lipid content (up to 70% of dry weight) (Zhao et al., 2011; Ling et al., 2013a; Santamauro et al., 2014; Zhang et al., 2014b; Yen et al., 2015).

Food industry wastewaters contain high levels of organic matters and nutrients (phosphorus and nitrogen in several chemical forms), which are often difficult to treat. The US Congressional Budget Office estimated that the annual cost investment for wastewater treatment would average between $13.0 and $20.9 billion for 2000–2019 (Congressional Budget Office, 2015). If organic matters could be recovered and converted into energy, not only could the cost of wastewater treatment be reduced, the waste could be turned into a resource developing environmentally sustainable waste-to-energy technologies (Mutanda et al., 2011).

The use of a mixed culture of microalgae and yeasts could be a strategy to improve both lipid production and the removal of organic matters and nutrients from wastewater, because mixed microbial cultures are common in natural ecological systems (Cheirsilp et al., 2011). Microalgae could produce oxygen for yeasts whereas yeasts could provide carbon dioxide to microalgae. Yeasts mainly uptake organic matters and microalgae essentially require nitrogen and phosphorus from wastewater. Cheirsilp et al. (2011) suggested that biomass production and lipid content for the mixed culture of *Rhodotorula glutinis* and *Chlorella vulgaris* in seafood-processing effluent would be beneficial in terms of the economic improvement of biodiesel. However, few studies have been performed to produce microbial lipid from high-strength real nonsterile wastewater without adding other nutrients for the mixed culture of yeast and microalgae. Our previous study (Ling et al., 2013a) showed that the increase in yeast initial cell density could make the remarkable lipid production possible even in real nonsterile wastewater. This chapter explores the potential of further improvement in lipid productivity of oleaginous yeast in a mixed culture with microalgae in real mixed wastewater obtained from distillery and local domestic wastewater treatment plants, and investigates the associated removal efficiencies for organic matters (chemical oxygen demand [COD]) and nutrients (nitrogen and phosphorus) present in wastewater under nonsterile conditions.

8.2 MATERIALS AND METHODS

8.2.1 MATERIALS

8.2.1.1 Strains

Yeast strain *Rhodosporidium toruloides* AS 2.1389 was purchased from the China General Microbiological Culture Collection Center, subcultured, and maintained on yeast extract/peptone/D-glucose (YPD) medium plates at 4°C. After the distillery wastewater was chosen for further studies, yeast was subcultured on the distillery wastewater agar medium plates and maintained on the distillery wastewater agar slants at 4°C. The composition of lipid produced by *R. toruloides* AS 2.1389 from the medium containing glucose as single carbon source and distillery wastewater was reported to be mainly palmitic acid, palmitoleic acid, stearic acid, oleic acid, and linoleic acid, which are suitable for biodiesel production (Li et al., 2005; Lin et al., 2010; Wu et al., 2010; Zhao et al., 2011).

Microalgal strain *Chlorella pyrenoidosa* FACHB-9 was purchased from the Institute of Hydrobiology (Chinese Academy of Sciences, Wuhan, China) and maintained on an SE (soil extract) medium agar plate (Song et al., 1999) at room temperature with a continuous light supply at 2.93 W m^{-2} (Su et al., 2011). Biodiesel produced by this microalgal strain mainly contains octadecenoic acid methyl ester (70.18%), octadecadienoic acid methyl ester, hexadecanoic acid methyl ester, and octadecanoic acid methyl ester (Li et al., 2011). This microalgal strain can be harvested by self-sedimentation (Su et al., 2011), which could save some expense in the harvesting process when applied on an industrial scale.

8.2.1.2 Medium

The YPD medium used for seed culture and storage purposes contained (per liter) D-glucose 20 g, yeast extract 10 g, and peptone 20 g (Ausubel et al., 1994; Han et al., 2003; Li et al., 2005). The distillery wastewater agar medium was made of rice wine distillery wastewater (vinasses) and 20 g L^{-1} agar with the pH adjusted to 5.5. The SE medium was prepared according to Song et al. (1999). All media were sterilized at 121°C for 20 min before use.

Table 8.1 Composition of Wastewater Samples for Initial Cell Density Optimization Experiment and Mixed Culture Experiment

Wastewater	SCOD (mg L^{-1})	TN (mg L^{-1})	TP (mg L^{-1})	NH_3-N (mg L^{-1})	pH
Distillery wastewater	39,800–59,950	1610–2680	378–873	140–206	3.6–3.7
Domestic wastewater	80–197	39–51	4–6.5	32–40	7.3–8.2
Mixed wastewater (1:1, v/v)	17,150–30,500	720–1315	181–349	91–95	3.7–3.8

SCOD, *soluble chemical oxygen demand;* TN, *total nitrogen;* TP, *total phosphorus.*

8.2.1.3 Wastewater

Domestic wastewater was obtained from "T," "C," and "M" wastewater treatment plants in Macau Special Administrative Region, China. Beer brewery wastewater and rice wine distillery wastewater were collected from "S1" and "S2" breweries in Foshan city, China. Milk candy wastewater was provided by the regional candy and biscuit factory in Guangdong province, China. Except for the wastewater sample from the "R" candy factory, all the wastewater samples were filtered (47 mm in diameter, 0.7-μm pore size glass fiber) and stored at 4°C before use. Distillery wastewater was mixed with domestic wastewater at a 1:1 ratio (v/v) (Table 8.1) for initial cell density optimization and mixed culture experiments. The main practical reason for the use of mixed wastewater in the current study was because the distillery wastewater would eventually be combined with domestic wastewater in the regional existing sewerage system. In addition, the competitiveness of the mixed culture of yeast and microalgae against indigenous organisms was further investigated while maintaining high lipid production with the high removal of organics and nutrients from the mixed wastewater.

8.2.2 EXPERIMENTAL SETUP

8.2.2.1 Orthogonal Experiments for Different Wastewater Samples

The oleaginous yeast strain was taken from YPD medium slants using an inoculating loop and cultured in YPD medium at 30°C and 200 rpm for 24 h and used as seed culture. Fifty milliliters of different kinds of wastewater was diluted with distilled water as designed and then added into 150-mL clean flasks sealed with foil and sterilized (121°C for 20 min) before use. The seed culture was inoculated into different wastewater samples at different temperatures (25°C, 30°C, and 35°C), inoculum sizes (5%, 10%, and 15%, v/v), and incubation periods (2–7 days) at pH 3.5–5.5 and 180 rpm, as designed. For the rice wine distillery wastewater, total volumes of 50 and 30 mL and shaking speeds of 180 and 200 rpm were tested. The orthogonal experimental design tables L_9 (3^4) and L_8 (4×2^4) were used in this experiment (Lazic and Mastorakis, 2008). After incubation, the samples were centrifuged at 4000 rpm for 10 min (Li et al., 2005). The COD of supernatant, biomass production, and lipid yield were analyzed at the end of the experiment. All the experiments were conducted in replicate.

8.2.2.2 Influence of Initial Cell Density

The oleaginous yeast *R. toruloides* strain grown in the distillery wastewater medium slant or plate was transferred to 150-mL flasks containing 50 mL YPD medium and cultivated at 30°C and 200 rpm for 36 h. The cultures were then centrifuged at 4000 rpm for 10 min to obtain the high cell density of

2.5×10^9 cells mL^{-1} (Lin et al., 2010) and used in the orthogonal experiment employing the mixture of distillery and domestic wastewaters. Cell density was measured and calculated using the cell chamber.

Thirty milliliters of distillery wastewater diluted with distilled water was added to 150-mL flasks without sterilization. The seed culture was inoculated into wastewater under the conditions of 200 rpm, different temperatures (25°C, 30°C, and 35°C), different initial cell densities in culture medium (0.5×10^8, 1.0×10^8, and 2.0×10^8 cells mL^{-1}), and different incubation periods (3–5 days). The orthogonal experimental design table L_9 (3^4) was used in this experiment (Lazic and Mastorakis, 2008). After incubation, the samples were centrifuged at 4000 rpm for 10 min (Li et al., 2005), and the COD and pH of supernatant and biomass production and lipid yield of the pellet were analyzed.

For the mixture of distillery and domestic wastewater (1:1 ratio) (Table 8.1) used as culture medium, 30 mL of nonsterile wastewater mixture was first added to 150-mL flasks. Then, the high–cell density seed culture (at 1.5×10^9 cells mL^{-1}) was inoculated into the mixture culture medium to bring the initial cell density in the wastewater to 5.0×10^7, 4.0×10^7, 3.0×10^7, and 2.0×10^7 cells mL^{-1} for optimization experiment I and 2.0×10^7, 1.0×10^7, 0.8×10^7, and 0.5×10^7 cells mL^{-1} for the optimization experiment II, and cultivated at 30°C and 200 rpm for 7 days. Samples were taken every 24 h for the analyses of COD, total nitrogen (TN), total phosphorus (TP), and pH in supernatants and biomass and lipid yield in pellets.

The yeast seed culture was prepared in 150-mL flasks containing 30 and 50 mL of sterilized YPD medium and cultivated at 30°C and 200 rpm for 24 and 36 h. All cell densities were measured using the cell chamber and all experiments were conducted in replicate.

8.2.2.3 Mixed Culture of Yeast and Microalgae

After the oleaginous yeast *R. toruloides* was cultivated in YPD medium, the seed culture was centrifuged at 4000 rpm for 10 min to obtain the high cell density of 1.2×10^9 cells mL^{-1}. *C. pyrenoidosa* was taken from the SE agar plate, cultivated in 150-mL conical flasks containing 30 mL sterile mixed (distillery and domestic, 1:1, v/v) wastewater (pH adjusted to 6.5) under the light intensity of 2.93 W m^{-2} (2000 lux, with continuous light) at 30°C and 80 rpm for 6–7 days, and used as seed culture. The cell density of microalgal seed culture was around 3.0×10^7 cells mL^{-1}.

Thirty milliliters (for the yeast pure culture experiment) and 25 mL (for the mixed culture experiment) of real mixed wastewater (Table 8.1) were first added to 150-mL flasks. Then, yeast cells with the initial cell density of 2.0×10^7 cells mL^{-1} (dry weight of inoculum biomass, 0.62 ± 0.07 g L^{-1}) were inoculated and cultivated at 30°C and 200 rpm for 7 days for the pure culture experiment. For the microalgal experiment, after the addition of 25 mL of real mixed wastewater with the pH adjusted to 6.5 in 150-mL flasks, 5 mL of seed culture with the cell density of 3.0×10^7 cells mL^{-1} was inoculated and cultivated at 30°C and 140 rpm for 7 days with a light supply of 2.93 W m^{-2} (2000 lux, 12:12 h light and dark cycles). In case of the mixed culture experiment I (Table 8.2), 5 mL of microalgal seed culture at the cell density of 3.0×10^7 cells mL^{-1} was added after 40 h of the initial yeast cultivation and incubated at 30°C and 200 rpm with a light supply of 2.93 W m^{-2} (2000 lux, 12:12 h light and dark cycles), without the pH adjustment. With ("Mix 1") or without ("Mix 2") part of biomass harvested at 72 h by sedimentation for 80 min using a split funnel, the production of lipid and biomass from days 4 to 7 was calculated as the sum of biomass harvested on the third day plus the accumulated sum of biomass harvested on each day ("Mix 1") or as the amount of biomass harvested on each day ("Mix 2"). In mixed culture experiment II (Table 8.2), 5 mL of microalgal seed culture at different cell densities (2.5×10^7, 1.5×10^7, and 1.0×10^7 cells mL^{-1}) was inoculated after 40 h of the initial yeast cultivation. Light was supplied from 40 h at 2.93 W m^{-2}

Table 8.2 Experimental Setup for Mixed Culture Experiment

Experiment	Test No.	Remark	Initial Cell Density of Yeast (cells mL^{-1})	Initial Cell Density of Microalgae Seed Culture (cells mL^{-1})	Initial Cell Density of Microalgae (cells mL^{-1})	With Microalgae Inoculation at 40 h (Y/N)	With Part of Biomass Harvested on Third Day (Y/N)	pH Adjustment
Mixed culture experiment I	1	Yeast pure culture	2.0×10^7	0	0	N	N	N
	2	Microalgae pure culture	0	3.0×10^7	5.0×10^6	N	N	pH adjusted to 6.5
	3	Mix 1	2.0×10^7	3.0×10^7	5.0×10^6	Y	Y	N
	4	Mix 2	2.0×10^7	3.0×10^7	5.0×10^6	Y	N	N
Mixed culture experiment II	5	Mixed culture	2.0×10^7	2.5×10^7	4.2×10^6	Y	Y	pH adjusted to 6.5–7.0 at 40 and 72 h
	6	Mixed culture	2.0×10^7	1.5×10^7	2.5×10^6	Y	Y	
	7	Mixed culture	2.0×10^7	1.0×10^7	1.7×10^6	Y	Y	
	8	Yeast pure culture	2.0×10^7	0	0	N	N	N

(2000 lux) with 12:12 h light and dark cycles. The pH of wastewater was adjusted to 6.5–7.0 with the addition of 0.1 mL HCl (4 mol L^{-1}) at 40 and 72 h. Part of biomass was harvested at 72 h by sedimentation for 80 min using a split funnel. The production of lipid and biomass during cultivation from days 4 to 7 was calculated as the sum of biomass harvested on the third day plus the accumulated sum of biomass harvested on each day (from days 4 to 7). Samples were collected in every 24 h for the analyses of COD, TN, TP, NH_3-N, and pH in supernatants and biomass and lipid yield in pellets. All experiments were performed in replicate.

8.2.2.4 Biodiesel Production Experiment

Approximately 2.5 g of fresh wet yeast biomass was placed into the 50-mL centrifuge tube and mixed with chloroform, methanol, and sulfuric acid as designed. The yeast biomass was heated via microwave irradiation and maintained at 60°C for the reaction time designed. After the treatment, an equal volume of distilled water was added to the mixture, which was then centrifuged. The organic layer was transferred into a preweighed glass tube. Then, after the chloroform in samples was evaporated, the crude bio-oil was obtained and then dissolved in 5 mL hexane for analysis by gas chromatography with flame ionization detection.

8.2.3 ANALYTICAL METHODS

COD, TN, and TP were measured using Hach's reagents with a Hach reactor DRB200 and spectrophotometer DR2800, following the HACH methods (HACH DR/2800, Hach Company, Loveland, Colorado) and standard methods (APHA/AWWA/WEF, 1995). For the suspended solids (SS), the well-mixed sample (25 mL) was first filtered through a filter paper (47 mm in diameter, 0.7-μm pore size glass fiber), and the increase in the weight of the filter paper represented SS after it was dried at 105°C for 1 h. The pH was measured using a HACH ONE laboratory pH meter.

After the culture was centrifuged at 4000 rpm for 10 min at ambient temperature, cell pellets were washed with distilled water twice and then dried at 80°C until the weight remained constant, and the cell dry weight was measured as biomass production. The total lipids were analyzed according to Bligh and Dyer (1959) with modifications described by Li et al. (2001) and Zhou et al. (2013). Six milliliters of 4 mol L^{-1} HCl solution per 1 g of cells was added to the cell pellet, heated at 100°C for 3 min using a water bath, and then frozen at −20°C overnight. The mixture of methanol and chloroform (1:1 ratio) at a volume of twice the sample was then added and mixed thoroughly. The mixture was centrifuged at 5000 rpm for 5 min and the aqueous phase layer was removed. Sodium chloride (0.1%) was added and mixed at the same volume of the organic solution phase layer. After the mixture was centrifuged at 5000 rpm for 5 min, the aqueous phase layer was removed again. Then, after chloroform evaporated, the weight of lipid extract was measured for the lipid yield. The residue of glucose remained was measured following the 3,5-dinitrosalicylic acid (DNS) method developed by Miller (1959) and modified by Zhao et al. (2008).

The software SPSS was applied in statistical analysis; data obtained were statistically analyzed by one-way analysis of variance (ANOVA) at 95% confidence. The *t* test (paired samples test) was applied to compare the lipid production data for mixed culture samples with and without harvesting part of biomass at 72 h of cultivation.

The yeast-based bio-oil was analyzed using a gas chromatograph (Agilent 6890N) equipped with a flame ionization detector and a capillary column (HP-88 series, 60 m × 0.25 mm × 0.20 μm).

The initial column temperature was 140°C for 5 min and programmed to a final temperature of 240°C at 4°C min^{-1}. The temperature was held at 240°C for 15 min (Shun and Yun, 2011). The bio-oil components were identified by comparing their retention times with those obtained by standard compounds.

8.3 RESULTS AND DISCUSSION

8.3.1 LIPID PRODUCTION FROM DIFFERENT WASTEWATERS BY OLEAGINOUS YEAST

8.3.1.1 Characterization of Wastewaters

The COD level of domestic wastewater was much lower than that of wastewater generated from food industries, with the maximum soluble COD of 503 mg L^{-1} (Table 8.3). The COD of rice wine distillery wastewater was several times higher than that of beer brewery wastewater. The wastewater from milk candy production showed the second highest COD, SCOD/TN/TP, and COD/TN ratios, whereas beer brewery wastewater had the highest COD/TN/TP and COD/TN ratios but the third highest COD. The highest COD was observed for rice wine distillery wastewater, with the third highest COD/TN and COD/TP ratios.

8.3.1.2 Domestic Wastewater

For domestic wastewater, the maximum COD removal efficiency and lowest COD of effluent achieved were 84.26% and 79.11 mg L^{-1}, respectively. Both were achieved under the conditions of initial SCOD 502.67 mg L^{-1}, inoculum size of 15%, room temperature of 23.5–24.5°C, and incubation time of 5 days. In general, the biomass production in domestic wastewater was low under all culture conditions, with the maximum of 0.436 g L^{-1}; the associated lipid yields were negligible.

8.3.1.3 Beer Brewery Wastewater

As shown in Table 8.4, in the case of beer brewery wastewater, the lipid yield was not remarkable; it ranged from 0.08 to 0.24 g L^{-1}, as well as a maximum lipid content of 16.72%. The highest COD removal efficiency and lipid yield achieved were 59.77% and 0.24 g L^{-1}, respectively, with the associated effluent COD of 5173.33 mg L^{-1} and lipid content of 9.49%. Both values were obtained under the conditions of initial SCOD 12,860 mg L^{-1}, inoculum size 10%, temperature 30°C, and incubation time of 5 days.

8.3.1.4 Milk Candy Wastewater

Table 8.5 shows that the high initial TCOD resulted in a high lipid yield and COD removal efficiency in milk candy wastewater although the lipid content obtained decreased slightly with the increase in initial COD. The highest COD removal efficiency of 68.25 ± 1.59% was obtained under the conditions of initial COD 5607 mg L^{-1}, inoculum size of 5%, 30°C, and incubation time of 7 days, with the respective high lipid content of 69.61% ± 4.15% and lipid yield of 1.16 ± 0.10 g L^{-1}. The highest lipid yield of 2.06 ± 0.37 g L^{-1} with the respective high lipid content of 51.65% ± 9.92% and COD removal efficiency of 59.68% ± 13.02% was obtained under the conditions of initial COD 29,600 mg L^{-1}, inoculum size of 5%, 25°C, and incubation time of 3 days.

Table 8.3 Characteristics of Wastewater Samples

	TCOD (mg L^{-1})	SCOD (mg L^{-1})	TN (mg L^{-1})	TP (mg L^{-1})	COD/TN/TP	COD/TN	SS (mg L^{-1})	pH
Domestic wastewater (T)	1060	503	30	4.9	103:6:1	17:1	300	7.0
Domestic wastewater (C)	1909	260	28	4.3	60:7:1	9:1	436	7.1
Domestic wastewater (M)	337	176	25	4.1	43:6:1	7:1	160	7.4
Beer brewery wastewater	–	12,860	54	31	415:2:1	240:1	55	5.2
Rice wine distillery wastewater	82,411	52,200–53,167	2080–2140	1192–208	44:2:1–256:10:1	25:1	11,527	3.5–3.7
Milk candy wastewater	29,600	–	190	22	555:3:1	197:1	4947	6.54

COD/TN/TP and COD/TN refer to soluble chemical oxygen demand to total nitrogen to total phosphorus and soluble chemical oxygen demand to total nitrogen except for milk candy wastewater, for which COD in COD/TN/TP and COD/TN refers to total chemical oxygen demand. SS, suspended solids; TCOD, *total chemical oxygen demand;* SCOD, *soluble chemical oxygen demand.*

Table 8.4 Lipid Production From Beer Brewery Wastewater

	Factor										
Test No.	Initial SCOD (mg L^{-1})	Inoculum (%, v/v)	Time (day)	T (°C)	COD Removal Efficiency (%)	COD of Effluent (mg L^{-1})	Standard Deviation of Effluent COD (mg L^{-1})	Biomass (g L^{-1})	Lipid Content (%)	Lipid Yield (g L^{-1})	Standard Deviation of Lipid Yield (g L^{-1})
1	12,860	5	3	25	17.83	10,567	448	0.73	15.18	0.11	<0.01
2	12,860	10	5	30	59.77	5173	160	2.53	9.49	0.24	0.02
3	12,860	15	7	35	21.85	6790	933	1.77	12.23	0.22	0.15
4	6430	5	5	35	42.66	5025	2934	0.62	16.72	0.10	0.07
5	6430	10	7	25	42.66	3687	155	1.82	7.27	0.13	0.01
6	6430	15	3	30	30.02	4500	226	2.23	8.13	0.18	0.01
7	3215	5	7	30	39.45	1947	71	3.95	2.92	0.12	0.11
8	3215	10	3	35	17.37	2657	68	0.88	9.44	0.08	0.01
9	3215	15	5	25	7.83	2963	185	2.00	8.86	0.18	0.06

Table 8.5 Lipid Production From Milk Candy Wastewater

Test No.	Factor										
	Initial TCOD (mg L^{-1})	Inoculum (%, v/v)	T (°C)	Time (day)	COD Removal Efficiency (%)	COD of Effluent (mg L^{-1})	Standard Deviation of Effluent COD (mg L^{-1})	Biomass (g L^{-1})	Lipid Content (%)	Lipid Yield (g L^{-1})	Standard Deviation of Lipid Yield (g L^{-1})
1	29,600	5	25	3	59.68	11,933	3855	4.43	51.65	2.06	0.37
2	29,600	10	35	5	58.24	12,360	216	4.94	34.69	1.91	0.15
3	29,600	15	35	7	65.56	10,193	290	6.55	26.52	1.73	0.06
4	5607	5	30	7	68.25	1780	28	1.67	69.61	1.16	0.10
5	5607	10	35	3	59.60	2200	87	2.62	45.89	1.20	0.15
6	5607	15	25	5	43.10	3067	93	3.33	32.73	1.09	0.06
7	2882	5	35	5	58.47	1197	16	4.18	62.31	0.76	0.21
8	2882	10	25	7	33.04	1930	90	1.63	71.36	1.16	0.16
9	2882	15	30	3	1.80	2830	115	2.33	42.19	0.97	0.08

8.3.1.5 Rice Wine Distillery Wastewater

As shown in Table 8.6, in the case of rice wine distillery wastewater, the highest lipid content was 32.68% ± 3.01% under the conditions of 15% inoculum, 30°C, COD of 52,200 mg L^{-1}, pH 3.5 (original pH), and cultivation time of 4 days, whereas the maximum lipid yield of 1.36 ± 0.08 g L^{-1} was obtained under the conditions of 15% inoculum, 25°C, COD of 52,200 mg L^{-1}, pH 5.5, and cultivation time of 6 days with a total volume of 50 mL. When the total volume was reduced to 30 mL to improve the aeration condition, the relatively high lipid yield of 1.81 ± 0.01 g L^{-1}, lipid content of 35.9% ± 1.43% (of cell dry weight), and COD removal efficiency of 74.43% ± 0.09% were observed for test 8 under the conditions of dilution factor 3, initial COD of 16,527 mg L^{-1}, inoculum size of 10% (v/v), 25°C, and cultivation time of 7 days. The initial COD level was the most important factor, with a P value = 0.000, 0.002, and 0.001 (all smaller than 0.05) for COD removal, biomass production, and lipid production, respectively. The yeast strain grew better when the wastewater COD was diluted from 53,167 to 16,527 mg L^{-1}, and the average COD removal efficiency increased from 5.5% to 71.0%.

8.3.1.6 Comparison of Different Wastewaters

Fig. 8.1 shows that the effects of wastewater type and initial COD were more significant than other factors on COD removal efficiency ($P = 0.000 < 0.05$ in ANOVA analysis). Samples with lower initial COD had higher COD removal efficiencies when the initial COD was higher than 15,000–25,000 mg L^{-1}. The effects of inoculum size, temperature, and incubation time were not significant on COD removal efficiency in all four kinds of wastewater, with P values > 0.05. The COD removal efficiency increased in milk candy wastewater whereas those in brewery and distillery wastewaters decreased when the temperature was higher than 30°C. In general, COD removal efficiencies increased with incubation time.

Tables 8.4, 8.5, and 8.6 and Fig. 8.2 show that the effects of initial COD, inoculum size, temperature, and incubation time and their associated optimal levels for lipid production in different kinds of wastewater were different. The type of wastewater or the initial COD showed the most significant influence on both lipid yield and lipid content ($P = 0.000 \ll 0.05$ in ANOVA analysis) (Fig. 8.1 and Tables 8.4, 8.5, and 8.6). Milk candy and distillery wastewaters were suitable for lipid production, with a remarkable lipid yield (maximum 2.06 g L^{-1} and 1.36–2.11 g L^{-1}, respectively), whereas lipid production from domestic and brewery wastewaters was not notable; the maximum lipid yield was lower than 0.25 g L^{-1} under the respective different optimal conditions. The initial soluble COD of beer brewery wastewater was 12,860 mg L^{-1}, with COD/TN and COD/TP ratios of 240:1 and 415:1, respectively, whereas the soluble COD for rice wine distillery wastewater was about four times higher (52,200 mg L^{-1}), with COD/TN and COD/TP ratios of 25:1 and 44:1, respectively. The highest lipid yield of 1.98 g L^{-1} was achieved in milk candy wastewater, with the second highest COD/TN and COD/TP ratios, and the second highest initial COD (29,600 mg L^{-1}). Results suggest that both initial COD and COD/TN and COD/TP ratios had the most important role in this study, and the initial COD seemed slightly more significant than COD/TN and COD/TP ratios, from the comparison of two different kinds of wastewater from the beer brewery and rice wine distillery. On the other hand, the reason for less lipid production in rice wine distillery wastewater under the same total volume compared with milk candy wastewater could be the high COD/TN and COD/TP ratios. Previous studies also reported that the limiting nitrogen and phosphorus would lead to a high amount of lipid accumulation in this yeast strain (Li et al., 2006; Wu et al., 2010).

Table 8.6 Lipid Production From Rice Wine Distillery Wastewater

Test No.	Factor												
	Time (day)	Initial SCOD (mg L^{-1})	Inoculum (%, v/v)	T (°C)	pH	COD Removal Efficiency (%)	COD of Effluent (mg L^{-1})	Standard Deviation of Effluent COD (mg L^{-1})	Biomass (g L^{-1})	Lipid Content (%)	Lipid Yield (g L^{-1})	Standard Deviation of Lipid Yield (g L^{-1})	Standard Deviation of Lipid Content (%)
Total Volume = 50 mL													
1	3	52,200	10	25	3.5	0.64	51,867	3166	1.63	14.74	0.24	0.18	10.20
2	3	26,100	15	30	5.5	60.28	10,367	751	7.35	15.17	1.11	0.14	2.11
3	4	52,200	15	30	3.5	18.30	42,650	5869	1.87	32.68	0.91	0.19	3.01
4	4	26,100	10	25	5.5	65.56	8990	410	6.46	18.57	1.20	0.11	1.89
5	5	52,200	10	30	5.5	53.38	24,333	202	7.99	15.91	1.29	0.11	2.22
6	5	26,100	15	25	3.5	67.04	8603	146	8.44	11.91	1.01	0.03	0.27
7	6	52,200	15	25	5.5	57.66	22,100	409	9.47	14.40	1.36	0.08	1.10
8	6	26,100	10	30	3.5	69.60	7933	205	7.13	14.11	1.01	0.17	2.10
Total Volume = 30 mL (No pH Adjustment)													
1	3	53,167	5	25	3.7	5.99	49,983	589	0.59	39.81	0.24	0.01	1.01
2	5	53,167	10	30	3.7	2.51	51,833	1084	0.99	37.72	0.36	0.21	24.32
3	7	53,167	15	35	3.7	7.99	48,917	684	0.99	23.40	0.22	0.14	14.87
4	7	35,444	5	30	3.7	81.06	6712	6	7.87	25.66	2.01	0.06	2.29
5	3	35,444	10	35	3.7	0.40	35,303	1297	0.67	36.79	0.25	0.06	7.36
6	5	35,444	15	25	3.7	60.09	14,146	2534	9.07	23.29	2.11	0.09	1.27
7	5	16,527	5	35	3.7	77.36	3741	103	2.84	34.40	0.99	0.30	3.57
8	7	16,527	10	25	3.7	74.43	4226	15	5.03	35.90	1.81	0.01	1.43
9	3	16,527	15	30	3.7	62.51	6197	563	6.68	30.07	2.00	0.13	2.98

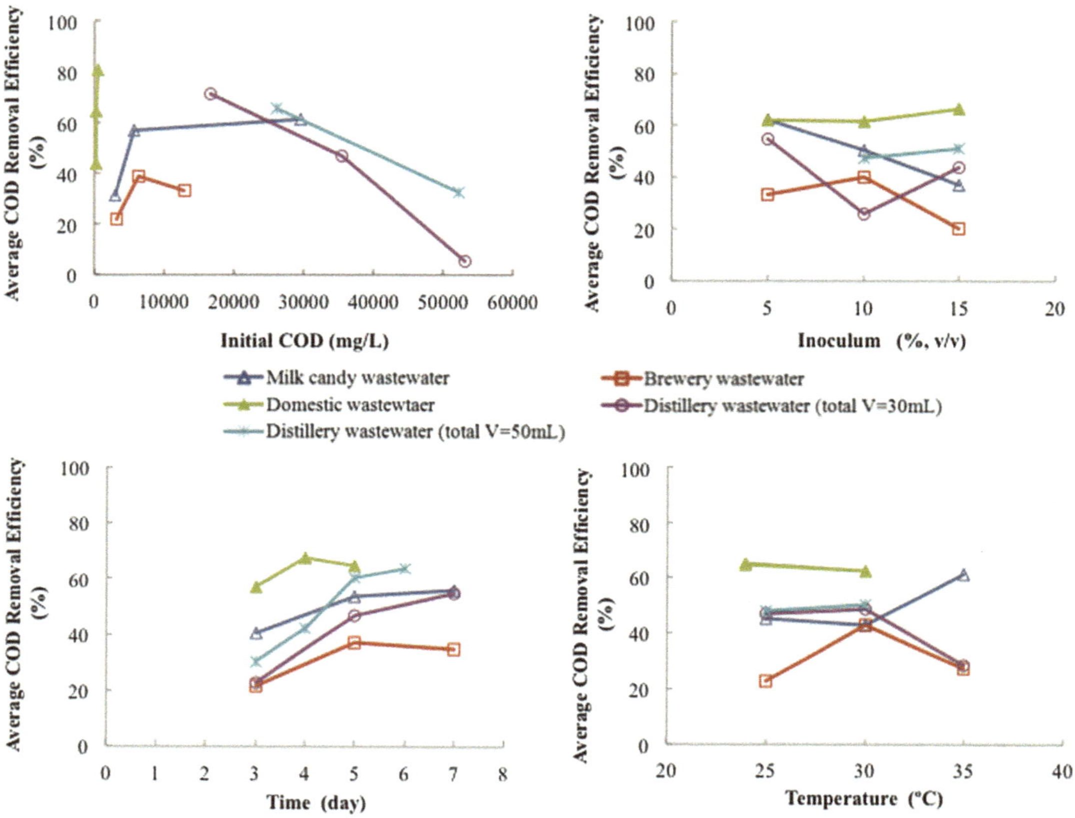

FIGURE 8.1

Effects of wastewater type, initial COD, inoculum size, temperature, and incubation time on COD removal efficiency in different types of wastewater. COD refers to soluble chemical oxygen demand except for milk candy wastewater, for which it refers to total chemical oxygen demand.

Inoculum size and incubation time also had important effects on biomass production ($P = 0.018$ and 0.000, both smaller than 0.05 in ANOVA analysis), although their effects on lipid production were less significant, with P values higher than 0.05. When the inoculum size was higher than 10%, lipid production from milk candy wastewater started to decrease, whereas the trends in beer brewery, rice wine distillery, and domestic wastewater were different (Fig. 8.1 and Tables 8.4, 8.5, and 8.6). In the case of incubation time, the highest lipid production in domestic and beer brewery wastewater was observed on the fourth and fifth days, respectively, whereas lipid production in rice wine distillery and milk candy wastewater increased on the sixth and seventh days, respectively. On the other hand, the effect of temperature was not significant for all four different types of wastewater, with P values >0.05 (Ling et al., 2013a).

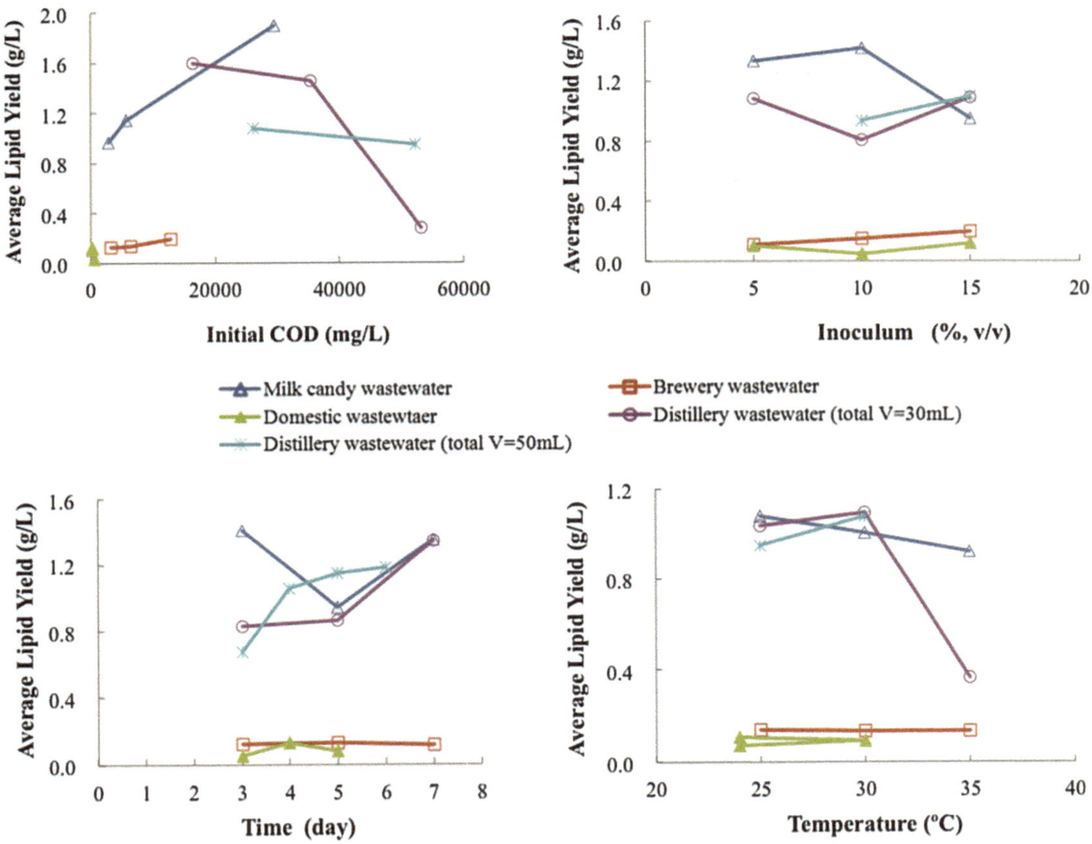

FIGURE 8.2

Effects of wastewater type, initial COD, inoculum size, temperature, and incubation time on lipid production (assuming the lipid content of biomass from domestic wastewater was 50%) from different types of wastewater. COD refers to soluble chemical oxygen demand except milk candy wastewater, for which it refers to total chemical oxygen demand.

8.3.2 INFLUENCE OF INITIAL CELL DENSITY IN RICE WINE DISTILLERY WASTEWATER UNDER NONSTERILE CONDITIONS

8.3.2.1 Orthogonal Experiment Using Nonsterile Distillery Wastewater

As shown in Table 8.7, at an initial cell density of $\geq 0.5 \times 10^8$ cells mL^{-1}, this yeast strain survived and dominated in real wastewater and was not much affected by indigenous microorganisms during the cultivation period of 5 days. It took 2 days less for the organism to produce more lipid and remove about 10% more organic matter from wastewater at the higher initial SCOD level and this relatively high initial cell density, compared with the traditional inoculation method (Table 8.6). For the real

Table 8.7 Results of Orthogonal Experiment Using Nonsterile Distillery Wastewater

Test No.	Factor									
	Initial SCOD ($mg\ L^{-1}$)	Initial Cell Density ($cells\ mL^{-1}$)	Temperature (°C)	Time (day)	COD Removal Efficiency (%)	COD of Effluent ($mg\ L^{-1}$)	Biomass ($g\ L^{-1}$)	Lipid Content (%)	Lipid Yield ($g\ L^{-1}$)	pH
1	69,067	0.5×10^8	25	3	15.22 ± 5.09	$58,556 \pm 3512$	0.79 ± 0.09	27.42 ± 3.27	0.22 ± 0.05	3.73
2	69,067	1.0×10^8	30	4	20.32 ± 1.93	$55,033 \pm 1333$	2.83 ± 0.28	16.54 ± 3.35	0.46 ± 0.10	4.38
3	69,067	2.0×10^8	35	5	18.79 ± 0.80	$56,089 \pm 551$	3.69 ± 0.21	17.35 ± 4.57	0.66 ± 0.41	3.79
4	45,433	0.5×10^8	30	5	54.05 ± 8.22	$20,878 \pm 3735$	8.42 ± 1.34	39.37 ± 7.68	3.28 ± 0.60	6.25
5	45,433	1.0×10^8	35	3	16.17 ± 5.58	$38,089 \pm 2533$	1.82 ± 0.08	37.67 ± 14.82	0.68 ± 0.27	3.78
6	45,433	2.0×10^8	25	4	19.16 ± 3.76	$36,728 \pm 1708$	4.35 ± 0.69	10.35 ± 9.14	0.42 ± 0.36	3.84
7	22,808	0.5×10^8	35	4	89.39 ± 0.66	2419 ± 152	4.75 ± 0.72	47.00 ± 11.95	2.22 ± 0.60	8.42
8	22,808	1.0×10^8	25	5	91.74 ± 0.54	1883 ± 123	6.73 ± 0.42	42.82 ± 5.78	2.89 ± 0.46	8.29
9	22,808	2.0×10^8	30	3	87.08 ± 0.62	2947 ± 142	9.81 ± 0.10	30.95 ± 11.64	3.03 ± 1.11	7.91

SCOD, *soluble chemical oxygen demand.*

distillery wastewater, the highest lipid content of 47.00% ± 11.95% and the second highest COD removal efficiency of 89.39% ± 0.66% were obtained in test 7 under the conditions of dilution factor 3, initial SCOD of 22,808 mg L^{-1}, initial cell density of 0.5×10^8 cells mL^{-1}, 35°C, and cultivation time of 4 days, and achieved a lipid yield of 2.22 ± 0.60 g L^{-1} (Table 8.7).

Either dilution factor or initial SCOD level had the most significant effect on COD removal efficiency ($P = 0.000 < 0.05$), lipid content ($P = 0.039 < 0.05$), and lipid yield ($P = 0.000 < 0.05$) among the four factors investigated for both sterile and nonsterile wastewater samples (Tables 8.6 and 8.7 and Figs. 8.2 and 8.3). The performance of this yeast strain was much better when the distillery wastewater was diluted to the initial SCOD below 25,000 mg L^{-1}. There might be some compounds present in the brewery or distillery wastewater, such as furfural, inhibiting the growth of yeast (Zhao et al., 2012), and dilution might have relieved the inhibition caused by these compounds.

The effect of temperature was not significant in terms of COD removal efficiency ($P = 0.771 > 0.05$), lipid yield ($P = 0.195 > 0.05$), and lipid content of dry biomass ($P = 0.653 > 0.05$); it varied little

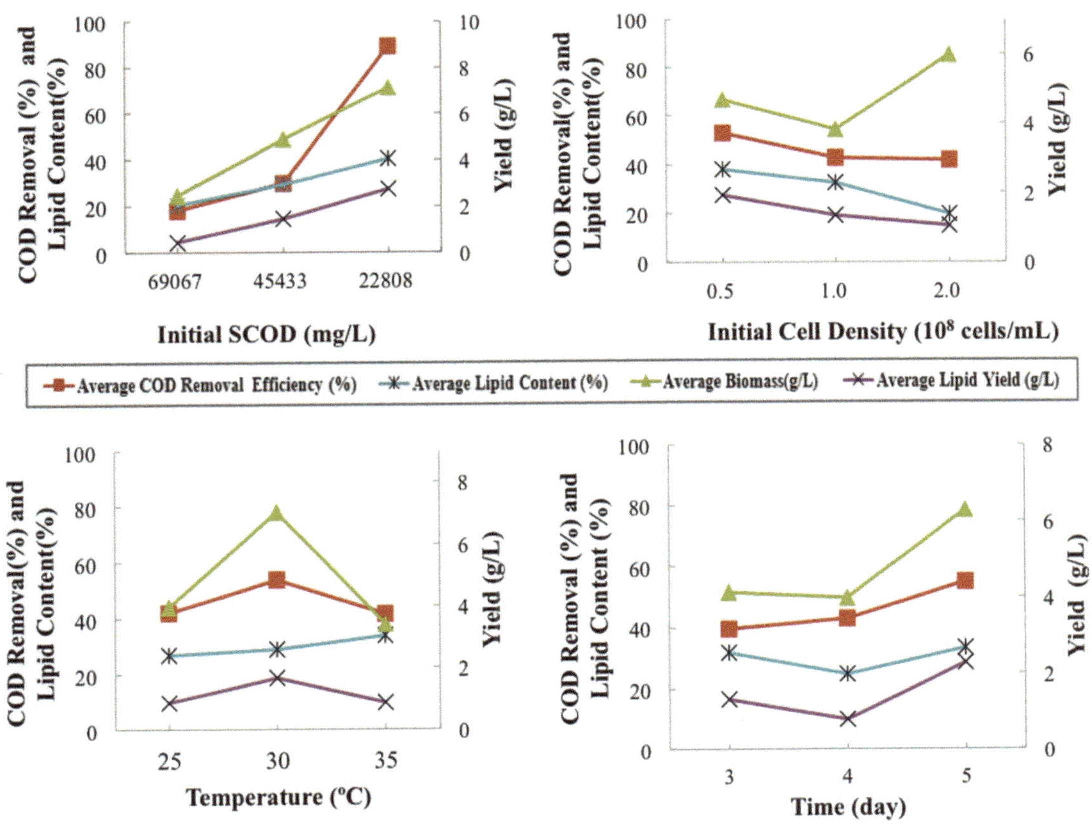

FIGURE 8.3

Effects of initial soluble chemical oxygen demand (SCOD), cell density, temperature, and cultivation time on COD removal efficiency, lipid content, and lipid yield.

when temperature changed from 25°C to 35°C. In comparison, its effect on biomass production was slightly more significant, with a P value $= 0.005 < 0.05$ (Tables 8.6 and 8.7 and Figs. 8.2 and 8.3). The optimal temperature for COD removal efficiency, lipid yield, and lipid content in distillery wastewater samples was 30°C. For the lipid content, the initial cell density was the second most important factor ($P = 0.041 < 0.05$), followed by the cultivation time and temperature. In terms of COD removal efficiency, lipid content, and lipid yield, the optimal conditions were the initial COD level of 22,808 mg L^{-1}, initial cell density of 0.5×10^8 cells mL^{-1}, temperature of 30°C, and cultivation time of 5 days.

8.3.2.2 Optimization Experiment

For the nonsterile 1:1 mixed (distillery and domestic) wastewater with the initial COD of 23,975 mg L^{-1}, the removal efficiencies for COD, TN, and TP increased rapidly during the first 3 days of cultivation, at around 90%, 50%, and 75%, respectively, regardless of the initial cell densities tested (2.0, 3.0, 4.0, and 5.0×10^7 cells mL^{-1}) (Fig. 8.4). Removal of COD as well as TN and TP did not show much difference with the change in initial cell densities. The highest removal of COD, TN, and TP was obtained on the fourth day and reached about 85%, 30%, and 80%, respectively, when the initial cell density changed from 0.5 to 2.0×10^7 cells mL^{-1}. These results suggest that the indigenous microorganisms in the distillery wastewater would also be able to degrade organic matters and nutrients in case yeast did not grow well in the wastewater. Therefore, the initial cell density of yeast did not have a significant effect on the removal of organic matter and nutrients from the mixed wastewater. The highest COD removal efficiency was $92.44 \pm 0.04\%$ at the cultivation time of 7 days and the initial cell density of 4×10^7 cells mL^{-1} (Fig. 8.4).

Figs. 8.4 and 8.5 show that the growth of yeast strain *R. toruloides* was not significantly affected by the indigenous microorganisms in nonsterile distillery and domestic mixed wastewaters when the initial cell density was $\geq 2.0 \times 10^7$ cells mL^{-1}. Lipid production in wastewater samples with the initial cell density of 2.0×10^7 cells mL^{-1} was higher than in samples with lower initial cell densities (0.5×10^7, 0.8×10^7, and 1.0×10^7 cells mL^{-1}) (Ling et al., 2014a). Lin et al. (2010) also reported when the cell density of yeast was $>0.3 \times 10^8$ cells mL^{-1}, the effect of contamination was negligible in the process of ethanol production. Because lipid production in samples with the initial cell density of 2.0×10^7 cells mL^{-1} was higher than or similar to samples with the initial cell densities of 20×10^7, 10×10^7, 5.0×10^7, 4.0×10^7, and 3.0×10^7 cells mL^{-1}, the initial cell density of 2.0×10^7 cells mL^{-1} was considered optimal for the lipid production from distillery wastewater by the yeast *R. toruloides* in the presence of indigenous microorganisms.

This yeast strain has also been reported able to survive over a wide pH range from 3 to 9 with better performance at pH 3–6 than pH 7–9 (Zhou et al., 2013). The pH of wastewater used in this study was not adjusted, with a value around 3.7. *R. toruloides* was previously shown to be more competitive than other microorganisms in domestic wastewater with an initial pH around 3.7 until its growth entered the stationary phase and kept its dominance when the pH was increased to around 7. On the other hand, the pH increased from 3.7 to about 8.2 on the fifth and sixth days of cultivation, and then decreased slightly, to about 8 on the seventh day (Fig. 8.4). The brewery or distillery wastewater contains several kinds of organic acid including malic acid, citric acid, lactic acid, acetolactic acid, and tartaric acid, which are produced by the metabolism of wine yeast during fermentation (Whiting, 1976; Saayman and Viljoen-Bloom, 2006). It has also been reported that yeasts could use such organic acids as a single carbon source to produce single-cell protein (Gao et al., 2008) and lipid (Zhou et al., 2010). Therefore,

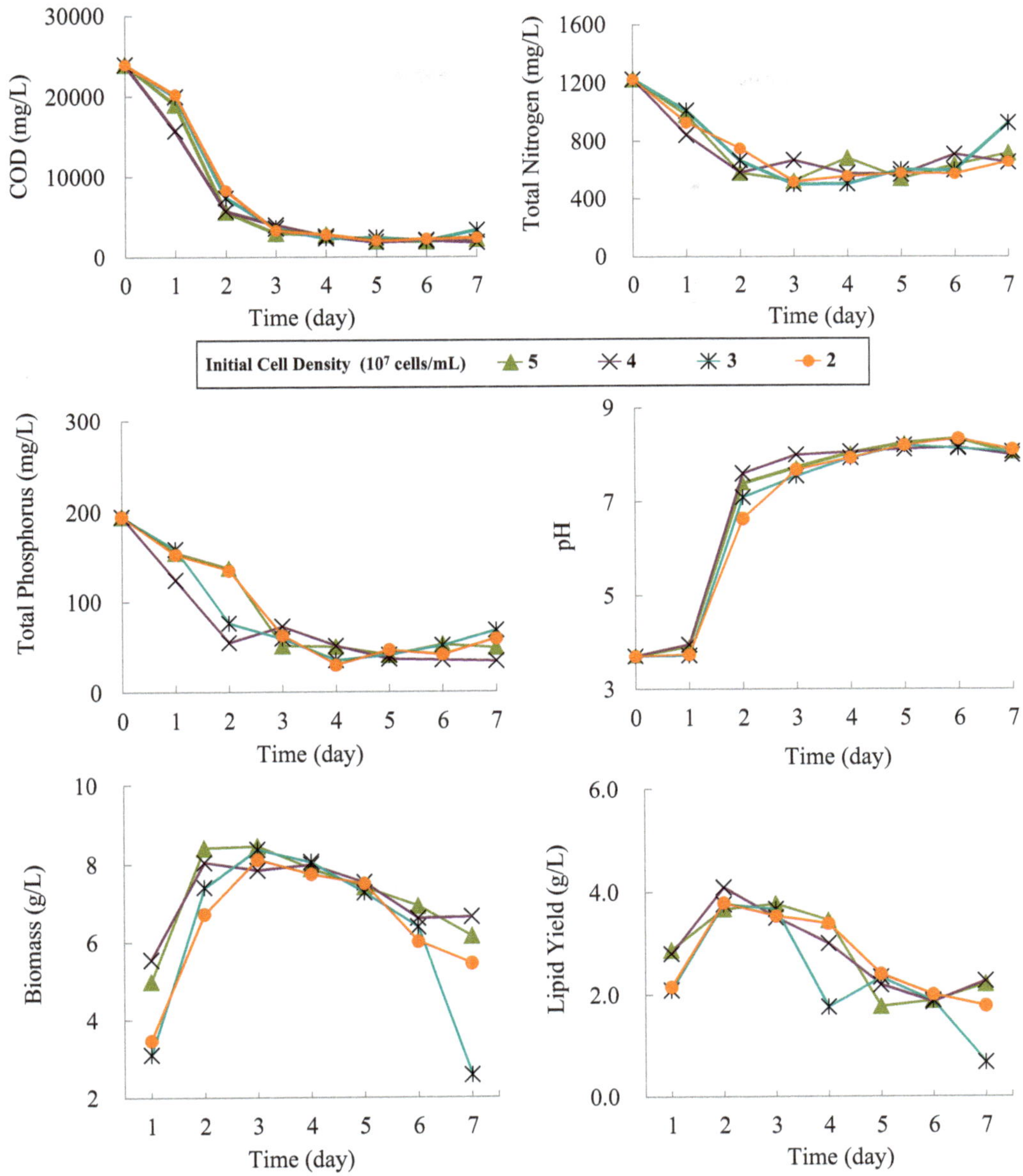

FIGURE 8.4

Changes in COD, TN, TP, pH, biomass production, and lipid production in nonsterile mixed (distillery and domestic) wastewater with initial SCOD of 23,975 mg L^{-1} and initial cell densities of 2.0–5.0 × 10^7 cells mL^{-1}.

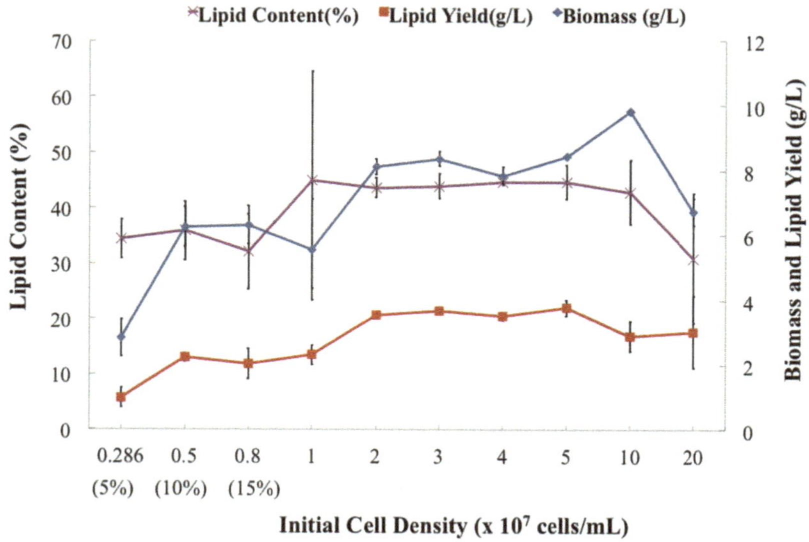

FIGURE 8.5

Effects of initial cell densities on lipid production in mixed (distillery and domestic) wastewater under nonsterile conditions in 3-day cultivation. The initial cell density of samples in an orthogonal experiment with an inoculum size of 5% (v/v) was estimated to be around 2.86×10^6 cells mL^{-1}, and those for 10% (v/v) and 15% (v/v) were 5.73×10^6 and 8.22×10^6 cells mL^{-1}, respectively (Ling et al., 2013b).

a possible reason for the pH increase after 48 h could be the uptake of organic acids by yeasts during cultivation.

For the initial cell densities from 5.0 to 2.0×10^7 cells mL^{-1}, the lipid content of dry biomass generally decreased up until the fifth day, whereas biomass production and lipid yield increased up until the second or third day and then decreased afterward. A similar tendency was observed when the initial cell density ranged from 2.0 to 0.5×10^7 cells mL^{-1} under nonsterile conditions (Ling et al., 2014a). Because the peak lipid yield was recorded on days 2–3 (Figs. 8.4 and 8.5), 3 days of cultivation was suggested to be optimal for lipid content and lipid yield while achieving the purposes of wastewater treatment and lipid production and minimizing energy consumption for aeration and temperature control.

After considering removal efficiencies for COD, TN, and TP, generation of NH_3-N, pH of effluent, lipid yield, lipid content of cell dry weight, cost for seed culture preparation, and energy consumption for aeration and temperature control, the optimal conditions were suggested to be an initial cell density of 2.0×10^7 cells mL^{-1} and cultivation time of 3 days, resulting in associated removal efficiencies for COD, TN, and TP at $86.11\% \pm 0.41\%$, $57.81\% \pm 0.29\%$, and $67.69\% \pm 0.73\%$, respectively, with the effluent COD, TN, TP, NH_3-N, and pH at 3331 ± 98 mg L^{-1}, 518 ± 4 mg L^{-1}, 63 ± 1 mg L^{-1}, 321 ± 10 mg L^{-1}, and pH 7.68, biomass production of 8.12 ± 0.23 g L^{-1}, lipid yield of 3.54 ± 0.04 g L^{-1}, and lipid content of dry biomass of $43.65\% \pm 1.74\%$ (Fig. 8.6). As shown in Tables 8.6 and 8.7 and Fig. 8.5, when the initial cell density was increased from 2.0×10^7 to 5.0×10^7 cells mL^{-1}, both COD removal and lipid production were improved without sterilization,

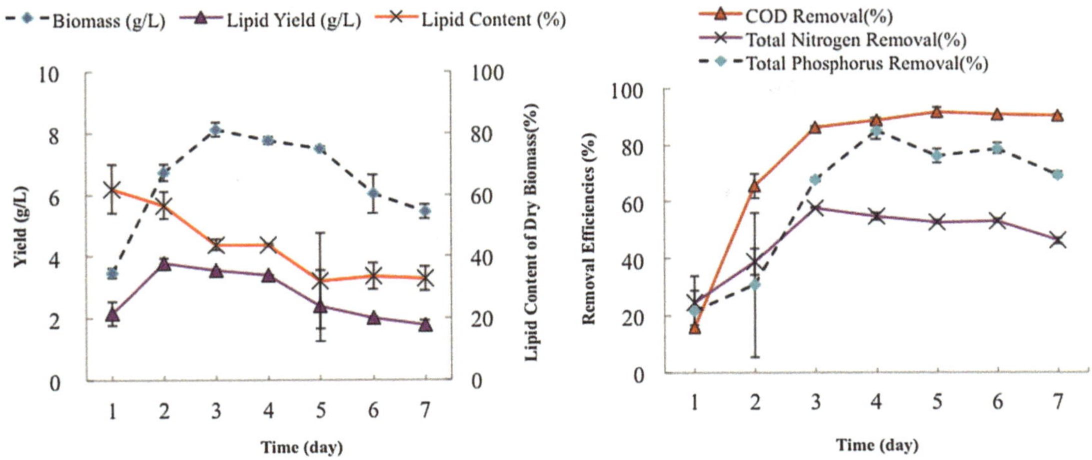

FIGURE 8.6

Lipid production and removal of organics and nutrients from rice wine distillery and domestic mixed wastewater at initial cell density of 2×10^7 cells mL^{-1}.

supplementation of external nutrient, and pH adjustment, which suggested that a lot of energy for sterilization as well as the reagent cost for pH adjustment could be saved. In addition, the relatively shorter time taken in this study compared with previous work using a traditional inoculation method and similar wastewater (Gonzalez-Garcia et al., 2013) or using a similar strain (Zhou et al., 2013) (Table 8.8) would be another advantage for biodiesel production.

8.3.3 MIXED CULTURE OF YEAST AND MICROALGAE

8.3.3.1 Comparison of Pure and Mixed Cultures

The highest lipid yields for the pure cultures of yeast and microalgae were obtained on the third (3.4 g L^{-1}) and seventh (3.0 g L^{-1}) days of cultivation, respectively, and the difference in optimal cultivation time was attributed to the faster growth of yeast compared with microalgae (Fig. 8.7). The highest lipid yield (4.0–4.6 g L^{-1}) for the mixed culture was obtained on the fifth day of cultivation, which suggests that the use of mixed culture improved lipid production. Cheirsilp et al. (2011, 2012) and Xue et al. (2010) also evaluated the use of mixed culture to enhance lipid production. The type of culture (pure or mixed) was highly significant, as evidenced by the low probability value ($P = 0.010 < 0.05$) for pure yeast and mixed cultures, as well as for lipid production data for mixed and pure microalgal cultures, at $P < 0.05$ (Ling et al., 2014b). One reason for the higher lipid and biomass production in mixed culture compared with the pure cultures may be related to the symbiotic relationship between yeast and microalgae. This relationship was also reported by Cheirsilp et al. (2011), who showed that the mixed culture mitigated or even eliminated the stresses caused by CO_2 for yeast and O_2 for microalgae. Zhang et al. (2014a) further reported that this synergistic effect was based on gas exchange but also dissolved oxygen, pH adjustment, and substance exchange.

Table 8.8 Lipid Production by Oleaginous Yeasts on Different Substrates

Microorganism	Substrate	Inoculum/ Initial Cell Density	Process	Time (h)	Biomass ($g L^{-1}$)	Lipid Content (%, w/w)	Lipid Yield ($g L^{-1}$)	COD Removal (%)	References
Lipomyces starkeyi	Potato starch wastewater with 20 $g L^{-1}$ glucose and 3.0 $g L^{-1}$ $(NH_4)_2SO_4$	10% (v/v)	Pure culture	96	2.59	8.88	0.23	–	Liu et al. (2013)
L. starkeyi	Fishmeal wastewater with 20 $g L^{-1}$ glucose	5% (v/v)	Pure culture	144	17.6	–	2.7	43.4	Huang et al. (2011)
Trichosporon dermatis	Wastewater from butanol fermentation	10% (v/v)	Nonsterile	120	7.4	13.5	–	68.2	Peng et al. (2013)
Cryptococcus curvatus	Distillery wastewater	16.67% (v/v)	Pure culture	144	5.19 ± 0.05	25.2 ± 1.98	1.49 ± 0.06	78.98 ± 1.38	Gonzalez-Garcia et al. (2013)
Rhodotorula glutinis	Distillery wastewater	16.67% (v/v)	Pure culture	144	6.06 ± 0.63	27.02 ± 2.36	1.86 ± 0.22	84.44 ± 3.52	Gonzalez-Garcia et al. (2013)
R. toruloides Y2	Bioethanol wastewater with glucose 1.2 $g L^{-1}$ d added from 48 h	5% (v/v)	Pure culture	120	3.8	34.9	–	72.3	Zhou et al. (2013)
R. toruloides AS2.1389	40 $g L^{-1}$ glucose solution	0.5×10^8 cells mL^{-1}	Nonsterile	48	11	55.4	6.1	–	Lin et al. (2010)
	Distillery wastewater	**0.2×10^8 cells mL^{-1}**	**Nonsterile**	**72**	**8.12 ± 0.23**	**43.65 ± 1.74**	**3.54 ± 0.04**	**86.11 ± 0.41**	**This work**

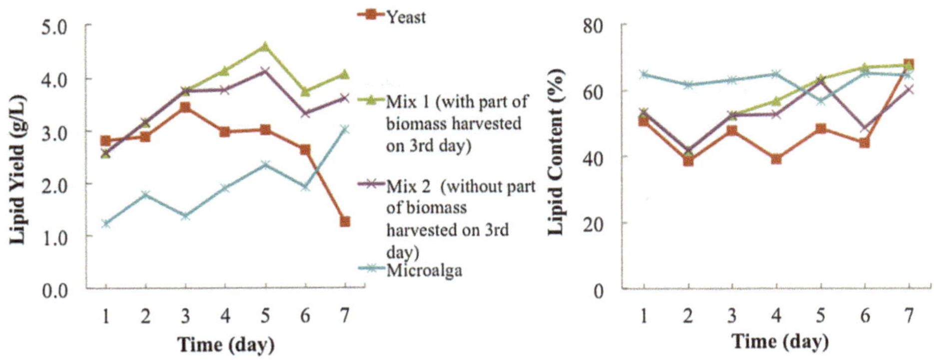

FIGURE 8.7

Lipid production by pure and mixed cultures of yeast and microalgae from rice wine distillery and domestic mixed wastewater. The initial cell density of yeast and microalgae was 2.0×10^7 and 0.5×10^7 cells mL^{-1}, respectively.

As shown in Figs. 8.7 and 8.8, the highest lipid yield (4.60 ± 0.36 g L^{-1}) and biomass production (7.25 ± 0.77 g L^{-1}) were achieved under the conditions with a yeast pure culture initial cell density of 2.0×10^7 cells mL^{-1}, mixed culture initial cell density of 5.0×10^6 cells mL^{-1}, without pH adjustment, after 5 days of cultivation, and with part of biomass harvested on the third day by sedimentation

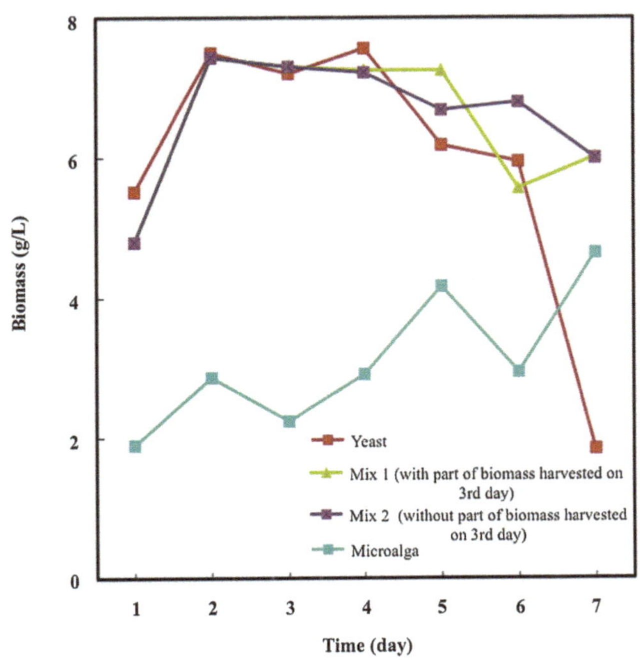

FIGURE 8.8

Biomass production by pure and mixed cultures of yeast and microalgae from rice wine distillery and domestic mixed wastewater. The initial cell density of yeast and microalgae was 2.0×10^7 and 0.5×10^7 cells mL^{-1}, respectively.

Table 8.9 Lipid Production and Removal of Organics and Nutrients on Fifth Day of Cultivation of Pure or Mixed Culture of Yeast and Microalgae in Distillery and Domestic Mixed Wastewater

Test No.	Part of Biomass Harvested on Third Day (Y/N)	pH Adjustment (Y/N)	Yeast Initial Cell Density ($\times 10^7$ cells mL^{-1})	Microalgae Initial Cell Density ($\times 10^6$ cells mL^{-1})	Initial SCOD ($mg\ L^{-1}$)	SCOD: TN:TP	Lipid Content (%, g lipid/g biomass)	Biomass ($g\ L^{-1}$)	Lipid Yield ($g\ L^{-1}$)
T1	N	N	2	0	30,500	169:7:1	48.48 ± 13.97	6.19 ± 0.31	3.02 ± 1.01
T2	N	Y	0	5.0			56.92 ± 7.38	4.19 ± 1.03	2.35 ± 0.28
T3	N	N	2	5.0			62.71 ± 9.54	6.68 ± 1.09	4.12 ± 0.11
T4	Y	N	2	5.0			63.45 ± 2.58	7.25 ± 0.77	4.60 ± 0.36
T5	Y	Y	2	4.2	17,150	49:2:1	65.24 ± 9.41	5.27 ± 0.76	3.46 ± 0.89
T6	Y	Y	2	2.5			64.14 ± 6.00	5.04 ± 1.65	3.21 ± 0.92
T7	Y	Y	2	1.7			58.44 ± 2.69	5.95 ± 2.20	3.50 ± 1.24
T8	N	N	2	0			44.77 ± 13.83	5.67 ± 0.14	2.53 ± 0.72

(Ling et al., 2014b). The associated removal efficiencies for SCOD, TN, and TP were 95.34% $\pm$ 0.07%, 51.18% $\pm$ 2.17%, and 89.29% $\pm$ 4.91%, respectively (Table 8.9) (test 4).

In general, the removal efficiencies for organics (COD) and nutrients (TN and TP) using mixed culture were about 4–20% higher than those for the pure cultures regardless of the initial cell density of microalgae in mixed culture. Table 8.9 shows the specific lipid yield and lipid content and the specific removal rates for COD, TN, and TP. The specific lipid yields of mixed culture samples (tests 3–4, 0.33–0.37 $\times$ 10^{-7} g/cell·day and tests 5–7, 0.29–0.32 $\times$ 10^{-7} g/cell·day) were similar and slightly higher than those of yeast pure culture (test 1, 0.30 $\times$ 10^{-7}g/cell·day and test 8, 0.25 $\times$ 10^{-7}g/cell·day) when grown in the same wastewater samples with higher total initial cell densities. The specific lipid yield and specific removal rates for organics and nutrients for mixed culture samples (tests 3–7) were lower compared with the microalgal pure culture (test 2) because of the higher microalgal initial cell density. However, the associated lipid yield and removal efficiencies for organics and nutrients were higher compared with the pure culture samples of yeast or microalgae.

The pH of mixed wastewater (Fig. 8.9) increased from about 3.7 to 7.5 after 48 h of incubation in the yeast pure culture (Ling et al., 2013b). This increase in pH is advantageous to the growth of microalgae; consequently, no further pH adjustment was necessary to grow this microalgal species in the mixed culture. Furthermore, the pH adjustment did not substantially improve lipid production (tests 5–7) compared with no pH adjustment (tests 3 and 4) (Table 8.9). After the pH was adjusted to 6.5–7.0 at 40 and 72 h by adding 0.1 mL HCl (4 mol L^{-1}), the pH further increased to 7.3–7.7 at 48 h and to 7.8–8.4 at 96 h (Fig. 8.9). This suggested that maintenance of the wastewater pH at 6.5–7.0 would require more chemicals and frequent pH adjustment, which may improve lipid production to some extent but also increase the cost of the process.

8.3.3.2 Effect of Harvesting Part of Biomass in Mixed Culture

The lipid yield increased (Fig. 8.7) when part of the yeast biomass was harvested during the mixed culture cultivation, which was attributable to the reduction in yeast biomass when its growth entered the late stationary or death phase, because the growth cycles for yeast and microalgae were different.

COD Removal Efficiency (%)	TN Removal Efficiency (%)	TP Removal Efficiency (%)	Specific Lipid Yield ($\times 10^{-7}$ g/cell·day)	Specific Lipid Content (lipid g/g biomass d)	Specific COD Removal Rate ($\times 10^{-2}$ mg/cell·day)	Specific TN Removal Rate ($\times 10^{-4}$ mg/cell·day)	Specific TP Removal Rate ($\times 10^{-4}$ mg/cell·day)
91.54 ± 0.66	46.31 ± 4.57	86.21 ± 1.59	0.30	9.70	2.79	6.09	1.56
61.75 ± 14.48	29.28 ± 17.74	75.14 ± 11.48	0.94	12.54	7.53	15.40	5.44
94.54 ± 0.36	51.84 ± 6.30	90.57 ± 1.07	0.33	11.38	2.31	5.45	1.31
95.34 ± 0.07	51.18 ± 2.17	89.29 ± 4.91	0.37	12.69	2.33	5.38	1.29
91.72 ± 2.02	26.39 ± 5.89	54.66 ± 8.87	0.29	13.05	1.30	1.57	1.58
92.61 ± 0.87	37.50 ± 0.98	59.86 ± 6.08	0.29	12.83	1.41	2.40	1.86
90.51 ± 1.20	33.33 ± 9.82	58.42 ± 15.71	0.32	11.69	1.43	2.21	1.88
87.87 ± 0.18	20.14 ± 1.96	54.48 ± 7.10	0.25	8.95	1.51	1.45	1.90

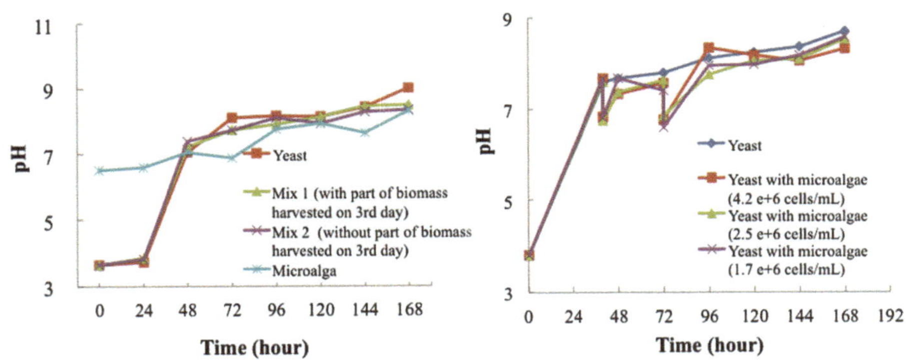

FIGURE 8.9

Changes in pH of rice wine distillery and domestic mixed wastewater with pure and mixed cultures.

After 3 days of cultivation, the biomass of yeast pure culture started to decrease, whereas the microalgal pure culture biomass increased after the fourth day, although with some fluctuations that may have been caused by the growth of indigenous microorganisms (Fig. 8.8). The generation time of yeast was about 90 min (Sherman, 2002) with a cell cycle time of 140–160 min (Abe et al., 1984; Lovrics et al., 2006), whereas for microalgae, the doubling time was 3.5–72 h (Sheehan et al., 1998; Chisti, 2007); this suggested that the best harvest time for yeast was earlier than the one for microalgae. There was no significant further increase in biomass after the inoculation of different microalgal cell densities ($1.7-4.2 \times 10^6$ cells mL^{-1}), which might be attributed to the suppression of yeast growth when the mixed culture reached an alkaline condition with a pH of 8.0 (Huang et al., 2013). Removal of part of the least active yeast biomass may also spare more resources for the growth of active yeast and microalgal cells in mixed culture and further improve the overall performance.

Table 8.9 shows that lipid production (4.60 ± 0.36 g L^{-1}) for the "Mix 1" samples with part of biomass harvested on the third day, with the specific lipid yield of 0.37×10^{-7} g/cell·day and the lipid content of 63.45% ± 2.58%, was slightly higher compared with the "Mix 2" samples, with the lipid production of 4.12 ± 0.11 g L^{-1}, the specific lipid yield of 0.33×10^{-7}g/cell·day, and the lipid content of 62.71% ± 9.54%, when part of biomass (lipid) was not harvested on the third day (Ling et al., 2014b). The *t* test (paired-samples test) was applied to the lipid production data for mixed culture samples and the factor with or without part of the biomass harvested on the third day showed a significant difference in lipid production with a very low probability value, P (two-tailed) = $0.025 < 0.05$.

8.3.3.3 Influence of Microalgal Initial Cell Density on Mixed Culture

The initial microalgal cell densities studied were 1.7×10^6, 2.5×10^6, 4.2×10^6, and 5.0×10^6 cells mL^{-1}, with the initial yeast cell density fixed at 2.0×10^7 cells mL^{-1}, corresponding to the ratio of yeast to microalgae at 12:1, 8:1, 5:1, and 4:1, respectively. Regardless of the initial microalgal cell density, the mixed culture generally had higher lipid yields compared with the yeast pure culture (Fig. 8.10 and Table 8.9). No significant improvements in both lipid yield and removal of organics and nutrients were obtained when the microalgal initial cell density increased from 1.7×10^6 to 4.2×10^6 cells mL^{-1}. Slight improvements in lipid production as well as organic and nutrient removal were obtained in samples with a microalgal initial cell density of 5.0×10^6 cells mL^{-1} (with the initial cell density of yeast fixed at 2.0×10^7 cells mL^{-1}); this behavior was partially attributed to the higher initial wastewater COD level. These results are in agreement with previous studies (Xue et al., 2010; Cheirsilp et al., 2011) which suggested that the mixed ratio of microalgae to yeast in mixed culture in the range of 1:1–3:1 may result in better overall performance. The increase in the initial cell density may reduce the time required for microalgae to exert a positive effect on mixed culture. However, it would take a longer time to obtain a higher microalgal cell density in seed culture (possibly 2–3 weeks), which would not be practical for a batch culture experiment (5–7 days), although it could be applicable to a continuous process (Ling et al., 2014b).

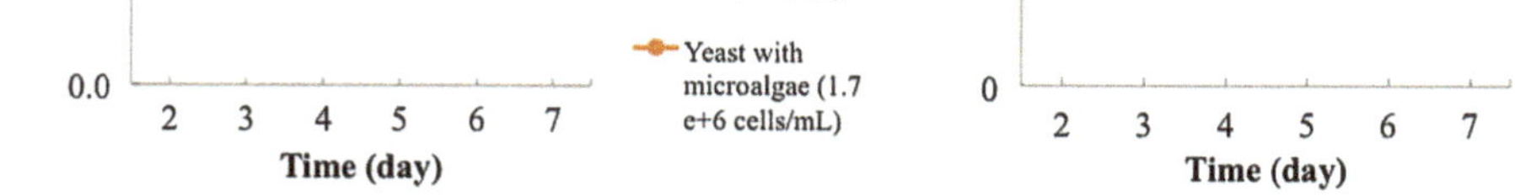

FIGURE 8.10

Lipid production by yeast pure culture (initial cell density, 2.0×10^7 cells mL^{-1}) and mixed culture of yeast (initial cell density, 2.0×10^7 cells mL^{-1}) and microalgae (three different initial cell densities) from rice wine distillery and domestic mixed wastewater.

A comparison of performance in the current study and previously published ones, especially in terms of lipid production together with the removal of organics and nutrients by pure or mixed cultures, is shown in Table 8.10. Lipid yield and lipid content in the current study were higher when mixed cultures were used (Xue et al., 2010; Cheirsilp et al., 2011, 2012). The specific lipid yield, lipid content, and COD removal rate obtained in the current study were lower compared with other studies, owing to the higher initial cell number used. The cost of using a high initial cell density under nonsterile conditions in the current study would not be significant, considering the higher cost of maintaining sterile conditions in most other studies, especially for a continuous process. The current study shows other advantages such as saving costs for chemicals because no external nutrients were added to the cultivation medium and a good performance was obtained under nonsterile conditions. Both lipid content and lipid yield using the mixed culture of yeast and microalgae were higher than those obtained when pure cultures of yeast (Huang et al., 2011; Zhou et al., 2013) and microalgae (Ren et al., 2014) were used.

The growth of mixed culture of yeast and microalgae may take a longer time, compared with a yeast pure culture, to achieve the highest lipid production, as shown in Table 8.10. The development of a continuous process could be a practical way to take full advantage of this mixed culture. A relatively high initial cell density may be another way to achieve remarkable lipid production under nonsterile conditions (Lin et al., 2010; Ling et al., 2013b), as well as at low temperature and pH (Santamauro et al., 2014). The cost for seed culture or preculture preparation may increase if a high cell density is used, although it could be reduced if the spent seed culture medium were recycled or reused. This cost would become much less important if the operation were a continuous process because the seed culture or preculture would only need to be prepared at the beginning (Ling et al., 2014b).

8.3.4 BIODIESEL PRODUCTION

The biodiesel (fatty acid methyl esters [FAMEs]) produced in the current study was mainly made of methyl palmitate (C16:0), methyl palmitoleate (C16:1), methyl stearate (C18:0), methyl oleate (C18:1), and methyl linoleate (C18:2), which corresponded to about 75% of the total amount of FAMEs generated under all conditions tested. Methyl oleate (C18:1) was the highest at over 40%, followed by methyl linoleate (C18:2) 15–20%, methylpalmitate (C16:0) 7–9%, methylpalmitoleate (16:1) 2–7%, and methyl stearate (C18:0) 2–4% (Table 8.11), similar to biodiesel produced from rapeseed (Rao et al., 2012).

The amount of catalyst H_2SO_4 was the most important parameter ($P = 0.021$ and $0.024 < 0.05$ for FAMEs and the conversion rate, respectively), followed by the mixture rates of methanol and chloroform, the ratio of solvent to wet biomass, and the reaction time (associated P values all >0.05) (Fig. 8.11 and Table 8.11). The FAMEs and lipid conversion rate showed a remarkable rise when the amount of H_2SO_4 was varied from 4% to 6%. The optimal mixture rate of methanol and chloroform was 1:1, similar to some previous studies on wet microalgal biomass (Cheng et al., 2013, 2014) but different from the conventional solvent mixture rate of 1:2 and from the previous study, which also used the wet biomass of yeast and microwave irradiation (Cui and Liang, 2014). Related to the ratio of solvent to wet biomass, the conversion rate increased when the ratio varied from 8 to 16 mL g^{-1} and remained constant after it further increased to 24 mL g^{-1}.

Table 8.10 Lipid Production by Pure or Mixed Culture of Oleaginous Yeasts and Microalgae Under Different Conditions

Organism	Substrate	Inoculum Size/Initial Cell Density (cells mL^{-1})	Sterile Condition (Y/N)	Time (day)	Biomass (g L^{-1})
Spirulina platensis and *Rhodotorula glutinis*	Monosodium glutamate wastewater (COD, 43,210 mg L^{-1}; reducing sugar, 2.9 ± 0.1 g L^{-1})	10% (v/v) yeast, 20% (v/v) microalgae	Y	5	1.6
R. glutinis and *Chlorella vulgaris*	Seafood processing effluent (pH, 5.3; COD, 34,680 mg L^{-1}; nitrate, 32 mg L^{-1}) with 1% molasses (62% total sugar contained)	2.8×10^5 yeast, 2.6×10^5 microalgae	Y	5	4.63
R. glutinis and *C. vulgaris*	3% Pure glycerol and urea with molar C/N = 32, pH = 6.0	1.0×10^6 yeast, 3.0×10^5 microalgae	Y	5	3.2
Lipomyce starkeyi	Fishmeal wastewater (COD, 78.4 g L^{-1} + 21.3 g L^{-1} (glucose) = 99.7 g L^{-1}; TN = 4.3 g L^{-1}; TP = 0.95 g L^{-1})	5% (v/v)	Y	5	3.8
Rhodosporidium toruloides Y2	Bioethanol wastewater (COD, 34 g L^{-1}) with glucose, 1.2 g L^{-1} d added from 48 h	5% (v/v)	Y	5	3.8
Scenedesmus sp.	Synthetic dark fermentation effluent (COD = 11.04 g L^{-1})	10% (v/v)	Y	5	1.8
R. toruloides and *Chlorella pyrenoidosa*	**Distillery wastewater (COD = 30.5 g L^{-1}; TN = 1315 mg L^{-1}; TP = 181 mg L^{-1})**	**2×10^7 (yeast), 0.5×10^7 (microalgae)**	**N**	**5**	**7.25 ± 0.77**

Table 8.11 Results of Direct Transesterification Experiment

Test No.	Solvent/ Biomass (mL g^{-1})	Methanol and Chloroform Mix Rate	H_2SO_4 (%)	React Time (min)	FAMEs/Lipid Conversion Rate (%)	Mean FAMEs (mg L^{-1})
1	8	1:0	2	10	31.53	204
2	8	1:1	4	20	37.33	241
3	8	2:1	6	30	55.99	362
4	16	1:0	4	30	37.19	240
5	16	1:1	6	10	71.12	460
6	16	2:1	2	20	42.39	274
7	24	1:0	6	20	53.06	343
8	24	1:1	2	30	53.36	345
9	24	2:1	4	10	44.60	288

Lipid (%, w/w)	Lipid ($g\ L^{-1}$)	COD Removal (%)	Nutrient Removal (%)	Specific Lipid Yield ($\times 10^{-10}$ $g\ L^{-1}$ cell d)	Specific Lipid Content (lipid ($g\ L^{-1}$)/($g\ L^{-1}$) biomass d)	Specific COD Removal Rate ($\times 10^{-7}$ $mg\ L^{-1}$ cell d)	References
13.75	0.22	73 (Reducing sugar removal, 94% ± 1%)	100 (Glutamic acid), 35 ($NH_4^+ - N$)	0.04	2.75	5.13	Xue et al. (2010)
62.2	2.88	79	N/A	9.63	12.68	70.64	Cheirsilp et al. (2011)
34.38	1.1	76.7 (Glycerol removal)	N/A	1.69	6.88	N/A	Cheirsilp et al. (2012)
20.79	0.79	35.83	N/A	0.50	4.16	22.69	Huang et al. (2011)
34.9	1.33	72.3	N/A	0.84	6.98	7.56	Zhou et al. (2013)
37	0.67	36.9	N/A	0.44	7.40	2.72	Ren et al. (2014)
63.45 ± 2.58	**4.60 ± 0.36**	**95.34 ± 0.07**	**51.18 (TN), 89.29 (TP)**	**0.37**	**12.69**	**2.33**	**Current work**

Methyl Palmitate (C16:0) (%)	Methyl Palmitoleate (C16:1) (%)	Methyl Stearate (C18:0) (%)	Methyl Oleate (C18:1) (%)	Methyl Linoleate (C18:2) (%)	Subtotal (%)	Other FAMEs (%)
9	7	3	42	18	79	21
8	3	3	43	17	74	26
8	6	2	43	16	76	24
8	4	3	43	16	74	26
8	5	4	43	16	76	24
8	2	4	39	18	70	30
7	5	3	43	17	75	25
8	4	3	36	15	65	35
8	7	4	42	20	81	19

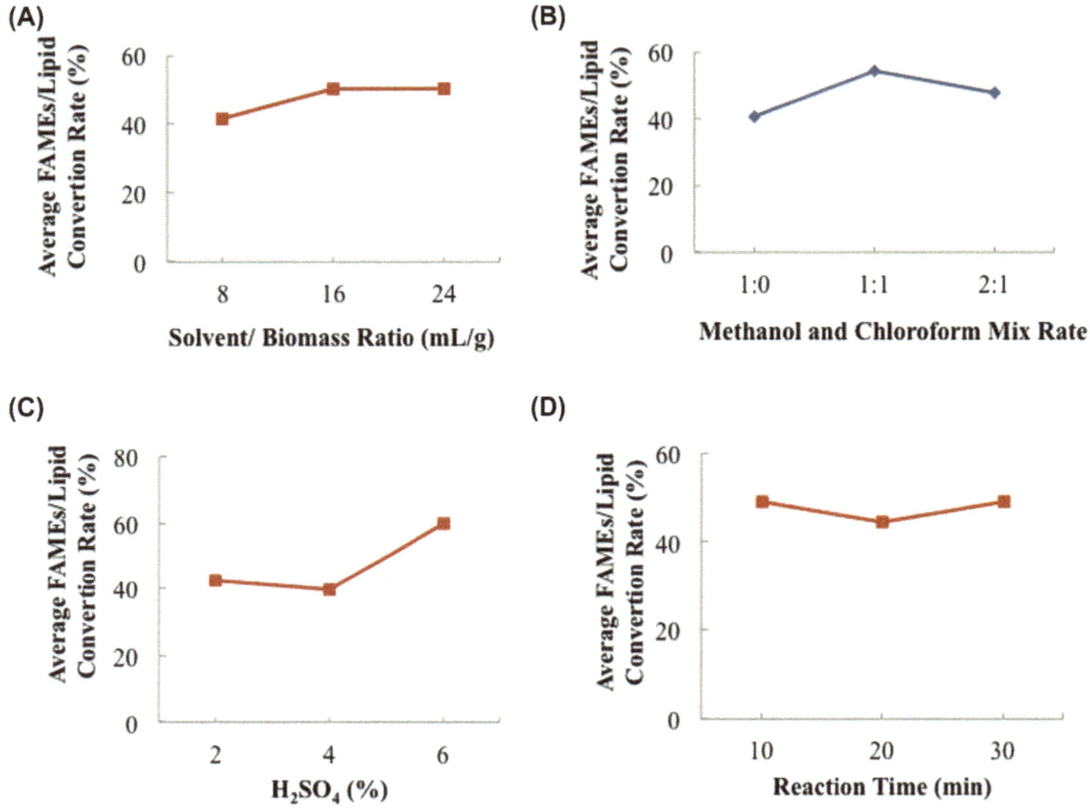

FIGURE 8.11

Effects of (A) solvent/biomass ratio, (B) methanol and chloroform mixture rate, (C) H_2SO_4, and (D) reaction time on FAMEs/lipid conversion rate.

No substantial difference in the conversion rate of FAMEs to lipid was observed when the reaction time changed from 10 to 30 min, which suggests that the reaction time could be shorter than in previous studies (Cheng et al., 2013, 2014), and consequently that energy could be saved. The highest conversion rate was 71% with the amount of FAMEs obtained from wastewater at 460 mg L^{-1}, lower than in some previous studies (over 90%) (Zhou et al., 2013; Cui and Liang, 2014). This lower conversion rate could be explained by the lack of mixing or agitation during the reaction process to ensure sufficient contact between solvent and biomass (Pinto et al., 2004) as well as the loss of FAMEs during chloroform evaporation. On the other hand, energy consumption for the biodiesel production process from dry biomass (20.1–94.8 MJ kg^{-1}) was much higher compared with wet biomass (0.9–3.8 MJ kg^{-1}) in the energy assessment (Ling et al., 2015), because the dewatering process takes a lot of energy, as high as 90.32 MJ kg^{-1} biodiesel including both heating and electricity (Lardon et al., 2009).

8.4 CONCLUSION

Brewery, rice wine distillery, and milk candy wastewater were suitable for use for lipid production by oleaginous yeast. Lipid production with increased initial cell density was assessed in distillery and domestic mixed (1:1) wastewater and showed a remarkable improvement in both lipid production and the removal of organic matter from real (nonsterile) wastewater in a relatively short time. The optimal initial cell density was 2.0×10^7 cells mL^{-1}. Lipid production by yeast in distillery and domestic mixed wastewater was further improved by mixed culture with microalgae after the pH of wastewater was increased from 3.7 to about 7 by the growth of yeast for 40 h under nonsterile conditions. Harvesting part of the inactive biomass in the middle of cultivation had a positive effect on lipid production. The approach of using a mixed culture is feasible for improving lipid production from real mixed wastewater under nonsterile conditions. The microwave-assisted process, which is applicable to the fresh wet biomass, is a promising technology in biodiesel production with lower energy consumption. Further improvements are still required especially for increasing conversion efficiency and reducing chemical consumption.

ACKNOWLEDGMENTS

This research was supported by a University of Macau Multi-Year Research Grant (MYRG2014-00112-FST) and by grants from the Macau Science and Technology Development Fund (FDCT/061/2013/A2 and FDCT/063/2013/A2).

REFERENCES

Abe, K., Fukui, S., Kusangi, A., 1984. Estimation of time lengths of cell-cycle phases from asynchronous cultures of the basidiomycetous yeast *Rhodosporidium toruloides*. The Journal of General and Applied Microbiology 30, 257–265.

APHA (American Public Health Administration)/AWWA (American Water Works Association)/WEF (Water Environment Federation), 1995. Standard Methods for the Examination of Water and Wastewater, 19th ed. Washington DC, USA.

Ausubel, F.M., Brent, R., Kingston, R.E., Moore, D.D., Seidman, J.G., Smith, J.A., Struh, K., 1994. Current Protocols in Molecular Biology, Current Protocols. Brooklyn, NY.

Bligh, E., Dyer, W., 1959. A rapid method of total lipid extraction and purification. Canadian Journal of Physiology and Pharmacology 37, 911–917.

Cheirsilp, B., Suwannarat, W., Niyomdecha, R., 2011. Mixed culture of oleaginous yeast *Rhodotorula glutinis* and microalga *Chlorella vulgaris* for lipid production from industrial wastes and its use as biodiesel feedstock. New Biotechnology 28, 362–368.

Cheirsilp, B., Kitcha, S., Torpee, S., 2012. Co-culture of an oleaginous yeast *Rhodotorula glutinis* and a microalga *Chlorella vulgaris* for biomass and lipid production using pure and crude glycerol as a sole carbon source. Annals of Microbiology 62, 987–993.

Cheng, J., Huang, R., Yu, T., Li, T., Zhou, J., Cen, K., 2013. Using wet microalgae for direct biodiesel production via microwave irradiation. Bioresource Technology 131, 531–535.

Cheng, J., Huang, R., Yu, T., Li, T., Zhou, J., Cen, K., 2014. Biodiesel production from lipids in wet microalgae with microwave irradiation and bio-crude production from algal residue through hydrothermal liquefaction. Bioresource Technology 151, 415–418.

Chisti, Y., 2007. Biodiesel from microalgae. Biotechnology Advances 25, 294–306.

Congressional Budget Office – The Congress of the United States. Future Investment in Drinking Water and Wastewater Infrastructure. http://www.cbo.gov/sites/default/files/11-18-watersystems.pdf (accessed 03.07.15.).

Cui, Y., Liang, Y., 2014. Direct transesterification of wet *Cryptococcus curvatus* cells to biodiesel through use of microwave irradiation. Applied Energy 119, 438–444.

Gao, W.W., Du, J.H., Han, W., Li, R.F., 2008. Studies on differential uptake of organic acids by *Candida utilis*. Food and Fermentation Industries 34, 117–121 (In Chinese).

Gonzalez-Garcia, Y., Hernandez, R., Zhang, G.C., Escalante, F.M.E., Holmes, W., French, W.T., 2013. Lipids accumulation in *Rhodotorula glutinis* and *Cryptococcus curvatus* growing on distillery wastewater as culture medium. Environmental Progress & Sustainable Energy 32, 69–74.

Han, C.H., Werder, M., Gustedt, W.V., Hauser, H., 2003. Heterologous expression of human scavenger receptor class B type I (SR-B1) in *Pichiapastoris*. In: Selinsky, B.S. (Ed.), Membrane Protein Protocols: Expression, Purification, and Characterization. Humana Press Inc., New Jersey, p. 50.

Huang, L.H., Zhang, B., Gao, B.Y., Sun, G.P., 2011. Application of fishmeal wastewater as a potential low-cost medium for lipid production by *Lipomyces starkeyi* HL. Environmental Technology 32, 1975–1981.

Huang, Q., Wang, Q., Gong, Z., Jin, G., Shen, H., Xiao, S., Xie, H., Ye, S., Wang, J., Zhao, Z.K., 2013. Effects of selected ionic liquids on lipid production by the oleaginous yeast *Rhodosporidium toruloides*. Bioresource Technology 130, 339–344.

International Energy Outlook, 2014. (U.S. Energy Information Administration). International Energy Outlook 2014 [PDF]. http://www.eia.gov/forecasts/ieo/pdf/0484(2014).pdf (accessed 12.06.15.).

Lardon, L., Helias, A., Sialve, B., Steyer, J.P., Bernard, O., 2009. Life-cycle assessment of biodiesel production from microalgae. Environmental Science & Technology 43, 6475–6481.

Lazic, L., Mastorakis, N., 2008. Orthogonal array application for optimal combination of software defect detection techniques choices. WSEAS Transactions on Computers 7, 1319–1336. http://wseas.org/cms.action?id=4026; http://www.wseas.us/e-library/transactions/computers/2008/27-646.pdf.

Li, P., Miao, X., Li, R., Zhong, J., 2011. In situ biodiesel production from fast-growing and high oil content *Chlorella pyrenoidosa* in rice straw hydrolysate. Journal of Biomedicine and Biotechnology 1–8.

Li, Y.H., Liu, B., Sun, Y., Zhao, Z.B., Bai, F.W., 2005. Screening of oleaginous yeast for broad spectrum carbohydrates assimilating capacity. Chinese Journal of Biotechnology 25, 39–44.

Li, Y.H., Liu, B., Zhao, Z.B., Bai, F.W., 2006. Optimization of culture conditions for lipid production by *Rhodosporidium toruloides*. Chinese Journal of Biotechnology 22, 650–656.

Li, Z.F., Ling, Z., Shen, X.J., Lai, B.S., Sun, S.Q., Wei, S., Wu, X., Tong, B., 2001. A comparative study on four method of fungi lipid extraction. Microbiology 28, 72–76 (In Chinese).

Lin, J.T., Shen, H.W., Zhang, Z.H., Hu, C.M., Jin, G.J., Tan, H.D., Zhao, Z.B., 2010. Microbial lipid production by *Rhodosporidium toruloides* in a two-stage culture mode. Chinese Journal of Biotechnology 26, 997–1002.

Ling, J., Nip, S., Shim, H., 2013a. Production of microbial lipid from food industry wastewater by *Rhodosporidium toruloides*. In: Proceedings of the 2013 International Conference on Materials Science and Environmental Engineering (MSEE 2013). Wuhan, China, August 17–18, pp. 171–177.

Ling, J., Nip, S., Shim, H., 2013b. Enhancement of lipid productivity of *Rhodosporidium toruloides* in distillery wastewater by increasing cell density. Bioresource Technology 146, 301–309.

Ling, J., Nip, S., de Toledo, R.A., Shim, H., 2014a. Optimization of initial cell density and usage efficiency of seed culture medium during lipid production from distillery wastewater by *Rhodosporidium toruloides*. Renewable Energies and Power Quality Journal 12.

Ling, J., Nip, S., Cheok, W.L., de Toledo, R.A., Shim, H., 2014b. Lipid production by a mixed culture of oleaginous yeast and microalga from distillery and domestic mixed wastewater. Bioresource Technology 173, 132–139.

Ling, J., Nip, S., de Toledo, R.A., Tian, Y., Shim, H., 2015. Evaluation of specific lipid production and nutrients removal from wastewater by *Rhodosporidium toruloides* and biodiesel production from wet biomass via microwave irradiation. Energy. http://dx.doi.org/10.1016/j.energy.2015.05.141 (in press).

Liu, J.X., Yue, Q.Y., Gao, B.Y., Wang, Y., Li, Q., Zhang, P.D., 2013. Research on microbial lipid production from potato starch wastewater as culture medium by *Lipomyces starkeyi*. Water Science and Technology 67, 1802–1808.

Lovrics, A., Csikasz-Nagy, A., Zsely, I.G., Zador, J., Turanyi, T., Novak, B., 2006. Time scale and dimension analysis of a budding yeast cell cycle model. BMC Bioinformatics 7, 494–504.

Miller, G.L., 1959. Use of dinitrosalicylic acid reagent for determination of reducing sugar. Analytical Chemistry 31, 426–428.

Mutanda, T., Ramesh, D., Karthikeyan, S., Kumari, S., Anandraj, A., Bux, F., 2011. Bioprospecting for hyper-lipid producing microalgal strains for sustainable biofuel production. Bioresource Technology 102, 57–70.

Olabi, A.G., 2013. State of the art on renewable and sustainable energy. Energy 61, 2–5.

Olabi, A.G., 2014. 100% sustainable energy. Energy 77, 1–5.

Peng, W.F., Huang, C., Chen, X.F., Xiong, L., Chen, X.D., Chen, Y., Ma, L.L., 2013. Microbial conversion of wastewater from butanol fermentation to microbial oil by oleaginous yeast *Trichosporon dermatis*. Renewable Energy 55, 31–34.

Pinto, G.A., Durao, F.O., Fiuza, A.M.A., Guimaraes, M.M.B.L., Madureira, C.M.N., 2004. Design optimization study of solvent extraction, chemical reaction, mass transfer and mixer-settler hydrodynamics. Hydrometallurgy 74, 131–147.

Raheem, A., Wan Azlina, W.A.K.G., Taufiq Yap, Y.H., Danquah, M.K., Harun, R., 2015. Thermochemical conversion of microalgal biomass for biofuel production. Renewable and Sustainable Energy Reviews 49, 990–999.

Rao, Y.R., Zubaidha, P.K., Reddy, J.N., Kondhare, D.D., Sushma, D.S., 2012. Crystalline manganese carbonate a green catalyst for biodiesel production. Green and Sustainable Chemistry 2, 14–20.

Raupach-Sumiya, J., Matsubara, H., Prahl, A., Aretz, A., Salecki, S., 2015. Regional economic effects of renewable energies-comparing Germany and Japan. Energy, Sustainability and Society 5, 1–17.

Ren, H.Y., Liu, B.F., Kong, F.Y., Zhao, L., Xing, D.F., Ren, N.Q., 2014. Enhanced energy conversion efficiency from high strength synthetic organic wastewater by sequential dark fermentative hydrogen production and algal lipid accumulation. Bioresource Technology 157, 355–359.

Saayman, M., Viljoen-Bloom, M., 2006. The biochemistry of malic acid metabolism by wine yeasts – a review. South African Journal for Enology and Viticulture 27, 113–122.

Santamauro, F., Whiffin, F.M., Scott, R.J., Chuck, C.J., 2014. Low-cost lipid production by an oleaginous yeast cultured in non-sterile conditions using model waste resources. Biotechnology for Biofuels 7, 34–43.

Sheehan, J., Dunahay, T., Benemann, J., Roessler, P., 1998. A Look Back at the U.S. Department of Energy's Aquatic Species Program—Biodiesel from Algae. National Renewable Energy Laboratory. NREL/TP-580–24190.

Sherman, F., 2002. Getting started with yeast. Methods in Enzymology 350, 3–41.

Shields, S.A., Marta Amirsadeghi, M., Sukhbaatar, B., Revellame, E., Hernandez, R., Donaldson, J.R., French, W.T., 2015. Lipid accumulation by *Rhodococcus rhodochrous* grown on glucose. Journal of Industrial Microbiology and Biotechnology 42, 693–699.

Shun, N., Yun, Z., 2011. Analysis of Fatty Acids in Infant Formulas Using an Agilent J&W HP-88 Capillary GC Column. Application Note. Agilent Technologies, Beijing, 5990-8429EN2011. http://www.chem.agilent.com/Library/applications/5990-8429EN.pdf (accessed 12.07.14.).

Song, L., Liu, Y., Gan, X., Zhu, Y., Yu, M., 1999. FACHB-collection, general introduction and the strain list. Acta Hydrobiologica Sinica 23 (Suppl.), 537–546.

Su, H.Y., Zhang, Y.L., Zhang, C.M., Zhou, X.F., Li, J.P., 2011. Cultivation of *Chlorella pyrenoidosa* in soybean processing wastewater. Bioresource Technology 102, 9884–9890.

Subramaniam, R., Dufreche, S., Zappi, M., Bajpai, R., 2010. Microbial lipids from renewable resources, production and characterization. Journal of Industrial Microbiology and Biotechnology 37, 1271–1287.

Whiting, G.C., 1976. Organic acid metabolism of yeasts during fermentation of alcoholic beverages – a review. Journal of the Institute of Brewing 82, 84–92.

Wu, S.G., Hu, C.M., Jin, G.J., Zhao, X., Zhao, K.Z.B., 2010. Phosphate-limitation mediated lipid production by *Rhodosporidium toruloides*. Bioresource Technology 101, 6124–6129.

Xue, F.Y., Miao, J.X., Zhang, X., Tan, T.W., 2010. A new strategy for lipid production by mix cultivation of *Spirulina platensis* and *Rhodotorula glutinis*. Applied Biochemistry and Biotechnology 160, 498–503.

Yen, H.W., Chan, J.T., Chang, J.S., 2015. The growth of oleaginous *Rhodotorula glutinis* in an internal-loop airlift bioreactor by using mixture substrates of rice straw hydrolysate and crude glycerol. Biomass and Bioenergy 80, 38–43.

Zhang, Z., Ji, H., Gong, G., Zhang, X., Tan, T., 2014a. Synergistic effects of oleaginous yeast *Rhodotorula glutinis* and microalga *Chlorella vulgaris* for enhancement of biomass and lipid yields. Bioresource Technology 164, 93–99.

Zhang, X., Yan, S., Tyagi, R.D., Surampalli, R.Y., Valéro, J.R., 2014b. Wastewater sludge as raw material for microbial oils production. Applied Energy 135, 192–201.

Zhao, K., Xue, P.J., Gu, G.Y., 2008. Study on determination of reducing sugar content using 3,5-dinitrosalicylic acid method. Food Science 29, 534–536.

Zhao, X., Hu, C.M., Wu, S.G., Shen, H.W., Zhao, Z.K., 2011. Lipid production by *Rhodosporidium toruloides* Y_4 using different substrate feeding strategies. Journal of Industrial Microbiology and Biotechnology 38, 627–632.

Zhao, X.B., Peng, F., Du, W., Liu, C.M., Liu, D.H., 2012. Effects of some inhibitors on the growth and lipid accumulation of oleaginous yeast *Rhodosporidium toruloides* and preparation of biodiesel by enzymatic transesterification of the lipid. Bioprocess and Biosystems Engineering 35, 993–1004.

Zhou, W.W., Zhang, Y.K., Zhou, W.T., Li, Y.H., 2010. Production of lipid from distillery wastewater by *Rhodosporidium toruloides*. Chinese Journal of Chemical Industry and Engineering Progress 29, 172–175 (In Chinese).

Zhou, W.W., Wang, W.R., Li, Y.H., Zhang, Y.K., 2013. Lipid production by *Rhodosporidium toruloides* Y2 in bioethanol wastewater and evaluation of biomass energetic yield. Bioresource Technology 127, 435–440.

UTILIZATION OF SLUDGE AS MANURE

9

V. Matichenkov[1], E. Bocharnikova[2]

Institute of Basic Biological Problems, Pushchino, Russia[1]; Institute of Physical-Chemical and Biological Problems in Soil Science, Pushchino, Russia[2]

CHAPTER OUTLINE

9.1 INTRODUCTION

Organization of the use of organic wastes in large cities has been realized for a long time. The main part of this organic material was transported out of cities, naturally composted, and used in agriculture (Ju et al., 2005). The use of organic wastes is closely related to the water supply. The idea of coagulating organic wastes using aluminum was discovered in Egypt after 1500 BCE (Crittenden et al., 2012). In Europe, sewage was used for irrigation in Prussia from the 16th to the 19th centuries, and sewage farms existed in England and outside Paris over 100 years ago. In 1804 the first actual municipal water treatment plant, designed by Robert Thom, was built in Scotland (Crittenden et al., 2012). In 1854 it was discovered that a cholera epidemic spread through water. The outbreak seemed less severe in areas where sand filters were installed. British scientist John Snow found that the direct cause of the outbreak was water pump contamination by sewage water. He applied chlorine to purify the water, which this paved the way for water disinfection (Crittenden et al., 2012).

At first, wastes from water treatment plants in Europe, the United States, and Australia were disposal in agricultural areas. However, industrialization and chemicalization amplified the level of the

Environmental Materials and Waste. http://dx.doi.org/10.1016/B978-0-12-803837-6.00009-3

pollutants in the city waste-water, which together with extremely high speed of population growing increases the risk for sewage contamination. In the result this waste without special treatment cannot be safety disposal to cultivated area.

Today municipal waste (solid, liquid, sludge, etc.) from water treatment plants usually has the following problems: metal contamination (Alvarez et al., 2002), including Al, because aluminum is used for organic matter precipitation (Xu et al., 2009); pathogen microorganisms (Wéry et al., 2008); and an unpleasant odor. As the result, numerous countries restrict the use of direct municipal wastes via agriculture. However, this type of the waste contains a high content of nutrients which are needed to increase the yield and sustainability of agriculture (Bhatnagar and Sillanpää, 2010). The use of this type of solid waste can provide an additional source for manufacturing highly effective organic fertilizers for agriculture and for greenery in large cities.

The use of active forms of Si, such as monosilicic and polysilicic acid can dramatically reduce the toxicity of Al and heavy metals (Exley et al., 2002; Matichenkov and Calvert, 2002; Matichenkov et al., 2013). This process is usually related to the formation of insoluble silicates of heavy metals (Matichenkov and Bocharnikova, 2001). Disinfection of the pathogen of microorganisms can be realized by using of several types of industrial by-products (Demin et al., 2006). Usually this process requires the use of by-products which contain metals or organic substances which are characterized as poisons for pathogenic microorganisms. As the result, such a procedure may have a negative influence on the environment. However, the combination of waste uses may provide an algorithm for the complete use of several types of by-products and ultimately make an environmentally friendly product (Matichenkov, 2008).

The main aim of this investigation was to determine the possibility of neutralizing and disinfecting municipal water treatment waste and obtaining an ecologically safe organic fertilizer or soil amendment.

9.2 MATERIALS AND METHODS

Sludge from the Santiago water treatment plant "La-Farfana" was used in the technology demonstration (Figs 9.1 and 9.2). The sludge contained 67% C (from dry material), pH 7.6, 25% moisture. Three steps were used to demonstrated the technology for treating sludge from the water treatment plant:

1. Preliminary test for disinfecting the sludge
2. Complete pathogen microorganisms and intestinal worm disinfection test of sludge
3. Complete pathogen microorganisms and intestinal worm disinfection test with complete Al and heavy metals detoxification test

During the preliminary disinfection test, tannery wastewater was used to disinfect the sludge (Demin et al., 2006). The sample of the wastewater was collected from a leather factory (Serpukhov, Moscow region, Russia), which contained a high level of soluble Cu (200 ± 5 mg L^{-1}) and organic substances (C, $1.5 \pm 0.3\%$). It was suggested that a specific amino acid–Cu complex is formed in this wastewater (Demin et al., 2006). This wastewater was used to disinfect the sludge from Chile. The following rates for amino acid–Cu substance were used: 2, 2.25, 2.5, 3, 3.5, 4, 5, 5.5, and 6 mmol kg^{-1} Cu to determine the pathogen microorganisms, and 3, 5, 6, 7, 7.25, 7.5, 7.75, 8, and 8.5 mmol kg^{-1} Cu to determine the intestinal worm eggs (helminth). Different concentrations for the amino acid–Cu complex were selected according to preliminary studies (Demin et al., 2006). The contents of the

FIGURE 9.1

Water treatment plant "La-Farfana," Santiago, Chile.

pathogen microorganisms and eggs of the intestinal worms were tested using the serological methods of diagnostics (Vainionpaa et al., 2008).

The test of complete disinfection of pathogen microorganisms and intestinal worms (helminth) from the sludge was conducted using 0.1%, 0.5% and 1% solution with amino acid–Cu complex. Disinfection of the sludge from Chile was conducted for 1 h at 24°C and 25% moisture. After incubation, the standard sanitary test for the availability of pathogenic microorganism colonies was conducted in the certified sanitary laboratory using serological diagnostic methods (Pushchino, Moscow region, Russian Federation).

The third test was conducted using complete pathogenic microorganisms and intestinal worm disinfection, and complete Al and heavy metal detoxification. Concentrated monosilicic acid was used

FIGURE 9.2

Temporary storage of sludge from water treatment plant processing before transportation to landfill, "La-Farfana," Santiago, Chile.

as a source of the active form of Si for Al and heavy metal detoxification. This material is produced by TerraThech, Inc. (Miami, Florida, United States). The colloidal Si had a pH of 12 and the content of the monosilicic acid was 15% Si. Three different combinations of the complete technology were tested: 0.1:1 (technology 1), 0.5:5 (technology 2), and 1:5 (technology 3), in which the numbers are the percentage of tannery wastewater (first number) and of colloidal Si (second number), which mix with sludge.

The sludge from Santiago water treatment plant was processed using several steps. The first step included the procedure, which described in the Russian patent, in which sludge was treated with tannery wastewater (Demin et al., 2006). This procedure makes it possible to eliminate the pathogenic microorganisms and eggs of the worms. After this, colloidal Si was used for the next stage of processing. The mixture of disinfected sludge and colloidal Si was incubated for 1 week at temperature 20°C to 24°C and 25–30% moisture. After incubation, the samples were collected for sanitary investigation. Other parts of the mixture were dried at 50°C to 60°C for 2 days. The dried material was ground, passed through a 0.2-mm sieve, and used in the greenhouse test.

Standard analysis of the pathogenic microorganisms and eggs of worms was conducted in a certified sanitary laboratory in Pushchino (Russia). A greenhouse test with barley (*Hordeum vulgare* L.) was conducted using washed quartz sand as a growth medium. Untreated and treated sludge (three different technologies, described previously) were used in the greenhouse test at 5 g kg^{-1}, which is equal to 5 ton ha^{-1}. The air temperature in the growth chamber was kept at 20 ± 2°C during daytime and 12 ± 2°C during nighttime. The light period was 12 h; light intensity was 950 μmol photons $m^{-2}\ s^{-1}$. The relative air humidity was 45 ± 5% during day and 70 ± 5% during night. Barley harvesting was conducted 1 month after seedlings sprouted. The biomass of roots and shoots was measured. The total content of the Al, Si, Cu, and Cd was tested in the shoots of the harvested barley using microwave digestion (Pougnet et al., 1992) and then using an ICP OES spectrometer (Perkin Elmer Optima 5300 DV). Each treatment had three replications.

9.3 RESULTS

9.3.1 FIRST TEST

The dynamics of the disinfection process is presented in Fig. 9.3. The data confirmed that the pathogenic microorganisms were less stable than the eggs of the intestinal worms. All pathogenic microorganisms were shown under application of 6 mmol kg^{-1} Cu as amino acid–Cu complex, whereas this concentration of Cu reduced the content of intestinal worm eggs to 20%.

9.3.2 SECOND TEST

The following groups of pathogen microorganisms were tested: eggs of helminth, intestinal bacillus (*Escherichia coli*), *Streptococcus*, *Micrococcus*, *Diplococcus*, *Staphylococcus Trichomonas*, intestinal amoeba, and *Clostridium perfringens* (Table 9.1). Using 0.5% and 1% of the weight of the sludge, all pathogenic microorganisms were completely eliminated. These results show that the technology tested can be used to completely disinfect municipal wastes. The treated sludge has no smell, which can be explained by the complete disinfection procedure.

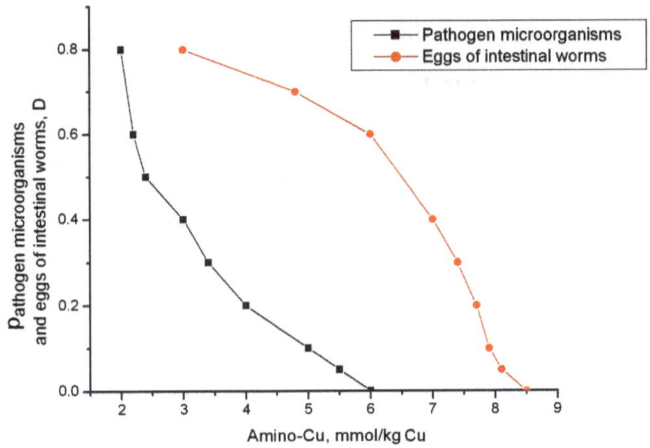

FIGURE 9.3

Effect of amino acid-Cu on the concentration of living pathogenic microorganisms and eggs of intestinal worms in treated municipal waste.

Table 9.1 Availability of Pathogenic Microorganism Colonies in the Control and Treated Municipal Sludge

		Solution With Amino Acid–Cu Complex		
Pathogen	**Control**	**0.1%**	**0.5%**	**1%**
Eggs of helminth	+++	++	–	–
Intestinal bacillus	+++	++	–	–
Streptococcus	+	++	–	–
Micrococcus	+++	–	–	–
Diplococcus	+++	+	–	–
Staphylococcus	++	+	–	–
Trichomonas	++	–	–	–
Intestinal amoeba	++	+	–	–
Clostridium perfringens	++	+	–	–

+++, many colonies of microorganisms; ++, medium colonies of microorganisms; +, low colonies of microorganisms; –, no colonies of microorganisms.

9.3.3 THIRD TEST

The greenhouse test showed that the using the technology to treat the sludge from water treatment plants significantly increased the biomass of the barley (Table 9.2). The application of disinfected untreated sludge increased the plant biomass 20% and 16% for roots and shoots, respectively. Treatment of the sludge by the suggested technologies, including both the disinfection and detoxification processes, increased the root biomass to 80% and the green part of the barley to 61%. Use of untreated disinfected sludge dramatically increased the total content of Al, Cu, and Cd in the leaves of the barley. Treatment of sludge by colloidal Si reduced the total content of Al, Cu, and Cd in leaves. By

Table 9.2 Biomass of Barley Roots and Shoots and Total Content of Al, Si, Cu, and Cd in Barley Leaves

	Treatment				
Parameter	Control (Sand)	Disinfected Untreated Sludge	Technology 1	Technology 2	Technology 3
Dry weight of 10 plant roots, g	0.15 ± 0.02	0.18 ± 0.02	0.22 ± 0.02	0.27 ± 0.02	0.26 ± 0.02
Dry weight of 10 plants shoots, g	0.18 ± 0.02	0.21 ± 0.02	0.25 ± 0.02	0.29 ± 0.02	0.29 ± 0.02
Total Al in leaves, %	0.02 ± 0.01	1.34 ± 0.10	0.12 ± 0.02	0.05 ± 0.01	0.06 ± 0.01
Total Si in leaves, %	0.67 ± 0.05	0.84 ± 0.07	1.19 ± 0.10	1.28 ± 0.08	1.35 ± 0.04
Total Cu in leaves, mg kg^{-1}	0.00	34.5 ± 2.5	3.5 ± 0.4	2.2 ± 0.2	2.3 ± 0.2
Total Cd in leaves, mg kg^{-1}	0.00	29.5 ± 1.3	0.54 ± 0.02	0.28 ± 0.02	0.18 ± 0.03

this means, the use of untreated disinfected sludge from municipal water treatment plants can protect plants against Al and heavy metal pollution.

9.3.4 DISCUSSION

The sludge smell is caused by microbial activity. Therefore, the unpleasant odor can be removed by a complete disinfection process. Use of other local by-products such as amino acid–Cu complexes from the leather industry makes it possible to employ both types of waste (from the leather industry and water treatment plants). However, such technology must be used carefully because the risk of secondary pollution by heavy metals is high. Over the past decade, several alternatives to water or suspension (including sludge) disinfection have been suggested. One is cavitation (Jyoti and Pandit, 2001; Badve et al., 2015). Officially, the cavitation process is the formation of vapor cavities in a liquid. It usually occurs when a liquid is subjected to rapid changes in pressure that cause the formation of cavities where the pressure is relatively low (Li et al., 2015). These microexplosives can physically crush and kill all microbes and worms in solution or sludge. By this means, the treated sludge will be completely disinfected and will have no unpleasant smell. However, the Al and heavy metal activities in the sludge will not be changed.

The use of concentrated monosilicic acid results in the precipitation and detoxification of Al and heavy metals (Matichenkov et al., 2013). The results obtained completely support this hypothesis. The following reactions can be initiated by monosilicic acid:

$$2Al^{3+} + 2H_4SiO_4 = Al_2Si_2O_5\downarrow + 2H^+ + 3H_2O$$

$$H_4SiO_4 + Cu^{2+} = CuSiO_4\downarrow + 4H^+$$

$$H_4SiO_4 + Pb^{2+} = PbSiO_4\downarrow + 4H^+$$

$$H_4SiO_4 + 2Cd(OH)_2 = Cd_2(SiO_4)\downarrow + 4H_2O$$

$$H_4SiO_4 + Cd^{2+} = CdSiO_3\downarrow + H_2O + H^+$$

$$Al_2O_3\cdot 2SiO_2 + CdCl_2 + H_2O = CdOAl_2O_32SiO_2\downarrow + 2HCl$$

The new formed silicates are characterized by extremely low solubility, which make Al and heavy metals immobile.

The Si-based technology that was tested and elaborated upon can be used for municipal wastes and to manufacture highly effective, environmentally friendly products for agriculture and greenery in large cities, and for the bioremediation of contaminated areas. This technology contain several stages: (1) removal of pathogenic microorganisms, which can be realized using special substances (specific industrial wastes); (2) detoxification of Al and heavy metals by active Si-rich substances; and (3) a granulation or composting process. Prospective investigations with hydrodynamic cavitation may modify this technology and make it a more effective and safe product.

9.4 CONCLUSIONS

The complex technology for the use of Al- and heavy metals—contaminated sludge from water treatment plants was elaborated upon and tested. The technology has several steps, including disinfection of municipal waste and then detoxification of Al and heavy metals. The final product has high efficiency as a fertilizer or complex organic fertilizer. The technology can be adapted to any conditions.

REFERENCES

Alvarez, E.A., Mochón, M.C., Sánchez, J.J., Rodríguez, M.T., 2002. Heavy metal extractable forms in sludge from wastewater treatment plants. Chemosphere 47 (7), 765—775.

Badve, M.P., Bhagat, M.N., Pandit, A.B., 2015. Microbial disinfection of seawater using hydrodynamic cavitation. Separation and Purification Technology 151 (4), 31—38.

Bhatnagar, A., Sillanpää, M., 2010. Utilization of agro-industrial and municipal waste materials as potential adsorbents for water treatment—a review. Chemical Engineering Journal 157 (2), 277—296.

Crittenden, J.C., Trussell, R.R., Hand, D.W., Howe, K.J., Tchobanoglous, G., 2012. MWH's Water Treatment: Principles and Design. John Wiley & Sons.

Demin, D.V., Sevostianov, S.M., Tatarkin, I.V., 2006. The Method for Waste-Water Treatment. Russian patent 2457909.

Exley, C., Schneider, C., Doucet, F.J., 2002. The reaction of aluminium with silicic acid in acidic solution: an important mechanism in controlling the biological availability of aluminium? Coordination Chemistry Reviews 228, 127—135.

Li, P., Song, Y., Yu, S., Park, H., 2015. The effect of hydrodynamic cavitation on *Microcystis aeruginosa*. Physical and chemical factors. Chemosphere 136, 245—251.

Ju, X., Zhang, F., Bao, X., Romheld, V., Roelcke, M., 2005. Utilization and management of organic wastes in Chinese agriculture: past, present and perspectives. Science in China Series C: Life Sciences 48 (2), 965—979. http://dx.doi.org/10.1007/BF03187135.

Jyoti, K.K., Pandit, A.B., 2001. Water disinfection by acoustic and hydrodynamic cavitation. Biochemical Engineering Journal 7 (3), 201—212.

Matichenkov, V.V., 2008. The Role of Soluble Silicon Compounds in Plant and Soil-Plant System (Doctoral dissertation). Institute Basic Biological Problems Russian Academy of Sciences, Pushchino.

Matichenkov, V.V., Bocharnikova, E.A., 2001. The relationship between silicon and soil physical and chemical properties. Studies in Plant Science 8, 209–219.

Matichenkov, V.V., Calvert, D.V., 2002. Silicon as a beneficial element for sugarcane. Journal American Society of Sugarcane Technologists 22 (2), 21–30.

Matichenkov, V.V., Wei, X., Liu, D., Bocharnikova, E.A., 2013. Theory, practice and prospection of Si fertilizer. Agricultural Science & Technology 14 (3), 498–502.

Pougnet, M.A.B., Schnautz, N.G., Walker, A.M., 1992. Development and application of a microwave digestion pressure vessel. South African Journal of Chemistry 25, 86.

Vainionpaa, R., Waris, M., Leinikki, P., 2008. Diagnostic techniques: serological and molecular approaches. Encyclopedia of Virology 29–37.

Wéry, N., Lhoutellier, C., Ducray, F., Delgenès, J.P., Godon, J.J., 2008. Behavior of pathogenic and indicator bacteria during urban wastewater treatment and sludge composting, as revealed by quantitative PCR. Water Research 42 (1), 53–62.

Xu, G.R., Yan, Z.C., Wang, Y.C., Wang, N., 2009. Recycle of alum recovered from water treatment sludge in chemically enhanced primary treatment. Journal of Hazardous Materials 161 (2), 663–669.

CHAPTER

10 ENERGY AND RESOURCE RECOVERY FROM SLUDGE: FULL-SCALE EXPERIENCES

V.K. Tyagi[1], **S.-L. Lo**[2]

The Energy and Resources Institute (TERI) University, New Delhi, India[1]*;*
National Taiwan University, Taipei, Taiwan, Republic of China (ROC)[2]

CHAPTER OUTLINE

Environmental Materials and Waste. http://dx.doi.org/10.1016/B978-0-12-803837-6.00010-X

10.1 INTRODUCTION

> Historically, it was common to see the schematics that showed the water treatment scheme in detail and an arrow at the end that simply said "sludge to disposal."
>
> **Neyens et al. (2004)**

In the current scenario, the growing global urbanization of society coupled with increasingly stringent sludge reuse and disposal regulations and increasing public pressure are forcing both public and private sludge generators to reevaluate sludge management strategies (Liu and Tay, 2001). Conventionally, waste sludge is disposed via incineration, landfills or ocean disposal or is reused as a soil conditioner in agriculture. With the banning of ocean disposal and new stringent European landfill criteria, much more sludge is now beneficially reused both in agriculture and via a variety of thermal technologies (Matthews, 1996). The selection of a sludge management strategy is of interest to a wide variety of groups including facility owners, engineering consultants, contract operators, equipment suppliers, politicians, regulators, environmental groups, and the general public. Selection of a sludge management strategy based on the actual needs of the community rather than on some artificial set of criteria is probably the single most important component in achieving long-term sustainability (Campbell, 2000). It is anticipated that upcoming sludge management efforts will accentuate the recovery and reuse of value-added crops from sludge (Rulkens, 2008). This interest in renewable energy has been driven by a combination of shrinking fossil fuel reserves caused by the rising demand for primary energy, fuel price spikes, climate change concerns, public awareness, and advancements in renewable energy technologies (Cao and Pawłowski, 2012; NACWA, 2010).

The two components in sludge that are technically and economically feasible to recycle are nutrients (primarily nitrogen and phosphorus) and energy (carbon) (Campbell, 2000). Several options are available for energy recovery from waste sludge. The most significant routes are anaerobic digestion of sludge with biogas recovery, co-digestion, incineration, and co-incineration with energy recovery, pyrolysis, gasification, supercritical (wet) oxidation, used in the production of construction materials, production of biofuels (hydrogen, syngas, and bio-oil), electricity generated using specific microbes, and the beneficial recovery of heavy metals, nutrient (nitrogen and phosphorus), protein, and enzymes (Tyagi and Lo, 2013).

10.2 SLUDGE CHARACTERIZATION

The sewage sludge is composed of a complex heterogeneous mixture of microorganisms, undigested organics such as paper, plant residues, oils, or fecal material, inorganic materials, and moisture (Degremont, 1979). The characteristics of primary and secondary activated sludge are summarized in Table 10.1.

Primary sludge is produced through a mechanical (screening, grit removal, and sedimentation) wastewater treatment process; it usually contains from 93% to 99.5% water and high ratios of suspended and dissolved organic matters. Waste-activated sludge (WAS), or secondary sludge, is produced during the biological treatment of wastewater and is composed mainly of microbial cells that are complex polymeric organic materials. The total solids (TS) concentration in secondary sludge ranges between 0.8% and 1.2%, which also depends on the type of biological treatment process employed (Tezel et al., 2011).

Table 10.1 Characteristics of Primary Sludge and Activated Sludge (Tchobanoglous et al., 2003)

Parameter	Primary Sludge	Activated Sludge	Composition
Total dry solids (total solids (TS)) %	5–9	0.8–1.2	• Nontoxic organic carbon compounds (approximately 60% on dry basis), Kjeldahl-N, phosphorus-containing components • Toxic pollutants: heavy metals (Zn, Pb, Cu, Cr, Ni, Cd, Hg, As: concentrations vary from 1000 mg/L to <1 mg/L), polychlorinated biphenyls, polyaromatic hydrocarbons, dioxins, pesticides, endocrine disrupters, nonyl-phenols • Pathogens and other microbiological pollutants • Inorganic compounds: silicates, aluminates, calcium, and magnesium-containing compounds • Water varying from a few percent to >95%
Volatile solids (VS) (% TS)	60–80	59–68	
Nitrogen (% TS)	1.5–4	2.4–5.0	
Phosphorus (% TS)	0.8–2.8	0.5–0.7	
Potash (K_2O %TS)	0–1	0.5–0.7	
Cellulose (%TS)	8–15	7–9.7	
Iron (Fe g/kg)	2–4	–	
Silica (SiO_2%TS)	15–20	–	
pH	5.0–8.0	6.5–8.0	
Grease and fats (%TS)	7–35	5–12	
Protein (%TS)	20–30	32–41	
Alkalinity (mg/L as $CaCO_3$)	500–1500	580–1100	
Organic acids (mg/L as acetate)	200–2000	1100–1700	
Energy content (kJ/kg TS)	23,000–29,000	19,000–23,000	

WAS contains 59–88% (w/v) organic matter, which is decomposable and produces unpleasant odors. Only a small part of the sludge is solid matter in which over 95% is water. The organic portion contains 50–55% carbon, 25–30% oxygen, 10–15% nitrogen, 6–10% hydrogen, 1–3% phosphorus, and 0.5–1.5% sulfur (Orhon and Artan, 1994). The ash from waste sludge contains mainly minerals such as quartz, calcite, or microcline. These minerals are formed by elements such as Fe, Ca, K, and Mg. Furthermore, some heavy metals such as Cr, Ni, Cu, Zn, Pb, Cd, and Hg can also be found in sludge (Fonts et al., 2009).

The potential for energy recovery from sludge is a function of their composition, which is a mixture of organic (volatile) matter, inorganic matter (inert material), and associated water. The energy content of sludge is laid in the volatile solids (VS), which is subdivided into two components: readily degradable (50% in primary sludge and 25% in WAS) and not readily degradable (30% in primary sludge and 55% in WAS) (NACWA, 2010). Considering the energy values, 1 lb of dry biosolids has the energy content of 6000–9000 British Thermal Units.

10.3 METHODS FOR ENERGY AND RESOURCE RECOVERY

By using the thermochemical, biochemical, and mechanical-chemical methods, several sludge-derived resources including biogas, fuel gas, electricity generation, production of construction materials, nutrients, biofuels (syngas, biodiesel, and bio-oil), hydrolytic enzymes, polyhydroxyalkanoates (PHA) (for bioplastic manufacturing), and biofertilizers, biosorbents can be recovered. Fig. 10.1 shows the techniques of energy and resource recovery from waste sludge.

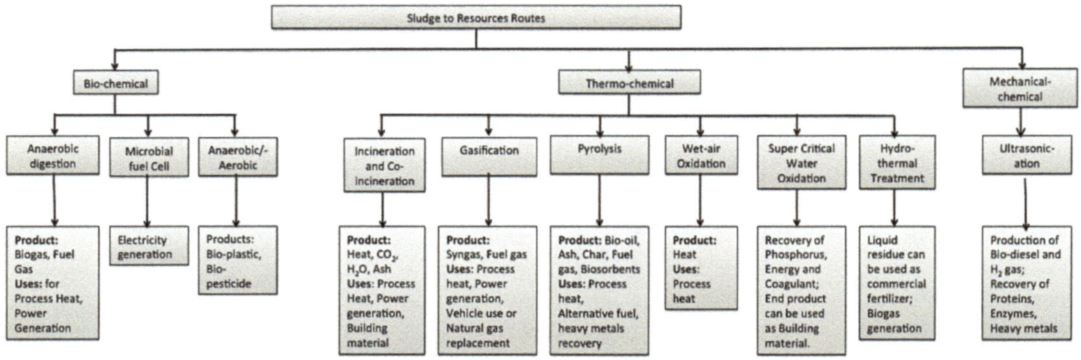

FIGURE 10.1

Routes of resource recovery from waste sludge (Tyagi and Lo, 2013).

10.3.1 ANAEROBIC DIGESTION

Anaerobic digestion is the most popular sludge stabilization technology currently on the market (Cao and Pawłowski, 2012). The process transforms sludge organic solids to biogas, which is a mixture of CH_4, CO_2, and traces of other gases, in an anaerobic environment (Reaction [1]) (Tezel et al., 2011):

$$C_cH_hO_oN_nS_s + yH_2O \rightarrow xCH_4 + nNH_3 + xH_2S + (c - x)CO_2$$
$$x = 1/8(4c + h - 2o - 3n - 2s) \qquad [10.1]$$
$$y = 1/4(4c - +h - 2o + 3n + 3s)$$

Biogas can be used as a source of energy to produce electricity and/or heat.

10.3.2 INCINERATION AND CO-INCINERATION

The main purpose of sludge incineration is the complete oxidation of organic compounds at a high temperature. In this process, biosolids are burned in a combustion chamber supplied with excess air (oxygen) to form mainly carbon dioxide and water, leaving only inert material (ash). This ash has to be disposed of or it can be used as a source to produce building materials (Guibelin, 2004).

10.3.3 GASIFICATION

Gasification involves the breakdown of dried sludge in an ash and in combustible gases at temperatures usually about 1000°C in an atmosphere with a reduced amount of oxygen (Jaeger and Mayer, 2000). Products of the process include heat (used to generate power and process heat) and syngas (synthetic gas). Gasification takes place at a high temperature (1400 to 1700°C) and high pressure (0.6−2.6 MPa) using pure oxygen as the oxidant (Spinosa, 2004).

10.3.4 PYROLYSIS

Sludge pyrolysis is an innovative method developed to manage sludge and energy, in which the sludge is thermally treated (350–500°C) under pressure and in an oxygen-deficient environment. In this method, the sludge is transformed into char, ash, pyrolysis oils, water vapor, and combustible gases (Stolarek and Ledakowicz, 2001; Khiari et al., 2004).

10.3.5 WET AIR OXIDATION

Wet oxidation is the chemical oxidation of sludge (by the addition of O_2) at a high temperature (150–330°C) and high pressure (6–20 MPa). The ZIMPRO process was the oldest process based on wet oxidation technology, developed in The Netherlands in the 1960s. The oxidation of sludge's organic matter produces water, carbon dioxide, and easily biodegradable organic compounds (acetic acid and fatty acid) (Weemaes and Verstraete, 1998).

10.3.6 SUPERCRITICAL WATER OXIDATION

Supercritical water oxidation (SCWO), which takes place at elevated temperatures and pressures (typically 25 MPa and 600°C), is a good solution for sludge disintegration. In this process, carbon and hydrogen from organic and biological constituents are oxidized to CO_2 and H_2O; nitrogen, sulfur, and phosphorus form N_2, SO_4^{2-}, and PO_4^{3-}, respectively; organic chlorides are converted to Cl^-; and heavy metals are oxidized to the corresponding oxides (Svanstrom et al., 2007). Although the cost of treatment is high, the value from the SCWO method in terms of sludge volume decrease together in excess of 90% recovery of energy, coagulants, and phosphate represent a value which will offset the cost of operation (Perez-Elvira et al., 2006).

10.3.7 HYDROTHERMAL TREATMENT

Hydrothermal treatment involves heating sludge in the water phase at temperatures between 150 and 450°C in the absence of oxygen or another oxidant. Sludge hydrolysis during hydrothermal treatment results in the production and accumulation of high concentrations of dissolved organic compounds in the liquid phase (Shanableh and Jomaa, 2001). Protein hydrolysis results in amino acids, lipid hydrolysis produces fatty acids, and the hydrolysis of fiber material and hydrocarbons leads to low-molecular hydrocarbon substances such as sugars. These compounds are particularly interesting as carbon resource not only in the production of biogas but also in the denitrification process and the biological removal of phosphorus from wastewater (Rulkens and Bien, 2004). The liquid residue of hydrothermally treated sewage sludge can be used as a marketable fertilizer because it is composed of three principal nutrients (N, P, and K) (Yoshikawa, 2010).

10.4 ENERGY AND RESOURCE RECOVERY

10.4.1 BIOGAS RECOVERY BY ANAEROBIC DIGESTION

Biogas produced during the anaerobic digestion of sewage sludge contains 60–70% methane, 30–40% carbon dioxide, and trace amounts of nitrogen, hydrogen, hydrogen sulfide, and water vapor.

Methane gas generated by anaerobic digesters is the main source of energy at a municipal wastewater treatment plant (WWTP). In most cases, the recovered methane is used to power gas engines or produce electrical and thermal energy for onsite use in the treatment plant. The cost of electricity for a treatment plant is about 80% of the total operational cost, and energy recovered through methane can cover about half of this cost (Deublein and Steinhauser, 2008). The Bio-terminator$^{24/85}$ is mesophilic anaerobic digestion technology developed by Total Solids Solution from research conducted at the University of Louisiana, United States. This process was found to be capable of destroying 85% of TS in 24 h at a reactor retention time of 24 h or less. A pilot-scale plant of 3.785-m^3 capacity was installed at Baton Rouge, Louisiana in 2005 and operated for 5 months (Burnett and Togna, 2007). The system was observed to remove 93% VS at 2 days' hydraulic retention time. Another commercial method, "Columbus Advanced Biosolids Flow-through Thermophilic Treatment (CBFT3), is a modification of thermophilic anaerobic digestion using a plug flow reactor. This process incorporates advanced reciprocating engines to produce electricity that supplies 40–50% of the plant electricity requirements. The overall energy efficiency of the process is 68–83% (Kalogo and Monteith, 2008).

A substantial increase in sludge mineralization and subsequent biogas generation can be achieved by employing a physical, chemical, thermal, mechanical, or biological pretreatment step such as microwave (MW) heating, ultrasonication, ozonation, enzymatic treatment, use of liquid jets, treatment with alkali or acids, high-performance pulse technique, or wet oxidation (Tyagi and Lo, 2011). With regard to the assessment of the feasibility of a pretreatment method, extra biogas production, total energy balance, final amount of sludge, and cost have to be taken into account and analyzed (Rulkens, 2008). Several pretreatment technologies such as Cambi (thermal), BioThelys (thermal), MicroSludge (physical-chemical), CROWN (ultrasonic), and Lysatec GmbH (mechanical) have been applied successfully at full scale in several countries (Kepp et al., 2000; Panter and Kleiven, 2005; Stephenson et al., 2005; Kruger and Hogan, 2005; Elliott and Mahmood, 2007).

The Cambi process was reported to increase net electricity production by 27% (Elliott and Mahmood, 2007). During full-scale municipal trials, Onyeche (2006) reported that high-pressure homogenization of WAS before anaerobic digestion increased gas production by 30%. Zabranska et al. (2006) reported long-term monitoring results from three full-scale installations of lysate thickening centrifuges. They reported 15–26% increases in biogas yield. A pilot-scale ozonation process (0.026 kg O_3/kg VS) to pretreat mixed primary and secondary sludge (weight by weight ratio of 1:3.5) was started in Japan by Kurita Water Industries. The process was capable of producing 36% more energy than the control anaerobic digester. The energy input (ozonation and pumping) and energy produced were estimated at 1923 kWh and 1736 kWh/dry metric ton (MT) sludge treated, respectively (Kalogo and Monteith, 2008; Goel et al., 2004). Xie et al. (2007) studied the effectiveness of ultrasonic treatment at a full-scale sludge digester in Singapore. The anaerobic digesters (4500 m^3 volume, sludge retention time (SRT), 30 days) were fed sonicated sludge (20 kHz; flow rate, 200 m^3/day). Over the 6-month study, the experimental system consistently produced a minimum of 200 m^3/day more gas than the control digester (methane production increased by 45%). Hogan et al. (2004) studied the feasibility of Sonix technology (20 kHz) for the pretreatment of WAS before anaerobic digestion at the demonstration and full-scale plants (Avonmouth, United Kingdom and Orange County, California, United States). They reported that Sonix technology was capable of improving biogas production (up to a 50% increase) and has a relatively short payback period of 2 years. The German company Hielscher (up to 48 kW) claimed to improve biogas yield by 25%. Another German company, Sonotronic, claimed that the integration of a high-output ultrasonic reactor (20 kHz) into existing biogas production systems increased biogas production (up to 50%) and the methane content of the biogas (to 70% of CH_4). Barber (2005) reported the outcomes of several full-scale part-stream ultrasound systems (Germany, Austria, Switzerland, Italy, and Japan). He observed a 22% increase in biogas production and VS reduction, and up to 7% improvement in sludge dewatering. The energy and mass balance study of the anaerobic digester (1200 m^3, 20 days SRT, flow rate 200 m^3/day, and 5% DS dry solids (DS)), treating the sludge sonicated at 2.5 W/m^2/K, shows that more energy is generated than consumed; ie, 1 kW of applied energy will generate seven times more electrical energy after losses. Moreover, the researchers suggested that a typical payback period for a full-scale ultrasound installation is 2−3 years.

Biogas is an excellent fuel for a large number of applications and it can be used more or less in all applications that were developed for natural gas. The biogas can be used for the production of heat and steam or electricity generation/cogeneration, as vehicle fuel, and for the production of chemicals.

Biogas can be used as a fuel to generate electrical power using engine generators, turbines, and fuel cells and as a fuel in gas vehicles (Bridger et al., 1962). An analysis accomplished by the Combined Heat and Power (CHP) partnership observed that if CHP were installed at all 544 wastewater treatment facilities in the United States (influent flow rates >5 million gallons/day and that operate anaerobic digesters), approximately 340 MW (340,000 kWh) of electricity could be generated, which is enough to power 261,000 homes (NACWA, 2010). According to the US Environmental Protection Agency, 2.3 million MT of carbon dioxide emissions annually (equivalent to 430,000 cars) could be offset if existing WWTP (with capacity over 5 million gallons/day) that employ anaerobic digestion installed energy recovery facilities. Harnessing the energy from biosolids offers energy security, reduced dependence on fossil fuels, and lowered greenhouse gas emissions (NACWA, 2010).

In several Swedish cities, use of sludge-derived biogas as biofuel in transportation sector is a well-established practice. The Henriksdal treatment plant produces and sells biogas to Stockholm's bus company. At least 30 buses in Stockholm are running on biogas (Salter, 2006). In the United States, various well-established energy recovery techniques are in use, including electricity and mechanical energy production and heat recovery through biogas generated from anaerobic digestion of waste sludge (Kalogo and Monteith, 2008). The use of methane as a source of hydrogen to generate energy with liquefied carbonate fuel has been effectively exhibited at King County, Washington's South treatment plant (Parry et al., 2004). Co-digestion of grease (from restaurant trap haulers) with sewage sludge is practiced in Watsonville, California, to improve biogas yield by over 50% (Cockrell, 2007). Grease is composed of energy-rich compounds such as fats, carbohydrates, and sugars (Bailey, 2007); thus, it is an appropriate substrate for biogas production during anaerobic digestion of sludge. The Sewerage Bureau of Tokyo Metropolitan Government (SBTMG) implemented several projects including the use of dewatered sewage sludge to produce fuel charcoal and sold for thermal power generation (Oda, 2007) and for electricity generation with a gas engine using syngas produced by the pyrolysis of sewage sludge (Takahashi, 2007). In China, biogas harvesting from sewage sludge is a common way to recover resources. The annual methane generation from all feedstocks including sewage sludge was assessed at 720 million cubic meters (Aalbers, 1999). In the United Kingdom, a new program for energy recovery was proposed by the government, including the generation of 20% of electricity from renewable sources by 2020 (Trumper, 2007). In 2005, 10.8% and 4.2% of all UK renewable energy was recovered by combustion and biogas production, respectively.

10.4.2 NUTRIENT RECOVERY

Sewage sludge contains considerable amounts of nutrients, especially phosphorus (0.5–0.7% TS) and nitrogen (2.4–5.0% TS) (Tchobanoglous et al., 2003); these nutrients exist mainly in the form of proteinaceous material. The breakdown and solubilization of sludge biomass and its subsequent conversion to ammonia and phosphates could be used to produce plant fertilizers such as magnesium ammonium phosphate (struvite), which can be applied directly to soil (SenterNovem, 2008; Liao et al., 2005a; Wong et al., 2006).

Much effort has been directed toward phosphorus recovery from sewage sludge via crystallization, which has been developed and implemented in Japan and The Netherlands (Stratful et al., 1999; Woods et al., 1999). Emerging commercial techniques for phosphorus recovery from waste sludge, including KREPO, Aqua-Reci, Kemicond, BioCon, SEPHOS, and SUSAN, are mainly based on physicochemical and thermal treatment to dissolve phosphorus and then recover it by precipitation. Phosphorus can be recovered from sludge as iron phosphate, calcium phosphate, phosphoric acid, and struvite (magnesium ammonium phosphate). The Aqua-Reci technology was developed in Sweden to recover both phosphorus and energy using combined SCWO following an extraction method (Stendahl and Jafverstrom, 2003). About 100% of phosphorus could be extracted with HCl or H_2SO_4 at 90°C and 2 h reaction time. Stendahl and Jafverstrom (2003) calculated that the total cost of the full-scale Aqua-Reci plant at Stockholm would be approximately US$946/dry MT sludge treated per year. The OSTARA process, another commercial technology to recover struvite from a phosphorus-rich sludge stream using magnesium chloride (80–85% P recovery), has been in operation (full-scale) at the City of Edmonton, Canada. The process has been in operation since May 2007, and is expected to produce between 200 and 250 MT of struvite per year (Kalogo and Monteith, 2008; Prasad et al., 2007). A full-scale process (45,000 m^3/day) for phosphorus recovery (>90% P recovery) as struvite (approximately 550 kg/day; equivalent to 0.01 kg struvite/m^3) was installed at the Lake Shinji East Clean Center in Japan (Ueno and Fujii, 2001). Seaborne technology, developed in Germany by Seaborne Environmental Research Laboratory (Berg and Schaum, 2005), a two-step acid–base leaching extraction method proposed by Kungl Tekniska Hogskolan, Swedish Royal Institute of Technology (Levlin and Hultman, 2004), and KREPO technique (pH 2, 100–110°C, 3.6 bars) (Hansen et al., 2000) and Kemicond technology (modified KREPO technique), are other efficient technologies applied at the pilot and full scale for phosphorus recovery. Crystalactor technology was applied at a full scale in The Netherlands. However, the cost of phosphorus recovery has been estimated at 22 times higher than the cost of mined phosphate rocks, and thus is not considered economical (Roeleveld et al., 2004). Phostrip technology recovers calcium phosphate from a phosphorus-enriched sludge processing side stream. A portion of the return sludge is pumped to an anaerobic tank where acetic acid may be dosed to the tank to increase the amount of phosphorus released to the liquid phase. The phosphorus-rich water is separated from the sludge and treated with lime to precipitate the phosphorus as calcium phosphate. In United States, at least four full-scale PhoStrip plants used the process in the mid 1980s. The cost of production is high compared with natural sources, but the value of the recovered calcium phosphate should increase as natural supplies decline (Kalogo and Monteith, 2008).

Since 1990, the Swedish government has emphasized the significance of nutrient recycling from sewage sludge. Thus use of sludge in agronomy is not a common practice despite the low concentration of contaminants in sludge compared with that in other countries (Bengtsson and Tillman, 2004). In 2000, the Swedish government investigation projected the target of recycling 75% of phosphorus from waste and sludge into productive soils by 2010 (SME, 2000). The Netherlands is well-known as one of the first countries to employ phosphorus recovery at a full scale (Stark, 2004). The Netherlands has aimed to replace 20% of its current phosphate rock consumption with recovered phosphate (Roeleveld et al., 2004). Around 32% of sewage sludge generated in The Netherlands is currently used in the cement industry and power stations (Uijterlinde, 2007). Germany is actively involved in advancing novel technologies for phosphorus recovery from sewage sludge; however, four pilot- or bench-scale technologies have been developed since 2002 (Berg and Schaum, 2005). Moreover, the dried biosolids products (OCEANGRO) are used as organic fertilizers by over 60 New Jersey golf courses and are registered with the New Jersey Department of Agriculture. The biosolids are used as a fertilizer to grow canola (used to make biodiesel fuel) at King County, Washington (Seattle) (Kalogo and Monteith, 2008). BioCon Technology was developed for phosphorus recovery as phosphoric acid. The sludge incinerated ash is leached with H_2SO_4, and resources including ferric chloride ($FeCl_3$), potassium bisulfate ($KHSO_4$), and phosphoric acid (H_3PO_4) are recovered by ion exchange (Hultman et al., 2003; Levlin and Hultman, 2004). However, BioCon Technology has been studied on a pilot-scale at a municipal WWTP near Aalborg, Denmark (Hultman et al., 2001).

The ammonia recovery process (ARP) is a technology developed in the United States and commercialized by the ThermoEnergy Corporation. In this process, an ion-exchange unit concentrates ammonia in the influent from approximately 1000 parts per million (ppm) to 15,000 ppm. The concentrated ammonia stream is then vaporized and the ammonia gas becomes in contact with sulfuric acid and crystallized as ammonium sulfate. The first ARP pilot plant was constructed at Oakwood Beach Water Pollution Control Plant, Staten Island, New York (Kalogo and Monteith, 2008).

10.4.3 HEAVY METALS RECOVERY

Heavy metals such as Zn, Cu, Ni, Cd, Pb, Hg, and Cr are principal elements restricting the use of sludge for land application owing to probable soil and ground water contamination, which ultimately affect human and animal health. Therefore, appropriate treatment of waste sludge is necessary before landfill disposal. Generally, metal-bearing sludges are treated to extract metal ions or stabilize the metals in solid form.

The effect of an ultrasonication-assisted acid leaching process on the separation and recovery of Cu and Fe from printed circuit board (PCB) waste sludge was studied by Xie et al. (2009). This technique was successfully implemented on an industrial scale in a heavy metal recovery plant in Huizhou city, China, from more than 2 years. The pilot-scale installation treated 5800 tons of waste sludge from PCB factories in 2007; 1000 tons of 98% copper sulfate and 3500 tons of 20% ferric chloride were produced. All of the copper sulfate was sold on the market and the ferric chloride was reused in the local PCB manufacturing industries. No secondary pollution was produced by the ultrasonication-assisted acid leaching method. This process has better separation and recovery efficiency, a low recovery cost, greater end-product quality, and zero process waste emissions (Xie et al., 2009). In another study (Li et al., 2010), a similar group reported significantly higher recovery rates of Cu (97.42%), Ni (98.46%), Zn (98.63%), Cr (98.32%), and Fe (100%) with a two-stage ultrasonically enhanced acid leaching process (pH 4.0, 100 min contact time, and 100 W power). The laboratory-verified process performance parameters (pH, ultrasonic power, and contact time) were successfully applied at a pilot-scale treatment and almost similar recovery rates were achieved for all of the studied metals as observed at a laboratory scale.

10.4.4 BIOFUEL PRODUCTION

Biofuels have gained worldwide attention because they have the potential to replace the nonrenewable petroleum fuels in the future. Emerging research is focused on biofuel recovery, especially bioethanol, biodiesel, syngas, biohydrogen, and bio-oil from biomass (Demirbas et al., 2011). Biofuels are produced from sources such as corn, soybeans, flaxseed, rapeseed, sugarcane, palm oil, sugar beet, raw sewage sludge, food scraps, animal parts, and rice (Singh, 2011). Using waste sludge as the substrate for biofuel production offers several advantages over the use of other biomass sources. It is a waste product and so is available at little or no cost, and the supply is plentiful because it is produced wherever there is a sizable human settlement (Massanet-Nicolau et al., 2010).

10.4.4.1 Hydrogen

Hydrogen is a promising alternative energy to fossil fuels. Hydrogen has high energy (122 kJ/g) that is 2.75 times greater than that of hydrocarbon fuel (Han and Shin, 2004). The possibility of producing hydrogen-rich fuel gas by thermochemical treatments of wet sewage sludge that include drying, pyrolysis, and gasification was studied by several researchers. Wet sewage sludge pyrolysis at high temperature (1000°C), combined with high heating rates, enhances the production of H_2-rich fuel gas (Dominguez et al., 2006). Moreover, a gaseous product of a much higher percentage of H_2 is produced from pyrolysis of wet rather than dry sludge (Manara and Zabaniotou, 2012). The extreme moisture content of sewage sludge generates high temperatures, and a steam-rich atmosphere consequently leads to in situ steam reforming of the volatile compounds and partial gasification of the solid char, which contributes to the production of hydrogen-rich fuel gas (Zhang et al., 2011). CO_2, CH_4, and H_2 concentration, low heating value of the produced gas, and aqueous yield are increased by the increasing moisture content, whereas CO concentration and tar yield are decreased (Manara and Zabaniotou, 2012).

10.4.4.2 Syngas ($H_2 + CO$)

Syngas, a mixture of carbon monoxide and hydrogen, may be used as a clean alternative to fossil fuels in generating electricity or for the production of liquid fuels such as synthetic diesel, dimethyl ether, and methanol (Lv et al., 2007). Syngas production is a two-step process. In the first step, the pyrolysis of sewage sludge at around 600°C in an oxygen-deficient atmosphere takes place, which leads to the production of carbon-rich char. In the second step, the char is gasified in the presence of oxygen or air and produces syngas. A commercial method, EBARA-fluidized bed gasification technology (Japan), co-treats municipal sludge with other solid wastes, including municipal solid waste, plastic waste, medical waste, and fly ash to recover the energy-rich syngas. As of September 2002, six TwinRec process lines were in operation and 14 more were under construction (Steiner et al., 2002). Another commercial method is KOPF gasification technology, including: a solar drying unit, a fluidized-bed gasification unit, a gas engine unit for energy recovery, and a postcombustion chamber (Kalogo and Monteith, 2008). The solar unit is used to dry the wet digested sludge to a solid content of between 70% and 85%. The fluidized-bed gasification unit operates at a temperature up to 900°C and a retention time of 30 min. The dried solids are converted into inert granules and combustible gas. The gas is recovered and cooled to a temperature below 150°C and dried. The gas engine unit generates about 70 kW of electricity. The postcombustion chamber is used to dispose of the surplus of gas that cannot be used. The gas is combusted at a temperature of 850°C. Most of the full-scale gasification installations are under operation in Europe, and mainly in Germany.

10.4.4.3 Bio-oil

Bio-oils, components of *n*-alkanes and 1-alkenes, aromatic compounds (that range from benzene derivatives to polyaromatic hydrocarbons, nitrogenated compounds, long-chain aliphatic carboxylic acids, ketones, esters, monoterpenes, and steroids), which are refined to high-quality hydrocarbon fuels, might have some advantages including facility of transport, storage, and combustion, and flexibility in marketing (Dominguez et al., 2003). Furthermore, the bio-oil is a possible source of light aromatics (for example, benzene, toluene, and xylene), which control a greater marketplace value than do raw oils (Tian et al., 2011).

Sludge pyrolysis (liquefaction) at intermediate temperatures of around 425–575°C (heating rate around 100°C/min) allows the dominant production of bio-oil, the yield of which amounts to 30–40 wt% of sewage sludge (Cao and Pawłowski, 2012). In 1980, a pilot-scale study was carried out to produce bio-oil from waste sludges using thermochemical liquefaction, in which liquid sludge (20% TS) was heated at 300°C and 10 MPa pressure for about 90 min, generating a heavy oil, char, gas, and reaction water. The technology was patented as the Sludge-to-Oil Reaction System (STORS). Typically, oil yields ranged from 10% to 20% and char from 5% to 30% by weight. Conventionally, liquefaction of sludge is always carried out in an electric or gas furnace (Dominguez et al., 2003).

The commercial EnerSludge demonstration plant was installed at Perth, Australia. Overall, 45% energy in the biosolids was converted to bio-oil; however, the plant was discontinued after 16 months of operation because it was not considered cost-effective (GVRD, 2005). In another STORS process, which was commercialized by ThermoEnergy (2007) at high pressure and elevated temperatures (275–315°C, 11,400–14,800 kPa, and 1–3 h), the sludge (20% DS) was converted into a fuel consisting of oil with 90% of the heating value of diesel and a solid (char) similar to coal. The recovered bio-oil could be used to generate electricity and/or heat using an engine. A general cost estimate of STORS technology for a small (10,000 p.e. (population equivalent), 292 dry MT/year), medium (100,000 p.e., 2920 dry MT/year), and large (1 million p.e., 29,200 dry MT/year) population was provided by Molton et al. (1986). They estimated that the energy used in the process was 1410, 1394, and 1394 kWh/dry MT, respectively, and the electricity from oil produced was 1898, 1480, and 1480 kWh/dry MT for the small, medium and large population, respectively. However, there is currently no full-scale installation in operation (Kalogo and Monteith. 2008).

10.4.4.4 Bio-diesel

Limited fossil fuel reserves and environmental benefits of biodiesel (decrease in acid rain and emissions of CO_2, SO_x and unburned hydrocarbons during the combustion process; easy biodegradability; less toxicity; and safety in storage and handling) increase the significance of biodiesel (Gogate and Kabadi, 2009). Biodiesel is the esters of simple alkyl fatty acids that can be produced from various lipid sources by a trans-esterification reaction with alcohol in the presence of a base, acid, enzyme, or solid catalyst. Municipal sewage sludge is gaining traction worldwide as a lipid feedstock for biodiesel production because it is available in abundance and contains significant amounts of lipids, which can make biodiesel production from sludge lucrative. These energy-rich lipids include phospholipids, monoglycerides, diglycerides, triglycerides, and free fatty acids in the oils and fats (Kargbo, 2010).

Studies (Dufreche et al., 2007) show that combining lipid extraction processes in 50% of all existing municipal WWTP in the United States and trans-esterification of the extracted lipids could produce approximately 1.8 billion gallons of biodiesel, which is about 0.5% of the yearly national petroleum diesel demand. Currently, the estimated cost of production is US$3.11/gallon of biodiesel.

Nevertheless, this cost should be reduced to levels that are at or below the current petro diesel costs of US$3.00/gallon (Kargbo, 2010). Currently, the estimated cost of production of biodiesel from dry sludge is US$3.11/gallon of biodiesel (Mondala et al., 2009; Susanne, 2010) compared with US$3.00/gallon for petro diesel (as of January 2010). To be more reasonable, it is necessary to reduce the cost to levels that are equal to or under petroleum diesel costs.

According to a case study in South Korea, the price of the lipids extracted from sewage sludge (SS) was approximately US$0.03/L, which was lower than that of all current biodiesel feedstocks (Kwon et al., 2012). The researchers revealed that the production of biodiesel using lipids extracted from SS could be economically feasible because of the remarkably high yield of oil and low cost of this feedstock compared with conventional biodiesel feedstocks. The yield of oil from SS, 980,000 L/ha per year, is higher than that from microalgal and soybean oils, 446 and 2200 L/ha per year, respectively. The disposal cost of 1 ton of wet SS paid by the Korean government is US$58.3 and the drying cost of 1 ton of wet SS is US$57.17. Dried SS is usually sent to the coal power plant and cement manufacturer and its cost (ie, 1 ton of dried SS) is US$10. Thus, biodiesel production from lipids extracted from SS would be the most economical. This economic feasibility would be synergetic with an inexpensive and reliable biodiesel conversion process (Kwon et al., 2012).

10.4.5 CONSTRUCTION MATERIAL

Reuse of waste sludge for construction materials can reduce the problems of disposal while offering a renewable substitute for the depletion of nonrenewable resources (Tay et al., 2004). Sewage sludge containing both organic carbon-containing complexes and inorganic composites represents a source of valuable materials. Substantial work has been carried out for the manufacture of valuable products by the thermal solidification of inorganic sludge composites, particularly in Japan. Dried sludge or incinerator ash is used as a primary material in manufacturing construction material. Solidification occurs at high temperatures up to 1000°C. Waste heat is available for the drying process. Depending on specific process modifications and operating conditions, various types of products can be made, such as artificial lightweight aggregates, slags, and bricks (DRRSS, 2002; Wang et al., 2008).

Upon mixing with clay or on their own, biosolids ash can be used to make bricks that are similar in appearance and physical properties to standard building bricks (Tay and Show, 1997; Wiebusch and Seyfried, 1997). In Japan, biosolids ash has been used to make bricks for over a decade (Okuno and Takahashi, 1997). The first full-scale sludge brick plant commenced in 1991 in Tokyo with a production capacity of 5500 bricks per day using 15,000 kg of incinerated sludge ash. More important, no heavy metal leaching was observed from the finished bricks, even in adverse environments with pH levels as low as 3 (Spinosa, 2004). Anderson et al. (1996) added sewage sludge ash from a fluidized bed incinerator to a series of common commercial brick; they reported encouraging results that directed a UK brick company to use this material as a potential substitute for the addition of sand to the bricks. At the University of Leeds, United Kingdom, Forth (2007) produced a building block, "Bitublock," made by mixing waste products such as recycled glass, sewage sludge, metal slag, and incinerator ash with a sticky binder called bitumen. Bitublock was considered to be almost six times stronger than traditional concrete block. Waste sludge can also be used to produce pumice, which consists of a highly vesicular rough texture (characterized by a rock being pitted with many cavities, known as vesicles, at its surface and inside). It can be used for the underlayer of athletic grounds because it has the characteristic of draining excess water and holds sufficient moisture.

Generally volcanic gravel is used for this purpose; thus, pumice can be a good alternative to less available volcanic gravel (Spinosa, 2004). Okuno and Takahashi (1997) reported the commercial-scale manufacture of bricks consisting of 100 wt% ISSA (Incinerated Sewage Sludge Ash). They stated that important parameters for the starting ash were an average particle size of <30 μm, loss on ignition of <1 wt%, CaO content <15 wt%, and an optimum maximum temperature of 1070−1080°C. The bricks complied with all relevant Japanese standards. However, during service life, problems with moss growth, efflorescence, and ice formation were observed which were linked to water absorption. A silicon-resin coating was applied that eliminated these performance problems. However, the coating resulted in a significant increase in manufacturing costs compared with conventional clay bricks. As landfill costs continue to rise, this will become a more significant factor in determining the economic viability of ISSA-based products. According to Wiesbusch et al. (1998), leaching tests on ISSA-containing bricks revealed concerns about the leaching of Cl^-, SO_4^{2-} and certain heavy metals, which led the authors to increase the brick-firing temperature from 1000−1060°C to 1100−1200°C.

The production of Portland cement using waste sludge is another way to use valuable inorganic and organic compounds of the sludge (Taruya et al., 2002). The waste sludge can be used in three different forms: as incinerated ash, dewatered sludge, or dried sludge powder. Of the three forms, use of dewatered sludge directly into Portland cement kilns seems to be the most appealing because this process does not require new incinerators or generate additional running costs. At a high operating temperature, toxic organic pollutants in the sludge are completely oxidized and heavy metals are immobilized in the cement (Rulkens, 2008). Thus the sludge can be a used as a raw material in Portland cement manufacturing, and subsequently can reduce the burden from natural resources such clay and limestone (sources of CaO, SiO_2, and F_2O_3) (Spinosa, 2004). According to previous studies (Paya et al., 2002; Tay et al., 2002; Lin et al., 2012), cement-like materials made from sludge can replace ordinary Portland cement for up to 20% by weight.

Waste sludge can also be reused to manufacture artificial lightweight aggregate (ALWA). ALWA can be used as planter soils, fillers for clearance between kerosene storage tanks and room walls, flowerpot additives, heat-proofing panels, replacement for water-infiltrating pavement, substitute of anthracite media in rapid sand filters as well as in walkways pavement (Spinosa, 2004). A 500-kg/h plant for producing ALWA commenced operations in 1996 at the Nambu plant in Tokyo. Compared with commercial lightweight aggregates, ALWA has better sphericity, lesser specific gravity, and low compressive strength. Concrete made with the sludge−clay aggregates had lower density and hence a higher strength to mass ratio compared with that produced from conventional granite aggregates. On the basis of their research findings, Tay et al. (2004) recommended that the sludge−clay aggregates with up to 20% clay content are suitable for structural applications, whereas aggregates with up to 50% clay content could be used for other general applications in which strength is not a critical requirement. Leaching test results showed that the peak levels of all health-based contaminants were within the respective safety limits specified in the World Health Organization (WHO) guidelines for drinking water. Compliance with WHO safety limits indicates that the use of these sludge−clay aggregates should not have a significant effect on human health or the environment.

Slag, a marble-like mineral of semicrystalline structure, is a possible solution to volume reduction and the immobilization of heavy metals in sludge. Water-cooled slag and air-cooled slag are two types which can be produce from waste sludge and can be used as an alternative to natural coarse aggregate, including ready-mixed concrete aggregate, road bed materials, concrete aggregate and back-filling material, interlocking tiles, permeable pavement, and other secondary concrete products

(Spinosa, 2004). GlassPack, a vitrification process (developed by Minergy Corporation, United States), uses the organic fraction of biosolids as a renewable fuel source to produce an inert glass aggregate product from the inorganic (ash) fraction. Wet sludge (approximately 20% solids) is pre-dried to <15% moisture; the dried solids are then subjected to temperatures between 1330 and 1500°C, at which the ash component melts into molten glass.

Japan may be considered a pioneer country in the production of construction materials from sewage sludge. Sludge is used to produce construction materials at full-scale plants (Ozaki et al., 1997). Incinerated sludge ash is also used in Columbus, Ohio, as a water-absorbent surface amendment in sports fields and horse arenas (USEPA, 1994). Ash from thermal oxidation installations are using to manufacture bricks by municipalities in Virginia and Georgia, and as a source of phosphorus by the city of Cincinnati, Ohio (Welp et al., 2002). In China, sewage sludge is also used in to manufacture of bricks and other building materials (Wang, 1997).

Thus a promising output was reported worldwide regarding the use of waste sludge as a construction material. Nevertheless, despite the development of technically feasible processes, most of the techniques are not economically viable because of the higher production cost (compared with the market price). Therefore, the commercialization of sewage sludge–based construction materials (ie, scale-up of economically feasible technologies and market development) are major challenges.

10.4.6 ELECTRICITY PRODUCTION FROM SLUDGE BY MICROBIAL FUEL CELLS

Electricity production from sludge using microbial fuel cells (MFC) has been considered a beneficial way to reuse sludge. A mixed bacterial population can be used in MFC to produce electricity while achieving the biodegradation of organic wastes. Many different bacterial species including *Escherichia*, *Shewanella*, *Clostridum*, and *Desulfovibrio* have been reported to reduce metallic ions (eg, manganese, ferric, uranium, and cupric) while oxidizing available carbon substrates by the redox mechanism (Lovely, 1993; Oh and Logan, 2005; Du et al., 2007; MFC, 2013). Rabaey and Verstraete (2005) described the mechanisms of waste treatment in MFC reactors. Microbes in the anodic chamber of an MFC oxidize added substrates and produce electrons and protons in the process. Electric current generation is made possible by keeping microbes separate from oxygen or any other end-terminal acceptor other than the anode, which requires an anaerobic anodic chamber. Typical electrode reactions are shown in Eqs. [10.2] and [10.3] using acetate as an example substrate.

$$\text{Anodic reaction: } CH_3COO^- + 2H_2O \xrightarrow{\text{microbes}} 2CO_2 + 7H^+ + 8e^- \qquad [10.2]$$

$$\text{Cathode reaction: } O_2 + 4e^- + 4H^+ \rightarrow 2H_2O \qquad [10.3]$$

The whole reaction is the decomposition of the organic substrate to carbon dioxide and water with simultaneous electricity generation as a by-product. Based on this electrode reaction pair, an MFC bioreactor can generate electricity from the electron flow from the anode to cathode in the external circuit (Du et al., 2007).

Various types of commercially available fuel cells include low-temperature (including phosphoric acid, proton exchange membrane, and alkaline types) and high-temperature (molten carbonate and solid oxide types) fuel cells. Overall fuel cell efficiency varies by type between 47% and 87%. The voltage of a biofuel cell is normally on the order of several hundred millivolts. Power densities have been reported on the order of a 50- to 100-W/m^3 reactor (Rabaey et al., 2005). Only phosphoric acid and molten carbonate fuel cells have been studied at full scale. Fuel cell capacities in North America range from 200 kW to 1 MW. Although the operating costs of fuel cells can be low, on the order of \$0.01/kWh, capital costs currently are high compared with other technologies (Kalogo and Monteith, 2008).

10.4.7 BIOPLASTIC

PHA is polyester of hydroxyalkanoic acid and is well known to be biodegradable and thus an environmentally friendly alternative to petroleum plastics (Lee and Yu, 1997; Akaraonye et al., 2010; Thomson et al., 2010). Poly-β-hydroxybutyric acid and its copolymer, poly (3-hydroxybutyrate-*co*-hydroxyvalerate), are the most widespread PHAs, whereas other forms are possible (Chua et al., 1999). PHA is produced in nature by the bacterial fermentation of sugar or lipids. PHA in microorganisms, particularly in bacteria, serves as a carbon and energy reserve and/or as a sink for redundant reducing power or electrons in stressful states (Anderson and Dawes, 1990; Lee, 1996). PHA-accumulating microorganisms are found in activated sludge. Polyphosphate-accumulating organisms accumulate PHA by taking up volatile fatty acids (VFA) under anaerobic conditions. PHA production by activated sludge has the following advantages (Satoh et al., 1999): (1) By using activated sludge, waste organic materials can be recovered and reused as biodegradable plastics; (2) the cost of PHA production can be reduced by using waste sludge as feedstock because the waste sludge is easily available in plenty; (3) PHA that is not produced by known pure culture can be obtained by using activated sludge instead of pure culture (Satoh et al., 1992).

PHAs are appealing as packaging films and disposable products (ie, utensils, diapers, cosmetic containers, bottles and cups) owing to their biodegradability (Reddy et al., 2003). In medicine, PHA can be used as functionalized nanobeads or microbeads for diagnostic and therapeutic applications, for soft- and hard-tissue repair and regeneration, and as conduits and carrier scaffolds for nerve repairs, drug delivery systems, devices for joints and wound dressings, drug eluting stents for cardiovascular applications, and heart valves in heart tissue engineering (Misra et al., 2006; Philip et al., 2007).

The current cost of producing microbial PHAs is about US\$4–6/kg, which is almost 10 times higher than petroleum plastic (Chua et al., 1999; Akaraonye et al., 2010). The cost of carbon sources has triggered slow growth experienced by the PHA industry. For example, the cost of substrate or carbon sources accounts for about 50% of the cost of microbial PHA (Law et al., 2001). Even with genetically engineered *Escherichia coli*, the carbon source is about 31% of PHA's production cost (Ma et al., 2000). According to Brar et al. (2009), the major technical difficulties with bioplastic production from wastewater sludge are optimizing the organic loading rate (OLR) and the economically lower yields. This requires extensive studies of enhancing OLR and efficacy in using mixed substrates (sewage sludge and agroindustry wastewater sludge) with mixed cultures.

Micromidas LLC is developing a biological process which will use carbon and other nutrients in sludge to generate biodegradable plastic. The resulting plastic will have a lifespan of months instead of the centuries needed to breakdown petroleum-based plastics. They are in the process of developing a trailer-mounted pilot unit that can be taken to a resource recovery facility to be tested on a larger scale (Moss et al., 2013).

10.4.8 BIOFERTILIZERS

Amino acid–chelated trace elements (AACTE) fertilizer is recognized as an environmentally friendly fertilizer for cotton, fruit, and other cash crops in China (Zhang et al., 2003); however, its production is restricted by limited protein sources such as animal hair, hooves, horn, and leather (Lu et al., 2007). Liu et al. (2009) developed a novel technology for sewage sludge use. Bacteria proteins in sewage sludge were extracted to generate AACTE fertilizer using several chemical methods. Initially, sewage sludge was hydrolyzed under hot HCl solution to obtain a protein solution. Then, the protein solution was further hydrolyzed into amino acids under hot acidic conditions. After that, the raw amino acid solution was purified through activated carbon decolorization and glacial acetic acid dissolution. Finally, the purified amino acids were used to produce AACTE fertilizer by chelation with trace elements.

Under optimal hydrolysis conditions, 78.5% of protein was extracted from the sewage sludge and the amino acids yield was 10–13 g/100 g dry sludge. The AACTE fertilizer produced was observed to be in agreement with China Standard for Amino Acids Foliar Fertilizer (GB/T 17,419-1998) (Lu et al., 2007).

10.5 OVERALL STATUS OF SLUDGE REUSE FOR ENERGY AND RESOURCE RECOVERY

There is general consensus among wastewater treatment experts worldwide that waste sludge is not a waste but a source of valuable resources. Major factors behind this consent are sustainability and environmental concerns (resource depletion, soil pollution, and global warming), the hike in energy prices, stringent directives for sludge disposal, and increasing pressure and protests from environmental authorities and the public domain (Kalogo and Monteith, 2008).

Japan may be considered a pioneer in the production of construction materials from sewage sludge. Sludge is used to produce construction materials at full-scale plants (Ozaki et al., 1997). The SBTMG implemented several projects including the use of dewatered sewage sludge to produce fuel charcoal and sold for thermal power generation (Oda, 2007) and for electricity generation with a gas engine using syngas produced by the pyrolysis of sewage sludge (Takahashi, 2007). Since 1990, the Swedish government has emphasized the significance of nutrient recycling from sewage sludge. Thus, the use of sludge in agronomy is not a common practice despite the low concentration of contaminants in sludge compared with other countries (Bengtsson and Tillman, 2004). In 2000, the Swedish government projected the target of recycling 75% of phosphorus from waste and sludge to productive soils by 2010 (SME, 2000). In several Swedish cities, use of sludge-derived biogas as a biofuel in the transportation sector is a well-established practice. The Henriksdal treatment plant produces and sells biogas to Stockholm's bus company. At least 30 buses in Stockholm are running on biogas (Salter, 2006).

In the United States, various well-established energy recovery techniques are in use, including electricity and mechanical energy production, and heat recovery through biogas generated from the anaerobic digestion of waste sludge (Kalogo and Monteith, 2008). The use of methane as a source of hydrogen to generate energy with liquefied carbonate fuel has been exhibited at King County, Washington's South treatment plant (Parry et al., 2004). Co-digestion of grease (from restaurant trap haulers) with sewage sludge is practiced in Watsonville, California, to improve the biogas yield by over 50% (Cockrell, 2007). Grease is composed of energy-rich compounds such as fats, carbohydrates, and sugars (Bailey, 2007); thus it is an appropriate substrate for biogas production during anaerobic digestion of sludge. Incinerated sludge ash is using in Columbus, Ohio, as a water-absorbent surface amendment in sports fields and horse arenas (USEPA, 1994). The ash from thermal oxidation installations is used for brick manufacture by the municipalities of Virginia and Georgia; and as a source of phosphorus by city of Cincinnati, Ohio (Welp et al., 2002). The dried biosolids products (OCEANGRO) are used as organic fertilizer by over 60 New Jersey golf courses and registered with the New Jersey Department of Agriculture. Moreover, the biosolids are used as a fertilizer to grow canola (used to make biodiesel fuel) at King County, Washington (Seattle) (Kalogo and Monteith, 2008).

In China, biogas harvesting from sewage sludge is a common method of resource recovery. Annual methane generation from all feedstock including sewage sludge was assessed at 720 million cubic meters (Aalbers, 1999). Moreover, sewage sludge is used in the manufacture of bricks and other building material (Wang, 1997). The Netherlands is well known to be one of the first countries to employ phosphorus recovery on a full scale (Stark, 2004). The Netherlands has aimed to replace 20% of its current phosphate rock consumption by recovered phosphate (Roeleveld et al., 2004). Around 32% of sewage sludge generated in The Netherlands is currently used in the cement industry and power stations (Uijterlinde, 2007). In the United Kingdom, a new program for energy recovery was proposed by the government, including the generation of 20% of electricity from renewable sources by 2020 (Trumper, 2007). In 2005, 10.8% and 4.2% of all UK renewable energy was recovered by combustion and biogas production, respectively. Germany is actively involved in advancing novel technologies for phosphorus recovery from sewage sludge; however, only four pilot or bench-scale technologies have been developed since 2002 (Berg and Schaum, 2005).

In New Zealand, the city of Napier is implementing the production of sludge compost using primary sludge from the Awatoto wastewater plant and wood chips (MacDonald et al., 2007). The main goal of this project is to use the sludge compost as a soil conditioner and a source of fuel for energy production. Furthermore, Malaysia is performing research to evaluate the potential of using dried sewage sludge as a raw material to produce clay-sludge bricks (Liew et al., 2004).

10.6 SUMMARY

Building sludge-derived resources recovery systems will help to produce environmentally benign products, reduce dependency on nonrenewable resources, thus facilitating the conservation of natural resources, decrease human health risks and environmental pollution (by complete eliminating pathogens, reducing the risk of soil and groundwater contamination from heavy metals and other emerging contaminants associated with landfills and land application of sludge, reducing the greenhouse gas emissions associated with burning petroleum fuels), and offer routes to the sustainable management of waste sludge (ie, environmentally friendly, economically feasible, and socially acceptable).

The major issue with resource recovery from waste sludge is related to the manufacturing cost of value-added products compared with the market price. Thus, the cost-effective production of value-added products with supply chain management and environmental compatibility must be ensured during the development of different processes. Several developments have failed mainly owing to the high capital and operation and maintenance costs (Kalogo and Monteith, 2008). Moreover, the method must generate a few or several low-volume and high-value chemical products as well as low-value and high-volume liquid transportation fuel, while producing electricity and process heat for its own use and which would be possibly adequate for the sale of electricity. The high-value products increase profitability, the high-volume fuel fulfills national energy demands, and the power generation decreases costs and avoids the release of greenhouse-gases (Speight, 2011).

REFERENCES

Aalbers, H., 1999. Resource recovery from faecal sludge using constructed wetlands – a survey of literature. In: UWEP Working Document, Waste, Nieuwehaven 201, 2801 CW Gouda, The Netherlands, 69 p.

Akaraonye, E., Keshavarz, T., Roy, I., 2010. Production of polyhydroxyalkanoates: the future green materials of choice. Journal of Chemical Technology and Biotechnology 85, 732–743.

Anderson, A.J., Dawes, E.A., 1990. Occurrence, metabolism, metabolic role, and industrial uses of bacterial polyhydroxyalkanoates. Microbiological Reviews 54, 450–472.

Anderson, M., Skerratt, R.G., Thomas, J.P., Clay, S.D., 1996. Case study involving using fluidised bed incinerator sludge ash as a partial clay substitute in brick manufacture. Water Science and Technology 34 (3–4), 507–515.

Bailey, R.S., 2007. Anaerobic digestion of restaurant grease wastewater to improve methane gas production and electrical power generation potential. In: Proceedings of the 80th Annual Technical Exhibition and Conference of the Water Environment Federation; October 13–17, San Diego, CA, pp. 6793–6805.

Barber, W.P., 2005. The effects of ultrasound on sludge digestion. The Chartered Institution of Water and Environmental Management 19, 2–7.

Bengtsson, M., Tillman, A.M., 2004. Actors and interpretations in an environmental controversy: the Swedish debate on sewage sludge use in agriculture. Resources, Conservation and Recycling 42, 65–82.

Berg, U., Schaum, C., 2005. Recovery of Phosphorus from Sewage Sludge and Sludge Ashes – Applications in Germany and Northern Europe. Dokuz Eylul Universitesi.

Brar, S.K., Verma, M., Tyagi, R.D., Surampalli, R.Y., 2009. Value addition of wastewater sludge: future course in sludge reutilization. Practice Periodical of Hazardous, Toxic, and Radioactive Waste Management 13 (1), 59–74.

Bridger, G.L., Starostka, R.W., Grace, W.R., 1962. Metal ammonium phosphate as fertilizers. Journal of Agricultural and Food Chemistry 10, 181–188.

Burnett, C.H., Togna, A.P., 2007. New high-rate plug flow anaerobic digester technology for small communities. In: Proceeding of the IWA Conference on Biosolids, Moving Forward Wastewater Biosolids Sustainability: Technical, Managerial, and Public Synergy; June 24–27, Moncton, New Brunswick, Canada, pp. 383–385.

Campbell, H.W., 2000. Sludge management – future issues and trends. Water Science and Technology 41 (8), 1–8.

Cao, Y., Pawłowski, A., 2012. Sewage sludge-to-energy approaches based on anaerobic digestion and pyrolysis: brief overview and energy efficiency assessment. Renewable and Sustainable Energy Reviews 16, 1657–1665.

Chua, H., Yu, P.H.F., Ma, C.K., 1999. Accumulation of biopolymers in activated sludge biomass. Applied Biochemistry and Biotechnology 77–79, 389–399.

Cockrell, P., 2007. Grease digestion to increase digester gas production – 4 years of operation. In: Proceedings of the 80th Annual Technical Exhibition and Conference of the Water Environment Federation, October 13–17, San Diego, CA, pp. 6706–6715.

Degremont, 1979. Manual tecnico del agua: Bilbao. Art. Graf. Grijelmo.

Demirbas, M.F., Balat, M., Balat, H., 2011. Biowastes-to-biofuels. Energy Conversion and Management 52, 1815–1828.

Deublein, D., Steinhauser, A., 2008. Biogas from Waste and Renewable Resources. Wiley-VCH, Weinheim, Germany.

Dominguez, A., Menendez, J.A., Inguanzo, M., Bernard, P.L., Pis, J.J., 2003. Gas chromatographic-mass spectrometric study of the oil fractions produced by microwave-assisted pyrolysis of different sewage sludges. Journal of Chromatography A 1012, 193–206.

Dominguez, A., Menéndez, J.A., Pis, J.J., 2006. Hydrogen rich fuel gas production from the pyrolysis of wet sewage sludge at high temperature. Journal of Analytical and Applied Pyrolysis 77, 127–132.

DRRSS, 2002. Disposal and Recycling Routes for Sewage Sludge. European Commission. Office for Official Publications of the European Communities, ISBN 92-894-1801-X.

Du, Z., Li, h., Gu, T., 2007. A state of the art review on microbial fuel cells: a promising technology for wastewater treatment and bioenergy. Biotechnology Advances 25, 464–482.

Dufreche, S., Hernandez, R., French, T., Sparks, D., Zappi, M., Alley, E., 2007. Extraction of lipids from municipal wastewater plant microorganisms for production of biodiesel. Journal of the American Oil Chemists' Society 84, 181–187.

Elliott, A., Mahmood, T., 2007. Pretreatment technologies for advancing anaerobic digestion of pulp and paper biotreatment residues. Water Research 41, 4273–4286.

Fonts, I., Azuara, M., Gea, G., Murillo, M.B., 2009. Study of the pyrolysis liquids obtained from different sewage sludge. Journal of Analytical and Applied Pyrolysis 85, 184–191.

Forth, J., 2007. New building material on the block. Water & Environmental Technologies 19 (7), 34.

Guibelin, E., 2004. Sludge thermal oxidation processes: mineral recycling, energy impact, and greenhouse effect gases release. Water Science and Technology 49 (10), 209–216.

Goel, R., Komatsu, K., Yasui, H., Harada, H., 2004. Process performance and change in characteristics during anaerobic digestion of sewage sludge with ozonation. Water Science and Technology 49 (10), 105–113.

Gogate, P.R., Kabadi, A.M., 2009. A review of applications of cavitation in biochemical engineering/biotechnology. Biochemical Engineering Journal 44, 60–72.

GVRD, September 2005. Review of Alternatives Technologies for Biosolids Management. Report of the Greater Vancouver Regional District.

Han, S.K., Shin, H.S., 2004. Bio-hydrogen production by anaerobic fermentation of food waste. International Journal of Hydrogen Energy 29, 569–577.

Hansen, B., Karlson, I., Cassidy, S., Pettersson, L., 2000. Operational experiences from a sludge recovery plant. Water Science and Technology 41 (8), 23–30.

Hogan, F., Mormede, S., Clark, P.B., Crane, M., 2004. Ultrasonic sludge treatment for enhanced anaerobic digestion. Water Science and Technology 50 (9), 25–32.

Hultman, B., Levlin, E., Stark, K., 2001. Phosphorus recovery from sewage sludges: research and experiences in Nordic countries. In: Scope Newsletter 41. Available at: www.ceepphosphates.org/Files/Newsletter/Scope%20Newsletter%2041.PDF.

Hultman, B., Levlin, E., Plaza, E., Stark, K., 2003. Phosphorus Recovery from Sludge in Sweden – Possibilities to Meet Proposed Goals in an Efficient, Sustainable and Economical Way. No. 10, pp. 19–28. Available at: www.lwr.kth.se/forskningsprojekt/Polishproject/JPS10s19.pdf.

Jaeger, M., Mayer, M., 2000. The Noell Conversion Process – a gasification process for the pollutant-free disposal of sewage sludge and the recovery of energy and materials. Water Science and Technology 41 (8), 37–44.

Kalogo, Y., Monteith, H., 2008. State of Science Report: Energy and Resource Recovery from Sludge. Global Water Research Coalition. Water Environment Research Foundation, Alexandria, VA, 238 p.

Kargbo, D.M., 2010. Biodiesel production from municipal sewage sludges. Energy & Fuels 24, 2791–2794.

Kepp, U., Machenbach, I., Weisz, N., Solheim, O.E., 2000. Enhanced stabilization of sewage sludge through thermal hydrolysis – three years of experience with full-scale plant. Water Science and Technology 42 (9), 89–96.

Khiari, B., Marias, F., Zagrouba, F., Vaxelaire, J., 2004. Analytical study of the pyrolysis process in a wastewater treatment pilot station. Desalination 167, 39–47.

Kruger, R., Hogan, F., 2005. Using sonix to enhance anaerobic digestion: an overview from different trials and installations. In: WEF Proceedings of the Residuals and Biosolids Conference, Covington, KY.

Kwon, E.E., Kim, S., Jeon, Y.J., Yi, H., 2012. Biodiesel production from sewage sludge: new paradigm for mining energy from municipal hazardous material. Environmental Science and Technology 46, 10222–10228.

Law, K.H., Leung, Y.C., Lawford, H., Chua, H., Hung, L.W., Yu, P.H., 2001. Production of polyhydroxybutyrate by *Bacillus* species isolated from municipal activated sludge. Applied Biochemistry and Biotechnology 91–93 (1–9), 515–524.

Lee, S.Y., 1996. Bacterial polyhydroxyalkanoates. Biotechnology and Bioengineering 49 (1), 1–14.

Lee, S., Yu, J., 1997. Production of biodegradable thermoplastics from municipal sludge by two stage bioprocess. Resources, Conservation and Recycling 19 (3), 151–164.

Levlin, E., Hultman, B., 2004. Phosphorus recovery from sewage sludge – ideas for further studies to improve leaching. In: Report No 12, Integration and Optimization of Urban Sanitation Systems, Stockholm June 5–9. Available at. www.lwr.kth.se/Personal/personer/levlin_erik/Pfurtherecovery.pdf (last access August 2007).

Li, C., Xie, F., Ma, Y., Cai, T., Li, H., Huang, Z., Yuan, G., 2010. Multiple heavy metals extraction and recovery from hazardous electroplating sludge waste via ultrasonically enhanced two-stage acid leaching. Journal of Hazardous Materials 178, 823–833.

Liao, P.H., Wong, W.T., Lo, K.V., 2005. Release of phosphorus from sewage sludge using microwave technology. Journal of Environmental Engineering and Science 4, 77–81.

Liew, A.G., Idris, A., Wong, C.H.K., Samad, A.A., Noor, M.J.M.M., Baki, A.M., 2004. Incorporation of sewage sludge in Clay Brick and its characterization. Waste Management & Research 22, 226–233.

Lin, Y., Zhou, S., Li, F., Lin, Y., 2012. Utilization of municipal sewage sludge as additives for the production of eco-cement. Journal of Hazardous Materials 213–214, 457–465.

Liu, Y., Kong, S., Li, Y., Zeng, H., 2009. Novel technology for sewage sludge utilization: preparation of amino acids chelated trace elements (AACTE) fertilizer. Journal of Hazardous Materials 171 (1–3), 1159–1167.

Liu, Y., Tay, J.H., 2001. Strategy for minimization of excess sludge production from the activated sludge process. Biotechnology Advances 19 (2), 97–107.

Lovley, D.R., 1993. Dissimilatory metal reduction. Annual Review of Microbiology 47, 297–299.

Lu, J.C., Zhao, Y.F., Gong, H.B., Li, D.T., Liu, J., 2007. New technology for production of the amino acid chelated trace metal fertilizers. Phosphate and Compound Fertilizer 6, 49–52 (in Chinese).

Lv, P., Yuan, Z., Wu, C., Ma, L., Chen, Y., Tsubaki, N., 2007. Bio-syngas production from biomass catalytic gasification. Energy Conversion and Management 48, 1132–1139.

Ma, C.K., Chua, H., Yu, P.H., Hong, K., 2000. Optimal production of polyhydroxyalkanoates in activated sludge biomass. Applied Biochemistry and Biotechnology 84–86, 981–989.

MacDonald, G., Dunford, P., Ehlers, J., 2007. Compost dried primary sludge – a renewable energy source? In: Proceedings of the 80th Annual Technical Exhibition and Conference of the Water Environment Federation, October 13–17, San Diego, CA, pp. 2952–2970.

Manara, P., Zabaniotou, A., 2012. Towards sewage sludge based biofuels via thermochemical conversion- A review. Renewable and Sustainable Energy Reviews 16 (5), 2566–2582.

Massanet-Nicolau, J., Guwy, A., Dinsdale, R., Premier, G., Esteves, S., 2010. Production of hydrogen from sewage biosolids in a continuously fed bioreactor: effect of hydraulic retention time and sparging. International Journal of Hydrogen Energy 35, 469–478.

Matthews, P.A., 1996. Global Atlas of Wastewater Sludge and Biosolids Use and Disposal. IAWQ Technical Report No 4.

MFC, 2013. Microbial Fuel Cells. School of Chemical Engineering and Advance Materials, New Castle University. http://research.ncl.ac.uk/mfc/background.htm (accessed April 2013).

Misra, S.K., Valappil, S.P., Roy, I., Boccaccini, A.R., 2006. Polyhydroxyalkanoate (PHA)/inorganic phase composites for tissue engineering applications. Biomacromolecules 7, 2249–2258.

Molton, P.M., Fassbender, A.G., Brown, M.D., 1986. STORS: The Sludge-to-Oil Reactor System. EPA/600/S2–86/034, Research and Development, Project summary.

Mondala, A., Liang, K., Toghiani, H., Hernandez, R., French, T., 2009. Biodiesel production by in situ trans-esterification of municipal primary and secondary sludges. Bioresource Technology 100, 1203–1210.

Moss, L.H., Donovan, J.F., Carr, S., Stone, L., Polo, C., Khunjar, W.O., Latimer, R., Jeyanayagam, S., Beecher, N., McFadden, L., 2013. Enabling the Future: Advancing Resource Recovery from Biosolids. Water Environment Federation, pp. 1–89.

NACWA, 2010. Renewable Energy Resources: Banking on Biosolids. National Association of Clean Water Agencies. www.nacwa.org.
Neyens, E., Baeyens, J., Dewil, R., De Heyder, B., 2004. Advanced sludge treatment affects extracellular polymeric substances to improve activated sludge dewatering. Journal of Hazardous Materials 106 (2–3), 83–92.
Oda, T., 2007. Making fuel charcoal from sewage sludge for thermal power generation plant – first in Japan. In: Proceedings of the 80th Annual Technical Exhibition and Conference of the Water Environment Federation, October 13–17, San Diego, CA, pp. 2943–2951.
Oh, S.E., Logan, B.E., 2005. Hydrogen and electricity production from a food processing wastewater using fermentation and microbial fuel cell technologies. Water Research 39, 4673–4682.
Okuno, N., Takahashi, S., 1997. Full scale application of manufacturing bricks from sewage. Water Science and Technology 36, 243–250.
Onyeche, T., 2006. Sewage sludge as source of energy. In: Proceedings of the IWA Specialized Conference on Sustainable Sludge Management: State-of-the-art, Challenges and Perspectives, Moscow, Russia; May 29–31, pp. 235–241.
Orhon, D., Artan, N., 1994. Modelling of Activated Sludge Systems. Technomic Publishing Co., Inc., Lancaster, PA, pp. 39–110. Ch. 2.
Ozaki, M., Watanabe, H., Weibusch, B., 1997. Characteristics of heavy metal release from incinerated ash, melted slag and their products. Water Science and Technology 36 (11), 267–274.
Panter, K., Kleiven, H., 2005. Ten years experience of full-scale thermal hydrolysis projects. In: Proceedings of the 10th European Biosolids and Bio-waste Conference, Wakefield, UK.
Parry, D., Bivins, J., Buehler, V., Coccaro, J., Deshpande, S., Forbes Jr., R., et al., 2004. High Performance Anaerobic Digestion. WEF White Paper. www.wef.org/NR/rdonlyres/4D51F55C-2CDD-42A0-AB9CD77C4139F51C/0/AnaerobicDigestion.pdf (last access October 2007).
Paya, J., Monzo, J., Borrachero, M.V., Amahjour, F., Girbe, I., Velagsquez, S., Ordognez, L.M., 2002. Advantages in the use of fly ashes in cements containing pozzolanic combustion residues: silica fume, sewage sludge ash, spent fluidized bed catalyst and rice husk ash. Journal of Chemical Technology and Biotechnology 77, 331–335.
Perez-Elvira, S.I., Diez, P.N., Fernandez-Polanco, F., 2006. Sludge minimization technologies. Reviews in Environmental Science and Biotechnology 5 (4), 375–398.
Philip, S., Keshavarz, T., Roy, I., 2007. Polyhydroxyalkanoates: biodegradable polymers with a range of applications. Journal of Chemical Technology and Biotechnology 82, 233–247.
Prasad, R., Britton, A., Balzer, B., Schafran, G., 2007. Nutrient recovery by crystallization process: Virginia experience. In: Proceedings of the 80th Annual Technical Exhibition and Conference of the Water Environment Federation, San Diego, CA; October 13–17, pp. 356–358.
Rabaey, K., Clauwaert, P., Aelterman, P., Verstraete, W., 2005. Tubular microbial fuel cells for efficient electricity generation. Environmental Science and Technology 39, 8077–8082.
Rabaey, K., Verstraete, W., 2005. Microbial fuel cells: novel biotechnology for energy generation. Trends in Biotechnology 23, 291–298.
Reddy, C.S.K., Ghai, R., Rashmi, Kalia, V.C., 2003. Polyhydroxyalkanoates: an overview. Bioresource Technology 87, 137–146.
Roeleveld, P., Loeffen, P., Temmink, H., Klapwijk, B., 2004. Dutch analysis for P-recovery from municipal wastewater. Water Science and Technology 49 (10), 191–199.
Rulkens, W., 2008. Sewage sludge as a biomass resource for the production of energy: overview and assessment of the various options. Energy & Fuels 22, 9–15.
Rulkens, W.H., Bien, J.D., 2004. Recovery of energy from sludge-Comparison of the various options. Water Science and Technology 50 (9), 213–221.
Salter, S., 2006. Treating Waste as a Resource, Inspiration from Sweden, 39 p. Available at: www.caw114.bc.ca/sewage/Treating%20Waste%20as%20a%20Resource-%20November%202006.pdf (last access August 2007).

Satoh, H., Mino, T., Matsuo, T., 1999. PHA production by activated sludge. International Journal of Biological Macromolecules 25 (1–3), 105–109.

Satoh, H., Mino, T., Matsuo, T., 1992. Uptake of organic substrates and accumulation of polyhydroxyalkanoates linked with glycolysis of intracellular carbohydrates under anaerobic conditions in the biological excess phosphate removal processes. Water Science and Technology 26 (5–6), 933–942.

SenterNovem, 2008. Biogas as Vehicle Fuel, a Trendsetter Report. Available at: http://www.novem.nl (accessed on January 2008).

Shanableh, A., Jomaa, S., 2001. Production and transformation of volatile fatty acids from sludge subjected to hydrothermal treatment. Water Science and Technology 44 (10), 129–136.

Singh, N.R., 2011. Biofuels. RSC Energy Series No. 5. In: Speight, J.G. (Ed.), The Biofuels Handbook. Royal Society of Chemistry, pp. 160–198. Ch.5.

SME, 2000. Swedish Ministry of the Environment. The Future Environment – Our Common Responsibility. Summary, Final Report of the Committee on Environmental Objectives. SOU, Stockholm, Sweden, p. 52.

Speight, J.G., 2011. A Biorefinery. Ch. 4, RSC Energy Series No. 5. In: Speight, James G. (Ed.), The Biofuels Handbook. Royal Society of Chemistry, pp. 118–159. http://dx.doi.org/10.1039/9781849731027-00118.

Spinosa, L., 2004. From sludge to resources through biosolids. Water Science and Technology 50 (9), 1–8.

Stark, K., 2004. Phosphorus recovery – experience from European countries. In: Plaza, E., Levlin, E., Hultman, B. (Eds.), Proceedings of Polish-swedish Seminars, Stockholm June 6–8, Integration and Optimization of Urban Sanitation Systems. Available at: www.lwr.kth.se/Forskningsprojekt/Polishproject/StarkSthlm19.pdf (last access August 2007).

Steiner, C., Kameda, O., Oshita, T., Sato, T., 2002. EBARA's fluidized bed gasification: atmospheric 2 × 225 t/d for shredding residues recycling and two-stage pressurized 30 t/d for ammonia synthesis from waste plastics. In: 2nd International Symposium on Feedstock Recycling of Plastics. Oostende, Belgium.

Stendahl, K., Jafverstrom, S., 2003. Phosphate recovery from sewage sludge in combination with supercritical water oxidation. Water Science and Technology 48, 185–190.

Stephenson, R., Rabinowitz, B., Laliberte, S., Elson, P., 2005. Teaching an old digester new tricks: full-scale demonstration of the MicroSludge process to liquefy municipal WAS. In: WEF Proceedings of the Residuals and Biosolids Management Conference, Covington, KY.

Stolarek, P., Ledakowicz, S., 2001. Thermal processing of sewage sludge by drying, pyrolysis, gasification and combustion. Water Science and Technology 44 (10), 333–339.

Stratful, I., Brett, S., Scrimshaw, M.B., Lester, J.N., 1999. Biological phosphorus removal, its role in phosphorus recycling. Environmental Technology 20, 681–695.

Susanne, R.S., 2010. Making the Best of Wastewater. Available at: www.biodieselmagazine.com/article-print.jsp?article_id=2776 (accessed April 2010).

Svanstrom, M., Patrick, T.N., Froling, M., Peterson, A.A., Tester, J.W., 2007. Choosing between green innovative technologies—hydrothermal processing of bio wastes. Journal of Advanced Oxidation Technologies 10, 177–185.

Takahashi, H., 2007. Study on sewage sludge gasification. In: Proceedings of the 80th Annual Technical Exhibition and Conference of the Water Environment Federation, San Diego, CA; October 13–17, pp. 4265–4272.

Taruya, T., Okuno, N., Kanaya, K., 2002. Reuse of sewage sludge as raw material of Portland cement in Japan. Water Science and Technology 46 (10), 255–258.

Tay, J.H., Show, K.Y., Hong, S.Y., Chien, C.Y., Lee, D.J., 2002. Potential reuse of wastewater sludge for innovative applications in construction industry. Bulletin of the College of Engineering, Nanyang Technological University 86, 103–112.

Tay, J.H., Show, K.Y., Lee, D.J., Hong, S.Y., 2004. Reuse of wastewater sludge with marine clay as a new resource of construction aggregates. Water Science and Technology 50 (9), 189–196.

Tay, J.H., Show, K.Y., 1997. Resource recovery of sludge as a building and construction material – a future trend in sludge management. Water Science and Technology 36, 259–266.

Tchobanoglous, G., Barton, F., Stensel, H., 2003. Wastewater Engineering Treatment and Reuse, fourth ed. Metcalfe and Eddy, New York.

Tezel, U., Tandukar, M., Pavlostathis, S.G., 2011. Anaerobic biotreatment of municipal sewage sludge. In: Moo-Young (Editor-in-Chief), M. (Ed.), Comprehensive Biotechnology. In: Agathos, S. (Ed.), Environmental Biotechnology and Safety, second ed., vol. 6. Elsevier, Amsterdam, The Netherlands.

ThermoEnergy, 2007. www.thermoenergy.com/munistors2.htm.

Thomson, N., Roy, I., Summers, D., Sivaniah, E., 2010. In vitro production of polyhydroxyalkanoates: achievements and applications. Journal of Chemical Technology and Biotechnology 85, 760–767.

Tian, Y., Zuo, W., Ren, Z., Chen, D., 2011. Estimation of a novel method to produce bio-oil from sewage sludge by microwave pyrolysis with the consideration of efficiency and safety. Bioresource Technology 102, 2053–2061.

Trumper, K., 2007. Energy and Sewage. Parliamentary Office of Science and Technology. Postnote. 282: 4 p. Available at: www.parliament.uk/documents/upload/postpn282.pdf (last access August 2007).

Tyagi, V.K., Lo, S.L., 2011. Application of physico-chemical pretreatment methods to enhance the sludge disintegration and subsequent anaerobic digestion: an up to date review. Reviews in Environmental Science and Biotechnology 10, 215–242.

Tyagi, V.K., Lo, S.L., 2013. Sludge: a waste or renewable source for energy and resources recovery? Renewable and Sustainable Energy Reviews 25C, 708–728.

U.S. EPA, 1994. A Plain English Guide to the EPA Part 503 Biosolids Rule. Report No. EPA/832/R-93/003. Office of Wastewater Management, Washington, DC.

Ueno, Y., Fujii, M., 2001. 3 years operating experience selling recovered struvite from full-scale plant. In: Proceedings of the 2nd International Conference on Recovery of Phosphates from Sewage and Animal Wastes, Holland, NL, March 12–13.

Uijterlinde, C., 2007. Sludge research STOWA. In: Presented at the WERF/GWRC Meeting, Alexandria, VA; July 16.

Wang, H., Brown, S.L., Magesan, G.N., Slade, A.H., Quintern, M., Clinton, P.W., et al., 2008. Technological options for the management of biosolids. Environmental Science and Pollution Research 15, 308–317.

Wang, M.J., 1997. Land application of sewage sludge in China. Science of the Total Environment 197, 149–160.

Weemaes, M., Verstraete, W.H., 1998. Evaluation of current wet sludge disintegration techniques. Journal of Chemical Technology and Biotechnology 73, 83–92.

Welp, J., Lundberg, L., Borghesi, J., Chattopadhyay, A., Copeland, B., D'acierno, J., et al., 2002. Thermal Oxidation of Sewage Solids. WEF white paper. 11 p. Available at: www.wef.org/NR/rdonlyres/979954FA-CAA0-45C2 91F94 D07107 F7A4B/0/ThermalOxidation.pdf (last access October 2007).

Wiebusch, B., Seyfried, C.F., 1997. Utilization of sewage sludge ashes in the brick and tile industry. Water Science and Technology 36, 251–258.

Wiesbusch, B., Ozaki, M., Watanabe, H., Seyfried, C.F., 1998. Assessment of leachingtests on construction materials made of incinerator ash (sewage sludge): investigations in Japan and Germany. Water Science and Technology 38, 195–205.

Wong, W.T., Chan, W.I., Liao, P.H., Lo, K.V., Mavinic, D.S., 2006. Exploring the role of hydrogen peroxide in the microwave advanced oxidation process: solubilisation of ammonia and phosphates. Journal of Environmental Engineering and Science 5, 459–465.

Woods, N.C., Sock, S.M., Daigger, G.T., 1999. Phosphorus recovery technology modeling and feasibility evaluation for municipal wastewater treatment plants. Environmental Technology 20, 663–679.

Xie, B., Liu, H., Yan, Y., 2009. Improvement of the activity of anaerobic sludge by low-intensity ultrasound. Journal of Environmental Management 90, 260–264.

Xie, R., Xing, Y., Ghani, Y.A., Ooi, K., Ng, S., 2007. Full-scale demonstration of an ultrasonic disintegration technology in enhancing anaerobic digestion of mixed primary and thickened secondary sewage sludge. Journal of Environmental Engineering and Science 6, 533–541.

Yoshikawa, K., 2010. Hydrothermal Treatment. WTERT Japan. http://www.wtert.jp/index.php?option=com_content&view=category&layout=blog&id=50&Itemid=187&lang=en.

Zabranska, J., Dohanyos, M., Jenicek, P., Kutil, J., Cejka, J., 2006. Mechanical and rapid thermal disintegration methods of enhancement of biogas production—full-scale applications. In: Proceedings of the IWA Specialized Conference on Sustainable Sludge Management: State-of-the-art, Challenges and Perspectives. Moscow, Russia; May 29–31, pp. 235–241.

Zhang, B., Xiong, S., Xiao, B., Yu, D., Jia, X., 2011. Mechanism of a wet sewage sludge pyrolysis in a tubular furnace. International Journal of Hydrogen Energy 36, 355–363.

Zhang, G.Y., Zhang, L.P., Liu, Z., 2003. Implementation methods and performance investigation for the amino acid chelated trace metal fertilizers. New Technologies in Applied Agriculture 3, 18–20 (in Chinese).

CHAPTER

CHROMITE

11

N. Koleli, A. Demir
Mersin University, Mersin, Turkey

CHAPTER OUTLINE

11.1 CHROMITE RESOURCES: GLOBAL AND IN TURKEY

Chromite (Cr) occurs exclusively in mafic and ultramafic igneous rocks (Papp and Lipin, 2006). Although some Cr-based minerals such as crocoite ($PbCrO_4$), vauquelinite ($Pb_2Cu[CrO_4][PO_4]$), uvarovite ($Ca_3Cr_2[SiO_4]_3$), and merumite ($4(Cr,Al)_2O_3 \cdot 3H_2O$) are well known, chromite ($FeCr_2O_4$) is the only one that is a commercially recoverable source of Cr (Sueker, 2006; Glastonbury et al., 2010; Murthy et al., 2011; van der Merwe et al., 2012; Sanchez-Segado et al., 2015). Chromite is an oxide mineral belonging to the spinel group, a complex mineral containing magnesium, iron, aluminum, and Cr in varying proportions, depending on the deposit, that can be represented as $FeCr_2O_4$ or $(Fe^{2+},Mg)[Cr,Al,Fe^{3+}]_2O_4$ (Murthy et al., 2011; Sanchez-Segado et al., 2015). Fig. 11.1 shows crystal structures of chromite spinel (Sanchez-Segado et al., 2015). Large variations in the total and relative amounts of Cr and iron in the lattice occur depending on the different geological and geographical distribution of chromite deposits (Rao et al., 2010).

Commercial chromite deposits that are large enough for mining are generally found in: (1) stratiform deposits, and (2) podiform deposits (Murthy et al., 2011; Papp and Lipin, 2001a). Other sources

Environmental Materials and Waste. http://dx.doi.org/10.1016/B978-0-12-803837-6.00011-1

FIGURE 11.1

Crystal structures of chromite spinel: *red*, O^{2-}; *blue*, Al^{3+}; *orange*, Fe^{2+}; *white*, Mg^{2+} (0.066 nm); *brown*, Cr(III) (0.076 nm).

From Sanchez-Segado, S., Makanyire, T., Escudero-Castejon, L., Hara, Y., Jha, A., 2015. Reclamation of reactive metal oxides from complex minerals using alkali roasting and leaching – an improved approach to process engineering. Green Chemistry 17, 2059–2080.

of chromite are laterites that are weathering products of peridotite. Laterites are more widely known as sources of nickel and cobalt (Papp, 2010). Laterites have historically been a minor source of chromite (Papp and Lipin, 2001a; Papp, 2010). Fig. 11.2 shows a map of locations, geological types, and the size of global chromite resources and reserves. As seen in Fig. 11.2, laterites are evaluated with podiform deposits. Most of the world's chromite resources occur as stratiform deposits in layered intrusions such as South Africa, Canada, Finland, and Madagascar (Papp and Lipin, 2001a,b; Papp, 2010). South Africa holds approximately three-quarters of the world's viable chromite reserves (Loock et al., 2014). The most important historic sources of chromite from podiform deposits are in Kazakhstan, Turkey, and Albania (Papp and Lipin, 2001a,b; Papp, 2010). Zimbabwe is the only country to exploit both stratiform and podiform deposits (Fig. 11.2). Mining of chromite is carried out by open-pit and underground mining. Whereas open-pit mining is generally applied to podiform deposits, underground mining is applied to stratiform deposits. Chromite ore is mined in over 20 countries, but about 80% of production originates from four countries: South Africa, India, Kazakhstan, and Turkey (Kleynhans et al., 2012).

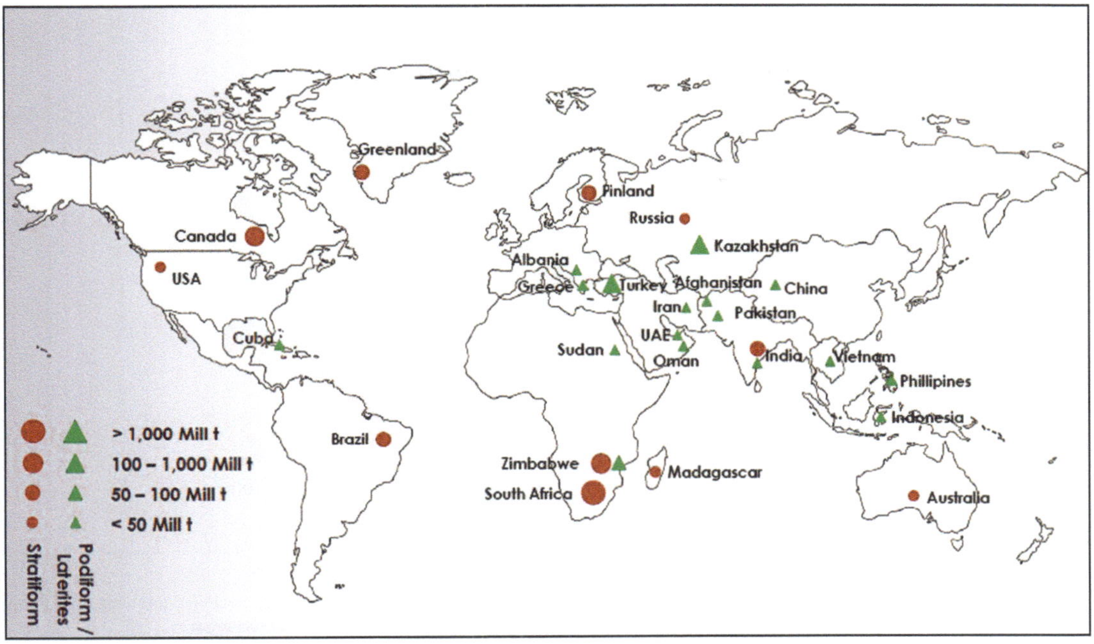

FIGURE 11.2

Map of locations, geological types, and size (millions of tons of Cr) of global chromite resources and reserves.

From Papp, J.F., 2007. Chromium. A National Mineral Commodity Perspective, Report: USGS Numbered Series, Open-File Report, 48 p.

After the discovery of elemental Cr by Nicolas-Louis Vauquelin in 1797, in the early 19th century the use of Cr was mainly limited to the textile industry, in which potassium dichromate was employed in the dying process. Use of Cr in leather tanning salts and paint pigments quickly followed. The first patent for the use of Cr in steel was granted in 1865, although the large-scale employment of Cr did not occur until the early 1900s when the electric arc furnace started widely being used and could smelt chromite into the master alloy (alloy was used as a feedstock to produce other alloys). By the early 1930s, Cr was commonly used in metal finishing and stainless steel. Since 1960, major changes have occurred in the Cr industry because of developments in steel-making technology (Papp and Lipin, 2001a; Papp, 2010).

According to data from the International Chromium Development Association (ICDA), global chromite ore and concentrate production totaled 29.4 metric tons in 2014 (Richard, 2015). Data from the US Geological Survey (USGS) estimated world production in 2014 at 29 metric tons, with South Africa at 15 metric tons, followed by Kazakhstan at 4 metric tons, India at 3 metric tons, and Turkey at 2.4 metric tons. Other countries producing chromite ore include Albania, Australia, Brazil, Finland, Iran, Madagascar, Oman, Pakistan, Philippines, Russia, and Zimbabwe (Fig. 11.3).

Timetric's "Global Chromium Mining to 2020" report comprehensively covers global reserves of chromite, the historical and forecast data on global chromite ores and concentrates production, and production by country and trade. According to this report, global reserves as of January 2015 were

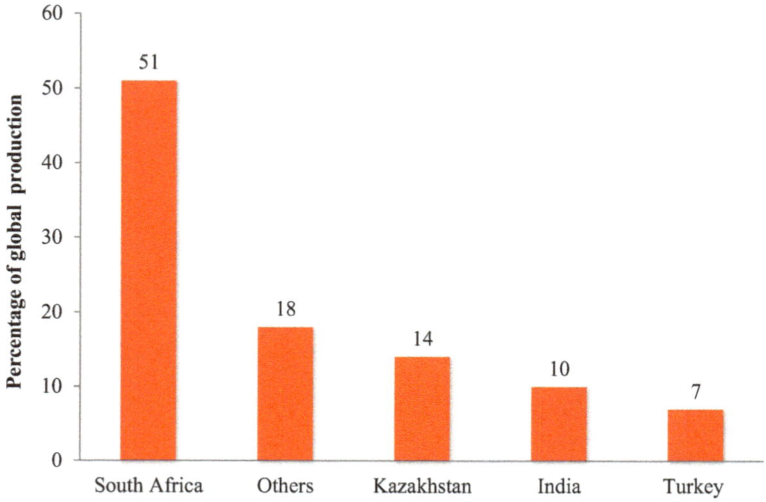

FIGURE 11.3

World chromite production capacity of major producers.

Source: Data from USGS; Papp, J.F., 2015. Chromium: U.S. Geological Survey (USGS) Mineral Commodity Summaries 2015, pp. 42–43. Available at: http://minerals.usgs.gov/minerals/pubs/mcs/2015/mcs2015.pdf (accessed 10.08.15.).

484.6 metric tons, with three countries accounting for 95% of the total: Kazakhstan at 230 metric tons, followed by South Africa (200 metric tons) and India (54 metric tons) (Global Chromium Mining to 2020). The remainder is found in the United States, Turkey, Finland, and Brazil. This report noted that the main area of application of chromite ore is in metallurgical processes, after conversion into ferrochrome; approximately 90% of all chromite ore is used to produce stainless steel. The steady growth in global chromite ore sand concentrates production resulted from a rise in demand from the ferrochrome and stainless-steel markets, with China accounting for 36% and 52% of global ferrochrome and stainless-steel production, respectively (Research and Markets, 2015).

Fig. 11.4 shows the chromite production capacity of major producers between 1994 and 2014 based on data from USGS across the world. Chromite ore is mined in over 20 countries, but about 80% of production originates from four countries; in 2007–2009, South Africa accounted for 37% of the world's production, with both India and Kazakhstan accounting for around 16% and Turkey for 8% (Papp, 2008, 2009; ICDA, 2010).

Turkey is the world's fourth largest producer of chromite after South Africa, Kazakhstan, and India. Turkey has podiform chromite deposits located in six major areas (Fig. 11.5). As seen in Fig. 11.5, in order of importance, the regions are: Guleman (Elazığ, East Turkey), Kopdag (Erzincan-Erzurum, East Turkey), Fethiye-Koycegiz-Denizli (South West Turkey), Bursa-Kutahya-Eskisehir (Nort West Turkey), Mersin-Aladag-Pinarbasi (South Turkey), and İskenderun-K.Maras (South Turkey) (Engin, 2002).

Chromite mining in Turkey started in 1850; since then, Turkey has had an important role in the world chromite market. Turkey produces chromite ore and ferrochrome. It is estimated that Turkey has produced about 45 metric tons of chrome ore. Chromite production is mainly export-oriented. Turkey

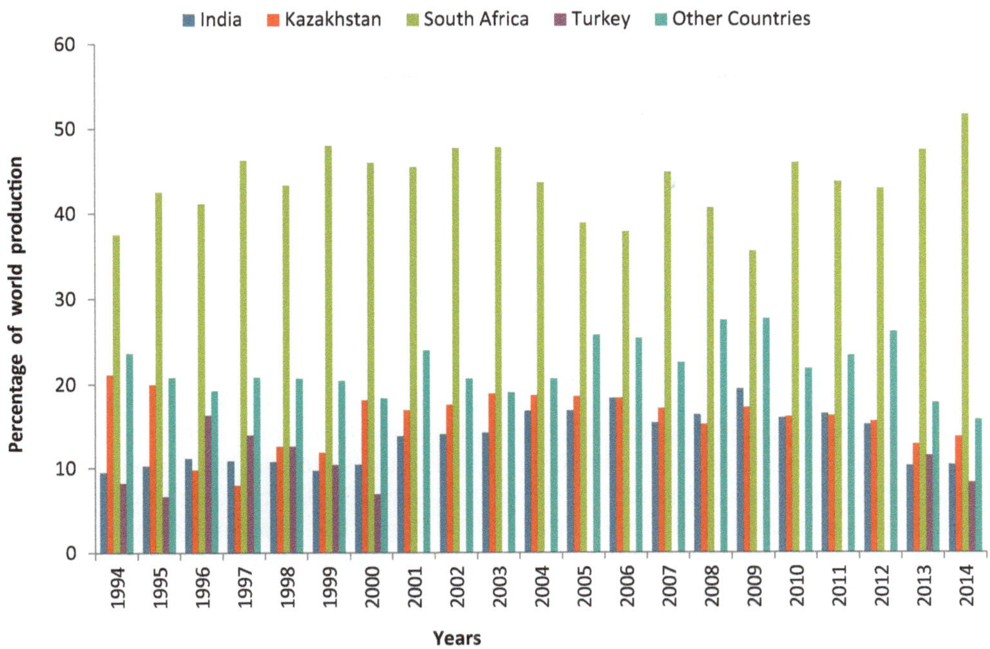

FIGURE 11.4

World's chromite production capacity of major producers between 1994 and 2014.

Source: Data from USGS; Papp, J.F., 2015. Chromium: U.S. Geological Survey (USGS) Mineral Commodity Summaries 2015, pp. 42–43. Available at: http://minerals.usgs.gov/minerals/pubs/mcs/2015/mcs2015.pdf (accessed 10.08.15.).

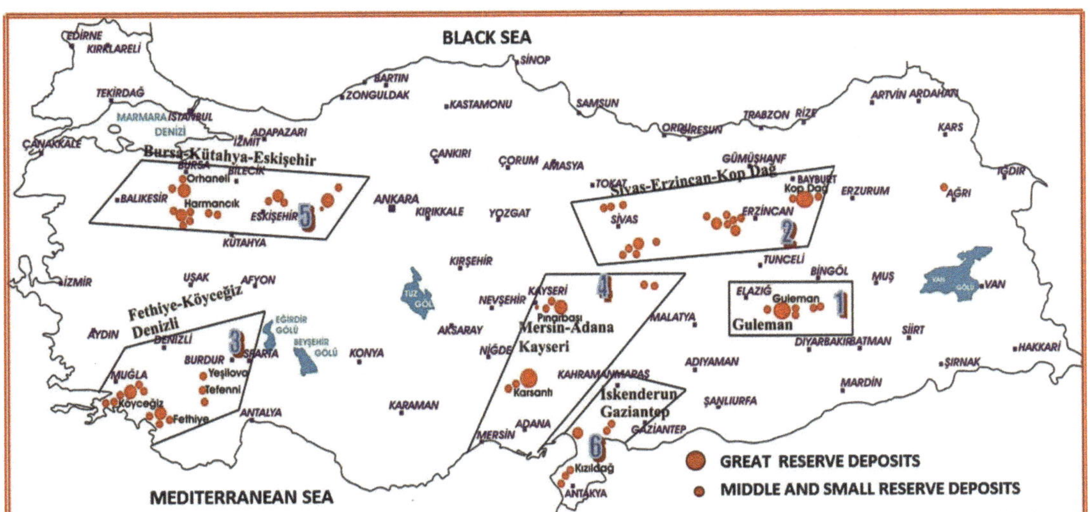

FIGURE 11.5

Chromite deposits located in six major areas in Turkey (*red points*).

Source: Data from Mineral Research & Exploration General Directorate (MTA); Chromite deposits in Turkey . Available at: http://www.mta.gov.tr/v2.0/images/turkiye_maden/maden_yataklari/b_h/krom.jpg (accessed 10.08.15.).

has one chrome and two ferrochrome chemical plants. A typical mineral composition of Turkish chromite ore is composed of Cr_2O_3 (48.5%), Al_2O_3 (10.01%), FeO (13.28%), MgO (18.83%), and CaO (9.38%) with a Cr:Fe ratio of 3:25 (Yalcin, 2003).

Turkey has a 6% share in the world's chromite mining and possess 25 million tons of reserves (Uyanik, 2010). In Aladag (Adana, South Turkey) a low-grade chromite deposit (5.42% Cr_2O_3 grade) with 200 metric tons reserve has yet to be exploited. Chromite demand in Turkey is mostly met by metallurgic-type ores. Chromite ore is one of Turkey's major income sources in primary mineral exports. Ferrochrome production of Turkey was different between 2007 and 2011, at 69.73–60.00 metric tons. The demand for chemical-type ores has also increased with the operation of sodium bichromite-producing factories. The annual demand for metallurgic ores is estimated to be 442,000 tons. The produced chromite ore is consumed mostly by the iron-steel (about 85%) and chemical industries (about 15%) (Yalcin, 2003). Etibank, a state-owned company, is Turkey's major chromite ore producer and the only producer of ferrochrome. Etibank (Eti Krom A.S.) operated a low-carbon ferrochrome smelter in Antalya and a high-carbon ferrochrome plant in Elazıg (Mobbs, 2013).

The ICDA reports world production of chromite by end-use sector and categorizes it as chemical, foundry sand, metallurgical, and refractory (Papp and Lipin, 2006). According to data from ICDA, global chromite use totaled 29.4 metric tons in 2014 (Fig. 11.6). Approximately 96% of mined chromite is consumed by the metallurgical industry for the production of different grades of ferrochrome, which is available in low-carbon, high-carbon, or charge-grade forms. The stainless-steel industry consumes 80–90% of ferrochrome, primarily as high-carbon or charge-grade ferrochrome (Murthy et al., 2011; Kleynhans et al., 2012; ICDA, 2010; Abubakre et al., 2007). Approximately 2.0% of the demand comes from the chemical industry, and the rest from refractory (0.20%) and foundry sand industries (2%) (Richard, 2015). Any change in the dynamics of the stainless-steel industry, which is the largest consumer of ferrochrome, affects overall chrome ore demand significantly.

Fig. 11.7 shows Cr chemicals used by industry in 2015. The largest nonmetallurgical market, chemicals, may further be subdivided into leather tanning, 32%; chrome plating, 20%; chrome metal, 20%; pigments, 15%; wood treatment, 9%; and others, 20% (Richard, 2015).

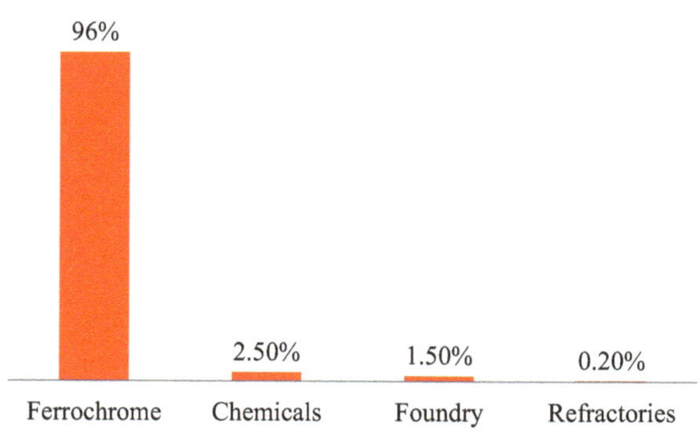

FIGURE 11.6

World chromite market in 2014 (total of 29.4 metric tons).

Source: Data from Richard, P., 2015. ICDA: overview of the global chrome market. In: 1st Indinox Stainless Steel Conference, Ahmedabad, India.

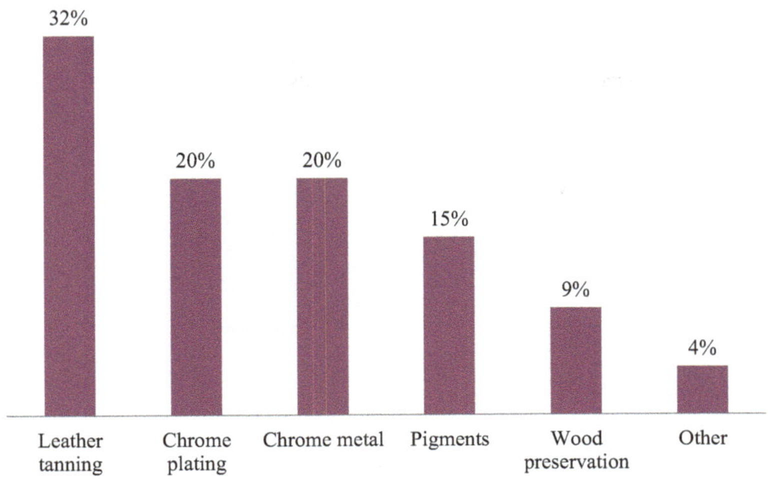

FIGURE 11.7

World chromite chemicals market in 2014 (total of 725,000 tons).

Source: Data from Richard, P., 2015. ICDA: overview of the global chrome market. In: 1st Indinox Stainless Steel Conference, Ahmedabad, India.

Chromium(VI) compounds are widely used as corrosion inhibitors, in the manufacturing of pigments, metal finishing and chrome plating, stainless-steel production, in leather tanning, and in wood preservatives (Costa, 1997; Agency for Toxic Substances and Disease Registry (ATSDR), 2000). The United States is one of the world's leading producers of Cr compounds.

11.2 RAW MATERIALS USED IN INDUSTRY AND WASTE GENERATED

This section focuses on threats directly associated with raw materials used in chromite industry and its wastes. To facilitate discussions about raw materials used in chromite industry and waste generated, a generalized Cr material flow diagram, from chromite ore mining to end use, is shown in Fig. 11.8. Fig. 11.8 represents a material flow diagram for Cr, from import to export, from manufacturing to final product, and from use to release. As seen in Fig. 11.8, mining includes screening and hand sorting; beneficiation includes crushing, grinding, and separation based on gravimetric, heavy media, and magnetic methods (Papp and Lipin, 2006; Papp, 1990). Residues from mining and processing chromite are often toxic not only because of the presence of Cr(VI) but also because of the agents and other waste materials in overburden and wastes (Mining Watch Canada, 2012a,b). The types of waste generated are explained in detail next.

11.2.1 WASTES OF CHROMITE MINING

Chromite mining wastes have the potential to cause serious environmental contamination. This occurs in places around the world where these wastes are inadequately managed.

Overburden is an important type of mine waste. Overburden includes the soil and rock that is removed to gain access to ore deposits at open cast mines. It is usually piled on the surface at mine sites

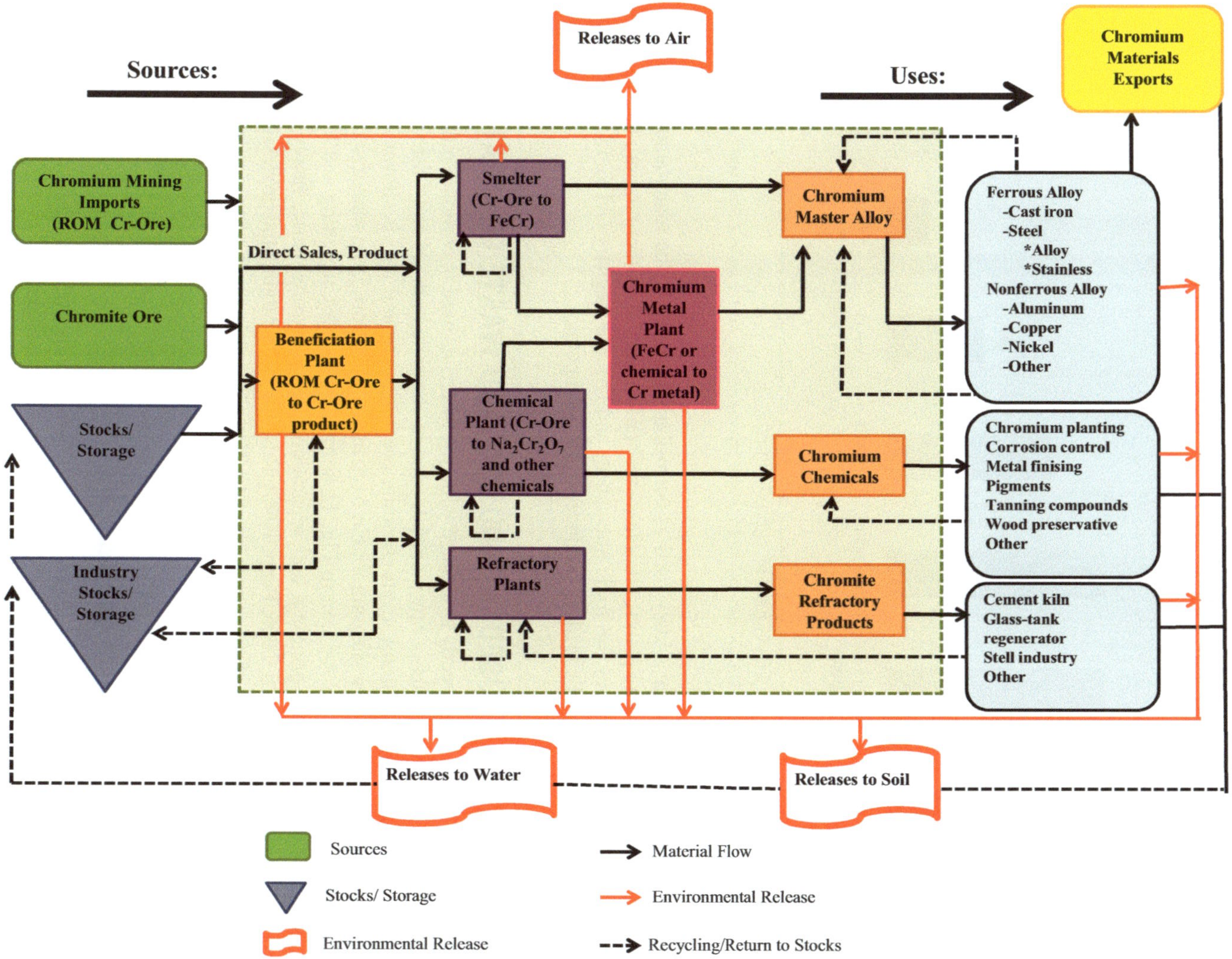

FIGURE 11.8

Chromium material flow diagram from chromite ore mining to end use. *Cr-ore*, chromite ore; *FeCr*, ferrochrome; *ROM*, run of mine.

Adapted from Papp, J.F., 1994. Chromium Life Cycle Study, Report, Federal Government Series, Information Circular 9411. U.S. Bureau of Mines, 94 p.; Papp, J.F., 2002. Chromium. United States Geological Survey (USGS).

where it will not impede further expansion of the mining operation, because moving large volumes of material is expensive. Sukinda Valley, located in India, is highly studied in the literature because the application of the open cast mining process since 1950 is a good sample of mining wastes of chromite ore. Mishra and Sahu (2013) reported that the environmental impacts of chromite mining in Sukinda Valley are caused by the presence of hexavalent Cr in mine drainage water, dust generation, overburden generation, and loss of biodiversity and forest. Open cast chromite mining generates a large quantity of overburden (Mishra and Sahu, 2013). If overburden is not protected or treated properly, the runoff from the overburden can pollute the nearby surface as well as groundwater bodies owing to leakage of hexavalent Cr. Overburden generally has a low potential for environmental contamination and often is used at mine sites for landscape contouring and revegetation during mine closure (Rankin, 2011).

A large amount of dust is generated during chromite mining activity, particularly from drilling, blasting, and transportation. Stacking and loading of both ores and chromite overburdens also generate a great amount of particulate matter. The dust is mostly chromite particles but sometimes may contain traces of hexavalent Cr, which has an adverse impact.

For example, in India, in the Sukinda mining area, around 7.6 metric tons of solid wastes have been generated in the form of rejected minerals, overburden material and waste rock, and subgrade ore that have caused environmental degradation, mainly creating lowering of the water table as well as deterioration of surface and groundwater quality (Mohanty and Patra, 2011; Tiwary et al., 2005). According to Indian standards, the permissible limit of hexavalent Cr for potable water is 0.05 mg/L, and for industrial discharge water is 0.1 mg/kg (Pattanaik et al., 2012). In the Sukinda area, the lateralization process involving oxidation and alternation of the serpentines creates alkaline pore water, which facilities the generation of Cr(VI) from inert chromites and causes hazardous Cr pollution of water (Das and Singh, 2011).

11.2.2 WASTES OF BENEFICIATION

As seen in Fig. 11.8, imported chromite is sent to smelters, chemical plants, and refractories, through either direct sales or domestic beneficiation plants. Beneficiation begins with the milling of mined run of mine (ROM) ore to make the ROM ore suitable for further activities to recover Cr values. The processing and beneficiation techniques differ and depend on the characteristics of the ore found in the region. Generally the process starts with milling operations (such as crushing and grinding) to produce liberated, uniform-sized particles followed by enrichment of the liberated ore by a combination of gravity separation techniques. Gravity concentration techniques have already been adopted in several mines worldwide (Gokarn et al., 2003). The recovery of Cr units from ferrochrome wastes up to now has been mainly associated with physical separation methods such as jigging, magnetic separation, dense media separation, flotation, shaking tables, and spirals. These processes have also predominantly been limited to the recovery of >75-μm particles; waste streams consisting of smaller particles have been somewhat neglected. This classification technique has the advantage of the specific gravity difference between chromite and gangue minerals.

11.2.3 WASTES OF SMELTERS

Most of the world's production of chromite (about 96%) is used in the metallurgical industry in the form of ferrochrome alloys. The alloys are produced by high-temperature reduction (smelting) of chromite.

They are essentially alloys of iron and Cr with much lower amounts of carbon and silicon; the amounts depend on the grade or type of alloy and impurities such as sulfur, phosphorus, and titanium. The conversion of chromite to ferrochrome alloys is dominated by electric submerged arc furnace smelting with carbonaceous reductants, predominantly coke, and fluxes to form the correct slag composition. Slag and furnace dusts (flue dust) are generated as waste in enormous quantities during ferrochrome production. These flue dust particles released into the environment during the ferrochrome production process are the source of high levels of toxic elements that pose ecological problems (Cohen and Petrie, 1997). As seen in Fig. 11.8, the increase in environmental concentrations of Cr has been linked to air, soil, groundwater, and wastewater release of Cr, mainly from metallurgical (steel, ferrous, and nonferrous alloys), refractories (chrome and chrome–magnesite), foundry sand, and chemical industries (pigments, electroplating, tanning, and others) (Stanin and Pirnie, 2004).

11.2.3.1 Waste of Refractory Industry

Refractory plants create products such as cement kiln, glass-tank regenerators, and other instruments for the steel industry. During these processes, a considerable amount of material is lost. As an example, if a Cr ink or dye is used and disposed of, it is dispersed into the environment and is no longer recoverable (Allen, 2003).

11.2.3.2 Waste of Metallurgical Industry

Approximately 96% of mined chromite is consumed by the metallurgical industry for the production of different grades of ferrochrome. Ferrochrome is made by smelting chromite ore in an electric arc furnace with flux materials (quartz, dolomite, limestone, and aluminosilicates) and a carbonaceous reductant (wood chips, coke, or charcoal) (Kleynhans et al., 2012; USEPA, 1998). The solid products obtained from smelting ferrochrome are metal, slag, and dust. In open top furnaces, the dust is collected in a bag filter plant, whereas in a closed top furnace, the dust is scrubbed in a venturi system and is produced as a slurry. Slags are nonmetallic and have been historically considered a waste product (Rankin, 2011).

Ferrochrome is currently mainly produced with (1) a conventional semiclosed or open submerged arc furnace operation, with bag filter off-gas treatment; (2) a closed submerged arc furnace operation that usually uses oxidative sintered pelletized feed, with venturi scrubbing of off-gas; (3) a closed submerged arc furnace operation consuming prereduced pelletized feed, with venturi scrubbing of off-gas; and (4) a closed direct current arc furnace operation, with venturi scrubbing of off-gas (Beukes et al., 2010). According to Beukes et al. (2010), the main waste products of the ferrochrome industry are slag, as well as bag filter dusts and scrubber sludge. Of these, bag filter dust is the most hazardous because it contains some Cr(VI) and is a fine dry material that is subject to wind dispersion if not contained properly (van der Merwe, 2011). Scrubber sludge is also a fine material, but because it originates from closed furnaces (Beukes et al., 2010) it does not contain as much Cr(VI) and is not subject to wind dispersion. By volume, slag is the most significant waste material produced by the ferrochrome industry. Beukes et al. (2010) estimated that approximately 5.3 metric tons of slag was produced in South Africa in 2007 alone (van der Merwe, 2011).

11.2.3.3 Waste of Chemical Industry

Sodium chromate and dichromate are produced at two facilities by a hydrometallurgical process during which ground chrome ore and soda ash are mixed (lime and/or leached calcine are sometimes added as well), roasted in an oxidizing atmosphere, and leached with weak chromate liquor or water (USEPA, 1998). Sodium dichromate can be converted into both anhydrous chromic oxide and hydrated chromic oxide. To produce anhydrous chromic oxide, sodium dichromate, sulfur, and wheat flour are blended with water, and the resultant slurry is heated in a kiln. The material recovered from the kiln is slurried with water, filtered, washed, dried, ground to size, screened, and packaged (USEPA, 1998). Chromium metal can be made either pyrometallurgically or electrolytically. In the pyrometallurgical method (not shown), chromium oxide (Cr_2O_3) reacts with aluminum powder in a refractory lined vessel after being ignited with barium peroxide and magnesium powder. Chromium metal may also be made from the oxide by reduction with silicon in an electric arc furnace (USEPA, 1998). Chromium in chemical industry wastes occurs predominantly in the hexavalent form. Treatment of Cr(VI) waste often involves reduction to Cr(III) and precipitation as hydrous oxide with lime or caustic soda. Chromium(III) waste can also be converted into hydrous oxide or may be incinerated to form an oxide before land disposal. Land filling appears to be the most important method for the disposal of Cr wastes generated by chemical industries. Of the total Cr released into the environment by the chemical industries, approximately 82.3% is released into land. An equally large amount of Cr waste is transferred off-site. It is anticipated that most of this off-site waste is disposed into landfills after proper treatment.

11.2.3.4 Waste of Foundry Sand Industry

The vast majority of chromite sand used in the foundry industry originates in South Africa (Amcol Metalcasting, 2013). South Africa is the reference standard for quality and has the most reserves to supply the global foundry industry. Chromite sand is the best sand medium for chilling metal. Chromite sand has the ability to draw heat from the molten metal, allowing the metal to skin over faster and minimizing burn-on. Excessive burn-on will need to be removed before heat treating, but it can also lead to other issues, causing castings to be scrapped. Foundry-grade chromite sand consists of 46% chromic oxide and 1.0% silica. Although many states have developed beneficial reuse regulations for industrial by-products, large quantities of foundry by-products are still being land filled in the United States (Abichou et al., 2004).

11.3 TOXIC EFFECT OF CHROMIUM WASTES ON HUMANS

Chromium is the 22nd most abundant element in Earth's crust, with an average concentration of 100 mg/kg (Emsley, 2003). Chromium is listed by the United State Environmental Protection Agency (USEPA) as one of the 126 priority pollutants (USEPA, 1992). In humans and animals, Cr(III) is an essential nutrient that has a role in glucose, fat, and protein metabolism by potentiating the action of insulin (Tchounwou et al., 2012). Most daily vitamin formulations contain between 50 and 200 μg Cr. Compared with Cr(III), Cr(VI) is extremely toxic and is considered by the World Health Organization and USEPA to be a human carcinogen. Hexavalent Cr(VI) can move more easily into the cells of plants, animals, and microorganisms, so it has a greater potential for toxicity (Mining Watch). Moreover, oxidation and reduction reactions can convert Cr(III) to Cr(VI), and vice versa (James, 1994; Kotas and Stasick, 2000). Even the metallic form of Cr, Cr(0), is oxidized to Cr(III) unless it is made passive by superficial oxidation (Beukes et al., 2012). Chromium is one of most dangerous heavy metals in terms of environmental health because of the interconversion of Cr forms.

The USEPA regulates Cr releases into the environment, and the Occupational Safety and Health Administration also regulates workplace exposure (Papp and Lipin, 2001b). People who live near industrial facilities that use Cr(VI) compounds or near Cr waste disposal sites have the greatest potential for exposure (Agency for Toxic Substances and Disease Registry (ATSDR), 2000; Niehs). Chromium is usually released into the environment from anthropogenic activity that occurs mainly in the hexavalent Cr(VI) form (Tchounwou et al., 2012; Abubakre et al., 2007). It is estimated that 33 tons of total Cr is released annually into the environment (ATSDR, 2012; Tchounwou et al., 2012). However, occupational exposure has been a major concern because of the high risk of Cr-induced diseases in industrial workers occupationally exposed to Cr(VI) (Guertin, 2005; Tchounwou et al., 2012).

Beginning in the mid-20th century, cohort studies were undertaken in these industries as well as in some other occupations and industries with potential exposure to Cr compounds, such as ferrochrome or stainless-steel production and leather tanning. By the 1980s, considerable evidence had been accumulated on cancer risks of Cr-exposed workers, which led to the identification of Cr(VI) compounds as a human carcinogen (IARC (International Agency for Research on Cancer), 1997). The strongest evidence presented at the time concerned the lungs. Lehmann (1932) was the first to suggest that a link between occupational exposure to chromates and increased risk of lung cancer. Inhalation and retention of Cr(VI)-containing materials can cause perforation of the nasal septum, asthma, bronchitis, pneumonitis, inflammation of the larynx and liver, and increased incidence of bronchogenic carcinoma (Gad et al., 1986; Lee et al., 1989; Bielicka et al., 2005). In a study of 97 workers from a chromate plant who were exposed to a mixture of insoluble chromite ore containing Cr(III) and soluble Cr(VI) as sodium chromate and dichromate, evaluation for respiratory effects revealed that 63% had perforations of the nasal septum, 86.6% had chemical rhinitis, 42.3% had chronic chemical pharyngitis, 10.35% had laryngitis, and 12.1% had sinus, nasal, or laryngeal polyps (ATSDR, 2012). Typical concentration ranges for airborne Cr(VI) in industries that use Cr(VI) compounds are as follows: stainless-steel welding, 50–400 $\mu g/m^3$; chromate production, 100–500 $\mu g/m^3$; chrome plating, 5–25 $\mu g/m^3$; ferrochrome alloy production, 10–140 $\mu g/m^3$; and chromate pigment production, 60–600 $\mu g/m^3$ (Agency for Toxic Substances and Disease Registry (ATSDR), 2000; Niehs; IARC (International Agency for Research on Cancer), 1997).

Skin contact of Cr(VI) compounds can induce skin allergies, dermatitis, dermal necrosis, and dermal corrosion (Gad et al., 1986; Lee et al., 1989). The toxic properties of chromates arise from the possibility of free diffusion across cell membranes and strong oxidative potential. The toxicological impact of Cr(VI) originates from the action of this form itself as an oxidizing agent, as well as from the formation of free radicals during the reduction of Cr(VI) to Cr(III) occurring inside the cell (Kotas and Stasick, 2000; Bielicka et al., 2005).

The principal route by which Cr(III) enters the body is the digestive system. Chromium in foods is present in both the inorganic form and inorganic complexes. The body has several systems for reducing Cr(VI) to Cr(III). Thus, Cr(VI) detoxification leads to increased levels of Cr(III). Too much Cr(III) can be toxic to plants and animals (Mining Watch Canada, 2012a,b). Trivalent Cr(III) formed there in a significant concentration can cause further adverse effects because of its high capability of coordinating various organic compounds, resulting in inhibition of some metalloenzyme systems (Kotas and Stasick, 2000). If Cr (VI) is ingested by humans, it will be absorbed into tissue 53 times faster and readily than Cr(III) (Yu et al., 2012; Freese et al., 2014). Hexavalent Cr(VI) is more mobile and toxic than the trivalent state. Hexavalent(VI) is also about 1000-fold more cytotoxic and mutagenic than Cr(III) (Freese et al., 2014; Kanchinadham et al., 2013).

The USEPA lists Cr(VI) as one of the top contaminants, posing a significant threat to human health and the environment (Freese et al., 2014). Therefore, the use of chromite ore-processing residue (COPR) as a fill material has been banned owing to its classification as a hazardous waste (Freese et al., 2014). Nevertheless, the old fill sites still pose a serious environmental and health concern because Cr continuously leaches from COPR over long periods of time (Freese et al., 2014).

11.4 RECOVERY OF USEFUL PRODUCTS FROM CHROMITE INDUSTRY

Although the intention of chromite mining is to remove chromite from the ore for the production of ferrochrome, the process is not 100% efficient, so mining and smelting wastes including dust, wastewater, tailings, waste rock, and slag (produced during ferrochrome separation from the ore) still can contain different forms of Cr, including Cr(VI). Other heavy metals and chemicals of concern may also be in the wastes (Mining Watch).

The wastes of chromite mining (especially open cast mining) results in the generation of waste materials containing unrecovered chromite. The huge volumes of these waste materials are substantial, although no reliable estimate currently exists (probably millions of tons per year). Chromite is relatively inert (Gu and Wills, 1988) and the Cr in these waste exists almost exclusively as Cr(III). They are generally transferred to stockpile or landfill without further treatment, because landfill is the simplest, cheapest, and most cost-effective method of disposing of waste (Barrett and Lawlor, 1995). Storage of waste in a disposal facility minimizes the effects of waste. Today, the USEPA lists Cr(VI) as one of the top contaminants posing a significant threat to human health and the environment (Freese et al., 2014). Therefore, the use of COPR as a fill material has been banned because of its classification as a hazardous waste (Freese et al., 2014). Nevertheless, the old fill sites still pose a serious environmental and health concern because Cr continuously leaches from COPR over long periods of time (Freese et al., 2014). Waste disposal by landfill has led to the pollution of groundwater resources. Therefore, these sites require long-term monitoring.

However, most discarded waste can be reused or recycled, according to one of the principles of most waste management philosophies (Taylor and Allen, 2006). The key strategy is to minimize and reuse the volume of such hazardous waste. Waste is a source of secondary raw materials (Rao et al., 2010). Rapid industrialization and the demand for heavy metals are increasing because the reserves of high-grade ores are diminishing each day. Therefore, there is a need to explore alternative sources of heavy metals. Management of huge volumes of overburden at chromite mines is a concern, and there is a need to make effective use of this material. The chromite mining overburden, however, has appreciable iron (45–55%), Cr (5–20%), and Ni content. Overburdens in world are generally disposed of safely by capping the landfill site. Kapure et al. (2011) studied the feasibility of the direct reduction process for low-grade chromite overburden on a bench scale and 40% and 90% Fe and 90% Ni were recovered in the rotary hearth furnace reduction process at a cycle time of 25 min. Thus, the volume of overburden waste at chromite mines was potentially reduced. Today, many pyrometallurgical and hydrometallurgical methods are studied mainly to extract valuable metals. Valuable metals (Ni, Cr, and Fe) can be recovered from these industrial wastes. Pyrometallurgical recovery consists of the thermal treatment of metal-containing wastes to bring about physical and chemical transformations (Jadhav and Hocheng, 2012). This enables the recovery of valuable metals. Calcining, roasting, smelting, and refining are pyrometallurgical processes used to recover metal. Hydrometallurgical recovery mainly employs the leaching process. It involves the use of aqueous solutions containing a lixiviant which is brought into contact with a material containing a valuable metal. Furthermore, the metals are concentrated and purified by using precipitation, cementation, solvent extraction, and ion exchange. The metals are finally recovered in pure form by using electrolysis and precipitation methods (Jadhav and Hocheng, 2012).

Industries with the largest contribution to Cr release include metal processing, tannery facilities, chromate production, stainless-steel welding, and ferrochrome and chrome pigment production (Kleynhans et al., 2012). In particular, Cr in the chemical industry wastes occurs predominantly in the hexavalent form. Land filling appears to be the most important method for the disposal of Cr wastes generated by the chemical industries. Of the total Cr released in the environment by the chemical industries, approximately 82.3% is released on land (Wilbur et al., 2012).

The best-known disposal method of waste in the literature is often reduction to Cr(III) of Cr(VI), because Cr in chemical industry wastes occurs predominantly in the hexavalent form. The treatment of Cr(VI) waste often involves reduction to Cr(III) and precipitation as hydrous oxide with lime or caustic soda. Chromium(III) waste can also be converted into hydrous oxide or may be incinerated to form oxide before land disposal. An equally large amount of Cr waste is transferred off-site. It is anticipated that most of this off-site waste will be disposed of in landfills after proper treatment (ATSDR, 2012).

Remediation, removal, immobilization, and reduction technologies are commonly used to reduce the environmental pollution potential of Cr. Some in situ or ex situ treatment technology such as pyrolysis, bioremediation, chemical reduction has been applied to convert hexavalent Cr to the less toxic trivalent form because COPR-contaminated sites still pose a serious environmental risk (Freese et al., 2014) compared with the some COPR treatment technologies. For ex situ treatments, pyrolysis of COPR soils was most effective in providing a long-term, permanent solution. Chemical reduction with sulfate-based materials or bioremediation was the most widely studied in situ treatment methods. Although it was effective, chemical reduction approaches could rebound to unacceptable Cr(VI) as a result of scavenging sulfate-based compounds or alteration of the pH. Bioremediation could be the most promising because the bacteria could help regulate the pH, thereby preventing the reduction of Cr(III) back to Cr(VI) (Freese et al., 2014).

The ferrochrome market demand is increasing daily, especially for global leading producers such as South Africa, Kazakhstan, Turkey, and India. In light of this, the ferrochrome industry is directing its efforts to use flue dust waste as a product to reuse and recycle to meet its environmental responsibility. So far, only sporadic attempts have been made to characterize and beneficiate such waste, and the data are still inadequate (Rao et al., 2010). Specific industry information is confidential, and reported data are suspect. Reported values of Cr consumption and disposal do not balance, and the USGS readily states that industries do not adequately report releases of Cr. The complex recycling of primary and secondary Cr sources further complicates an understanding of the flow of Cr (Allen, 2003).

Typical Cr recovery during ferrochrome smelting is 70–93%, depending on the specific smelting technology applied (Naiker and Riley, 2006; Cramer et al., 2004). Owing to increased cost, efficiency, and environmental pressures, all ferrochrome producers strive to reduce waste generation with associated lower Cr unit losses. The loss of Cr units in ferrochrome production wastes, ie, slag, bag filter dust and venturi sludge, is a significant contributor to the loss of revenue.

The flue dust generated from the ferrochrome industry is one such type of waste containing chromite minerals. The toxic nature of trivalent and hexavalent Cr makes flue dust from the ferrochrome industry hazardous. About 400–500 kg residue is generated per ton of ferrochrome metal produced; this residue includes slag, flue dust, and sludge (Rao et al., 2010). Increasing costs of waste disposal and strict environmental rules and regulations suggest that priority should first be given to minimizing, recycling, and/or metal recovery of waste from the ferrochrome industry (Rao et al., 2010; Mishra and Sahu, 2013).

However, the disposal method of waste cited in the literature is often oxidation of Cr(III) to Cr(VI) (van Staden et al., 2014). The liberation of Cr units from ferrochrome bag filter dust and the ultrafine fraction of slag with aqueous ozonation and the advanced oxidation process were investigated to convert Cr(III) to Cr(VI). Results indicated that limited Cr liberation could be obtained from waste materials using aqueous ozonation. For bag filter dust, only a maximum of 4.2% of total Cr liberation was achieved. Application of the advanced oxidation process by adding H_2O_2 during ozonation increased Cr liberation dramatically. More than 21% of total Cr liberation could be achieved for both of the waste materials used in this investigation. From the studies presented by van der Merwe et al. (2012) and Rodman et al. (2006), it seems possible that the conversion of insoluble Cr(III) to soluble Cr(VI) could be considered an alternative method (the advanced oxidation process) for Cr recovery from wastes. Although this route for Cr recovery seems counterintuitive, because Cr(VI) is generally considered to be carcinogenic, the risks associated with aqueous Cr(VI) are certainly much less than airborne Cr(VI), and these risks can be mitigated (Beukes et al., 2010).

Rao (2010) studied on recovery of chromite values from ferrochrome industry flue dust. Ferrochrome industry waste such as flue dust contains chromite minerals that are considered hazardous materials if left untreated, stockpiled, or land filled. Recovery of chromite values by physical beneficiation can be applied to support recycling; the remaining waste is transformed into nontoxic materials for safe disposal. It is beneficial with regard to saving raw material resources as well as in reducing Cr pollution (Rao et al., 2010).

Increasing costs of waste disposal and strict environmental rules and regulations suggest that priority should first be given to minimizing the flue dust from the ferrochrome industry. Then the possibilities for partial or full recycling and/or metal recovery should be investigated (Rao et al., 2010). In this research, gravity and magnetic separation wet methods were applied because the flue dust sample contains quartz, carbon and useful chromite of significant value, which have different phase densities and magnetic properties (Rao et al., 2010). However, toxic substances such as Cr(VI) can leach out during the wet mineral separation process (Rao et al., 2010), which necessitates further treatment of the wastewater from the process (Rao et al., 2010). However, wet beneficiation processes have specific advantages as well as disadvantages over dry beneficiation. For example, wet processes can be effective with regard to sharp separation of the constituent phases and produce a high-grade concentrate, whereas dry methods appear to be less sharp with regard to separating the phases. In the current case, better results were achieved: Cr_2O_3 content was enhanced from 25.6% to 40.14% with a weight recovery of 38.5% with the gravity method compared with the magnetic separation method (Rao et al., 2010). Gericke (1998) reported that heavy metals in the ferrochrome flue dust could be immobilized by mixing 20% ferrochrome dust with 80% or 50% clay with 30% ferrochrome dust and 20% ferrochrome slag. Maine et al. (2005) also used 50% clay and 50% ferrochrome bag filter dust and sintered it at 1200°C to stabilize the Cr(VI) and salts in the ferrochrome dusts. The results showed that more than 99% Cr(VI) and 90% salt in the ferrochrome dusts were stabilized. The Cr(VI) concentration in the leachate also decreased with increasing sintering temperature. These tests can be performed for tailings of the ferrochrome dust samples after sufficient Cr values are extracted by physical means for the stabilization of Cr(VI) values, if any (Rao et al., 2010).

Waste is a source of secondary raw materials but it can negatively affect the natural environment and public health (Rao et al., 2010). Flue dust generated from the ferrochrome industry is one such type of waste containing chromite minerals. The toxic nature of trivalent and hexavalent Cr makes flue dust from the ferrochrome industry hazardous.

ACKNOWLEDGMENTS

The authors wish to thank their colleague, Professor Dr. Cetin Kantar, University of Canakkale Onsekiz Mart, Canakkale, Turkey, and their friend, Dr. Abdi Kurt, University of Mersin, Mersin, Turkey, for their valuable suggestions and correction of the English text; and their doctoral student Zeynep Gorkem Dogaroglu for help with the figures.

REFERENCES

Abichou, T., Edil, T.B., Benson, C.H., Bahia, H., 2004. Beneficial use of foundry by-products in highway construction. In: Geotechnical Engineering for Transportation Projects: Proceedings of Geo-trans 2004, July 27–31, 2004. Los Angeles, CA, United States. American Society of Civil Engineers, Reston, VA 20191-4400, United States.

Abubakre, O.K., Muriana, R.A., Nwokike, P.N., 2007. Characterization and beneficiation of Anka chromite ore using magnetic separation process. Journal of Minerals & Materials Characterization & Engineering 6 (2), 143–150.

Agency for Toxic Substances and Disease Registry (ATSDR), 2000. Toxicological Profile for Chromium. US Department of Health and Human Services, Atlanta, US.

Allen, D., 2003. Material Flows and Waste Streams: Industrial Ecology. ESM 282. The Bren School at UCSB.

ATSDR, 2012. Toxicological Profile for Chromium. US Department of Health and Human Services, Agency for Toxic Substances and Disease Registry.

Barrett, A., Lawlor, J., 1995. The Economics of Waste Management in Ireland. Economic and Social Research Institute, Dublin, p. 129.

Beukes, J.P., Dawson, N.F., van Zyl, P.G., 2010. Theoretical and practical aspects of Cr(VI) in the South African ferrochrome industry. The Journal of the Southern African Institute of Mining and Metallurgy 110, 53–62.

Beukes, J.P., Van Zyl, P.G., Ras, M., 2012. Treatment of Cr(VI) containing wastes in the South African ferrochrome industry – a review of currently applied methods. The Journal of the Southern African Institute of Mining and Metallurgy 112, 347–352.

Bielicka, A., Bojanowska, I., Wiśniewski, A., 2005. Two faces of chromium – pollutant and bioelement. Polish Journal Environmental Studies 14, 5–10.

Cohen, B., Petrie, J.G., 1997. Containment of chromium and zinc in ferrochromium flue dusts by cement-based solidification. Canadian Metallurgical Quarterly 36, 251–260.

Costa, M., 1997. Toxicity and carcinogenicity of Cr(VI) in animal models and humans. Critical Reviews in Toxicology 27 (5), 431–442.

Cramer, L.A., Basson, J., Nelson, L.R., 2004. The impact of platinum production from UG2 ore on ferrochrome production in South Africa. The Journal of the South African Institute of Mining and Metallurgy 104 (9), 517–527.

Das, A.P., Singh, S., 2011. Occupational health assessment of chromite toxicity among Indian miners. Indian Journal of Occupational & Environmental Medicine 15 (1), 6–13.

Emsley, J., 2003. Chromium. In: Nature's Building Blocks: An A-Z Guide to the Elements. Oxford University Press, New York, p. 538.

Freese, K., Miller, R., Cutright, T.J., Senko, J., 2014. Review of chromite ore processing residue (COPR): past practices, environmental impact and potential remediation methods. Current Environmental Engineering 1, 82–90.

Gad, S.C., Powers, W.J., Dunn, B.J., Hoffman, G.M., Siino, K.M., Walsh, R.D., 1986. Acute toxicity of four chromate salts. In: Serrone, D.M. (Ed.), Chromium Symposium 1986: An Update. Industrial Health Foundation Inc, Pittsburgh, PA, pp. 43–58.

Gericke, W.A., 1998. Environmental solutions to waste products from ferrochrome production. In: The Proceedings of INFACON 8, Beijing, China, pp. 51–58.

Glastonbury, R.A., van Der Merwe, W., Beukes, J.P., Van Zyl, P.G., Lachmann, G., Steenkamp, C.J.H., Dawson, N.F., Stewart, H.M., 2010. Cr(VI) generation during sample preparation of solid samples – a chromite ore case study. Water SA 36 (1), 105–109.

Global Chromium Mining to 2020. Timetric Reports. Available at: https://timetric.com/research/report/mmgcr001mr–global-chromium-mining-to-2020/ (accessed 10.07.15.).

Gokarn, P., Ranjana, A., Nayaka, C.R., Murty, Y.R., 2003. Sustainable Development Through Beneficiation of Low Grade Chromite Ores. Available at: http://www.slideshare.net/prabhashgokarn/paper-sustainable-development-through-beneficiation-of-low-grade-chromite-ores (accessed 05.08.15.).

Gu, F., Wills, B.A., 1988. Chromite- mineralogy and processing. Minerals Engineering 1, 235–240.

Guertin, J., 2005. Toxicity and health effects of chromium (all oxidation states). In: Guertin, J., Jacobs, J.A., Avakian, C.P. (Eds.), Chromium(VI) Handbook. CRC Press, Boca Raton, pp. 216–234.

IARC (International Agency for Research on Cancer), 1997. Monographs on the evaluation of carcinogenic risks to humans. In: Chromium, Nickel and Welding, vol. 49.

ICDA, 2010. Statistical Bulletin. The International Chromium Development Association, Paris, France.

Jadhav, U.U., Hocheng, H., 2012. A review of recovery of metals from industrial waste. Journal of of Achievements in Materials and Manufacturing Engineering 154 (2), 159–162.

James, B.R., 1994. Hexavalent chromium solubility and reduction in alkaline soils enriched with chromite ore processing residue. Journal of Environmental Quality 23, 227–233.

Kanchinadham, S.B.K., Loganathan, V.D., Kalyanaraman, C.A., 2013. A preliminary study on the leachability of chromium from a contaminated site. Environmental Progress & Sustainable Energy 32, 1096–1110.

Kapure, G.U., Rao, C.B., Tathavadkar, V.D., Sen, R., 2011. Direct Reduction of Low Grade Chromite Overburden for Recovery of Metals. Institute of Materials, Minerals and Mining, Published by Maney on Behalf of the Institute.

Kleynhans, E.L.J., Beukes, J.P., Van Zyl, P.G., Kestens, P.H.I., Langa, J.M., 2012. Unique challenges of clay binders in a pelletised chromite pre-reduction process. Minerals Engineering 34, 55–62.

Kotas, A.J., Stasick, Z., 2000. Chromium occurrence in the environment and methods of its speciation. Environmental Pollution 107, 263–283.

Lee, K.P., Ulrich, C.E., Geil, R.G., Trochimowicz, H.J., 1989. Inhalation toxicity of chromium dioxide dust to rats after two years exposure. Science Total Environment 86, 83–108.

Lehmann, K.B., 1932. 1st Grund zu einer besonderen Beunruhigung wegen des Auftretens von Lungen krebs bei Chromat arbeitern vorhanden? Zentralblatt für Gewerbehygiene 19, 168–170.

Loock, M.M., Beukes, J.P., van Zyl, P.G., 2014. A survey of Cr(VI) contamination of surface water in the proximity of ferrochromium smelters in South Africa. Water SA 40 (4), 709–716.

Maine, C.F., Smit, J.P., Giesekke, E.W., 2005. The Solid Stabilization of Soluble Wastes Generated in the South African Ferrochrome Industry, Final Report to the Water Research Commission. WRC Report No 942/1/05.

van der Merwe, W., Beukes, J.P., van Zyl, P.G., 2012. Cr(VI) formation during ozonation of Cr-containing materials in aqueous suspension – implications for water treatment. Water SA 38, 505–510.

van der Merwe, W., 2011. Ozone Treatment of Chromium Waste Materials (Master thesis). Science in Chemistry at the Potchefstroom Campus of the North-West University.

Engin, T., 2002. Mineral Deposits in Turkey, Mineral resource base of the Southern Caucasus and systems for its management in the XXI century. In: Tvalchrelidze, A.G., Morizot, G. (Eds.), Proceedings of the NATO Advanced Research Workshop on Mineral Resource Base of the Southern Caucasus and Systems for Its

Management in the XXI Century Tbilisi, Georgia 3–6 April 2001, NATO Science Series. Series IV, Earth and Environmental Sciences, vol. 17. Kluwer Academic Publisher, Netherlands, pp. 81–104.
Mining Watch Canada, 2012a. Potential Toxic Effects of Chromium, Chromite Mining and Ferrochrome Production: A Literature Review, 56 p. Available at: http://www.miningwatch.ca/sites/www.miningwatch.ca/files/Chromite%20Review.pdf (accessed 10.07.15.).
Mining Watch Canada, 2012b. An Overview of Chromium, Chromite and Toxicity. Mining Watch Canada, Mines Alerte, 56 p. Available at: http://www.miningwatch.ca/sites/www.miningwatch.ca/files/Chromite%20Review.pdf (accessed 10.07.15.).
Potential Environmental of Impacts Mining and Processing. 5 p. Available at: http://www.miningwatch.ca/sites/www.miningwatch.ca/files/PotentialEnvironmental.pdf (accessed 10.09.15.).
Mishra, H., Sahu, H.B., 2013. Environmental scenario of chromite mining at Sukinda valley – a review. International Journal of Environmental Engineering and Management 4 (4), 287–292.
Mobbs, M., 2013. Advance release the Mineral Industry of Turkey. In: USGS 2011 Minerals Yearbook U.S. Department of the Interior U.S. Geological Survey Turkey. Available at: http://minerals.usgs.gov/minerals/pubs/country/2011/myb3-2011-tu.pdf (accessed 10.07.15.).
Mohanty, M., Patra, H.K., 2011. Attenuation of chromium toxicity in mine waste water hyacinth. Journal of Stress Physiology & Biochemistry 7 (4), 336–346.
Murthy, Y.R., Tripathy, S.K., Kumar, C.R., 2011. Chrome ore beneficiation challenges & opportunities – a review. Minerals Engineering 24 (5), 375–380.
Naiker, O., Riley, T., 2006. Xstrata alloys in profile. In: Southern African Pyrometallurgy 2006 Conference. South African Institute of Mining and Metallurgy, Johannesburg, pp. 297–306.
Chromium Hexavalent Compounds. Report on Carcinogens, thirteenth ed. Available at: http://ntp.niehs.nih.gov/go/roc13 (accessed 10.07.15.).
Papp, J.F., Lipin, B.R., 2001a. Chromium. Report: USGS Numbered Series, Open-File Report 2001–381.
Papp, J.F., Lipin, B.R., 2001b. Chromium and Chromium Alloys, Kirk-Othmer Encyclopedia of Chemical Technology. John Wiley & Sons, Inc. http://dx.doi.org/10.1002/0471238961.0308181523051920.a01.pub2. Published Online: September 14, 2001.
Papp, J.F., Lipin, B.R., 2006. Chromite. In: Kogel, J.E., Trivedi, N.C., Barker, J.M., Drukowski, S.T. (Eds.), Industrial Minerals and Rocks-commodities, Markets, and Uses, seventh ed. AIME, Society for Mining, Metallurgy, and Exploration, Inc., Littleton, CO, pp. 309–333.
Papp, J.F., 1990. Chromium. United States Geological Survey (USGS), pp. 259–292. Available at: http://images.library.wisc.edu/EcoNatRes/EFacs2/MineralsYearBk/MinYB1990v1/reference/econatres.minyb1990v1.jpapp.pdf (accessed 10.07.15.).
Papp, J.F., 1994. Chromium Life Cycle Study, Report, Federal Government Series, Information Circular 9411. U.S. Bureau of Mines, 94 p.
Papp, J.F., 2002. Chromium. United States Geological Survey (USGS).
Papp, J.F., 2007. Chromium. A National Mineral Commodity Perspective, Report: USGS Numbered Series, Open-File Report, 48 p.
Papp, J.F., 2008. Chromium. In: Metals and minerals: U.S. Geological Survey Minerals Yearbook 2008, vol. I, pp. 171–181.
Papp, J.F., 2009. 2009 Minerals Yearbook Chromium (Advance Release).
Papp, J.F., 2010. Mineral of the Month-Chromium 2009.
Papp, J.F., 2015. Chromium: U.S. Geological Survey (USGS) Mineral Commodity Summaries 2015, pp. 42–43. Available at: http://minerals.usgs.gov/minerals/pubs/mcs/2015/mcs2015.pdf (accessed 10.08.15.).
Pattanaik, S., Pattanaik, D.K., Das, M., Panda, R.B., 2012. Environmental scenario of chromite ore mining at Sukinda valley beyond 2030. Discovery Science 1 (2), 35–39.
Amcol Metalcasting. 2013. Hevi-Sand Process: Is the new technology a good reason to consider changing to foundry chromite sand chromite assurance parameters?. Technical Reference. Available at:http://www.amcolmetalcasting.com/DesktopModules/Bring2mind/DMX/Download.aspx?EntryId=12569&Command=Core_Download≈language=en-US&PortalId=43&TabId=4484 (accessed 10.07.15.)

Rankin, W.J., 2011. Minerals, Metals and Sustainability: Meeting Future Material Needs. CSIRO Pub, Collingwood, VIC.
Rao, D.S., Angadi, S.I., Muduli, S., Nayak, B.D., 2010. Valuable waste. Recovery of chromite values from ferrochrome industry flue dust. Aufbereitungs-Technik/Mineral Processing 51, 56−63.
Global Chromium Mining to 2020, 2015, 21 p. Available at: http://www.researchandmarkets.com/research/7wkbsm/global_chromium (accessed 10.08.15.).
Richard, P., 2015. ICDA: overview of the global chrome market. In: 1st Indinox Stainless Steel Conference, Ahmedabad, India.
Rodman, D.L., Carrington, N.A., Xue, Z., 2006. Conversion of chromium(III) propionate to chromium(VI) by the advanced oxidation process. Pretreatment of a biomimetic complex for metal analysis. Talanta 70, 668−675.
Sanchez-Segado, S., Makanyire, T., Escudero-Castejon, L., Hara, Y., Jha, A., 2015. Reclamation of reactive metal oxides from complex minerals using alkali roasting and leaching − an improved approach to process engineering. Green Chemistry 17, 2059−2080.
van Staden, Y., Beukes, J.P., van Zyl, P.G., du Toit, J.S., Dawson, N.F., 2014. Characterisation and liberation of chromium from fine ferrochrome waste materials. Minerals Engineering 56, 112−120.
Stanin, F.T., Pirnie, M., 2004. The transport and fate of Cr(VI) in the environment. In: Guertin, J., Jacobs, J.A., Avakian, C.P. (Eds.), Chromium(VI) Handbook. CRC Press, Florida, pp. 165−214.
Sueker, J.K., 2006. Chromium. In: Morrison, R.D., L.Murph, B. (Eds.), Environmental Forensics: Contaminant Specific Guide. Academic Press, Burlington, pp. 81−95.
Taylor, R., Allen, A., 2006. Waste disposal and landfill: information needs. In: Schmoll, O., Howard, G., Chilton, J., Jand Chorus, I. (Eds.), Protecting Groundwater for Health: Managing the Quality of Drinking-Water Sources. IWA, London.
Tchounwou, P.B., Yedjou, C.G., Patlolla, A.K., Sutton, D.J., 2012. Heavy metal toxicity and the environment. In: Luch, A. (Ed.), Molecular, Clinical and Environmental Toxicology, Experientia Supplementum, vol. 101. Springer, Basel, pp. 133−164.
Tiwary, R.K., Dhakate, R., Rao, V.A., Singh, V.S., 2005. Assessment and prediction of contaminant migration in ground water from chromite waste dump. Environmental Geology 48, 420−429.
USEPA, 1992. Environmental Criteria and Assessment Office. United States Environmental Protection Agency; Integrated Risk Information System (IRIS), Cincinnati, OH.
USEPA. 1998. Chromium, Ferrochromium, and Ferrochromium-Silicon. In: Identification and Description of Mineral Processing Sectors; Technical Background Document; Final. pp. 207−220. Available at: http://nepis.epa.gov/Exe/ZyPDF.cgi/P100FVHR.PDF?Dockey=P100FVHR.PDF (accessed 10.07.15.).
Uyanik, T., 2010. Mining, Export Promotion Center of Turkey, 9 p. Available at: http://www.fenimining.com/upload/dosyalar/m_rapor_1359493103.pdf (accessed 10.07.15.).
Wilbur, S., Abadin, H., Fay, M., Yu, D., Tencz, B., Ingerman, L., Julie, K., Klotzbach, J., James, S., 2012. Production, Import/Export, Use, and Disposal. Toxicological Profile for Chromium. Agency for Toxic Substances and Disease Registry (US), Atlanta, GA. Available at: http://www.ncbi.nlm.nih.gov/books/NBK158858/http://www.lexissecuritiesmosaic.com/gateway/FedReg/toxprofiles_tp7.pdf (accessed 10.07.15.).
Yalcin, S., 2003. Modelling Chromium Leaching From Chromite Ore Processing Waste. The Degree of Master of Science, the Department of Environmental Engineering, the Graduate School of Natural and Applied Sciences of the Middle East Technical University.
Yu, S., Du, J., Luo, T., Huang, Y., Jing, C., 2012. Evaluation of chromium bioaccessibility in chromite ore processing residue using *in vitro* gastrointestinal method. Journal of Hazardous Materials 209−210, 250−255.

CHAPTER

12

DETOXIFICATION AND RESOURCE RECOVERY OF CHROMIUM-CONTAINING WASTES

Y. Fei, C. Liu

Guangdong Institute of Eco-Environment and Soil Sciences, Guangzhou, People's Republic of China; Chinese Academy of Sciences, Guiyang, People's Republic of China

CHAPTER OUTLINE

Environmental Materials and Waste. http://dx.doi.org/10.1016/B978-0-12-803837-6.00012-3

12.1 CHROMIUM WASTES

12.1.1 CHROMIUM AND POLLUTION SOURCES

Chromium (Cr) is one of the most essential elements and import raw materials in modern metallurgical, chemical, and refractory industries (Darrie, 2001; Murthy et al., 2011). It is widely used in all kinds of chemical industries as pigments, reagents, and catalysts (Unceta et al., 2010). As a result, Cr-containing wastes can be generated by the metallurgical, refractory, chemical, electroplating, leather tanning, pigment, welding, and glass industries (Ma and Garbers-Craig, 2006). Up to 130,000 mg kg^{-1} of Cr contamination may be found in reports in the literature (Wang and Vipulanadan, 2000). Among these industries, about 90% of mined chromite ore is converted into ferrochrome, 80% of which is consumed by the stainless-steel industry (Murthy et al., 2011). According to data from 2008, 95% of the total volume was produced for metallurgical use, 2% for chemical-grade use, and 3% for refractory- and foundry-grade use (Murthy et al., 2011).

It was estimated that the world annual consumption of chromite was 12.5 million ton in 1999 (Darrie, 2001), whereas the world total chromite ore production reached 24 million tons in 2008 (Murthy et al., 2011). South Africa holds the most chromite reserves, with about 3.1 billion tons, followed by Zimbabwe, with reserves of 140 million tons and resources of 1 billion tons (Murthy et al., 2011). South Africa, India, and Kazakhstan are the major chromite ore- and concentrate-producing countries, together contributing 70% of the world's production in 2008 (Fig. 12.1) (Murthy et al., 2011).

Those countries and areas are often developing and underdeveloped countries and areas in which serious environmental pollution issues accompany intensive mass production. In China, up to 6 million tons of Cr-containing slag had been accumulating in chromate and metallic chromium manufacturing factories until 2005, and only 2 million tons was treated whereas 4 million tons did not receive safe disposal, which created a big environmental risk (Wu et al., 2010). For the galvanic wastes containing

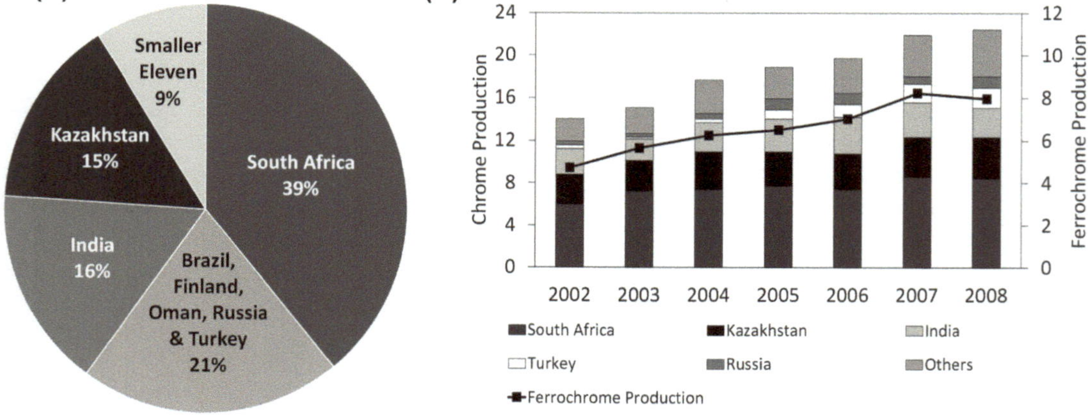

FIGURE 12.1

(A) World chromite ore production. (B) World chromite and ferrochrome production (2002–08).

Adapted from Murthy, Y.R., Tripathy, S.K., Kumar, C.R., 2011. Chrome ore beneficiation challenges & opportunities – a review. Minerals Engineering 24 (5), 375–380.

not only Cr but also other heavy metals such as Ni, Zn, Cu, and Pb, it is estimated that over 10^6 ton year^{-1} is produced by electroplating processes across the world (Reinosa et al., 2010).

12.1.2 CHARACTERISTICS OF CHROMIUM SPECIATIONS

Cr can exist in several chemical forms between 0 and VI, whereas trivalent Cr(III) and hexavalent Cr(VI) are the most stable forms in the natural environment (Unceta et al., 2010). For Cr(III), which is less soluble and less mobile, it is one of the essential micronutrients for human body, although excessive concentration may cause eczema or even DNA damage (Unceta et al., 2010). Main concerns regarding Cr-containing waste often come from Cr(VI) (Wang and Vipulanadan, 2000), which is highly soluble and mobile in the environment, and also is much more toxic than Cr(III) for both acute and chronic exposures (Darrie, 2001; Unceta et al., 2010). It is recognized as carcinogenic specifically to the lungs and in some cases to the sinonasal cavity (Ma and Garbers-Craig, 2006), and thus it is a highly toxic and is classified as human carcinogen by the US Environmental Protection Agency (EPA) and as a class I human carcinogen by the International Agency for Research on Cancer (Unceta et al., 2010). The threat to human health even resulted in a lawsuit to the owner of a 34-acre site on the Hackensack River in Jersey City, New Jersey, United States, and resulted in the large project of digging and hauling 1.5 million tons of chromite ore-processing residue (COPR) (Chrysochoou et al., 2009a).

Naturally, because they are affected by redox potential and pH, species between Cr(III) and Cr(VI) can transform into each other. Based on the natural behavior of Cr(III) and Cr(VI) as presented in Fig. 12.2, Cr(III) stabilization favors acidic media with low redox potential whereas Cr(VI) is stable under alkaline conditions with higher redox potential (Unceta et al., 2010).

Regarding to the extremely high toxicity of Cr(VI), reduction of Cr(VI) to more stable and less mobile and bioavailable Cr(III) is usually preferred during waste treatment and detoxification (Chrysochoou et al., 2009a). Different reducing agents such as Fe(II), phosphate, and sulfide, and organic matter such as humic acid can reduce $HCrO_4^-$ and CrO_4^{2-} to Cr(III) (Unceta et al., 2010).

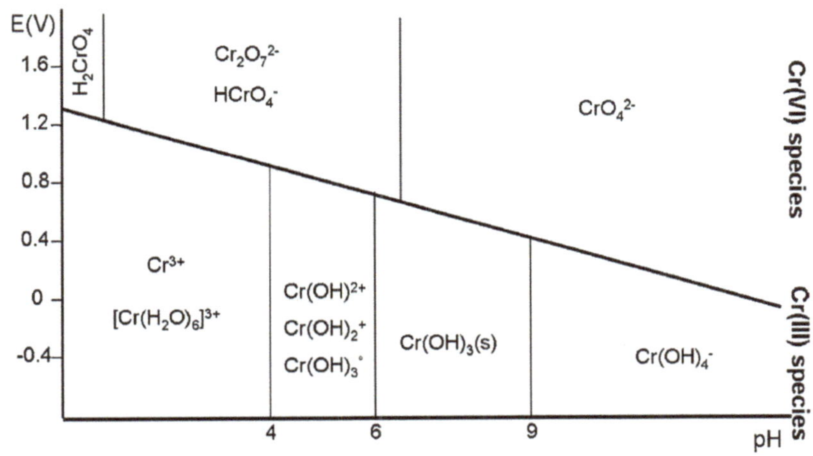

FIGURE 12.2

Chemical species of chromium as a function of pH and potential.

Adapted from Unceta, N., et al., 2010. Chromium speciation in solid matrices and regulation: a review. Analytical and Bioanalytical Chemistry 397 (3), 1097–1111.

12.1.3 CHROMITE ORE-PROCESSING RESIDUE

Chromite ore, (Mg, Fe)(Cr, Al, Fe)$_2$O$_4$, contains 15–65% of chromic oxide content, mostly in the form of spinel or corundum oxides (Darrie, 2001). Chromite ore processing refers to converting the chromium into commercial products such as sodium chromate (Na_2CrO_4), or more typically dichromate ($Cr_2O_7^{2-}$), to give hexavalent or trivalent chromium compounds for further application in various industries (Darrie, 2001; Chrysochoou et al., 2009a). A typical flow sheet for chromium chemicals production is shown as Fig. 12.3.

Until the late 1950s, lime-based processes were exclusively employed and produced large volumes of Cr-rich solid wastes (Chrysochoou et al., 2009a) until more environmentally friendly lime-free extraction technology was developed. However, some countries in Eastern Europe, India, Pakistan, and China may still use lime-based processes (Darrie, 2001). Compared with modern lime-free processes, high-lime process waste contains 30–40% calcium oxide equivalent as well as significant (>0.5%) levels of calcium chromate, which makes Cr compounds dissolve slowly; Cr(VI) can hardly be reduced by common reducing agents, which causes a greater challenge for treatment and disposal (Darrie, 2001).

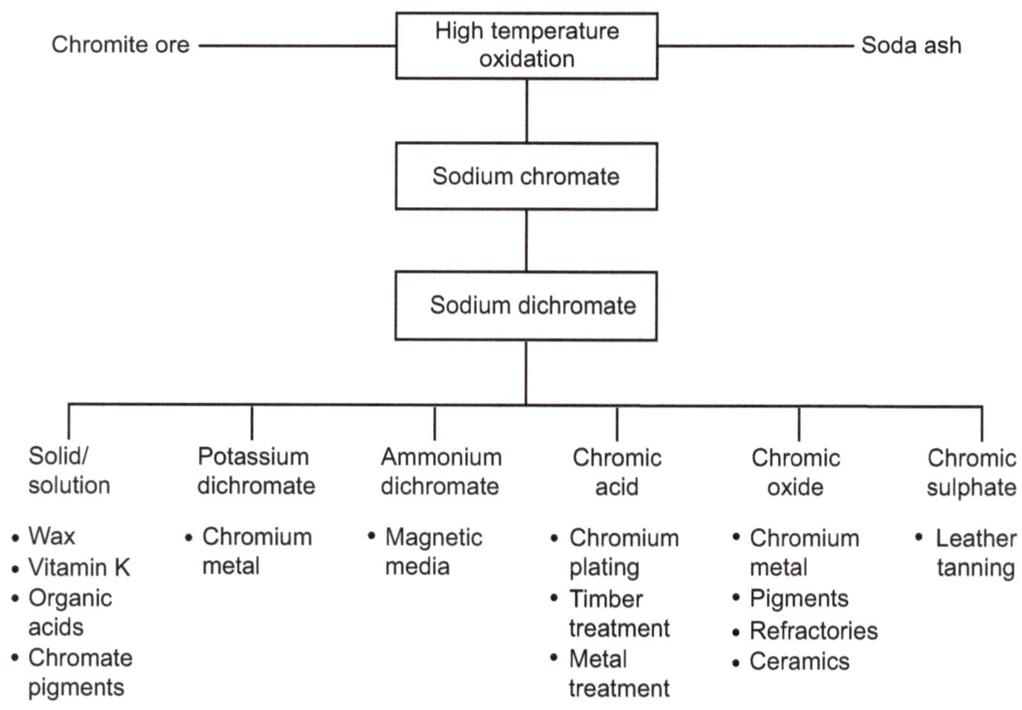

FIGURE 12.3

Typical flow sheet for chromium chemical production.

Adapted from Darrie, G., 2001. Commercial extraction technology and process waste disposal in the manufacture of chromium chemicals from ore. Environmental Geochemistry and Health 23 (3), 187–193.

The residual content of Cr in COPR may reach 50,000 mg kg^{-1}, in which about half is Cr(VI) (Chrysochoou et al., 2009a). As suggested by microstructural analysis, most of the Cr(VI) in COPR may exist in the form of calcium aluminum chromium oxide hydrate (CAC) ($Ca_4Al_2[CrO_4][OH]_{12}\cdot nH_2O$) ($n = 3, 6, 8$) (Chrysochoou et al., 2009a). Hydrocalumite ($Ca_4Al_2[CrO_4][OH]_{12}\cdot 6H_2O$), an isostructure with the CACs, and hydrogarnet ($Ca_3Cr_2[SiO_4]_3$) may also be important host phases for chromate in COPR from other reports (Hillier et al., 2007).

12.1.4 FERROCHROME AND STAINLESS-STEEL PLANTS: DUST AND WASTE ACID TREATMENT SLUDGE

Inside the electric furnace employed in these plants, dust is produced through the vaporization of elements or compounds from high-temperature zones, as well as the ejection of slag and metal; then it is captured in the off-gas by cyclone separators or by bag house filters (Ma and Garbers-Craig, 2006). Cr(III) is oxidized to Cr(VI) under high temperature in the presence of oxygen (Ma and Garbers-Craig, 2006). Typical crystalline phases of chromite (Fe, Mg)Cr_2O_4, chromates such as $CaCrO_4$, K_2CrO_4, and Na_2CrO_4, as well as oxides including CrO and CrO_3 are identified in these dusts (Ma and Garbers-Craig, 2006).

The acid pickling process is necessary when the surface quality of stainless steel is improved, and highly acidic wastewater with high concentrations of Cr is produced. After the treatment process of neutralization, reduction, and precipitation, sludge or filter cake is formed, containing lots of $Cr(OH)_3$ (Ma and Garbers-Craig, 2006).

12.1.5 LEATHER INDUSTRY: TANNING AND POSTTANNING WASTES

The leather industry is another heavy-polluting industry which usually located in developing economies (Thanikaivelan et al., 2005), in which environmental protection often yields to economic development, causing huge troubles to the natural environment. About 90% of the light leather industry was reported to be using Cr-based compounds in its processes (Pinto et al., 2005). The tanning and posttanning process involves permanently stabilizing leather-making proteins against heat, degradation, and thermomechanical stress and improving the intrinsic properties of leather. In commercial practices, chrome tanning is the most commonly employed approach, which uses a large quantity of basic chromium salts that become the source of Cr pollution in this industry. As reported, for each ton of processed leather, more than 0.12 kg of Cr pollution is generated (Thanikaivelan et al., 2005).

12.2 REGULATIONS FOR CHROMIUM IN WASTES AND ENVIRONMENT MATRIX

To restrict the casual disposal of Cr-containing wastes, many countries and organizations have strict limits regarding the Cr(VI) and total Cr content in disposed waste or receiving environmental compartments including atmosphere, drinking water, and soils, as listed in Table 12.1. As summarized, less than 0.5 mg L^{-1} Cr(VI) is acceptable in general water samples, whereas the total Cr content should be less than 0.5 mg L^{-1}. For some countries such as China and some areas such as Maryland in the United States, the regulation is more strict: The total Cr limit is set at 0.05 mg L^{-1} for drinking water.

As for solid samples, the concern is more with the soluble or leachable content than with the total concentration. One of the most accepted criteria is the Toxicity Characteristic Leaching Procedure (TCLP) recommended by the US EPA, that Cr(VI) concentration in the leachate should be less than 5 mg L^{-1}.

Table 12.1 Regulations for Chromium in Environmental Matrices (mg L^{-1} or mg kg^{-1})

Country	Year	Matrix	Cr(VI) Limit	Total Chromium Limit
World Health Organization	2008	Drinking water		0.05 (Rakhunde et al., 2012)
European Union	2003	Cement	(Soluble) 2 (Unceta et al., 2010)	
France	2001	General water	0.1 (Ma and Garbers-Craig, 2006)	0.5 (Ma and Garbers-Craig, 2006)
Germany		Drinking water	–	0.05 (Ma and Garbers-Craig, 2006)
		Metal/chemical industry	0.1 (Ma and Garbers-Craig, 2006)	0.5 (Ma and Garbers-Craig, 2006)
Japan		Drinking water	0.05 (Ma and Garbers-Craig, 2006)	–
		Public water systems	0.5 (Ma and Garbers-Craig, 2006)	2 (Ma and Garbers-Craig, 2006)
Kazakhstan		Drinking water	–	0.0031 (Ma and Garbers-Craig, 2006)
		Liquid effluent emissions	0.005–0.03 (Ma and Garbers-Craig, 2006)	–
South Africa	1998	Effluent discharge	–	0.5 (Ma and Garbers-Craig, 2006)
		Drinking water	0.05 (Ma and Garbers-Craig, 2006)	0.1 (Ma and Garbers-Craig, 2006)
United States		Drinking water	0.1 (Ma and Garbers-Craig, 2006)	0.1 (Ma and Garbers-Craig, 2006)
New Jersey	2007	Soil	20 (Chrysochoou et al., 2009a)	
Maryland	2001	Soil (residential)	23 (Chrysochoou et al., 2010)	
		Soil (nonresidential)	610	
United Kingdom		Drinking water	–	0.05 (Ma and Garbers-Craig, 2006)
China	1984	Chromium slag	(Soluble) 5	
		Wastewater	0.3	
		Drinking water		0.05

12.3 SIMPLE DETOXIFICATION AND DISPOSAL

Thus, to meet regulations, Cr-containing wastes cannot be stockpiled or land filled directly, but must first be treated (Ma and Garbers-Craig, 2006). Otherwise, the land filling sites would require a lot of remediation work in the future. The water-insoluble postextraction from no-lime plants is usually treated first to reduce the Cr(VI) concentration to around 0.1−0.2% and then land filled (Darrie, 2001). Reducing reagents such as $FeSO_4$ or CaS_5 are often employed in these posttreatments (Darrie, 2001; Chrysochoou et al., 2009a,b). However, it was reported that the reduction may not always be efficient enough to detoxify the Cr(VI), despite high stoichiometric doses from twice to eight times, possibly because Cr(VI) was found to locate with nodules and usually bound to hydrogarnet and hydrotalcite which is acid-assisted and not easily accessible to dissolved reductants (Chrysochoou et al., 2009b). Another problem is that Fe(II) often precipitates easily because of the high pH of Cr-containing waste such as COPR (Wang et al., 2007). $FeSO_4$ might even increase Cr(VI) leaching as a result of the anion exchange of sulfate for chromate (Wang et al., 2007).

Before better solutions can be developed, land filling is sometimes the simplest and most cost-effective way to dispose waste and is an age-old method in actual practice (Omotoso et al., 1998). In the United Kingdom, a large proportion of tannery waste is land filled mainly for economic reasons (Basegio et al., 2009). Owing to its sand-like properties, COPR was widely used as a backfill for wetlands, sewer line, roadway construction, construction of tank berms, and foundation material in the United States by the millions of tons (Chrysochoou et al., 2009b, 2010). However, in time, the toxic elements in the wastes are exposed to soil, water, and air through leaching and evaporation. For instance, ferrochrome plant and stainless-steel plant dusts and sludge can easily leach out large volumes of Cr(VI), whereas the aging process does not help, and may even oxidize Cr(III) species into Cr(VI) in the presence of lime at room temperature (Ma, 2005), increasing the potential of Cr mobility and toxicity.

12.4 STABILIZATION/SOLIDIFICATION PROCESSES

Regarding the ineffectiveness of simple detoxification and landfill in dealing with Cr-containing waste, stabilization/solidification (S/S) is a more important technique from the viewpoint of environmental safety. In the S/S process, hazardous wastes can be chemically or thermally transformed into a nontoxic form or additives can bond with the waste so that the mobility of heavy metals is reduced (Ma and Garbers-Craig, 2006). The S/S strategy is considered to be one of the best demonstrated available technologies by the US EPA for the treatment of industrial wastes and contaminated soils (Rodriguez-Pinero et al., 1998). Clay, lime, ordinary Portland cement (OPC), and other wastes (such as coal fly ash, blast furnace slag, and ferrochrome slag) can be found in the literature as the main binders, with different degrees of success (Ma and Garbers-Craig, 2006). The resulting nontoxic material can usually be used in land filling or as a raw material for other uses (Ma and Garbers-Craig, 2006).

12.4.1 BASIC CEMENTATION PROCESSES

Cementation is commonly used because of its low cost and adaptability to wet wastes. Besides, because the process could be conducted on-site, the transport costs and secondary emissions are largely minimized (Ma and Garbers-Craig, 2006). Since 1989, the super-detoxification process has been commercialized by Bethlehem Steel Corporation to mix electric furnace dust with lime and alumina-silicates as well as other additives to make a concrete-like material that can efficiently immobilize heavy metals in the matrix by precipitating or oxidizing/reducing heavy metals into the least soluble components (Zunkel, 1997).

Because of its porosity and permeability, cement has a major role in the performance of S/S processes (Ma and Garbers-Craig, 2006; Omotoso et al., 1998). OPC is one of the most commonly used raw materials for S/S treatment. Through cement—water reactions, a high-strength monolithic cement—waste matrix can be developed, and the heavy metal ion, ie, Cr^{3+}, $CrO_4{}^{2-}$, etc., can be stabilized via chemisorption, precipitation, chemical incorporation in the hydrated cement components, and/or physical entrapment (Omotoso et al., 1998). According to Cohen and Petrie's report (Cohen and Petrie, 1997, 2005), ferrochromium dust can be stabilized with OPC. Prewashing the dust or longer curing times and increased OPC contents are helpful to improve Cr retention.

The main mechanism is supposed to include (Dermatas and Moon, 2006): (1) precipitation of Cr(III) in the alkaline pore solution as hydroxide; (2) physical or/and chemical inclusion, or incorporation into the hydration products of the cement; and (3) sorption to clay and a new product formed during the reactions. Cement hydration and pozzolanic reaction (Eqs. [12.1] and [12.2]) take place in the cementation process when inorganic agents are used as binders, and a calcium silicate and aluminate hydrate polymeric net is obtained based for similar reactions which has the main role of solidifying heavy metals in the wastes (Rodriguez-Pinero et al., 1998; Dermatas and Meng, 2003):

$$\mathrm{CH + SH \rightarrow CSH} \qquad [12.1]$$

$$\mathrm{CH + AH \rightarrow CAH} \qquad [12.2]$$

where $C = CaO$, $S = SiO_2$, and $A = Al_2O_3$, $H = H_2O$ in cement chemist notation.

However, most published reports demonstrated only the effectiveness of treating Cr(III), whereas the simulated stabilization of Cr(VI)-containing waste by OPC suggested a failure (Dermatas and Moon, 2006), because the formed $CaCrO_4 \cdot 2H_2O$ and $Ca_2CrO_5 \cdot 3H_2O$ were not stable enough and Cr(VI) was still mobile in the leaching test (Omotoso et al., 1998). The presence of Cr(VI) inhibited the cement hydration process owing to the reaction with Ca^{2+} and increased Cr(VI) content reduced the compressive strength of the produced cement (Wang and Vipulanadan, 2000). To meet the TCLP limit of 5 mg L^{-1}, the treating ratio of Cr(VI) to cement has to be strictly limited.

12.4.2 MODIFIED CEMENTATION PROCESSES

Clays may be applied in the cementation process as additives to reinforce the S/S effects. As Pinto et al. reported, tannery waste could be stabilized by OPC with a ratio of 10—15%, and the addition of 5% more bentonite or modified organoclay would significantly improve the efficiency and largely reduce the leachable Cr concentration (Pinto et al., 2005).

Various industrial wastes are also employed simultaneously as additives in the S/S processes, including blast furnace slag, ferrochrome slag, and fly ash. Approximately 95% of Cr(VI) was found to be stabilized when ferrochrome dust and electric arc furnace slag were mixed with OPC (Giesekke et al., 2000). Spent pickling acid slag and lime were used to treat electric furnace at Atlas Steel; the result was success in reducing the leachable concentration of Cr(VI) to below the Ontario drinking water standard and US EPA criteria (Stone and Marasco, 1986).

Regarding the Cr(VI) speciations of most concern, which can hardly be immobilized by simple S/S processes, a reducing pretreatment is usually required (Rodriguez-Pinero et al., 1998). Ammonium iron sulfate was reported to be capable of reducing Cr(VI) without pH adjustment when OPC or cement kiln dust was employed as the binder (Jacobs, 1992). Metabisulfite was also suggested to be able to reduce more than 90% of Cr(VI) from a contaminated soil site (Eid et al., 1996).

Ferrous Fe(II) salts are more often employed as a reducing agent. $FeCl_2$ was demonstrated to be helpful in improving the stability of Cr(VI) in electric arc furnace slag (Ma and Garbers-Craig, 2006). Despite the pH sensitivity and dependence of Fe(II) reagents, it provides a reductant to Cr(VI) but also can provide Fe(III), the reaction product, as the co-precipitator and adsorbent of Cr(III) or other heavy metals (Rodriguez-Pinero et al., 1998). Thus, a modified cementation process including chemical pretreatment by Fe(II), the addition of lime to neutralize the acid used in the pretreatment, and finally a cementation process was developed (Rodriguez-Pinero et al., 1998). A similar process using CaS_5 as the reductant and fly ash as the cementation binder was reported to be successful in treating COPR dump sites (Kameswari et al., 2015).

As a newly emerging reductant, nanoscale zero-valent iron (nZVI) was found to be effective in reducing Cr(VI) from COPR for the S/S process. As Du et al. (2012) reported, with the presence of an adequate water content, the incubation of COPR with nZVI could reach nearly 100% of Cr(VI) reduction and immobilization, which suggests potential application in the actual treatment and remediation processes.

12.4.3 LIME-BASED STABILIZATION/SOLIDIFICATION PROCESSES

Because lime is more economical than cement and the heat generated by lime during hydration can potentially accelerate the rate of reaction, lime or quicklime is also employed in S/S processes. The addition of lime can result in the effective immobilization of Cr, mainly Cr(III), through the simple S/S process (Dermatas and Moon, 2006), probably owing to the high alkaline condition formed that can precipitate the metal ions.

However, because it has the same problem as basic cement-based S/S processes, the addition of lime alone cannot achieve good performance on Cr(VI) stabilization. A possible cause might be that the high alkaline environment is not favored by Cr(VI) reduction. Research found that the additional use of fly ash or blast furnace slag, which contains some extent of sulfates, is capable of achieving the aim of Cr(VI) immobilization (Dermatas and Moon, 2006). When quicklime and fly ash were added to Cr-contaminated soils, one of the pozzolanic compounds, $3CaO \cdot Al_2O_3 \cdot 0.5CaCrO_4 \cdot 0.5CaSO_4 \cdot nH_2O$, was identified as the precipitate which might contribute the most to the S/S of Cr(VI) (Dermatas and Moon, 2006).

12.4.4 THERMAL STABILIZATION PROCESSES

As for the inefficiency of the chemical reduction of Cr(VI) in the solid wastes, thermal techniques were attempted to improve the performance of such a treatment. Actually, incineration with no pretreatment may promote the oxidation of Cr(III) to Cr(VI), rather than reverse (Kavouras et al., 2015). Thus, reagents as reductants or stabilizers are usually required during thermal treatment.

In a trial of CaO treatment of Cr-containing sludge (Mao et al., 2015), it was found that different types of Cr(VI) products were generated under different temperatures ranging from 200°C to 1100°C.

Only when the temperature exceeded 400°C was it possible that Cr(VI) might be reduced to Cr(III). Comparing with CaO, MgO might be a better additive to inhibit the formation of Cr(VI) during oxic heating treatment.

Wei et al. (2002) used kaolin as the sorbent for Cr(VI) immobilization and then thermally treated the slurry at 500−900°C. It was found that CrO_3 transformed to Cr_2O_3 after 4 h of heating, and chromium octahedral were identified as bridged to silica tetrahedral and incorporated into the minerals such as mullite or sillimanite when heated to 1100°C. By a similar process, bentonite was reported to be a suitable material for this sintering process, and with a temperature over 500°C, more than 87% of Cr(VI) could be converted to Cr(III); this transforming ratio increased to almost 100% along with an increase in temperature (Wei et al., 2010).

Cr(VI) containing stainless-steel plant dust was thermally stabilized through sintering under 1100°C for 5 h using silica-rich clay, and heavy metals including not only Cr but also Zn and Pb were successfully immobilized to below environmental limits (Ma and Garbers-Craig, 2009). As Confirmed by X-ray diffraction analysis, Cr_2O_3 was stabilized in spinel ([Fe, Zn, Ni, Mg, Mn][Fe, Cr, Al]$_2O_4$), hematite (Cr_2O_3), augite ([Ca(Mg, Fe, Al][Si, Cr, Al]$_2O_6$), and glassy phases.

Wang et al. (2007) suggested a method using sucrose, starch, or flour as the reductants and calcinate in the mixture of COPR and reagent in inertial gas, which could complete reduce Cr(VI) to Cr(III) at above 500°C with more than twice the stoichiometric ratio of the reductants to Cr(VI).

Based on the reduction of Cr(VI) to Cr(III) by organic matter, a novel treatment was invented using sewage sludge as the reductant during thermal treatment (Wu et al., 2010). After mixture with sewage sludge, Cr(VI)-containing slag was heated to 200−900°C under nitrogen gas. It was reported that the organic matter from the sewage sludge could effectively reduce Cr(VI) by 60−100%, although most could easily be leached whereas heating could strengthen the reduction and further stabilize the reduced product.

12.5 SECONDARY RECYCLING

Besides S/S treatment, it is usually more economically favorable if new usable products can be obtained from wastes, which is called secondary recycling (Reinosa et al., 2010; Silva et al., 2008). The building industry is one of the most technological active sectors and often can absorb large volumes of solid wastes as well as recycled materials (Reinosa et al., 2010; Pioro and Pioro, 2004). For Cr-containing wastes, cement, glass, medium-grade abrasive, bricks, pigments, etc., with added value made by vitrification and other procedures then can be reused. Some typical slag reprocessing technology flows for slag-waste recycling into different uses are illustrated in Figs. 12.4−12.6.

12.5.1 CEMENT

Regarding the previous discussion of S/S processes, cement and clay-like matrix is good choice for the stabilization of Cr wastes. Besides using cement to solidify wastes in landfill sites or contaminated soils, it is more product-oriented if Cr waste can be incorporated into the composition of the cement. Similar to S/S processes, the use of industrial waste in cement composites has a long history (Riganti et al., 1986).

The working procedure and main mechanism are more or less the same as the cementation processes already described. According to the assessment of Sturm et al. (2009), the leachates of both

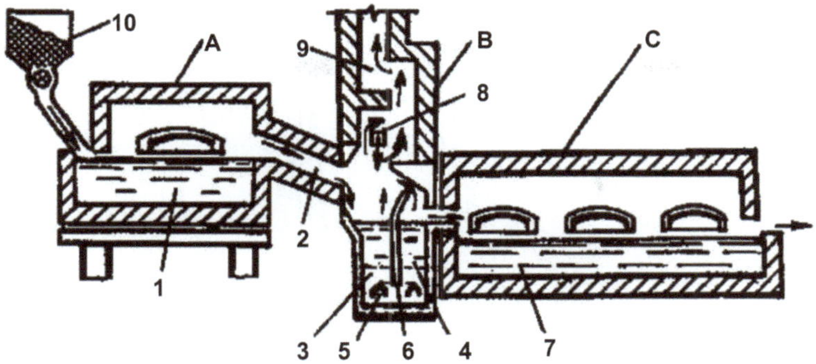

FIGURE 12.4

Experimental metallurgical slag reprocessing line: (A) Slag-melting furnace, (B) melting converter, (C) melt-clarification furnace: (1) melted slag, (2) slag chute, (3) primary melting chamber, (4) secondary melting chamber, (5) submerged gas–air burners, (6) partition, (7) melted glass-ceramic, (8) loading window, (9) flue gas duct, (10) crushed cold slag.

Adapted from Pioro, L.S., Pioro, I.L., 2004. Reprocessing of metallurgical slag into materials for the building industry. Waste Management 24 (4), 371–379.

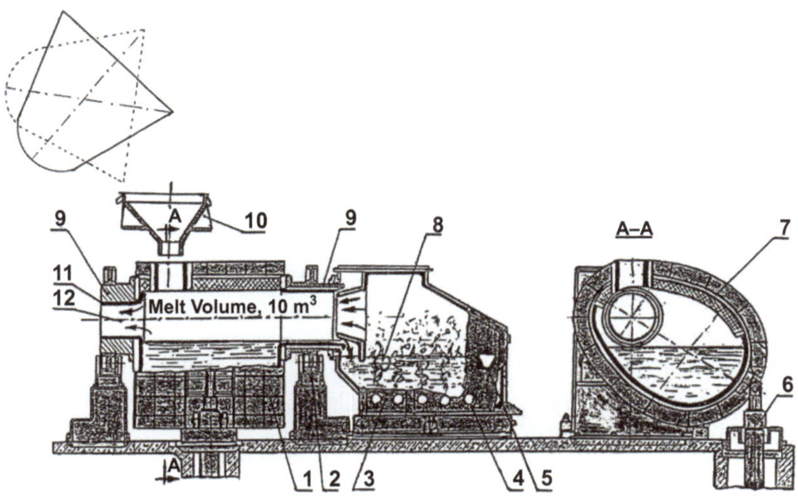

FIGURE 12.5

Melting converter for reprocessing melted slag and charge into building materials with a capacity of 250 ton day^{-1}: (1) slag receiver, (2) supporting rollers, (3) converter, (4) openings for lances, (5) tap hole, (6) hydraulic jack, (7) refractory, (8) charge loading, (9) receiver trunnions, (10) interim funnel, (11) melt lining, (12) flue gases.

Adapted from Pioro, L.S., Pioro, I.L., 2004. Reprocessing of metallurgical slag into materials for the building industry. Waste Management 24 (4), 371–379.

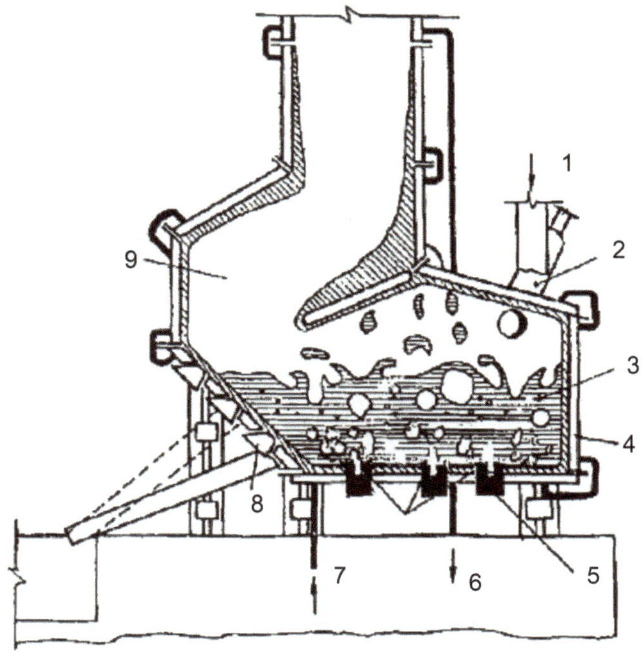

FIGURE 12.6

Melting converter for glass-ceramic filler production: (1) charging materials loading, (2) feeder, (3) melt, (4) converter body (water jackets), (5) burners, (6) and (7) cooling water inlet and outlet, (8) tap hole for melt draining, (9) flue gas duct.

Adapted from Pioro, L.S., Pioro, I.L., 2004. Reprocessing of metallurgical slag into materials for the building industry. Waste Management 24 (4), 371–379.

compact and ground cement with or without the addition of steel plant electric arc filter dust are below the environmental limits of Cr and other metals. With the optimized process, in adding wastes it was possible to improve strength and corrosion resistance to the cement and even saved part of the energy consumption for producing normal Portland cement (Riganti et al., 1986; Shi and Qian, 2000). From the physicomechanical and environmental aspects, Cr waste can be an option for the cement component.

12.5.2 GLASS-CERAMIC MATERIALS

Among the value-added processes, vitrification or glassification might be the most studied and reported. As a well-established technology, vitrification is an attractive option that can converse waste into a stable and homogeneous (silicate) glass (Colombo et al., 2003). It is flexible enough to deal with various types of waste and is efficient in reducing the large volume of wastes (Ma and Garbers-Craig, 2006; Basegio et al., 2009) as well as the inertization of heavy metal ions (Basegio et al., 2009; Colombo et al., 2003). However, because of high energy consumption during the process, vitrification is more expensive than simple disposal in landfills or the S/S processes (Colombo et al., 2003). The cost is estimated to be US$110–135/ton of dust (Gress, 1993) or US$80/ton of glass (Ek and Schlobohm, 1993). It may be helpful to cover part of the expenditure if a high value–added product can be made.

Characterized by a high content of glass-forming and network-modifying oxides such as SiO_2, Al_2O_3, and CaO, Cr-containing wastes including electric arc furnace dust from stainless-steel plants, sludge, or mud generated during tanning processes, for example, are good raw materials for vitrification. The final glass is generally in the system of Al_2O_3-SiO_2 or CaO-MgO-Al_2O_3-SiO_2, and the presence of a sufficient amount of Fe_2O_3, P_2O_5 and TiO_2 would have a role in controlling the crystallization process (Colombo et al., 2003), whereas Cr may be functional in coloring the final glassy products.

In the literature, numerous examples exist of turning all kinds of Cr-containing wastes to glass through vitrification processes (Silva et al., 2008; Kuo et al., 2007; Celary and Sobik-Szoltysek, 2014; Sobiecka and Szymanski, 2014; Li et al., 2007). Stainless-steel dust was reported in the vitrification process with clay, and both the stabilization of heavy metals and the production of stable vitrification product were achieved successfully (Ma and Garbers-Craig, 2006). The melting procedure could reduce mobile Cr(VI) to immobilized Cr(III) (Kuo et al., 2003; Huang et al., 2004), and Cr_2O_3 could be crystallized as Cr-rich spinel and augite (Huang et al., 2004).

Medium-grade abrasive and roofing granule products can be developed from furnace dust and foundry sand (Huang et al., 2004). In simulated Cr_2O_3 melting under 1500°C, the produced basaltic glass had exceptional resistance against corrosion and increased hardness and fracture toughness; based on the related indices, the annealed products could be classified as medium-grade abrasives (Huang et al., 2004). Furnace dusts mixed with silicates such as silica sand, clay, or cullet and then melted in the vitrification furnace at 816–1371°C can make environmentally friendly abrasives applied in architectural applications (Gress, 1993). Glassification International developed a similar process using all kinds of pyrometallurgic wastes including electric furnace dust, slag, spent refractory, and mill scale, to produce nonhazardous glassy products such as tiles and sheets for architectural applications (Ek and Schlobohm, 1993).

Other glassy products such as stoneware tiles could be produced by incorporating Cr waste into industrial compositions (Barbieri et al., 2000). After a calcination step to eliminate the gas-forming species, the galvanic waste may be mixed with kaolinite, feldspar, quartz, and recycled domestic glass and then go through the industrial process. After being thermally treated, well-faceted particles can be obtained from the mixture of galvanic waste with the minerals (Fig. 12.7). The finalized porcelain stoneware can satisfy both mechanical and the chemical standards as a promising product (Reinosa et al., 2010).

Through optimized plasma vitrification, tannery sewage sludge could also be employed to produce products with negligible Cr leaching levels as well as hardness similar to commercial glass, which might be further modified and used as construction aggregate substitutes (Celary and Sobik-Szoltysek, 2014). However, better performances was found in metal leaching and product hardness if the sludge was incinerated to ash before vitrification (Celary and Sobik-Szoltysek, 2014; Sobiecka and Szymanski, 2014). A similar result was found for Cr-containing electroplating sludge: that the addition of ash or cullet during the vitrification of sludge would improve glassy conditions as well as the compressive strength of the produced slags (Li et al., 2007).

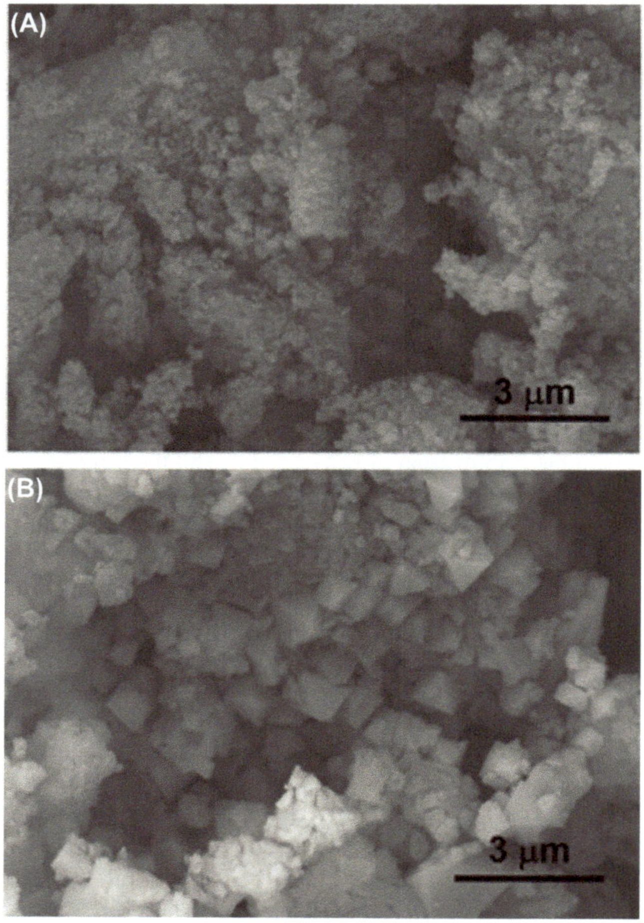

FIGURE 12.7

Scanning electron micrographs corresponding to galvanic waste (A) as received and (B) thermally treated. After the calcination process, well-faceted particles appeared that were 1–2 μm in size and highly agglomerated.

Adapted from Reinosa, J.J., et al., 2010. High chemical stability of stoneware tiles containing waste metals. Journal of the European Ceramic Society 30 (14), 2997–3004.

12.5.3 BRICKS

Another type of reusable product via thermal stabilization of Cr wastes is sintered or fired bricks. A lot of work has been done and reviewed in recycling different types of waste into those eco-friendly bricks (Velasco et al., 2014). It was shown that reuse of waste in brick manufacture might benefit both waste management and brick production, because it solves the problem of waste treatment and disposal and may also enhance the brick's properties. Typical procedures for brick making from Cr-containing wastes is illustrated in Fig. 12.8.

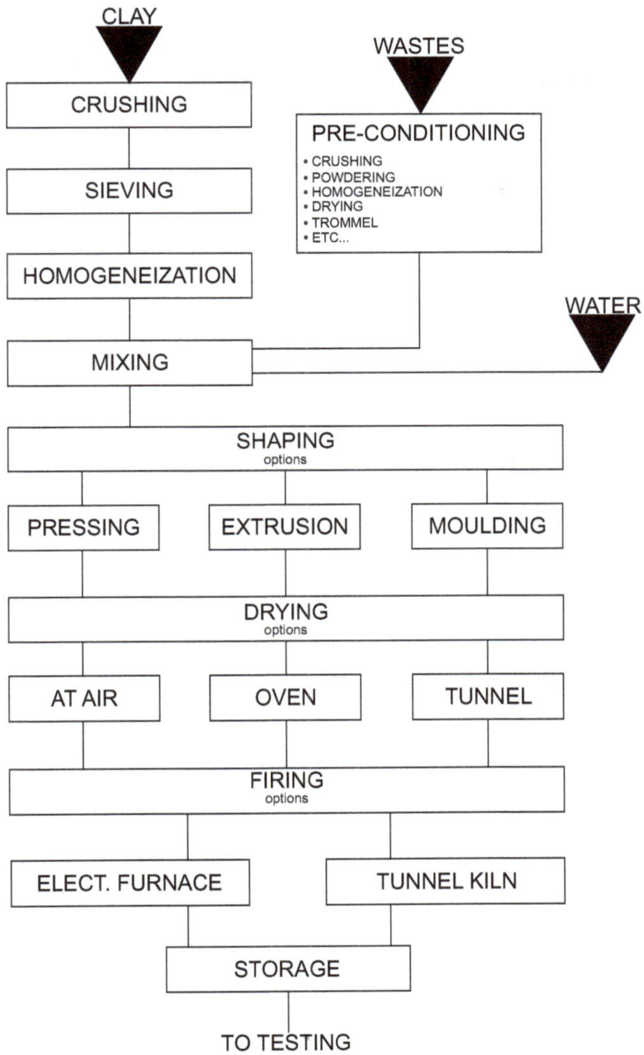

FIGURE 12.8

Flow diagram illustrating procedures for brick making from wastes. *ELECT*, electrical.

Adapted from Velasco, P.M., et al., 2014. Fired clay bricks manufactured by adding wastes as sustainable construction material—a review. Construction and Building Materials 63, 97–107.

According to reports, a mixture of stainless-steel dust and clay could effectively stabilize Cr(VI) in the waste after sintering between 1000°C and 1200°C and also produce bricks for further application (Ma, 2005; Geldenhuis and Horne, 2002). Up to 99% of Cr(VI) was demonstrated to be stabilized by this method (Ma and Garbers-Craig, 2006). Also, the addition of Cr wastes to the composites may change the final fired bricks, including decreasing linear shrinkage and bulk density, as well as closing some of the open porosity, resulting in better mechanical properties in environmental resistance (Perez-Villarejo et al., 2015).

12.5.4 PIGMENTS

A large amount of work was also done to develop pigment-based material using Cr waste as the precursor (Sreeram et al., 2011). For that, Cr is always important in the pigment industry, Cr-rich waste is certainly one of the cheaper choices as a color precursor compared with natural salts or metallic oxides.

As Berry et al. (2002) reported, Cr(III) could be extracted from tanned leather shavings by alkaline reagents and then be used in the same way that Cr_2O_3 was used in commercial pigment production. Cobalt-chromite green and chrome-tin pink pigments were successfully obtained in their trial. Cr(III) was stabilized and oxidation to Cr(VI) was precluded through high-temperature (1000−1400°C) treatment.

The Cr-containing pigments pink $CaCr_{0.04}Sn_{0.97}SiO_5$ and green $Ca_3Cr_2(SiO_4)_3$ were also synthesized using dried Cr sludge as a chromium oxide precursor via solid-state reactions by other researchers (Fig. 12.9) (Andreola et al., 2008). Black pigment based on the spinel structures $NiCr_2O_4$ and

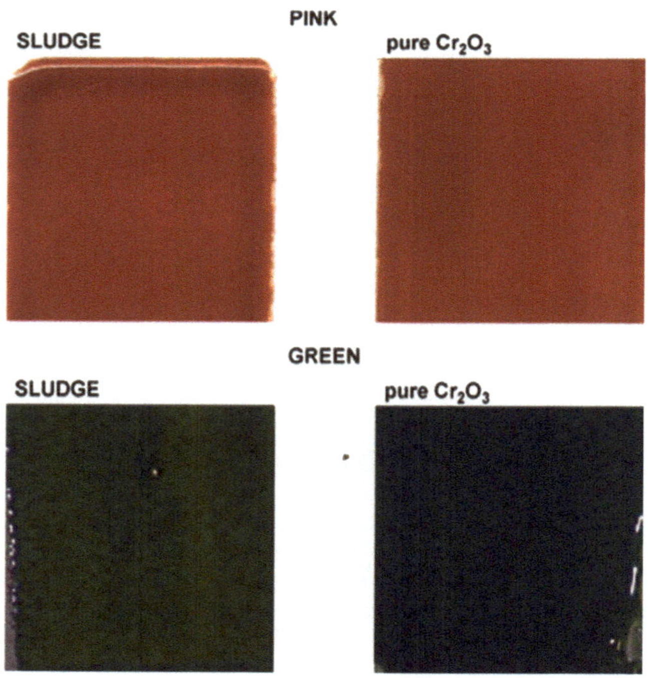

FIGURE 12.9

Pink and green glazed tiles obtained with synthesized pigments from waste compared with those starting from pure Cr_2O_3.

Adapted from Andreola, F., et al., 2008. Synthesis of chromium containing pigments from chromium galvanic sludges. Journal of Hazardous Materials 156 (1–3), 466–471.

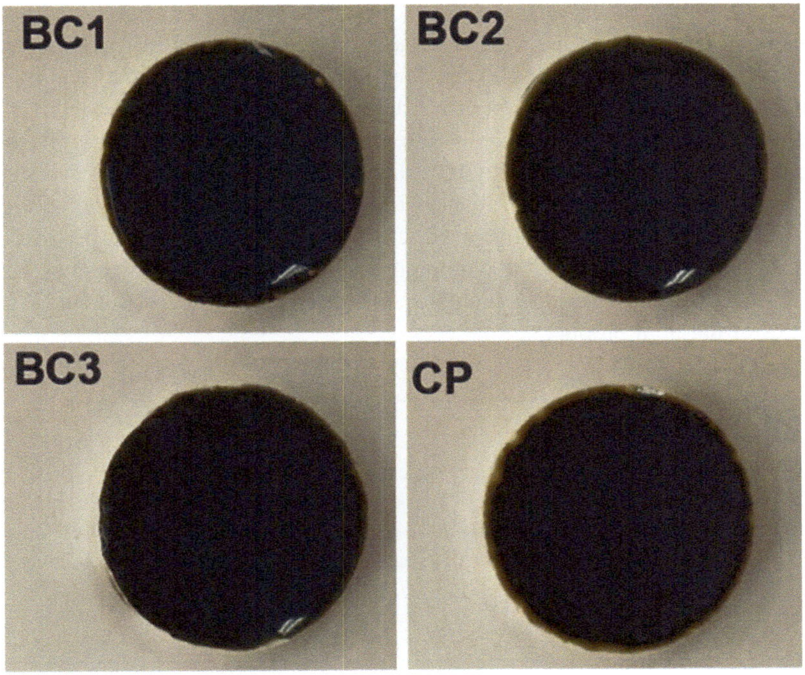

FIGURE 12.10

Black glazed tiles obtained with synthesized pigments from waste (BC1, BC2, and BC3) compared with commercial pigment (CP).

Adapted from Chen, Z.T., et al., 2015. Synthesis of black pigments containing chromium from leather sludge. Ceramics International 41 (8), 9455–9460.

$CuCr_2O_4$ could also be produced by Cr-rich leather sludge after they were calcined at high temperature with other metallic oxides (Fig. 12.10) (Chen et al., 2015). The color developed in ceramic glaze and was found to be similar to that produced from pure Cr_2O_3.

12.6 SUMMARY

- Cr is currently one of the most essential elements in industrial production. Most intensive exploitation, extraction, production, and use concentrate on developing countries and areas where economic development is usually accompanied by heavy environmental pollution. Typical Cr-rich wastes can be found from ore processing, stainless-steel plants, the electroplating industry, and leather tanning processes.
- Among various forms of Cr components, Cr(III) and Cr(VI) are two of the most stable valence states, and Cr(VI) often receives more concern regarding its high mobility and extreme toxicity. Thus, for the detoxification of Cr-containing waste, reductants such as ferrous salts ($FeSO_4$) and CaS_5 are used as the first-step treatment to reduce Cr(VI) to Cr(III) for easier stabilization before disposal.

- The S/S process is an important approach for the safe treatment and disposal of industrial wastes. By using cement, clay or/and lime with or without other additives, metal ions can be effectively immobilized. Thermal treatment can also be employed to achieve success in stabilizing metals of concern. It was found that stabilization of Cr(III) is more effective, whereas for Cr(VI), pretreatment such as a reduction process is often required to obtain better performance.
- Regarding economic benefits, secondary recycling is increasingly focused on waste treatment. Because Cr is originally the important material in most industrial production, the reuse of Cr components recycled from Cr-containing waste is promising. Value-added products such as cements, abrasives, glass, tiles, bricks, and pigments can be generated by chemical or/thermal treatment of Cr-rich wastes. Besides efficient stabilization of Cr, the presence of Cr may enhance the properties of the final product.

ACKNOWLEDGMENTS

This work was supported by the National Natural Science Foundation of China (41171364), the Guangdong Natural Science Foundation of China (S2013050014266 and S2011030002882), the Science and Technology Planning Project of Guangzhou, China (2013J4500024), the One Hundred Talents Program of the Chinese Academy of Sciences, and the Platform Project of Guangdong Academy of Sciences, China (2016GDASPT-0212).

REFERENCES

Andreola, F., et al., 2008. Synthesis of chromium containing pigments from chromium galvanic sludges. Journal of Hazardous Materials 156 (1–3), 466–471.

Barbieri, L., Bonamartini, A.C., Lancellotti, I., 2000. Alkaline and alkaline-earth silicate glasses and glass-ceramics from municipal and industrial wastes. Journal of the European Ceramic Society 20 (14–15), 2477–2483.

Basegio, T., et al., 2009. Vitrification: an alternative to minimize environmental impact caused by leather industry wastes. Journal of Hazardous Materials 165 (1–3), 604–611.

Berry, F.J., Costantini, N., Smart, L.E., 2002. Synthesis of chromium-containing pigments from chromium recovered from leather waste. Waste Management 22 (7), 761–772.

Celary, P., Sobik-Szoltysek, J., 2014. Vitrification as an alternative to landfilling of tannery sewage sludge. Waste Management 34 (12), 2520–2527.

Chen, Z.T., et al., 2015. Synthesis of black pigments containing chromium from leather sludge. Ceramics International 41 (8), 9455–9460.

Chrysochoou, M., et al., 2009a. Use of micro X-ray absorption spectroscopy and diffraction to delineate Cr(Vi) speciation in COPR. Global Nest Journal 11 (3), 318–324.

Chrysochoou, M., et al., 2009b. Microstructural analyses of Cr(VI) speciation in chromite ore processing residue (COPR). Environmental Science & Technology 43 (14), 5461–5466.

Chrysochoou, M., et al., 2010. Importance of mineralogy in the geoenvironmental characterization and treatment of chromite ore processing residue. Journal of Geotechnical and Geoenvironmental Engineering 136 (3), 510–521.

Cohen, B., Petrie, J.G., 1997. Containment of chromium and zinc in ferrochromium flue dusts by cement-based solidification. Canadian Metallurgical Quarterly 36 (4), 251–260.

Cohen, B., Petrie, J.G., 2005. The application of batch extraction tests for the characterisation of solidified ferro alloy waste products. Journal of Environmental Management 76 (3), 221–229.

Colombo, P., et al., 2003. Inertization and reuse of waste materials by vitrification and fabrication of glass-based products. Current Opinion in Solid State & Materials Science 7 (3), 225–239.

Darrie, G., 2001. Commercial extraction technology and process waste disposal in the manufacture of chromium chemicals from ore. Environmental Geochemistry and Health 23 (3), 187–193.

Dermatas, D., Meng, X.G., 2003. Utilization of fly ash for stabilization/solidification of heavy metal contaminated soils. Engineering Geology 70 (3–4), 377–394.

Dermatas, D., Moon, D.F., 2006. Chromium leaching and immobilization in treated soils. Environmental Engineering Science 23 (1), 77–87.

Du, J.J., et al., 2012. Reduction and immobilization of chromate in chromite ore processing residue with nanoscale zero-valent iron. Journal of Hazardous Materials 215, 152–158.

Eid, A.H., Keshtkar, H., Zahir, K.O., 1996. Characterization and treatment of the soil of an industrial site contaminated with chromium(VI). Journal of Environmental Science and Health Part A, Environmental Science and Engineering & Toxic and Hazardous Substance Control 31 (1), 227–247.

Ek, R.B., Schlobohm, J.E., 1993. Glassification of electric arc furnace dust. Iron and Steel Engineer 70 (4), 82–84.

Geldenhuis, J.M.A., Horne, A.W., 2002. Recovery of valuables and stabilisation of chromium (VI) in stainless steel flue dust fines. In: Steelmaking Conference Proceedings. Nashville, USA.

Giesekke, E.W., et al., 2000. Evaluation of solid-stabilized products made from Cr(VI)-containing ferrochrome bag-filter dust. In: Waste Materials in Construction Conference. Harrogate, England, UK.

Gress, L., 1993. Recycling vitrification process for electric arc furnace dust. Iron and Steel Engineer 70 (8).

Hillier, S., et al., 2007. Hydrogarnet: a host phase for Cr(VI) in chromite ore processing residue (COPR) and other high pH wastes. Environmental Science & Technology 41 (6), 1921–1927.

Huang, D., et al., 2004. Incorporation of chromium(VI) and chromium(VI) oxides in a simulated basaltic, industrial waste glass-ceramic. Journal of the American Ceramic Society 87 (11), 2047–2052.

Jacobs, J.H., 1992. Treatment and stabilization of a hexavalent chromium containing waste material. Environmental Progress 11 (2), 123–126.

Kameswari, K.S.B., et al., 2015. Diffusion and leachability index studies on stabilization of chromium contaminated soil using fly ash. Journal of Hazardous Materials 297, 52–58.

Kavouras, P., et al., 2015. Incineration of tannery sludge under oxic and anoxic conditions: study of chromium speciation. Journal of Hazardous Materials 283, 672–679.

Kuo, Y.M., Lin, T.C., Tsai, P.J., 2003. Effect of SiO_2 on immobilization of metals and encapsulation of a glass network in slag. Journal of the Air & Waste Management Association 53 (11), 1412–1416.

Kuo, Y.M., Wang, J.W., Tsai, C.H., 2007. Encapsulation behaviors of metals in slags containing various amorphous volume fractions. Journal of the Air & Waste Management Association 57 (7), 820–827.

Li, C.T., et al., 2007. Vitrification of chromium electroplating sludge. Environmental Science & Technology 41 (8), 2950–2956.

Ma, G., Garbers-Craig, A.M., 2006. A review on the characteristics, formation mechanisms and treatment processes of Cr (VI)-containing pyrometallurgical wastes. Journal of the South African Institute of Mining and Metallurgy 106 (11), 753–763.

Ma, G.J., Garbers-Craig, A.M., 2009. Stabilisation of Cr(VI) in stainless steel plant dust through sintering using silica-rich clay. Journal of Hazardous Materials 169 (1–3), 210–216.

Ma, G., 2005. Cr(VI)-Containing Electric Furnace Dust and Filter Cake: Characteristics, Formation, Leachability and Stabilization. University of Pretoria.

Mao, L.Q., et al., 2015. The role of temperature on Cr(VI) formation and reduction during heating of chromium-containing sludge in the presence of CaO. Chemosphere 138, 197–204.

Murthy, Y.R., Tripathy, S.K., Kumar, C.R., 2011. Chrome ore beneficiation challenges & opportunities – a review. Minerals Engineering 24 (5), 375–380.

Omotoso, O.E., Ivey, D.G., Mikula, R., 1998. Hexavalent chromium in tricalcium silicate – Part I – quantitative X-ray diffraction analysis of crystalline hydration products. Journal of Materials Science 33 (2), 507–513.

Perez-Villarejo, L., et al., 2015. Valorization and inertization of galvanic sludge waste in clay bricks. Applied Clay Science 105, 89–99.

Pinto, C.A., et al., 2005. Hazardous wastes disposal: stabilization/solidification of tannery residue in the presence of chromium. Advanced Powder Technology IV 498–499, 697–703.

Pioro, L.S., Pioro, I.L., 2004. Reprocessing of metallurgical slag into materials for the building industry. Waste Management 24 (4), 371–379.

Rakhunde, R., Deshpande, L., Juneja, H.D., 2012. Chemical speciation of chromium in water: a review. Critical Reviews in Environmental Science and Technology 42 (7), 776–810.

Reinosa, J.J., et al., 2010. High chemical stability of stoneware tiles containing waste metals. Journal of the European Ceramic Society 30 (14), 2997–3004.

Riganti, V., Fiumara, A., Odobez, G.B., 1986. The use of industrial sludges as raw-materials in the cement industry. Waste Management & Research 4 (3), 293–302.

Rodriguez-Pinero, M., et al., 1998. Stabilization of a chromium-containing solid waste: immobilization of hexavalent chromium. Journal of the Air & Waste Management Association 48 (11), 1093–1099.

Shi, C.J., Qian, J.S., 2000. High performance cementing materials from industrial slags – a review. Resources Conservation and Recycling 29 (3), 195–207.

Silva, A.C., et al., 2008. Incorporation of galvanic waste (Cr, Ni, Cu, Zn, Pb) in a soda-lime-borosilicate glass. Journal of the American Ceramic Society 91 (4), 1300–1305.

Sobiecka, E., Szymanski, L., 2014. Thermal plasma vitrification process as an effective technology for fly ash and chromium-rich sewage sludge utilization. Journal of Chemical Technology and Biotechnology 89 (7), 1115–1117.

Sreeram, K.J., Rao, J.R., Nair, B.U., 2011. Chromium(III) pigments: use of leather wastes as alternative starting material. Journal of the American Leather Chemists Association 106 (7), 219–225.

Stone, J.N., Marasco, G.F., 1986. Integrated treatment of wastes at Atlas Steels. Iron and Steel Engineer 63 (4), 43–47.

Sturm, T., et al., 2009. The use of EAF dust in cement composites: assessment of environmental impact. Journal of Hazardous Materials 166 (1), 277–283.

Thanikaivelan, P., et al., 2005. Recent trends in leather making: processes, problems, and pathways. Critical Reviews in Environmental Science and Technology 35 (1), 37–79.

Unceta, N., et al., 2010. Chromium speciation in solid matrices and regulation: a review. Analytical and Bioanalytical Chemistry 397 (3), 1097–1111.

Velasco, P.M., et al., 2014. Fired clay bricks manufactured by adding wastes as sustainable construction material – a review. Construction and Building Materials 63, 97–107.

Wang, S., Vipulanadan, C., 2000. Solidification/stabilization of Cr(VI) with cement – leachability and XRD analyses. Cement and Concrete Research 30 (3), 385–389.

Wang, T.G., He, M.L., Pan, Q., 2007. A new method for the treatment of chromite ore processing residues. Journal of Hazardous Materials 149 (2), 440–444.

Wei, Y.L., et al., 2002. Thermal stabilization of chromium (VI) in kaolin. Environmental Science & Technology 36 (21), 4633–4641.

Wei, Y.L., et al., 2010. Thermal detoxification and bloating of chromium(VI) with bentonite. Nuclear Instruments & Methods in Physics Research Section A, Accelerators Spectrometers Detectors and Associated Equipment 619 (1–3), 108–111.

Wu, C.L., et al., 2010. Thermal stabilization of chromium slag by sewage sludge: effects of sludge quantity and temperature. Journal of Environmental Sciences (China) 22 (7), 1110–1115.

Zunkel, A.D., 1997. Electric arc furnace dust management: a review of technologies. Iron and Steel Engineer 74 (3), 33–38.

CHAPTER

ASBESTOS: RESOURCE RECOVERY AND ITS WASTE MANAGEMENT

13

A. Kumar[1], M.N.V. Prasad[1,2], S.K. Maiti[3], Tripti[1]

Ural Federal University Named After the First President of Russia B.N. Yeltsin, Ekaterinburg, Russia[1]; University of Hyderabad, Hyderabad, Telangana, India[2]; Indian School of Mines, Dhanbad, Jharkhand, India[3]

CHAPTER OUTLINE

13.1 INTRODUCTION

Active and abandoned asbestos mines and waste cause severe environmental pollution to their surroundings. Despite its toxic nature, the production and consumption of asbestos is still proliferating. The world production of asbestos was estimated to be 1.98 million metric tons (Mt) in 2014, which is a slight decline from 2.02 Mt in 2013. Russia is the leading asbestos producer, contributing more than half of the total world production in 2014, followed by China, Brazil, and Kazakhstan (USGS, 2012 and 2013). Asbestos has average tensile strength and resistance to heat, fire, electrical,

Environmental Materials and Waste. http://dx.doi.org/10.1016/B978-0-12-803837-6.00013-5

and chemical damage. Because of its numerous vital properties, it became the first choice for almost all sectors of industry (textile, chemical, electrical, automotive, roofing, ceramics, cement, and paper). Because of its lethal effects, it has been banned in some developed countries. However, its huge production is still in progress, and every year its production generates enormous amounts of potentially toxic waste. These wastes contain minute asbestos fibers which can easily transfer from one place to another by air and other means, causing a huge impact on the environment (soil, air, and water) and human health. Asbestos is highly carcinogenic and exposure to it may result in lung cancer, mesothelioma, pneumoconiosis, and other respiratory problems (OSHA, 2006; Whalley et al., 1999; Pham et al., 2013; Dave and Beckett, 2005). In addition, leaching of heavy metals from asbestos waste in water bodies may destroy the aquatic ecosystem; in addition, high metal contamination and a lack of nutrients in the soil results in vegetation and habitat loss. Heavy runoff during monsoons causes the deposition of a thick layer of asbestos waste in nearby agricultural fields, fallow lands, and vegetative areas, simultaneously affecting vegetation as well. Overall use in different industrial sectors, health impacts, and the environmental impact of asbestos are presented in Fig. 13.1.

Bioremediation/phytoremediation is a green, sustainable, microbial, cost-effective, eco-friendly and solar-driven technology which has gained special interest worldwide to stabilize contaminated derelict sites. Bioremediation has rarely been used by researchers for asbestos. However, because of the lethal effect of asbestos, the use of fungal studies in association with plants which can transform asbestos chemistry is gaining importance compared with physical and chemical methods.

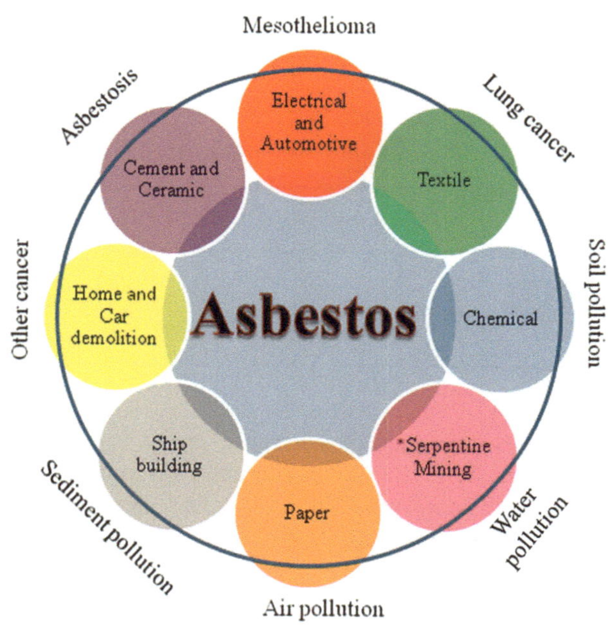

FIGURE 13.1

Pollution and use of asbestos in different industrial sectors and its health and environmental impacts.

13.2 WORLD SCENARIO OF ASBESTOS

Asbestos is the generic commercial designation for a group of naturally occurring mineral silicate fibers which can be divided into two classes: serpentine (chrysotile) and amphibole asbestos (amosite: brown asbestos; crocidolite: blue asbestos, anthophyllite, actinolyte, and tremolite) (Schüz et al., 2013) (Fig. 13.2). Asbestos has average tensile strength and resistance to heat, fire, electrical, and chemical damage. Because of its numerous vital properties, it became the first choice for almost all sectors of industry.

The mining of asbestos started with minimum effort almost 4000 years ago. However, slowly with the passage of time and the needs of development, its demand increased. During the mid 18th century, asbestos became popular when different industries started manufacturing asbestos-based materials and products. In 1850–80, asbestos was mainly used for packing and flat-sealing purposes. During these 3 decades, different products such as molded and bonded heat-insulated products, textiles, roofing felt, and cement products were manufactured. However in the early 20th century, because of its higher demand, asbestos production increased tremendously. Industries started using asbestos on a large scale in the textile, chemical, electrical, automotive, roofing, ceramic, cement, and paper industries (Zaremba and Peszko, 2008).

13.2.1 WORLDWIDE PRODUCTION AND CONSUMPTION

Despite the extreme harmful nature of asbestos fibers, the production of asbestos is still in progress. The world production of asbestos was estimated to be 1.98 Mt in 2014, which is a slight decline from 2.02 Mt in 2013 (Table 13.1). Four countries accounted for over 99% of world asbestos production. In consumption, China was the leading consumer of asbestos, followed by India, Russia, Brazil, Kazakhstan, Indonesia, and Thailand in 2012 (Table 13.2). These seven countries (85% of total

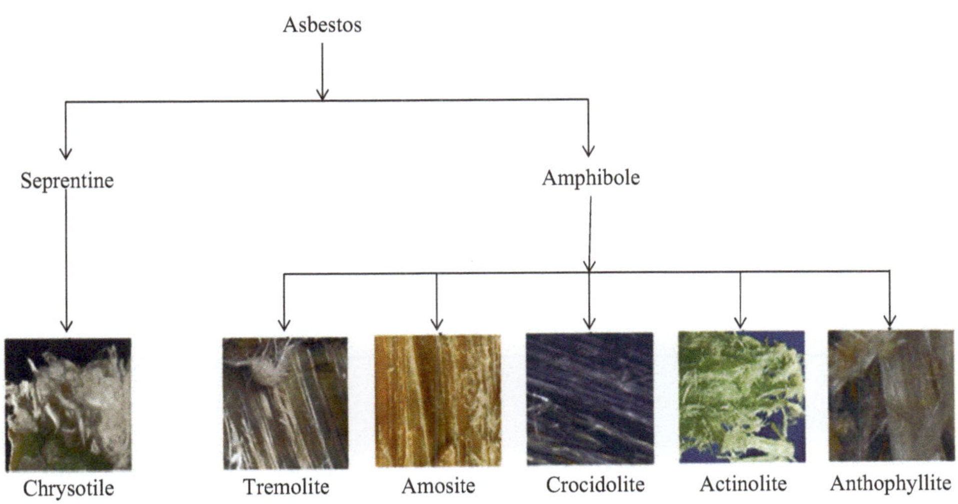

FIGURE 13.2

Different types of asbestos found throughout the world.

Table 13.1 World Total Asbestos Production (2008–14) by Different Countries (in metric tons)

Country[a]	2008	2009	2010	2011	2012	2013	2014[e]
Russia[e]	1,017,000	1,000,000[e]	995,174[r]	1,031,880[r]	1,041,000[r]	1,050,000	1,050,000
China[e]	380,000	440,000	400,000	440,000	420,000	420,000	400,000
Brazil	287,673	288,452	302,257	306,321	304,569[r]	307,000	291,000
Kazakhstan	230,100	230,000	214,100	223,100	241,200	242,000	240,000
Argentina	298	322	341	105	100[e]	100[e]	ns
Canada[e]	160,000	150,000	100,000	50,000	–	–	ns
Zimbabwe	11,489	4971	2400[e]	–	–	–	ns
India[e]	304	261[b]	254[b]	250	245	240	ns
Total	*2,090,000*	*2,110,000*	*2,010,000*[r]	*2,050,000*[r]	*2,010,000*[r]	*2,020,000*	*1,981,000*

[e], Estimated; [r], revised, –, zero.
World totals and estimated data are rounded to no more than three significant digits; may not add to totals shown.
Marketable fiber production. Table includes data available through May 2, 2014.
[a], In addition to the countries listed, Afghanistan, North Korea, Romania, and Slovakia also produced asbestos but output was not officially reported and available general information was inadequate to formulate reliable estimates of output levels.
[b], Reported figure.
Source: USGS, 2012 and 2013. Mineral Year Book. United States Geological Survey. Asbestos 2011 and 2012 (advanced release).

Table 13.2 Estimated Asbestos Consumption, by Major Regions, 2005–12[a,b,c]

Location	2005	2006	2007	2009	2010	2011	2012
Africa	63,400	44,400	61,700	12,000[r]	16,800[r]	14,500	10,400
Asia and Middle East	1,180,000	1,100,000	1,220,000	1,730,000[r]	1,750,000[r]	1,770,000[r]	1,690,000
Central and North America	87,300	50,200	67,700	24,700	−12,800	−6260[r]	28,300
Europe	759,000	634,000	621,000	93,400[r]	83,800[r]	81,000[r]	50,500
South America	170,000	167,000	107,000	159,000[r]	194,000[r]	223,000[r]	204,000
Other areas (unspecified)	2610	2410	1720	−588	2000	2000	120
Grand total	*2,110,000*	*2,100,000*	*2,260,000*	*2,020,000*[r]	*2,040,000*[r]	*2,080,000*[r]	*1,980,000*

[a], Calculated as country production plus imports minus exports. r, reserved.
[b], Previously, Kazakhstan, Kyrgyzstan, Russia, Turkmenistan, and Uzbekistan were included under Europe, to be consistent with country groupings in the early to middle 1900s. The table was revised to be consistent with current conventions.
[c], Negative values are net exports.

consumption) along with Vietnam, Sri Lanka, and Ukraine accounted for about 94% of global asbestos consumption (Table 13.3). Only China and India consumes more than half of the total asbestos (Fig. 13.3). Russia produces nearly half of the total asbestos; however, it consumes a low quantity of asbestos; most is exported worldwide. India contributes less than 1% of total asbestos, although the requirement and demand for asbestos is high and its consumption is over 18% from 2009 to 2012.

Table 13.3 World Total Consumption (2004–12) by Different Countries (in metric tons)

Country	2004	2005	2006	2007	2008	2009	2010	2011	2012
China	537,000	515,000	541,000	626,000	665,000	565,000	614,000	638,000	531,000
India	190,000	255,000	240,000	302,000	349,000	341,000	426,000	322,000	473,000
Russia	321,000	315,000	293,000	280,000	360,000	277,000	263,000	251,000	196,000
Brazil	66,900	139,000	134,000	93,800	131,000	140,000	171,000	185,000	168,000
Kazakhstan	289,000	150,000	96,200	109,000	186,000	40,000	−1370	155,000	5290
Indonesia	51,000	23,300	36,300	46,200	78,000	82,000	112,000	124,000	162,000
**Total*	*91,000*	*170,000*	*167,000*	*107,000*	*2,200,000*	*1,980,000*	*2,060,000*	*2,070,000*	*1,980,000*

**, Including other countries;*
The world totals and estimated data are rounded to no more than three significant digits; may not add to totals shown.
Table includes data available through July 1, 2013.
Calculated as country production plus imports minus exports.
Source: USGS, 2012 and 2013. Mineral Year Book. United States Geological Survey. Asbestos 2011 and 2012 (advanced release).

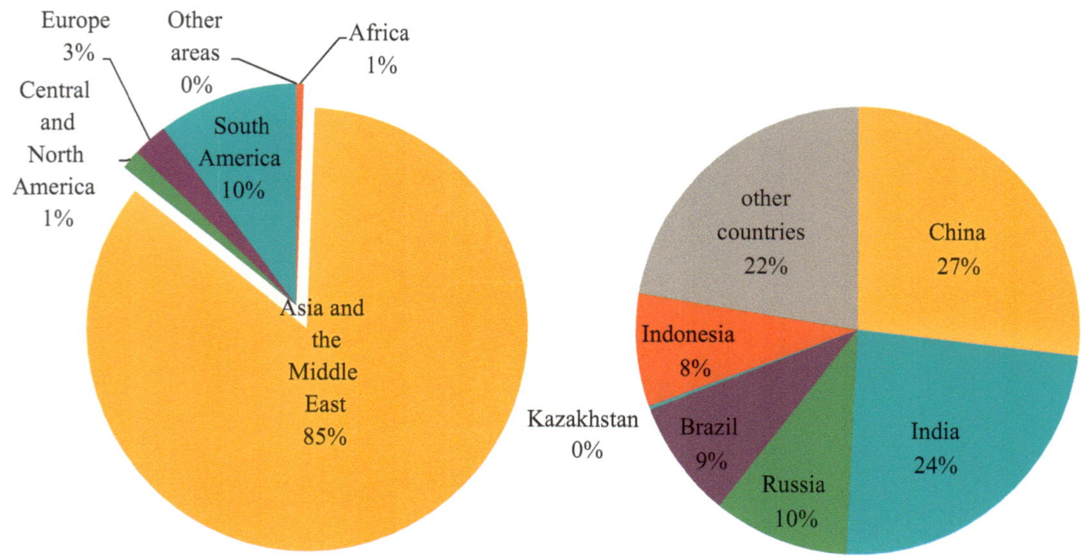

FIGURE 13.3

Total global percent asbestos consumption (2012) by (A) different regions and (B) different countries.

13.2.2 USE OF ASBESTOS AND ITS WASTE

Asbestos is one of the most important materials used by different industrial sectors. It is used mainly in the textile, chemical, electrical, automotive, roofing, ceramics, cement, and paper industries. However, because of the adverse effects of asbestos, it has been banned in almost 52 countries and its use has been reduced (Fig. 13.4). The United States stopped producing asbestos in 2002. However, the use of asbestos is not restricted in the United States and it imports asbestos to make asbestos-cement products, footwear, paper, mill board and felt, compressed asbestos fiber jointing, yarn and thread, cord and string, woven or knitted fabric, civil aircraft, gaskets, packing and seals, fabricated asbestos fiber, brake lining and pads, other fabricated asbestos fibers, clothing, pads, and friction materials. Apart from these uses, asbestos is used mainly for cement pipes, cement sheets, coatings and compounds, electrical and thermal insulation, and plastic flooring.

Attention has been paid to using asbestos waste generated from mine waste. Materials which require minute quantities of asbestos for production with minimal effect on the environment and human health are gaining in importance. Experimental studies are ongoing using asbestos to making carbon nanotubes, semiconductors, and so forth. Some advancements in the use of asbestos for manufacturing new products are listed in Table 13.4.

13.2.3 SUBSTITUTES OF ASBESTOS

Asbestos is considered to be one of the most toxic materials in the world. To reduce the commercial use of asbestos, new alternatives material should be introduced on the market with both nontoxic and commercial value (Table 13.5). In 1938, fiberglass insulation was invented as an alternative to asbestos

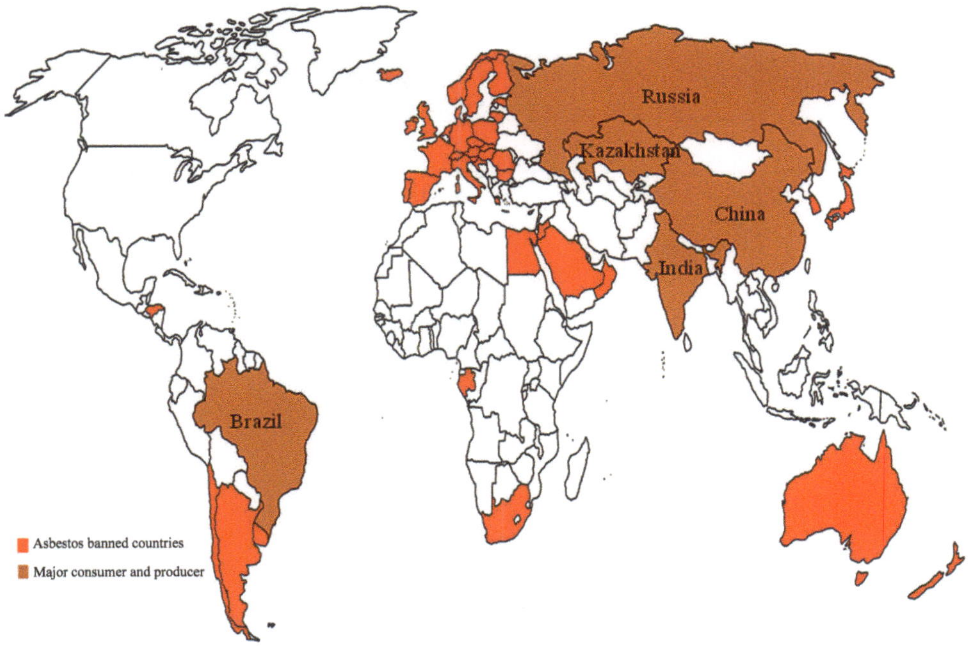

FIGURE 13.4

World map showing major asbestos producers, consumers, and countries banning asbestos. Asbestos banned countries: Algeria, Argentina, Australia, Austria, Bahrain, Belgium, Bulgaria, Chile, Croatia, Cyprus, Czech Republic, Denmark, Egypt, Estonia, Finland, France, Gabon, Germany, Greece, Hungary, Iceland, Israel, Italy, Japan, Jordan, Korea, South, Kuwait, Latvia, Lithuania, Luxembourg, Malta, Mozambique, Netherlands, New Caledonia, Norway, Oman, Poland, Portugal, Qatar, Romania, Saudi Arabia, Serbia, Seychelles, Slovakia, Slovenia, South Africa, Spain, Sweden, Switzerland, Turkey, United Kingdom, Uruguay.

Table 13.4 List of Different Materials Prepared From Asbestos Waste

Products	References
Precipitated silicon dioxide (SiO_2)	Gaoxiang (2007), Shuilin et al. (2007), Jiang et al. (2011), and Xinyan (1998)
Magnesium hydroxide ($Mg[OH]_2$)	Gaoxiang (2007), Shuilin et al. (2007), and
Clinker ceramic wall bricks	Kusiorowski et al. (2014)
Glass ceramics decorative slabs and materials	Qilong et al. (1997), and Zanzhong et al. (1998)
Reinforcing fillers in polymers	Zhai et al. (2014)
Heat-resistant concrete and other construction materials	Zhu (2009)
Highway concrete	Zhu (2009)
Sintered clinker wall bricks	Kusiorowski et al. (2014)

Table 13.5 List of Possible Substitutes for Asbestos

Substitutes for asbestos	General properties/ characteristics	Major uses	References
Mineral wool			
Fiberglass (glass wool)	Hand resistance, abrasion resistance, high seam strength	Construction, automotive parts and spares, Packaging material, furniture industry, laminates in aircraft, boats	Lee et al. (1981) and Costa and Orriols (2012)
Stone and slag wool	Heat resistant, sound resistant, high seam strength	Thermal and acoustic insulation, buildings, automotive industries and appliances, inflammable materials and flame-retardant protection, hydroponics growth media, filtration	Costa and Orriols (2012) and Kim and Jeong (2014)
High-temperature insulation wool (rock wool)	Made of alkaline-earth silicates, less biopersistent than refractory ceramic fibers	Thermal insulation (lining of blast furnace)	Fyles et al., 1996 and Costa and Orriols (2012)
Cement-bonded wood fibers	Acoustics based, low, medium and high density, high toughness index	Construction industry (sliding doors, wallboard, insulating concrete forms)	Wolfe and Gjinolli (1999) and Karade (2010)
Biological materials			
Banana peel	High compressive strength, hardness, and specific gravity	Automotive brake pads	Idris et al. (2013)
Palm kernel fibers	High coefficient of friction, compressive strength, hardness, and specific gravity	Automotive brake pads	Ikpambese et al. (2014) and Ibhadode and Dagwa (2008)
Bagasse	High coefficient of friction and wear resistant	Automotive brake pads	Aigbodion et al. (2010)
Cashew nutshell liquid (Kevlar pulp)	Hardness reduced using carbon fibers as reinforcement and graphite powder as friction modifier	Brake linings and clutch facings in automotive industry	Mathur et al. (2004)
Palm slag	High thermal properties, compressive strength, and wear behavior	Automotive brake pads	Ruzaidia et al. (2011)
Kenaf fibers (*Hibiscus cannabinus*)	High thermal conductivity, wear resistant	Automotive brake pads	Namessan et al. (2013)

Table 13.5 List of Possible Substitutes for Asbestos—cont'd

Substitutes for asbestos	General properties/ characteristics	Major uses	References
Cashew dust with metal fiber–steel fiber	Stable frictional performance, high wear resistance	Automotive brake pads	Liew and Nirmal (2013)
Refractory ceramic fibers		Thermal insulation, electronics and construction industries	Walker et al. (2012) and Mukherjee (2013)
Polybenzimidazole fiber	High melting point, thermal, chemical, and oxidative stability	High-temperature welders' apparel, protective gloves, aircraft wall fabrics, astronaut space suits, firefighter gear	Powers and Serad (1986) and Costa and Orriols (2012)
Nanofibers (nanotubes)	$\leq$100 nm, made of inorganic nonmetallic C, Si, B, great electrical conductivity, mechanical resistance, and thermal stability	Electronics, medical industry, and other industrial applications.	Costa and Orriols (2012)
Hydrogenerator	High friction resistance	Lining material and brake pad for heavy machines	Chand et al. (2004)
Continuous filament glass fiber	Rectilinear, with uniform diameters, noninhalable	Insulation, electronics, and construction industries	Costa and Orriols (2012)
Carbon nanotubes			

material; it is now the most commonly used type of insulation material. Because of its great similarity to asbestos's structure, issues have been raised. However, the International Agency for Research on Cancer removed fiberglass from its list of possible human carcinogens in 2001.

A new highly textured fiberglass fabric called "Zetex" was discovered in 1978. Because of its miraculous properties (it is lighter and more resistant to abrasion compared with asbestos), it has received much attention. It offers the same bulk, thickness, hand, and feel as asbestos. The fabric was textured to eliminate some of the problems that arise with fiberglass, such as poor abrasion resistance and poor seam strength. It is an excellent alternative to asbestos fiberglass and is used mainly for protection from heat and fire and for thermal and acoustic insulation.

Mineral wool or glass wool, also called man-made vitreous fiber or cotton fiber, is well-known mostly in European countries and is used mainly in insulators in houses (as both structural and pipe insulation), filtration, soundproofing, and as a hydroponics growth medium. Because of its high surface area and fibrous nature, it holds a huge amount of water and air; for these reasons it is used in hydroponics plant studies.

Many companies that produced asbestos-cement products reinforced with asbestos fibers developed products incorporating organic fibers. Product such as "Eternit" and "Everite" use "Nutec" fibers which consist of organic fibers, Portland cement, and silica. Cement-bonded wood fiber is another substitute. Stone fibers are used in gaskets and friction materials.

Another potential synthetic fiber is polybenzimidazole fiber (PBI), which has exceptional thermal and chemical stability. Important places where PBI is used are fire departments and space agencies. Asbestos alternatives for industrial use include sleeves, rope, tape, fabric, textiles, and insulation batt materials made from fiberglass and silica.

13.3 ENVIRONMENTAL AND HEALTH IMPACT OF ASBESTOS

13.3.1 ENVIRONMENTAL IMPACT

Surface and underground mining results in the huge production of mine waste. Active and abandoned mines dumped with waste create a severe adverse impact on the environment. Major sources of introducing asbestos into natural waters and the atmosphere are through the disposal of industrial asbestos wastes, the erosion of asbestos-rich rock formations, and wind erosion. Generated waste slowly and continuously contaminates the water, air, and soil.

Rivers are used as the main sources of drinking water for large cities. Fiber-contaminated water bodies located close to these asbestos mines impart toxic fibers and metals into the soil and vegetation (Karkanas, 1995; Kumar and Maiti, 2015b). Furthermore, these fibers contain significant concentrations of Cr and Ni along with other metals, which increase the toxicity of the drinking water. Epidemiological studies revealed that asbestos-contaminated drinking water has both positive and negative associations with asbestos exposure for the population (Berndt and Brice, 2008).

Atmospheric pollution also contributes to the pollution of natural waters. In India, Andhra Pradesh, Jharkhand, and Rajasthan face the problem of fluorosis because naturally occurring silicate fiber often occurs in high-fluoride host rock, which occurs as a result of the enrichment of fluoride in the groundwater, which is the only source of water for the local population (Subramanian and Madhavan, 2005).

Another possible source of asbestos pollution is the milling and crushing of host rocks, which results in the removal of small asbestos fibers containing heavy metals. These fibers remain suspended in air for a long time and become deposited in the soil which is leached and percolated through water into the surface and groundwater. Furthermore, increases in toxic Cr, Ni, and other metals result in sparse vegetation growth at mine sites (Kumar and Maiti, 2015a). Kumar and Maiti (2014) reported the presence of asbestos fibers in the agricultural fields of Roroburu, Chaibasa, India (Fig. 13.5). Mine waste dumped on the Roro hills is a mixture of asbestos and chromite; because of it, the local population faces the toxic effects of Cr and other metals. Crops and plants growing on the asbestos-chromite–contaminated fields have accumulated a significant concentration of heavy metals; their regular uptake may lead to different health problems and carcinogenic diseases.

13.3.2 HEALTH IMPACT

Asbestos production and the use of asbestos materials result in severe diseases and health issues. Asbestos creates a major impact on the human body and sometimes on animals exposed to the mineral.

FIGURE 13.5

Scanning electron microscope results of (A) raw asbestos fibers associated with rock; (B) presence of asbestos fibers in chromite-asbestos mine waste–contaminated agricultural soil.

Because of the very fine size of asbestos fibers, it is easily inhaled through the nostrils and enters the human body. Asbestosis, lung cancer, and mesothelioma are three major primary diseases associated with asbestos exposure and its inhalation. The terms "asbestosis" and "asbestos bodies" were first used in English in 1927; in 1950, the relation between exposure to asbestos and lung cancer and mesothelioma became known (Cooke, 1927; Wagner et al., 1960). Selikoff et al. (1979) and Doll and Peto (1985) documented that asbestos is the occupational cause of asbestosis, pleural plaques, lung cancer, pleura and peritoneal mesothelioma, larynx cancer, and other cancers. When the fine fibers enter through the nostrils or gastrointestinal, they become attached to the lungs, resulting in cancer and other related disorders. Diseases caused as a result of asbestos fibers can be seen after 20–40 years from the time of exposure to asbestos and its products (INSERM, 2005). A study showed that between 1982 and 1996, the chances of acquiring pleural mesothelioma widely increased, with an average annual rate of increase for men of 5%, and for women, 3%.

Wagner et al. (1960) was the first to discover mesothelioma of pleura and peritoneum, which is a fatal disease with a long latent period. It forms a hard white tumor that characteristically encases the lung and causes lung cancer. In the United States (CDC, 2009) and worldwide, it was found that of all types of vital disease resulting from asbestos and asbestos fibers, mesothelioma is a very rare cancer. The global 15-year (1994–2008) magnitude of mesothelioma was estimated to be 174,300, according to a group of 56 countries that reported the disease, and 38,900 for the same 15-year period by extrapolation to a group of 33 countries that did not systematically report the disease (Park et al., 2011).

The disease "asbestosis" is an occupational disease caused by exposure to the fine fibers of asbestos and its deposition into the lungs of workers working in asbestos mines. Clinical features of asbestosis include breathlessness, inspiratory crackles, and pulmonary function abnormalities or pulmonary gas exchange disturbances during exertion and inflammation of lungs pleura which can be seen after 20 or more years from the time of exposure. Asbestos exposure occurs only when the mineral is disturbed or shaken by mechanical or natural means. The fine particles remain suspended in air for a long time and result in exposure of the fibers to humans.

The American Thoracic Society (2004) first considered asbestos-related lung cancers to be scar cancers, which changed as more data became available about the fiber burden in workers with other types of asbestos-related disease. A survey on the workers of Quebec working in a chrysotile asbestos mine showed that the fiber concentration in lungs of chrysotile miners and millers dying of asbestosis were almost over twofold higher than in those dying of lung cancer (30 vs 13 million chrysotile fibers per gram of dry lung tissue) (Cowie et al., 2005). All lung cancer cell types have been associated with asbestos exposure, which further supports the concept of exposure having a direct role in their genesis, independent of smoking (Gong and Christiani, 2002).

13.4 CLEANUP TECHNOLOGIES

Different methods are available to clean up mine waste generated from waste (Fig. 13.6). However, physical and chemical methods are expensive and sometimes cause other environmental pollution. Currently, attention is given to biological methods because of their high efficiency, cost-effectiveness and minimal toxicity to the environment. Bioremediation and phytoremediation are the two of the most important approaches available today to clean up hazardous sites. Planting transformed sites with some energy plants, timber trees, and other economy-generating plants further helps develop rural areas and populations (Fig. 13.7).

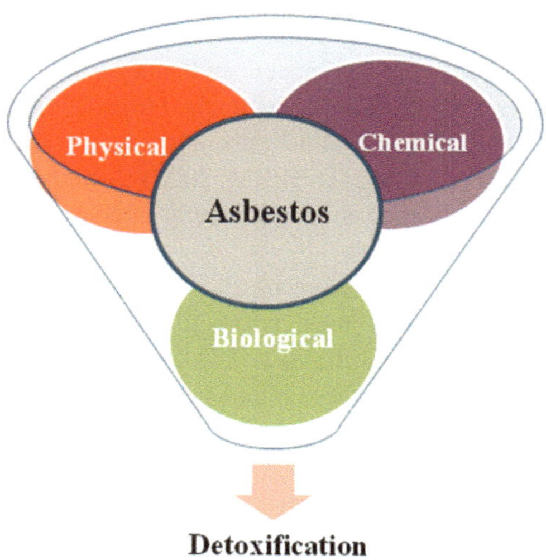

FIGURE 13.6

Different cleanup methods available or applied alone or coupled to detoxify the effect of asbestos.

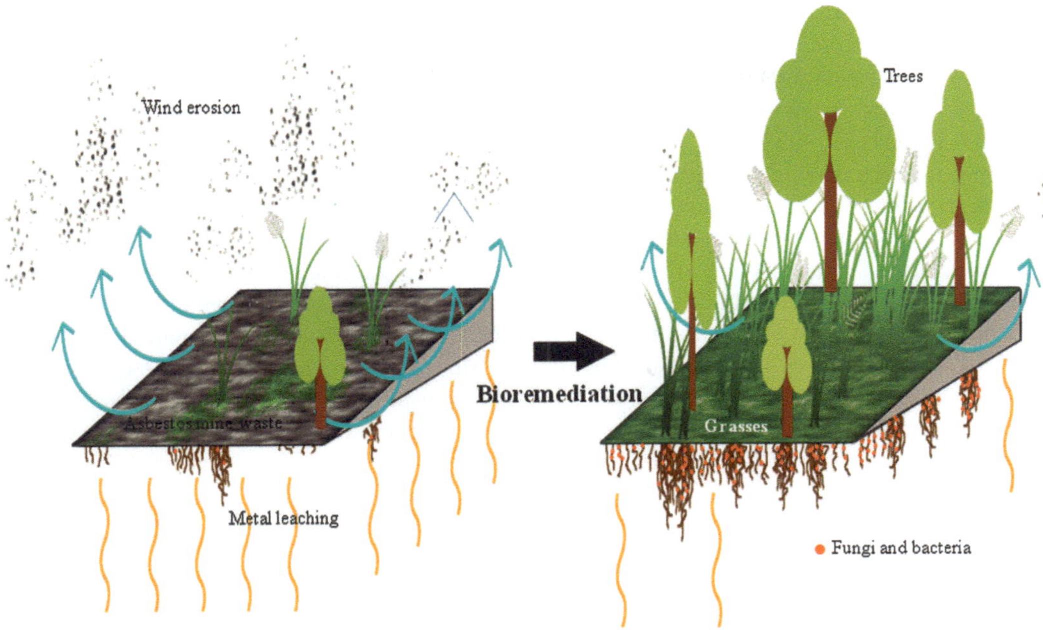

FIGURE 13.7

Bioremediation and phytoremediation of asbestos mine waste using a variety of grasses and trees.

13.4.1 PHYTOREMEDIATION

13.4.1.1 Plant and Lichen Colonization

Exposure to asbestos fibers has adverse effects on human health, such as lung cancer, asbestosis, and mesothelioma. Despite the restricted use of asbestos, it is still used in many countries. After mining operations, loose waste or soil materials are left or dumped on hills or in the open environment and are gradually driven away by environmental means (water and wind), causing pollution. Plants including grasses, herbs, shrubs, and trees are better approaches for remediating and stabilizing derelict or degraded land sites. The mining of asbestos results in the generation of millions of tons of waste associated with asbestos fibers. Abandoned asbestos mines represent potential sources of airborne mineral fibers with adverse effects on human health, and thus are considered hazardous environmental sites (Lee et al., 2008; Favero-Longo et al., 2009). These wastes are dumped directly into the environment, which requires revegetation. Revegetation of asbestos mine spoils has been proposed as a way to stabilize asbestos-bearing substrate, thereby reducing fiber dispersion into the air (O'Dell and Claassen, 2009). Favero-Longo et al. (2009) evaluated the effect of natural revegetation on the air dispersion of asbestos fibers from asbestos-rich serpentine lithosoils at an abandoned chrysotile mine. Different air samples were collected from vegetated and barren plots. The plots with natural plant growth had reduced asbestos dispersion (50% reduction with 15–40% vegetative cover) compared with barren plots. Several natural locally growing plants were also collected, including *Minuartia* and *Thymus* species, to revegetate the mine spoil. *Thymus* sp., a mat-forming, serpentine endemic plant with high transplantation survival, growth rates, and reproductive output, was highly effective in revegetating the asbestos mine spoil.

Because higher plants are generally poor in colonizing the surface of the soil and rocks, the use of lichens was studied by Favero-Longo et al. (2005, 2007). Saxicolous lichens significantly colonized asbestos-rich surfaces, including asbestos cement tiles and asbestos-bearing serpentinite rock. Sometimes lichen metabolites determine dissolution processes associated with a significant reduction in the toxicity-relevant surface reactivity of asbestos (Turci et al., 2007; Favero-Longo et al., 2009). However, higher plants had a higher growth rate compared with lichen colonization, which limits lichens as agents of spontaneous bioremediation. The mine walls of the most important asbestos mine in Western Europe (the chrysotile mine of Balangero, northwestern Italy), which was abandoned 20–55 years ago, still show only scattered and poor lichen colonization, whereas natural outcrops of analogous serpentinite rock in the surrounding area show an average lichen cover of over 75%. Balangero mine walls were cultivated with isidia and transplanted adult thalli of *Xanthoparmelia tinctina* (Maheu & Gilet) hale. This foliose species is common on natural serpentinite outcrops outside the mining basin, but only a few young thalli occur on mine walls abandoned 40–55 years ago (Favero-Longo et al., 2006).

13.4.2 BIOREMEDIATION

Currently, there is no good way to decontaminate asbestos-contaminated sites. However, a new approach to remediate asbestos-contaminated sites is bioremediation. Simply put, bioremediation is the decontamination or reduction of contaminants from the substrate using microorganisms. Asbestos is widely used in different industries to manufacture asbestos products. An increase in asbestos mine waste and its legacy can cause potentially fatal diseases. One application of bioremediation is the use of fungi, which can be employed for the cleanup of asbestos. However, cleaning contaminated soil around asbestos mines and factories is much harder than removing asbestos sheets from buildings. The widely dispersed fibers can easily become airborne if disturbed.

Microorganisms are already used to clean oil spills and are being studied for the decontamination of soil and water containing toxic heavy metals. Asbestos fibers contain iron, which is partially toxic. It can cause the formation of highly reactive chemicals called free radicals, which can damage DNA and trigger cancer (Daghino et al., 2006; Ball, 2003). By removing iron from asbestos, its hazardous percentage can be reduced to a great extent. Many microorganisms require iron to generate energy, and they grab hold of iron atoms in soil minerals and concentrate it using chemical hooks called siderophores. Iron-reduced fibers cannot generate carcinogenic free radicals. Fungus such as *Fusarium oxysporum* (a common cause of plant rot), *Mortierella hyalina*, and *Oidiodendron maius* (a fungus that attacks trees and shrubs) can withstand multiple heavy metal pollution. These fungi sucks iron and weave a network around their hyphae; the fibers remained adhered to that, resulting in the nondispersal of fibers even after soil disturbance. The far-reaching hyphae networks that fungi spin throughout large areas of soil could immobilize asbestos fibers. To implement this technique, fungi cultured under artificial conditions can be applied to asbestos-contaminated sites with suitable nutrients which can spread hyphae to stabilize the number of asbestos fibers (Ball, 2003).

Martino et al. (2003) demonstrated that modifications at the surface of the fibers could result in a decrease in their carcinogenic potential and sometimes full inactivation of asbestos fibers dispersed in the soil or substrate. Siderophores were released from the plants, depriving them of the free radical–generating sites crucial to triggering carcinogenic mechanisms. Soil fungi display a huge functional and genetic biodiversity, and suitable strains with high iron-mobilizing abilities can be selected from natural sources. Further research can be done to identify genes involved in siderophore production, which may be the starting point for genetically engineered organisms with increased iron mobilization from asbestos.

Daghino et al. (2006) reported that few genera of soil fungi growing on asbestos fibers release significant chelators and antioxidants. *Verticillium* sp. and *Paecilomyces* sp. were isolated from chrysotile-bearing rock fragments, as the most potent in iron extraction, and compared with previously studied *F. oxysporum*. All fungi extracted iron from fibers (7.3% from crocidolite and 33.6% from chrysotile by *Verticillium* sp.), releasing it into the medium. *Fusarium oxysporum* and *Paecilomyces* sp. suppressed the potential of the fibers to release hydroxyl radical, whereas *Verticillium* sp. suppressed it on crocidolite but enhanced it on chrysotile, a hallmark of ongoing mobilization of reactive iron. Fibers incubated in the growth medium, but in the absence of fungi, exhibited a remarkable potential to damage DNA in vitro, as measured by the generation of 8-oxo-7,8-dihydro-2′-deoxyguanosine, whereas all of the fungi reduced such an effect. Fungi may thus be regarded as appropriate candidates for the bioremediation of asbestos-rich soils whereby the reactive iron ions responsible for DNA damage are progressively removed from the fibers.

On the basis of the results described by different researchers, it can be concluded that fungi could be possible agents for the decontamination of asbestos-polluted soil and substrate.

13.5 CASE STUDIES

13.5.1 INDIA

India, which is a developing country, is second in the consumption of asbestos, with an overall production of 268 tons in 2010–11 (USGS, 2012 and 2013; IBM, 2013). Table 13.6 represents the occurrences, mineralogical variety, reserves, remaining resources, and production during 2009–12 by different asbestos-producing states in India. The reported consumption of asbestos in 2011–12 was about 104,000 tons and was used almost entirely to manufacture asbestos-cement and asbestos-based products.

Five reporting mines, three in Andhra Pradesh (Cuddapah district) and two in Rajasthan, are the major producers of the chrysotile variety of asbestos. A total of 69% of the chrysotile variety of asbestos is entirely produced by Padma Minerals (P) Ltd and Baba Minerals Corporation of Andhra Pradesh. The major mineralogical varieties of asbestos found in India are chrysotile, tremolite, and chrysotile mixed with other amphibole minerals. To date, Andhra Pradesh is the only asbestos-producing state in India. Exports and import of asbestos in 2011–12 were 1296 and 378,122 tons, respectively, compared with 231 and 365,795 tons in 2010–11.

Roro mine in Chaibasa, West Singhbhum, Jharkhand, India, was continuously mined for almost 30 years and abandoned with a huge amount of asbestos waste (Fig. 13.8). The mine waste containing minute asbestos fibers is slowly rolling down from the hill owing to heavy runoff and is easily carried away by wind, creating numerous gullies and causing a serious adverse effect on human health and the environment (Figs. 13.9 and 13.10).

Kumar and Maiti (2015a) attempted to remediate the asbestos mine waste site by implementing the phytostabilization technique of phytoremediation. Before implementation phytostabilization in the

Table 13.6 Occurrences, Reserves, Remaining Resources, and Production During 2009–12 by Different Asbestos-Producing States in India (in tons)

State	District	Mineralogical variety	No. of mines	Reserves	Remaining resources	Production		
						2009–10	2010–11	2011–12^P
Andhra Pradesh	Cuddapah	Chrysotile	3	14,782	41,749	243	268	280
Jharkhand	Singhbhum	Chrysotile, tremolite, chrysotile mixed with other minerals	–	–	154,893	–	–	–
Karnataka	Chickmagalur	Amosite	–	–	8,282,457	–	–	–
	Hassan	Anthophyllite						
	Mandya	Mixed amphibole minerals						
	Mysore	Chrysotile						
	Shimoga	Amosite						
Rajasthan	Ajmer	Mixed amphibole minerals	2	2,496,059	11,119,651	–	–	–
	Bhilwara	Mixed amphibole minerals						
	Dungarpur	Mixed amphibole minerals						
	Pali	Tremolite, chrysotile mixed with other amphibole minerals						
	Rajsamand	Tremolite, actinolite and mixed amphibole minerals						
	Udaipur	Chrysotile, tremolite, and mixed amphibole minerals						
Uttarakhand	Chamoli	Others	–	–	311	–	–	–

P, Provisional; reserves and resources as of April 2010.

Source: IBM, 2012 and 2013. Indian Mineral Year Book, 2011 & 2012, fiftieth and fifty-first ed. Indian Bureau of Mines. Asbestos (advanced release).

FIGURE 13.8

Distant view of asbestos mine in Roro, Chaibasa, Jharkhand, India.

FIGURE 13.9

(A) Gully erosion, (B) water erosion, (C) wind erosion, and (D) loose asbestos mine waste.

FIGURE 13.10

(A) Children playing with their skin exposed and feet bare; (B) cattle grazing on toxic asbestos waste.

FIGURE 13.11

(A) Growth of *Cymbopogon citratus* (lemongrass) and *Chrysopogon zizanioides* (vetiver) in control and differently amended pots after 120 days for phytostabilization of asbestos mine waste; (B) extensive roots of vetiver bind the top layer of garden soil and amended asbestos mine waste; (C) extensive roots of lemongrass bind the entire asbestos mine waste.

field, a pot-scale study was conducted for 4 months in a greenhouse using two aromatic grasses (*Cymbopogon citrates* and *Chrysopogon zizanioides*) and organic manure (farmyard and chicken manure) along with some garden soil. Different treatments were prepared to check the best treatment for the phytostabilization of asbestos mine waste. Both grasses were growing luxuriantly on mine waste with amendments with a high root–shoot ratio. Because of the extensive root system, both grasses were capable of bind very fine loose asbestos mine waste in the pots (Fig. 13.11). Both small pot and long pot experiments were conducted to study the root growth system; it was found that these grasses can grow up to 60 cm in 5 months and help stabilize asbestos waste and reduce the dispersal of asbestos fibers.

REFERENCES

Aigbodion, V.S., Akadike, U., Hassan, S.B., Asuke, F., Agunsoye, J.O., 2010. Development of asbestos – free brake pad using bagasse. Tribology in Industry 32 (1), 45–50.

American Thoracic Society, 2004. Diagnosis and initial management of nonmalignant diseases related to asbestos. American Journal of Respiratory and Critical Care Medicine 170, 691–715.

Ball, P., 2003. Fungi iron-out asbestos pollution. Nature. http://dx.doi.org/10.1038/news030120-2.

Berndt, M.E., Brice, W.C., 2008. The origins of public concern with taconite and human health: reserve mining and the asbestos case. Regulatory Toxicology and Pharmacology 52, S31–S39.

CDC, 2009. Malignant mesothelioma mortality – United States, 1999–2005. MMWR 58, 393–396.

Chand, N., Hashmi, S.A.R., Lomash, S., Naik, A., 2004. Development of asbestos free brake pad. Journal of the Institution of Engineers (India): Mechanical Engineering Division 85 (1), 13–16.

Cooke, W.E., 1927. Pulmonary asbestosis. British Medical Journal 2, 1024–1025.

Cowie, R.L., Murray, J., Becklake, M.R., 2005. Pneumoconioses. Chapter 61. In: Mason, R.J., Broaddus, V.C., Murray, J.T., Nadel, J.A. (Eds.), Murray and Nadel's Textbook of Respiratory Medicine, , fourth ed.vol. 2. Elsevier Saunders, Philadelphia, PA, USA, pp. 1748–1782.

Costa, R., Orriols, R., 2012. Man-made mineral fibers and the respiratory tract. Archivos de Bronconeumología 48 (12), 460–468.

Daghino, S., Turci, F., Tomatis, M., Favier, A., Perotto, S., Douki, T., Fubini, B., 2006. Soil fungi reduce the iron content and the DNA damaging effects of asbestos fibers. Environmental Science & Technology 40, 5793–5798.

Dave, S.K., Beckett, W.S., 2005. Occupational asbestos exposure and predictable asbestos-related diseases in India. American Journal of Industrial Medicine 48 (2), 137–143.

Doll, R., Peto, J., 1985. Asbestos. Effects on Health of Exposure to Asbestos. Health and Safety Commission, Her Majesty's Safety Office, London.

Favero-Longo, et al., 2009. Plant colonization limits dispersion in the air of asbestos fibers in an abandoned asbestos mine. Northeastern Naturalist 16 (sp5), 163–177.

Favero-Longo, S.E., Girlanda, M., Honegger, R., Fubini, B., Piervittori, R., 2007. Interactions of sterile-cultured lichen-forming ascomycetes with asbestos fibres. Mycological Research 111, 473–481.

Favero-Longo, S.E., Siniscalco, C., Piervittori, R., 2006. Plant and lichen colonization in an asbestos mine: spontaneous bioattenuation limits air dispersion of fibres. Plant Biosystems 140 (2), 190–205.

Favero-Longo, S.E., Turci, F., Tomatis, M., Castelli, D., Bonfante, P., Hochella, M.F., Piervittori, R., Fubini, B., 2005. Chrysotile asbestos is progressively converted into a non-fibrous amorphous material by the chelating action of lichen metabolites. Journal of Environmental Monitoring 7, 764–766.

Fyles, K.M., Shorrock, P., Davidson, M.R., 1996. Compositions for High Temperature Fiberisation. US 5962354 A.

Gaoxiang, D., 2007. Advancement of the Research on the Synthesis Utility of Asbestos Tailing. China Non-Metallic Mining Industry Herald 2, 003.

Gong, M.N., Christiani, D., 2002. Lung cancer. In: Hendrick, D.J., Burge, P.S., Beckett, W.S., Churg, A. (Eds.), Occupational Disorders of the Lung: Recognition, Management and Prevention. WB Saunders. Elsevier Science, Edinburgh, UK, pp. 305–326.

Ibhadode, A.O.A., Dagwa, I.M., April–June 2008. Development of asbestos-free friction lining material from palm kernel shell. Journal of the Brazilian Society of Mechanical Sciences and Engineering 2, 166–173.

IBM, 2012 and 2013. Indian Mineral Year Book, 2011 & 2012, fiftieth and fifty-first ed. Indian Bureau of Mines. Asbestos (advanced release).

Idris, U.D., Aigbodion, V.S., Abubakar, I.J., Nwoye, C.I., 2013. Eco-friendly asbestos free brake-pad: using banana peels. Journal of King Saud University – Engineering Sciences 27 (2), 185–192.

Ikpambese, K.K., Gundu, D.T., Tuleun, L.T., 2014. Evaluation of palm kernel fibers (PKFs) for production of asbestos-free automotive brake pads. Journal of King Saud University – Engineering Sciences 28 (1), 110–118.

INSERM, 2005. The use of chrysotile asbestos in quebec, direction risqué biologiques, environnementaux et occupationnels.

Jiang, Y.P., Peng, T.J., Sun, H.J., 2011. March. Preparation of Acicular Basic Magnesium Carbonate by the Activation Product of Chrysotile Asbestos Tailing. Advanced Materials Research 178, 230–235.

Karade, S.R., 2010. Cement-bonded composites from lignocellulosic wastes. Construction and Building Materials 24, 1323–1330.

Karkanas, 1995. The slip-fiberchrysotile asbestos deposit in the Zidani area, Northern Greece. Ore Geology Reviews 10, 19–29.

Kim, K.-W., Jeong, Y.-S., 2014. Experimental study on the comparison of the material properties of glass wool used as building materials. Materials Science 20 (1), 103–107. http://dx.doi.org/10.5755/j01.ms.20.1.3714.

Kumar, A., Maiti, S.K., 2014. Translocation and bioaccumulation of metals in *Oryza sativa* and *Zea mays* growing in chromite-asbestos contaminated agricultural fields, Jharkhand, India. Bulletin of Environmental Contamination and Toxicology 93 (4), 434–441.

Kumar, A., Maiti, S.K., 2015a. Effect of organic manures on the growth of *Cymbopogon citratus* and *Chrysopogon zizanioides* for the phytoremediation of chromite-asbestos mine waste: a pot scale experiment. International Journal of Phytoremediation 17 (5), 437–447.

Kumar, A., Maiti, S.K., 2015b. Assessment of potentially toxic heavy metal contamination in agricultural fields, sediment, and water from an abandoned chromite-asbestos mine waste of Roro hill, Chaibasa, India. Environmental Earth Sciences 74 (3), 2617–2633.

Kusiorowski, R., Zaremba, T., Piotrowski, J., 2014. The potential use of cement–asbestos waste in the ceramic masses destined for sintered wall clay brick manufacture. Ceramics International 40, 11995–12002.

Lee, K.P., Barras, C.E., Griffith, F.D., Waritz, R.S., Lapin, C.A., 1981. Comparative pulmonary responses to inhaled inorganic fibers with asbestos and fiberglass. Environmental Research 24 (1), 167–191.

Lee, R.J., Strohmeier, B.R., Bunker, K.L., Van Orden, D.R., 2008. Naturally occurring asbestos – a recurring public policy challenge. Journal of Hazardous Materials 153, 1–21.

Liew, K.W., Nirmal, U., 2013. Frictional performance evaluation of newly designed brake pad materials. Materials and Design 48, 25–33.

Martino, E., Prandi, L., Fenoglio, I., Bonfante, P., Perotto, S., Fubini, B., 2003. Soil fungal hyphae bind and attack asbestos fibers. Angewandte Chemie International Edition 42, 219–222.

Mathur, R.B., Thiyagarajan, P., Dhami, T.L., 2004. Controlling the hardness and tribological behavior of non-asbestos brake lining materials for automobiles. Journal of Carbon Science 5 (1), 6–11.

Mukherjee, S., 2013. The Science of Clays. Traditional and Modern Uses of Ceramics, Glass and Refractories, pp. 123–150.

Namessan, N.O., Maduako, J.N., Iya, S. Comparative study of the effects of treatment techniques on the thermal and frictional properties of kenaf (*Hibiscus cannabinus*) fibre reinforced brake pads. African Journal of Science and Technology (AJST) Science and Engineering Series 12 (2), 44–54.

O'Dell, R.E., Claassen, V.P., 2009. Serpentine revegetation: a review. Northeastern Naturalist 16, 253–271.
OSHA, 2006. Occupational Safety and Health Administration: Health Effects of Hexavalent Chromium, vol. 800. US Department of Labor, p. 321.
Park, E.-K., Takahashi, K., Hoshuyama, T., et al., 2011. Global magnitude of reported and unreported mesothelioma. Environmental Health Perspectives 119, 514–518.
Pham, V.H., Tran, T.N.L., Le, G.V., Movahed, M., Jiang, Y., Pham, N.H., Ogawa, H., Takahashi, K., 2013. Asbestos and asbestos-related diseases in Vietnam: in reference to the International Labor Organization/World Health Organization National Asbestos Profile. Safety and Health at Work 4 (2), 117–121.
Powers, E.J., Serad, G.A., 1986. High Performance Polymers: Their Origin and Development in Book: History and Development of Polybenzimidazoles, pp. 355–373.
Qilong, L., Zhongyuan, L., Kefeng, T., 1997. Investigation on manufacturing class-ceramic decorative slabs from asbestos tailings. Multipurpose Utilization of Mineral Resources 4, 008.
Ruzaidia, C.M., Kamarudin, H., Shamsul, J.B., Abdullah, M.M.A., 2011. Comparative study on thermal, compressive, and wear properties of palm slag brake pad composite with other fillers. Advanced Materials Research 328–330, 1636–1641.
Schüz, J., Schonfeld, S.J., Kromhout, H., Straif, K., Kashanskiy, S.V., Kovalevskiy, E.V., Bukhtiyarov, I.V., McCormack, V., August 2013. A retrospective cohort study of cancer mortality in employees of a Russian chrysotile asbestos mine and mills: study rationale and key features. Cancer Epidemiology 37 (4), 440–445. http://dx.doi.org/10.1016/j.canep.2013.03.001.
Selikoff, I.J., Hammond, E.C., Seidman, H., 1979. Hammond, E.C., Seidman, H. 1979. Mortality experience of insulation workers in the United States and Canada. Ann NY Acad Sci 330, 91–116.
Shuilin, Z., Yang, L., Gaoxiang, L.F.D., Duogong, Z., 2007. Study on The Pilot Technique of Comprehensive Utilization of Asbestos Tailing. Non-Metallic Mines 05.
Subramanian, V., Madhavan, N., 2005. Asbestos problem in India. Lung Cancer 49S1, S1–S9.
Turci, F., Favero-Longo, S.E., Tomatis, M., Martra, G., Castelli, D., Piervittori, R., Fubini, B., 2007. A biomimetic approach to the chemical inactivation of chrysotile fibres by lichen metabolites. Chemistry – A European Journal 13, 4081–4093.
USGS, 2012 and 2013. Mineral Year Book. United States Geological Survey. Asbestos 2011 and 2012 (advanced release).
Wagner, J.C., Sleggs, C.A., Marchand, P., 1960. Diffuse pleural mesothelioma and asbestos exposure in the Northwestern Cape Province. British Journal of Industrial Medicine 17, 260–271.
Walker, A.M., Maxim, L.D., Utell, M.J., 2012. Are airborne refractory ceramic fibers similar to asbestos in their carcinogenicity? Inhalation Toxicology 24 (7), 416–424. http://dx.doi.org/10.3109/08958378.2012.683892.
Whalley, C., Hursthouse, A., Rowlat, S., Iqbal-Zahid, P., Vaughan, H., Durant, R., 1999. Chromium Speciation in Natural Waters Draining Contaminated Land, Glasgow, UK.
Wolfe, R.W., Gjinolli, A., 1999. Durability and strength of cement bonded wood particle composites made from construction waste. Forest Products Journal 49 (2), 24–31.
Xinyan, Z., Jianping, F., 1998. Production of Precipitated Silicon Dioxide from Asbestos Tailings. Chemical Production and Technology 4, 38.
Zanzhong, Y., Fengzhi, D., Yujin, L., 1998. Tailings from asbestos ores: application to the ceramic industry. Conservation and Utilization of Mineral Resources 4.
Zaremba, T., Peszko, M., 2008. Investigation of the thermal modification of asbestos wastes for potential use in ceramic formulation. Journal of Thermal Analysis and Calorimetry 92, 873–877.
Zhai, W., Wang, Y., Deng, Y., Gao, H., Lin, Z., Li, M., 2014. Recycling of asbestos tailings used as reinforcing fillers in polypropylene based composites. Journal of hazardous materials 270, 137–143.
Zhu, W., 2009. Application research of asbestos tailing in highway concrete. Shanxi Architecture 08.

CHAPTER

14 RESOURCE POTENTIAL OF NATURAL AND SYNTHETIC GYPSUM WASTE

M.N.V. Prasad

University of Hyderabad, Hyderabad, Telangana, India

CHAPTER OUTLINE

Environmental Materials and Waste. http://dx.doi.org/10.1016/B978-0-12-803837-6.00014-7

14.1 INTRODUCTION

About 100–200 million years ago, the mineral gypsum (calcium sulfate dihydrate [$CaSO_4 \cdot 2H_2O$]) became precipitated as a result of the evaporation of seawater. It is deposited in sedimentary layers in the marine environment. Under high pressure and temperature, gypsum is transformed into anhydrite ($CaSO_4$). In nature, gypsum and anhydrite occur as beds or nodular masses up to a few meters thick. Gypsum is formed by the hydration of anhydrite. The depth of hydration can range from the surface of the deposit down to 300 m, depending on the temperature and pressure, topography, and structure of the deposit. Anhydrite is often mined in conjunction with gypsum, but it is comparatively limited in its technical applications. The content of gypsum in sedimentary rock varies from 75% to 95%; the rest is clay and chalk.

Gypsum and anhydrite are produced from open-cast mines or underground mines using pillar and stall mining methods that give extraction rates of up to 75% (Singh, 2015). Gypsum is crushed and finely ground for cement manufacture for further fine grinding with cement clinker. Anhydrite has limited uses, although large quantities of a mixture of anhydrite and gypsum are blended with cement clinker and are finely ground to produce Portland cement. Addition of about 5% is used to control the initial rate of reaction with water and to retard the setting time of the cement.

Gypsum is used mainly to manufacture building products such as plaster, plasterboard, gypsum fiberboard, and plaster blocks. Demand is principally driven by activity in the construction sector. The value of construction output continues to increase in real terms. Demand for new and refurbished housing is increasing, and in conjunction with the need for new schools, hospitals, offices, and shops, there is likely to be increasing demand for gypsum building products in the foreseeable future.

When gypsum ($CaSO_4 \cdot 2H_2O$) is ground into a powder and heated at 150 to 165°C, three-quarters of its combined water is removed, producing hemihydrate plaster ($CaSO_4$, 0.5 H_2O), commonly known as "plaster of paris." When this powder is mixed with water, the resulting paste sets hard as the water recombines to produce gypsum again. This process can be repeated almost indefinitely, with important implications for recycling. Gypsum has a variety of names that are widely used in the mineral trade: selenite, satin spar, alabaster, and plaster of paris.

Gypsum-based products are in considerable demand in industries for construction, decoration, medicine and dentistry, architecture, chemical product manufacturer, cosmetics, and even agriculture (http://www.fertilizer.org/Phosphogypsum; Abril et al., 2009). Gypsum is a naturally occurring mineral that is mined in various parts of the world. The word is derived from the Greek *gypsum*, meaning "to cook," which refers to burned or calcined mineral. Its chemical formula is $CaSO_4 \cdot 2H_2O$ ie, calcium sulfate dehydrate. World reserves of natural gypsum are large and adequate to meet demand. The total reported production of gypsum in 2011 was about 141.3 million tons. China is the largest producer, accounting for 26%, followed by Iran, Iraq, and Thailand (8% each), the United States (7%), Mexico and Spain (5% each), Brazil (3%) Australia, Canada, Egypt, France, India, and Russia (2% each) (Fig. 14.1).

14.2 NATURAL GYPSUM APPLICATIONS AND REUSE

The most important applications of gypsum are in the production of plaster, plasterboard, gypsum fiberboard, and gypsum blocks. The mineral forms the basis of a large industry processing a wide range

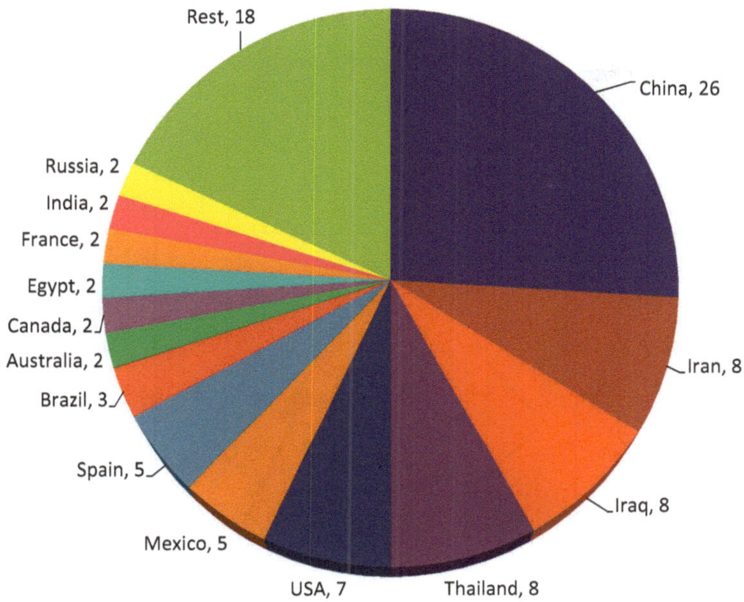

FIGURE 14.1

World reserves of natural gypsum and production shares as percentages.

of building products. Synthetic gypsum is now more widely used in the manufacture of plasterboard and gypsum fiberboards. Natural gypsum is especially suitable for the manufacture of building plasters because it contains clays that improve the workability of the plaster. High-purity natural gypsum is also used to produce special plasters: for example, for use as plaster molds in the pottery industry and for surgical and dental work. Small quantities of high-purity gypsum are also used in confectionary, food, the brewing industry, pharmaceuticals, and sugar beet refining, and as cat litter, and oil absorbent. Finally, but not least, gypsum has been used for more than 200 years as a soil amendment and fertilizer (Xu et al., 2012). Indeed, it improves water penetration and workability of impermeable sodic "alkali" soils, it softens soils with a high clay content, it helps neutralize soil acidity, and it adds the plant nutrients calcium and sulfur.

Gypsum products are among the few construction materials for which "closed-loop" recycling is possible, ie, for which the waste can be used to make the same product again and not merely recovered for use in other "down-cycling" applications, as is the case with some other construction materials, eg, waste concrete and bricks used for aggregates in road construction.

The resource potential of gypsum and its products is shown in Fig. 14.2 and Table 14.1.

Synthetic gypsum is calcium sulfate from sulfuric acid–processing fluorspar, datolite, and other boron-containing concentrates from boric acid and other processes, based on the calcium-sulfuric acid method.

Soil used for a construction purpose may naturally contain gypsum components and be referred to as gypseous, gypsiferous, or sulfate-bearing soils, or the gypsum may be added to the nongypsiferous

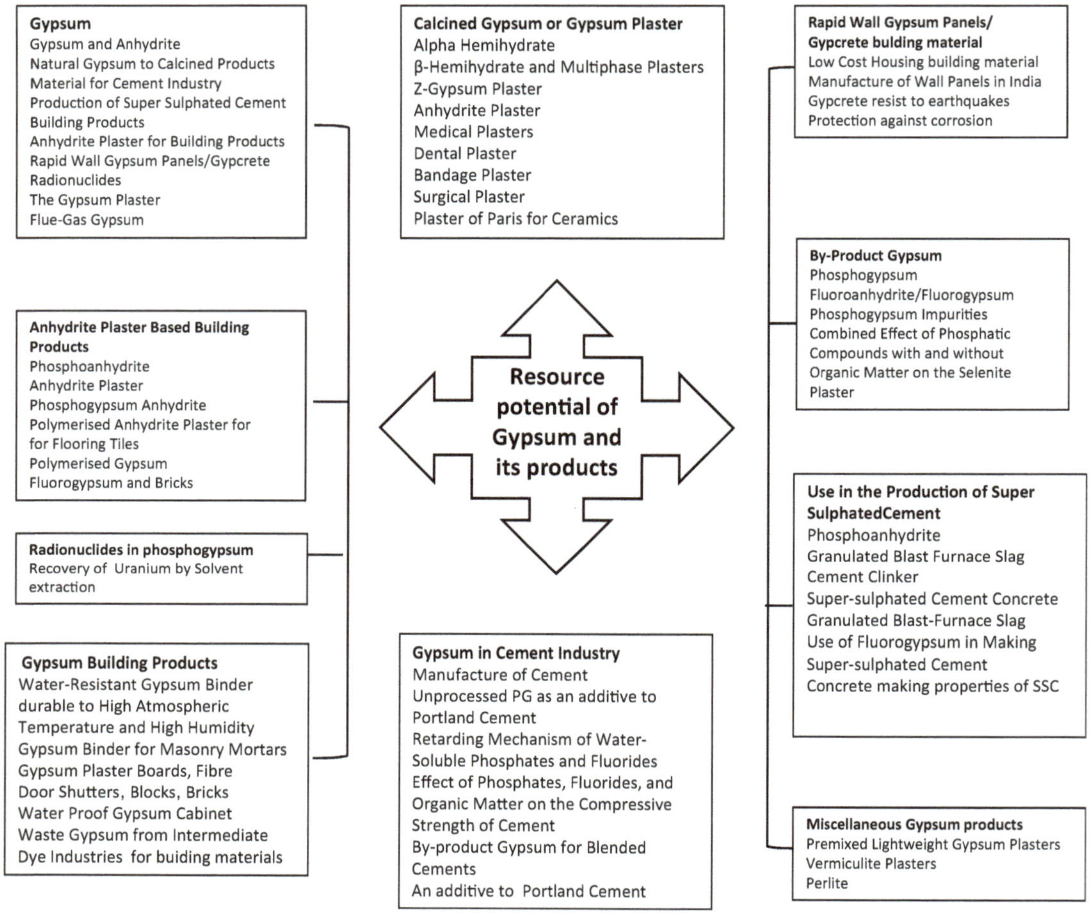

FIGURE 14.2

Resource potential of gypsum and its products (Singh, 2015; Valkov et al., 2014).

soil in small quantities to improve its properties. Therefore, the presence of gypsum in soils used as construction materials and its effect on the different soil properties have been studied independently by researchers all over the world (Kuttah and Sato, 2015).

Phosphogypsum (PG) is an industrial waste derived from phosphoric acid and phosphate fertilizer manufactures using the wet method. This process is economical but results in the production of a large amount of PG (Carbonell-Barrachina et al., 2002) in which every ton of phosphoric acid produced about 5 tons of PG (Bhawan, 2012). World PG production is variously estimated to be in the region of 100–280 megatons per year, so it may cause serious storage and environmental problems (Carbonell-Barrachina et al., 2002). PG is used partially as a soil conditioner (Giavarini et al., 2000; Hanna et al., 1999); however, PG my contain radioactive elements and heavy metals, which would hinder use of the waste (Hanna et al., 2000). PG can be converted into calcium

Table 14.1 Resource Potential of Natural and Synthetic Gypsum Waste Production of Beneficial Materials and Pollution Prevention

Research Observations About Natural/Synthetic Gypsum	References
Removal of lead ions using hydroxyapatite nano-material prepared from phosphogypsum waste	Mousa et al. (2015)
Health risk assessment of heavy metals contamination in tomato and green pepper plants grown in soils amended with phosphogypsum	Al-Hwaiti and Al-Khashman (2015)
Assessment of the radiological impacts of phosphogypsum used as main constituent of building materials in Jordan	Al-Hwaiti (2015)
Radiological assessment of Iraqi phosphate rock and phosphate fertilizers	Ali and Awad (2015)
Phosphogypsum as soil stabilizer of pavement layers	Al-Zaidyeen and Al-Qadi (2015)
Effect of phosphogypsum amendment on two *Arabidopsis thaliana*: eco type growth and development	Ayadi et al. (2015)
A kinetic study on cadmium adsorption from aqueous solutions by preconditioned phosphogypsum	Balkaya and Cesur (2015)
Phosphogypsum-enriched urea on soil quality, nitrogen use efficiency, and quality of basmati rice (*Oryza sativa*) and their residual effect on succeeding wheat (*Triticum aestivum*)	Bana et al. (2015)
Impact of harbor, industry, and sewage on phosphorus geochemistry of subtropical estuary in Brazil	Berbel et al. (2015)
Valorization of rare earth–containing industrial process residues toward zero waste	Binnemans et al. (2015)
Radioactivity in phosphogypsum in Romanian industrial areas	Calin et al. (2015)
Biogeochemical aspects of uranium mineralization, mining, milling, and remediation	Campbell et al. (2015)
Purification of calcium oxide in phosphogypsum decomposition residue based on sucrose–CO_2 method	
Effect of phosphogypsum on growth, physiology, and antioxidative defense system in sunflower seedlings	Elloumi et al. (2015)
Viability of adding gypsum and calcite for remediation of metal-contaminated soil: laboratory and pilot plant scales	González-Núñez et al. (2015)
Optimized and validated wavelength-dispersive X-ray fluorescence spectrometry method for determination of uranium in phosphogypsum	Guitouni (2015)
Phosphogypsum as soil fertilizer	Hentati et al. (2015)
Factors affecting release of radon^{222} from phosphogypsum	Hilal et al. (2015)
Potential of bentonite–lime mix modified with phosphogypsum and reinforced with sisal fibers	Kumar et al. (2015)
Experimental investigation of physical properties of soil in permeation process of phosphogypsum leachate.	
Uranium (^{234}U, ^{235}U, and ^{238}U) contamination of environment surrounding phosphogypsum waste heap in northern Poland	Olszewski et al. (2015)
Potential use of phosphogypsum in alkali-activated fly ash under effects of elevated temperatures and thermal shock cycles	Rashad (2015)
Soil stabilization with phosphogypsum and fly ash: microlevel study	Ravichandran et al. (2015)
Immobilization of lead in anthropogenic contaminated soils using phosphates with and without oxalic acid	Su et al. (2015)
Effects of phosphogypsum and superphosphate on compost maturity and gaseous emissions during kitchen waste composting	Yang et al. (2015)

Continued

Table 14.1 Resource Potential of Natural and Synthetic Gypsum Waste Production of Beneficial Materials and Pollution Prevention—cont'd

Research Observations About Natural/Synthetic Gypsum	References
Effects of kaolinite addition on melting characteristics of reaction between phosphogypsum and CaS	Zhao et al. (2015)
Preparation of β-hemihydrate gypsum with phosphogypsum	Zheng et al. (2015)
Influence of constant magnetic field on properties of waste phosphogypsum and fly ash composites	Zieliński (2015)
Alkaline slag is more effective than phosphogypsum in amelioration of subsoil acidity in Ultisol profile	Li et al. (2015a,b)
Radionuclides in food	Bhat and Gómez-López (2014)
radiological impact from a Brazilian phosphate facility	Glória dos Reis and Da Costa Lauria (2014)
Biological sulfate reduction for treatment of gypsum-contaminated soils	Kijjanapanich et al. (2014)
Strengthening effect of mycorrhizal technology on application of phosphogypsum in agriculture	Gu et al. (2013)
Use of phosphogypsum for preparation of nonautoclaved aerated concrete	Yang et al. (2013a)
Thermodynamic study of phosphogypsum decomposition by sulfur	Yang et al. (2013b)
Valorization of PG for manufacturing sulfur polymer concrete	Garcia-Diaz et al. (2013)
Physicochemical approaches to control phosphogypsum-heavy metal releases in aquatic environment	Ammar et al. (2013)
Restoration of uncontrolled phosphogypsum dump site and impacts	Kobbi-Rebai et al. (2013)
Increasing arsenic sorption on red mud by phosphogypsum addition	Lopes et al. (2013)
Heavy metal immobilization in contaminated soils using phosphogypsum and rice straw compost	Mahmoud and Abd El-Kader (2013)
Phytoremediation potential of *Arabidopsis thaliana*, ectopically expressing a vacuolar proton pump, for industrial waste phosphogypsum	Khoudi et al. (2013)
Food safety issues for mineral and organic fertilizers	Chaney (2012)
Reclamation of saline-sodic soil properties and improvement of rice (*O. sativa* L.) growth and yield using desulfurized gypsum in China	Chi et al. (2012)
Comparative study on use of fly ash and phosphogypsum in brick production.	Selcuk and Emrah (2012)
Technology using waste phosphogypsum to produce gypsum block	Shi (2012)
Method for producing baking-free brick by doping phosphogypsum and fly ash	Wang et al. (2012)
Use of waste phosphogypsum to prepare nonfired bricks by novel hydration–recrystallization process	Zhou et al. (2012)
Evaluation of organic substrates to enhance sulfate-reducing activity in phosphogypsum	Castillo et al. (2012)
Remediation of phosphate plant sinkhole in Florida	Garlanger and Schmall (2012)
Remediation of bauxite residue amended with poultry manure and phosphogypsum	Jones et al. (2012)
Mobility of trace elements in Mediterranean red soil amended with phosphogypsum	Kassir et al. (2012)
Amelioration of soda saline–alkali with gypsum application	Xu et al. (2012)
Use of phosphogypsum industrial waste for mineral CO_2 sequestration	Cardenas-Escudero et al. (2011)
Effects of different phosphate fertilizers on copper sorption in red soils	Song et al. (2011)
Microencapsulation of phosphogypsum into sulfur polymer matrix	López et al. (2011)

Table 14.1 Resource Potential of Natural and Synthetic Gypsum Waste Production of Beneficial Materials and Pollution Prevention—cont'd

Research Observations About Natural/Synthetic Gypsum	References
Valorization of phosphogypsum as building material	Tayibi et al. (2011)
Use of phosphogypsum industrial waste for mineral CO_2 sequestration	Cárdenas-Escudero et al. (2011)
Evaluation of PG as soil conditioners to reduce potential aluminum toxicity and acidity in western Spain	Garrido et al. (2011)
Use of fly ash, phosphogypsum, and red mud as liner material for disposal of hazardous zinc leach residue waste.	Çoruh and Ergun (2010)
Phosphogypsum−steel slag−granulated blast furnace slag−limestone cement.	Huang and Lin (2010)
Responses of rice (*O. sativa* L.) growth and yield to phosphogypsum amendment in saline-sodic soils of North−East China	Liu et al. (2010a)
Impacts of sand amendment on rice (*O. sativa* L.) growth and yield in saline-sodic soils of northeast China	Liu et al. (2010b)
Redox chemistry of sulfate and uranium in phosphogypsum tailings dump	Papanicolaou et al. (2010)
Method of producing bricks of autoclaved phosphogypsum. State Intellectual Property Office of People's Republic of China	Pu et al. (2010)
Complex nanominerals and ultrafine particles assemblages in phosphogypsum of fertilizer industry and implications on human exposure	Silva et al. (2010)
Discussion of ways to use phosphogypsum in China	Zhong et al. (2010)
Use of phosphogypsum, fly ash, and red mud as linear material for disposal of hazardous zinc leach residue waste	Coruh and Ergun (2010)
Waste gypsum from intermediate dye industries for production of building materials	Garg and Jain (2010)
Phosphogypsum for sulfur concrete production for construction industry	Mohamed and Gamal (2010)
Occupational dosimetric assessment (inhalation pathway) of phosphogypsum in agriculture in southwest Spain	Abril et al. (2009)
Method of producing nonfired bricks for industrial by-production of phosphogypsum	Chi and Lin (2009)
High-strength phosphogypsum wall material and method for making bricks using material	Tan and Chen (2009)
Fluxes of natural radionuclides in production process of phosphoric acid plant	Bolívar et al. (2009)
Method of manufacturing nonfired bricks of phosphogypsum	Chen (2008)
Use of phosphogypsum as raw and calcined material in manufacturing building products	Degirmenci (2008a)
Use of waste phosphogypsum and natural gypsum in adobe stabilization	Degirmenci (2008b)
Phosphogypsum use in polyphosphoric acid and bitumen	Masson (2008)
Application of phosphogypsum in soil stabilization	Degirmenci et al. (2007)
Method of manufacturing high-strength building bricks of phosphogypsum	Yang (2007)
Method of manufacturing high-strength and proof-water and proof-fire nonfired bricks of industrial by-production of phosphogypsum	Yin (2007)
Application of steel slag fly-ash phosphogypsum-solidified material as road base material	Shen et al. (2007)
Phosphogypsum for soil stabilization	Degirmenci et al. (2007)
Flue gas desulfurization products used in agricultural and other land applications	Anonymous (2006a,b)

Continued

Table 14.1 Resource Potential of Natural and Synthetic Gypsum Waste Production of Beneficial Materials and Pollution Prevention—cont'd

Research Observations About Natural/Synthetic Gypsum	References
Application of phosphogypsum in agriculture and radiological impact	Papastefanou et al. (2006)
Win–win solution for flue gas desulfurization gypsum: researchers discover beneficial applications for by-product in agriculture	Ramsier and Norton (2006)
Reduction of clinkerization temperature using phosphogypsum	Kacimi et al. (2006)
Flue gas desulfurization products as sulfur sources for alfalfa and soybean	Chen et al. (2005)
Method of manufacturing environmental and high-pressure nonfired bricks of phosphogypsum	
Immobilization of radioactive wastes in bitumen	Ojovan and Lee (2005)
Use of weathered phosphogypsum as set retarder in Portland cement	Altun and Sert (2004)
Use of phosphogypsum for cement and plaster manufacture	Singh (2002)
Use of lime, gypsum, alum, and potassium permanganate in water quality management	Wilkinson (2002)
Sodic soils reclaimed with by-product from flue gas desulfurization: corn production and soil quality	Chun et al. (2001)
Stabilized phosphogypsum: class C fly ash: Portland type II cement composites for potential marine application	Guo et al. (2001)
Radiochemicals in phosphogypsum for engineering use	Tayibi et al. (2001)
Beneficial uses of flue gas desulfurization by-products	Dick et al. (2000)
Properties and examples of beneficial use of gypsum-like by-products.	Ritchey et al. (2000)
Production of nonfired bricks of phosphogypsum	Zhang and Chi (2000)
Phosphate waste in mixtures to improve deformation	Katamine (2000)
Long-term effects of gypsum on crop yield and subsoil chemical properties	Toma et al. (1999)
Phosphogypsum: cement: lime composites	Guo et al. (1999)
Sodic soils: distribution, properties, management, and environmental consequences	Sumner and Naidu (1998)
Radioactive materials in phosphogypsum	Bolívar et al. (1998)
High-gypsum by-products as soil amendments for horticultural crops	Korcak (1997)
Ammonium sulfate fertilizer as by-product in flue gas desulfurization	Wallach (1997)
High-gypsum by-products as soil amendments for horticultural crops	Korcak (1997)
Literature review on gypsum as calcium and sulfur source for crops and soils in southeastern United States	Sumner (1995)
Improving aging resistance of straight-run bitumens by addition of phosphorus compounds	de Filippis et al. (1995)
Environmental impacts of phosphogypsum	Rutherford et al. (1994)
Amelioration of subsoil acidity with minimum disturbance.	Sumner (1994)
Gypsum amendment to soil can reduce selenium uptake by alfalfa in presence of coal fly ash	Arthur et al. (1993)
Gypsum for agronomic investigations	Sumner (1993)
Ammonia-based flue gas desulfurization waste solution as nitrogen fertilizer	Gissel-Nielsen and Bertelsen (1989)
Use of gypsum to improve physical properties and water relations in soils.	Miller (1988)
Sulfur fertilization of winter clovers, coastal Bermuda grass, and corn on north and west Florida soils	Thompson and Neller (1963)

hydroxyapatite ($Ca_{10}[PO_4]_6\ [OH]_2$) and ammonium sulfate, in which hydroxyapatite (HAP), is one of the most important constituents of the biocompatible inorganic materials used in human hard tissues (bone and teeth) (Hayati et al., 2011; Mohamed et al., 2014). The chemical species constituting of HAP crystals are calcium, phosphorus, oxygen, and hydrogen, which are nontoxic (Murugan and Ramakrishnan, 2005). Synthetic hydroxyapatite is particularly an attractive material which has received considerable attention over the past 2 decades for bone and tooth implants because of its chemical similarity to natural bone (Jarcho et al., 1976). It is also important in many industrial applications such as catalysis, ion exchange, and metal removal (Baton et al., 2007; Mortimer and Lemaitre, 1989; Baillez et al., 2007; Hadioui et al., 2008). Several methods have been used to synthesize HAP (Huang et al., 2011; Jokic et al., 2011; Minh et al., 2013). Therefore various laboratories are planning to produce nanomaterials from PG waste (Mousa and Hanna, 2013), which help remove Pb(II) from aqueous solutions.

Hammas et al. (2013) studied the solubility of PG in different aqueous solutions and chose the one with the highest value of soluble material to produce useful products. The researchers explained that the reaction process leads to the formation of calcium carbonate and hydrated sodium hydrogenosulfate rods. After decarbonation with heat at 40°C and the addition of sulfuric acid, evaporation of the solution at 100°C resulted in rod crystals. These were identified as $NaHSO_4 \cdot H_2O$. Thus, Hammas et al. (2013) succeeded in converting a salt solution of PG (considered a waste product in some processes) into calcium carbonate and sodium hydrogenosulfate. The latter is an interesting compound because it can be transformed into ammonium sulfate.

14.3 SYNTHETIC GYPSUM

The gypsum industry is working to build value for society by offering safe, economical, and recyclable products for consumers using substitutes for natural gypsum to reduce pressure on natural resources.

14.4 FLUE GAS DESULFURIZATION GYPSUM

In coal-fired electricity-generating stations, flue gases are cleaned using finely ground limestone (or lime), which reacts with sulfur dioxide emissions to produce a very pure gypsum (ACAA, 2006). This synthetic gypsum has a higher purity, 96% gypsum, than natural gypsum, typically 80% gypsum. The primary objective of this process is to reduce acid precipitation from the atmosphere. For details of the different types of cleaning processes, reference may be made to Clean Coal Technologies flue gas desulfurization (FGD) for SO_2 control (http://www.iea-oal.org.uk/content/default.asp?PageId=981).

The desulfurization process produces carbon dioxide. The reaction takes place during wet scrubbing using a $CaCO_3$ (limestone) slurry, producing $CaSO_3$ (calcium sulfite) and can be expressed as:

$$CaCO_3\ (\text{solid}) + SO_2\ (\text{gas}) \rightarrow CaSO_3\ (\text{solid}) + CO_2\ (\text{gas})$$

Other by-products of gypsum are titanogypsum and fluorogypsum.

FGD gypsum is an important supplement to the supply of natural gypsum. This synthetic gypsum has a high purity (gypsum content of 96%).

14.5 PROCESSING FLUE GAS DESULFURIZATION GYPSUM

Of the available FGD processes , limestone-based scrubbing processes have proved the most popular. Desulfurization takes place in scrubbing towers in which the flue gases are brought into contact with an aqueous suspension containing powdered limestone or slaked quicklime as the alkaline component. The SO_2 is washed out with water oxidized to sulfates, SO_3, in the aqueous solution, and precipitated with calcium from the limestone/quicklime into dihydrous calcium sulfate ($CaSO_4 \cdot 2H_2O$) gypsum. The gypsum crystals are separated out of the suspension as a moist, fine crystalline powder with the aid of centrifuges or filters.

Recovering gypsum from FGD plants is becoming more important, as the technical feasibility of substituting FGD gypsum for natural gypsum in traditional applications such as wallboard, cement, and soil conditioners has demonstrated. It is estimated that there are several hundreds of new coal-fired plants and over a few thousands of upgrades of existing coal-fired plants in various stages around the world.

14.6 PHOSPHATE MINERAL FERTILIZERS: P_2O_5

The production of phosphate mineral fertilizers is one of the most profitable and financially stable among manufacturing industries. Phosphate ore (apatite) is converted into fertilizer with sulfuric acid and phosphogypsum is a by-product of this process. Approximately 5 tons per ton of P_2O_5 is produced. Rock phosphate is a naturally occurring source for P_2O_5. Production of rock phosphate in India is shown in Fig. 14.3.

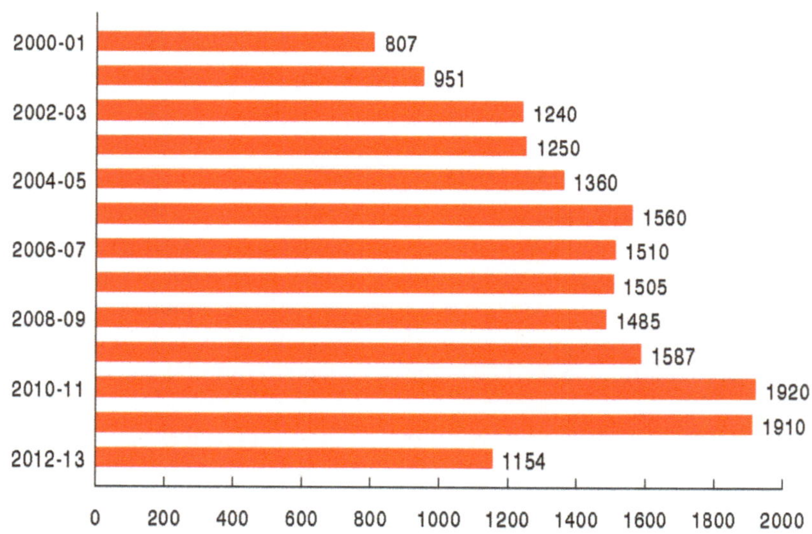

FIGURE 14.3

Production of rock phosphate in India.

Data source: Fertilizer Association of India and Central Pollution Control Board.

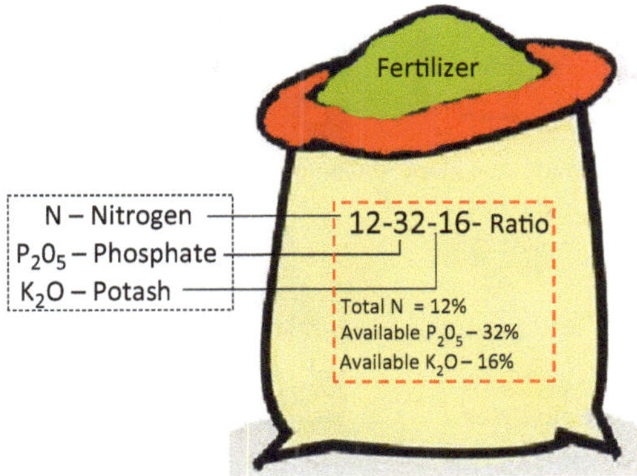

FIGURE 14.4

Fertilizer 12–32–16 grade, meaning that it is 12% N, 32% P_2O_5, and 16% K_2O by weight.

- 1 lb of P = 2.3 lb of P_2O_5, or
- 1 lb P_2O_5 = 0.44 lb P

Fertilizers are designed to have a known "grade" or concentration of nutrients which is listed on the fertilizer bag in that order (Fig. 14.4).

For example, if 100 lb of the fertilizer mentioned previously were applied evenly over an acre, you would apply:

12% N × 100 lb fertilizer/acre = 12 lb N/acre
32% P_2O_5 × 100 lb fertilizer/acre = 32 lb P_2O_5/acre
16% K_2O × 100 lb fertilizer/acre = 16 lb K_2O/acre

Production of phosphate fertilizers involves the dissolution of phosphate rock (mainly apatite, $Ca_5[PO_4]3OH$) with sulfuric acid (H_2SO_4), generating a gypsum by-product known as PG ($CaSO_4 \cdot nH_2O$) (Fig. 14.5).

For the acidulation of finely ground rock phosphate with sulfuric acid to form phosphoric acid and gypsum, the chemical reaction is detailed:

$$\underset{\text{Rock Phosphate}}{Ca_{10}F_2(PO_4)_6} + \underset{\text{Sulphuric acid}}{10H_2SO_4} + \underset{\text{Water}}{10nH_2O} \rightarrow \underset{\text{Phosphoric acid}}{6H_3PO_4} + \underset{\text{Gypsum}}{10(CaSO_4 \cdot nH_2O)} + \underset{\text{Hydrogen Fluoride}}{2HF}$$

where $n = 0$, 0.5, or 2, depending on the hydrate form in which the calcium sulfate is crystallized.

14.7 PRODUCTION OF PHOSPHORIC ACID

The content of phosphorus in the form of calcium phosphate in phosphate rock cannot be taken up directly by plants. Hence, the water-soluble form of phosphorus is derived from rock phosphate by converting calcium phosphate to phosphoric acid or to single super phosphate (SSP). In the production of SSP, calcium sulfate (gypsum) remains in the product.

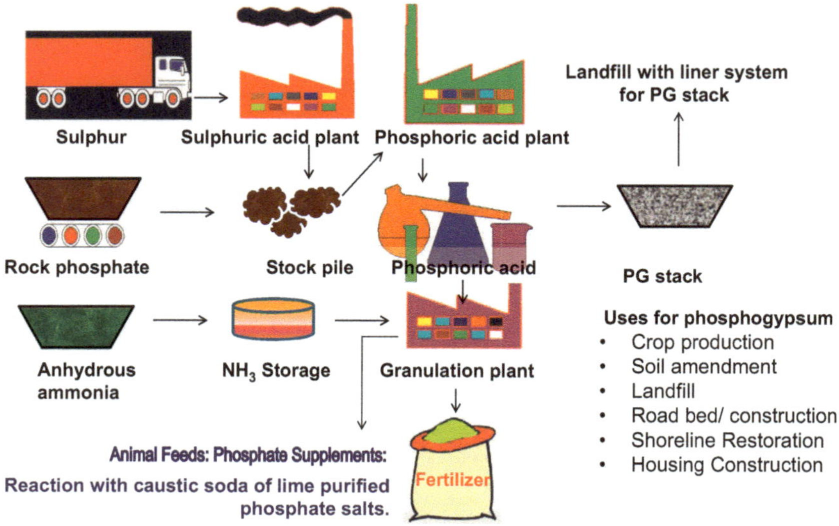

FIGURE 14.5

Manufacturing process of phosphoric acid and the production of fertilizers and PG.

14.7.1 MANUFACTURING PROCESS OF PHOSPHORIC ACID

Production figures for phosphoric acid are expressed in terms of their P_2O_5 content. For example, 100 tons per day of P_2O_5 corresponds to 333 tons per day of phosphoric acid at 30% concentration as P_2O_5.

There are two basic methods for the commercial production of phosphoric acid.

14.7.1.1 Wet Process

The basic reaction for the production of wet process phosphoric acid, simplified as:

Reaction										
$3CaO \cdot P_2O_5$	$+$	$3\ H_2SO_4$	$+$	$6\ H_2O$	$\rightarrow$	$3\ CaSO_4 \cdot 2\ H_2O$	$+$	$2\ H_3PO_4$		
State solid	$+$	liquid			$\rightarrow$	solid	$+$	liquid		[14.1]
Molecular weight										
310	$+$	294	$+$	108	$\rightarrow$	516	$+$	196		

It is thus a reaction in a heterogeneous medium between a solid and a liquid, whose products are a liquid (the aim of the operation) and a solid by-product. The latter is more voluminous than the solid rock from which it is derived. If the impurities present in the rock are taken into account, whether they are elements associated with the calcium phosphate in the apatite molecule or present in the exo-gangue, the following rounded figures are obtained:

Weight of gypsum/weight of rock: about 1.5
Weight of gypsum/weight of P_2O_5: about 5

In India, the wet process is used mostly for the production of phosphoric acid to make phosphatic fertilizer.

PG is generated from the filtration process in phosphoric acid plants, in which insoluble gypsum and other insolubles are separated from the product, ie, phosphoric acid, as efficiently as possible. Depending on the source of rock phosphate, about 4.5–5 tons (dry basis) of PG is generated per ton of phosphoric acid (as P_2O_5) recovered.

The quality and quantum of PG generation depend on the quality of the phosphate rock and the process used to produce phosphoric acid: calcium sulfate generated either in dihydrate ($CaSO_4 \cdot 2H_2O$) or hemihydrate ($CaSO_4 \cdot 1/2H_2O$) form.

PG generated from phosphoric acid plants is currently stacked; some is sold to the other industries for beneficial use, especially in cement manufacturing as a raw material substitute for mineral gypsum, plasterboard manufacturing, and alkali soil amendment (Al-Hwaiti and Al-Khashman, 2015) or reconditioning. Handling of PG by trucks or railways involves loading, unloading, and temporary storage at yards, which is likely to have an impact on the environment especially during the rainy season if proper measures are not taken. To have a minimal impact on the environment, it is necessary to devise guidelines for safe handling, including the transportation, storage, use, and disposal of PG. Safe handling, transportation, and storage of PG within phosphoric acid plants, safe disposal of PG through stacking, and the beneficial end use of PG are encouraged to solve problems associated with its disposal.

14.8 CHARACTERISTICS OF PHOSPHOGYPSUM

PG is a gray, damp, fine-grained, powdered silt or silty-sand material with a maximum size range between 0.5 mm (number 40 mesh sieve) and 1.0 mm (number 20 mesh sieve); most particles (50–75%) are finer than 0.075 mm (number 200 mesh sieve). The specific gravity of PG ranges from 2.3 to 2.6. The maximum dry bulk density is likely to range from 1470 to 1670 kg m^{-3} (92–104 lb ft^{-3}), based on Standard Proctor Compaction. After filtration, the gypsum cake usually has a free moisture content between 18% and 35%. Hemihydrate in the presence of free water will rapidly convert to dihydrate, and in the process, if left undisturbed will set into a relatively hard cemented mass and will not cause dust problems.

PG consists primarily of calcium sulfate dihydrate with small amounts of silica, usually as quartz and unreacted phosphate rock, traces of radioactive material (such as radium and uranium) (Al-Hwaiti, 2015; Ali and Awad, 2015), traces of heavy metals such as arsenic, cadmium, chromium, and lead (Ammar et al., 2013), and fluoride. The concentration of the metals depends on the composition of the phosphate rock. Following are main concerns with respect to the management of PG:

1. High fluoride concentration (in the range of 0.5–1.5%) which may leach and contaminate groundwater if not stored and handled properly
2. May contain heavy metals (Cd, Cr, Pb, etc.) that may enter the food chain through potable water and agricultural products
3. Presence of radionuclide radium^{226}, which upon decay may emit harmful alpha particles
4. Acidic runoff from PG owing to the presence of residual phosphoric acid may cause fish mortality in receiving water bodies such as ponds or lakes (because of low pH)

14.9 MANAGEMENT AND HANDLING OF PHOSPHOGYPSUM

PG retains 18–35% moisture depending on the type and degree of filtration and handling facilities installed for further disposal. In most cases, washed filter cake is received into a hopper and mixed with process or return or makeup water to make slurry so that it can be pumped to the PG disposal stack. In some cases, PG is removed from the filter as a dry cake and transported to the disposal area by a conveyor belt or any other means of solids handling system. However, this method is not widely practiced because slurry handling is simpler and less expensive.

Most phosphoric acid plants in India and else where are based on the wet process (dihydrate) for the production of phosphoric acid. A certain degree of difference may be seen in the process or technology adopted which influences the characteristics and type of handling of PG within the plant. The handling of PG is first done by the industry itself, in which washed PG is shifted to the PG stack, intermediate storage, or direct disposal for use, loading point for sale, or disposal.

14.10 PHOSPHOGYPSUM IN PHOSPHORIC ACID PLANTS

Slurry pumping for transporting PG offers significant operating, economic, and environmental advantages over dry stacking although it requires more makeup water, which might upset the water balance, especially for lower-capacity plants. Handling of PG is detailed next.

In dry stacking, the filtered cake of PG is transported to the disposal area by trucks or a belt conveyor, in which bulldozers with or without a mechanical stacker and movable lateral conveyors are used to spread the PG evenly. PG dry stacks which are in operation in the country have a height of 30–60 m. The top surface of the stack is continuously distributed evenly using bulldozer equipment. Except in very high rainfall locations, where the surface of the stack never gets a chance to dry out, dust generated by construction equipment is a problem which can be controlled by limiting the operational area of PG handling and keeping surfaces damp by spraying water or recycling wastewater as needed.

In wet stacking, PG slurry from an agitated surge tank is conveyed by gravity or pumped to the PG tack through a rubber-lined steel or high-density polyethylene pipeline. Additional pipeline is required to pump decanted water back. With this method, the PG slurry settles on the top portion of the stack with slurry flow distribution through peripheral rim ditches. The decanted water flows down by gravity to the surge tank or holding pond from where it is pumped back for reuse in a phosphoric acid plant. Internal dike walls are built up from the dewatered PG to give a steadily elevating pond. This water cannot be discharged as such because it may have a low pH, fluorine, and phosphate as P_2O_5. The outer dikes are generally earthen dikes designed to prevent the escape of contaminated water into nearby streams. PG dry stacks which are in operation in the country are designed to have a height of 60–80 m. A stabilized PG stack may not cause dust problems; however, dust emissions can be seen during excavation, shaping, and movement of vehicles as well as during the maintenance of access roads (Fig. 14.6).

14.11 USE OF PHOSPHOGYPSUM

As much as possible, phosphoric acid plants attempt to direct PG for beneficial use such as plasterboard manufacturing, cement manufacturing as a substitute for natural gypsum, recovery of

FIGURE 14.6

Use of Neutralised, stabilised or suitably treated phosphogypsum (i.e., blending with suitable stabilising material, binding material and bentonite) for road construction. Courtesy: Dr. Boris Levin, Phosagro, Russia.

by-products such as ammonium sulfate and sulfuric acid, reclamation of alkali or saline–alkali soils, controlled use as a fertilizer, and road making. Use of PG can be achieved as per these guidelines:

1. Used by installing captive plaster/gypsum board manufacturing unit:
   Phosphoric acid plants may explore the possibility of recycling PG by installing captive plaster/gypsum board manufacturing units.
2. Used by plaster, block, or gypsum board manufacturing industry:
   The plaster, block, or gypsum board manufacturing industry should use PG as a raw material to manufacture plaster, putty blocks, or gypsum boards after a reduction in impurities according to the specifications as given in Indian Standards IS: 12679-1989, Reaffirmed 2010 (Requirement of By-product Gypsum for Use in Plaster, Blocks and Gypsum Boards).
3. Used by cement manufacturing units:
   As of March 31, 2014, the installed capacity of the cement industry is about 360 million tons and produces about 255 million tons. About 4–5% of natural gypsum is used in cement

production, ie, about 10 million tons. If natural gypsum is replaced with PG generated from phosphoric acid plants, all of PG generated can be consumed by the cement industry alone in the country.

As much as possible, cement manufacturing unit(s) should use PG in place of natural gypsum as a regulator for the setting time of cement, at the rate of 4–5% in its composition.

4. Used to recover or manufacture sulfuric acid:
Any entrepreneur who wishes to recover or manufacture sulfuric acid from PG may approach the concerned state pollution control board or pollution control committee to obtain consent as is required to establish such a recycling industry.
5. Used to reclaim alkali soils and saline–alkali soils and as fertilizer in agriculture:
PG may be used to reclaim alkali soils and saline–alkali soils with high sodium ion concentration. The fertilizer industry may maximize the use of PG as "fertilizer" (Silva et al., 2010). For this purpose, the Indian Standard Specifications for By-product Gypsum, as stipulated under IS: 10170-1982, needs to be followed. To condition alkaline soil, as a manure in agriculture mainly to correct black alkali soil, BIS specifications for low-grade gypsum for agriculture use should be followed (ie, IS:6046-1982, First Revision; reaffirmed 2008).
6. Used for road making:
Neutralized, stabilized or suitably treated PG (ie, blending with suitable stabilizing material, binding material, and bentonite) and free from water-leachable contaminants such as fluoride, cadmium, lead, arsenic, and mercury content (as per the limits suggested subsequently) and with a coefficient of permeability not more than 10–7 cm s^{-1} is permitted for road construction (Fig. 14.6).

PG waste is used extensively to replace natural gypsum because of the similarities of their chemical composition and binding features (Gaidučis et al., 2009). Currently, PG is widely applied as agricultural fertilizers (Papastefanou et al., 2006; Abril et al., 2009), soil stabilization amendments (Degirmenci et al., 2007; Degirmenci, 2008a), cement retarders (Singh, 2002; Altun and Sert, 2004), cementitious binders (Degirmenci, 2008a), and building bricks and blocks (Lokshin et al., 2013; Selcuk and Emrah, 2012; Pu et al., 2010). In particular, much attention has been paid to the production of PG nonfired bricks and blocks (Yang et al., 2009; Shi, 2012; Pu et al., 2010; Tan and Chen, 2009; Wang et al., 2012; Yin, 2007; Yang, 2007; Li et al., 2005; Zhang and Chi, 2000; Chi and Lin, 2009).

Phosphorus in the form of phosphate (expressed as P_2O_5) is a nutrient for plants and a building block in food production. The agriculture sector uses a large amount of chemical fertilizers to replenish and supplement nutrients that growing plants take from the soil. Modern intensive agriculture heavily demands phosphate fertilizers. PG, essentially "calcium sulfate," is generated as a waste from phosphoric acid plants by the reaction of rock phosphate with sulfuric acid. If PG is disposed of in open yards, it may pose a threat to the environment. Handling and management of PG are a major problem in phosphoric acid plants because of the large volume if generated and the large area required, as well as the potential for the release of dust, fluoride, and heavy metals.

PG waste generated in fertilizer industries is stacked. PG stacks can be found in more than 50 countries around the world (eg, Engis, Belgium; Kavale, Greece; Siilinjärvi, Finland; Florida, United States). As a matter of fact, 85% of worldwide production currently remains in regulated stacks which occupy considerable land resources and lead to serious contamination (Tayibi et al., 2009) (Figs. 14.7 and 14.8).

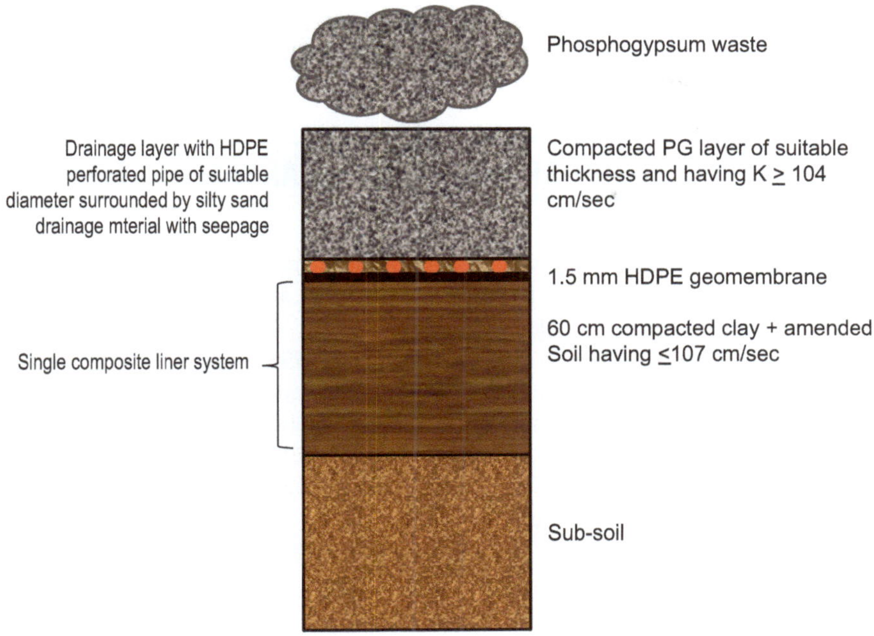

FIGURE 14.7

Suggested capping technology for PG stack.

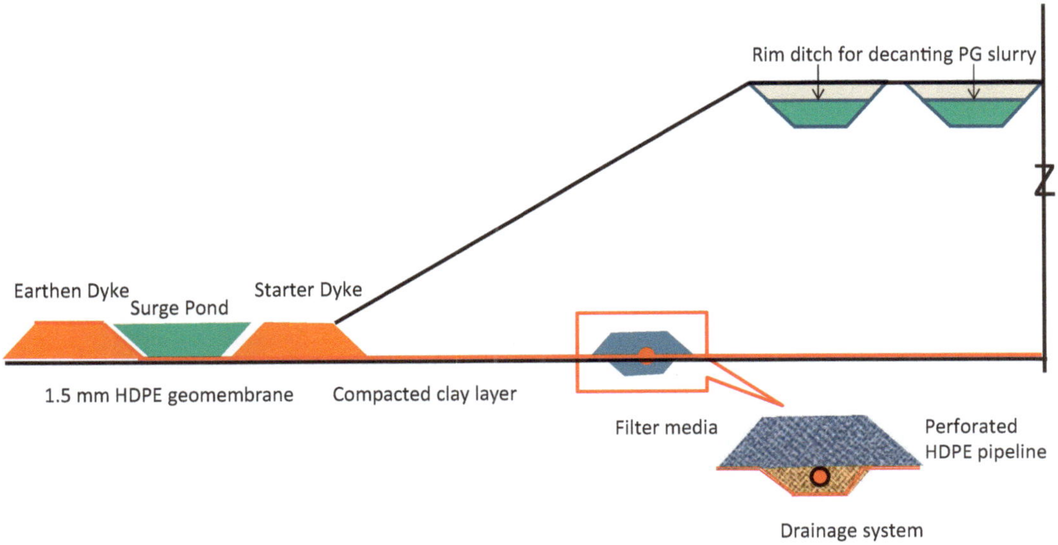

FIGURE 14.8

Sectional view of PG stack.

PG generated from phosphoric acid plants contains three types of impurities that are considered to be potentially harmful: residual acidity, fluorine compounds, and trace elements including radioactivity.

14.12 APPLICATION OF PHOSPHOGYPSUM IN AREAS OF ENVIRONMENT AND MATERIAL SCIENCE

1. Agriculture (Abril et al., 2009; Al-Hwaiti and Al-Khashman, 2015; Bana et al., 2015; Gu et al., 2013) to improve plant growth and development (Ayadi et al., 2015; Elloumi et al., 2015; Hentati et al., 2015; Liu et al., 2010a,b)
2. Source of radionuclides (Al-Hwaiti, 2015; Bolivar et al., 1998; Borges et al., 2013; Calin et al., 2015; Hilal et al., 2015; Olszewski et al., 2015; Papanicolaou et al., 2009)
3. Physicochemical approaches are being considered and applied to control heavy metal release/immobilization (Ammar et al., 2013; Mahmoud and Abd El-Kader, 2014)
4. Use in cement industry and construction and building material production (Altun and Sert, 2004; Singh, 2002; Zhao et al., 2015; Zheng et al., 2015; Zhong et al., 2010; Zieliński, 2015)
5. Soil stabilizer (Al-Zaidyeen and Al-Qadi, 2015; Degirmenci, 2008; Degirmenci et al., 2007; Ravichandran et al., 2015; Perez-López et al., 2010; Perez-López et al., 2010; Rashad, 2015; Rusch et al., 2002)
6. Used in mineral CO_2 sequestration (Cardenas-Escudero et al., 2011)
7. Its chemistry can be exploited for environmental benefits (Carbonell-Barrachina et al., 2002; Dai et al., 2015; Hanna et al., 1999)
8. Use of fly ash, PG, and red mud as a liner material for the disposal of hazardous leach residue waste (Çoruh and Ergun, 2010)
9. Production of different types of bricks for specific applications (Demyanenko et al., 2012; Selcuk and Emrah, 2012; Tan and Chen, 2009; Wanf et al., 2012; Zhou et al., 2012)
10. Used as ameliorant for phytoremediation of industrial waste (Khoudi et al., 2013; Kobbi-Rebai, 2013; Li et al., 2015a,b; Lopes et al., 2013)
11. Fabrication and production of biocomposites (Kumar et al., 2015) polymer matrix (Lopez et al., 2011), and nanomaterial (Mohamed et al., 2014; Silva et al., 2010)
12. Used for road construction (Shen et al., 2007)
13. Used as a building material (Tayibi et al., 2011)
14. Used to prepare concrete (Yang et al., 2013a,b)
15. Used to prepare compost (Yang et al., 2015)

14.13 DIFFERENT TYPES OF BY-PRODUCTS OF GYPSUM

14.13.1 PHOSPHOGYPSUM

A large quantity of phosphoric acid is used to produce fertilizers and detergents. Phosphoric acid is obtained by processes based on the decomposition of rock phosphate with sulfuric acid. Tricalcium phosphate reacts with sulfuric acid to form phosphoric acid and calcium sulfate. Usually, less soluble calcium sulfate is separated from phosphoric acid by filtration. Calcium sulfate generally appears as hemihydrate or dihydrate, depending on the temperature conditions of the process and sulfuric and phosphoric acid concentrations.

PG is used in agriculture in countries such as Australia, Bangladesh, Brazil, India, Kazakhstan, Pakistan, Spain, Syria, and the United States. In Brazil, 40% of PG that is generated is used in agriculture. In California, United States, PG stock derived from Idaho and Wyoming phosphate rock was fully used for agricultural purposes, which owing to strict regulations was banned until 1989. A small amount of PG with low radium content from Florida, United States (about 0.03% of Florida's total production) is used as a soil-improving agent in peanut production. In the European Union PG is approved for use as an additive for soil improvement, called "calcium sulfate."

PG improves soil in the following cases:

- bog drainage and subsequent use in the agricultural sector
- restoration of alkali and sodium soils
- to increase soil water retention
- use as a fertilizer for crops and pastures
- Usually after PG application, soil does not need plowing. Fresh PG does not erode. It is humid, and in this state it can be applied directly to the soil surface by spraying.

Benefits of using PG on alkali and sodium are:

- reduction in the negative impact of sodium and aluminum on soil
- increase in calcium and sulfur obtained from PG
- increase in ammonia retention by soil
- increase in soil water retention through better composition
- increase in efficiency of irrigation

The use of PG as a fertilizer at 40–200 kg per acre increases the yield of the following cultures: alfalfa, apples, avocados, buckwheat, beans, zucchini, carrots, coffee, wheat, cotton, lime, maize, zucchini, onions, oranges, peaches, peanuts, pineapple, rice, soybeans, sugar beets, tea, tobacco, tomatoes, vegetables, and corn, among others.

PG is often used in its original form without being processed. Depending on the culture, it may require the removal of acid water and even some impurities. For example, PG from the North Florida dump is kept in the rain for about a year. Rainwater saturates it while replacing acid water and increase the pH to more than 5. Generally the removal of impurities from PG is minimized to preserve its economic attractiveness compared with the cost of natural gypsum.

In addition, the practice has developed of using PG as an insulating material for the upper protective layer of radioactive waste. Long-term status observations on radioactive waste disposal sites in the Southern Federal District, Russia, showed no impact on the environment of radioisotopes as well as PG.

During sulfuric acid processing of phosphate raw materials with radioactive isotopes, there is an uneven transition to liquid and solid phases with uranium (^{238}U) enrichment of phosphoric acid, and thorium (^{230}Th and ^{232}Th) and radium (^{226}Ra) enrichment of PG. For nonradioactive phosphate raw materials such as Kola apatite concentrate, its derivatives also are nonradioactive.

PG that is processed from Kola apatite concentrate contains no hazardous radioactive isotopes of ^{238}U, ^{230}Th, ^{228}Th, ^{228}Ra, and ^{210}Po, which determines its full radiological safety and the lack of requirements for storage and use in various fields. There are similar radioactivity characteristics for

Brazil and South African apatite concentrate and PG. However, phosphorite itself contains a significant amount of radioactive impurities such as Th, U, Ra, Po, and Pb, which require special security measures in the storage and handling of PG.

Considering the complex nature of processes at the location of synthetic gypsum landfill facility because of external and internal factors to reduce and ultimately exclude the impact on the environment, environmental impact evaluations based on the hazard class definition of the material are not sufficient and do not reflect the real situation.

14.13.2 TITANOGYPSUM

Titanium dioxide is a white pigment and is by far the most important pigment in terms of quantity. The world production of titanium dioxide is approximately 4 million tons per year, of which about half is produced by the sulfate process and the other half by the chloride process. Only the sulfate process results in gypsum, of which a maximum of about 50% is so-called "white gypsum," which can be used by the gypsum industry.

14.13.3 CITROGYPSUM

Citric acid is a component found in almost all plant and animal species and is pivotal in the energy cycle of living organisms. Large quantities of citric acid for use in the food, pharmaceutical, and detergent industries are produced by mycological fermentation of crude sugar solutions such as molasses. To free citric acid from impurities such as proteins and sugars, it is precipitated with lime to calcium citrate and washed. Pure citric acid is then recovered by acidification with sulfuric acid and filtered off from the formed gypsum. Technically, this gypsum can be purified to a grade which can be processed by the gypsum industry. Part of this gypsum, however, is used as a filter aid to remove complex trace metals. It has a blue color owing to the absorption of ferric hexacyanoferrate complexes which can be removed by only high-temperature calcination producing anhydrite (usable by the cement industry).

14.13.4 FLUOROANHYDRITE

To produce hydrofluoric acid, the mineral fluorspar or fluorite is heated with sulfuric acid. Because the reaction is normally conducted under dry conditions at elevated temperatures, the resulting calcium sulfate is formed in the anhydrous form, anhydrite.

14.13.5 OTHER SYNTHETIC GYPSUM

Small amounts of gypsum are recovered by the production of some organic acids such as tartaric, lactic, formic, and oxalic acid. In addition, all processes which end with the subproduct sulfuric acid are potential gypsum producers. Neutralization of acidic effluents with lime or limestone yields gypsum, for which the potential use depends on the remaining impurities. Titanogypsum is a classic example. The removal of sulfates from brines with the help of lime produces gypsum; the high magnesium content of this "salt gypsum" is one of the main obstacles to its use.

The Hazardous Waste Rules (Management, Handling and Transboundary Movement), 2008 (HWM Rules 2008) as notified by the MoEF&CC under the Environment (Protection) Act 1986, has excluded PG produced from "Phosphoric Acid Plants" under Schedule-1. PG is considered a "High volume low effect waste" and the regulation stipulates that PG management should be carried out in accordance with the provisions of the Central Pollution Control Board (CPCB) India.

In India, about 11 phosphoric acid plants produce abut 6.5 million tons per year of PG. About 4.5–5 tons (dry basis) of PG is produced per ton of phosphoric acid (as P_2O_5) recovered from phosphate rock. PG is currently stacked within the PA fertilizer plants (Fig. 14.7) and some is sold for use in cement as well as for plasterboard manufacturers. However, improper handling of PG is likely have an environmental impact owing to the low pH and fluoride and phosphates present in the PG.

Considering the technological innovations, the CPCB suggests that the guidelines for Management, Handling, Utilization, and Disposal of Phosphogypsum generated from phosphoric acid plants need to be reviewed or reexamined at least once every 5 years to update developments with respect to the possible beneficial use of PG in due course.

14.14 MANUFACTURE OF AMMONIUM SULFATE

Manufacture of ammonium sulfate from the PG is possible according to the following reaction (Fig. 14.9) (Abbas, 2011):

$$\underset{\text{Gypsum}}{CaSO_4 \cdot 2H_2O} + \underset{\text{Ammonium carbonate}}{(NH_4)_2CO_3} \rightarrow \underset{\text{Ammonium sulfate}}{(NH_4)_2SO_4} + \underset{\text{Calcium carbonate}}{CaCO_3} + 2H_2O$$

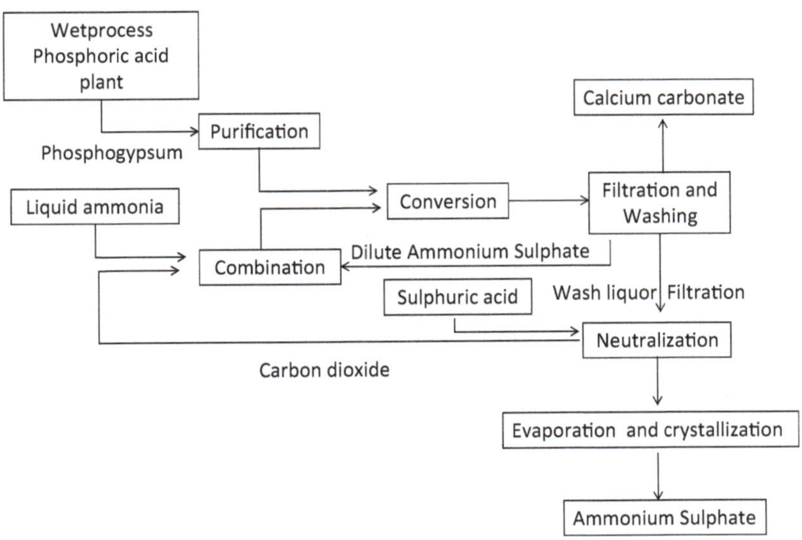

FIGURE 14.9

Scheme showing manufacture of ammonium sulfate from PG.

14.15 FLUE GAS DESULFURIZATION GYPSUM AS A SOURCE OF CALCIUM AND SULFUR FOR CROPS

All crops require calcium (Ca) and sulfur (S) for adequate growth, but certain crops have a specific or higher requirement, particularly for Ca. Peanuts (*Arachis hypogea*) are the best known example. Application of 1 T gypsum ha^{-1} resulted in a yield increase of 984 kg of peanuts ha^{-1} with a concomitant improvement in peanut quality, measured as the percentage of sound mature kernels in the harvest. Application of FGD gypsum is profitable to farmers (Anonymous, 2006a,b; Arthur et al., 1993). In addition, the seed Ca content, which is important in germination of the seeds and emergence of the seedlings, was also increased by gypsum application.

Another crop that responds well to applications of gypsum is tomatoes (*Lycopersicon esculentum*), which can have a Ca deficiency under certain circumstances that results in a condition known as blossom end rot. When soluble soil Ca levels are low, the gradient in tomato skin Ca from the stem end of the fruit to the blossom end is such that the skin at the blossom end is not as robust, which allows pathogens to gain entry and rot the fruit. This represents the worst-case scenario, but marginal levels of Ca in the tomato skin can also result in premature rotting of the fruit during transport and storage.

Melons (*Cucumis* spp.) and cantaloupes (*Cucumis melo*) in particular have been suspected of responding to increased levels of soil Ca, but little evidence has been published in the literature. The fruit are sometimes soaked in a Ca solution to promote shelf life, and in some instances yield responses have been obtained to foliar Ca sprays Fig. 14.10. Addition of Ca to the soil is recommended, but little supporting evidence has been published. The improved Ca nutrition contributed positively to shelf life, as reflected in the rate of fruit rejection after storage. Similar responses can be expected in other members of the curcurbit family, such as watermelons and honeydew melons.

Many other crops (peaches [*Prunus persica*], coffee [*Coffea arabica*], sorghum, soybean, wheat, and sugarcane [*Saccharum officinarum*]), have responded well to the amelioration of subsoil acidity by gypsum applications (Shainberg, 1989; Sumner, 1993; Dick et al., 2000; Ritchey et al., 2000).

14.16 URANIUM RECOVERY FROM PHOSPHORIC ACID VIA HYDROMETALLURGY AND SOLVENT EXTRACTION

Although gypsum ($CaPO_4 \cdot 2H_2O$) is the major component of PG, it also contains metal impurities and some radionuclides that have potential health and environmental impacts (Rutherford et al., 1994). Thus, their effects consist primarily of groundwater contamination with trace amounts of heavy metals and air pollution for radionuclides, the most threatening of which are ^{226}Ra and its daughter product, ^{222}Rn (Guo et al., 2001) Consequently, the presence of some U-series radionuclides in phosphate rock are called NORM (naturally occurring radioactive material) and its impact on the environment require alternative routes of disposal for this kind of residue (Al-Hwaiti, 2015; Ali and Awad, 2015). Numerous investigations have focused on searching for uses for PG waste. Among them, PG has been used as an agricultural fertilizer, in soil stabilization amendments, in the cement industry (Altun and Sert, 2004), as a setting regulator in place of natural gypsum, and as a raw mineralizer in a burning Portland cement clinker. Unfortunately, the presence of impurities can negatively affect the performance of PG in some of these applications. Currently, a widely accepted and environmentally friendly

FIGURE 14.10

FGD gypsum as a source of calcium and sulfur for tomatoes and melons to increase shelf life.

option for reusing waste materials is its incorporation in asphaltic pavements. The easiest way to dispose of PG would be its direct addition as a fraction of mineral aggregates used in asphalt mixtures for road construction. However, we propose an alternative method consisting of preblending asphaltic bitumen with PG before it is used in the asphalt mixture. This alternative constitutes a way to obtain a bituminous binder which enhances high-temperature performance compared with the original bitumen. In fact, addition of phosphorus compounds to bitumen is well known to increase bitumen stiffness significantly (Cuadri et al., 2014). Regarding environmental concerns, bitumen constitutes a way to immobilize the NORM radioactive waste.

Phosphate rocks are characterized by the high content of natural alpha radioactive elements, especially from uranium decay series (^{210}Po, ^{226}Ra, ^{234}U, and ^{238}U) and beta emitter (^{210}Pb). In the process of phosphoric acid production, about 80% of uranium is associated with the phosphoric acid fraction, whereas about 90% of the ^{210}Po and ^{210}Pb is bound to the PG fraction.

ACKNOWLEDGMENTS

Thanks are due to the Central Pollution Control Board, Ministry of Environment, Forests and Climate Change, Delhi, for providing information about the management, handling, use, and disposal of PG generated from phosphoric acid plants in India and elsewhere. The author is grateful to Dr. Boris Levin, Phosagro, Russia (Leninsky prospect, Moscow) for providing useful literature and illustrations (Fig. 14.6) about PG and its use in road construction.

REFERENCES

Abbas, K.K., 2011. Study on the production of ammonium sulfate fertilizer from phosphogypsum. Journal of Engineering and Technology 29, 814–821.

Abril, J.M., García-Tenorio, R., Periáñez, R., 2009. Occupational dosimetric assessment (inhalation pathway) from the application of phosphogypsum in agriculture in South West Spain. Journal of Environmental Radioactivity 100 (1), 29–34.

Anonymous, 2006a. Coal Combustion Product (CCP) Production and Use Survey. ACAA. www.acaa-usa.org.

Anonymous, 2006b. A Review of Agricultural and Other Land Application Uses of Flue Gas Desulfurization Products. Electric Power Research Institute, Palo Alto, CA, 010385.

Al-Hwaiti, M., Al-Khashman, O., 2015. Health risk assessment of heavy metals contamination in tomato and green pepper plants grown in soils amended with phosphogypsum waste materials. Environmental Geochemistry and Health 37 (2), 287–304.

Al-Hwaiti, M.S., 2015. Assessment of the radiological impacts of treated phosphogypsum used as the main constituent of building materials in Jordan. Environmental Earth Sciences 74, 3159–3169.

Ali, K.K., Awad, Y.D., 2015. Radiological assessment of Iraqi phosphate rock and phosphate fertilizers. Arabian Journal of Geosciences 8 (11), 9481–9488.

Altun, I.A., Sert, Y., 2004. Utilization of weathered phosphogypsum as set retarder in Portland cement. Cement and Concrete Research 34, 667–680.

Al-Zaidyeen, S.M., Al-Qadi, A.N.S., 2015. Effect of phosphogypsum as a waste material in soil stabilization of pavement layers. Jordan Journal of Civil Engineering 9, 1–7.

Ammar, R., El Samrani, A.G., Kazpard, V., Bassil, J., Lartiges, B., Saad, Z., Chou, L., 2013. Applying physicochemical approaches to control phosphogypsum heavy metal releases in aquatic environment. Environmental Science and Pollution Research 20 (12), 9014–9025.

Arthur, M.A., Rubin, G., Woodbury, P.B., Weinstein, L.H., 1993. Gypsum amendment to soil can reduce selenium uptake by alfalfa in the presence of coal fly ash. Plant Soil 148, 83–90.

Ayadi, A., Chorriba, A., Fourati, A., Gargouri-Bouzid, R., 2015. Investigation of the effect of phosphogypsum amendment on two *Arabidopsis thaliana* ecotype growth and development. Environmental Technology (United Kingdom) 36, 1547–1555.

Baillez, S., Nzihou, A., Bernache-Assolant, D., Champion, E., Sharrock, P., 2007. Removal of aqueous lead ions by hydroxyapatites: equilibria and kinetic processes. Journal of Hazardous Materials 139, 443–446.

Balkaya, N., Cesur, H., 2015. A kinetic study on cadmium adsorption from aqueous solutions by pre-conditioned phosphogypsum. Desalination and Water Treatment 7.

Bana, R.S., Shivay, Y.S., Tyagi, V.K., 2015. Effect of summer forage crops and phosphogypsum-enriched urea on soil quality, nitrogen-use efficiency and quality of basmati rice (*Oryza sativa*) and their residual effect on succeeding wheat (*Triticum aestivum*). The Indian Journal of Agricultural Sciences 85, 531–538.

Baton, J., Kadaksham, A.J., Nzihou, A., Singh, P., Aubry, N., 2007. Trapping heavy metals by using calcium hydroxyapatite and dielectrophoresis. Journal of Hazardous Materials 139, 461–466.

Berbel, G.B.B., Favaro, D.I.T., Braga, E.S., 2015. Impact of harbour, industry and sewage on the phosphorus geochemistry of a subtropical estuary in Brazil. Marine Pollution Bulletin 93, 44–52.

Bhat, R., Gómez-López, V.M., 2014. Radionuclides in food: past, present and future. In: Practical Food Safety: Contemporary Issues and Future Directions, pp. 281–309.

Bhawan, P., 2012. Guidelines for Management and Handling of Phosphogypsum Generated from Phosphoric Acid Plants. CPCB. Available at: www.cpcb.in (accessed 19.06.12.).

Binnemans, K., Jones, P.T., Blanpain, B., Van Gerven, T., Pontikes, Y., 2015. Towards zero-waste valorisation of rare-earth-containing industrial process residues: a critical review. Journal of Cleaner Production 99, 17–38.

Bolívar, J.P., García-Tenorio, R., Más, J.L., 1998. Radioactivity of phosphogypsum in the southwest of Spain. Radiation Protection Dosimetry 76, 185–189.

Bolívar, J.P., Martín, J.E., García-Tenorio, R., Pérez-Moreno, J.P., Mas, J.L., 2009. Behaviour and fluxes of natural radionuclides in the production process of a phosphoric acid plant. Applied Radiation and Isotopes 67, 345–356.

Borges, R.C., Ribeiro, F.C.A., Lauria, D.C., Bernedo, A.V.B., 2013. Radioactive characterization of phosphogypsum from Imbituba, Brazil. Journal of Environmental Radioactivity 126, 188–195.

Calin, M.R., Radulescu, I., Calin, M.A., 2015. Measurement and evaluation of natural radioactivity in phosphogypsum in industrial areas from Romania. The Journal of Radioanalytical and Nuclear Chemistry, 10 p. http://dx.doi.org/10.1007/s10967-015-3970-3.

Campbell, K.M., Gallegos, T.J., Landa, E.R., 2015. Biogeochemical aspects of uranium mineralization, mining, milling, and remediation. Applied Geochemistry 57, 206–235.

Carbonell-Barrachina, A., Delaune, R.D., Jugsujinda, A., 2002. Phosphogypsum chemistry under highly anoxic conditions. Waste Management 22, 657–665.

Cardenas-Escudero, C., Morales-Florez, V., Perez-Lopez, R., 2011. Procedure to use phosphogypsum industrial waste for mineral CO_2 sequestration. Journal of Hazardous Materials 196, 431–435.

Castillo, J., Pérez-López, R., Sarmiento, A.M., Nieto, J.M., 2012. Evaluation of organic substrates to enhance the sulfate-reducing activity in phosphogypsum. Science of the Total Environment 439, 106–113.

Chaney, R.L., 2012. Food safety issues for mineral and organic fertilizers. Advances in Agronomy 117, 51–116.

Chen, Z.W., 2008. The Method of Manufacturing the Non-fired Bricks of Phosphogypsum. State Intellectual Property Office of the People's Republic of China. CN200710053782 (in Chinese with English abstract).

Chen, L., Dick, W.A., Nelson Jr., S., 2005. Flue gas desuilfurization products as sulfur sources for alfalfa and soybean. Agronomy Journal 97, 265–271.

Chi, J.H., Lin, X.J., 2009. The Method of Producing the Non-fired Bricks of the Industrial By-production Phosphogypsum. State Intellectual Property Office of the People's Republic of China. CN200910058460 (in Chinese with English abstract).

Chi, C.M., Zhao, C.W., Sun, X.J., Wang, Z.C., 2012. Reclamation of saline-sodic soil properties and improvement of rice (*Oryza sativa* L.) growth and yield using desulfurized gypsum in the west of Songnen Plain, northeast China. Geoderma 187–188, 24–30.

Chun, S., Nishiyama, M., Matsumoto, S., 2001. Sodic soils reclaimed with by-product from flue gas desulfurization: corn production and soil quality. Environmental Pollution 114, 453–460.

Çoruh, S., Ergun, O.N., 2010. Use of fly ash, phosphogypsum and red mud as a liner material for the disposal of hazardous zinc leach residue waste. Journal of Hazardous Materials 173 (1), 468–473.

Cuadri, A.A., Navarro, F.J., García-Morales, M., Bolívar, J.P., 2014. Valorization of phosphogypsum waste as a sphaltic bitumen modifier. Journal of Hazardous Materials 279, 11–16.

Dai, Q.-X., Ma, L.-P., Yan, B., Xie, L.-G., Mao, Y., Zhang, H., Zi, Z.-C., 2015. Purification of calcium oxide in phosphogypsum decomposition residue based on the sucrose-CO_2 method. Separation Science and Technology (Philadelphia) 50, 479–486.

Degirmenci, N., Okucu, A., Turabi, A., 2007. Application of phosphogypsum in soil stabilization. Building and Environment 42, 3393–3398.

Degirmenci, N., 2008a. Utilization of phosphogypsum as raw and calcined material in manufacturing of building products. Construction and Building Materials 22, 1857–1862.

Degirmenci, N., 2008b. The using of waste phosphogypsum and natural gypsum in adobe stabilization. Construction and Building Materials 22, 1220–1224.
Demyanenko, N.V., Wang, P.X., Quan, M., Zhang, Q.J., 2012. Method for Producing Baking-Free Brick by Doping Phosphogypsum and Fly Ash. State Intellectual Property Office of the People's Republic of China. CN201210289778 (in Chinese with English abstract).
Dick, W.A., Hao, Y., Stehouwer, R.C., Bigham, J.M., Wolfe, W.E., Adriano, D., Beeghly, J.H., Haefner, R.J., 2000. Beneficial uses of flue gas desulfurization by-products: examples and case studies of land application. In: Powers, J.F., Dick, W.P. (Eds.), Land Application of Agricultural, Industrial and Municipal By-products. Soil Science Society of America, Madison, WI, pp. 505–536.
Elloumi, N., Zouari, M., Chaari, L., Abdallah, F.B., Woodward, S., Kallel, M., 2015. Effect of phosphogypsum on growth, physiology, and the antioxidative defense system in sunflower seedlings. Environmental Science and Pollution Research 22 (19), 14829–14840.
de Filippis, P., Giavarini, C., Scarsella, M., 1995. Improving the ageing resistance of straight-run bitumens by addition of phosphorus compounds. Fuel 74, 836–841.
Gaidučis, S., Mačiulaitis, R., Kaminskas, A., 2009. Eco-balance features and significance of hemihydrate phosphogypsum reprocessing into gypsum binding materials. Journal of Civil Engineering and Management 15, 205–213.
Garcia-Diaz, I., Lopez, F.J., Alguacil, F.J., Bolivar, J.P., Gazquez, M., 2013. Valorisation of two inorganic industrial wastes for manufacturing sulphur polymer concrete. Chemical Engineering Transactions 34, 115–120.
Garg, M., Jain, N., 2010. Waste gypsum from intermediate dye industries for production of building materials. Construction and Building Materials 24, 1632–1637.
Garlanger, J.E., Schmall, P.C., 2012. Investigation and Remediation of a Phosphate Plant Sinkhole in Florida. Geotechnical Special Publication, pp. 972–982, 228 GSP.
Garrido, F., Aguilar-Carrillo, J., García-González, M.T., 2011. Evaluation of two industrial by-products as soil conditioners for reducing potential aluminium toxicity and acidity in a palexerult of Western Spain. Communications in Soil Science and Plant Analysis 42, 2604–2614.
Gissel-Nielsen, G., Bertelsen, F., 1989. Ammonia-based flue gas desulfurization waste solution as a nitrogen fertilizer. Environmental Geochemistry and Health 11, 54–56.
Giavarini, C., Mastrofini, D., Scarsella, M., Barré, L., Espinat, D., 2000. Macrostructure and rheological properties of chemically modified residues and bitumens. Energy Fuels 14, 495–502.
Glória dos Reis, R., Lauria, D., 2014. The potential radiological impact from a Brazilian phosphate facility. Journal of Environmental Radioactivity 136, 188–194.
González-Núñez, R., Alba, M.D., Vidal, M., Rigol, A., 2015. Viability of adding gypsum and calcite for remediation of metal-contaminated soil: laboratory and pilot plant scales. International Journal of Environmental Science and Technology 12, 2697–2710.
Gu, L., Bai, L., Zhang, N., Zhang, S., Yue, X., Chen, Y., Xia, Y., 2013. Strengthening effect of mycorrhizal technology on application of phosphogypsum in agriculture. Nongye Gongcheng Xuebao/Transactions of the Chinese Society of Agricultural Engineering 29, 152–159.
Guitouni, M., 2015. Optimized and validated wavelength-dispersive X-ray fluorescence spectrometry (WDXRF) method for the determination of uranium in phosphogypsum. Journal of Radioanalytical and Nuclear Chemistry 303 (3), 1649–1657.
Guo, P.T., Hawke, A.S., Rusch, K.A., 1999. Determination of calcium diffusion coefficients as an estimator of the long-term dissolution potential of phosphogypsum: cement: lime composites. Environmental Science & Technology 33, 3185–3192.
Guo, T.Z., Malone, R.F., Rusch, K.A., 2001. Stabilized phosphogypsum: class C fly ash: portland type II cement composites for potential marine application. Environmental Science & Technology 35, 3967–3973.

Hadioui, M., Sharrock, P., Mecherri, M.-O., Brumas, V., Fiallo, M., 2008. Reaction of lead ions with hydroxylapatite granules. Chemical Papers 62, 516–521.

Hammas, I., Horchani-Naifer, K., Férid, M., 2013. Solubility study and valorization of phosphor-gypsum salt solution. International Journal of Mineral Processing 123, 87–93.

Hanna, A.A., Akarish, A.I.M., Mousa, S.M., 1999. Phosphogypsum 1-mineralogical, thermo-gravimetric, chemical and infrared characterization. Journal of Materials Science and Technology 15, 431–434.

Hanna, A.A., Abu-Ayana, Y.M., Mousa, S.M., 2000. Phosphogypsum utilization as adhesive feller and composite materials. Journal of Materials Science and Technology 16, 439–444.

Hayati, A.N., Rezaie, H.R., Hosseinalipour, S.M., 2011. Preparation of poly(3-hydroxybutyrate)-nano-hydroxyapatite composite scaffolds for bone tissue engineering. Materials Letters 65, 736–739.

Hentati, O., Abrantes, N., Caetano, A.L., Bouguerra, S., Gonçalves, F., Römbke, J., Pereira, R., 2015. Phosphogypsum as a soil fertilizer: ecotoxicity of amended soil and elutriates to bacteria, invertebrates, algae and plants. Journal of Hazardous Materials 294, 80–89.

Hilal, M.A., El Afifi, E.M., Nayl, A.A., 2015. Investigation of some factors affecting on release of radon^{222} from phosphogypsum waste associated with phosphate ore processing. Journal of Environmental Radioactivity 145, 40–47.

Huang, I.C., Zhou, Y.B., Tang, Z.M., Guo, X., Qian, Z.Y., Zhou, S.B., 2011. Synthesis of multifunctional Fe_3O_4 core/hydroxyapatiteshell nanocomposites by biomineralization. Dalton Transactions 40, 5026–5031.

Huang, Y., Lin, Z.S., 2010. Investigation on phosphogypsum–steel slag–granulated blast-furnace slag-limestone cement. Construction and Building Materials 24, 1296–1301.

Jarcho, M., Bolen, C.H., Thomas, M.B., Kay, J.F., Doremus, R.H., 1976. Hydroxyapatite synthesis and characterization in dense polycrystalline form. Journal of Materials Science 11, 2027–2035.

Jokic, B., Mitered, M., Radmilovid, V., Drmanic, S., Petrovic, R., Janackovic, D., 2011. Synthesis and characterization of monetite andhydroxyapatite whiskers obtained by a hydrothermal method. Ceramics International 37, 167–173.

Jones, B.E.H., Haynes, R.J., Phillips, I.R., 2012. Cation and anion leaching and growth of *Acacia saligna* in bauxite residue sand amended with residue mud, poultry manure and phosphogypsum. Environmental Science and Pollution Research 19, 835–846.

Kacimi, L., Simon-Masseron, A., Ghomari, A., Derriche, Z., 2006. Reduction of clinkerization temperature by using phosphogypsum. Journal of Hazardous Materials 137, 129–137.

Kassir, L.N., Darwish, T., Shaban, A., Lartiges, B., Ouaini, N., 2012. Mobility of selected trace elements in Mediterranean red soil amended with phosphogypsum: experimental study. Environmental Monitoring and Assessment 184, 4397–4412.

Katamine, N.M., 2000. Phosphate waste in mixtures to improve their deformation. Journal of Transportation Engineering 126, 382–389.

Khoudi, H., Maatar, Y., Brini, F., Fourati, A., Ammar, N., Masmoudi, K., 2013. Phytoremediation potential of Arabidopsis thaliana, expressing ectopically a vacuolar proton pump, for the industrial waste phosphogypsum. Environmental Science and Pollution Research 20 (1), 270–280.

Kijjanapanich, P., Annachhatre, A.P., Lens, P.N.L., 2014. Biological sulfate reduction for treatment of gypsum contaminated soils, sediments, and solid wastes. Critical Reviews in Environmental Science and Technology 44, 1037–1070.

Kobbi-Rebai, R., Annabi-Trabelsi, N., Khemakhem, H., Ayadi, H., Aleya, L., 2013. Impacts of restoration of an uncontrolled phosphogypsum dumpsite on the seasonal distribution of abiotic variables, phytoplankton, copepods, and ciliates in a man-made solar saltern. Environmental Monitoring and Assessment 185 (3), 2139–2155.

Korcak, R.F., 1997. High-gypsum by-products as soil amendments for horticultural crops. HortTechnology 3, 156–160.

Kumar, S., Dutta, R.K., Mohanty, B., 2015. Potential of bentonite-lime-mix modified with phosphogypsum and reinforced with sisal fibres. Periodica Polytechnica Civil Engineering 59, 143–154.

Kuttah, D., Sato, K., 2015. Review on the effect of gypsum content on soil behavior. Transportation Geotechnics 4, 28–37.

Li, Z.J., Shen, T., Xie, Q., 2005. The Method of Manufacturing the Environmental and High-Pressure Non-fired Bricks of Phosphogypsum. State Intellectual Property Office of the People's Republic of China. CN200510020444 [in Chinese with English abstract].

Li, G.-Y., Bai, R.-P., Mi, Z.-K., Wang, F., 2015a. Experimental investigation on physical properties of soil in permeation process of phosphogypsum leachate. Yantu Gongcheng Xuebao/Chinese Journal of Geotechnical Engineering 37, 419–425.

Li, J.-Y., Liu, Z.-D., Zhao, W.-Z., Masud, M.M., Xu, R.-K., 2015b. Alkaline slag is more effective than phosphogypsum in the amelioration of subsoil acidity in an Ultisol profile. Soil and Tillage Research 149, 21–32.

Liu, M., Liang, Z., Ma, H.-Y., Huang, L.-H., Wang, M.-M., 2010a. Responses of rice (*Oryza sativa* L.) growth and yield to phosphogypsum amendment in saline-sodic soils of North-East China. Journal of Food Agriculture and Environment 8, 827–833.

Liu, M., Liang, Z., Yang, F., Ma, H., Huang, L., Wang, M., 2010b. Impacts of sand amendment on rice (*Oryza sativa* L.) growth and yield in saline-sodic soils of North-East China. Journal of Food Agriculture and Environment 8, 412–418.

Lokshin, E.P., Tareeva, O.A., Elizarova, I.R., 2013. On integrated processing of phosphogypsum. Russian Journal of Applied Chemistry 86, 463–468.

Lopes, G., Guilherme, L.R.G., Costa, E.T.S., Curi, N., Penha, H.G.V., 2013. Increasing arsenic sorption on red mud by phosphogypsum addition. Journal of Hazardous Materials 262, 1196–1203.

López, F.A., Gázquez, M., Alguacil, F.J., Bolívar, J.P., García-Díaz, I., López-Coto, I., 2011. Microencapsulation of phosphogypsum into a sulfur polymer matrix: physico-chemical and radiological characterization. Journal of Hazardous Materials 192, 234–245.

Mahmoud, E., Abd El-Kader, N., 2014. Heavy metal immobilization in contaminated soils using phosphogypsum and rice straw compost. Land Degradation & Development 26 (8), 819–824.

Mahmoud, E.K., El-Kader, N.A., 2013. Soil immobilization of heavy metal using soil amendments in a greenhouse study. Compost Science & Utilization 21, 156–163.

Masson, J.F., 2008. Brief review of the chemistry of polyphosphoric acid (PPA) and bitumen. Energy Fuels 22, 2637–2640.

Miller, W.P., 1988. Use of Gypsum to Improve Physical Properties and Water Relations in South Eastern Soils. FL. Inst. Phos. Res. Proj. Rep. 83-01-020.

Minh, D.P., Tran, N.D., Nzihou, A., Sharrock, P., 2013. Hydroxyapatitegel for the improved removal of Pb^{2+} ions from aqueous solution. Chemical Engineering Journal 232, 128–138.

Mohamed, A., El Gamal, M.M., 2010. Sulfur Concrete for the Construction Industry – A Sustainable Development Approach. J. Ross Publishing, USA.

Mohamed, K.R., Mousa, S.M., El Bassyouni, G.T., 2014. Fabrication of nano structural biphasic materials from phosphogypsum waste and their in vitro applications. Materials Research Bulletin 50, 432–439.

Mortimer, A., Lemaitre, J., 1989. Synthesis and thermal behavior of well-crystal size calcium-deficient phosphateapatite. Journal of Solid State Chemistry 78, 215–219.

Mousa, S.M., Ammar, N.S., Ibrahim, H.A., 2015. Removal of lead ions using hydroxyapatite nano-material prepared from phosphogypsum waste. Journal of Saudi Chemical Society. http://dx.doi.org/10.1016/j.jscs.2014.12.006.

Mousa, S.M., Hanna, A.A., 2013. Synthesis of nano-crystalline hydroxyapatite and ammonium sulfate from phosphogypsum waste. Materials Research Bulletin 48, 823–828.

Murugan, R., Ramakrishna, S., 2005. Development of nanocomposites for bone grafting. Composites Science and Technology 65, 2385–2405.

Ojovan, M.I., Lee, W.E., 2005. Immobilisation of radioactive wastes in bitumen. In: An Introduction to Nuclear Waste Immobilisation. Elsevier Science Ltd., Amsterdam, pp. 201–212.

Olszewski, G., Boryło, A., Skwarzec, B., 2015. Uranium (^{234}U, ^{235}U and ^{238}U) contamination of the environment surrounding phosphogypsum waste heap in wiślinka (northern Poland). Journal of Environmental Radioactivity 146, 56–66.

Papanicolaou, F., Antoniou, S., Pashalidis, I., 2009. Experimental and theoretical studies on physico-chemical parameters affecting the solubility of phosphogypsum. Journal of Environmental Radioactivity 100 (10), 854–857.

Papanicolaou, F., Antoniou, S., Pashalidis, I., 2010. Redox chemistry of sulphate and uranium in a phosphogypsum tailings dump. Journal of Environmental Radioactivity 101, 601–605.

Papastefanou, C., Stoulos, S., Ioannidou, A., Manolopoulou, M., 2006. The application of phosphogypsum in agriculture and the radiological impact. Journal of Environmental Radioactivity 89, 188–198.

Perez-López, R., Nieto, J.M., López-Coto, I., Aguado, J.L., Bolivar, J.P., Santisteban, M., 2010. Dynamics of contaminants in phosphogypsum of the fertilizer industry of Huelva (SW Sap). From phosphate rock ore to the environment. Applied Geochemistry 25, 705–715.

Pu, Z.Y., Chen, W.G., Ma, G.F., Yang, S.M., Luo, W.L., Zhong, X.G., 2010. The Method of Producing Bricks of the Autoclaved Phosphogypsum. State Intellectual Property Office of the People's Republic of China. CN200910305788 [in Chinese with English abstract].

Ramsier, C., Norton, D., 2006. A win/win solution for FGD gypsum: researchers discover beneficial applications for by-product in agriculture. Ash at Work (I), 10–15.

Rashad, A.M., 2015. Potential use of phosphogypsum in alkali-activated fly ash under the effects of elevated temperatures and thermal shock cycles. Journal of Cleaner Production 87, 717–725.

Ravichandran, P.T., Divya Krishnan, K., Janani, V., Annadurai, R., Gunturi, M., 2015. Soil stabilization with phosphogypsum and fly ash – a micro level study. International Journal of ChemTech Research 7, 622–628.

Ritchey, K.D., Clark, R.B., Elrashidi, M.A., Baligar, V.C., 2000. Properties and examples of beneficial use of gypsum like by-products. In: Powers, J.F., Dick, W.P. (Eds.), Land Application of Agricultural, Industrial and Municipal By-products. Soil Science Society of America, Madison, WI, pp. 537–565.

Rusch, K.A., Guo, T., Roger, K., Seals, R.K., 2002. Stabilization of phosphogypsum using class C fly ash and lime: assessment of the potential for marine applications. Journal of Hazardous Materials B93, 167–186.

Rutherford, P.M., Dudas, M.J., Samek, R.A., 1994. Environmental impacts of phosphogypsum. Science of the Total Environment 99, 1–38.

Selcuk, T., Emrah, A., 2012. A comparative study on the use of fly ash and phosphogypsum in the brick production. Sadhana-Academy Proceedings in Engineering Sciences 37, 595–607.

Shainberg, I., Sumner, M.E., Miller, W.P., Farina, M.P.W., Pavan, M.A., Fey, M.V., 1989. Use of gypsum on soils: A review. Adv. Soil Sci. 9, 1–111.

Shen, W., Zhou, M., Ma, W., Hu, J., Cai, Z., 2007. Investigation on the application of steel slag fly-ash phosphogypsum solidified material as road base material. Journal of Hazardous Materials 164, 99–104.

Shi, L., 2012. Technology Making Use of Waste Phosphogypsum to Produce Gypsum Block. State Intellectual Property Office of the People's Republic of China. CN201110074895 [in Chinese with English abstract].

Silva, L.F.O., Hower, J.C., Izquierdo, M., 2010. Complex nanominerals and ultrafine particles assemblages in phosphogypsum of the fertilizer industry and implications on human exposure. Science of the Total Environment 408, 5117–5122.

Singh, M., 2002. Treating waste phosphogypsum for cement and plaster manufacture. Cement and Concrete Research 32, 1033–1038.

Singh, M., 2015. Gypsum and Gypsum Products – Their Science and Technology. Manakin Press Pvt Ltd, New Delhi, p. 352.

Song, Z., Tang, S., Ding, Y., Yang, Y., Shi, L., 2011. Effects of different phosphate fertilizers on copper sorption in red soils. Journal of Basic Science and Engineering 19 (5), 783–791.

Su, X., Zhu, J., Fu, Q., Zuo, J., Liu, Y., Hu, H., 2015. Immobilization of lead in anthropogenic contaminated soils using phosphates with/without oxalic acid. Journal of Environmental Sciences (China) 28, 64–73.

Sumner, M.E., 1993. Gypsum and acid soils: the world scene. Advances in Agronomy 51, 1–32.

Sumner, M.E., 1994. Amelioration of subsoil acidity with minimum disturbance. In: Jayawardane, N.S., Stewart, B.A. (Eds.), Subsoil Management Techniques. Lewis Publishers, Boca Raton, FL, pp. 147–185.

Sumner, M.E., 1995. Literature Review on Gypsum as a Calcium and Sulfur Source for Crops and Soils in the Southeastern United States. FIPR Publication #01-118-118. Florida Institute of Phosphate Research, Bartow, FL.

Sumner, M.E., Naidu, R., 1998. Sodic Soils: Distribution, Properties, Management, and Environmental Consequences. Oxford University Press, Oxford, England.

Tan, K.F., Chen, D.Y., 2009. High-Strength Phosphogypsum Wall Material and Method for Making Brick Using the Material. State Intellectual Property Office of the People's Republic of China. CN200810045312 [in Chinese with English abstract].

Tayibi, H., Choura, M., López, F.A., Alguacil, F.J., López-Delgado, A., 2009. Environmental impact and management of phosphogypsum. Journal of Environmental Management 90, 2377–2386.

Tayibi, H., Gascó, C., Navarro, N., López-Delgado, A., Alvarez, A., Alguacil, F.J., López, F.A., 2011. Valorisation of phosphogypsum as building material: radiological aspects. Materiales de Construcción 61 (304), 503–515.

Tayibi, H., Gascó, C., Navarro, N., López-Delgado, A., Choura, M., Alguacil, F.J., López, F.A., 2001. Radiochemical characterization of phosphogypsum for engieering use. Journal of Environmental Protection 2, 168–174.

Thompson, L.G., Neller, J.R., 1963. Sulfur Fertilization of Winter Clovers, Coastal Bermuda Grass and Corn on North and West Florida Soils. FL. Ag. Expt. Sta. Bull., p. 656

Toma, M., Sumner, M.E., Weeks, G., Saigusa, M., 1999. Long-term effects of gypsum on crop yield and subsoil chemical properties. Soil Science Society of America Journal 39, 891–895.

Valkov, A.V., Andreev, V.A., Anufrieva, A.V., Makaseev, Y.N., Bezrukova, S.A., 2014. Phospho-gypsum technology with the extraction of valuable components. Procedia Chemistry 11, 176–181.

Wallach, D.L., 1997. Ammonium sulfate fertilizer as a by-product in flue gas desulfurization: the Dakota Gasification Company experience. In: Rechcigl, J.E., MacKinnon, H.C. (Eds.), Agricultural Uses of By-products and Wastes. American Chemical Society, Washington, D.C, pp. 240–254.

Wang, P.X., Quan, M., Zhang, Q.J., et al., 2012. Method for Producing Baking-Free Brick by Doping Phosphogypsum and Fly Ash. State Intellectual Property Office of the People's Republic of China. CN201210289778 [in Chinese with English abstract].

Wilkinson, S., 2002. Use of lime, gypsum, alum and potassium permanganate in water quality management. Aquaculture Asia 7, 12–14.

Xu, L., Wang, Z.-C., Zhao, C.-W., Wang, Y.-H., Sun, X.-J., Wei, B.-W., Cao, L.-L., Zhu, M.-W., 2012. Amelioration of soda saline-alkali soil through equidistant slotting in combining with gypsum application. Chinese Journal of Applied Ecology 31, 1179–1185.

Yang, J.K., Liu, W.C., Zhang, L.L., Xiao, B., 2009. Preparation of load-bearing building materials from autoclaved phosphogypsum. Construction and Building Materials 23, 687–693.

Yang, J.P., 2007. The Method of Manufacturing the High-Strength Building Bricks of Phosphogypsum. State Intellectual Property Office of the People's Republic of China. CN200610054521 [in Chinese with English abstract].

Yang, L., Yan, Y., Hu, Z., 2013a. Utilization of phosphogypsum for the preparation of non-autoclaved aerated concrete. Construction and Building Materials 44, 600–606.

Yang, X., Zhang, Z., Wang, X., 2013b. Thermodynamic study of phosphogypsum decomposition by sulfur. Journal of Chemical Thermodynamics 57, 39–45.

Yang, F., Li, G., Shi, H., Wang, Y., 2015. Effects of phosphogypsum and superphosphate on compost maturity and gaseous emissions during kitchen waste composting. Waste Management 36, 70–76.

Yin, X.L., 2007. The Method of Manufacturing the High Strength and Proof-Water and Proof-Fire Non-Fired Bricks of the Industrial By-Production Phosphogypsum. State Intellectual Property Office of the People's Republic of China. CN200510032313 [in Chinese with English abstract].

Zhang, T.B., Chi, J.H., 2000. The Method of Producing the Non-Fired Bricks of Phosphogypsum. State Intellectual Property Office of the People's Republic of China. CN00109901 [in Chinese with English abstract].

Zhao, L., Wan, T., Yang, X., Yang, L., Kong, X., Zhang, Z., Wang, X., 2015. Effects of kaolinite addition on the melting characteristics of the reaction between phosphogypsum and CaS. Journal of Thermal Analysis and Calorimetry 119, 2119–2126.

Zheng, S.-C., Yu, Q., Ning, P., Ma, J.-B., Zheng, Y.-L., 2015. Preparation of β-hemihydrate gypsum with phosphogypsum. Xiandai Huagong/Modern Chemical Industry 35, 60–63.

Zhong, B.H., Wang, X.L., Zhang, Z.Y., Yang, X.S., 2010. Discussion on utilization ways of phosphogypsum in China. Phosphate & Compound Fertilizer 25, 61–63 [in Chinese with English abstract].

Zhou, J., Gao, H., Shu, Z., 2012. Utilization of waste phosphogypsum to prepare non-fired bricks by a novel hydration–recrystallization process. Construction and Building Materials 34, 114–119.

Zieliński, M., 2015. Influence of constant magnetic field on the properties of waste phosphogypsum and fly ash composites. Construction and Building Materials 89, 13–24.

CHAPTER

METALLIFEROUS WASTE IN INDIA AND KNOWLEDGE EXPLOSION IN METAL RECOVERY TECHNIQUES AND PROCESSES FOR THE PREVENTION OF POLLUTION

15

J. Koelmel[1,2], M.N.V. Prasad[1], G. Velvizhi[3], S.K. Butti[3], S. Venkata Mohan[3]
University of Hyderabad, Hyderabad, Telangana, India[1]; USIEF Fulbright-Nehru Scholar[2]; CSIR-Indian Institute of Chemical Technology (CSIR-IICT), Hyderabad, Telangana, India[3]

J. Koelmel present address: University of Florida, Gainesville, FL, USA

CHAPTER OUTLINE

Environmental Materials and Waste. http://dx.doi.org/10.1016/B978-0-12-803837-6.00015-9

15.1 INTRODUCTION

Metal contamination is a global problem because metals are widely used in everything from building materials to information technology, and the bioavailable fraction of metals is increasing. Metals are reaching toxic concentrations in many parts of the world. Although toxic at high doses, some metals are important components of many organisms; for example, Fe is a necessary component of hemoglobin and Zn is found in over 200 enzymes in humans and is critical to cell growth and division, and in genetic expression (Sandstead, 1994). Copper is involved in a range of physiological functions such as immune function, red and white cell maturation, cholesterol and glucose metabolism, Fe transport, and brain development (Olivares and Uauy, 1996). Other metals such as Pb are toxic at even trace levels and have no known biological function.

Development of infrastructure is highly dependent on metals because of their useful properties. Their electrons are easily removed and hence they are great conductors of electricity. Most metals are ductile, which means they can easily be drawn into wires; they are malleable, which means that they can easily be shaped. Because of their high melting points most metals will preserve their form after being shaped under high temperatures. In addition, metal crystal structures are tightly packed (high atomic density), which makes them extremely strong materials to build with. Cu, Pb, and Zn are found in the most tightly packed structures, called closest packed structures, with 74% of space in the crystal taken by the range of their atomic orbitals.

Because of metals' necessity for infrastructure, industrialization causes a huge increase in metal use and waste. Before industrialization the work necessary for societal functions was mainly not mechanized and consisted primarily of agriculture. Once mechanization began, in most cases the use of metals increased tremendously and then fell off at a later time or at a higher per-capita gross domestic product (GDP), as represented by the environmental Kuznets curve (Bernardini and Galli, 1993). The inverted U-shaped environmental Kuznets curve best models the amount of pollution emitted as a country develops, and shows a drastic increase followed by a gradual decline owing to regulatory policy. After 1974 there was a significant decrease in demand for most metals except Al, which may correlate with the slower growth in GDP (Cleveland and Ruth, 1998).

Today with newly developing economic giants such as India and China, the trend is reversing. A twofold increase in metal production (mainly based on Cu and Al) is needed in the next 2 decades to meet the expected demand. Half of the increase in demand is from consumption in India and China (Streifel, 2006). With the increased production of metals comes an increase in environmental hazards posed by certain metals' toxicities. Metals are naturally found in soils but usually in concentrations below those posing significant health risks. Problems arise when metals are highly concentrated and bioavailable, both of which are a direct consequence of anthropogenic extraction and the use of these metals. In soil, metal bioavailability decreases with time as metals bind to humic acids, fulvic acids, and other organic compounds' active sites containing sulfur, nitrogen, and oxygen (Martin and Kaplan, 1998). Therefore, the slow, naturally occurring release of metals from ores will have a low fraction of bioavailable metals whereas drastic increases in metals from anthropogenic sources may have a high fraction of bioavailable metals. Because the development of infrastructure currently depends on the use of toxic metals, the possible rate of development while maintaining a healthy environment depends on developing remediation technologies for metal contaminants.

This is especially true for India, where high population density and enormous growth over the past decades have led to widespread metal contamination. India is entering into the "developed phase" and it is the ideal time to develop cost-effective strategies to meet the environmental challenges of India's rapid growth.

The purposes of this review are to:

1. Use currently available data to predict major sources of metal contamination in the water, soil, and air in India
2. Predict how these sources may change in the near future
3. Determine how these sources pose a threat by entering human food and water, as well as the air we breathe
4. Review the prospects for recovery and preventing pollution.

15.2 MINERALS AND METALS MASS CONSUMPTION

15.2.1 FUEL COMBUSTION

15.2.1.1 Coal-Fired Power Plants

Coal has trace amounts of toxic metals and minerals which are released into the biosphere when coal is burned in power plants. Indian coal has an estimated 8500, 3.72, 0.06, 2.29, 6.28, 3.67, 6.1, and 11.89 parts per million (ppm) Al, As, Cd, Co, Cu, Pb, Ni, and Zn, respectively (Khandekar et al. in Sajwan et al., 1999), although concentrations vary significantly depending on the source and type of coal. The two major types of coal mined in India are bituminous and lignite, which have been mined in similar percentages from 2000 to 2010 at 93.2 ± 0.4% and 6.8 ± 0.4% of mined coal (calculated using British Geological Survey (BGS), 2012 data). The trace metal content in coal tends to decrease with increasing grade of coal; of the three grades of coal, bituminous and lignite are the second and third, respectively, from highest to lowest grade. Although concentrations of toxic metals are not alarmingly high in coal, the enormous quantity of coal burned in India (and globally) makes it a major factor in the accumulation of metals in the biosphere. This section provides the first model of anthropogenic inputs of environmentally relevant toxic metals and minerals to the biosphere from both fly ash and emissions.

India is the third largest producer and consumer of coal, preceded by the United States and then China. Production of coal in India has increased dramatically over the past 30 years (Fig. 15.1), as has consumption of coal in India following the global trend (Fig. 15.2). Between 2003 and 2010 the percentage of total energy used in India coming from coal (averaged annually) ranged from 68% to 69% (according to BGS data). Estimated coal production varies depending on the organization that compiles data: For example, using BGS (2012) data, the increase from 2008 to 2010 in coal production was 1.7%, whereas using World Coal Association WCA, (2012) data, the increase was 9.8%, but general trends remain the same. Coal consumption continues to increase in India and with about 61 billion metric tons of coal reserves remaining in India as of 2008 (EIA, 2012), even at a rate of a billion tons consumed per year coal reserves would last for another 60 years as of 2008. The WCA estimated that about 118 years' worth of coal remain worldwide (WCA, 2012). Production rates have not kept off; consumption rates in India and the percentage of coal from imports about doubled from 2000 to 2010 (Fig. 15.3). This increase in imports means that fly ash from coal burning may have increasingly greater amounts of toxic metals, because India's fly ash has much less toxic metal load than the global average (Sushil and Batra, 2006).

Fly ash is an important source of toxic metals. The metal content of fly ash varies significantly between plants because of the type of coal used and the combustion conditions. Generally Group I elements (eg, Al) collect in the bottom and fly ash, Group II elements (eg, As, Cd, Cu, Pb, and Zn) are often found in fugitive emissions unless captured in fly ash with electrostatic precipitator technology, and Group III elements (ie, Hg) are found mainly in gas phase. Generally the levels of metals in Indian

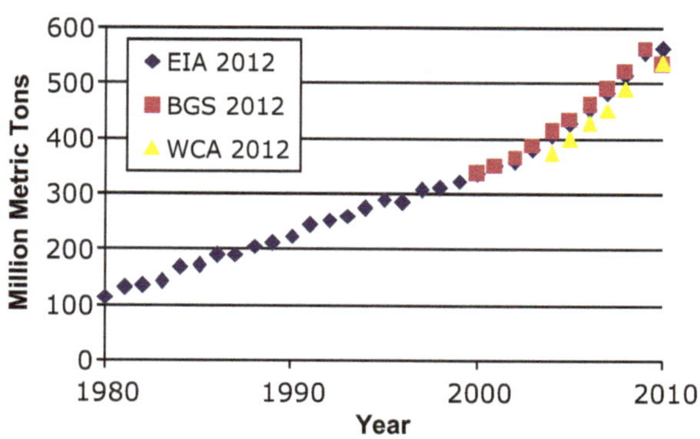

FIGURE 15.1

Scenario of coal production in India. *EIA*, Environmental Impact Assessment; *BGS*, British Geological Survey; *WCA*, World Coal Association.

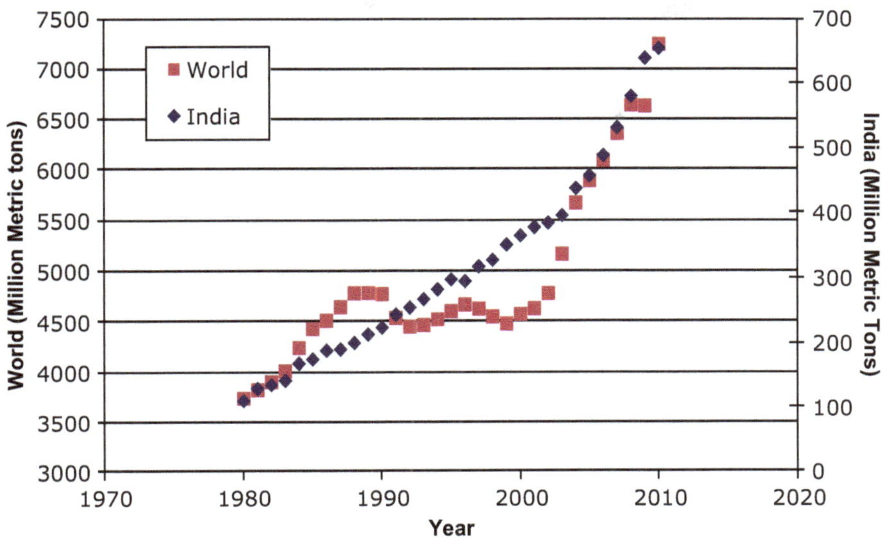

FIGURE 15.2

Scenario of coal consumption in India and the world.

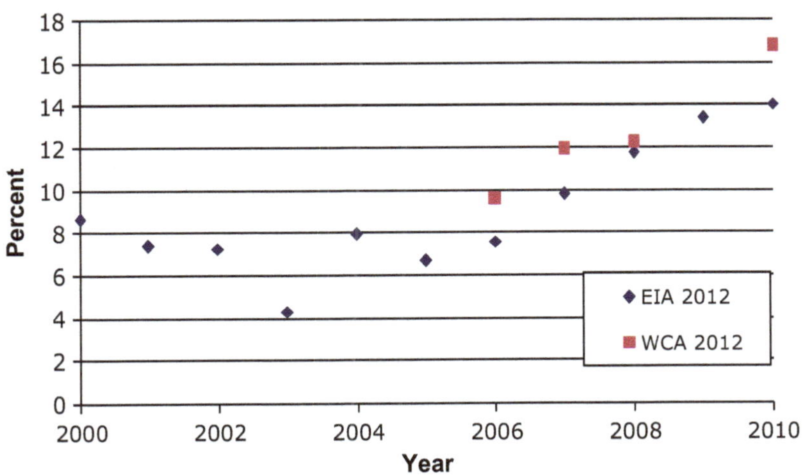

FIGURE 15.3

Percentage of coal consumption from imports.

fly ash are less than in other parts of the world. This is especially true for Pb: 56, 20, 34, and 55 ppm Pb in fly ash was found in four plants in India whereas the range of Pb in Chinese fly ash is 843–847 ppm, and in Greece is 123–143 ppm (Sushil and Batra, 2006). This lower concentration could be because Indian coal has high ash content (35–45%) compared with other coals (Sushi and Batra, 2006) and hence the metal may be "diluted." Average fly ash metal and mineral content in India varies significantly between authors, and hence there is currently no consensus on fly ash metal content in India. Average Cr, Co, Cu, Pb, Mn, and Ni in fly ash from Indian plants is 120, 23.6, 100, 35, 338.91, and 150 ppm, respectively, in one review (Sivakumar and Datta, 1996). In another review, ranges for Cd and Sb are from 1 to 10 ppm and for As, Cr, Ni, Pb, and Zn are 10−100 ppm; for Cu it is 100−1000 ppm in fly ash from Indian coal plants (Dhadse et al., 2008). Average Hg is about 1 ppm in fly ash according to Dhadse et al. (2008) and 0.007−0.28 ppm compared with the world average of 0.62 ppm according to Pirrone et al. (2010). As with the general burning of coal, the quantity of fly ash generated in India leads to significant toxic metal loads to the environment, with, for example, 110 million metric tons of coal ash generated in 2005 (Sushil and Batra, 2006). Globally, China is the largest producer, and then the United States and Russia.

Fly ash is usually stored as compacted fly ash in a controlled fill or dumped in a landfill, or as a slurry in an ash pond which is then discharged elsewhere. Most fly ash is either land filled or dumped in an ash pond (Singh et al., 2012). The metals can then be leached from rainwater in the case of the landfill or added directly into the ash slurry and then into the groundwater. Fugitive ash emissions can also travel in the atmosphere and deposit the heavy metal load. One major problem in India is that ash ponds often do not follow correct procedure and are unlined, which magnifies leaching into groundwater, as was observed in two power plants in North India (Sushil and Batra, 2006) and a power plant in New Delhi (Singh et al., 2012). In addition to lacking lining, the two ash ponds from North India discharge directly into a river or drain (Sushil and Batra, 2006). The amount of metals leached from fly ash varies highly, depending not only on the characteristics of the environment such as pH and dilution, but also on the characteristics of fly ash from different sources. For example, in fly ash from two coal mines in Andhra Pradesh, percentages of trace metals in fly ash leached into water under the same environmental parameters were ordered from highest to lowest as: Zn, Cr, Pb, Cu, Ni, and Cd for fly ash from Chandrapura, and Zn, Cu, Cr, Ni, Pb, and Cd for fly ash from Ramagundam (Prasad and Mondal, 2008).

In groundwater or drinking water from wells near ash ponds in India, trace metals and the minerals As, Cd, Cr, Cu, Ni, Pb, and Zn have been shown to reach levels above drinking water standards in certain instances, often with strong evidence pointing to ash ponds as the source of the metals. The elements Cu, Pb, and Zn leach easily from the surface of the fly ash into water because they condense on the fly ash surface after heating and cooling during combustion. In Orissa, India, a fly ash pond was found to contaminate nearby tube wells and open well water with Cu, Pb, and Zn, and Pb was above drinking water standards. As, Cr, Co, and Ni were at very low concentrations within safe limits. Cu and Zn were predicted to reach levels above drinking water standards owing to cumulative effects. Most metals were found at highest concentrations in tube wells within about 600 m of the ash pond, and again at around 1000 m where the leachate mixed with the groundwater (Praharaj et al., 2002). On the other hand, in a study by Agrawal et al. (2011), As, Pb, Cd, Ni, and other trace metals all were above World Health Organization (WHO) standards in the drinking water surrounding ash ponds in Uttar Pradesh (Agrawal et al., 2011). Yet another study in New Delhi showed that nearby groundwater samples by an ash pond near a plant commissioned in 1967 contained Cr, Cd, Ni, and Pb above drinking water standards, but not Cu (Singh et al., 2012).

A model of the amount of metal released into the biosphere within fly ash or leached from fly ash was developed using the average metal content in fly ash from Sivakumar and Datta (1996), and when data on metals were not available, from EKTA International (2012). In that review, the fly ash percentage of 32% of the coal's weight was used based on 40% of the ash content for coal, 80–85% of which was predicted to be fly ash (Prasad and Mondal, 2008). The total consumption of coal was multiplied by 0.32 and the percent metal content in the fly ash to obtain the overall metal released. Table 15.1 shows the estimated amount in metric tons of various metals in coal consumed, and in predicted fly ash generated for 2010 as well as cumulative inputs for 1980–2010. Fig. 15.4 shows cumulative inputs between 1980 and 2010 for some trace metals studied.

Fly ash is the major by-product of burning coal and has many potential uses. India's Ministry of Environment and Forests (MoEF) has set laws to push for a higher percentage of fly ash use: For example, until 2018 plants must make fly ash available free of charge, and within 50 km of a plant, most clay construction materials must contain at least 25% fly ash (Dhadse et al., 2008). In addition the

Table 15.1 Predicted Anthropogenic Inputs of Trace Metals From Coal Fly Ash in India

Element	As	Cd	Cr	Co	Cu	Pb	Ni	Zn	Hg	F
2010[a]	9	0.71	29	5	21	7	31	37	0.2	6
Cumulative[b]	137	11	429	74	315	110	473	561	3	91

[a]*Total metal content in fly ash generated in 2010; values in 1000 metric tons.*
[b]*Total metal content in 1000 metric tons in fly ash generated from 1980 to 2010.*

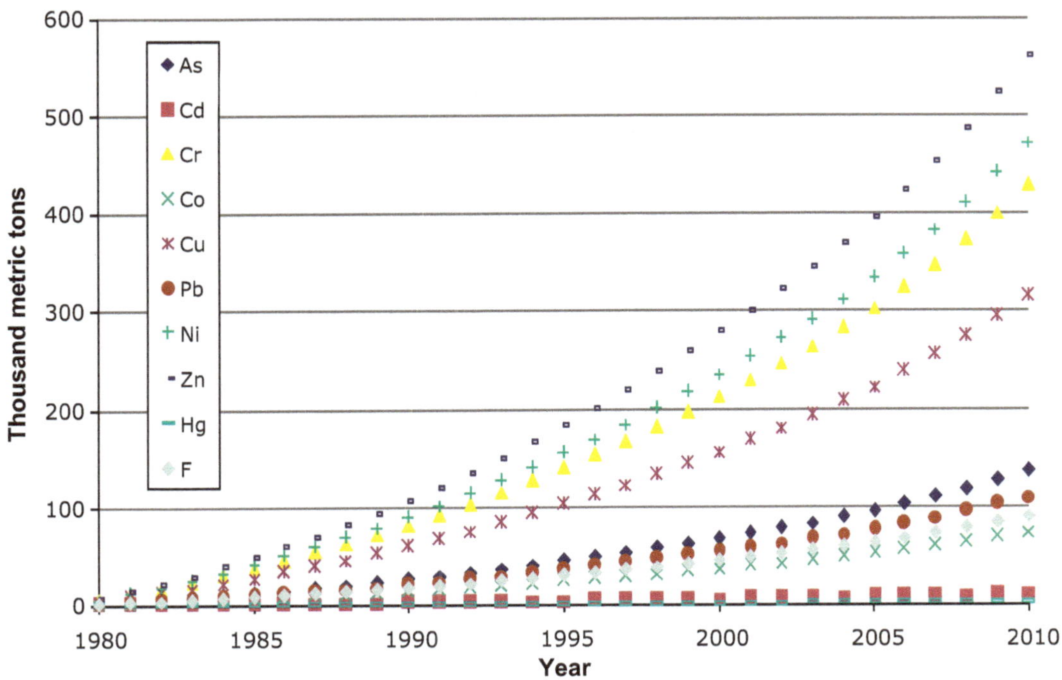

FIGURE 15.4

Cumulative inputs of metals from coal fly ash.

Fly Ash Mission Department of Science and Technology, Government of India, was started to develop and promote new ways to use fly ash (TIFAC, 2001). These initiatives are important because India's use of fly ash (13%) is about half the global average (25%) (Iyer and Scott, 2001). Other measures to reduce contamination and increase reuse of fly ash in India include Asia's first 100% dry fly ash storage, in which mounds of fly ash are sprayed with a polymer and vegetation is planted on top or the fly ash is used (Sushil and Batra, 2006).

The Fly Ash Mission Department of Science and Technology has pushed for and developed the use of fly ash mainly in roads and embankments, concrete dams, and soil amendments for various purposes, from waste sites to agriculture, and has initiated densification of ash ponds and the reclamation of ash ponds. The initiatives have led to a 45-fold increase in fly ash production from 1995 to 2005 (Kumar et al., 2005). As of 2001, 1% of reused fly ash is used for agriculture (Dhadse et al., 2008). Fly ash is a good source of nutrients and organic matter, and can amend acidic soils (TIFAC, 2001). Because of the high concentration of toxic metals in some fly ash, it is interesting that it is being pushed for agricultural use. Hazardous metal contents can potentially become less bioavailable in fly ash used for building materials, with the only bioavailable contents being those leaching from the concrete, and leaching from under the structural fills (or into the groundwater below). Therefore, it is often both environmentally sound and sustainable to reuse fly ash, especially for construction materials.

Other by-products from coal burning include volatized components, bottom ash, and boiler slag. Volatized metals and fugitive emissions of trace metals from the burning of coal are other major anthropogenic sources of these metals into the biosphere that have global implications. To model emissions from coal consumption in India, emission factors for Asia proposed by Pacyna and Pacyna (2001) were used (Table A1a). The amount emitted in 2010 and final cumulative anthropogenic inputs from 1980 to 2010 are shown in Table 15.2.

Table 15.2 Emissions of Metal into the Atmosphere From Burning Coal (1000 metric tons)

Element	As	Cd	Cr	Cu	Hg	Ni	Pb	Sb	V	Zn
2010[a]	0.1	0.1	1.1	0.9	0.3	1.3	0.7	0.1	0.7	1.0
Cumulative[b]	2.0	1.0	16.8	13.8	4.9	19.7	9.9	2.0	9.9	14.8

[a]*Total metal emissions in 2010; values in 1000 metric tons.*
[b]*Total cumulative emissions in 1000 metric tons from 1980 to 2010.*

15.2.1.2 Wood Fuel Combustion

Based on United Nations statistics division data between 1990 and 2005, India was the top consumer of wood fuel by country, except in 1994, when China had higher consumption. Fig. 15.5 shows the consumption of fuel wood in India from 1995 to 2005. Using these data, elemental emission factors developed by the Eastern Research Group (2001) (Table A1e), and a conversion factor for wood volume to energy from the biomass energy center (2011), a model was developed estimating total emissions of trace metals from wood burning on an annual basis in India. Emission data are presented in Table 15.3 for 1990 and 2005.

Trace metal emissions from wood will continue to be dwarfed by coal, especially as coal begins to replace wood for personal energy use. As of 2002, wood fuel still remained the major fuel in India. In

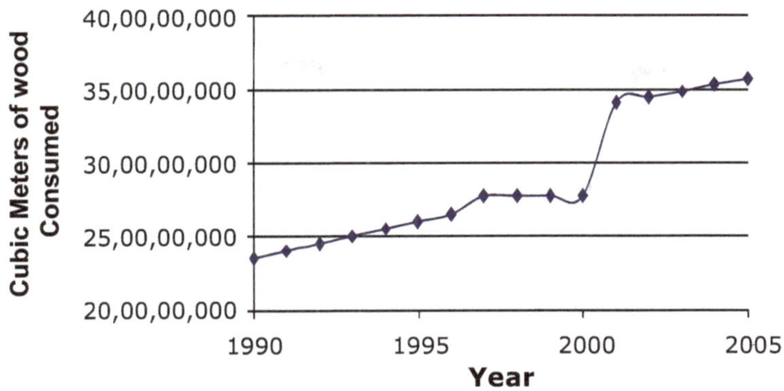

FIGURE 15.5

Wood consumption as fuel in India (in cubic meters).

Table 15.3 Emission Data for Wood Combustion (metric tons)

Element	Sb	As	Cd	Cr	Cu	Pb	Hg	Ni	Zn
1990	5	14	3	13	31	30	2	21	267
2005	8	21	4	20	47	46	3	32	406

urban sectors coal is replacing wood for fuel consumption and energy needs, which releases a much larger fraction of toxic metals, but reduces the ecological burden of deforestation (Pandey, 2002). Therefore it can be predicted that with the growth of GDP and more equitable distribution of wealth, coal consumption will continue to increase significantly if no alternative source of electricity takes the main stage.

15.2.2 WASTE RECYCLING OPERATIONS

15.2.2.1 Metal Content in Waste: Indian Scenario

India generates a large amount of waste, and concentrations were directly correlated with the population densities as of 2001. Using data from the National Solid Waste Association of India (NSWAI, 2012) based on data from six states and 15 cities, a correlation coefficient of 0.95 was obtained from the metric tons of waste generated daily and the population (Fig. 15.6). The metal content of waste was calculated as the percent metal content multiplied by the tons of waste. The percent metal content did not correlate with the percent domestic waste, percent industrial waste, or percent commercial waste, nor did the total metal content correlate with the population or total waste production. The percentage of metals in waste and total metal waste generated by all cities which reported percent metal content in waste and total waste generated as of 2001 is shown in Table 15.4, which demonstrates highly heterogeneous levels of metal in waste. The average percentage of metal in municipal waste from all of India did not grow with the further development and population growth of 0.49% in 1996 and 0.50% in 2005.

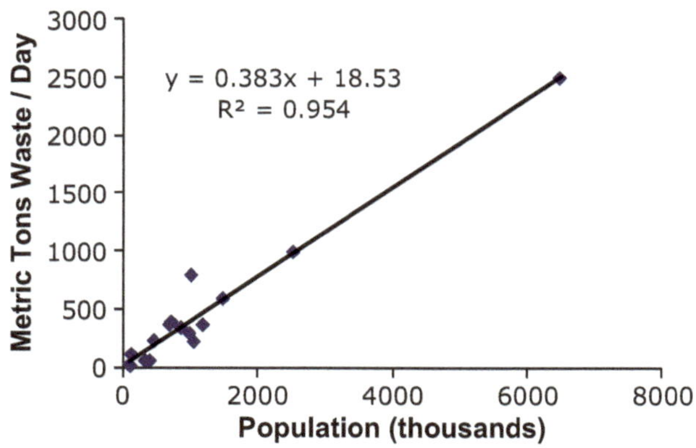

FIGURE 15.6

Waste as a factor of population.

Table 15.4 Emissions of Toxic Metals Resulting From Incineration of Municipal Waste in Major Cities in India

District/City	State	Metal %	Tons Metal Year^{-1}
Bangalore	Karnataka	1.00	9125
Cochin	Kerala	0.70	958
Bhopal	Madhya Pradesh	1.50	3285
Dhule	Maharashtra	0.01	1
Pune	Maharashtra	0.40	1460
Yavatmal	Maharashtra	0.50	44
Nashik	Maharashtra	0.50	420
Solapur	Maharashtra	0.50	639
Ulhasnagar	Maharashtra	0.70	603
Barshi	Maharashtra	1.00	110
Malegaon	Maharashtra	1.00	237
Navi Mumbai	Maharashtra	2.16	3154
Akola	Maharashtra	8.00	3504
Aurangabad	Maharashtra	15.00	16,425
Ajmer	Rajasthan	0.13	60
Madurai	Tamil Nadu	0.10	137

Municipal waste emissions from incineration and in leachate in India are a potential major national source of toxic metals. Data on a national scale of municipal waste generation, imports, and incineration are minimal, and hence a model using emission factors developed by Pacyna and Pacyna (2001) (Table A1c) could not be generated. In addition, a large database of trace metal contents of waste and leachate is missing.

15.2.2.2 Recycling

One of the most polluting industries with regard to toxic metals as well as organics is the recycling industry. Small-scale industries do not have the financial and technological capacity to invest in pollution control experts and technology, so that regulations force them out of the market or they move operations and continue illegally. Also, by saving costs by operating without pollution control measures, illegal or unregulated operations can out-compete those with measures by outbidding providers of recyclable materials. Some major recyclables in India with metal contaminants as bi-products are waste of electrical and electronic equipment (WEEE), steel, Cu, Pb, and Zn (Kojima and Jain, 2008). Steel, Cu, Pb, and Zn contamination from secondary production is further discussed in Section 15.2.3 and in respective sections on individual elements.

15.2.2.2.1 Waste of Electrical and Electronic Equipment

WEEE is one of the fastest types of waste to increase in generation; the disposal and recycling of WEEE can lead to severe metal contamination. Global waste of WEEE reached 40 million metric tons as of 2010. The fastest-growing markets as well as large generators of electronic waste are developing nations such as India, whereas countries such as North America and Europe illegally export WEEE to India (Sepúlveda et al., 2010). India generated 330,000 metric tons of WEEE per year as of 2006–2007, with an additional 50,000 metric tons of WEEE illegally imported (Kojima and Jain, 2008). The growth of WEEE generation as of 2008 was predicted to be 34% per year (Kukday, 2007). The top 10 states from highest to lowest generation of WEEE are Maharashtra, Andhra Pradesh, Tamil Nadu, Uttar Pradesh, West Bengal, Delhi, Karnataka, Gujarat, Madhya Pradesh, and Punjab, with Maharashtra generating about 20,270 metric tons (Pinto, 2008). Recycling of this waste can be even more profitable than mining ores because metals are often more concentrated. For example, Cu is 10 to 50 times more concentrated in WEEE, whereas gold is five to 10 times higher in cell phones than in the metals' respective ores (Kukday, 2007). A total of 19,000 tons of WEEE was recycled by mainly unregulated informal recycling operations in 2006–2007 (Kojima and Jain, 2008). Although there is national legislation led mainly by European countries and the European Directive on WEEE, in India no specific law regulated WEEE as of 2010 (Fig. 15.7), although regulations were being drafted (Sepúlveda et al., 2010).

Metal inputs into the biosphere from WEEE in India were modeled. Although there are extensive data on the WEEE content of economically profitable metals to mine, such as Au, Cu, Pb, Pd, and Ni, the trace metal content of mixed WEEE is scarce. Table 15.5 lists a compilation of metal content in mixed WEEE at various times. Average concentrations in (mg/kg) for Cu, Pb, Ni, Zn, Sb, Cd, Cr, and Hg were $138{,}625 \pm 55{,}765$, $61{,}983 \pm 124{,}137$, $13{,}025 \pm 7595$, $33{,}033 \pm 40{,}917$, 1300 ± 566, 330 ± 212, 5300 ± 6505, and 1.29 ± 0.86, respectively. Based on the standard deviations of the samples, and as reported in numerous articles, one can see that there is great variation. Metal content depends on the sample location, because the mixture of electronic equipment can vary significantly, as well as the time of sampling. This is because ratios of electronic change over time as well, and the amount of metals used in electronics also changes. For example, the content of economically important metals such as Au and Cu has declined in electronic overtime as technology reduces the amount needed. This can be seen in WEEE samples for Cu in the data in Table 15.5 and Fig. 15.8.

Based on the metal content average in WEEE, and 330,000 metric tons generated of WEEE in 2006 with a 34% increase each year in WEEE generation, a model of metal inputs into the biosphere was developed (Figs. 15.9 and 15.10). Copper averages were taken only for 2007 based on a reduction

FIGURE 15.7

Electronic circuit board at a flea market in Hyderabad. Waste of Electrical and Electronic Equipment (WEEE) in India is often first sold to for secondary use at flea markets. What is not sold is then brought to small recycling operations to extract the precious metals.

Table 15.5 Toxic Metal Content in Mixed Waste of Electrical and Electronic Equipment, 1947–2007

Year/ References	Cu (%)	Pb (%)	Ni (%)	Zn (%)	Sb ppm	Cd ppm	Cr ppm	Hg ppm
1947[a]	14.1							
1972[a]	18.6							
1991[b]	18.2	1.6						
1998[c]	11.0	0.4	0.2	1.4	900	480	700	1.90
1999[d]	20.0	2.0	2.0					
1999[d]	16.4	1.4						
2007[e]	8.5	31.5	2.0	8.0				
2007[f]	4.1	0.3	1.0	0.5	1700	180	9900	0.68

ppm, *parts per million.*
[a]*Hoffmann (1992).*
[b]*Sum (1991).*
[c]*Lower Saxony Environmental Protection Agency, 1998 in Morf et al. (2007).*
[d]*Busselle et al. (1999).*
[e]*Ilyas et al. (2007).*
[f]*Morf et al. (2007).*

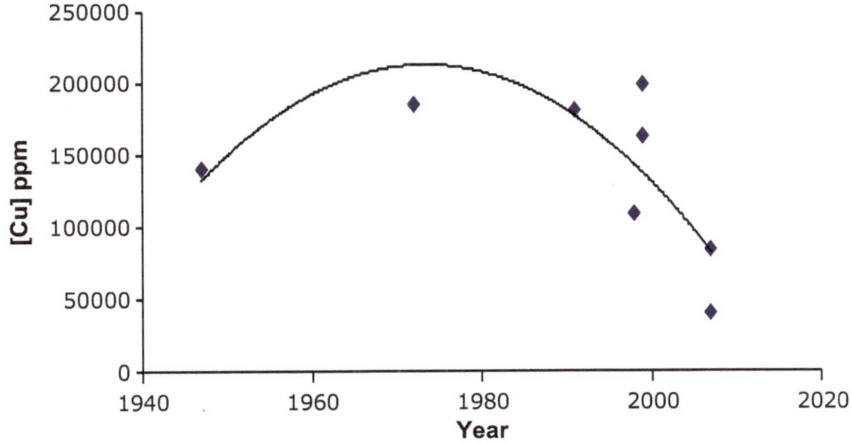

FIGURE 15.8

Copper in mixed WEEE versus year of sampling.

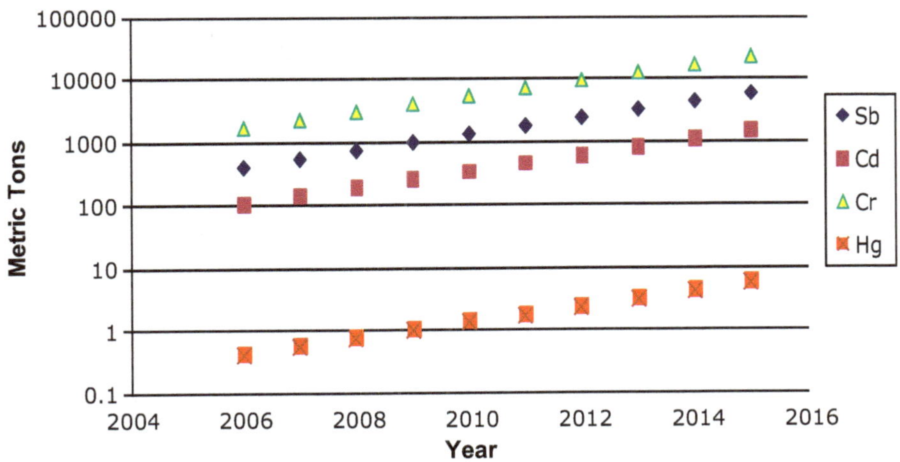

FIGURE 15.9

Predicted metals from WEEE (logarithmic scale).

in copper content. Predicted metric tons of Cu, Pb, Ni, Zn, Sb, Cd, Cr, and Hg in generated WEEE in 2015 is 418,310 $\pm$ 182,025, 284,926 $\pm$ 570,634, 59,874 $\pm$ 34,912, 151,848 $\pm$ 188,088, 5976 $\pm$ 2600, 1516 $\pm$ 975, 24,363 $\pm$ 29,904, and 5.9 $\pm$ 4.0, respectively.

For recycling industries in which small and medium enterprises cannot afford clean technologies, they can form an industry association. Some current industry associations are the lead/nonferrous recycler's association and the industry association, which have been able to provide multiple combined effluent treatment plants and hazardous landfill facilities. Other possibilities for industry associations

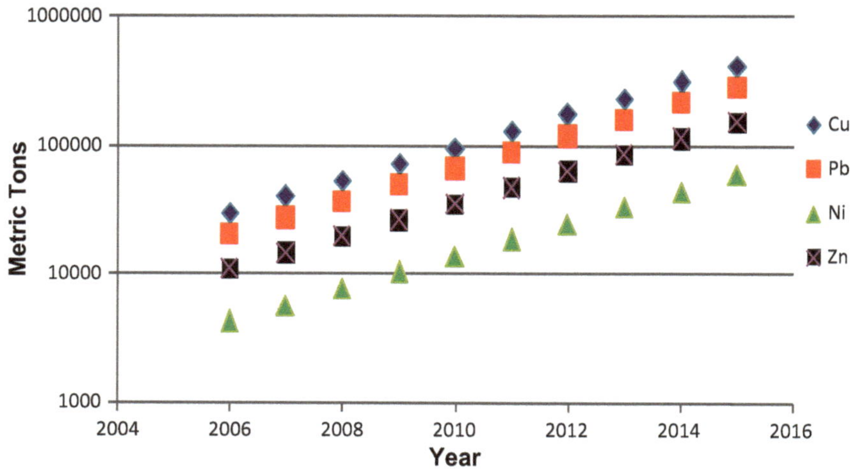

FIGURE 15.10

Predicted major metals from WEE (logarithmic scale).

are to research and provide pollution controls expertise, low-interest loans, and technology. One example of a group providing industrial expertise is the "Waste Minimization Circle" in India. India lacks recycling industrial parks which could allow for common pollution control facilities and easier oversight of operations, especially for small and medium enterprises for which regulating them is difficult. As of 2007, India had 115 common effluent treatment plants and 22 common treatment, storage, and disposal facilities (Kojima and Jain, 2008).

15.2.3 MINING, SMELTING, AND SECONDARY PRODUCTION

15.2.3.1 Copper

India consumes significantly more copper than it produces, which provides sufficient consumer discarded copper for a secondary market. The secondary market is controlled by price fluctuations in the primary copper market. Fig. 15.11 shows the production (primary and secondary) of copper from 2000 to 2010; there was an increase of almost 300% in primary production over this period. Copper secondary production seems to have a negative correlation with primary production, although data are limited. This could be because consumables with copper and copper ore may compete for limited production capacity at copper smelters, which means that whichever is more cost-effective replaces the other. This trend is also seen with zinc (Figs. 15.12 and 15.13; Table 15.6).

15.2.3.2 Zinc

India produces about 2–3% of global zinc, and India is rapidly increasing production. China and India are the major contributors to current increases in global zinc production, with global production increased by 4.5% from 2010 to 2015 (International Lead and Zinc Study Group, 2016). India currently contains the largest open pit zinc mine (Rampura Agucha Mine, Rajasthan) in the world. Increases in zinc primary production directly led to less zinc secondary production as can be seen in Fig. 15.12 and Table 15.7.

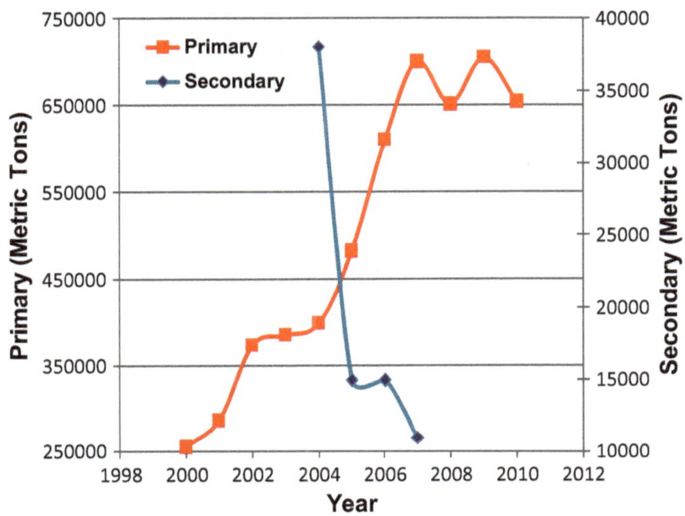

FIGURE 15.11

Copper primary and secondary production.

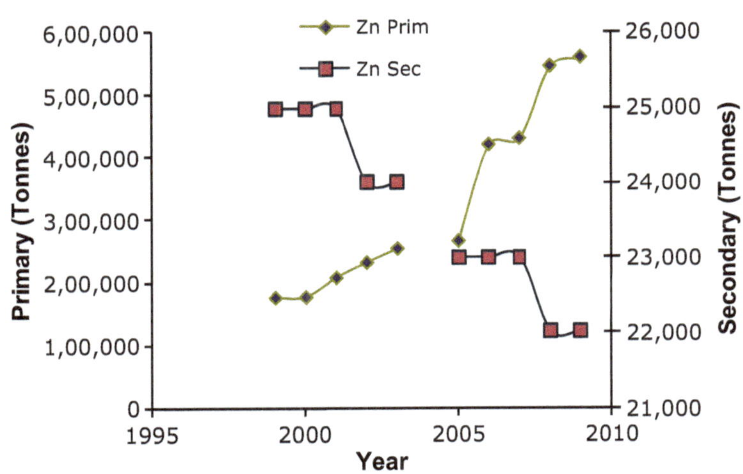

FIGURE 15.12

Zinc primary and secondary production across time.

15.2.3.3 Lead

Data on lead smelting in India are scarce. In 2006, lead primary production was 92,000 metric tons and secondary production was 60,000 metric tons (Kojima and Jain, 2008), whereas mined lead was 67,363 metric tons (BGS, 2012). In India, lead demand outstrips production, which may be why lead ore is imported into India for smelting (24,637 metric tons, or 27% calculated from data for 2006) as

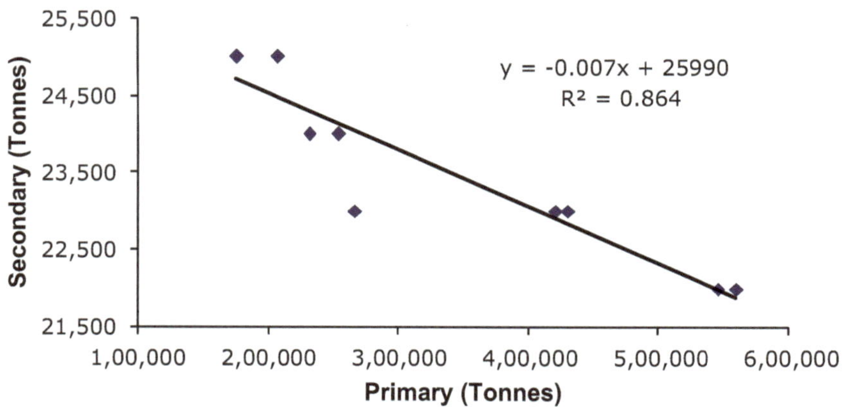

FIGURE 15.13

Zinc primary and secondary production, linear correlation.

Table 15.6 Toxic Metal Emissions From Primary Production

Year	As	Cd	Cu	Hg	Ni	Pb	Sb	Zn
2000	128	51	768	3	384	256	13	128
2010	327	131	1962	7	981	654	33	327
Cumulative[a]	2753	1101	16,517	55	8258	5506	275	2753

[a] *Cumulative emissions from 2000 to 2010.*

Table 15.7 Metal Emissions From Primary Zinc Smelting

Element	As	Cd	Cu	Hg	Pb	Sb	Zn
1999	9	35	9	4	88	3	1750
2009	28	112	28	11	280	8	5600
Cumulative[a]	176	705	176	71	1764	53	35,271

[a] *From 1999 to 2009. 2004 production data are missing; an average of 2003 and 2005 production data were used.*

well as why secondary production is so high. Hindustan Zinc Ltd. (HZL) (Udaipar) is the major producer of lead in India and has constructed a new zinc and lead smelter in Rajpura Dariba, which had a lead smelting capacity of 85,000 tons as 2010, and as of 2011 was constructing a second lead smelter with an expected production capacity of 100,000 tons (Dhanuka and Prakash, 2012). Toxic metal emissions from primary lead production in 2006 and HZL's primary lead production potential in 2012 are shown in Table 15.8 using emission factors from Pacyna and Pacyna (2001) in Table A1b.

Table 15.8 Emissions of Toxic Metals (metric tons) From Primary Lead Production in 2006 and 2012

Element	As	Cd	Cu	Hg	Ni	Pb	Sb	Zn
2006	0.92	1.38	0.92	0.28	0.92	184.00	6.90	7.36
2012[a]	1.85	2.78	1.85	0.56	1.85	370.00	13.88	14.80

[a]*Data only for Hindustan Zinc Ltd., based on predicted production capacity, not actual smelting.*

15.2.4 MAJOR BUILDING MATERIALS: CEMENT AND STEEL

15.2.4.1 Steel

Crude steel and pig iron are close in total annual production in India. The amount of crude steel produced globally is enormous because it is the cheapest metal suitable for construction. Metals of environmental concern are not emitted in high concentrations compared with other smelting operations, but the magnitude of production makes crude steel and pig iron production a major source of metals in the atmosphere. Crude steel production based on BGS (2012) data is shown in Fig. 15.14. Predicted emissions of various metals from steel production are shown in Table 15.9.

15.2.5 ARSENIC

Arsenic contamination in India is mainly from natural ores that leach arsenic into the groundwater. Endemic arsenic contamination is most severe in West Bengal, Bihar, Uttar Pradesh, Assam, and Chhattisgarh, with West Bengal being the most severe case (Sahni, 2011). The other effected states in the Bengal Delta Plains are Arunachal Pradesh, Nagaland, Manipur, Mizoram, Meghalaya, and

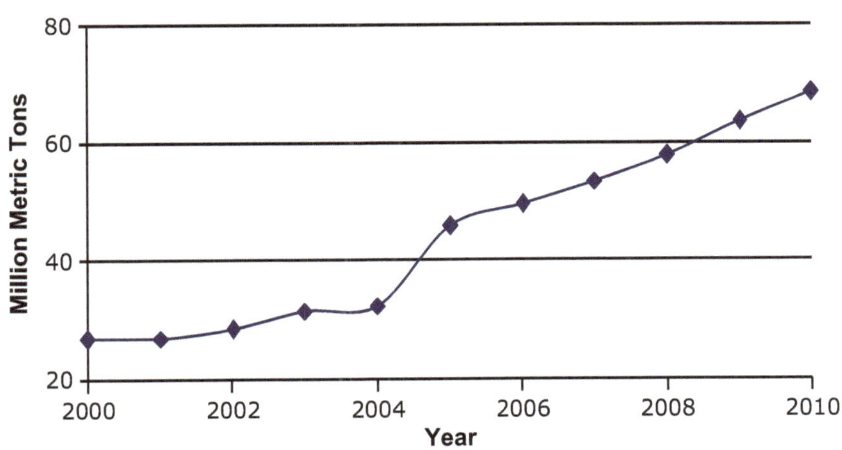

FIGURE 15.14

Crude steel production.

Table 15.9 Steel Production Emissions, 2000 and 2010, and Accumulation by 2010

Element	As	Cd	Cr	Cu	Hg	Ni	Pb	Sb	Zn
2000	13	3	108	5	1	1	40	0.3	81
2010	34	7	273	14	3	3	102	0.7	205
Cumulative[a]	243	49	1943	97	19	24	729	4.9	1457

[a]*Cumulative metal emissions from 2000 to 2010.*

Tripura. In West Bengal, preliminary sampling of 58,166 water samples in nine districts showed that 34% of samples had As levels greater than 50 μg L^{-1} (Chowdhury et al., 2000). At first, As contamination of groundwater was thought to be mainly confined to the Lower Ganga Plain, but soon research such as that reported in Chakraborti et al. (2003) showed toxic As levels in the Middle and Lower Ganga Plain. In Bihar, of 206 tube wells sampled, 56.8% had As levels greater than 50 μg L^{-1}, and 19.9% had levels greater than 300 μg L^{-1} (Chakraborti et al., 2003). In Chhattisgarh, the source of As has been determined to be leaching of As from altered realgar (As_4S_4), para realgar (As_4S_4), and tennantite ($Cu_{12}As_4S_{13}$) into the groundwater, with greater than 250 μg L^{-1} commonly observed (Shukla et al., 2010). In Chhattisgarh, nontube wells (10–25 M), tube wells (30–60 M), surface water, soil, and sediment As was 336 ± 158 μg L^{-1}, 187 ± 223 μg L^{-1}, 62 ± 52 μg L^{-1}, 61 ± 88 mg kg^{-1}, and 105 ± 130 mg kg^{-1}, respectively (Patel et al., 2005).

As(III) is the most toxic form, followed by As(V); all inorganic forms are generally more toxic than organic forms, and there is a decrease in toxicity with an increase in methylation for organic forms. In wells in Chattisgarh, As(V) was the dominant species at low levels of As whereas As(III) was the dominant species at high levels, with almost no detection of organic forms (Patel et al., 2005). In West Bengal, out of thousands of hair, nail, and urine samples, 77% of samples were at abnormal or toxic levels, and out of 29,035 participants 15.02% had skin lesions. As the authors put it, "We feel that we have only seen the tip of the iceberg" (Chowdhury et al., 2000). In Bihar, out of 390 adults, sampled 13% had skin lesions and 63% had arsenic typical neuropathy; out of 160 children, 6.3% had skin lesions (Chakraborti et al., 2003).

Because of the cost of large pipelines for a centralized solution to As contamination in groundwater, interim low-cost solutions have been implemented. Examples include harvesting rainwater by the Nongovernmental Organization Barefoot College of Rajasthan, use of pond sand filters and modified dug wells as implemented by Dhaka Community Hospital in Bangladesh, and shallow modified chlorinated wells installed by Project Well, West Bengal (Hira-Smith et al., 2007). In tube wells, the concentration of As was directly proportional to the well depth, presumably because the farther water travels down, the more As accumulates from leaching (Hira-Smith et al., 2007; Patel et al., 2005). Therefore, shallow wells serve as an effective interim solution. In 90% of 700 traditionally dug shallow wells sampled in Bangladesh and West Bengal, As levels were below 50 μg L^{-1}, with a total average of 15 μg L^{-1}. Furthermore, a study in West Bengal showed average As levels of 11 modified shallow wells (average depth of 18 feet) to be 29 μg L^{-1}, whereas the original 20 wells these wells replaced had average As levels of 473 μg L^{-1} (Hira-Smith et al., 2007). There is still an urgent need to invest in long-term solutions for most areas affected by endemic As contamination of water and food.

Although most arsenic enrichment in the water and soil of India is from natural sources, anthropogenic sources can have a role as well. As is often found in metal ore deposits, arsenides are mainly of Cu, Ni, and Fe. Ores such as these can release large amounts of As into the surrounding region. For example, in the state of Chhattisgarh there is a severe endemic of As contamination as well as in 20% of India's iron ore. Concentrations of As at 15 ppm have been found in the iron ore. Steel plants are a major source of As to already naturally contaminated areas. For example, the Bhilai Steel Plant produces about 5 million tons of crude steel per year, with an estimated 75 tons of As emitted per year from these operations (Patel et al., 2005).

Coal-fired power plants can be another source of arsenic, especially for workers within the plants. Contamination of As in the air in coal-fired power plant workplaces can be above the Occupational Safety and Health Administration (OSHA) standards of 10 $\mu g\ m^{-3}$ in the air. Whether workplaces contain dangerous levels of As depends on working conditions, the type of coal, and how the coal is burned. In a study of four Georgia power plants, out of 128 samples all arsenic concentrations were below 0.8 $\mu g\ m^{-3}$, far below OSHA standards (Bird et al., 2004). On the other hand, in a coal fire plant in the Slovak Republic, average airborne As levels were 48.3 $\mu g\ m^{-3}$, with boiler makers being exposed to 67.7 $\mu g\ m^{-3}$ and cleaners to 138.9 $\mu g\ m^{-3}$ (Yager et al., 1997). However, OSHA standards may be too high, because in this same power plant in the Slovak Republic, the researchers found that urine levels of arsenic were on average about three times less than corresponding urine levels in other studies in which workers were exposed to arsenic trioxide vapors, dust, and arsine gas, which means that arsenic in fly ash is much less bioavailable (Yager et al., 1997).

Other major sources of arsenic include pesticides and possibly emissions from China's production of white arsenic. White arsenic production in China as of 2010 was 25,000 metric tons, greater than the next 11 top producers combined (BGS, 2012).

15.2.6 MERCURY

Natural sources of Hg emission into the atmosphere constitute about 5207 tons $year^{-1}$ whereas anthropogenic sources contribute about 2320 tons $year^{-1}$ (Pirrone et al., 2010), which means that as of 2010 anthropogenic sources contribute about 31% of atmospheric Hg. The value should be much greater if you calculate Hg flux into the atmosphere, because oceans, vegetation, and soils can also be a major sink of Hg. Of natural sources, oceans contribute 52% of Hg emissions because Hg enters the atmosphere as the result of temperature or concentration gradients and photoreduction at the water surface, whereas terrestrial sources contribute 47%. Of natural sources, 13% of all emissions come from biomass burning. Of anthropogenic sources, the four major contributors are coal from stationary combustion (35%), gold production (17%), nonferrous metal production (13%), and cement production (10%) (Pirrone et al., 2010). In cement production, coal is often used to heat cement kilns and Hg comes from burning this coal. Although mining is a relatively low contributor, the exception is artisanal and small-scale gold mining, which often directly uses Hg to extract Au, and Hg is often also in the ores (Pacyna et al., 2010). If current regulations and technologies do not change by 2020, anthropogenic mercury emissions may increase by 25% from 2005 levels (Pacyna et al., 2010) (Fig. 15.15; Table 15.10).

India is the second largest polluter after China and preceding the United States. In 2005 China emitted about 800 tons of Hg and India emitted 180 tons of Hg according to Pacyna et al. (2010). Hg has a relatively long atmospheric lifetime and hence it can be transported over long distances, which

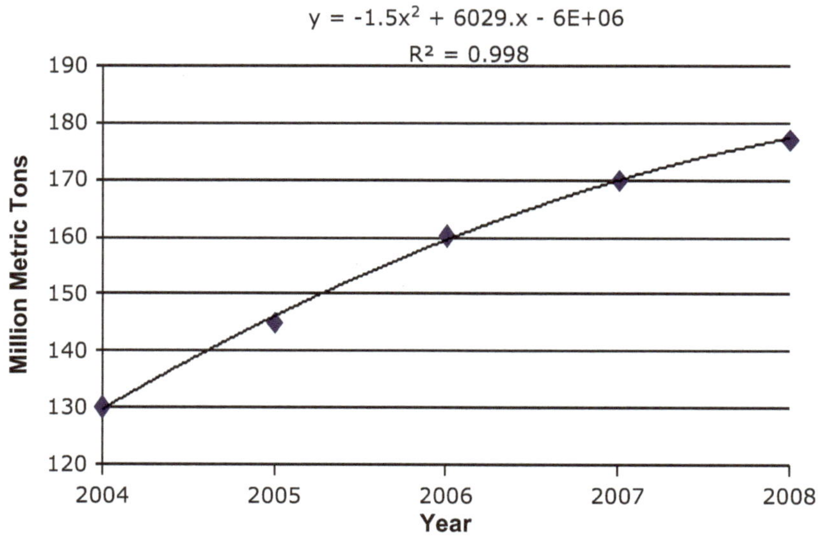

FIGURE 15.15

Hydraulic cement production.

Table 15.10 Metal Emissions From Hydraulic Cement Production

Element	As	Cd	Cr	Hg	Ni	Pb	Zn
2004	26	1.3	130	13	13	26	260
2008	35	1.8	177	18	18	35	354

makes China a major probable source of India's atmospheric mercury. Local Hg emissions in India come mainly from stationary combustion consisting of an estimated 52% of total emissions and waste disposal (especially incineration) consisting of 32% of total emissions (Pirrone et al., 2010). According to other estimates, fossil fuel combustion for heating and power contribute three times more than all other anthropogenic sources combined (Pacyna et al., 2010). Another potentially major source of Hg in India that has not been included in models because of the lack of data is Hg emissions from the burning of wood and wood waste, because these materials are major sources of heat in both the industrial and residential sectors (Pirrone et al., 2010). According to our estimates (Section 15.2.1.2), only about 3 metric tons of Hg was emitted in 2005 owing to the consumption of fuel wood, and hence fuel wood is not a major anthropogenic source of Hg.

Hg emissions have been reduced significantly in India as a result of the conversion of Hg-cell chlorine plants to membrane processes in chlor-alkali plants, which can be seen as major contributing factors to Hg emissions in India in 2004 versus 2000 (see Figs. 15.16 and 15.17). In 2008, the MoEF stated that 86% of chlor-alkali plants were converted and that by 2012 100% would be Hg-free (Mukherjee et al., 2009).

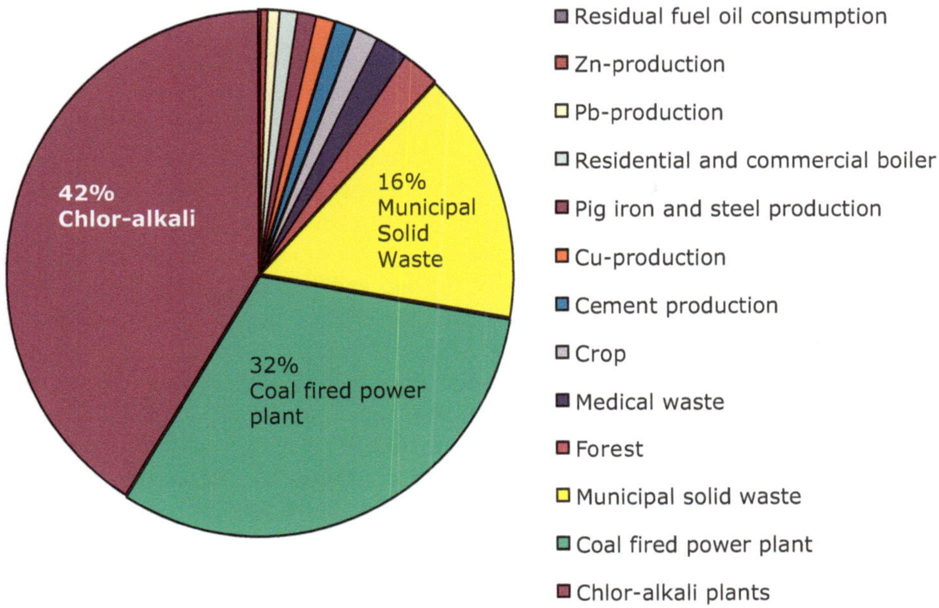

FIGURE 15.16

Major sources of mercury in 2000.

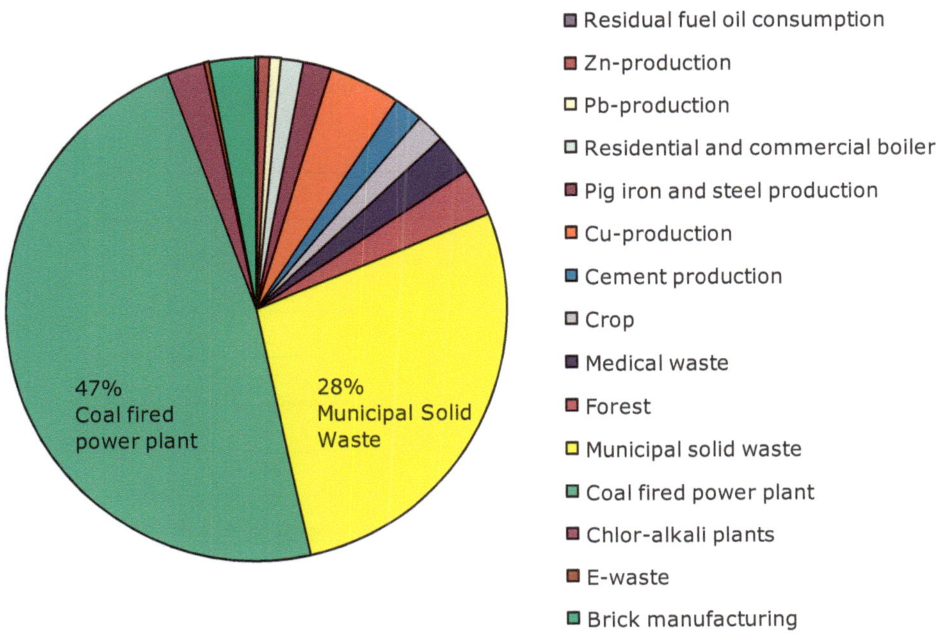

FIGURE 15.17

Major sources of mercury in 2004.

15.2.7 LEAD

One of the most severe instances of seemingly irreversible damage to life on earth is the anthropogenic increase in the fraction of bioavailable Pb. About 95% of Pb in the biosphere in 1995 was from anthropogenic sources, which was linked to a 100- to 500-fold increase in Pb content in humans since preindustrial levels (Smith and Flegal, 1995). Because there is no lower limit to Pb toxicity (US Department of Health and Human Services, 2007), any increase is detrimental, especially on a global scale. In the late 1970s, global emissions of Pb decreased significantly when leaded gasoline was phased out of use in the United States. The consumption of Pb in leaded gasoline exceeded all other categories combined in the United States before this decline, and the world consumption of Pb strongly reflected this decrease in use in the United States. After the 1980s, however, Pb consumption began to rise again from use in different sectors (Smith and Flegal, 1995).

Anthropogenic inputs of lead into the atmosphere drastically increased soon after as developing nations such as India and China advanced in the automobile sector and continued to use leaded gasoline. India began to phase out the use of leaded gasoline in 1994; by 1999, petroleum provided to the entire nation was mandated at 0.013 $g\ L^{-1}$ compared with 0.56 $g\ L^{-1}$ before 1994 (ENVIS, 1999). Data provided by ENVIS (1999) was used to calculate emission factors for the consumption of petroleum over various stages of the phase-out of lead campaign (for more information on emission factors, see Table A1d). Emission factors along with motor fuel consumption in India (data from EIA (2012)) were used to estimate emissions of lead per year and cumulative emissions from 1986 to 2008 (Figs. 15.18 and 15.19, respectively). These figures show that emissions of lead into the atmosphere each year decreased by over an order of magnitude since the phase-out of leaded gasoline.

This has had beneficial impacts on the environment and health. From November 2003 to 2004, there was a steady decrease in Pb concentration in particulate matter ranging from 2.5 to 10 micrometers (PM10) which may be the result of the increased use of unleaded gasoline (Karar et al., 2006). A study on the phase-out of lead showed that estimated lead concentrations in Ganga River water, Lucknow

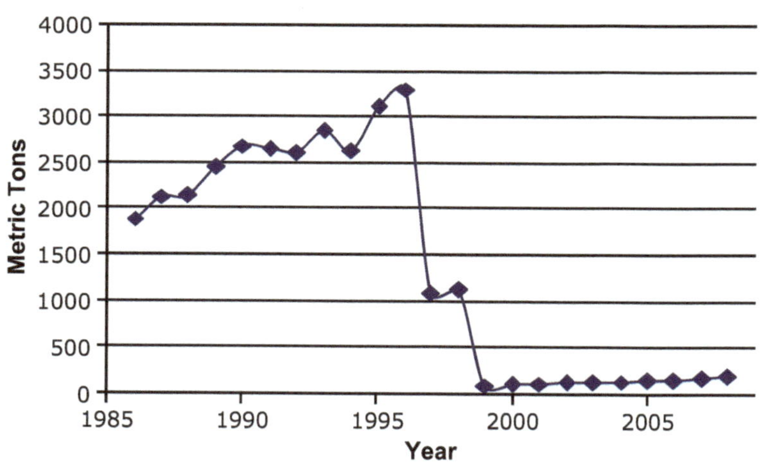

FIGURE 15.18

Emissions of lead from petroleum.

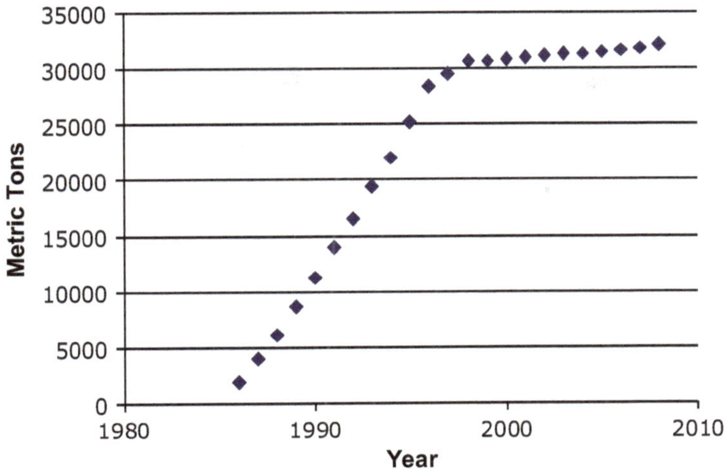

FIGURE 15.19

Cumulative biosphere lead from petroleum.

urban air, and urban children's blood decreased by 82.8%, 87.5%, and 33.5%, respectively, from samples during the leaded-petrol phase (before 1996) and the unleaded petrol phase (after 1999) (Singh and Singh, 2006). However, the cumulative effects of leaded gasoline in the biosphere, especially in urban areas, are most likely still a health hazard. A study in Sweden correlating the distance from roads with Pb blood levels in children from 1978 to 2007 using geographic information systems showed that blood levels of Pb can still be higher owing to leaded gasoline even more than a decade after leaded gasoline was phased out, although major impacts on blood levels were reduced almost immediately after the phase-out. Twenty-nine years after lead was phased out, children's blood Pb level continued to decline at about 5% per year (Stroh et al., 2009).

There is currently a drastic increase in petroleum consumption each year in India that generally follows the global trend (Fig. 15.20). The consumption of petroleum in 2010 was more than 20% of proven crude oil reserves in India, which supports the fact that India may shift to even a higher percentage of coal consumption as part of the entire energy demand. On the other hand, India continues to determine new sources of domestic crude oil production, which means that proven oil reserves fluctuate and have actually grown in the past 15 years. Therefore, it is difficult to predict when economically accessible oil reserves will diminish in India, and hence Pb emissions from this source will further diminish. Diminished oil reserves could amplify metal emissions into the atmosphere because readily available large coal reserves in India may make coal partly replace oil consumption.

Urban centers are not always the major Pb-contaminated areas. Comparing blood levels in suburban fishing and farming villages with a village with a large informal dumping ground in South India and the city of Chennai, Agusa et al. (2006) found that blood levels were highest by the dumping sight, with an average of 10.5 $\mu g\ dL^{-1}$, and that participants in Chennai had lower blood levels than did participants in the villages of Chidambaram, Palaverkadu, and Parangipettai. A major source of Pb in a dumping site is often car batteries and WEEE, and these dangers are magnified when these items are recycled, especially on a small scale. A major concentrated source of Pb in the environment is

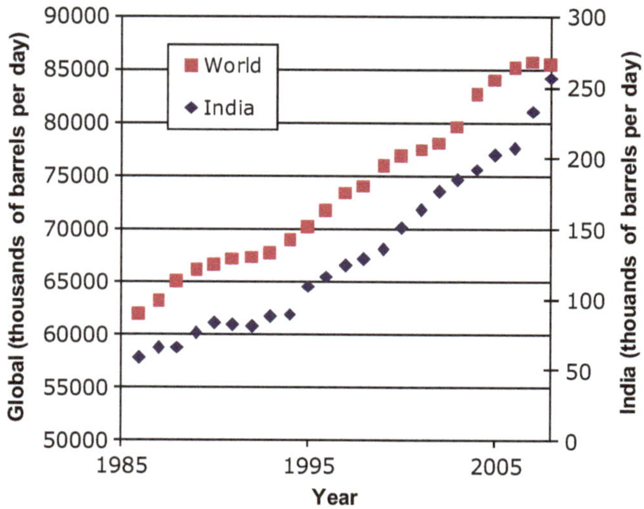

FIGURE 15.20

Petroleum consumption in India and the world.

backyard, small, and medium recycling enterprises for lead batteries. Seventy-five percent of Pb consumption involves these batteries and most recycled Pb (60,000 tons in 2007) is from Pb batteries. Lead batteries consist of 70–80% Pb and PbO_2 by weight; after inefficient recycling practices, that often leaves 30–40% Pb in the waste, or 300,000 to 400,000 ppm. In addition, a rotary kiln furnace can lead to concentrated Pb fumes (Kojima and Jain, 2008). Another concentrated source of Pb is from WEEE recycling. For example, in and near WEEE plants in New Delhi, 3560–6450 ppm Pb was found in bottom ash, 150–8815 ppm Pb in separation workshop dust, 31–1300 ppm in street dust, and 20.5 mg L^{-1} in wastewater from leaching. In China, Pb concentrations in river sediment near a shredding recycling plant reached 44,300 ppm (Greenpeace, 2005).

Lead smelting can also lead to significant environmental contamination with Pb. One reason is that Pb emissions from smelting operations are high (see Section 1.3.3 and Tables 15.8 and A1b). As with Pb from the environment, because of leaded petrol, Pb from smelting operations may continue to be a hazard once operations have stopped or emissions are reduced. By correlating children's distances from a Pb smelter (calculations including time spent at school versus home and those distances from the smelter) with children's Pb levels, Stroh et al. (2009) showed that children were exposed to significant levels of Pb from the smelter even after a reduction in emissions. The researchers proposed that this is potentially due to soil contamination in the area.

15.3 HUMAN EXPOSURE TO METALS THROUGH FOOD AND COSMETICS

15.3.1 SEAFOOD

Contamination of waterways by toxic metals is an extreme hazard to the population of India, which relies on marine systems for food and drinking water. In the Ganges the concentration of the potentially toxic elements Co, Ni, Cu, and Zn in sediments is below the global average, whereas Cr is far above, at 147 ppm versus 100 ppm. Higher levels of the lithophile element Cr potentially result from leaching and corrosion of basic rocks in the crystalline zone of the Himalayan region. Cr and Ni (46 ppm) are on average above permissible exposure limit (PEL) levels, posing significant harm to the ecosystem, whereas Cu (55 ppm) may pose harm because it is above the lowest effect level (LEL). The general concentration of Co, Ni, Cu, Cr, Zn, Pb, and Cd is unpolluted or at background levels according to geo-accumulation indices, although near urban centers metal concentrations can be enriched compared with natural levels (Singh et al., 2003).

The Gomti River is a major tributary in the Ganga River system and is one of the most polluted rivers in India. Between 1994 and 1998, Cd, Cu, and Ni levels along various sampling locations of the river were generally above the LEL for ecosystems (0.6, 16, and 16 ppm, respectively), which means that they were above the threshold at which no negative effect on the ecosystem could be observed. For Cd and Ni, four samples were above 3.53 and 36 ppm, respectively, posing significant harm to the ecosystem. At times, in portions of the river Cr, Pb, and Zn were above LEL (Singh et al., 2005). The Ganges feeds into the Bay of Bengal, where some sediments and metals are deposited. Selvaraj et al. (2004) found that sediments in the Bay of Bengal are naturally enriched with Zn and enriched with Cd and Pb from anthropogenic sources, and potentially Co, Cr, and Ni as well.

Often water systems serve as a dumping ground for industrial, commercial, and public waste, and hence can become severely contaminated. Industrial areas can have a dangerous mixture of toxic metal contamination. The Central Pollution Control Board, based in New Delhi, has identified major contaminated sites of concern including the Patencheru industrial area in Andhra Pradesh. Concentrations of Cd, Cr, Cu, and Zn in soils surrounding this area have been measured at 200, 240, 500, and 600 ppm, respectively, which means that soils should be taken out of agricultural production, and leaching into aquifers poses a hazard to the major drinking water supply (Krishna and Govil, 2001). The Nakkavagu stream crosses the Patencheru industrial area and is a tributary to the Manjira, which provides drinking water for a major portion of Hyderabad and Secunderabad city residences and the districts of Medak and Nizamabad. Another example is Lake Hussein Sagar in Hyderabad, which serves as dumping ground for over 400 industrial units; it was found that Cd, Pb, Ni, and Zn levels were on average 0.0046, 0.421, 0.024, and 0.1813 ppm, respectively, in the lake. Ground and surface water is used by over half a million residents. Groundwater closer to the lake was more contaminated than groundwater that was farther away (Srikanth et al., 1993).

Further research should be done on the bioaccumulation of these metals in aquatic organisms and plants living within or along polluted river systems such as these, to determine health consequences for human consumers. There could be significant impacts on coastal fisheries of the Bay of Bengal, where the Ganges river has deposited 109,000, 35,000, and 41,000 tons of Cr, Ni, and Cu, respectively, per year as of 2001 (Singh et al., 2003). India is ranked third in fisheries and second in aquaculture globally, with about seven million metric tons produced in 2007; it makes up 4.7% of the agricultural GDP (NFDB, 2007). Nutrient elements such as Fe, Zn, Cu, and Mn are higher in fish species whereas nonessential elements such as Cd, Hg, and As are often lower (Kumar et al., 2012). The levels of selected nonessential elements (Table 15.11) and nutrient elements (Table 15.12) found in seafood in various reports in the literature were compiled and compared with various global standards.

Table 15.11 Nonessential Trace Elements As, Cd, Hg, and Pb in Seafood (From Compilation of Literature)

Location	Group	As (ppm)	Cd (ppm)	Hg (ppm)	Pb (ppm)
NE and SE_1	Cephalopod		3.759 ± 4.856**	0.040 ± 0.034	0.035 ± 0.029
NE_2 (f)	Fish		0.62–1.20**		12.40–19.96**
SW_3	Mollusks	Nd–0.69	Nd–0.98	Nd–0.33	Nd–0.98*
SW_3	Cephalopod	Nd	Nd–0.47	Nd–0.6*	0.07–0.76*
SW_3	Crustaceans	Nd–2.98*	Nd–0.12	Nd–0.47	Nd–0.94
SW_3	Fish	Nd–4.14*	Nd–1.32*	Nd–2.31*	Nd–1.32*
SE_4	Fish		0.18–0.54**		
SE_5	Fish		0.008–0.114		0.062–1.569**
NE_6	Crustaceans		0.2–0.4		0.8–1.9*
NE_7 (f)	Fish		0.184–0.968**		0.005–0.017
NE_8 (f)	Fish	0.02–2.37*	0.01–1.10*	0.05-1.60**	
NE_9	Fish				1.9–2.5**
Fish limit		1_{17}	1_{10}, 0.05_{12},0.2_{17}	0.5_{13}, 0.5_{14}	1.5_{17}, $0.5–6_{11}$, 0.5_{14}
Crustacea limit		76_{15},1_{17}	3_{15}, 0.5_{16}		1.5_{15}
Mollusks	Limit	1_{17}	2_{17}		2.5_{17}

NE, *northeast;* ppm, *parts per million;* SE, *southeast;* Nd, *not detected;* (f), *from a freshwater location.*

Table 15.12 Nutrient Elements in Seafood (Cr, Cu, Ni, and Zn) (From Compilation of Literature)

Location	Group	Cr	Cu	Ni	Zn
NE and SE_1	Cephalopod		10.99 ± 8.60	NA	16.54 ± 2.32
NE_2 (f)	Fish	0–3.89**	16.22–47.97**	2.20–3.69**	12.13–44.74*
SW_3	Mollusks	0.18–3.65**	1.17–24.1	Nd–0.89	3.8–165*
SW_3	Cephalopod	Nd–1.33**	0.65–10.7	Nd–0.94	3.99–16.3
SW_3	Crustaceans	0.1–1.87	1.88–18.9	Nd–0.71	5.14–49.6*
SW_3	Fish	Nd–1.87*	0.13–14.7*	Nd–0.69	0.66–39.2
SE_4	Fish	0.65–0.92	0.12–0.31	0.38–1.54**	14.1–33.5
SE_5	Fish	0.415–1.562**			0.103–0.807
NE_6	Crustaceans	1.7–3.1	12.2–35.7*	2.9–7	17.6–64.5*
NE_7(f)	Fish	0.335–3.687**	0.345–5.421		4.496–12.213
NE_8 (f)	Fish		0.5–28.2*		3.0–99.1*
NE_9	Fish	3.2–13.7**	13.5–15**		27.9–46.9*
Safety limit fish		50_{10}, 1_{11}	30_{10}, 10_{17}, 20_{14}	$0.5–1_{10}$	100_{10}, 40_{13}
Crustacea limit		12_{15}		70_{15}	
Mollusks			30_{17}		$100,0_{17}$

NE, *northeast;* SE, *southeast;* SW, *southwest;* Nd, *not detected.*
FAO, 2011. Fish and Fishery Products Hazards and Controls Guide > Appendix 5-FDA & EPA Safety Levels in Regulations and Guidance, third ed.

Reference Number of Species (if given)	Location	Sample Size (if given)
1. Panutrakul et al. (2008)	Bay of Bengal, India	1
2. De et al. (2010)	Hooghly River, Bay of Bengal	62
3. Sivaperumal et al. (2007)	Kochi area	20
4. Raja et al. (2009)	Parangipettai coast	60
5. Lakshmanan et al. (2009)	Parangipettai coast	5
6. Hossain and Khan (2001)	Bangladesh coast	60
7. Bhattacharya et al. (2008)	Upper stretch of Gangetic West Bengal	6
8. Kumar et al. (2012)	Landing station of Digha	54
9. Aich et al. (2011)	East Calcutta wetland in West Bengal	27
10. WHO, (1989), in Mokhtar et al. (2009)		
11. maximum limits for fish, in Mokhtar et al. (2009)		
12. Figueroa (2008)		
13. Australia and New Zealand Food Authority Amendment No. 53, 2004		
14. Ministry of Public Health, Thailand, 1986		
15. USFDA (1993)		
16. EC, 2001, in Raja et al. (2009)		
17. FAO (1983) in Raja et al. (2009)		
18. FAO 2011		

***Average at dangerous levels or levels above minimum standard in more than 50%;*
**Some samples above minimum standards (fish standards used if lacking for group; cephalopod is a mollusk), (f): From a freshwater location.*

Of the nonessential elements (Table 15.11), As and Hg were found only in a minor portion of samples to be above standards of 1 and 0.5 ppm, respectively, although only two studies were compiled for As. In a study by Aich et al. (2011), Hg levels in carp from ponds fed by municipal and industrial waste sewage had the most specimens above the 0.5-ppm threshold. Cadmium and lead, on the other hand, were commonly above 0.2 ppm in most fish specimens, at least in the northeast and southeast coastal waters (the Bay of Bengal). As mentioned previously, the Bay of Bengal has been shown to have enriched Cd and Pb levels from anthropogenic inputs (Selvaraj et al., 2004). Therefore, more work should be done to monitor trace metal sources, and when they are anthropogenic, to mitigate the problem to the Bay of Bengal, because there can be major health and economic losses owing to Pb and Cd contamination.

Cadmium levels above 0.05 ppm or even lower are also considered unsafe in fish by standards in multiple countries including Norway, European communities (for some species), New Zealand, and Hong Kong (in this case, 0.002 ppm) (Figueroa, 2008). Although these low levels are found naturally in fish at many shores owing to different geological conditions, regardless of whether there will be health hazards, these standards have severe consequences for Indian fisheries. All of the studies in Table 15.11 show Cd levels above 0.05 ppm in most or all fish samples. The European Rapid Alert System for Food and Feed is set up to be able to respond rapidly to food contaminants above the given threshold, with 4.7% of 927 alerts in 2005 resulting from toxic metals and most metal contamination alerts being Cd in seafood (Figueroa, 2008). Therefore, Indian fish exporters, especially from the Bay of Bengal, are commonly at risk of losing shipping, handling, and export value when shipping to countries with strict standards. This limits India's markets, especially if other countries in Asia follow suit. It is important to take into account species, class, and genus when looking at trace element distribution. For example, purpleback squid (*Sthenoteuthis oualaniensis*) from the Bay of Bengal has concentrated Cd levels in edible tissues almost an order of magnitude above most other samples species shown in Table 15.11. The purpleback squid's concentration of Cd is in the same range for both sample locations in the northeast and southeast of India, which shows that location is not a major factor in Cd levels (Panutrakul et al., 2008).

Of the nutrient elements, zinc is not of concern in any of the literature; only a few fish if any are above the standard of 40 ppm. Because zinc deficiency is often a problem in Indian citizens and Zn can help children with diarrhea (Sazawal et al., 1995), higher values may often be beneficial. In most of the literature, Cu was within the limits of 10 ppm for fish and 30 ppm for mollusks, except in the Hooghly River (De et al., 2010), which is part of one of the most polluted estuaries in the world, and in carp ponds fed by industrial and municipal waste (Aich et al., 2011). Concentrations of Ni were also commonly above standards for the Hooghly River, as well as for the Parangipettai Coast (Raja et al., 2009), which is an important fishing and trading area. Chromium contamination of fish was the most severe, with four out of the six research papers where chromium was measured in fish showing levels above 1 ppm. In part, natural corrosion of basic rocks from the Himalayan region naturally leads to enriched Cr in the Ganges, that is finally deposited in the Bay of Bengal (Singh et al., 2003). The only studies not finding Cr above 1 ppm in most samples was a study done in the southwest and far southeast (Parangipettai Coast), both of which are a fair distance from the Ganges River delta.

A mixture of anthropogenic and natural trace metal elevations in coastal and freshwater bodies has led to excessive trace metal contamination in most seafood. Only a study in the southwest Kochi area (Sivaperumal et al., 2007) found most fish to be within safe limits for all metal contaminants. The two most common trace metal pollutants above standards are Cd and Cr, even when 0.2 ppm is used as an upper limit for Cd. Excessive Cr levels in the Bay of Bengal could be mainly the result of natural sources, whereas enriched Cd levels have been shown to be at least partly due to anthropogenic sources. Freshwater and coastal areas where industrial and municipal waste drains to have sometimes been shown to be highly productive areas for fisheries (De et al., 2010), but research shows that monitoring of these sites for toxic metals should be mandatory.

15.3.2 CONTAMINATION OF MAJOR FOOD SOURCES IN INDIA

Major food products of India in 2005, by production in dollars, are rice, buffalo milk, wheat, cows milk, and vegetables, with about 27, 25, 11, 10, and 7 billion dollars' worth of production, respectively. For 2005, total production was 129, 51, 72, 39, and 35 million metric tons. In 2005, India was the largest producer of buffalo milk and the second largest producer of rice, wheat, cow's milk, and vegetables by country (FAO, 2012), which made India an important factor in food security. For most people, the greatest exposure pathway to metals in humans is through food (Sharma et al., 2008).

Food crops are an important staple of the Indian diet, as well as crops grown for feedstock to raise livestock. Most metals enter vegetables through irrigation using sewage or via industrial effluents.

Treated wastewater in areas such as Titagarh, West Bengal, often still have metals significantly above recommended standards. Even those below the safety limit can pose risks if they are bioconcentrated in food crops. This bioconcentration is exemplified in Titagarh, where lead, copper, and nickel were above safe limits in treated wastewater whereas zinc, cadmium, and chromium were not. On the contrary, only cadmium and nickel were above safe standards in wastewater-irrigated agricultural soil. Finally of the nine vegetables tested that were growing on these wastewater-irrigated soils, Pb, Zn, Cd, Cr, and Ni were far above safe limits for all nine whereas copper was within safe limits for most species (Gupta et al., 2008). Sharma et al. (2007) found that in Varanasi, using treated or untreated wastewater elevated metal content in *Beta Vulgaris* and that even when only Cd was above Indian irrigation standards in the wastewater, Cd, Pb, and Ni were above Indian permissible levels in *B. vulgaris*, as was Cr in Lohta.

Rice is a major staple of India and is grown in rice paddies, which makes water contamination a major concern. In a study done in Chattisgarh in areas with endemic As contamination, rice grains from 10 species were found to have 0.274 ± 0.15, 0.5, and 71.2 ± 80.6 ppm As in grains, husks, and growing soils, respectively, and As concentration in rice was not explained by the As concentration in soil (Patel et al., 2005). Factors such as pH, oxidation reduction potential, effective cation exchange capacity, and soil organic matter can have a much larger role in determining metal content in plants than the metal concentration in the soil. Sometimes metal concentrations in the soil are insignificant variables in models explaining plant metal content (Koelmel and Amarasiriwardena, unpublished data). Therefore, future regulations regarding trace metals in fertilizer and water for crops must take into account varied soil conditions.

Most metal contamination in food can come from atmospheric deposition in the marketplace. Sharma et al. (2008) found that 23–68% of the concentration of Cu, Cd, Pb, and Zn in three different vegetables samples from urban markets in Varanasi was removed by washing, compared with 10% at the site of production; this showed that deposition of metals on produce in urban markets can be as significant as, or even more significant than, metal uptake by the plant before harvest. Cu and Cd levels in vegetables being sold at market in Varanasi are above safe levels, whereas Zn and Pb levels are not (Sharma et al., 2008). Not only will the amount of metals accumulated depend on the crop species (Gupta et al., 2008) but deposition of metals onto vegetables can be based on the species because of varying surface area properties and the way in which the produce is packed and displayed (Sharma et al., 2008).

Much research has been done on the accumulation of metals, especially Cd and Pb, in buffalo and cow's milk in India (Rajaganapathy et al., 2011), which shows that toxic metals in soil, water, fodder, and air accumulate in these livestock and their milk, with levels often above acceptable standards. National standards are at a maximum of 0.02 mg mL^{-1}. Examples include findings of 1.13 ppm Pb in buffalo milk from polluted sites (industrial or urban) in Delhi and 0.24 ppm in rural areas (Dey et al., 1996), and findings are 0.75 ppm Pb and 0.05 ppm Cd in buffalo milk from buffalos near a Pb and Zn smelter (Dwivedi et al., 2001).

Sahayaraj and Ayyadurai (2009) found elevated levels of lead in buffalo milk from buffalos raised by River Cooum, which passes through the city of Chennai and feeds into the Bay of Bengal. In 216 samples of cattle feed, 216 samples of buffalo milk, 83 samples of well water, and 124 samples of bore well water, Pb average levels were 8.26 mg kg^{-1}, 0.06 mg mL^{-1}, 0.37 μg mL^{-1}, and 0.27 μg mL^{-1}, respectively. A highly significant model predicting the milk content of Pb was developed as: $\text{MilkPb} = 31.03 + 77.38\ (\text{WaterPb}) + 37.77\ (\text{FeedPb})$, with the water and feed lead content explaining about 9.2% of the milk lead levels ($R^2 = 0.092$). The major contributors are expected to be from air, and other possible contributors are from posters the buffalo were observed to lick. Feed material was grown in sewage-fertilized soils. Another major finding was that a government operation of buffalo farming was found to have only 0.013 mg mL^{-1}, presumably because they were using corporation water with 0.008 μg mL^{-1} Pb. Hg was also found to be increased in buffalo milk owing to industrial contamination in Haryana, with Hg ranging from 0 to 2.78 ppb in an industrial zone and 0 to 1.3 ppb in a nonindustrial zone (Roy, 2009).

South India produces about 230 million kg of tea annually, with about 50% exported. Teas were found to have 24.1 ± 2.3, 4.7 ± 1.3, 2.5 ± 1.0, 0.1 ± 0.1, and 0.8 ± 0.3 ppm Cu, Cr, Ni, Cd, and Pb, respectively (Seenivasan et al., 2008). Trace element standards for tea are scarce and vary among countries; south India Cu levels are far below all standards whereas Pb is far above the upper limits of Pb in teas of 1 ppb in Iran and 5 ppb in China, but below the 10 ppm standards of India (standards from Karimi et al., 2008).

Chocolates and other sweets are important, especially because of high consumption by children. Nickel is used as a catalyst to hydrogenate unsaturated fats. In this way, it is used to harden chocolates and hence can be found in high concentrations in chocolate and cocoa butter. Of 69 chocolates analyzed from Mumbai, those that were cocoa based had the highest concentration of Pb, Ni, and Cd, with average levels of Pb and Cd above recommended levels according to the Polish National Standard. Most metal contamination was from raw materials (Dahiya et al., 2005).

Although food products have been studied extensively, there is less regulation and research on cosmetics. Herbal cosmetics for application on skin are not regulated by the WHO, although Health Canada has set up regulatory limits. In an analysis of 20 herbal cosmetics sold in Indian markets, it was found that all of them were significantly above WHO Cd standards for herbal preparations (>0.3 ppm), ranging from 0.6 to 1.9 ppm. Two samples were found to have Hg above WHO regulatory limits (>1 ppm) and six were found to have Pb above the regulatory limit (>10 ppm); four were above Health Canada limits (>3 ppm). Therefore, regulation is essential, especially because the health effects of toxic metals on skin include skin cancer, skin allergies, and osteoporosis (Sukender et al., 2012).

15.4 METAL RECOVERY TECHNIQUES AND PROCESSES FOR THE PREVENTION OF POLLUTION

Rapid increase in population, urbanization, and high standards of living increase the demand for metals. Metals have wide applications such as mining, metallurgical operations, electroplating, leather tanning, anticorrosive agents, vehicle emissions, lead-acid batteries, paints, treated woods, smelting of metalliferous material, surface-finishing industries, and construction (Fig. 15.21). Apart from their applications these metals have a detrimental effect on human health, which creates the need for waste treatment before disposal into the ecosystem. Moreover, the reserves of high-grade metal ores are diminishing; hence, in addition to treatment there is a need to recover metals from large volumes of metal-laden wastewaters (Galanis et al., 2009; Wu et al., 2012). There are various ways to recover metal ions, such as physical, chemical, and biological methods. Advanced biological methods such as bioelectrochemical systems (BESs) have been initiated for the recovery of metals (Fig. 15.22) (Lefebvre et al., 2012; Nancharaiah et al., 2015; Lu et al., 2015).

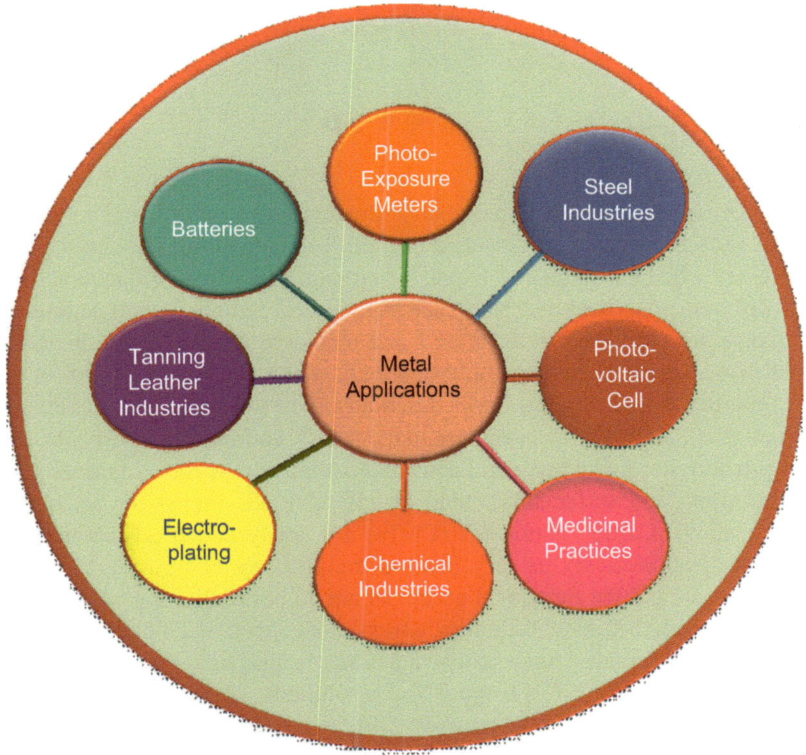

FIGURE 15.21

Application of different types of metal.

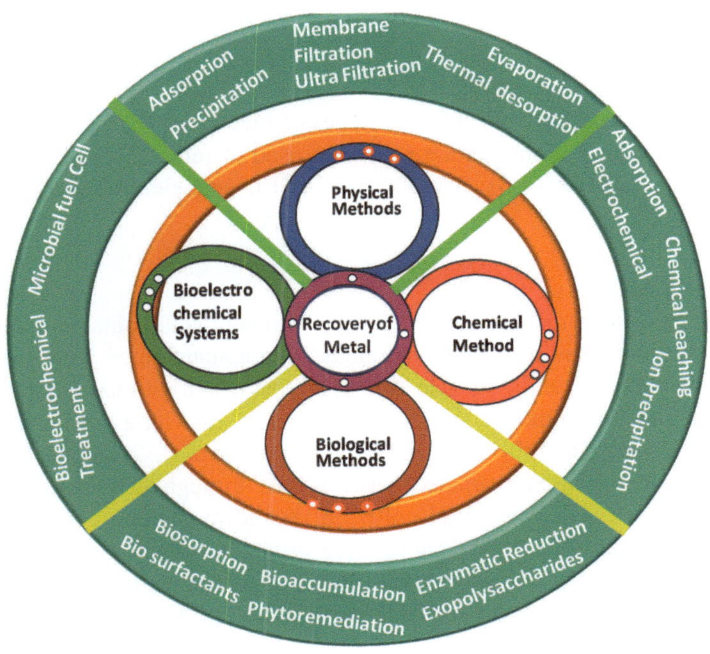

FIGURE 15.22

Detailed description of methods used to recover metals.

15.4.1 PHYSICAL METHODS

15.4.1.1 Adsorption

Adsorption is a process in which substances or pollutants (adsorbates) bind to the surface of the adsorbing material (adsorbent) by a physical and/or chemical interaction. The surface properties of the adsorbent provide sites of binding heavy metals or contaminants from an aqueous solution (Wang et al., 2008a,b). Different adsorbent materials such as activated carbon, carbon nanotubes (CNTs) and combined absorbents are used for metal removal and recovery. Activated carbon–based absorbents have a porous surface area which has large micropore and mesopore volumes used for the recovery of metals. Studies reported that eucalyptus bark–based adsorbents showed good removal of Cu+2 and Pb+2 (Kongsuwan et al., 2009). With a decrease in the natural sources for adsorbents, there is a search for an alternative material as an additive and activated carbon composite. Additives such as alginate, tannic acid, magnesium, and surfactants, could be effective adsorbents for heavy metal recovery (Kurniawan et al., 2005). Carbon nanotubes (CNTs) are newly added adsorbents which have the capability of heavy metal recovery as they possess the property of forming electrostatic linkages between the heavy metals and the functional groups of nanotubes (Salipira et al., 2008). Using CNTs, recovery of heavy metals such as lead, cadmium, chromium, copper, and nickel was reported (Li et al., 2004). Although CNTs inherently possess low adsorbent properties, they can be made active by oxidizing treatment using HNO_3, NaClO, and $KMnO_4$ solutions; among them, acidified, multiwalled CNTs showed good performance for lead recovery (Srivastava, 2013). Widespread use of CNTs will eventually pose a risk to humans because they are discharged into the water environment. To resolve this problem, an environmental friendly adsorbent, CNTs immobilized by calcium alginate (CNTs/CA), was prepared and tested to remove copper (Li et al., 2004). Considering the specificity, economical value, and the degree of adsorption for the cumulative removal and recovery of metals, combined adsorbents (carbon, charcoal, CNTs, or activated carbon) are designed.

15.4.1.2 Membrane Filtration Techniques

Membrane filtration technologies with different types of membranes have high efficiency, are easy to operate, and save space. The membrane processes used to remove metals from wastewater are ultrafiltration, reverse osmosis (RO), nanofiltration (NF), and electrodialysis (ED) (Kurniawan et al., 2006).

15.4.1.2.1 Ultrafiltration

Ultrafiltration uses permeable membranes to remove inorganic contaminants such as heavy metals along with suspended solids and organic compounds from contaminated waste sources (Juang and Shiau, 2000). Based on the size of the particle, different types of ultrafiltration membranes can be used to separate heavy metals, macromolecules, and suspended solids with a pore size varying from 5 to 20 nm and a molecular weight of 100–100,000 Da (Vigneswaran et al., 2004). The operational process involves passage of the metal-contaminated waste source through the membranes; the filtrate is collected separately. The residual material collected on the membrane contains recovered heavy metals. The main advantages of this process are the low driving force and of smaller space requirement owing to its high packing density. However, because of its high cost and the problems associated with fouling, its application is limited. Metals such as copper, cadmium, and nickel can be remediated using this technique (Saffaj et al., 2004).

15.4.1.2.2 Reverse Osmosis

Reverse Osmosis is used to remove heavy metals with a pressure-driven membrane separation process with the function of a concentration gradient. The contaminant water is pumped across the membrane with pressure, which aids in the removal of heavy metals. A semipermeable membrane allows fluids to pass through and filters out contaminated heavy metal ions, enabling metal recovery. Membranes with a pore size of 10−4 μm can filter all types of metal contamination in addition to other contaminants (Bohdziewicz et al., 1999). However, high pressure needs high-energy inputs and most wastewater is drained out as reject water, which makes the process expensive. Studies have reported that Cu^{2+} and Ni^{2+} ions are separated using an Na^2-ethylenediamenetetraacetic acid−RO process and the rejection efficiency of the two ions increased to 99.5% (Mohsen-Nia et al., 2007).

15.4.1.2.3 Nanofiltration

Nanofiltration involves sieving and the Donnan effect during separation processes. The electric potential developed between the charged anions in the membrane helps to separate the positively charged metal ions during filtration (Al-Rashdi et al., 2013). The small pores and membrane surface charge allows small and charged solutes to pass through it. However, it also depends on temperature, pressure, cross-flow velocity, pH, salinity, etc. The NF membranes have pore diameters varying from 0.5 to 2 nm; the molecular weight of the solute ranges from 200 to 1000 Da. NF membranes are used in purification plants operating between UF and RO to enable effective and holistic removal and recovery of heavy metals. A study reported that thin-film composite polyamide NF membrane is used for nickel ion removal from aqueous wastewater (Murthy and Chaudhari, 2008).

15.4.1.2.4 Electrodialysis

Electrodialysis is a charge separation techniques in which the positively charged heavy metals move across the corresponding electrodes through a cation exchange membrane. By applying an electric potential, the anions migrate toward the anode and the cations toward the cathode, crossing the anion exchange and cation exchange membranes for metal recovery, respectively (Chen, 2004). Effects of flow rate, temperature, and voltage at different concentrations affect the treatment process. Increased metal recovery can be achieved by increasing the voltage and temperature; however, the percentage of separation decreases with an increasing in flow rate (Mohammadi et al., 2005). ED offers advantages for the treatment of wastewater laden with heavy metals, such as the ability to produce a highly concentrated stream for recovery and the rejection of undesirable impurities from water. Nataraj et al. (2007) studied the removal of hexavalent chromium ions using an electrodialysis pilot plant with a combination of ion exchange membranes (Nataraj et al., 2007).

15.4.1.3 Flotation

Flotation involves the separation of heavy metals by forming a hydrophobic adjunct in liquid phase. The flotation process is generally of three types: dissolved air flotation (DAF), ion flotation, and precipitation flotation that are employed for the removal of metal ions. In DAF metal agglomerates are formed when the suspended solids become attached to the micro air bubbles; these agglomerates have a density lower than water, which allows them to float and later be recovered (Lundh et al., 2000). This process works by imparting hydrophobic ends to charged metal species in wastewater using surfactants of different sources. These formed hydrophobic complexes are then floated using air bubbles formed by adding frothers such as sodium dodecyl sulfate, hexadecyltrimethyl ammonium bromide, ethanol, and methyl isobutyl carbinol (Polat and Erdogan, 2007). Precipitate flotation works by forming a precipitate which is either the metal hydroxide or a metal salt, depending on the concentration of the metal solution. The formed precipitate is then floated using an air bubble attachment (Capponi et al., 2006). Nenov et al. (2008) reported recovery of toxic metals such as copper, nickel, and zinc with an efficiency of 96% using integrated flotation methods (Nenov et al., 2008).

15.4.2 CHEMICAL METHODS

15.4.2.1 Chemical Leaching

Chemical leaching uses chemicals that solubilize and provide an extractable form of metals that are inseparably attached to organic solids, minerals, or soil complexes. Strong acids and alkalis are administered to break down the linkages and leach out metals, enabling recovery and removal. Every metal reacts differently and needs a specific removal technique. The most common solubilization process involves acidification using strong concentrated acids (Dermont et al., 2008). At the optimum redox condition, metals are leached out. This is followed by other separation methods such as centrifugation to collect the metals. Studies reported removal efficiencies of 72% and 79% Zn, 70% and 75% Cu, 69% and 73% Ni, 57% and 70% Pb, 55% and 65% Cd, and 11% and 22% Cr using chemical leaching methods (Bayat and Sari, 2010). However, this process has the drawback of being uneconomical because it uses strong and expensive chemicals (Bayat and Sari, 2010).

15.4.2.2 Chemical Precipitation

Chemical precipitation operates by forming insoluble precipitates of different heavy metals in the form of hydroxides, sulfides, chitosans, etc. (Ku and Jung, 2001). The formed precipitates can be separated later from the water by sedimentation or filtration. Conventional chemical precipitation processes include precipitation methods such as hydroxide precipitation, sulfide precipitation, and heavy metal chelating precipitation etc.

15.4.2.2.1 Hydroxide Precipitation

Hydroxide precipitation involves reacting the metals with hydroxides such as $Ca(OH)_2$ or NaOH to form their respective hydroxides. The redox conditions are maintained in the alkaline range of 8–11 where most of the metal hydroxides are insoluble precipitates which are then removed by flocculation or sedimentation. This method is used for almost all kinds of metal contaminations; however, hydroxide precipitation also has drawbacks in terms of producing low-density upper precipitates which require a lot of effort to separate from the reactors. Aziz et al. (2008) studied the recovery of heavy metals with concentrations higher than 1000 mg L^{-1} using lime water precipitation (Aziz et al., 2008). For industrial-scale operations, hydroxide precipitation methods provide a cost-effective process with ease of handling (Baltpurvins et al., 1997).

15.4.2.2.2 Sulfide Precipitation

Sulfide precipitation involves sulfide/hydrogen sulfide reactions with heavy metals to form insoluble precipitates. The metal sulfides are not amphoteric and precipitate at a broad range of pH, which has an advantage over hydroxide precipitation. These metal sulfide precipitates are easier to remove because they form thicker sludge and have good dewatering properties. Limitations of this process involve the production of toxic hydrogen sulfide fumes, which pose a health hazard. Ozverdi and Erdem (2006) investigated pyrite and synthetic iron sulfide to remove Cu^{2+}, Cd^{2+}, and Pb^{2+}.

15.4.2.2.3 Heavy Metal Chelating Precipitation

Chelating agents provide a specific binding site for metals to bind and form a precipitate (Fei et al., 2005). Commonly used chelating agents are trimercaptotriazine, potassium/sodiumthiocarbonate, and sodiumdimethyldithiocarbamate. There are specific chelating agents developed for specific metals: 1,3-benzenediamidoethanethiol for mercury contamination, *N,N0-bis*-(dithiocarboxy)piperazine and 1,3,5-hexahydrotriazinedithiocarbamate for complex heavy metals, potassium ethyl xanthate to remove copper ions from wastewater, and dipropyl dithiophosphate for remediating lead, cadmium, copper, and mercury (Chang and Yen, 2006). This precipitation method involves adding various chemical agents that produce toxic gases and need to maintain stringent redox conditions. To avoid these issues, metal chelating agents are used to precipitate the heavy metals directly in the aqueous system.

15.4.3 ELECTROCHEMICAL

Electrochemical methods remove and recover heavy metals based on the principle that metals become precipitated in their elemental form on the solid electrodes when the potential or electricity is applied on the electrode. The basic reaction occurring is the reduction of metals in different oxidation states into a zero oxidation state (elemental state of metal) at the cathode when there is a flow of electrons through the circuit from the anode (Nancharaiah et al., 2015). These reductions are of three types based on their functions: electrocoagulation (EC), electroflotation (EF), and electrodeposition. EC involves the use of electrodes that release of coagulating ions that react with the metals, and these coagulates can be separated later by centrifugation (Chen, 2004). EF involves hydrogen gas that is generated in the electrochemical cell at the cathode to float the suspended floccules of heavy metals in the catholyte to the surface, which are then separated out. Electrodeposition is the process in which there is removal and recovery at the same time owing to the applied potential or current. The heavy metals that are in different oxidation states become reduced at the cathode and are plated onto it. Electrochemical methods have been employed to remove Cr, Zn, As, Cu, Ni, Mn, and Ag. Heidmann and Calmano (2008) studied specific removal of Zn^{2+}, Cu^{2+}, Ni^{2+}, Ag^{+}, and Cr^2O^7 in an electrochemical system with aluminum electrodes (Heidmann and Calmano, 2008). Electrochemical treatment methods involve investments in the form of an electricity supply, which makes them unusable in large-scale applications. However, electrochemical technologies have regained significance for application over the past 2 decades (Wang et al., 2007).

15.4.4 BIOLOGICAL METHODS

15.4.4.1 Biosorption

Biosorption sequesters dissolved metal ions or metalloid species using various natural materials of biological origin such as bacteria, fungi, yeast, and algae, through active and passive modes (Gadd et al., 1998). The active mode is metabolism dependent, in which metal ions are transported across the cell membrane through intracellular uptake processes. The passive mode is metabolism independent, in which binding occurs between metals and ligands present in the cell wall on dead biomass. Living cells showed a toxic effect at a higher concentration; hence the passive mode was widely used as biosorbents (Zabochnicka–Swiatek and Krzywonos, 2014). The passive mode involves physico-chemical interactions which binds metal ions on the cell membrane and cell wall, which act as a complex ion exchanger that has similar properties as commercial available resins. The surface of the cell is often negatively charged for the passive binding of cationic metals (Chojnacka, 2010). A sequence of different mechanisms occur in this process, such as ion exchange (exchanging mobile ions), complexation (ions of heavy metals bound to the functional group present in the cell membrane), and physical adsorption (triggered by intermolecular interaction) to recovery the metals (Zabochnicka Swiatek and Krzywonos, 2014). Studies reported that lead was recovered with a capacity of 92.3 mg g^{-1} using *Bacillus* sp. (Tunali et al., 2006), zinc with a capacity of 17.7 mg g^{-1} using *Pseudomonas* putida (Pardo et al., 2003), and copper with a capacity of 381 mg g^{-1} using *Bacillus firmus* (Salehizadeh and Shojaosadati, 2003). The main advantage of the passive mode is it does not require nutrients and energy for the growth of biomass (Ahalya et al., 2004). Biosorption of heavy metals from aqueous solutions has been proved to be promising; it offers significant advantages such as low cost, availability, profitability, ease of operation, and high efficiency, especially when dealing with low concentrations (Kratochvil and Volesky, 1989). However, metal reduction (elemental state), metal desorption, and early saturation are its limitations (Goyal, 2007).

15.4.4.2 Bioaccumulation

Bioaccumulation occurs in living organisms through the transport of metals to cells and the accumulation of metals inside cells by various metabolic activities (Luoma and Rainbow, 2005). Cells absorb metal ions through the same routes by which nutrients are supplied during metabolic functions (Chojnacka, 2010). Metal influx across bacterial membranes includes ion pumps, ion channels, carrier-mediated transport, endocytosis, complex permeation, and lipid permeation (Issazadeh et al., 2013). *Pseudomonas putida* 62BN is used to accumulate cadmium (Pandey et al., 2013), *Citrobacter* sp. for lead and cadmium (Rani and Goel, 2009), *Thiobacillus ferrooxidans* for silver (Ahmed and Khan, 2011), *Bacillus subtilis* for chromium, *Pseudomonas aeruginosa* for uranium, and *Micrococcus luteus* for strontium (Umrania, 2006). Studies also reported that species such as *Mytillus edulis*, *Corbicula fluminea*, and *Macoma balthica* are effective in bioaccumulating metals such as chromium, copper and silver respectively (Wang et al., 1997; Croteau and Luoma, 2005; Griscom et al., 2002). The accumulation of heavy metals depends on many factors such as temperature, pH, salinity, dissolved oxygen, and hardness, as well as biological factors (Sanjay et al., 2011). Recovery of metal ions from bioaccumulated cells is effluxed by the activity of membrane-bound proteins in the cell (Yu et al., 2013).

15.4.4.3 Phytoextraction and Agro-mining

Phytoremediation is an effective technique involving plants, soil, and microbes to treat a broad range of soils contaminated with metals and radionuclides. Certain plants are able to remove metals from the soil through their roots by absorption along with water and nutrients from contaminated sites (Banuelos, 2000). Studies reported that plants such as *Brassica napus* cv. Westar (canola), *Hibiscus cannabinus* L. cv. Indian (kenaf), and *Festuca arundinacea* Schreb were effective in reducing selenium in contaminated sites (Banuelos et al., 2001). *Populus nigra* was the best accumulator of different metals such as Mn, Zn, and Cd (Parizanganeh et al., 2015). Several techniques are involved in phytoremediation, of which phytoextraction is especially notable for the recovery of metals (Ali and Khan, 2013). Plants solubilize heavy metals by secreting metal-mobilizing substances in the rhizosphere.

Root exudates can lower the rhizosphere soil pH and increase the concentration of heavy metals in solution by promoting desorption (Thangavel and Subbhuraam, 2004).

Phytoextraction may be economical when applied to expensive metals such as gold, palladium, and platinum (Pt). Antarctic snow and crustal enrichment of Pt has been reported. This suggests that global transportation may have contributed to Pt pollution in the atmosphere over past decades (Soyol-Erdence et al., 2011). Pt is used to make automobile and industrial catalytic converters. It is more expensive than gold. It can be beneficial to remove Pt from wastewater by biosorption (Lalhruaitluanga and Prasad, 2015). Phytoextraction and phytomining using metal-accumulating ecotypes of plants have attained global attention (Chaney et al., 2004; Losfeld et al., 2014; Sheoran et al., 2013; van der Ent et al., 2015). There are blue and dark skies for the application of metallophytes as follows:

1. Blue skies: The grand success of the Phytomining Workshop, July 21–22, 2014, Brisbane, Australia, "From discovery to full-scale application," Held at the University of Queensland, St. Lucia Campus, Brisbane, Queensland," Organized by the Center for Mined Land Rehabilitation, Sustainable Minerals Institute
2. Dark skies: Expressions such as "Phytoextraction: Where's the action?" (Robinson et al., 2015).

Despite some reservations, promising bioeconomic potential exists for metallophytes in the following areas:

1. Biofortified nutraceuticals for human health
2. Biofortified animal feed/poultry feed
3. Ecocatalysts for green chemistry (Grison, 2015)
4. Bioremediation-based bioeconomy (Prasad, 2016)

15.4.4.4 Biosurfactants

Biosurfactants are amphiphilic compounds produced mostly on microbial cell surfaces, which contain hydrophobic and hydrophilic moieties. They are produced extracellularly or as part of the cell membrane by bacteria, yeast, and fungi (Mulligan, 2005). The ability of biosurfactants to form complexes with metals is the main reason for their use in the remediation of metal-contaminated soil. Biosurfactants are grouped as glycolipids, lipopeptides, phospholipids, fatty acids, neutral lipids, and polymeric and particulate compounds which are either anionic or neutral. These biosurfactants are less toxic and more biodegradable than chemical surfactants that are commonly used for metal-contaminated soil (Biermann et al., 1987). Biosurfactant metal binding includes ion exchange, precipitation dissolution, counter-ion association, and electrostatic interaction (Rufino et al., 2012). Mechanisms involved in removing metal are sorption at the soil interface and metal complexation, followed by desorption of the metal by lowering soil–water interfacial tension and fluid forces, and finally complexation of the metal with micelles (Mulligan, 2005). A wide variety of microorganisms can produce these compounds, such as *Arthrobacterparaffineus*, *Corynebacterium* spp., *Mycobacterium* spp., *Rhodococuserythropolis*, and *Nocardiasp* (Mulligan and Gibbs, 1993). Studies have reported that copper and zinc were evaluated using surfactin from *B. subtilis*, rhamnolipids from *P. aeruginosa*, and sophorolipid from *Torulopsisbombicola* (Mulligan et al., 2001). Biosurfactants effectively enhance remediation of contaminants owing to their biodegradability and low toxicity.

15.4.5 ADVANCED BIOLOGICAL METHODS: BIOELECTROCHEMICAL SYSTEMS

Advanced biological techniques such as bioelectrochemical systems (BES) are effective techniques for the removing and recovering metals. BES are microbially catalyzed electrochemical systems which use bacteria as a catalyst to carry out redox reactions for metal recovery (Venkata Mohan et al., 2014a). BES work under similar principles of electrochemical cells, with the difference of using bacteria as a biocatalyst. Basically, microbial fuel cells are configured as two-chambered systems consisting of an anode and a cathode electrode assembly. At the anode, the protons and electrons are generated by oxidizing the organic substrate; protons migrate to the cathode through a proton exchange membrane, creating a potential difference between the anode and cathode, against which the electrons flow through the circuit, where they become reduced in the presence of terminal electron acceptor-generating power and the recovery of metal or any pollutants that act as a terminal electron acceptor (Venkata Mohan et al., 2009). The cathodic chamber could be biotic or abiotic, which is determined by the presence or absence of a biocatalyst (Fig. 15.23) (Venkata Mohan et al., 2014b). Biological conversions of substrates at the anode are the source of electrons which are driven to the cathode through the external circuit, reduces the metals to form precipitates (Nancharaiah et al., 2015). Studies have reported that metal ions such as Cu(II), V(V), Pb(II), Zn(II), Cd(II), Cr(VI), Ag(I), and Hg(II) have been removed through BES (Li et al., 2008; Wang et al., 2008a,b; Zhang et al., 2012; Li et al., 2008). In BES, metal ions can be recovered by using the in situ−generated potential and the ex situ or externally applied potential (Wang et al., 2008a,b; Li et al., 2008; Tandukar et al., 2009; Huang et al., 2010).

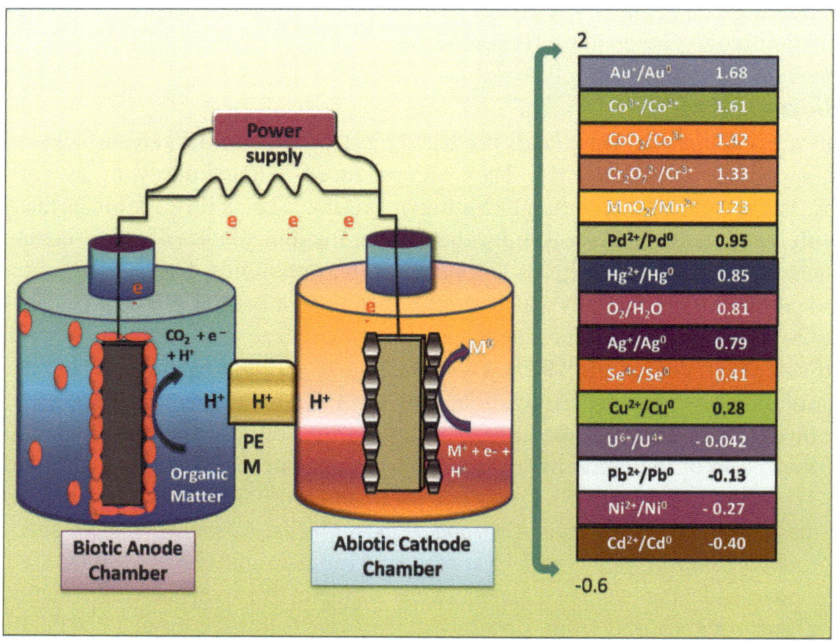

FIGURE 15.23

Schematic representation of BES in double-chamber systems with a biotic anode and abiotic cathode.

15.4.5.1 In Situ Potential Influence on Metal Recovery in Anodic and Cathodic Chamber

BES provide a flexible platform to reduce metal elements or ions with a lower chemical valency in microbial fuel cells as a result of the in situ potential developed in the system. These ions have positive standard potentials, which create a large difference between the anodic and cathodic potentials which become reduced based on the potential developed in the system without requiring additional potential (Wang et al., 2008a,b, 2011). Metal recovery is composed of an abiotic or biotic cathode. In a biotic system, bacteria have a crucial role by using the organic substrate and thereby liberating reducing equivalents which reduce metal species. Most metal reduction takes place in the cathode chamber which aids in removing and recovering metals, degrading oxidized pollutants etc. (Nancharaiah et al., 2015). Studies reported that microbial fuel cells with a power generation of 3.6 W m^{-2} resulted in Cd and Zn removal of 90% and 97%, respectively (Abourached et al., 2014). The metal compounds act as good electron acceptors owing to their high electron affinity and can be reduced to a different form which can be recovered easily (Fig. 15.24) (Kim et al., 2002; Modin et al., 2012; Tao et al., 2011). Researchers have started using biocathodes for metal recovery, pollutant degradation, and the wastewater treatment because of the advantages offered, such as economic viability and ease of operation (He and Angenant, 2006). Biocathodes are effective in degrading specific pollutants such as chloramphenicol and have the capability of reducing metal compounds (Sun et al., 2013; Terheijne et al., 2006). More specifically, metal speciation occurs from its complex form to a simpler one (Gangadharan and Nambi, 2015). Highly toxic Cr(VI) is reduced to less soluble, less toxic Cr(III) by diverse microorganisms (Nancharaiah et al., 2010). Because not all the metals can be recovered through in situ potential, some specific metals require an external potential to drive electrons for metal reduction.

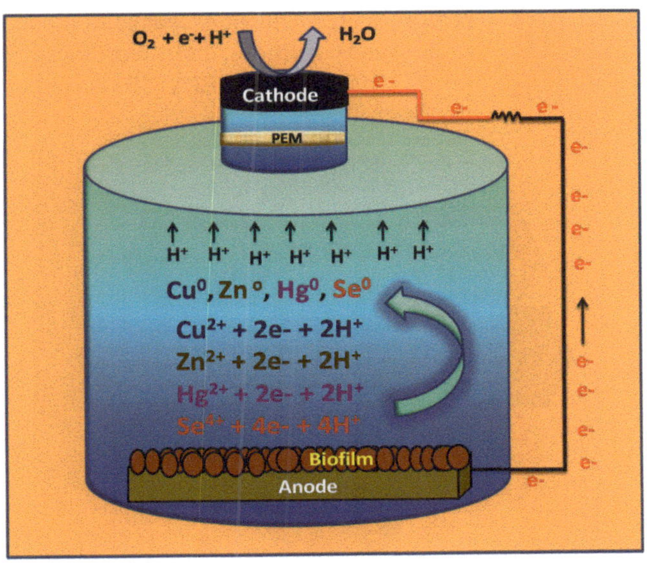

FIGURE 15.24

Schematic representation of a BES single-chamber, open-air cathode system for the removal of metals.

15.4.5.2 Ex Situ Potential for Recovery of Metals in Bioelectrochemical Systems

To enhance metal reduction in the cathodic chamber, strategies such as applied potential or current can be used for effective metal removal and recovery of metals. The in situ potential generated by the inherent capabilities of the system is insufficient to drive electrons from the anode to the cathode; hence, they need some external potential or current to drive the electrons (Wang and Ren, 2014). Studies reported that the in situ potential is too small to drive electrons transferred from the anode to the cathode to reduce Ni^{2+} to Ni^0. According to the Nernst equation, about −0.54 V (versus SCE) of cathodic potential is required to reduce 1000 mg L^{-1} of Ni^{2+} to Ni^0 (at 30 °C); hence, it requires an additional external potential to drive electrons to reduce the metals to an elemental form (Qin et al., 2012). Studies also reported that with sequential application of a set potential such as 0.34, 0.52, and 1.7 V, it is possible to recover mixed metals or metal ions selectively, such as Cu(II), Pb(II), Cd(II), and Zn(II), from a simulated municipal solid waste (MSW) (Modin et al., 2012). Cu^{2+}, Ni^{2+}, and Fe^{2+} were also observed to be removed from artificial acid mine drainage with an external potential of 1.0 V (Luo et al., 2014). Hence, applying an external potential in BES helps to recover metals from metallurgical waste, MSW, acid mine waste, etc., with a limited energy requirement.

15.5 POLLUTION PREVENTION

Heavy metals come under the environmental pollutant category because of their toxic effects, which cause significant hazards to humans, animals, and ecosystems. The source of pollution is mostly anthropogenic industrial activity or natural means. Pollution through metals can be handled effectively by reducing upstream sources of pollutants with the initial steps of processing metals rather than by using extensive treatment processes downstream (ConnTAP, 1992). Pollution prevention encourages metal industries to realize waste minimization strategies to reduce the disposal of hazardous substances and other harmful substances by source reduction, change in products, process modification, change in technologies, improved operating conditions, direct recycling, and good housekeeping, and to follow a closed-loop approach (Fig. 15.25) (Ohio EPA, 1999). Implementing an effective

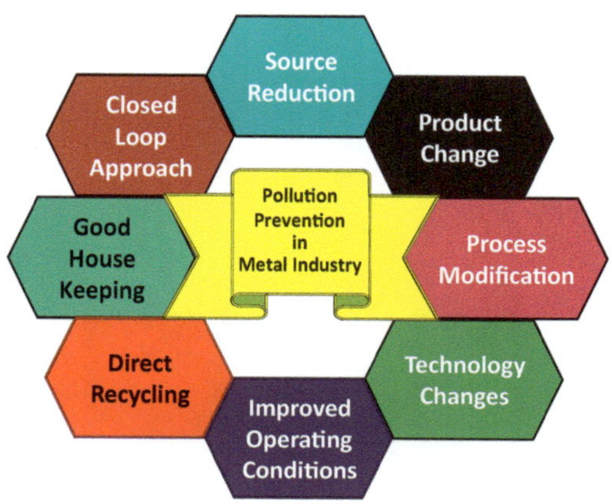

FIGURE 15.25

Strategic plan of pollution prevention in metal processing industries.

inventory control system and segregating waste streams, allowing wastes to be recycled or reused, will help to reduce pollution. Technology changes will determine technical alternatives using a cost-effective approach. Improving operating conditions by planning effective methods reduces operating time and waste. Direct recycling will use the waste product effectively; good housekeeping prevents injuries and maintains equipment. Following a closed-loop approach completely reduces waste and effectively reuses the waste products with a proper balance between input and output (Venkata Mohan et al., 2016). Hence, a pollution prevention strategy is a widespread process which emphasizes the concept of sustainable development.

ACKNOWLEDGMENTS

World Mineral Statistics contributed by permission of the British Geological Survey. JK thankfully acknowledges the awards of USIEF and US Nehru Fulbright Fellowship for 8 months. Thanks are also due to the University of Hyderabad, Hyderabad, for hosting the tenure of fellowship. GV and SKB acknowledge CSIR and UGC, respectively, for providing a research fellowship. SVM acknowledges funding from DBT in the form of a National Bioscience Award (No. BT/HRD/NBA/34/01/2012(vi)).

REFERENCES

Abourached, C., Catal, A., Liu, H., 2014. Efficacy of single chamber microbial fuel cell for removal of cadmium and zinc with simultaneous power generation. Water Research 51, 228–233.

Ahmed, M., Khan, M.S., 2011. Pesticide interactions with soil microflora: importance in bioremediation. In: Ahmad, I., Ahmad, F., Pichtel, J. (Eds.), Microbes and Microbial Technology: Agricultural and Environmental Applications. Springer, New York, pp. 393–413.

Agrawal, P., Mittal, A., Prakash, R., Kumar, M., Tripathi, S.K., 2011. Contamination of drinking water due to coal-based thermal power plants in India. Environmental Forensics 12, 92–97.

Agusa, T., Kunito, T., Ramu, K., Chamnan, C., Trang, P.T.K., Minh, T.B., Subramanian, A., Iwata, H., Viet, P.H., Tana, T.S., Tanabe, S., 2006. Lead contamination and its human health effects in India, Vietnam and Cambodia. Biomedical Research on Trace Elements 17, 413–416.

Ahalya, N., Ramachandra, T.V., Kanamadi, R.D., 2004. Biosorption of heavy metals. Research Journal of Chemistry and Environment 7, 71–79.

Aich, A., Chakraborty, A., Sudarshan, M., Chattopadhyay, B., Mukhopadhyay, S.K., 2011. Study of trace metals in Indian major carp species from wastewater-fed fishponds of East Calcutta Wetlands. Aquaculture Research 43, 53–65.

Ali, H., Khan, E., 2013. Phytoremediation of heavy metals – Concepts and applications. Chemosphere 91, 869–881.

Al-Rashdi, B.A.M., Johnson, D.J., Hilal, N., 2013. Removal of heavy metal ions by nanofiltration. Desalination 315, 2–17.

Australia and New Zealand Food Authority Provisional Regulation Limitation of Contamination in Food, April 2004. Specifications and Standards for Foods, Food Additives, etc. Under the Food Sanitation Laws. Standard 1.4.1, Contamination and Natural Toxicants, Maximum level of metal contaminates in food.

Aziz, H.A., Adlan, M.N., Ariffin, K.S., 2008. Heavy metals Cd, Pb, Zn, Ni, Cu and Cr(III) removal from water in Malaysia: post treatment by high quality limestone. Bioresource Technology 99 (6), 1578–1583.

Baltpurvins, K.A., Burns, R.C., Lawrance, G.A., Stuart, A.D., 1997. Effect of electrolyte composition on zinc hydroxide precipitation by lime. Water Research 31, 973–980.

Banuelos, G.S., 2000. Phytoextraction of selenium from soils irrigated with selenium –laden effluent. Plant and Soil 224, 251–258.

Banuelos, G.S., Ajwa, H.A., Mackey, B., Wu, L., Cook, C., Akohoue, S., Zambruzuski, S., 2001. The green technology of selenium phytoremediation. Journal of Environmental Quality 26, 639–646.

Bayat, B., Sari, B., 2010. Comparative evaluation of microbial and chemical leaching processes for heavy metal removal from dewatered metal plating sludge. Journal of Hazardous Materials 174, 763–769.

Bernardini, O., Galli, R., 1993. Dematerialization: long-term trends in the intensity of use of materials and energy. Futures 25, 431–448.

BGS, 2012. World Mineral Statistics.

Bhattacharya, A.K., Mandal, S.N., Das, S.K., 2008. Heavy metals accumulation in water, sediment and tissues of different edible fishes in upper stretch of Gangetic West Bengal. Trends in Applied Sciences Research 3, 61–68.

Biermann, M., Lange, F., Piorr, R., Ploog, U., Rutzen, H., Schindler, J., Schmidt, R., 1987. Surfactants in consumer products. In: Falbe, J. (Ed.), Theory, Technology and Application. Springer-Verlag, Heidelberg, pp. 86–106.

Bird, M.J., MacIntosh, D.L., Williams, P.L., 2004. Occupational exposures during routine activities in coal-fueled power plants. Journal of Occupational and Environmental Hygiene 1, 403–413.

Bohdziewicz, J., Bodzek, M., Wasik, E., 1999. The application of reverse osmosis and nanofiltration to the removal of nitrates from groundwater. Desalination 1221, 139–147.

Busselle, L.D., Moore, T.A., Shoemaker, J.M., Allred, R.E., 1999. Separation processes and economic evaluation of tertiary recycling of electronic scrap. In: Electronics and the Environment. ISEE –1999. Proceedings of the 1999 IEEE International Symposium on 1999, pp. 192–197.

Capponi, F., Sartori, M., Souza, M.L., Rubio, J., 2006. Modified column flotation of adsorbing iron hydroxide colloidal precipitates. International Journal of Mineral Processing 79, 167–173.

Chakraborti, D., Mukherjee, S.C., Pati, S., Sengupta, M.K., Rahman, M.M., Chowdhury, U.K., Lodh, D., Chanda, C.R., Chakraborti, A.K., Basu, G.K., 2003. Arsenic groundwater contamination in Middle Ganga Plain, Bihar, India: a future danger? Environmental Health Perspectives 111, 1194–1201.

Chaney, R.L., Reeves, P.G., Ryan, J.A., Simmons, R.W., Welch, R.M., Scott Angle, J., 2004. An improved understanding of soil Cd risk to humans and low cost methods to phytoextract Cd from contaminated soils to prevent soil Cd risks. Biometals 17, 549–553.

Chang, T.C., Yen, J.H., 2006. On-site mercury-contaminated soils remediation by using thermal desorption technology. Journal of Hazardow Materials 128, 208–217.

Chen, G.H., 2004. Electrochemical technologies in wastewater treatment. Separation and Purification Technology. Journal of Hazardous Materials 177, 876–880.

Chojnacka, K., 2010. Biosorption and bioaccumulation – the prospects for practical applications. Environment International 36, 299–307.

Chowdhury, U.K., Biswas, B.K., Chowdhury, T.R., Samanta, G., Mandal, B.K., Basu, G.C., Chanda, C.R., Lodh, D., Saha, K.C., Mukherjee, S.K., Roy, S., Kabir, S., Quamruzzaman, Q., Chakraborti, D., 2000. Groundwater arsenic contamination in Bangladesh and West Bengal, India. Environmental Health Perspectives 108, 393–397.

Cleveland, C.J., Ruth, M., 1998. Indicators of dematerialization and the materials intensity of use. Journal of Industrial Ecology 2, 15–50.

ConnTAP Quarterly/Summer, 1992. Connecticut Technical Assistance Program 5 (3). http://infohouse.p2ric.org/ref/18/17167.pdf.

Croteau, M.N., Luoma, S.N., 2005. Delineating copper accumulation pathways for the freshwater bivalve Corbicula using stable copper isotopes. Environmental Toxicology and Chemistry 24, 2871–2878.

Dahiya, S., Karpe, R., Hegde, A.G., Sharma, R.M., 2005. Lead, cadmium and nickel in chocolates and candies from suburban areas of Mumbai, India. Journal of Food Composition and Analysis 18, 517–522.

Dermont, G., Bergeron, M., Mercier, G., Richer-Laflèche, M., 2008. Soil washing for metal removal: a review of physical/chemical technologies and field applications. Journal of Hazardous Materials 152, 1–31.

Dey, S., Dwivedi, S.K., Swarup, D., 1996. Lead concentration in blood, milk and feed of lactating buffalo after acute lead poisoning. Veterinary Record 138, 336.

Dhadse, S., Kumari, P., Bhagia, L.J., 2008. Fly ash characterization, utilization and government initiatives in India – a review. Journal of Scientific and Industrial Research 67, 11–18.

Dhanuka, T., Prakash, V., 2012. Surface and Underground Installations. Department of Mechanical Engineering of Poornima College of Engineering.

De, T.K., De, M., Das, S., Ray, R., Ghosh, P.B., 2010. Level of heavy metals in some edible marine fishes of mangrove dominated tropical estuarine areas of Hooghly River, north east coast of Bay of Bengal, India. Bulletin of Environmental Contamination and Toxicology 85, 385–390.

Dwivedi, S.K., Swarup, D., Dey, S., Patra, R.C., 2001. Lead poisoning in cattle and buffalo near primary lead-zinc smelter in India. Veterinary and Human Toxicology 43 (2), 93–94.

Eastern research group background document report on revisions to 5th edition AP-42 section 1.6 wood residue combustion. In: Boilers Department of Mechanical Engineering of Poornima College of Engineering: Morrisville, North Carolina, 2001.

EIA, 2012. International Energy Statistics. http://205.254.135.7/cfapps/ipdbproject/IEDIndex3.cfm (accessed 31.05.12.).

EKTA International, 2012. Fly- Ash. http://ektainternational.tradeindia.com/fly-ash-838525.html (accessed 31.05.12.).

ENVIS, 1999. Auto Emissions. http://cpcbenvis.nic.in/newsletter/autoemissionjun1999/jun1999step.htm (accessed 31.05.12.).

EPA, 1999. A Citizen's Guide to Phytoremediation. EPA 542-F-12-016.

FAO, 1983. Compilation of Legal Limits for Hazardous Substances in Fish and Fishery Products. FAO Fishery Circular, Washington, DC, pp. 5–100.

FAO, 2011. Fish and Fishery Products Hazards and Controls Guide > Appendix 5-FDA & EPA Safety Levels in Regulations and Guidance, third ed.

FAO, 2012. Major Food and Agricultural Commodities and Producers. http://www.fao.org/es/ess/top/country.html (accessed 31.05.12.).

Fei, C., Sheng, L.G., Wei, Y.W., Jun, W.Y., 2005. Preparation and adsorption ability of poly-sulfone microcapsules containing modifiedchitosan gel. Tsinghua Science & Technology 10, 535–541.

Figueroa, B.E., 2008. Are more restrictive food cadmium standards justifiable health safety measures or opportunistic barriers to trade? An answer from economics and public health. Science of the Total Environment 389, 1–9.

Gadd, G.M., White, C., De Rome, L., 1998. Heavy metal and radionuclide uptake by fungi and yeasts. In: Norris, P.R., Kelly, D.P., Rowe, A. (Eds.), Biohydrometallurgy. Chippenham, Wilts, United Kingdom.

Galanis, A., Karapetsas, A., Sandaltzopoulos, R., 2009. Metal-induced carcinogenesis, oxidative stress and hypoxia signalling. Mutation Research. Netherlands 674 (1–2), 31–35.

Gangadharan, P., Nambi, I.M., 2015. Hexavalent chromium reduction and energy recovery by using dual-chambered microbial fuel cell. Water Sci Technol 71, 353–358.

Goyal, A.S.S.D., 2007. Microbial and plant derived biomass for removal of heavy metals from wastewater. Bioresource Technology 98, 2243–2257.

Greenpeace, 2005. Greenpeace Report. In: Brigden, K., Labanska, I., Sanyillo, D., Allsopp, M. (Eds.), Recycling of Electronic Waste in China and India: Workplace and Environmental Contamination. Greenpeace International.

Griscom, S.B., Fisher, N.S., Luoma, S.N., 2002. Kinetic modeling of Ag, Cd and Co bioaccumulation in the clam Macomabalthica: quantifying dietary and dissolved sources. Marine Ecology Progress Series 240, 127–141.

Grison, C., 2015. Combining phytoextraction and ecocatalysis: a novel concept for greener chemistry, an opportunity for remediation. Environmental Science and Pollution Research 22, 5589–5591.

Gupta, N., Khan, D., Santra, S., 2008. An assessment of heavy metal contamination in vegetables grown in wastewater-irrigated areas of Titagarh, West Bengal, India. Bulletin of Environmental Contamination and Toxicology 80, 115–118.

He, Z., Angenent, L.T., 2006. Application of bacterial biocathodes in microbial fuel cells. Electroanalysis 18, 2009–2015.

Heidmann, I., Calmano, W., 2008. Removal of Zn(II), Cu(II), Ni(II), Ag(I) and Cr(VI) present in aqueous solutions by aluminium electrocoagulation. Journal of Hazardous Materials 152, 934–941.

Heijne, Ter, A., Liu, F., Weijden, R., Van der Weijma, J., Buisman, C.J.N., Hamelers, H.V.M., 2010. Copper recovery combined with electricity production in a microbial fuel cell. Environmental Science & Technology 44, 4376–4381.

Hira-Smith, M., Yuan, Y., Savarimuthu, X., Liaw, J., Hira, A., Green, C., Hore, T., Chakraborty, P., von Ehrenstein, O., Smith, A., 2007. Arsenic concentrations and bacterial contamination in a pilot shallow dugwell program in West Bengal, India. Journal of Environmental Science and Health, Part A 42, 89–95.

Hoffmann, J., 1992. Recovering precious metals from electronic scrap. JOM: The Journal of the Minerals, Metals & Materials Society 44, 43–48.

Hossain, M.S., Khan, Y.S.A., 2001. Trace metals in Penaeid shrimp and Spiny lobster from the Bay of Bengal. ScienceAsia 27, 165–168.

Huang, L., Chen, J., Quan, X., Yang, F., 2010. Enhancement of hexavalent chromium reduction and electricity production from a biocathode microbial fuel cell. Bioprocess and Biosystems Engineering 33, 937–945.

Ter Heijne, A., Hamelers, H.V.M., de Wilde, V., Rozendal, R.A., Buisman, C.J.N., 2006. A bipolar membrane combined with ferric iron reduction as an efficient cathode system in microbial fuel cells. Environmental Science & Technology 40, 5200–5205.

International Lead and Zinc Study Group, 2016. Press Release: 18 January 2016. ILZSG. United Nations.

Ilyas, S., Anwar, M.A., Niazi, S.B., Afzal Ghauri, M., 2007. Bioleaching of metals from electronic scrap by moderately thermophilic acidophilic bacteria. Hydrometallurgy 88, 180–188.

Issazadeh, K., Jahanpour, N., Pourghorbanali, F., Raeisi, G., Faekhondeh, J., 2013. Heavy metal resistance by bacterial strains. Annals of Biological Research 4, 60–63.

Iyer, R., Scott, J., 2001. Power station fly ash - a review of value-added utilization outside of the construction industry. Resources, Conservation and Recycling 31, 217–228.

Juang, R.S., Shiau, R.C., 2000. Metal removal from aqueous solutions using chitosan – enhanced membrane filtration. Journal of Membrane Science 165, 159–167.

Karar, K., Gupta, A.K., Kumar, A., Biswas, A.K., 2006. Characterization and identification of the sources of chromium, zinc, lead, cadmium, nickel, manganese and iron in PM10 particulates at the two sites of Kolkata, India. Environmental Monitoring and Assessment 120, 347–360.

Karimi, G., Hasanzadeh, M.K., Nili, A., Khashayarmanesh, Z., Samiei, Z., Nazari, F., Teimuri, M., 2008. Concentrations and health risk of heavy metals in tea samples marketed in Iran. Pharmacologyonline 3, 164–174.

Khasim, D.I., Kumar, N.V., Hussain, R.C., 1989. Environmental contamination of chromium in agricultural and animal products near a chromate industry. Bulletin of Environmental Contamination and Toxicology 43, 742–746.

Kim, H.J., Park, H.S., Hyun, M.S., Chang, I.S., Kim, M., Kim, B.H., 2002. A mediator-less microbial fuel cell using a metal reducing bacterium, *Shewanella putrefaciens*. Enzyme and Microbial Technology 30 (2), 145–152.

Kojima, M., Jain, A., 2008. Controlling pollution in small-scale recycling industries: experiences in India and Japan. In: Promoting 3Rs in Developing Countries: Lessons From the Japanese Experience IDE-JETRO: Chiba, vol. 30, pp. 36–58.

Kongsuwan, A., Patnukao, P., Pavasant, P., 2009. Binary component sorption of Cu (II) and Pb (II) with activated carbon from *Eucalyptus camaldulensis* Dehn bark. Journal of Industrial and Engineering Chemistry 15, 465–470.

Kratochvil, D., Volesky, B., 1989. Advances in the biosorption of heavy metals. Trends in Biotechnology 16, 291–300.

Krishna, A.K., Govil, P.K., 2001. Heavy metal distribution and contamination in soils of Thane–Belapur industrial development area, Mumbai, Western India. Environmental Geology 47, 1054–1061.

Ku, Y., Jung, I.L., 2001. Photocatalytic reduction of Cr (VI) in aqueous solutions by UV irradiation with the presence of titanium dioxide. Water Research 35, 135–142.

Kukday, K., 2007. Making profit from mining of e-waste – http://timesofindia.indiatimes.com/business/india-business/Making-profit-from-mining-of-e-waste/articleshow/2107581.cms (accessed 28.05.12.).

Kumar, B., Sajwan, K.S., Mukherjee, D.P., 2012. Distribution of heavy metals in valuable coastal fishes from North East coast of India. Turkish Journal of Fisheries and Aquatic Sciences 12, 81–88.

Kumar, V., Mathur, M., Sinha, S.A., 2005. Case Study: Manifold Increase in Fly Ash Utilisation in India Fly Ash Utilization Programme (FAUP). TIFAC, DST, New Delhi 110016.

Kurniawan, T.A., Chan, G.Y.S., Lo, W.H., Babel, S., 2006. Physicochemical treatment techniques for wastewater laden with heavy metals. Chemical Engineering Journal 118, 83–98.

Kurniawan, T.A., Chan, G.Y.S., Lo, W.H., Babel, S., 2005. Comparisons of low-cost adsorbents for treating wastewaters laden with heavy metals. Science of the Total Environment 366 (2–3), 409–426.

Lalhruaitluanga, H., Prasad, M.N.V., 2015. Removal of heavy metals from aqueous solutions through biosorption. In: Dündar, M., Bruschi, F., Gartland, K., Magni, M.V., Gahan, P., Deeni, Y. (Eds.), Current Applications of Biotechnology. Erciyes University, Kayseri, Turkey, pp. 87–102.

Lefebvre, O., Neculita, C.M., Yue, X., Ng, H.Y., 2012. Bioelectrochemical treatment of acid mine drainage dominated with iron. Journal of Hazardous Materials 241–242, 411–417.

Losfeld, G., Mathieu, R., L'Huillier, L., Fogliani, B., Jaffré, T., Grison, C., 2015. Phytoextraction From Mine Spoils: Insights From New Caledonia. Environmental Science and Pollution Research. http://dx.doi.org/10.1007/s11356-014-3866-8, 12 p.

Lu, Z., Chang, D., Ma, J., Huang, G., Cai, L., Zhang, L., 2015. Behavior of metal ions in bioelectrochemical systems: a review. Journal of Power Sources 275, 243–260.

Lundh, M., Jonson, L., Dahlquist, J., 2000. Experimental studies of the fluid dynamics in the separation zone in dissolved air flotation. Water Research 34, 21–30.

Luo, H., Liu, G., Zhang, R., Bai, Y., Fu, S., Hou, Y., 2014. Heavy metal recovery combined with H_2 production from artificial acid mine drainage using the microbial electrolysis cell. Journal of Hazardous Materials 270, 153–1159.

Luoma, Rainbow, 2005. Why is metal bioaccumulation so variable biodynamics as a unifying concept. Environmental Science & Technology 39, 1921–1931.

Lakshmanan, R., Kesavan, K., Vijayanand, P., Rajaram, V., Rajagopal, S., 2009. Heavy metals accumulation in five commercially important fishes of parangipettai, Southeast coast of India. Advance Journal of Food Science and Technology 1 (1), 63–65.

Li, Y., Di, Z., Luan, Z., Ding, J., Zuo, H., Wu, X., Xu, C., Wu, D., 2004. Removal of heavy metals from aqueous solution by carbon nanotubes: adsorption equilibrium and kinetics. Journal of Environmental Sciences (China) 16, 208–211.

Li, Z.J., Zhang, X.W., Lei, L.C., 2008. Electricity production during the treatment of real electroplating wastewater containing Cr6+ using microbial fuel cell. Process Biochemistry 43, 1352–1358.

Martin, H.W., Kaplan, D.I., 1998. Temporal changes in Cadmium, Thallium, and Vanadium removal from water in Malaysia: post treatment by high quality limestone. Bioresource Technology 99, 1578–1583.

Modin, O., Wang, X., Wu, X., Rauch, S., Fedje, K.K., 2012. Bioelectrochemical recovery of Cu, Pb, Cd, and Zn from dilute solutions. Journal of Hazardous Materials 235–236, 291–297.

Mohammadi, T., Mohebb, A., Sadrzadeh, M., Razmi, A., 2005. Modeling of metal ion removal from wastewater by electrodialysis. Separation and Purification Technology 41, 73–82.

Mohsen-Nia, P., Montazeri, Modarress, H., 2007. Removal of Cu^{2+} and Ni^{2+} from wastewater with a chelating agent and reverse osmosis processes. Desalination, 217, 276–281.

Mokhtar, M.B., Aris, A.Z., Munusamy, V., Praveena, S.M., 2009. Assessment level of heavy metals in Penaeus monodon and *Oreochromis* spp in selected aquaculture ponds of high densities development area. European Journal of Scientific Research 30 (3), 348–360.

Morf, L.S., Tremp, J., Gloor, R., Schuppisser, F., Stengele, M., Taverna, R., 2007. Metals, non-metals and PCB in electrical and electronic waste – actual levels in Switzerland. Waste Management 27, 1306–1316.

Mukherjee, A.B., Bhattacharya, P., Sarkar, A., Zevenhoven, R., 2009. Mercury emissions from industrial sources in India and its effects in the environment. In: Mason, R., Pirrone, N. (Eds.), Mercury Fate and Transport in the Global Atmosphere. Springer, US, pp. 81–112.

Mulligan, C.N., 2005. Environmental applications for biosurfactants. Environmental Pollution 133, 183–198.

Mulligan, C.N., Gibbs, B.F., 1993. Factors influencing the economics of biosurfactants. In: Kosaric, N. (Ed.), Biosurfactants, Production, Properties, Applications. Marcel Dekker, New York.

Murthy, Z.V.P., Chaudhari, L.B., 2008. Separation of cadmium ions and estimation of membrane transport parameters of a nanofiltration membrane. Indian Journal of Chemical Technology 15, 107–112.

Nancharaiah, Y.V., Venkata Mohan, S., Lens, P.N.L., 2015. Metals removal and recovery in bioelectrochemical systems?: a review. Bioresource Technology 195, 102–114.

Nancharaiah, Y.V., Dodge, C., Venugopalan, V.P., Narasimhan, S.V., Francis, A.J., 2010. Immobilization of Cr (VI) and its reduction to Cr(III) phosphate by granular biofilms comprising a mixture of microbes. Applied and Environmental Microbiology 76, 2433–2438.

Nataraj, S.K., Hosamani, K.M., Aminabhavi, T.M., 2007. Potential application of an electrodialysis pilot plant containing ion-exchange membranes in chromium removal Desalination, 27, 181–190.

NFDB, 2007. National Fisheries Development Board: About Indian Fisheries. http://nfdb.ap.nic.in/html/aboutus.htm (accessed 31.05.12.).

Nenov, V., Lazaridis, N.K., Blocher, C., Bonev, B., Matis, K.A., 2008. Metal recovery from a copper mine effluent by a hybrid process. Chemical Engineering Process. 47, 596–602.

NSWAI, 2012. National Solid Waste Association of India Data Bank. http://www.nswai.com/databank.php (accessed 31.05.12.).

Olivares, M., Uauy, R., 1996. Copper as an essential nutrient. The American Journal of Clinical Nutrition 63, 791S–796S.

Ozverdi, A., Erdem, M., 2006. Cu^{2+}, Cd^{2+} and Pb^{2+} adsorption from aqueous solutions by pyrite and synthetic iron sulphide. Journal of Hazardous Materials 137, 626–632.

Pacyna, J.M., Pacyna, E.G., 2001. An assessment of global and regional emissions of trace metals to the atmosphere from anthropogenic sources worldwide. Environmental Reviews 9, 269–298.

Pacyna, E.G., Pacyna, J.M., Sundseth, K., Munthe, J., Kindbom, K., Wilson, S., Steenhuisen, F., Maxson, P., 2010. Global emission of mercury to the atmosphere from anthropogenic sources in 2005 and projections to 2020. Atmospheric Environment 44, 2487–2499.

Pandey, R., Chavan, P.N., Walokar, N.M., Sharma, N., Tripathi, V., Khetmalas, M.B., 2013. Pseudomonas stutzeri RP1: A versatile plant growth promoting endorhizospheric bacteria inhabiting sunflower (Helianthus annus). Res. J. Biotech 8, 48–55.

Panutrakul, S., Jayasinghe, R.P.P.K., Chookong, C., 2008. Heavy metal contents in purpleback squid *(Sthenoteuthis oualaniensis)* from the Bay of Bengal. The Ecosystem-Based Fishery Management in the Bay of Bengal.

Pardo, R., Herguedas, M., Barrado, E., Vega, M., 2003. Biosorption of cadmium, copper, lead, and zinc by inactive biomass of *Pseudomonas putida*. Analytical and Bioanalytical Chemistry 376, 26−32.

Parizanganeh, A.H., Zamani, A., Saba, J., 2015. Phytoremediation of heavy metals contaminated environments: screening for native accumulator plants in Zanjan-Iran. International Journal of Environmental Research 9, 309−316.

Patel, K.S., Shrivas, K., Brandt, R., Jakubowski, N., Corns, W., Hoffmann, P., 2005. Arsenic contamination in water, soil, sediment and rice of central India. Environmental Geochemistry and Health 27, 131−145.

Pinto, V.N., 2008. E-waste hazard: the impending challenge. Indian Journal of Occupational and Environmental Medicine 12, 65−70.

Pirrone, N., Cinnirella, S., Feng, X., Finkelman, R.B., Friedli, H.R., Leaner, J., Mason, R., Mukherjee, A.B., Stracher, G.B., Streets, D.G., Telmer, K., 2010. Global mercury emissions to the atmosphere from anthropogenic and natural sources. Atmospheric Chemistry and Physics. Discussions 10, 4719−4752.

Polat, H., Erdogan, D., 2007. Heavy metal removal from waste waters by ion flotation. Journal of Hazardous Materials 148, 267−273.

Praharaj, T., Swain, S.P., Powell, M.A., Hart, B.R., Tripathy, S., 2002. Delineation of groundwater contamination around an ash pond: geochemical and GIS approach. Environment International 27, 631−638.

Prasad, M.N.V., 2016. In: Bioremediation and Bioeconomy. Elsevier, USA, ISBN 978-0-12-802830-8.

Prasad, B., Mondal, K.K., 2008. Heavy metals leaching in Indian fly ash. Environmental Science & Engineering 50, 127−132.

Pandey, D., 2002. Fuelwood studies in India: myth and reality. Center for International Forestry Research, Bogor Barat, Indonesia.

Qin, B., Luo, H., Liu, G., Zhang, R., Chen, S., Hou, Y., Luo, Y., 2012. Nickel ion removal from wastewater using the microbial electrolysis cell. Bioresource Technology 121, 458−461.

Raja, P., Veerasingam, S., Suresh, G., Marichamy, G., Venkatachalapathy, R., 2009. Heavy metals concentrations in four commercially valuable marine edible fish species from Parangipettai Coast, South East Coast of India. International Journal of Animal and Veterinary Advances 1, 10−14.

Rajaganapathy, V., Xavier, F., Sreekumar, D., Mandal, P.K., 2011. Heavy metal contamination in soil, water and fodder and their presence in livestock and products? A review. Journal of Environmental Science and Technology.

Rani, A., Goel, R., 2009. Strategies for crop improvement in contaminated soils using metal-tolerant bioinoculants. In: Khan, M.S., Zaidi, A., Musarrat, J. (Eds.), Microbial Strategies for Crop Improvement. Springer, Berlin, pp. 105−132.

Roy, D., 2009. Heavy metal contents in cow and buffalo milk samples from Haryana. Indian Journal of Animal Nutrition 26, 29−33.

Robinson, B.H., Anderson, C.W.N., Dickinson, N.M., 2015. Phytoextraction: Where's the action? Journal of Geochemical Exploration 151, 34−40.

Rufino, R.D., Luna, J.M., Campos-Takkai, G.M., Ferreir, S.R.M., Sarubbo, L.A., 2012. Applications of the biosurfactant produced by *Candida lipolytica* in the remediation of heavy metals. Chemical Engineering Transactions 27, 61−66.

Sepúlveda, A., Schluep, M., Renaud, F.G., Streicher, M., Kuehr, R., Hagelüken, C., Andreas, C., Gerecke, A.C., 2010. A review of the environmental fate and effects of hazardous substances released from electrical and electronic equipments during recycling: examples from China and India. Environmental Impact Assessment Review 30, 28−41.

Sharma, R.K., Agrawal, M., Marshall, F., 2007. Heavy metal contamination of soil and vegetables in suburban areas of Varanasi, India. Ecotoxicology and Environmental Safety 66, 258−266.

Saffaj, N., Loukil, H., Younssi, S.A., Albizane, A., Bouhria, M., Persin, M., Larbot, A., 2004. Filtration of solution containing heavy metals and dyes by means of ultrafiltration membranes deposited on support made of Morrocan clay. Desalination 168, 301−306.

Sahayaraj, P.A., Ayyadurai, K., 2009. Bioaccumulation of lead in milk of buffaloes from Cooum River Belt in Chennai. Journal of Environmental Biology 30, 651–654.

Sajwan, K.S., Alva, A.K., Keefer, R.F., 1999. Biogeochemistry of Trace Elements in Coal and Coal Combustion Byproducts. Springer.

Salehizadeh, Shojaosadati, 2003. Removal of metal ions from aqueous solution by polysaccharide produced from *Bacillus firmus*. Water Research 37, 4231–4235.

Salipira, K.L., Krause, R.W., Mamba, B.B., Malefetse, T.J., Cele, L.M., Durbach, S.H., 2008. Cyclodextrin polyurethanes polymerized with multi-walled carbon nanotubes: synthesis and characterization. Materials Chemistry and Physics 111 (2–3), 218–224.

Sandstead, H.H., 1994. Understanding zinc: recent observations and interpretations. Journal of Laboratory and Clinical Medicine 124, 322–327.

Sanjay Kumar Dubey, J., Dubey, A.J., Viswas, Tiwari, P., 2011. Studies on Cyanobacterial biodiversity in paper mill and pharmaceutical industrial effluents. British Biotechnology Journal 1, 61–67.

Sazawal, S., Black, R.E., Bhan, M.K., Bhandari, N., Sinha, A., Jalla, S., 1995. Zinc supplementation in young children with acute diarrhea in India. New England Journal of Medicine 333, 839–844.

Seenivasan, S., Manikandan, N., Muraleedharan, N.N., Selvasundaram, R., 2008. Heavy metal content of black teas from south India. Food Control 19, 746–749.

Selvaraj, K., Ram Mohan, V., Szefer, P., 2004. Evaluation of metal contamination in coastal sediments of the Bay of Bengal, India: geochemical and statistical approaches. Marine Pollution Bulletin 49, 174–185.

Sharma, R.K., Agrawal, M., Marshall, F.M., 2008. Heavy metal (Cu, Zn, Cd and Pb) contamination of vegetables in urban India: a case study in Varanasi. Environmental Pollution 154, 254–263.

Shukla, D.P., Dubey, C.S., Singh, N.P., Tajbakhsh, M., Chaudhry, M., 2010. Sources and controls of Arsenic contamination in groundwater of Rajnandgaon and Kanker District, Chattisgarh Central India. Journal of Hydrology 395, 49–66.

Singh, A.K., Singh, M., 2006. Lead decline in the Indian environment resulting from the petrol-lead phase-out programme. Science of the Total Environment 368, 686–694.

Singh, K.P., Malik, A., Sinha, S., Singh, V.K., Murthy, R.C., 2005. Estimation of source of heavy metal contamination in sediments of Gomti River (India) using principal component analysis. Water Air and Soil Pollution 166, 321–341.

Singh, M., Müller, G., Singh, I., 2003. Geogenic distribution and baseline concentration of heavy metals in sediments of the Ganges River, India. Journal of Geochemical Exploration 80, 1–17.

Singh, R.K., Gupta, N.C., Guha, B.K., 2012. The leaching characteristics of trace elements in coal fly ash and an ash disposal system of thermal power plants. Energy Sources, Part A: Recovery, Utilization, Environmental Effects 34, 602–608.

Singh, S., 2005. Review of Urban Transportation in India. World Transit Research.

Sivaperumal, P., Sankar, T.V., Viswanathan Nair, P.G., 2007. Heavy metal concentrations in fish, shellfish and fish products from internal markets of India vis-a-vis international standards. Food Chemistry 102, 612–620.

Smith, D.R., Flegal, A.R., 1995. Lead in the biosphere: recent trends. Ambio 24, 21–23.

Soyol-Erdene, T., Huh, Y., Hong, S., Hur, S.D., 2011. A 50-Year Record of Platinum, Iridium, and Rhodium in Antarctic Snow: Volcanic and Anthropogenic Sources. Environmental Science and Technology 45, 5929–5935.

Srikanth, R., Rao, A.M., Kumar, C.S., Khanum, A., 1993. Lead, cadmium, nickel, and zinc contamination of ground water around Hussain Sagar lake, Hyderabad, India. Bulletin of Environmental Contamination and Toxicology 50, 138–143.

Srivastava, S., 2013. Sorption of divalent metal ions from aqueous solution by oxidized carbon nanotubes and nanocages: a review. Advanced Materials Letters 4, 2–8.

Streifel, S., 2006. Impact of China and India on global commodity markets: focus on metals & minerals and petroleum. Development Prospects Group. World Bank.

Stroh, E., Lundh, T., Oudin, A., Skerfving, S., Strömberg, U., 2009. Geographical patterns in blood lead in relation to industrial emissions and traffic in Swedish children, 1978–2007. BMC Public Health 9, 225.

Sukender, K., Jaspreet, S., Sneha, D., Munish, G., 2012. AAS estimation of heavy metals and trace elements in indian herbal cosmetic preparations. Research Journal of Chemical Sciences 2, 46–65.

Sum, E., 1991. The recovery of metals from electronic scrap. JOM Journal of the Minerals, Metals and Materials Society 43, 53–61.

Sun, F., Liu, H., Liang, B., Song, R., Yan, Q., Wang, A., 2013. Reductive degradation of chloramphenicol using bioelectrochemical system (BES): a comparative study of abiotic cathode and biocathode. Bioresource Technology 143, 699–702.

Sushil, S., Batra, V.S., 2006. Analysis of fly ash heavy metal content and disposal in three thermal power plants in India. Fuel 85, 2676–2679.

Sivakumar, D.S., Datta, M., 1996. Assessment of groundwater contamination potential around ash pond through field sampling: a review. In: Raju, V.S. (Ed.), Ash ponds and ash disposal systems. Narosa Publishing House, New Delhi, pp. 311–325.

Sahni, S.K., 2011. Hazardous metals and minerals pollution in India. A position paper: Indian National Science Academy, Bahadurshah Zafar Marg, New Delhi.

Tandukar, M., Huber, S.J., Onodera, T., Pavlostathis, S.G., 2009. Biological chromium (VI) reduction in the cathode of a microbial fuel cell. Environmental Science and Technology 43, 8159–8165.

Tao, H.-C., Liang, M., Li, W., Zhang, L.-J., Ni, J.-R., Wu, W.-M., 2011. Removal of copper from aqueous solution by electrodeposition in cathode chamber of microbial fuel cell. Journal of Hazardous Materials 189 (1–2), 186–192.

Thangavel, P., Subbhuraam, C., 2004. Phytoextraction: role of hyperaccumulators in metal contaminated soils. Proceedings of the Indian National Science Academy, Part B: Biological Sciences 70, 109–130.

TIFAC, 2001. Non Conventional Sources of Plant Nutrient & Soil Conditioners to Enhance Agricultural Productivity. http://tifac.org.in/index.php?option=com_content&view=article&id=714&Itemid=205 (accessed 31.05.12.).

Tunali, S., Cabuk, A., Akar, T., 2006. Removal of lead and copper ions from aqueous solutions by bacterial strains isolated from soil. Chemical Engineering Journal 115, 203–211.

Umrania, V.V., 2006. Umrania, Bioremediation of toxic heavy metals using acidothermophilic autotrophes. Bioresour. Technol. 97, 1237–1242.

USFDA, 1993. Food and Drug Administration Guidance Document for [various Elements] in Shellfish. Washington, DC.

US Department of Health and Human Services, 2007. Toxicological profile for lead. Public health service, Agency for toxic substance and disease registry. Georgia US, Atlanta.

van der Ent, A., Baker, A.J., Reeves, R.D., Chaney, R.L., Anderson, C.W., Meech, J.A., Erskine, P.D., Simonnot, M.O., Vaughan, J., Morel, J.L., Echevarria, G., Fogliani, B., Rongliang, Q., Mulligan, D.R., 2015. Agromining: farming for metals in the future? Environmental Science & Technology 49, 4773–4780. http://dx.doi.org/10.1021/es506031u.

Venkata Mohan, S., Veer Raghavulu, S., Dinakar, P., Sarma, P.N., 2009. Integrated function of microbial fuel cell (MFC) as bio-electrochemical treatment system associated with bioelectricity generation under higher substrate load. Biosensors & Bioelectronics 24, 2021–2027.

Venkata Mohan, S., Velvizhi, G., Annie Modestra, J., Srikanth, S., 2014a. Microbial fuel cell: critical factors regulating bio-catalyzed electrochemical process and recent advancements. Renewable & Sustainable Energy Reviews 40, 779–797.

Venkata Mohan, S., Velvizhi, G., Vamshi Krishna, K., Lenin Babu, M., 2014b. Microbial catalyzed electrochemical systems: a bio-factory with multi-facet applications. Bioresource Technology 165, 355–364.

Venkata Mohan, S., Annie, J., Amulya, K., Butti, S.K., Velvizhi, G., 2016. A circular bioeconomy with biobased products from CO_2 Sequestration. Journal of Trends in Biotechnology. http://dx.doi.org/10.10.1016/j.tibtech.2016.02.012 in press.

Vigneswaran, S., Ngo, H.H., Chaudhary, D.S., Hung, Y.T., 2004. Physico-chemical treatment processes for water reuse. In: Wang, L.K., Hung, Y.T., Shammas, N.K. (Eds.), Physicochemical Treatment Processes, 3. Humana Press, New Jersey, pp. 635–676.

Wang, G., Huang, L., Zhang, Y., 2008a. Cathodic reduction of hexavalent chromium [Cr(VI)] coupled with electricity generation in microbial fuel cells. Biotechnology Letters 30, 1959–1966.

Wang, L.K., Hung, Y.T., Shammas, N.K., 2007. Advanced physicochemical treatment technologies. In: Handbook of Environmental Engineering, 5. Humana press, New Jersey, p. 710.

Wang, S., Ang, H.M., Tade, M.O., 2008b. Novel applications of red mud as coagulant, adsorbent and catalyst for environmentally benign processes. Chemosphere 72, 1621–1635.

Wang, Z., Lim, B., Choi, C., 2011. Removal of Hg^{2+} as an electron acceptor coupled with power generation using a microbial fuel cell. Bioresource Technology 102 (10), 6304–6307.

Wang, H., Ren, Z.S., 2014. Bioelectrochemical metal recovery from wastewater: a review. Water Research 66, 219–232.

Wang, W.X., Griscom, S.B., Fisher, N.S., 1997. Bioavailability of Cr-(III) and Cr(VI) to marine mussels from solute and particulate pathways. Environmental Science & Technology 31, 603–611.

WCA, 2012. Coal Statistics – World Coal Association. http://www.worldcoal.org/resources/coal-statistics/ (accessed 31.05.12).

Wu, G., Wu, J., Shao, H., 2012. Hazardous heavy metal distribution in Dahuofang catchment, Fushun, Liaoning, an important industry city in China: a case study. CLEAN – Soil Air Water 40, 1372–1375.

Yu, W., Zhang, Q., Brown, M.H., 2013. Microbial Efflux Pumps Current Research, ISBN 978-1-908230-21-8.

Yager, J.W., Hicks, J.B., Fabianova, E., 1997. Airborne arsenic and urinary excretion of arsenic metabolites during boiler cleaning operations in a slovak coal-fired power plant. Environmental Health Perspectives 105 (8), 836–842.

Zabochnicka Swiatek, Krzywonos, 2014. Potentials of biosorption and bioaccumulation processes for heavy metal removal. Polish Journal of Environmental Studies 23, 551–561.

Zhang, L.J., Tao, H.C., Wei, X.Y., Lei, T., Li, J.B., Wang, A.J., Wuc, W.J., 2012. Bioelectrochemical recovery of ammonia–copper (II) complexes from wastewater using a dual chamber microbial fuel cell. Chemosphere 89, 1177–1182.

APPENDIX 1: EMISSION FACTORS OF TRACE METALS

Table A1a Emission Factors From Combustion (g ton^{-1}) (Pacyna and Pacyna, 2001)

Element	Coal	Oil	Cement	Pig Iron/Steel
As	0.2	0.02	0.2	0.5
Cd	0.1	0.05	0.01	0.1
Cr	1.7	1	1	4

Table A1a Emission Factors From Combustion (g ton^{-1}) (Pacyna and Pacyna, 2001)—cont'd

Element	Coal	Oil	Cement	Pig Iron/Steel
Cu	1.4	0.5	—	0.2
Hg	0.5	0.06	0.1	0.04
Mn	1.5	1	—	1.5
Mo	0.4	0.3	—	—
Ni	2	20	0.1	0.05
Pb	1	2	0.2	1.5
Sb	0.2	—	—	0.01
Se	0.8	0.3	0.002	0.01
Sn	0.1	0.8	—	—
Tl	0.5	—	—	—
V	1	60	—	0.1
Zn	1.5	1	2	3

Table A1b Emission Factors From Production (g ton^{-1}) (Pacyna and Pacyna, 2001)

Element	Primary Cu	Secondary Cu	Primary Pb	Secondary Pb	Primary Zn
As	500	2	10	0.5	50
Cd	200	3	15	2.5	200
Cu	3000	100	10	1	50
Hg	10	—	3	—	20
In	4	—	—	—	3
Mn	10	—	—	—	—
Ni	1500	1	10	—	—
Pb	1000	90	2000	100	500
Sb	50	3	75	—	15
Se	50	—	35	—	15
Sn	50	—	—	—	—
Zn	500	200	80	20	10,000
V	10	—	—	—	—

Table A1c Emission Factors for Waste Incineration (g ton^{-1}) (Pacyna and Pacyna, 2001)

Elements	Municipal	Sewage
As	1.1	5
Cd	0.4	1
Cr	0.7	50

Continued

Table A1c Emission Factors for Waste Incineration ($g\ ton^{-1}$) (Pacyna and Pacyna, 2001)—cont'd

Elements	Municipal	Sewage
Cu	7	10
Hg	1	5
Mn	1.8	50
Ni	0.7	10
Pb	10	80
Sb	3	5
Se	0.2	1
Sn	1	5
Zn	20	50
V	3	

Table A1d Emission Factors of Pb at Different Phases in India's Shift From Leaded Petroleum

Year	Petroleum Pb ($g\ L^{-1}$)
1980–93	0.560
1994	0.508
1995	0.491
1997	0.150
1998	0.150
1999	0.013
1999–2012	0.013

Note: During 1994 only Delhi, Mumbai, Calcutta, and Chennai shifted to 0.15 $g\ L^{-1}$ Pb rather than in 1995 to 0.013. Using the fact that an estimated 12.7% of vehicles are in this metropolitan area (Singh, 2005) and assuming that this carries over to 12.7% of petroleum use, the factors for 1994 and 1995 were calculated by summing the metropolitan petroleum (Pb) and the rest of India's petroleum (Pb) multiplied by their respective percentages.
Adapted from data in ENVIS, 1999 Auto Emissions http://cpcbenvis.nic.in/newsletter/autoemissionjun1999/jun1999step.htm (accessed 31.05.12.).

Table A1e Emission Factors for Wood Combustion

Element	Factor (lb/MM BTU)
Sb	7.9E-06
As	2.2E-05
Cd	4.1E-06
Cr	2.1E-05
Cu	4.9E-05
Pb	4.8E-05
Hg	3.5E-06
Ni	3.3E-05
Zn	0.00042

CHAPTER

RESOURCES RECOVERY FROM WASTEWATER BASED ON EXTRACELLULAR ELECTRON TRANSFER

16

F. Zhao, D.M. Gurumurthy

Institute of Urban Environment, Chinese Academy of Sciences, Xiamen, China

CHAPTER OUTLINE

16.1 INTRODUCTION TO EXTRACELLULAR ELECTRON TRANSFER

16.1.1 BACKGROUND

It has been a century since M.C. Potter first reported his seminal work entitled "Electrical effects accompanying the decomposition of organic compounds" in 1911 (Potter, 1911). He recognized that eukaryotic yeast, *Saccharomyces cerevisiae* and certain species of bacteria created electrical energy while degrading glucose in platinum-based electrodes (Potter, 1910, 1911). Over 2 decades later, Barnet Cohen built "bacterial batteries" which produced a frail current for a short span. In his research, he noticed that actively growing microbial cultures generated the great potential of 35 V at a current of

Environmental Materials and Waste. http://dx.doi.org/10.1016/B978-0-12-803837-6.00016-0

0.2 mA when they were connected in series (Cohen, 1931). The efforts of Potter and Cohen remained hidden until early 1960, when they were revived as "biofuel cells" with the interest of converting biowaste to energy in spacecraft (Canfield et al., 1963; Bean et al., 1964). The first patent to describe microbial fuel cell (MFC) technology was issued to John Davis in 1967; he used external mediators to transfer electrons to produce alcohol, aldehyde, and carboxylic acid from hydrocarbon-oxidizing *Nocardia salmonicolor* (Biffinger and Ringeisen, 2008). However, interest gradually diminished owing to rapid advances in other energy technologies such as photovoltaics and lower prices for mineral oils. In the 1980s to 2000, many works including that by Bennetto extended the previous idea of MFCs for power generation to self-sustainable applications (Bennetto, 1987; Wilkinson, 2000a,b; Wilkinson and Campbell, 1996).

Although it is recent, interest in extracellular electron transfer (EET) is slowly emerging because rapidly growing industrialization and human activities have created waste that is damaging to the environment. The development of sustainable microbial technologies for the recovery of useful products from wastes and wastewaters is urgently needed. Despite the challenges in waste and wastewater treatment, producing bioenergy and recovering useful resources from microbial-mediated transformation have been key issues. After the recovery of metals, it is highly difficult to recover nutrients, which are secondary useful products, mainly owing to lack of knowledge about microbial respiration and the early state of technology. Hence, it is necessary to understand microbial physiology and technology development with great economic incentives. This chapter on resources recovery based on the extracellular electrons of microorganisms provides new insight into microbial respiration, which has excellent performance in removing pollutants and recovering resources from different wastes and wastewaters.

16.1.2 MICROBIAL "RESPIRATION"

It has long been known that energy generation from metabolism is a fundamental process of all living organisms (Lehninger, 1971). The enigma of energy generation in cells is solved by electron resilience to or from extracellular substrates linked to proteins, metabolites, and reduction-oxidation (redox) reactions in the cell (Lehninger, 1971; Nicholls and Ferguson, 2002; Green and Zande, 1981). Interestingly, the term "microbial respiration" or "breath" was used in the early work of Antoine Lavoisier (1789), in which it was stated that most aerobic organisms derive energy needed for their life process by burning foodstuffs with molecular oxygen (Duveen and Klickstein, 1954). In contrast to prokaryotes, eukaryotic metabolic pathways to extract energy depend on firm groups of organic compounds as electron donors, whereas prokaryotes have evolved as dexterous scavengers of diverse organic and inorganic electron donors and acceptors (Gottschalk, 1986). Today it is well known that prokaryotes create energy by substrate-level phosphorylation of molecules (electron donor) and later delaminate into protons (H; when charged, H^+) and electrons (e; when charged, e^-), in which protons are consumed inside and liberated outside the membrane. Fig. 16.1 illustrates the generation of an electrochemical proton gradient in the inner membrane and periplasmic space for the synthesis of adenosine triphosphate (ATP) in the case of microorganisms. With the help of redox chain protein complexes and carrier molecules on the cell surface, electrons are then transported to an electron acceptor (oxidants) to reduce molecular oxygen to water and produce an additional proton. This redox reaction pumps additional protons to the same membrane side, creating an electrochemical potential across the membrane that itself has a difference in pH (alkaline inside) and an electric potential

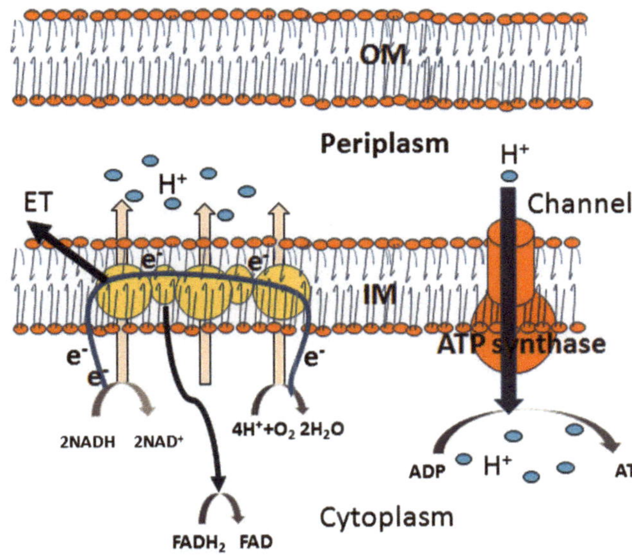

FIGURE 16.1

Generation of electrochemical proton gradient inside membrane and in periplasmic space for the synthesis of ATP.

(negative side). The electrochemical proton gradient across the membrane can be transformed into ATP molecules for cellular functions with the help of ATP synthase located in the membrane (Gottschalk, 1986; Oren, 2002).

16.2 EXTRACELLULAR ELECTRON TRANSFER

Scientists are still attempting to understand details of how electrons transfer across the cell membrane, including what the prerequisites are for electron donors and acceptors and how to arrive at appropriate newer biofuel designs. Over the past decades, environmental microbiologists and biogeochemists have devoted study to microbial respiration that involves electron transfer between cell-insoluble "mineral" substrates and cell–electrode surfaces in bioelectrochemical systems. However, the fundamental quest is still unclear because:

- It is difficult to study microbe–mineral interactions, because it involves complex sets of parameters (eg, various soluble and insoluble electron acceptors and metabolic processes)
- It involves a species-specific electron transfer strategy

Understanding microbial interactions with mineral surfaces is a ideal criterion for electrochemical systems, biogeochemical processes, and environmental remediation. Studies on dissimilar metal-reducing microbes that use minerals as terminal electron acceptors are the prime focus of current research because these microbes evolve as a result of EET that reduces minerals (primarily iron, magnesium oxide, and possibly all other minerals) (Lovley, 1993, 2006b). A number of mechanistic, genetic, and molecular observations for EET involved prokaryotic microorganisms. Breakthroughs in genome-scale metabolic modeling in *Geobacter* spp., a gram-negative, iron-reducing, anaerobe, provided in-depth understanding of the course of electron transfer capability (Reguera et al., 2005; Lovley, 2006b, 2008a,b; Butler et al., 2010). This group of bacteria has direct contact with iron's surface to reduce in the absence of exogenous electron shuttles without forming a thick biofilm owing to its nutrient limitation. On other hand, *Shewanella* is a common gram-negative facultative anaerobe mainly used as a redox mediator for EET (Lies et al., 2005; Myers and Myers, 1992, 1993, 2003). Although both *Geobacter* and *Shewanella* can make outer-membrane proteins and cytochromes in electron transfer, at least in the case of *Shewanella*, more than one pathway for mineral respiration may exist: a pathway that requires direct contact and one that does not (Reguera et al., 2005; Lovley, 2006b, 2008a,b; Butler et al., 2010; Lies et al., 2005; Myers and Myers, 1992, 1993, 2003).

16.2.1 EXTRACELLULAR ELECTRON TRANSFER TYPE AND MECHANISM

Like *Geobacter* and *Shewanella*, commonly used groups for these studies including *Acidotheobacillus, Pseudomonas*, and gram-positive *Bacillus* spp. pointed to the role of these outer-membrane proteins and cytochromes in catalyzing electron transfer (Yarzabal et al., 2002; Hernandez and Newman, 2001). To give an idea, electrons transfer ranges from a short distance (<2 nm) to several centimeters. Long-distance electron transfer involves cellular appendages or redox carriers and other small molecules (Sunil et al., 2012). Among many bacteria, electron tunneling has a crucial role in short-distance electron transfer, ie, between respiratory chain redox sites of membranes separated by less than 2 nm (Yarzabal et al., 2002; Hernandez and Newman, 2001; Sunil et al., 2012). However, long-distance electron travel in various microbial systems is explained by two major categories: indirect/mediated EET and direct EET.

16.2.2 INDIRECT ELECTRON TRANSFER

In indirect electron transfer, cells organize themselves by means of small molecules that are capable of redox cycling (ie, as electron mediators), to chelate metals and deliver them to an intracellular metal oxidoreductase, particularly in biofilms, in which cells cannot establish direct contact with the mineral surface. However, these molecules also require some outer membrane proteins for reduction. In the case of redox cycling, a single shuttle molecule can cycle (back and forth between cells and minerals) as a terminal electron acceptor (ie, reduced form) to become oxidized and transfer its electrons to iron minerals. Exogenous shuttles generally are chelators, which are well known to enhance the rate of reduction by stimulating and making Fe(III) or other minerals by more accessibility to cells.

Some artificial redox organic compounds such as phenazines, phenothiazines, phenoxazines, and viologen (Lovley et al., 1993, 1996; Coates et al., 1998) with their structural notations are presented in Fig. 16.2. Organic quinone-containing compounds have been shown to stimulate Fe(III) reduction effectively. Compounds such as anthraquinone-2,6-disulfonate has gained much interest as a redox-active shuttle because its highly soluble nature could be accessible to all classes of bacteria as a terminal electron acceptor and its bioavailability provides the cell with comparatively more energy gained per unit time. Certain ligands such as citrate and nitrilotriacetic acid are able to reduce extracellularly; however, their function is contradictory (Lovley et al., 1993, 1996; Coates et al., 1998). In addition to organic molecules, there have been some studies on inorganic molecules such as sulfide and reduced uranium reacted with iron minerals, which are not discussed further here. Generally these

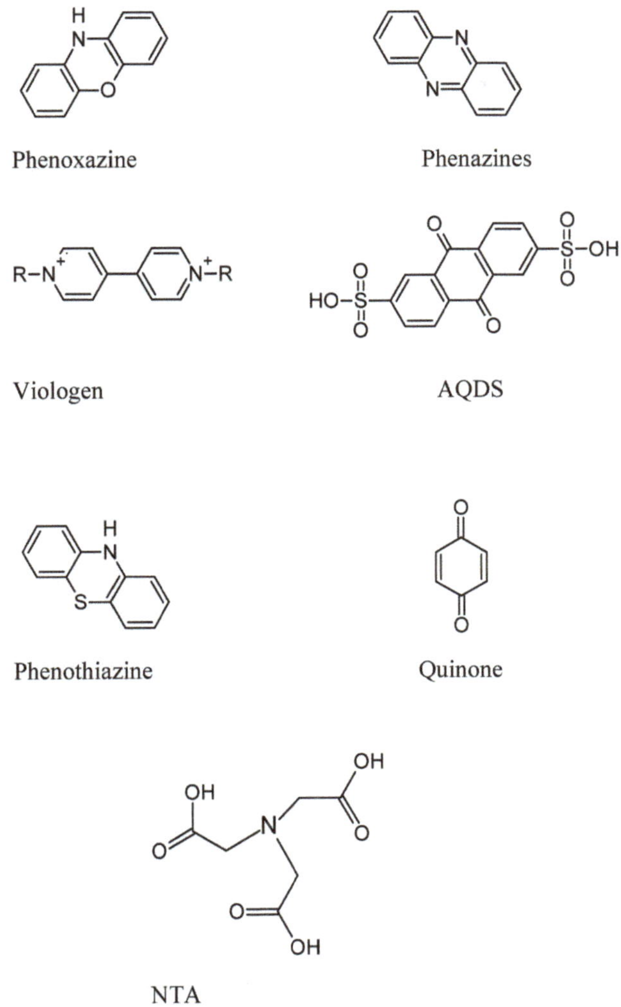

FIGURE 16.2

Examples of artificial redox mediators used in bioelectrochemical systems.

processes are regulated by oxidoreductase enzymes, but their respective subcellular locations have not yet been determined to have known shuttles.

Endogenous pathways are similar to exogenous electron transfer, in which shuttles are produced by the microbes themselves. Fig. 16.3 lists some examples of exogenous mediators produced by microorganisms. Some extracellular quinone-like molecule (menaquinone) is recovered from wild-type *Shewanella oneidensis* MR-1, which is known for electron shuttles (described previously). The MenC mutant strain of *S. oneidensis* H2 is unable to produce a quinone-like substance, in which MenC encodes *o*-succinylbenzoic acid synthase, an enzyme involved in the biosynthesis of

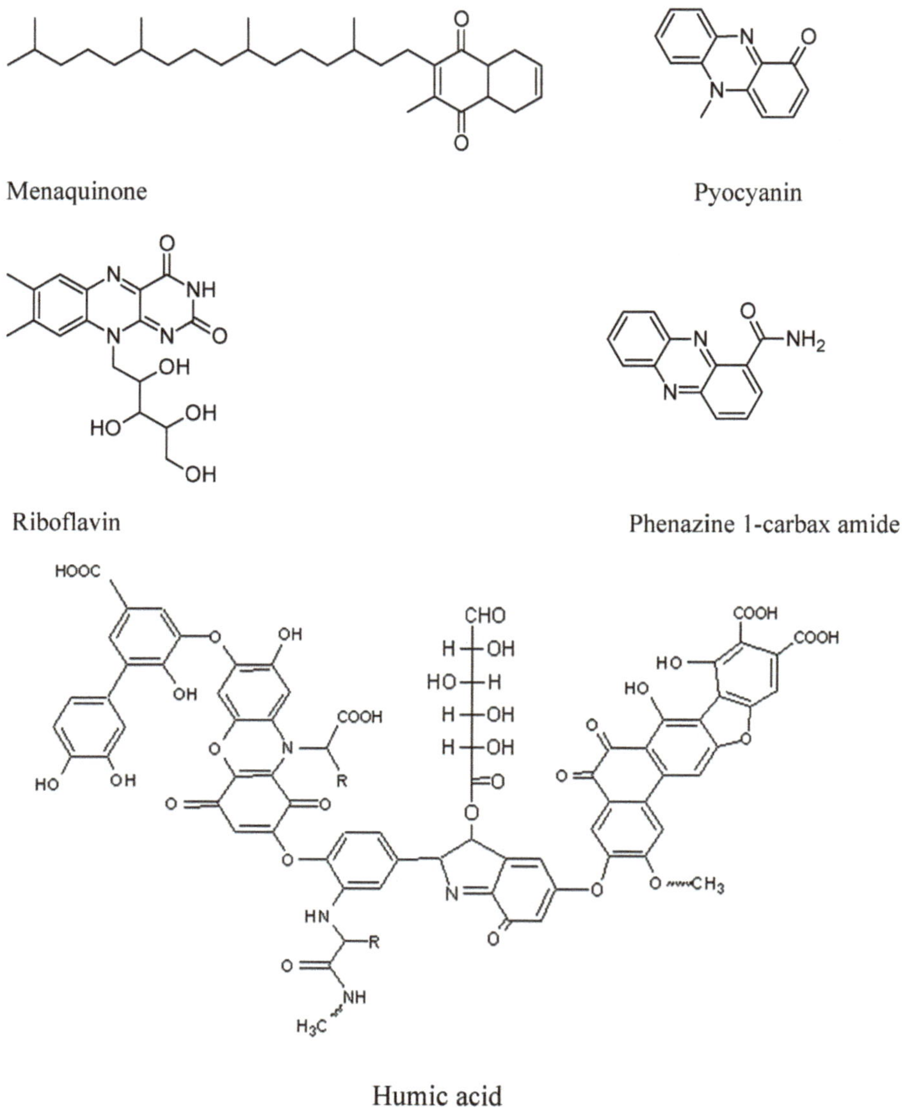

FIGURE 16.3

Endogenous mediators for electron transfer.

membrane-diffusible menaquinone, which acts as an electron mediator between proteins. That the MenC mutant is unable to make the extracellular factor shows that menaquinone might also be an interspecies cycling shuttles. The role of flavin in exogenous electron shuttling in *S. oneidensis* in mineral oxide reduction demonstrates that electron shuttling accounts for about 75% of the substrate respiratory capacity. Flavin electron shuttles may be important for the ability of *S. oneidensis* to

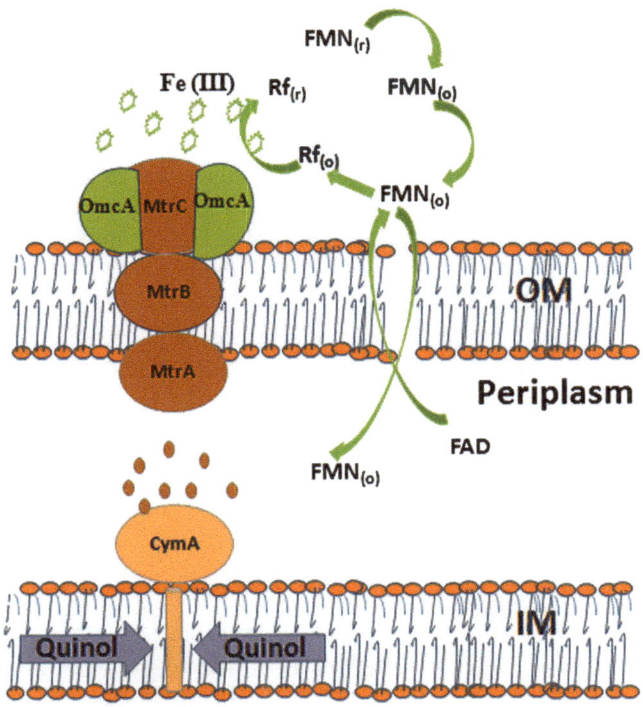

FIGURE 16.4

Mechanism of indirect electron transfer in *Shewanella*, including flavins and C type cytochromes.

Drawn based on information available from VonCanstein, H., Ogawa, J., Shimizu, S., Lloyd, J.R., 2008. Secretion of flavins by Shewanella *species and their role in extracellular electron transfer. Applied and Environmental Microbiology 74, 615–623.*

respire to other insoluble substrates (Fig. 16.4) (Myers and Myers, 1993; Saffarini et al., 2002; Newman and Kolter, 2000; VonCanstein et al., 2008; Brutinel and Gralnick, 2012; Velasquez-Orta et al., 2010).

Humic substances are an example of exogenous shuttles, which are large, complex, and poorly defined organic compounds. Because they are a weak chelator, they cannot account for the degree of stimulation to reduce iron minerals. In the humic-stimulated environment, cells (hydroquinone state of humic) abiotically reduced Fe(III) by cyclic reduction (Gralnick and Newman, 2007). However, it is known that the quinone moiety in the humic substance acts as an electron acceptor in microbial respiration.

16.2.3 DIRECT ELECTRON TRANSFER

The direct pathway is generally assumed to be a direct interaction between minerals and cells involved in membrane-bound redox active *c*-type cytochromes and multicopper oxidases (outer membrane or exosporium) (Yarzabal et al., 2002; Hernandez and Newman, 2001; Sunil et al., 2012). However, it is unclear how these microorganisms establish contact with the solid electron acceptor and what physical force acts to build contact from the cell or membrane. The answer to these questions was partially given when it was discovered that OmcA, MtrC, and MtrB were outer membrane *c*-type cytochromes from *S. oneidensis* (Lower et al., 2007; Meitl et al., 2009; Myers and Myers, 2002, 2003; Ross et al., 2009; Reardon et al., 2010). Fig. 16.5 depicts the entire direct electron transfer in *S. oneidensis* MR-1. Proteins OmcA and MtrC can bind hematite with high affinity and transfer electrons directly to it but not to goethite (Shi et al., 2006). In addition, for redox proteins, electron transfer also occurs by means of extracellular polymeric substances, which may affect adherence to mineral surfaces and mediate metal reactivity, because these are sites for the reduction of minerals (Reardon et al., 2010; Lower et al., 2009; Marshall et al., 2006). A proteomic study showed that homologs of MtrC and OmcA are key components of bound and loosely associated extracellular polymeric substances isolated from the biofilm of the metal-reducing bacterium *Shewanella* sp. HRCR-1 (Cao et al., 2011). These results suggest that after their release from the bacterial cell surface, MtrB, MtrC, and OmcA may not be in association with MtrA. Although their roles in Fe(III) oxide reduction and their relationship with *Shewanella* nanowires and the outer-membrane vesicles are not currently characterized, it is proposed that extracellular polymeric substance-associated MtrC, OmcA, and probably MtrB may be part of

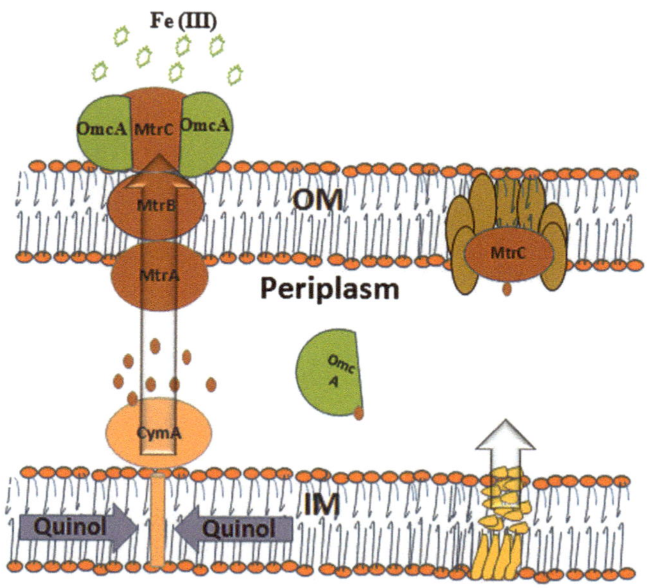

FIGURE 16.5

Proposed Mtr- and Omc-based direct EET pathway of *S. oneidensis* MR-1.

Drawn based on information available from Lower, B.H., Shi, L., Yongsunthon, R., Droubay, T.C., McCready, D.E., Lower, S.K., 2007. Specific bonds between an iron oxide surface and outer membrane cytochromes MtrC and OmcA from Shewanella oneidensis *MR-1. Journal of Bacteriology 189, 4944–4952; Meitl, L.A., Eggleston, C.M., Col-berg, P.J.S., Khare, N., Reardon, C.L., Shi, L., 2009. Electro-chemical interaction of* Shewanella oneidensis *MR-1 and its outer membrane cytochromes OmcA and MtrC with hematite electrodes. Geochimica et Cosmochimica Acta 73, 5292–5307; Myers, C.R., Myers, J.M., 2002. MtrB is required for proper incorporation of the cytochromes OmcA and OmcB into the outer membrane of* Shewanella putrefaciens *MR-1. Applied Environmental Microbiology 68, 5585–5594; Myers, C.R., Myers, J.M., 2003. Cell surface exposure of the outer membrane cytochromes of* Shewanella oneidensis *MR-1. Letters in Applied Microbiology 37, 254–258.*

nonlocal electron transfer strategy used by *S. oneidensis* MR-1, acidophilic iron-oxidizing bacteria; and *Bacillus* spp. for reduction of Fe(III) oxide minerals are distant from the bacterial cell surface (Lower et al., 2009; Marshall et al., 2006; Cao et al., 2011; Rosso et al., 2003; Gorby et al., 2008; Bose et al., 2009).

"Conductive nanowire" is another pathway of direct electron transfer; it generally refers to the "pili" of the microorganism, especially those studied in *G. sulfurreducens* and *S. oneidensis* for Fe(III) oxide reduction (Gorby et al., 2006; Reguera et al., 2005, 2006). Fig. 16.6 depicts *G. sulfurreducens* electron transfer through long and thin pili acting like a sort of biological nanowire. The basis behind this high affinity toward minerals has yet to be investigated; some studies on the electrical conductivity of pili suggested that they are involved in direct electron conduits between the cell and the mineral surface. In the terminal step of *G. sulfurreducens*, electrons are required to pass through C-type cytochrome OmcS, which acts as a barrier for electron transfer to Fe(III) oxide reduction. It is proposed that electron conduction is along the length of the pili through the conjugated orbital of aromatic amino acids in the pilin constituent. Like *G. sulfurreducens*, *S. oneidensis* produces electrically conductive nanowires, which are regulated by outer-surface C-type cytochromes OmcB (MtrC) and OmcA in Fe(III) reduction (Reguera et al., 2005, 2006).

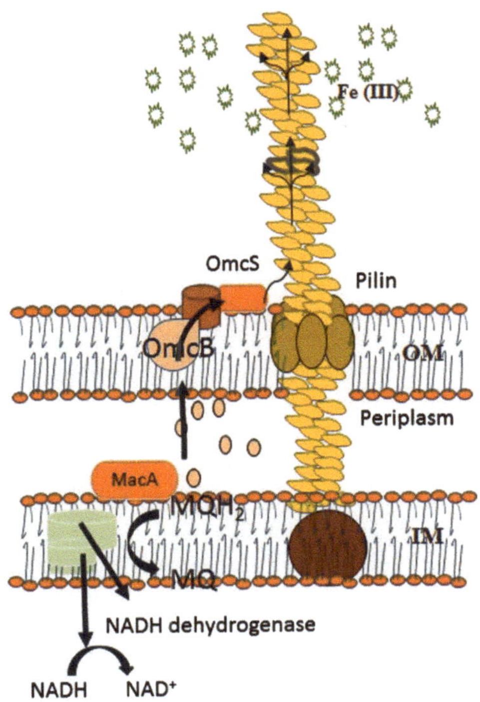

FIGURE 16.6

Electron transfer mechanism in *G. sulfurreducens* biological nanowires.

Modified from Lovley, D.R., 2006. Bug juice: harvesting electricity with microorganisms. Nature Review of Microbiology 4, 497–508.

16.3 RESOURCES RECOVERY BASED ON EXTRACELLULAR ELECTRON TRANSFER

16.3.1 METALS

Bio-inspired technology for metal recovery is revolutionary for modern remediation and a sustainable environment. The technology was originally inspired by the "bioleaching" ability of microorganisms, in which microbial reduction and oxidation mechanisms and EET have an important role in recovering various metals from solid substrates (minerals and electronic waster [e-waste]) and metal-laden wastewater. Many of the initial studies were concerned with copper (Cu^0) recovery based on the bioleaching ability of *Acidithiobacillus* spp., in which EET to Fe(II) and sulfur in chalcopyrite are oxidized to yield Cu^0 (Ilyas et al., 2007). Other studies were concerned with various microorganisms with an EET ability, including *Geobactor*, *Shewanella*, and *Desulfovibrio*, which were previously studied for their Fe(III) and Mn(IV) oxidation-reduction potential and were also used to recover precious metals including Pt(IV), Pd(II), Rd(III), Ag(I), and Au(III) (Brandl et al., 2001). Some studies showed that the accumulation of bacteria on elemental metal directs the reduction process through redox machineries on the cell surface or in periplasmic spaces. Several reports are available on adding acid such as sulfuric acids to the leachate mixture to enhance the reduction process. In the case of Cu, it was observed that upon adding sulfuric acid to *Acidithiobacillus* spp. leachate, recovery of Cu was greatly enhanced several fold, which may be a result of the influence of acids on the redox machinery in *acidophiles* (Brandl et al., 2001; Hutchins et al., 1986; Lloyd, 2003). Iron—sulfur-oxidizing bacteria participation in the recovery of precious metals such as such as Ag, Au, Zn, and Ni from e-waste, catalyst batteries, and electroplate wastewater has been reported (Brandl et al., 2001; Hutchins et al., 1986; Lloyd, 2003; Choi and Hu, 2013). These iron—sulfur bacteria have the same potential for metal recovery as Fe(III) and Mn(IV) oxidation reduction, as previously discussed.

At least four different mechanisms have been proposed for microbial electrochemical-assisted metal recovery from waste and wastewater (Fig. 16.7). These mechanisms are characterized by the redox potential for direct metal reduction at the cathodic environment. In the first, "direct metals recovery using abiotic cathode," in which the cathode chamber directly influences the reduction of soluble metal oxides. In a typical microbial electrochemical system, in the reduction of metal oxides by electrochemically active bacteria using an organic electron donor, the anode potential would be generally constant and is influenced by the redox potential of bacteria and a solution. The cathode potential can vary depending on the electron acceptors used spontaneously. Their redox potentials need to be higher than the anode potential in the given condition. The reduction takes place spontaneously and is thermodynamically favorable, and the metals can be used directly as the electron acceptor with no external power consumption. The reduction and recovery (60—99%) of metals such as Au(III), Ag(I), Cu(II), V(V), Cr(VI), Fe(III), and Hg(II) with simultaneous power generation was determined using a two-chambered MFC (Choi and Hu, 2013; Zhang et al., 2009, 2012; Li et al., 2008; Wang et al., 2008, 2010, 2011; Heijne et al., 2010; Tandukar et al., 2009; Tao et al., 2011a,b,c, 2012). Several studies revealed that reduction and separation of metals is not feasible for wastewater treatment because metals and organic matter coexist in the same water stream and the power output of the two-chamber MFC was low as a result of its high internal resistance, but the reduction of metals in a reactor is capable of generating its co-products, which enables commercial application (Wang and Ren, 2014). As an example, the reduction Cu(II) to Cu^0 also generates co-products such as Cu_2O, CuCl,

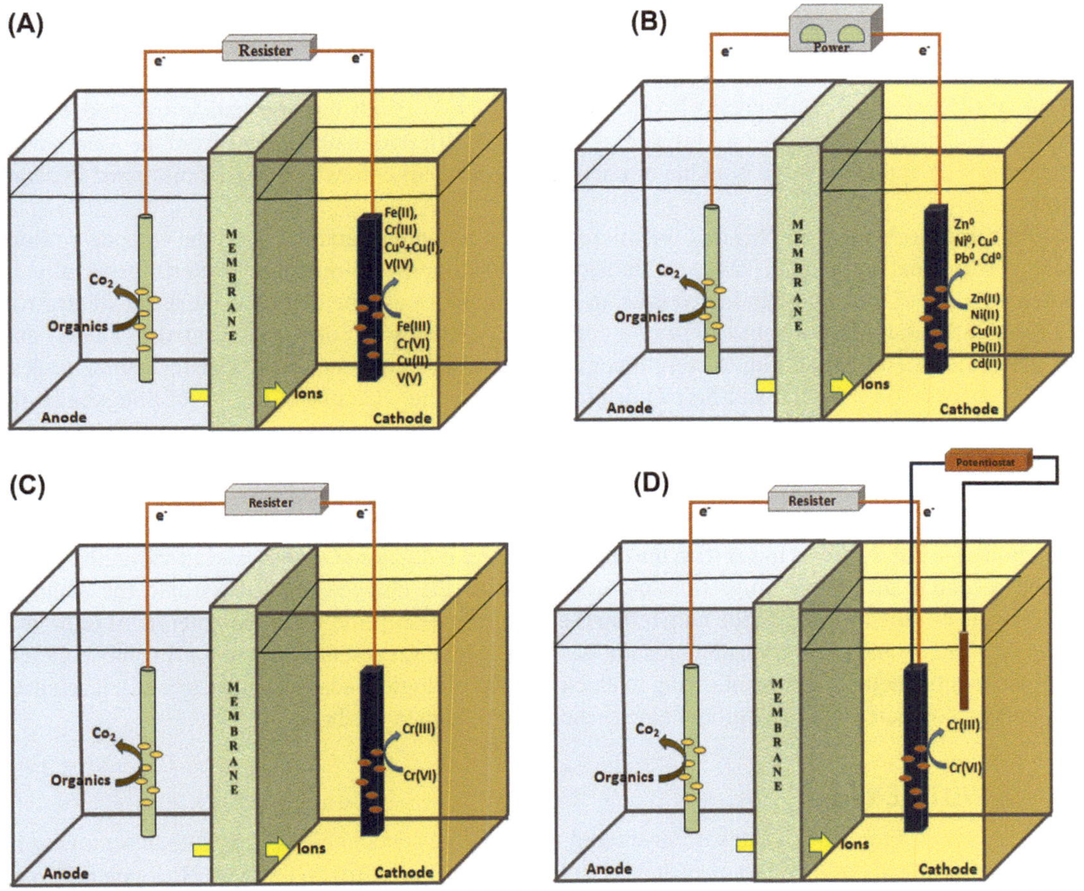

FIGURE 16.7

Mechanism of bioelectrically assisted metal recovery in a microbial electrochemical system. (A) Direct recovery using abiotic cathode. (B) Abiotic cathode supplemented with external power. (C) Metal conversion using a biocathode. (D) Metal conversion using a biocathode supplemented with external power.

Redrawn with reference from Wang, H., Ren, Z.J., 2014. Bioelectrochemical metal recovery from wastewater: a review. Water Research 66, 219–232.

$Cu_4(OH)_6SO_4$, or $CuSO_4 \cdot 3Cu(OH)_2$. Like Cu(II), Cr(VI) will become reduced to Cr^2O_3, Cr^{3+}, and $Cr(OH)_3$, whereas V(V) is reduced to V(IV) and reoxidized as the deposits of Na_3VO_4 and V_2O_5. Approximately 99% of gold (Au) and Ag(I) was recovered in a two-chamber MFC with the maximum power production of 6.58 and 4.25 W m^{-2}, respectively, in the presence of a catholyte (Choi and Hu, 2013; Choi and Cui, 2012). Research considering the use of single-chamber, three-chambered, and strip-chambered reactors has been developed for wastewater treatment along with the removal of Cd, Zn, Se(IV), and As(III), which are not given importance here.

In the second method, metals with lower redox potentials than anode potentials can be reduced at the cathode using external power. Such metals include Ni(II), Pb(II), Cd(II), and Zn(II), which generally require less than 2 V for reduction. Approximately 51 ± 4.6% and 67 ± 5.3% of Ni(II) was reduced at the externally applied 0.5−1.1 V but changing the voltages resulted in a sequential reduction of mixed metals in one catholyte, with the sequence of Cu(II), Pb(II), Cd(II), and Zn(II) (0, 0.34, 0.52, and 1.7 V, respectively) (Modin et al., 2012). With this mechanism of recovery by applying a variable voltage, metal ions were reduced with different removal rates from the wastewater. The setup with applied voltage is a kind of microbial electrolysis cell, which will be discussed in detail subsequently.

The third mechanism emphasizes the microbial conversion of metal oxide on the cathode without using an external potential. This was studied in dissimilar metal-reducing bacteria especially in bioremediation of metal-contaminated sites, in which microbes use metal ions as the terminal electron acceptors and assimilate within as enzyme cofactors. A study was conducted related to Cr(VI) and Se(VI) reduction by dissimilar metal-reducing bacteria and conversion or transformation to less mobile and less toxic Cr(III) and Se(IV) or precipitation to Se^0 (Combs et al., 1996; Shugaba et al., 2012; Tandukar et al., 2009; Garbisu et al., 1996; He and Yao, 2011).

Similar to the second mechanism, the fourth reduction mechanism requires external power required for low-potential metals. An example is the case of sedimentary recovery of U(VI) reduced by *G. sulfurreducens*, in which the external potential of −500 mV was applied to the cathode for the reduction of metal, which is lower than the electrochemical reduction (Gregory and Lovley, 2005). The poised potential attracts the migration and adsorption of the charged metal ions onto the cathode, whereas the biofilm on the cathode help retain the ions from desorption when the potential is removed. Metal conversion and recovery using biocathodes with a small set potential can quickly remove metals from a contaminated environment owing to electrical and biological adsorption; as a result, it requires less energy compared with traditional electrochemical reduction methods.

16.3.2 NUTRIENTS

Recovery of nitrogen from wastes is considered to be a more sustainable approach than removing it, because of depleted natural resources and the significant cost of nitrogen fixation. Nitrogen recovery in a microbial electrochemical system is mainly through the migration and recovery of ammonia. Methods have been developed for ammonia recovery in which the conversion of nitrogen to ammonium in a microbial electrochemical system takes place with simultaneous nitrification and denitrification, and then the produced nitrate can be further reduced using a cathode. Nitrates and nitrites can be used interchangeably as a electron acceptor for autotrophic denitrification at the cathode, but nitrite is oxidized via biological or electrochemical processes if oxygen is present (Desloover et al., 2012).

In MFCs, 100% ammonia recovery efficiency and reduced ammonia inhibition upon anaerobic digestion can be observed (Haddadi et al., 2013). A higher current density can greatly improve ammonia recovery in an MFC, with 61% of ammonium transportation by electricity-driven migration (Wu and Modin, 2013). A key factor in ammonia recovery is the high pH of the catholyte, which can change ammonium to ammonia gas. This feature alone has been used to recover ammonia from some special wastewaters such as reject water, which contains a high concentration of ammonium (1000 mg L^{-1}) but a low amount of organic compounds. The concept was examined in a microbial electrolysis cell with simultaneous hydrogen production and ammonia recovery. The catholyte pH increased to above 12 as a result of current generation; the microbial electrolysis

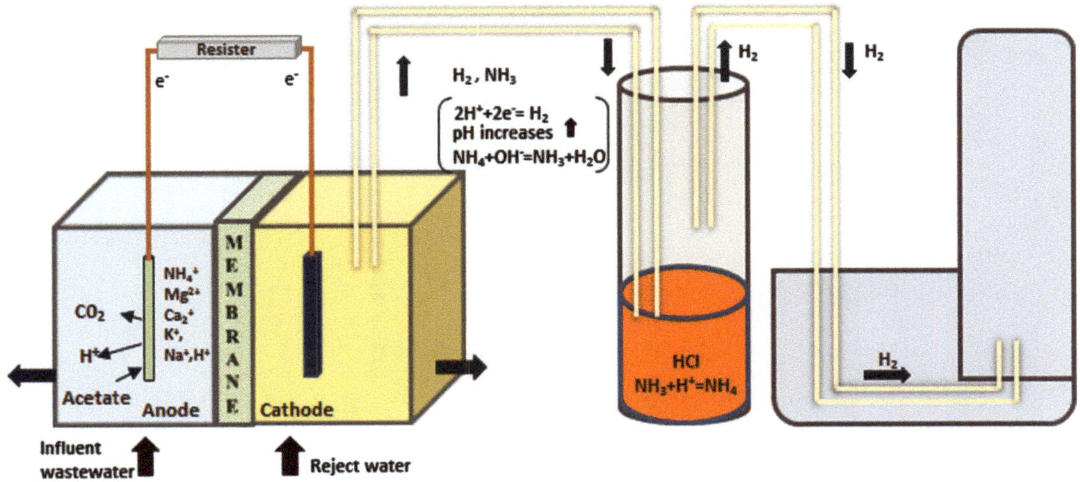

FIGURE 16.8

Ammonia recovery with hydrogen production using a microbial electrochemical system.

Modified from Wu, X., Modin, O., 2013. Ammonium recovery from reject water combined with hydrogen production in a bioelectrochemical reactor. Bioresource Technology 146, 530–536.

cell recovered 96% of ammonia in synthetic reject water and 79% from real reject water (Fig. 16.8) (Wu and Modin, 2013).

Research has been reported on nitrogen recovery from synthetic wastewater using three-chamber MFCs (Fig. 16.9). In a startup process, simple two-chamber MFCs were operated with potassium ferricyanide as the catholyte for stable power output. The two-chamber MFCs were then modified to three-chamber MFCs by integrating both cation exchange membrane and anion exchange membrane, which were used to recover the NO_3^-—N and NH_4^+—N from synthetic wastewater into value-added nitrogenous fertilizer from both the cathode and anode chambers. Ferric nitrate was used as the sole electron acceptor in the cathode, which also was used to evaluate the NO_3^-—N recovery efficiency in the case of the major anion of NO_3^- in the cathode. The output voltage of these MFCs was about 600—700 mV at an external load of 500 Ω. About 47% of NH_4^+—N in the anode chamber and 83% of NO_3^-—N in the cathode chamber was recovered. This method of nitrogen recovery shows that higher current density can selectively improve the recovery efficiency of both NO_3^-—N and NH_4^+—N (Xiao et al., 2014).

Conventional enhanced biological phosphorus removal is accomplished by polyphosphate-accumulating organisms, which release phosphate in anaerobic phase and take in and store excess phosphate during the following aerobic stage. Phosphorus removal and recovery in microbial electrochemical systems can be processed through electrochemical precipitation as struvite or magnesium ammonium phosphate hexahydrate ($MgNH_4PO_4 \cdot 6H_2O$), which is a slow-release fertilizer. It was reported that phosphate was extracted by reducing iron phosphate ($FePO_4$) in digested sewage sludge, and then the phosphate-containing liquid was stoichiometrically reacted with $MgCl_2$ and NH_4OH to precipitate struvite (Fig. 16.10) (Fischer et al., 2011). Struvite precipitation can be enhanced under high pH or boosted by an external power supply. In an integrated photo-bioelectrochemical system,

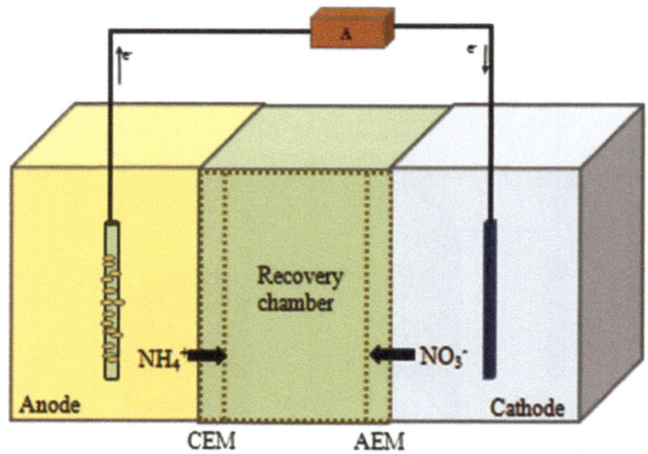

FIGURE 16.9

Nitrogen recovery in three-chambered microbial fuel cells.

Modified from Xiao, Y., Zheng, Y., Wu, S., Yang, Z.-H., Zhao, F., 2014. Nitrogen recovery from waste water using microbial fuel cell. Frontiers of Environmental Science and Technology, 1–7.

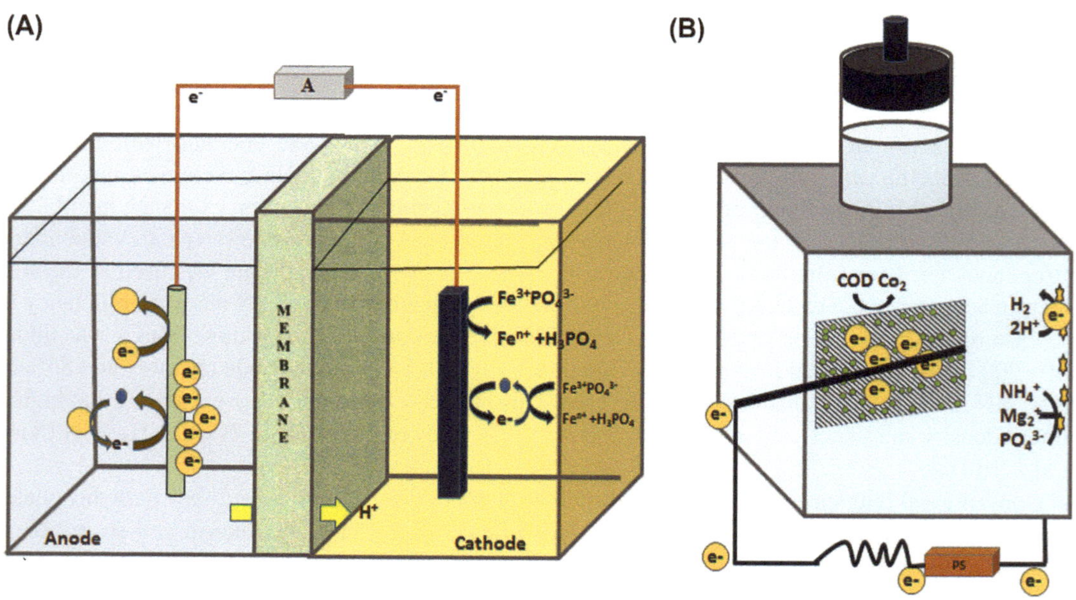

FIGURE 16.10

Phosphorus recovery as precipitate using a microbial electrochemical system: (A) two-chamber system; (B) single-chamber system.

Modified from Fischer, F., Bastian, C., Happe, M., Mabillard, E., Schmidt, N., 2011. Microbial fuel cell enables phosphate recovery from digested sewage sludge as struvite. Bioresource Technology 102, 5824–5830; Cusick, R.D., Logan, B.E., 2012. Phosphate recovery as struvite within a single chamber microbial electrolysis cell. Bioresource Technology 107, 110–115.

phosphate can be removed in the cathode compartment by algal growth (Cusick and Logan, 2012; Xiao et al., 2012). Although there have been different strategies to remove nitrogen or phosphate partially or completely from contaminated water, so far no process has been considered viable and scalable, partially owing to the complex removal mechanisms of these nutrients.

Sulfur-containing wastes from industrial and domestic sources, for instance, composed of H_2SO_4, SO_3^{2-} (eg, gypsum), and other oxyanion species such as thiosulfate $S_2O_3^{2-}$, sulfite SO_3^{2-}, and dithionate $S_2O_4^{2-}$, represent major environmental pollutants that can be removed and recovered before discharging into the environment. A number of removal strategies for sulfur species have been developed and implemented using microbial oxidation. The methods involve a two-step process. First, sulfate is reduced to sulfide, which is subsequently oxidized biologically to elemental sulfur (S^0) or recovered as metal sulfide. Studies were conducted on *Alcaligenes* sp. or *Paracoccus* sp.; the latter was capable of oxidizing sulfide using nitrate as an electron acceptor (Rabaey et al., 2009). Other sulfide-oxidizing bacteria including *Pseudomonas* and *Acinetobacter* were present in greater abundance on the anode than in the solution, whereas the sulfate-reducing bacteria preferably lived in the solution (dominant genera *Comamonas* and *Acinetobacter*). Biological factors did not uncouple from abiotic spontaneous electrochemical sulfide oxidation, which predicted that sulfide-oxidizing bacteria could donate electrons directly to the electrode (anode) and thus generate electricity. SO_4^{2-}, $S_2O_3^{2-}$, polythionates, S^0, and sulfide $H_2S/HS_2/S_2^{2-}$ were the potential sulfur compounds present in the anode and microbe-assisted production of $S_2O_3^{2-}$ and SO_4^{2} resulted in a persistent current (115 mA m^{-2}), in which the former, 91% and 86% sulfite and thiosulfate, were removed and recovered, respectively. At a given anode potential of −0.24 V versus Ag/AgCl reference electrode, the sulfide was rapidly oxidized to elemental sulfur, which can be recovered from wastewater (Zhao et al., 2009).

16.3.3 HYDROGEN

MFCs can be modified as a microbial electrolysis cell to produce gaseous molecules such as hydrogen (Logan et al., 2008). Under normal operating conditions, protons released by the anodic reaction migrate to the cathode to combine with oxygen to form water. Hydrogen generation from protons and electrons produced by the metabolism of microbes in an MFC is thermodynamically unfavorable. Applying an external potential to increase the cathode potential in an MFC circuit overcomes the thermodynamic barrier. In this mode, protons and electrons produced by the anodic reaction are combined at the cathode to form hydrogen. The required external potential for a microbial electrolysis cell is theoretically 110 mV, much lower than the 1210 mV required for direct electrolysis of water at neutral pH, because some energy comes from the biomass oxidation process in the anodic chamber. Microbial electrolysis cell can potentially produce about 8−9 mol H_2 per mol glucose compared with the typical 4 mol H_2 per mol glucose achieved in conventional fermentation (Liu et al., 2005). In biohydrogen production using microbial electrolysis cells, oxygen is no longer needed in the cathodic chamber. Thus, efficiency is improved because oxygen leaking into the anodic chamber is no longer an issue.

16.3.4 METHANE

Methane can be sequentially produced and recovered at the cathode chamber of a microbial electrolysis cell. Provided with the electron donor, a small external potential is applied for the favorable production of methane in the cathode chamber along with power generation. Methane production becomes energetically unfavorable at applied voltages higher than 1.05 V; hence a lower potential should be considered for application. In microbial electrolysis cells, carbon dioxide is reduced to methane at the biocathode as electromethanogenesis by electron transfer interference to *Methanobacterium palustre* (Cheng et al., 2009). However, some reports suggest that acetate via hydrogen gas also have the ability to produce methane in microbial electrolysis cells (Lovley, 2011; Clauwaert et al., 2008). The main disadvantages of methane production using a microbial electrochemical system are high solubility and the discharge of methane from the effluent into the environment, which needs to be avoided. Also, for methane production at higher temperatures, there is the need to remove methane from digester effluent (Deepak et al., 2012).

16.3.5 ETHANOL

Similar to methane production in microbial electrolysis cells, biological conversion of acetate can produce ethanol at the cathode chamber (Steinbusch et al., 2010). The process relies on the mediated electron transfer mechanism, in which external mediators (eg, methyl viologen) act as irreversible inhibitors for methanogenesis (induced by butyrate, when methyl viologen was depleted) and promote catalytic conversion to ethanol. Along with ethanol, hydrogen and *n*-butyrate were also produced (Steinbusch et al., 2008; Aulenta et al., 2007; Rabaey and Rozendal, 2010). To tackle methanogenesis, the effective conversion of butyrate to butanol as an end product is effective for ethanol production. To improve ethanol production, further research should focus on unmediated reduction of acetate at the cathode by microorganisms or the immobilization of methyl viologen on the electrode.

16.3.6 DIOXIDANE AND HYDROGEN PEROXIDE

The production of a strong oxidizer such as hydrogen peroxide in a microbial electrochemical system is coupled with the reduction of oxygen on the cathode at an applied voltage of 0.5 V. This process produces approximately 1.9 kg H_2O_2 per m^3 from acetate at an overall efficiency of 83.1% (Rozendal et al., 2009). Because most of the required energy was derived from acetate, the system had a low energy requirement of around 0.93 kWh per kg H_2O_2. However, H_2O_2 was produced at low concentrations (0.13%), which makes any useful recovery difficult. In one MFC study, a spectrographically pure graphite rod was used as a cathode electrode (Fu et al., 2010); the oxygen reduction reaction on the electrode mainly followed the two-electron pathway yielding H_2O_2, whereas the optimal condition for H_2O_2 production was in the 0.1-mol L^{-1} Na_2SO_4 electrolyte. When the graphite rods were used as the cathodic electrode in the MFC, the H_2O_2 concentration reached 78.85 mg L^{-1} after 12-h operation with an external resistance of 20 Ω (the H_2O_2 production rate was 6.57 mg L^{-1} h^{-1}). Coulombic efficiency, current efficiency, and chemical oxygen demand conversion efficiency were 12.26%, 69.47%, and 8.51%, respectively.

16.4 SUMMARY AND FUTURE PERSPECTIVE

In this chapter we describe the action of the microbial respiration mechanism in recovering various resources from different wastewaters and wastes. These methods have been discussed against the backdrop of current practices in various types of research across the globe. We are still developing and understanding new methods; it is uncertain to what extent these "new" methods will complement existing ones to replace fossil fuels.

If we considered EET to be a mechanistic phenomenon for microorganisms for energy generation, much needs to be understood about factors responsible regarding EET. We have made tremendous advances in understanding the role of outer membrane proteins, redox mediators, extracellular polymeric substances, and conductive pili in both gram-positive and gram-negative bacteria and various yeasts. Despite energy and power generation, much effort has been also directed at isolating novel molecules and recovering resources in waste and wastewaters using microbial reactors. It is as an attractive option for reducing treatment costs with the simultaneous production of power and recovery of useful products. Unlocking the knowledge of EET, which involves various molecules, novel natural mediators, and advances in cost-effective fuel cells, has given us a powerful tool for many applications.

Several challenges affect the recovery process, include the alkalinity of wastewaters, electrical losses, process configurations, metabolic diversity, substrate competition with bacteria or Archaea, exposure to high concentrations of recalcitrant compounds, and biofilm pH gradients. The production and recovery of various metals and biofuels such as hydrogen, methane, ethanol, and even bulk chemicals such as caustic and hydrogen peroxide have already been scaled up in industrial processes. In addition, a wide range of other products theoretically can be produced using wastewater as a source of redox equivalents. For example, the biocatalyzed reduction of butyrate to butanol and the reduction of glycerol to 1,3-propanediol, mainly used as a building block in the production of polymers, have gained attention. Glycerol is a by-product of biodiesel production; it is generated in increasingly large quantities, often with impurities. Therefore, the use and "upgrading" of such a low-value by-product using a microbial electrochemical system is attractive. Conventional methods for removing recalcitrant compounds such as dyes, pesticides, heavy metals, and phenolic compounds normally involve physical and chemical techniques with certain limitations. These disadvantages could theoretically be avoided in a microbial electrochemical system. The removal and recovery of heavy metals and metalloids such as chromium, copper, uranium, and selenium using either biotic or abiotic cathodes has been demonstrated.

REFERENCES

Aulenta, F., Catervi, A., Majone, M., Panero, S., Reale, P., Rossetti, S., 2007. Electron transfer from a solid-state electrode assisted by methyl viologen sustains efficient microbial reductive dechlorination of TCE. Environmental Science and Technology 41, 2554–2559.

Bean, R.C., Inami, Y.H., Basford, P.R., et al., 1964. Study of the fundamental principles of bio-electrochemistry. In: NASA Final Technical Report Research Laboratories. Philco Corporation, p. 107.

Biffinger, J.C., Ringeisen, B.R., 2008. Engineering microbial fuels cells: recent patents and new directions. Recent Patents in Biotechnology 2, 150–155.

Bennetto, H.P., 1987. Microbes come to power. New Scientist 114, 36–40.

Butler, J.E., Young, N.D., Lovley, D.R., 2010. Evolution of electron transfer out of the cell: comparative genomics of six geobacter genomes. BMC Genomics 11, 40.

Brutinel, E.D., Gralnick, J.A., 2012. Shuttling happens: soluble flavin mediators of extracellular electron transfer in *Shewanella*. Applied Environmental Microbiology 93, 41–48.

Bose, S., Hochella, M.F., Gorby, Y., Kennedy, D.W., McCready, D.E., Madded, A.S., Lower, B.H., 2009. Bio reduction of hematite nanoparticles by the dissimilatory iron reducing bacterium *Shewanella oneidensis* MR-1. Geochimica et Cosmochimica Acta 73, 962–976.

Brandl, H., Bosshard, R., Wegmann, M., 2001. Computer munching microbes: metal leaching from electronic scrape by bacteria and fungi. Hydrometallurgy 59, 319–326.

Cohen, B., 1931. The bacterial culture as an electrical half-cell. Journal of Bacteriology 21, 18–19.

Canfield, J.H., Goldner, B.H., Lutwack, R., 1963. Utilization of human wastes as electrochemical fuels. NASA Technical Report 63. Magna Corporation, Anaheim.

Coates, J.D., Ellis, D.J., Blunt-Harris, E.L., Gaw, C.V., Roden, E.E., Lovley, D.R., 1998. Recovery of humic-reducing bacteria from a diversity of environments. Applied Environmental Microbiology 64, 1504–1509.

Cao, B., Shi, L., Brown, R.N., Xiong, Y., Fredrickson, J.K., et al., 2011. Extra cellular polymeric substances from *Shewanella* sp. HRCR-1 biofilms: characterization by infrared spectroscopy and proteomics. Environmental Microbiology 13, 1018–1031.

Choi, C., Hu, N., 2013. The modeling of gold recovery from tetra chloroaurate wastewater using a microbial fuel cell. Bioresource Technology 133, 589–598.

Choi, C., Cui, Y., 2012. Recovery of silver from wastewater coupled with power generation using a microbial fuel cell. Bioresource Technology 107, 522–525.

Combs, G.F., Garbisu, C., Yee, B.C., Yee, A., Carlson, D.E., et al., 1996. Bioavailability of selenium accumulated by selenite reducing bacteria. Biology of Trace Element Research 52, 209–225.

Cusick, R.D., Logan, B.E., 2012. Phosphate recovery as struvite within a single chamber microbial electrolysis cell. Bioresource Technology 107, 110–115.

Cheng, S., Xing, D., Call, D.F., Logan, B.E., 2009. Direct biological conversion of electrical current into methane by electromethanogenesis. Environmental Science and Technology 43, 3953–3958.

Clauwaert, P., Toledo, R., van der Ha, D., Crab, R., Verstraete, W., Hu, H., Udert, K.M., Rabaey, K., 2008. Combining biocatalyzed electrolysis with anaerobic digestion. Water Science and Technology 57, 575–579.

Duveen, D.I., Klickstein, H.S., 1954. A Bibliography of the Works of Antoine Laurent Lavoisier, pp. 1743–1794. London.

Desloover, J., Woldeyohannis, A.A., Verstraete, W., Boon, N., Rabaey, K., 2012. Electrochemical resource recovery from digestate to prevent ammonia toxicity during anaerobic digestion. Environmental Science and Technology 46, 12209–12216.

Deepak, P., Anoop, S., Gilbert, V.B., Stig, I.O., Singh, N.P., Dielsa, L., Vanbroekhoven, K., 2012. Bio-electrochemical systems (BES) for sustainable energy production and product recovery from organic wastes and industrial wastewaters. RSC Advances 2, 1248–1263.

Fischer, F., Bastian, C., Happe, M., Mabillard, E., Schmidt, N., 2011. Microbial fuel cell enables phosphate recovery from digested sewage sludge as struvite. Bioresource Technology 102, 5824–5830.

Fu, L., You, S.J., Yang, F.l., Gao, M.-M., Fang, X.-H., Zhang, G.-Q., 2010. Synthesis of hydrogen peroxide in microbial fuel cell. Journal of Chemical Technology and Biotechnology 85, 715–719.

Green, D.E., Zande, H.D., 1981. Universal energy principle of biological systems and the unity of bioenergetics. Proceedings of National Academy Sciences 78, 5344–5347.

Gottschalk, G., 1986. An overview of the dissimilatory and assimilatory metabolism of prokaryotes, emphasizing metabolic diversity in the bacterial world. In: Bacterial Metabolism, second ed. Springer-verlag, New York.

Gralnick, A.J., Newman, D.K., 2007. Extracellular respiration. Molecular Microbiology 65 (1), 1–11.

Gorby, Y., McLean, J., Korenevsky, A., Rosso, K., El-Naggar, M.Y., Beveridge, T.J., 2008. Redox reactive membrane vesicles produced by *Shewanella*. Geobiology 6, 232–241.

Gorby, Y.A., Yanina, S., McLean, J.S., Rosso, K.M., Moyles, D., Dohnalkova, A., et al., 2006. Electrically conductive bacterial nanowires produced by *Shewanella oneidensis* strain MR-1 and other microorganisms. Proceedings of National Academy Sciences 103, 11358–11363.

Garbisu, C., Ishii, T., Leighton, T., Buchanan, B.B., 1996. Bacterial reduction of selenite to elemental selenium. Chemical Geology 132, 199–204.

Gregory, K.B., Lovley, D.R., 2005. Remediation and recovery of uranium from contaminated subsurface environments with electrodes. Environmental Science and Technology 39, 8943–8947.

Hernandez, M.E., Newman, D.K., 2001. Extracellular electron transfer. Cellular and Molecular Life Science 58, 1562–1571.

Hutchins, S.R., Davidson, M.S., Brierley, J.A., Brierley, C.L., 1986. Microorganisms in reclamation of metals. Annual Review of Microbiology 40, 311–336.

Heijne, A.T., Liu, F., Weijden, R.V.D., Weijma, J., Buisman, C.J.N., Hamelers, H.V.M., 2010. Copper recovery combined with electricity production in a microbial fuel cell. Environmental Science and Technology 44, 4376–4381.

He, Q., Yao, K., 2011. Impact of alternative electron acceptors on selenium(IV) reduction by *Anaeromyxobacter dehalogenans*. Bioresource Technology 102 (3), 3578–3580.

Haddadi, S., Elbeshbishy, E., Lee, H.S., 2013. Implication of diffusion and significance of anodic pH in nitrogen-recovering microbial electrochemical cells. Bioresource Technology 142, 562–569.

Ilyas, S., Anwar, M.A., Niazi, S.B., Chauri, M.A., 2007. Bioleaching of metals from electronic scrap by moderately thermophilic acidophilic bacteria. Hydrometallurgy 88, 180–188.

Li, Z., Zhang, X., Lei, L., 2008. Electricity production during the treatment of real electroplating wastewater containing Cr using microbial fuel cell. Process Biochemistry 43, 1352–1358.

Lehninger, A.L., 1971. Bioenergetics: The Molecular Basis of Biological Energy Transformations, second ed. Addison-Wesley.

Lovley, D.R., 1993. Dissimilatory metal reduction. Annual Review of Microbiology 47, 263–290.

Lovley, D.R., Giovannoni, S.J., White, D.C., Champine, J.E., Phillips, E.J., Gorby, Y.A., Goodwin, S., 1993. *Geobacter metallireducens* gen. nov. sp. nov., a microorganism capable of coupling the complete oxidation of organic compounds to the reduction of iron and other metals. Archives of Microbiology 159, 336–344.

Lovley, D.R., Woodward, J.C., Chapelle, F.H., 1996. Rapid anaerobic benzene oxidation with a variety of chelated Fe(III) forms. Applied Environmental Microbiology 62, 288–291.

Lovley, D.R., 2006a. Bug juice: harvesting electricity with microorganisms. Nature Review of Microbiology 4, 497–508.

Lovley, D.R., 2006b. Microbial energizer: fuel cells that keep on going. Microbe 1, 323–329.

Lovley, D.R., 2008a. Extracellular electron transfer: wires, capacitors, iron lungs, and more. Geobiology 6, 225–231.

Lovley, D.R., 2008b. The microbe electric: conversion of organic matter to electricity. Current Opinion in Biotechnology 19, 564–571.

Lies, D.P., Hernandez, M.E., Kappler, A., Mielke, R.E., Gralnick, J.A., Newman, D.K., 2005. *Shewanella oneidensis* MR-1 uses overlapping pathways for iron reduction at a distance and by direct contact under conditions relevant for biofilms. Applied Environmental Microbiology 71, 4414–4426.

Lower, B.H., Shi, L., Yongsunthon, R., Droubay, T.C., McCready, D.E., Lower, S.K., 2007. Specific bonds between an iron oxide surface and outer membrane cytochromes MtrC and OmcA from *Shewanella oneidensis* MR-1. Journal of Bacteriology 189, 4944–4952.

Lower, B.H., Yongsunthon, R., Shi, L., Wildling, L., Gruber, H.J., et al., 2009. Antibody recognition force microscopy shows that outer membrane cytochromes OmcA and MtrC are expressed on the exterior surface of *Shewanella oneidensis* MR-1. Applied Environmental Microbiology 75, 2931–2935.

Lloyd, J.R., 2003. Microbial reduction of metals and radionuclides. FEMS Microbiology Reviews 27, 411–425.

Logan, B.E., Call, D., Cheng, S., Hamelers, H.V.M., et al., 2008. Microbial electrolysis cells for high yield hydrogen gas production from organic matter. Environmental Science and Technology 42, 8630–8640.

Liu, H., Grot, S., Logan, B.E., 2005. Electrochemically assisted microbial production of hydrogen from acetate. Environmental Science and Technology 39, 4317–4320.

Lovley, D.R., 2011. Powering microbes with electricity: direct electron transfer from electrodes to microbes. Environmental Microbiology Reports 3, 27–35.

Myers, C.R., Myers, J.M., 1992. Localization of cytochromes to the outer-membrane of anaerobically grown *Shewanella putrefaciens* MR-1. Journal of Bacteriology 174, 3429–3438.

Myers, C.R., Myers, J.M., 1993. Role of menaquinone in the reduction of fumarate, nitrate, iron(III) and manganese(IV) by *Shewanella putrefaciens* MR-1. FEMS Microbiology Letters 114, 215–222.

Myers, C.R., Myers, J.M., 2003. Cell surface exposure of the outer membrane cytochromes of *Shewanella oneidensis* MR-1. Letters in Applied Microbiology 37, 254–258.

Meitl, L.A., Eggleston, C.M., Col-berg, P.J.S., Khare, N., Reardon, C.L., Shi, L., 2009. Electro-chemical interaction of *Shewanella oneidensis* MR-1 and its outer membrane cytochromes OmcA and MtrC with hematite electrodes. Geochimica et Cosmochimica Acta 73, 5292–5307.

Myers, C.R., Myers, J.M., 2002. MtrB is required for proper incorporation of the cytochromes OmcA and OmcB into the outer membrane of *Shewanella putrefaciens* MR-1. Applied Environmental Microbiology 68, 5585–5594.

Marshall, M.J., Beliaev, A.S., Dohnalkova, A.C., Kennedy, D.W., Shi, L., et al., 2006. *c*-Type cytochrome dependent formation of U(IV) nanoparticles by *Shewanella oneidensis*. PLoS Biology 4, 1325–1333.

Modin, O., Wang, X., Wu, X., Rauch, S., Fedje, K.K., 2012. Bioelectrochemical recovery of Cu, Pb, Cd, and Zn from dilute solutions. Journal of Hazardous Materials 235–236, 291–297.

Nicholls, D.G., Ferguson, S.J., 2002. Bioenergetics, third ed. Academic Press.

Newman, D.K., Kolter, R., 2000. A role for excreted quinones in extracellular electron transfer. Nature 405, 94–97.

Oren, A., 2002. Metabolic diversity of prokaryotes and eukaryotes. In: Contrafatto, G. (Ed.), Biological Science Fundamentals and Systematics, vol. II. Encyclopedia of Life Support Systems, EOLSS, Publishers, Oxford.

Potter, M.C., 1911. Electrical effects accompanying the decomposition of organic compounds. Proceedings of the Royal Society of London. Series B 84, 260–276.

Potter, M.C., 1910. On the difference of potential due to the vital activity of microorganisms. Proceedings of University Durham Philosophical Society 3, 245–249.

Rabaey, K., Angenent, L., Schorder, U., Keller, J. (Eds.), 2009. Bio-electrochemical Systems: From Extracellular Electron Transfer to Biotechnological Application. IWA publishing, London.

Reguera, G., McCarthy, K.D., Mehta, T., Nicoll, J.S., Tuominen, M.T., Lovley, D.R., 2005. .Extracellular electron transfer via microbial nanowires. Nature 435, 1098–1101.

Ross, D.E., Brantley, S.L., Tien, M., 2009. Kinetic characterization of OmcA and MtrC, terminal reductases involved in respiratory electron transfer for dissimilatory iron reduction in *Shewanella oneidensis* MR-1. Applied Environmental Microbiology 75, 5218–5226.

Reardon, C.L., Dohnalkova, A.C., Nachimuthu, P., Kennedy, D.W., et al., 2010. Role of outer-membrane cytochromes MtrC and OmcA in the bio mineralization of ferrihydrite by *Shewanella oneidensis* MR-1. Geobiology 8, 56–68.

Rosso, K.M., Zachara, J.M., Fredrickson, J.K., Gorby, Y.A., Smith, S.C., 2003. Non local bacterial electron transfer to hematite surface. Geochimica et Cosmochimica Acta 67, 1081–1087.

Reguera, G., Nevin, K.P., Nicoll, J.S., Covalla, S.F., Woodard, T.L., Lovley, D.R., 2006. Biofilm and nanowire production leads to increased current in *Geobacter sulfurreducens* fuel cells. Applied Environmental Microbiology 72, 7345–7348.

Rabaey, K., Rozendal, R.A., 2010. Microbial electrosynthesis – revisiting the electrical route for microbial production. Nature Review of Microbiology 8, 706–716.

Rozendal, R.A., Leone, E., Keller, J., Rabaey, K., 2009. Efficient hydrogen peroxide generation from organic matter in a bioelectrochemical system. Electrochemical Communication 11, 1752–1755.

Sunil, A.P., Cecilia, H., Lo, G., 2012. Electron transfer mechanism between microorganisms and electrodes in bioelectrochemical system. Bioanalytical Reviews 4, 159–192.

Saffarini, D.A., Blumerman, S.L., Mansoorabadi, K.J., 2002. Role of menaquinones in Fe(III) reduction by membrane fractions of *Shewanella putrefaciens*. Journal of Bacteriology 184, 846–848.

Shi, L., Chen, B., Wang, Z., Elias, D.A., Mayer, M.U., et al., 2006. Isolation of a high-affinity functional protein complex between OmcA and MtrC: two outer membrane decaheme *c*-type cytochromes of *Shewanella oneidensis* MR-1. Journal of Bacteriology 188, 4705–4714.

Shugaba, A., Buba, F., Kolo, B.G., Nok, A.J., Ameh, D.A., Lori, J.A., 2012. Uptake and reduction of hexavalent chromium by *Aspergillus niger* and *Aspergillus parasiticus*. Journal of Petroleum and Environmental Biotechnology 3 (3), 119–126.

Steinbusch, K.J.J., Hamelers, H.V.M., Schaap, J.D., Kampman, C., Buisman, C.J.N., 2010. Bioelectrochemical ethanol production through mediated acetate reduction by mixed cultures. Environmental Science and Technology 44, 513–517.

Steinbusch, K.J.J., Hamelers, H.V.M., Buisman, C.J.N., 2008. Alcohol production through volatile fatty acids reduction with hydrogen as electron donor by mixed cultures. Water Research 42, 4059–4066.

Tandukar, M., Huber, S.J., Onodera, T., Pavlostathis, S.G., 2009. Biological chromium(VI) reduction in the cathode of a microbial fuel cell. Environmental Science and Technology 43, 8159–8165.

Tao, H., Li, W., Liang, M., Xu, N., Ni, J., Wu, W., 2011a. A membrane-free baffled microbial fuel cell for cathodic reduction of Cu(II) with electricity generation. Bioresource Technology 102, 4774–4778.

Tao, H., Liang, M., Li, W., Zhang, L., Ni, J., Wu, W., 2011b. Removal of copper from aqueous solution by electrodeposition in cathode chamber of microbial fuel cell. Journal of Hazardous Materials 189, 186–192.

Tao, H., Zhang, L., Gao, Z., Wu, W., 2011c. Copper reduction in a pilot-scale membrane free bioelectrochemical reactor. Bioresource Technology 102, 10334–10339.

Tao, H., Gao, Z., Ding, H., Xu, N., Wu, W., 2012. Recovery of silver from silver(I)-containing solutions in bioelectrochemical reactors. Bioresource Technology 111, 92–97.

VonCanstein, H., Ogawa, J., Shimizu, S., Lloyd, J.R., 2008. Secretion of flavins by *Shewanella* species and their role in extracellular electron transfer. Applied and Environmental Microbiology 74, 615–623.

Velasquez-Orta, S.B., Head, I.M., et al., 2010. The effect of flavin electron shuttles in microbial fuel cells current production. Applied Environmental Microbiology 85, 1373–1381.

Wilkinson, S., 2000a. Gastrobots-benefits and challenges of microbial fuel cells in food powered robot applications. Autonomous Robots 9, 99–111.

Wilkinson, S., 2000b. Gastronome – a pioneering food powered mobile robot. In: Proc. IASTED 2000 (8th) Int. Conf. on Robotics and Applications, pp. 318–037. Honolulu, Hawaii.

Wilkinson, A.G., Campbell, C., 1996. Green bug robots-renewable environmental power for miniature robots. In: Proceedings of 4th IASTED International Conference, Robotics and Manufacturing, pp. 275–278. Honolulu, Hawaii.

Wang, G., Huang, L., Zhang, Y., 2008. Cathodic reduction of hexavalent chromium [Cr(VI)] coupled with electricity generation in microbial fuel cells. Biotechnology Letters 30, 1959–1966.

Wang, Z., Lim, B., Lu, H., Fan, J., Choi, C., 2010. Cathodic reduction of Cu^{2+} and electric power generation using a microbial fuel cell. Bulletin of Korean Chemical Society 31, 2025–2030.

Wang, Z., Lim, B., Choi, C., 2011. Removal of Hg^{2+} as an electron acceptor coupled with power generation using a microbial fuel cell. Bioresource Technology 102, 6304–6307.

Wang, H., Ren, Z.J., 2014. Bioelectrochemical metal recovery from wastewater: a review. Water Research 66, 219–232.

Wu, X., Modin, O., 2013. Ammonium recovery from reject water combined with hydrogen production in a bioelectrochemical reactor. Bioresource Technology 146, 530–536.

Xiao, Y., Zheng, Y., Wu, S., Yang, Z.-H., Zhao, F., 2014. Nitrogen recovery from waste water using microbial fuel cell. Frontiers of Environmental Science and Technology 1–7.

Xiao, L., Young, E.B., Berges, J.A., He, Z., 2012. Integrated photo-bioelectrochemical system for contaminants removal and bioenergy production. Environmental Science and Technology 46, 11459–11466.

Yarzabal, A., Brasseur, G., Ratouchniak, J., Lund, K., Lemesle-Meunier, D., DeMoss, J.A., Bonnefoy, V., 2002. The high molecular weight cytochrome *c*-cyc2 of *Acidithiobacillus ferrooxidans* is an outer membrane protein. Journal of Bacteriology 184, 313–317.

Zhang, B., Zhao, H., Shi, C., Zhou, S., 2009. Simultaneous removal of sulfide and organics with vanadium(V) reduction in microbial fuel cells. Journal of Chemical Technology and Biotechnology 84, 1780–1786.

Zhang, B., Feng, C., Ni, J., Zhang, J., Huang, W., 2012. Simultaneous reduction of vanadium(V) and chromium(VI) with enhanced energy recovery based on microbial fuel cell technology. Journal of Power Sources 204, 34–39.

Zhao, F., Rahunen, N., Varcoe, J., Roberts, A., Avignone-Rossa, C., Thumser, A., Slade, R., 2009. Factor affecting the performing of microbial fuel cells for sulfur pollutants removal. Biosensors and Bioelectronics 24, 1931–1936.

CHAPTER

17

ACID MINE DRAINAGES FROM ABANDONED MINES: HYDROCHEMISTRY, ENVIRONMENTAL IMPACT, RESOURCE RECOVERY, AND PREVENTION OF POLLUTION

P.J.C. Favas[1,2], S.K. Sarkar[3], D. Rakshit[3], P. Venkatachalam[4], M.N.V. Prasad[5]

University of Trás-os-Montes e Alto Douro, Vila Real, Portugal[1]; MARE – Marine and Environmental Sciences Centre, University of Coimbra[2]; University of Calcutta, Calcutta, West Bengal, India[3]; Periyar University, Salem, Tamil Nadu, India[4]; University of Hyderabad, Hyderabad, Telangana, India[5]

CHAPTER OUTLINE

Environmental Materials and Waste. http://dx.doi.org/10.1016/B978-0-12-803837-6.00017-2

17.1 INTRODUCTION

The mining exploration of metallic mineral deposits has as a consequence the exposure of large quantities of sulfides to weathering processes (oxygen, water, and bacteria) that generate acidic environments which promote metal(loid) solubilization. The water acts as a reagent, as the medium in which the chemical reactions proceed, and as a transport vector of the formed products. These waters constitutes the so-called "acid mine drainage" or "acid and metalliferous drainage" (AMD) (Schmiermund and Drozd, 1997; Cánovas et al., 2007; Seal and Shanks, 2008; Koski et al., 2008; Foster et al., 2008; Romero et al., 2011; Davies et al., 2011; Favas et al., 2012; Soltani et al., 2014; Li et al., 2015).

According to Schmiermund and Drozd (1997), the term "acid mine drainage" was originally used to describe effluents resulting from underground mining operations, which were characterized by low pH values. However, those authors assumed a critical position regarding the use of the expression to classify most mining effluents. They considered the term inappropriate because of its limited scope, which often led to misconceptions. The justification for this position is made essentially based on two premises. On the one hand, the authors report that low pH values are not a universal characteristic of mining effluents. In fact, effluents generated in a mine may also be neutral or alkaline, depending on the paragenesis of the mineral deposit. On the other hand, the most worrying aspect in relation to the environmental effects of mining effluents may not be the low pH but rather the high content of heavy metals, sulfates, suspended solids, etc.

Thus, Schmiermund and Drozd (1997) considered it more accurate to use the expression "mining-influenced waters" (MIW). The authors also made reference to an another proposal, "acid rock drainage" (ARD), which, however, was considered valid only in cases in which water–rock interaction result in acid drainage in the absence of mining, such as in the reported situation in Antarctica (Dold et al., 2013). Similarly, acid sulfate soils (ASS) are rich in metal sulfides which can oxidize to generate acidity upon exposure to the atmosphere and moisture. ASS and ARD are widely distributed around the world and can cause severe agricultural, engineering, and environmental problems if they are not managed properly.

However, throughout this chapter we will use the expression "acid mine drainage," considering that the effluents generated in the studied mines are in some cases significantly acidic and always have a pH value below 7.

Most minerals that constitute ores, especially sulfides and sulfosalts, are formed under reducing environments and under temperatures significantly higher than atmospheric. Thus, when these minerals are exposed to surface environmental conditions, they tend to disintegrate chemically to form new compounds until equilibrium it is restored (Park and MacDiarmid, 1975; García-Lorenzo et al., 2012; Wei and Wolfe, 2013; Li et al., 2015).

Before the development of mining activities, the amount of sulfides exposed to the favorable condition of this oxidative weathering is reduced. As a consequence of mining exploration, large quantities of sulfides are exposed to oxygen, water, and bacteria that generate acidic environments which promote the metal(loid) solubilization, forming AMD.

The abandonment of a mine as the final episode of mining activities leads in many cases to the maintenance of or even an increase in acid drainage—generating processes, which is the results of potential pollution sources more than its activity period (Fig. 17.1). Acid drainage from abandoned mines can originate from inside as well as from the surface by the oxidation of sulfides contained in the galleries, in the slope excavation, and in the tailings.

The oxidative weathering of metal sulfides is a complex process involving many types of chemical reactions (oxidation—reduction, hydrolysis, ion complex formation, dissolution/precipitation, etc.) which result in various products, including oxidized forms of Fe and/or other metals, sulfates, and strong acidity. Among the sulfides, pyrite is the most abundant and the one whose weathering process is best studied. For this reason, it will be used as an illustrative example of AMD formation.

Many studies have been conducted around the world in mineralized and mining areas to perform a hydrochemical characterization of AMD (or ARD) and surface waters to evaluate the subsequent mining environmental impact, and in the context of metal recovery as well as to prevent pollution (Table 17.1).

The current work was undertaken for an in-depth study of AMD situated in the northern part of Portugal to consider two important aspects. The first part has been prepared to highlight basic features dealing with: (1) the origin of AMD; (2) characteristics of AMD; (3) factors controlling the formation of AMD; (4) environmental impacts of AMD; (5) resource recovery from AMD; and (6) prevention, mitigation, and treatment of AMD. The second part accentuates in-depth case studies of mine drainage water collected bimonthly from abandoned mines situated in northern Portugal over 1 year.

17.2 ORIGINS OF ACID MINE DRAINAGE

17.2.1 PYRITE OXIDATION MECHANISMS

It is accepted that the natural process of pyrite weathering is not just chemical but biogeochemical. Some bacteria species may act as catalysts to increase the oxidation rate up to 100 times (Kelly, 1991). Ahonen and Tuovinen (1992) showed a significant increase in the rate of Ni, Co, and Zn leaching from various sulfides in the presence of bacteria. These bacteria are characteristic of acidic aqueous environments with pH ranging from 2.8 to 3.2 (acidophilic organisms) that become inactive (or absent) in water with a pH above 5.5 (Caruccio et al., 1980; Johnson and Hallberg, 2003).

FIGURE 17.1

View of an old mine gallery with evidence of the presence of acidic water (Vale das Gatas mine, Vila Real, Northern Portugal).

Reactions involved in the pyrite dissolution process are well known and result in the production of sulfuric acid (H_2SO_4), estimated at 2 mol acid per mole of pyrite (Urrutia et al., 1987), which in turn creates conditions for Fe solubilization (Kelly, 1991; Gray, 1997; Banks et al., 1997; Elberling et al., 2003; Jönsson et al., 2006; Chou et al., 2013):

$$FeS_2 + \frac{7}{2}O_2 + H_2O \rightarrow Fe^{2+} + 2SO_4{}^{2-} + 2H^+ \quad [17.1]$$

$$Fe^{2+} + \frac{1}{4}O_2 + H^+ \rightarrow Fe^{3+} + \frac{1}{2}H_2O \quad [17.2]$$

$$Fe^{3+} + 3H_2O \rightarrow Fe(OH)_3 + 3H^+ \quad [17.3]$$

$$FeS_2 + 14Fe^{3+} + 8H_2O \rightarrow 15Fe^{2+} + 2SO_4{}^{2-} + 16H^+ \quad [17.4]$$

Table 17.1 Annotated Synopsis of Knowledge of Acid Mine Drainages From Abandoned Mines: Hydrochemistry, Environmental Impact, Resource Recovery, and Pollution Prevention

Main Subjects and Findings	Reference(s)
Production of bio-oil and enhanced metal recovery from microalgae cultivated in acid mine drainage	Raikova et al. (2016)
Composition of acid mine water from Iberian Pyrite Belt (southwest Spain)	Olías et al. (2016)
Efflorescent sulfates from Baia Sprie mining area (Romania)	Buzatu et al. (2016)
Novel method of using permeable reactive kiddle for remediation of acid mine drainage	Lee et al. (2016)
Water acidification trends in reservoir of Iberian Pyrite Belt (southwest Spain)	Cánovas et al. (2016)
Environmental risk assessment of acid rock drainage under uncertainty	Betrie et al. (2016)
Potential of coal mine waste rock for generating acid mine drainage	Qureshi et al. (2016)
Long-term fluctuations of groundwater mine pollution in sulfide mining district with dry Mediterranean climate: implications for water resources management and remediation	Caraballo et al. (2016)
Recovery of Zn from acid mine water and electric arc furnace dust in integrated process	Carranza et al. (2016)
Gold and silver in sulfide tailings	Myagkaya et al. (2016)
Manipulating air-cathode fuel cell toward recovering highly active heterogeneous electro-Fenton catalyst from Fe(II) in acid mine drainage	Sun et al. (2015)
Comprehensive use of lead-zinc tailings: resource recovery of sulfur	Lei et al. (2015)
Life-cycle assessment of desulfurization flotation process to prevent acid rock drainage	Broadhurst et al. (2015)
Environmental aspects of metals removal from waters and gold recovery	Salminen et al. (2015)
Reactive transport modeling of early diagenesis in reservoir lake affected by acid mine drainage: trace metals, lake overturn, benthic fluxes, and remediation	Torres et al. (2015)
Ecological assessment of coal mine and metal mine drainage in South Korea using *Daphnia magna* bioassay	Lee et al. (2015)
Influence of pH and temperature on alunite dissolution: rates, products, and insights into mechanisms from atomistic simulation	Acero et al. (2015)
Potential role of aluminum hydroxysulfates in removal of contaminants in acid mine drainage	Carrero et al. (2015)
Treatment of open-cast lignite mining–induced acid mine drainage using rotating microfiltration system	Meschke et al. (2015)
Sulfate-reducing bioreactor dependence on organic substrates for remediation of coal-generated acid mine drainage in field	Lefticariu et al. (2015)
Efficiency of ball-milled South African bentonite clay for remediation of acid mine drainage	Masindi et al. (2015)
Suppression of pyrite oxidation in acid mine drainage by carrier microencapsulation using liquid product of hydrothermal treatment of low-rank coal, and electrochemical behavior of resultant encapsulating coatings	Yuniati et al. (2015)
Acidithiobacillus ferrooxidans DLC-5, an extremely acidophilic bacterium isolated from acid mine drainage in northeast China	Chen et al. (2015b)
Effect of pH on Cu, Ni, and Zn removal by biogenic sulfide precipitation in inversed fluidized-bed bioreactor	Janyasuthiwong et al. (2015)

Continued

Table 17.1 Annotated Synopsis of Knowledge of Acid Mine Drainages From Abandoned Mines: Hydrochemistry, Environmental Impact, Resource Recovery, and Pollution Prevention—cont'd

Main Subjects and Findings	Reference(s)
Investigation into use of cement kiln dust in high-density sludge treatment of acid mine water	Mackie and Walsh (2015)
pH and Eh influence in acid mine drainage system using fuzzy qualitative tool (Iberian Pyrite Belt, southwest Spain)	de la Torre et al. (2015)
Metal speciation in sediments and soils associated with acid mine drainage in Mount Morgan (Queensland, Australia)	Vicente-Beckett et al. (2015)
Use of residue gas sludge to remove heavy metals from acid mine drainage	Jafaripour et al. (2015)
Temporal evolution of bacterial communities associated with in situ wetland-based remediation of marine shore porphyry copper deposit	Diaby et al. (2015)
Environmental management in North American mining sector	Asif and Chen (2015)
Design of photovoltaic systems to power aerators for natural purification of acid mine drainage	Song and Choi (2015)
Selective recovery of dissolved metals from mine drainage using electrochemical reactions	Park et al. (2015)
Temporal geochemical variations in above- and below-drainage coal mine discharge	Burrows et al. (2015)
Geochemical and microbiological characterization of drainage from low-sulfide waste rock: active zone field experiments of Diavik Waste Rock Project	Bailey et al. (2015)
Suitability of different growth substrates as source of nitrogen for sulfate-reducing bacteria	Dev et al. (2015)
Investigation of acidic headwater streams in Judith Mountains (Montana, United States)	Williams et al. (2015)
Hydrogeochemical effects of bulkhead in Dinero mine tunnel, Sugar Loaf mining district, near Leadville (Colorado, United States)	Walton-Day and Mills (2015)
Study of acid mine drainage production potential in flotation tailings of Sarcheshmeh porphyry copper mine	Shoja and Salari (2015)
Prediction of effects of acid mine drainage on variations in hydrogeochemical environment at sulfide-bearing ore deposits	Limantseva et al. (2015)
Tailings weathering and arsenic mobility at abandoned silver mine (Morocco)	El Adnani et al. (2015)
Inhibition of pyrite oxidation by surface coating: long-term field study	Kang et al. (2015)
Treatment of acid mine drainage using fly ash zeolite column	Prasad and Kumar (2015)
Grain-size effects on mine water quality and acid–neutral rock drainage production in kinetic testing (Recsk porphyry skarn Cu–Zn deposit, Hungary)	Embile et al. (2015)
Controls on Zn concentrations in acidic and neutral mine drainage from New Zealand's bituminous coal and epithermal mineral deposits	Pope and Trumm (2015)
Critical review of acid rock drainage prediction methods and practices	Parbhakar-Fox and Lottermoser (2015)
Identifying catchment-scale predictors of coal mining impacts on New Zealand stream communities	Clapcott et al. (2016)
Noninvasive flow path characterization in mining-impacted wetland	Bethune et al. (2015)
Sustainable remediation of mine drainage	Emili et al. (2015)
Determination of carbonate minerals responsible for alkaline mine drainage at Xikuangshan antimony mine (China)	Nyirenda et al. (2015)

Abandoned mine tailings in cultural itineraries (Don Quixote Route, Spain)	Martín-Crespo et al. (2015)
Environmental tracers for elucidating weathering process in phosphogypsum disposal site: implications for restoration	Pérez-López et al. (2015)
Influence of freeze–thaw dynamics on internal geochemical evolution of low-sulfide waste rock	Sinclair et al. (2015)
Hydrogeochemical and isotopic studies of Kettara mine watershed (Morocco)	Toughzaoui et al. (2015)
Protocol improvements for optimal sulfide tailings reactivity (ASTM-normalized humidity cell kinetic test)	Bouzahzah et al. (2015)
Particular river-whiting phenomenon caused by discharge of hypolimnetic water from stratified reservoir (Xiaoche River, China)	Chen et al. (2015a)
Impact of acid mine drainages on surficial waters of abandoned mining site	García-Lorenzo et al. (2015)
Synthesis and characterization of magnetic nanoparticles and study of their removal capacity of metals from acid mine drainage	Kefeni et al. (2015)
Electrochemical approaches to preventing and remediating acidity and toxic metals of acid mine drainage	Bejan and Bunce (2015)
Mobility of As, Cu, Cr, and Zn from tailings covered with sealing materials using alkaline industrial residues: comparison of two leaching methods	Jia et al. (2015)
Conversion of coal mine drainage ochre to water treatment reagent: production, characterization, and application for P and Zn removal	Sapsford et al. (2015)
Identification of processes controlling chemical composition of pit lake waters located in eastern part of Muskau Arch (Polish–German borderland)	Lutyńska and Labus (2015)
Inorganic constituents of conductivity in five central Appalachian watersheds with mixed source–driven pollutants	Cook et al. (2015)
Overview of resources from acid drainage and postmining opportunities	Jacobs (2014b)
Acid mine drainage, rock drainage, and acid sulfate soils: causes, assessment, prediction, prevention, and remediation	Jacobs (2014a)
Pollution control and metal resource recovery for acid mine drainage	Chen et al. (2014)
Purification techniques for recovery of valuable compounds from acid mine drainage and cyanide tailings: application of green engineering principles	García et al. (2014)
Recovery of iron from acid mine drainage in form of oxides	Mačingová and Luptáková (2014)
Biogeometallurgical premining characterization of ore deposits: approach to increase sustainability in mining process	Dold and Weibel (2013)
Sulfate removal from sodium sulfate–rich brine and recovery of barium as barium salt mixture	Vadapalli et al. (2013)
New microbiological strategies that enable selective recovery and recycling of metals from acid mine drainage and mine process waters	Ňancucheo et al. (2012)
Comparison of sulfidogenic up-flow and down-flow fluidized-bed reactors for biotreatment of acidic metal-containing wastewater	Sahinkaya and Gungor (2010)
Metals recovered from mine water provide new revenue stream	Gleason et al. (2009)
Treatment of acid mine drainage in South Brazil: comparative active processes and water reuse	Silveira et al. (2009)
Wine wastes as carbon source for biological treatment of acid mine drainage	Costa et al. (2009)

Continued

Table 17.1 Annotated Synopsis of Knowledge of Acid Mine Drainages From Abandoned Mines: Hydrochemistry, Environmental Impact, Resource Recovery, and Pollution Prevention—cont'd

Main Subjects and Findings	Reference(s)
Natural attenuation processes in two water reservoirs receiving acid mine drainage	Sarmiento et al. (2009)
Bioremediation of sulfate-rich mine effluents using grass cuttings and rumen fluid microorganisms	Greben et al. (2009)
Bioleaching of chalcopyrite concentrate by moderately thermophilic culture in stirred tank reactor	Zhou et al. (2009)
Sustainability in metal mining: from exploration, over processing to mine waste management	Dold (2008)
Bioremediation of acid mine drainage in uranium deposit	Groudev et al. (2008)
In situ and laboratory studies of natural attenuation of acid mine drainage	López-Luna (2007)
Comparison of bioleaching effect of mesophilic (35°C) and thermophilic (45°C) bacteria on TIZAPA tailings (Mexico)	Rivera-Santillán and Becerril-Reyes (2007)
Biohydrometallurgy and environment	Johnson (2006)
Identifying refractory nickel in acid mine drainage lime treatment sludge	Robertson et al. (2006)
Recovery of metal resources from passive treatment systems and sustainable development initiatives associated with large mining projects	Gusek and Clarke-Whistler (2005)
Proteogenomic approaches for molecular characterization of natural microbial communities	Banfield et al. (2005)
Depauperate macroinvertebrates in mine-affected stream: clean water may be key to recovery	Battaglia et al. (2005)
Chemical and physical properties of iron hydroxide precipitates associated with passively treated coal mine drainage in bituminous region of Pennsylvania and Maryland (United States)	Kairies et al. (2005)
Advances in biotreatment of acid mine drainage and biorecovery of metals	Tabak and Govind (2003); Tabak et al. (2003)
Conversion of pollutants to fertilizers: ion exchange synthesis of potassium sulfate from acidic mine waters	Muraviev (2003)
Testing and performance of newly constructed full-scale passive treatment system (Whittle Colliery, Northumberland, United Kingdom)	Nuttall (2003)
Mining reactive shale at BHP Bulliton's Mt Whaleback mine (Western Australia)	Waters (2002)
Assessment of contamination of surface water and groundwater by acid mine drainage	Gray (1996)
Management of acidic effluents from tailing dams in metalliferous mines	Saharan et al. (1995)
Mining waste research in United States	Shea-Albin and Fitch (1991)
Resource recovery: economic approach to remediation	Beck et al. (1990)

Hence the biogeochemical model can be defined in three major steps, as described subsequently, in the oxidative weathering of pyrite (Ferguson and Erickson, 1988; Kim et al., 1982; Sengupta, 1993; Johnson and Hallberg, 2003; López-Archilla, 2005; Wei and Wolfe, 2013; Chen et al., 2014; Nordstrom et al., 2015):

1. In the first stage, pH is near neutral and the rate of pyrite oxidation caused by oxygen (inorganic mechanism) (Eq. [17.1]) and by microbial action, in particular by the species *Thiobacillus ferrooxidans*, *Leptospirillum ferrooxidans*, and *Acidithiobacillus ferrooxidans* (biological mechanism), is comparable. The oxidation process is relatively slow, even when the concentration of pyrite is too high. The natural alkalinity of the water is often sufficient to neutralize the acidity produced. Iron hydroxide formation (Eqs. [17.2] and [17.3]), which includes the hydrolysis of ferric ion, is processed abiotically, decreasing the reaction rate with a decrease in pH. In this step, pH is maintained approximately above 4.5, the sulfate concentration is high (unlike the Fe concentration), and the production of acidity is reduced.
2. In the second stage, when the system neutralizing capacity is exceeded, the formed acid accumulates, causing a decrease in pH. In response to this acidification, the oxidizing ability of microorganisms increases and the amount of pyrite oxidized by oxygen decreases. Oxidation of pyrite continues as described in the first step, contrary to the oxidation of ferrous ion (Eq. [17.2]), whose rate becomes controlled mainly by the microbial activity. The sulfate concentration remains high and the total concentration of Fe increases with a concurrent increase in acidity. Fe^{3+}/Fe^{2+} is low and the pH of the medium is approximately between 2.5 and 4.5.
3. In the third stage, the pH in the vicinity of pyrite grains is less than 3 and the production rate of acid varies owing to increased solubility of Fe as well as reduced precipitation of Fe oxides and hydroxides. The oxidation of the ferrous ion is uniquely determined by microorganisms, which indirectly influence the oxidation of pyrite by ferric ion (Eq. [17.4]). Therefore, the steady-state activity of ferric ion is controlled by the combined effects of bacterial oxidation of ferrous ion, reduction of ferric ion by pyrite, and formation of Fe sulfate and complex hydroxides. In this step, the pH of the medium is less than 2.5. The acidity and concentrations of sulfate and total Fe are high, as well as the Fe^{3+}/Fe^{2+} ratio.

Thus, biogeochemical pyrite oxidation occurs by means of two oxidation mechanisms: inorganic oxidation and biological oxidation. Both mechanisms may occur simultaneously with the predominance of one or another determined primarily by the pH and Eh of the medium and by the limiting factors of bacterial growth.

In turn, the inorganic mechanism can be developed according to two mechanisms (Reed, 1986): a direct mechanism in which oxygen is the oxidizing agent, and an indirect mechanism in which the oxidizing agent is ferric ion. These two mechanisms will be processed simultaneously but at different rates, depending on the pH.

Again, the biological oxidation mechanism can occur in two ways (Urrutia et al., 1987; Kelley and Tuovinen, 1988; Wei and Wolfe, 2013): (1) a direct mechanism, in which bacteria act on the sulfide surface, oxidizing it enzymatically; and (2) through an indirect mechanism, in which the bacteria catalyze the oxidation of ferrous ion to ferric ion, in this way intermediating the inorganic indirect mechanism.

In the case of the weathering of other sulfides, the produced acid is the same but the metals (Me) in solution (aq) depend on the composition of such sulfides. Obviously, the chemical reactions will differ from those referred to pyrite. The sulfide oxidation process can be schematized by the following general reactions (Cidu et al., 1997) (Eqs. [17.5] and [17.6]):

$$(\mathrm{Fe, Me})\mathrm{S_2} + \frac{7}{2}\mathrm{O_2} + \mathrm{H_2O} \rightarrow \mathrm{Fe^{2+}}_{aq}, \mathrm{Me^{2+}}_{aq} + 2\mathrm{SO_4}^{2-} + 2\mathrm{H^+} \quad [17.5]$$

$$2\mathrm{Fe^{2+}}_{aq} + \frac{1}{2}\mathrm{O_2} + 2\mathrm{H^+} \rightarrow 2\mathrm{Fe^{3+}}_{aq} + \mathrm{H_2O} \quad [17.6]$$

17.2.2 SULFIDE WEATHERING PRODUCTS

Oxidative weathering of pyrite (and other sulfides) can be described using a variety of equations that nevertheless are approximate representations of a complex natural process which reflects the following results (Seal and Shanks, 2008; Koski et al., 2008; Kim, 2015): (1) solubilization of metallic ions or their precipitation in stable and insoluble forms under surface conditions (hydrated oxides, carbonates, sulfates, arsenates, phosphates, silicates, and heavy metals); (2) converting the sulfur to sulfate ions; and (3) produce acidic solutions.

Most nonsulfide gangue minerals are also dissolved during the weathering process. Quartz is one of the few original minerals that remain intact. The others are represented only by their alteration products (Fe and Mn oxides, clays, and other stable compounds) or in some cases may resist and remain in variable amounts (Sherlock et al., 1995; Kim et al., 2002; Murciego et al., 2011).

The precipitation of metals in stable and insoluble forms depends on several factors, including the current anionic species and the solubility of the metal compounds (and therefore on environmental conditions).

According to Krauskopf and Bird (1994), the behavior of metals is essentially determined by the anions with which they form stable insoluble compounds (as a precipitate form) rather than by the anions with which they form soluble salts. The insoluble compounds establish limits on the amounts of the various metals that may be carried in solution, moving away from the source and finally dispersing into the various environmental systems.

For example, the frequent occurrences of cerargyrite (AgCl) and native silver (Ag) in oxidized silver deposits should confirm that silver easily precipitates as chloride or be reduced to metal. Copper (Cu) can be partially retained in the oxidation zone as basic carbonate (malachite and azurite), oxide, or native metal. However, at less acidic conditions (pH 6 or 7), Cu can also be deposited as sulfates (eg, brochantite) and basic chlorides (eg, atacamite). Zinc (Zn) can remain in the oxidized zone essentially as carbonate (eg, smithsonite). Iron (Fe) and manganese (Mn) tend to remain in outcrops, owing to the insolubility of their oxides and hydroxides of higher valencies and carbonates. However, in more reducing conditions as well as acidic solutions, Fe and Mn neither co-precipitate with common anions nor reduce to native metal; they are transported in solution in significant amounts. Lead (Pb) forms a carbonate (cerussite) and a sulfate (anglesite). The similar insoluble forms were also applicable to Cd (forming otavite), Ni and Co (forming arsenates), and Ni (forming silicate). Zinc, Cu, and Cd stand out as mobile elements compared with other metals because the anions normally allow significant quantities of these metals to remain in solution.

Therefore, the mobility of the metal ions is determined by a wide range of factors, from the composition of water and rocks to the conditions of pH and Eh of the medium, and to interactions among the various elements and compounds in solution.

17.2.2.1 Pyrite Weathering Products

As result of pyrite oxidative weathering, several products can result, including a diversity of precipitates and their derivatives with varying crystallinities (Table 17.2). The most common are fundamentally composed of combinations of Fe with oxygen or sulfate (Bloomfield, 1973; Calvo de Anta and Perez Otero, 1993; Jackson and Bistricki, 1995; McGregor et al., 1998; Kirby et al., 1999; Sánchez España et al., 2006a, 2006b; Seal and Shanks, 2008; Koski et al., 2008; Frau et al., 2009; Kim, 2015).

From laboratory experiments, Brady et al. (1986) showed that the concentration of sulfate in AMD has a strong influence on the type and proportion of the final oxides which compose the precipitates. Stumm and Morgan (1996) suggested a competition mechanism between hydroxyl and sulfate anions for surface adsorption sites. Moreover, the presence of transition metals also influences the degree of crystallinity of the precipitates (Barral Silva, 1989) as having an inhibitory effect on the crystallization of Fe amorphous compounds.

Acid production is continuous throughout the pyrite oxidation process. Its amount varies according to the degree of Fe oxidation, the oxidation mechanism, and the resulting products (Breemen, 1973; Urrutia et al., 1987; Barral Silva, 1989; Wei and Wolfe, 2013). Thus, we highlight the following probable consequences: (1) If the medium is highly acidic and oxidizing (Eh $>$ 500 mV), the entire Fe is oxidized and remains in solution as ferric ion, eg, each mole of pyrite produces an equivalent mole of protons; (2) if the medium is not sufficiently oxidizing (Eh $<$ 500 mV), Fe is oxidized and remains in solution as ferrous ion, eg, originates a double amount of protons; (3) if the medium is highly acidic (pH $<$ 4) and oxidant (Eh $>$ 400 mV), and with a high concentration of sulfates, all of the oxidized Fe precipitates as jarosite, eg, each mole of pyrite produces 2.66 mol of protons; and (4) if the Eh conditions are higher than 200 mV and the pH is greater than 4, the oxidized Fe hydrolyses to iron hydroxide, eg, each mole of pyrite produces 4 mol of protons, which corresponds to the highest production of acid.

These reactions occur in nature in a combined form with a predominance of the latter two, establishing an average production of acid varying from 1.2 to 2 mol of protons per mole of oxidized pyrite (Urrutia et al., 1987). The prevalence of the last two reactions mainly results from the strong tendency that the ferric ion has to hydrolyze and precipitate, which confirms the oxidation of pyrite among the most acidic weathering reactions (Stumm and Morgan, 1996).

17.3 CHARACTERISTICS OF ACID MINE DRAINAGE

The AMD quality can be characterized by the physical and chemical properties. Physical properties commonly considered are temperature, electrical conductivity, and the suspended and dissolved solids. As significant chemical properties we can cite pH, alkalinity and acidity, hardness, metal ions in solution, silica content, salt content, ammonium ion content, hydrocarbon content, and radioactivity.

These parameters depend on many factors, from those which influence the sulfide oxidation process, to those which influence the migration and dilution of AMD. Thus, it becomes clear that each mine is a specific case and should be carefully considered and studied. In the case of metal mining

Table 17.2 Examples of Pyrite Oxidation Products: Precipitates

Iron Combinations	Pyrite Oxidation Products: Precipitates[a]
Iron–oxygen combinations	Iron oxides and hydroxides: • Goethite (α-FeO(OH)) • Lepidocrocite [γ-FeO(OH)] • Hematite (α-Fe_2O_3) • Maghemite (γ-Fe_2O_3) • Magnetite (Fe_3O_4) Amorphous and poorly crystallized iron oxyhydroxides: • Ferrihydrite ($Fe_2O_3 \cdot 2FeOOH \cdot 2, 6H_2O$) • Feroxyhite ($\delta'$-FeOOH) • Amorphous $Fe(OH)_2$ • Amorphous $Fe(OH)_3$

Iron–sulfate combinations (and other metals that occur in medium)	Jarosite [$XFe_3(OH)_6(SO_4)_2$]
	(with X = NH_4^+, K^+, Na^+ or H_3O^+)
	Schwertmannite [$Fe_8O_8(OH)_6(SO_4)$]
	Hydrated sulfates:
	• Efflorescences:
	  ◦ Melanterite ($SO_4Fe \cdot 7H_2O$)
	  ◦ Rozenite ($SO_4Fe \cdot 4H_2O$)
	  ◦ Szomolnokite ($SO_4Fe \cdot H_2O$)
	  ◦ Roemerite [$(SO_4)_4Fe^{2+}Fe^{3+} \cdot 12H_2O$]
	  ◦ Gypsum ($CaSO_4 \cdot 2H_2O$)
	  ◦ Alunogen [$Al_2(SO_4)_3 \cdot 18H_2O$]
	  ◦ Meta-alunogen [$Al_2(SO_4)_3 \cdot 12H_2O$]
	  ◦ Khademite [$Al_2(SO_4)F \cdot 5H_2O$]
	  ◦ Pickeringite [$MgAl_2(SO_4)_4 \cdot 22H_2O$]
	• Copiapite [$Fe^{3+}Fe^{2+}(OH)_2(SO_4)_6 \cdot 20H_2O$]
	• Voltaite [$K_2Fe^{3+}Fe^{2+}(SO_4)_{12} \cdot 18H_2O$]
	• Rhomboclase [$FeH(SO_4)_2 \cdot 4H_2O$]
	• Coquimbite [$Fe_2(SO_4)_3 \cdot 9H_2O$]
	• Tamarugite [$NaAl(SO_4)_2 \cdot 20H_2O$]
	• Sodium alunite [$NaAl(SO_4)_2 \cdot 12H_2O$]
	• Sulfates of Fe and K, Cu, Pb, Ca, Al (…)

Iron–carbonate combinations	Siderite ($FeCO_3$)
Iron–silica combinations	Nontronite [$Fe_2(Si,Al)_4O_{10}(OH)_2$] Cronstedtite [$(Fe^{3+};Fe^{2+})_3(Fe^{3+},Si)_2O_5(OH)_4$] Chamosite [$(Fe_4^{2+}Al_2)(Si_2Al_2)O_{10}(OH)_8$]

[a]*According to Breemen, 1973; Park and MacDiarmid, 1975; Thornber and Wildman, 1984; Thornber, 1985; Guilbert and Park, 1986; Urrutia et al., 1987; Barral Silva, 1989; Doner and Lynn, 1989; Schwertmann and Taylor, 1989; Ferris et al., 1989; Krauskopf and Bird, 1994; Bigham et al., 1996; Nordstrom and Alpers, 1999a, 1999b; Bigham and Nordstrom, 2000; Jönsson et al., 2006; García-Lorenzo et al., 2012; Chou et al., 2013; Kim, 2015; and Carrero et al., 2015.*

operations, the most commonly discussed characteristics are pH, metal ion concentration (Fe, Al, Mn, Zn, Cu, Pb, etc.), concentrations of dissolved anions (sulfates, nitrates, chlorides, arsenates, etc.), hardness, and suspended solids.

To facilitate AMD characterization (and also the characterization of waters affected by AMD), Schmiermund and Drozd (1997) summarized five common features of these waters: (1) pH, acidity, and alkalinity; (2) heavy metal concentrations; (3) Fe and Al concentrations; (4) sulfate and arsenate concentrations; and (5) turbidity and suspended solids.

17.3.1 pH, ACIDITY, AND ALKALINITY

The pH of a solution is a measure of the hydrogen ion concentration; formally, it can be defined as the negative logarithm of the hydrogen ion concentration ($pH = -\log [H^+]$). Absolutely pure water is weakly dissociated and H^+ and OH^- ions are present in the same concentrations, ie, are neutral in their reaction. If the concentration of H^+ is higher than that of pure water, the solution is acidic; if it is less, the solution is alkaline. In nature, most observed pH values are between 4 and 9, depending on factors such as the water source, the bedrock geology, and the vegetation cover. The lowest pH values found in nature are found in solutions in contact with oxidized sulfides (Krauskopf and Bird, 1994; Schiermund and Drozd, 1997; Koski et al., 2008).

One of the most significant aspects related to the pH of the medium consists of its strong influence on the solution of many substances. In the case of heavy metals, the low pH increases its solubility and hence its toxicity.

Water is commonly called "acidic" or "basic" when the pH is lower or higher than neutral, respectively. However, the terms "acid" and "basic" in this context should not be confused with the terms "acidity" and "alkalinity" (Schiermund and Drozd, 1997).

Alkalinity measures the total concentration of bases present in the sample and is measured by the solution's ability to consume protons, ie, by the neutralization capacity of a strong acid. In superficial waters, alkalinity depends mainly on constituents which are related to the CO_2 content and to the mineralization of water, ie, bicarbonates, carbonates, and hydroxyl ions. If present, borates, phosphates, and silicates are included in the determinations. Thus, the total alkalinity of water is the sum of OH^-, CO_3^{-2}, HCO_3^-, PO_4^{-3}, $SiO(OH)_3^-$, $H_2BO_3^-$, NH_3^-, HS^-, and organic anions.

Acidity measures the total concentration of acids present in the sample and is measured by the solution's ability to consume hydroxyl ions, ie, by the neutralizing capacity of a strong base.

Thus, systems with low pH may have different values of total acidity, depending on the most abundant acids in solution. However, Kelly (1991) showed that there is a good linear correlation ($r^2 = 0.68$) between pH and the logarithm of the total acidity in AMD. At pH values greater than 7, high total acidity rarely exists because the concentration of sulfuric acid is low. In fact, one of the consequences of high total acidity is the ability of these media to maintain a very low pH. This capacity results from the destruction of the bicarbonate buffer system. For pH values below 4.2, all of the carbonate and bicarbonate are converted to carbonic acid, which in turn rapidly dissociate into water and carbon dioxide (Kelly, 1991). The release of CO_2 into the atmosphere ensures maintenance of the gradient.

Therefore, total acidity is a key concept for establishing the difference between the AMD and other acid systems such as, for example, sites affected by acid rain.

17.3.2 HEAVY METAL CONCENTRATIONS

High concentrations of one or more heavy metals are a common feature of AMD. In principle, any metal present in the mineral deposit can be incorporated into the AMD under certain geochemical conditions. Another aspect which highlights the importance of this characteristic relates to the threat of toxicity that significant levels of heavy metals represent to aquatic life forms downstream of the pollution source, as well as for human beings.

In fact, heavy metal concentrations and the pH are two most important characteristics commonly discussed. Accordingly, Ficklin et al. (1992) developed an empirical classification for mine and natural drainages in mineralized areas of Colorado (Colorado Mineral Belt). This classification is based on the pH and the combined contents of the main metals (Zn + Cu + Cd + Co + Ni + Pb) detected in the studied samples (besides Mn, Al, and Fe, which are not considered in this classification because they have very different geochemical behaviors).

With the proposed classification, the authors summarized the drainages of a wide range of geologic environments and explained the ranges in pH and the metal concentrations in accordance with the geologic and geochemical controls (Plumlee et al., 1992; Ficklin et al., 1994).

The control mechanisms of metal concentrations in AMD can interact in a complex way and are highly specific to the metal and the site. At the start of the AMD formation process, heavy metal concentrations are essentially controlled by the amount of metal-bearing minerals as well as their solubilities and dissolution rates. With the development of the process, as the effluent evolve in response to contact with the regional rocks, atmosphere, and other waters, there are changes in the metal complexation which favor precipitation and adsorption, thus controlling and mitigating AMD metal flow (Schmiermund and Drozd, 1997; Larocque and Rasmussen, 1998; Koski et al., 2008; Keleprtzis et al., 2012; Schaider et al., 2014; Bakar et al., 2015).

17.3.3 IRON AND ALUMINUM CONCENTRATIONS

Although not included in the classification of Ficklin et al. (1992), because the geochemical behavior is markedly distinct from that of the other metals, Fe and Al are two elements worth serious consideration in studies of AMD. They generally occur in significant concentrations in AMD, which are higher than the concentrations of the metals mentioned earlier.

Fe and Al are important in the study of AMD because in certain cases they can affect the behavior of other metals and often lead to the formation of coatings along water streams: a remarkable accumulation of yellow-orange Fe oxy-hydroxides (colloquially known as yellow-boy), and of white Al oxy-hydroxides, both with a pronounced visual impact (Thornber, 1985; Schmiermund and Drozd, 1997; Banks et al., 1997; Kim, 2015).

The extremely low solubility of these compounds under near-neutral conditions results in the persistence of these colorations for a long time.

The concentration of dissolved Fe is controlled by the precipitation of ferric oxy-hydroxides or oxy-hydroxy-sulfates, among other solid phases, whose formation and stability greatly depend on Eh and pH conditions. Besides this, Fe (Fe^{3+}) precipitation is of considerable importance as a potential control of other metal concentrations, owing to adsorption and co-precipitation phenomena (Krauskopf and Bird, 1994; Schmiermund and Drozd, 1997; Jönsson et al., 2006).

Aluminum is also considered to be an important characteristic of AMD along with Fe for several reasons including a similar trend in forming oxy-hydroxides and oxy-hydroxy-sulfates in "acid-neutralized" environments, and the potential to cause aesthetic degradation of affected river systems (Kim, 2015; Carrero et al., 2015).

17.3.4 SULFATE AND ARSENATE CONCENTRATIONS

The high concentration of sulfate is probably the most consistent feature of the AMD, because in most cases its origin depends on sulfide oxidation (Schmiermund and Drozd, 1997). Researchers have demonstrated that sulfate concentrations reflect the effectiveness of the sulfide oxidation, as well as the effluent flow rate from the oxidation zone and any subsequent dilution.

The authors also considered that the conservative nature of sulfate in the AMD provides a basis with which to evaluate and locate the destination of nonconservative components, especially metals. For example, a decrease in the concentration of sulfates in solution, downstream of the source, often can be attributed to dilution phenomena. Other components whose concentration decreases over the same water stream section may have a similar explanation, using the sulfate behavior as a reference, or on the contrary, to consider other mechanisms (eg, precipitation or adsorption).

The importance of sulfate ions in characterizing AMD is reinforced by the strong influence it seems to have on the type and propagation of the iron oxides that compose the precipitates ("coatings") (Brady et al., 1986; Jönsson et al., 2006).

Typically, arsenic does not occur in AMD at such high concentrations as sulfate, nor does it behave in an equally conservative way. Generally, dissolved As concentrations are similar to those observed for metals. However, this parameter is important because of the high toxicity of this metalloid.

17.3.5 TURBIDITY AND SUSPENDED SOLIDS

Turbidity is the capacity of water to disperse and absorb light, preventing its transmission in depth. It depends on suspended and dissolved solids including pelagic organisms (plankton). Turbidity is a parameter measured in situ, and is directly related to the visual appearance of water.

The total suspended solids (TSS) is a parameter measured in the laboratory and has the disadvantage, relative to turbidity, of needing to be sampled and stored. Indeed, during storage reactions may occur such as precipitation and flocculation, which irreversibly alter the sample relative to field conditions (Schmiermund and Drozd, 1997).

Particles that can constitute suspended solids, relevant to AMD, include macromolecular colloidal particles of Fe and Al oxy-hydroxides, macroscopic particles, and other compounds (clay and silt).

TSS (and turbidity) are an important feature of AMD because they can have great relevance to the transport of As and heavy metals owing to adsorption phenomena, in addition to adverse effects on water quality and the impact on aquatic ecosystems.

17.4 FACTORS CONTROLLING THE FORMATION OF ACID MINE DRAINAGE

The quality and quantity of AMD produced in pyritic environments depend on several factors, which may be categorized as primary, secondary, and tertiary (Ferguson and Erickson, 1988). Primary factors are those that are directly involved in acid production processes, ie, factors which directly determine sulfide weathering (eg, mineralogical characteristics of pyrite and other sulfides, oxygen, water, bacteria); whereas secondary factors control the "consumer" or modify the products of reactions involved in AMD formation (eg, acid neutralization by the presence of buffer minerals); and finally, tertiary factors include physical and structural aspects concerning the deposit, from mining to the location, which somehow influence the production, migration, and dilution of AMD (eg, characteristics of the material deposited on tailings, hydrological regime of the site) (Fig. 17.2).

Primary factors are mainly concerned with the mineralogical characteristics of sulfides, oxygen, water, temperature, pH, Eh, and microorganisms (Ferguson and Erickson, 1988; Kelley and Tuovinen, 1988; Plumlee et al., 1992; Sasaki, 1994; Banwart et al., 2002; Saaltink et al., 2002; Bowell, 2002; Seal and Shanks, 2008; Koski et al., 2008; Foster et al., 2008; Favas et al., 2012).

The mineralogical characteristics of sulfides are composed of a set of intrinsic factors which determine the sulfide's weatherability: namely, its ability to weather a function of time. The main features of a mineralogical nature which determine the weatherability of sulfides under superficial conditions are the size and shape of the mineral grains and the crystalline structures, crystallochemical properties, and structural defects (Caruccio, 1978; Urrutia et al., 1987; Ferguson and Erickson, 1988; Sasaki, 1994; Valente, 2004; Murciego et al., 2011; Wei and Wolfe, 2013).

Within the primary factors, those which influence the activity of the microorganisms are also considered. The pH and temperature, for example, determine the optimal range of activity. In addition, the availability of oxygen, nutrients, and carbon dioxide also interferes with the activity of microorganisms, depending on the physiology of each species.

As mentioned, both pH and Eh have a great influence on the sulfide's weathering and on the solubility and mobility of several chemical elements and compounds.

Among the secondary factors, the existence of minerals capable of neutralizing acid resulting from decomposition of the sulfides is one of the most important mechanisms in controlling AMD characteristics. The most common of these minerals is calcite ($CaCO_3$), which consumes hydrogen ions (H^+) producing HCO_3^- and H_2CO_3, but there are also other minerals. In the neutralization process, part of the acidity and Fe (and other metals) are removed from the solution and the pH rises to values close to neutral (Sengupta, 1993; Bowell, 2002; Huminicki and Rimstidt, 2008; Wei and Wolfe, 2013; Wu et al., 2014).

Therefore, this "buffering effect" induced by the dissolution of calcium carbonate as a result of pyrite oxidation (and sulfides in general) controls the concentration of Ca and S in solution. This raises the pH of the medium and increases the removal of dissolved metals (Holmströn et al., 1999; Cravotta and Trahan, 1999; Gunsinger et al., 2006; García-Lorenzo et al., 2012). This effect of "acid neutralization" can be described by the following reaction, assuming that the Fe precipitates as Fe hydroxide (Holmströn et al., 1999) (Eq. [17.7]):

$$FeS_2 + 3.75O_2 + 3.5H_2O + 4CaCO_3 \rightarrow Fe(OH)_3 + 4HCO_3^- + 2SO_4^{2-} + 4Ca^{2+} \qquad [17.7]$$

The tertiary factors are the "last" that control the quality and quantity of acid drainage. Typically, the acidity of the effluents decreases downstream as it increases the distance to the pyritic environments. On the one hand, dilutions occur along the course, which are related to the hydrography of the area, which may be of greater or lesser importance. On the other hand, the effluents can come into contact with neutralizing sources: for example, organic matter and/or basic minerals. Thus, the consequent increase in pH values is accompanied by significant changes in water column composition

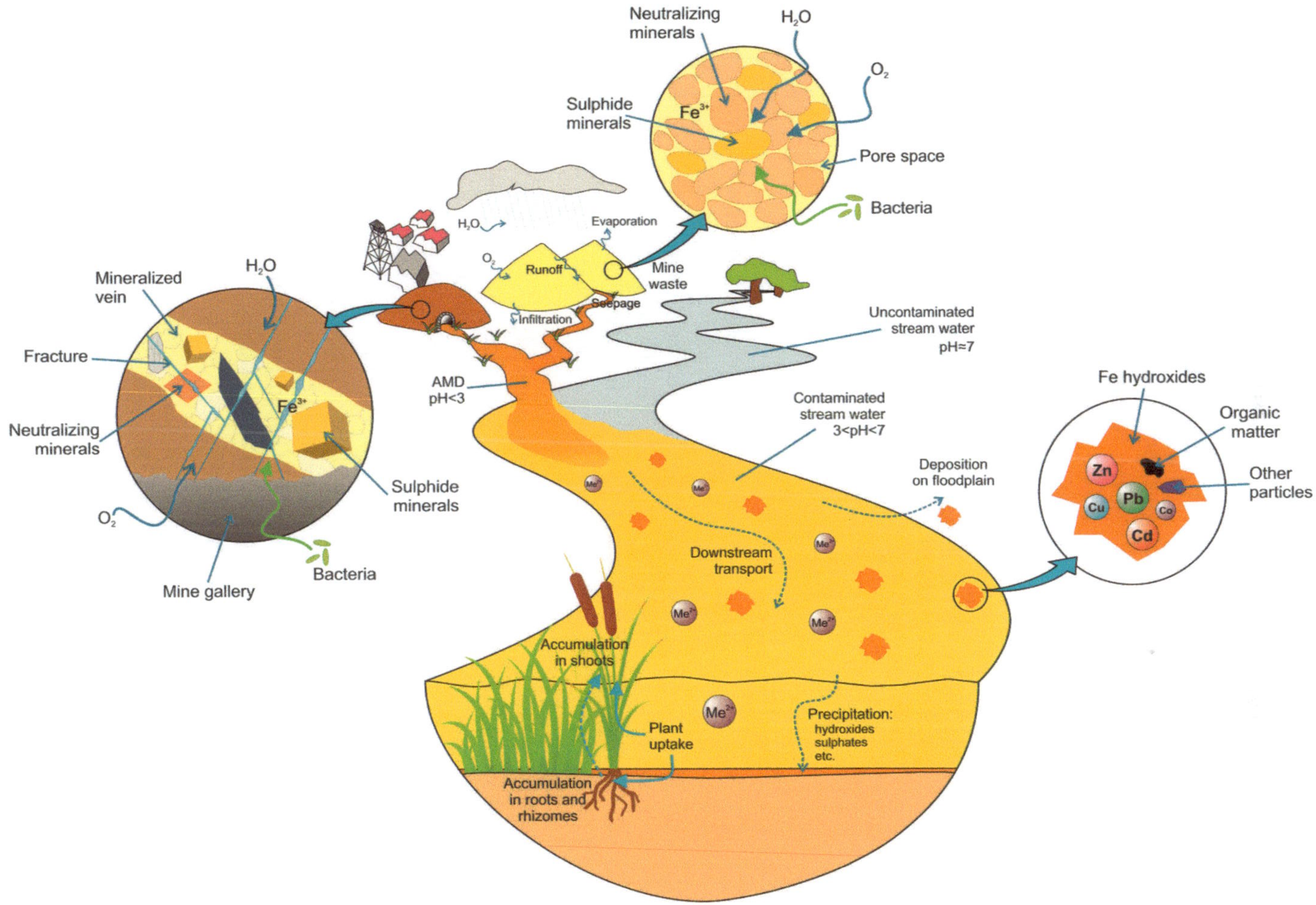

FIGURE 17.2

Schematic illustration of factors that affect sulfide weathering and modify the production, migration, and dilution of AMD.

as a result of the decreased solubility of many elements (Bowell, 2002; Rees et al., 2002; Robles-Arenas and Candela, 2010; Soltani et al., 2014; Shim et al., 2015).

17.5 ENVIRONMENTAL IMPACT OF ACID MINE DRAINAGE

As previously mentioned, mining effluents are characterized by high concentrations of heavy metals and high acidity, a combination that causes severe environmental problems in most cases. The migration of untreated AMD can contaminate surface stream water and groundwater, compromising the quality of water resources and affecting the health of aquatic life (Azapagic, 2004; Johnson and Hallberg, 2005). The environmental, social, and economic problems associated with AMD from abandoned mine sites are significant and global. The environmental impacts of AMD can occur in both active and abandoned mine lands. Water circulation inside mine galleries and in mine tailings promotes the oxidative weathering of sulfides and the formation of effluents with high pollution potential, which easily migrate to various natural systems (groundwater, fluvial systems, and soil). Some of the most problematic situations caused by acid and metalliferous drainage are the environmental impact on water quality, the impact on the fluvial ecosystem, and the visual impact.

17.5.1 IMPACT ON WATER QUALITY

AMD can percolate into the aquifers or form effluents that converge into superficial water streams, contaminating both ground and surface water (Fig. 17.3). In general, waters affected by AMD are not potable and have high levels of various undesirable substances such as heavy metals and sulfate, which determine their use, either domestic or industrial (Banks et al., 1997; Sangita et al., 2010; Robles-Arenas and Candela, 2010; Santisteban et al., 2015; Bakar et al., 2015). Thus, degradation of the water quality may lead to the impossibility of using them, requiring an alternate supply or expensive water purification processes.

As a consequence of the abandonment of underground mines, serious situations can occur involving groundwater contamination. In these cases, with the progressive flooding of the galleries, the water table is restored to original levels; therefore, water withdrawing in the surroundings may become contaminated. Also, the existence of many ventilation shafts, which were abandoned with no sealing or waterproofing, allows the flow of water contained in underground mines, which can flow to underlying aquifers.

The impact of AMD on water quality also raises questions about the sustainable use of such waters for the irrigation of agricultural crops (Strosnider et al., 2011; Yesilnacar and Kadiragagil, 2013; Modoi et al., 2014). For example, sugarcane is cultivated with iron mine leachate water in the Dabaoshan Iron Mine area, about 230 km south of Guangzhou, China (Fig. 17.3). The harvested sugarcane may be is a suitable substrate for the production of a variety of biorefinery products (Aguilar et al., 2012; Zahid et al., 2013; Xie et al., 2014; Bahurudeen et al., 2015; Moubarik and Grimi, 2015).

17.5.2 IMPACT ON THE FLUVIAL ECOSYSTEM

AMD can lead to significant changes in the ecological conditions of receptors streams and lakes, especially when the dilution of AMD is not effective (Fig. 17.3). This ecological disruption results from the acidification of water because of the content of toxic substances.

FIGURE 17.3

Dabaoshan Iron Mine area, Guangzhou, South China. (A) Mining for iron and other types of metal recovery; (B) red-colored river as a result of iron contamination; (C) concrete dam to control downstream flow; (D) sugar cane cultivation with iron-contaminated water.

Photos by M.N.V. Prasad.

As suggested by several authors (Williamson et al., 1982; Kelly, 1991; Phillips and Rainbow, 1994; Salomons, 1995; Arnekleiv and Storset, 1995; Gray, 1997; Sangita et al., 2010; Wei and Wolfe, 2013), the negative effects are as follows:

1. The oxidation of ferrous cations leads to a decrease in the amount of available oxygen for aquatic organisms to consume. The formation of iron oxy-hydroxides can also cause a decrease in nutrients such as phosphate, which is adsorbed.
2. The formation of Fe precipitate coatings in the water stream can prevent the oviposition of numerous fish species, modifying their reproductive cycle. This aspect, which is associated with a decreased amount of available food, can lead to the destruction of sensitive fish populations such as trout and salmon. Acidification of water also may have an important role because, for example,

salmonids cannot spawn in water with a pH below 5. When their aqueous suspension increases water turbidity, these precipitates interfere with photosynthesis and reduce visibility for consumers.

3. Benthic organisms or benthos, is usually sensitive to variations in water quality; they cannot adapt to these conditions, so their disappearance could seriously affect the food chain.
4. Metal cations accumulate in the tissues of several organisms; they are transferred by bioaccumulation along the food chain and concentrate in higher-level consumers, posing risks to human health. In addition, heavy metal concentrations in solution may have more immediate implications related to high toxicity expressed by some elements (eg, copper).
5. Heavy metal concentrations in the tissues of various organisms, from benthic invertebrates to large predators, can have particularly serious consequences in lacustrine environments, which because of its closed ecosystem nature, tends to be "poisoned" ("slow poisoning of the lakes").
6. Some species of algae and invertebrates can adapt to conditions created by AMD, and in the absence of competition can become seriously invasive.

17.5.3 VISUAL IMPACT

One symptom of contamination by acid drainage is the presence of strong ferruginous chromatization ("yellow-boy") on river margins and the river bed as a result of Fe oxy-hydroxide precipitation. These precipitates are brightly colored, between yellow and red, with a predominance of ocher tones; they providing a visually aggressive element and produce a generally unpleasant aesthetic effect (Fig. 17.3).

17.6 RESOURCE RECOVERY FROM ACID MINE DRAINAGE

Considering the adverse impact on the environment, removal and recycling of metal(loid) contaminants and other by-products from AMD is the subject of global research. Critical literature surveys and investigations into several selected alternatives such as chemical treatment (modifications to lime neutralization), solvent extraction, ion exchange, bioadsorption, and electromining aim to develop a viable treatment process. Different techniques are efficiently applied, as discussed subsequently, to explore novel ways in which to produce high-quality effluents to secure satisfactory revenue and minimize or eliminate hazardous sludge production.

17.6.1 APPLICATION OF FILTER MATERIALS

Filter materials are a novel option for treating AMD ($pH \leq 6$) because they are inexpensive and available on a local scale, have high functionality, and demand a minimum cost for maintenance. They are available in a wide variety in the context of metal sorption capacity and are used extensively in passive treatment systems. A perusal of the literature reveals that the composition of mining effluents varies to a large extent, depending on the mining site and the type of mining. Hence a large number of filter materials, both organic and inorganic, has been used to remove heavy metals for mine draining both in controlled laboratory setups as well as in natural conditions. The filter materials have the capacity to remove heavy metals through diverse mechanisms, eg, adsorption, ion exchange, sulfate reduction (Robinson-Lora and Brennan, 2009) and precipitation (Feng et al., 2004; Ríos et al., 2008).

The organic materials are composed of bark, chitin, chitosan, commercial ion exchangers, lignite, peat, rice husks, and yeast, whereas biocarbons, calcareous shale, dolomite, fly ash, limestone, olivine, steel slag materials, and zeolites are potential examples of inorganic materials (Westholm et al., 2014). However, the selective filter materials tested have performed well in both laboratory and field conditions, which provides a certain basis for judging the potential of the filter materials to remove metals. Generally, Cu, Fe, Zn, and Pb are easiest to remove among the metals with the application of filter materials such as steel and iron slag, lignite, chitosan, natural zeolite, and yeast (Feng et al., 2004; Mohan and Chander, 2006; Laus et al., 2007; Mamba et al., 2009; Ramirez-Paredes et al., 2011; Westholm et al., 2014). For example, in the case of Fe, this has been attributed to the precipitation mechanism when the removal of Fe is almost unaffected by the presence of other species in the solution (Motsi et al., 2009; Westholm et al., 2014). Filter materials have the capacity to accumulate metals in a finite and accessible volume; afterward, these may be replaced when metal saturation is achieved, as endorsed by Westholm et al. (2014).

17.6.1.1 Use of Lignite

Lignite is a brown, fibrous, low-grade coal with a relatively low carbon and high moisture content. This acts as a potential cation exchanger and may be used efficiently to remove and recover metal ions from AMD, as shown experimentally by Mohan and Chander (2006). Lignite has been characterized and used for AMD treatment. Sorption of ferrous, ferric, manganese, zinc, and calcium in multicomponent aqueous systems was investigated at different pH conditions to find the optimum pH.

17.6.1.2 Uses of Fly Ash

Fly ash is generally composed of fine particles produced from coal combustion and has multiple uses. However, the composition varies according to the sources and composition of the coal being used. Fly ash generated from the burning of coal in power plants poses widespread environmental and economic problems for many countries. Pérez-López et al. (2007) used fly ash to improve the quality of the AMD generated by the oxidation of sulfide-rich mining waste. The generation of AMD was prevented by adding fly ash to sulfide-rich residues in unsaturated column experiments. A column experiment filled with a pyrite-rich sludge with artificial irrigation leached acid drainages (pH $\approx$ 2) containing high concentrations of sulfate, iron, and other metals. However, unsaturated column experiments filled with pyritic-rich sludge and fly ash drained leachates characterized by alkaline pH (pH up to 10), low sulfate concentration, and a lack of iron and other metals in solution. As a consequence of the leaching of fly ash, the pyrite oxidative dissolution at high pH favors metal precipitation inside the column (mainly iron), coating of pyrite grains, and attenuation of the oxidation process, resulting in great improvement in the quality of the leachates. Prasad and Mortimer (2011) converted Indian fly ash into zeolites, and both the raw fly ash and zeolite were used to treat British acidic mine waters. The results demonstrated that fly ash zeolites are more effective at removing all metals than raw fly ash for the treatment of AMD.

Ríos et al. (2008) also used low-cost sorbents such as coal fly ash, natural clinker, and synthetic zeolites to clean up AMD generated at the Parys Mountain copper—lead—zinc deposit, Anglesey (North Wales) and to remove heavy metals and ammonium from AMD. pH had a very important role in the sorption and removal of contaminants, and a higher adsorbent ratio in the treatment of AMD promoted an increase in the pH, particularly using natural clinker-based faujasite (7.70—9.43) and the reduction in metal concentration. Na-phillipsite had a lower efficiency compared with that of faujasite. Selectivity of faujasite for metal removal was in the decreasing order: Fe $>$ As $>$ Pb $>$ Zn $>$ Cu $>$ Ni $>$ Cr. Based on these results, use of these materials has the potential to provide improved methods for treating AMD.

17.6.1.3 Uses of Chitin

Robinson-Lora and Brennan (2010) used sustainable composite waste material, crab-shell chitin complex (CC), which has proven to be a promising substrate for AMD remediation. Remediation rates and metal removal mechanisms promoted by CC were evaluated and found to be more effective than others. CC increased the pH from 3.0 to near neutral in 3 days. In microcosms containing CC, steady alkalinity generation and acidity removal were observed at average rates of 26.5−25.2 mg $CaCO_3$/ L^{-d}, respectively. Aluminum removal (100%) was eventually observed with all substrates, but occurred much faster with CC. Results from thermodynamic geochemical modeling indicate that Al removal was consistent with the precipitation of hydroxides and/or alunite. Iron removal was consistent with precipitation of Fe(III) oxides and Fe(II) sulfides, as well as sorption onto CC and SMC. The addition of Na lactate interfered with such mechanisms owing to complexation effects. Chitin complex has been found to be a good absorbent for Mn (>73%) (Robinson-Lora and Brennan, 2010), likely because of the formation of rhodochrosite and strong base anion exchangers for U (Ladeira and Gonçalves, 2007). The results of this study indicate that CC is an attractive substrate for treating metal-laden waste streams, especially under controlled laboratory conditions.

Drawbacks to using filter materials may be summarized as follows: (1) Application of specific filter materials for mine drainage is difficult mainly owing to heterogenous characteristics and their subsequent leaching of hazardous toxic chemicals from filter materials. In addition, disposal of filter materials at landfill sites and their regeneration are problematic and costly. (2) Proper care should be taken to select suitable filter materials in the context of their characteristics (eg, local availability, cost, metal reduction capacity) and application of specific doses, which is a matter of great concern. An increase in fly ash for AMD treatment has no additional benefit and is harmful because it releases many metals from its surface into mine water.

Most filter materials have been observed to be effective only under controlled laboratory conditions and only a few have been proven to have a true potential under natural conditions; these could be considered for commercial applications for future use.

17.6.2 USING MODULAR BIOREACTORS

Bioreactor systems are used to mediate the selective biomineralization of transition metals (such as Fe and Zn) present in acidic mine waters, which represents a more eco-friendly and sustainable approach to mine water treatment. Ňancucheo and Johnson (2012) demonstrated two novel bioreactor systems in which acidophilic bacteria are used, indigenous to mine-impacted environments. One is an aerobic system that promotes the selective removal of iron using a combination of biological oxidation of ferrous iron and pH-controlled precipitation of ferric iron, and was shown to generate pure-grade schwertmannite (idealized composition, $Fe_8O_8[SO_4][OH]_6$) from synthetic mine waters.

The second is an anaerobic system that uses a consortium of acidophilic sulfate-reducing bacteria to precipitate stable metal sulfides selectively in in-line reactor vessels from synthetic mine waters containing Cu, Zn, Fe, and Al (Stumm and Morgan, 1996; Ňancucheo and Johnson, 2012). Hedrich and Jhonson (2014) developed novel techniques that promote the improvement of acidic waste waters and selective recovery of metals. With these techniques, zinc and iron were removed via controlled biomineralization: zinc as ZnS in a sulfidogenic bioreactor and iron as schwertmannite by microbial iron oxidation and precipitation of ferric iron. This demonstrated that modular biomineralization units can be effective in processing complex mine water and generating metal products which may be recycled.

17.6.3 USING PERMEABLE REACTIVE BARRIERS

Sasaki et al. (2008), evaluated the efficiency of granulated blast furnace slag (GBFS) not containing iron oxides, as a reactive material for arsenic in a permeable reactive barrier (PRB) and discussed the mechanism in immobilizing arsenic and for potential use as an alternative material instead of iron granules. In this work, the immobilization of arsenate in groundwater affected by AMD was investigated using a PRB column bearing GBFS compared with iron granules, which are commonly used. Results presented by the authors showed that 15 mg L^{-1} As decreased to be less than 0.4 mg L^{-1} for more than 18 pore volumes (pv) in the GBFS-PRB by sorption, co-precipitation, and presumably formation of hydrated calcium arsenate, and less than 0.04 mg L^{-1} for more than 17 pv in the iron-bearing PRB, probably by co-precipitation with iron (oxy-hydro)oxides. According to Sasaki et al. (2008), GBFS has the advantages of compensating low reactivity with high porosity, facilitating industrial handling with low density, and using industrial wastes for more valuable applications.

17.7 PREVENTION, MITIGATION, AND TREATMENT OF ACID MINE DRAINAGE

Pollution caused by AMD can be minimized by adopting appropriate strategies and approaches to mitigate AMD or prevent its generation and migration (Fig. 17.4). The generation of AMD can be prevented or minimized by controlling the chemical reactions and microbiological activity which cause sulfide oxidative weathering, and thus the production of these acid and metalliferous effluents. The techniques applied to these objectives are known as "source control" measures (Johnson and Hallberg, 2005). For example, one way to prevent dissolved oxygen supply to the system is to flood and seal abandoned deep mines, and thus to prevent or minimize AMD production (Johnson and Hallberg, 2005; Sangita et al., 2010). Another viable option to fulfill the same objective is underwater storage for disposal and storage of mine tailings that potentially produce acid (Li et al., 1997). In this case, shallow water covers can be used or they can be combined with a layer of sediment or organic material covering the tailings. Alternatively, dry covers may be used for surface storage of reactive mineral spoils, with a "sealing layer" usually constructed from clay (Swanson et al., 1997; Johnson and Hallberg, 2005). Another approach to minimizing AMD production is to promote acid–neutralization reactions by adding acid-consuming materials. A variation on this theme is to add phosphate to acidic water or to pyritic mine waste to form insoluble ferric phosphate that reduces its potential to act as an oxidant of sulfides (Johnson and Hallberg, 2005; Sangita et al., 2010). Finally, a third type of strategy consists of adding anionic surfactants, such as benzoic acid, sorbic acid, and sodium dodecyl sulfate, to control or inhibiting the activity of lithotrophic iron- and sulfur-oxidizing bacteria (Johnson and Hallberg, 2005; Sangita et al., 2010).

In addition to "source control," "migration control" measures can be applied. The aim is to prevent or minimize the dispersion of AMD into environmental systems and minimize its impact as a source of contamination. The dispersion of AMD can be prevented by managing water flow, diverting surface water flowing toward the site of AMD production, minimizing water seepage into the mine area, and properly managing potentially acid-producing waste (Akcil and Koldas, 2006; Sangita et al., 2010). The reduction of potential AMD contamination can be performed by a set of chemical, physical, and biological mechanisms for the treatment of AMD (Fig. 17.4).

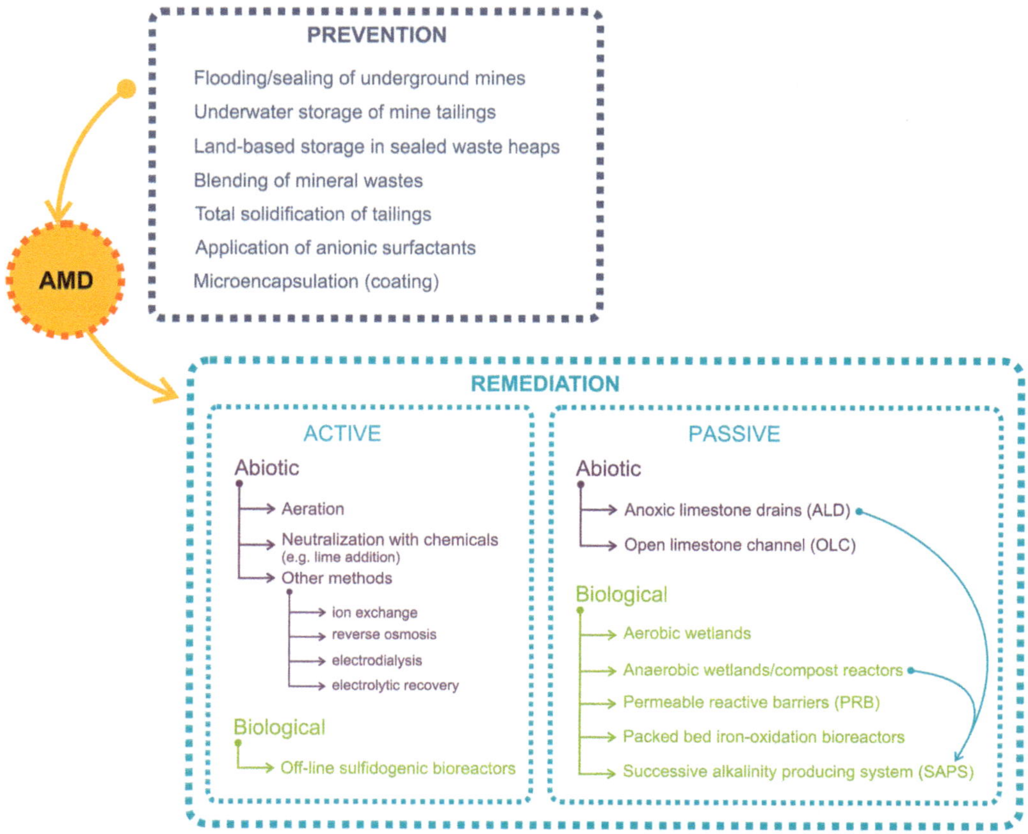

FIGURE 17.4

Examples of strategies to prevent or minimize the AMD generation and different treatment techniques for AMD waters (active or passive methods and biological or abiotic methods).

According to Johnson, D.B., Hallberg, K.B., 2005. Acid mine drainage remediation options: a review. Science of the Total Environment 338, 3–14; Taylor, J., Pape, S., Murphy, N., 2005. A summary of passive and active treatment technologies for acid and metalliferous drainage (AMD). Fifth Australian Workshop on Acid Drainage, 29–31 August 2005, Fremantle, Western Australia. Australian Centre for Minerals Extension and Research (ACMER), Earth Systems. 49 p and Sangita, Udayabhanu, G., Prasad, B., 2010. Studies on environmental impact of acid mine drainage generation and its treatment: an appraisal. Indian Journal of Environmental Protection 30(11), 953–967.

AMD treatment technologies are commonly categorized as "passive" or "active," both of which combine physical, chemical, and biological approaches (Taylor et al., 2005). Active treatment methods are generally more cost-effective because of regular maintenance fees (eg, chemical reagents, mechanical systems, labor input) for continuous operation (Costello, 2003; Taylor et al., 2005). Active systems include neutralization with chemicals (eg, adding lime), aeration, off-line sulfidogenic bioreactors, etc.

Compared with active treatment methods, passive treatment methods require only occasional maintenance (Taylor et al., 2005). Passive systems include open limestone channels, anoxic limestone drains, aerobic wetlands, compost reactors, permeable reactive barriers, packed-bed iron-oxidation bioreactors, etc. Although economically attractive, the methods have significant limitations. Thus, they are best suited to treating mine effluents with low acidity (<800 mg $CaCO_3$ L^{-1}) and low flow rates (<50 L s^{-1}) and therefore low acidity loads (Taylor et al., 2005).

However, for both treatment methods, the main purpose is to lower acidity and toxic metal concentrations, raise pH, and often lower sulfate concentrations and salinity (Taylor et al., 2005).

According to Taylor et al. (2005), neutralization (or pH control) is the most commonly used AMD treatment for both active and passive treatment systems. By creating alkaline conditions (eg, pH $\geq$ 9.5) by increasing the pH, the solubility of most metals can be significantly decreased by precipitation.

However, beyond the classification as an "active" or "passive" method, according to Johnson and Hallberg (2005), a more useful categorization is between treatment technologies that rely on biological activities ("biological strategies") and those that do not ("abiotic strategies"). Nevertheless, within these groups some processes may be described as either "active" or "passive."

Hence we will give more emphasis to passive treatment methods, specifically the biological approaches, because they have the main advantages of low construction costs, easy operation and maintenance, and good efficiency of treatment.

17.7.1 PASSIVE TREATMENT METHODS

The most commonly used methods for passive treatment are aerobic wetlands, anaerobic "wetlands"/ compost reactors, anoxic limestone drains (ALD), successive alkalinity-producing systems (SAPS), and, when contamination may affect groundwater, PRB.

In practice, these methods are used either alone or combined depending on the AMD characteristics and treatment requirements (Fig. 17.5). In general, alkaline materials are used in passive systems to neutralize acidity, organic substrates are used to create reducing environments, and bacteria are employed to catalyze the reactions and accelerate precipitation processes.

In a way, passive treatment methods are based on similar physical, chemical, and biological processes that occur in natural wetlands. In these systems the Eh and pH conditions of the water to be treated are modified, leading to pH neutralization in favor of the formation of insoluble chemical species that precipitate and retain the contaminant load (heavy metals and metalloids).

Similarly to natural wetlands, constructed wetlands are ecosystems consisting of organic soils, microorganisms, algae, and vascular aquatic plants in areas where the water level is at or near the surface, at least in part of the year. All of the components work together to treat effluents, through the combined actions of filtration, ion exchange, adsorption, and precipitation (Frers, 2009; Vymazal, 2009; Fonder and Headley, 2013; Favas et al., 2014). It is the oldest method of wastewater treatment and is not regarded as proper phytoremediation, because it is based on the contributions of the entire system (Horne, 2000; Prasad, 2004). Depending on how it processes the water circulation, the wetland is classified as horizontal or surface flow and vertical or subsurface flow. The latter case ensures greater reactivity of the influent with the substrate. Good cleaning efficiency, low cost of construction and easy operation and maintenance are the main advantages. It is widely applied in the treatment of domestic, agricultural, and industrial wastewater (Kadlec and Wallace, 2008; Vymazal and Kröpfelová, 2008), but it has also proved to be suitable for treating AMD (Sobolewski, 1999; López Pamo et al., 2002; Nyquist and Greger, 2009; Adams et al., 2013).

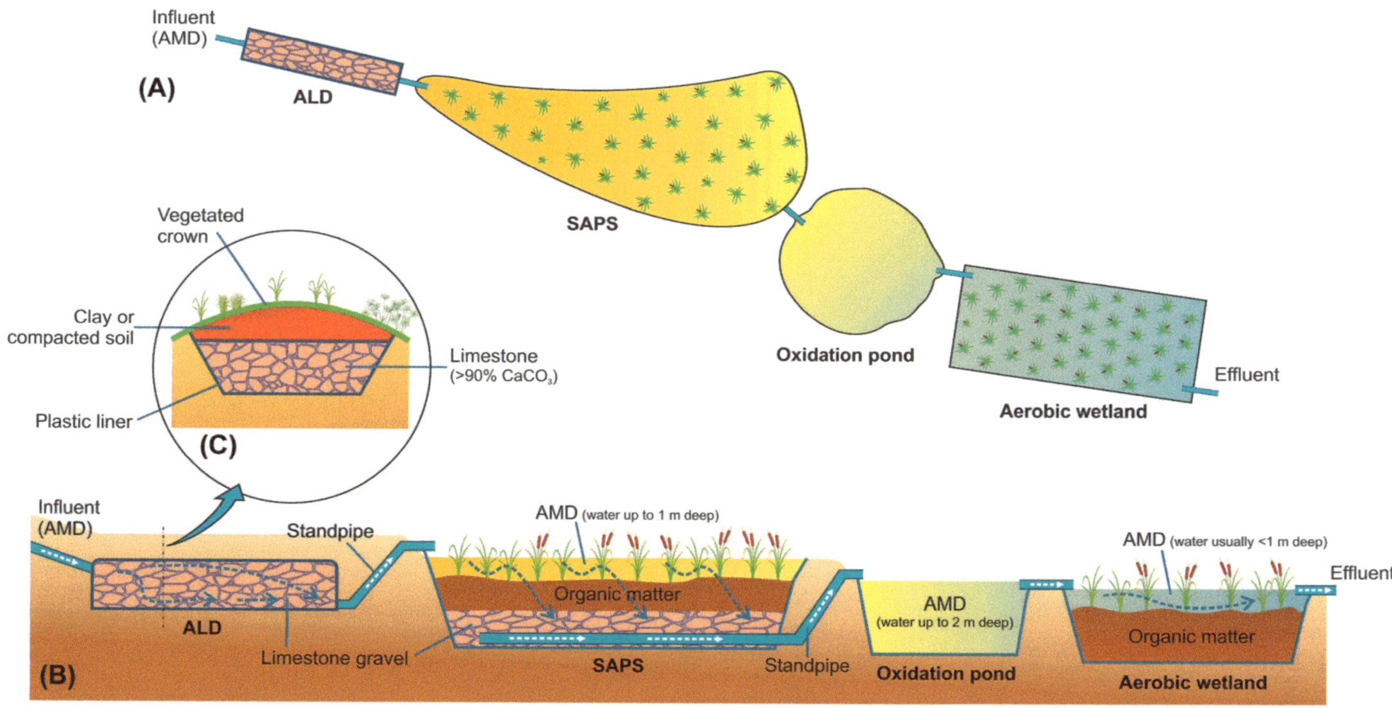

FIGURE 17.5

Schematic illustration of a hypothetical passive treatment facility with an ALD, SAPS, oxidation pond, and aerobic wetland. (A) Map view; (B) simplified general cross-sectional view; (C) simplified cross-section of the ALD.

17.7.1.1 Aerobic Wetlands

According to Kadlec and Knight (1996), this method aims to reproduce the phenomena and processes that occur in natural wetlands (swamps, salt marshes, peat bogs, etc.), creating an enabling environment for the development of plants (eg, *Typha*, *Equisetum*, *Juncus*), communities of organisms (algae, protozoa, and bacteria), and mosses (eg, *Sphagnum*), which participate in water purification (Fig. 17.5). These wetlands occupy a large surface and are characterized by a shallow water layer flooding the substrate on which vegetation grows. The slow water flow enables the retention time needed for water purification processes.

The use of aquatic plants which release oxygen by the roots and rhizomes favors oxygenation of the substrate. This promotes the formation of a suitable habitat for the development of bacteria colonies that act as catalysts in the oxidation reaction of contaminants, transforming, in the case of iron, Fe^{2+} into Fe^{3+}, which precipitates as hydroxide (López Pamo et al., 2002).

Among the several processes that occur in aerobic wetlands, we highlight the following: filtration of suspended matter, adsorption of metals and ion exchange in the substrate materials, bioaccumulation of metals in the roots and aerial parts of the plants, and precipitation of metal oxides and hydroxides. However, this type of system is more suitable for the treatment of mine waters that have the ability to self-generate alkalinity able to neutralize the acidity generated in sulfide oxidation.

An aerobic wetland system may consist of one or several cells connected to each other through which the water to be treated flows slowly and by gravity, establishing surface horizontal flow. In some cases, designed systems including cascades, winding beds, and shallow pools with large surfaces, to facilitate water oxygenation and improve the efficiency of treatment.

17.7.1.2 Anaerobic Wetlands/Compost Bioreactors

In this type of system, anoxic conditions are required for proper operation, in which the height of the water layer must be greater than 30 cm. This water layer cover a permeable substrate with a thickness of 30–60 cm, formed mainly of organic material (70–90% compost, peat, hay, sawdust, etc.), which in turn is mixed or arranged on a limestone layer (López Pamo et al., 2002). The purpose of the organic substrate is to eliminate dissolved oxygen, reduce Fe^{3+} to Fe^{2+}, and generate alkalinity by chemical processes or with the intervention of anaerobic microorganisms (sulfur-reducing bacteria). With this substrate it may develop the typical vegetation of these environments, which helps to stabilize the substrate and introduces additional organic matter. Anaerobic wetlands with generated alkalinity can receive mine drainage with a pH below 4.

However, in some installations these systems are enclosed entirely below ground level and do not support macrophytes (Johnson and Hallberg, 2005). In fact, the growth of plant roots in depth can cause the ingress of oxygen into the anaerobic zones, influencing the reductive processes. Often, the use of macrophytes in these systems is merely for aesthetic reasons. Thus, Johnson and Hallberg (2005) argued that the term "bioreactor compost" is a preferable generic term to describe these systems.

17.7.1.3 Anoxic Limestone Drains

This system consists of a drain filled with limestone gravel or other carbonated materials and sealed (impervious to both air and water) by a plastic bottom liner and a clay cover (Fig. 17.5). The drain is installed 1 or 2 m deep (typically about 1.5 m), to maintain anoxic conditions (López Pamo et al., 2002; Kleinmann et al., 1998; Johnson and Hallberg, 2005). Because of the anoxic conditions of the system, the precipitation of oxides and hydroxides is prevented. According to Johnson and Hallberg (2005), the dimensions of the drain typically vary from narrow (0.6–1.0 m) to wide (10–20 m) in their diameter, and about 30 m in length. Acid mine water is constrained to circulate through the drain causing the dissolution of limestone, which generates alkalinity ($HCO_3^- + OH^-$) and raises the water pH.

Thus, the only purpose of the ALD system is to convert acidic water into water with an excess of alkalinity, which flows into a sedimentation pond and/or an aerobic wetland or other aerobic system to precipitate and retain metallic oxy-hydroxides.

17.7.1.4 Successive Alkalinity-Producing System

A successive alkalinity-producing system (Kepler and McCleary, 1994) can be considered a treatment system which combines an anaerobic wetland and an ALD (Fig. 17.5). This system is also referred to as a reducing and alkalinity producing system (Younger et al., 2003). Thus, the installation of an SAPS has the objectives of an anaerobic wetland (sulfate reduction and metal retention) and an ALD (increase alkalinity). An SAPS consists of a tank in which two different substrates are deposited (alkaline material and organic compounds), which are submerged in the water to be treated, to a depth between 1 m and 3 m. The tank is drained from the bottom through a series of tubes. The lower substrate is usually limestone (0.5–1 m thick) and serves to neutralize the pH of the mine water. The upper substrate consists of an organic material (0.1–0.5 m thick) and promotes the removal of dissolved oxygen, sulfate reduction, and the transformation of Fe^{3+} into Fe^{2+}, avoiding the precipitation of Fe^{3+} hydroxide on the limestone bed (Skousen et al., 1998). Thus, this system can solve the problems of the large surface area suitable for anaerobic wetlands and the precipitation of Fe and Al hydroxides in ALD systems. Generally, the treatment of acidic water in an SAPS is complemented with an oxidation/decantation tank or/and an aerobic wetland.

17.7.1.5 Permeable Reactive Barriers

When AMD affect the groundwater or manifest as an underground flow, the solution may involve installing a PRB disposed perpendicular to the direction of the flow. The construction of this system consists of opening a trench transverse to the flow, which is then filled with different types of reactive materials (limestone, organic matter, swarf, etc.). In a simplified way, a PRB can be considered an SAPS with a vertical and underground disposition. Thus, the purification processes occurring within the barrier are the bacterial reduction of sulfates, the retention of metals which precipitate as sulfides, and the increase in pH by the dissolution of calcium carbonate.

17.8 HYDROCHEMISTRY OF ACID MINE DRAINAGES AND SUPERFICIAL WATERS FROM ABANDONED MINES OF NORTH PORTUGAL: ENVIRONMENTAL IMPLICATIONS

Several studies have been performed in Portugal, mostly to survey the hydrochemical characterization of AMD and their impact on surface waters (eg, Antunes et al., 2002; Gomes and Favas, 2006; Silva et al., 2009; Valente and Gomes, 2009; Abreu et al., 2011; Luís et al., 2011; Maia et al., 2012; Favas et al., 2012; Candeias et al., 2014). In this chapter, we present important findings from several studies we have performed in several abandoned tin/tungsten (Sn/W) mining areas of Northern Portugal.

17.8.1 STUDY AREAS

To obtain a panoramic view of the most representative types of Sn and W mines in Portugal, the following areas were selected according to mineralization and environmental factors (Figs. 17.6 and 17.7): Ervedosa mine with cassiterite and diverse sulfides (arsenopyrite, pyrite, sphalerite, pyrrhotite, chalcopyrite, and stannite); Rio de Frades mine with wolframite, cassiterite, scheelite, and sulfides (arsenopyrite, pyrite, sphalerite, marcasite, pyrrhotite, and stannite); Regoufe mine with wolframite, cassiterite, and sulfides (arsenopyrite, pyrite, and sphalerite); Adoria mine with volframite, cassiterite, scheelite, and sulfides (pyrite, sphalerite, chalcopyrite, galena, pyrrhotite, and stannite); and Tarouca mine with scheelite, cassiterite, and sulfides (pyrite, arsenopyrite, chalcopyrite, sphalerite, and pyrrhotite).

17.8.2 METHODS

To characterize the mine effluents (AMD), samples were collected from the galleries and tailings drainages. To evaluate the impact of the AMD in the fluvial system, reference points were selected upstream of the confluences with mine effluents. Downstream of those confluences, sampling was spaced longitudinally to examine the self-purification characteristics of the receiving medium.

The defined sample stations were sampled for a year between January and November in six bimonthly sampling campaigns. Samples were collected in clean, sterilized plastic bottles. The water samples were filtered through 0.45-μm Millipore cellulose membrane filters and cooled to 4°C immediately after collection. For the determination of metals, samples were acidified to $pH < 2$ with 65% HNO_3 (vol vol^{-1}).

The pH, Eh, temperature, and electrical conductivity (EC) of the water was determined using a pH meter (HI 982,804-02; Hanna Instruments) in the field. Bicarbonate (HCO_3) was also measured in situ using the titration method.

Collected samples were processed and analyzed later at the Chemistry Laboratory of the Earth Science Department, University of Coimbra (Portugal) using current analytical methods (atomic absorption spectrometry [AAS] for Ca, Mg, Na, and K; coupled graphite furnace AAS for Fe, Mn, Cu, Zn, Cd, Co, Ni, Pb, and As; the spectrophotometric method for Cl; SO_4 was analyzed by gravimetry). Water data quality control was performed by inserting reagent blanks and duplicate samples into each batch. Analytical precision, defined as the percent relative variation at the 95% confidence level, ranged from 2% to 6%, depending on the concentration levels. Certified reference material from the National Water Research Institute of Canada (reference TMDA-62) was used to maintain precision and quality control. The detection limits for major and trace elements in water samples were as follows: 0.025 mg L^{-1} SO_4, Cl; 0.5 mg L^{-1} Ca; 0.1 mg L^{-1} Mg; 1 mg L^{-1} Na, K; 10 μg L^{-1} Fe; 2.5 μg L^{-1} Mn; 6 μg L^{-1} Zn; 0.1 μg L^{-1} Cu, Pb; 0.5 μg L^{-1} As; 0.05 μg L^{-1} Cd; 0.02 μg L^{-1} Co; and 5 μg L^{-1} Ni.

Principal component analysis (PCA) was applied to interpret data using AnDad software (version 7.10). The aim of PCA is to reduce data dimensionality while preserving the variance–covariance structure of the data. The analysis starts with p random variables ($X1$, $X2$, … , Xp), where no assumption of multivariate normality is required. If the principal components are derived from multivariate normal random variables, they have a noteworthy geometric property. The axes of the constant density ellipsoids correspond to the principal components.

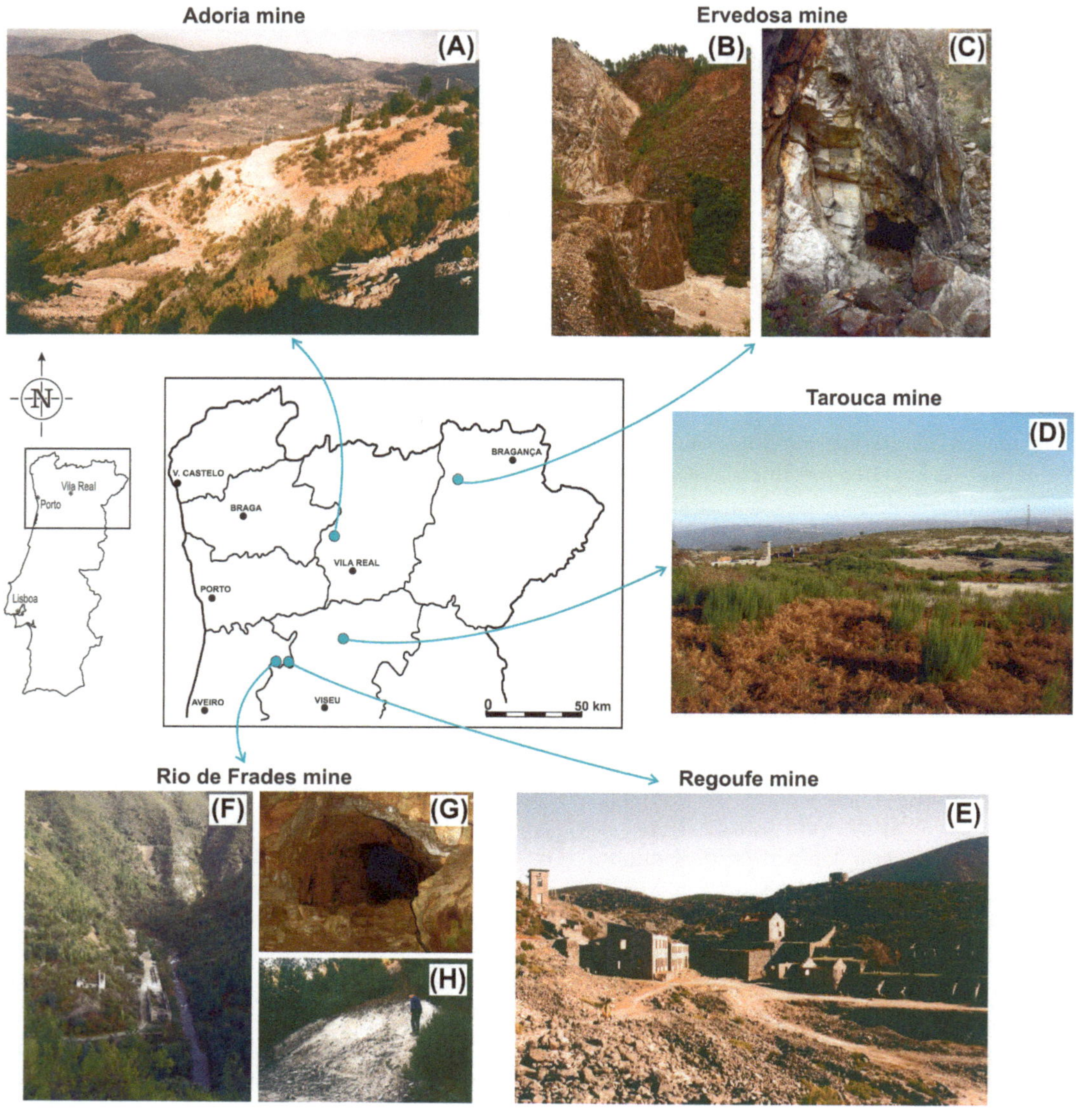

FIGURE 17.6

Location of the studied mining areas on the map of Portugal. Adoria mine: (A) general view of the waste pile; Ervedosa mine: (B) main open pit and (C) gallery; Tarouca: (D) panoramic view of old industrial installations and waste pile and tailings; Regoufe: (E) view of old industrial installations; Rio de Frades: (F) panoramic view of old industrial installations and waste piles and (G) gallery, and (H) waste dumps containing sulfides.

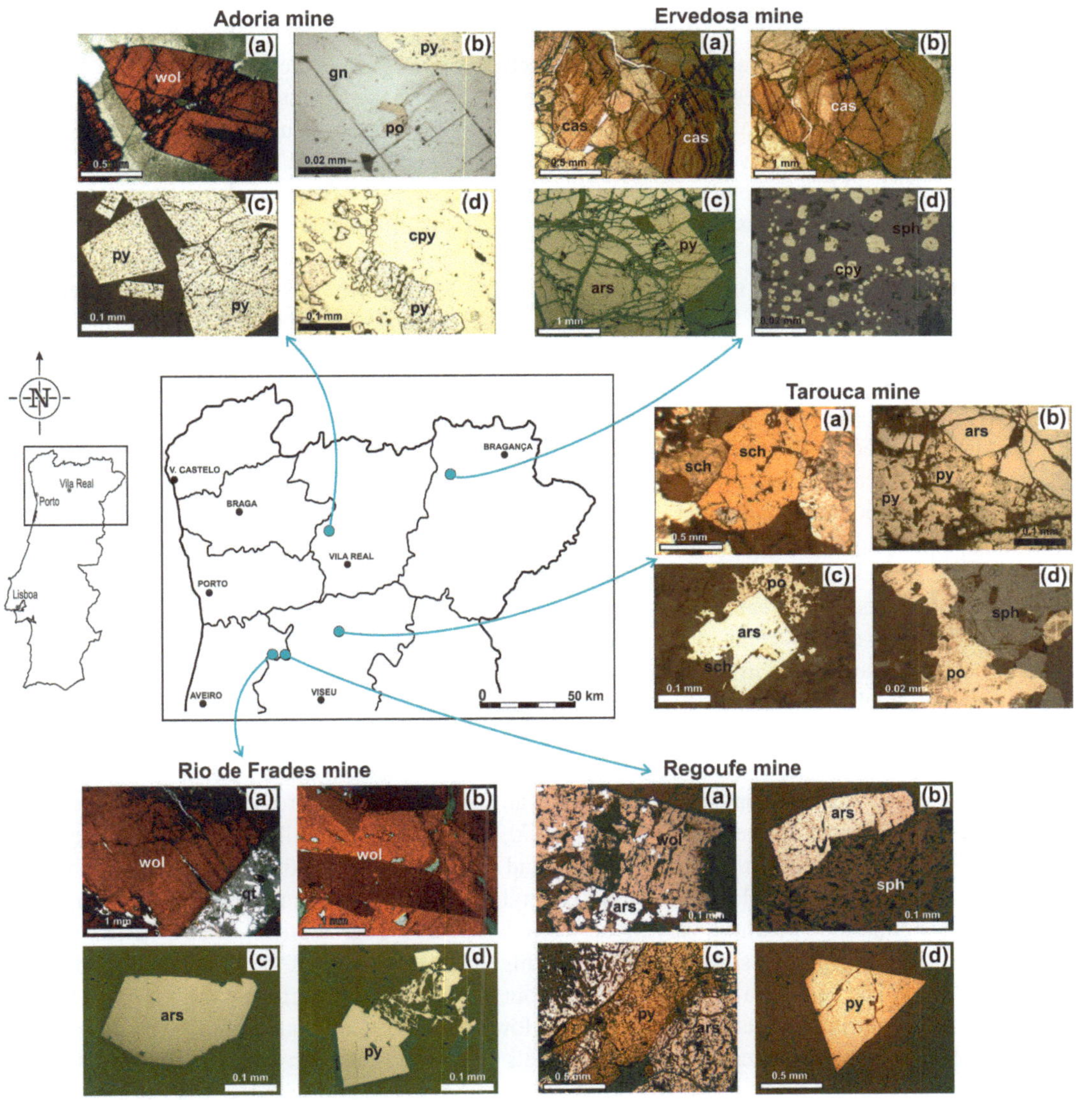

FIGURE 17.7

Location of the studied mining areas on the map of Portugal and mineralogy of the mine veins. *cas*, Cassiterite; *wol*, wolframite; *sch*, scheelite; *py*, pyrite; *cpy*, chalcopyrite; *ars*, arsenopyrite; *sph*, sphalerite; *gn*, galena; *po*, pyrrhotite.

17.8.3 RESULTS AND DISCUSSION

In general, the AMD of studied mines exhibited chemical characteristics that reflect the leaching processes of mineralized masses that contain sulfides. Therefore, these waters are acidic, with low pH and significant concentrations of SO_4 and metal(loid)s (As, Fe, Mn, Zn, Cu, Pb, Cd, and Ni). While in the superficial natural stream waters outside the mines' influence, the pH is close to neutral, characterized by low EC and lower metal concentrations. Thus, the stream waters inside the mines' influence shows an intermediate composition between AMD and natural stream waters outside the mines' influence.

The waters under mining influence reveal a sulfated composition, whereas the representative samples of hydrogeochemical background have a predominantly bicarbonated or chlorinated and sulfated typology. It is evident in most samples of AMD, and stream waters under the mining influence the prevalence of the SO_4^{2-} ion. The highest levels of SO_4 were found in the Ervedosa mine, followed by the Regoufe, Rio de Frades, Adoria, and Tarouca mines.

The stream waters under the mining influence, including the higher—flow rate streams, are characterized by significant increases in SO_4^{2-} (conservative ion), whereas only the lower—flow rate water streams show significant increases in EC values and metal levels and decreases in pH. Thus, the waters of rivers with the highest flow rate, as in the case of the Tuela River (Ervedosa mine) and the Frades River (Rio de Frades mine), are less sensitive to decreases in pH, increases in EC, and increases in metal concentrations, but reveal SO_4^{2-} contamination, owing to the conservative behavior of this ion.

This behavior is also evident in the results of the Tarouca mine, in which the presence of calc-silicate rocks and carbonates in the skarns eventually lead to the "acid-neutralization" phenomenon. This interpretation also explains the high levels of SO_4 in AMD with low metal loading and pH values close to neutral. The significantly higher concentrations of HCO_3 and Ca in AMD reinforce this interpretation.

Using the classification proposed by Ficklin et al. (1992) for water from mines and mineralized areas (based on variations in the levels of Zn, Cu, Cd, Ni, Co, and Pb in the solution as a function of pH), it is possible to determine the AMD typology and evaluate their effects on receptors river systems (Fig. 17.8). According to this classification, in general the stream waters outside the mines' influence has a low metal content in solution, varying between "acid" and "near neutral." The AMD samples are divided into different typologies, varying between "high metals" and "acid" to "low metal" and "acid" or "near neutral." The Ervedosa mine waters stand out as the most contaminated, followed by the Rio de Frades mine, Adoria mine, and Regoufe mine. However, Ni higher contents were found in Rio de Frades and Regoufe mines, whereas the highest levels of Pb were found in Adoria mine samples. The Tarouca mine waters had a significantly lower pollution load compared with the other mines studied. Generally the stream waters inside the mines' influence exhibit intermediate characteristics and are classified mostly as "acid" with "high metal."

The results clearly show the influence of sulfated facies of the studied water, particularly the AMD and the stream waters inside the mine (Fig. 17.9). These are characterized by significant sulfate addition even when significant increases in the EC and metal concentrations are not verified, or declining pH values.

The influence of seasonal fluctuations on the composition of AMD and surrounding surface waters is also evident in most of the results achieved, to the extent that there are significant variations in most of the analyzed parameters. The lower pH and higher EC and metal concentrations were recorded in a

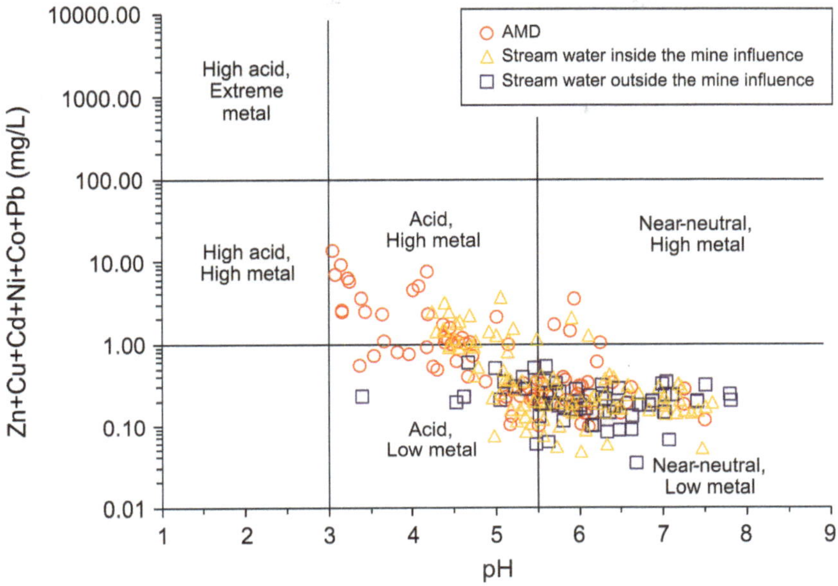

FIGURE 17.8

Water samples from the five studied mine areas plotted on the diagram of Ficklin et al. (1992) based on pH and metal concentrations (Zn + Cu + Cd + Ni + Co + Pb).

dry period. This can be attributed to the decrease in the dilution capacity as a consequence of reduced flow rates and enhancement of the evaporation phenomena. During the wet periods, because of the increased flow rate, the dilution effect causes a considerable reduction in the concentration of metals as well as the EC and an increasing in pH. Thus, in general the data collected verify that the water quality (Portuguese law, 2007, Decree No. 306/2007) varies with the seasonal regime, with contamination levels more significant in dry periods than in humid ones.

The hydrochemical data matrix used in the PCA consisted of 243 samples and 17 variables (pH, HCO_3, SO_4, Ca, Mg, Na, K, Cl, Fe, Mn, Zn, Cu, Pb, As, Cd, Co, and Ni). The first four principal components explain more than 84% of the total variance in the data (Favas, 2008). The first principal component (axis 1), which explains 51.5% of the total data variance, defines a HCO_3–Ca–K–Cl–Na–Mg–Co–Zn–Cu–Cd association (Fig. 17.10). The second component (axis 2), explaining 16.2% of the total data variance, reveals an Mn–Fe–SO_4 association. The third component (axis 3), explaining 10.3% of the total data variance, defines an As–Ni association. Lead alone is explained only by axis 4. These four associations demonstrate a behavior independent of each other. However, the second group (Mn–Fe–SO_4) reveals a close proximity with the variables Co–Zn–Cu–Cd of the first group.

Therefore, the PCA results verify in the first factorial plan a representative association of variables that indicate the geological control of the water matrix. Thus, the geochemical signature of the mineralization (eg, sulfates) and other minerals from paragenesis, in particular silicates, clearly appears in the composition of the waters, though their own geochemical behavior determining the different groups.

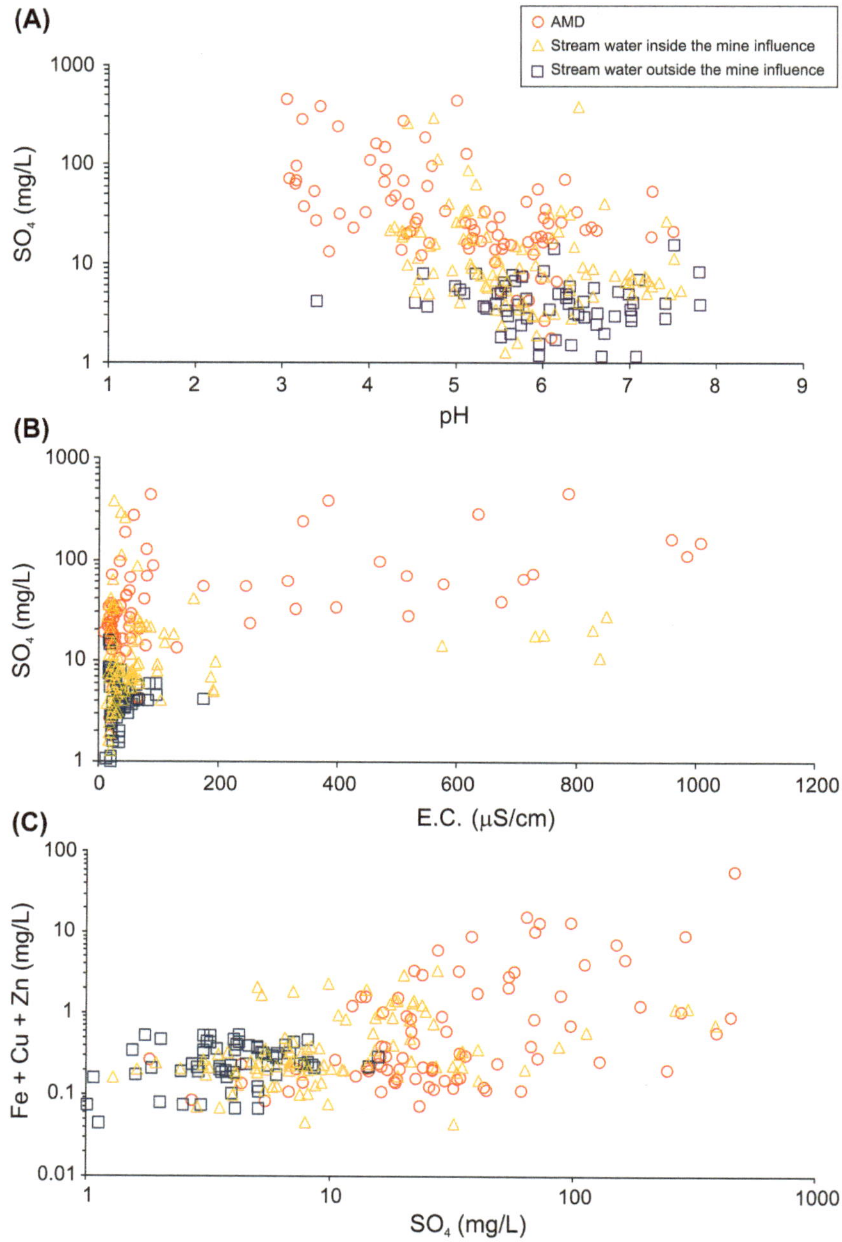

FIGURE 17.9

(A) SO_4 versus pH diagram; (B) SO_4 versus EC diagram; (C) (Fe + Cu + Zn) versus SO_4 diagram for AMD and stream waters inside and outside the mines' influence from the five studied mine areas.

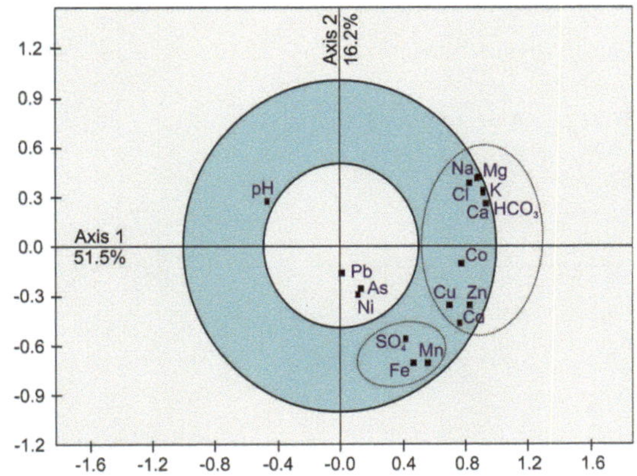

FIGURE 17.10

Projection of variables in the two first factorial plans (axis 1 versus axis 2 and axis 1 versus axis 3) of the PCA.

The variables As–Ni and Pb are explained separately in the second factorial plan and in the third factorial plan, respectively. This may be the result of anomalous concentrations found mainly in the Regoufe and Rio de Frades mines (As–Ni) and in the Adoria mine (Pb).

The projection of the samples in the two first factorial plans also highlights the separation into four differentiated "populations" in the first factorial plan and into three "populations" in the second factorial plan (Fig. 17.11). Considering the variables represented in the first factorial plan, it appears that: (1) "Population A" gathers the lesser mineralized waters, and is essentially composed of samples belonging to the regional hydrochemical background (including most of the samples of Tarouca, Adoria, Regoufe, and Rio de Frades mines); (2) "Population B," transiting progressively from the previous one, consists essentially of AMD samples of the driest periods from Regoufe and Rio de Frades mines and the AMD samples of wet periods from Ervedosa mine. Thus, these samples are more contaminated than "Population A," particularly by SO_4, Mn, Fe, Cd, Co, Cu, Zn, HCO_3, Ca, Na, Mg, Cl, and K, because they are under the influence of these variables' association; (3) "Population C" is composed of samples under the influence of the HCO_3–Ca–Na–Cl–K–Mg association (samples of tailings runoff from Ervedosa mine), which reflects the importance of silicate hydrolysis for the composition of the AMD generated in tailings; and (4) "Population D" results from the association of the samples most contaminated with SO_4, Mn, Fe, Cd, Co, Cu, and Zn, ie, samples of AMD from galleries of the Ervedosa mine, reflecting the oxidation of sulfides.

In the second factorial plan, the following features are observed: (1) "Population E," which is associated with the first factorial axis, results from the approximation of the "populations" C, D, and partly B of the first factorial plan. Namely, this "population" is formed by the more contaminated samples with SO_4, Mn, Fe, Cd, Co, Cu, Zn, HCO_3, Ca, Na, Mg, Cl, and K, which are the majority of AMD from the Ervedosa mine; (2) "Population F" (associated with the third factorial axis) is composed of some AMD samples from the Rio de Frades mine and the Regoufe mine. These samples, located in As–Ni association's influence zone, are the most contaminated by these elements considering all five mines studied.

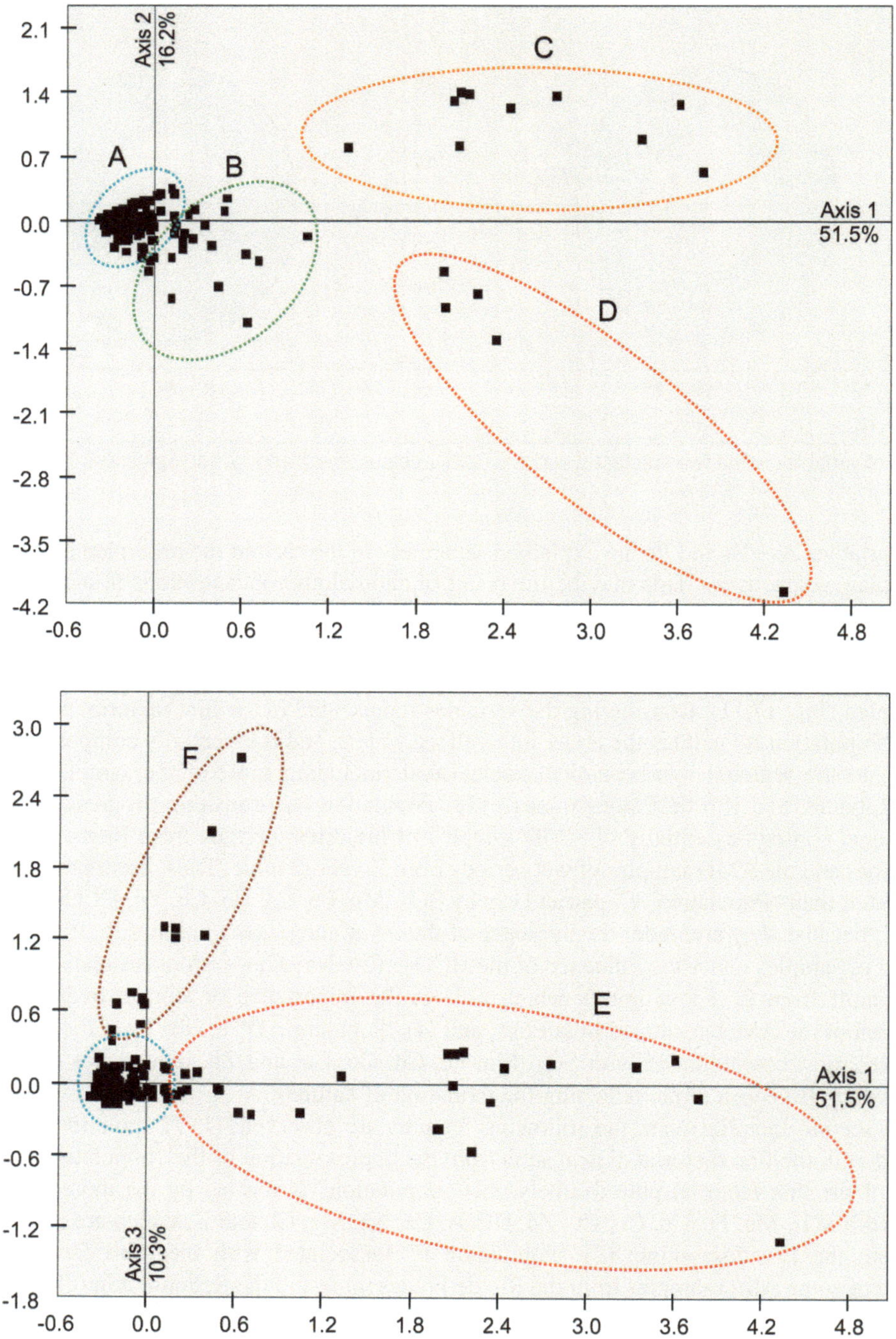

FIGURE 17.11

Projection of samples in the two first factorial plans (axis 1 versus axis 2 and axis 1 versus axis 3) of the PCA.

17.8.4 CONCLUSIONS

The current observations reveal that AMD is characterized by a low pH and significant concentrations of SO_4, and metal(loid)s (As, Fe, Mn, Zn, Cu, Pb, Cd, and Ni). In contrast, in the superficial natural stream waters outside the mines' influence, the pH is close to neutral, with low conductivity and lower metal concentrations. However, the stream waters inside the mines have an intermediate composition between AMD and natural stream waters outside the mines' influence. Sulfate is a conservative ion and exhibits a positive correlation with metals (Fe + Cu + Zn) and EC, and a negative correlation with pH. All AMD samples inside the mining areas, including the samples collected in major flow-rate rivers, have SO_4 contamination.

Contamination of surrounding river systems of the five mining areas studied is the most significant manifestation of the potential environmental impact associated with AMD. The nature and extent of the contamination and depend on the physicochemical characteristics of those waters, particularly on pH and metal content. The characterization of AMD generated both inside the galleries and in the tailings, as well as the characterization of receptor stream waters, allowed the detection of a seasonal influence on the metal concentration of AMD and consequently on its potential to disperse contaminants.

Meanwhile, the environmental impact on the quality of the superficial waters is marked only in the closest surroundings of the pollution source (1–2 km). Several coherent and integrated processes contribute to this situation, such as: (1) dilution by other groundwater and surface waters that flow toward stream waters; (2) precipitation or co-precipitation of metallic cations by hydroxides and sulfates; and (3) adsorption of the metallic cations by organic and inorganic sediments and aquatic plants. Hydrochemical processes and the environmental impact on the study sites point out to possible issues for potential resource recovery from AMD (eg, iron or other metals) and appropriate strategies to mitigate AMD and prevent pollution.

REFERENCES

Abreu, M.M., Batista, M.J., Magalhães, M.C.F., Matos, J.X., 2011. Acid mine drainage in the Portuguese Iberian Pyrite Belt. In: Robinson, B.C. (Ed.), Mine Drainage and Related Problems. Nova Science Publishers, New York, pp. 71–118.

Acero, P., Hudson-Edwards, K.A., Gale, J.D., 2015. Influence of pH and temperature on alunite dissolution: rates, products and insights on mechanisms from atomistic simulation. Chemical Geology 419, 1–9.

Adams, A., Raman, A., Hodgkins, D., 2013. How do the plants used in phytoremediation in constructed wetlands, a sustainable remediation strategy, perform in heavy-metal contaminated mine sites? Water and Environment Journal 27 (3), 373–386.

Aguilar, R.N., Castillo, M.A., Herrera, S.A., Rodríguez, L.D.A., Murguia, G.J., 2012. Productive diversification from sugarcane lignocellulosic byproducts. International Sugar Journal 114 (1359), 179–182.

Ahonen, L., Tuovinen, O., 1992. Mineral transformations during bacterial leaching of a complex sulfide ore. In: Kharaka, M. (Ed.), Water-Rock Interaction – Volume 1: Low Temperature Environments. Balkema, Rotterdam, pp. 253–256.

Akcil, A., Koldas, S., 2006. Acid mine drainage (AMD): causes, treatment and case studies. Journal of Cleaner Production 14, 1139–1145.

Antunes, I.M.H.R., Neiva, A.M.R., Silva, M.M.V.G., 2002. The mineralized veins and the impact of old mine workings on the environment at Segura, central Portugal. Chemical Geology 190, 417–431.

Arnekleiv, J.V., Storset, L., 1995. Downstream effects of mine drainage on benthos and fish in a Norwegian river: a comparison of the situation before and after river rehabilitation. Journal of Geochemical Exploration 52, 35–43.

Asif, Z., Chen, Z., 2015. Environmental management in North American mining sector. Environmental Science and Pollution Research. http://dx.doi.org/10.1007/s11356-015-5651-8.

Azapagic, A., 2004. Developing a framework for sustainable development indicators for the mining and minerals industry. Journal of Cleaner Production 12 (6), 639–662.

El Adnani, M., Plante, B., Benzaazoua, M., Hakkou, R., Bouzahzah, H., 2015. Tailings weathering and arsenic mobility at the abandoned Zgounder Silver Mine, Morocco. Mine Water and the Environment. http://dx.doi.org/10.1007/s10230-015-0370-4.

Bahurudeen, A., Vaisakh, K.S., Santhanam, M., 2015. Availability of sugarcane bagasse ash and potential for use as a supplementary cementitious material in concrete. Indian Concrete Journal 89 (6), 41–50.

Bailey, B.L., Blowes, D.W., Smith, L., Sego, D.C., 2015. The Diavik Waste Rock Project: geochemical and microbiological characterization of drainage from low-sulfide waste rock: active zone field experiments. Applied Geochemistry 62, 18–34.

Bakar, A.F.A., Yusoff, I., Ng, T.F., Ashraf, M.A., 2015. Cumulative impacts of dissolved ionic metals on the chemical characteristics of river water affected by alkaline mine drainage from the Kuala Lipis gold mine, Pahang, Malaysia. Chemistry and Ecology 31 (1), 22–33.

Banfield, J.F., Verberkmoes, N.C., Hettich, R.L., Thelen, M.P., 2005. Proteogenomic approaches for the molecular characterization of natural microbial communities. OMICS: A Journal of Integrative Biology 9 (4), 301–333.

Banks, D., Younger, P., Arnesen, R., Iversen, E., Banks, S., 1997. Mine-water chemistry: the good, the bad and the ugly. Environmental Geology 32 (3), 157–174.

Banwart, S., Evans, K., Croxford, S., 2002. Predicting mineral weathering rates at field scale for mine water risk assessment. In: Younger, P.L., Robins, N.S. (Eds.), Mine Water Hydrogeology and Geochemistry. The Geological Society, London, pp. 137–157. Special Publication, 198.

Barral Silva, M., 1989. Precipitados de hierro en medios sulfato-ácidos que resultan de la alteración de anfibolitas ricas en sulfuros. Cuadernos do Laboratorio Xeoloxico de Laxe 14, 29–42.

Battaglia, M., Hose, G.C., Turak, E., Warden, B., 2005. Depauperate macroinvertebrates in a mine affected stream: clean water may be the key to recovery. Environmental Pollution 138 (1), 132–141.

Beck, J.M., Engelking, J.F., Elder, R.L., 1990. Resource recovery. An economic approach to remediation. In: Mining and Mineral Processing Wastes. Littleton, CO. Proceedings of the Western Regional Symposium on Mining and Mineral Processing Wastes, Berkeley, CA, May 30–June 1. Society for Mining, Metallurgy, and Exploration, Inc, pp. 243–248.

Bejan, D., Bunce, N.J., 2015. Acid mine drainage: electrochemical approaches to prevention and remediation of acidity and toxic metals. Journal of Applied Electrochemistry 45, 1239–1254.

Bethune, J., Randell, J., Runkel, R.L., Singha, K., 2015. Non-invasive flow path characterization in a mining-impacted wetland. Journal of Contaminant Hydrology 183, 29–39.

Betrie, G.D., Sadiq, R., Nichol, C., Morin, K.A., Tesfamariam, S., 2016. Environmental risk assessment of acid rock drainage under uncertainty: the probability bounds and PHREEQC approach. Journal of Hazardous Materials 301, 187–196.

Bigham, J., Nordstrom, D., 2000. Iron and aluminium hydroxysulfates from acid sulfate waters. In: Alpers, C.N., Jambor, J.L., Nordstrom, D.K. (Eds.), Sulfate Minerals: Crystallography, Geochemistry, and Environmental Significance, Reviews in Mineralogy & Geochemistry, 40, pp. 351–403.

Bigham, J., Schwertmann, U., Traina, S., Winland, R., Wolf, M., 1996. Schwertmannite and the chemical modeling of iron in acid sulfate waters. Geochimica et Cosmochimica Acta 60 (12), 2111–2121.

Bloomfield, C., 1973. Acidification and ochre formation in pyritic soils. In: Dost, H. (Ed.), Acid Sulphate Soils – Publication 18, vol. II. Wageningen: International Institute for Land Reclamation and Improvement, pp. 40–51.

Bouzahzah, H., Benzaazoua, M., Bussière, B., Plante, B., 2015. ASTM normalized humidity cell kinetic test: protocol improvements for optimal sulfide tailings reactivity. Mine Water and the Environment 34 (3), 242–257.

Bowell, R., 2002. The hydrogeochemical dynamics of mine pit lakes. In: Younger, P.L., Robins, N.S. (Eds.), Mine Water Hydrogeology and Geochemistry. The Geological Society, London, pp. 159–185. Special Publication, 198.

Brady, K., Bigham, J., Jaynes, W., Logan, T., 1986. Influence of sulfate on Fe-oxide formation: comparisons with a stream receiving acid mine drainage. Clays and Clay Minerals 34 (3), 266–274.

Breemen, V., 1973. Soil forming processes in acid sulphate soils. In: Dost, H. (Ed.), Acid Sulphate Soils – Publication 18, vol. I. International Institute for Land Reclamation and Improvement, Wageningen, pp. 66–129.

Broadhurst, J.L., Kunene, M.C., von Blottnitz, H., Franzidis, J.-P., 2015. Life cycle assessment of the desulfurisation flotation process to prevent acid rock drainage: a base metal case study. Minerals Engineering 76, 126–134.

Burrows, J.E., Peters, S.C., Cravotta, C.A., 2015. Temporal geochemical variations in above- and below-drainage coal mine discharge. Applied Geochemistry 62, 84–95.

Buzatu, A., Dill, H.G., Buzgar, N., Damian, G., Maftei, A.E., Apopei, A.I., 2016. Efflorescent sulfates from Baia Sprie mining area (Romania): acid mine drainage and climatological approach. Science of the Total Environment 542, 629–641.

Calvo de Anta, R., Perez Otero, A., 1993. Evolución mineralógica en medios afectados por contaminación ácida. Cuadernos do Laboratorio Xeoloxico de Laxe 18, 337–343.

Candeias, C., Ávila, P.F., Silva, E.F., Ferreira, A., Salgueiro, A.R., Teixeira, J.P., 2014. Acid mine drainage from the Panasqueira mine and its influence on Zêzere river (Central Portugal). Journal of African Earth Sciences 99, 705–712.

Cánovas, C.R., Olías, M., Macias, F., Torres, E., San Miguel, E.G., Galván, L., Ayora, C., Nieto, J.M., 2016. Water acidification trends in a reservoir of the Iberian Pyrite Belt (SW Spain). Science of the Total Environment 541, 400–411.

Cánovas, C.R., Olías, M., Nieto, J.M., Sarmiento, A.M., Cerón, J.C., 2007. Hydrogeochemical characteristics of the Tinto and Odiel Rivers (SW Spain). Factors controlling metal contents. Science of the Total Environment 373, 363–382.

Caraballo, M.A., Macías, F., Nieto, J.M., Ayora, C., 2016. Long term fluctuations of groundwater mine pollution in a sulfide mining district with dry Mediterranean climate: Implications for water resources management and remediation. Science of the Total Environment 539, 427–435.

Carranza, F., Romero, R., Mazuelos, A., Iglesias, N., 2016. Recovery of Zn from acid mine water and electric arc furnace dust in an integrated process. Journal of Environmental Management 165, 175–183.

Carrero, S., Pérez-López, R., Fernandez-Martinez, A., Cruz-Hernández, P., Ayora, C., Poulain, A., 2015. The potential role of aluminium hydroxysulphates in the removal of contaminants in acid mine drainage. Chemical Geology 417, 414–423.

Caruccio, F., 1978. Depositional environment of carboniferous sediments: a predictor of coal mine problems. In: Goodman, C. (Ed.), Environmental Management of Mineral Wastes. Sijthoff & Noordhoff, pp. 127–139.

Caruccio, F.T., Geidel, G., Pelletier, A., 1980. The assessment of a stratum's capability to produce acid drainage. Paper in:. In: Grave, D.H. (Ed.), Proceedings Symposium on Surface Mining, Hydrology, Sedimentology and Reclamation. University of Kentucky, Lexington, Kentucky, pp. 437–443.

Chen, J., Yang, H., Zhang, D.D., Xu, D., Luo, J., Wang, J., 2015a. A particular river-whiting phenomenon caused by discharge of hypolimnetic water from a stratified reservoir. PLoS One 10 (9), e0137860. http://dx.doi.org/10.1371/journal.pone.0137860.

Chen, P., Yan, L., Wu, Z., Xu, R., Li, S., Wang, N., Liang, N., Li, H., 2015b. Draft genome sequence of extremely acidophilic bacterium *Acidithiobacillus ferrooxidans* DLC-5 isolated from acid mine drainage in Northeast China. Genomics Data 6, 267–268.

Chen, T., Yan, B., Lei, C., Xiao, X., 2014. Pollution control and metal resource recovery for acid mine drainage. Hydrometallurgy 147-148, 112–119.

Chen, Y.T., Li, J.T., Chen, L.X., Hua, Z.X., Huang, L.N., Liu, J., Xu, B.B., Liao, B., Shu, W.S., 2014. Biogeochemical processes governing natural pyrite oxidation and release of acid metalliferous drainage. Environmental Science & Technology 48, 5537–5545.

Chou, I.-M., Seal II, R.R., Wang, A., 2013. The stability of sulfate and hydrated sulfate minerals near ambient conditions and their significance in environmental and planetary sciences. Journal of Asian Earth Sciences 62, 734–758.

Cidu, R., Caboi, R., Fanfani, L., Frau, F., 1997. Acid drainage from sulfides hosting gold mineralization (Furtei, Sardinia). Environmental Geology 30 (3/4), 231–237.

Clapcott, J.E., Goodwin, E.O., Harding, J.S., 2016. Identifying catchment-scale predictors of coal mining impacts on New Zealand stream communities. Environmental Management. http://dx.doi.org/10.1007/s00267-015-0627-5.

Cook, N.A., Krometis, L.H., Sarver, E.A., Huang, J., 2015. Inorganic constituents of conductivity in five central Appalachian watersheds with mixed source-driven pollutants. Ecological Engineering 82, 175–183.

Costa, M.C., Santos, E.S., Barros, R.J., Pires, C., Martins, M., 2009. Wine wastes as carbon source for biological treatment of acid mine drainage. Chemosphere 75 (6), 831–836.

Costello, C., 2003. Acid Mine Drainage: Innovative Treatment Technologies. U.S. Environmental Protection Agency, Office of Solid Waste and Emergency Response Technology Innovation Office, Washington, DC.

Cravotta, C.A., Trahan, M.K., 1999. Limestone drains to increase pH and remove dissolved metals from acidic mine drainage. Applied Geochemistry 14, 581–606.

Davies, H., Weber, P., Lindsay, P., Craw, D., Pope, J., 2011. Characterisation of acid mine drainage in a high rainfall mountain environment, New Zealand. Science of the Total Environment 409, 2971–2980.

Dev, S., Patra, A.K., Mukherjee, A., Bhattacharya, J., 2015. Suitability of different growth substrates as source of nitrogen for sulfate reducing bacteria. Biodegradation 26 (6), 415–430.

Diaby, N., Dold, B., Rohrbach, E., Holliger, C., Rossi, P., 2015. Temporal evolution of bacterial communities associated with the in situ wetland-based remediation of a marine shore porphyry copper deposit. Science of the Total Environment 533, 110–121.

Dold, B., 2008. Sustainability in metal mining: from exploration, over processing to mine waste management. Reviews in Environmental Science and Biotechnology 7 (4), 275–285.

Dold, B., Gonzalez-Toril, E., Aguilera, A., Lopez-Pamo, E., Cisternas, M.E., Bucchi, F., Amils, R., 2013. Acid rock drainage and rock weathering in Antarctica: Important sources for iron cycling in the southern ocean. Environmental Science & Technology 47, 6129–6136.

Dold, B., Weibel, L., 2013. Biogeometallurgical pre-mining characterization of ore deposits: an approach to increase sustainability in the mining process. Environmental Science and Pollution Research 20 (11), 7777–7786.

Doner, H., Lynn, W., 1989. Carbonate, halide, sulfate, and sulfide minerals. In: Dixon, J.B., Weed, S.B. (Eds.), Minerals in Soil Environments, second ed. Soil Science Society of America, Madison, Wisconsin (Chapter 6).

Elberling, B., Balić-Žunić, T., Edsberg, A., 2003. Spatial variations and controls of acid mine drainage generation. Environmental Geology 43, 806–813.

Embile, R., Walder, I., Madai, F., Móricz, F., Rzepka, P., Walder, P., 2015. Grain size effects on mine water quality and acid/neutral rock drainage production in kinetic testing using Recsk porphyry skarn Cu–Zn deposit rocks. Mine Water and the Environment. http://dx.doi.org/10.1007/s10230-015-0369-x.

Emili, L.A., Pizarchik, J., Mahan, C.G., 2015. Sustainable remediation of legacy mine drainage: a case study of the Flight 93 National Memorial. Environmental Management. http://dx.doi.org/10.1007/s00267-015-0625-7.

Favas, P.J.C., 2008. Biogeochemistry in Tin-Tungsten Mining Areas (In Portuguese) (Ph.D. thesis). University of Trás-os-Montes e Alto Douro, Vila Real, 805 p.

Favas, P.J.C., Pratas, J., Varun, M., D'Souza, R., Paul, M.S., 2014. Phytoremediation of soils contaminated with metals and metalloids at mining areas: potential of native flora. In: Hernández-Soriano, M.C. (Ed.), Environmental Risk Assessment of Soil Contamination. Intech, Rijeka, pp. 485–517.

Favas, P.J.C., Pratas, J., Gomes, M.E.P., 2012. Hydrochemistry of superficial waters in the Adoria mine area (Northern Portugal): environmental implications. Environmental Earth Sciences 65, 363–372.

Feng, D., van Deventer, J.S.J., Aldrich, C., 2004. Removal of pollutants from acid mine wastewater using metallurgical by-product slags. Separation and Purification Technology 40 (1), 61–67.

Ferguson, K., Erickson, P., 1988. Pre-mine prediction of acid mine drainage. In: Salomons, W., Förstner, U. (Eds.), Environmental Management of Solid Waste. Dredged Material and Mine Tailings. Springer-Verlag, Berlin, pp. 24–43.

Ferris, F., Tazaki, K., Fyfe, W., 1989. Iron oxides in acid mine drainage environments and their association with bacteria. Chemical Geology 74 (3–4), 321–330.

Ficklin, W., Plumlee, G., Smith, K., 1994. Geologic and geochemical controls on the composition of water draining from diverse mineral deposits. In: Carter, L.M.H., Toth, M.I., Day, W.C. (Eds.), USGS Research on Mineral Resources. Part a – Program and Abstracts. U.S. Geological Survey Circular 1103-A. U.S. Geological Servey, pp. 34–35.

Ficklin, W., Plumlee, G., Smith, K., McHugh, J., 1992. Geochemical classification of mine drainages and natural drainages in mineralized areas. In: Kharaka, M. (Ed.), Water-rock Interaction, Low Temperature Environments, vol. 1. Balkema, Rotterdam, pp. 381–384.

Fonder, N., Headley, T., 2013. The taxonomy of treatment wetlands: a proposed classification and nomenclature system. Ecological Engineering 51, 203–211.

Foster, A.L., Munk, L., Koski, R.A., Shanks III, W.C., Stillings, L.L., 2008. Relationships between microbial communities and environmental parameters at sites impacted by mining of volcanogenic massive sulfide deposits, Prince William Sound, Alaska. Applied Geochemistry 23 (2), 279–307.

Frau, F., Ardau, C., Fanfani, L., 2009. Environmental geochemistry and mineralogy of lead at the old mine area of Baccu Locci (south-east Sardinia, Italy). Journal of Geochemical Exploration 100 (2–3), 105–115.

Frers, C., 2009. El uso de plantas acuáticas en el tratamiento de aguas residuales. El Planeta Azul, Carmen de Areco, Argentina.

García, V., Häyrynen, P., Landaburu-Aguirre, J., Pirilä, M., Keiski, R.L., Urtiaga, A., 2014. Purification techniques for the recovery of valuable compounds from acid mine drainage and cyanide tailings: application of green engineering principles. Journal of Chemical Technology and Biotechnology 89 (6), 803–813.

García-Lorenzo, M.L., Pérez-Sirvent, C., Martínez-Sánchez, M.J., Molina-Ruiz, J., 2012. Trace elements contamination in an abandoned mining site in a semiarid zone. Journal of Geochemical Exploration 113, 23–35.

García-Lorenzo, M.L., Marimón, J., Navarro-Hervás, M.C., Pérez-Sirvent, C., Martínez-Sánchez, M.J., Molina-Ruiz, J., 2015. Impact of acid mine drainages on surficial waters of an abandoned mining site. Environmental Science and Pollution Research. http://dx.doi.org/10.1007/s11356-015-5337-2.

Gleason, W., 2009. Metals recovered from mine water provide new revenue stream. Mining Engineering 61 (10), 22–24.

Gomes, M.E.P., Favas, P.J.C., 2006. Mineralogical controls on mine drainage of the abandoned Ervedosa tin mine in north-eastern Portugal. Applied Geochemistry 21, 1322–1334.

Gray, N., 1997. Environmental impact and remediation of acid mine drainage: a management problem. Environmental Geology 30 (1/2), 62–71.

Gray, N.F., 1996. The use of an objective index for the assessment of the contamination of surface wafer and groundwater by acid mine drainage. Journal of the Chartered Institution of Water and Environmental Management 10 (5), 332–340.

Greben, H.A., Baloyi, J., Sigama, J., Venter, S.N., 2009. Bioremediation of sulphate rich mine effluents using grass cuttings and rumen fluid microorganisms. Journal of Geochemical Exploration 100 (2–3), 163–168.

Groudev, S., Georgiev, P., Spasova, I., Nicolova, M., 2008. Bioremediation of acid mine drainage in a uranium deposit. Hydrometallurgy 94 (1–4), 93–99.

Guilbert, J., Park, C., 1986. The Geology of Ore Deposits. W. H. Freeman and Company, New York.

Gunsinger, M.R., Ptacek, C.J., Blowes, D.W., Jambor, J.L., Moncur, M.C., 2006. Mechanisms controlling acid neutralization and metal mobility within a Ni-rich tailings impoundment. Applied Geochemistry 21, 1301–1321.

Gusek, J., Clarke-Whistler, K., 2005. Where does the recovery of metal resources from passive treatment systems fit in sustainable development initiatives associated with large mining projects?. In: 22nd American Society of Mining and Reclamation Annual National Conference 2005, 1, pp. 454–469.

Hedrich, S., Johnson, D.B., 2014. Remediation and selective recovery of metals from acidic mine waters using novel modular bioreactors. Environmental Science & Technology 48, 12206–12212.

Holmströn, H., Ljungberg, J., Öhlander, B., 1999. Role of carbonates in mitigation of metal release from mining waste. Evidence from humidity cells tests. Environmental Geology 37 (4), 267–280.

Horne, A.J., 2000. Phytoremediation by constructed wetlands. In: Terry, N., Bañuelos, G. (Eds.), Phytoremediation of Contaminated Soil and Water. Lewis Publishers, New York, pp. 13–39.

Huminicki, D.M.C., Rimstidt, J.D., 2008. Neutralization of sulfuric acid solutions by calcite dissolution and the application to anoxic limestone drain design. Applied Geochemistry 23 (2), 148–165.

Jackson, T., Bistricki, T., 1995. Selective scavenging of copper, zinc, lead, and arsenic by iron and manganese oxyhydroxide coatings on plankton in lakes polluted with mine and smelter wastes: results of energy dispersive X-ray micro-analysis. Journal of Geochemical Exploration 52, 97–125.

Jacobs, J.A., 2014a. Case studies for acid mine drainage. In: Jacobs, J.A., Lehr, J.H., Testa, S.M. (Eds.), Acid Mine Drainage, Rock Drainage, and Acid Sulfate Soils: Causes, Assessment, Prediction, Prevention, and Remediation. Wiley, New Jersey, pp. 407–418.

Jacobs, J.A., 2014b. Overview of resources from acid drainage and postmining opportunities. In: Jacobs, J.A., Lehr, J.H., Testa, S.M. (Eds.), Acid Mine Drainage, Rock Drainage, and Acid Sulfate Soils: Causes, Assessment, Prediction, Prevention, and Remediation. Wiley, New Jersey, pp. 361–375.

Jafaripour, A., Rowson, N.A., Ghataora, G.S., 2015. Utilisation of residue gas sludge (BOS sludge) for removal of heavy metals from acid mine drainage (AMD). International Journal of Mineral Processing 144, 90–96.

Janyasuthiwong, S., Rene, E.R., Esposito, G., Lens, P.N.L., 2015. Effect of pH on Cu, Ni and Zn removal by biogenic sulfide precipitation in an inversed fluidized bed bioreactor. Hydrometallurgy 158, 94–100.

Jia, Y., Maurice, C., Öhlander, B., 2015. Mobility of As, Cu, Cr, and Zn from tailings covered with sealing materials using alkaline industrial residues: a comparison between two leaching methods. Environmental Science and Pollution Research. http://dx.doi.org/10.1007/s11356-015-5300-2.

Johnson, D., Hallberg, K., 2003. The microbiology of acidic mine waters. Research in Microbiology 154, 466–473.

Johnson, D.B., 2006. Biohydrometallurgy and the environment: intimate and important interplay. Hydrometallurgy 83 (1–4), 153–166.

Johnson, D.B., Hallberg, K.B., 2005. Acid mine drainage remediation options: a review. Science of the Total Environment 338, 3–14.

Jönsson, J., Jönsson, J., Lövgren, L., 2006. Precipitation of secondary Fe(III) minerals from acid mine drainage. Applied Geochemistry 21, 437–445.

Kadlec, R., Knight, R., 1996. Treatment Wetlands. CRC/Lewis Publishers, Boca Raton, FL. USA, 436.

Kadlec, R.H., Wallace, S.D., 2008. Treatment Wetlands, second ed. CRC Press, Boca Raton, FL.

Kairies, C.L., Capo, R.C., Watzlaf, G.R., 2005. Chemical and physical properties of iron hydroxide precipitates associated with passively treated coal mine drainage in the Bituminous Region of Pennsylvania and Maryland. Applied Geochemistry 20 (8), 1445–1460.

Kang, C.-U., Jeon, B.-H., Park, S.-S., Kang, J.-S., Kim, K.-H., Kim, D.-K., Choi, U.-K., Kim, S.-J., 2015. Inhibition of pyrite oxidation by surface coating: a long-term field study. Environmental Geochemistry and Health. http://dx.doi.org/10.1007/s10653-015-9778-9.

Kefeni, K.K., Msagati, T.M., Mamba, B.B., 2015. Synthesis and characterization of magnetic nanoparticles and study their removal capacity of metals from acid mine drainage. Chemical Engineering Journal 276, 222–231.

Kelepertzis, E., Argyraki, A., Daftsis, E., 2012. Geochemical signature of surface water and stream sediments of a mineralized drainage basin at NE Chalkidiki, Greece: a pre-mining survey. Journal of Geochemical Exploration 114, 70–81.

Kelley, B., Tuovinen, O., 1988. Microbiological oxidations of minerals and mine tailings. In: Salomons, W., Förstner, U. (Eds.), Chemistry and Biology of Solid Waste. Dredged Material and Mine Tailings. Springer-Verlag, Berlin, pp. 33–53.

Kelly, M., 1991. Mining and the Freshwater Environment, second ed. Elsevier Science Publishers Ltd, London, New York, p. 231.

Kepler, D.A., McCleary, E.C., 1994. Successive alkalinity producing systems (SAPS) for the treatment of acidic mine drainage. In: Proceedings of the International Land Reclamation and Mine Drainage Conference and the 3rd International Conference on the Abatement of Acidic Drainage, April, 1994, Pittsburgh, PA, vol. 1, pp. 195–204.

Kim, A.G., Heisey, B.S., Kleinmann, R.L.P., Deul, M., 1982. Acid Mine Drainage : Control and Abatement Research. Information Circular 8905. U.S. Department of the Interior, Bureau of Mines, Washington, DC, p. 22.

Kim, J.J., Kim, S.J., Tazaki, K., 2002. Mineralogical characterization of microbial ferrihydrite and schwertmannite, and non-biogenic Al-sulfate precipitates from acid mine drainage in the Donghae mine area, Korea. Environmental Geology 42, 19–31.

Kim, Y., 2015. Mineral phases and mobility of trace metals in white aluminum precipitates found in acid mine drainage. Chemosphere 119, 803–811.

Kirby, C., Decker, S., Macander, N., 1999. Comparison of color, chemical and mineralogical compositions of mine drainage sediments to pigment. Environmental Geology 37 (3), 243–252.

Kleinmann, R.L.P., Hedin, R.S., Nairn, R.W., 1998. Treatment of mine drainage by anoxic limestone drains and constructed wetlands. In: Geller, A., Klapper, H., Salomons, W. (Eds.), Acidic Mining Lakes: Acid Mine Drainage, Limnology and Reclamation. Springer, Berlin, pp. 303–319.

Koski, R.A., Munk, L.A., Foster, A.L., Shanks III, W.C., Stillings, L.L., 2008. Sulfide oxidation and distribution of metals near abandoned copper mines in coastal environments, Prince William Sound, Alaska, USA. Applied Geochemistry 23 (2), 227–254.

Krauskopf, K.B., Bird, D.K., 1994. Introduction to Geochemistry, third ed. McGraw-Hill. 640 pp.

Ladeira, A.C.Q., Gonçalves, C.R., 2007. Influence of anionic species on uranium separation from acid mine water using strong base resins. Journal of Hazardous Materials 148 (3), 499–504.

Larocque, A., Rasmussen, P., 1998. An overview of trace metals in the environment, from mobilization to remediation. Environmental Geology 33 (2/3), 85–91.

Laus, R., Geremias, R., Vasconcelos, H.L., Laranjeira, M.C.M., Favere, V.T., 2007. Reduction of acidity and removal of metal ions from coal mining effluents using chitosan microspheres. Journal of Hazardous Materials 149 (2), 471–474.

Lee, S.-H., Kim, I., Kim, K.-W., Lee, B.-T., 2015. Ecological assessment of coal mine and metal mine drainage in South Korea using *Daphnia magna* bioassay. SpringerPlus 4, 518. http://dx.doi.org/10.1186/s40064-015-1311-1.

Lee, W.-C., Lee, S.-W., Yun, S.-T., Lee, P.-K., Hwang, Y.S., Kim, S.-O., 2016. A novel method of utilizing permeable reactive kiddle (PRK) for the remediation of acid mine drainage. Journal of Hazardous Materials 301, 332–341.

Lefticariu, L., Walters, E.R., Pugh, C.W., Bender, K.S., 2015. Sulfate reducing bioreactor dependence on organic substrates for remediation of coal-generated acid mine drainage: field experiments. Applied Geochemistry 63, 70–82.

Lei, C., Yan, B., Chen, T., Quan, S.-X., Xiao, X.-M., 2015. Comprehensive utilization of lead-zinc tailings, part 1: pollution characteristics and resource recovery of sulphur. Journal of Environmental Chemical Engineering 3 (2), 862–869.

Li, M.G., Aube, B.C., St-Arnaud, L.C., 1997. Considerations in the use of shallow water covers for decommissioning reactive tailings. In: Proceedings of the Fourth International Conference on Acid Rock Drainage, May 30–June 6, Vancouver, BC, vol. I, pp. 115–130.

Li, Y., Qian, G., Li, J., Gerson, A.R., 2015. Kinetics and roles of solution and surface species of chalcopyrite dissolution at 650 mV. Geochimica et Cosmochimica Acta 161, 188–202.

Limantseva, O.A., Ryzhenko, B.N., Cherkasova, E.V., 2015. Prediction of the effects of acid mine drainage on variations in the hydrogeochemical environment at sulfide-bearing ore deposits. Geochemistry International 53 (10), 922–936.

López Pamo, E., Aduvire, O., Barettino, D., 2002. Tratamientos pasivos de drenajes ácidos de mina: estado actual y perspectivas de futuro. Boletín Geológico y Minero 113 (1), 3–21.

López-Archilla, A., 2005. Riotinto: un universo de mundos microbianos. Ecosistemas, 2005/2. http://www.revistaecosistemas.net/articulo.asp?Id=110&Id_Categoria=2&tipo=portada. Accessed on June 2015.

López-Luna, E.L., Ergas, S.J., Yuretich, R.F., 2007. In situ and laboratory studies of natural attenuation of acid mine drainage. In: 9th International In Situ and On-Site Bioremediation Symposium 2007, 1. Battelle Press, pp. 7–15.

Luís, A.T., Teixeira, P., Almeida, S.F.P., Matos, J.X., Silva, E.F., 2011. Environmental impact of mining activities in the Lousal area (Portugal): chemical and diatom characterization of metal-contaminated stream sediments and surface water of Corona stream. Science of the Total Environment 409 (20), 4312–4325.

Lutyńska, S., Labus, K., 2015. Identification of processes controlling chemical composition of pit lakes waters located in the eastern part of Muskau Arch (Polish-German borderland). Archives of Environmental Protection 41 (3), 60–69.

Mačingová, E., Luptáková, A., 2014. Recovery of iron from acid mine drainage in the form of oxides. Inzynieria Mineralna 15 (2), 193–198.

Mackie, A.L., Walsh, M.E., 2015. Investigation into the use of cement kiln dust in high density sludge (HDS) treatment of acid mine water. Water Research 85, 443–450.

Maia, F., Pinto, C., Waerenborgh, J.C., Gonçalves, M.A., Prazeres, C., Carreira, O., Sério, S., 2012. Metal partitioning in sediments and mineralogical controls on the acid mine drainage in Ribeira da Água Forte (Aljustrel, Iberian Pyrite Belt, Southern Portugal). Applied Geochemistry 27 (6), 1063–1080.

Mamba, B.B., Dlamini, N.P., Nyembe, D.W., Mulaba-Bafubiandi, A.F., 2009. Metal adsorption capabilities of clinoptilolite and selected strains of bacteria from mine water. Physics and Chemistry of the Earth 34 (13–16), 830–840.

Martín-Crespo, T., Gómez-Ortiz, D., Martín-Velázquez, S., Esbrí, J.M., de Ignacio-San José, C., Sánchez-García, M.J., Montoya-Montes, I., Martín-González, F., 2015. Abandoned mine tailings in cultural itineraries: Don Quixote Route (Spain). Engineering Geology 197, 82–93.

Masindi, V., Gitari, M.W., Tutu, H., DeBeer, M., 2015. Efficiency of ball milled South African bentonite clay for remediation of acid mine drainage. Journal of Water Process Engineering 8, 227–240.

McGregor, R.G., Blowes, D.W., Jambor, J.L., Robertson, W.D., 1998. Mobilization and attenuation of heavy metals within a nickel mine tailings impoundment near Sudbury, Ontário, Canada. Environmental Geology 36 (3/4), 305–319.

Meschke, K., Herdegen, V., Aubel, T., Janneck, E., Repke, J.-U., 2015. Treatment of opencast lignite mining induced acid mine drainage (AMD) using a rotating microfiltration system. Journal of Environmental Chemical Engineering 3 (4), 2848–2856.

Modoi, O.-C., Roba, C., Török, Z., Ozunu, A., 2014. Environmental risks due to heavy metal pollution of water resulted from mining wastes in NW Romania. Environmental Engineering and Management Journal 13 (9), 2325–2336.

Mohan, D., Chander, S., 2006. Removal and recovery of metal ions from acid mine drainage using lignite—A low cost sorbent. Journal of Hazardous Materials 137 (3), 1545–1553.

Motsi, T., Rowson, N.A., Simmons, M.J.H., 2009. Adsorption of heavymetals from acid mine drainage by natural zeolite. International Journal of Mineral Processing 92 (1–2), 42–48.

Moubarik, A., Grimi, N., 2015. Valorization of olive stone and sugar cane bagasse by-products as biosorbents for the removal of cadmium from aqueous solution. Food Research International 73, 169–175.

Muraviev, D., 2003. Conversion of pollutants to fertilisers: ion exchange synthesis of potassium sulphate from acidic mine waters. Water Science and Technology 48 (1), 199–206.

Murciego, A., Alvarez-Ayuso, E., Pellitero, E., Rodriguez, M.A., Garcia-Sanchez, A., Tamayo, A., Rubio, J., Rubio, F., Rubin, J., 2011. Study of arsenopyrite weathering products in mine wastes from abandoned tungsten and tin exploitations. Journal of Hazardous Materials 186, 590–601.

Myagkaya, I.N., Lazareva, E.V., Gustaytis, M.A., Zhmodik, S.M., 2016. Gold and silver in a system of sulfide tailings. Part 1: migration in water flow. Journal of Geochemical Exploration 160, 16–30.

Nordstrom, D.K., Blowes, D.W., Ptacek, C.J., 2015. Hydrogeochemistry and microbiology of mine drainage: an update. Applied Geochemistry 57, 3–16.

Nordstrom, D.K., Alpers, C.N., 1999a. Negative pH, efflorescent mineralogy, and consequences for environmental restoration at the Iron Mountain Superfund site, California. Proceedings of the National Academy of Sciences of the United States of America 96, 3455–3462.

Nordstrom, D.K., Alpers, C.N., 1999b. Geochemistry of acid mine waters. In: Plumlee, G.S., Logsdon, M.J. (Eds.), The Environmental Geochemistry of Mineral Deposits, Part A. Processes, Techniques, and Health Issues: Society of Economic Geologists, Reviews in Economic Geology, 6A, pp. 133–156.

Nuttall, C.A., 2003. Testing and performance of a newly constructed full-scale passive treatment system at Whittle Colliery, Northumberland. Land Contamination and Reclamation 11 (2), 105–112.

Nyirenda, T.M., Zhou, J., Xie, L., Pan, X., Li, Y., 2015. Determination of carbonate minerals responsible for alkaline mine drainage at Xikuangshan antimony mine, China: using thermodynamic chemical equilibrium model. Journal of Earth Science 26 (5), 755–762.

Nyquist, J., Greger, M., 2009. A field study of constructed wetlands for preventing and treating acid mine drainage. Ecological Engineering 35, 630–642.

Olías, M., Nieto, J.M., Pérez-López, R., Cánovas, C.R., Macías, F., Sarmiento, A.M., Galván, L., 2016. Controls on acid mine water composition from the Iberian Pyrite Belt (SW Spain). Catena 137, 12–23.

Parbhakar-Fox, A., Lottermoser, B.G., 2015. A critical review of acid rock drainage prediction methods and practices. Minerals Engineering 82, 107–124.

Park, C.F., MacDiarmid, R.A., 1975. Ore Deposits, third ed. Freeman, San Francisco. 529 pp.

Park, S.-M., Shin, S.-Y., Yang, J.-S., Ji, S.-W., Baek, K., 2015. Selective recovery of dissolved metals from mine drainage using electrochemical reactions. Electrochimica Acta 181, 248–254.

Pérez-López, R., Nieto, J.M., Almodóvar, G.R., 2007. Utilization of fly ash to improve the quality of the acid mine drainage generated by oxidation of a sulphide-rich mining waste: column experiments. Chemosphere 67 (8), 1637–1646.

Pérez-López, R., Nieto, J.M., de la Rosa, J.D., Bolívar, J.P., 2015. Environmental tracers for elucidating the weathering process in a phosphogypsum disposal site: implications for restoration. Journal of Hydrology 529, 1313–1323.

Phillips, D.J.H., Rainbow, P.S., 1994. Biomonitoring of Trace Aquatic Contaminants. Chapman & Hall, London.

Plumlee, G.S., Smith, K.S., Ficklin, W.H., Briggs, P.H., 1992. Geological and geochemical controls on the composition of mine drainages and natural drainages in mineralized areas. In: Kharaka, M. (Ed.), Water-rock Interaction, Low Temperature Environments, vol. 1. Balkema, Rotterdam, pp. 419–422.

Pope, J., Trumm, D., 2015. Controls on Zn concentrations in acidic and neutral mine drainage from New Zealand's bituminous coal and epithermal mineral deposits. Mine Water and the Environment 34 (4), 455–463.

Prasad, B., Mortimer, R.J.G., 2011. Treatment of acid mine drainage using fly ash zeolite. Water Air and Soil Pollution 218 (1–4), 667–679.

Prasad, M.N.V., 2004. Phytoremediation of metals and radionuclides in the environment: the case for natural hyperaccumulators, metal transporters, soil-amending chelators and transgenic plants. In: Prasad, M.N.V. (Ed.), Heavy Metal Stress in Plants: From Biomolecules to Ecosystems, second ed. Springer, Berlin, pp. 345–391.

Prasad, B., Kumar, H., 2015. Treatment of acid mine drainage using a fly ash zeolite column. Mine Water and the Environment. http://dx.doi.org/10.1007/s10230-015-0373-1.

Qureshi, A., Maurice, C., Öhlander, B., 2016. Potential of coal mine waste rock for generating acid mine drainage. Journal of Geochemical Exploration 160, 44–54.

Raikova, S., Smith-Baedorf, H., Bransgrove, R., Barlow, O., Santomauro, F., Wagner, J.L., Allen, M.J., Bryan, C.G., Sapsford, D., Chuck, C.J., 2016. Assessing hydrothermal liquefaction for the production of bio-oil and enhanced metal recovery from microalgae cultivated on acid mine drainage. Fuel Processing Technology 142, 219–227.

Ramirez-Paredes, F.I., Manzano-Munoz, T., Garcia-Prieto, J.C., Zhadan, G.G., Shnyrov, V.L., Kennedy, J.F., Roig, M.G., 2011. Biosorption of heavy metals from acid mine drainage onto biopolymers (chitin and alfa (1,3) beta-D-glucan) from industrial biowaste exhausted brewer's yeasts (*Saccharomyces cerevisiae* L.). Biotechnology and Bioprocess Engineering 16 (6), 1262–1272.

Reed, S.M., 1986. Groundwater Recovery Problems Associated with Opencast Mine Backfills (Ph.D. thesis). University of Nottingham.

Rees, S.B., Bowell, R.J., Wiseman, I., 2002. Influence of mine hydrogeology on mine water discharge chemistry. In: Younger, P.L., Robins, N.S. (Eds.), Mine Water Hydrogeology and Geochemistry. The Geological Society, London, pp. 379–390. Special Publication, 198.

Ríos, C.A., Williams, C.D., Roberts, C.L., 2008. Removal of heavy metals from acid mine drainage (AMD) using coal fly ash, natural clinker and synthetic zeolites. Journal of Hazardous Materials 156 (1–3), 23–35.

Rivera-Santillán, R.E., Becerril-Reyes, V., 2007. Comparison of the bioleaching effect of mesophilic (35°C) and thermophilic (45°C) bacteria on the TIZAPA tailings. Advanced Materials Research 20-21, 34–37.

Robertson, K., Gauvin, R., Rao, S.R., Finch, J., 2006. Identifying refractory nickel in acid mine drainage lime treatment sludge. Canadian Metallurgical Quarterly 45 (1), 95–102.

Robinson-Lora, M.A., Brennan, R.A., 2009. Efficient metal removal and neutralization of acid mine drainage by crab-shell chitin under batch and continuous-flow conditions. Bioresour Technol 100 (21), 5063–5071.

Robinson-Lora, M.A., Brennan, R.A., 2010. Chitin complex for the remediation of mine impacted water: geochemistry of metal removal and comparison with other common substrates. Applied Geochemistry 25 (3), 336–344.

Robles-Arenas, V.M., Candela, L., 2010. Hydrological conceptual model characterisation of an abandoned mine site in semiarid climate. The Sierra de Cartagena-La Unión (SE Spain). Geologica Acta 8 (3), 235–248.

Romero, A., González, I., Galán, E., 2011. Stream water geochemistry from mine wastes in Peña de Hierro, Riotinto area, SW Spain: a case of extreme acid mine drainage. Environmental Earth Sciences 62 (3), 645–656.

Saaltink, M.W., Domenech, C., Ayora, C., Carrera, J., 2002. Modelling the oxidation of sulphides in an unsaturated soil. In: Younger, P.L., Robins, N.S. (Eds.), Mine Water Hydrogeology and Geochemistry. The Geological Society, London, pp. 187–204. Special Publication, 198.

Saharan, M.R., Gupta, K.K., Jamal, A., Sheoran, A.S., 1995. Management of acidic effluents from tailing dams in metalliferous mines. Mine Water and the Environment 14 (1), 85–93.

Sahinkaya, E., Gungor, M., 2010. Comparison of sulfidogenic up-flow and down-flow fluidized-bed reactors for the biotreatment of acidic metal-containing wastewater. Bioresource Technology 101 (24), 9508–9514.

Salminen, J., Blomberg, P., Mäkinen, J., Räsänen, L., 2015. Environmental aspects of metals removal from waters and gold recovery. AIChE Journal 61 (9), 2739–2748.

Salomons, W., 1995. Environmental impact of metals derived from mining activities: processes, predictions, prevention. Journal of Geochemical Exploration 52, 5–23.

Sánchez España, J., López Pamo, E., Santofimia Pastor, E., Reyes Andrés, J., Martín Rubí, J.A., 2006a. The impact of acid mine drainage on the water quality of the Odiel river (Huelva, Spain): evolution of precipitate mineralogy and aqueous geochemistry along the Concepción-Tintillo segment. Water, Air, and Soil Pollution 173, 121–149.

Sánchez España, J., López Pamo, E., Santofimia Pastor, E., Reyes Andrés, J., Martín Rubí, J.A., 2006b. The removal of dissolved metals by hydroxysulphate precipitates during oxidation and neutralization of acid mine waters, Iberian Pyrite Belt. Aquatic Geochemistry 12, 269–298.

Sangita, Udayabhanu, G., Prasad, B., 2010. Studies on environmental impact of acid mine drainage generation and its treatment: an appraisal. Indian Journal of Environmental Protection 30 (11), 953–967.

Santisteban, M., Grande, J.A., de la Torre, M.L., Valente, T., Ceron, J.C., 2015. Acid mine drainage in semi-arid regions: the extent of the problem in the waters of reservoirs in the Iberian Pyrite Belt (SW Spain). Hydrology Research 46 (1), 156–167.

Sapsford, D., Santonastaso, M., Thorn, P., Kershaw, S., 2015. Conversion of coal mine drainage ochre to water treatment reagent: production, characterisation and application for P and Zn removal. Journal of Environmental Management 160, 7–15.

Sarmiento, A.M., Olías, M., Nieto, J.M., Cánovas, C.R., Delgado, J., 2009. Natural attenuation processes in two water reservoirs receiving acid mine drainage. Science of the Total Environment 407 (6), 2051–2062.

Sasaki, K., 1994. Effect of grinding on the rate of oxidation of pyrite by oxygen in acid solutions. Geochimica et Cosmochimica Acta 58 (21), 4649–4655.

Sasaki, K., Nukina, S., Wilopo, W., Hirajima, T., 2008. Removal of arsenate in acid mine drainage by a permeable reactive barrier bearing granulated blast furnace slag: column study. Materials Transactions 49 (4), 835–844.

Schaider, L.A., Senn, D.B., Estes, E.R., Brabander, D.J., Shine, J.P., 2014. Sources and fates of heavy metals in a mining-impacted stream: temporal variability and the role of iron oxides. Science of the Total Environment 490, 456–466.

Schmiermund, R.L., Drozd, M.A., 1997. Acid mine drainage and other mining-influenced waters (MIW). In: Marcus, J. (Ed.), Mining Environmental Handbook: Effects of Mining on the Environment and American Environmental Controls on Mining. Imperial College Press, London, pp. 599–617.

Schwertmann, U., Taylor, R.M., 1989. Iron oxides. In: Dixon, J.B., Weed, S.B. (Eds.), Minerals in Soil Environments, second ed. Soil Science Society of America, Madison, Wisconsin (Chapter 8).

Seal II, R.R., Shanks III, W.C., 2008. Sulfide oxidation: insights from experimental, theoretical, stable isotope, and predictive studies in the field and laboratory. Applied Geochemistry 23 (2), 101–342.

Sengupta, M., 1993. Environmental Impacts of Mining, Monitoring, Restoration, and Control. Lewis Publishers, London.

Shea-Albin, V.R., Fitch, W.N., 1991. Mining waste research in the U.S. Bureau of Mines 199–205.

Sherlock, E.J., Lawrence, R.W., Poulin, R., 1995. On the neutralization of acid rock drainage by carbonate and silicate minerals. Environmental Geology 25, 43–54.

Shim, M.J., Choi, B.Y., Lee, G., Hwang, Y.H., Yang, J.S., O'Loughlin, E.J., Kwon, M.J., 2015. Water quality changes in acid mine drainage streams in Gangneung, Korea, 10 years after treatment with limestone. Journal of Geochemical Exploration 159, 234–242.

Shoja, S.E., Salari, M.M., 2015. Study of acid mine drainage production potential in flotation tailings of Sarcheshmeh porphyry copper mine. Arabian Journal of Geosciences 8 (10), 8229–8236.

Silva, E.F., Bobos, I., Matos, J.X., Patinha, C., Reis, A.P., Fonseca, E.C., 2009. Mineralogy and geochemistry of trace metals and REE in volcanic massive sulfide host rocks, stream sediments, stream waters and acid mine drainage from the Lousal mine area (Iberian Pyrite Belt, Portugal). Applied Geochemistry 24, 383–401.

Silveira, A.N., Silva, R., Rubio, J., 2009. Treatment of acid mine drainage (AMD) in South Brazil. Comparative active processes and water reuse. International Journal of Mineral Processing 93 (2), 103–109.

Sinclair, S.A., Pham, N., Amos, R.T., Sego, D.C., Smith, L., Blowes, D.W., 2015. Influence of freeze-thaw dynamics on internal geochemical evolution of low sulfide waste rock. Applied Geochemistry 61, 160–174.

Skousen, J., Rose, A., Geidel, G., Foreman, J., Evans, R., Hellier, W., 1998. Handbook of Technologies for Avoidance and Remediation of Acid Mine Drainage. Published by the National Mine Land Reclamation Center, West Virginia University, Morgantown, West Virginia, 123 pp.

Sobolewski, A., 1999. A review of processes responsible for metal removal in wetlands treating contaminated mine drainage. International Journal of Phytoremediation 1 (1), 19–51.

Soltani, N., Moore, F., Keshavarzi, B., Sharifi, R., 2014. Geochemistry of trace metals and rare earth elements in stream water, stream sediments and acid mine drainage from Darrehzar Copper Mine, Kerman, Iran. Water Quality, Exposure and Health 6, 97–114.

Song, J., Choi, Y., 2015. Design of photovoltaic systems to power aerators for natural purification of acid mine drainage. Renewable Energy 83, 759–766.

Strosnider, W.H.J., Llanos López, F.S., Nairn, R.W., 2011. Acid mine drainage at Cerro Rico de Potosí II: severe degradation of the Upper Rio Pilcomayo watershed. Environmental Earth Sciences 64 (4), 911–923.

Stumm, W., Morgan, J.J., 1996. Aquatic Chemistry: Chemical Equilibria and Rates in Natural Waters. John Wiley & Sons, New York.

Sun, M., Wu, N.-N., Zhai, L.-F., Ru, X.-R., 2015. Manipulate an air-cathode fuel cell toward recovering highly active heterogeneous electro-Fenton catalyst from the Fe(II) in acid mine drainage. Minerals Engineering 84, 1–7.

Swanson, D.A., Barbour, S.L., Wilson, G.W., 1997. Dry-site versus wet-site cover design. In: Proceedings of the Fourth International Conference on Acid Rock Drainage, May 30–June 6, Vancouver, BC, vol. IV, pp. 1595–1610.

de la Torre, M.L., Grande, J.A., Valente, T., Perez-Ostalé, E., Santisteban, M., Aroba, J., Ramos, I., 2015. Definition of redox and pH influence in the AMD mine system using a fuzzy qualitative tool (Iberian Pyrite Belt, SW Spain). Environmental Science and Pollution Research. http://dx.doi.org/10.1007/s11356-015-5718-6.

Tabak, H.H., Govind, R., 2003. Advances in biotreatment of acid mine drainage and biorecovery of metals: 2. Membrane bioreactor system for sulfate reduction. Biodegradation 14 (6), 437–452.

Tabak, H.H., Scharp, R., Burckle, J., Kawahara, F.K., Govind, R., 2003. Advances in biotreatment of acid mine drainage and biorecovery of metals: 1. Metal precipitation for recovery and recycle. Biodegradation 14 (6), 423–436.

Taylor, J., Pape, S., Murphy, N., 2005. A summary of passive and active treatment technologies for acid and metalliferous drainage (AMD). In: Fifth Australian Workshop on Acid Drainage, 29–31 August 2005. Australian Centre for Minerals Extension and Research (ACMER), Earth Systems, Fremantle, Western Australia, 49 p.

Thornber, M.R., 1985. Supergene alteration of sulphides, VII: distribution of elements during the gossan-forming process. Chemical Geology 53 (3/4), 279–301.

Thornber, M.R., Wildman, J.E., 1984. Supergene alteration of sulphides, VI: the binding of Cu, Ni, Zn, Co and Pb with gossan (iron-bearing) minerals. Chemical Geology 44 (4), 399–434.

Torres, E., Couture, R.M., Shafei, B., Nardi, A., Ayora, C., Van Cappellen, P., 2015. Reactive transport modeling of early diagenesis in a reservoir lake affected by acid mine drainage: trace metals, lake overturn, benthic fluxes and remediation. Chemical Geology 419, 75–91.

Toughzaoui, S., Amari, K.E., Benkaddour, A., Hibti, M., Essarraj, S., 2015. Hydrogeochemical and isotopic studies of the Kettara mine watershed, Morocco. Mine Water and the Environment 34 (3), 308–319.

Urrutia, M., Graña, J., Garcia-Rodeja, R., Macias, F., 1987. Procesos de oxidación de pirita en medios superficiales: potential acidificante e interés para la recuperación de suelos de mina. Cuaderno do Laboratorio Xeolóxico de Laxe 11, 131–145.

Vadapalli, V.R.K., Zvimba, J.N., Mulopo, J., Motaung, S., 2013. Sulphate removal from sodium sulphate-rich brine and recovery of barium as a barium salt mixture. Journal of Environmental Science and Health – Part A Toxic/Hazardous Substances and Environmental Engineering 48 (8), 933–938.

Valente, T.M., Gomes, C.L., 2009. Fuzzy modelling of acid mine drainage environments using geochemical, ecological and mineralogical indicators. Environmental Geology 57, 653–663.

Valente, T.M.F., 2004. Modelos de Caracterização de Impacte Ambiental para Escombreiras Reactivas – equilíbrio e evolução de resíduos de actividade extractiva (Ph.D. thesis). Universidade do Minho, Braga, 301 p.

Vicente-Beckett, V.A., McCauley, G.J.T., Duivenvoorden, L.J., 2015. Metal speciation in sediments and soils associated with acid-mine drainage in Mount Morgan (Queensland, Australia). Journal of Environmental Science and Health – Part A Toxic/Hazardous Substances and Environmental Engineering. http://dx.doi.org/10.1080/10934529.2015.1087738.

Vymazal, J., 2009. The use constructed wetlands with horizontal sub-surface flow for various types of wastewater. Ecological Engineering 35, 1–17.

Vymazal, J., Kröpfelová, L., 2008. Wastewater Treatment in Constructed Wetlands with Horizontal Sub-surface Flow. Springer, Dordrecht.

Walton-Day, K., Mills, T.J., 2015. Hydrogeochemical effects of a bulkhead in the Dinero mine tunnel, Sugar Loaf mining district, near Leadville, Colorado. Applied Geochemistry 62, 61–74.

Waters, P., 2002. Mining reactive shale at BHP Bulliton's Mt Whaleback mine – Challenges and solutions. Australasian Institute of Mining and Metallurgy Publication Series 7, 219–221.

Wei, X., Wolfe, F.A., 2013. Minerals and mine drainage. Water Environment Research 85 (10), 1515–1547.

Westholm, L.J., Repo, E., Sillanpää, M., 2014. Filter materials for metal removal from mine drainage—a review. Environmental Science and Pollution Research 21, 9109–9128.

Williams, G.P., Petteys, K., Gammons, C.H., Parker, S.R., 2015. An investigation of acidic head-water streams in the Judith Mountains, Montana, USA. Applied Geochemistry 62, 48–60.

Williamson, N.A., Johnson, M.S., Bradshaw, A.D., 1982. Mine Wastes Reclamation. The Establishment of Vegetation on Metal Mine Wastes. Mining Journal Books, London.

Wu, J., Li, P., Qian, H., Duan, Z., Zhang, X., 2014. Using correlation and multivariate statistical analysis to identify hydrogeochemical processes affecting the major ion chemistry of waters: a case study in Laoheba phosphorite mine in Sichuan, China. Arabian Journal of Earth Geosciences 7, 3973–3982.

Xie, J., Weng, Q., Ye, G., Luo, S., Zhu, R., Zhang, A., Chen, X., Lin, C., 2014. Bioethanol production from sugarcane grown in heavy metal-contaminated soils. BioResources 9 (2), 2509–2520.

Yesilnacar, M.I., Kadiragagil, Z., 2013. Effects of acid mine drainage on groundwater quality: a case study from an open-pit copper mine in eastern Turkey. Bulletin of Engineering Geology and the Environment 72, 485–493.

Younger, P.L., Jayaweera, A., Elliot, A., Wood, R., Amos, P., Daugherty, A.J., et al., 2003. Passive treatment of acidic mine waters in subsurfaceflow systems: exploring RAPS and permeable reactive barriers. Land Contamination & Reclamation 11, 127–135.

Yuniati, M.D., Kitagawa, K., Hirajima, T., Miki, H., Okibe, N., Sasaki, K., 2015. Suppression of pyrite oxidation in acid mine drainage by carrier microencapsulation using liquid product of hydrothermal treatment of low-rank coal, and electrochemical behavior of resultant encapsulating coatings. Hydrometallurgy 158, 83–93.

Zahid, L., Malek, F., Nornikman, H., Mohd Affendi, N.A., Ali, A., Hussin, N., Ahmad, B.H., Abd Aziz, M.Z.A., 2013. Development of pyramidal microwave absorber using sugar cane bagasse (SCB). Progress in Electromagnetics Research 137, 687−702.

Zhou, H.-B., Zeng, W.-M., Yang, Z.-F., Xie, Y.-J., Qiu, G.-Z., 2009. Bioleaching of chalcopyrite concentrate by a moderately thermophilic culture in a stirred tank reactor. Bioresource Technology 100 (2), 515−520.

Ňancucheo, I., Johnson, D.B., 2012. Selective removal of transition metals from acidic mine waters by novel consortia of acidophilic sulfidogenic bacteria. Microbial Biotechnology 5, 34−44.

Ňancucheo, I., Hedrich, S., Johnson, D.B., 2012. New microbiological strategies that enable the selective recovery and recycling of metals from acid mine drainage and mine process waters. Mineralogical Magazine 76 (7), 2683−2692.

CHAPTER

18 RESTORATION OF SMELTER INDUSTRIAL BARRENS FOLLOWING POLLUTION REDUCTION DRIVES ECONOMIC RECOVERY

G. Koptsik[1], S. Koptsik[1], G. Spiers[1,2], P. Beckett[2]

M.V. Lomonosov Moscow State University, Moscow, Russia[1]; Laurentian University, Sudbury, ON, Canada[2]

CHAPTER OUTLINE

18.1 INTRODUCTION

The resolution of environmental problems created by technogenic pollution of ecosystems induced by the mining and metallurgical processing industry is a priority requiring fundamental science-based knowledge to implement methodological approaches. Lifecycle assessment (LCA) documents the environmental impacts of products, processes, or services through production, use, and disposal (Ayres et al., 2002; Landner and Reuther, 2004; Norgate et al., 2007). From an LCA perspective, the environmental consequences of mining and processing activities have highlighted ecosystem restoration as an issue of enormous importance for industrial regions.

The creation of metal-contaminated, phytotoxic, barren landscapes by atmospheric smelter fallout has occurred in many parts of the world, including Australia, Canada, Chile, the United States, and Russia (Kozlov and Zvereva, 2007; Mudd, 2010), with three such landscapes being located in the environmentally sensitive Kola Peninsula region of northern Russia. Technogenic barren lands have been, or are being, successfully regreened; Sudbury, Ontario, Canada, has achieved international recognition for successful community-focused land restoration initiatives (Lautenbach et al., 1995; Winterhalder, 1996). After a dramatic reduction in gaseous and aerosolic emissions in the 1970s, the

Environmental Materials and Waste. http://dx.doi.org/10.1016/B978-0-12-803837-6.00018-4

development of a sustainable forest ecosystem within damaged watersheds led to the recovery of regional lakes, allowing the development of a sports fishery and an aesthetically pleasing habitat that has attracted an entrepreneurial base critical for economic diversification. However, the land system remediation methodology and practice to manage specific pollutants and environmental impacts must be optimized. Detailed research is critical to understand and manage soil fertility issues to develop environmentally benign soil amendments to minimize contaminant metal bioavailability and identify key indigenous species that will colonize in a recovering polluted environment (Bolan and Duraisamy, 2003; Adriano et al., 2004; Guo et al., 2006).

An understanding of the critical role that soil health has in the structure and sustainable functioning of terrestrial and human ecosystems after extensive regional anthropogenic pollution determines the relevance of focused research themes. Soil mineral and organic components, with their pH and nutrient buffer properties, control not only the degree of internal change in the soils themselves as a consequence of anthropogenic contaminant impact but also the amount of contaminant potentially released into the other compartments of the impacted ecosystems: living organisms, groundwater, surface waters such as rivers and lakes, and, potentially, oceans. The accumulation of contaminants in soils is commonly accompanied by a long-term negative impact on the entire ecosystem, with problems often exacerbated by unfavorable natural and anthropogenic factors leading to potentially toxic levels of contaminants to both soils and aquatic biota. The development and implementation of science-based approaches to remediation of contaminated soils is therefore an important theoretical and practical challenge facing modern society. Ecosystem productivity limitations within soil and plant resources in the harsh climatic conditions of the Arctic make the development of solutions to the challenge acute.

The increasing anthropogenic impact on the environment has focused attention on developing a basic understanding of the rehabilitation requirements of industrial emission–contaminated soils in industrial, agricultural, and urban territories (Choi et al., 2008; Hudson-Edwards et al., 2011; Koptsik, 2014a,b). The technologies of phytoremediation (green technologies) as an approach to the in situ remediation of contaminant metal–impacted soils has attracted considerable attention (Cunningham and Ow, 1996; Prasad and Freitas, 2003; Robinson et al., 2003; Nowack et al., 2006; Mench et al., 2009; Vangronsveld et al., 2009; Prasad, 2011); phytoextraction and phytostabilization are the foci as widespread and alternative methods of soil phytoremediation. The efficiency of phytoremediation is related to the natural capacity of select plants for the accumulation and translocation of metals, with an associated tolerance to a high content of metals in the plant biomass (Bolan et al., 2011; Lottermoser et al., 2011) and to the surface amendments applied. The successful phytoextraction and phytomining for remediation of contaminated soils in Rasteburg (South Africa) (Robinson et al., 1997) and the chemo-phytostabilization of technogenic barrens near the copper–nickel smelters in Sudbury (Ontario, Canada) (Lautenbach et al., 1995; Winterhalder, 1996) are well-known examples. Chemo-phytostabilization of barren lands has also been initiated in the Kola Subarctic, where atmospheric emissions over many years have led to the degradation of large areas of the ecosystems near the copper–nickel smelters (Kryuchkov, 1985; Tsvetkov and Cherkizov, 1987; Kapel'kina, 1993; Evdokimova, 1995; Nikonov et al., 2005). The current approaches to remediation, however, require enhancement to reduce the impact of continuing air emissions, of extremely high levels of soil contamination, with associated landscape erosion leading to a loss of organic matter from soil surface horizons under harsh seasonal climatic conditions (Koptsik, 2014c; Koptsik et al., 2015a).

The aim of the current overview is to compare the efficiency of remediation practices of technogenic barrens, to evaluate opportunities to enhance the landscape restoration practices and reduce environmental impacts of nickel-copper mining and smelting in two northern industrial regions: Sudbury, Ontario, Canada and Kola Peninsula, Russia.

18.2 THE SUDBURY EXPERIENCE

Long before the advent of mining, the Anishinaabe people lived among the pines, spruce, maples, and oaks growing on the shallow Podzolic soils developed in glacial sediments overlying Precambrian Shield rocks of the Sudbury region. European economic activity was initiated in the 18th century through fur trading, with subsequent development of a lumber industry for ship building, urban construction, and construction of the transcontinental railway system. The first Cu-Ni rediscovery was made in 1856, with mining developing after completion of the transcontinental railway in the mid-1880s. Open heap roasting using local wood began in 1888 and was phased out in 1929 with operational consolidation into three large smelters. Smelter atmospheric releases of SO_2 were 2.5×10^6 tons $year^{-1}$ in the late 1960s; associated with a peak production of about 2.1×10^4 tons $year^{-1}$ of Ni in 1974 were atmospheric releases of approximately 5.0×10^4 tons $year^{-1}$ of fine Fe-, Ni-, Cu-, As-, and Pb-containing aerosols (Wren, 2012).

The sandy soils of the Sudbury landscape were affected by a century of interacting factors, such as logging, wildfires, wind and water erosion, SO_2 fumigations, acidification, and metal deposition, resulting in 17,000 ha of highly damaged watersheds and another 64,000 ha of stunted woodlands. Contaminant reductions commencing in 1972 (97% reduction in SO_2 emissions and large smelter particulates by 2016) created the possibility of reestablishing vegetation. The local government formed the Vegetation Enhancement Technical Advisory Committee (VETAC) in 1973. The volunteer action-orientated advisory committee with members from academic, government, business, and citizen sectors works together for the purpose of developing community pride in restoring healthy landscapes that support healthy, resilient, and sustainable communities through the successful recovery of industrial damage in regional watersheds to attract industrial partners, businesses, institutions, and people.

VETAC guided the translation of the Laurentian University research and mining company successes into vegetation establishment on the research scale to implement the internationally recognized Sudbury Method for technogenic barren landscape restoration on a regional scale. The original Sudbury Method (Lautenbach et al., 1995), used from 1978 to 2010, spread 10 tons ha^{-1} dolomitic limestone of various particle sizes: 400 kg ha^{-1} of 5–20–20 fertilizer; a 40-kg ha^{-1} nursery cover of five to seven agronomic grasses and two legumes; followed by planting native trees including pines (Figs. 18.1 and 18.2).

Following a major ecological risk assessment study (Wren, 2012), the resulting Biodiversity Action Plan in 2009 directed that regreening activities move from mere reclamation to a more complete biodiverse restoration strategy. The current Sudbury Protocol dictates liming, fertilizing where needed, and use of native grasses with less dependence on legumes. The planting of a wide array of native trees helps establish a forest structure with an emphasis on introducing forest shrubs and understory components (about 50 species). By 2015, 3460 ha have received limestone, fertilizer, and seed and nearly 10 million trees and shrubs have been planted for approximately Can$30 million.

The development of a sustainable regional forest ecosystem has led to the recovery of regional streams and lakes, allowing dramatic improvement in the overall quality of life within an aesthetically pleasant habitat and a city undergoing economic diversification. The evolving Sudbury Protocol, supported for over 40 years, will continue restoration activities that are now about 50% accomplished, with ongoing landscape change monitoring as restoration completion criteria are established. Sudbury has achieved international recognition for success in community-focused land restoration initiatives. Communities elsewhere have adopted derivative methods based on the Sudbury Protocol and the developing novel ecosystem which hosts scientific investigations and educational tours focused on ecosystem restoration.

FIGURE 18.1

Vegetation changes near Martindale Road, Sudbury, 1981 to 2015, after application of the Sudbury Method.

FIGURE 18.2

Local community landscape changes over 30 years, from 1975 to 2005, after application of the Sudbury Method.

18.3 THE KOLA STORY

Mining and metallurgy activities in the Kola Peninsula, northwest Russia, have contributed significantly to the economy for more than 70 years, with a dramatic impact on the fragile natural environment of the extreme north region. The Kola Mining and Metallurgical Company (Kola MMC) is one of the two main Russian production units of Norilsk Nickel Group, the world leader in nickel and palladium production with strong positions in platinum and copper (http://www.nornik.ru). The Pechenganikel and Severonikel copper–nickel industrial complexes, the main subdivisions of the Kola MMC, are the largest emitters of sulfur and potentially environmentally toxic elements, predominantly Ni and Cu, in the European north. The detailed eco-geochemical investigations in the early 1990s (Reimann et al., 1998) and in 2000–2001 (Salminen et al., 2004) showed that the Severonikel complex is also a source for emissions of a wide spectrum of contaminant elements, namely Ag, Al, As, Ca, Cd, Co, Cr, Fe, Hg, Mg, Mn, Pb, Sb, Sr, Th, Tl, V, Zn, and others (Kashulina et al., 2014).

The mining and smelting capacities of the Pechenganikel industrial complex are located at two sites, at the Nikel settlement and Zapolyarnyy town in the northwest region of the Kola Peninsula. Three mines, the nickel smelter and copper–nickel ore roasting plant comprise the core of the industrial complex. The output of the industrial complex is a converter matte, with further refining producing nickel, copper, cobalt, precious metals, and sulfuric acid. The Severonikel industrial complex, located in Monchegorsk in the central region of the Kola Peninsula, contains nickel electrolysis and copper metallurgical refineries, and associated slag heaps, with many sources of ore feed materials.

Until the early 1970s, both Pechenganikel and Severonikel complexes used local ore with comparatively low sulfur content (6.5%) as smelter feed producing annual SO_2 emissions not exceeding 100,000 tons at each complex, resulting in insignificant local damage to soils and vegetation of the regional watersheds. During the 1970s to 1980s, the feed for the industrial complexes was enriched with concentrate from Norilsk, which contained high concentrations of sulfidic minerals (up to 30%). This change in smelter feed increased the volume of SO_2 and metal-rich aerosolic emissions which severely impacted the regional ecology. Kola MMC data (Kola MMC, 2008) indicate that SO_2 emissions at the Pechenganikel complex reached a maximum in 1979–1980 (up to 400,000 tons) and at the Severonikel complex in 1974–1986 (up to 280,000 tons), with a decrease to 257,500 and 232 tons, respectively in 1990. Currently, the Pechenganikel complex emits about 100,000 tons of SO_2 annually, with the dominant amount of SO_2 emitted from high stacks (140–160 m), potentially contributing to long-range transport. However, short-term, high-atmospheric concentrations of sulfur dioxide emission events are still directly damaging vegetation at varying distances from the smelter (Arctic Pollution, 2006). The ongoing upgrading and installation of emission control technology at the Severonikel industrial complex have resulted in a significant decrease in SO_2 emissions during the past 20 years (to 31,000 tons of SO_2, 346 tons of Ni, and 461 tons of Cu in 2011). Between 1990 and 2014, overall emissions of SO_2 and aerosolic metal-rich dusts by Kola MMC decreased by 3.3 and 2 times, up to 150,200 and 11,760 tons per year, respectively (Delivering on Strategy - Unlocking the Potential of Norilsk, 2015). Despite the decrease in emissions during the past 20 years, Ni and Cu concentrations in snow pack remain extremely high near the Severonikel plant, at 2500 and 1500 times background values, respectively. Rainwater currently has Ni and Cu concentrations 150 and 80 times background levels, respectively (Kashulina et al., 2014), a five- to sixfold decrease in concentrations since 1994.

Production facilities of Kola MMC are all located within or close to the Arctic eco-zone; long-term emissions of SO_2 and contaminant metals are causing severe environmental damage in the region, leading to the technogenic destruction of forest soils and ecosystems to produce vast technogenic barrens around the contamination sources (Kozlov and Zvereva, 2007; Kashulina et al., 1997; Chernenkova and Kuperman, 1999; Koptsik et al., 2003). The severe climatic conditions aggravate the ecological situation of globally northernmost coniferous boreal forests in this region. With the strong natural climatic stresses, even minor atmospheric loads of pollutants may have severe effects on forest vitality. The high concentrations of sulfur dioxide have caused severe damage to trees and understory vegetation owing to direct tissue damage (Aamlid, 1992; Aamlid and Venn, 1993). The indirect impact caused by soil acidification and associated release of soil-solution aluminum, other potentially toxic elements, or induced nutrient deficiency in the root zone is also possible (Kashulina et al., 1997; Koptsik et al., 2003; Aamlid and Venn, 1993; Koptsik et al., 2001). Long-term ecosystem contamination has resulted in low ecosystem productivity as a result of the destruction of the tree canopy layer, of low biodiversity within the phytocenoses, with low soil biological activity induced by soil contamination and associated disturbance of biogeochemical element cycles (Evdokimova, 1995; Koptsik et al., 2003; Alekseev, 1990; Lukina and Nikonov, 1998; Evdokimova et al., 2014; Koptsik et al., 2015b). Furthermore, the dispersal of large amounts of toxic waste in the vicinity of industrial complexes has resulted in the risk of secondary contamination. The annual compensation costs for the annual damage to human health and the environment from air pollution in the Kola Subarctic may be estimated at least €1.5 billion, based on estimates of €9792–28,576 per ton of SO_2 and €3800 per ton of Ni (European Environment Agency, 2014) for emissions in the Kola region.

The threefold reduction of quantity of air emissions by the Kola MMC in the past 20 years has resulted in some improvement in vegetation health (Derome et al., 2008; Lyanguzova, 2010; Chernen'kova et al., 2014). However, as in the Sudbury area (Lautenbach et al., 1995; Winterhalder, 1996), unfavorable soil properties suppressed the revegetation recovery of the technogenic barrens despite the improvement in air quality. The decline in metal emission will not result in an immediate decline in plant nickel and copper concentrations because of the continuing release of bioavailable metals from accumulations in upper soil horizons for plant uptake. Secondary pollution of the region by dusts ablated from eroded soil surfaces is also significant. The high concentrations of contaminant metals accumulated in the soil horizons, coupled with the depletion of both nutrients and organic matter, the physical loss of the upper soil layers through erosion, the severe microclimate, the resultant decrease in the soil biological activity, the suppression of the root systems development, and the depletion of the seed reserves have all combined to suppress the recovery of the natural vegetation.

In 2003–2004, large-scale remediation efforts were initiated using two main methods (Koptsik, 2014c; Koptsik et al., 2015a) for the technogenic barrens of Monchegorsk and Nikel with the support of the Kola MMC. The chemo-phytostabilization method employed on limed sites included fall planting of willow and birch trees with simultaneous sowing of grasses (40–50 kg ha^{-1}) and the application of nitrogen, phosphate, and potassium fertilizer. The lime application rate was 2 tons per hectare on most sites, with 1 ton per hectare of agricultural NPK fertilizer. Tree seedlings were transplanted from abandoned agricultural lands.

The alternate remediation method involved the development and even placement of a fertile substrate cover layer on contaminated soils, followed by the application of fertilizers and liming materials, sowing of grass seed mixtures, and planting of tree and shrub species. The substrate cover layer in the Nikel area was manufactured using soil materials for the former arable horizon collected from abandoned agricultural lands in valley bottoms of the region. The introduced surface Technosol cover substrate for sites in the Monchegorsk area consisted of an admixture of peat, sand, sawdust, and wastewater sludge blended in varying ratios. Both crushed limestone (2 tons per hectare) and mineral fertilizers (1 ton NPK per hectare) were applied, with subsequent sowing of grasses and planting of trees and shrubs (Koptsik et al., 2015a). The tree and shrub species consisted mainly of willows (*Salix* sp.) and birches (*Betula* sp.); the grass mixture was composed of timothy (*Phleum pratense*), tall fescue (*Festuca arundinacea*), rump (*Bromopsis* sp.), and winter ryegrass (*Lolium perenne*) (Figs. 18.3 and 18.4).

The relict surface horizons of soils in the technogenic barrens surrounding the Pechenganikel and Severonikel industrial complexes in the Kola Peninsula are acidic, depleted in organic matter and nutrients, contaminated with heavy metals, and eroded (Table 18.1). Remediation of the surface soils of the technogenic barrens has resulted in medium-term improvement of the selected soil properties, especially in a notable decrease in acidity and enrichment in key plant nutrients such as phosphorus and nitrogen. However, effectiveness of the remediation depends on the initial soil conditions and the actual treatment method applied.

Most of the treated soils remained strongly acid after the chemo-phytostabilization treatments, with measured concentrations of plant available calcium, magnesium, potassium, and phosphorus being much lower than background levels and plant nutrient requirements. Plant available manganese and zinc in the remediated soils were also still depleted. Most of the treated soils, however, contained equivalent (or higher) concentrations of available nickel and copper than their untreated analogues.

The application of a natural soil material or manufactured organo-mineral coating layer together with liming and fertilization to cover the polluted soils promotes a drastic and lasting decrease in soil acidity and an improvement in plant nutrient supply. Compared with the soils of technogenic barrens, remediation decreased actual soil acidity by almost 2 pH units, composed of an 80% reduction in exchangeable acidity and a 50% drop in total acidity. The land remediation approaches led to the enrichment of soils with organic carbon and key plant nutrients such as nitrogen, phosphorus, potassium, calcium, magnesium, manganese, and zinc; the impact on the behavior of contaminant metals was more ambiguous. Under current industrial emissions conditions, the manufactured surface layers in half of the soils showed ongoing accumulation of nickel and copper, despite the observed reduction of aerosolic emissions. The sole decrease in levels of plant available or mobile nickel was observed in the chemozems which have organic-rich surface horizons. An accumulation of copper occurred in the surface layer of all treated soils, with the most visibly fertile soil having a thick carbon-rich anthropogenic cover layer composed of wastewater sludge, sawdust, and sand.

The remediation of technogenic barren lands has encouraged an increase in biological activity of the soils (Koptsik et al., 2015b), best documented as a greater measured total soil respiration rate on the fertile technogenic soil covers than in the contaminated soils after chemo-phytostabilization. Total soil respiration rate is effectively the sum of both microbial and root respiration rates.

Although plant survival in the willow and birch plantations on the remediated plots was satisfactory, there was considerable necrosis of basal shoots on many of the young trees, with active coppice shoots developing in most trees. The development of willow and birch trees in plantations near the industrial complexes and on the crests of slopes exposed to the strong winds of the region, especially in the chemo-phytostabilization sites, was extremely low (Fig. 18.5). The leaves of the willow and birch in these sites had visible ubiquitous red spots and single black spots (necrosis).

FIGURE 18.3

(A) General view of the Pechenganikel smelter in the Kola Peninsula. (B) Remediation plot in the vicinity of the smelter; in the background is Norwegian territory. (C) Chemo-phytostabilization and (D) fertile soil cover layer applied over contaminated soils in the vicinity of the smelter in 2010. (E) Willow plantings 6 years later after chemo-phytostabilization and (F) vegetation development in the fertile cover layer over the contaminated soils.

Table 18.1 Acidity, Concentrations of Total Carbon, Nitrogen, and Sulfur, and Available Elements in Topsoil in the Industrial Barren and Remediation Plots in Nikel and Monchegorsk Areas in the Kola Peninsula ($n = 4$ Composite Samples)

Para-meters[a]	pH_{H2O}	pH_{KCl}	EA[b]	C	N	S	P	Ca	Mg	K	Fe	Mn	Zn	Ni	Cu
			cmol kg^{-1}	%			mg kg^{-1}								
Industrial Barren (Nikel Area)															
M	3.67	3.35	0.94	3.1	0.12	0.081	5.0	5.4	8.1	6.1	250	0.05	0.47	21	90
Conf	0.01	0.02	0.03	0.2	0.006	0.006	0.24	0.45	0.36	0.28	14	0.01	0.03	0.66	5.3
Chemo-phytostabilization (Nikel Area)															
M	4.36	3.85	0.82	3.1	0.13	0.089	21	170	7.80	8.8	198	0.59	1.10	75	158
Conf	0.04	0.01	0.03	0.2	0.009	0.003	2.2	20	0.59	0.44	12	0.12	0.06	4.1	3.3
P_{diff}	<0.01	<0.001	–	–	–	–	<0.2	<0.05	–	<0.2	–	<0.2	<0.01	<0.01	<0.01
Coating of Polluted Soils With Fertile Former Arable Layer (Nikel Area)															
M	6.10	5.75	0.043	4.22	0.18	0.056	98	1950	174	55	47	3.1	1.07	36	16
Conf	0.03	0.10	0.006	0.10	0.004	0.001	9.1	51	3.5	1.6	2.4	0.24	0.10	4.2	1.5
P_{diff}	<0.001	<0.01	<0.01	<0.05	<0.05	–	<0.05	<0.001	<0.001	<0.001	<0.01	<0.05	<0.2	–	<0.01
Industrial Barren (Monchegorsk Area)															
M	4.46	3.98	11.6	21.2	0.71	0.21	7.0	127	49	70	320	4.8	3.1	220	805
Conf	0.07	0.03	0.89	1.6	0.06	0.02	1.0	59	13	18	33	1.1	0.6	35	58
Chemo-phytostabilization (Monchegorsk Area)															
M	4.77	4.27	7.22	13.4	0.54	0.18	8.3	54.2	28	43	480	3.6	1.5	103	560
Conf	0.03	0.05	0.59	0.9	0.04	0.01	0.2	7.1	3.5	15	81	0.81	1.0	8.3	59
P_{diff}	<0.05	<0.05	<0.05	<0.05	<0.02	–	–	–	–	–	–	–	–	<0.02	<0.02
Coating of Polluted Soils With Constructed Organo-mineral Layer (Monchegorsk Area)															
M	5.90	5.67	0.32	6.5	0.33	0.11	8.8	950	170	58	150	8.7	6.4	102	170
Conf	0.15	0.14	0.02	0.5	0.02	0.01	0.6	115	25	3.8	36	3.2	1.8	8.0	50
P_{diff}	<0.01	<0.01	<0.001	<0.01	<0.01	<0.01	–	<0.05	<0.05	–	<0.05	–	–	<0.02	<0.001

Conf, *95% confidence interval for the mean;* P_{diff}, *significance level of difference between properties in remediation plot and adjoining barren; –, insignificant difference. Available elements were extracted with 1 M* CH_3COONH_4 *(pH 4.65).*

[a] M = *mean value.*

[b] *EA = exchangeable acidity.*

FIGURE 18.4

(A) Industrial barren landscapes surrounding the Severonikel smelter in the Kola Peninsula. (B) Natural regeneration of vegetation in moist depressional areas of the landscape. (C) Willow and (D) birch plantings in the plots covered with the manufactured organo-mineral layer. (E, F) Green landscapes in the vicinity of the Severonikel industrial complex (2015).

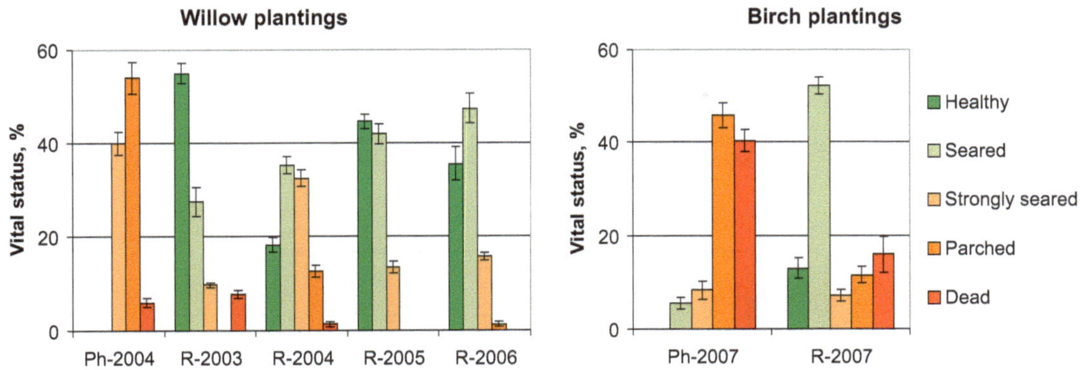

FIGURE 18.5

Vital health status of willow and birch plantings (mean ± standard deviation).

The maintenance of optimal soil conditions for plant nutrition is a key factor for productivity; the nutrient status of plants is often assessed by chemical analysis of storage organs or by leaf sampling for diagnosis. The assessment approach requires a comparison of measured nutrient concentrations in the plants with optimal levels obtained by long-term research experiments (Kopinga and van der Burg, 1995). The accuracy of an assessment of nutrient status through the chemical analysis of leaf tissue may not be reliable in areas experiencing active atmospheric deposition containing the nutrient elements of interest, because fine dust may adhere to the leaves being sampled for nutrient diagnosis.

Elevated concentrations of nickel and copper in leaves sampled from the new willow plantations (Table 18.2) may reflect either high soil metal levels or airfall contamination by industrial emissions. However, in the Nikel area, measured concentrations of calcium, magnesium, and potassium in willow leaves in all of the reclaimed plantations were much lower than background levels from pristine sites, whereas the concentrations of phosphorus were close to background levels. In the Monchegorsk area, in all of the reclaimed plantations, the concentrations of calcium in willow leaves exceeded background levels, the concentrations of magnesium and potassium were similar to background levels, and the concentrations of phosphorus were much lower than the background level. The willows in the remediated areas all exhibited a deficiency of both manganese and zinc, perhaps triggered by antagonism with iron and copper. The vegetation establishment in the manufactured soil cover over the polluted soils was only partial and possibly not self-sustainable.

The sharp increase in metal uptake by the plant roots in the first and/or second years after liming may be induced by the release of metals from the soil exchange complex into the soil solution through ion-exchange reactions with the added soluble Ca and Mg cations. This observation implies that willow planting should be delayed for 2 to 3 years after liming the metal-contaminated soils to avoid excess uptake of metals displaced into the soil solution by the added calcium and magnesium.

The success of chemo-phytostabilization as a land remediation treatment may require repeated liming of the reclaimed soils, with the initial liming of the soils of the technogenic barrens with dolomite ($CaMgCO_3$) at rates of up to 10 tons per hectare being necessary to decrease soil acidity, immobilize accumulated metals, and increase nutrient availability. Furthermore, the dolomite applied should have a broad range of particle sizes from coarse silt to medium and coarse sand, to facilitate a

Table 18.2 Elemental Content of Willow Leaves in Barents Region and in Willow Plantings in Nikel and Monchegorsk Areas in Kola Peninsula (n = 4) (mg kg^{-1})

Location	Parameters	S	P	Ca	Mg	K	Fe	Mn	Zn	Ni	Cu
Barents region (Reimann et al., 2001)	Median	2930	3600	11,000	3010	16,500	79	310	125	9.0	7.7
	min	1580	1080	5290	–	8250	44	40	15	1.9	4.5
	max	4680	7690	23,100	–	23,900	246	1440	262	207	64
Chemo-phytostabilization (Nikel area)	*M*	–	3460	6400	1320	6300	490	160	50	180	77
	std		560	1000	41	1500	100	17	9.0	22	6.7
	Conf		890	1700	66	2400	160	28	14	35	11
Coating of polluted soils with fertile humus layer (Nikel area)	*M*	–	4230	6800	1240	11,500	330	120	136	110	59
	std		330	930	16	810	33	74	17	13	7.1
	Conf		530	1500	25	1300	52	120	27	21	11
Chemo-phytostabilization (Monchegorsk area)	*M*	4340	1080	14,000	2020	7700	420	220	27	640	270
	std	470	110	3400	370	1100	180	80	8.7	110	42
	Conf	740	180	5400	590	1800	280	130	14	180	67
Coating of polluted soils with constructed organo-mineral layer (Monchegorsk area)	*M*	5410	920	14,400	3500	14,600	370	120	72	360	170
	std	760	130	2600	420	1800	110	41	2.8	70	47
	Conf	1200	210	4200	670	2900	170	65	4.5	110	75

slower dissolution rate to supply calcium and magnesium to the soil–plant system to minimize the potential for loss of the soluble components to groundwater in the low clay content soils of the region, as well as discourage loss by overland flow or by wind erosion in sandy soils.

18.4 COMMON SOLUTIONS

During the lifetime of the nickel- and copper-producing industries in both Sudbury and the Kola Peninsula regions, where metal-rich sulfidic ores were mined for more than half a century, large amounts of mining and smelting waste materials were emitted into the surrounding environment. Enormous tonnages of sulfur dioxide and contaminant metals were released into the atmosphere and discharged into forest soils and surface waters, causing a dramatic impact on the environment and on human health. However, a detailed ecological risk assessment revealed that the soils and terrestrial ecosystems in the Sudbury region have recovered to a remarkable degree in the past 40 years after a series of emission-reduction measures allowed the successful introduction of a regional regreening program (Wren, 2012). The regreening of the land systems within the Sudbury watersheds, coupled with chemical pretreatment of fugitive waters from mines and smelting complexes within water treatment plants, has led to amelioration of streams, rivers, and lakes that were previously exposed to acidic, metal-laden waters. The watersheds now exhibit clear signs of recovery.

The initial state of the 81,000 hectares of the affected Sudbury environment when the Sudbury Protocol was being developed bear amazing similarities to those currently observed in the Kola region. There is abundant evidence of similar vegetation and landscape damage from anthropogenic influences in Kola, with the erosion appearing less intense in Kola than in Sudbury, a reflection of less impact from forestry operations and wildfires. Thus there is a greater regional cover of the relict coarse-textured Podzolic soils remaining, with obvious metal toxicity, depressed pH, and fertility issues to be addressed with additional detailed field and laboratory research studies. The current plots in the Kola region will provide an excellent series on which to implement some secondary matched-pair fertility and, eventually, local species survival experiments.

As shown by experiences in Sudbury (Lautenbach et al., 1995; Winterhalder, 1996; Nkongolo et al., 2013), Kola (Tsvetkov and Cherkizov, 1987; Evdokimova, 1995; Koptsik, 2014c), and worldwide (Bolan and Duraisamy, 2003; Adriano et al., 2004), the application of liming materials can lower the mobility, biological availability, and toxicity of metals in contaminated soils. The impact of liming material application is not universal; the efficiency of liming is determined by the extent of the soil contamination, the initial acidity and buffer capacity of the soils, and the chemistry and mineralogical natures of the contaminant metals, as well as the resilience of the plants. The application of liming materials thus should not be considered a reliable method of soil remediation, especially under conditions of contamination with multiple metals, because successful immobilization may require the use of a combination of both mineral and organic soil ameliorants.

A common aspect of degraded soils is the lack of either surface or mineral-complexed organic matter compared with adjacent undisturbed areas. Organic amendments such as livestock manure, biosolids, pulp and paper mill by-products, wood residuals, and crop residues are produced in abundance in both Canada and Russia and could be widely used in land system reclamation (Larney and Angers, 2012). Moreover, the use of organic amendments for reclamation is mutually beneficial because the waste products from agriculture, forestry, and urban areas may assist other sectors in meeting their land reclamation goals. Nevertheless, the application in soil remediation of organic residual materials traditionally used as sources of nutrients, as means to increase the cation exchange capacity, to intensify microbiological activity, and to improve the physical properties and fertility of soils in temperate and agricultural regions requires detailed study in sub-Arctic environments such as exist in the Kola region of Russia. These organic residual materials, if low in organic compounds and metallic contaminant content, can be applied to decrease the biological availability of metals and metalloids in contaminated soils and sediments by enhancing adsorption, complexation, restoration, and evaporation processes (Bolan and Duraisamy, 2003; Park et al., 2011). Humic substances bind metal ions into stable complexes to reduce their bioavailability in soils and natural waters. Solid sorbents based on humic acids have been produced and are already used (Perminova, 2008). Humates are excellent adsorbents which contain elements essential for plant nutrition and growth, are involved in the structure and formation of aggregates, thereby improving the humus content, water, and air regime of soils. However, the possibility of metal retention and mobility within the soil depends on a complex interacting series of factors including the chemical and physical properties of the element such as ionic radius and strength, and properties of the soil matrix such as pH, oxidation–reduction conditions, the amount and composition of organic matter, and particle size distribution, which are not completely understood.

The use of metal-tolerant native plant species adapted to the local bioclimatic conditions (eg, *Deschampsia caespitosa*, *Poa pratensis*, *Agrostis gigantea*, *Calamagrostis epigeios*, C. *arundinacea*, *Phleum pratense*), along with site liming and fertilization, is important. Canadian reclamation practice and experience (Lautenbach et al., 1995; Winterhalder, 1996) demonstrates the need to add legumes into herbage mixtures, with optimal pH conditions between 5.5 and 7.5.

The use of a combination of trees and grasses for reclamation promotes effective phytostabilization. Grasses with branching root systems promote surface soil aggregation and encourage porosity development, which in turn reduces wind erosion and surface runoff. Planting of trees improves the physical stabilization of soils by decreasing surface wind turbulence and intercepts precipitation, thus reducing the potential for rilling and water erosion, especially on slopes. The development of massive root systems promotes soil stabilization, and the release of soluble organic substances in litter fall on the surface of the soil organic horizons (or humus forms) helps in the development organo-metal complexes for forest soil genesis. Worldwide data (Pulford and Watson, 2003) support the Sudbury (Winterhalder, 1996) and Kola (Nikonov et al., 2005; Koptsik, 2014c; Koptsik et al., 2015a; Kozlov et al., 1999) experiences, suggesting that willows (*Salix* spp.) have a greater potential for phytoremediation because of rapid effective territorial expansion, high productivity, effective evapotranspiration ability, high tolerance to metals, and demand for available nutrients. Birch spp. are known to be colonists in the harsh environmental conditions of deforested habitats in both primary and human-induced secondary successional systems in boreal forests. Seedlings of mountain birch (*Betula pubescens* ssp. *czerepanovii*) can be thus also used successfully for vegetation recovery. In addition to liming material applications, the revegetation of industrial barrens can be enhanced significantly by inexpensive treatments such as wind sheltering and watering (Kozlov and Haukioja, 1999).

Ongoing research initiatives must include studies of the interactions between plants and microorganisms. The long-term monitoring of the successes of remediation measures for technogenic barrens should include regular measurements of soil acidity and nutrient status to guide planned correction measures to prevent the development of nutrient deficiencies or remobilization of contaminant metals.

18.5 LESSONS TO BE LEARNED

Sustainable recovery to create green landscapes from barren lands in mining and processing industrial areas offers tremendous scope for process innovation. The mitigation of environment and human health hazards arising from industrial contamination impacts can be addressed by restoring functions and services of contaminated soils and damaged ecosystems. The use of plants and associated microorganisms could be a sustainable, environmentally friendly, and cost-effective solution to reduce contaminant exposure. Basic remediation options such as chemo-phytostabilization can provide rapid risk management through containment and stabilization of the contaminants while potentially providing a range of additional economic (eg, biomass generation as renewable energy source), social (eg, leisure and recreation, educational value) and environmental (eg, CO_2 sequestration, secondary pollution prevention) benefits (Mench et al., 2009; Hargreaves et al., 2012; Cundy et al., 2013).

Studies in Sudbury (Hargreaves et al., 2012) demonstrated the feasibility of managing wastes from the forest resource industries to manufacture organic soil covers on mining brownfields for the growth of biomass crops for use in the production of biofuels. Corn, canola, switchgrass and willow crops can be grown successfully using agricultural and horticultural techniques on acid-generating tailings that have deep covers (over 50 cm) of organic residuals such as paper mill sludges or municipal composts. These studies are an important preliminary phase for a novel reclamation practice that has environmental and economic benefits including the use of one industrial waste (biosolids) to reclaim the wastes (tailings) of another industry; a contribution by the mining industry to sustainable development and community growth; and a contribution to greenhouse gas reductions through the addition of carbon to otherwise barren or less productive land. An example of successful application of organic residual cover materials for Technosol manufacture and biomass production on acid-generating tailings is shown in Fig. 18.6.

Another adaptive type of restoration option, ecological restoration, or phytostabilization was successfully used for the biological recovery of various mine industry–damaged sites in the Middle Urals, one of the oldest mining regions of Russia. This approach involved establishing ecosystem structure and function, thus promoting biodiversity (Chibrik et al., 2016).

Impressive examples of natural landscape protection on biotopes of industrial heritage are presented by Lemoine (2016) for Nord-Pas de Calais, one of the older industrial regions of Western Europe. The first large-scale example was described for the protection of slag of heaps (coal mining wastes) by developing vegetative cover. The second small but complex local-scale example involves the protection of the most interesting and polluted calaminarian grassland, which allows the development of rich vegetation cover, with growth and survival of a rare species of metallophytes.

Furthermore, the surroundings of copper–nickel smelters in Sudbury and in the Kola Peninsula provide excellent natural or living laboratories for environmental studies and remediation practitioner education. They represent two of the most intensively studied industrial regions in the world. As consequence of the long-term detrimental impact of the nickel- and copper-producing industry in the Sudbury area (Lautenbach et al., 1995; Winterhalder, 1996; Nkongolo et al., 2013; Hargreaves et al., 2012) and in the Kola Peninsula (Kryuchkov, 1985; Tsvetkov and Cherkizov, 1987; Kapel'kina, 1993; Evdokimova, 1995), the two regions have been the focus of environment studies since the 1960s and 1970s. In the late 1980s, increased international contact with the Scandinavian countries, especially with bordering Norway and Finland, stimulated research initiation in the field of environmental

FIGURE 18.6

Reclamation activities at a tailings management area in the Sudbury Basin. (A) Tailings impoundment site. (B) Preparation of organic residual cover materials from wood product industry residuals. (C) Application of manufactured biosolids. (D) Biomass production of biosolids. (E) Organic Technosols created from biosolids on tailings. (F) Fast-growing willows planted for possible use as biofuel crop.

protection in the Kola Subarctic (Chernenkova and Kuperman, 1999; Koptsik et al., 2003; Aamlid, 1992; Aamlid and Venn, 1993; Koptsik et al., 2001; Lukina and Nikonov, 1998; Alekseev, 1990), which continued through a series of national and international research projects (Koptsik, 2014c; Koptsik et al., 2015a; Reimann et al., 1998; Salminen et al., 2004; Kashulina et al., 1997, 2014; Koptsik et al., 2015b; Derome et al., 2008; Chernen'kova et al., 2014; Reimann et al., 2001).

As with Laurentian University based field and academic programs in the Sudbury region (Juckers et al., 2013), the Kola Peninsula is also an important field region for educational activities. For over 20 years, students from Moscow State University and other universities have completed summer internships, and conducted field studies to fulfill course work requirements and complete both master's and doctoral degree research. These students are engaged in ongoing research programs involving monitoring and remediation of the extensively contaminated sites. Involvement of both undergraduate and postgraduate students in the monitoring and assessment process is an important learning opportunity which allows students to obtain first-hand experience outside the classroom, and also helps to bring new creative ideas into the long-term remediation programs. The introduction of new summer training courses will raise environmental education to a new level.

The ongoing major challenge is to refine, develop, and scale up remediation technologies to ameliorate both the environmental and human health successfully in a manner that will lead to the recovery of the natural ecological resources in northern mining and processing industrial regions. Three stages were previously indicated in the Sudbury Protocol; the challenge is now to transfer the approach to assist in the adaptation and development of the Kola Method of community-led landscape change initiatives. Observations in the Nikel–Zapolyarnyy and Monchegorsk areas (Koptsik, 2014c; Koptsik et al., 2015a) indicate a series of well-developed large-scale field experiments exhibiting various stages of successful vegetation establishment. The goal will be to refine these current experiments to solve ongoing and complex issues of fertility imbalance, organic matter shortage, and suitable indigenous species selection to produce to crucial sustainable experimental results for transfer to a community-led regional regreening program. In addition to environmental issues, the experimental demonstration sites can make a significant contribution to community and stakeholder engagement by providing visual evidence of the effectiveness of selected remediation approaches under varying site conditions.

To date, 105 hectares of barren lands have been treated, with approximately 150,000 trees and shrubs planted in the Monchegorsk area at a total cost of 75 million rubles (http://kolagmk.ru). The successful results of future modified field experiments in the Kola region will lead to the development of a field-applicable technology suitable for the ongoing development of a large-scale ecosystem recovery and biodiversity enhancement program. Effective and sustained ecosystem recovery will require the implementation of an operational long-term monitoring and management program.

Successful recovery from accumulated environmental damage to allow the sustainable use of natural resources on the Kola Peninsula, the largest industrial region of the Russian Arctic, obviously requires the joint efforts of society, government at the federal, regional, and municipal levels, and industries. For remediation success under the continued metal production program in the harsh climate, important science-based technology selection decisions are required, with the probable modification of current technologies being implemented to provide adaption to the conditions of Kola. The development and implementation of intensive erosion control and remediation processes will require legislative encouragement for implementation in the industrial territories. The combination of activities will assist in reducing secondary pollution, improving the environment, expanding recreational areas, and enhancing the attractiveness of places of residence and providing employment for the local population. Ultimately, the proposed measures will increase investment attractiveness and lead to a qualitative growth in the gross domestic product of the region. Sustainable ecosystem recovery and landscape restoration of the fragile region from industrial impacts are seen as crucial for both economic and community sustainability in the Kola region, with a possible extension to the Norilsk industrial area of the Siberian sub-Arctic.

ACKNOWLEDGMENTS

This work was supported by the Russian Science Foundation (Project No. 15-14-20021).

REFERENCES

Aamlid, D., 1992. Forest Pathological Research in Northern Forests with Special Reference to Abiotic Stress Factors. Finnish Forest Research Institute. Research Papers 451.

Aamlid, D., Venn, K., 1993. Methods of monitoring the effects of air pollution on forest and vegetation of eastern Finnmark, Norway. Norwegian Journal of Agricultural Sciences 7, 71.

Adriano, D.C., Wenzel, W.W., Vangronsveld, J., Bolan, N.S., 2004. Role of assisted natural remediation in environmental cleanup. Geoderma 122, 121–142.

Annual report 2014 Delivering on Strategy – Unlocking the Potential of Norilsk, 2015. OJSC MMC Norilsk Nickel, p. 104.

Arctic Pollution, 2006. Acidification and Arctic Haze. AMAP, Arctic Monitoring and Assessment Programme, Oslo, pp. 15–20.

Alekseev, V.A. (Ed.), 1990. Lesnye Ekosistemy I Atmosfernoe Zagryaznenie (Forest Ecosystems and Atmospheric Pollution). Nauka, Leningrad (In Russian).

Ayres, R.U., Ayres, L.W., Råde, I., 2002. The Life Cycle of Copper, Its Co-products and By-products, Report Commissioned by the MMSD (Mining, Minerals and Sustainable Development) Project of IIED (International Institute for Environment and Development). World Business Council for Sustainable Development (WBCSD).

Bolan, N.S., Duraisamy, V.P., 2003. Role of inorganic and organic soil amendments on immobilization and phytoavailability of heavy metals: a review involving specific case studies. Australian Journal of Soil Research 41, 533–555.

Bolan, N.S., Park, J.H., Robinson, B., Naidu, R., Huh, K.Y., 2011. Phytostabilization: a green approach to contaminant containment. Advances in Agronomy 112, 145–204 (Chapter 4).

Choi, Y.D., Temperton, V.M., Allen, E.B., Grootjans, A.P., Halassy, M., Hobbs, R.J., Naeth, M.A., Torok, K., 2008. Ecological restoration for future sustainability in a changing environment. Ecoscience 15, 53–64.

Cunningham, S.D., Ow, D.W., 1996. Promises and prospects of phytoremediation. Plant Physiology 110, 715–719.

Chernenkova, T.W., Kuperman, P.G., 1999. Changes in the spruce forest communities along a heavy metal deposition gradient on Kola Peninsula. Water, Air, & Soil Pollution 111, 187–200.

Chernen'kova, T.V., Bochkarev, Y.N., Friedrich, M., Boettger, T., 2014. The impact of natural and anthropogenic factors on radial tree growth on the northern Kola Peninsula. Contemporary Problems of Ecology 7 (7), 759–769.

Cundy, A.B., Bardos, R.P., Church, A., Puschenreiter, M., Friesl-Hanl, W., Müller, I., Neu, S., Mench, M., Witters, N., Vangronsveld, J., 2013. Developing principles of sustainability and stakeholder engagement for "gentle" remediation approaches: the European context. The Journal of Environmental Management 129, 283–291.

Chibrik, T.S., Lukina, N.V., Filimonova, E.I., Glazyrina, M.A., Rakov, E.A., Maleva, M.G., Prasad, M.N.V., 2016. Biological recultivation of mine industry deserts: facilitating the formation of phytocoenosis in the Middle Ural Region, Russia. In: Prasad, M.N.V. (Ed.), Bioremediation and Bioeconomy. Elsevier, Amsterdam, pp. 389–418 (Chapter 16).

Derome, J., Myking, T., Aarrestad, P.A. (Eds.), 2008. Current State of Terrestrial Ecosystems in the Joint Norwegian, Russian and Finnish Border Area in Northern Fennoscandia, vol. 85. Working Papers of the Finnish Forest Research Institute, 98 p.

Evdokimova, G.A., 1995. Ecological and Microbiological Principles of Soil Protection in the Extreme North. Kola Science Centre, Russian Academy of Science, Apatity (In Russian).

Evdokimova, G.A., Mozgova, N.P., Korneikova, M.V., 2014. The content and toxicity of heavy metals in soils affected by aerial emissions from the Pechenganikel plant. Eurasian Journal of Soil Science 47 (5), 504–510.

European Environment Agency, 2014. Costs of Air Pollution from European Industrial Facilities 2008–2012 – An Updated Assessment. EEA Technical Report No 20/2014. Publications Office of the European Union, Luxembourg, 74 p.

Guo, G., Zhou, Q., Ma, L.Q., 2006. Availability and assessment of fixing additives for the in situ remediation of heavy metal contaminated soils: a review. Environmental Monitoring and Assessment 116, 513–528.

Hudson-Edwards, K.A., Jamieson, H.E., Lottermoser, B.G., 2011. Mine wastes: past, present, future. Elements 7 (6), 375–380.

Hargreaves, J., Lock, A., Beckett, P.J., Spiers, G.A., Tisch, B., Lanteigne, L., Posadowski, T., Soenens, M., 2012. Suitability of an organic residual cover on tailings for bioenergy crop production: a preliminary assessment. Canadian Journal of Soil Science 92 (1), 203–211.

Juckers, M., Watkinson, A., D'Andrea, L., Beckett, P., Spiers, G., 2013. Sudbury, Ontario: an outdoor laboratory for experiential learning and land reclamation. Canadian Reclamation 13 (1), 43–45.

Kozlov, M.V., Zvereva, E.L., 2007. Industrial barrens: extreme habitats created by non-ferrous metallurgy. Reviews in Environmental Science and Biotechnology 6, 231–259.

Koptsik, G.N., 2014a. Modern approaches to remediation of heavy metal polluted soils: a review. Eurasian Journal of Soil Science 47 (7), 707–722.

Koptsik, G.N., 2014b. Problems and prospects concerning the phytoremediation of heavy metal polluted soils: a review. Eurasian Journal of Soil Science 47 (9), 923–939.

Koptsik, G.N., 2014c. Efficiency of remediation of technogenic barrens around the Pechenganikel works in the Kola Subarctic. Eurasian Journal of Soil Science 47 (5), 519–528.

Kryuchkov, V.V., 1985. Reclamation of disturbed lands in the North. Priroda 7, 68–77 (In Russian).

Kapel'kina, L.P., 1993. Ecological Aspects of the Optimization of Technogenic Landscapes. Nauka, St. Petersburg (In Russian).

Koptsik, G.N., Smirnova, I.E., Koptsik, S.V., Zakharenko, A.I., Turbaevskaya, V.V., 2015a. Efficiency of remediation of technogenic barrens around the Severonikel factory on the Kola Peninsula. Moscow University Soil Science Bulletin 70 (2), 78–84.

Koptsik, G.N., Kadulin, M.S., Zakharova, A.I., 2015b. The effect of technogenic contamination on carbon dioxide emission by soils in the Kola Subarctic. Biology Bulletin Reviews 5 (5), 480–492.

Kashulina, G., de Caritat, P., Reimann, C., 2014. Snow and rain chemistry around the "Severonikel" industrial complex, NW Russia: current status and retrospective analysis. Atmospheric Environment 89, 672–682.

Kola MMC Year-Book, 2008. No. 6, pp. 25–26 (In Russian).

Kashulina, G., Reimann, C., Finne, T.E., Halleraker, J.H., Äyräs, M., Chekushin, V.A., 1997. The state of the ecosystems in the central Barents region: scale, factors and mechanism of disturbance. Science of the Total Environment 206, 203–225.

Koptsik, S., Koptsik, G., Livantsova, S., Eruslankina, L., Zhmelkova, T., Vologdina, Z., 2003. Heavy metals in soils near the nickel smelter: chemistry, spatial variation, and impacts on plant diversity. Journal of Environmental Monitoring 5, 441–450.

Koptsik, G., Koptsik, S., Aamlid, D., 2001. Pine needle chemistry near a large point SO_2 source in Northernmost Europe. Water, Air, & Soil Pollution 130, 929–934.

Kopinga, J., van der Burg, J., 1995. Using soil and foliar analysis to diagnose the nutritional status of urban trees. Journal of Arboriculture 21 (1), 17–24.

Kozlov, M.V., Zvereva, E.L., Niemelä, P., 1999. Effects of soil quality and air pollution on the rooting and survival of *Salix borealis* cuttings. Boreal Environment Research 4, 67–76.

Kozlov, M.V., Haukioja, E., 1999. Performance of birch seedlings replanted in heavily polluted industrial barrens of the Kola Peninsula, NW Russia. Restoration Ecology 7, 145–154.

Landner, L., Reuther, R., 2004. Heavy Metals in Society and in the Environment: A Critical Review of Current Knowledge on Fluxes, Speciation, Bioavailability and Risk for Adverse: Effects of Copper, Chromium, Nickel and Zinc. Kluwer Academic Publishers, Dordrecht.

Lautenbach, W.E., Miller, J., Beckett, P.J., Negusanti, J.J., Winterhalder, K., 1995. Municipal land restoration program: the regreening process. In: Gunn, J.M. (Ed.), Restoration and Recovery of an Industrial Region: Progress in Restoring the Smelter-Damaged Landscape Near Sudbury, Canada. Springer-Verlag, New York, pp. 109–122.

Lottermoser, B.G., Glass, H.J., Page, C.N., 2011. Sustainable natural remediation of abandoned tailings by metal-excluding heather (*Calluna vulgaris*) and gorse (*Ulex europaeus*), Carnon Valley, Cornwall, UK. Ecological Engineering 37, 1249–1253.

Lukina, N.V., Nikonov, V.V., 1998. Nutrient Status of North Taiga Forests (Natural Regularities and Pollution-Induced Changes). Kola Science Centre, Russian Academy of Science, Apatity (In Russian).

Lyanguzova, I.V., 2010. Tolerance of the Components in Forest Ecosystems of the Russian North to the Environmental Contamination. Abstract of diss. for the degree of doctor of biol. sciences. St. Petersburg. 39 p.

Larney, F.J., Angers, D.A., 2012. The role of organic amendments in soil reclamation: a review. Canadian Journal of Soil Science 92, 19–38.

Lemoine, G., 2016. Brownfield restoration as a smart economic growth option for promoting ecotourism, biodiversity, and leisure: two case studies in Nord-Pas de Calais. In: Prasad, M.N.V. (Ed.), Bioremediation and Bioeconomy. Elsevier, Amsterdam, pp. 361–388 (Chapter 15).

Mudd, G.M., 2010. Global trends and environmental issues in nickel mining: sulfides versus laterites. Ore Geology Reviews 38 (1–2), 9–26.

Mench, M., Schwitzguébel, J.-P., Schroeder, P., Bert, V., Gawronski, S., Gupta, S., 2009. Assessment of successful experiments and limitations of phytotechnologies: contaminant uptake, detoxification and sequestration, and consequences for food safety. Environmental Science and Pollution Research 16, 876–900.

Norgate, T.E., Jahanshahi, S., Rankin, W.J., 2007. Assessing the environmental impact of metal production processes. The Journal of Cleaner Production 15 (8–9), 838–848.

Nowack, E., Schulin, R., Robinson, B.H., 2006. Critical assessment of chelant-enhanced metal phytoextraction. Environmental Science and Technology 40 (17), 5225–5232.

Nikonov, V.V., Lukina, N.V., Isaeva, L.G., Gorbacheva, T.T., Belova, E.A., 2005. Restoration of the territory, damaged by air pollution from copper-nickel processing industry in the Kola Peninsula. In: Innovation Potential of the Kola Science, vol. 2. Kola Science Centre, Russian Academy of Science, Apatity, pp. 284–288 (In Russian).

Nkongolo, K.K., Spiers, G., Beckett, P., Narendrula, R., Theriault, G., Tran, A., Kalubi, K.N., 2013. Long-term effects of liming on soil chemistry in stable and eroded upland areas in a mining region. Water, Air, & Soil Pollution 224, 1618–1632.

Prasad, M.N.V., Freitas, H.M.O., 2003. Metal hyperaccumulation in plants – biodiversity prospecting for phytoremediation technology. Electronic Journal of Biotechnology 6 (3), 285–321.

Prasad, M.N.V., 2011. A State-of-the-Art Report on Bioremediation, Its Applications to Contaminated Sites in India. Ministry of Environment and Forests. Government of India, New Delhi, 90 p.

Park, J.H., Lamb, D., Paneerselvam, P., Choppala, G., Bolan, N., Chungd, J.-W., 2011. Role of organic amendments on enhanced bioremediation of heavy metal(loid) contaminated soils. Journal of Hazardous Materials 185, 549–574.

Perminova, I.V., 2008. Humic substances – a challenge to chemists of the 21st century. Khimiya i Zhizn' 1, 50–55 (In Russian).

Pulford, I.D., Watson, C., 2003. Phytoremediation of heavy metal-contaminated land by trees – a review. Environment International 29, 529–540.

Robinson, B.H., Fernández, J.-E., Madejón, P., Marañón, T., Murillo, J.M., Green, S., Clothier, B., 2003. Phytoextraction: an assessment of biogeochemical and economic viability. Plant and Soil 249, 117–125.

Robinson, B.H., Brooks, R.R., Howes, A.W., Kirkman, J.H., Gregg, P.E.H., 1997. The potential of the high-biomass nickel hyperaccumulator *Berkheya coddii* for phytoremediation and phytomining. Journal of Geochemical Exploration 60, 115–126.

Reimann, C., Äyräs, M., Chekushin, V.A., Bogatyrev, I., Boyd, R., de Caritat, P., Dutter, R., Finne, T.E., Halleraker, J.H., Jæger, Ø., Kashulina, G., Niskavaara, H., Lehto, O., Pavlov, V., Räisänen, M.L., Strand, T., Volden, T., 1998. Environmental Geochemical Atlas of the Central Barents Region. NGU-GTK-CKE Special Publication, Trondheim, Norway.

Reimann, C., Koller, F., Frengstad, B., Kashulina, G., Niskavaara, H., Englmaier, P., 2001. Comparison of the element composition in several plant species and their substrate from a 1,500,000 km^2-area in Northern Europe. Science of the Total Environment 278, 87–112.

Salminen, R., Chekushin, V., Tenhola, M., Bogatyrev, I., Glavatskikh, S.P., Fedotova, E., Gregorauskiene, V., Kashulina, G., Niskavaara, H., Polischuk, A., Rissanen, K., Selenok, L., Tomilina, O., Zhdanova, L., 2004. Geochemical atlas of Eastern Barents region. Journal of Geochemical Exploration 83 (1–3, special issue), 1–530.

Tsvetkov, V.F., Cherkizov, E.A., 1987. An experience in forest reclamation of land in the impact zone of industrial emissions in the Kola Peninsula. In: Impact of Industrial Enterprises on the Environment. Nauka, Moscow, pp. 112–119 (In Russian).

Vangronsveld, J., Herzig, R., Weyens, N., Boulet, J., Adriaensen, K., Ruttens, A., Thewys, T., Vassilev, A., Meers, E., Nehnevajova, E., van der Lelie, D., Mench, M., 2009. Phytoremediation of contaminated soils and groundwater: lessons from the field. Environmental Science and Pollution Research 16 (7), 765–794.

Winterhalder, K., 1996. Environmental degradation and rehabilitation of the landscape around Sudbury, a major mining and smelting area. Environmental Reviews 4 (3), 185–224.

Wren, C.D., 2012. Risk Assessment and Environmental Management: A Case Study in Sudbury, Ontario, Canada. Maralte B.V., Leiden, The Netherlands.

CHAPTER

19 METHODS FOR UTILIZATION OF RED MUD AND ITS MANAGEMENT

M.K. Sahu, R.K. Patel
National Institute of Technology, Rourkela, Odisha, India

CHAPTER OUTLINE

Environmental Materials and Waste. http://dx.doi.org/10.1016/B978-0-12-803837-6.00019-6

19.1 INTRODUCTION

Red mud (RM) is a solid caustic waste or a by-product formed after the digestion of bauxite ore with sodium hydroxide in a solution of concentrated NaOH at 150–230°C under pressure in Bayer's process for the extraction of alumina (Kurniawan et al., 2006; Zhang et al., 2011). During the digestion process, aluminum reacts with NaOH to form soluble sodium aluminate, leaving RM slurry (Newson et al., 2006). RM slurry is highly alkaline, with a pH > 12 owing to the presence of NaOH and Na_2CO_3 (1–6%, w/w); this is expressed in terms of Na_2O (Genç-Fuhrman et al., 2004; Fois et al., 2007). The composition of the RM is a mixture of solid and metallic oxide–bearing impurities. The red color is the result of the oxidized iron present in the RM; hence it is called red mud. Approximately 60% of the mass of the RM is due to the oxidized iron in the RM (Schmitz, 2006; Chandra, 1996). In addition to iron oxide, the other principal constituents of RM are silica, unbleached residual aluminum, and titanium oxide (Mohan and Pittman, 2007). The main constituents of RM (% w/w) are: Fe_2O_3 (30–60%), Al_2O_3 (10–20%), SiO_2 (3–50%), Na_2O (2–10%), CaO (2–8%), and TiO_2 (trace to 10%). The variation in the constituents of the RM is caused by the composition of bauxite from where it is generated. The amount of RM generated, per ton of alumina extraction, varies greatly depending on the type of bauxite ore used: about 0.3 tons of high-grade bauxite to 2.5 tons of very low-grade bauxite (Newson et al., 2006; Collazo et al., 2006). Around 201 million tons of bauxite ores is mined per year. This many tons of bauxite is digested globally to release a huge amount of RM into the environment (Wagh and Desai, 1987; Piga et al., 1993; Prasad et al., 1985). Both because of its chemical composition and because of the amount generated every year, RM is a serious problem for the environment. It cannot be disposed of easily because of its high alkali nature, and hence the slurry is stored in a large pond or reservoir, but the process of storage is also not free of problems such as seepage into underground water in a slurry form or alkaline air that allows dust to settle onto plant life in dry conditions. Because of the instability of storage, vast areas of land consumed by RM storage can neither be built on nor farmed, even when dry. There are many problems in managing RM slurry; a few important problems are listed below:

1. Costly maintenance of large RM pond areas
2. Risk of caustic pollutants for all living organisms
3. Leakage of alkaline contaminants into groundwater resources
4. Overflow of materials during rainy season causes harmful effects on surrounding environment

Dusting of dry surfaces interferes with nearby rehabilitation and plant life (Enick et al., 2001; Brunori et al., 2005). The quantity of generation of RM varies from 55% to 65%, depending on the quality of the bauxite. Generally RM is dumped near the plant site in RM ponds, but, in some countries such as France, England, Germany, and Japan, RM is discharged into the sea, because the availability of land for dumping is less and the sea is nearby. Furthermore, research on the RM use on a laboratory scale has been increasing worldwide because the technology is changing quickly. A lot of research and developmental activities are ongoing globally to find an effective use for RM, which involves various types of product development.

19.2 RED MUD CHEMISTRY

The Bayer process is an economical process for the extraction of alumina from bauxite ore using a concentrated NaOH solution (caustic soda) at high temperature and pressure. Austrian chemist Karl Bayer was the inventor of the Bayer process, which was patented in 1887. When Bayer was working in Saint Petersburg, Russia, he developed a method for supplying alumina to the textile industry (it was used as a mordant in dyeing cotton). He discovered that the aluminum hydroxide that precipitated from the alkaline solution was crystalline and could easily be isolated. The NaOH selectively dissolves Al_2O_3 from bauxite ore and produces a sodium-aluminate solution from which pure $Al(OH)_3$ precipitated in alkaline conditions, which is then calcined to produce Al_2O_3 used subsequently for the extraction of metallic aluminum.

The Bayer process is the only industrial method used to refine bauxite to produce alumina. Bauxite is the most important ore of aluminum which contains only 30–54% alumina, Al_2O_3; the rest is a mixture of silica, various iron oxides, and titanium dioxide along with trace amounts of zinc, phosphorous, nickel, vanadium etc., as indicated earlier. The alumina must be purified before it is refined to aluminum metal.

After alumina extraction from bauxite, the other components of bauxite, Fe_2O_3, SiO_2, TiO_2, etc., do not dissolve in the basic medium; however some SiO_2 dissolves as silicate and is filtered from the solution as solid impurities (clarification). For various reasons, most alumina producers add lime at some point in the process and the lime forms a number of compounds that end up with the bauxite residue. The solid impurities are called RM. RM is composed of a mixture of solid and metallic oxides. The red color arises from iron oxides, which comprise up to 60% of the mass of the RM. The mud is highly basic. In addition to iron, the other dominant components include silica, unbleached residual alumina, and titanium oxide. The RM causes disposal problems owing to its caustic nature.

19.2.1 RED MUD GENERATION

The world's 10 biggest aluminum companies, ranked by production and compiled by London-based researcher CRU in 2011, are represented in Table 19.1.

The approximate amount of RM generation in India (Chaddha, 2009) is represented in Table 19.2. The worldwide alumina production is around 58 million tons, in which India accounts for 2.7 million tons (Indian Aluminum Industry, 2009). India is the fourth largest deposit of bauxite in the world, whereas more than 50% of bauxite deposits in India are present in the state of Orissa. The important alumina refineries in India are shown in Fig. 19.1.

Table 19.1 World's 10 Largest Aluminum Industries in 2011, Ranked by Production and Compiled by London-Based Researcher CRU

Rank	Company	Output in Thousands of Metric Tons
1.	United Co. Rusal	4127
2.	Rio Tinto Group	3829
3.	Alcoa Inc.	3669
4.	Aluminum Corp. of China	3127
5.	Norsk Hydro ASA	1705
6.	Dubai Aluminum Co.	1386
7.	China Power Investment Co.	1381
8.	BHP Billiton Ltd.	1249
9.	Shandong Xinfa Aluminum and Electricity Group	1016
10.	Aluminum Bahrain BSC	881

Table 19.2 Status of Red Mud Generation (100,000 tons per year) in India in 2010

Company	2008–09	Additional Generations by 2010
NALCO	20.33	6.50
HINDALCO	16.54	4.55
BALCO	2.97	–
MALCO	0.95	–
VEDANTA	3.66	18.20
UTKAL	–	19.50
RAYKAL	–	18.20
ADITYA	–	18.20
JSW	–	18.20
Total	~44.45	~103.35
Grand total = 44.45 + 103.35 = ~147.8		

19.2.2 CONSTITUENT OF RED MUD

RM is a by-product of the Bayer process. The chemical and physical properties of RM depend primarily on the bauxite used and, to a lesser extent, the manner in which it is processed. RM is the main solid waste product of alumina industry. The production of 1 ton of alumina generates up to 2 tons of toxic waste, which translates into more than 100,000 metric ton per year of RM produced for a typical

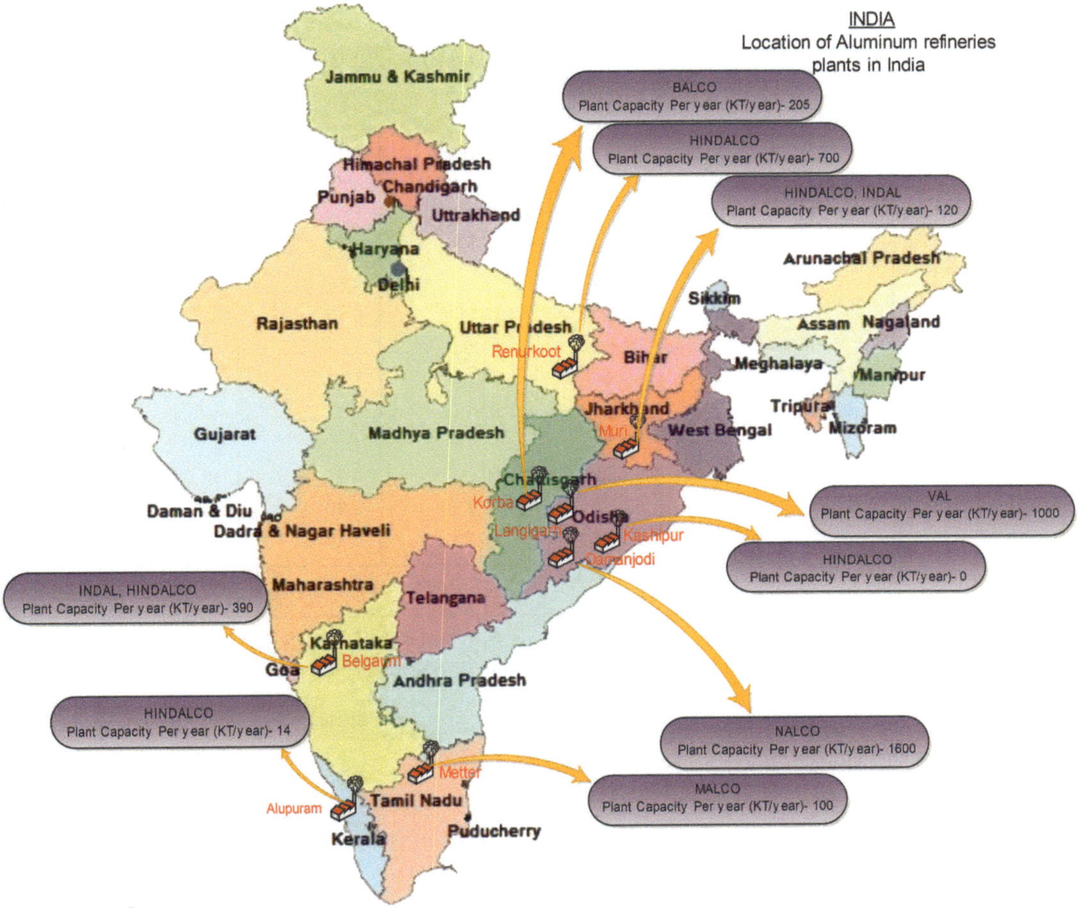

FIGURE 19.1

Important alumina refineries in India.

plant. The worldwide annual production of RM is 120 million tons (Power et al., 2011). Its disposal remains an issue of great importance with environmental implications.

Table 19.3 shows the main compositions of RM generated in plants worldwide (Wang et al., 2008). Iron oxide is the major component, followed by alumina and silica, except for some RMs from India where the titania content may be higher than alumina. The following data give some idea of the worldwide chemical composition of RM found in different bauxites.

Chemical analysis showed that RM contains Fe_2O_3, Al_2O_3, SiO_2, CaO, Na_2O, TiO_2, K_2O, and MgO as well as some minor constituents, namely: Na, K, V, Ni, Ba, Cu, Mn, Cr, Pb, and Zn. The difference in chemical composition between different RMs worldwide is high. The typical chemical composition range (%) of bauxite residue for main components is shown in Table 19.4 (Paramguru et al., 2004).

Table 19.3 Main Composition of Red Mud Generated in Plants Worldwide

Country		Major Chemical Composition (wt%)				
		Fe_2O_3	Al_2O_3	TiO_2	SiO_2	Na_2O
India	MALCO	45.17	27.00	5.12	5.70	3.64
	HINDALCO	35.46	23.00	17.20	5.00	4.85
	BALCO	33.80	15.58	22.50	6.84	5.20
	NALCO	52.39	14.73	3.30	8.44	4.00
Hungary		38.45	15.20	4.60	10.15	8.12
Jamaica		50.9	14.20	6.87	3.40	3.18
Surinam		24.81	19.00	12.15	11.90	9.29
United States	ALCOA Mobile	30.40	16.20	10.11	11.14	2
	Arkansas	55.6	12.15	4.5	4.5	1.5–5.0
	Sherwon	50.54	11.13	Traces	2.56	9.00
FRG Baudart		38.75	20.00	5.5	13.00	8.16
Taiwan		41.30	20.21	2.90	17.93	3.8
Australia		40.50	27.70	3.50	19.90	1–2

Table 19.4 Chemical Composition (wt%) of Red Mud in Different Alumina Industries Worldwide

Constituents	Chemical Composition
Fe_2O_3	30–60%
Al_2O_3	10–20%
SiO_2	3–50%
Na_2O	2–10%
CaO	2–8%
TiO_2	Trace to 10%

In general, RMs are a very fine material (Tian et al., 1998; Gu et al., 2012). The particle size of RM is in the range of 0.1–160 μm. The specific surface area of RM is around 10–31 $m^2\ g^{-1}$, which indicates that RM has a high degree of mineral particle dispersion. RM has a large water content, up to 700–1000 kg m^{-3}, accounting for 79–93% of the total weight. This water will be desorbed when the RM is shocked, which may lead to a decrease in the mechanical properties of RM.

The crystalline phases (Fig. 19.2) of RM obtained by powder X-ray diffraction (XRD) are: hematite (α-Fe_2O_3), goethite (α-FeO[OH]), gibbsite (γ-Al[OH]$_3$), calcite ($CaCO_3$), rutile/anatase (TiO_2), sodalite: zeolite(I) ($1.08Na_2OAl_2O_31.68SiO_21.8H_2O$), quartz ($SiO_2$), sodium aluminum silicate (Na[$AlSiO_4$]), and magnetite (Fe_3O_4).

The major constituents of RM are the oxides of Fe, Al, Si, Ti, Na, and Ca, along with a large number of minor constituents. The alumina, silica, and soda account for about 40% of the total RM; the rest are mostly in the form of iron and titanium oxides. The chemical compositions of Indian RM from different sides are given in Table 19.5 (Chaddha, 2009).

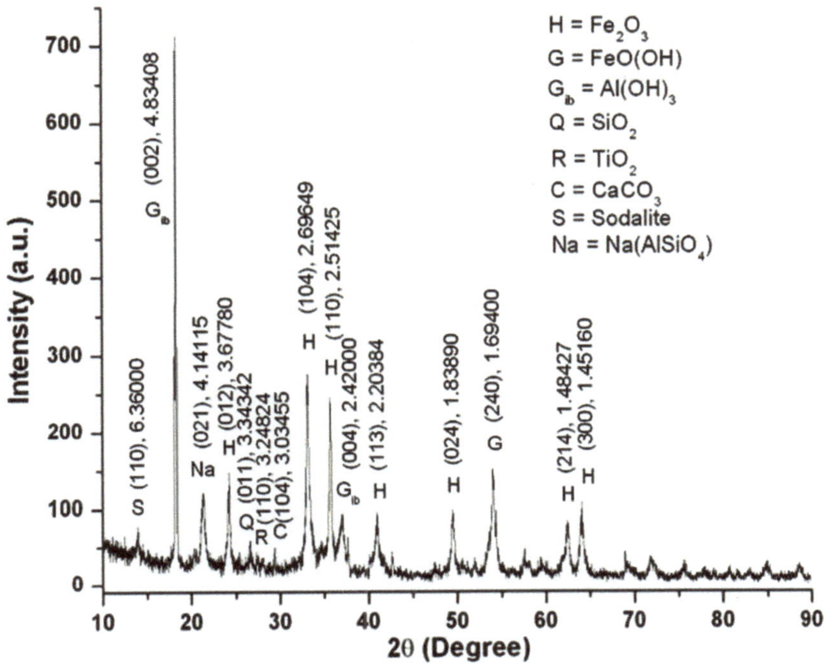

FIGURE 19.2

XRD pattern of RM.

Table 19.5 Major Constituents and Their Chemical Compositions of Red Mud in India

Composition	East Coast Bauxite (High Fe and Low Ti)	Central India Bauxite (Low Fe and High Ti)
Al_2O_3	18–20	18–20
Fe_2O_3	50–55	35–37
TiO_2	4–5	18–20
SiO_2	5–6	7–9
Na_2O	4.5–6	5–6
CaO	0.1–0.6	1–2
LOI	11–12	10–12

19.2.3 INTERNATIONAL STATUS OF RED MUD MANAGEMENT

Because of the highly caustic nature of RM, there are reasonable problems associated with its disposal methods. The major method of disposal currently in use worldwide of huge amounts of RM includes seawater discharge, lagooning, dry stacking, and dry disposal.

In seawater discharge, after the washing and thickening process of RM, the slurry is directly disposed into the deep sea via pipeline. This process decreases the environmental effect of land disposal but it may bring toxic metals into the marine environment and increase the turbidity of the sea owing to the fine mud and the formation of colloidal magnesium and aluminum compounds. French and Japanese practices have favored disposal at sea as the best option on economic and environmental grounds. In Japan, alumina plants are restricted to the available land area for disposal of residues, and so have discharged the residue into the deep sea.

Lagooning is an orthodox disposal method in which the residue slurry is pumped directly into land-based ponds. This consists of the construction of clay-lined dams into which bauxite residue slurry is simply pumped and allowed to dry naturally (Hind et al., 1999). Most of the alumina refineries until 1975 used lagooning for RM disposal, but some such as Pinjarra, Kwinana, and Wagerup in Australia moved to the dry stacking method. Queensland Alumina Ltd. (Australia), after treating its RM with seawater, and CVG Bauxilium (Venezuela), still use the wet disposal method by disposing their RM in lagoons (Emmett and Klepper, 1991). In the dry stacking method, the residue slurry is thickened to 48–55% solids and discharged in thin layers, dewatered, and air dried before discharge of the next layer on it. It can be safely stacked. This decreases the area of disposal but may increase dust generation and requires funds for its long-term closure. These methods have been successfully applied at the MOTIM plant in Hungary (Solymar et al., 2002).

Dry disposal is a method in which the residue is filtered to a dry cake (>65% solids) and the material is washed on the filter with water or steam to recover soda and minimize the alkalinity of residue. Without further treatment, the dry residue is carried by truck or conveyor to the disposal site. This reduces the storage area but requires installation and operation of the filtration plant. Solids contents of greater than 75% have been achieved with Bokela hyperbaric filtration technology at the Stade plant in Germany (Bott et al., 2002). Even with the excellent washing performance offered by hyperbaric steam filtration, significant alkalinity remains associated with the solids because of the complex nature of RM. Hence these hazards associated with alkalinity may be further reduced by employing suitable methods for neutralizing the RM slurry. As per the Guidelines of Australian and New Zealand Environment and Conservation Council and Agriculture and Resource Management Council of Australia and New Zealand, the liquor, which is strongly alkaline with a high pH, requires neutralization to a pH below 9 with an optimum value of 8.5–8.9 before becoming environmentally benign (Hanahan et al., 2004).

19.2.4 CURRENT STATUS OF RED MUD MANAGEMENT IN INDIA

Generally two types of RM disposal (dry and wet) have been followed by the alumina industries in India. RM disposal in alumina industry areas causes seepage of alkaline liquid into groundwater, which might contaminate industrial, domestic, and agricultural water supplies; spillage from damaged pipelines or from retaining-dyke failure, reduction in the availability of arable land, requirement of huge areas (about 2 million square kilometers per year), dust pollution in arid regions, aesthetic impacts, etc., are other problems of the alumina industry (Agrawal et al., 2004a). With growing concerns about environmental protection as well as land conservation, a systematic effort has been made to modify and improve on the methods of disposal. Details of RM disposal practices at Indian alumina plants are given in Table 19.6 (Agrawal et al., 2004a).

Table 19.6 Red Mud Disposal Practices at Indian Alumina Plants

Name of Plant	Red Mud (t/t) of Alumina	Dumping Process
Indal, Muri	1.35–1.45	This refinery adopted the closed-cycle (wet slurry disposal) system (CCD). The disposal ponds have not been provided with a liner.
Nalco, Damonjodi	1.2	A modified CCD method is used for wet disposal. Subjected to six stages of countercurrent washing by pond-returned water (0.5 g per Na_2O) and condensate from evaporators. Washed red mud is wrapped and sent to disposal sites. The bottom and sites of the pond are covered by impervious and semipervious clay with base filters.
Vedanta, Lanjigarh	1.3	Wet disposal
Balco, Korba	1.3	Residue after setting, countercurrent washing is in four stages and filtered. The filtered cake is wrapped with the pond-returned water and dumped in pond. Uses modified close cycle disposal (MCCD) system of wet disposal. Dykes of the currently used pond have stone masonry and well-protected polythene liners and clay layer.
HINDALCO, Renukoot	1.4	Traditional CCD method of impoundment was used. In late 1979, dry disposal method was implemented. After five stages, countercurrent washing the solid is filtered (70% solids) and disposed of into the pond.
Indal, Belgaum	1.16	Plant switched over to dry disposal mode from wet slurry disposal mode in 1985. Mud after clarification process through six-stage countercurrent washing and after filtration (65% solids) is disposed of by dumpers at pond site. Dry portion of pond is covered with 15-cm black cotton soil to promote green vegetation.

19.2.5 PROBLEM RELATED TO RED MUD MANAGEMENT

There are many problems in managing RM disposal.

1. Soda content: RM contains mainly leachable soda, around 1–2%, and bound soda, around 3–6%, in the form of sodium aluminum silicate. A high amount of soda acts as a preliminary barrier in its use as a raw material for industrial applications such as cement and clinker production, the steel industry, constructional bricks, and blocks.
2. Composition variability: The chemical and mineralogical composition of RM generated depends on the nature of bauxite and processing parameters.
3. Environmental pollution: Alumina refineries are mineral-based chemical plants involved in refining crushed bauxite ore. Alumina plant operations include various physical operations such as crushing, grinding, conveying, loading, and transporting, which generate fine particles. They can cause serious health hazards upon inhalation, depending on the size, shape, constituents, and duration of exposure of the particles. Of these parameters, the concentration of the respirable fraction of airborne dust (0.5–5.0 μm) and its free silica content have been reported to cause lung fibrosis and occupational disorders. Expose to dust causes a lung-related disease called pneumoconiosis. Silicosis is the most common form of pneumoconiosis, which is caused by

exposure to free silica (Pattajoshi, 2006). Furthermore, captive power plants (coal-based) are also responsible for air and dust pollution. Noise pollution is found in industrial and mining areas. Planting and forestation around conveyer belts can effectively reduce noise and air pollution.

4. Water pollution: Leakage and overflow of alkaline slurry from RM ponds into water bodies and rivers cause cattle deaths and crop loss. Villagers are mostly affected owing to the discharge of effluents from effluent treatment plants. Domestic animals and humans who are exposed to such polluted water are severely affected with various types of skin disease owing to the high alkaline pH of the water. Groundwater pollution near the refinery area results from leakage from fly ash ponds and RM ponds. During heavy rains, the ponds overflow and their toxic materials spill into nearby streams and rivers (Pattajoshi, 2006). Moreover, coal-power plants release large amounts of CO_2 into the atmosphere, which causes air pollution.

19.3 METHODS FOR THE USE OF RED MUD

The uses of RM are numerous but it requires neutralization before it is used. The storage of large amounts of RM in landfills generally creates serious environmental problems for the local population because of its high level of alkalinity (usually near pH 12−13) but also economic problems because of maintenance. The use of these mineral deposits as a secondary raw material for a variety of applications would produce economic benefits and solve a great environmental problem. For the disposal (Rai et al., 2012), storage, and use of RM, a huge amount of investigation and development is being carried out all over the world. Hence the use of RM cannot be uniform because it depends on its components (Chaddha, 2009). The feasible reuse of RM for any application has the following criteria:

1. Volume: The application should have high-volume use.
2. Performance: It should be a low-cost substitute for other material and its performance should be the same.
3. Cost: It should be cost-effective.
4. Risk: There should be no environmental risk associated with its use regarding health and safety (Chaddha, 2009).

Many efforts are being made globally to find suitable uses for RM so that the alumina industry may end up with no residue at all. Different uses for of RM have been carried out to minimize RM-related issues:

1. Removal of toxic heavy metals such as fluoride (Tor et al., 2009a), Cr(VI) (Costa et al., 2010), arsenic (Genç-Fuhrman et al., 2004; Genç-Fuhrman et al., 2005), vanadate and molybdate (Palmer et al., 2010), copper (Nadaroglu et al., 2010), phosphate (Yue et al., 2010; Zhao et al., 2009), and nickel (Smiljanic et al., 2010) from water and wastewater
2. Used as coagulant, adsorbent, and catalyst (Wang et al., 2008; Karimi et al., 2010)
3. Additives for production of cements (Vangelatos et al., 2009; Zhang et al., 2009; Despland et al., 2010)
4. Used as soil conditioner/fertilizer and revegetation (Wehr et al., 2006)
5. Preparation of ceramic materials, tiles, and glass ceramics (Yang et al., 2009)
6. Constructional bricks as building material and road bed material (Yang and Xiao, 2008)
7. Synthesis of inorganic polymeric materials (Dimas et al., 2009)

8. Development of RM building block, brick, and hollow bricks, using an admixture of RM, blast furnace slag, and fly ash
9. Development of fiber-reinforced RM composite doors
10. Acid mine drainage remediation
11. Red oxide and primer is produced using an admixture of RM
12. Recovery of metals iron, alumina, zirconia, titanium, etc. from RM and production of ferro-titanium (Hind et al., 1999; Kumar et al., 1998; Liu et al., 2009a)

To solve RM storage is to develop a complete utilization technology that consumes RM or converts it into a secondary resource. Since the 1950s, scientists have carried out research projects that explore the disposal and use of RM. According to the exceptional physical and chemical properties of RM, use can be divided into different categories as reported subsequently.

19.3.1 RECOVERY OF VALUABLE ELEMENTS FROM RED MUD

RM mainly contains elemental compositions such as Fe_2O_3, Al_2O_3, SiO_2, CaO, Na_2O, and K_2O. It also contains other compositions such as Li_2O, V_2O_5, TiO_2, and ZrO_2. For instance, the content of TiO_2 in RM produced in India can be as much as 24%. There are usually many elements remaining in bauxite residues such as titanium, scandium, and gallium. Therefore, it is greatly important to recover metals, especially rare earth elements, from RM. The process and technology have not yet been developed to recover all of the components from RM and active research is ongoing to find a suitable solution for its management.

19.3.1.1 Recovery of Divalent Metals From Red Mud

Because of the high iron content of RM, widespread research into the recovery of iron from the Bayer process RM has been carried out by scientists all over the world. The recovering process of iron from RM can be divided into the following categories (Kurniawan et al., 2006): roasting magnetic recovery (Newson et al., 2006), reducing smelting method (Genç-Fuhrman et al., 2004), direct magnetic separation method, and leaching-extraction method (Fois et al., 2007).

Liu et al. (2009b) extracted iron from RM by direct reduction roasting followed by magnetic separation. The process was effective with optimal reaction parameters such as a ratio of carbon powder to RM of 18:100, a ratio of additives to RM of 6:100, and roasting at a temperature of 1300°C for 110 min. Total iron content in the concentrate materials was 88.77% with a metallization ratio of 97.69% and recovery ratio of 81.40%. Iron was recovered from RM by using sodium carbonate through reduction roasting-magnetic separation by Zhu et al. (2012). This process was carried out by adding 8% sodium carbonate and roasting at 1050°C for 80 min followed by magnetic separation of the reduced pellets. The pellets were ground up to 90% at a magnetic field intensity of 0.08 T. The final product of the process contained Fe at 90.87% and Al_2O_3 at 0.95% with an overall iron recovery of 95.76%.

Li et al. (2009) recovered alumina up to 89.71%, Fe up to 60.67% and magnetite up to 61.78% by reduction-sintering, leaching, and then magnetic separation under optimized conditions. Aluminum and sodium were extracted by sintering the RM with magnesium oxide and sodium carbonate at a ratio of 1:0.1:0.25 at a temperature of 900°C for 4 h (Meher et al., 2010). Zhong et al. (2009) reported that at a low temperature and pressure, Al_2O_3 and Na_2O were extracted from RM as sodium aluminate hydrate by a mild hydrochemical process. The optimal conditions of Al_2O_3 extraction were: a 45%

concentration of NaOH solution, a CaO-to-RM mass ratio of 0.25, and a liquid-to-solid ratio of 0.9 under 0.8 MPa at 200°C for 3.5 h. Subsequently, Na_2O was extracted from the residue after the extraction of alumina with a 7% NaOH solution and a liquid-to-solid ratio of 3.8, under 0.9 MPa at 170°C for 2 h. By this overall process, 87.8% of Al_2O_3 and 96.4% of Na_2O were extracted from RM. Alumina was recovered from RM by a hydrothermal process by Zhang et al. (2009). Batch experiments were used to evaluate the main factors influencing the recovery process, namely a reaction temperature in the range of 200–280°C, a caustic ratio of 20 (molar ratio of Na_2O to Al_2O_3 in sodium solution), and a total time of 2 h.

Palmer et al. (2010) used thermally activated hydrotalcite materials, RM, and seawater-neutralized RM to remove arsenate, vanadate, and molybdate from individual and mixed solutions. The percentage of anion removal increased by increasing the Mg:Al ratio to 4:1. Qu et al. (2013a) recovered heavy metals from RM by leaching with fungus *Aspergillus niger*. Bioleaching experiments were carried out with the RM at various pulp densities (1–5%, w/v) under various bioleaching conditions (one-step, two-step, and spent medium bioleaching). In spent medium bioleaching, the highest leaching ratio was achieved and was probably the best condition for leaching heavy metals from RM.

19.3.1.2 Recovery of Rare Earth Metals From Red Mud

Some rare earth metal such as scandium (60–120 g t^{-1}), gallium (60–80 g t^{-1}), and yttrium (60–150 g t^{-1}) are present in small quantities in RM (Srnirnov and Molchanova, 1997). Srnirnov and Molchanova (1997) extracted scandium and uranium from RM with the nitrogen-phosphorus–containing ampholite resins, AFI-21 and AFL22. Wang et al. (2011) recovered scandium from RM by emulsion liquid membrane (ELM) with various compositions (P507 + span80 + sulfonated kerosene + HCl). A total of 98% Sc^{3+} was extracted from RM under optimal conditions of ELM. A pilot-plant process was established for the recovery of scandium (Ochsenku et al., 2002). Zhou et al. (2008) selectively extracted scandium from RM by activated carbon modified by tri-butyl phosphate. At the optimal condition of adsorbent dosage, 6.25 g L^{-1}, 308 K, and 40 min, the highest percentage of scandium was extracted and the adsorption process followed pseudo–second-order kinetics. Titanium oxide (TiO_2) nanopowders were extracted from RM by leaching with dilute sulfuric acid (3 mol L^{-1}) at 60°C with a solid-to-liquid ratio of 5% (Tsakiridis et al., 2011). A filamentous acid-producing fungus named RM-10, identified as *Penicillium tricolor*, was used for biological leaching rare earth elements and radioactive elements from RM (Qu et al., 2013b).

19.3.2 USE OF RED MUD AS BUILDING MATERIALS

19.3.2.1 Cement

RM mainly contains Fe_2O_3 and Al_2O_3 as well as some silica, and thus it should be appropriate as a raw material for the preparation of cement clinkers. Tsakiridis et al. (2004) prepared Portland cement clinkers by mixing 3.5% RM with 74.8% limestone, 11.4% schist, 7.3% Milos sand, and 3% bauxite and sintering the mixture at 1450°C. Physicomechanical tests of the cement such as the grindability, setting time, compressive strength, and expansibility clinker showed that the addition of RM did not negatively affect the quality of the cement. Ordinary Portland cement (OPC) was prepared by Vangelatos et al. (2009) by mixing 5 wt% ferroalumina (FA) with dried RM, limestone, and sandstone. The authors studied the compressive strength of OPC changes from 26.5 to 30.8 MPa after 2 days and 57.5–62.9 MPa after 28 days. The sulfoaluminate cement was prepared by Wang et al. (2013) using

RM and desulfurized gypsum. The cement prepared by this process had excellent mechanical strength and was an efficient way to consume industrial solid waste. The prepared cement was much less expensive than Portland cement owing to low material costs, low material pretreatment costs, and a low calcination temperature.

Singh et al. (1997) performed a detailed investigation for the preparation of calcium sulfoaluminoferrite and calcium aluminoferrite (C_4AF)-based cements using RM, lime, bauxite, and gypsum. The properties of cements strongly depend on the proportion of ingredients and the firing temperature. The authors studied the compressive strength of the prepared cement, which was found to be higher than that of OPC. The hydration temperature of Portland cement increased with the addition of RM up to 30 wt% (Ribeiro et al., 2012). The hydration process of the cement was also accelerated. Pan et al. (2002) developed cementitious material using a mixture of solid water glass (a composite solid alkali activator), sodium aluminate clinkers, and RM by a blast-furnace slag-sintering process. The authors reported that the material possesses high early and ultimate strength and good resistance against chemical attacks.

Zhang et al. (2009) conducted an investigation for the preparation of cement by mixing sintering-process RM with coal gangue. To improve cementitious activity, the RM and coal gangue were mixed together at a ratio of 3:2 and made small spheres with water and calcined at 600°C for 2 h. The prepared composite cement had excellent physical properties such as setting time, soundness, and mechanical properties compared with the OPC. Gordon et al. (1996) prepared RM composite materials in which the strength of RM was increased by incorporating hydrated lime, condensed silica fume, and limestone. The RM composite was used as a construction material without applying Portland cement as a binder with a compressive strength in the range of 15–18 MPa at 28 days and 18–22 MPa at 122 days, which was caused by the formation of stratlingite (C_2ASH_8) and complex carbonates such as hydrogrossular ($Ca_3Al_2(SiO_4,CO_3,OH)_3$). Kumar et al. (1989) prepared blended cements by replacing OPC with acid-neutralized RM in different proportions of 0%, 5%, 10%, and 15%. The blended cement with a RM content of 5–10% had higher compressive strength than that of OPC.

19.3.2.2 Concrete Preparation

De Oliveira and Rossi used RM to prepare different types of new RM coarse aggregate (RMFS80) by taking different amounts of recycled waste material by a mass percentage of 70% or 80% of RM, 30% or 20% of fine sand, 2% of burned oil, and 30% of water, respectively, and improving the workability of the composite (de Oliveira and Rossi, 2012). Hydrochloric acid neutralized RM was used by Sawant et al. (2012) to prepare concrete for partially replacing Portland cement. The authors prepared different grades of concrete blended cement (M30, M40, and M50) using 0, 5, 10, 15, 20, and 25 wt% of neutralized RM by replacing Portland cement.

19.3.2.3 Brick Production

An alternative use of RM in brick production can decrease the cost of raw materials and also have great environmental importance. A process for manufacturing fired bricks was patented in which raw bricks were prepared by mixing 50–90 wt% wet RM with a water-fixing substance and clay (Bayer and Haerter, 1975). The raw bricks were dehydrated by heating below 70°C and then fired at a temperature between 900 and 1100°C. The bricks had better properties than German bricks. Yang and Xiao (2008) prepared unsintered bricks by mixing RM (25–40 wt%) with sand (30–35 wt%), fly ash (18–28 wt%), gypsum (1−3 wt%), lime (8−10 wt%), and Portland cement (1 wt%). RM bricks were prepared using RM with 5, 10, and 15 wt% boron waste and fired at 700, 800, and 900°C separately (Kavas, 2006). Bricks with 15% boron waste had good mechanical performance.

19.3.2.4 Glass Ceramics

RM usually contains a selected amount of CaO, TiO_2, Al_2O_3, and SiO_2 (Peng et al., 2005). The chemical composition of RM is appropriate for making glass ceramics. Romero and Rincon (1999, 2000), Zhang and Yan (2000), and Bhat et al. (2002) produced glass and glass ceramics with RM. Yang et al. (2008) prepared $CaO-SiO_2-Al_2O_3$ glass ceramics and glass from CaO rich slag RM and fly ash. The authors reported that the total weight of RM and fly ash could be up to 85% for better-produced glass ceramics.

19.3.2.5 Corrosion-Resistant Materials

Many researchers have used RM waste for surface modification. Plasma-sprayed RM was used to modify surfaces by coating metal substrates (Satapathy et al., 2007). The authors used sieved RM waste with a particle size of about 60−100 μm. The coating was made by plasma spraying over aluminum, copper, mild steel, and stainless-steel substrates with various applied torch input powers. With this method, the maximum hardness achieved was 560 HV. Collazo et al. (2012) investigated ways to improve corrosion protection properties on AA2024-T3 aluminum alloy using a mixture of 5% (w/w) dried and calcined RM powder with hybrid sol−gel films made from tetra-*n*-propoxyzirconium as the inorganic precursor sol and 3-glycidoxypropyltrimethoxysilane as the organic one. Rebar corrosion-inhibiting concrete was prepared using highly alkaline RM (Ribeiro et al., 2012).

19.3.3 USE OF RED MUD AS PIGMENT

Many researchers have used red mud as a pigment. Pozzolanic pigment was prepared from RM only by calcining at a temperature of 600−800°C and used to color concrete (Pera et al., 1997). During this operation, aluminum hydroxides and clay minerals are transformed into a pozzolanic admixture which consumes the calcium hydroxide produced by cement hydration. A uniform and durable colored concrete was obtained by using white cement interground with 11% burned RM. Karasu et al. (2004) synthesized different colored pigments such as black, brown, and dark green from a mixture of RM, CaO, Cr_2O_3, and MnO by following standard pigment preparation routes and incorporating porcelain tiles, which does not affect the physical properties of the final product.

19.3.4 USE OF RED MUD AS FILLING MATERIAL

Researchers have attempted to recycle RM for filling materials such as landfill cover, road base, wall materials, and filling materials for mined or quarry areas that may lead to the large-scale consumption of RM. A high-grade road base material was prepared by using RM (Qi, 2005). A 15-m-wide, 4-km-long highway was constructed using RM as a base material in Zibo, Shandong Province. The stability and strength of the road constructed met level I standards. RM was also used to prepare roadbed materials (Brown and Kirkpatrick, 1999). The road remained in good condition even after long use and heavy rains.

19.3.5 USE OF RED MUD IN WASTEWATER TREATMENT

Application of RM on a large scale provides important benefits in terms of the environment and economics by reducing landfill volume, contamination of soil and groundwater, and the release of land for alternative uses. Also high-alkalinity RM typically has a large specific surface area and high ion-exchange capacity. Investigations have also been protracted to improve bauxite residues as an adsorbent to remove toxic heavy metals, metalloid ions, and inorganic anions, as well as organic compounds. The leaching test should be conducted before it is measured for water treatment. The removal of inorganic anions which are potentially harmful in drinking water is an unindustrialized issue. RM was used to remove different anions from water and wastewater.

19.3.5.1 Removal of Phosphorus

Phosphorus is a preventive nutrient leading to productivity in most inland, freshwater systems. Surface waters receiving excess inputs of nutrients through wastewater discharges are liable to eutrophication, leading to algal and hydrophytic boom and rapid deterioration of water quality. There are several treatment options for reducing the levels of phosphate in water streams, including metal precipitation, constructed wetland systems, adsorption by various microorganisms either in a free state or immobilized in polysaccharide gels, enhanced biological phosphorus removal using activated sludge systems, and several innovative engineering solutions (de-Bashan & Bashan, 2004). Many researchers have used RM and activated RM to remove phosphorus from water and wastewater.

Shiao and Akashi (1977) prepared hydrochloric acid–activated RM for the removal of phosphate in range 1–5 mg L^{-1} from aqueous solution. Heat-activated and acid-heat–activated RM was synthesized by Liu et al. (2007) to enhance the phosphate sorption capacity. The RM was calcined at 700°C for 2 h to make heat-activated RM, and the RM was activated at 80°C with 0.25 mol L^{-1} HCl for 2 h in an acid-heat–active process. The optimum pH for phosphate removal was at pH 7, in which 31.0 and 30.7 mg P g^{-1} of phosphate was removed from solution by acid-heat–activated RM and heat-activated RM, respectively, with an initial phosphate concentration of 155 mg P L^{-1}. The Langmuir isotherm model fitted well and indicated that the maximum sorption capacity of phosphate by the acid-heat–activated RM and heat-activated RM was 202.9 and 155.2 mg P g^{-1}, respectively. Acid and acid-thermal performance was employed to activate the raw RM using different acids such as HCl and HNO_3 (Huang et al., 2008). The HCl-treated RM had the highest adsorption capacity (0.58 mg P g^{-1} at pH 5.5 and 40°C) because of the high surface area for phosphate removal. The adsorption capacity of the RM decreased with an increasing in pH. The HCl-treated RM at pH 2 had an adsorption of 0.8 mg P g^{-1}, whereas the adsorption reduced to 0.05 mg P g^{-1} at pH 10. The adsorption was increased by 25% with an increase in temperature from 30 to 40°C. Because of the existence of two acidic phosphorus species, $H_2PO_4^-$ and HPO_4^{2-}, the adsorption of phosphorous on the RM showed first-order kinetics. Koumanova et al. (1997) prepared sulfuric acid–treated RM and used it for phosphate removal from an aqueous solution of potassium orthophosphate and sodium tripolyphosphate. The activated RM had a higher adsorption capacity than raw RM. The degree of phosphate removal depends on the ratio of acid to RM and the contact time between them. Pradhan et al. (1998) reported on phosphate adsorption on activated RM prepared by refluxing RM in 20% HCl for 2 h. Phosphate adsorption efficiency was reported to be 80–90% at room temperature at an initial concentration of 30–100 mg L^{-1}. The percentage of phosphate removal decreases with an increase in pH.

The adsorption data followed both the Langmuir and Freundlich isotherm. Two activated RM samples were prepared (1) by stirring RM with 0.25 M HCl for 2 h (RM 0.25) and (2) by heating RM at 700°C for 2 h (RM 700); they were used to remove phosphate from aqueous solution (Li et al., 2006). Both samples showed maximum removal of phosphate (99%) at pH 7.0 and 25°C with an initial phosphate concentration of 155 mg P L^{-1}. The acid-sintered RM (RM 0.25) had a higher phosphate removal capacity, ie, 24.67 mg P g^{-1} than the heat-treated RM (RM 700).

The activated RM was prepared by treatment with sulfuric acid and used to adsorb phosphorus from a dilute aqueous solution (Mohanty et al., 2004). Optimal removal of phosphate was achieved at pH 4.5 at an equilibrium time of 60 min. Lower adsorbent dose and higher initial phosphorus concentration resulted in a preferred higher loading capacity. The equilibrium adsorption process fitted well with the Freundlich isotherm and Lagergren first-order kinetics. The effect of different anions on phosphorus removal was explained on the basis of the changing affinity of anions for the surface and their relative concentrations. Zhao et al. (2011) prepared a new composite coagulant using RM and hydrochloric pickle liquor and used it for phosphate removal from wastewater. The authors compared the composite with commercial polyaluminum chloride coagulant and reported that it could promote phosphate removal efficiency from 4.9% to 10.4% and that it exhibited better coagulation performance than commercial polyaluminum chloride coagulant. Furthermore, it was inferred that the coagulant is not hazardous.

The combination of RM, aluminum, and iron doping agents with commercial peat was used to treat phosphorus from municipal wastewater (Roberge et al., 1999). In this investigation, phosphorus removal was found to be 95% with RM-treated peat in a column study. The use of RM decreased the phosphorus concentration in wastewater below 0.15 mg L^{-1}. The RM and 8% (w/w) $CaSO_4$ were used to prepare aggregates which were used to adsorb phosphorus from wastewater in both batch and column experiments, and 6.8–58.1 mg g^{-1} of phosphorus was adsorbed by the aggregates (Lopez et al., 1998). Fortin and Karam (2001) achieved the maximum sorption of phosphorus using RM residue in batch sorption experiments. The authors reported that the ligand-exchange mechanism was involved in rapid phosphorus sorption and the sorption capacity was 40–45 mg g^{-1}. The sorption of phosphorus on RM residue was concentration dependent, which may due to the concentration gradient and the change in the range of diffusion coefficients. An RM-granulated adsorbent (RMGA) was prepared from main raw materials such as RM, bentonite, and starch at different weight ratios and used to remove phosphate from an aqueous solution (Yue et al., 2010). In another investigation, RMGA was been prepared by Zhao et al. (2012) using dewatered RM as the main raw material, with bentonite and starch and some natural additives to remove phosphate from an aqueous solution. Initially the phosphate removal rate decreased with an increase in the initial pH. The maximum phosphate removal capacity was reported to be 6.64 mg g^{-1} at pH 5. McConchie et al. (1996) synthesized Bauxsol by treating RM with seawater and used it for phosphate removal. Similarly, Akhurst et al. (2006) used Bauxsol and acid-treated Bauxsol for phosphate removal from an aqueous solution. The authors reported that the adsorption of PO_4^{3-} increased with a decrease in pH, and the maximum adsorption efficiency was obtained at pH 5.2 ± 0.1. The adsorption capacity of Bauxsol and acid-treated Bauxsol was 0.21 and 0.48 mmol g^{-1} at pH 9.0 and 5.2, respectively. The PO_4^{3-} adsorption by Bauxsol decreased in the presence of HCO_3^- ions, whereas SO_4^{2-} and Cl^- ions had little effect and the adsorption efficiency increased in the case of Ca^{2+} and Mg^{2+} ions. The adsorption process followed a ligand-exchange mechanism.

19.3.5.2 Removal of Fluoride

With the increase in industrial accomplishments, water bodies with excess levels of fluoride are becoming a matter of great concern. High fluoride concentrations in groundwater, up to more than 30 mg L^{-1}, occur in many parts of the world such as the United States, Africa, and Asia. More than 260 million people worldwide consume drinking water with a fluoride content of >1.0 mg L^{-1}. Most of these people live in tropical countries where the problem is exacerbated by the need to drink more water because of the heat. It is thus absolutely essential to bring down fluoride levels to acceptable limits; because of this, tremendous research and development efforts are being put forward all over the world (Mohapatra et al., 2009). Many researchers have used RM to remove fluoride from water and wastewater.

The removal of fluoride from aqueous solutions by RM and HCl-activated RM has been investigated (Cengeloglu et al., 2002). HCl-activated RM was found to have higher fluoride adsorption ability than the original RM. The fluoride removal mechanism could be described on the basis of the chemical nature and a specific interaction with metal oxide surfaces. Tor et al. (2009a) synthesized granular RM (GRM) for the removal of fluoride from water by mixing fly ash (2 g), sodium carbonate (1 g), powdered quick lime (0.8 g), and sodium silicate (1.2 g) with RM (15 g) and roasted the mixture at 400°C for 2 h followed by calcination at 900°C for 0.5 h. The maximum fluoride removal of 0.644 mg g^{-1} was obtained at pH 4.7 with an equilibrium time of 6 h. RM modified with $AlCl_3$ (MRMA) and heat activation (MRMAH) was used to remove fluoride from water (Wei et al., 2009). The adsorption capability was reported to be 68.07 mg g^{-1} for MRMA and 91.28 mg g^{-1} for MRMAH, which were much higher than that of RM (13.46 mg g^{-1}). The Langmuir isotherm was fitted to the adsorption isotherm model. The highest removal efficiency was achieved at pH 7–8.

19.3.5.3 Removal of Nitrate

Nitrate concentrations in groundwater have increased comprehensively. Wastewater, fertilizers, and livestock farming are main sources of nitrate in groundwater supplies. Groundwater in many locations is used as a supply for drinking water, and high nitrate concentrations present a prospective risk to public health, particular to infants. Use of the RM in the original or activated form as an adsorbent to remove nitrate from drinking water was reported by various researchers.

RM and HCl-activated RM was synthesized by Cengeloglu et al. (2006) and used to remove nitrate from aqueous solutions in a batch adsorption technique. The authors reported that the adsorption capacity of the RM and activated RM was 1.86 and 5.86 mmol g^{-1}, respectively, at pH 7. The adsorption mechanism was described on the basis of the chemical nature and specific interaction with metal oxide surfaces and nitrate ions. Nitrate removal was enhanced from groundwater samples containing 2.52 mM nitrate by using untreated and acid-treated RM combined with zero-valent iron (Fe[0]) (Cho et al., 2011). Nitrate removal increased with an increase in the dose of acid-treated RM, resulting in a 164% increase in the first-order rate constant (k_{obs}) compared with Fe(0). A summary of the adsorption capacity of RM for different anions is presented in Table 19.7.

19.3.5.4 Removal of Other Metal Ions

Most heavy metals are considered toxic beyond permissible limits and usually are discharged from industrial processes even at a low concentration. Precipitation is the most commonly used method for the removal of heavy metals from wastewater. Most researchers have investigated fine RM powder as an adsorbent for metal ions directly because powder adsorbents have a high specific surface area and other properties suitable for adsorption.

Table 19.7 Adsorption Capacity of Different Anions by Red Mud

Adsorbent	Adsorbate	Temperature (°C)	pH	Amount Adsorbed	References
HCl-activated red mud	Phosphorous	–	–	1–5 mg L^{-1}	Shiao and Akashi (1977)
Red mud activated by 0.25 mol L^{-1} HCl at 80°C	Phosphorous	–	7	202.9 mg P g^{-1}	Liu et al. (2007)
Acid (HCl/HNO_3)-treated red mud	Phosphorous	40	5.5	0.58 mg P g^{-1}	Huang et al. (2008)
Red mud activated by 0.25 mol L^{-1} HCl at 80°C	Phosphorous	25	7	24.67 mg P g^{-1}	Li et al. (2006)
Red mud activated by H_2SO_4	Phosphorous	40	4.5	7.4 mg G^{-1}	Mohanty et al. (2004)
Red mud + $CaSO_4$	Phosphorous	–	–	6.8–58.1 mg G^{-1}	Lopez et al. (1998)
Red mud	Phosphorous	–	–	40–45 mg G^{-1}	Fortin and Karam (2001)
Bauxsol acid-treated	Phosphorous	23	9.0	0.21 mmol G^{-1}	Akhurst et al. (2006)
Bauxsol	Phosphorous	23	5.2	0.48 mmol G^{-1}	Akhurst et al. (2006)
Red mud–granular	Phosphorous	–	5	6.64 mg G^{-1}	Zhao et al. (2012)
Red mud–granular	Fluoride	–	4.7	0.644 mg G^{-1}	Tor et al. (2009a)
Red mud	Fluoride	–	7–8	13.46 mg G^{-1}	Wei et al. (2009)
Red mud–$AlCl_3$	Fluoride	–	7–8	68.07 mg G^{-1}	Wei et al. (2009)
Heated red mud	Fluoride	–	7–8	91.28 mg G^{-1}	Wei et al. (2009)
Red mud	Nitrate	–	7	1.86 mmol G^{-1}	Cengeloglu et al. (2006)
HCl-activated red mud	Nitrate	–	7	5.86 mmol G^{-1}	Cengeloglu et al. (2006)

Altundoğan et al. (2000) used RM to remove As(III) and As(V) from water. The uptake capacity of the 20 g L^{-1} RM was reported to be 8.86 and 6.86 μmol g^{-1} within 45 and 90 min for As(III) and As(V), respectively, at 25°C with an initial concentration of 133.5 μmol L^{-1} (10 mg L^{-1}). The authors reported that As(III) adsorption was exothermic and physical in nature, whereas As(V) adsorption was endothermic and chemical in nature. Furthermore, the same group used heat-treated and acid-treated RM to increase arsenic adsorption capacity from water (Altundoğan et al., 2002). The adsorption efficiency of RM was reported to increase by acid treatment and the adsorption was higher at a lower pH and failed at pH >4 for As(V), whereas As(III) adsorption was favorable below pH 7.5. Later, liquid-phase RM (LPRM) was used to remove As(V) by two different routes, ie, neutralization of an LPRM–arsenic solution mixture with an acid solution accompanied by air agitation (Kurniawan et al., 2006); and neutralization of that mixture by CO_2 gas (Newson et al., 2006;

Altundoğan and Tümen, 2003). LPRM with a volumetric LPRM:As(V) solution ratio of 0.1 removed As(V) successfully, as reported in the literature.

Fe_3O_4-nanoparticles were synthesized from RM in an ammonia solution to remove As(V) from synthetic and natural groundwater (Akin et al., 2012). The sorption of As(V) increased with an increase in temperature and maximal As(V) was removed at pH 2.5 in the solution, an initial concentration of 50 $\mu g\ L^{-1}$, a contact time of 45 min, and an adsorbent dose of 8 $g\ L^{-1}$. Sahu et al. (2010) synthesized activated CO_2 neutralized RM and used it to remove arsenate from aqueous solutions on a laboratory scale. The percent removal of arsenate increased gradually with a decrease in pH and maximum removal was found to be 55.55 $mg\ g^{-1}$ at pH ~4, as reported by the authors. Arsenic removal from an aqueous solution in seawater-neutralized RM (Bauxsol) was investigated by Genç-Fuhrman et al. (2003). The adsorption capability of Bauxsol for arsenate varied at 6.08–14.43 $\mu mol\ g^{-1}$ at pH 6.3–10.0, respectively. The adsorption of arsenate on Bauxsol may be due to an inner-sphere complex, as reported in the literature.

Activated seawater-neutralized RM (activated Bauxsol [AB]) was used to remove inorganic arsenic from water; 100% removal of As(V) was achieved at pH 4.5 (Genç-Fuhrman et al., 2004). The adsorption process was spontaneous and endothermic in nature and followed the Langmuir isotherm. Again, the authors investigated the removal of arsenic from water using a number of different sorbents: seawater-neutralized RM (Bauxsol), acid-treated Bauxsol (ATB), AB, Bauxsol-coated sand (BCS), and activated Bauxsol-coated sand (ABCS) (Genç-Fuhrman et al., 2005). The adsorption of arsenic on the developed sorbents decreased in the order AB > ATB > ABCS > BCS > Bauxsol. As per the authors, the adsorption of arsenate on seawater-neutralized RM was favored by a slightly alkaline pH of 8.5. Furthermore, the same group developed seawater-neutralized RM (Bauxsol) through acid treatment, combining acid, heat treatment, and the addition of ferric sulfate ($Fe_2(SO_4)_3 \cdot 7H_2O$) or aluminum sulfate ($Al_2[SO_4]_3 \cdot 18H_2O$) and used it for arsenate adsorption (Genç-Fuhrman et al., 2004). They reported that the adsorption of arsenate from water on Bauxsol was 89%, which increased considerably to 95% by acid-treated RM and further increased to roughly 100% by combined acid and heat treatment. The addition of aluminum sulfate and ferric sulfate suppressed the efficiency of removal.

Genç-Fuhrman et al. (2005) modified Bauxsol and AB with sand and prepared BCS and ABCS grains, respectively, and studied the adsorption of arsenate from water in both batch and column experiments. The maximum adsorption capacities of arsenate for ABCS and BCS were 2.14 and 1.64 $mg\ g^{-1}$, respectively, at pH 7.1. Modified RM was developed with $FeCl_3$ and used to remove arsenate from water (Zhang et al., 2008). The adsorption capability of modified RM was reported to be 68.5, 50.6, and 23.2 $mg\ g^{-1}$ at pH 6, 7, and 9, respectively. According to the authors, the NO_3^- ion had little effect on adsorption, whereas Ca^{2+} enhanced the adsorption and HCO_3^- decreased it. RM was used for arsenate adsorption at different pH values of 4, 7, and 10 and the adsorption of arsenate on RM increased with a decrease in pH (Castaldi et al., 2010). Arsenate sorption on RM samples followed a ligand exchange mechanism which involved the formation of inner-sphere complexes. A different composition of RM was prepared by mixing RM with hematite, china clay, and fly ash and was used to remove As(III) in a batch process by Singh et al. (2008). Ferrous-based RM sludge was synthesized by dissolving hydrated $FeCl_2$ in deionized water and RM and was used to remove low-concentration arsenic from a rural area (Li et al., 2010). The authors reported that ferrous-based RM sludge was more effective in reducing the aqueous arsenic concentration and the removal efficiency of As(V) was higher than As(III). Lopes et al. (2012) reported the sorption of arsenic on RM mixed with phosphogypsum. The authors prepared an RM mixture by adding phosphogypsum to RM at 0%, 1%, 2%, 5%, 10%, and 25% by weight. The adsorption capacity of As (As_{MAC}) increased with an increase in the percentages of phosphogypsum in the RM owing to the presence of Ca^{2+} in phosphogypsum, which altered the charge balance of the adsorbent, leading to the formation of ternary complexes.

Cr(VI) is a highly toxic metal considered to be an aquatic pollutant. Researchers used RM as an adsorbent to remove chromium from an aqueous solution. Pradhan et al. (1999) prepared activated RM (ARM) by treating RM with acid followed by ammonia precipitation and used it to remove Cr(VI) from a synthetic solution and industrial effluents. The authors reported that the adsorption increased with an increase in the initial concentration of Cr(VI), whereas a negative trend was detected with an increase in temperature. Dursun et al. (2008) used HCl-treated RM for chromate removal and the removal efficiency of chromate was about 70% at optimal conditions. RM was used to remove trivalent chromium Cr(III) from synthetic wastewater and the maximum adsorption efficiency was found to be 99.9% from a 150 mg Cr(III)/100 mL synthetic solution at pH 6 with an agitation time of 10 h (Rajkumar et al., 2001).

The presence of RM effectively removes nickel from aqueous solutions (Zouboulis and Kydros, 1993). According to the authors, RM acts as an alkalinity controller and precipitates nickel to an insoluble hydroxide. Smiljanić et al. (2010) used rinsed Bosnian RM (RBRM) to adsorb Ni^{2+} ions from an aqueous solution. The mineral mixture had a high acid-neutralizing capability and pH 2–8 the mixture had the highest cation immobilization. The accumulation of 5 g L^{-1} RBRM effected 100% removal from solutions of 1×10^{-4} to 5×10^{-4} mol L^{-1}, which decreased to 26% with an increase in Ni^{2+} concentration to 8×10^{-3} mol L^{-1}. The maximum sorption capacity by the Langmuir model was 0.372 mmol g^{-1} at pH 5. RM powder was used to remove Ni(II) from an aqueous solution in the presence and absence of citrate ligand by Smičiklas et al. (2013). The authors reported that a small amount of Ni(II) was removed at pH 3 with a lower initial metal concentration (10^{-3} mol L^{-1}), but 100% removal of Ni(II) was obtained at a higher concentration (2×10^{-3} mol L^{-1}) at pH >8. The sorption capacities decreased in the order: 27.4, 25, 21, 7.6, and 2.5 mg g^{-1}, with an increase in Ni/citrate molar ratio of 1:0, 1:0.25, 1:0.5, 1:1, and 1:2, respectively. Ni(II) removal followed sorption and precipitation mechanisms depending on the solution pH, metal, and citrate content. Hannachi et al. (2010) used RM to remove Ni(II) from an aqueous solution and reported that the maximum sorption of Ni(II) was at pH 5, a temperature of 30°C, and a contact time of 4 h. The adsorption process followed a first-order rate mechanism with an adsorption capacity of 13.69 mg g^{-1}.

Agrawal et al. (2004b) removed copper from an aqueous solution by using RM, sea nodule residue, and fly ash separately. The maximum adsorption capacity of copper on RM was 2.28 mg g^{-1} at pH 5.5. RM and ARM were employed for the sorption of copper from water (Nadaroglu et al., 2010). The maximum adsorption of copper occurred at an initial solution concentration of 3 mmol, pH of 5.5, temperature of 30°C, and adsorbent dose of 1 g mL^{-1}. The ARM had a higher level of adsorption capacity for Cu^{2+} than did normal RM.

Zhu et al. (2007) prepared GRM ceramics for the removal of cadmium ions from aqueous solutions by mixing fly ash (2 g), sodium carbonate (1 g), powdered quick lime (0.8 g), and sodium silicate (1.2 g) with RM (15 g) and roasted at 400°C for 2 h followed by calcination at 900°C for 0.5 h. The removal of cadmium from aqueous solutions on RM, acidified RM, and ball-milling nanoparticle RM was also reported (Luo et al., 2011). The maximum sorption capacity of cadmium was 0.16, 0.19, and 0.21 mol kg^{-1} for the original, acidified, and nanoparticle RM, respectively, at pH 6.5. Both acidification and ball-milling treatments considerably enhanced cadmium sorption and facilitated the conversion of cadmium into less extractable fractions. The formation of inner-sphere complexes of cadmium similar to XCdOH (X represents surface groups on RM) on the RM surfaces is the mechanism of cadmium sorption, although outer-sphere complexes of cadmium were the primary species.

Sahu et al. (2013) prepared acid-activated RM to investigate the removal efficiency of Pb(II) from aqueous solutions. The percent removal increased gradually with a decrease in pH and maximum removal was achieved at pH 4. The authors reported that the material had a high adsorption capacity for the lead ion and was also capable of lowering the metal ion concentration to a great extent. The metal ion concentration gradient is a fundamental force that transfers metal ions from the solution to an adsorbent surface and diffuses metal ions into the inside of the adsorbent. ARM was prepared by an acid dilution and ammonia precipitation process and used to remove Cd(II) from an aqueous solution by a batch mode experiment (Sahu et al., 2015). The maximum adsorption capacity of Cd(II) on ARM was reported to be 12.046 and 12.548 mg g^{-1} at temperatures 293 and 303 K, respectively. The mass transfer study revealed that the adsorption of cadmium ions was faster at a lower temperature. The adsorption of Cu(II), Pb(II), and Cd(II) from contaminated water on bauxite waste RM and coal fly ash was reported (Apak et al., 1998). Lopez et al. (1998) prepared RM aggregate for the adsorption of phosphorus and other heavy metals by using moistened RM and 8% (w/w) $CaSO_4$. The authors reported that the maximum adsorption capacity of RM aggregates based on the Langmuir isotherm for Cu^{2+}, Zn^{2+}, Ni^{2+}, and Cd^{2+} was 19.72, 12.59, 10.95, and 10.57 mg g^{-1}, respectively, with a contact time of 48 h and pH of 4.7–7.9. In a continuous removal process, secondary effluents from an urban sewage treatment plant were allowed to percolate through the RM aggregate-packed column. Gupta and Sharma (2002) prepared ARM by treating RM with hydrogen peroxide and used it to adsorb cadmium and zinc from an aqueous solution in a batch and column process. RM removed lead and chromium from aqueous solutions, as reported by Gupta et al. (2001). The adsorption capacity of RM for lead was 64.79 mg g^{-1} at pH 4–6, and for chromium was 35.66 mg g^{-1} at pH 2. A pellet-type RM adsorbent was developed by mixing 58.7 wt% RM, 25.2 wt% kaolin, 11.7 wt% sodium silicate solutions, 2.9 wt% fly ash, and 1.5 wt% magnesium chloride at 600°C and was used for the adsorption of Pb^{2+}, Cu^{2+}, and Cd^{2+} from aqueous solutions (Han et al., 2002). The removal efficiency was more than 95% for Pb^{2+}, Cu^{2+}, and Cd^{2+} after 24 h of operation. The response surface methodology (RSM) and artificial neural network (ANN) prediction models were used for lead removal from industrial sludge by using RM (Geyikçi et al., 2012). The authors reported that the ANN results were more reliable then RSM results. Vaclavikovaa et al. (2005) used RM to remove cadmium, zinc, copper, and lead from aqueous solutions in the presence of 0.01 M $NaNO_3$. The maximum uptake capacity for cadmium was 68 mg g^{-1} at pH 6 and 133 mg g^{-1} for zinc at pH 7. The removal was caused by precipitation of the respective insoluble hydroxides. Heavy metals such as Pb^{2+}, Cd^{2+}, and Zn^{2+} were removed using untreated RM (RM_{nt}) and acid-treated RM (RM_a) by Santona et al. (2006). The adsorption capacity of RM_{nt} for these metals was in the order of zinc $\geq$ lead > cadmium. Industrial wastes such as fly ash, RM, and scale arrears were used to remove heavy metals from mine waters (Zoumis et al., 2000). The fly ash and RM had better adsorption capacities for zinc, cadmium, nickel, and manganese. The main mechanism proposed for these removals were sorption and precipitation. The calcined RM was used as an adsorbent to remove phosphate and heavy metals from the sewage of swine wastewater in a sequencing batch reactor (Shi et al., 2009). The adsorption process was favorable at a higher pH for the removal of phosphate, copper, zinc, and arsenic. The adsorption mechanism was resulted from surface complexation reactions and the mechanism of phosphate and arsenic removal was from co-precipitation. Komnitsas et al. (2004) used RM and limestone to prepare permeable reactive barriers (PRBs) on a laboratory scale to remove inorganic metal ions such as iron, zinc, manganese, aluminum, nickel, copper, cobalt, and cadmium from water. The adsorption efficiency of the RM PRBs was reported to be 8.47 for iron, 0.45 for aluminum, 0.005 for manganese, 0.066 for zinc, 0.025 for copper, 0.019 for cadmium, 0.014 for cobalt, 0.012 for nickel, and 36.1 for SO_4^{2-} (in mg g^{-1} per week). Main mechanisms for the removal of these metal ions were precipitation, co-precipitation, and adsorption. Seawater-neutralized RM (SWN-RM) with a hydrotalcite structure was prepared by using magnesium and aluminum with a ratio between 3.5:1 and 4:1 and employed for the adsorption of arsenate, vanadate, and molybdate in individual and mixed solutions; it was compared with several synthetic hydrotalcite minerals (Palmer et al., 2010). The authors reported that the percentage of anion removal from the solution increased by increasing the Mg:Al ratio to 4:1 and the increasing affinity adsorption was in the order: arsenate > vanadate > molybdate. Removal of boron from an aqueous solution using neutralized RM by batch equilibration technique was reported by Cengeloglu et al. (2007). The RM was neutralized by suspension in distilled water with a liquid-to-solid ratio of 2:1 on a weight basis and stirring until it attained equilibrium at pH 8–8.5. The authors reported that the maximum adsorption capacity of the material was 5.996 mg g^{-1} at pH 7, an adsorbent dose of 4 g L^{-1}, initial boron concentration of 43 mg L^{-1}, and temperature of $25 \pm 1°C$. In addition to these studies, a number of other researchers discovered the potential use of RM as an adsorbent to remove toxic metals from aqueous solutions. A summary of the adsorption capacity of RM for different metal ions is presented in Table 19.8.

19.3.5.5 Removal of Dyes

Colored dye wastes are important type of pollutant discharged from textile, paper, printing, and leather industries, which are highly toxic to aquatic animals. Various health-related problems, such as allergy, skin irritation, dermatitis, cancer, and mutations in humans, are connected with dye pollution in water. Several investigations have been carried out to discover the applicability of RM for dye removal from wastewater.

Sahu and Patel (2015) evaluated the efficiency of safranin-O dye removal from aqueous solutions with the application of RM waste modified with sodium dodecyl sulfate. The maximum adsorption capacity of safranin-O takes place at pH 4. The authors reported that at acidic pH the surfactant molecule adsorbed on the surface of the RM by hydrophobic bonding and created surface aggregates (hemi-micelles, mixed hemi-micelles, or ad-micelles) on the hydrophilic adsorbent surface. Waste RM was used to adsorb congo red (an anionic dye) from aqueous solution by Namasivayam and Arasi (1997). The adsorption capacity was reported to be 4.05 mg g^{-1} at pH 2.0 and 30°C. The mechanism of adsorption was mainly ion exchange. The percentage of desorption increased with an increase in the pH of the aqueous solution. Acid violet dye removal of from wastewater by using RM was reported (Namasivayam et al., 2001). The adsorption capacity of the adsorbent was 1.37 mg g^{-1} at pH 4.1 and followed both Freundlich and Langmuir models. The adsorption of dye was mainly due to the ion-exchange mechanism. Furthermore, the same group reported the adsorption of Procion orange dye on waste RM (Namasivayam et al., 2002). The authors optimized different initial dye concentrations, adsorbent dosages, agitation times, and pH as controlling parameters. There was a decrease in the percentage of adsorption from 82 to 0 with an increase in pH from 2.0 to 11.0 as a result of the formation of an aqua complex and acid–base dissociation at the solid–solution boundary. H_2O_2-activated RM has been used to remove some basic dyes such as rhodamine B, fast green, and methylene blue from wastewater (Gupta et al., 2004a). The percent removals were 92.5%, 94.0%, and

Table 19.8 Adsorption Capacity of Red Mud for Different Metal Ions

Adsorbent	Adsorbate	Temperature (°C)	pH	Amount Adsorbed	References
Red mud	As(III)	25	9.5	8.86 μmol g^{-1}	Altundoğan et al. (2000)
Red mud	As(V)	25	3.2	6.86 μmol g^{-1}	Altundoğan et al. (2000)
Heat-activated red mud	As(III)	25	5.8–7.5	11.80 μmol g^{-1}	Altundoğan et al. (2002)
Acid-activated red mud	As(V)	25	1.8–3.5	12.57 μmol g^{-1}	Altundoğan et al. (2002)
Red mud	As(V)	25–65	2.5	15.3 μg L^{-1}	Akin et al. (2012)
CO_2-neutralized red mud	As(V)	–	4	55.55 mg G^{-1}	Sahu et al. (2010)
Seawater-neutralized red mud (Bauxsol)	As(V)	–	6.3–10.0	6.08–14.43 μmol g^{-1}	Genç-Fuhrman et al. (2003)
Bauxsol-coated sand	As(V)	–	4.5	3.32 mg G^{-1}	Genç-Fuhrman et al. (2005)
Activated Bauxsol-coated sand	As(V)	–	7.1	1.64–2.14 mg G^{-1}	Genç-Fuhrman et al. (2005)
$FeCl_3$-coated red mud	As(V)	–	6–9	23.2–68.5 mg G^{-1}	Zhang et al. (2008)
Activated red mud	Cr(VI)	–	7.06	30.74 mmol G^{-1}	Pradhan et al. (1999)
Rinsed Bosnian red mud	Ni(II)	–	5	0.372 mmol G^{-1}	Smiljanić et al. (2010)
Red mud	Ni(II)	30	5	13.69 mg G^{-1}	Hannachi et al. (2010)
Red mud	Cu(II)	–	5.5	2.28 mg G^{-1}	Agrawal et al. (2004b)
Granular red mud	Cd(II)	20–40	3–6	38.2–52.1 mg G^{-1}	Zhu et al. (2007)
Red mud	Cd(II)	–	6.5	0.16 mol kg^{-1}	Luo et al. (2011)
Acidified red mud	Cd(II)	–	6.5	0.19 mol kg^{-1}	Luo et al. (2011)
Nanoparticle red mud	Cd(II)	–	6.5	0.21 mol kg^{-1}	Luo et al. (2011)
Treated red mud	Cd(II)	–	–	46.9–66.8 mg G^{-1}	Apak et al. (1998)
Treated red mud	Cu(II)	–	–	35.2–75.2 mg G^{-1}	Apak et al. (1998)

Continued

Table 19.8 Adsorption Capacity of Red Mud for Different Metal Ions—cont'd

Adsorbent	Adsorbate	Temperature (°C)	pH	Amount Adsorbed	References
Treated red mud	Pb(II)	–	–	117.3–165.8 mg G^{-1}	Apak et al. (1998)
Red mud	Cu(II)	–	5.6–6.2	19.72 mg G^{-1}	Lopez et al. (1998)
Red mud	Zn(II)	–	6.9–7.8	12.59 mg G^{-1}	Lopez et al. (1998)
Red mud	Ni(II)	–	7.5–7.9	10.59 mg G^{-1}	Lopez et al. (1998)
Red mud	Cd(II)	–	4.7–6.2	10.57 mg G^{-1}	Lopez et al. (1998)
Activated red mud	Cd(II)	30	4	1.16×10^4 mol g^{-1}	Gupta and Sharma (2002)
Activated red mud	Zn(II)	30	4	2.22×10^4 mol g^{-1}	Gupta and Sharma (2002)
Red mud	Pb(II)	–	4–6	64.79 mg G^{-1}	Gupta et al. (2001)
Red mud	Cr(VI)	–	2	35.66 mg G^{-1}	Gupta et al. (2001)
Red mud	Cd(II)	–	6	68 mg G^{-1}	Vaclavikovaa et al. (2005)
Red mud	Zn(II)	–	7	133 mg G^{-1}	Vaclavikovaa et al. (2005)
Calcined red mud	Copper	–	–	18.18–65.17 mg G^{-1}	Shi et al. (2009)
Calcined red mud	Zinc	–	–	15.45–99.20 mg G^{-1}	Shi et al. (2009)
Calcined red mud	Arsenic	–	–	18.83–27.51 mg G^{-1}	Shi et al. (2009)
Calcined red mud	Phosphate	–	–	46.26–149.00 mg G^{-1}	Shi et al. (2009)
Neutralized red mud	Boron	–	2–7	30.12 mg G^{-1}	Cengeloglu et al. (2007)

75.2% for rhodamine B, fast green, and methylene blue dyes at an optimum pH of 1.0, 7.0, and 8.0, respectively. The adsorption of the three dyes on H_2O_2-activated RM was reported to be exothermic in nature. The authors described the suitability of acetone for desorption of rhodamine B, fast green, and methylene blue dyes. Arias et al. (1999) used RM for the adsorption of methylene blue by both batch and column techniques. The process was followed two kinetic processes: a rapid process followed by a slow kinetic reaction. The optimum adsorption was 0.74 mg g^{-1} for a reaction time of 192 h. Wang et al. (2005) reported the removal of methylene blue dye from an aqueous solution on heat-treated (800°C overnight) and acid-treated (1 N HNO_3 solution for 24 h) RM and fly ash. As per the authors, heat treatment reduced the adsorption ability for both fly ash and RM but acid treatment by HNO_3

increased the adsorption capacity of fly ash, whereas it decreased the adsorption capacity of RM (3.2×10^{-6} mol g^{-1}) owing to the neutralization of hydroxide ions on the basic surface of RM, leading to the lower adsorption capacity. The adsorption of methylene blue on RM was endothermic in nature and fitted well with the Redlich–Peterson model. Tor and Cengeloglu used HCl-activated RM to remove congo red dye from water in batch adsorption experiments (Tor and Cengeloglu, 2006). The maximum monolayer sorption capacity of ARM was described to be 7.08 mg g^{-1} at pH 7.0 with an equilibrium time of 90 min. SWN-RM was thermally treated at 400 and 500°C to increase the adsorption capacity and was used for the adsorption of reactive blue dye 19 (RB 19) (de Souza et al., 2013). The adsorption of RB 19 dye significantly depends on the pH of the solution, and the highest percentage removal of RB 19 occurred for SWN-RM heated to 400°C at pH 4. The Langmuir isotherm is more appropriate to describe the phenomenon of the removal of RB 19 using SWN-RM, SWN-RM 400, and SWN-RM 500, with a maximum adsorption capacity of 250.0, 416.7, and 384.6 mg g^{-1}, respectively. A summary of the adsorption capacity for different dyes on RM is presented in Table 19.9.

19.3.5.6 Removal of Organic Pollutants

Phenol and its derivatives are highly carcinogenic and priority classes of pollutants of water. RM has also been explored as a potential adsorbent for the removal of phenol and its derivative from water and wastewater.

Gupta et al. (2004b) studied the removal of phenol, 2-chlorophenol, 4-chlorophenol, and 2,4-dichlorophenol from wastewater by RM. The maximum adsorption of phenol and 2-chlorophenol was reported to be 50–81% at pH 6.0, whereas the maximum adsorption of 4-chlorophenol and 2,4-dichlorophenol was 94–97% at pH 5.0 and 4.0, respectively. The authors described the removal of phenol and its derivatives by particle diffusion; the order of removal was: 2,4-dichlorophenol > 4-chlorophenol > 2-chlorophenol > phenol. The adsorption process was endothermic in nature and fitted well with both Langmuir and Freundlich models. RM has also been used to remove 2,4-dinitrophenol (Gupta and Ali, 2006). About 95% of 2,4-dinitrophenol was sorbed by the developed adsorbent and the removal efficiency was reached 96% in column experiments at a flow rate of 0.5 mL min^{-1}. Tor et al. (2006) studied the removal of phenol from an aqueous solution on neutralized RM by a batch process. The maximum monolayer sorption ability for the adsorption of phenol by neutralized RM was found to be 4.12 mg g^{-1} at a pH of 1–9 with an equilibrium of 10 h. The authors reported that the adsorption of phenol decreased from 66% to 32% and 15% in the presence of nitrate and sulfate, respectively, but there was no negative effect on adsorption by the chloride ion. HCl-activated RM was used to remove phenol from an aqueous solution by Tor et al. (2009b). The maximum adsorption capacity of activated RM was 8.156 mg g^{-1} at pH 6 and 25 ± 1°C with an equilibrium of 10 h. The adsorption kinetics followed a pseudo–second-order reaction and both Langmuir isotherms. A summary of adsorption capacity of RM for the removal of different phenolic pollutants from water is presented in Table 19.10.

19.3.5.7 Removal of Bacteria and Virus

RM was mixed with amended sand and used to remove *Escherichia coli*, *Salmonella adelaide*, and poliovirus type 1 from a minor effluent in a wastewater treatment plant by 65-cm-long soil packed with amended sands (Ho et al., 1990). The authors developed a medium by neutralizing the RM using 5% gypsum and incorporated it to form 30% of the amended sands. After 72 days moving over 10 pore

Table 19.9 Adsorption Capacity of Red Mud for the Removal of Different Dyes From Water

Adsorbent	Adsorbate	Temperature (°C)	pH	Amount Adsorbed	References
Red mud	Congo red	30	2	4.05 mg G^{-1}	Namasivayam and Arasi (1997)
Red mud	Acid violet	–	4.1	1.37 mg G^{-1}	Namasivayam et al. (2001)
H_2O_2-activated red mud	Rhodamine B	30–50	1	$(1.01-1.16) \times 10^{-5}$ mol g^{-1}	Gupta et al. (2004a)
H_2O_2-activated red mud	Fast green	30–50	7	$(7.25-9.35) \times 10^{-6}$ mol g^{-1}	Gupta et al. (2004a)
H_2O_2-activated red mud	Methylene blue	30–50	8	$(4.35-5.23) \times 10^{-5}$ mol g^{-1}	Gupta et al. (2004a)
Red mud	Methylene blue	–	–	0.74 mg G^{-1}	Arias et al. (1999)
Treated red mud	Mcthylene blue	–	–	7.8×10^{-6} mol g^{-1}	Wang et al. (2005)
HCl-activated red mud	Congo red	–	7	7.08 mg G^{-1}	Tor and Cengeloglu (2006)
Seawater-neutralized red mud	Reactive blue dye 19	–	–	250.0 mg G^{-1}	de Souza et al. (2013)
Seawater-neutralized red mud thermally treated with 400°C	Reactive blue dye 19	–	–	416.7 mg G^{-1}	de Souza et al. (2013)
Seawater-neutralized red mud thermally treated with 500°C	Reactive blue dye 19	–	–	384.6 mg G^{-1}	de Souza et al. (2013)

Table 19.10 Adsorption Capacity of Red Mud for Removal of Phenolic Pollutants From Water

Adsorbent	Adsorbate	Temperature (°C)	pH	Amount Adsorbed	References
Red mud	Phenol	30–50	6	0.63–0.74 mol G^{-1}	Gupta et al. (2004b)
Red mud	2-Chlorophenol	30–50	6	0.72–0.79 mol G^{-1}	Gupta et al. (2004b)
Red mud	4-Chlorophenol	30–50	5	0.78–0.82 mol G^{-1}	Gupta et al. (2004b)
Red mud	2,4-Dichlorophenol	30–50	4	0.80–0.85 mol G^{-1}	Gupta et al. (2004b)
Neutralized red mud	Phenol	–	–	4.12 mg G^{-1}	Tor et al. (2006)
HCl-activated red mud	Phenol	25 ± 1	6	8.156 mg G^{-1}	Tor et al. (2009b)

volumes of wastewater, no or little *E. coli* was detected; *S. adelaide* and poliovirus type 1 was not detected in the outlet samples after 9 days of moving over 2.4 pore volumes. The removal efficiency was in the order: poliovirus type1 > *E. coli* > S. *adelaide*.

19.3.6 USE OF RED MUD TO ADSORB AND PURIFY TOXIC WASTE GASES

Toxic waste gases such as H_2S, SO_x, NO_x, CO_2, and CO obtained from fossil fuel and different industrial processes cause environmental problems. However, it was beneficial to use bauxite residues as a low-cost raw material for the absorption and purification of toxic waste gases.

Yadav et al. (2010) used RM for the sequestration of CO_2 as an important greenhouse gas. The authors prepared three different size fractions of RM (RM I, RM II, and RM III) of varying densities (1.5–2.2 g cm^{-3}) and used them for carbonation in a stainless-steel reaction chamber at a fixed pressure of 3.5 bar. Carbonation of RM was studied as a function of reaction time (0.5–12 h) and a liquid-to-solid ratio of 0.2–0.6. The authors reported that the main calcium-bearing mineral, chantalite, and cancrinite phases are responsible for the carbonation of RM. Maximum carbonation was shown by the fraction RM II having a higher concentration of cancrinite. The carbonation capacity was calculated to be 5.3 g of CO_2/100 g of RM II. Raw RM and neutralized RM (Bauxsol) was used for carbon dioxide capture by carbonation (Jones et al., 2006). Hydroxide alkalinity ingestion was 85% on raw RM and 89% on Bauxsol after only 5 min of carbonation. The carbonation of RM with CO_2 was described by the following reactions (Eqs. [19.1]–[19.5]):

$$NaAl(OH)_4 + CO_2 \rightarrow NaAlCO_3(OH)_2 + H_2O \quad [19.1]$$

$$NaOH + CO_2 \rightarrow NaHCO_3 \quad [19.2]$$

$$Na_2CO_3 + CO_2 + H_2O \rightarrow 2NaHCO_3 \quad [19.3]$$

$$3Ca(OH)_2 \cdot 2Al(OH)_3 + CO_2 \rightarrow 3CaCO_3 + Al_2O_3 \cdot 3H_2O + 3H_2O \quad [19.4]$$

$$Na_6[AlSiO_4]_6 \cdot 2NaOH + CO_2 \rightarrow Na_6[AlSiO_4]_6 + 2NaHCO_3 \quad [19.5]$$

Sahu et al. (2011) studied the removal of hydrogen sulfide (H_2S) gas on RM under ambient conditions on a laboratory scale. The authors reported that hydrogen sulfide gas was removed in the form of FeS_2, FeS, S, and sulfide raw materials. The color of the RM changed from red to black during the sulfidation reaction, which indicated the conversion of iron oxide to iron sulfide. The removal capacity of RM was 2.1 g H_2S/100 g RM. Acid-activated RM was developed for the oxidation of carbon monoxide (CO) in the temperature range 100–500°C (Sushil and Batra, 2012). The maximum conversion of >90% was reported at a temperature above 400°C. RM was also evaluated for the absorption and purification of the waste gas of SO_2 owing to its high alkalinity, large absorption capacity, and simple production process (Yi et al., 2007). The absorption rate of SO_2 on RM was more than 95%. As per the authors, the mechanism for the process was physical adsorption, followed by chemical neutralization reactions. Nan et al. (2012) described the absorption of low-concentration SO_2 on RM and roasted RM. The amount of SO_2 fixation by original and roasted RM was reported to be 25.1 and 30.2 mL g^{-1}, respectively, which was increased using ultrasonic waves. The optimum amount of SO_2 fixation was 36.7 mL g^{-1} under the optimal condition of a liquid-to-solid ratio of 9:1. Fois et al. (2007) evaluated the absorption of SO_2 on RM by a model based on film theory to calculate the gas and liquid–side mass transfer coefficients. Fan et al. (2005) prepared a fixed-bed reactor using mixed clay with an RM sorbent for a laboratory-scale and bench-scale operation for coal gas desulfurization. CuO impregnated on alumina, bauxite, and RM was used for the sorption of low-concentration SO_2 in a fluidized-bed reactor 5 cm in diameter (Uysal et al., 1988). The rate of sorption and sorption capacities increased with an increase in temperature. The desulfurization rate was initially 100% and was as high as 94% after 10 cycles.

19.3.7 USE OF RED MUD AS CATALYST

In many industries, catalytic processes are widely used for various industrial activities. Because RM contains important components such as iron oxide, titanium oxide, and silicon oxide, it can be employed as a metal precursor to act as a catalyst.

19.3.7.1 Use of Red Mud as Hydrogenation Catalyst

In a hydrogenation reaction there is an addition of H_2 directly or indirectly, usually to unsaturated organic compounds. The reaction between the compound and the H_2 is carried out in the presence of a catalyst, which is required to activate the H_2. Thus, RM was investigated as a hydrogenation catalyst for coal, petroleum residue, and certain organic residues.

Pratt and Christoverson (1982) reported that activated RM prepared by acid digestion and reprecipitation acts as a primary catalyst for the hydrogenation of naphthalene to tetralin. The authors reported that upon activation, the surface area of the RM increased to 155 m^2 g^{-1}, and thus conversion of naphthalene by activated RM increased to 49% whereas conversion with untreated RM was 3.55%. Furthermore, upon addition of 20% TiO_2, the conversion percentage increased to 58%. In this same process, ARM was prepared for the hydrogenation of naphthalene, phenanthrene, and pyrene

(Eamsiri et al., 1992). The conversion efficiency over activated RM was naphthalene (80%), phenanthrene (54%), and pyrene (52%) at a pressure of 6 MPa, temperature of 405°C, and time of 1 h. Liano et al. prepared RM and ARM using the method of Pratt and Christoverson (1982) and used them as catalysts for the hydrogenation of anthracene oil in a trickle-bed reactor at 350°C, 10 MPa, and H_2 at a flow rate of 4 mL s^{-1} at a standard temperature and pressure (Liano et al., 1994). The authors reported that the catalytic activity of RM was enhanced by the activation. Mastral et al. (1993) synthesized three different types of iron-based catalyst precursors, such as RM (Fe_2O_3), iron sulfide, and iron pentacarbonyl, for the hydrogenation of high sulfur—low-rank coals under mild conditions (350°C, 10 MPa, H_2, cold, and 30 min). As per the authors, the RM seemed to have higher conversion percentage and tetrahydrofuran (THF)-soluble production compared with other iron-based catalysts. Hydrotreatment of diphenylether and diphenylmethane was carried out through hydroliquefication of coal with sulfur-promoted RM and other iron-containing catalysts (FeS_2, Fe_2O_3/S, Fe_3O_4/S, and $FeSO_4$) (Yokoyama et al., 1989).

19.3.7.2 Use of Red Mud as Catalyst in Dechlorination and Hydrodechlorination Reaction

Ordóñez et al. (2001a) investigated the use of RM as a catalyst for hydrodechlorination of tetrachloroethylene (TTCE) in a continuous fixed-bed reactor made of stainless steel. The authors reported that the conversion of TTCE by sulfided RM increased with an increase in pressure and temperature. The sulfided RM was an active hydrodechlorination catalyst owing to its iron sulfide content. The same authors used sulfided RM as a catalyst for hydrodechlorination of TTCE at 100 bar pressure and 350°C (Ordóñez et al., 2001b) and noted an increase in the life of the catalyst in the presence of carbon sulfide. Furthermore, the same authors prepared ARM by dissolving the RM in an acid solution (HCl or a mixture of HCl/H_3PO_4) followed by ammonia precipitation, and subsequently used it for the hydrodechlorination of TTCE in an organic medium; they compared with that of unmodified RM (Ordóñez et al., 2001c). The modified RM catalysts had better activity and resistance to deactivation than unmodified RM. Dechlorination of different organochlorinated compounds such as 1,2-dichloroethane (DCE), 1,1,2-trichloroethane, 1,1,2,2-tetrachloroethane, trichloroethane, and tetrachloroethane was catalyzed over pretreated RM and compared with other catalysts such as Fe_2O_3, Fe_2O_3/Al_2O_3 and iron-containing modified zeolite Fe-ZSM5 (Halász et al., 2005). The surface area of calcined RM was 37 m^2 g^{-1} and the conversion percentage of DCE was 39% at 300°C, which was superior to the other tested catalyst.

19.3.7.3 Use of Red Mud as Catalyst in Oxidation of Hydrocarbon

RM was used (Muhammad et al., 2012) as a support to prepare Co-oxide—based catalysts (Co/RM). The heterogeneous catalytic activity was calculated in aqueous phenol degradation using an oxidizing agent peroxymonosulfate. The authors reported that the Co_3O_4 particles were highly dispersed on the RM surface and that pretreatment of RM (RM-T) significantly influence the catalytic activity. The phenol degradation followed first-order kinetics and activation energies on Co/RM-T and Co/RM-NT were 46.2 and 47.0 kJ mol^{-1}, respectively. Karimi et al. (2012) reported that RM was an effective catalyst for the conversion of carboxylic acids. The RM was prereduced using formic and/or a blend of formic acid and acetic acid, or by reduction with H_2 (g) at $T > 350$°C in situ to give a reduced RM with a nonalkaline magnetic solid. The catalyst converted biomass-derived levulinic acid into a blend of C9 alkenes and alkanes up to 76% (w/w) yield in the presence of H_2 (g) and could be reused as a catalyst without loss of activity. Sushil et al. (2011) used RM as a source of catalyst along with other metal oxides for the oxidation of propane, a volatile organic compound. The enhancement of catalytic activity of RM was achieved by adding other active components such as chromium, copper, and nickel oxides. The treated RM with the highest Fe_2O_3 content and highest surface area provided the highest conversion in the temperature range 100–500°C with a gas hourly space velocity of 5000 h^{-1}. A 50% conversion in the range 320–380°C was attained. Paredes et al. (2004) used RM to prepare three iron-based catalysts by dissolution-precipitation for the catalytic combustion of methane and to study the activity and stability of the catalyst. The first RM was activated by the method of Pratt and Christoverson (1982). The second RM was activated by dissolving RM in a mixture of ortho-phosphoric acid and hydrochloric acid followed by precipitation and calcination. The activity and stability of the RM materials were compared with commercial catalyst of iron oxide (hematite) and Cu-Cr-Ti. The RM showed the least activity whereas RM activated by Pratt and Christoverson (1982) exhibited higher catalytic activity and good thermal stability. Lamonier et al. (2005) evaluated the catalytic potential of RM for the oxidation of VOCs in particular toluene. The major components of RM Al_2O_3 and Fe_2O_3 were active and CO_2 selective in the toluene oxidation reaction. The authors correlated the RM activity directly to the amount of iron in the sampling, and the chemical reaction–specific rate (by grams of Fe_2O_3) in the presence of RM was similar to that observed in the presence of a pure Fe_2O_3 iron oxide. RM and fly ash are used as supports for the synthesis of Co-oxide–based catalysts Co/RM and Co/FA, which are also used in oxone activation for phenol degradation for wastewater treatment (Saputra et al., 2012). It was reported that Co/R has a much higher efficiency in phenol degradation than Co/FA. The Co oxide on the catalysts was present in the Co_3O_4 phase and the dispersion was found to be higher on RM. The phenol degradation followed first-order kinetics for both catalysts and the activation energies for Co/RM and Co/FA were 66.3 and 47.0 kJ mol^{-1}, respectively.

19.4 CONCLUSION

In this chapter, we emphasized various proposals for the use of RM in different industrial applications to solve the problem of RM as a waste material. Each constituent of RM is useful in isolation but is treated as waste in combination. In bulk, RM is used as building materials in the form of cement, concrete, bricks, and glass ceramics. Because of the superfine particle properties of RM, the cement produced exhibits good physical and excellent mechanical properties. The comprehensive strength was found to increase from 26.5 to 30.8 MPa after 2 days and 57.5–62.9 MPa after 28 days compared with normal cement. RM can be used from 40% to 90% by weight to make concrete with varied proportions of other materials which exhibit more strength than normal concrete. Researchers have reported that 40–90 wt% of RM is used to produce bricks which have demonstrated the properties of good bricks. Similarly, the RM are used to produce glass ceramics by adding varying quantities of RM, fly ash, and other materials which have better mechanical properties. Because RM consists of fine particles and contains many metal oxides with porous structures, researchers have widely investigated the adsorption of contaminants with RM in both the liquid and solid phases and have found them to be efficient for the removal of anions, heavy metals, organic (dyes and phenols), and bacteria from water and wastewater. However, after activation by acidification, heat, and modification with other materials, the RM was found to have a high sorption capacity. Because of the presence of activity-enhancing constituents (Fe and Ti), sintering agents (Na and Ca), and phases of metal oxide present (anatase

TiO_2, rutile TiO_2 etc.), the modified RM was widely used by researchers for catalytic reactions such as hydrogenation, dechlorination, and hydrodechlorination, as well as oxidation of hydrocarbon. It has also been observed that the batch adsorption process is preferred by many researchers because adsorption appears to be an attractive method for removing contaminants from water. It is an eco-friendly, simple, and cost-effective option for the average person using groundwater as a main source of water. Although RM is capable of being employed for many value-added uses, the problem of RM remains, indicating the need for more intensive research for the practical use of such an important material.

ABBREVIATIONS

ABCS	Activated Bauxsol-coated sand
ANN	Artificial neural network
ANRM	Activated CO_2-neutralized red mud
ARM	Activated red mud
ATB	Acid treated Bauxsol
BCS	Bauxsol-coated sand
C_2ASH_8	Stratlingite
C4AF	Calcium aluminoferrite
CCD	Close cycle disposal
DCE	1,2-dichloroethane
DPE	Diphenylether
DPM	Diphenylmethane
ELM	Emulsion liquid membrane
ETP	Effluent treatment plant
FA	Ferroalumina
GHSV	Gas hourly space velocity
GPTMS	3-Glycidoxypropyltrimethoxysilane
GRM	Granular red mud
GRM	Granular red mud ceramics
K_{obs}	First-order rate constant
OPC	Ordinary Portland cement
OPh	Potassium orthophosphate
PACl	Polyaluminum chloride
PMS	Peroxymonosulphate
PRBs	Permeable reactive barriers
RB 19	Blue dye 19
RB	Reactive barrier
RBRM	Rinsed Bosnian red mud
RM	Red mud
RM_a	Acid-treated red mud

Continued

RMGA	Red mud granulated adsorbent
RM_{nt}	Untreated red mud
RM-T	Pretreatment red mud
RSM	Response surface methodology
SBR	Sequencing batch reactor
SDS/RM	Sodium dodecyl sulfate modified red mud
SNR	Sea nodule residue
SRM	Unmodified red mud
SWN-RM	Seawater-neutralized red mud
SWRM	Sea water neutralized red mud
SWRM400	Sea water neutralized red mud heated with 400°C
SWRM500	Sea water neutralized red mud heated with 500°C
TBP	Tri-butyl phosphate
TCEE	Tetrachloroethane
TPOZ	Tetra-n-propoxyzirconium
TPPh	Sodium tripolyphosphate
TRCEE	Trichloroethane
TTCE	Tetrachloroethylene

ACKNOWLEDGMENTS

The authors are thankful to Prof. S.K. Sarangi, Director, National Institute of Technology, Rourkela, for providing necessary facilities and necessary help in carrying out this work. The authors are also thankful to Vedanta Alumina Limited, Langigarh, Odisha, India, for necessary funds to carry out the red mud research.

REFERENCES

Agrawal, A., Sahu, K.K., Pandey, B.D., 2004a. Solid waste management in non-ferrous industries in India. Resources, Conservation and Recycling 42, 99–120.

Agrawal, A., Sahu, K.K., Pandey, B.D., 2004b. A comparative adsorption study of copper on various industrial solid wastes. AIChE Journal 50, 2430–2438.

Akhurst, D.J., Jones, G.B., Clark, M., McConchie, D., 2006. Phosphate removal from aqueous solutions using neutralised bauxite refinery residues (Bauxsol™). Environmental Chemistry 3, 65–74.

Altundoğan, H.S., Altundoğan, S., Tümen, F., Bildik, M., 2000. Arsenic removal from aqueous solutions by adsorption on red mud. Waste Management 20, 761–767.

Altundoğan, H.S., Altundoğan, S., Tümen, F., Bildik, M., 2002. Arsenic adsorption from aqueous solutions by activated red mud. Waste Management 22, 357–363.

Altundoğan, H.S., Tümen, F.F., 2003. As(V) removal from aqueous solutions by coagulation with liquid phase of red mud. Journal of Environmental Science and Health – Part A Toxic/Hazardous Substances and Environmental Engineering 38, 1247–1258.

Akin, I., Arslan, G., Tor, A., Ersoz, M., Cengeloglu, Y., 2012. Arsenic(V) removal from underground water by magnetic nanoparticles synthesized from waste red mud. The Journal of Hazardous Materials 235–236, 62–68.

Apak, R., Tutem, E., Hugul, M., Hizal, J., 1998. Heavy metal cation retention by unconventional sorbents (red mud and fly ashes). Water Research 32, 430–440.

Arias, M., López, E., Nuñez, A., Rubinos, D., Soto, B., Barral, M.T., Díaz-Fierros, F., 1999. In: Berthelin, J., Huang, P.M., Bollag, J.-M., Andreux, F. (Eds.), Adsorption of Methylene Blue by Red Mud, an Oxide-Rich Byproduct of Bauxite Refining.

de-Bashan, L.E., Bashan, Y., 2004. Recent advances in removing phosphorus from wastewater and its future use as fertilizer (1997–2003). Water Research 38, 4222–4246.

Brunori, C., Crmisini, C., Massanisso, P., Pinto, V., Torricelli, L., 2005. Reuse of treated red bauxite waste: studies on environmental compatibility. Journal of Hazardous Materials 117, 55–63.

Bott, R., Langeloh, T., Hahn, J., 2002. Dry bauxite residue by hi-bar steam pressure filtration. In: Chandrashekar, S. (Ed.), 6th International Alumina Quality Workshop, pp. 24–32. Brisbane.

Bayer, G., Haerter, M., 1975. Method for producing bricks from red mud, US Patent 3886244.

Bhat, P.N., Ghosh, D.K., Desai, M.V.M., 2002. Immobilisation of beryllium in solid waste (red mud) by fixation and vitrification. Waste Management 22, 549–556.

Brown, S.O., Kirkpatrick, D.B., 1999. Red mud product development. In: Light Metals, pp. 25–30.

Chandra, S., 1996. Waste Materials Used in Concrete Manufacturing.

Collazo, A., Cristobal, M.J., Novoa, X.R., Pena, G., Perez, M.C., 2006. Electrochemical impedance spectroscopy as a tool for studying steel corrosion inhibition in simulated concrete environments-red mud used as rebar corrosion. Journal of ASTM International 3, 1–10.

Chaddha, M.J., 2009. Asia pacific partnership of clean development & climate on industry perspectives-an overview, Jawaharlal Nehru Aluminum Research Development and Design Centre.

Costa, R.C.C., Moura, F.C.C., Oliveira, P.E.F., Magalhaes, F., Ardisson, J.D., Lago, R.M., 2010. Controlled reduction of red mud waste to produce active systems for environmental applications: heterogeneous Fenton reaction and reduction of Cr(VI). Chemosphere 78, 1116–1120.

Collazo, A., Covelo, A., Nóvoa, X.R., Pérez, C., 2012. Corrosion protection performance of sol–gel coatings doped with red mud applied on AA2024-T3. Progress in Organic Coatings 74, 334–342.

Cengeloglu, Y., Kir, E., Ersoz, M., 2002. Removal of fluoride from aqueous solution by using red mud. Separation and Purification Technology 28, 81–86.

Cengeloglu, Y., Tor, A., Ersoz, M., Arslan, G., 2006. Removal of nitrate from aqueous solution by using red mud. Separation and Purification Technology 51, 374–378.

Cho, D.W., Abou-Shnab, R.A.I., Kim, Y., Jeon, B.H., Song, H., 2011. Enhanced reduction of nitrate in groundwater by zero-valent iron with activated red mud. Geosystem Engineering 14, 65–70.

Castaldi, P., Silvetti, M., Enzo, S., Melis, P., 2010. Study of sorption processes and FT-IR analysis of arsenate sorbed onto red muds (a bauxite ore processing waste). Journal of Hazardous Materials 175, 172–178.

Cengeloglu, Y., Tor, A., Arslan, G., Ersoz, M., Gezgin, S., 2007. Removal of boron from aqueous solution by using neutralized red mud. Journal of Hazardous Materials 142, 412–417.

Despland, L.M., Clark, M.W., Aragno, M., Vancov, T., 2010. Minimising alkalinity and pH spikes from Portland cement-bound Bauxsol (seawater-neutralized red mud) pellets for pH circum-neutral waters. Environmental Science & Technology 44, 2119–2125.

Dimas, D.D., Giannopoulou, I.P., Panias, D., 2009. Utilization of alumina red mud for synthesis of inorganic polymeric materials. Mineral Processing and Extractive Metallurgy Review 30, 211–239.

Dursun, S., Guclu, D., Berktay, A., Guner, T., 2008. Removal of chromate from aqueous system by activated red-mud. Asian Journal of Chemistry 20, 6473–6478.

Enick, R.M., Beckman, E.J., Shi, C., Xu, J., 2001. Remediation of metal-bearing aqueous waste streams via direct carbonation. Energy & Fuels 15, 256–262.

Emmett, R.C., Klepper, R.P., 1991. High density red mud thickeners. In: Rooy, E.L. (Ed.), New Orleans Light Metals, pp. 229–233.

Eamsiri, A., Jackson, W.R., Pratt, K.C., Christov, V., Marshall, M., 1992. Activated red mud as a catalyst for the hydrogenation of coals and of aromatic compounds. Fuel 71, 449–453.

Fois, E., Lallai, A., Mura, G., 2007. Sulfur dioxide absorption in a bubbling reactor with suspensions of Bayer red mud. Industrial & Engineering Chemistry Research 46, 6770–6776.

Fortin, J., Karam, A., 2001. Phosphorus sorption by red mud residue as affected by concentration and reaction time. Agrochimica 45, 55–66.

Fan, H.L., Li, C.H., Xie, K.C., 2005. Testing of iron oxide sorbent for high-temperature coal gas desulfurization. Energy Sources 27, 245–250.

Genç-Fuhrman, H., Tjell, J.C., McConchie, D.M., Schuiling, O., 2003. Adsorption of arsenate from water using neutralized red mud. Journal of Colloid and Interface Science 264, 327–334.

Genç-Fuhrman, H., Tjell, J.C., McConchie, D.M., 2004. Increasing the arsenate adsorption capacity of neutralized red mud (Bauxsol). Journal of Colloid and Interface Science 271, 313–320.

Genç-Fuhrman, H., Bregnhøj, H., McConchie, D., 2005. Arsenate removal from water using sand–red mud columns. Water Research 39, 2944–2954.

Gu, H.N., Wang, N., Liu, S.R., Tian, Y.J., 2012. Composition and particle characteristics of sintering process red mud. Rock and Mineral Analysis 31, 312–317.

Gordon, J.N., Pinnockh, W.R., Mooreh, M.M., 1996. A preliminary investigation of strength development in Jamaican red mud composites. Cement and Concrete Composites 18, 371–379.

Gupta, V.K., Sharma, S., 2002. Removal of cadmium and zinc from aqueous solutions using red mud. Environmental Science and Technology 36, 3612–3617.

Gupta, V.K., Gupta, M., Sharma, S., 2001. Process development for the removal of lead and chromium from aqueous solutions using red mud—an aluminium industry waste. Water Research 35, 1125–1134.

Geyikçi, F., Kılıç, E., Çoruh, S., Elevli, S., 2012. Modelling of lead adsorption from industrial sludge leachate on red mud by using RSM and ANN. Chemical Engineering Journal 183, 53–59.

Gupta, V.K., Suhas, Ali, I., Saini, V.K., 2004a. Removal of rhodamine B, fast green, and methylene blue from wastewater using red mud, an aluminum industry waste. Industrial & Engineering Chemistry Research 43, 1740–1747.

Gupta, V.K., Ali, I., Saini, V.K., 2004b. Removal of chlorophenols from wastewater using red mud: an aluminum industry waste. Environmental Science & Technology 38, 4012–4018.

Gupta, V.K., Ali, I., 2006. Removal of 2,4-dinitrophenol from wastewater by adsorption technology: a batch and column study. International Journal of Environment and Pollution 27, 104–120.

Hind, A.R., Bhargava, K.S., Grocott, S.C., 1999. The surface chemistry of Bayer process solids: a review. Colloids and Surfaces A: Physicochemical and Engineering Aspects 146, 359–374.

Hanahan, C., McConchie, D., Pohl, J., Creelman, R., Clark, M., Stocksiek, C., 2004. Chemistry of seawater neutralization of bauxite refining residues (red mud). Environmental Engineering science 21, 125–138.

Huang, W., Wang, S., Zhu, Z., Li, L., Yao, X., Rudolph, V., Haghseresht, F., 2008. Phosphate removal from wastewater using red mud. Journal of Hazardous Materials 158, 35–42.

Hannachi, Y., Shapovalov, N.A., Hannachi, A., 2010. Adsorption of nickel from aqueous solution by the use of low-cost adsorbents. Korean Journal of Chemical Engineering 27, 152–158.

Han, S.W., Kim, D.K., Hwang, I.G., Bae, J.H., 2002. Development of pellet-type adsorbents for removal of heavy metal ions from aqueous solutions using red mud. Journal of Industrial and Engineering Chemistry 8, 120–125.

Ho, G.E., Gibbs, R.A., Mathew, K., 1990. Bacteria and virus removal from secondary effluent in sand and red mud columns. Water Science and Technology 23, 261–270.

Halász, J., Hodos, M., Hannus, I., Tasi, G., Kiricsi, I., 2005. Catalytic detoxification of C_2-chlorohydrocarbons over iron-containing oxide and zeolite catalysts. Colloids and Surfaces A: Physicochemical and Engineering Aspects 265, 171–177.

Indian Aluminium Industry. 2009.

Jones, G., Joshi, G., Clark, M., McConchie, D., 2006. Carbon capture and the aluminium industry: preliminary studies. Environmental Chemistry 3, 297.

Kurniawan, T.A., Chan, G.Y., Lo, W.H., Babel, S., 2006. Comparisons of low-cost adsorbents for treating wastewaters laden with heavy metals. Science of the Total Environment 366, 409–426.

Karimi, E., Gomez, A., Kycia, S.W., Schlaf, M., 2010. The thermal decomposition of acetic and formic acid catalyzed by red mud—implications for the potential use of red mud as a pyrolysis bio-oil upgrading catalyst. Energy & Fuels 24, 2747–2757.

Kumar, R., Srivastava, J.P., Premchand, 1998. Utilization of iron values of red mud for metallurgical applications. In: Bandopadhyay, A., Goswami, N.G., Rao, P.R. (Eds.), Environmental and Waste Management, Jamshedpur. National Metallurgical Laboratory.

Kumar, V., Nautiyal, B.D., Jha, A.K., 1989. Use of neutralized red mud in concrete. Indian Concrete Journal 63, 505–507.

Kavas, T., 2006. Use of boron waste as a fluxing agent in production of red mud brick. Building and Environment 41, 1779–1783.

Karasu, B., Caki, C., Akgun, E., Kaya, G., 2004. Effects of the red mud based pigment addition on the physical and micro-structural properties of porcelain tiles. Key Engineering Materials 264–268, 1613–1616.

Koumanova, B., Drame, M., Popangelova, M., 1997. Phosphate removal from aqueous solutions using red mud wasted in bauxite Bayer's process. Resources, Conservation and Recycling 19, 11–20.

Komnitsas, K., Bartzas, G., Paspaliaris, I., 2004. Efficiency of limestone and red mud barriers: laboratory column studies. Minerals Engineering 17, 183–194.

Karimi, E., Teixeira, I.F., Ribeiro, L.P., Gomez, A., Lago, R.M., Penner, G., Kycia, S.W., Schlaf, M., 2012. Ketonization and deoxygenation of alkanoic acids and conversion of levulinic acid to hydrocarbons using a red mud bauxite mining waste as the catalyst. Catalysis Today 190, 73–88.

Li, X., Xiao, W., Liu, W., Liu, G., Peng, Z., Zhou, Q., Qi, T., 2009. Recovery of alumina and ferric oxide from Bayer red mud rich in iron by reduction sintering. Transactions of Nonferrous Metals Society of China 19, 1342–1347.

Li, Y., Liu, C., Luan, Z., Peng, X., Zhu, C., Chen, Z., Zhang, Z., Fan, J., Jia, Z., 2006. Phosphate removal from aqueous solutions using raw and activated red mud and fly ash. Journal of Hazardous Materials 137, 374–383.

Li, Y., Wang, J., Luan, Z., Liang, Z., 2010. Arsenic removal from aqueous solution using ferrous based red mud sludge. Journal of Hazardous Materials 177, 131–137.

Liu, W., Yang, J., Xiao, B., 2009a. Review on treatment and utilization of bauxite residues in China. International Journal of Mineral Processing 93, 220–231.

Liu, W., Yang, J., Xiao, B., 2009b. Application of Bayer red mud for iron recovery and building material production from alumosilicate residues. Journal of Hazardous Materials 161, 474–478.

Liu, C., Li, Y., Luan, Z., Chen, Z., Zhang, Z., Jia, Z., 2007. Adsorption removal of phosphate from aqueous solution by active red mud. Journal of Environmental Sciences 19, 1166–1170.

Lopez, E., Soto, B., Arias, M., Nuñez, A., Rubinos, D., Barral, M.T., 1998. Adsorbent properties of red mud and its use for waste water treatment. Water Research 32, 1314–1322.

Lopes, G., Guilherme, L.R.G., Costa, E.T.S., Curi, N., Penha, H.G.V., 2012. Increasing arsenic sorption on red mud by phosphogypsum addition. Journal of Hazardous Materials. http://dx.doi.org/10.1016/j.jhazmat.2012.06.051.

Luo, L., Ma, C., Ma, Y., Zhang, S., Lv, J., Cui, M., 2011. New insights into the sorption mechanism of cadmium on red mud. Environmental Pollution 159, 1108–1113.

Liano, J.J., Rosal, R., Sastre, H., Diez, F.V., 1994. Catalytic hydrogenation with red mud. Fuel 73, 688–694.

Lamonier, J.F., Wyralski, F., Leclercq, G., Aboukais, A., 2005. Recycling of waste, red mud, as a catalyser for the elimination of volatile organic compounds. The Canadian Journal of Chemical Engineering 83, 737–741.

Mohan, D., Pittman, C.U., 2007. Arsenic removal from water/wastewater using adsorbents – a critical review. Journal of Hazardous Materials 142, 1–53.

Meher, S.N., Rout, A.K., Padhi, B.K., 2010. Extraction of Al and Na from red mud by magnesium oxide sodium carbonate sinter process. African Journal of Environmental Science and Technology 4, 897–902.

Mohanty, S., Pradhan, J., Das, S.N., Thakur, R.S., 2004. Removal of phosphorus from aqueous solution using alumized red mud. International Journal of Environmental Studies 61, 687–697.

McConchie, D., Saenger, P., Fawkes, R., 1996. An environmental assessment of the use of seawater to neutralise bauxite refinery wastes. In: Proceedings of 2nd International Symposium on Extraction and Processing for the Treatment and Minimisation of Wastes, Minerals, Metals and Materials Society, pp. 407–416.

Mohapatra, M., Anand, S., Mishra, B.K., Giles, D.E., Singh, P., 2009. Review of fluoride removal from drinking water. Journal of Environmental Management 91, 67–77.

Mastral, A.M., Mayoral, C., Izquierdo, M.T., Pardos, C., 1993. Iron catalyzed hydrogenation of high sulphur content coals. Fuel Processing Technology 36, 177–184.

Muhammad, S., Saputra, E., Sun, H., Ang, H., Tade, M.O., Wang, S., 2012. Heterogeneous catalytic oxidation of aqueous phenol on red mud-supported cobalt catalysts. Industrial & Engineering Chemistry Research 51, 15351–15359.

Newson, T., Dyer, T., Adam, C., Sharp, S., 2006. Effect of structure on the geotechnical properties of bauxite residue. Journal of Geotechnical Engineering 132, 143–151.

Nadaroglu, H., Kalkan, E., Demir, N., 2010. Removal of copper from aqueous solution using red mud. Desalination 251, 90–95.

Namasivayam, C., Arasi, D.J.S.E., 1997. Removal of congo red from wastewater by adsorption onto waste red mud. Chemosphere 34, 401–417.

Namasivayam, C., Yamuna, R.T., Arasi, D.J.S.E., 2001. Removal of acid violet from wastewater by adsorption on waste red mud. Environmental Geology 41, 269–273.

Namasivayam, C., Yamuna, R.T., Arasi, D.J.S.E., 2002. Removal of procion orange from wastewater by adsorption on waste red mud. Separation Science and Technology 37, 2421–2431.

Nan, X., Zhang, T., Zhang, L., 2012. Study on absorption of low-concentration SO_2 with basic slag intensified by ultrasonic wave. In: Suarez, C.E. (Ed.), Light Metals.

Ochsenku, M.T., Hatzilyberis, K.S., Mendrinos, L.N., Salmas, C.E., 2002. Pilot-plant investigation of the leaching process for the recovery of scandium from red mud. Industrial & Engineering Chemistry Research 41, 5794–5801.

de Oliveira, D.R.C., Rossi, C.R.C., 2012. Concretes with red mud coarse aggregates. Materials Research 15, 333–340.

Ordóñez, S., Sastre, H., Díez, F.V., 2001a. Catalytic hydrodechlorination of tetrachloroethylene over red mud. Journal of Hazardous Materials 81, 103–114.

Ordóñez, S., Sastre, H., Dıez, F.V., 2001b. Characterisation and deactivation studies of sulfided red mud used as catalyst for the hydrodechlorination of tetrachloroethylene. Applied Catalysis B: Environmental 29, 263–273.

Ordóñez, S., Sastre, H., Dıez, F.V., 2001c. Hydrodechlorination of tetrachloroethylene over modified red mud: deactivation studies and kinetics. Applied Catalysis B: Environmental 34, 213–226.

Piga, L., Pochetti, F., Stoppa, L., 1993. Recovering metals from red mud generated during alumina production. Journal of the Minerals, Metals, and Materials Society 45, 54–59.

Prasad, P.M., Kachawha, J.S., Gupta, R.C., Mankhand, T.R., Sharma, J.M., 1985. Light Metals Science and Technology. Trans. Tech. Publications Ltd, Switzerland.

Power, G., Gräfe, M., Klauber, C., 2011. Bauxite residue issues: I. Current management, disposal and storage practices. Hydrometallurgy 108, 33–45.

Paramguru, R.K., Rath, P.C., Misra, V.N., 2004. Trends in red mud utilization – a review. Mineral Processing and Extractive Metallurgy Review 26, 1–29.

Pattajoshi, P.K., 2006. Assessment of airborne dust associated with chemical plant: a case study. Indian Journal of Occupational and Environmental Medicine 10, 32–34.

Palmer, S.J., Nothling, M., Bakon, K.H., Frost, R.L., 2010. Thermally activated seawater neutralized red mud used for the removal of arsenate, vanadate and molybdate from aqueous solution. Journal of Colloid and Interface Science 342, 147–154.

Pan, Z., Cheng, L., Lu, Y., Yang, N., 2002. Hydration products of alkali-activated slag–red mud cementitious material. Cement and Concrete Research 32, 357–362.

Peng, F., Liang, K.M., Shao, H., Hu, A.M., 2005. Nano-crystal glass-ceramics obtained by crystallization of vitrified red mud. Chemosphere 59, 899–903.

Pera, J., Boumaza, R., Ambroise, J., 1997. Development of a pozzolonic pigment from red mud. Cement and Concrete Research 27, 1513–1522.

Pradhan, J., Das, J., Das, S., Thakur, R.S., 1998. Adsorption of phosphate from aqueous solution using activated red mud. Journal of Colloid and Interface Science 204, 169–172.

Pradhan, J., Das, S.N., Thakur, R.S., 1999. Adsorption of hexavalent chromium from aqueous solution by using activated red mud. Journal of Colloid and Interface Science 217, 137–141.

Pratt, K.C., Christoverson, V., 1982. Hydrogenation of a model hydrogen-donor system using activated red mud catalyst. Fuel 61, 460–462.

Paredes, J.R., Ordóñez, S., Vega, A., Dıez, F.V., 2004. Catalytic combustion of methane over red mud-based catalysts. Applied Catalysis B: Environmental 47, 37–45.

Queensland Aluminium Limited. http://www.qal.com.au/Process.html.

Qu, Y., Lian, B., Mo, B., Liu, C., 2013a. Bioleaching of heavy metals from red mud using *Aspergillus niger*. Hydrometallurgy 136, 71–77.

Qu, Y., Lian, B., 2013b. Bioleaching of rare earth and radioactive elements from red mud using *Penicillium tricolor* RM-10. Bioresource Technology 136, 16–23.

Qi, J.Z., 2005. Experimental research on road materials of red mud. University of Huazhong Science and Technology, Wuhan, China.

Rai, S., Wasewar, K.L., Mukhopadhyay, J., Yoo, C.K., Uslu, H., 2012. Neutralization and utilization of red mud for its better waste management. Archives of Environmental Science 6, 13–33.

Ribeiro, D.V., Labrincha, J., Morelli, M.R., 2012. Effect of calcined red mud addition on the hydration of Portland cement. Materials Science Forum 727–728, 1408–1411.

Romero, M., Rincon, J.M., 2000. The controlled vitrification/crystallisation process applied to the recycling of inorganic industrial wastes. Boletín de la Sociedad Española de Cerámica y Vidrio 39, 155–163.

Romero, M., Rincon, J.M., 1999. Surface and bulk crystallization of glass-ceramic in the $Na_2O-CaO-ZnO-PbO-Fe_2O_3-Al_2O_3-SiO_2$ system derived from a goethite waste. The Journal of the American Ceramic Society 82, 1313–1317.

Ribeiro, D.V., Labrincha, J.A., Morelli, M.R., 2012. Cement and concrete research effect of the addition of red mud on the corrosion parameters of reinforced concrete. Cement and Concrete Research 42, 124–133.

Roberge, G., Blais, J.F., Mercier, G., 1999. Phosphorus removal from wastewater treated with red mud-doped peat. The Canadian Journal of Chemical Engineering 77, 1185–1194.

Rajkumar, M., Nagendran, N., Sasikumar, S.S., 2001. Removal of trivalent chromium from wastewater using red mud. Indian Journal of Environmental Protection 21, 97–100.

Schmitz, C., 2006. 'Red Mud Disposal' Handbook of Aluminium Recycling.
Solymar, K., Ferenczi, T., Papanastassiou, D., 2002. Digestion alternatives of the Greek diasporic bauxite. In: Schneider, W. (Ed.), Seattle Light Metals, pp. 75–81.
Smiljanic, S., Smiciklas, I., Peric-Grujic, A., Loncar, B., Mitric, M., 2010. Rinsed and thermally treated red mud sorbents for aqueous Ni^{2+} ions. Chemical Engineering Journal 162, 75–83.
Srnirnov, D.I., Molchanova, T.V., 1997. The investigation of sulphuric acid sorption recovery of scandium and uranium from the red mud of alumina production. Hydrometallurgy 45, 249–259.
Singh, M., Upadhayay, S.N., Prasad, P.M., 1997. Preparation of iron rich cements using red mud. Cement and Concrete Research 27, 1037–1046.
Sawant, A.B., Kumthekar, M.B., Diwan, V.V., Hiraskar, K.G., 2012. Experimental study on partial replacement of cement by neutralized red mud in concrete. International Journal of Engineering & Advanced Technology 2, 282–286.
Satapathy, A., Mishra, S.C., Ananthapadmanabhan, P.V., Sreekumar, K.P., 2007. Development of ceramic coatings using red mud – a solid waste of alumina plants. The Journal of Solid Waste Technology and Management 33, 48–53.
Shiao, S.J., Akashi, K., 1977. Phosphate removal from aqueous solution from activated red mud. Water Pollution Control Federation 49, 280–285.
Sahu, R.C., Patel, R., Ray, B.C., 2010. Utilization of activated CO_2-neutralized red mud for removal of arsenate from aqueous solutions. Journal of Hazardous Materials 179, 1007–1013.
Singh, A.P., Srivastava, K.K., Shekhar, H., 2008. Effect of pH on removal of arsenic(III) by mixed adsorbents. The Proceedings of the National Academy of Sciences 78, 283–286.
Smičiklas, I., Smiljanić, S., Perić-Grujić, A., Šljivić-Ivanović, M., Antonović, D., 2013. The influence of citrate anion on Ni(II) removal by raw red mud from aluminum industry. Chemical Engineering Journal 214, 327–335.
Sahu, M.K., Mandal, S., Dash, S.S., Badhai, P., Patel, R.K., 2013. Removal of Pb(II) from aqueous solution by acid activated red mud. The Journal of Environmental Chemical Engineering 1, 1315–1324.
Sahu, M.K., Mandal, S., Yadav, L.S., Dash, S.S., Patel, R.K., 2015. Equilibrium and kinetic studies of Cd(II) ion adsorption from aqueous solution by activated red mud. Desalination and Water Treatment 2, 1–15.
Santona, L., Castaldi, P., Melis, P., 2006. Evaluation of the interaction mechanisms between red muds and heavy metals. Journal of Hazardous Materials 136, 324–329.
Shi, L., Peng, X., Luan, Z., Wei, N., Wang, Q., Zhao, Y., 2009. Use of activated red mud to remove phosphate and heavy metals from the effluent of biologically treated swine wastewater. Acta Scientiae Circumstantiae 29, 2282–2288.
Sahu, M.K., Patel, R.K., 2015. Removal of safranin-O dye from aqueous solution using modified red mud: kinetics and equilibrium studies. RSC Advances 5, 78491–78501.
de Souza, K.C., Antunes, M.L.P., Couperthwaite, S.J., da Conceição, F.T., de Barros, T.R., Frost, R., 2013. Adsorption of reactive dye on seawater-neutralised bauxite refinery residue. Journal of Colloid and Interface Science 396, 210–214.
Sahu, R.C., Patel, R., Ray, B.C., 2011. Removal of hydrogen sulfide using red mud at ambient conditions. Fuel Processing Technology 92, 1587–1592.
Sushil, S., Batra, V.S., 2012. Modification of red mud by acid treatment and its application for CO removal. Journal of Hazardous Materials 203–204, 264–273.
Sushil, S., Scholz, P., Pollok, K., Ondruschka, B., Batra, V.S., 2011. Application of industrial waste based catalysts for total oxidation of propane. Chemical Engineering Journal 166, 568–578.
Saputra, E., Muhammad, S., Sun, H., Ang, H.M., Tadé, M.O., Wang, S., 2012. Red mud and fly ash supported Co catalysts for phenol oxidation. Catalysis Today 190, 68–72.

Tian, Y., Wang, F.X., Ma, S.C., 1998. The mechanical properties of red mud stockpiling. In: Light Metals, vol. 2, pp. 32–34 (In Chinese).

Tor, A., Danaoglu, N., Arslan, G., Cengeloglu, Y., 2009a. Removal of fluoride from water by using granular red mud: batch and column studies. Journal of Hazardous Materials 164, 271–278.

Tor, A., Cengeloglu, Y., Ersoz, M., 2009b. Increasing the phenol adsorption capacity of neutralized red mud by application of acid activation procedure. Desalination 242, 19–28.

Tsakiridis, P.E., Oustadakis, P., Katsiapi, A., Perraki, M., Agatzini-Leonardou, S., 2011. Synthesis of TiO_2 nano-powders prepared from purified sulphate leach liquor of red mud. Journal of Hazardous Materials 194, 42–47.

Tsakiridis, P.E., Agatzini-Leonardou, S., Oustadakis, P., 2004. Red mud addition in the raw meal for the production of Portland cement clinker. Journal of Hazardous Materials 116, 103–110.

Tor, A., Cengeloglu, Y., 2006. Removal of congo red from aqueous solution by adsorption onto acid activated red mud. Journal of Hazardous Materials 138, 409–415.

Tor, A., Cengeloglu, Y., Aydin, M.E., Ersoz, M., 2006. Removal of phenol from aqueous phase by using neutralized red mud. Journal of Colloid and Interface Science 300, 498–503.

Uysal, B.Z., Aksahin, I., Yucel, H., 1988. Sorption of SO_2 on metal oxides in a fluidized bed. Industrial & Engineering Chemistry Research 27, 434–439.

Vangelatos, I., Angelopoulos, G.N., Boufounos, D., 2009. Utilization of ferroalumina as raw material in the production of ordinary Portland cement. Journal of Hazardous Materials 168, 473–478.

Vaclavikovaa, M., Misaelidesb, P., Galliosb, G., Jakabskya, S., Hredzaka, S., 2005. Removal of cadmium, zinc, copper and lead by red mud, an iron oxides containing hydrometallurgical waste. Studies in Surface Science and Catalysis 155, 517–525.

Wagh, A.S., Desai, P., 1987. Bauxite Tailings: 'Red Mud'.

Wang, S., Ang, H.M., Tade, M.O., 2008. Novel applications of red mud as coagulant, adsorbent and catalyst for environmentally benign processes. Chemosphere 72, 1621–1635.

Wehr, J.B., Fulton, I., Menzies, N.W., 2006. Revegetation strategies for bauxite refinery residue: a case study of Alcan in Northern Territory, Australia. Environmental Management 37, 297–306.

Wang, C.Z., Zhou, G.Z., Kou, X.Y., Zheng, Z.L., 2011. Recovery of Sc^{3+} from red mud leaching solution by emulsion liquid membrane. Advanced Materials Research 335–336, 1465–1468.

Wang, W., Wang, X., Zhu, J., Wang, P., Ma, C., 2013. Experimental investigation and modeling of sulfoaluminate cement preparation using desulfurization gypsum and red mud. Industrial & Engineering Chemistry Research 52, 1261–1266.

Wei, N., Luan, Z.K., Wang, J., Shi, L., Zhao, Y., Wu, J.W., 2009. Preparation of modified red mud with aluminum and its adsorption characteristics on fluoride removal. Chinese Journal of Inorganic Chemistry 25, 849–854.

Wang, S., Boyjoo, Y., Choueib, A., Zhu, Z.H., 2005. Removal of dyes from aqueous solution using fly ash and red mud. Water Research 39, 129–138.

Yue, Q., Zhao, Y., Li, Q., Li, W., Gao, B., Han, S., Qi, Y., Yu, H., 2010. Research on the characteristics of red mud granular adsorbents (RMGA) for phosphate removal. Journal of Hazardous Materials 176, 741–748.

Yang, H., Chen, C., Pan, L., Lu, H., Sun, H., Hu, X., 2009. Preparation of double-layer glass-ceramic/ceramic tile from bauxite tailing and red mud. Journal of the European Ceramic Society 29, 1887–1894.

Yang, J., Xiao, B., 2008. Development of unsintered construction materials from red mud wastes produced in the sintering alumina process. Construction and Building Materials 22, 2299–2307.

Yang, J., Zhang, D., Hou, J., He, B., Xiao, B., 2008. Preparation of glass-ceramics from red mud in the aluminium industries. Ceramics International 34, 125–130.

Yadav, V.S., Prasad, M., Khan, J., Amritphale, S.S., Singh, M., Raju, C.B., 2010. Sequestration of carbon dioxide (CO_2) using red mud. Journal of Hazardous Materials 176, 1044–1050.

Yi, C., Jun-qi, L., Fang, H., Jun, Z., Den-feng, Z., Wei, L., 2007. The Performance Research on Absorbing SO2 Waste Gas with Bayer Red Mud (In Chinese). J. Guizhou Univ. Technol. Nat. Sci. Ed., 36, 30–37.

Yokoyama, S., Yamamoto, M., Maekawa, Y., Kotanigawa, T., 1989. Catalytic activity of sulphate for hydroliquefaction of coal by using diphenylether and diphenylmethane. Fuel 68, 531–533.

Zhao, Y., Wang, J., Luan, Z., Peng, X., Liang, Z., Shi, L., 2009. Removal of phosphate from aqueous solution by red mud using a factorial design. Journal of Hazardous Materials 165, 1193–1199.

Zhang, N., Sun, H., Liu, X., Zhang, J., 2009. Early-age characteristics of red mud–coal gangue cementitious material. Journal of Hazardous Materials 167, 927–932.

Zhu, D., Chun, T., Pan, J., He, Z., 2012. Recovery of iron from high-iron red mud by reduction roasting with adding sodium salt. Journal of Iron and Steel Research International 19, 1–5.

Zhong, L., Zhang, Y., Zhang, Y., 2009. Extraction of alumina and sodium oxide from red mud by a mild hydrochemical process. Journal of Hazardous Materials 172, 1629–1634.

Zhang, R., Zheng, S., Ma, S., Zhang, Y., 2011. Recovery of alumina and alkali in Bayer red mud by the formation of andradite-grossular hydrogarnet in hydrothermal process. Journal of Hazardous Materials 189, 827–835.

Zhou, H., Li, D., Tian, Y., Chen, Y., 2008. Extraction of scandium from red mud by modified activated carbon and kinetics study. Rare Metals 27, 223–227.

Zhang, P.X., Yan, J.Q., 2000. Infrared spectra of crystallization of glasses using red mud as raw materials. Journal of Inorganic Materials 15, 751–755.

Zhao, Y., Zhang, L., Ni, F., Xi, B., Xia, X., Peng, X., Luan, Z., 2011. Evaluation of a novel composite inorganic coagulant prepared by red mud for phosphate removal. Desalination 273, 414–420.

Zhao, Y., Yue, Q., Li, Q., Xu, X., Yang, Z., Wang, X., Gao, B., Yu, H., 2012. Characterization of red mud granular adsorbent (RMGA) and its performance on phosphate removal from aqueous solution. Chemical Engineering Journal 193–194, 161–168.

Zhang, S., Liu, C., Luan, Z., Peng, X., Ren, H., Wang, J., 2008. Arsenate removal from aqueous solutions using modified red mud. Journal of Hazardous Materials 152, 486–492.

Zouboulis, A.I., Kydros, K.A., 1993. Use of red mud for toxic metals removal: the case of nickel. Journal of Chemical Technology and Biotechnology 58, 95–101.

Zhu, C., Luan, Z., Wang, Y., Shan, X., 2007. Removal of cadmium from aqueous solutions by adsorption on granular red mud (GRM). Separation and Purification Technology 57, 161–169.

Zoumis, T., Calmano, W., Förstner, U., 2000. Demobilization of heavy metals from mine waters. Acta Hydrochimica et Hydrobiologica 28, 212–218.

CHAPTER

THERMAL BEHAVIOR OF RED MUD AND ITS BENEFICIAL USE IN GLASS-CERAMIC PRODUCTION

20

C.-Z. Liao, K. Shih

University of Hong Kong, Hong Kong, Hong Kong SAR, China

CHAPTER OUTLINE

20.1 INTRODUCTION

The largest ecological disaster in Hungary's history occurred on October 4, 2010. This accident, which killed 10 people and contaminated about 40 km^2 of agricultural land along the Torma stream and the Marcal River (Ruyters et al., 2011; Gelencser et al., 2011), was caused by red mud stored in the reservoir of an alumina plant in Ajka. This accident drew government attention to the urgent problem of red mud. Red mud is the by-product created when the Bayer process is used to extract alumina from bauxite ore. The amount of red mud generated per ton of alumina varies greatly depending on the type of bauxite used, from 0.3 tons for high-grade bauxite to 2.5 tons for very low-grade bauxite (Pascual et al., 2008). Nearly 120 years of alumina extraction from bauxite ore has generated approximately 2.7 billion tons of red mud, and this amount is increasing by approximately 120 million tons per year (Power et al., 2011). The accident just mentioned shows that this red mud is dangerous to humans and

Environmental Materials and Waste. http://dx.doi.org/10.1016/B978-0-12-803837-6.00020-2

harmful to the environment. In addition, red mud is an industrial by-product with a high concentration of sodium (because NaOH is used as a digested solution), high pH, high alkalinity, radioactive components, and heavy metals. The Brazilian Standards Association classifies red mud as a dangerous waste (Rivas Mercury et al., 2010).

Currently, the main practical disposal methods for red mud are marine disposal, lagooning, "dry" stacking, and dry cake disposal (Power et al., 2011). Because of the increasing production of red mud and its effects on the environment, these methods are becoming increasingly unacceptable. Finding an economical and environmental method for the reclamation of red mud so that it can be used as a secondary raw material in other useful applications is an important consideration. In this regard, three types of application have been considered: construction and chemical, environmental and agronomic, and metallurgical. In the construction and chemical applications category, researchers have proposed using the mud in civil and building construction, to produce ceramic, plastics, coatings, and pigments, and as catalysts and adsorbents (Klauber et al., 2011; Wang et al., 2008; Singh et al., 1996; Liu and Zhang, 2011; Pera et al., 1997). In environmental and agronomic applications, red mud can be used in the disposal of wastewater and waste gas, and in agronomic acidification (Klauber et al., 2011; Wang et al., 2008). In metallurgical applications, there is the potential to recover many valuable metals, such as rare earth elements, Ti, Fe, and so on (Liu and Li, 2015; Liu and Naidu, 2014). Although many studies have reported the possibility of such applications, limited practices have been implemented owing to a lack of community acceptance.

One promising large-scale use of red mud is as a secondary raw material for the ceramic industry. In this potential application, red mud can make up to 50 wt% of the final products (Sglavo et al., 2000) and the demand for ceramic products is huge. In the ceramic industry, most of the energy is consumed during the sintering process.

Previous studies have focused on managing red mud and its potential applications but few have considered the thermal behavior of red mud. In this chapter we first discuss how red mud is generated and then summarize the chemical compositions and phase components of red mud from different areas. We then review the thermal behavior, phase transformations, and processes of red mud from different sources. For a case study of the thermal process, we examine the use of red mud to stabilize hazardous metal-containing wastes and make glass-ceramic products.

20.2 CHARACTERIZATION OF RED MUD

20.2.1 GENERATION OF RED MUD

Fig. 20.1 is a simplified flowchart of the Bayer process, which generates red mud. After grinding, the bauxite slurry is transferred to digestion tanks; then, an NaOH solution and lime (in the form of CaO or $Ca(OH)_2$) are added. The lime can significantly reduce the amount of NaOH consumption. During digestion, the hydrated alumina (gibbsite/boehmite/diaspore) is dissolved by the NaOH solution (Liu et al., 2009) as follows:

$$Al_2O_3\cdot 3H_2O\ \text{(gibbsite)} + 2OH^- \rightarrow 2[Al(OH)_4]^-$$

or

$$Al_2O_3\cdot H_2O\ \text{(boehmite/diaspore)} + 2OH^- + 2H_2O \rightarrow 2[Al(OH)_4]^-$$

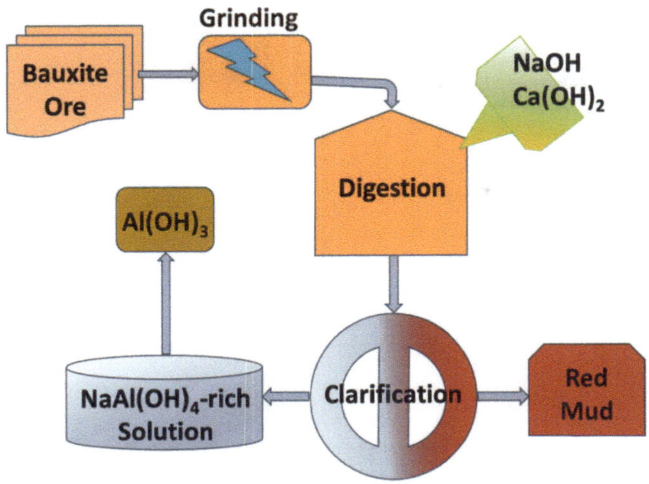

FIGURE 20.1

Schematic representation of the Bayer process, illustrating red mud generation.

Digestion conditions such as the temperature, pressure, and concentration of the NaOH solution depend on the composition of the bauxite ore (Gräfe et al., 2011). For example, the concentration of the NaOH solution varies from 3 M for gibbsitic bauxite to 7 M for diasporic bauxite (Jones and Haynes, 2011). Among these two types of bauxite ore, diasporic bauxite needs a higher pressure and temperature to extract alumina, and therefore it consumes more energy. In addition to the dissolution of hydrated alumina, silicate minerals are dissolved during the digestion process, which leads to the formation of a desilication product (DSP). After the digestion process, a clarification process is carried out in which the $NaAl(OH)_4$-rich solution is separated from the DSP and other undissolved components in the bauxite ore. The $NaAl(OH)_4$-rich solution is then precipitated as $Al(OH)_3$ and finally to alumina (Al_2O_3); the separated solids are known as red mud. The amount of red mud generated per ton of alumina produced varies greatly with the type of bauxite used, from 0.3 tons for high-grade bauxite to 2.5 tons for very low-grade bauxite (Pascual et al., 2008).

20.2.2 CHEMICAL AND PHASE COMPOSITIONS OF RED MUD

To handle, dispose of, and reuse red mud appropriately, its chemical and mineralogical compositions should be known. In general, the chemical and mineralogical compositions of red mud are complex and depend on the type (gibbsite/boehmite/diaspore) and grade (Al_2O_3/SiO_2 ratio) of bauxite ore and the production processes used to extract the alumina.

Chemical analyses have revealed that red mud is composed of six major constituents in terms of oxides, ie, Fe_2O_3, CaO, Al_2O_3, SiO_2, TiO_2, and Na_2O, together with a number of trace hazardous elements such as V, Cr, Ni, Cu, Zn, Pb, Ra, Th, and U. Table 20.1 shows the contents of the six major constituents in red mud samples from different parts of the world. The Fe_2O_3 content of the samples, originating in the bauxite ore and giving the mud its characteristic red color, had a wide range

Table 20.1 Chemical Compositions of Red Mud Samples From Different Areas

Source	Fe_2O_3	CaO	Al_2O_3	SiO_2	TiO_2	Na_2O	Others	LOI	References
China-1	35.73	16.05	14.26	11.13	8.08	4.12	2.03	8.6	Liao et al. (2015)
China-2	11.77	13.97	25.48	20.58	4.14	6.55	3.61	–	Liu et al. (2014)
China-3	6.75	42.25	10.5	22.2	2.55	3.00	3.31	–	Liu et al. (2014)
China-4	32.14	8.21	17.85	20.48	3.57	6.75	–	–	Liu et al. (2014)
Turkey-1	36.94	2.23	20.39	15.74	4.98	10.10	0.05	8.19	Sushil and Batra (2008) and Altundoğan et al. (2000)
Turkey-2	35.04	5.30	20.20	13.50	4.00	9.40	0.72	8.44	Sushil and Batra (2008) and Kavas (2006)
India-1	28.1	10.2	21.9	7.5	15.6	4.5	–	12.2	Gräfe et al. (2011)
India-2	27.9	11.8	19.4	7.3	16.4	3.3	–	12.6	Gräfe et al. (2011) and Prasad and Subramanian (1997)
India-3	54.8	2.5	14.8	6.4	3.7	4.8	–	9.5	Gräfe et al. (2011) and Prasad and Subramanian (1997)
India-4	44.5	0.8	19.2	7.0	13.5	4.0	-	10.0	Gräfe et al. (2011) and Prasad and Subramanian (1997)
Jamaica-1	53.5	14.8	8.16	0.8	10.9	0.3	–	11.6	Gräfe et al. (2011) and Li and Rutherford (1996)
Jamaica-2	62.1	15.9	2.34	1.3	12.9	0.8	–	4.6	Gräfe et al. (2011) and Li (1998)
Jamaica-3	64.2	14.2	2.12	0.7	13.3	0.6	–	5	Gräfe et al. (2011) and Li (1998)
Jamaica-4	59.4	13.2	5.98	0.6	12.9	0.4	–	7.6	Gräfe et al. (2011) and Li (1998)
Brazil-1	61.7	4.2	9.4	3.4	10.4	1.5	–	12.5	Gräfe et al. (2011) and Li and Rutherford (1996)
Brazil-2	45.6	1.16	15.1	15.6	4.29	7.50	10.75	–	Liu and Li (2015)
United States-1	30.40	–	16.20	11.14	10.11	2.00	30.15	–	Paramguru et al. (2005)
United States-2	55.60	–	12.15	4.50	4.50	1.5	21.75	–	Paramguru et al. (2005)
United States-3	50.54	–	11.13	2.56	–	9.00	26.77	–	Paramguru et al. (2005)
Australia-1	40.50	–	27.70	19.90	3.50	2.00	6.4	–	Paramguru et al. (2005)
Australia-1	34.05	3.69	25.45	17.06	4.90	2.74	2.06	–	Sushil and Batra (2008) and Genç et al. (2003)
Greece-1	40.80	12.60	19.95	6.80	5.80	2.70	0.81	10.54	Sushil and Batra (2008) and Tsakiridis et al. (2004)
Greece-2	44.34	11.64	17.04	7.79	5.12	3.17	1.13	9.77	Sushil and Batra (2008) and Pontikes et al. (2007)
Italy	30.45	7.77	17.19	9.58	8.61	12.06	1.96	12.38	Sushil and Batra (2008) and Bertocchi et al. (2006)
Hungary	33–48	0.5–3.5	16–18	9–15	4–6	8–12	–	–	Sushil and Batra (2008) and Halász et al. (2005)

–, data not available; LOI, loss of ignition.

(6.75–64.2 wt%). The amount of calcium in the samples ranged from 0.5 to 42.25 wt%. As an indicator of the alumina extraction efficiency in a refinery, the Al_2O_3 contents left in the red mud varied from 2.12 wt% in the Jamaican sample to 27.70 wt% in the Australian sample. The soda loss in some refineries was very low, as reflected in the samples with an Na_2O content lower than 1 wt% (Jamaican). However, the sodium level in the Italian red mud was up to 12.06 wt% (in terms of the oxide).

Powder X-ray diffraction (XRD) can determine the mineral composition of a sample. Typically, approximately 70 wt% of red mud is composed of crystalline phases, and amorphous materials make up the remaining 30 wt% (Gräfe and Klauber, 2011). Some of the crystalline phases originate from the bauxite ore and others are formed during the Bayer process. The mineralogical constituents in the red samples from different parts of the world are summarized in Table 20.2. As indicated by Gräfe and Klauber (2011) and Paramguru et al. (2005), the mineral components vary with the original bauxite ore, parameters of the digestion process, deposition management, and aging processes. Fig. 20.2 shows the Rietveld quantitative XRD analysis of a sample of red mud from the Pingguo refinery in China. In this sample, the amorphous content was estimated to be around 15.9 wt%. Hematite, which is found in all red mud samples (Table 20.2), made up about 27.5 wt% of the Pingguo sample. Most of this crystalline phase came from the original bauxite ore, although some may have been converted from goethite during the digestion process. In the Pingguo red mud, TiO_2 (rutile or anatase) was not observed. Because the Pingguo alumina refinery used a higher level of CaO and a higher digestion temperature (≥250°C) for alumina extraction (Zhao et al., 2002), the $CaTiO_3$ and andradite (or the $Ca_3(FeAlTi)_2(SiO_4)_3 \cdot nH_2O$ if further incorporated with Fe, Al, and Si) probably formed instead of the TiO_2 seen in the other samples. Thus, the Pingguo red mud sample had about 5.6 wt% of perovskite ($CaTiO_3$), which was not found in the other red mud samples. In contrast, TiO_2 (rutile or anatase) was observed in the other red mud samples but not detected in the Pingguo red mud. Perovskite, rutile, and anatase coexisted in the Renukoot red mud sample (India). Quartz was not observed in the Pingguo red mud sample but it was observed in other red mud samples, probably because the high temperature used in the digestion process (≥250°C) at the Pingguo alumina refinery leads to the dissolution of quartz during the Bayer process (Zhao et al., 2002). The dissolved silicon eventually precipitates in the Pingguo red mud, forming silicon-containing phases (such as cancrinite $3(Na_2O \cdot Al_2O_3 \cdot 2SiO_2 \cdot nH_2O) \cdot 2NaX$, in which $n = 0–2$ and $X = OH^-, Cl^-, CO_3^{2-}$, and SO_4^{2-}), as was observed in the XRD result (Fig. 20.2). Andradite, cancrinite, kaolinite, gibbsite, and undissolved diaspore were also found in the Pingguo red mud. These phases contain most of the unextracted Al. Owing to the addition of lime in the digestion process, a calcite phase was also detected in the Pingguo red mud sample and in red mud samples from some other regions (Table 20.2).

20.3 THERMAL ANALYSIS

20.3.1 ANALYSIS OF THERMOGRAVIMETRY, DERIVATIVE THERMOGRAVIMETRY, DIFFERENTIAL SCANNING CALORIMETRY, AND DIFFERENTIAL THERMAL ANALYSIS

Several researchers have conducted thermal analyses of red mud from various sources. Papp et al. (1977) investigated the effects of pressure on the shape of thermogravimetry (TG), derivative thermogravimetry (DTG), and differential thermal analysis (DTA) curves in Hungarian red mud and found that the decomposition rates of the phases change with changes in pressure, making it possible to

Table 20.2 Phase Compositions of Red Mud Samples From Different Areas

Phases (wt%)	China Pingguo (Liao et al., 2015)	Spain San Ciprian (Pascual et al., 2008)	Brazil Para state (Liu and Li, 2015)	Italy Cagliari (Sglavo et al., 2000)	United Kingdom (Liu and Li, 2015)	Australia Eurallumia (Liu and Li, 2015)	India Renukoot (Liu and Li, 2015)
Hematite	27.3	√	√	√	13.5	29	22.2
Perovskite	5.6	–	–	–	–	–	1.1
Andradite	26.1	√	–	–	–	–	–
Cancrinite	11.7	–	–	√	–	33	4.7
Kaolinite	5.4	–	–	–	–	–	–
Diaspore	3.4	–	–	–	–	5	0.6
Gibbsite	2.4	–	√	√	–	–	3
Calcite	2.2	√	√	–	0.7	–	1
Chantalite	–	–	–	√	–	–	–
$Na_5Al_3CSi_3O_{15}$	–	–	–	√	–	–	–
Bayerite	–	–	–	√	–	–	–
Boehmite	–	√	–	√	–	6	1
Rutile	–	√	–	√	4.6	–	1.8
Antase	–	–	–	√	–	5	3.8
Quartz	–	√	√	√	1.2	–	–
Goethite	–	–	√	–	21.8	–	10.9
Sodalite	–	–	√	–	17.5	16	3.7
Ilmenite	–	–	–	–	–	–	–
Muscovite	–	–	–	–	2.4		
Hydrohematite	–	√	–	–	–	–	–
Aluminogoethite	–	√	–	–	–	–	–
Ferrihydrite	–	√	–	–	–	–	–
Natrodavyne	–	√	–	–	–	–	–
Plazolite	–	√	–	–	–	–	–
Sodium Carbarnate hydrate	–	√	–	–	–	–	–
Amorphous	15.9	√	–	√	38.2	–	–

√, Detected; –, not detectable.

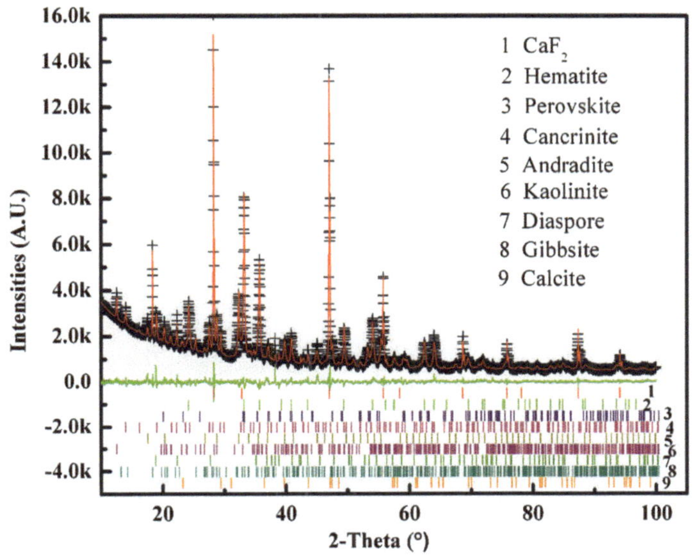

FIGURE 20.2

Rietveld quantitative XRD analysis of a Pingguo red mud sample. CaF_2 was used as an internal standard. Crosses indicate the observed data and are fitted with the calculated pattern. The bottom curve indicates the difference; vertical bars indicate the Bragg reflection positions of the corresponding phases (Liao et al., 2015).

separate overlapping curves easily. Alp and Goral (2003), Atasoy (2005, 2007), Pascual et al. (2008), and Palmer and Frost (2009) studied the thermal behavior of various red muds, including the dehydration, decomposition, possible solid-state reaction, and solid–liquid phase transitions, using TG and DTA at temperature ranges from room temperature to 1400°C. Below 700°C the physical water loss and gibbsite and calcite decomposition were similar. However, the behaviors of compounds containing Fe, Al, Si, Na, Ca, or Ti were different above this temperature. These studies demonstrated that because red mud from various sources has different chemical and mineralogical compositions, its thermal behavior is different. Researchers (Rivas Mercury et al., 2010; Sglavo et al., 2000) have not only studied red mud using TG and DTA, they have examined the phase transformations at different stages of the heating process. In the raw and heated samples of red mud, there are crystalline and amorphous phases, and the proportion of these two components affects the solid reaction during processing. Therefore, one study (Rivas Mercury et al., 2010) quantified the proportions of amorphous content in a raw sample and in a sample heated to 1250°C using quantitative XRD analysis.

Fig. 20.3 shows a typical TG and differential scanning calorimetry (DSC) skeleton of a Pingguo red mud sample. The total mass loss was about 8.65% up to 1253°C, and the sample melted above 1253°C. A total of 5.7% of the mass loss before 478°C may be related to the loss of physical water and water molecules located in gibbsite (Atasoy, 2007), goethite (Laskou et al., 2006), and cancrinite (Castaldi et al., 2008). An endothermic peak at 374°C in the DSC curve can be explained by the decomposition of goethite (Laskou et al., 2006). The mass loss (1%) between 478°C and 580°C may be attributed to the decomposition of diaspore and kaolinite. The decomposition of diaspore and kaolinite can be

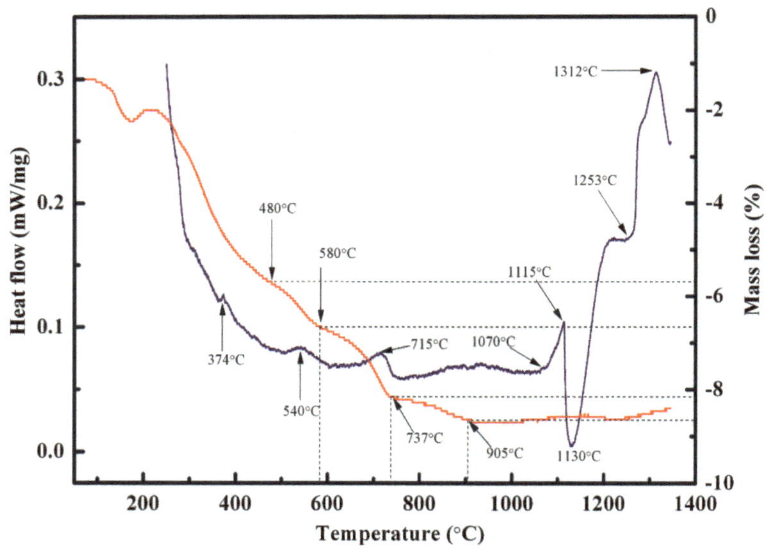

FIGURE 20.3

Results of thermogravimetry and DSC for the Pingguo red mud sample (Liao et al., 2015).

deduced from the endothermic peak at 543°C, which had a large range in the DSC curve (Alp and Goral, 2003; Kloprogge et al., 2002). The mass loss of about 1.5% in the temperature range of 580–737°C may result from the decomposition of calcite and the dehydration of andradite. The endothermic peak at 715°C, which also had a large range in the DSC curve, may be due to the decomposition of calcite and andradite (Rivas Mercury et al., 2010; Atasoy, 2007). From 737°C to 905°C, there was a small mass loss (0.45%) that could be attributed to the release of CO_2 from cancrinite (Hassan, 1996; Hassan et al., 2006). The peak beginning at 1070°C with a maximum at 1115°C may reflect the decomposition of an amorphous phase that contained Ca, Al, and Si elements. At 1115°C, there was a sharp drop, and there was an exothermic peak at 1131°C, which may be attributed to the formation of a new phase. The sharp drop at 1115°C may mean that there was an overlap between the decomposition of the amorphous phase and the formation of a new phase. In other words, the SFCA phase may have formed at temperatures below 1115°C. An endothermic peak in the DSC curve, starting at about 1253°C with a maximum of 1312°C, revealed that the sample began to melt at 1253°C.

20.3.2 THERMALLY PHASE TRANSFORMATIONS AND PROCESSES

To explore the recycling of red mud as a raw material for the ceramic industry, it is necessary to study the phase components and phase reactivity of red mud when it is sintered at high temperatures. Only three research groups have focused on the phase components and phase transformation mechanisms of their corresponding red muds at a series of temperatures ranging from room temperature to high temperatures (up to 1400°C).

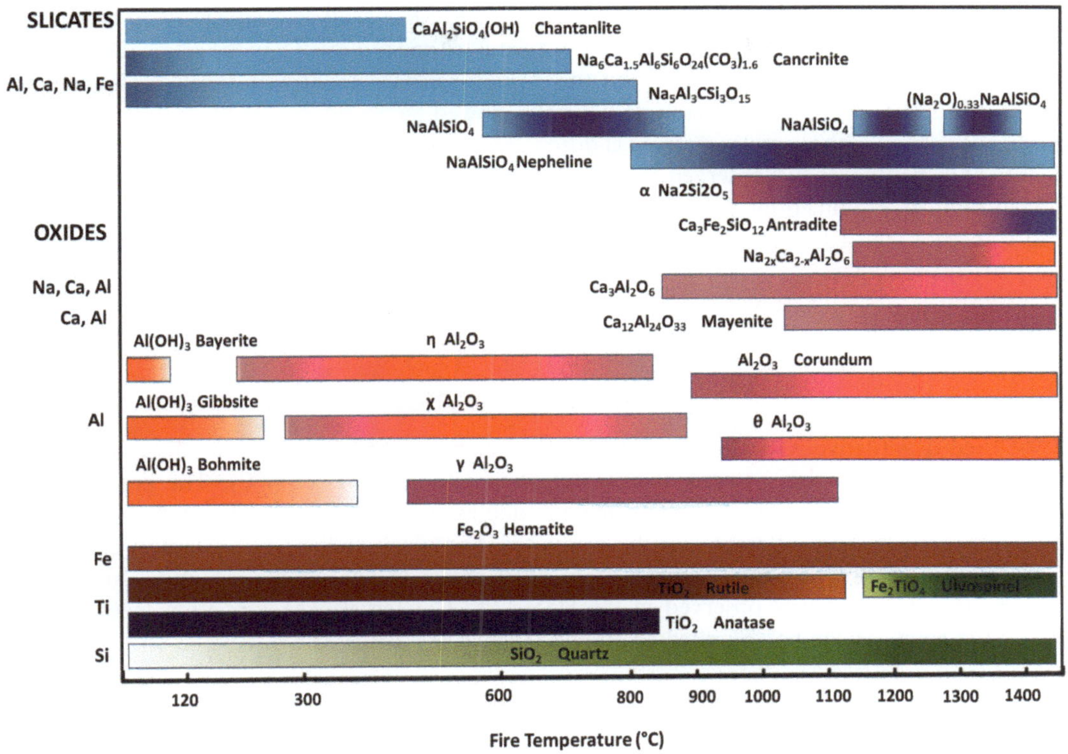

FIGURE 20.4

Evolution of crystalline phases in red mud samples sintered at different temperatures (Sglavo et al., 2000).

Sglavo et al. (2000) analyzed the solid-state transformation and solid–liquid phase transitions of red mud over a range of temperatures from 120 to 1400°C. XRD analysis was used to identify the crystalline phases in the thermally heated samples. Their results are summarized in Fig. 20.4. The authors pointed out that the main reactions in a red mud sample that was sintered below 900°C was the loss of H_2O and CO_2 from the minerals in which they coexisted. In the 900–1100°C interval, new crystalline phases such as $NaAlSiO_4$, $Na_2Si_2O_5$, and $Ca_3Al_2O_6$ were formed. Quartz and hematite had limited reactivity with other components in the red mud at temperatures below 1100°C. When the sintering temperature was higher than 1100°C, the following reactions occur (Sglavo et al., 2000):

$$Fe_2O_3 \rightarrow 2FeO + 0.5O_2,$$

$$2FeO + TiO_2 \rightarrow Fe_2TiO_4,$$

and

$$3CaO + Fe_2O_3 + 3SiO_2 \rightarrow Ca_3Fe_2Si_3O_{12}$$

Rivas Mercury et al. (2010) used Rietveld quantitative XRD analysis to quantify crystalline and amorphous phases in a red mud sample at room temperature and at 1250°C. In the original sample, the amorphous content made up 24.9 wt%, but this value decreased to 11 wt% when the sample was sintered at 1250°C for 1 h. XRD analysis not only showed an abundance of crystalline phases and amorphous content in the samples sintered at different temperatures, it also showed that a considerable quantity of hematite was still observed at 1050°C. However, anatase (or rutile), which remained in the sintered samples examined by Sglavo et al. (2000), was not detectable at temperatures higher than 1050°C, because the TiO_2 reacted with CaO and Fe_2O_3 to form $CaTiO_3$ and Fe_2TiO_5 (Rivas Mercury et al., 2010) as follows:

$$CaO + TiO_2 \rightarrow CaTiO_3 \text{ and}$$

$$Fe_2O_3 + TiO_2 \rightarrow Fe_2TiO_5$$

We used quantitative XRD to examine the phase compositions and transformation processes of Pingguo red mud samples sintered at different temperatures (700–1200°C). The proportions of crystalline and amorphous phases for the sintered samples are summarized in Table 20.3 (Liao et al., 2015). These results were different from those reported in previous studies (Rivas Mercury et al., 2010; Sglavo et al., 2000) of red mud samples from other areas. In our sample, an intermediate-phase gehlenite ($Ca_2Al_2SiO_7$) was observed in the 800–1100°C interval. The formation mechanism of this crystalline phase can be described as follows (Liao et al., 2015):

$$Al_2Si_2O_5(OH)_4 \text{ (kaolinite)} \rightarrow Al_2Si_2O_7$$

$$CaCO_3 \rightarrow CaO + CO_2$$

$$Al_2Si_2O_7 + 4CaO \rightarrow Ca_2Al_2SiO_7 \text{ (gehlenite)} + Ca_2SiO_4$$

Table 20.3 X-ray Diffraction Quantitative Analysis of Crystalline and Amorphous Phases in Red Mud Samples Sintered at 700–1200°C (Liao et al., 2015)

Phases (wt%)	700°C	800°C	900°C	1000°C	1100°C	1200°C
Hematite	30.6	30.8	31.9	29	13	3.6
Perovskite	6.2	7.9	10.1	11.2	13.1	15.1
Andradite	11.9	12.4	12.3	9.4	11.1	–
Cancrinite	10.7	4.5	–	–	–	–
Gehlenite	–	9.7	16.2	17.5	8.5	–
Ca_2SiO_4	–	3.8	3.6	–	–	–
Sodalite	–	–	4.6	4	6.9	6.9
$NaAlSiO_4$	–	–	8.3	8.9	7	6.9
SFCA[a]	–	–	–	–	37.3	59.1
Amorphous	40.7	30.9	13	20	3.1	8.4

[a]SFCA, *silico-ferrite of calcium and aluminum;* –, *not detectable.*

It is interesting that the amount of hematite decreased dramatically at temperatures higher than 1100°C. This result was different from that reported by Sglavo et al. (2000) and Rivas Mercury et al. (2010), in which the hematite phase in sintered samples was still abundant. It was also reported that a new silico-ferrite of calcium and aluminum (SFCA) phase formed above 1100°C in their treated red mud system. Based on quantitative XRD results, we demonstrated two formation processes for this SFCA phase (Liao et al., 2015):

$$\text{Ca-Al-Si}-\text{containing amorphous content} + \text{hematite} + \text{gehlenite} \rightarrow \text{SFCA}$$

$$\text{Andradite} + \text{gehlenite} + \text{hematite} \rightarrow \text{SFCA}$$

Accordingly, these studies showed that the phase compositions, phase transformations, and reaction processes of our thermally treated red mud sample varied significantly from those of other investigated red muds. Thus, before red mud can be recycled in the ceramic industry, it is important to investigate the possible range of reactions to thermal processes at different temperatures.

20.4 THERMALLY IMMOBILIZING HAZARDOUS METALS

Previous studies found that spinel phases such as $NiFe_2O_4$, $CuFe_2O_4$, and $ZnFe_2O_4$ have an important role in stabilizing hazardous metals (Ni, Cu, and Zn) (Shih et al., 2006a,b; Tang et al., 2011a,b). There are considerable amounts of Fe_2O_3 in red mud. Thus, we examined the use of red mud to stabilize hazardous metals into spinel phases. To investigate how heavy metals are incorporated into red mud, we used metal oxides (such as NiO, CuO, and ZnO) to simulate the solid wastes that contain heavy metals, and then mixed them into red mud samples, which were then sintered at different temperatures and for different holding times. XRD analysis showed that the heavy metals were mainly incorporated into the ferrite spinel phase, as shown in Fig. 20.5. The maximum incorporation quality of metal oxides was evaluated and results showed that up to 25 wt% of the metal oxides (NiO, CuO, and ZnO) was captured by the red mud, depending on the type of hazardous oxides.

Of all types of electronic waste, cathode ray tubes (CRTs) are the most hazardous to the environment (Nnorom et al., 2011) because they contain large amounts of lead, which pollutes the environment if management strategies are inappropriate. Thus, CRTs are classified as "hazardous materials" and cannot be deposited directly in landfills (Xu et al., 2012). The funnel glass in a CRT has a lead content of about 20–25 wt% (in the form of metal oxide). The crystalline phase $PbFe_{12}O_{19}$ immobilizes such lead content better than other Pb-containing phases (Mao et al., 2014; Lu, 2013). To immobilize the lead content appropriately in CRT funnel glass, we proposed a strategy that not only recycled red mud but also captured the Pb and transformed it into a $PbFe_{12}O_{19}$ phase. In this strategy, different weight ratios of CRT funnel glass and red mud were sintered at various temperatures to obtain $PbFe_{12}O_{19}$. Fig. 20.6 shows the formation of the $PbFe_{12}O_{19}$ phase in samples with different fractions of CRT funnel glass and red mud (5 wt% CRT + 95 wt% red mud; 10 wt% CRT + 90 wt% red mud; 15 wt% CRT + 85 wt% red mud; 20 wt% CRT + 80 wt% red mud; 25 wt% CRT + 75 wt% red mud; and 30 wt% CRT + 70 wt% red mud) sintered at 1000°C for 5 hours. The $PbFe_{12}O_{19}$ phase (magnetoplumbite) formed in all mixtures of CRT funnel glass and red mud sintered at 1000°C for 5 hours. Except for magnetoplumbite, all of the crystalline phases, such as hematite, gehlenite, andradite, and perovskite, originated from the sintered red mud. This further indicates the importance

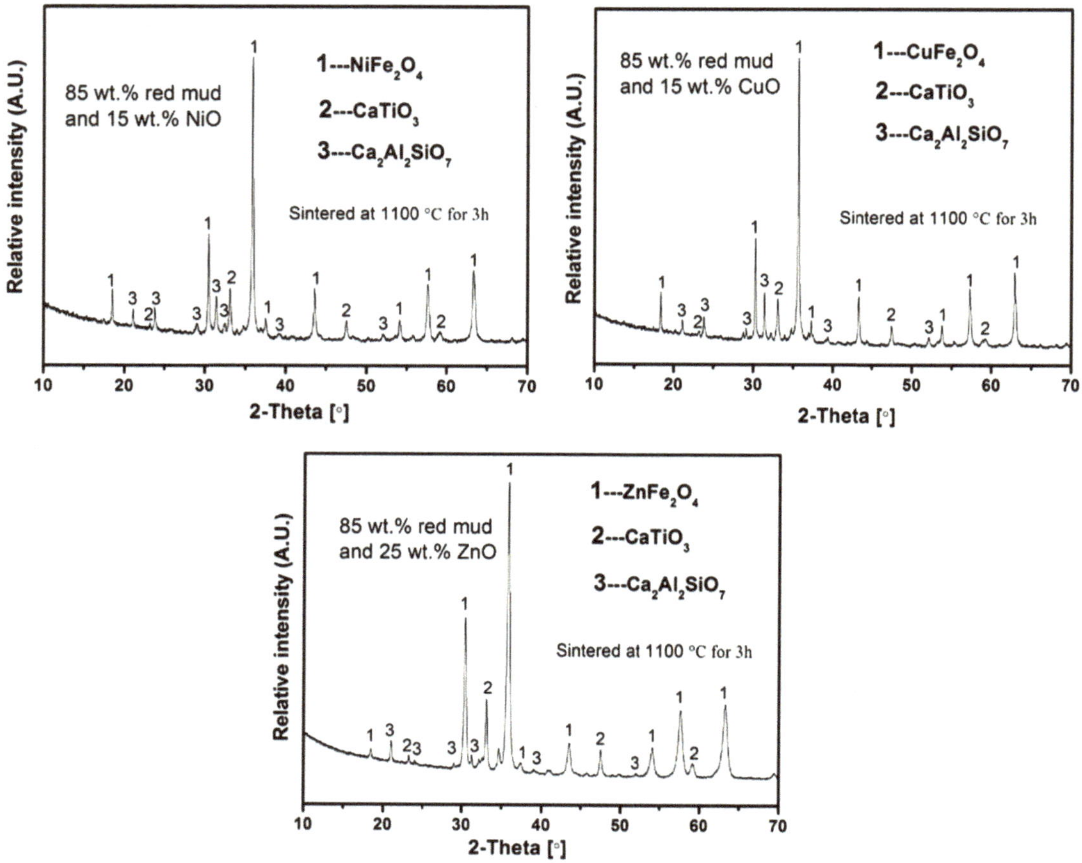

FIGURE 20.5

XRD results showing the major crystalline phases of the red mud and hazardous metal oxide (NiO, CuO, or ZnO) mixtures sintered at 1100°C for 3 hours.

of studying the thermal processes of red mud at different temperatures. It is also interesting that the sintering should be lower than 1000°C. This is a moderate temperature for the ceramic industry.

20.5 MAKING GLASS-CERAMIC

Glass-ceramic is an excellent product that combines the advantages of ceramic with the distinctive characteristics of glass. The disordered structure of the glass components in glass-ceramic can incorporate a wide range of elements, and thus it can stabilize different hazardous wastes (such as hazardous metals). As a result, glass-ceramic may be a useful way to recycle industrial waste. Many researchers have investigated different methods for mixing red mud with different solid wastes to make the raw material for glass-ceramic.

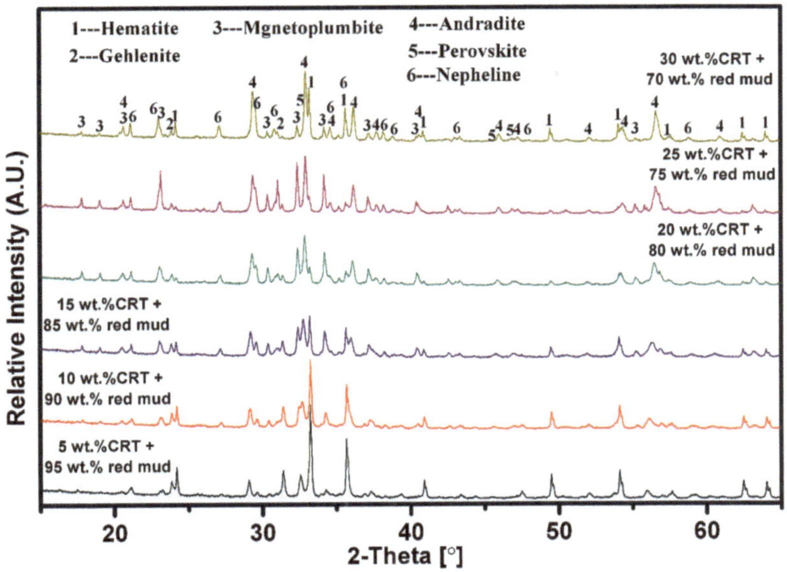

FIGURE 20.6

Phase identification results for samples with different weight fractions of CRT funnel glass and red mud (5 wt% CRT + 95 wt% red mud; 10 wt% CRT + 90 wt% red mud; 15 wt% CRT + 85 wt% red mud; 20 wt% CRT + 80 wt% red mud; 25 wt% CRT + 75 wt% red mud; and 30 wt% CRT + 70 wt% red mud) sintered at 1000°C for 5 hours. The formula of the magnetoplumbite is $PbFe_{12}O_{19}$.

Erol (2004) first mixed fly ash from a thermal power plant with red mud to make glass at 1600°C, and then developed two methods to produce glass-ceramic products. XRD results revealed that the major crystalline phase in the obtained glass-ceramic was diopside $Ca(Mg,Al)(Si,Al)_2O_6$. Glass-ceramic products with a large diopside phase were also obtained by treating a mixture of red mud, coal fly ashes, and silica fume with controlled nucleation and crystallization (Erol et al., 2007). However, another crystalline phase, aluminum augite ($Ca(Mg,Fe^{3+},Al)(Si,Al)_2O_6$), was found to coexist in this type of glass-ceramic product. The results of a toxicity characteristic leaching procedure (TCLP) test showed that the produced glass-ceramic samples are nontoxic. Yang et al. (2008) also attempted to recycle both red mud and fly ash. In their studies, a mixture of these two industrial wastes (52 wt% red mud and 33 wt% fly ash) with small amounts (15 wt% in total) of silica sand, TiO_2, and Na_2CO_3 was first converted into glass at 1360°C. This glass was then heated with a two-stage nucleation-crystallization process (nucleation at 697°C for 2 hours followed by crystallization at various temperatures) to produce glass-ceramic products. After crystallization, two major crystalline phases, gehlenite ($Ca_2Al_2SiO_7$) and augite ($Ca(Fe,Mg)Si_2O_6$), were observed in the products. The augite phase transformed into gehlenite when the crystallization temperature was increased from 850 to 1100°C. A chemically durable glass-ceramic product with crystalline phases of esseneite, wollastonite, and plagioclase can be prepared from a vitrified mixture of red mud, fly ash, and porcelain stoneware tiles (Bernardo et al., 2009). The bending strength and Vickers micro-hardness of the obtained glass-ceramic exceed 100 MPa and 7 GPa, respectively. To avoid the limitations (cost and limited use of wastes) of using red mud for ceramic tile or glass-ceramic, Yang et al. (2009) developed a method for producing double-layer glass-ceramic/ceramic tiles. In this double-layer tile, red mud and bauxite tailings can make up to 56 wt% of the tile. Additives such as quartz sand, sodium caubonate, and lime were used to improve the properties of the double-layer tiles.

Because of the high porosity and low thermal conductivity of foamed glass-ceramic, it is a promising material for thermal insulation in buildings. Red mud was used as a secondary raw material to produce foamed glass-ceramic along with other additives such as $CaCO_3$ or other solid wastes. In another study (Ye et al., 2014), a mixture of red mud, waste glass, and dolomite was used to produce foamed glass-ceramic. The optimum foaming temperature was about 1130°C, much higher than the decomposition temperature of the dolomite that acted as the foaming agent. Glass-ceramic obtained from a mixture of red mud and other solid wastes is usually green in color, and the main crystalline phase is augite. With $CaCO_3$ as a foaming agent, a mixture of red mud, fly ash, $Na_2B_4O_7 \cdot 10H_2O$, and Na_3PO_4 was sintered at different temperatures with various holding times to produce glass-ceramic foams (Guo et al., 2014). As indicated by Guo et al. (2014), when the red mud–fly ash ratio was lower than 40:60 and the total weight percentage was 60%, low density glass-ceramic foams were obtained at low temperatures (760–840°C). Foamed glass-ceramic products with homogenous pore distributions (as shown in Fig. 20.7) were also synthesized from a mixture of red mud, lead-zinc mine tailings, and silica sand (Liu et al., 2016). The thermal conductivity, mechanical properties, and

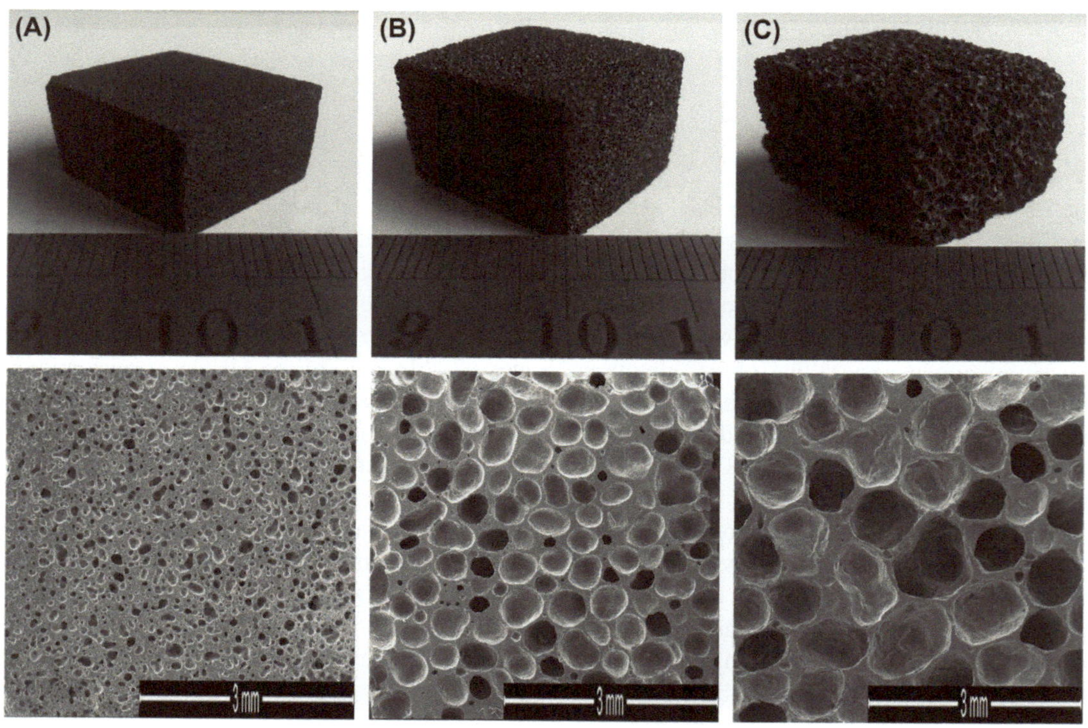

FIGURE 20.7

Microstructure evolution as a function of temperature: (A) 900°C, (B) 950°C, (C) 1000°C (Liu et al., 2016).

chemical stability of the obtained foamed glass-ceramic products were highly affected by the sintering temperature. Liu et al. suggested that the integrated performance (porosity, mechanical strength, and thermal conductivity) of the glass-ceramic foam obtained at 970°C made it a potential candidate for thermal insulation material (Liu et al., 2016). It has also been noted that the sintering temperature (970°C) is a moderate temperature in the ceramic industry, which would make the products cost-effective.

20.6 FUTURE RESEARCH

As discussed, red mud has been used successfully as a secondary raw material in many useful applications. When red mud is combined with other solid wastes, especially hazardous waste, leaching of hazardous metals from the products must be strictly monitored. The results of chemical tests (such as TCLP) and the analysis of its physical properties (such as mechanics) show that the obtained products have excellent potential for use as industrial and construction materials. However, the studies discussed here did not investigate which crystalline phases contained hazardous components such as heavy metals. This is an important consideration for the long-term use of such products and for public acceptance of their use. Therefore, the element distribution of hazardous metals in the final products should be examined in future studies.

In addition, although red mud has been employed successfully to prepare many useful products when loaded with other solid wastes, there are still many questions to be answered. For example, the Rietveld quantitative XRD technique could be used to quantify the crystalline phases and amorphous contents of glass-ceramics. With this technique, the sintering temperature and processing time needed to obtain excellent integrated properties for the products could be optimized. When loading additives such as fly ash, Cr slag, bottom ash, and Cu mining tailors are used, immobilization of hazardous metals is the greatest concern. However, the concentrations of hazardous metals in the final products are low (<1 wt%), which makes many techniques for measuring or immobilizing them unworkable. X-ray absorption spectroscopy can be used to detect a hazardous metal concentrations of less than 1 wt%. It is also useful for obtaining information about oxidation states, bonding lengths, and surrounding environments. Such information is needed to analyze the hazardous metals-bearing species in the final products and the metal incorporation mechanisms.

ACKNOWLEDGMENTS

The authors gratefully acknowledge the funding for this research provided by the General Research Fund (17206714 and 17212015) and Collaborative Research Fund (C7044-14G) schemes of the Research Grants Council of Hong Kong, and the HKU strategic research themes on Clear Energy and Earth as a Habitable Planet.

REFERENCES

Altundoğan, H.S., Altundoğan, S., Tümen, F., Bildik, M., 2000. Arsenic removal from aqueous solutions by adsorption on red mud. Waste Management 20 (8), 761–767.

Alp, A., Goral, M., 2003. The influence of soda additive on the thermal properties of red mud. Journal of Thermal Analysis and Calorimetry 73 (1), 201–207.

Atasoy, A., 2005. An investigation on characterization and thermal analysis of the Aughinish red mud. Journal of Thermal Analysis and Calorimetry 81 (2), 357–361.

Atasoy, A., 2007. The comparison of the Bayer process wastes on the base of chemical and physical properties. Journal of Thermal Analysis and Calorimetry 90 (1), 153–158.

Bernardo, E., Esposito, L., Rambaldi, E., Tucci, A., Pontikes, Y., Angelopoulos, G.N., 2009. Sintered esseneite–wollastonite–plagioclase glass–ceramics from vitrified waste. Journal of the European Ceramic Society 29 (14), 2921–2927.

Bertocchi, A.F., Ghiani, M., Peretti, R., Zucca, A., 2006. Red mud and fly ash for remediation of mine sites contaminated with As, Cd, Cu, Pb and Zn. Journal of Hazardous Materials 134 (1), 112–119.

Castaldi, P., Silvetti, M., Santona, L., Enzo, S., Melis, P., 2008. XRD, FTIR, and thermal analysis of bauxite ore-processing waste (red mud) exchanged with heavy metals. Clays and Clay Minerals 56 (4), 461–469.

Erol, M., 2004. Characterization of glass-ceramics produced from waste materials. In: Key Engineering Materials, vol. 264, pp. 1883–1886.

Erol, M., Küçükbayrak, S., Ersoy-Mericboyu, A., 2007. Production of glass-ceramics obtained from industrial wastes by means of controlled nucleation and crystallization. Chemical Engineering Journal 132 (1), 335–343.

Gelencsér, A., Kováts, N., Turóczi, B., Rostási, A., Hoffer, A., Imre, K., et al., 2011. The red mud accident in Ajka (Hungary): characterization and potential health effects of fugitive dust. Environmental Science and Technology 45 (4), 1608–1615.

Genç, H., Tjell, J.C., McConchie, D., Schuiling, O., 2003. Adsorption of arsenate from water using neutralized red mud. Journal of Colloid and Interface Science 264 (2), 327–334.

Gräfe, M., Power, G., Klauber, C., 2011. Bauxite residue issues: III. Alkalinity and associated chemistry. Hydrometallurgy 108 (1), 60–79.

Gräfe, M., Klauber, C., 2011. Bauxite residue issues: IV. Old obstacles and new pathways for in situ residue bioremediation. Hydrometallurgy 108 (1), 46–59.

Guo, Y., Zhang, Y., Huang, H., Meng, K., Hu, K., Hu, P., et al., 2014. Novel glass ceramic foams materials based on red mud. Ceramics International 40 (5), 6677–6683.

Halász, J., Hodos, M., Hannus, I., Tasi, G., Kiricsi, I., 2005. Catalytic detoxification of C_2-chlorohydrocarbons over iron-containing oxide and zeolite catalysts. Colloids and Surfaces A: Physicochemical and Engineering Aspects 265 (1), 171–177.

Hassan, I., 1996. The thermal behavior of cancrinite. Canadian Mineralogist 34, 893–900.

Hassan, I., Antao, S.M., Parise, J.B., 2006. Cancrinite: crystal structure, phase transitions, and dehydration behavior with temperature. American Mineralogist 91 (7), 1117–1124.

Jones, B.E., Haynes, R.J., 2011. Bauxite processing residue: a critical review of its formation, properties, storage, and revegetation. Critical Reviews in Environmental Science and Technology 41 (3), 271–315.

Kavas, T., 2006. Use of boron waste as a fluxing agent in production of red mud brick. Building and Environment 41 (12), 1779–1783.

Klauber, C., Gräfe, M., Power, G., 2011. Bauxite residue issues: II. options for residue utilization. Hydrometallurgy 108 (1), 11–32.

Kloprogge, J.T., Ruan, H.D., Frost, R.L., 2002. Thermal decomposition of bauxite minerals: infrared emission spectroscopy of gibbsite, boehmite and diaspore. Journal of Materials Science 37 (6), 1121–1129.

Laskou, M., Margomenou-Leonidopoulou, G., Balek, V., 2006. Thermal characterization of bauxite samples. Journal of Thermal Analysis and Calorimetry 84 (1), 141–146.

Li, L.Y., 1998. Properties of red mud tailings produced under varying process conditions. Journal of Environmental Engineering 124 (3), 254–264.

Li, L.Y., Rutherford, G.K., 1996. Effect of bauxite properties on the settling of red mud. International Journal of Mineral Processing 48 (3), 169–182.

Liao, C.Z., Zeng, L., Shih, K., 2015. Quantitative X-ray diffraction (QXRD) analysis for revealing thermal transformations of red mud. Chemosphere 131, 171–177.

Liu, T., Li, X., Guan, L., Liu, P., Wu, T., Li, Z., Lu, A., 2016. Low-cost and environment-friendly ceramic foams made from lead–zinc mine tailings and red mud: foaming mechanism, physical, mechanical and chemical properties. Ceramics International 42 (1), 1733–1739.

Liu, W., Yang, J., Xiao, B., 2009. Review on treatment and utilization of bauxite residues in China. International Journal of Mineral Processing 93 (3), 220–231.

Liu, W., Chen, X., Li, W., Yu, Y., Yan, K., 2014. Environmental assessment, management and utilization of red mud in China. Journal of Cleaner Production 84, 606–610.

Liu, X., Zhang, N., 2011. Utilization of red mud in cement production: a review. Waste Management and Research 29 (10), 1053–1063.

Liu, Y., Naidu, R., 2014. Hidden values in bauxite residue (red mud): recovery of metals. Waste Management 34 (12), 2662–2673.

Liu, Z., Li, H., 2015. Metallurgical process for valuable elements recovery from red mud—A review. Hydrometallurgy 155, 29–43.

Lu, X., 2013. Application of X-ray Diffraction to Identify the Phases Formed during Lead Stabilization and Resource Recovery (Ph.D. thesis). The University of Hong Kong, Hong Kong.

Mao, L., Cui, H., An, H., Wang, B., Zhai, J., Zhao, Y., Li, Q., 2014. Stabilization of simulated lead sludge with iron sludge via formation of $PbFe_{12}O_{19}$ by thermal treatment. Chemosphere 117, 745–752.

Nnorom, I.C., Osibanjo, O., Ogwuegbu, M.O.C., 2011. Global disposal strategies for waste cathode ray tubes. Resources, Conservation and Recycling 55 (3), 275–290.

Palmer, S., Frost, R., 2009. Thermal decomposition of Bayer precipitates formed at varying temperatures. Journal of Thermal Analysis and Calorimetry 100 (1), 27–32.

Paramguru, R.K., Rath, P.C., Misra, V.N., 2005. Trends in red mud utilization—A review. Mineral Processing and Extractive Metallurgy Review 26 (1), 1–29.

Pascual, J., Corpas, F.A., López-Beceiro, J., Benítez-Guerrero, M., Artiaga, R., 2008. Thermal characterization of Spanish red mud. Journal of Thermal Analysis and Calorimetry 96, 407–412.

Papp, K., Kocsardy, E., Kiss, A., Lakatos, T., 1977. Influence of pressure reduction on the shapes of thermoanalytical curves of red muds. Journal of Thermal Analysis and Calorimetry 11 (2), 249–256.

Pera, J., Boumaza, R., Ambroise, J., 1997. Development of a pozzolanic pigment from red mud. Cement and Concrete Research 27 (10), 1513–1522.

Pontikes, Y., Nikolopoulos, P., Angelopoulos, G.N., 2007. Thermal behaviour of clay mixtures with bauxite residue for the production of heavy-clay ceramics. Journal of the European Ceramic Society 27 (2), 1645–1649.

Power, G., Gräfe, M., Klauber, C., 2011. Bauxite residue issues: I. current management, disposal and storage practices. Hydrometallurgy 108 (1), 33–45.

Prasad, P.M., Subramanian, S., 1997. Problems and prospects of red mud utilization. Transactions of the Indian Institute of Metals 50 (5), 427–442.

Rivas Mercury, J.M., Cabral, A.A., Paiva, A.E.M., Angélica, R.S., Neves, R.F., Scheller, T., 2010. Thermal behavior and evolution of the mineral phases of Brazilian red mud. Journal of Thermal Analysis and Calorimetry 104 (2), 635–643.

Ruyters, S., Mertens, J., Vassilieva, E., Dehandschutter, B., Poffijn, A., Smolders, E., 2011. The red mud accident in Ajka (Hungary): plant toxicity and trace metal bioavailability in red mud contaminated soil. Environmental Science and Technology 45 (4), 1616–1622.

Shih, K., White, T., Leckie, J.O., 2006a. Spinel formation for stabilizing simulated nickel-laden sludge with aluminum-rich ceramic precursors. Environmental Science and Technology 40 (16), 5077–5083.

Shih, K., White, T., Leckie, J.O., 2006b. Nickel stabilization efficiency of aluminate and ferrite spinels and their leaching behavior. Environmental Science and Technology 40 (17), 5520–5526.

Sglavo, V.M., Campostrini, R., Maurina, S., Carturan, G., Monagheddu, M., Budroni, G., Cocco, G., 2000. Bauxite 'red mud' in the ceramic industry. Part 1: thermal behaviour. Journal of the European Ceramic Society 20 (3), 235–244.

Singh, M., Upadhayay, S.N., Prasad, P.M., 1996. Preparation of special cements from red mud. Waste Management 16 (8), 665–670.

Sushil, S., Batra, V.S., 2008. Catalytic applications of red mud, an aluminium industry waste: a review. Applied Catalysis B: Environmental 81 (1), 64–77.

Tang, Y., Chui, S.S.Y., Shih, K., Zhang, L., 2011a. Copper stabilization via spinel formation during the sintering of simulated copper-laden sludge with aluminum-rich ceramic precursors. Environmental Science and Technology 45 (8), 3598–3604.

Tang, Y., Shih, K., Wang, Y., Chong, T.C., 2011b. Zinc stabilization efficiency of aluminate spinel structure and its leaching behavior. Environmental Science and Technology 45 (24), 10544–10550.

Tsakiridis, P.E., Agatzini-Leonardou, S., Oustadakis, P., 2004. Red mud addition in the raw meal for the production of Portland cement clinker. Journal of Hazardous Materials 116 (1), 103–110.

Wang, S., Ang, H.M., Tadé, M.O., 2008. Novel applications of red mud as coagulant, adsorbent and catalyst for environmentally benign processes. Chemosphere 72 (11), 1621–1635.

Xu, Q., Li, G., He, W., Huang, J., Shi, X., 2012. Cathode ray tube (CRT) recycling: current capabilities in China and research progress. Waste Management 32 (8), 1566–1574.

Yang, H., Chen, C., Pan, L., Lu, H., Sun, H., Hu, X., 2009. Preparation of double-layer glass-ceramic/ceramic tile from bauxite tailings and red mud. Journal of the European Ceramic Society 29 (10), 1887–1894.

Yang, J., Zhang, D., Hou, J., He, B., Xiao, B., 2008. Preparation of glass-ceramics from red mud in the aluminium industries. Ceramics International 34 (1), 125–130.

Ye, Z.N., Wang, Y.W., Jiang, H., Li, N., Liu, S.Q., January 2014. Foamed glass-ceramics made from red-mud. In: Key Engineering Materials, vol. 575, pp. 461–464.

Zhao, H., Hu, H., Jin, M., Li, J., Li, Q., Liu, Y., 2002. Digestion of diasporic bauxite with mass ratio of Al_2O_3/SiO_2 no greater than 7 by Bayer process with an excessive addition of lime. In: Light Metals-Warrendale-proceedings. TMS, pp. 101–106.

CHAPTER

CLAY MINERALS: STRUCTURE, CHEMISTRY, AND SIGNIFICANCE IN CONTAMINATED ENVIRONMENTS AND GEOLOGICAL CO_2 SEQUESTRATION

21

I. Bibi[1,2], J. Icenhower, N.K. Niazi[1,2], T. Naz[1], M. Shahid[3], S. Bashir[1]

University of Agriculture Faisalabad, Faisalabad, Punjab, Pakistan[1]; Southern Cross University, East Lismore, NSW, Australia[2]; COMSATS Institute of Information and Technology, Vehari, Pakistan[3]

CHAPTER OUTLINE

Environmental Materials and Waste. http://dx.doi.org/10.1016/B978-0-12-803837-6.00021-4

21.1 INTRODUCTION

Clay minerals include the phyllosilicate or sheet silicate family of minerals, which are characterized by their layered structures composed of polymeric sheets of silica (SiO_4) tetrahedra attached with octahedra sheets (containing aluminum (Al), magnesium (Mg), iron (Fe)$(O,OH)_6$). These minerals are layered-type aluminosilicates which are formed as a result of chemical weathering of other silicate minerals present on the Earth's surface (Ismadji et al., 2015).

The definition of what constitutes a "clay mineral" is complicated by multiple meanings of the word "clay." Clay can denote a family of minerals or it can pertain to a size fraction (3.9–0.98 μm) in soils and sediments. This latter meaning is sometimes intertwined with clay minerals; however, in the connotation of size, clay-size particles, including Fe and Al oxyhydroxides, micas, silica polymorphs, and even feldspars, are common in soil and sediments. Generally, clay minerals are less than 1 μm in diameter. Therefore, they have been described as micro- to nanocrystalline materials. Clay minerals have a platy morphology owing to the arrangement of atoms in their layered structure. To understand the behavior of clay minerals and their remarkable properties, we will need a thorough comprehension of their structures: how the tetrahedral and octahedral sheets join and the variety of ways in which bonding between tetrahedral–octahedral (T–O) sheets results in the various forms of clay minerals observed in various environments (Murray, 2007; Cygan et al., 2012; Balan et al., 2014; Detellier and Schoonheydt, 2014; Ismadji et al., 2015).

In this chapter, we will mainly focus on the four major groups of clay minerals (kandite, illite, smectite, and vermiculite) and discuss the chemical and structural differences between and within the groups; Fe–Al oxide minerals will not be a focus of this review. Clay minerals are crucially important industrial minerals; they have been in use as raw materials for hundreds of industrial and some environmental applications owing to their abundance, availability, and inexpensive nature. For example, clay minerals have been used in agricultural applications, engineering and construction applications, environmental remediation, geology, pharmaceuticals, food processing, and many other industrial applications (Detellier and Schoonheydt, 2014; Ismadji et al., 2015). This review will summarize the types, structural chemistry, and characteristics of clay minerals; elucidate their role as cleaning agents in various environmental settings (immobilization of hazardous heavy metals and organic contaminants, eg, dyes, antibiotics, in water and wastewater); highlight their significance in natural and engineered environments (eg, clay barriers) and in leaching/dissolution of elements; and describe the fundamentally important, although partially investigated, role of clay minerals for sequestration of carbon dioxide (CO_2) at geological carbon sequestration (GCS) sites.

21.2 TYPES, STRUCTURE, CLASSIFICATION, AND CHARACTERISTICS OF CLAY MINERALS

21.2.1 CLAY MINERALS: STRUCTURE AND CHEMISTRY

The structure of clay minerals has fascinated investigators for years and has invited studies from scientific luminaries (Pauling, 1930; Brindley, 1951; Yoder and Eugster, 1955). A thorough review of the structure and chemistry of clay minerals is beyond the scope of this chapter, but good general summaries can be found in Langmuir (1997) and Drever (1997). For a more detailed discussion of clay mineralogy, we encourage the interested scholar to consult Deer et al. (1962).

Like all phyllosilicate minerals, clay mineralogy is defined by sheet structures based on the Si_4O_{10} unit cell (Fig. 21.1). To distinguish clay minerals from the other phyllosilicates, such as biotite, muscovite, chlorite, etc., a general definition is needed.

As described earlier, clay can denote a family of minerals or it can pertain to a size fraction (3.9−0.98 μm) in sediments. This latter meaning is sometimes intertwined with clay minerals, but in the connotation of size, clay-size particles, including Fe and Al oxyhydroxides, micas, silica polymorphs, and even feldspars, are common in soil and sediments. Some definitions exclude colloidal size fractions, but the size of clay minerals places them in a range that overlaps with colloidal material. A general definition of clay includes:

1. A phyllosilicate with an Si_4O_{10} tetrahedral base structure. Tetrahedral sheets are connected to octahedral layers that can be conceptualized as $Mg(OH)_2$ (brucite) or $Al(OH)_3$ (gibbsite) layers (see subsequent discussion). The composite T−O layers may contain interlayer sites.
2. "Hydrous" silicate containing water and/or hydroxyl molecules. In many clay minerals, H_2O, CO_2, or organic molecules can be reversibly intercalated into the interlayer sites. Exchange of CO_2 for H_2O (and vice versa) has important implications for carbon dioxide sequestration, as more fully explicated later. The addition of cations or molecules into the interlayer sites results in volume changes, both positive and negative, and changes in volume of these "swelling clays" also have important implications for permeability in rocks containing significant fractions of clays.

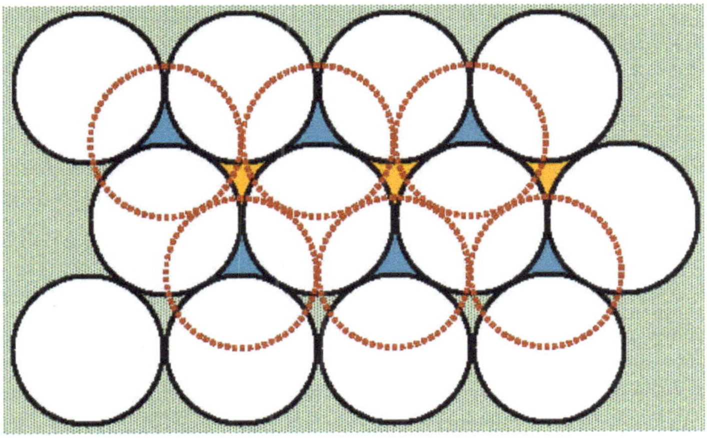

FIGURE 21.1

Two layers of closest-packed hydroxyl ions (*solid lines represent bottom layer; dashed red represent top layer*) in the brucite- and gibbsite-like structure. Tetrahedral voids are shown in blue and octahedral in gold. In the gibbsite-like layer, every third void is unoccupied.

Modified from Drever, J.I., 1997. The Geochemistry of Natural Waters. Prentice Hall, Upper Saddle River, New Jersey, USA.

3. Micro- to nanocrystalline size, requiring identification by X-ray diffraction (XRD). Because of their small crystal size, the large surface area of clays renders clays as reactive substrates for the sorption of cations and anions.
4. Plate-like hexagonal crystals (although tubular and fibrous types are also known).
5. When mixed with water, the materials display various degrees of plasticity.
6. Clays are the result of low-temperature aqueous or hydrothermal interactions with preexisting phases, including feldspars, ferromagnesian phases, volcanic glass, and micas. As discussed later, the major divisions in the clays reflect the source material from which clays are derived.

Key to understanding the behavior of clay minerals and their remarkable properties is to understand their structure—how the tetrahedral and octahedral sheets join and the variety of ways that bonding between T—O sheets gives rise to the various forms of clay minerals observed in the environment. In the sections below, we will describe the four major groups of clay minerals (kandite, illite, smectite and vermiculite) and discuss the chemical and structural differences between and within the groups.

A fundamental division in clay mineralogy arises from differences in the occupancy of sites within the octahedral sheet. The structure of octahedral sheets is largely dictated by the arrangement of hydroxyl anions and the sites between the relatively large anions. In a closely packed structure of hydroxyl anions, there are both tetrahedral and octahedral "holes" or sites between anions (Fig. 21.1). For the structure to remain electrically neutral, every third octahedral site must be unoccupied. On this basis, fully occupied sites are designated "trioctahedral" phyllosilicates. In brucite-like octahedral layers, Mg^{2+} cations are sixfold coordinated and each OH^- anion has three Mg^{2+} closest neighbors, resulting in an electrically neutral $Mg(OH)_2$ structure.

In contrast, every third octahedral site in the gibbsite-like octahedral layer is unoccupied, and this structure is called "dioctahedral." Within each group of clay minerals, both dioctahedral and trioctahedral phases can exist.

The simplest structural type of clay mineral, from a structural and chemical standpoint, is those which make up the kandite family. The general chemical formula of the kandites is $Al_4Si_4O_{10}(OH)_8$, represented by the polymorphs kaolinite, nacrite, halloysite, and dickite. In this structural family, a sheet of tetrahedrally coordinated SiO_4 tetrahedra is connected to a sheet of octahedrally coordinated $Al(OH)_3$ through the apical oxygen in the tetrahedral sheet. The tetrahedral sheets are made up of six-member rings of silica tetrahedra and the octahedral sheets are similar in geometry to the mineral gibbsite. An illustration of the unit cell of kaolinite is shown in Fig. 21.2.

Where the tetrahedral and octahedral units connect, the OH^- in the $Al(OH)_3$ sheets is replaced by oxygen. The basal spacing between T—O layers is 7.1 Å (10 Å in halloysite). Fig. 21.3 illustrates the T—O structure of kaolinite. Because the T and O sheets share oxygen through common bonds, and because there is a variety of ways in which these structural units can connect, the change in structure is manifested as the polymorphs. In the case of halloysite, the mismatch in the ideal repeat distance between the T and O sheets is large enough that the tetrahedral sheet has to curl, resulting in tubular forms of the mineral.

Because of the variety of ways in which T—O layers can stack, there can be mixtures of polymorphs with a seemingly homogeneous sample. Imaging by high-resolution transmission electron microscopy has shown that dickite domains can exist in dominant kaolinite structures, and vice versa (Kogure and Inoue, 2005).

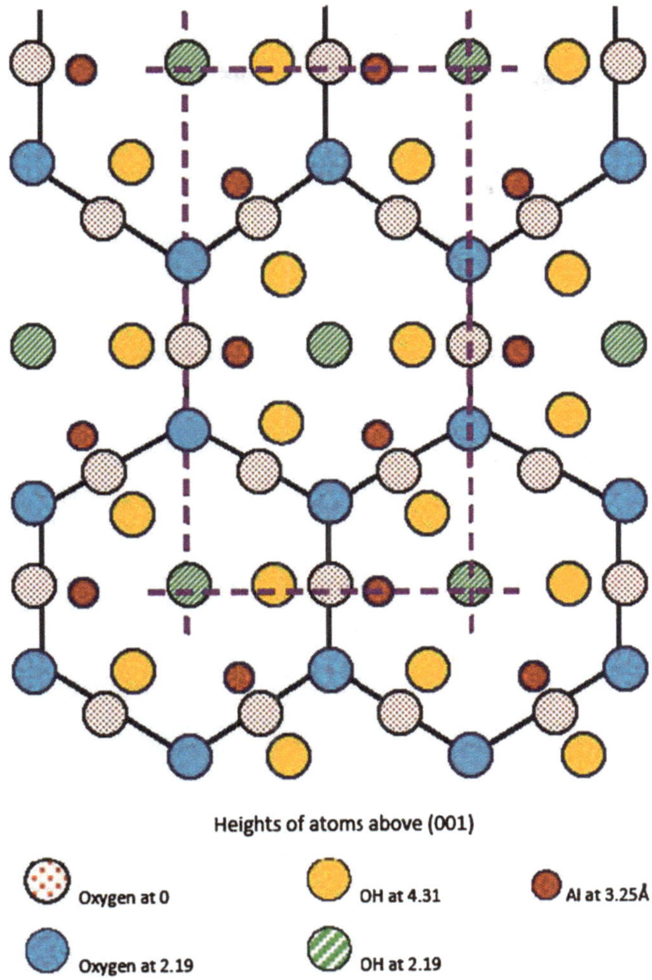

FIGURE 21.2

Idealized kaolinite structure projected onto (001). The *dashed purple line* represents the unit cell.

Modified from Deer, W.A., Howie, R.A., Zussman, J., 1962. Rock-Forming Minerals, vol. 3. Sheet Silicates, Longmans, Green and Co. Ltd, London, UK.

Because kandites contain the gibbsite-like octahedral sheet, they are dioctahedral phyllosilicates. Bonds between T—O layers are relatively strong and do not accommodate interlayer cations or anions. Accordingly, these clays do not swell when contacted by fluids. In addition, silicon (Si) and Al in the tetrahedral and octahedral sites do not accommodate substitution by other cations, which means that the cation exchange capacity (CEC) and surface charge are relatively limited. In contrast to the typical

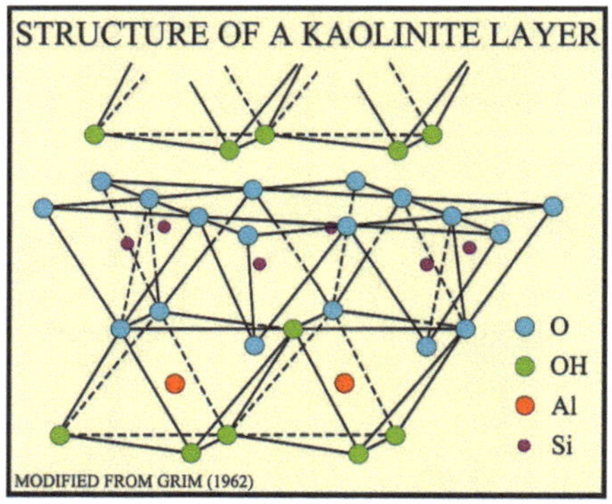

FIGURE 21.3

Structure of the tetrahedral–octahedral layer of kaolinite.

kandite characteristic of not accommodating water into the interlayer site, halloysite can contain small amounts of water in its interlayer site.

For more information on the structure of kaolinite and its use in industry, environmental management, health sciences, etc., see Cygan et al. (2012), Balan et al. (2014), Detellier and Schoonheydt (2014), and Schroeder and Erickson (2014).

Similar to the kandites, members of the illite family also do not typically accommodate water. Thus, like kandites, heating to 200°C does not result in major structural changes with the loss of water. Because their structures allow for some substitution of Al onto the tetrahedral site, the excess negative charge can be accommodated by the addition of potassium (K) into the interlayer site. Accordingly, the general form of the illite chemistry is $K_xAl_4[Si_{8-x}Al_xO_{20}](OH)_4$ where $x = 1.0-1.5$. There are a number of chemical variations within the illite family not captured by this general formula, and both dioctahedral and trioctahedral forms have been documented. The fundamental structure of illite minerals is two tetrahedral (T) sheets with an octahedral (O) sheet sandwiched between to form a T–O–T composite. The apices of the T sheets point inward toward the O sheets and the basal spacing is 10 Å, which helps to distinguish illites from other clay families by use of XRD.

The strong basal reflection from the d_{001} plane also helps to distinguish the smectite or montmorillonite group of clays from other families. Because smectites can accommodate high degrees of interlayer molecules, including water, organics, and carbon dioxide (see subsequent discussion), a characteristic feature of smectites is expansion when treated with ethylene glycol, in which the basal spacing increases to 17 Å. Although originally the name "montmorillonite" was used to denote a group of clay minerals with common physical and chemical properties, the name "smectite" has come to encompass the entire family of clays of which montmorillonite is a member. As in the case of illite, the smectites are characterized by a T–O–T structure. Unlike the other clay minerals, however, the number of possible substitutions documented for the smectites is large both on the tetrahedral and octahedral sheets. Substitution of Al^{3+} and possibly Fe^{3+} can occur widely for Si^{4+} on the tetrahedral site whereas substitution of Al^{3+}, Fe^{2+}, Fe^{3+}, Mg^{2+}, Zn^{2+}, and Li^{+} occurs on the octahedral layer. Accordingly, both dioctahedral and trioctahedral types exist within this family. To balance the excess negative charge that results from substitutions (Al^{3+} for Si^{4+} on the tetrahedral site, for example),

Table 21.1 Summary of Smectites Derived From Base Compositions by Cation Substitutions

	Z (Tetrahedral)	Y (Octahedral)	X (Interlayer)
Dioctahedral			
Pyrophyllite	Si_8	Al_4	–
Montmorillonite	Si_8	$Al_{3.34}Mg_{0.66}$	$(½Ca, Na)_{0.66}$
Beidellite	$Si_{7.34}Al_{0.66}$	Al_4	$(½Ca, Na)_{0.66}$
Nontronite	$Si_{7.34}Al_{0.66}$	Fe_4^{3+}	$(½Ca, Na)_{0.66}$
Trioctahedral			
Talc	Si_8	Mg_6	–
Saponite	$Si_{7.34}Al_{0.66}$	Mg_6	$(½Ca, Na)_{0.66}$
Hectorite	Si_8	$Mg_{5.34}Li_{0.66}$	$(½Ca, Na)_{0.66}$
Sauconite	$Si_{6.7}Al_{1.3}$	$Zn_{4-5}(Mg,Al,Fe^{3+})_{2-1}$	$(½Ca, Na)_{0.66}$

Modified from Deer, W.A., Howie, R.A., Zussman, J., 1962. Rock-Forming Minerals, vol. 3. Sheet Silicates, Longmans, Green and Co. Ltd, London, UK.

widespread substitution on the interlayer site is possible. The most common interlayer charge-balancing cations are Na^+, K^+, and Ca^{2+}. Deer et al. (1962) summarized the smectite family and the possible substitutions that result in individual members (Table 21.1).

In this table, substitutions from the base compositions, pyrophyllite and talc, yield the various smectite compositions.

The structure of montmorillonite is illustrated in Fig. 21.4. Of special significance in the smectite family is bentonite, which is derived from the weathering of volcanic glass (Christidis and Huff, 2009).

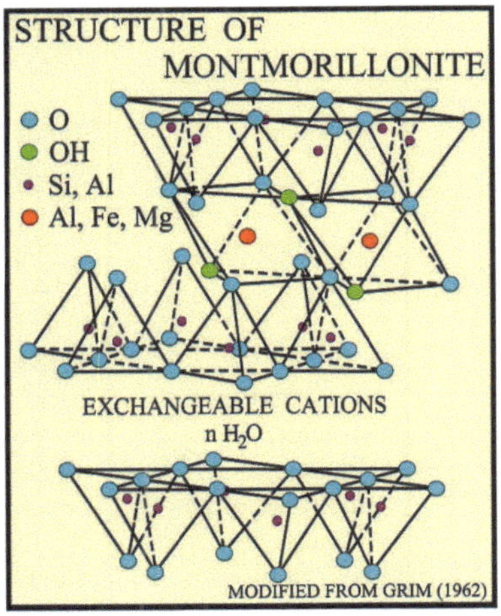

FIGURE 21.4

Structure of the smectite mineral, montmorillonite.

Bentonite appears to be a mixture of montmorillonite and beidellite (Table 21.1), but some K-rich bentonites are illite-dominated or contain kaolinite. Bentonite is characterized by a high degree of water uptake and CEC. These properties make bentonite ideal for waste disposal issues, in which there is a reduction in porosity owing to swelling and uptake of contaminant ions owing to cation exchange. Thus, the use of bentonite clays in waste pond lining and backfill in nuclear waste disposal vaults is viewed as advantageous.

Mixed clay minerals such as bentonite are often referred to as mixed-layer illite–smectite (I/S) minerals and are characterized by individual layers of illite within a dominant smectite structure, or vice versa. With burial time and temperature, I/S clays transform slowly into more illite-rich compositions (Huang et al., 1993) (Table 21.2).

For further discussions of bentonite and its use in waste disposal, polymer-clay nanocomposites, mining, paper processing, medicine, etc., see Carrado and Komadel (2009), Eisenhour and Brown (2009), Güven (2009), and Williams et al. (2009).

Vermiculites are chemically complex clay minerals that typically form as a result of biotite and other originally ferromagnesian minerals. They are typically trioctahedral and are 2:1-layered phyllosilicates that have the characteristic of modest expansion, although not to the degree manifested by

Table 21.2 Principal Characteristics of Clay Minerals

	Kandites	Illites	Smectites	Vermiculites
Structure type	1:1 T–O Dioctahedral	2:1 T–O–T Mostly dioctahedral	2:1 T–O–T Both di- and trioctahedral	2:1 T–O–T Mostly trioctahedral
Main interlayer cations	Nil	K	Ca, Na	Mg
Interlayer water	Only halloysite	Only in hydromuscovite	0 to 3 layers	2 layers
Basal spacing	7.1 Å (10 Å in halloysite)	10 Å	Variable; up to 19 Å (3 layer)	Variable; 14.4 Å when fully hydrated
Glycol	Taken up by halloysite only	No effect	Takes 2 layers of glycol, expands to 17 Å	Takes 1 layer glycol, expands to 14 Å
Heating to 200°C	Halloysite collapses to 7.4 Å, others unchanged	No marked change	Collapse to about 10 Å	Exfoliation; shrinkage of layer spacing
Paragenesis	Alteration of granitic rocks, feldspars under acidic conditions	Alteration of micas, feldspars, especially under alkaline conditions. High Al and K concentrations favorable	Alteration of mafic rocks and volcanic material under alkaline conditions. Availability of Mg and Ca, deficiency of K	Alteration of biotite and other ferromagnesian phases and volcanic material

Modified from Deer, W.A., Howie, R.A., Zussman, J., 1962. Rock-Forming Minerals, vol. 3. Sheet Silicates, Longmans, Green and Co. Ltd, London, UK.

smectite. Magnesium-rich vermiculites typically have two layers of water in their interlayer site, leading to an expansion of about 14 Å. In contrast, K-rich vermiculites have a basal spacing of only about 10 Å and contain no interlayer water. The CEC values of different clay minerals are presented in Table 21.3.

Table 21.4 presents the point of zero charge (pH_{pzc}) of various clay minerals, which is considered to be an important characteristic because of their significance in various environmental applications (as discussed subsequently).

Table 21.3 Cation Exchange Capacity (CEC) Values of Different Clay Minerals

Clay Mineral	CEC (mEq/100 g)
Allophane	70
Kaolinite	3–15
Halloysite ($2H_2O$)	5–10
Na-montmorillonite	80–130
Ca-montmorillonite	40–70
Hectorite	80–130
Palygorskite	30–40
Sepiolite	30–40
Illite	10–40
Vermiculite	100–150

Modified from Ismadji, S., Soetaredjo, F.E., Ayucitra, A., 2015. Natural clay minerals as environmental cleaning agents. In: Clay Materials for Environmental Remediation. Springer International Publishing, Switzerland, pp. 5–37.

Table 21.4 Point of Zero Charge (pH_{pzc}) Values for Various Clay Minerals From Different Origins in the World

Clay Mineral	Origin	pH_{pze}
Kaolin	Georgia, United States	3.00
Illite	Zhejiang, China	2.50
Kaolinite	Nigeria	4.40
Bentonite	Fluka	3.00
Montmorrilonite	Milos, Greece	9.40
Ca-bentonite	Pacitan, Indonesia	3.85
Montmorillonite	Sigma Aldrich	3.40
Montmorillonite	Sipovo, Bosnia	7.70
Smectite	Tunisia	8.20
Kaolinite	Thailand	4.00–5.00

Modified from Ismadji, S., Soetaredjo, F.E., Ayucitra, A., 2015. Natural clay minerals as environmental cleaning agents. In: Clay Materials for Environmental Remediation. Springer International Publishing, Switzerland, pp. 5–37.

21.3 CLAY MINERALS AND ENVIRONMENTAL APPLICATIONS

Clay minerals are highly important minerals which have been used as raw materials for hundreds of industrial and environmental applications because of their abundance and low cost. They are used in agricultural, engineering, and construction applications, in environmental remediation, in geology, as agents for carbon dioxide sequestration, in food processing, and in many other industrial and environmental applications (Murray, 2007). Fig. 21.5 shows a schematic overview of the major environmental significance of clay minerals.

21.3.1 CLAY MINERALS AS SORBENTS FOR TOXIC GASEOUS CONTAMINANTS

Several industrial operations have led to the release of different toxic gaseous contaminants, such as volatile organic compounds (VOCs). A reduction in the generation of VOCs using thermal oxidation is costly because of the high energy required during the process. Although activated carbons are considered to be good commercial sorbents for the depletion of toxic gaseous contaminants, the major disadvantage in their application is that they are combustible, which can lead to difficulties during regeneration and reuse. Clay minerals (eg, smectite, kaolinite, bentonite, zeolite) have also been used as sorbents to remove various toxic gaseous contaminants (Ismadji et al., 2015). For example, pillared clays, which are prepared by cation exchange between alkali cations (generally Na^+) of the original

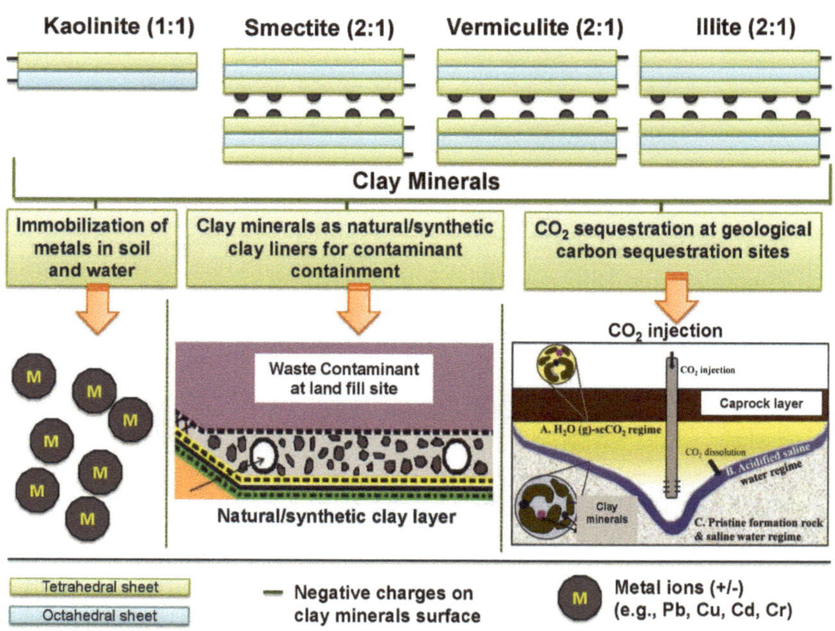

FIGURE 21.5

Schematic representation of different clay minerals and their significance in the immobilization of environmental contaminants, as clay barriers/liners, and in the sequestration of geological CO_2.

clay and hydroxy metal cations from a solution in which the clay is dispersed, could be alternative sorbents for toxic gaseous contaminant abatement (Pires and Pinto, 2010).

Hydrogen sulfide (H_2S) is a colorless gas with corrosive, flammable, poisonous, and explosive properties; it is classified as a noxious gaseous contaminant. Previous studies reported the use of clay minerals for the removal of H_2S gas from environmental media (Thanh et al., 2005; Stepova et al., 2009; Batista et al., 2014; Zhang et al., 2014; Ismadji et al., 2015).

Thanh et al. (2005) modified Na-bentonite with Fe to add active sites to enhance H_2S sorption. Iron-impregnated Na-bentonite demonstrated a significant improvement in the removal of H_2S, despite a noticeable decrease in microporosity compared with the original pillared clay material (Na-bentonite). Differences in sorption efficiency could result from variations in the chemistry of Fe species, the dispersion pattern of Fe on the clay surface, and the accessibility of small pores for the H_2S. Similarly, the activated carbon-sepiollite composite was used for sorbent removal in NH_3 gas removal. The specific interactions between NH_3 with the acidic surface functional groups of sepiollite resulted in strong binding energy, leading to NH_3 removal.

21.3.2 CLAY MINERALS FOR IMMOBILIZATION OF ENVIRONMENTAL CONTAMINANTS

21.3.2.1 Heavy Metal Immobilization

The contamination of soil, sediments, and water with elevated concentrations of heavy metals has become a global health and environmental concern owing to their toxic and carcinogenic nature. The major anthropogenic sources of heavy metals are waste and wastewater from various industries, municipalities, agrochemicals, wastewater treatment systems, runoff from agricultural lands, and mining and leather tanning operations, thus leading to the contamination of water and land resources of ecosystems (Niazi et al., 2011; Austruy et al., 2014; Niazi et al., 2015; Shakoor et al., 2015). Several heavy metals are known to have toxic impacts on humans, animals, plants, and microbes, such as mercury (Hg), arsenic (As, metalloid), lead (Pb), cadmium (Cd), chromium (Cr), copper (Cu), cobalt (Co), manganese (Mn), nickel (Ni), and tin (Sn). Some heavy metals including iron (Fe), selenium (Se), copper (Cu), zinc (Zn), and molybdenum (Mo) in trace concentrations are considered essential for humans.

Previous research has shown that clay minerals have been explored as efficient sorbents for the removal of heavy metals (particularly cationic metals, eg, Pb, Zn, Cd, Fe, Cu) from aqueous solutions (Fig. 21.5). The merits of employing clay minerals as sorbents for the immobilization of heavy metals include high ion sorption/exchange capacity, huge CEC, low permeability, swelling/shrinking ability, chemical and mechanical stability, and large specific surface area. Table 21.5 summarizes some selected studies of clay minerals as sorbents for heavy metal removal.

The sorption of heavy metals in aqueous environments using clay minerals is affected by many factors such as the pH, temperature, the presence of other ions, and the initial analyte concentration.

For instance, Dos Santos et al. (2015) studied the effect of pH (2−8) and ionic strength on the removal of Hg(II) by montmorillonite as a sorbent. At a pH below the pH_{pzc} of montmorillonite, the surface of the clay mineral became positively charged; thus, ion exchange between Hg(II) and Ca^{2+}, Na^+, or K^+ on the surface of the investigated montmorillonites led to the sorption of Hg(II). However, at a pH above 8 (pH_{pzc}), the formation of metal hydroxides or intermediate species occurred, inhibiting an estimate of the decline of the ion concentration in solution (Dos Santos et al., 2015).

Table 21.5 Selected Studies of Heavy Metal Removal Using Clay Minerals

Clay Mineral	Heavy Metal	Remarks
Bentonite	Zn	Bentonite modified with HCl
Bentonite	Pb	Bentonite modified with HCl
Bentonite	Ni and Cd	Chitosan-clay cross-linked to epichlorohydrin
Bentonite	Pb	Bentonite pillared by $MnCl_2$
Kaolin	Pb	Influences of initial concentration, pH, equilibration time, dosage, and temperature in adsorption performance were studied
Kaolin	Cr	Kaolin-supported bacterial biofilm of *Bacillus* sp.
Kaolinite and montmorillonite	Cu	Clay minerals modified with H_2SO_4
Kaolinite	Pb	Kaolinite modified with polyvinylalcohol
Kaolinite	Cd	Kaolinite pillared using MnO_2
Kaolinite	Pb and Cd	Cd and Pb sorption isotherms performed at different pH values
Montmorillonite	Cu	Bentonite pillared by polyhydroxyl ferric
Montmorillonite	Cr	Effect of activation and modification on absorption performance was studied
Montmorillonite	As(III) and As(IV)	Montmorillonite pillared by $TiCl_4$
Montmorillonite	Cu	Chitosan-montmorillonites prepared by cross-linking with pentasodium tripolyphosphate
Montmorillonite	Co	Montmorillonite modified with chitosan
Montmorillonite	Hg	Influence of pH, ionic strength, and treatment with different cations
Palygorskite	Cd	Palygorskite also modified with HCl
Smectite, Illite, calcite	Cu, Zn	Evaluated absorptive capacity of 3 different natural clays
Vermiculite	Hg	Studies influence of humic acid on adsorption performance

Modified from Ismadji, S., Soetaredjo, F.E., Ayucitra, A., 2015. Natural clay minerals as environmental cleaning agents. In: Clay Materials for Environmental Remediation. Springer International Publishing, Switzerland, pp. 5–37.

Temperature is also a crucial process parameter in the adsorption of heavy metals onto clays. Depending on the structure of clay minerals, temperature has an impact on the sorption capacity to an extent. In the case of physical sorption, temperature is known to have a negative effect on the metal uptake by sorbent: metal uptake decreases with an increase in temperature. However, if chemisorption is the control mechanism, increasing the temperature also increases the uptake of metals by the sorbent (clay minerals).

The sorption of cationic heavy metals on the surface of clay minerals could be classified depending on the bonding mechanisms by which the heavy metal cations bind with the mineral. In the case of transition heavy metal cations such as lanthanides and actinides, sorption on most clay minerals at low pH depends on the ionic strength, whereas at pH values >6, it is relatively less ionic strength and pH dependent (Borisover and Davis, 2015). This has been ascribed to the type of surface charge present on the clay mineral surface: fixed charge sites (also termed constant potential charge sites), which predominate sorption at low pH values. The heavy metal (cations) sorption to such fixed charge sites is a result of outer-sphere surface complexation, which is driven by electrostatic attraction and hydrogen-bonding. For this type of (physical) sorption process, the bonding strength is influenced by ionic strength. At high pH values (pH > 7.5), edge sites on the clay mineral surface are more negatively charged and as such are dominant contributors to metal sorption because of strong covalent bonding (inner-sphere surface complexation). These typical sorption behaviors have been found for transition metal cations, lanthanides, and actinides for a range of clay minerals and cationic sorbates (Gu et al., 2010; Dahn et al., 2011; Borisover and Davis, 2015).

Anionic heavy metal/metalloid sorption on surface of clay minerals is relatively weaker compared with cationic heavy metals, as a result of charge repulsion from the fixed negative charge. However, some anions (eg, arsenate [As(V)]) are capable of forming strong covalent bonds with the edge sites of clay minerals (Goldberg, 2002).

Interlayer Sorption of Heavy Metals

The clay mineral montmorillonite carries a permanent negative charge owing to the substitution of lattice cations in a mineral structure with lower charge cations. The negative surface charge (ie, CEC) in its structure is neutralized in the interlayer space by outer-sphere complexed (electrostatically bound) cations (Na and Ca) which could exchange with other heavy metals cations present in an aqueous solution. The cation exchange process is generally driven by pH-independent sorption which can be highly significant at low pH values and at low ionic strength (Dahn et al., 2011). For instance, strontium (Sr) sorption to montmorillonite depends on the ionic strength; however, it is independent of pH (Missana et al., 2008).

Mica mineral layers are bound together by interlayer K^+, which is weakly hydrated and fits into the di-trigonal cavities on the surfaces of adjacent layers. McKinley et al. (2004) reported that micas are a special case for the sorption of cesium (Cs^+) that is a vital component of low-level nuclear wastes. The authors observed that Cs^+ was preferentially sorbed to mica edges, steps on mica surfaces, or fractured regions in sediments from the Hanford (US) nuclear waste site.

Edge Site Sorption of Heavy Metals

In addition to the electrostatic cation exchange sites described earlier, supplementary sorption sites are the surface edge sites, located at the edges of clay mineral layers. Generally, they are present as surface hydroxyl groups bound to the metal ions of octahedral and tetrahedral sheets. These surface functional groups can be protonated or deprotonated, and form surface complexes such as (Fe/Al) metal oxide surface hydroxyl groups. Surface complexation of heavy metal cations and anions is briefly described in following sections.

Surface Complexation of Heavy Metal Cations. Over the past many decades, studies have investigated heavy metal sorption on clay minerals such as montmorillonite, kaolinite, and bentonite. For example, Gu et al. (2010) provided a comparison of sorption of five different heavy metal cations (Cd, Cu, Ni, Pb, and Zn) onto montmorillonite as a function of solution pH and ionic strength. Their results demonstrated that sorption of Cd, Cu, Ni, Pb, and Zn was affected by ionic strength in an acidic pH range (3–6), whereas in a high (pH > 6) pH range no or minimal influence was found on heavy metal sorption. The sorption of uranium(VI) (U[VI]) and Eu was examined on montmorillonite, illite, and kaolinite under a range of environmentally relevant conditions (Gaskova and Bukaty, 2008). These authors described that both ion exchange (outer-sphere) and surface complexation (inner-sphere) processes were responsible for the sorption processes (Bradbury and Baeyens, 2005).

A study by Soltermann et al. (2014) demonstrated the sorption of Fe^{2+} and Zn in anaerobic systems with an ionic strength of 0.1 M at pH 6.4. Their results showed that Fe^{2+} was a highly competing cation at micromolar concentrations when Zn was present in trace concentrations. Under anoxic systems, Fe^{2+} can often be present at concentrations as high as 1 mM; as such, it could be an important cation occupying edge sites on the mineral surface. However, when Fe^{2+} was at trace concentrations, Zn was not an important competing cation for sorption. Spectroscopic investigations suggested partial oxidation of the sorbed Fe^{2+} to Fe^{3+}, showing that some specific edge sites on mineral surface catalyzed oxidation to Fe^{3+}. Further research is warranted to resolve the catalytic properties of these clay minerals to facilitate detoxification reactions of heavy metals/metalloids (eg, Cr[VI] to Cr[III] and As[III] to As[V]) at the clay mineral—water interface in aqueous environments.

A fundamentally important application of sorption of clay minerals is in the management of many nuclear waste programs through clay rocks. For example, illite, montmorillonite, I/S mixed layers, and kaolinite are vital components in such clay rock types and often make up about 50% of the total rock mass (Borisover and Davis, 2015).

The sorption properties of clay minerals could be significantly changed as a result of the presence of inorganic and organic coatings which are generally found in soils and sediments; thus, they have an important role in the mobility and availability of metals in natural environments. For instance, Bellenger and Staunton (2008) investigated the sorption of low amounts of Cs on kaolinite, montmorillonite, and illite, with varying concentrations of humic acid or fulvic acid. The sorption of Cs on illite was decreased in the presence of organic coatings, with smaller effects found for montmorillonite and kaolinite minerals. Fan et al. (2014) demonstrated that Cs sorption on vermiculite (intermediate expansion) and illite (nonexpanding) was reduced to some extent in the ternary clay-HA-Cs system owing to (1) blocked access of Cs by humic acid, and (2) low availability of the interlayer site in vermiculite.

Sepiolite is a natural magnesium (Mg) hydrosilicate clay mineral with a unique fibrous structure that allows absorption of organic and inorganic ions, and because of its low-cost, sepiolite has great potential as a cleaning agent for removing heavy metals from water and wastewaters (Borisover and Davis, 2015). Future research should be directed to comparing and exploring the potential of these clay minerals (sepiolite, illite, and montmorillonite) for immobilizing heavy metals under varying temperatures and pH and in the presence of different naturally occurring ions. This represents a critically important research gap in our scientific ability.

Surface Complexation of Heavy Metal Anions. There have been some studies detailing the nature of heavy metal (anion) surface complexes formed on clay mineral surfaces. Charlet et al. (2012) determined the structure of As(V) surface complexes formed at the edge sites of Ca-montmorillonite mineral. The authors indicated that AsO_3^{4-} oxyanions could share two oxygen atoms with the apical oxygen atoms of Al (and Mg) edge sites in octahedra, thus forming a binuclear, bidentate inner-sphere complex.

The competitive sorption and surface complexation mechanism of As(V), arsenite (As[III]), phosphate, and molybdate was examined on the surface of different clay minerals (ie, kaolinite, montmorillonite, and illite). Each of these heavy metal anions is known to make surface complexes on the edge sites of montmorillonite. At the arsenic concentration of 20 mM and high clay mineral contents used in those experiments, there was little sorptive competition between As(V) and As(III) owing to their different pK values. The sorptive competition between As(V) and phosphate led to enhanced As(V) transport through clay barriers.

In a study by Manning and Goldberg (1997), the authors reported oxidation of sorbed As(III) to As(V) on kaolinite and illite mineral surfaces at a pH of about 7. Similarly, Lin and Puls (2000) observed the transformation of sorbed As(III) to As(V) by five various clay mineral samples (ie, halloysite, kaolinite, illite, illite-montmorillonite, chlorite) which were aged for two months. Lenoble et al. (2002) examined the efficiency of Fe(III)-pillared montmorillonite and suggested that it was as strong as goethite as a sorbent for As(V) and As(III), which proving that it was an efficient reactive material for clay barriers where arsenic immobilization is critically important. Although partially understood, investigations are warranted to resolve the role of various modification methods to develop modified clay minerals to immobilize redox-sensitive heavy metal/metalloid species.

21.4 SORPTION OF ORGANIC CONTAMINANTS

21.4.1 SORPTION OF DYES, ANTIBIOTICS, AND OTHER ORGANIC CONTAMINANTS

Synthetic dyes are classified into 20–30 different groups: for example, azo dye, anthraquinone, phthalocyanine, triarylmethane, azine, oxazine. Dyes with reactive groups are known as reactive dyes and can form covalent bonds with OH–, NH–, or SH– groups in fibers. Most of the dyes are visible in aqueous solutions even at very low concentrations (1 mg L^{-1}).

The dyes are highly persistent under natural environments because they are synthesized to be stable, both chemically and photolytically. The release of dyes may therefore present an eco-toxic hazard and introduces the potential danger of bioaccumulation that may eventually affect humans by transport through the food chain. Compared with many physical and chemical removal techniques, sorption using various sorbents has been found to be the best choice for treating water and wastewater (Ismadji et al., 2015).

Several studies have been directed to explore efficient and cheap sorbents for removing dyes from water and wastewater (Table 21.6 shows some selective studies). Clay minerals possess a high sorption capacity toward several classes of dyes, and their sorption capabilities are comparable to those of activated carbons. The sorption performance of clay minerals and its modified forms also depends strongly on the class of dye. Many natural clay minerals have a high sorption capacity for binding basic (cationic) dyes (Borisover and Davis, 2015; Ismadji et al., 2015); however, often they have low capacity to remove dyes from other groups or classes of dyes. Similar to the other liquid phase adsorption systems, factors that affect the adsorption of dyes onto clay minerals and its modified forms are solution pH, temperature, initial dye concentration, the presence of other substances, etc.

Variation in the solution pH results in a change in the surface charge of the sorbent and the degree of ionization of the absorbing ion. For instance, Bhattacharyya et al. (2014) conducted sorption experiments at a pH of 2–12 to remove rhodamine B from water using kaolinite and montmorillonite. They found that the surface of kaolinite was positive under an acidic pH range; thus, its sorption capacity was low (maximum uptake was around 24 mg g^{-1}) for rhodamine B species (ie, positively charged). The surface of solid became repulsive to $RhBH^+$ and thus led to reduced uptake of rhodamine B by kaolinite and its modified forms in acidic solutions. In the case of montmorillonite, uptake of rhodamine B was much higher than kaolinite and its acid-modified forms, owing to the negatively charge of the montmorillonite surface under the entire range of pH studied.

Temperature is another significant physicochemical process parameter in the sorption system. In most of cases for dye–clay mineral sorption systems, temperature has a positive effect on the removal

Table 21.6 Selected Studies of Sorption Capacity of Clay Minerals for Dyes

Clay Mineral	Dye Type	Performance
Bentonite	Methylene blue	Maximum sorption capacity of bentonite 2.22 mmol g^{-1}; after acid activation, sorption capacity decreased to 1.56 mmol g^{-1}
Kaolin	Malachite green	Adsorption capacity of kaolin 0.509 mmol g^{-1}; rarasponin modified kaolin to 0.919 mmol g^{-1}
Alunite	Acid red 88	Sorption capacity of calcinedalunite 832.31 mg g^{-1}
Na-montmorillonite	Methyl orange	Adsorption capacity Na-montmorillonite 22.83 mg g^{-1}; modified by cethyltrimethyl ammonium bromide 42.04 mg g^{-1} anionic surfactants sodium stearate 121.97 mg g^{-1}
Attapulgite	Congo red	Maximum adsorption capacity of HTMA-bentonite 189.39 mg g^{-1}
Bentonite	Evans blue	Adsorption capacity 0.263 mmol g^{-1}; treated with rarasaponin 0.516 mmol g^{-1}
Bentonite	Methylene blue	Adsorption capacity of bentonite 194 mg g^{-1}, and for rarasaponin-bentonite 256 mg g^{-1}
Bentonite	Methylene blue	Adsorption capacity of chitosan-bentonite composite 142.86 mg g^{-1}
Bentonite	Reactive red 120	Adsorption capacity of cetylpyridinium-bentonite 81.97 mg g^{-1}
Kaolinite	Coomassie brilliant blue R 250	Adsorption capacity 22.89 mg g^{-1}; acid-treated kaolinite 30.08 mg g^{-1}
Bentonite	Amido black 10B	Adsorption capacity of cross-linked chitosan–bentonite composite 323.6 mg g^{-1}
Bentonite	Orange II	Adsorption capacity of alkyltriphenyl phosphonium-bentonite 53.78 mg g^{-1}
Rectorite	Methylene blue	Adsorption capacity of acid-modified rectorite 37 mg g^{-1}
Bentonite	Acid blue 12	Adsorption capacity of bentonite 0.76 mmol g^{-1}, and for CTA (cetyl trimethylammonium bromide)-bentonite 2.76 mmol g^{-1}
Na-bentonite	Congo red	Adsorption capacity of natural bentonite 19.5 mg g^{-1}, acid activation 69.44 mg g^{-1}, thermal activation 54.64 mg g^{-1}, combination of acid and thermal 75.75 mg g^{-1}

Modified from Ismadji, S., Soetaredjo, F.E., Ayucitra, A., 2015. Natural clay minerals as environmental cleaning agents. In: Clay Materials for Environmental Remediation. Springer International Publishing, Switzerland, pp. 5–37.

of (dye) solute with an increase in temperature (Bhattacharyya and Ray, 2014; Liu et al., 2014, 2015; Ismadji et al., 2015). However, in several systems temperature has negative effect on the sorption capacity of dyes (Vimonses et al., 2009; Suwandi et al., 2012; Chandra et al., 2013; Ozturk and Malkoc, 2014). Further research is warranted to resolve this confusion by investigating clay mineral–dye interaction under environmentally significant conditions with varying temperatures and pH (Pentrak et al., 2012).

In addition to dyes, excessive use of antibiotics has created serious health problems for the environment. Over 30–90% of the given dose can remain undegradable in the human or animal body, and thus is excreted as an active compound (Putra et al., 2009). As a result of overuse, bacteria have developed antibiotic resistance. The removal of antibiotics from aqueous solutions has been conducted effectively by sorption using clay minerals (Hamilton et al., 2014; Sturini et al., 2015).

Sorption and intercalation of ciprofloxacin on montmorillonite were studied by Wu et al. (2010). Quantitative desorption of exchangeable cations confirmed that cation exchange was the most important mechanism of ciprofloxacin sorption on the surface of montmorillonite (Lin and Puls, 2000; Li et al., 2010; Wu et al., 2010). To our understanding, it is critically important to examine the ability of clay minerals to remove inorganic and organic contaminants in real wastewater simultaneously, although this is a partially explored area of research.

21.5 HEAVY METAL AND METALLOID INTERACTIONS WITH CLAY MINERALS UNDER NATURAL AND ENGINEERED ENVIRONMENTAL CONDITIONS

In environmental applications, clay mineral liners are commonly applied to seal landfills and obstruct the transport of hazardous metals/metalloids through these barriers (Fig. 21.5). This is important to stop the leaching of mobile and toxic elements into groundwater. For instance, As(V), As(III), and Se(IV) sorption from landfill leachates onto montmorillonite and kaolinite was studied (Borisover and Davis, 2015) owing to the occurrence of these minerals in clay barriers.

Missana et al. (2009) investigated the immobilization of Se(IV) on FEBEX Bent; on an illite sample it was high, in the slightly acidic pH range ($pH < 7$). Although Cr(VI) is relatively more weakly sorbed than As(V), As(III), or phosphate, and is only efficiently removed by illite or kaolinite under more acidic environments (pH 2–5) than Se(IV), removal of Cr(VI) is still high enough to allow natural clays to be used as landfill liners to attenuate Cr(VI) as a contaminant. The Fe(II)-bearing clay minerals (eg, montmorillonite, illite, vermiculite), were effective in reducing Cr(VI) to Cr(III) at the mineral surface, which greatly increased the retention of Cr and indicated that the redox potential within clay barriers is important in its mobility (Borisover and Davis, 2015; Ismadji et al., 2015). To our understanding, research is required to determine the reduction-coupled sorption of Cr(VI) and other oxyanionic metal/metalloid species, to explore the natural potential of clay minerals for detoxification and immobilization of these elements under a wide range of environmentally relevant situations.

The speciation of Ni and Zn in a clay layer of a paddy soil (Taiwan) was investigated using extended X-ray absorption fine structure spectroscopy (Manceau et al., 2007). Nickel substituted for Mn in the Mn layer of a lithiophorite in soil, whereas Zn led to filling of vacant sites in the gibbsitic layer of the lithiophorite. The metal substitution reactions in soils and sediments can have a significant role in the sequestration of some heavy metals and govern their mobility and availability in soil pore-water solutions (Vespa et al., 2010).

21.6 CLAY MINERALS AND CO_2 SEQUESTRATION

There are a number of ways in which the presence of clays can affect the success of subsurface geologic sequestration of CO_2 (GSC). Although the CO_2 pumped underground is relatively dry (ie, contains minor concentrations of dissolved H_2O), supercritical carbon dioxide ($scCO_2$) will eventually mix with and dissolve into formation waters to form a single-phase solution (Fig. 21.5). The presence of H_2CO_3 (in addition to low concentrations of sulfuric acid) will decrease the solution pH, thereby increasing the dissolution of the porous matrix. The positive aspect of this dissolution is that divalent cations (primarily Mg^{2+}, Ca^{2+}, and Fe^{2+}) will be released, and over time they will combine with the carbonate anion to form carbonate minerals, thus locking up CO_2 long-term. However, the presence of clays, especially smectites, may exert a strong influence on the amount of divalent cations available for carbonate mineral formation. Either by sorption onto clay surfaces or by cation exchange, the activities of divalent cations may be decreased if the surface area of clays contacting the solution is large. Although this scenario has not been fully investigated, the results of some field tests (Apps et al., 2011) appear to suggest that this mechanism could be important, especially for Ca^{2+}.

A second way in which the presence of clays can have a strong bearing on the GSC strategy is in the uptake of CO_2 from solution to clay surfaces or interlayer sites. Chen and Lu (2015) showed that CO_2 can be adsorbed onto kaolinite surfaces to a large extent, thereby immobilizing CO_2. They showed that the amount of CO_2 sorbed onto a kaolinite surface is proportional to the surface area, which they manipulated in their samples by contacting kaolinite with strong acid. However, an even more important effect is the intercalation of CO_2 into the interlayer sites of smectites. This mechanism can work to produce swelling or shrinkage of the clay, an effect that clearly has implications for sealing pathways through which CO_2 might escape from its emplacement formation. Accordingly, there is increasing interest in determining the conditions in which CO_2 can replace H_2O in the interlayer site in smectite clays, the degree of expansion or contraction, and the impact this may have on cap-rock integrity.

The ability of smectite minerals to accommodate CO_2 into the interlayer site is based on the ease with which the interlayer thickness can expand, as discussed earlier. Up to three layers of water can be accommodated by smectites, and the uptake of H_2O into the interlayer site should be visible by stepwise changes in the interlayer distance, which in XRD patterns is manifested as changes in the d_{001} spacing (Ferrage et al., 2005). In addition, the d_{001} distance should also be composition-dependent, with slight swelling differences between Na- and Ca-rich compositions. The 0, 1, 2, and 3 hydration states are designated 0W, 1W, 2W, and 3W, respectively, and have the *d*-spacings (Table 21.7).

However, the expected stepwise increase as the hydration state in smectites increases is not always seen. The possible reasons for this include a lack of access of water to the smectites, the presence of other phases, and heterogeneous uptake of water. In fact, Ferrange et al. (2005) stated: "Heterogeneity (*of interlayer expansion*) rather than homogeneity seems to be the rule for smectite hydration" (the words in italics are the current author's).

Both molecular simulations (Cygan et al., 2012; Myshakin et al., 2013; Lee et al., 2014; Schaef et al., 2015) and experiments (Giesting et al., 2012a,b; Ilton et al., 2012; Loring et al., 2012, 2014; Schaef et al., 2012, 2015; Romanov 2013; Rother et al., 2013; De Jong et al., 2014) indicate that carbon

Table 21.7 Interlayer Spacing (Indexed by d_{001} in X-Ray Diffraction Patterns) for Na- or Ca-Smectites as a Function of State of Hydration

Dominant Interlayer Cation in Smectite	Interlayer Spacing, Indexed by d_{001} (in Angstroms)			
	0W	1W	2W	3W
Na[a]	9.6–10.1	12.3–12.7	15.0–15.8	18.5–19.0
Ca[b]	9.6–9.7	11.9–12.5	15.0–15.5	18.0–19.1

[a]*Ferrange et al. (2005).*
[b]*Quoted in Schaef et al. (2012).*

dioxide can substitute into the interlayer site of hydrated smectites. A variety of techniques were employed to prove that carbon dioxide was actually intercalated into the interlayer site, and not simply sorbed onto clay surfaces, and these methods include in situ high-pressure XRD, in situ infrared spectroscopy (Loring et al., 2014; Schaef et al., 2015), diffuse reflectance or attenuated transmission reflectance infrared (Romanov 2013; Loring et al., 2014), nuclear magnetic resonance (Loring et al., 2012), neutron diffraction (Rother et al., 2013), and quartz crystal microbalance (Schaef et al., 2015). Most of the experiments were conducted at temperatures and pressures appropriate for discussion of GSC, and measurements were carried out in situ rather than ex situ. A variety of chemical compositions were made by saturating the smectites in Na, K, Ca, or Mg, or using natural samples, and various starting hydration states were used. Many of the experiments used pure CO_2, but others used CO_2 variably saturated in H_2O, which greatly increased the applicability of the results. In detail, there are some relatively minor differences among these studies, which will not be highlighted here, but the general conclusions will be summarized subsequently.

The investigations showed that the extent to which CO_2 can be intercalated into the smectite structure depends on the degree of hydration, the identity of the dominant interlayer cation, the pressure (or partial pressure) of CO_2, temperature, and the extent to which the $scCO_2$ is saturated in H_2O. In general, all studies agreed that some amount of initial water is necessary in the smectite for CO_2 to enter into the interlayer site. Smectites with a 0W hydration state showed little or no swelling when contacted by dry $scCO_2$. However, even dehydrated smectites can serve as a substrate for significant CO_2 sorption (Loring et al., 2014; Schaef et al., 2015). The amount of sorption increases with an increase in CO_2 pressure but decreases with an increase in sorbed H_2O. The effect of an increase in temperature is to cause shrinkage in the hydrated smectites (De Jong et al., 2014). In contrast to the behavior of dehydrated smectites in the presence of "dry" CO_2, hydration and swelling occur in 0W smectites when water is dissolved in CO_2 (Ilton et al., 2012). A fairly large degree of interlayer expansion occurred when approximately 1W smectite was exposed to $scCO_2$ at temperatures and pressures typical for geologic disposal. For example, Giesting et al. (2012b) showed that K- and Ca-exchanged smectite expanded from their initial states to maximum values of 12.2 and 15.1 Å, respectively, and this expansion occurred rapidly. On the other hand, some investigations have shown that smectites with initial hydrations states of 2W or 3W shrink when in contact with $scCO_2$ (Ilton et al., 2012; Schaef et al., 2015). The smectites that displayed evidence of shrinkage from these relatively high initial hydration states did so at or below 100% saturation with H_2O in CO_2. When the contacting fluid was at a higher saturation state (ie, two-phase liquid), the 3W smectites underwent a small degree of swelling. These data underscore the importance of the initial hydration state but also the proportion of water to carbon dioxide in the system.

Dehydrated smectites (0W) showed little or no response to the presence of dry $scCO_2$, but a small amount of water dissolved in $scCO_2$ may cause a small degree swelling. The greatest change in hydration state occurs for smectites with higher initial hydration states (1W to 3W). In general, smectites with an amount of interlayer water up to 1W show evidence for swelling when in contact with CO_2. On the other hand, 2W and 3W initial hydration state smectites have significant shrinkage when in contact with $scCO_2$ at conditions below saturation with H_2O. Small degrees of swelling occur for 3W smectites in contact with two-phase ($H_2O + scCO_2$) fluids, with the swelling likely caused by an increase in water, not carbon dioxide. Both stepwise (Ilton et al., 2012; Schaef et al., 2015) and more linear (De Jong et al., 2014) hydration or dehydration steps can occur.

The geologic implications of these data are that swelling of clays during internment of CO_2 in the subsurface will close off potential pathways in cap rocks, increasing safety for GCS. On the other hand, shrinkage of clays may result in an increase in permeability, resulting in negative effects for GCS. Which of these scenarios (overall expansion or shrinkage of clays) will prevail will depend on a number of factors already described, such as the initial hydration state, the identity of the dominant cation in the interlayer site, the degree of water saturation in the $scCO_2$, and the pressure and temperature of the injection site. Potentially, the worst scenario would be relatively shallow injection, in which the initial hydration state of the smectites is 2W to 3W, such that the dry $scCO_2$ would likely cause significant shrinkage in the expandable clays. Deeper injection into rocks containing smectites with about 1W hydration will likely result in clay swelling, especially in areas distal to the point of injection, in which mixing of $scCO_2$ and water will result in partial saturation of the initially dry CO_2. As pointed out by Schaef et al. (2015), there is potential overlap between shales (rocks rich in clays) that are targeted for hydraulic fracturing and those that serve as a cap rock for storage and sequestration integrity. It is therefore vital to conduct detailed site investigations to determine the risks associated with GSC.

The reversibility of CO_2 intercalation or sorption is also an important consideration. Even rocks that lack smectites but are rich in kaolinite can serve to accommodate large amounts of CO_2 through sorption. For intercalated CO_2, there is still controversy surrounding the permanence of interlayer CO_2 once the pressure has been reduced. Although Romanov (2013) proposed that CO_2 is "irreversibly" intercalated in smectites, other investigators (Giesting et al., 2012a) suggested that the reversibility of intercalation depends on "aging" of the clays (ie, time of contact with CO_2). This is an area of investigation that warrants further attention.

21.7 CLAY MINERAL WEATHERING AND METAL CATION LEACHING IN THE SOIL AND SEDIMENT ENVIRONMENTS

Clay minerals are central to our natural environment. Reactions occurring at the mineral–water interface affect a wide range of natural processes, all of which involve geochemical elemental cycling. Chemical weathering of clay minerals is a naturally occurring process in all soil and sediment environments. Clay mineral dissolution process under acidic aqueous environments is particularly important because of the rapid consumption of protons and enhanced leaching of metal cations from mineral breakdown (Bibi et al., 2014). In the past few decades, significant advances have been made in understanding the effect of acidic conditions prevailing under a wide range of environments (eg, acid sulfate soils, acid mine drainage) on clay mineral dissolution and elemental leaching.

Investigation of the effect of acidic conditions on clay dissolution has been carried out in batch and flow-through systems using pure minerals and soil clay powders by means of laboratory experiments (Bibi et al., 2011a,b; 2014). Both 1:1 (kaolinite) and 2:1 (mica minerals, smectite, vermiculite, chlorite) type clay mineral dissolution kinetics have been examined in the presence of a wide range of organic and inorganic acids (Brantley, 2007). In a set of experiments, Bibi et al. (2011a,b, 2014) determined the effects of saline acidic conditions on dissolution kinetics of the most commonly and abundantly occurring clay minerals (kaolinite, illite, and montmorillonite) using pure clay samples as well as clays derived from soils. The authors reported that both pure and soil clay minerals demonstrated the greatest leaching of cations (Si, Al, K, Fe, and Mg) at a highly acidic pH (pH 1), and vice versa.

Acid sulfate conditions investigated by Bibi et al. (2011a, 2014) did not seem to affect the dissolution kinetics of pure minerals and soil clays, as evaluated from leaching of Si (major structural cation in clay minerals) under steady-state conditions compared with other sources of acidity (eg, HCl, HNO_3) previously considered by other investigators. However, the leached amount of Al (major octahedral cation) was significantly reduced in lower–ionic strength solutions compared with higher–ionic strength solutions, mainly because of the adsorption of dissolved Al at the lower ionic strength. Hence, it could be concluded that the enhanced concentration of salts present in the reacting solution had deteriorative effects on the health of soils and associated water resources owing to enhanced leaching and the release of Al from clay mineral weathering under saline acidic conditions.

21.8 SUMMARY AND CONCLUSIONS

Clay minerals constitute a group of highly reactive, naturally occurring minerals that are of prime importance in the natural environment as well as for industrial and environmental applications. Most processes governed by clay minerals involve geochemical elemental cycling, such as clay mineral weathering, inorganic and organic contaminant sorption on clay minerals, and CO_2 sequestration in the subsurface geological settings. Clay minerals have wide applications in industry, agriculture, engineering, construction, environmental remediation, geology, etc. Because of the dependence of these processes on the structure and properties of clay minerals, a thorough comprehension of the crystal structure, chemistry, and composition of four major groups (kandite, illite, smectite, and vermiculite) has been provided in this chapter.

Clay minerals have been widely used to immobilize heavy metals in water and wastewater environments. However, the presence of clay minerals at geologic CO_2 sequestration sites may affect the success of this strategy in several different ways. The major source of modification of clays as a result of clay mineral–sCO_2 interaction is brought about by reduction in the solution pH owing to the production of H_2CO_3 and small amounts of H_2SO_4 under high-pressure environments. Some of the most common impacts of the interaction between clay minerals and sCO_2 at GSC sites are that:

- Clay mineral dissolution kinetics may be enhanced as a result of acidic pH conditions produced at the GSC sites associated with enhanced leaching of cations such as Ca^{2+}, Mg^{2+}, and Fe^{2+}. Over time, these cations will combine with carbonate anion to form carbonate minerals, thus locking up CO_2 for a long time.
- Another method of interaction could be the uptake of CO_2 onto the surfaces or interlayer sites of clay minerals. For example, the uptake of CO_2 into the interlayer of smectites could lead to changes in the swelling of smectite layers, depending on the amount of hydrated layers originally present in the smectite.
- The geologic implications of these results are that swelling of clays during the internment of CO_2 in the subsurface will close off potential pathways in cap rocks, thereby increasing the safety for GCS. Also, the reversible nature of this process should be considered in deciding the long-term sustainability of GSC sites.

ACKNOWLEDGMENTS

The authors are thankful to the Grand Challenges Canada—Stars in Global Health (GCC Grant No. 0433-01) and the University of Agriculture Faisalabad for providing financial assistance.

REFERENCES

Apps, J.A., Zheng, L.G., Spycher, N., Birkholzer, J.T., Kharaka, Y., Thordsen, J., Kakouros, E., Trautz, R., 2011. Transient changes in shallow groundwater chemistry during the MSU ZERT CO_2 injection experiment. In: Gale, J., Hendriks, C., Turkenberg, W. (Eds.), 10th International Conference on Greenhouse Gas Control Technologies, pp. 3231–3238.

Austruy, A., Shahid, M., Xiong, T.T., Castrec, M., Payre, V., Niazi, N.K., Sabir, M., Dumat, C., 2014. Mechanisms of metal-phosphates formation in the rhizosphere soils of pea and tomato: environmental and sanitary consequences. Journal of Soils and Sediments 14, 666–678.

Balan, E., Calas, G., Bish, D.L., 2014. Kaolin-group minerals: from hydrogen-bonded layers to environmental recorders. Elements 10, 183–188.

Batista, L.C.D., Dantas, D.D., de Farias, R.F., 2014. Dye adsorption on inorganic matrices as a new strategy to gas capture: hydrogen sulfide adsorption on Rodhamine B modified kaolinite. Synthesis and Reactivity in Inorganic Metal-Organic and Nano-Metal Chemistry 44, 1398–1400.

Bellenger, J.P., Staunton, S., 2008. Adsorption and desorption of (85)Sr and (137)CS on reference minerals, with and without inorganic and organic surface coatings. Journal of Environmental Radioactivity 99, 831–840.

Bhattacharyya, R., Ray, S.K., 2014. Micro- and nano-sized bentonite filled composite superabsorbents of chitosan and acrylic copolymer for removal of synthetic dyes from water. Applied Clay Science 101, 510–520.

Bhattacharyya, K.G., SenGupta, S., Sarma, G.K., 2014. Interactions of the dye, Rhodamine B with kaolinite and montmorillonite in water. Applied Clay Science 99, 7–17.

Bibi, I., Singh, B., Silvester, E., 2011a. Dissolution of illite in saline-acidic solutions at 25°C. Geochimica et Cosmochimica Acta 75, 3237–3249.

Bibi, I., Singh, B., Silvester, E., 2011b. Dissolution of common phyllosilicates in acid sulfate systems. In: Proceedings of the ASA, CSSA, SSSA International Annual Meetings. Fundamental for Life: Soil, Crop and Environmental Sciences, 16–19 October, San Antonio, Texas, USA.

Bibi, I., Singh, B., Silvester, E., 2014. Dissolution kinetics of soil clays in sulfuric acid solutions: ionic strength and temperature effects. Applied Geochemistry 51, 170–183.

Borisover, M., Davis, J.A., 2015. Adsorption of inorganic and organic solutes by clay minerals. In: Christophe Tournassat, C.I.S.I.C.B., Fa□za, B. (Eds.), Developments in Clay Science. Elsevier, pp. 33–70.

Bradbury, M.H., Baeyens, B., 2005. Experimental measurements and modeling of sorption competition on montmorillonite (vol. 69, p. 4187, 2005). Geochimica et Cosmochimica Acta 69, 5863–5864.

Brantley, S.L., 2007. Kinetics of mineral dissolution. In: Brantley, S.L., White, A.F., Kubicki, J.D. (Eds.), Kinetics of Water-Rock Interaction. Springer, New York.

Brindley, G.W., 1951. X-Ray Identification and Crystal Structures of Clay Minerals. Mineralogical Society, London, UK.

Carrado, K.A., Komadel, P., 2009. Acid activation of bentonites and polymer—clay nanocomposites. Elements 5, 111–116.

Chandra, I.K., Ju, Y.H., Ayucitra, A., Ismadji, S., 2013. Evans blue removal from wastewater by rarasaponin-bentonite. International Journal of Environmental Science and Technology 10, 359–370.

Charlet, L., Bardelli, F., Parsons, C., He, J., Chakraborty, S., Gailer, J., 2012. Arsenic binding onto phyllosilicates and glutathione: soil immobilisation and human excretion mechanisms. In: Ng, J.C., Noller, B.N., Naidu, R., Bundschuh, J., Bhattacharya, P. (Eds.), Understanding the Geological and Medical Interface of Arsenic. Taylor & Francis Group, London, UK, pp. 59–62.

Chen, Y.-H., Lu, D.-L., 2015. CO_2 capture by kaolinite and its adsorption mechanism. Applied Clay Science 104, 221–228.
Christidis, G.E., Huff, W.D., 2009. Geological aspects and genesis of bentonites. Elements 5, 93–98.
Cygan, R.T., Romanov, V.N., Myshakin, E.M., 2012. Molecular simulation of carbon dioxide capture by montmorillonite using an accurate and flexible force field. Journal of Physical Chemistry C 116, 13079–13091.
Dahn, R., Baeyens, B., Bradbury, M.H., 2011. Investigation of the different binding edge sites for Zn on montmorillonite using P-EXAFS – the strong/weak site concept in the 2SPNE SC/CE sorption model. Geochimica et Cosmochimica Acta 75, 5154–5168.
De Jong, S.M., Spiers, C.J., Busch, A., 2014. Development of swelling strain in smectite clays through exposure to carbon dioxide. International Journal of Greenhouse Gas Control 24, 149–161.
Deer, W.A., Howie, R.A., Zussman, J., 1962. In: Rock-Forming Minerals, vol. 3. Sheet Silicates, Longmans, Green and Co. Ltd, London, UK.
Detellier, C., Schoonheydt, R.A., 2014. From platy kaolinite to nanorolls. Elements 10, 201–206.
Dos Santos, V.C.G., Grassi, M.T., Abate, G., 2015. Sorption of Hg(II) by modified K10 montmorillonite: influence of pH, ionic strength and the treatment with different cations. Geoderma 237, 129–136.
Drever, J.I., 1997. The Geochemistry of Natural Waters. Prentice Hall, Upper Saddle River, New Jersey, USA.
Eisenhour, D.D., Brown, R.K., 2009. Bentonite and its impact on modern life. Elements 5, 83–88.
Fan, Q.H., Tanaka, M., Tanaka, K., Sakaguchi, A., Takahashi, Y., 2014. An EXAFS study on the effects of natural organic matter and the expandability of clay minerals on cesium adsorption and mobility. Geochimica et Cosmochimica Acta 135, 49–65.
Ferrage, E., Lanson, B., Sakharov, B.A., Drits, V.A., 2005. Investigation of smectite hydration properties by modeling experimental X-ray diffraction patterns: Part I. Montmorillonite hydration properties. American Mineralogist 90, 1358–1374.
Gaskova, O.L., Bukaty, M.B., 2008. Sorption of different cations onto clay minerals: modelling approach with ion exchange and surface complexation. Physics and Chemistry of the Earth 33, 1050–1055.
Giesting, P., Guggenheim, S., van Groos, A.F.K., Busch, A., 2012a. Interaction of carbon dioxide with Na-exchanged montmorillonite at pressures to 640 bars: implications for CO_2 sequestration. International Journal of Greenhouse Gas Control 8, 73–81.
Giesting, P., Guggenheim, S., van Groos, A.F.K., Busch, A., 2012b. X-ray diffraction study of K- and Ca-exchanged montmorillonites in CO_2 atmospheres. Environmental Science and Technology 46, 5623–5630.
Goldberg, S., 2002. Competitive adsorption of arsenate and arsenite on oxides and clay minerals. Soil Science Society of America Journal 66, 413–421.
Gu, X.Y., Evans, L.J., Barabash, S.J., 2010. Modeling the adsorption of Cd (II), Cu (II), Ni (II), Pb (II) and Zn (II) onto montmorillonite. Geochimica et Cosmochimica Acta 74, 5718–5728.
Güven, N., 2009. Bentonites – clays for molecular engineering. Elements 5, 89–92.
Hamilton, A.R., Hutcheon, G.A., Roberts, M., Gaskell, E.E., 2014. Formulation and antibacterial profiles of clay-ciprofloxacin composites. Applied Clay Science 87, 129–135.
Huang, W.L., Longo, J.M., Pevear, D.R., 1993. An experimentally derived kinetic model for smectite-to-illite conversion and its use as a geothermometer. Clays and Clay Minerals 41, 162–177.
Ilton, E.S., Schaef, H.T., Qafoku, O., Rosso, K.M., Felmy, A.R., 2012. In situ X-ray diffraction study of Na^+ saturated montmorillonite exposed to variably wet super critical CO_2. Environmental Science and Technology 46, 4241–4248.
Ismadji, S., Soetaredjo, F.E., Ayucitra, A., 2015. Natural clay minerals as environmental cleaning agents. In: Clay Materials for Environmental Remediation. Springer International Publishing, Switzerland, pp. 5–37.
Kogure, T., Inoue, A., 2005. Stacking defects and long-period polytypes in kaolin minerals from a hydrothermal deposit. European Journal of Mineralogy 17, 465–473.
Langmuir, D., 1997. Aqueous Environmental Geochemistry. Prentice Hall, Upper Saddle River, New Jersey, USA.
Lee, M.S., McGrail, B.P., Glezakou, V.A., 2014. Microstructural response of variably hydrated Ca-rich montmorillonite to supercritical CO_2. Environmental Science and Technology 48, 8612–8619.

Lenoble, V., Bouras, O., Deluchat, V., Serpaud, B., Bollinger, J.C., 2002. Arsenic adsorption onto pillared clays and iron oxides. Journal of Colloid and Interface Science 255, 52–58.

Li, Z.H., Schulz, L., Ackley, C., Fenske, N., 2010. Adsorption of tetracycline on kaolinite with pH-dependent surface charges. Journal of Colloid and Interface Science 351, 254–260.

Lin, Z., Puls, R.W., 2000. Adsorption, desorption and oxidation of arsenic affected by clay minerals and aging process. Environmental Geology 39, 753–759.

Liu, B., Lu, J.X., Xie, Y., Yang, B., Wang, X.Y., Sun, R.C., 2014. Microwave-assisted modification on montmorillonite with ester containing Gemini surfactant and its adsorption behavior for triclosan. Journal of Colloid and Interface Science 418, 311–316.

Liu, Q., Yang, B.C., Zhang, L.J., Huang, R.H., 2015. Adsorption of an anionic azo dye by cross-linked chitosan/bentonite composite. International Journal of Biological Macromolecules 72, 1129–1135.

Loring, J.S., Schaef, H.T., Turcu, R.V.F., Thompson, C.J., Miller, Q.R.S., Martin, P.F., Hu, J.Z., Hoyt, D.W., Qafoku, O., Ilton, E.S., Felmy, A.R., Rosso, K.M., 2012. In situ molecular spectroscopic evidence for CO_2 intercalation into montmorillonite in supercritical carbon dioxide. Langmuir 28, 7125–7128.

Loring, J.S., Ilton, E.S., Chen, J., Thompson, C.J., Martin, P.F., Benezeth, P., Rosso, K.M., Felmy, A.R., Schaef, H.T., 2014. In situ study of CO_2 and H_2O partitioning between Na-montmorillonite and variably wet supercritical carbon dioxide. Langmuir 30, 6120–6128.

Manceau, A., Tommaseo, C., Rihs, S., Geoffroy, N., Chateigner, D., Schlegel, M., Tisserand, D., Marcus, M.A., Tamura, N., Chen, Z.S., 2007. Natural speciation of Mn, Ni, and Zn at the micrometer scale in a clayey paddy soil using X-ray fluorescence, absorption, and diffraction (vol. 69, p. 4007, 2005). Geochimica et Cosmochimica Acta 71, 526–527.

Manning, B.A., Goldberg, S., 1997. Adsorption and stability of arsenic(III) at the clay mineral-water interface. Environmental Science and Technology 31, 2005–2011.

McKinley, J.P., Zachara, J.M., Heald, S.M., Dohnalkova, A., Newville, M.G., Sutton, S.R., 2004. Microscale distribution of cesium sorbed to biotite and muscovite. Environmental Science and Technology 38, 1017–1023.

Missana, T., Garcia-Gutierrez, M., Alonso, U., 2008. Sorption of strontium onto illite/smectite mixed clays. Physics and Chemistry of the Earth 33, S156–S162.

Missana, T., Alonso, U., Garcia-Gutierrez, M., 2009. Experimental study and modelling of selenite sorption onto illite and smectite clays. Journal of Colloid and Interface Science 334, 132–138.

Murray, H.H., 2007. Applied Clay Mineralogy: Occurrences, Processing and Application of Kaolins, Bentonites, Palygorskite-Sepiolite, and Common Clays. Elsevier Amsterdam, the Netherlands.

Myshakin, E.M., Saidi, W.A., Romanov, V.N., Cygan, R.T., Jordan, K.D., 2013. Molecular dynamics simulations of carbon dioxide intercalation in hydrated Na-montmorillonite. Journal of Physical Chemistry C 117, 11028–11039.

Niazi, N.K., Bishop, T.F.A., Singh, B., 2011. Evaluation of spatial variability of soil arsenic adjacent to a disused cattle-dip site, using model-based geostatistics. Environmental Science and Technology 45, 10463–10470.

Niazi, N.K., Singh, B., Minasny, B., 2015. Mid-infrared spectroscopy and partial least-squares regression to estimate soil arsenic at a highly variable arsenic-contaminated site. International Journal of Environmental Science and Technology 12, 1965–1974.

Ozturk, A., Malkoc, E., 2014. Adsorptive potential of cationic basic yellow 2 (BY2) dye onto natural untreated clay (NUC) from aqueous phase: mass transfer analysis, kinetic and equilibrium profile. Applied Surface Science 299, 105–115.

Pauling, L., 1930. The structure of chlorites. In: Proceedings of the National Academy of Sciences, vol. 16. National Academy of Sciences, USA, p. 578.

Pentrak, M., Czimerova, A., Madejova, J., Komadel, P., 2012. Changes in layer charge of clay minerals upon acid treatment as obtained from their interactions with methylene blue. Applied Clay Science 55, 100–107.

Pires, J., Pinto, M.L., 2010. Pillared interlayered clays as adsorbents of gases and vapors. In: Gil, A., Korili, S.A., Trujillano, R., Vicente, M.A. (Eds.), Pillared Clays and Related Catalyst. Springer, Heidelberg.

Putra, E.K., Pranowo, R., Sunarso, J., Indraswati, N., Ismadji, S., 2009. Performance of activated carbon and bentonite for adsorption of amoxicillin from wastewater: mechanisms, isotherms and kinetics. Water Research 43, 2419–2430.

Romanov, V.N., 2013. Evidence of irreversible CO_2 intercalation in montmorillonite. International Journal of Greenhouse Gas Control 14, 220–226.

Rother, G., Ilton, E.S., Wallacher, D., Hauss, T., Schaef, H.T., Qafoku, O., Rosso, K.M., Felmy, A.R., Krukowski, E.G., Stack, A.G., Grimm, N., Bodnar, R.J., 2013. CO_2 sorption to subsingle hydration layer montmorillonite clay studied by excess sorption and neutron diffraction measurements. Environmental Science and Technology 47, 205–211.

Schaef, H.T., Ilton, E.S., Qafoku, O., Martin, P.F., Felmy, A.R., Rosso, K.M., 2012. In situ XRD study of Ca^{2+} saturated montmorillonite (STX-1) exposed to anhydrous and wet supercritical carbon dioxide. International Journal of Greenhouse Gas Control 6, 220–229.

Schaef, H.T., Loring, J.S., Glezakou, V.A., Miller, Q.R.S., Chen, J., Owen, A.T., Lee, M.S., Ilton, E.S., Felmy, A.R., McGrail, B.P., Thompson, C.J., 2015. Competitive sorption of CO_2 and H_2O in 2:1 layer phyllosilicates. Geochimica et Cosmochimica Acta 161, 248–257.

Schroeder, P.A., Erickson, G., 2014. Kaolin: from ancient porcelains to nanocomposites. Elements 10, 177–182.

Shakoor, M., Niazi, N.K., Bibi, I., Rahman, M., Naidu, R., Dong, Z., Shahid, M., Arshad, M., 2015. Unraveling health risk and speciation of arsenic from groundwater in rural areas of Punjab, Pakistan. International Journal of Environmental Research and Public Health 12, 12371.

Soltermann, D., Fernandes, M.M., Baeyens, B., Miehe-Brendle, J., Dahn, R., 2014. Competitive Fe(II)-Zn(II) uptake on a synthetic montmorillonite. Environmental Science and Technology 48, 190–198.

Stepova, K.V., Macquarrie, D.J., Krip, I.M., 2009. Modified bentonites as adsorbents of hydrogen sulfide gases. Applied Clay Science 42, 625–628.

Sturini, M., Speltini, A., Maraschi, F., Rivagli, E., Pretali, L., Malavasi, L., Profumo, A., Fasani, E., Albini, A., 2015. Sunlight photodegradation of marbofloxacin and enrofloxacin adsorbed on clay minerals. Journal of Photochemistry and Photobiology A: Chemistry 299, 103–109.

Suwandi, A.C., Indraswati, N., Ismadji, S., 2012. Adsorption of *N*-methylated diaminotriphenilmethane dye (malachite green) on natural rarasaponin modified kaolin. Desalination and Water Treatment 41, 342–355.

Thanh, D., Block, K., Bandosz, T.J., 2005. Adsorption of hydrogen sulfide on montmorillonites modified with iron. Chemosphere 59, 343–353.

Vespa, M., Lanson, M., Manceau, A., 2010. Natural attenuation of zinc pollution in smelter-affected soil. Environmental Science and Technology 44, 7814–7820.

Vimonses, V., Lei, S.M., Jin, B., Chowd, C.W.K., Saint, C., 2009. Kinetic study and equilibrium isotherm analysis of Congo Red adsorption by clay materials. Chemical Engineering Journal 148, 354–364.

Williams, L.B., Haydel, S.E., Ferrell, R.E., 2009. Bentonite, bandaids, and borborygmi. Elements 5, 99–104.

Wu, Q.F., Li, Z.H., Hong, H.L., Yin, K., Tie, L.Y., 2010. Adsorption and intercalation of ciprofloxacin on montmorillonite. Applied Clay Science 50, 204–211.

Yoder, H.S., Eugster, H.P., 1955. Synthetic and natural muscovites. Geochimica et Cosmochimica Acta 8, 225.

Zhang, Z.H., Liang, H.J., Du, X.G., Zuo, X.L., 2014. H_2S removal: correlation between performance and loading species of Zn-Fe/attapulgite. Environmental Progress & Sustainable Energy 33, 378–384.

CHAPTER

22 ZEOLITE FOR NUTRIENT STRIPPING FROM FARM EFFLUENTS

S. Chowdhury[1], N. Khan[2], G.-H. Kim[1], J. Harris[3], P. Longhurst[3], N.S. Bolan[4]

Hannam University, Daejeon, Republic of Korea[1]; *University of South Australia, Mawson Lakes, SA, Australia*[2]; *Cranfield University, Cranfield, Bedford, United Kingdom*[3]; *University of Newcastle, Callaghan, NSW, Australia*[4]

CHAPTER OUTLINE

22.1 INTRODUCTION

Various industrial sectors produce a significant amount of wastewater, which is often discharged into land and water systems. Wastewaters usually contain various pollutants such as ionic chemicals, oils, and organics, some of which may have toxic effects on ecosystems (Asl et al., 2013; Li et al., 2013). However, wastewater effluents may also contain a large amount of plant nutrients (Bolan et al., 2009; Li et al., 2013). For example, in New Zealand dairy and piggery farm wastewaters can supply enough nitrogen (N) to meet the requirements of 40,500 hectares (ha) of corn field or enough phosphorus (P) for 62,500 ha of pasture. Worldwide, such wastewaters could be used to fertilize 6.7 million ha of corn field with N, or 10.5 million ha of pasture with P (Bolan et al., 2009).

Environmental Materials and Waste. http://dx.doi.org/10.1016/B978-0-12-803837-6.00022-6

These types of farm waste are often treated biologically using a two-stage system (Bolan et al., 2004; Guo et al., 2013). The first stage involves a settlement or sedimentation pond; and the second aerobic stage is usually stratified with an aerobic top layer over an anaerobic base layer. Aerobic pond treatment is followed by discharge of the wastewater to either land or surface water. In many countries including New Zealand and Australia, discharge of farm wastewaters to surface waters currently requires legal consent; the concentration of effluent nutrients needs to be minimized before entering surface waters. Various approaches involving either disposal to land or tertiary treatments are commonly used (Bolan et al., 2004, 2009).

Some countries such as New Zealand encourages land disposal of effluents, because this is generally less harmful to water quality than discharge directly into water bodies. However, land disposal of farm wastewaters has some challenges (Bolan et al., 2004; Guo et al., 2013):

1. soil moisture and climatic conditions are critical because they may not be favorable for operating the treatment processes
2. free land is essential for holding wastewater when land application is not possible owing to unfavorable environmental conditions
3. land can be waterlogged, which makes treatment operations difficult
4. groundwater and surface water can be contaminated by leaching and runoff, respectively
5. odor nuisance may occur when spreading the effluent (eg, piggery wastewater)

Removal of contaminants should engage cost-effective technologies. A variety of techniques have been developed in past decades for wastewater treatment. Although biologically based pond systems, ie, biological treatment, may be effective in removing suspended solids and carbon, there has been some debate about their efficiency in removing macro- and micronutrients (Huang et al., 2014). Currently, adsorption of nutrients is believed to be an effective technique for wastewater treatment when efficient adsorbents are involved. Porous materials can be used to adsorb nutrients from wastewaters. For example, activated carbon, clay minerals, biomaterials, and zeolites can be used to retain cations and anions from wastewaters (Bolan et al., 2009). Contaminant-stripped effluents can then be discharged safely into waterways, and the nutrient-enriched adsorbent materials can be used as soil conditioners or nutrient sources.

Zeolitic tuffs have been found in many parts of the world after the original discovery of zeolitic minerals in volcanogenic sedimentary rock in the 1750s. Natural zeolites have found applications in adsorption, catalysis, agriculture, aquaculture, soil remediation, medicine, and energy (Muller and Arlt, 2013; Primo and Garcia, 2014; Laurino and Palmieri, 2015). It has been estimated that global natural zeolite consumption was 3.98 million tons in 2009, and has increased every year with industrialization (Wang and Peng, 2010).

The use of natural zeolites for environmental applications is gaining renewed interest owing to their abundance, ease of extraction, and unique properties, such as adsorption and ion exchange. The employment of natural zeolites for the treatment of water and wastewaters is already well established and expanding (Guo et al., 2013; Huang et al., 2014). Modified forms of natural zeolite have also been used to remove contaminants from water systems (Larsen, 2007; Camacho et al., 2011). However, little information is available on the postuse of nutrient-enriched zeolites as a potential source of plant nutrients. In this chapter, we review work on nutrient stripping from farm wastewaters using natural and modified zeolites, and the potential use of nutrient-enriched zeolites as a source of nutrients and soil amendment.

22.2 PROPERTIES OF NATURAL ZEOLITE

Natural zeolites are crystalline aluminosilicates, compositionally similar to clay minerals but differing in three-dimensional nano- and microporous structure (Auerback et al., 2003) with useful physico-chemical properties such as ion exchange, molecular sieving, catalysis, and sorption (Larsen, 2007; Szerement et al., 2014). Many forms of natural zeolites are available. Clinoptilolite, analcime, chabazite, mordenite, phillipsite, stilbite, and laumontite are common forms of natural zeolites, whereas offretite, paulingite, barrerite, and mazzite are rare. Among these various natural forms, clinoptilolite is the most abundant and is widely used (Rehakova et al., 2004).

The properties of natural zeolites depend on several factors, including the structure, size, and shape, the charge density of the anionic framework, and the ionic charge and concentration of the external electrolyte solution (Zygadlo et al., 2010). The structure of natural zeolite is complex (Table 22.1); SiO_4 and AlO_4 tetrahedra are the primary building units. These tetrahedrals attach through oxygen ions to form secondary building units, which then associate to form a three-dimensional crystalline structure. The negative charge of the zeolite framework appears to result from the substitution of silicon (Si) by aluminum (Al), which is compensated by alkaline and/or alkaline metal cations (Margeta et al., 2013). The zeolites act as cation exchangers owing to the presence of a negative charge on the surfaces. Zeolites have varying chemical and physical properties. Tables 22.1 and 22.2 show some selected properties of various natural zeolites including cation exchange capacity (CEC). The CEC of zeolites ranges between 52 and 268 cmol(+) kg^{-1} (Table 22.2).

Table 22.1 Structural Properties of Various Forms of Natural Zeolites (Burzo, 2013a,b; Guo et al., 2007; Margeta et al., 2013; Vaughan, 1966; Wang and Peng, 2010)

Zeolite	Chemical Formula	Framework Type	Structural Type
Analclime	$Na_{16}(Al_{16}Si_{32}O_{96})\cdot16H_2O$	ANA	Cubic
Laumonite	$Ca_4(Al_8Si_{36}O_{48})\cdot16H_2O$	LAU	Monoclinic
Phillipsite	$K_2(Ca_{0.5},Na)_4\cdot(Al_6Si_{10}O_{32})\cdot12H_2O$	PHI	Monoclinic
Erionite	$NaK_2MgCa_{1.5}\cdot(Al_8Si_{28}O_{72})\cdot28H_2O$ $(Na_2K_2MgCa_{1.5})_4\cdot Al_8Si_{28}O_{72}\cdot28H_2O$	ERI	Hexagonal
Ferrierite	$Na_{1.5}Mg_2Si_{30.}5Al_{5\cdot5}O_{72}\cdot18H_2O$ $(Na_2,K_2,Ca, Mg)_3Al_6\ Si_{30}O_{72}\cdot20H_2O$	FER	Orthorhombic, Monoclinic
Chabazite	$Ca_2(Al_4Si_8O_{24})\cdot12H_2O$ $(Ca,Na_2,K_2)_2Al_4Si_8O_{24}\cdot12H_2O$	CHA	Hexagonal Rhombohedral or triclinic
Natrolite	$Na_{16}[(AlO_2)_{16}(SiO_2)_{24}]\cdot16H_2O$	NAT	Orthorhombic
Mordenite	$Na_3KCa_2(Al_8Si_{40}O_{96})\cdot28H_2O$ $(Na_2,K_2,Ca)_4(Al_8Si_{40}O_{96})\cdot24H_2O$	MOR	Orthorhombic
Heulandite	$(Na,K)Ca_4(Al_9Si_{27}O_{72})\cdot24H_2O$ $Ca_4(Al_8Si_{28}O_{72})\cdot24H_2O$	HEU	Monoclinic
Clinoptilolite	$(Na,K)_6(Al_6Si_{30}O_{72})\cdot20H_2O$ $(Na,K,Ca)_{2-3}Al_3(Al,Si)_2Si_{13}O\cdot12H_2O$	CLI	Monoclinic
Stilbite	$Na_4Ca_8(Al_{20}Si_{52}O_{144})\cdot56H_2O$	STI	Monoclinic

Table 22.2 Physical and Chemical Properties of Zeolites

Zeolite	CEC (cmol kg^{-1})	Specific Surface Area (m^2 g^{-1})	References
Clinoptilolite	102	20.3	Alvarez-Ayuso et al. (2003)
Synthesized zeolite–NaPI (from fly ash)	195	27.7	Alvarez-Ayuso et al. (2003)
New Zealand clinoptilolite	88.2–113		Bolan et al. (2003)
New Zealand clinoptilolite: chemically modified using 0.5 M NaOH	159–198		Bolan et al. (2003)
New Zealand clinoptilolite: chemically modified using 0.5 M $NaHCO_3$	154–182		Bolan et al. (2003)
New Zealand clinoptilolite: chemically modified using 0.5 M HCl	120–152		Bolan et al. (2003)
Iranian clinoptilolite + mordenite	120		Ghiaci et al. (2004)
Croatian clinoptilolite	145		Farkas et al. (2005)
Chilean clinoptilolite	105	177	Englert and Rubio (2005)
Italian phillipsite + chabazite	212		Capasso et al. (2005)
Turkish clinoptilolite	95.1–141		Karadag et al. (2006)
Synthetic zeolite: made from coal fly ash using NaOH	88.2–213	29.0–90.8	Chen et al. (2006)
Clinoptilolite	52.2	13.2	Sprynskyy et al. (2006)
Modified clinoptilolite: made with 2 M HCl acid		78.9	Sprynskyy et al. (2006)
Clinoptilolite		6.64	Wen et al. (2006)
Clinoptilolite	160–180		Guo et al. (2007)
Synthetic zeolite: made from fly ash (ZFA)	268	37.4	Li et al. (2013)
Synthetic zeolite–Ca^{2+} amended ZFA	252	21.5	Li et al. (2013)
Synthetic zeolite–Mg^{2+} amended ZFA	257	75.5	Li et al. (2013)
Chinese clinoptilolite		23.0	Huang et al. (2014)
Chinese clinoptilolite: chemically modified using $MgCl_2$		29.0	Huang et al. (2014)
Clinoptilolite: commercial	111		Omar et al. (2015)

22.3 MODIFICATION OF NATURAL ZEOLITE

Natural zeolite can be modified by single or combined treatment including heat and chemical agents such as acids, bases, and inorganic salts (Tables 22.2, 22.4 and 22.5). Chemical and heat treatment of zeolite may result in cation relocation and pore opening. Pore manipulation may lead to changes in the sorbent properties of zeolite. Acid or base treatment, and surfactant impregnation by ion exchange, are commonly used to alter the hydrophilic or hydrophobic properties for adsorption of various ions or organics (Wang and Peng, 2010; Guo et al., 2013).

22.3.1 ACID–BASE TREATMENT

Zeolite structure and its physical and chemical properties can be modified with either inorganic base solutions such as NaOH and $Ca(OH)_2$ or acid solutions such as HCl and HNO_3. Acid treatment is a simple method and the most commonly used one for zeolite modification. The effectiveness of acid treatment depends on some factors such as the chemical composition, structure, mineral purity, and reaction conditions. Dissolution of natural zeolites occurs as a result of the acidic/basic behavior of the aluminosilicate structure in the presence of H^+ or OH^- ions in the acid solution. Interactions of natural high-silica zeolites such as clinoptilolite, heulandite, mordenite, and erionite with acidic and basic aqueous media commonly occur at low dissolution rates and are sufficiently acid-resistant (Filippidis and Kantiranis, 2007). Aluminum ions can be increasingly removed from the aluminosilicate structure by the dealumination process. The reactions are carried on favorably at lower pH and the $AlOH^{2+}$ species formed are detached because of their high degree of surface protonation. However, acid washing of natural zeolite may remove impurities that block the pores (Wang and Peng, 2010). Dissolution of some amorphous materials that block the pores of natural zeolites is another consequence of acid modification.

Hydrochloric acid solution treatment leads to decationization (obtaining the so-called "hydrogen form" zeolites), dealuminization, and sometimes destruction of the crystal lattice. The effects of HCl on various zeolites vary. As an example, HCl acid modification of the natural materials mordenite and erionite conducted under similar conditions led to weak decationization and almost no dealuminization of mordenite, whereas the extent of alkaline and alkaline earth metal uptake as well as of aluminum uptake from erionite was more than 90% (Allen et al., 2009). Changes in the chemical composition and structure of zeolites as a result of the decationization and dealumination steps led to changes in zeolite's properties.

The nature of the mineral and exchangeable cation content, as well as impurities, have a significant influence on acid modification of structurally identical natural zeolites. Zeolite/solution contact time, heating before and after modification, and pretreatment with water or other solutions such as NH_4Cl, influence the efficiency of modification as well. Low siliceous zeolites are unstable in acid and their decationization is conducted by other methods. Ion exchange with more soluble ammonium (NH_4^+) salts (usually NH_4Cl) is the initial stage, followed by heating of samples rich in NH_4^+ to eliminate NH_4^+ and hydrogen. This method of decationization is also applied to highly siliceous zeolites. The influence of the concentration of the modifying solution and the reaction time are also taken into consideration.

22.3.2 INORGANIC SALT OR SURFACTANT MODIFICATION

Chemical modification with inorganic salts (NaCl, $CaCl_2$, $BaCl_2$, NH_4Cl, and $FeCl_3$) or a cationic surfactant (hexsadecyltrimethylammonium [HDTMA] bromide) may improve zeolite's properties and thus increase its efficiency in wastewater treatment (Cerjan-Stefanovic et al., 2007; Kumar and Hayashi, 2009). A high concentration solution of inorganic salts on the surface of zeolite is significant for successful modification. Under normal conditions, large cavities and entries to the channels inside the zeolite framework are filled with water-forming hydration spheres around exchangeable cations. After the contact of zeolite with an inorganic salt solution such as NaCl, the exchange of cations (hydrogen ion [H^+] or sodium ion [Na^+]) from a solution with exchangeable cations (Na^+, potassium ions [K^+], calcium ions [Ca^{2+}], or magnesium ions [Mg^{2+}]) from the zeolite framework occurs. To remove anions from the aqueous system, the zeolite surface has to be modified with a solution of inorganic salts (for example, $FeCl_3$) whose adsorption on the zeolite surface leads to the formation of oxihydroxides, which then form stable complexes with anions in solution. This modification can result in the creation of an adsorption layer on the zeolite surface and modification of the surface charge on the zeolite surface (from negative to positive) (Bosinceanu and Sulitanu, 2008; Stefanakis et al., 2009). The specific surface area of some natural and modified zeolites after pretreatment with inorganic salts is shown in Table 22.2.

22.4 APPLICATION OF ZEOLITE IN WASTEWATER TREATMENT

Zeolites can be used in various sectors including agriculture, household products, radioactive waste, and wastewater treatment (Table 22.3). Wastewater contains many different types of pollutants

Table 22.3 Potential Applications of Zeolites

Sector	Application	References
Agriculture	Nutrient retainer and soil conditioner	Bolan et al. (2009) and Lim et al. (2015)
	Odor control	Lenntech (2015)
	Livestock feed additives	Lenntech (2015)
	Ammonia filtration in fish hatcheries	Lenntech (2015)
	Medium for hydroponic growing	Markou et al. (2014, 2015)
Household products	Household odor control	Lenntech (2015)
	Pet odor control	Lenntech (2015)
Industrial products	Adsorbents for oil and spills	Lenntech (2015)
	Gas separation	Lenntech (2015)
Radioactive waste	Site remediation or decontamination	Lenntech (2015)
Water treatment	Water filtration	Ramesh et al. (2011)
	Heavy metal removal	Ramesh et al. (2011)
	Antibiotic and pathogen removal	Ramesh et al. (2011) and Liu et al. (2013)
	Swimming pools	Lenntech (2015)
Wastewater treatment	Ammonia and phosphate removal	Huang et al. (2014)
	Heavy metal removal	Dal Bosco et al. (2005)
	Antibiotic and pathogen removal	Ramesh et al. (2011) and Liu et al. (2013)

including inorganic and organic compounds, which are more or less hazardous to humans, animals, and plants. In the following section, applications of natural and modified zeolites to remove of contaminants from wastewaters are discussed in more detail.

22.4.1 NUTRIENTS

Nitrogen $(NH_4^+ - N)$ and phosphorus $(PO_4^{3-} - P)$ are the most important (by amount) nutrients for plant growth and development. However, NH_4^+ and phosphate (PO_4^{3-}) accelerate the eutrophication of water bodies such as lakes and rivers, which receive wastewaters (Zhang et al., 2007). A higher concentration of NH_4^+ and PO_4^{3-} causes a sharp decrease in dissolved oxygen and obvious toxicity to aquatic organisms. Because a high amount of nutrients may cause pollution to the environment, it is important to remove the nutrients from wastewaters before discharge (Wang and Peng, 2010). In the past, various methods of nutrient removal from waters have been developed; among these, biological treatments and adsorption are the most widely used methods. In the adsorption process, natural zeolites as well as their modified or synthesized forms were explored as an effective adsorbent for nutrient removal from wastewaters. Table 22.4 summarizes NH_4^+ and PO_4^{3-} adsorption capacity of various sourced natural zeolites as well as their modified or synthesized forms.

In a laboratory study, Markou et al. (2014) used natural clinoptilolite to adsorb NH_4^+ from poultry wastewater, in which the removal efficiency was about 9 mg N g^{-1} of zeolite. Cooney et al. (1999) also used natural clinoptilolite to remove NH_4^+ from water, but the highest NH_4^+ removal efficiency was achieved when zeolite's exchange sites were converted to the sodium form. Rozic et al. (2000) also investigated the removal of NH_4^+ from aqueous solutions using clinoptilolite and found the highest (about 61 wt%) removal efficiency at the lowest initial concentration (ie, 100 mg N L^{-1}). A modified clinoptilolite (treated with alkali solution) was evaluated to remove NH_4^+ from a fellmongery wastewater stream (Bolan et al., 2003). It was observed that after treatment, the CEC of the natural zeolite increased significantly (up to 74%) and the zeolite could hold a potential maximum of 18.7–20.1 mg NH_4^+ g^{-1}. Guo et al. (2013) reported that synthesized zeolite from coal fly ash and chemically modified zeolite (treated with MgO), respectively, were effective (up to 95% and 87%, respectively) for NH_4^+ removal of swine wastewater.

Lebedynets et al. (2004) used clinoptilolite in a study of NH_4^+ sorption from synthetic aqueous solutions and concluded that NH_4^+ sorption by zeolites could be completed within 24 h for all fraction sizes and initial NH_4^+ ion concentrations. NH_4^+ adsorption increased with an increase in initial NH_4^+ concentration, with the maximum sorption capacity of 11.5 mg g^{-1}; this opposed the findings of Rozic et al. (2000). Sprynskyy et al. (2005) also investigated NH_4^+ uptake from synthetic aqueous solutions onto raw and pretreated forms of the same clinoptilolite under dynamic conditions. They found that the maximum sorption capacity of the clinoptilolite toward NH_4^+, estimated under dynamic conditions, was significantly higher than that under static conditions. NH_4^+ ion sorption onto the zeolite occurred by exchange with Na^+, Ca^{2+}, and K^+ ions in which Na^+ ions were observed to be more easily exchanged for both hydrogen and NH_4^+. However, the role of Ca^{2+} ions increased with zeolite saturation by NH_4^+. Pretreatment of the zeolite by NaCl, heat, and acid confirmed the predominant importance of the ion-exchange mechanism. Pretreatment of clinoptilolite with NaCl showed an improvement in NH_4^+ removal over other treatments.

Farkas et al. (2005) carried out a study of adsorption of NH_4^+ on a natural clinoptilolite in the treatment of leakage water from waste dumps. With and without pretreatment of leakage water by

Table 22.4 Selected Literature on Sorption of Ammonium and Phosphate Onto Zeolites

Material	Natural/Modified	Zeolite Dose (mg L^{-1})	Sorption Capacity (mg g^{-1})		References
			NH_4^+	PO_4^{3-}	
Canadian clinoptilolite	Natural	Batch kinetic study of sorption of NH_4^+ from wastewater (190.31 mg L^{-1}) or model solutions (200 mg L^{-1}) onto zeolite (2 g per 200 mL solution)	18.4 or 22.9		Liu and Lo (2001)
New Zealand clinoptilolite	Natural (commercial)	Column study of sorption of NH_4^+ from wastewater of varying concentrations (715–1210 mg L^{-1}) onto various fractions of zeolite	6.69		Bolan et al. (2003)
New Zealand clinoptilolite	Modified using 0.5 M NaOH.	As above	6.45		Bolan et al. (2003)
New Zealand clinoptilolite	Modified using 0.5 M $NaHCO_3$	As above	7.24		Bolan et al. (2003)
New Zealand clinoptilolite	Modified using 0.5 M HCl	As above	13.05		Bolan et al. (2003)
New Zealand clinoptilolite	Chemically modified using NaCl	Batch equilibrium study of sorption of NH_4^+ from solutions of varying concentrations (10–200 mg L^{-1}) onto zeolite (0.5 g per 50-mL solution)	~8.9		Weatherley and Miladinovic (2004)
New Zealand mordenite	Chemically modified using NaCl	As above	~10.7		Weatherley and Miladinovic (2004)
Croatian clinoptilolite	Natural	Batch kinetic study of sorption of NH_4^+ from wastewater and model solutions (800 mg L^{-1}) at designated time intervals for 6 h onto zeolite (2 g per 100-mL solution)	17.7		Farkas et al. (2005)
Chinese clinoptilolite	Natural	Batch kinetic study of sorption of NH_4^+ from wastewater (11.12–115.16 mg L^{-1}) and model solution (12.92–119. mg L^{-1}) at designated time intervals onto zeolite (5 g per 100–mL solution)	Wastewater: 1.74 Model solution: 1.926		Wang et al. (2006a)

Chinese clinoptilolite	Natural	Batch equilibrium study of sorption of NH_4^+ from solutions of varying concentrations (50–250 mg L^{-1}) onto zeolite (1 g per 100-mL solution)	11.2		Wang et al. (2007)
Chinese Na–clinoptilolite	Natural	As above	22.6		Wang et al. (2007)
American clinoptilolite	Natural	Batch equilibrium study of sorption of NH_4^+ from solutions (100–600 mg L^{-1}) onto zeolite (1.0 g per 50-mL solution)	18.6	20–21	Guo et al. (2007)
American clinoptilolite	Chemically modified using NaCl	As above	20.6	20–21	Guo et al. (2007)
Chinese clinoptilolite	Natural	Batch kinetic study of sorption of NH_4^+ from solution (5.5 mEq L^{-1} or 99.2 g L^{-1}) onto zeolite (400 g per 100-mL solution)	12.3–13.0		Wen et al. (2006)
Chinese clinoptilolite	Chemically modified using NaOH	As above	9.6–13.2		Wen et al. (2006)
Chinese clinoptilolite	Chemically modified using NaOH	As above	7.9–10.1		Wen et al. (2006)
Chinese clinoptilolite	Chemically modified using NaCl under microwave irradiation	Batch equilibrium study of sorption of NH_4^+ from solutions of varying concentrations (0–200 mg L^{-1}) onto zeolite (0.15 g per 50-mL solution)	23.8		Lei et al. (2008)
Turkish clinoptilolite	Natural	Batch equilibrium study of sorption of NH_4^+ from solutions of varying concentrations (25–150 mg L^{-1}) onto zeolite (0.5 g per 50-mL solution)	5.2–8.1		Karadag et al. (2006)
Turkish clinoptilolite	Natural	Batch equilibrium study of sorption of NH_4^+ from solutions of varying concentrations (20–100 mg L^{-1}) onto zeolite (1 g per 50-mL solution)	7.3		Karadag et al. (2007)

Continued

Table 22.4 Selected Literature on Sorption of Ammonium and Phosphate Onto Zeolites—cont'd

Material	Natural/Modified	Zeolite Dose (mg L^{-1})	Sorption Capacity (mg g^{-1})		References
			NH_4^+	PO_4^{3-}	
Turkish clinoptilolite	Natural	Batch equilibrium study of sorption of NH_4^+ from solutions of varying concentrations (5–150 mg L^{-1}) onto zeolite (0.4 g per 100-mL solution)	14.5–15.2		Gunay (2007)
Ukrainian clinoptilolite	Natural	Column study of sorption of NH_4^+ from solutions of varying concentrations (10–100 mg L^{-1}) onto various fractions of zeolite (bed height 8 cm)	13.6–21.5		Sprynskyy et al. (2005)
Chilean zeolite (clinoptilolite and mordenite)	Natural	Batch kinetic study of sorption of NH_4^+ from solutions (100 mg L^{-1}) onto zeolite (0.25 g per 100-mL solution)	12.302		Englert and Rubio (2005)
Synthesized zeolite from 15 China fly ashes	Synthesized zeolite from coal fly ash using NaOH solution	Batch study of sorption of PO_4^{3-} from solutions (1000 mg P L^{-1}) onto zeolite (0.75 g per 40-mL solution)	–	11.8–47.2	Chen et al. (2006)
Clinoptilolite	Natural	Batch study of sorption of PO_4^{3-} from artificial solutions composed by various nutrient solutions onto zeolite (20 g per L solution)	–	<0.2	Markou et al. (2015)
Clinoptilolite	Natural	Batch study of sorption of PO^{4-} from artificial solutions ranging in concentration from 6 to 80 mg P L^{-1} onto zeolite (40 g per 100-mL solution)	–	0.043	Wang et al. (2013)
Synthesized zeolite	Synthesized zeolite from fly ash treated with 0.01 M H_2SO_4	Batch study of sorption of NH_4^+ and PO_4^{3-} from solutions (2.5–1200 mg N L^{-1} and 0.5–1000 mg P L^{-1}) onto zeolite (0.4 g per 40-mL solution)	60.2	62.7	Zhang et al. (2007)
Clinoptilolite	Chemically modified using $Ca(OH)_2$	Batch study of sorption of PO_4^{3-} from artificial solutions composed by various nutrient solutions onto zeolite (20 g per L solution)	8.5	10.3	Markou et al. (2015)

activated carbon, NH_4^+ sorption on the zeolite was 5.7 and 4.2 mg NH_4^+ g^{-1}, respectively. In the artificial NH_4^+ solution, NH_4^+ sorption was 17.70 mg NH_4^+ g^{-1}, which indicated that K^+ and Ca^{2+} and organic pollutants in leakage water prevented NH_4^+ adsorption. Study revealed that for a multicomponent system there was a competitive effect between NH_4^+ and other cations such as Ca^{2+}, Mg^{2+}, and K^+ which caused a decrease in NH_4^+ removal (Cooney et al., 1999). Jorgensen and Weatherley (2006) also reported variable effects of organic compounds on NH_4^+ uptake by investigating different absorbents (natural clinoptilolite, a gel resin [Dowex50wx8], and a macronet resin [Purolite MN-500]) for the removal of NH_4^+ from wastewaters after a secondary biological treatment. However, among the exchangers and absorbents, natural clinoptilolite was found to be the highest NH_4^+ exchanger.

The influences of contact time, pH, NH_4^+ concentration, and the presence of other cations and anions were investigated using a natural zeolite tuff and synthetic zeolites (5A, 13X, and ZSM-5) (Maranon et al., 2006). The researchers observed that adsorption capacities were affected by different cations present in multicomponent solutions (containing NH_4^+, zinc ion [Zn^{2+}], cadmium ion [Cd^{2+}], Ca^{2+}, and Na^+), although the effect of Zn and Cd at low concentrations was small. In a comparison of natural and synthetic zeolites, the removal efficiencies of NH_4^+ by zeolite tuff were higher than those of synthetic zeolites at a low initial NH_4^+ concentration. Karadag et al. (2006) also conducted a batch experiment study of kinetics and equilibrium isotherms of NH_4^+ on a natural clinoptilolite, in which the effects of temperature, contact time, and initial NH_4^+ concentration were examined. It was concluded, according to thermodynamic parameters, that the exchange of NH_4^+ by the clinoptilolite is a spontaneous and exothermic process. Some natural clinoptilolites were also tested for NH_4^+ removal from the leachate solution of sewage sludge (Wang et al., 2006a). The researchers assessed the effect of the particle size of zeolite to be more pronounced than other relevant parameters such as contact time and initial NH_4^+ concentration. It was found that ion exchange between NH_4^+ and cations in the zeolite increased with a decrease in the particle size of the zeolite (Bolan et al., 2004). Wen et al. (2006) also reported that even after modification (covered by biofilm), zeolite did not affect ion exchange for the smaller particle size but decreased the ion-exchange capacity by 22% for the larger particle size.

Similar to N, farm wastewater contains a large amount of P (Bolan et al., 2004; Huang et al., 2014). Because a high content of P can cause pollution to ecosystems, it should be removed from wastewaters before discharging to the land and water bodies (Chen et al., 2006; Wang and Peng, 2010). Natural zeolites usually have little or no affinity for anions such as PO_4^{3-} because of the net negative charge on its framework (Huang et al., 2014). Therefore, natural zeolite may demonstrate a low adsorption capacity for PO_4^{3-} (<0.2 mg P g^{-1}) in an aqueous environment (Markou et al., 2015). In general, the surface properties of natural zeolites are modified to remove P from wastewaters (Wang and Peng, 2010). The modification of natural zeolites was discussed previously in detail.

In general, two types of reagents must be used for the simultaneous removal of NH_4^+ and PO_4^{3-} from aqueous systems: for example, the combination of a cation and an anion exchanger, or that of a cation exchanger and an inorganic cohesion precipitant (Zhang et al., 2007). These procedures of combining two materials are expensive and complicated. Magnesium salts can be used to crystallize Mg^+, NH_4^+, and PO_4^{3-} present in farm wastewaters, and form struvite ($MgNH_4PO_4 \cdot 6H_2O$) (Huang et al., 2014), thereby removing NH_4^+ and PO_4^{3-} simultaneously. This procedure may involve simpler technologies and can be less expensive compared with the previously noted procedures in which two reagents are required. However, additional PO_4^{3-} salt may be required in the struvite precipitation procedure to increase NH_4^+ removal efficiency, which can increase the treatment cost (Zhang et al., 2007; Huang et al., 2014).

Calcium(II) ion-amended zeolite synthesized from fly ash was used to treat swine wastewater by Li et al. (2013); they achieved 74% N and 96% P removal from the wastewater. Zhang et al. (2007) also reported that acid (0.01 M H_2SO_4)-treated zeolite synthesized from coal fly ash was an important material to remove NH_4^+ and PO_4^{3-} simultaneously from the synthetic effluent at low nutrient concentrations. They explained that although the CEC of the treated zeolite decreased slightly (ie, NH_4^+ removal was not significantly affected), P immobilization by the treated zeolite increased significantly owing to the transformation of $CaCO_3$ into more soluble forms (eg, gypsum and anhydrite) which favor the precipitation of Ca phosphate.

It has been shown that the main mechanism that accounts for the adsorption of cations such as NH_4^+ and K^+ ions and anions such as PO_4^{3-} onto natural zeolites is an ion-exchange process (Bolan et al., 2004). Bolan et al. (2004) observed that the adsorption maxima for NH_4^+ and K^+ were close to the CEC of zeolites, which indicated a simple ion-exchange mechanism for adsorption. The results in Table 22.3 show that the adsorption capacity of clinoptilolite ranges from 5.2 to 23.8 mg NH_4^+ g^{-1}. PO_4^{3-} adsorption increased from 0.043 to 62.7 mg g^{-1} for synthesized or modified zeolites (Table 22.4). However, other components in solutions such as heavy metal ions and organics may reduce NH_4^+ and PO_4^{3-} exchange capacity owing to competitive adsorption (Bolan et al., 2004; Huang et al., 2014).

22.4.2 HEAVY METALS

Because of rising industrial activity, contamination of heavy metals in the water supply has gradually increased over the past decades. The presence of heavy metals in the environment is a serious concern because of their toxic effects on human health, plants, and animals. Most of these metals, such as lead (Pb), copper (Cu), Cd, Zn, nickel (Ni), manganese (Mn), mercury (Hg), and chromium (Cr) are present in farm and industrial wastewaters. Before discharging these wastewaters, heavy metals contents should be minimized. Natural zeolites and their modified forms have been used in many investigations to adsorb heavy metals from the wastewaters (Table 22.5).

A study was carried out using a natural zeolite and its modified forms (treated with NaOH, NaCl, HCl, and CH_3COONa) to remove metal ions such as Ni^{2+}, Cu^{2+}, Cd^{2+}, Zn^{2+}, and Pb^{2+} from synthetic wastewater (Panayotova, 2001). The researchers observed that metal adsorption from multicomponent solutions decreased, probably as a result of competition between the metals (Sprynskyy et al., 2006). Like Panayotova and Velikov (2002), the researchers concluded that modified zeolites (by NaCl, CH_3COONa, and NaOH) were much more effective (by 25–30%) in removing metals from wastewater than natural zeolites. Synthetic zeolite was also effective in metal removal from wastewaters. Alvarez-Ayuso et al. (2003) reported that the synthetic zeolite (NaP1) was about 10 times more efficient at adsorbing metals than natural zeolite. Asl et al. (2013) suggested that zeolite prepared from fly ash can be a suitable adsorbent for removing Cr (VI) from wastewaters because they found greater than 80% of Cr (VI) removal by zeolite in their batch experiments. Cardoso et al. (2015) also effectively removed heavy metals (arsenic [As], Ni, Cu, and Mn) from acid mine wastewater by using synthesized zeolite from fly ash.

The metal removal efficiency of zeolites can be influenced by temperature, solution pH, particle size, and contact time (Asl et al., 2013). Inglezakis et al. (2007) reported that clinoptilolite dust rather

Table 22.5 Selected Literature on the Sorption of Heavy Metals Onto Zeolites

Material	Natural/Modified/ Synthesized	Zeolite Dose (mg L^{-1})	Metal	Sorption Capacity (mEq g^{-1})	References
Turkish clinoptilolite	Natural	Batch equilibrium study of sorption of single- or multiple-metal solutions of varying concentrations (50–3000 mg L^{-1}or 5 mEq L^{-1}) onto zeolite (single-metal solution: 1 g zeolite per 20 mL Cu^{2+}, Zn^{2+}, or Ni^{2+}: solution, and 1 g zeolite per 100 mL Pb^{2+}: solution)	Cu^{2+} Ni^{2+} Pb^{2+} Zn^{2+}	0.022[a]; 0.227[b] 0.017[a]; 0.173[b] 0.299[a]; 0.730[b] 0.108[a]; 0.251[b]	Oter and Akcay (2007)
Greek clinoptilolite	Natural	Batch equilibrium study of sorption of metals from solutions of varying concentrations (10–200 mg L^{-1}) onto zeolite (2.5–10 g L^{-1} solution)	Cd^{2+} Cu^{2+} Cr^{3+} Ni^{2+} Zn^{2+}	0.082 0.186 0.238 0.067 0.106	Alvarez-Ayuso et al. (2003)
NaPI	Synthetic zeolite made from fly ash via hydrothermal alkaline activation	As above	Cd^{2+} Cu^{2+} Cr^{3+} Ni^{2+} Zn^{2+}	0.904 1.589 2.516 0.685 0.997	Alvarez-Ayuso et al. (2003)
Scolecite	Natural	Batch equilibrium study of sorption of metals from multiple-metal solutions of varying concentrations (5–30 mg L^{-1}) onto zeolite (0.5 g L^{-1} solution)	Cd^{2+} Co^{2+} Cu^{2+} Ni^{2+} Pb^{2+} Zn^{2+}.	0.0032 0.0078 0.130 0.031 0.056 0.064	Bosso and Enzweiler (2002)
Ukraine clinoptilolite	Modified using HCl and NH_4Cl	Batch equilibrium study of sorption of single-metal solutions of varying concentrations (Pb^{2+}, Cu^{2+}, Ni^{2+}: 10–800 mg L^{-1}; Cd^{2+}: 1–80 mg L^{-1}) onto zeolite (0.5 g per 200-mL solution)	Cd^{2+} Cu^{2+} Ni^{2+} Pb^{2+}	0.087 0.809 0.530 0.259	Sprynskyy et al. (2006)

Continued

Table 22.5 Selected Literature on the Sorption of Heavy Metals Onto Zeolites—cont'd

Material	Natural/Modified/ Synthesized	Zeolite Dose (mg L^{-1})	Metal	Sorption Capacity (mEq g^{-1})	References
Turkish clinoptilolite	Natural	Batch equilibrium study of sorption of metals from single-metal solutions of varying concentrations (100–400 mg L^{-1}) onto zeolite (10 g per 500-mL solution)	Cu^{2+} Co^{2+} Mn^{2+} Zn^{2+}	0.282 0.448 0.154 0.268	Erdem et al. (2004)
Brazilian scolecite	Natural	Batch equilibrium study of sorption of single metal from solutions of varying concentrations (10–5000 mg L^{-1}) onto zeolite (1 g per 60-mL solution)	Cd^{2+} Cr^{3+} Ni^{2+} Mn^{2+}	1.779 6.139 4.157 4.001	Dal Bosco et al. (2005)
Clinoptilolite (CLI)	Modified using Na, Fe, and Ag (CLINa, CLIFe, and CLIAg, respectively)	Batch equilibrium study of sorption of Mn from solutions of varying concentrations (2.5–300 mg L^{-1}) onto zeolite (150 g per 15-mL solution)	Mn^{2+}: CLINa, CLIFe, and CLIAg	0.232 0.315 0.537	Rojas-Pavón et al. (2015)
Mordenite (MOR)	Modified using Na, Fe, and Ag (MORNa, MORFe, and MORAg, respectively)	As above	Mn^{2+}: MORNa, MORFe, and MORAg	0.324 0.336 0.383	Rojas-Pavón et al. (2015)
CLI	Natural	Batch equilibrium study of sorption of metals from solutions of varying concentrations (10 mEq L^{-1}) onto zeolite (0.5–18 g per 100-mL solution)	Cu^{2+} Cr^{3+} Fe^{3+} Pb^{2+}	0.37 0.39 0.30 0.98	Inglezakis et al. (2002)

[a] *Sorption from multiple-metal solution.*
[b] *Sorption from single-metal solution.*

than granules was more efficient in removing Pb. The finer clinoptilolite fractions adsorbed higher amounts of the metals owing to relative enrichment of the adsorption surface and higher cleavage (Sprynskyy et al., 2006). Agitation and temperature also affected Pb^{2+} adsorption, especially in the case of granular clinoptilolite (2.5−5.0 mm). The adsorption of Pb^{2+} increased with an increase in the pH of the solution from 1 to 4. For instance, Llanes-Monter et al. (2007) observed maximal Pb^{2+} adsorption (141 cmol kg^{-1}) by a clinoptilolite-rich tuff at pH 3. However, Camacho et al. (2011) did not observed a pH effect on As^{5+} removal of groundwater by MnO^{2-}-modified natural clinoptilolite.

Activated (Gordes zeolite) and nonactivated (Bigadic) natural clinoptilolite were used by Turkman et al. (2004) to remove heavy metals (Pb^{2+}, Cd^{2+}, Ni^{2+}, and Zn^{2+}) in synthetic and real wastewaters. After 90 min of contact in batch experiments, 46% removal efficiency was achieved on Gordes zeolite whereas Cd^{2+} was not removed successfully by Bigadic zeolite. Oren and Kaya (2006) also observed similar results for Zn^{2+} adsorption characteristics of Gordes and Bigadic zeolites, in which the removal efficiency of Gordes zeolite was two times higher than that of Bigadic zeolite.

It has been shown that the metal removal efficiency of zeolites may depend on their physical and chemical properties, and the temperature of the environment, concentration and pH of wastewater, contact time, and presence of other metals. Therefore, the factors which affect the removal efficiency should be taken into account during the application of zeolite to remove heavy metals from wastewaters.

22.4.3 ORGANIC CONTAMINANTS

Organic contaminants including dye, humic substances, phenolic compounds, petroleum, surfactants, pesticides, and pharmaceuticals are important pollutants in wastewaters. The presence of organic contaminants in water may produce toxic chemicals during disinfection. Among the contaminants, humic substances such as humic acid, fulvic acid, or humin, which result from the decay of organic matter and pharmaceuticals such as antibiotics, are most abundant in farm wastewaters; and these should be removed before discharge. Adsorption by zeolites has been found to be an effective technique for removing organic contaminants from wastewaters. However, modified zeolite (surfactant modification) is commonly used to remove organic contaminants from wastewaters (Wang and Peng, 2010).

The study conducted by Wang et al. (2006b) reported that surfactant modification of clinoptilolite (treated with HDTMA) largely increased fulvic acid adsorption on zeolite surfaces owing to hydrophobic interaction and hydrogen bonding. However, Capasso et al. (2005) investigated the interaction of humic acid with untreated zeolite in the presence of various cations such as Na^+, K^+, $NH_4{}^+$, and Ca^{2+}. Their results revealed that the effect of cations on the removal of humic acid by zeolite was significant: divalent cations enhanced efficiency and monovalent cations decreased adsorption. It was also found that the sorption capacity increased with ionic strength and was highest at pH 7.4. The humic acid sorption capacity of Ca-enriched zeolite was also higher than that of normal zeolite (Capasso et al., 2007a,b). Wang et al. (2008) reported that the adsorption capacity of humic acid by natural zeolite was highest (68 mg g^{-1}) at 30°C and pH 5.0. They also found that Pb^{2+} and Cu^{2+} exhibit competitive adsorption with humic acid on natural zeolite, resulting in reduced adsorption of humic acid. Those results indicated that humic acid sorption on a zeolite surface is a complex phenomenon in which the efficiency can depend on the ability of cations to form stable bonds with organics and/or their specific selective sorption by the zeolite (Wang and Peng, 2010).

Few studies on the removal of antibiotics from wastewaters and other materials by zeolites have been reported in the literature. Liu et al. (2013) compared two vertical flow–constructed wetlands containing either volcanic rock or zeolite. They found that both systems were effective in removing three commonly deployed antibiotics (ciprofloxacin HCl, oxytetracycline HCl, and sulfamethazine) to 10% of influent concentration, and that zeolite was the more effective of the two, although this was a marginal but consistent difference.

22.5 BIOMASS PRODUCTION

Nutrient-enriched zeolites can be used to produce microbial biomass. For instance, Markou et al. (2014) used natural zeolite as a medium for the sorption of ammonia from wastewater, and subsequently as nitrogen releaser in cultures of *Arthrospira platensis*. Markou et al. (2015) reported that modified zeolite (treated with $Ca[OH]_2$) was a significant adsorbent for PO_4^{3-} in synthetic wastewater, and subsequently a P source for the cultivation of green microalga (*Chlorella vulgaris*) and cyanobacterium (*A. platensis*).

22.6 REVEGETATION POTENTIAL OF NUTRIENT-ENRICHED ZEOLITE

Zeolites are traditionally used in agriculture as a soil conditioner and as a nitrogen-retaining medium to increase N efficiency (Polat et al., 2004). As a result, they promote better plant growth and consequently enhance yield. For example, Torii (1978) reported that apple yield was significantly increased (by 13–38%) after application of zeolite at the rate of 4–8 tons per acre (10–20 tons ha^{-1}). Sepaskhah and Barzegar (2010), and Szerement et al. (2014) reported that zeolites can be used to cultivate a wide variety of crops including vegetables, cereals, grapes, and other fruits.

Because zeolites help to retain nutrients for long time, application of zeolite to soil may be helpful to sustain soil fertility. It concerns the most important (by amount) plant nutrients such as N and also deals with other nutrients such as P (particularly for modified or synthetic zeolites), K, Ca, Mg, and also micronutrients. Zeolites can hold these nutrients in the root zone, and plants use these when necessary; this increases the efficiency of nutrient use and yield. Large losses of nutrient by leaching are a common phenomenon in soil, particularly sandy soil, which has little capability of retaining a high amount of nutrients for better plant growth (Flanigen and Mumpton, 1981; Polat et al., 2004; Sepaskhah and Barzegar, 2010). Therefore, application of zeolites into soil is helpful for better plant growth by reducing leaching and increasing the efficiency of nutrient use.

Zeolites can also be used as soil amendments because they do not break down over time like other soil amendments (eg, lime) but help soil to be more productive (Wang and Peng, 2010). Irrigation and fertilizer costs can be reduced significantly by applying zeolites in the field. The porous structure of zeolites helps to keep soil aerated and moist, thereby improving biological activity for a long period (Ghanbari and Ariafar, 2013). Zeolites are alkaline by nature, which can help buffer the soil pH level (Bish and Ming, 2001) after fertilization and reduces the need to apply lime.

Because zeolite alone can act as an effective soil amendments and/or nutrient retainer, nutrient-enriched zeolite (after nutrient stripping from wastewaters) can be an important soil conditioner and a potential source of plant nutrients. Although many studies have found increased crop yield owing to zeolite application (Kavoosi, 2007; Sepaskhah and Barzegar, 2010), a study conducted by Lim et al. (2015)

reported that application of zeolite on a rice field reduced plant growth as well as yield, which was attributed to the effect of the low nutrient level of the soil. Therefore, it can be assumed that the revegetation potential of nutrient-enriched zeolites depends on the soil fertility status and the rate of nutrient desorption from zeolites, although information about the application of nutrient-enriched zeolites in agriculture is limited. However, Ippolito et al. (2011) reported that the application of urea fertilizer mixed with zeolite improved the status of soil water, reduced nitrate leaching loss, and consequently increased corn yield. The application of clinoptilolite zeolite mixed with urea was also found to be effective in reducing N loss by controlling NH_4^+ and nitrate-N released from urea (Omar et al., 2015).

Because the need for global food security increases in the face of soil degradation (Amundson et al., 2015; McLaughlin and Kinzelbach, 2015), there are significant opportunities to explore the use of zeolites as a soil conditioner and nutrient delivery systems. The act of enriching zeolites with nutrients from wastewater treatment and applying them to land transforms nutrient-containing wastewater to a resource. However, increased use of zeolites in wastewater treatment may require the further demonstration of treatment efficacy and the development of regulation. Moreover, a comprehensive analysis of treatment efficacy and cost needs to be understood to ensure a viable economic rationale.

22.7 CONCLUSIONS

Zeolite structural configuration and physicochemical properties have been well studied and characterized. They are widely used for their catalytic abilities and storage potential in a diverse range of industries. They are used to treat contaminated wastewaters and as soil conditioners. Natural zeolites, which have CEC, are effective and low-cost materials for wastewater treatment, especially for positively charged solutes such as NH_4^+, K^+, and heavy metals. However, modification of natural zeolites using cationic surfactant enables them to adsorb negatively charged solutes such PO_4^{3-} and other organics such as humic acid and antibiotics from farm wastewaters. Therefore, base-enriched modified zeolites can be used for the simultaneous removal of both positively and negatively charged nutrients from farm wastewaters. Because zeolite acts as an effective soil conditioner and nutrient retainer, application of nutrient-enriched zeolite may be a suitable option as a source of nutrients for plants. However, the effectiveness of field application of nutrient-enriched zeolites in agriculture will depend on soil nutrient levels, desorption rates of nutrients from zeolites, and the plant uptake of nutrients. The use of zeolites is likely to increase significantly if their primary use for nutrient adsorption is followed by their secondary use in agriculture for nutrients.

ACKNOWLEDGMENT

This work was supported by the Ministry of Environment through "The GAIA Project" in the Korea Republic.

REFERENCES

Allen, S.J., Ivanova, E., Koumanova, B., 2009. Adsorption of sulfur dioxide on chemically modified natural clinoptilolite. Acid modification. Chemical Engineering Journal 152, 389–395.

Alvarez-Ayuso, E., Garcia-Sanchez, A., Querol, X., 2003. Purification of metal electroplating waste waters using zeolites. Water Research 37, 4855–4862.

Amundson, R., Berhe, A.A., Hopmans, J.W., Olson, C., Sztein, A.E., Sparks, D.L., 2015. Soil and human security in the 21st century. Science 348. http://dx.doi.org/10.1126/science.1261071.

Asl, S.M.H., Ahmadi, M., Ghiasvand, M., Tardast, A., Katal, R., 2013. Artificial neural network (ANN) approach for modeling of Cr(VI) adsorption from aqueous solution by zeolite prepared from raw fly ash (ZFA). Journal of Industrial and Engineering Chemistry 19, 1044–1055.

Auerback, S.M., Carrado, K.M., Dutta, P.K., 2003. Handbook of Zeolite Science and Technology. Marcel Dekker, Inc., New York.

Bish, L.W., Ming, D.W., 2001. Applications of natural zeolites in water and wastewater treatment. Natural Zeolites: Occurrence, Properties, Applications 45, 519–550.

Bolan, N., Laurenson, S., Luo, J., Sukias, J., 2009. Integrated treatment of farm effluents in New Zealand's dairy operations. Bioresource Technology 100, 5490–5497.

Bolan, N., Wong, L., Adriano, D.C., 2004. Nutrient removal from farm effluents. Bioresource Technology 94, 251–260.

Bolan, N.S., Mowatt, C., Adriano, D.C., Blennerhassett, J.D., 2003. Removal of ammonium ions from fellmongery effluent by zeolite. Communications in Soil Science and Plant Analysis 34, 1861–1872.

Bosinceanu, R., Sulitanu, N., 2008. Synthesis and characterization of FeO(OH)/Fe3O4 nanoparticles encapsulated in zeolite matrix. Journal of Optoelectronics and Advanced Materials 10, 3482–3486.

Bosso, S., Enzweiler, J., 2002. Evaluation of heavy metal removal from aqueous solution onto scolecite. Water Research 36, 4795–4800.

Burzo, E., 2013a. 8.1. 6.7 Analcime and related silicates. In: Tectosilicates. Springer, pp. 129–305.

Burzo, E., 2013b. 8.1. 6.8 Zeolites having mordenite-, ferrierite-, boggsite-, epistilbite-, and terranovaite-type framework. In: Tectosilicates. Springer, pp. 1–170.

Camacho, L.M., Parra, R.R., Deng, S., 2011. Arsenic removal from groundwater by MnO_2-modified natural clinoptilolite zeolite: effects of pH and initial feed concentration. Journal of Hazardous Materials 189, 286–293.

Capasso, S., Colella, C., Coppola, E., Iovino, P., Salvestrini, S., 2007a. Removal of humic substances from water by means of calcium-ion-enriched natural zeolites. Water Environment Research 79, 305–309.

Capasso, S., Coppola, E., Iovino, P., Salvestrini, S., Colella, C., 2007b. Sorption of humic acids on zeolitic tuffs. Microporous and Mesoporous Materials 105, 324–328.

Capasso, S., Salvestrini, S., Coppola, E., Buondonno, A., Colella, C., 2005. Sorption of humic acid on zeolitic tuff: a preliminary investigation. Applied Clay Science 28, 159–165.

Cardoso, A.M., Paprock, A., Ferret, L.S., Azevedo, C.M.N., Pires, M., 2015. Synthesis of zeolite Na-P1 under mild conditions using Brazilian coal fly ash and its application in wastewater treatment. Fuel 139, 59–67.

Cerjan-Stefanovic, S., Zabukovec Logar, N., Margeta, K., Novak Tusar, N., Arcon, I., Maver, K., Kovac, J., Kaucic, V., 2007. Structural investigation of Zn^{2+} sorption on clinoptilolite tuff from the Vranjska Banja deposit in Serbia. Microporous and Mesoporous Materials 105, 251–259.

Chen, J., Kong, H., Wu, D., Hu, Z., Wang, Z., Wang, Y., 2006. Removal of phosphate from aqueous solution by zeolite synthesized from fly ash. Journal of Colloid and Interface Science 300, 491–497.

Cooney, E.L., Booker, N.A., Shallcross, D.C., Stevens, G.W., 1999. Ammonia removal from wastewaters using natural Australian zeolite. I. Characterization of the zeolite. Separation Science and Technology 34, 2307–2327.

Dal Bosco, S.M., Jimenez, R.S., Carvalho, W.A., 2005. Removal of toxic metals from wastewater by Brazilian natural scolecite. Journal of Colloid and Interface Science 281, 424–431.

Englert, A., Rubio, J., 2005. Characterization and environmental application of a Chilean natural zeolite. International Journal of Mineral Processing 75, 21–29.

Erdem, E., Karapinar, N., Donat, R., 2004. The removal of heavy metal cations by natural zeolites. Journal of Colloid and Interface Science 280, 309–314.

Farkas, A., Rozic, M., Barbaric-Mikocevic, Z., 2005. Ammonium exchange in leakage waters of waste dumps using natural zeolite from the Krapina region, Croatia. Journal of Hazardous Materials B117, 25–33.

Filippidis, A., Kantiranis, N., 2007. Experimental neutralization of lake and stream waters from N. Greece using domestic HEU-type rich natural zeolitic material. Desalination 213, 47–55.

Flanigen, M., Mumpton, F.A., 1981. Commercial properties of natural zeolites. In: Mumpton, F.A. (Ed.), Mineralogy and Geology of Natural Zeolites, Reviews in Mineralogy. Mineralogical Society of America 4, 165–175.

Ghanbari, M., Ariafar, S., 2013. The effect of water deficit and zeolite application on growth traits and oil yield of medicinal peppermint (*Mentha piperita* L.). International Journal of Medicinal and Aromatic Plants 3, 33–39.

Ghiaci, M., Kia, R., Abbaspur, A., Seyedeyn-Azad, F., 2004. Adsorption of chromate by surfactant-modified zeolites and MCM-41 molecular sieve. Separation and Purification Technology 40, 285–295.

Gunay, A., 2007. Application of nonlinear regression analysis for ammonium exchange by natural (Bigadic) clinoptilolite. Journal of Hazardous Materials 148, 708–713.

Guo, J., Yang, C., Zeng, G., 2013. Treatment of swine wastewater using chemically modified zeolite and bioflocculant from activated sludge. Bioresource Technology 143, 289–297.

Guo, X., Zeng, L., Li, X., Park, H.S., 2007. Removal of ammonium from RO permeate of anaerobically digested wastewater by natural zeolite. Separation and Purification Technology 42, 3169–3185.

Huang, H., Xiao, D., Pang, R., Han, C., Ding, L., 2014. Simultaneous removal of nutrients from simulated swine wastewater by adsorption of modified zeolite combine with struvite crystallization. Chemical Engineering Journal 256, 431–438.

Inglezakis, V., Loizidou, M., Grigoropoulou, H., 2002. Equilibrium and kinetic ion exchange studies of Pb^{2+}, Cr^{3+}, Fe^{3+} and Cu^{2+} on natural clinoptilolite. Water Research 36, 2784–2792.

Inglezakis, V.J., Stylianou, M.A., Gkantzou, D., Loizidou, M.D., 2007. Removal of Pb(II) from aqueous solutions by using clinoptilolite and bentonite as adsorbents. Desalination 210, 248–256.

Ippolito, J.A., Tarkalson, D.D., Lehrsch, G.A., 2011. Zeolite soil application method affects inorganic nitrogen, moisture, and corn growth. Soil Science 176, 136–142.

Jorgensen, T.C., Weatherley, L.R., 2006. Continuous removal of ammonium ion by ion exchange in the presence of organic compounds in packed columns. Journal of Chemical Technology and Biotechnology 81, 1151–1158.

Karadag, D., Koc, Y., Turan, M., Ozturk, M., 2007. A comparative study of linear and non-linear regression analysis for ammonium exchange by clinoptilolite zeolite. Journal of Hazardous Materials 144, 432–437.

Karadag, D., Koc, Y., Turan, M., Armagan, B., 2006. Removal of ammonium ion from aqueous solution using natural Turkish clinoptilolite. Journal of Hazardous Materials 136, 604–609.

Kavoosi, M., 2007. Effects of zeolite application on rice yield, nitrogen recovery, and nitrogen use efficiency. Communications in Soil Science and Plant Analysis 38, 69–76.

Kumar, V., Hayashi, S., 2009. Modification on natural clinoptilolite zeolite for its NH4+ retention capacity. Journal of Hazardous Materials 169, 29–35.

Larsen, S.C., 2007. Nanocrystalline zeolites and zeolite structures: synthesis, characterization, and applications. The Journal of Physical Chemistry C 111, 18464–18474.

Laurino, C., Palmieri, B., 2015. Zeolite: "the magic stone"; main nutritional, environmental, experimental and clinical field of application. Nutrición Hospitalaria 32, 573–581.

Lebedynets, M., Sprynskyy, M., Sakhnyuk, I., Zbytniewski, R., Golembiewski, R., Buszewski, B., 2004. Adsorption of ammonium ions onto a natural zeolite: Transcarpathian clinoptilolite. Adsorption Science and Technology 22, 731–741.

Lei, L., Li, X., Zhang, X., 2008. Ammonium removal from aqueous solutions using microwave-treated natural Chinese zeolite. Separation and Purification Technology 58, 359–366.

Lenntech, B.V., 2015. Zeolite Application (accessed October 2015). http://www.lenntech.pl/zeolites-applications.htm.

Liu, C.H., Lo, K.V., 2001. Ammonia removal from composting leachate using zeolite. I. Characterization of the zeolite. Journal of Environmental Science and Health A 36, 1671–1688.

Li, J., Zhang, Z., Khunjar, W., Zhao, K., 2013. Enhanced nutrient sequestration from swine wastewater using zeolite synthesized from fly ash integrated with surface amendment technique. Fuel 111, 57–65.

Lim, S.S., Lee, D.S., Kwak, J.H., Park, H.J., Kim, H.Y., Choi, W.J., 2015. Fly ash and zeolite amendments increase soil nutrient retention but decrease paddy rice growth. Journal of Soils and Sediments 16.

Liu, L., Liu, C., Zheng, J., Huang, X., Wang, Z., Liu, Y., Zhu, G., 2013. Elimination of veterinary antibiotics and antibiotic resistance genes from swine wastewater in the vertical flow constructed wetlands. Chemosphere 91, 1088–1093.

Llanes-Monter, M.M., Olguin, M.T., Solache-Rios, M.J., 2007. Lead sorption by a Mexican, clinoptilolite-rich tuff. Environmental Science and Pollution Research 14, 397–403.

Maranon, E., Ulmanu, M., Fernandez, Y., Anger, I., Castrillon, L., 2006. Removal of ammonium from aqueous solutions with volcanic tuff. Journal of Hazardous Materials 137, 1402–1409.

Margeta, K., Farkas, A., Siljeg, M., Logar, N.Z., 2013. Natural zeolites in water treatment – how effective is their use. In: Elshorbagy, W., Chowdhury, R.K. (Eds.), Water Treatment. InTech, Rijeka, Croatia, pp. 81–112.

Markou, G., Depraetere, O., Vandamme, D., Muylaert, K., 2015. Cultivation of *Chlorella vulgaris* and *Arthrospira platensis* with recovered phosphorus from wastewater by means of zeolite sorption. International Journal of Molecular Sciences 16, 4250–4264.

Markou, G., Vandamme, D., Muylaert, K., 2014. Using natural zeolite for ammonia sorption from wastewater and as nitrogen releaser for the cultivation of *Arthrospira platensis*. Bioresource Technology 155, 373–378.

McLaughlin, D., Kinzelbach, W., 2015. Food security and sustainable resource management. Water Resources Research 51, 4966–4985.

Muller, K., Arlt, W., 2013. Status and development in hydrogen transport and storage for energy solutions. Energy Technology 1, 501–511.

Omar, L., Ahmed, O.H., Majid, N.M.A., 2015. Improving ammonium and nitrate release from urea using clinoptilolite zeolite and compost produced from agricultural wastes. The Scientific World Journal. http://dx.doi.org/10.1155/2015/574201.

Oren, A.H., Kaya, A., 2006. Factors affecting adsorption characteristics of Zn^{2+} on two natural zeolites. Journal of Hazardous Materials 131, 59–65.

Oter, O., Akcay, H., 2007. Use of natural clinoptilolite to improve water quality: sorption and selectivity studies of lead (II), copper (II), zinc (II), and nickel (II). Water Environment Research 79, 329–335.

Panayotova, M., 2001. Kinetics and thermodynamics of removal of nickel ions from wastewater by use of natural and modified zeolite. Fresenius Environmental Bulletin 10, 267–272.

Panayotova, M., Velikov, B., 2002. Kinetics of heavy metal ions removal by use of natural zeolite. Journal of Environmental Science and Health A 37, 139–147.

Polat, E., Karaca, M., Demir, H., Naci-Onus, A., 2004. Use of natural zeolite (clinoptilolite) in agriculture. Journal of Fruit and Ornamental Plant Research 12, 183–189.

Primo, A., Garcia, H., 2014. Zeolites as catalysts in oil refining. Chemical Society Reviews 43, 7548–7561.

Ramesh, K., Reddy, D.D., Biswas, A.K., Rao, A.S., 2011. Zeolites and their potential uses in agriculture. Advances in Agronomy 113, 215–236.

Rehakova, M., Cuvanova, S., Dzivak, M., Rimar, J., Gavalova, Z., 2004. Agricultural and agrochemical uses of natural zeolite of the clinoptilolite type. Current Opinion in Solid State and Materials Science 8, 397–404.

Rojas-Pavon, C.X., Olguin, M.T., Jimenez-Cedillo, M.J., Maubert, A., 2015. Sorption properties of modified clinoptilolite-and mordenite-rich tuffs for manganese removal from aqueous systems. Research and Reviews in Materials Science and Chemistry 5, 29–61.

Rozic, M., Cerjan-Stefanovic, S., Kurajica, S., Vancina, V., Hodzic, E., 2000. Ammoniacal nitrogen removal from water by treatment with clays and zeolites. Water Research 34, 3675–3681.

Sepaskhah, A.R., Barzegar, M., 2010. Yield, water and nitrogen-use responses of rice to zeolite and nitrogen fertilization in a semi-arid environment. Agricultural Water Management 98, 38–44.

Sprynskyy, M., Buszewski, B., Terzyk, A.P., Namiesnik, J., 2006. Study of the selection mechanism of heavy metal (Pb^{2+}, Cu^{2+}, Ni^{2+}, and Cd^{2+}) adsorption on clinoptilolite. Journal of Colloid and Interface Science 304, 21–28.

Stefanakis, A.I., Akratos, C.S., Gikas, G.D., Tsihrintzis, V.A., 2009. Effluent quality improvement of two pilot-scale, horizontal subsurface flow constructed wetlands using natural zeolite. Microporous and Mesoporous Materials 124, 131–143.

Sprynskyy, M., Lebedynets, M., Terzyk, A.P., Kowalczyk, P., Namiesnik, J., Buszewski, B., 2005. Ammonium sorption from aqueous solutions by the natural zeolite Transcarpathian clinoptilolite studied under dynamic conditions. Journal of Colloid and Interface Science 284, 408–415.

Szerement, J., Ambrozewicz-Nita, A., Kedziora, K., Piasek, J., 2014. Use of zeolite in agriculture and environmental protection. A short review. Institute of Agrophysics PAS, Lublin, Poland. Home page: http://zeolity.wszia.edu.pl/documents/2014/12/use-of-zeolite-in-agriculture-and-environmental-protection-a-short-review-in-english.pdf.

Torii, K., 1978. Utilization of natural zeolites in Japan. In: Sand, L.B., Mumpton, F.A. (Eds.), Natural Zeolites: Occurrence, Properties, Use. Pergamon Press, Elmsford, New York, pp. 441–450.

Turkman, A.E., Aslan, S., Ege, I., 2004. Treatment of metal containing wastewaters by natural zeolites. Fresenius Environmental Bulletin 13, 574–580.

Vaughan, P.A., 1966. The crystal structure of the zeolite ferrierite. Acta Crystallographica 21, 983–990.

Wang, Y.F., Lin, F., Pang, W.Q., 2007. Ammonium exchange in aqueous solution using Chinese natural clinoptilolite and modified zeolite. Journal of Hazardous Materials 142, 160–164.

Wang, S., Peng, Y., 2010. Natural zeolites as effective adsorbents in water and wastewater treatment. Chemical Engineering Journal 156, 11–24.

Wang, S., Terdkiatburana, T., Tade, M.O., 2008. Adsorption of Cu(II), Pb(II) and humic acid on natural zeolite tuff in single and binary systems. Separation and Purification Technology 62, 64–70.

Wang, Y.Q., Liu, S.J., Xu, Z., Han, T.W., Chuan, S., Zhu, T., 2006a. Ammonia removal from leachate solution using natural Chinese clinoptilolite. Journal of Hazardous Materials 136, 735–740.

Wang, S.B., Li, H.T., Xie, S.J., Liu, S.L., Xu, L.Y., 2006b. Physical and chemical regeneration of zeolitic adsorbents for dye removal in wastewater treatment. Chemosphere 65, 82–87.

Wang, Z., Dong, J., Liu, L., Zhu, G., Liu, C., 2013. Screening of phosphate-removing substrates for use in constructed wetlands treating swine wastewater. Ecological Engineering 54, 57–65.

Weatherley, L., Miladinovic, N., 2004. Comparison of the ion exchange uptake of ammonium ion onto New Zealand clinoptilolite and mordenite. Water Research 38, 4305–4312.

Wen, D., Ho, Y.S., Tang, X., 2006. Comparative sorption kinetic studies of ammonium onto zeolite. Journal of Hazardous Materials 133, 252–256.

Zhang, B.H., Wu, D.Y., Wang, C., He, S.B., Zhang, Z.J., Kong, H.N., 2007. Simultaneous removal of ammonium and phosphate by zeolite synthesized from coal fly ash influenced by acid treatment. Journal of Environmental Sciences 19, 540–545.

Zygadlo, M., Seweryn, A., Wozniak, M., 2010. The synthesis of zeolites based on fly ash from selected heat recovery. Archives of Waste Management and Environment Protection (Arch. Gosp. Odpad. Ochr. Srod.) 12, 15–26.

CHAPTER

NATURAL AND SURFACTANT-MODIFIED ZEOLITE FOR THE REMOVAL OF POLLUTANTS (MAINLY INORGANIC) FROM NATURAL WATERS AND WASTEWATERS

23

N.S. Dionisiou, T. Matsi

Aristotle University of Thessaloniki, Thessaloniki, Greece

CHAPTER OUTLINE

23.1 INTRODUCTION

Natural zeolites are hydrated aluminosilicate minerals with high cation exchange capacity (CEC) owing to extended isomorphic substitution of Si^{4+} by Al^{3+} in their tetrahedra and large specific surface area, among other properties. Based on their high CEC, zeolites can serve as cation exchangers and thus as sorbents for cations, but their ability to sorb anions is negligible (Mumpton, 1999; Widiastuti et al., 2008; Wang and Peng, 2010). However, their sorption capacity for anions can substantially increase upon modifying their surface, which results in a change in the unsatisfied charge of their external surface from negative to positive. For the surface modification of zeolites, certain surfactants can be used and the produced zeolite is called a surfactant-modified zeolite (SMZ). SMZ can serve as sorbent for both anions because of the positive charge of its external surface, but also for cations, mainly owing to the negative charge of its interior (Wang and Peng, 2010; Bowman, 2003).

Environmental Materials and Waste. http://dx.doi.org/10.1016/B978-0-12-803837-6.00023-8

The objective of this chapter was to compile and evaluate existing research and knowledge concerning natural zeolite and SMZ with the perspective of using them to remove inorganic pollutants mainly from natural waters and wastewaters. Specifically, the structure, properties, and sources of natural zeolites and their ability to retain cations are reported briefly. Emphasis is given to the surface modification of natural zeolites by cationic surfactants and the ability of SMZ to sorb anions. In addition, SMZ sorption capacity for cations and organic pollutants is reported.

23.2 STRUCTURE, PROPERTIES, AND SOURCES OF NATURAL ZEOLITE

Natural zeolites are crystalline, tectosilicate minerals composed of silica and aluminum tetrahedra linked to each other through shared oxygen atoms. Because of the latter, zeolites have a ring-style, open, three-dimensional, infinite crystal structure consisting of interconnected cavities and channels (Fig. 23.1). Because of extended isomorphic substitution of Si^{4+} by Al^{3+} in tetrahedra, a large unsatisfied negative charge is produced at the zeolites' lattice. The negative charge is balanced by cations, eg, Ca^{+2}, Mg^{+2}, K^{+}, Na^{+}, NH_4^{+}, etc., which are retained by electrostatic bonds and forces onto the zeolites' surface or into their crystal structure. Many of the zeolites' counter-cations (ie, exchangeable cations) are weakly held and thus are free to exchange with others in solution (Widiastuti et al., 2008; Wang and Peng, 2010). The structural cavities and entry channels leading into them are large enough to retain hydrated inorganic cations selectively, as well as certain organic molecules and water molecules (which can be easily lost upon heating at high temperatures but also easily regained at room temperatures). Because of their high selectivity, zeolites can be used as catalysts. Furthermore, upon removal of water molecules from zeolites, small substances can pass through entry channels, whereas large substances are excluded; in this sense, zeolites can be used as chemical sieves (Mumpton, 1999; Widiastuti et al., 2008). Certain properties of zeolites are reported in Table 23.1.

The CEC of zeolites is based on their unsatisfied negative charge and consequently is strongly related to the extent of isomorphic substitution of Si^{4+} by Al^{3+}. Natural zeolites have values of CEC usually ranging from 200 to 400 $cmol_c\ kg^{-1}$, almost twice the CEC of bentonite clay. Unlike most noncrystalline ion exchangers, eg, organic resins and aluminosilicate gels (mislabeled in the trade as "zeolites"), zeolites selectively sorb cations. Because of their hydration spheres, certain cations cannot approach

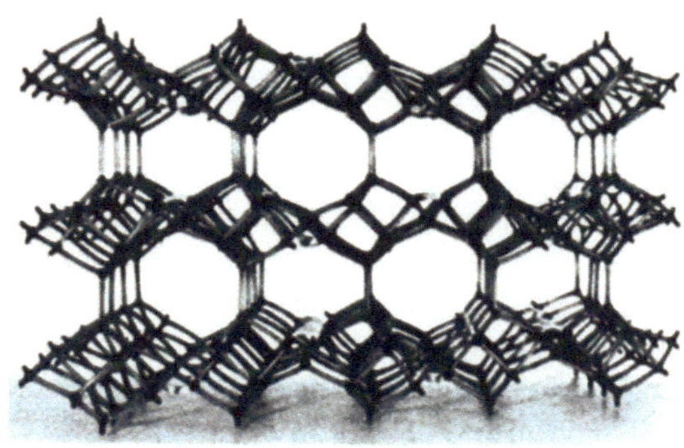

FIGURE 23.1

Open three-dimensional structure of natural zeolite. From Ackley et al. (1992).

Table 23.1 Certain Properties of Natural Zeolite

Property	Common Values
Channels	0.22–0.80 nm
Cavities	0.66–1.18 nm
Thermal stability	500–1000°C
Cation exchange capacity	≤400 $cmol_c$ kg^{-1}

the zeolites' framework, whereas certain others are generally more tightly held by zeolites and are more selectively exchanged than other cations. For example, although clinoptilolite has a relatively small CEC (225 $cmol_c$ kg^{-1}), it exhibits cation selectivity which was found to follow the decreasing order: $Cs^+ > Rb^+ > K^+ > NH_4{}^+ > Ba^{2+} > Sr^{2+} > Na^+ > Ca^{2+} > Fe^{2+} > Al^{3+} > Mg^{2+} > Li^+$ (Mumpton, 1999).

About 50 different kinds of natural zeolites are known, and more than 150 have been synthesized for different applications (Ozaydin et al., 2006). The most common natural zeolites are analcime, chabasite, clinoptilolite, erionite, ferrierite, heulandite, laumontite, mordenite, phillipsite, and stilbite; clinoptilolite is the most studied and used (Wang and Peng, 2010; Ozaydin et al., 2006). Natural zeolites have been found in the United States, Bulgaria, Russia (Mumpton, 1999), Australia, Armenia, Canada, Chile, China, Cuba, Georgia, Italy, New Zealand, Slovakia, Romania (Wang and Peng, 2010), Brazil (Campos and Buchler, 2007), Croatia (Curkovic et al., 1997), Greece (Misaelides et al., 1995; Perraki and Orfanoudaki, 2004), Hungary, the United Kingdom (Widiastuti et al., 2008), Iran (Faghihian et al., 1999), Japan (Takagi, 1978), Mexico (Llanes-Monter et al., 2007), Serbia (Lemic et al., 2006), Turkey (Ozaydin et al., 2006), Ukraine (Sprynskyy et al., 2006), and elsewhere.

Because of their structure and the derived chemical properties, such as increased CEC, dehydration and rehydration, sorption capacity, catalytic role, and biological activity, but also because of certain physical properties such as a large specific surface area, natural zeolites have numerous applications in various fields. Owing to their unique cation exchange behavior, zeolites have been extensively recommended for environmental applications such as purification and softening of natural waters, wastewater treatment, and rehabilitation of soils polluted with heavy metals and radionuclides (Mumpton, 1999).

23.3 APPLICATIONS OF NATURAL ZEOLITE FOR ENVIRONMENTAL PURPOSES

Surface and groundwater, irrigation and drinking water, and wastewater (industrial, municipal, agricultural) may contain cations more or less hazardous for human beings, animals, and plants and in general for the environment. Because of the previously mentioned properties, especially the high CEC, zeolites can exchange their harmless counter-cations with hazardous cations (especially those of heavy metals), retaining and thus removing them from water and aqueous media. Numerous studies with respect to the ability of natural zeolites to retain cationic pollutants are reported in the literature, and selected studies are reviewed in the current section.

Natural zeolites can efficiently and selectively remove several heavy metals from solution. This ability is related, among other parameters, to the crystal size and the kind of counter-cations of zeolites, the hydration ionic radii and concentration of metals in solution, the solution pH, and the contact time and temperature (Llanes-Monter et al., 2007; Sprynskyy et al., 2006).

Zeolites' affinity for heavy metals, especially of clinoptilolite for Pb^{2+}, has been demonstrated by several studies. It is reported that clinoptilolite's affinity for certain heavy metals follows the decreasing order: $Pb^{2+} > Cd^{2+} > Cs^{+} > Cu^{2+} > Co^{2+} > Cr^{3+} > Zn^{2+} > Ni^{2+} > Hg^{2+}$ (Zamzow et al., 1990). The efficiency of the same kind of zeolite to retain Pb^{2+}, Cd^{2+}, Cu^{2+}, and Ni^{2+} follows the decreasing order: $Pb^{2+} > Cu^{2+} > Cd^{2+} > Ni^{2+}$ and increases when both zeolite's particle size and metal concentration in solution decreases (Sprynskyy et al., 2006). In another study of Pb^{2+}, Cu^{2+}, Fe^{3+}, and Cr^{3+} retention by clinoptilolite, the mineral's selectivity for Pb^{2+} was the maximum (Inglezakis et al., 2004). Similar findings are reported regarding Pb^{2+} retention compared with Ba^{2+}, Cd^{2+}, Ni^{2+}, Sr^{2+}, and Cs^{+} (Faghihian et al., 1999). Zeolites retained Pb^{2+} more selectively than Cd^{2+}, although retention of both metals decreased drastically upon an increase in their concentrations in solution (Curkovic et al., 1997).

In other studies, clinoptilolite showed strong affinity for Pb^{2+} and large amounts of the metal were retained by the mineral regardless of the kind and concentration of competing ions in solution (Zamzow et al., 1990; Bailey et al., 1999). Furthermore, a natural zeolitic tuff containing 50% clinoptilolite retained Pb^{2+}, Zn^{2+}, and Cu^{2+} to a higher degree at a lower concentrations of metals in solution and its maximum retention capacity for Pb^{2+} but also for Cu^{2+} was double than that for Zn^{2+} (Peric et al., 2004). The ability of natural zeolites to retain these heavy metals has also been studied by other researchers (Cincotti et al., 2006; Oter and Akcay, 2007).

In addition, treatment of wastewater of an abandoned copper mine with clinoptilolite reduced Cu^{2+}, Zn^{2+}, and Fe^{3+} below drinking water standards (Zamzow et al., 1990) and its selectivity for certain heavy metals followed the decreasing order: $Co^{2+} > Cu^{2+} > Zn^{2+} > Mn^{2+}$ (Erdem et al., 2004).

From the literature, it seems that zeolites retain heavy metals mainly by electrostatic forces after exchange with their counter-cations. However, precipitation and specific sorption as additional retention mechanisms cannot be excluded. It is reported that clinoptilolite removed Pb^{2+} efficiently from solution at pH 2–5, which was attributed to Pb^{2+} retention by the mineral via a cation exchange process at all pH values, especially the most acidic. However, $Pb(OH)_2$ precipitation onto zeolite at pH 5 could not be excluded (Llanes-Monter et al., 2007). Except for Pb^{2+}, the cation exchange process is reported as the main mechanism of Cd^{2+} (Curkovic et al., 1997), Zn^{2+}, and Cu^{2+} retention (Peric et al., 2004), although precipitation and sorption are also mentioned as other possible mechanisms in the latter case.

Raw and pretreated (with HCl and NH_4Cl) clinoptilolite were proven to be effective means for Pb^{2+}, Cd^{2+}, Cu^{2+}, and Ni^{2+} retention (Sprynskyy et al., 2006). The retention mechanism for the cation exchange process is accomplished in three stages. During the first 30 min, metals are retained on the microcrystals' surface of clinoptilolite. Then, for a certain period of time, desorption is prevalent over cation exchange. Finally, retention of the metals occurs in the microcrystals' interior at a slow rate. Almost 25% of Pb^{2+}, 40% of Cd^{2+} and Cu^{2+} and 90% of Ni^{2+} are retained during the first stage of the process (Sprynskyy et al., 2006).

Several researchers connected Na^{+} enrichment of natural zeolites with enhancement of their ability to remove heavy metals from solution (Curkovic et al., 1997; Faghihian et al., 1999; Llanes-Monter et al., 2007; Zamzow et al., 1990; Inglezakis et al., 2004). Probably the main reasons for this are both the low charge and the large hydration ionic radius of Na^+. Because of these properties, Na^+ is expected to be weakly bound to zeolite, and thus it can be more easily exchanged with heavy metals in solution than other common exchangeable cations of zeolite, ie, Ca^{2+}, Mg^{2+}, K^+, and NH_4^+. However, an undesirable increase in pH in aqueous medium at a strongly alkaline range cannot be excluded in the case of zeolite's enrichment with Na^+. Similarly, an undesirable decrease in solution pH at a strongly acidic range cannot be excluded in the case of zeolite's treatment with HCl, as previously mentioned.

Except for heavy metals, several studies report the ability of natural zeolites to remove NH_4^+ from aqueous systems, the most common cation in polluted natural waters and wastewaters that negatively affects human and animal health, aquatic organisms, and eutrophication in rivers and lakes. As for heavy metals, also in this case the most studied kind of natural zeolite is clinoptilolite.

Clinoptilolite's capacity for NH_4^+ is about 5–10 times higher than that of sepiolite (Bernal and Lopez-Real, 1993) and greater than that of bentonite clay (Rozic et al., 2000). Ammonium retention was unaffected by the clinoptilolite's particle size but it increased with an increase in the surface charge density of the mineral, which was attributed to the increase in readily available exchange sites Consequently, the cation exchange process was considered to be the main mechanism of NH_4^+ retention by zeolite (Bernal and Lopez-Real, 1993). Similar findings were reported by other researchers, who also found that NH_4^+ removal by clinoptilolite was completed within the first 15 min of equilibration time and depended on the solution pH at the range 5.5–7.6 (Ashrafizadeh et al., 2008), but it decreased at concentrations higher than 100 mg NH_4^+ L^{-1} in solution (Rozic et al., 2000; Ashrafizadeh et al., 2008).

Similarly to heavy metals, Na^+ treatment of clinoptilolite can increase its retention capacity for NH_4^+ because of its higher affinity for NH_4^+ over Na^+. Zeolite's affinity for alkali and alkaline earth metals and NH_4^+ is reported to follow the decreasing order: $Rb^+ > K^+ > NH_4^+ > Ba^{2+} > Na^+ > Ca^{2+} > Li^+$ (Wang and Peng, 2010). In addition, NH_4^+ exchange selectivity for zeolite's counter-cations followed the decreasing order: $Na^+ > Ca^{2+} > K^+$ (Sprynskyy et al., 2005) and although a competitive effect was evident between NH_4^+ and other cations in solution, eg, Ca^{2+}, Mg^{2+}, and K^+, selectivity of Na-rich zeolite for NH_4^+ was the highest (Cooney et al., 1999). However, other researchers reported a reduction in NH_4^+ retention by zeolite in the presence of Na^+ in solution (Nguyen and Tanner, 1998).

Zeolite was studied as a means for NH_4^+ removal from different kinds of wastewater. Clinoptilolite and mordenite showed considerable ability in removing NH_4^+ from domestic (dairy and piggery) wastewater. This ability depended on the mineral's particle size, the contact time between zeolite and wastewater, and the wastewater loading flow rate and composition (Nguyen and Tanner, 1998). Similarly to the latter finding, results of a study of zeolite's ability to retain NH_4^+ from leakage water of waste dumps containing high amounts of NH_4^+ (820 mg L^{-1}) and organic substances (1033 mg C L^{-1}) indicated that organic molecules present in wastewaters may reduce zeolite's capacity for NH_4^+ owing to their sorption onto the mineral (Farkas et al., 2005). In a study of NH_4^+ removal from wastewater by zeolite conducted with columns, it was found that the maximum capacity of zeolite for NH_4^+ was observed at pH 4 or after acid pretreatment of the mineral (Sarioglu, 2005).

In addition to this, natural zeolites have the ability to retain radionuclides, especially those of Cs, from aqueous medium. Clinoptilolite selectively removed ^{137}Cs, ^{90}Sr, ^{60}Co, and ^{110m}Ag from liquid radioactive waste, especially the two former isotopes (Osmanlioglu, 2006). Three natural zeolites (clinoptilolite, chabazite, and mordenite) were tested as sorbents for ^{137}Cs. All the three zeolites showed appreciable capacities for ^{137}Cs; that of chabazite was the highest. In all cases, ^{137}Cs retention was unaffected by solution pH but significantly decreased in the presence of K^+ in solution (Borai et al., 2009). The radioactive nuclides ^{134}Cs, ^{85}Sr, and ^{131}I were added to river and seawater to prepare artificially contaminated liquid samples, and their retention by natural zeolite (mordenite) but also by bentonite clay and activated carbon was investigated (Kubota et al., 2013). In the river water samples, mordenite successfully removed only ^{134}Cs and ^{85}Sr at levels comparable to those of bentonite clay, whereas activated carbon partly removed only ^{134}Cs and ^{131}I. In the seawater samples, only ^{134}Cs was removed by mordenite and bentonite clay to a lesser degree than the river water samples, whereas ^{85}Sr and ^{131}I were hardly removed by all three substances (Kubota et al., 2013).

23.4 SURFACE MODIFICATION OF NATURAL ZEOLITE

As mentioned, zeolites have a permanent negative charge caused by isomorphic substitution of Si^{4+} by Al^{3+} in the tetrahedra of their crystal lattice. Because of the negative charge, zeolites can retain cations but their affinity for anions is negligible. Their ability to retain anions, and thus to remove them from natural waters and wastewaters, can substantially increase by altering their surface charge from negative to positive, via sorption of certain substances onto their external surface.

Such substances are $Al_2(SO_4)_3$ (Campos and Buchler, 2007), long-chain primary amines such as oleylamine (Vujakovic et al., 2000), and cationic surfactants, usually quaternary ammonium salts with chloride or bromide. Such surfactants are tetraethylammonium (Sullivan et al., 1998), hexadecyltrimethylammonium (HDTMA) (Sullivan et al., 1998; Haggerty and Bowman, 1994; Schulze-Makuch et al., 2003), ethylhexadecyldimethylammonium (Campos et al., 2007), dodecyltrimethyammonium, octyltrimethyammonium (Li et al., 2002), benzyldimethyltetradecylammonium (Kuleyin, 2007), benzyltrimethylammonium, stearyldimethylbenzylammonium (Lemic et al., 2006), dodecypyridinium (Li, 2006), cetylpyridinium (Zhan et al., 2011), and 4-methylpyridinium (Widiastuti et al., 2008).

Cationic surfactants contain both hydrophobic and hydrophilic groups in their chemical formula, and because of the latter groups they are water-soluble substances. Their molecules consist of a positively charged head group, usually quaternary ammonium, and a counter-anion, commonly Cl^- or Br^-. In addition, their molecules are larger than the diameter of zeolite's channels, and thus they cannot penetrate into the interior (internal surface) of zeolite. However, they can be sorbed onto the external surface of zeolite, altering in this way the external charge of the mineral from negative to positive. The strength and stability of surfactant's retention onto zeolite's surface increase as the length of the surfactant's carbon chain increases. Probably for this reason, HDTMA salts are commonly used for surface modification of natural zeolites (Li et al., 1998a).

At very low concentrations of surfactant in aqueous solution, its molecules exist as monomers. Under this condition, the molecules can be sorbed onto zeolite's surface at a monolayer. When the surfactant's concentration becomes greater than a critical value (critical micelle concentration), the molecules tend to associate with each other to form micelles in addition to monomers in solution. At this point, the surfactant's molecules can be sorbed onto zeolite by forming a bilayer; the first layer is retained by electrostatic forces and the second layer, by hydrophobic interactions (eg, van der Waals forces), with the positive charge of the second layer (ie, the positively charged heads of the surfactant's molecules) being neutralized by its counter-anions (Fig. 23.2) (Li et al., 1998a).

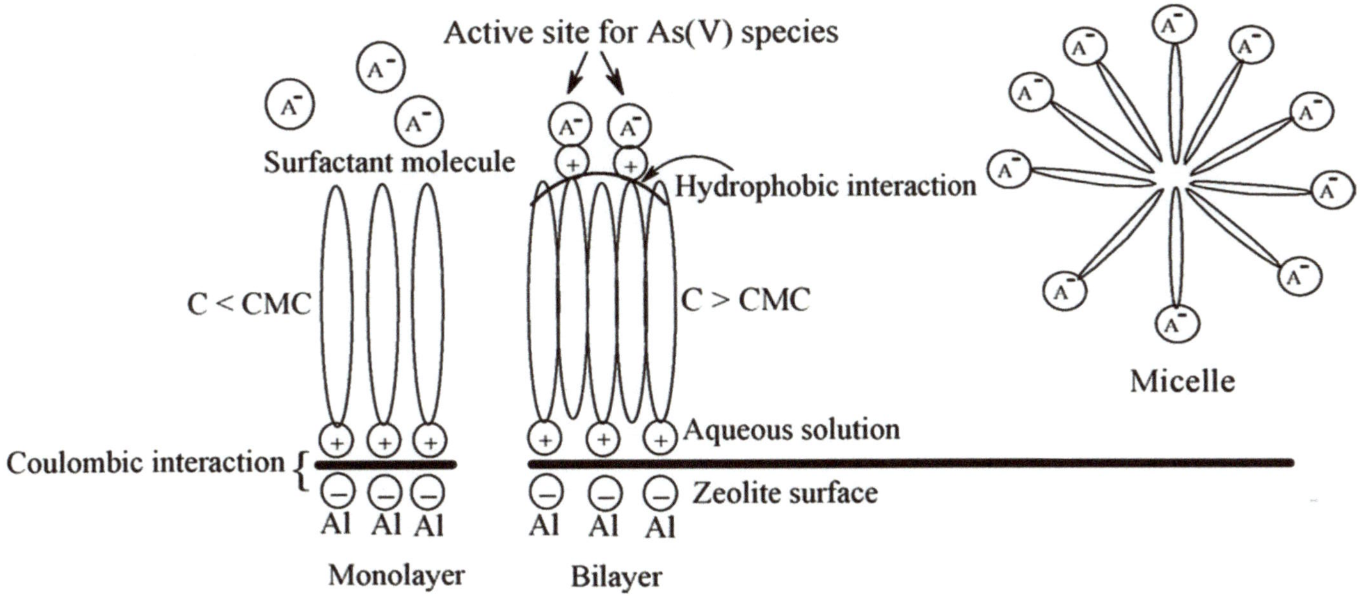

FIGURE 23.2
Surfactant modification of zeolite's external surface. From Chutia et al. (2009).

In addition to the critical micelle concentration, surfactant's concentration in relation to the CEC of zeolite's external surface (ECEC) controls also the formation of a monolayer or a bilayer. Specifically, large surfactant molecules such as HDTMA salts are sorbed only onto the external surface of zeolite, and consequently the capacity of the mineral to retain and exchange the surfactant depends on its ECEC. When the surfactant's concentration is below or equivalent to the zeolite's ECEC, the positively charged head groups of the surfactant's molecules can substitute the counter-cations of the zeolite's surface selectively, forming a monolayer. At concentrations higher than the ECEC of the zeolite, the surfactant's molecules may also be sorbed on the organic monolayer, in this way forming a bilayer via hydrophobic interactions between the organic tails of the two layers. The bilayer can be patchy or complete when in the latter case the surfactant's concentration is twice the zeolite's ECEC (Bowman, 2003; Sullivan et al., 1998). However, results of stability studies of SMZ question the bilayer formation. It was shown that upon addition of HDTMA-Br to zeolite at rates equal to 100% and 200% of its ECEC, HDTMA-Br was irreversibly bound to zeolite only in the former case (Haggerty and Bowman, 1994).

In either case of patchy or complete bilayer formation, the external charge of SMZ becomes positive and the mineral obtains anion exchange capacity. Furthermore, the surfactant's bilayer is a solvent-like medium into which organic compounds can be dissolved (Bowman, 2003; Sullivan et al., 1998) and SMZ is stable in water and concentrated chemical solutions. Based on thermodynamics measurements, clinoptilolite, modified by surfactants whose concentrations were higher than the critical micelle concentration and the mineral's ECEC, was found to be the most stable (Sullivan et al., 1998). Except for the critical micelle concentration and the relative concentration of surfactant to zeolite's ECEC, counter-anions of the surfactant's molecules also affect the kind of layer. HDTMA-Br and HDTMA-Cl can be sorbed onto zeolite's surface in a complete bilayer, whereas HDTMA-HSO_4 can be sorbed only onto an incomplete one (Li et al., 1998b).

After its surface modification, the resultant SMZ is capable of simultaneous sorption from water and other aqueous systems of: (1) anions (by electrostatic forces with the positively charged heads of the surfactant's molecules onto the external surface of SMZ, ie, anion exchange with the counter-anions of the surfactant), (2) cations (by electrostatic forces with the negatively charged interior of SMZ or the remaining unsatisfied by the surfactant's molecules negative charge of the SMZ's external surface, ie. in either case, cation exchange with the zeolite's cations), and (3) organic molecules (by hydrophobic attraction with the surfactant's carbon tails onto the external surface of SMZ). In addition, the cost of the zeolite's surface modification is relatively low and makes SMZ a viable alternative sorbent for different pollutants with respect to other reactive substances, such as activated C and ion-exchange resins (Bowman, 2003).

23.5 APPLICATIONS OF SURFACTANT-MODIFIED ZEOLITE FOR ENVIRONMENTAL PURPOSES

As was previously mentioned, SMZ is capable of sorbing three different kinds of pollutants: anions (mainly by an anion-exchange process), cations (mainly by a cation-exchange process) and organic compounds (by hydrophobic attraction). Consequently, SMZ becomes a material that could be used effectively to remove anionic but also cationic inorganic pollutants and organic contaminants from drinking and irrigation water, other aqueous media, soil solution, and wastewater. According to the literature, the most studied kind of SMZ is clinoptilolite modified by HDTMA-Br or HDTMA-Cl.

23.5.1 REMOVAL OF INORGANIC ANIONIC POLLUTANTS

The main environmental application of SMZ is based on its ability to sorb, retain, and thus remove anionic pollutants from natural waters and wastewaters. Such anions are mainly toxic chromates and arsenates, and the phosphates and nitrates, which are responsible for eutrophication (especially the former) and enrichment (especially the latter) of natural waters. All of them are oxy-anions; the main part of this subsection summarizes research into their sorption onto SMZ. In addition, SMZ sorption capacity for other oxy-anions (selenates, perchlorates, sulfates, and borates) is mentioned.

In a study of chromates removal from a hexavalent chromium solution by SMZ (96% mordenite), the maximum sorption capacity of SMZ was 93% and 67% in the case of using HDTMA-HSO_4 and ethylhexadecyldimethylammonium for zeolite's modification, respectively (Campos et al., 2007). SMZ modified with cetylpyridinium was proven to be a superior sorbent for chromates to SMZ modified with ethylhexadecyldimethylammonium (Bailey et al., 1999). Certain researchers reported that SMZ modified by HDTMA-Br had significant sorption capacity for chromates, but also for nitrates and sulfates (Li et al., 1998a). This capacity was highest when the initial HDTMA-Br concentration was much greater than the critical micelle concentration and also the HDTMA-Br loading was twice the zeolite's ECEC, ensuring in this way the formation of a complete bilayer on zeolite's surface (Li et al., 1998a). Based on this, and because chromates sorption was inhibited in the presence of nitrates and sulfates, it was concluded that oxy-anions sorption onto SMZ is primarily a surface anion-exchange process (Li et al., 1998a), which is in agreement with other researchers' conclusions (Li, 2006; Li and Hong, 2009). However, a different mechanism of oxy-anions sorption onto SMZ is also reported in the literature (Haggerty and Bowman, 1994). SMZ substantially reduced the concentration of chromates, but also of selenates and sulfates, in aqueous solutions, when the ECEC of zeolite had been fully satisfied by HDTMA-Br (ie, formation of a monolayer). It was concluded that the most likely mechanism of anions sorption by SMZ is surface precipitation of an HDTMA−anion complex (Haggerty and Bowman, 1994).

Especially for chromates, their sorption on SMZ decreased as the solution pH and ionic strength increased (Li, 2004). Based on the consideration that the mechanism of chromates sorption on SMZ is a surface anion-exchange process, this decrease was attributed to increased competition between the chromates and the background electrolyte anions for the same sorption sites (Li, 2004).

Sorption of chromates by SMZ modified with HDTMA-Br was investigated with column experiments. Although SMZ efficiently sorbed chromates, a slow and persistence release of HDTMA-Br but also of chromates was evident throughout the experiments. It was concluded that a change in the surfactant surface configuration from a bilayer to a monolayer formation had occurred which resulted in the loss of the SMZ ability to sorb and thus immobilize chromates (Li, 2006; Li and Hong, 2009). The latter was confirmed by the results of a subsequent column experiment which showed that SMZ had lost almost 90% of its sorption capacity for chromates. In addition, chromates' sorption was related to the size of SMZ's aggregations: the smaller the size, the higher was chromates' sorption (Li and Hong, 2009).

In general, arsenic sorption on SMZ depends on the zeolitic material, the arsenic chemical species, the chemical properties of aqueous medium, as well as the characteristics of SMZ (Macedo-Miranda and Olguin, 2007). It was found that sorption capacity for arsenates of SMZ modified with HDTMA-HSO_4 was higher than that of SMZ modified with ethylhexadecyldimethylammonium and both SMZs exhibited a higher capacity than that of natural zeolite modified with $Al_2(SO_4)_3$ (Campos and Buchler, 2007). SMZ (clinoptilolite modified with HDTMA-Br) sorption capacity for arsenates significantly increased when the loading level of the surfactant onto zeolite exceeded the monolayer coverage; this was attributed to anion exchange of the SMZ's Br^- with the As^{5+} ionic species of solution, which was $H_2AsO_4^-$ and $HAsO_4^{2-}$ at common pH values (Li et al., 2007). In addition, increased arsenite sorption on the particular SMZ was found, although it was less than that of arsenates. This was attributed to arsenite retention on SMZ by mechanisms other than the anion-exchange process, such as surface complexation, because As^{3+} species at common pH values is neutral H_3AsO_3. The ionic strength of solution significantly reduced only the arsenate sorption, probably because of competition of other anions with arsenates for the same exchange sites of SMZ. However, in both cases of As^{5+} and As^{3+} species, solution pH had a slight effect on their sorption (Li et al., 2007).

Others researchers observed a significant effect of solution pH on arsenate sorption onto SMZ and concluded that $HAsO_4^{2-}$ were sorbed more than $H_2AsO_4^-$ (Mendoza-Barron et al., 2010). Inconclusive findings are reported with respect to the reversibility of arsenate sorption on SMZ (Mendoza-Barron et al., 2010; Chutia et al., 2009). Furthermore, mordernite modified by HDTMA-Br was more efficient than clinoptilolite modified by the same surfactant for reducing arsenic concentrations in water below the World Health Organization guideline value of 10 $\mu g\ L^{-1}$ (Chutia et al., 2009).

Research on the use of SMZ for phosphates removal from waters was also been carried out. Almost 50% of $H_2PO_4^-$ was sorbed by SMZ modified with oleylamine during the first 10 min of equilibration (Vujakovic et al., 2000). In another study, SMZ modified by cetylpyridinium chloride removed phosphates at 50–90%, depending on the concentration of the surfactant used to produce SMZ, the initial concentration of phosphates, the contact time, and the initial pH (Widiastuti et al., 2008). In a study of P sorption-desorption on SMZ with batch experiments, SMZ exhibited high sorption capacity for P, which was comparable to a kaolinite–goethite mixture and kaolinite clay. SMZ sorption capacity increased with pH, and although it decreased with temperature and in the presence of a relative high concentration of nitrates, it remained appreciable in all cases (Dionisiou et al., 2013a). In addition, SMZ retained the greatest amount of added P, and P concentrations desorbed from SMZ ranged at levels below than those above which environmental risks are likely. The possibility of using P-laden SMZ as a slow-release P fertilizer after its use for removing P from waters was also mentioned (Dionisiou et al., 2013a).

Nitrates sorption on SMZ has been adequately investigated. The influence of various parameters of the solution on nitrates sorption onto SMZ was investigated with batch sorption experiments (Schick et al., 2010). The equilibration time was short (0.5–1 h), the final percentage of nitrates removal from the solution was higher than 80%, the presence of competing anions at equivalent concentrations did not significantly reduce nitrates sorption onto SMZ, and leaching of the cationic surfactant from the SMZ was negligible. Especially for anions, their affinity for SMZ followed the increasing order: $Cl^- << HCO_3^- \leq SO_4^{2-} < NO_3^-$ (Schick et al., 2010). Results of both batch and column sorption and desorption experiments confirmed the high sorption ability of SMZ for nitrates (Li, 2003; Schick et al., 2011).

As for the other oxy-anions previously mentioned, the increase in nitrates sorption corresponded to the increase in HDTMA-Br loading onto zeolite (100%, 150%, and 200% of zeolite's ECEC), which indicated the formation of a surfactant bilayer and of nitrates retention by anion exchange. The sorbed nitrates were desorbed instantaneously but slowly from SMZ, under different ionic strength conditions, probably owing to competition with other anions of the background electrolyte solution (Li, 2003). Based on the slow rate of nitrate release from the SMZ, as well as the results of a pot experiment in which SMZ loaded with nitrates was used as N fertilizer for corn, it was concluded that SMZ could be used as a slow-release fertilizer for NO_3-N after its use for nitrates removal from aqueous systems (Li, 2003). Similar findings with respect to the formation of a surfactant bilayer, the effect of solution ionic strength, and the mechanism of nitrates retention by SMZ are also reported by other researchers who investigated zeolite modified with cetylpyridinium bromide. They also reported that the presence of chlorides, sulfates, or bicarbonates in solution only slightly reduced SMZ efficiency for nitrates (Zhan et al., 2011).

In addition, it was reported that SMZ can be easily and fully regenerated, and consequently it could be used continuously to purify aqueous systems enriched with nitrates by using two series of SMZ-loaded columns. However, contrary to the batch experiments previously mentioned, continuous leaching of HDTMA from SMZ was evident during column experiments and a reduction in HDTMA concentrations in the effluent was achieved by subsequent filtration through an activated carbon bed (Schick et al., 2011).

As far as other anions are concerned, SMZ effectively sorbed selenates, sulfates (Haggerty and Bowman, 1994; Li et al., 1998a), and perchlorates (Zhang and Pathan, 2007; Zhang et al., 2007). Particularly for perchlorates, their affinity for SMZ was higher than that of other anions in solution and followed the increasing order: $Cl^- < HCO_3^- < SO_4^{2-} < NO_3^- << ClO_4^-$ (Zhang et al., 2007). Boron sorption on SMZ was studied at different pH values, at a preliminary level (Dionisiou et al., 2013b). In all pH values the relative amounts of sorbed B were much lower than other anions reported in the literature. This was expected for the strongly acid and neutral pH, where B species is the neutral H_3BO_3, whereas it was unexpected for the strongly alkaline pH, where $B(OH)_4^-$ are formed at appreciable amounts. In addition to the low sorption capacity of SMZ, B concentration in the equilibrium solution was above the critical limit of 3 mg L^{-1} in irrigation waters with respect to the risk of B phytotoxicity (Dionisiou et al., 2013b).

23.5.2 REMOVAL OF INORGANIC CATIONIC POLLUTANTS

Except for anions, SMZ has the ability to retain cations, mainly because of the negative charge of its interior, and to a much smaller degree because of certain negatively charged sites of its external surface, which remained unmodified. Consequently, SMZ could also be used as a sorbent for cationic heavy metals from waters.

In a study of Cd^{2+} sorption to natural zeolite and SMZ, both materials were effective sorbents of Cd^{2+}. The natural zeolite was slightly superior to SMZ (Cortes-Martinez et al., 2004). However, in another study of Sr^{2+} and Pb^{2+} sorption by both kinds of zeolite, Sr^{2+} was strongly retained by the natural zeolite but not by the SMZ (Bowman et al., 2000). This was attributed to the saturation of the SMZ's external exchange sites by the surfactant's molecules and to the low penetration of Sr^{2+} into the SMZ's interior. Contrary to Sr^{2+}, Pb^{2+} sorption by natural zeolite and SMZ was similar. This difference in SMZ's sorption capacity for the two cations was attributed to different retention mechanisms. Pb^{2+} may be bound to the remaining positively charged external surfaces of SMZ by mechanisms in addition to electrostatic attraction, such as precipitation onto the exterior or retention by the interior of SMZ (Bowman et al., 2000).

The influence of quaternary ammonium surfactants (HDTMA, dodecytrimethyammonium, and octyltrimethyammonium) on the retention of Cs^+, Sr^{2+}, Pb^{2+}, Zn^{2+}, and La^{3+} by a modified clinoptilolite was investigated (Li et al., 2002). Generally, the SMZ retention capacity for the cations decreased compared with clinoptilolite, probably because the exchange sites were blocked by the surfactant's molecules. However, Cs^+ and Pb^{2+} retention by the SMZ was higher than for the other cations, probably because of selective sorption of Cs^+ and specific sorption of Pb^{2+}, in addition to the cation-exchange process. Upon an increase in the surfactant's carbon chain length, metal desorption from clinoptilolite increased, especially that of Sr^{2+} and La^{3+}, because of competition of the surfactants' molecules for the same exchange sites (Li et al., 2002).

23.5.3 REMOVAL OF ORGANIC POLLUTANTS

Apart from inorganic anionic and cationic pollutants, SMZ has also the ability to sorb and thus to remove organic compounds (polar and nonpolar) from aqueous systems. This ability is probably the result of hydrophobic attraction of the organic pollutants (especially nonpolar) to surfactant molecules attached to SMZ. However, retention of the polar organics by electrostatic attraction also cannot be excluded.

In a study of perchloroethylene sorption on SMZ, perchloroethylene was most efficiently sorbed by the SMZ, in the case of a monolayer formation of the surfactant onto zeolite's surface. At higher surfactant loading levels, perchloroethylene sorption was limited (Li and Bowman, 1998). It was concluded that the perchloroethylene was retained in the organic phase formed by the surfactant onto the zeolite's surface. In a similar study, perchloroethylene concentrations in contaminated water were reduced by two orders of magnitude owing to sorption by SMZ (Bowman, 2003).

Sorption of benzene and its ionizable products, phenol and aniline, on a clinoptilolite-rich zeolitic tuff modified with HDTMA-Br was also studied (Li et al., 2000). The results showed that SMZ was effective in sorbing ionizable organic compounds as well as the hydrophobic nonionic. Similarly, SMZ efficiently sorbed volatile organic compounds benzene, toluene, ethylbenzene, and xylene (Bowman, 2003; Ranck et al., 2005; Altare et al., 2007) and the ionizable phenol and *p*-chlorophenol (Kuleyin, 2007); *p*-chlorophenol had a higher degree of sorption than phenol, probably because of the more hydrophobic nature of *p*-chlorophenol.

SMZ modified with stearyldimethylbenzylammonium chloride was an effective sorbent for the pesticides atrazine, lindane, and diazinone, especially when a monolayer surfactant cover had formed on its surface (Lemic et al., 2006). The sorption ability of the same SMZ for polycyclic aromatic hydrocarbons (PAHs), ie, fluorene, phenanthrene, benz[*a*]anthracene, fluoranthene, and pyrene, was studied. SMZ with monolayer coverage was the most effective for sorption of PAHs, with the more hydrophobic PAHs replacing the less hydrophobic compounds on the SMZ surface (Lemic et al., 2007).

Furthermore, SMZ was an efficient sorbent for reactive azo dyes, because of their retention by electrostatic attraction on the surfactant's bilayer formed on zeolite's surface (Benkli et al., 2005). The same was reported for clinoptilolite-heulandite—rich tuff modified with octadecyldimethylbenzyl ammonium as a sorbent for ochratoxin A (Dakovic et al., 2003).

SMZ removed fulvic acids from solution. The optimum HDTMA loading level was 120% of the zeolite's ECEC, at which 98% of fulvic acids were sorbed by the SMZ. The sorption of fulvic acids increased in the presence of Ca^{2+} and Mg^{2+}, whereas it decreased in the presence of other organic substances such as phenols (Wang et al., 2006). In addition, SMZ totally removed bacteria (*Escherichia coli*) and viruses from sewage effluents and groundwater (Bowman, 2003; Schulze-Makuch et al., 2003).

23.6 CONCLUSIONS

Natural zeolite has a high ability to retain heavy metals (especially Pb^{2+}), ammonium, and radionuclides (especially Cs^{+}); consequently, it can be used effectively as a sorbent to decontaminate natural waters and wastewater. However, after using natural zeolite for water decontamination, the fate of the loaded zeolite with heavy metals and/or radionuclides needs further investigation. The same stands for NH_4^{+}-loaded zeolite regarding its efficiency as slow-release N fertilizer, and the financial feasibility of such use.

SMZ has a high sorption capacity for anionic inorganic pollutants (specifically oxy-anions). Its sorption capacity for organic compounds and certain cations (eg, Pb^{2+}) is also appreciable. However, the sorption mechanism in relation to the arrangement of the surfactant molecules on zeolite's surface (ie, monolayer, patchy, or complete bilayer) and the stability of the arrangement need further clarification.

In addition, multiple-component (inorganic anions and cations and organic compounds) sorption studies are rare in the literature. As for natural zeolite, the fate of SMZ loaded with anions (ie, regeneration and reuse) after it is used for water decontamination needs further investigation. The same stands for the possibility of using SMZ loaded with nitrates and/or phosphates as slow-release N and/or P fertilizer. However, in that case, the potential adverse environmental impacts of released surfactant molecules from SMZ after mixing with soil should be taken seriously.

REFERENCES

Ackley, M.W., Giese, R.F., Yang, R.T., 1992. Clinoptilolite: untapped potential for kinetic gas separations. Zeolites 12, 780–788.

Altare, C.R., Bowman, R.S., Katz, L.E., Kinney, K.A., Sullivan, E.J., 2007. Microporous and Mesoporous Materials 105, 305–316.

Ashrafizadeh, S.N., Khorasani, Z., Gorjiara, M., 2008. Ammonia removal from aqueous solutions by Iranian natural zeolite. Separation Science and Technology 43, 960–978.

Bailey, S.E., Olin, T.J., Bricka, R.M., Adrian, D.D., 1999. A review of potentially low-cost sorbents for heavy metals. Water Research 33, 2469–2479.

Benkli, Y.E., Can, M.F., Turan, M., Celik, M.S., 2005. Modification of organo-zeolite surface for the removal of azo dyes in fixed-bed reactors. Water Research 39, 487–493.

Bernal, M.P., Lopez-Real, J.M., 1993. Natural zeolites and sepiolite as ammonium and ammonia adsorbent materials. Bioresource Technology 43, 27–33.

Borai, E.H., Harjula, R., Malinen, L., Paajanen, A., 2009. Efficient removal of cesium from low-level radioactive liquid waste using natural and impregnated zeolite. Journal of Hazardous Materials 172, 416–422.

Bowman, R.S., Sullivan, E.J., Li, Z., 2000. Uptake of cations, anions and nonpolar organic molecules by surfactant-modified clinoptilolite-rich tuff. In: Collela, C., Mumpton, A. (Eds.), Natural Zeolites for the Third Millennium. De Frede, Napoli, Italy, pp. 287–297.

Bowman, R.S., 2003. Applications of surfactant-modified zeolites to environmental remediation. Microporous and Mesoporous Materials 61, 43–56.

Campos, V., Buchler, P.M., 2007. Anionic sorption onto modified natural zeolites using chemical activation. Environmental Geology 52, 1187–1192.

Campos, V., Morais, L.C., Buchler, P.M., 2007. Removal of chromate from aqueous solution using treated natural zeolite. Environmental Geology 52, 1521–1525.

Chutia, P., Kato, S., Kojima, T., Satokawa, S., 2009. Adsorption of As(V) on surfactant-modified natural zeolites. Journal of Hazardous Materials 162, 204–211.

Cincotti, A., Mameli, A., Locci, A.M., Orru, R., Cao, G., 2006. Heavy metals uptake by Sardinian natural zeolites: experiment and modeling. Industrial and Engineering Chemical Research 45, 1074–1084.

Cooney, E.L., Booker, N.A., Shallcross, D.C., Stevens, G.W., 1999. Ammonia removal from wastewaters using natural Australian zeolite. II. Pilot-scale study using continuous packed column process. Separation Science and Technology 34, 2741–2760.

Cortes-Martinez, R., Martinez-Miranda, V., Solache-Rios, M., Garcia-Sosa, I., 2004. Evaluation of natural and surfactant-modified zeolites in the removal of cadmium from aqueous solutions. Separation Science and Technology 39, 2711–2730.

Curkovic, L., Cerjan-Stefanovic, S., Filipan, T., 1997. Metal ion exchange by natural and modified zeolites. Water Research 31, 1379–1382.

Dakovic, A., Tomasevic-Canovic, M., Rottinghaus, G., Dondur, V., Masic, Z., 2003. Adsorption of ochratoxin A on octadecyldimethylbenzyl ammonium exchanged-clinoptilolite-heulandite tuff. Colloids and Surfaces B: Biointerfaces 30, 157–165.

Dionisiou, N.S., Matsi, T., Misopolinos, N.D., 2013a. Phosphorus adsorption-desorption on a surfactant-modified natural zeolite: a laboratory study. Water Air and Soil Pollution 224, 1362–1371.

Dionisiou, N.S., Matsi, T., Misopolinos, N.D., 2013b. Removal of boron by surfactant modified zeolotic tuff from Northeastern Greece. Journal of Agricultural Science 5, 94–99.

Erdem, E., Karapinar, N., Donat, R., 2004. The removal of heavy metal cations by natural zeolites. Journal of Colloid and Interface Science 280, 309–314.

Faghihian, H., Ghannadi Maragheh, M., Kazemian, H., 1999. The use of clinoptilolite and its sodium form for removal of radioactive cesium, and strontium from nuclear wastewater and Pb^{2+}, Ni^{2+}, Cd^{2+}, Ba^{2+} from municipal wastewater. Applied Radiation and Isotopes 50, 655–660.

Farkas, A., Rozi, M., Barbari-Mikocevic, Z., 2005. Ammonium exchange in leakage waters of waste dumps using natural zeolite from the Krapina region, Croatia. Journal of Hazardous Materials B117, 25–33.

Haggerty, G.M., Bowman, R.S., 1994. Sorption of other inorganic anions by organo-zeolite. Environmental Science and Technology 28, 452–458.

Inglezakis, V.J., Loizidou, M.M., Grigoropoulou, H.P., 2004. Ion exchange studies on natural and modified zeolites and the concept of exchange site accessibility. Journal of Colloid and Interface Science 275, 570–576.

Kubota, T., Fukutani, S., Ohta, T., Mahara, Y., 2013. Removal of radioactive cesium, strontium, and iodine from natural waters using bentonite, zeolite, and activated carbon. Journal of Radioanalytical and Nuclear Chemistry 296, 981–984.

Kuleyin, A., 2007. Removal of phenol and 4-chlorophenol by surfactant-modified natural zeolite. Journal of Hazardous Materials 144, 307–315.

Lemic, J., Kovacevic, D., Tomasevic-Canovic, M., Kovacevic, D., Stanic, T., Pfend, R., 2006. Removal of atrazine, lindane and diazinone from water by organo-zeolites. Water Research 40, 1079–1085.

Lemic, J., Tomasevic-Canovic, M., Adamovic, M., Kovacevic, D., Milicevic, S., 2007. Competitive adsorption of polycyclic aromatic hydrocarbons on organo-zeolites. Microporous and Mesoporous Materials 105, 317–323.

Li, Z., Hong, H., 2009. Retardation of chromate through packed columns of surfactant-modified zeolite. Journal of Hazardous Materials 162, 1487–1493.

Li, Z., Anghel, I., Bowman, R.S., 1998a. Sorption of oxyanions by surfactant modified zeolite. Journal of Dispersion Science and Technology 19, 843–857.
Li, Z., Roy, S.I., Zou, Y., Bowman, R.S., 1998b. Long-term chemical and biological stability of surfactant modified zeolite. Environmental Science and Technology 17, 2628–2632.
Li, Z., Burt, T., Bowman, R.S., 2000. Sorption of ionisable organic solutes by surfactant-modified zeolite. Environmental Science and Technology 34, 3756–3760.
Li, Z., Alessi, D., Allen, L., 2002. Influence of quaternary ammonium on sorption of selected metal cations onto clinoptilolite zeolite. Journal of Environmental Quality 31, 1106–1114.
Li, Z., Beachner, R., McManama, Z., Hong, H., 2007. Sorption of arsenic by surfactant-modified zeolite and kaolinite. Microporous and Mesoporous Materials 105, 291–297.
Li, Z., 2003. Use of surfactant-modified zeolite as fertilizer carriers to control nitrate release. Microporous and Mesoporous Materials 61, 181–188.
Li, Z., 2004. Influence of solution pH and ionic strength on chromate uptake by surfactant-modified zeolite. Journal of Environmental Engineering 130, 205–208.
Li, Z., 2006. Chromate transport through surfactant-modified zeolite columns. Ground Water Monitoring and Remediation 26, 117–124.
Li, Z., Bowman, R.S., 1998. Sorption of perchloroethylene by surfactant-modified zeolite as controlled by surfactant loading. Environmental Science Technology 32, 2278–2282.
Llanes-Monter, M.M., Olguin, M.T., Solache-Rios, M.J., 2007. Lead sorption by a Mexican, clinoptilolite-rich tuff. Environmental Science and Pollution Research 14, 397–403.
Macedo-Miranda, M.G., Olguin, M.T., 2007. Arsenic sorption by modified clinoptilolite-heulandite rich tuffs. Journal of Inclusion Phenomena and Macrocycilc Chemistry 59, 131–142.
Mendoza-Barron, J., Jacobo-Azuara, A., Leyva-Ramos, R., Berber-Mendoza, M.S., Guerrero-Coronado, R.M., Fuentes-Rubio, L., Martinez-Rosales, J.M., 2010. Adsorption of arsenic (V) from a water solution onto a surfactant modified zeolite. Adsorption 17, 489–496.
Misaelides, P., Godelitsas, A., Filippidis, A., Charistos, D., Anousis, I., 1995. Thorium and uranium uptake by natural zeolitic materials. The Science of the Total Environment 173–174, 237–246.
Mumpton, F., 1999. La roca magica: uses of natural zeolites in agriculture and industry. Proceedings of the National Academy of Sciences of the USA 96, 3463–3470.
Nguyen, M.L., Tanner, C.C., 1998. Ammonium removal from wastewaters using natural New Zealand zeolites. New Zealand Journal of Agricultural Research 41, 427–446.
Osmanlioglu, A.E., 2006. Treatment of radioactive liquid waste by sorption on natural zeolite in Turkey. Journal of Hazardous Materials B137, 332–335.
Oter, O., Akcay, H., 2007. Use of natural clinoptilolite to improve water quality: sorption and selectivity studies of lead(II), copper(II), zinc(II), and nickel(II). Water Environment Research 79, 329–335.
Ozaydin, S., Kocer, G., Hepbasli, A., 2006. Natural zeolites and energy applications. Energy Sources, Part A: Recovery, Utilization, and Environmental Effects 28, 1425–1431.
Peric, J., Trgo, M., Vukojevic-Medvidovic, N., 2004. Removal of zinc, copper and lead by natural zeolite—a comparison of adsorption isotherms. Water Research 38, 1893–1899.
Perraki, T., Orfanoudaki, A., 2004. Mineralogical study of zeolites from Pentalofos area. Applied Clay Science 25, 9–16.
Ranck, J.M., Bowman, R.S., Weeber, J.L., Katz, L.E., Sullivan, E.J., 2005. BTEX removal from produced water using surfactant-modified zeolite. Journal of Environmental Engineering – ASCE 131, 434–442.
Rozic, M., Cerhan-Stefanovic, S., Kurajica, S., Vancina, V., Hodzic, E., 2000. Ammoniacal nitrogen removal from water by treatment with clays and zeolites. Water Research 34, 3675–3681.
Sarioglu, M., 2005. Removal of ammonium from municipal wastewater using natural Turkish (Dogantepe) zeolite. Separation and Purification Technology 41, 1–11.

Schick, J., Caullet, P., Paillaud, J.L., Patarin, J., Mangold-Callarec, C., 2010. Batch-wise nitrate removal from water on a surfactant-modified zeolite. Microporous and Mesoporous Materials 132, 395–400.

Schick, J., Caullet, P., Paillaud, J.L., Patarin, J., Mangold-Callarec, C., 2011. Nitrate sorption from water on a surfactant-modified Zeolite. Fixed-bed column experiments. Microporous and Mesoporous Materials 142, 549–556.

Schulze-Makuch, D., Bowman, R.S., Pillai, S.D., Guan, H., 2003. Field evaluation of the effectiveness of surfactant modified zeolite and iron-oxide-coated sand for removing viruses and bacteria from ground water. Ground Water Monitoring and Remediation 23, 68–75.

Sprynskyy, M., Lebedynets, M., Terzyk, A.P., Kowalczyk, P., Namiesnik, J., Buszewski, B., 2005. Ammonium sorpion from aqueous solutions by the natural zeolite Transcarpathian clinoptilolite studied under dynamic conditions. Journal of Colloid and Interface Science 284, 408–415.

Sprynskyy, M., Buszewski, B., Terzyk, A.P., Namiesnik, J., 2006. Study of selection mechanism of heavy metal (Pb^{2+}, Cu^{2+}, Ni^{2+}, and Cd^{2+}) adsorption on clinoptilolite. Journal of Colloid and Interface Science 304, 21–28.

Sullivan, E.J., Carey, J.W., Bowman, R.S., 1998. Thermodynamics of cationic surfactant sorption onto natural clinoptilolite. Journal of Colloid and Interface Science 206, 369–380.

Takagi, S., 1978. Preliminary study on use of zeolite for isolation and removal of long-lived cesium in liquid waste from nuclear power stations. Journal of Nuclear Science and Technology 15, 213–221.

Vujakovic, A.D., Tomasevic-Canovic, M.R., Dakovic, A.S., Dondur, V.T., 2000. The adsorption of sulphate, hydrogenchromate and dihydrogenphosphate anions on surfactant-modified clinoptilolite. Applied Clay Science 17, 265–277.

Wang, S., Peng, Y., 2010. Natural zeolites as effective adsorbents in water and wastewater treatment. Chemical Engineering Journal 156, 11–24.

Wang, S.G., Gong, W.X., Kiu, X.W., Gao, B.Y., Yue, Q.Y., Zhang, D.H., 2006. Removal of fulvic acids from aqueous solutions via surfactant modified zeolite. Chemical Research in Chinese Universities 22, 566–570.

Widiastuti, N., Wu, H., Ang, M., Zang, D.K., 2008. The potential application of natural zeolite for greywater treatment. Desalination 218, 271–280.

Zamzow, M.J., Eichbaum, B.R., Sandgren, K.R., Shanks, D.E., 1990. Removal of heavy metals and other cations from wastewater using zeolites. Separation Science and Technology 25, 1555–1569.

Zhan, Y., Lin, J., Zhu, Z., 2011. Removal of nitrate from aqueous solution using cetylpyridinium bromide (CPB) modified zeolite as adsorbent. Journal of Hazardous Materials 186, 1972–1978.

Zhang, P., Pathan, A.B.M.B.U., 2007. Sorption and desorption of perchlorate on surfactant-modified zeolite. Studies in Surface Science and Catalysis 170, 2086–2091.

Zhang, P., Avudzega, D., Bowman, R.S., 2007. Removal of perchlorate from contaminated waters using surfactant-modified zeolite. Journal of Environmental Quality 36, 1069–1075.

CHAPTER

TREATMENT AND REUSE OF INCINERATION BOTTOM ASH

24

T. Astrup[1], A. Muntoni[2], A. Polettini[3], R. Pomi[3], T. Van Gerven[4], A. Van Zomeren[5]

Technical University of Denmark, Lyngby, Denmark[1]; University of Cagliari, Cagliari, Italy[2]; University of Rome "La Sapienza," Rome, Italy[3]; Katholieke Universiteit Leuven, Leuven, Belgium[4]; Energy Research Centre of The Netherlands (ECN), Petten, The Netherlands[5]

CHAPTER OUTLINE

Environmental Materials and Waste. http://dx.doi.org/10.1016/B978-0-12-803837-6.00024-X

24.1 INTRODUCTION

Waste incineration based on mass-burn technology is currently the most commonly adopted method for the thermal treatment of municipal solid waste. Waste incineration further allows the recovery of energy from the waste, thereby providing significant savings with respect to environmental impacts (Fruergaard and Astrup, 2011; Vehlow, 2012, 2015). Modern municipal solid waste incinerators are typically equipped with effective flue gas–cleaning systems able to comply with the stringent air emission criteria employed in most countries (European Commission, 2006). As flue gas cleaning technologies improve, allowed air emissions can be assumed to be further lowered in the future. This renders the solid residues from the incineration processes the primary emission route for many contaminants out of the incineration plants. Consequently, the overall environmental profile of a waste incineration plant is to a large extent related to the quality and management of solid residues such as bottom ashes (BAs), fly ashes (FAs), and air pollution control (APC) residues. Of these residues, BAs are generated in the largest amounts (150–300 kg t^{-1} of waste incinerated, with an average of 200 kg t^{-1}) and are at the same time the residues with technical properties most suited for use. From an environmental and management point of view, it is therefore essential to optimize BA processing to reduce landfilling of the residues and enhance utilization.

Over the past 2 decades, numerous studies have addressed aspects related to understanding the technical and environmental properties of incinerator BA as well as defining the suitable conditions for recycling. Investigations have focused on the leaching behavior (Meima and Comans, 1999, 1998; Meima et al., 2002; Polettini and Pomi, 2004; van Zomeren and Comans, 2004; Johnson et al., 1995; Dijkstra et al., 2006a; Freyssinet et al., 2002; Dijkstra et al., 2008), geochemical reactions (Meima and Comans, 1997; Eusden et al., 1999; Piantone et al., 2004; Astrup et al., 2006), weathering aspects (Meima and Comans, 1999, 1998; Polettini and Pomi, 2004; Freyssinet et al., 2002; Meima and Comans, 1997a; Piantone et al., 2004; Chimenos et al., 2000; Zevenbergen et al., 1996; Zevenbergen and Comans, 1994; Zevenbergen et al., 1998), upgrading and stabilization methods (Meima et al., 2002; Meima and Comans, 1998; Cheeseman et al., 2003; Bethanis, 2004; Arickx et al., 2006; van der Sloot et al., 2001; Crannell et al., 2000; Van Gerven et al., 2007), utilization options (Aberg et al., 2006; Onori et al., 2011a; Filipponi et al., 2003; Polettini et al., 2009, 2005a; Giampaolo et al., 2002; Pera et al., 1997; Pecqueur et al., 2001; Dabo et al., 2009; Olsson et al., 2006; Vegas et al., 2008; Cioffi et al., 2011; Triffault-Bouchet et al., 2005; Tang et al., 2015; Forteza et al., 2004) and pilot/full-scale experiences (Aberg et al., 2006; Dabo et al., 2009; De Windt et al., 2011; Hjelmar et al., 2007; François and Pierson, 2009). Internationally, the individual investigations to some extent reflect varying concerns and priorities in individual countries, but also provide an important knowledge base for improving BA management. Worldwide, current management of BA varies to a considerable extent: In some countries ashes are primarily landfilled without treatment and in other countries the ashes are treated extensively before use in construction works. The basic characteristics and problems related to BA and its processing are, however, generally similar across countries and regions. In this sense, huge potentials for building on existing knowledge and research experiences exist with a potential to provide improved solutions for BA management and use in the future.

The current chapter focuses on both state-of-the-art and innovative techniques for the treatment and reuse/recycling of incinerator BA. Based on a review of existing research, an outline of common characteristics of waste incineration BA as well as a discussion of important BA processing options with focus on effects on leaching are presented. The chapter further outlines important environmental aspects to be considered in relation to BA management and use.

24.2 BOTTOM ASH CHARACTERISTICS

This section provides an overview of BA characteristics, mineralogy and leaching behavior, and also includes identification of the constituents generally considered most critical from an environmental point of view.

24.2.1 PHYSICAL CHARACTERISTICS

Mass-burn incinerator BA typically consists of noncombustible or unburned materials collected at the outlet of the combustion chamber. BA is typically collected in a quenching tank with water but may also be collected dry. Water quenching of BA serves two main purposes: cooling of the material while preventing tertiary air from entering the combustion chamber. Compared with dry discharge, water quenching has the major drawbacks of causing the agglomeration of particles and the formation of glassy constituents along with a high water content of the ash discharged, which are both believed to affect the potential for metal recovery negatively (Meylan and Spoerri, 2014; Berkhout et al., 2011; Rem et al., 2004; De Vries et al., 2012) (see subsequent discussion for further details on practical implications).

For practical reasons, incinerators often combine BA with grate siftings (the fine materials passing through the grate) and in some cases also with the boiler ashes (the coarse fraction of particles carried over by the flue gas from the combustion chamber, which is collected in the heat recovery section). Although waste incinerators receive varying mixtures of waste input and the ash-handling options may differ, BAs often share common characteristics. Ash characteristics are affected considerably by the incinerator technology (eg, fluidized-bed incineration versus grate-fired incineration). Because grate-fired mass-burn incineration is the most widespread and most robust technology for mixed municipal solid waste, only this technology is discussed in the current chapter.

Fresh, untreated BA contains all of the solids remaining after the incineration process (including in some cases grate siftings and boiler ashes, as mentioned earlier). The constituents of BA can be classified as noncombustible materials (waste glass, soil minerals, metals, and metal alloys) and melt products (glasses, silicate minerals, and oxide minerals) (Meima and Comans, 1997b). The mineral fraction of the BA is typically light to dark gray and is a granular material (although it may also contain large fused lumps). The particle size distribution is typical of well-graded materials and generally conforms with that of sandy gravel, with a content of 40-mm oversize particles commonly below 5% by total mass (Tang et al., 2015; Forteza et al., 2004; Heinrichs et al., 2012; Izquierdo et al., 2011; Arm, 2004; Allegrini et al., 2014), as well as a low portion of fines (<63 μm) (Izquierdo et al., 2011).

In addition to the mineral fraction (which usually accounts for 50−75% of the total mass Berkhout et al., 2011; Šyc et al., 2015; Chimenos et al., 1999; Hu and Bakker, 2015; Sormunen and Rantsi, 2015), BA also contains a number of other components including ferrous metals (FeM) (iron and steel, 5−13% on average), nonferrous metals (NFeM) (mainly aluminum and stainless steel), and heavy nonferrous metals (HNFeM) (mainly Cu and Zn, 2–5%), relict glass and ceramic particles (15–30%), and unburned organic matter (0.2–5%) (Berkhout et al., 2011; Heinrichs et al., 2012; Izquierdo et al., 2011; Šyc et al., 2015; Chimenos et al., 1999; Muchová and Rem, 2006; Holm et al., 2015). It has been reported in numerous studies that the content of such major components is largely variable concerning the grain size of the BA particles (Berkhout et al., 2011; Allegrini et al., 2014; Muchova et al., 2008), which also strongly affects the recovery potential of valuable fractions.

With regard to the recycling potential of BA, both the mineral and the metal fractions are potentially valuable. The former can be reused as an aggregate or be embedded in other materials such as cement, concrete or asphalt in a number of engineering applications, whereas the latter can be reused in the metal industry. Metals can be present in BA as metal scraps, either as metal particles physically adhering to the mineral fraction or as metallic species incorporated into the mineral structure. Obviously, the recovery of the different metal constituents requires increasingly larger efforts depending on the degree of their interaction with the mineral fraction.

24.2.2 CHEMICAL CHARACTERISTICS

24.2.2.1 Inorganic Content

The inorganic content of BA varies among incinerators, depending on the type of waste received, the incinerator technology, and the operating conditions of the incinerator (Hyks and Astrup, 2009). However, the contents of alkali metals, alkaline earth metals, and heavy metals typically vary within certain ranges regardless of the origin. Table 24.1 shows the typical elemental composition ranges as derived from various literature sources. In the table, the elemental constituents are grouped into the following categories: (1) major elements; (2) elements of potential environmental concern (ie, recognized as toxic or hazardous elements); (3) rare earth metals; (4) platinum group metals; and (5) other precious and critical elements. The last three groups of constituents have attracted considerable attention in the scientific literature (Allegrini et al., 2014, 2015; Muchová and Rem, 2006; Muchova et al., 2008; Funari et al., 2015; Morf et al., 2013) from the perspective of defining the overall resource recovery potential of incinerator BA.

Quantitatively, Al, Ca, Cl, Cu, Fe, K, Mg, Na, P, Pb, S, Si, and Zn are important inorganic constituents. As explained later in relation to ash mineralogy, in particular Al, Ba, Ca, K, Mg, S, and Si are important constituents of major solid phases controlling ash chemistry and pH. On the other hand, it is generally acknowledged that the total metal content of the ash is not related to the potential environmental impact exerted by the material in their use or the disposal site. Metals are typically considered of concern once they are released into the environment by leaching (see later discussion regarding the leaching properties).

As for the metal recovery potential from BA, the total economic value of the main recoverable metals estimated on the basis of their respective concentration ranges (Table 24.1); market prices are reported in Fig. 24.1. The reported values indicate that, as observed in previous investigations (Allegrini et al., 2014; Funari et al., 2015; Morf et al., 2013), the concentration of most rare earth elements as well as platinum group and other precious metals, although often of the same order of magnitude as their content in the upper continental crust, is generally much lower than that typical of concentrated ores and therefore is too low to justify any potential recovery option. Also, the real economic value of BA would be affected by the actual recovery yield, selectivity, and degree of

Table 24.1 Observed Ranges for Elemental Content of Bottom Ash From Different Literature Sources (Allegrini et al., 2014; Funari et al., 2015; Morf et al., 2013; Chandler et al., 1997; Sabbas et al., 2003; Bayuseno and Schmahl, 2010; Chung et al., 2007; Fujimori et al., 2004, 2005; Jung and Osako, 2007; Kida et al., 1996; Zhang et al., 2001; Zhao et al., 2008)

Element	Concentration Range (mg/kg) (Minimum to Maximum)
Major elements	
Al	14,000–79,000
Ca	8600–170,000
Fe	3100–150,000
K	660–16,000
Mg	240–26,000
Mn	7.7–3200
Na	2200–42,000
P	440–10,500
Si	4300–308,000
Elements of potential environmental concern	
As	0.12–190
Ba	69–5700
Cd	0.3–70
Cu	190–25,000
Cr	20–3400
Mo	2.5–280
Ni	7–4300
Pb	75–14,000
Se	0.05–10
Sn	2–470
Tl	0.0077–0.23
V	16–120
Zn	10–20,000
Rare earth elements	
Sc	1.3–22
La	2–30
Ce	11–51
Pr	1.1–10
Nd	4.0–37
Sm	0.93–5
Eu	0.25–2.6
Gd	0.88–5

Continued

Table 24.1 Observed Ranges for Elemental Content of Bottom Ash From Different Literature Sources (Allegrini et al., 2014; Funari et al., 2015; Morf et al., 2013; Chandler et al., 1997; Sabbas et al., 2003; Bayuseno and Schmahl, 2010; Chung et al., 2007; Fujimori et al., 2004, 2005; Jung and Osako, 2007; Kida et al., 1996; Zhang et al., 2001; Zhao et al., 2008)—cont'd

Element	Concentration Range (mg/kg) (Minimum to Maximum)
Tb	0.18−3
Dy	0.54−3
Ho	0.11−0.45
Er	0.31−2
Tm	0.01−0.18
Yb	0.31−5
Lu	0.02−0.23
Platinum group metals	
Pt	0.074−0.53
Pd	0.03−1.8
Ir	0.00072−0.0075
Rh	<0.030
Ru	<0.0175
Other precious and critical elements	
Ag	<0.29−36.9
Au	<0.11−2.2
Sb	7.6−430
Be	1.2−6
Co	6−350
Ga	7.8−24
Ge	0.78−2.7
In	<1.7
Nb	2−14
Ta	2.5−14
W	10−320

purity of the extracted metals that can be attained through hydrometallurgical extraction processes (Allegrini et al., 2014). The data in Fig. 24.1 show that, as similarly pointed out by Morf et al. (2013), most (about 85%) of the total economic value of BA is in fact associated with few elements, including Al, Si, Cu, Cr, Ni, and Au (assuming for this metal the upper value for the concentration range in Table 24.1). This provides some useful indications about the fact that there is little economic

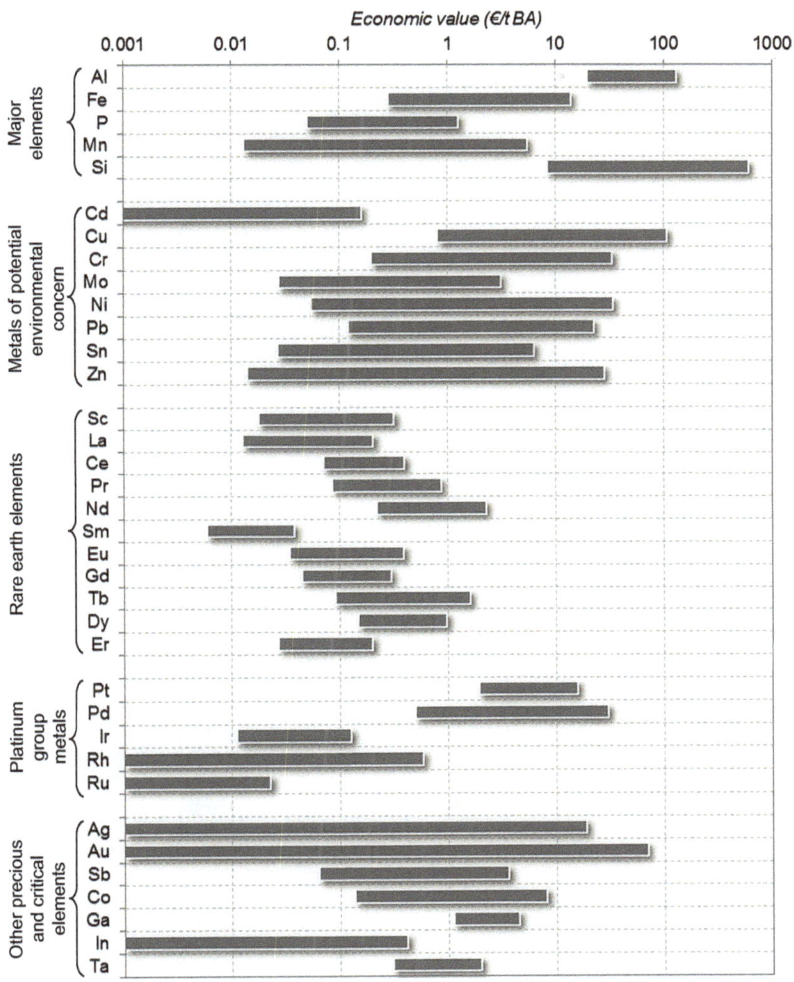

FIGURE 24.1

Estimated total economic value of metals in BA.

incentive to recover other metals than the ones mentioned, unless the recovery methods can be applied at very low additional costs.

24.2.2.2 Organic Content

Organic carbon is a relevant constituent of BA, because it has been extensively reported to have the capability, even at low concentrations, of affecting the leaching behavior of a number of inorganic components. According to van Zomeren and Comans (van Zomeren and Comans, 2009), carbon may be present in BA in different forms (Fig. 24.2). Specifically, total organic carbon is speciated between

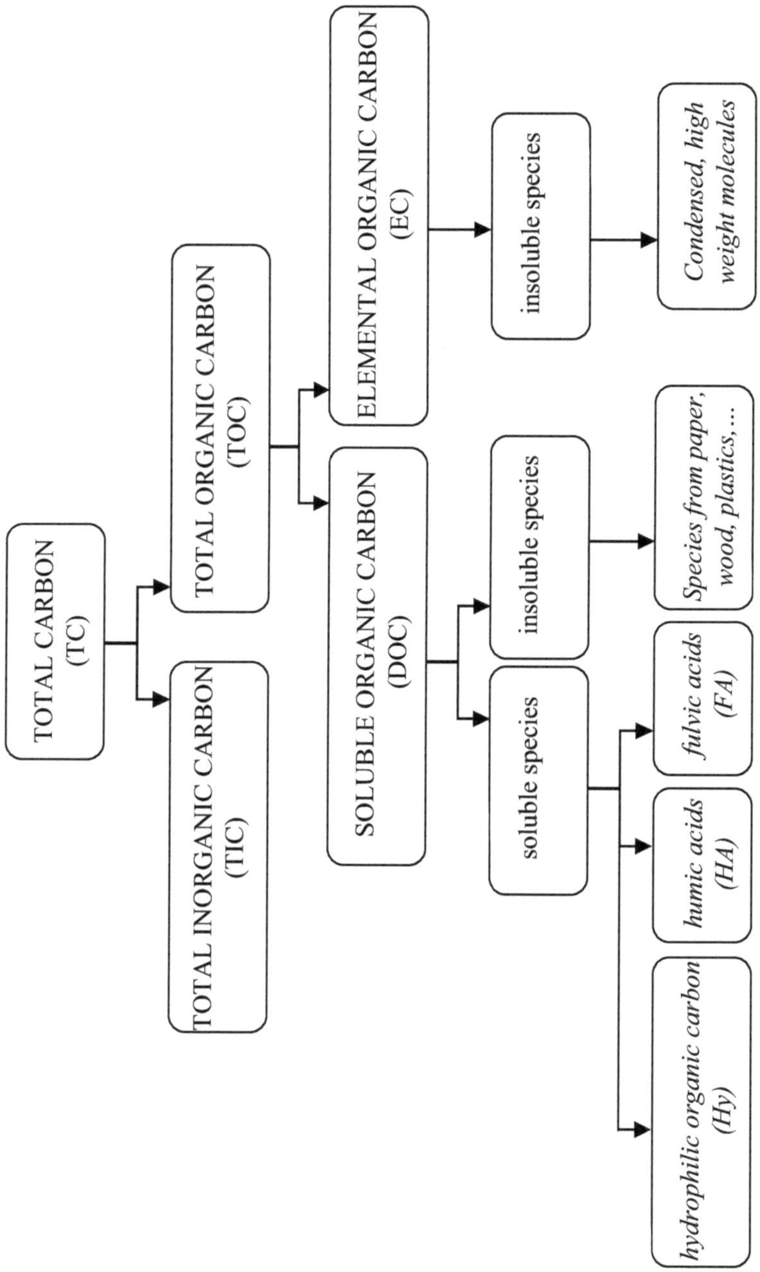

FIGURE 24.2
Carbon speciation in MSWI BA.

Adapted from van Zomeren and Comans (2009).

elemental and organic carbon species (van Zomeren and Comans, 2009; Rocca et al., 2013), which exhibit different properties in terms of water solubility and stability. Owing to its chemical structure, elemental carbon has an extremely low solubility and high stability, and may contribute to adsorption between organic micropollutants and particle surfaces (van Zomeren and Comans, 2009; Cornelissen et al., 2005). Conversely, organic carbon includes soluble, and less biologically stable, organic compounds (humic and fulvic acids, along with the so-called "hydrophilic organic compounds" such as carbohydrates) as well as an insoluble fraction, which cannot be extracted even under acidic or basic conditions. The portion of humic and fulvic acids, along with the hydrophilic organic compounds, that can be leached out from the solid matrix is found in solution in the form of dissolved organic carbon (DOC), which thus represents an aggregate parameter. Several studies have pointed to the role of DOC in the enhanced leaching of heavy metals from municipal solid waste incineration (MSWI) BA (van Zomeren and Comans, 2004; Chandler et al., 1997; Meima et al., 1999). It has been demonstrated that humic and fulvic acids are present in incinerator BA (van Zomeren and Comans, 2004; Dijkstra et al., 2008; Arickx et al., 2007) and possess a high complexing capacity for heavy metals, including, eg, Cu above all (Arickx et al., 2007, 2010; Milne et al., 2003; Yao et al., 2010; Hyks et al., 2009; Olsson et al., 2007).

To highlight the short- and long-term behavior of BA, the potential microbial activity also should be taken into account. By measuring the biodegradable organic carbon (BOC) through respiration tests, some researchers (Rendek et al., 2006a) found that this was linearly correlated with DOC, and suggested that a link between leachable and biodegradable organic carbon exists, although evidence of correlation between total organic carbon (TOC) and BOC or DOC was not evidenced.

24.2.2.3 Mineralogy and Geochemical Characteristics

The mineralogical composition of BA is well established as being capable of dictating the chemical and leaching properties of the material. BA mineralogy is also known to change with time as a result of the chemical and mineralogical transformations occurring during weathering, which comprise numerous interrelated processes such as hydrolysis, hydration, dissolution/precipitation, carbonation, complexation with organic and inorganic ligands, oxidation/reduction, surface complexation, surface (co)precipitation, sorption, ion exchange, formation of solid solutions and others (for further details see, eg, Sabbas et al. (2003)). As a result of mineralogical alterations, the basic properties of BA, including pH and acid neutralization capacity, are also expected to change, in turn affecting leaching.

Because of the importance of these issues, a number of studies have focused on the mineralogical characteristics of incinerator BA (Eusden et al., 1999; Kirby and Rimstidt, 1993), in some cases also investigating the influence that aging may exert on them (Piantone et al., 2004; Zevenbergen et al., 1996, 1998; Bayuseno and Schmahl, 2010; Wei et al., 2011). From a mineralogical point of view, incinerator BA is recognized to be a multicomponent material with a partially amorphous character. The commonly detected components have different characteristics in terms of chemical composition and mineralogical association, and typically include a low-density slag or melt phase which often also incorporates metal impurities, as well as a vitreous phase with variable quantities of crystals and glassy material. A rough distinction is often made between refractory components (including high−melting point solids contained in the original material, which remain unchanged or only partially melted upon combustion) and newly formed glassy and crystalline products. As reported in Table 24.2, silicates (quartz), aluminosilicates of Ca and Na (melilite minerals such as gehlenite and akermanite, feldspars,

Table 24.2 Overview of Main Mineralogical Composition of Fresh and Weathered Bottom Ash

Mineral	Formula	Presence in Bottom Ash		References
		Fresh	Weathered	
Refractory components				
Quartz	SiO_2	×	×	Eusden et al. (1999), Zevenbergen et al. (1998), Kirby and Rimstidt (1993), Bayuseno and Schmahl (2010), Baciocchi et al. (2010)
Feldspar minerals	$(K,Ca,Na)(Al,Si)_4O_8$	×		Eusden et al. (1999)
Biotite	$K(Mg,Fe)_3AlSi_3O_{10}O_{22}(OH)_2$	×		Eusden et al. (1999)
Olivine	$(Mg,Fe)_2SiO_4$	×		Eusden et al. (1999), Bayuseno and Schmahl (2010)
Pyroxene	$Ca(Mg,Fe)Si_2O_6$	×		Eusden et al. (1999), Bayuseno and Schmahl (2010)
Newly formed products				
Melilite-group minerals:				
Ghelenite	$Ca_2Al_2SiO_7$	×	×	Eusden et al. (1999), Zevenbergen et al. (1998), Kirby and Rimstidt (1993), Wei et al. (2011), Baciocchi et al. (2010)
Akermanite	$Ca_2(Mg,Fe)Si_2O_7$			
Spinel-group minerals:				
Magnetite	Fe_3O_4	×	×	Eusden et al. (1999), Zevenbergen et al. (1998), Bayuseno and Schmahl (2010), Wei et al. (2011)
Hercynite	$FeAl_2O_4$			
Hematite	Fe_2O_3	×	×	Piantone et al. (2004), Zevenbergen et al. (1998), Kirby and Rimstidt (1993), Baciocchi et al. (2010)
Feldspar minerals	$(Ca,Na)(Al,Si)_4O_8$	×		Eusden et al. (1999)

Wollastonite	$CaSiO_3$	×		Eusden et al. (1999), Wei et al. (2011)
Chromite	$FeCr_2O_4$	×		Eusden et al. (1999)
Lime	CaO	×		Eusden et al. (1999)
Portlandite	$Ca(OH)_2$	×		Wei et al. (2011), Baciocchi et al. (2010)
Ca carbonates:				
calcite, vaterite	$CaCO_3$	×	×	Piantone et al. (2004), Zevenbergen et al. (1998), Kirby and Rimstidt (1993), Baciocchi et al. (2010)
Sulfates:				
Anhydrite	$CaSO_4$	×	×	Polettini and Pomi (2004), Piantone et al. (2004), Zevenbergen et al. (1998)
Gypsum	$CaSO_4 \cdot 2H_2O$			
Aluminocopiapite	$Al_{2/3}Fe_4(OH)_2(SO_4)_6 \cdot 20H_2O$			
Ettringite	$Ca_6Al_2(SO_4)_3(OH)_{12} \cdot 26H_2O$	×	×	Polettini and Pomi (2004), Piantone et al. (2004), Zevenbergen et al. (1998), Bayuseno and Schmahl (2010)
Weddellite	$CaC_2O_4 \cdot 2H_2O$	×		Piantone et al. (2004)
Hydroxides:				
Goethite	$FeOOH$	×	×	Piantone et al. (2004)
Gibbsite	$AlOOH$			
Ca silicate hydrates	C-S-H–type phases		×	Piantone et al. (2004), Bayuseno and Schmahl (2010)
Zeolites			×	Piantone et al. (2004), Bayuseno and Schmahl (2010)
Clay-group minerals			×	Piantone et al. (2004), Zevenbergen et al. (1998), Bayuseno and Schmahl (2010)
Corundum	Al_2O_3	×	×	Polettini and Pomi (2004), Bayuseno and Schmahl (2010)
Muscovite-group minerals		×		Bayuseno and Schmahl (2010)

pyroxene minerals, olivine, pseudowollastonite, alite, and belite), metal oxides (magnetite and corundum), hydroxides (portlandite), sulfates (anhydrite, gypsum, and ettringite), carbonates (calcite and siderite), as well as metallic inclusions encapsulated in the melt phase are recognized as the most common phases in BA.

It has also been reported (Zevenbergen and Comans, 1994; Zevenbergen et al., 1998; Bayuseno and Schmahl, 2010; Wei et al., 2011) that weathering causes mineralogical alterations on different time scales. At the initial stages of the process, changes in mineralogy are mainly related to the alteration, at a relatively fast rate, of Ca-rich solids in the strongly alkaline environment which is established, with consequent precipitation of carbonate minerals (see subsequent discussion for additional details) and Al (oxy)hydroxides such as ferrihydrite and gibbsite, as well as dissolution of ettringite with gypsum formation. At later stages of the process, mineralogical changes are instead dominated by slow alteration of the silicate-based glassy components, which are comparatively more resistant to the external environmental conditions. The alteration of the glassy phase is also believed to result in mainly new amorphous or poorly crystalline phases, which are formed as Ca, Al, and other elements are released during the dissolution reactions. Reported products of glass weathering include hydrated silicates, aluminates, and sulfates (hydrocalumite, C-S-H–type phases, gypsum, clay minerals, and zeolites).

24.2.2.4 Leaching Behavior

Risks associated with the transfer of potentially hazardous components from BA to groundwater, surface water, and soil are recognized to be determined by their leaching potential rather than by their total content in the solid material (Dijkstra et al., 2006a). As a consequence, comprehensive information on the leaching behavior of BA is of paramount importance to evaluate the potential environmental impacts of the material correctly under the intended use scenarios or during final disposal. At the European level, it is commonly acknowledged that the estimation of contaminant release from waste materials should be based on the results of multiple standardized leaching tests, the objective of which is to address different aspects of contaminant release under a wide range of environmental conditions. To this aim, specific standard leaching procedures have been developed in the different countries; in Europe, testing procedures have been defined by the standardization committees CEN/TC 292: "Characterisation of waste" and CEN/TC 351: "Construction products: Assessment of the release of dangerous substances." Of course, the amount of information required to estimate the extent of contaminant leaching from the waste matrix is a function of the target one wishes to attain through a leaching test (characterization testing with an in-depth evaluation of the underlying mechanisms governing leaching, compliance testing to verify whether a waste material is suitable for certain purposes, and quality control testing to evaluate, eg, the variability in the waste characteristics).

It is well established that no correlation exists between the leaching of elements (except for highly soluble salts) from predominantly inorganic waste materials such as incinerator BA and their respective total content in the ash (Hyks and Astrup, 2009; Saveyn et al., 2014). Leaching has also been found to be poorly dependent on both the incinerator operating conditions and the composition of the waste input to the incinerator (Hyks and Astrup, 2009). Thus, with the exception of elements associated with highly soluble salts such as Na, K, and Cl, for which leaching is determined by their availability and is therefore directly related to their total content in the material (Hjelmar, 1996), the leaching behavior of major and trace elements in BA is largely dominated by the dissolution/precipitation reactions of common mineral phases (Meima and Comans, 1997, 1999; Meima et al., 1999) as well as (particularly for weathered BA) by the interaction of elements with reactive surfaces (mainly Fe and Al [hydr]oxides) and complexing components (organic matter) (Meima and Comans, 1998; Dijkstra et al., 2006b, 2008; Meima and Comans, 1998; Dijkstra et al., 2002). A brief overview of the main geochemical processes governing the leaching of individual elements/components from the BA matrix is provided in Table 24.3.

Several countries have developed criteria or regulations to define the conditions for reuse of waste materials (including incinerator BA) in different applications. In some cases, both unrestricted and restricted reuse scenarios are identified on the basis of both the total content of contaminants in the ash and their leaching behavior. Because the leaching test procedures required to evaluate the suitability of waste materials for use generally differ among countries, the limit values are likewise different depending on the specific evaluation approach adopted (Table 24.4). However, irrespective of the individual criteria adopted, there is some general agreement that, among the contaminants of potential environmental concern, Cu, Mo, Sb, Cl, and SO_4 may represent critical substances with respect to the

Table 24.3 Overview of Geochemical Processes Reported to Control Leaching of Major and Minor Elements or Components From Quenched Noncarbonated Municipal Solid Waste Incineration Bottom Ash (Meima and Comans, 1999, 1998; Dijkstra et al., 2006a; Dijkstra et al., 2008; Meima and Comans, 1997a, 1998; Meima et al., 1999; Hjelmar, 1996; Dijkstra et al., 2002; Cornelis et al., 2006, 2008, 2012)

Element	Controlling Mechanism
Ca, SO_4, CO_3	Solubility control by ettringite and gypsum/anhydrite Solubility control by calcite/monohydrocalcite
Al	Solubility control by gibbsite, amorphous $Al(OH)_3$ or amorphous Al silicates
Fe	Solubility control by ferrihydrite
Mg	Solubility control by brucite or magnesite
Si	Solubility control by wairakite, amorphous SiO_2 or amorphous Ca-Al-silicates
Na, K, Cl	Solubility control by availability of corresponding salts
Mo	Solubility control by powellite
Sb	Solubility control by Ca antimonates Incorporation into ettringite
Cu	Organic complexation by fulvic acids. Solubility control by tenorite/$Cu(OH)_2$ Sorption control by amorphous Al minerals
Pb	Solubility control by cerussite or $Pb(OH)_2$. Sorption control by amorphous Al minerals. Organic complexation
Zn	Solubility control by zincite, $Zn(OH)_2$ or $ZnSiO_3$. Sorption control by amorphous Al minerals
Cd	Solubility control by otavite. Sorption by amorphous Al silicates
Others (Mn, Ba, V)	Solubility control by $MnO(OH)$, barite and $Pb_2V_2O_7$

Table 24.4 Summary of Criteria and Regulations of Different European Countries for Assessment of Suitability of Waste Materials to Use (Saveyn et al., 2014)

Country	Applied Criteria	Leaching Test Procedure Required
Austria	Total content + leaching	EN 12457-4
Belgium (Flanders)	Total content + leaching	CEN/TS 14405
Czech Republic	Total content + leaching	EN 12457-4
Denmark	Total content + leaching	EN 12457-1
Finland	Total content + leaching	EN 12457-3 CEN/TS 14405
France	Total content + leaching	EN 12457-2 and 4 CEN/TS 14405
Germany	Total content + leaching	EN 12457-2 DIN 19528
Italy	Leaching	EN 12457-2
The Netherlands	Total content + leaching	CEN/TS 14405
Spain (selected regions)	Leaching	EN 12457-4 DIN 38414-S4
Sweden	Total content + leaching	CEN/TS 14405

potential environmental impacts from BA use (Sormunen and Rantsi, 2015; Saveyn et al., 2014; De Wijs and Cleven, 2008). For this reason, if BA is intended to be reused in civil engineering applications, dedicated efforts should be devoted, eg, through appropriate BA pretreatment, to improve the environmental profile of the material with major regard to those specific contaminants. The specific effects of the individual treatment options on the leaching behavior of critical contaminants are outlined in the subsequent sections. As will be noted later in the chapter, the various treatment processes available act differently on modifying the leaching of the various potentially harmful components; therefore, careful selection of the type of treatment applied should be conducted with a view to meeting the standards required for use.

24.3 BOTTOM ASH PROCESSING

The processing methods that have been proposed and are in some cases routinely applied to incinerator BA have two main purposes: the separation of valuable fractions (basically, the mineral and metal fractions) to be reused in different applications and improvement in their technical and environmental behavior to meet the requirements set by technical standards for the use and regulatory thresholds for the reduction of potential environmental impacts.

The following provides an overview of the most important processing and treatment techniques applied to BA. The techniques are discussed with respect to their recycling potential as well as their potential effects on the leaching behavior of the material.

24.3.1 EXTRACTION AND SEPARATION

24.3.1.1 Integrated Scrubbing

The common use of quenching tanks for BA cooling provides an opportunity to separate a large portion of the readily soluble components (mainly chloride and, to a lesser extent, sulfate salts) through integrated scrubbing. The relatively high temperatures (about 70°C) in the quenching tank contribute to create the conditions for a high rate of dissolution. If relatively clean surplus water is added and/or the contact time between water and the incineration residues is increased, a reduction of more than 50% in readily soluble salts can be achieved (Zwahr, 2004). However, regarding the leaching of metals, integrated scrubbing is not expected to result in a significant improvement with a view to use, because as mentioned earlier, most metal contaminants display solubility- rather than availability-controlled leaching.

24.3.1.2 Mechanical Separation

Mechanical separation of BA components has become a routine practice in several incineration plants, especially in northern Europe, and together with aging represents the most widely adopted treatment method of incineration BA. As outlined in the previous sections, BA is increasingly recognized as a source for high-value materials, which justifies the considerable economic drive to enhanced recovery of BA fractions. It has been shown (Berkhout et al., 2011) that the value of NFeM recovered from BA alone is sufficient to cover the costs of the entire treatment sequence for the production of the recycled mineral fraction. The environmental benefits of recycling of BA fractions are obviously associated with the reduced consumption of natural resources and avoided emissions from the primary industrial production processes. For instance, the production of secondary Al from recycled metal scraps requires approximately 6% of the energy demand of Al production from virgin materials, which amounts to about 45 $kWh\ kg^{-1}$ of metal (Hu and Rem, 2009). Lifecycle assessment (LCA) studies of BA recovery showed that the related savings in greenhouse gas emissions are on the order of 100–400 kg $CO_2\ t^{-1}$ BA processed, depending on the assumptions adopted and the specific conditions considered (Allegrini et al., 2015; Boesch et al., 2014), with a major contribution of Al (>50%) and Fe (35%) recovery. However, potential toxic impacts of BA recycling may be critical, mainly owing to Fe scrap recovery (Allegrini et al., 2015), although it was also observed that this may be the result of an overestimation of contaminant release by the adopted inventory of process emissions.

Enhanced recovery of metals from BA is also expected to result in benefits in terms of the technical performance of the separated mineral fraction for use as an aggregate or mineral addition in cement, concrete or asphalt mixtures in construction applications, because diverting metallic constituents may avoid documented detrimental effects in the usage site (Allegrini et al., 2015), including the oxidation of metallic Al with associated H_2 production and expansion or cracking phenomena in concrete products caused by the formation of ettringite.

Separation of the metal and mineral fractions of BA is usually accomplished through an often complex treatment sequence which combines traditional processing and classification units (sieve separation, crushing, impact crushing, attrition liberation, magnetic separation, eddy current separation, air classification, hydraulic separation, and aging) with more advanced classification techniques (optical separation, magnetic density separation, X-ray classification, and electrodynamic fragmentation) (Berkhout et al., 2011; Holm et al., 2015). Most commonly, dry processing is applied (Holm et al., 2015; Hu et al., 2009), although wet-processing techniques have been proposed

(Berkhout et al., 2011; Hu and Bakker, 2015; Muchová and Rem, 2006; Muchova et al., 2008; Hu and Rem, 2009) to overcome the negative effects of the high moisture content in the ash that is thought to impair the recovery of NFeM and improve the separation of the light organic fraction which tends to adhere to the metal particles. Application of combined dry and wet separation techniques has led to full-scale implementations, particularly in The Netherlands (e.g. Keulen et al., 2016).

The need to combine multiple processing units in BA treatment plants stems from the fact that although recovery yields for ferrous scraps in excess of 80% are commonly achieved (Meylan and Spoerri, 2014; Allegrini et al., 2014; Muchová and Rem, 2006), efficient recovery of NFeM is considerably more sensitive to the feed material characteristics, requiring in particular an accurate control of the grain size range of the material streams fed to the metal separation units (European Commission, 2006; Heinrichs et al., 2012; Allegrini et al., 2014; Holm et al., 2015; Hu and Rem, 2009; Hu et al., 2009). In particular, the finer fractions are known to contribute significantly to the inventory of HNFeM in BA (Allegrini et al., 2014; Hu and Bakker, 2015; Muchová and Rem, 2006; Hu et al., 2009; Biganzoli et al., 2013; Muchova, 2010; Grosso et al., 2011). To this aim, in modern BA separation plants, multiple size fractions are usually processed in parallel (Rem et al., 2004; De Vries et al., 2012; Heinrichs et al., 2012; Allegrini et al., 2014; Šyc et al., 2015; Hu and Bakker, 2015; Holm et al., 2015; Muchova et al., 2008; Hu and Rem, 2009; Hu et al., 2009; Biganzoli et al., 2013; Muchova, 2010). It has been suggested (Heinrichs et al., 2012) that an optimized feed to eddy-current separation should have a ratio of 3 between the upper and lower diameters of the particle size range.

An overview of the treatment layout of different BA processing plants aimed at separating various fractions (typically, mineral fraction, ferrous scraps, and NFeM, either combined or individually), with details (when available) on the related recovery rates, is provided in Table 24.5.

24.3.1.3 Extraction With Water, Acids, and Chelating Agents

Slurry-phase extraction of metal contaminants from BA using various extracting agents has been proposed as a means to improve the leaching behavior of the mineral fraction in view of its use as an aggregate material. Washing with water is the simplest process that can be applied to remove soluble constituents from BA. The L/S ratio and residence time in the BA quenching tank are such that the thermodynamic equilibrium of the dissolution processes is seldom attained. Thus, after quenching BA will still contain residual soluble components which can be further extracted through a water washing treatment. Because of the alkaline nature of BA, the pH of fresh BA−water suspensions is typically in the range 9.5−12, which corresponds to the solubility minimum for most metal species. For this reason and through the fact that the leaching of most metals is solubility-controlled (leaching is determined by the solubility of certain minerals present), the efficiency of trace metal removal from BA through water washing is typically relatively low. Hence, this and the treatment is mainly applied to remove major components that are soluble, such as chloride, Na, and sulfate. However, the less soluble sulfate minerals may not dissolve completely upon water washing, which makes the treatment generally unsuitable to improve the leaching characteristics of BA at the level required by regulation for reuse.

Enhanced sulfate solubilization yields have been reported using $NaHCO_3$ or CO_2 in the washing solution, owing to the precipitation of Ca as carbonate in place of sulfate forms. However, it has also been reported that assisted water washing produced a permanent effect on sulfate leaching in only a few cases (Astrup, 2007).

In some cases, water washing displays appreciable dissolution yields for Cu, and to some extent Cr and Pb, although it was also observed that the attained removal is inadequate to reduce leaching below the regulatory limits (Astrup, 2007). The application of assisted water washing using, eg, CO_2 to lower the pH has been reported to allow for the extraction of a number of trace metals including Cu, Ni, and

Table 24.5 Overview of Processing Schemes Applied for Bottom Ash Recovery and Information on Related Recovery Yields (When Available)

Type of Technology	Type and Number of Separation Units	Separated Size Fractions	Target Streams (and Recovery Efficiencies)	References
Dry	Sieving (2) Manual sorting (1) Magnetic separation (3) Eddy-current separation (1)	>250 mm 45–250 mm <45 mm	FeM NFeM (29%)	Heinrichs et al. (2012)
Dry	Sieving (4) Magnetic separation (6) Air classification (1) Eddy-current separation (3)	>32 mm 16–32 mm 6–16 mm <6 mm (discarded)	Mineral fraction FeM NFeM (75%)	Heinrichs et al. (2012)
Wet	Sieving (3) Hydrocycloning (2) Magnetic separation (1) Eddy-current separation (3)	>45 mm 16–45 mm 4–16 mm 0.25–4 mm <0.25 mm	Mineral fraction	Holm et al. (2015)
Dry	Impact crushing (1) Manual sorting (1) Attrition liberation (1) Sieving (4) Magnetic separation (5) Eddy-current separation (4)	>60 mm 18–60 mm 5–18 mm 2–5 mm <2 mm (discarded)	Mineral fraction FeM NFeM Stainless steel Metal oxides	Holm et al. (2015)
Dry	Sieving (4) Manual sorting (1) Attrition liberation (1) Magnetic separation (5) Eddy-current separation (4)	>300 mm 45–300 mm 8–45 mm <8 mm	Mineral fraction FeM NFeM	Holm et al. (2015)
Wet	Wet screening (4) Hydrocycloning (1) Washing separation (1) Thickening (1) Magnetic separation (1) Eddy-current separation (1) Magnus separation (1)	>2 mm 0.045–2 mm <0.045 mm	Mineral fractions (sand, gravel) NFeM (58%) Al (67%) Cu/Zn (38%)	Rem et al. (2004)
Dry	Sieving Impact separation (Advanced dry recovery, ADR) (1) Wind sifting (1) Magnetic separation Eddy-current separation	>20 mm Others <2 mm	Mineral fraction FeM Al (66%) NFeM	De Vries et al. (2012), Hu et al. (2009)
Dry	Sieving (4) Manual sorting (1) Magnetic separation (3) Eddy-current separation (4) Air classification (2)	>150 mm 50–150 mm 10–150 mm 50–100 mm 20–50 mm 4–20 mm <10 mm <4 mm	Mineral fraction FeM NFeM	Biganzoli et al. (2013)

Continued

Table 24.5 Overview of Processing Schemes Applied for Bottom Ash Recovery and Information on Related Recovery Yields (When Available)—cont'd

Type of Technology	Type and Number of Separation Units	Separated Size Fractions	Target Streams (and Recovery Efficiencies)	References
Wet	Screening (4) Wind sifting (1) Magnetic separation (3) Eddy-current separation (2) Magnetic density separation (1) Jigging (1) Shaking table separation (1) Dewatering (1)	>20 mm 6–20 mm 2–6 mm 0.1–2 mm <0.1 mm	Mineral fractions (sand, gravel) FeM (83%) Al (83%) NFeM (73%) HNFeM Precious metals Light organics	Hu and Bakker (2015), Muchová and Rem (2006), Muchova et al. (2008), Hu and Rem (2009)
Wet	Wet screening (3) Hydrocycloning (1) Washing separation (1) Dewatering (1) Eddy-current separation (1) Wet eddy-current separation (2) Kinetic gravity separation (1) Magnus separation (1) Magnetic separation (1) Density separation (1)	>20 mm 6–20 mm 2–6 mm 0.1–2 mm <0.1 mm	Mineral fractions (sand, gravel; 76%) FeM (77%) Al (29%) NFeM (82%) HNFeM Precious metals Light organics	Muchova (2010)
Dry	Screening (6) Manual sorting (1) Magnetic separation (4) Crushing (1) Eddy-current separation (3) Inductive sorting system (1)	>50 mm 16–50 mm 8–16 mm 2–8 mm <2 mm	Mineral fraction FeM (85%) NFeM (61%) Al (62%) Cu (61%)	Allegrini et al. (2014)
Dry	Screening (3) Magnetic separation (3) Eddy-current separation (5) Table separation (3) Cyclone separation (1)	>5 mm 0.7–5 mm 0.1–0.7 mm 2–8 mm <2 mm	Mineral fraction FeM (85%) NFeM (61%)	Meylan and Spoerri (2014)

FEM, *ferrous metals;* NFEM, *nonferrous metals.*

Zn (Polettini et al., 2005b). On the other hand, although some removal was attained for selected contaminants, CO_2-assisted water washing has been observed to have a negative impact on the leaching characteristics, enhancing the release of Cr, Cu, Mo, and Ni (Polettini et al., 2005a,b), likely owing to an indirect effect exerted by a decrease in pH.

Other chemical extraction processes have been proposed which involve the use of inorganic acids including hydrochloric, nitric, or sulfuric acid as well as aqua regia, chelating agents including nitrilotriacetate, ethylenediamine-tetraacetate, diethylentriamine-pentaacetate, and saponins, but these have been mainly proposed for fly ash and APC residues for heavy metal recovery and detoxification of the material. Van Gerven et al. (2007) investigated different organic solutions to upgrade BA quality and found that 0.2 M ammonium citrate gave the best results in terms of metal removal and leaching of contaminants from the treated material. The extraction process involved one to three extraction steps at L/S = 5 for 1 h followed by one to three distilled water washing steps at L/S = 5 for 1 h each to remove the excess of chelating agent. Although three-stage extraction enhanced metal removal from BA, metal leaching from the final residue was not always positively affected by the increase in the extraction steps. Increasing the number of washing stages after extraction was more effective toward improving the leaching behavior of treated BA, and the combination of a single-step extraction with three washing stages using water reduced the leaching of regulated metals below the corresponding Flemish limits.

Other authors (Anjum et al., 2013) studied bioleaching to extract metals from BA using *Aspergillus niger* metabolites under the combined effect of ultrasonic radiation. The acidic metabolites of the microbial species (particularly citric acid and oxalic acid) promoted the leaching of a number of metals (Al, Fe, and Zn) from the ash matrix.

The environmental burdens of chemical extraction are evident (Reijnders, 2005). Substantial inputs of chemicals are needed for alkaline, acid, and organic extraction. Acid leaching of BA to use the metals is not recommended because substantial amounts of acid are needed to gain a low pH and high $L\ S^{-1}$, and the leaching/precipitation is a time-consuming, multistage procedure. Furthermore, subsequent selective extraction of the liquid solution for potential recovery of the extracted metals is energy-consuming. When nondegradable chelating agents are applied, the treatment of the residual effluent will be problematic.

24.3.2 CHEMICAL PROCESSES

A number of BA processing options involve some level of chemical alteration of the solid matrix. The chemical reactions involved may bind and immobilize specific elements, but may also solubilize others. These chemical reactions may either occur naturally upon interaction with the environment or be purposely induced as a part of an industrial process.

The principle of chemical treatment of BA stems from consideration of the material's chemical reactivity (particularly under alkaline conditions), which has been well documented by several studies (Meima and Comans, 1999; Johnson et al., 1995; Meima and Comans, 1997; Zevenbergen et al., 1996, 1998; Zevenbergen and Comans, 1994; Comans and Meima, 1994). This results from the fact that the high-temperature solids formed during the combustion process become thermodynamically unstable as soon as BA is readily cooled from the furnace temperatures down to ambient conditions during quenching. The complex series of chemical and mineralogical transformations that arise is referred to as natural weathering and is composed of numerous interrelated processes such as hydrolysis, hydration, dissolution/precipitation, carbonation, complexation with organic and inorganic ligands, oxidation/reduction, surface complexation, surface (co)precipitation, sorption, ion exchange, formation of solid solutions, and others (Sabbas et al., 2003).

The purpose of chemical processing of BA is to intentionally induce and accelerate the natural alteration processes of the major mineral phases of the material simultaneously so that adequate stabilization is attained within technically reasonable time frames of hours to weeks or months. Chemical processing may be applied as a pretreatment before use or final disposal. When use is the intended destination for BA, chemical treatments are applied to allow for the development of engineering properties and leaching characteristics that render BA more suited for the application in mind. In the case of landfill disposal, chemical processing of BA may still be applied for the material to attain the "final storage quality" (Hjelmar, 1996) before landfilling, thus minimizing the technical requirements for leachate collection and treatment (Sabbas et al., 2003).

In the following, the mechanisms involved in the main chemical processes available for BA are critically reviewed, to derive a comparison of the results that can be obtained from application of the different treatment methods.

24.3.2.1 Natural Aging and Weathering

Natural aging of BA by means of stockpiling under atmospheric conditions is considered to be a cost-effective and simple stabilization treatment step to be applied, possibly in combination with additional processing, before use or disposal. For this reason, a stockpiling period of a couple of weeks to several months before use is typical practice in several countries (see e.g. Astrup, 2007).

Weathering reactions proceed naturally in incinerator BA upon contact with atmospheric agents including water, O_2 and CO_2, and basically mimic the mineralogical alteration processes documented for soils of volcanic origin (Zevenbergen and Comans, 1994; Zevenbergen et al., 1998). Although it is well established that natural weathering leads to the formation of thermodynamically stable phases or phase assemblages (Meima and Comans, 1999; Johnson et al., 1995; Meima and Comans, 1997; Zevenbergen et al., 1996, 1998; Zevenbergen and Comans, 1994; Meima and Comans, 1998; Comans and Meima, 1994; Pfrang-Stotz et al., 2000), completion of the chemical and mineralogical transformations is expected to occur within time frames on the order of hundreds to thousands of years. Nevertheless, it has been shown that a certain degree of BA stabilization can be attained when the material is stockpiled for periods ranging from some weeks to a few months. Within such relatively short periods, mineralogical alterations and pH changes may modify the leaching-controlling mechanisms altering the release of major ions and trace elements from BA.

However, a number of studies still indicate that stockpiling is not always adequate for the quality standards required for BA use to be attained, which implies the need for additional treatment for specific contaminants (eg, Cu, Cr, Mo).

Natural weathering may also induce the formation of cementitious phases, which act as binders and confer BA improved mechanical properties (compressive strength and elasticity modulus). This has been observed to occur particularly in the short term (first 2 weeks of aging) and has been ascribed to the presence of portlandite (Chimenos et al., 2005). For longer weathering periods, when ettringite and amorphous Al (hydr)oxides become the most relevant neo-formed phases along with minor amounts of gypsum, calcite, and Ca silicate aluminate hydrates, mechanical properties are only slightly improved. Compaction of freshly quenched BA before natural weathering also can affect mechanical behavior, improving stiffness and compressive strength (Chimenos et al., 2005). However, detrimental effects, including the formation of gel from the oxidation of metallic aluminum at high pH and concurrent hydrogen gas evolution, as well as ettringite formation and hydration of Ca and Mg oxides, may cause BA expansion phenomena. Typically, the largest volume increase is observed during the first days of weathering and is mainly due to lime hydration.

A number of authors (Freyssinet et al., 2002; Piantone et al., 2004; Pfrang-Stotz et al., 2000) investigated the fate of major elements during short-term natural aging of BA and concluded that the main mineralogical transformations include hydration of oxides (CaO, MgO, and Al_2O_3), precipitation of Ca carbonates such as calcite and vaterite, formation of Ca silicate hydrates, conversion of anhydrite into various sulfate forms (bassanite, ettringite, and gypsum), as well as progressive transformation of crystalline Fe species (from magnetite to maghemite to hematite). Additional mineral phases have also been identified during natural weathering of BA, including metal oxides, oxy-hydroxides, quartz, and oxalates, as well as zeolite species and aluminum hydroxides polymorphs (bayerite-gibbsite-nordstrandite) (Meima and Comans, 1999, 1997a; Piantone et al., 2004; Zevenbergen et al., 1996). The relative amount of these phases is related both to the duration of the aging period and the extent of contact between solid particles and atmospheric agents.

The occurrence of exothermic reactions during weathering results in temperature changes of BA that may also affect the kinetics of other processes. During BA maturation in heaps, temperature has been observed to range from 15 to 20°C higher than ambient temperature to even higher temperature differentials, up to 40°C (Freyssinet et al., 2002).

BA maturation also involves organic matter degradation, but neither methane nor organo-halogenated compounds or volatile organic compounds (VOCs) were identified, probably because of the low TOC content of BA. However, the occurrence of organic matter biodegradation is indicated by the O_2 depletion over time, as observed by a number of authors (Freyssinet et al., 2002).

Investigation of leaching of major ions indicates a rapid washout of chlorides, a reduction in Ca leaching owing to the changes in the corresponding solubility-controlling species, as well as prolonged leaching of sulfates (Meima and Comans, 1999; Freyssinet et al., 2002; Baranger et al., 2002). An initial increase in the release of Na, K, and Mg may also be observed, which is likely related to chloride mobilization, cation exchange mechanisms, or dissolution of aluminosilicate minerals (Zevenbergen and Comans, 1994). In experiments by Freyssinet and colleagues (Freyssinet et al., 2002), dissolution of NaCl, KCl, and $MgCl_2$ was relatively rapid, so that about 80% of the chloride species was dissolved within a stockpiling period of 5 months, whereas Meima and Comans (1999) observed low leaching of Na, K, and chlorides for aging periods longer than 6 weeks. However, the results from Pfrang-Stotz and coworkers (Pfrang-Stotz et al., 2000) disagreed with such findings, because they indicated no significant changes in K and chloride leaching after 3-month weathering compared with fresh BA.

Leaching of Ca was reported to be controlled initially by portlandite (Freyssinet et al., 2002; Chimenos et al., 2003); for longer aging periods, gypsum, and ettringite controlled Ca leaching, with a decrease in Ca release resulting in the lower solubility of such phases compared with portlandite; at later stages of weathering, calcite becomes the solubility-controlling mineral for Ca (Meima and Comans, 1997).

Changes in Ca-controlling phases are reflected by pH evolution over time, with a decrease in pH as soon as portlandite is converted into other minerals. However, pH changes also depend on particle size. Chimenos et al. (2003) observed that whereas larger gravel particles (>6 mm), which typically have lower Ca contents, had an initial pH of <11, which was not appreciably affected by weathering, smaller gravel particles (2–6 mm) had an initial pH of about 12 and decreased to approximately 11.3 within the first 4 weeks of aging, and then further to about 10 after 3 months. Surprisingly, the study also showed that the 2-mm undersize fraction displayed some slight decrease in pH after only 4 months of stockpiling; this disagrees with other findings of Arickx et al. (2006), which indicated a decrease in pH from 12.1 to 10.7 for the BA sand fraction (0.1–2 mm) after 3 months of natural weathering.

Sulfates are reported to be retained over longer time scales compared with other species, and leaching has been observed to be governed by either ettringite or gypsum, depending on the degree of weathering. The increase in sulfate leaching over time observed by some investigators is likely related to the progressive dissolution of ettringite caused by carbonation, because sulfate concentrations as high as 1000 mg L^{-1} were measured in the leachate from 15- to 18-month stockpiled BA (Freyssinet et al., 2002; Baranger et al., 2002).

The kinetics of Si minerals dissolution from weathered BA is slowed by the precipitation of secondary silicate phases at alkaline pH; Si leaching decreases over time during natural weathering, likely because of the transition of unstable minerals to more stable (crystalline) phases (Freyssinet et al., 2002). However, it was reported (Chimenos et al., 2003) that for appreciable changes in Si (and Al) mobility from BA to occur, weathering periods of more than 2 weeks are required.

Changes in mineralogical composition and major ions leaching from BA during natural weathering have variously been reported to alter heavy metal mobility. Reductions in Zn, Cu, Pb, Cr, and Ni release were observed during 3-month stockpiling (Meima and Comans, 1997a; Arickx et al., 2006; Pfrang-Stotz et al., 2000). Some authors (Freyssinet et al., 2002) indicated that among various mineral phases, carbonates are responsible for Pb and Zn entrapment; other neo-formed phases such as oxides, phosphates, and sulfates may also display some metal-bearing capacity with respect to Cr, As, and Pb. The occurrence of heavy metals in solid solutions with newly formed minerals (such as calcite) may also have a role (Piantone et al., 2004).

Natural weathering has been found to reduce the leaching of Pb and Zn below the limits established by the Flemish regulation for the use of granular materials in construction applications (Arickx et al., 2006), and some studies even report leachate concentrations below the analytical detection limits (Meima and Comans, 1999).

Although Cu leaching is positively affected by natural aging, Cu has been reported to be a critical element in view of BA use. Changes in Cu leaching over time have been related to both pH changes and organic matter degradation (Meima and Comans, 1997; Arickx et al., 2006), which also occurs to various degrees depending on the particle size as a function of the initial TOC content. It was shown that, although after 3-month aging the gravel fraction of BA complied with the Flemish limits for Cu in view for use, the sand fraction still displayed unacceptable Cu release (Arickx et al., 2006), presumably due to the high residual DOC concentrations in the material.

For typical oxyanion-forming metals (Cr, Mo, and Sb), natural weathering appears to exert mainly a mobilization effect, although the governing mechanisms have not been yet fully elucidated.

24.3.2.2 Forced Carbonation

Carbonation of BA has a major role in modifying the mineralogical characteristics and leaching behavior of BA. Whereas carbonation occurs naturally during weathering, accelerated carbonation has also been studied as an efficient method to control the leaching behavior of some metal contaminants, although it may also have some contrasting effects, depending on the contaminant of concern (see subsequent discussion).

Macroscopically, the effect of carbonation is a general decrease in pH from the initial alkaline range typical of fresh BA (about 10.5–11.5) to values (about 8.3) dictated by pH control by calcite (Meima et al., 2002; Polettini and Pomi, 2004; Arickx et al., 2006; Polettini et al., 2005b; Baciocchi et al., 2010; Cornelis et al., 2006, 2012). Changes in pH reflect a number of mineralogical alterations, with new mineral phases being formed and others disappearing. Calcite precipitation leads to decreased Ca leaching owing to the lower solubility of Ca carbonate forms compared with ettringite, which typically governs the leaching behavior of Ca from freshly quenched BA. Ettringite dissolution upon carbonation has also been proposed as the reason for the increased release of sulfate from carbonated BA (Meima et al., 2002; Polettini and Pomi, 2004; Bodénan et al., 2000; Fernández Bertos et al., 2004).

As for the other major elements, Al is known to precipitate mainly as amorphous $Al(OH)_3$, Si as amorphous aluminosilicates, and sulfate as gypsum or anhydrite (Meima and Comans, 1999, 1997a; Meima et al., 2002; Dijkstra et al., 2006b).

The influence of accelerated carbonation on trace contaminant leaching from MSWI BA has been the subject of several literature studies. In general, the overall effect of carbonation is the result of multiple mechanisms, including pH changes, precipitation of metal carbonates, and interaction with newly formed minerals, which in turn depend on both carbonation conditions and the mineralogy of the starting material. In the examined literature studies, different leaching test methods were applied, with the consequence that the mechanisms affecting leaching are accounted for differently, which complicates the interpretation of the experimental results in general terms. A schematic overview of the ranking of the effects of carbonation on contaminant mobility from BA as observed in various experimental studies is provided in Table 24.6.

24.3.2.3 Chemical Binding

Most chemical processes studied for BA treatment aim to decrease the mobility of (mainly) trace metals and are therefore regarded as stabilization processes. The principle of chemical stabilization is to induce the formation of low-solubility mineral phases which possess thermodynamic and geochemical stability under either use or disposal conditions.

Among the chemical binding processes which deserve major attention, the addition of chemical additives capable of increasing the sorptive properties of BA promotes the immobilization of trace metals.

The use of Al(III) and Fe(III) salts and other sorbing compounds (steel slag, apatite, allophane, and bauxite) was tested with the aim of reducing the leaching of trace metals from BA (Meima and Comans, 1998; Polettini et al., 2005b; Comans et al., 2000), because of their recognized role in trace contaminant binding, especially for weathered BA (see earlier discussion). Pure Al(III) and Fe(III) salts were capable of immobilizing cationic metals (Cu) (Comans et al., 2000) and oxyanion-forming metals (Mo, Cr, and Sb) (Meima and Comans, 1998; Polettini et al., 2005b). As expected, metal binding was found to be pH dependent; neutral to alkaline pHs favored cations whereas lower values favored (oxy)anions. However, a concomitant mobilization effect was observed for other trace metals such as Ni and Zn, and also for a number of major cations including Ca, Mg, and Na (Polettini et al., 2005b). It should be noted that the addition of Al(III) and Fe(III) salts does also imply an increased leaching of chlorides or sulphates since these salt are mainly added in the form of Al or Fe chloride or sulphate.

Other authors (Crannell et al., 2000; Polettini et al., 2005b) studied the addition of soluble phosphate to reduce trace metal leaching from BA. Stabilization treatment produced a strong reduction in leaching of Pb and Cu and had a positive (although less relevant) effect on Cd and Zn (Polettini et al., 2005b). Leaching curves as a function of pH for phosphate-treated BA were below those of the untreated material for Cd, Cu, Ni, Zn, and Mo, confirming the stability of the corresponding solubility-controlling minerals under varying pH conditions. However, a mobilization effect of the applied process toward Pb was observed (Polettini et al., 2005b).

24.3.3 THERMAL PROCESSES

Thermal techniques involve heating the ashes to about 1000-1500°C to promote changes in the physical and chemical characteristics. In most cases, thermal processing give the treated material improved characteristics in view of both use and final disposal. Final products have a reduced volume

Table 24.6 Reported Effects of Carbonation on Contaminant Mobility From Bottom Ash

Contaminant	Leaching Test Method	Effect	References
Metals			
Ba	EN 12457 (LS 10)	~	Bodénan et al. (2000)
	EN 12457 (LS 10)	↓	Arickx et al. (2006), Van Gerven et al. (2005)
	NEN 7343 (column)	↓	Arickx et al. (2006)
Cd	EN 12457 (LS 10)	↓	Rendek et al. (2006)
Cr	EN 12457 (LS 10)	↑	Bodénan et al. (2000), Van Gerven et al. (2005), Rendek et al. (2006b)
	EN 12457 (LS 2)	↑	Polettini et al. (2005b)
	EN 12457 (LS 10)	~	Polettini et al. (2005b)
	NT Enviro 003 (availability)	↓	Todorovic and Ecke (2006)
	NT Enviro 003 (availability)	~	Todorovic and Ecke (2006)
	NEN 7345 (diffusion)	↓	Todorovic and Ecke (2006)
Cu	EN 12457 (LS 10)	↓	Arickx et al. (2006), Van Gerven et al. (2005)
	EN 12457 (LS 10)	~	Bodénan et al. (2000)
	NEN 7343	↓	Arickx et al. (2006)
	EN 12457 (LS 2)	↑	Polettini et al. (2005b)
	EN 12457 (LS 10)	↑	Polettini et al., 2005b
	WTC ANC	~	Polettini and Pomi (2004)
	NT Enviro 003 (availability)	~	Todorovic and Ecke (2006)
	NT Enviro 003 (availability)	~	Todorovic and Ecke (2006)
	NEN 7345 (diffusion)	↓	Todorovic and Ecke (2006)
	NEN 7345 (diffusion)	↑	Todorovic and Ecke (2006)
	—	↓	Meima et al. (2002)
	TS 14497 (pH-stat)	↓	Dijkstra et al. (2006b)
Mo	EN 12457 (LS 10)	~	Bodénan et al. (2000), Van Gerven et al. (2005)
	EN 12457 (LS 10)	↓	Arickx et al. (2006)
	NEN 7343 (column)	↓	Arickx et al. (2006)
	EN 12457 (LS 2)	~	Polettini et al. (2005b)
	EN 12457 (LS 10)	↑	Polettini et al. (2005b)
	NT Enviro 003 (availability)	↓	Todorovic and Ecke (2006)
	NT Enviro 003 (availability)	~	Todorovic and Ecke (2006)
	NEN 7345 (diffusion)	↓	Todorovic and Ecke (2006)
	—	↑	Meima et al. (2002)
	TS 14497 (pH-stat)	↓	Dijkstra et al. (2006b)
Ni	EN 12457 (LS 10)	~	Arickx et al. (2006)
	NEN 7343	↑	Arickx et al. (2006)
	EN 12457 (LS 2)	↑	Polettini et al. (2005b)
	EN 12457 (LS 10)	↑	Polettini et al. (2005b)

Table 24.6 Reported Effects of Carbonation on Contaminant Mobility From Bottom Ash—cont'd

Contaminant	Leaching Test Method	Effect	References
Pb	EN 12457 (LS 10)	↓↓	Bodénan et al. (2000), Fernández Bertos et al. (2004), Bone et al. (2003)
	EN 12457 (LS 10)	↓	Arickx et al. (2006), Rendek et al. (2006b)
	NEN 7343 (column)	↓	Arickx et al. (2006)
	WTC ANC	↓↓	Polettini et al. (2005b)
	NT Enviro 003 (availability)	↑	Todorovic and Ecke (2006)
	NEN 7345 (diffusion)	↑	Todorovic and Ecke (2006)
Sb	EN 12457 (LS 10)	↑	Arickx et al. (2006), Cornelis et al. (2006), Van Gerven et al. (2005)
	NEN 7343 (column)	↑	Arickx et al. (2006)
	NT Enviro 003 (availability)	↓	Todorovic and Ecke (2006)
	NT Enviro 003 (availability)	~	Todorovic and Ecke (2006)
	NEN 7345 (diffusion)	↓	Todorovic and Ecke (2006)
V	EN 12457 (LS 10)	↓	Arickx et al. (2006)
	NEN 7343 (column)	↓	Arickx et al. (2006)
Zn	EN 12457 (LS 10)	↓	Arickx et al. (2006), Polettini et al. (2005b), Bodénan et al. (2000), Fernández Bertos et al. (2004)
	NEN 7343 (column)	↓	Arickx et al. (2006)
	EN 12457 (LS 2)	↑	Polettini et al. (2005b)
	WTC ANC	↓	Polettini and Pomi (2004)
	NT Enviro 003 (availability)	↑	Todorovic and Ecke (2006)
	NT Enviro 003 (availability)	~	Todorovic and Ecke (2006)
	NEN 7345 (diffusion)	↑	Todorovic and Ecke (2006)
Anions			
Chloride	EN 12457 (LS 10)	~	Fernández Bertos et al. (2004)
	NT Enviro 003 (availability)	~	Todorovic and Ecke (2006)
	NEN 7345 (diffusion)	~	Todorovic and Ecke (2006)
Sulfate	EN 12457 (LS 10)	↑	Meima and Comans (1999), Bodénan et al. (2000), Fernández Bertos et al. (2004)
	EN 12457 (LS 10)	↑	
	NT Enviro 003 (availability)	~	Todorovic and Ecke (2006)
	NT Enviro 003 (availability)	~↑	Todorovic and Ecke (2006)
	NEN 7345 (diffusion)	↑	Todorovic and Ecke (2006)
	—	↑	Meima et al. (2002)
Organics			
DOC	TS 14497 (pH-stat)	↓↓	Dijkstra et al. (2006b)

and mechanical properties suitable for engineering applications together with negligible leaching of constituents compared with the original residues (Chandler et al., 1997). Reduced leaching, in particular, is the result of the immobilization of lithophilic metals by means of chemical or physical mechanisms and the removal of volatile compounds as a consequence of vaporization.

The possibility of achieving these goals relies on process variables such as temperature, treatment duration, and the chemical composition of the material to be treated; in particular, this needs to be optimized in terms of the content of glass-formers (SiO_2 and Al_2O_3), melting (Na_2O and K_2O), and stabilizing agents (CaO, MgO, ZnO, and PbO), which can be achieved using appropriate additives. The presence of interfering substances such as chlorides, sulfates, and alkalis, is of concern, as well.

Vitrification, melting, and sintering are the major options for thermal treatment, and they differ principally in the characteristics of the treated material rather than the process itself (Chandler et al., 1997; Sabbas et al., 2003).

24.3.3.1 Vitrification/Melting

During vitrification, a mixture of ashes with glass-forming materials and melting/stabilizing oxides is processed at 1000−1500°C to form a homogeneous liquid phase. The molten material is then cooled to generate a single-phase glassy product. Virgin materials such as silica and calcium carbonate, or glass scraps may be used as process additives, and their quantity may vary from 20% to 70% by weight of the total feed.

Immobilization of contaminants is achieved via chemical bonding and encapsulation mechanisms. Chemical bonding is of concern when considering certain metals (network formers) able to substitute Si-forming covalent bonds with oxygen atoms in the silica network, or ions (network modifiers) able to bind elements of the glass network ionically. Encapsulation occurs during cooling of the molten material, when constituents are surrounded by a glassy layer without chemical interaction with the silica networks (Chandler et al., 1997).

Melting processes are operatively similar to vitrification, but without the addition of glass-forming materials. As a consequence, products from melting are not homogeneous and may contain multiple crystal phases, a separate molten metal phase, or inclusions of unmolten feed. Specific metal phases thus can be separated from the molten product and possibly recycled.

BA vitrification is already a proven treatment technology, mainly owing to its suitable chemical composition. While BA may be thermally treated together with fly ash and APC residues, combined vitrification and melting of both residue types may require previous removal of soluble salts and further treatment of the flue gases.

24.3.3.2 Sintering

Sintering involves heating the ash, either alone or mixed with additives, to temperatures below the melting points of the main constituents, close to those achieved during incineration in the combustion chamber (typically around 900°C). This results in the diffusion transport of material across particle surfaces so that a number of reactions occur between different chemical phases and physical bonding of particles. As a consequence of particle bonding and chemical phase reconfigurations, many constituents are incorporated into a stable matrix.

Thermal treatment at relatively low operating temperatures has the multiple objectives of preventing the loss of volatile species from the material, limiting energy consumption, and reducing treatment costs. For such reasons, sintering is probably the most important thermal process for BA. A number of studies have shown application of the process for the production of ceramic materials (Cheeseman et al., 2003; Bethanis, 2004). The perspectives of reusing sintered ash are based on evidence of improvement gained in different important properties upon treatment: The material gains higher strength and density as well as lower porosity, and heavy metal mobility can drastically be reduced.

The process can be implemented in a dedicated unit, eg, in a rotary kiln placed downstream of the combustion chamber or integrated in the combustion unit itself.

24.3.4 UTILIZATION OPTIONS

BA may be used in construction either directly as a filler, embankment or foundation material or as an aggregate in other materials such as cement, concrete, or asphalt mixtures. Most research regarding the use of binders to treat BA has been carried out in relation to solidification and stabilization with the purpose of immobilizing the contaminants in a monolithic matrix through chemical stabilization and physical encapsulation. The focus was therefore originally to minimize leaching, but attention has increasing been addressed to the potential use of BA-containing products for construction purposes.

In the following, the most common options and applications for BA use are described.

24.3.4.1 Use as Unbound/Bound Aggregate

BA may be used in construction for different applications mainly involving an aggregate in either unbound or bound forms (ie, added in mixtures with a binder such as cement or asphalt) for the construction of layers of roads, harbor areas, parking lots, and others (Aberg et al., 2006; Dabo et al., 2009; Vegas et al., 2008; Forteza et al., 2004; De Windt et al., 2011; Hjelmar et al., 2007; François and Pierson, 2009; Astrup, 2007) and for the formulation of structural cement or concrete products (Filipponi et al., 2003; Giampaolo et al., 2002; Pecqueur et al., 2001; Cioffi et al., 2011; van der Wegen et al., 2013; Crignon et al., 1999). While unbound applications are well established in several countries such as Denmark, Belgium, Germany, Sweden, Taiwan, The Netherlands, and Japan, the application of BA as an aggregate in concrete products has also gained interest, although the final technical properties of the products may be of concern.

A road is typically constructed of a number of layers of compacted materials, with progressively increasing quality from bottom to top. The layers generally include (from the surface to the lower course) a surface layer (usually made of asphalt), two subsurface layers (a base and a subbase course) made up of either unbound granular compacted materials or cement-bound materials, and a subgrade course, which is the native soil underneath the constructed road. The upper pavement layers are designed to support high, localized loads and are subject to greater wear than the underlying layers. As a result, the use of BA and other recycled aggregates in road construction is usually meant for the lower layers of road pavement (Hill et al., 2001) (the base, or preferably, the subbase courses).

Whereas BA conforms to traditional technical requirements prescribed for the construction of granular road base and subbases courses (Forteza et al., 2004; Becquart et al., 2009), it may fail in fulfilling the requirements prescribed for aggregate in concrete base layers, mainly because of to its poor abrasion resistance (Forteza et al., 2004).

Concerning the environmental behavior of BA in the constructed road site, the predominant effects are related to the dissolution of salts and associated leaching mainly of Na, K, and chloride (Aberg et al., 2006; François and Pierson, 2009; Izquierdo et al., 2008; Birgisdóttir et al., 2006). As for metal contaminants, whereas some studies suggest that they are not of environmental concern (François and Pierson, 2009) especially after the initial stage of road operation (Dabo et al., 2009), other researchers (Aberg et al., 2006; Izquierdo et al., 2008) indicated that Cu and/or Cr may be leached in higher amounts compared with the use of natural aggregates, and may be problematic in terms of compliance with environmental protection criteria, although it was also noted (De Windt et al., 2011) that the environmental performance of the road is strongly site-specific and may considerably vary from one installation to another. LCA studies on the use of BA in road construction (Birgisdóttir et al., 2006) indicate that the environmental impacts with and without BA are comparable to one another.

As far as BA reuse in bound cement-based forms is concerned, a major challenge is the potential for swelling of structures containing BA, with associated extensive cracking or formation of pop-outs (Pecqueur et al., 2001; Crignon et al., 1999; Alkemade et al., 1994; Müller and Rübner, 2006; Saikia et al., 2015). Such expansion has been associated with the oxidation of metallic aluminum (resulting in the generation of hydrogen gas and an increase in volume) and the formation of ettringite (through the reaction of calcium sulfates, calcium oxides, aluminum oxides, and water, resulting in a mineral of increased volume), hydration of lime and magnesium oxide (resulting in an increase in volume), and alkali-silica reactions (reaction of alkali hydroxides from cement with silica from the glassy particles in the BA, generating a swelling gel). Aluminum oxidation appears to be more important with respect to the expansion of BA mixed in cement and concrete applications (Pecqueur et al., 2001; Crignon et al., 1999; Müller and Rübner, 2006; Saikia et al., 2015). Removal or inactivation of metallic aluminum has been suggested as a potential option to overcome these swelling effects. Metallic aluminum may be oxidized by immersing the BA in an alkaline solution (Pera et al., 1997; Alkemade et al., 1994; Saikia et al., 2015) or by wet grinding (Bertolini et al., 2004). Improved removal of Al particles by eddy current separation is also a possibility. Aging and exposure to atmospheric air have been shown to decrease the reactivity of metallic Al in incineration ashes (Laenen et al., 2002; Ginés et al., 2009).

Besides the problem of expansion, the overall physical and mechanical quality of cement/concrete mixtures containing BA is lower than that of cement/concrete mixtures with natural aggregate: lower compressive and flexural strength (Pera et al., 1997; Dyer et al., 2000; Paine et al., 2000; Juric et al., 2006), higher water absorption (Pera et al., 1997; Dyer et al., 2000), and lower performance with respect to freeze–thaw cycles (Dyer et al., 2000). In many cases, however, the technical properties of the cementitious products still meet the required prescriptions. On the other hand, by appropriately controlling BA properties including particle density, metallic Al content, sulfate and alkali concentration, and loss on ignition, high-performance products can be obtained (van der Wegen et al., 2013; Keulen et al., 2016). In other cases, it may be possible to find less demanding applications for the produced construction materials, eg, specific lightweight concrete and other low-strength materials (Halliday and Dhir, 2002; Qiao et al., 2008).

Although the use of BA in cement-based applications is essentially a solidification/stabilization technique, leaching of heavy metals often remains a problem. Cai et al. (2004) reported that leaching from concrete-bound BA increased, compared with the reference concrete sample, by a factor of 50 for Cu and a factor of five for Cd, Pb, and Zn, whereas no increase was observed for Cr and Ni. Other authors (Tang et al., 2015; Saikia et al., 2015; Jansegers, 1997) reported that whereas the leaching of most contaminants is maintained well below the regulatory limits, Cu, Cr, Mo, Sb, Se, chloride, and sulfate may be problematic, probably because of the negative influence of the fine BA fraction on the hydration process of cement (Tang et al., 2015). Soluble salts have another leaching problem: K, Na, chloride, and sulfate cannot be immobilized by the hydrate system and are readily released by leaching (Dyer et al., 2000; Cai et al., 2004). Leaching of salts during the life cycle of the product is unwanted because it increases porosity, decreasing strength, and it leads to salt precipitates on the monolith surface, which often poses aesthetic problems. Again, for dedicated purposes the leaching problem should be remediable through pretreatment.

24.3.4.2 Use as a Pozzolanic Admixture

In principle, considering the typical relative content of the major oxides SiO_2, CaO, and Al_2O_3, BA may display some pozzolanic behavior in the presence of $Ca(OH)_2$. However, the pozzolanic activity of raw BA is low compared with other typical pozzolanic materials including coal fly ash and silica fume. This is likely the result of the mineralogical characteristics of the amorphous silicate and aluminosilicate phases in BA, the structure of which is probably not easily broken up by the OH^- ions released by portlandite. Some researchers have focused on activating BA with different methods (mechanical, chemical, or thermal activation) to improve the pozzolanic properties, to use BA as a substitute for other pozzolanic admixtures (Polettini et al., 2009, 2005; Bertolini et al., 2004; Lin et al., 2008; Lin and Lin, 2006). Among the chemical activation methods investigated, it appears that the use of 2% $CaCl_2$ results in improved mechanical properties of Portland cement–BA mixtures, whereas Na- and K-based activators do not seem to be capable of producing similar strength enhancement (Polettini et al., 2005b, 2009). However, it is highly probable that the efficacy of a given chemical activation method depends on the specific mineralogical properties of BA. Thermal activation also enhanced the chemical reactivity of BA in cementitious systems, with sintering and vitrification pretreatments promoting pozzolanic reactions (Lin et al., 2008; Lin and Lin, 2006). Another way to improve the pozzolanic properties of BA is to combine it with other incinerator ashes or mineral additions (Bertolini et al., 2004; Cai et al., 2004).

24.3.4.3 Use as Aggregate in Asphalt Mixtures

In the United States, solidification in bituminous structures has been extensively investigated in demonstration projects (Eighmy et al., 1997). In one project, for instance, 7% of bitumen was blended with an aggregate mix of 50% natural aggregate and 50% BA (Musselman et al., 1994). The BA-based pavement was covered by a 2.5-cm wearing surface. After 2 years of field monitoring combined with laboratory testing, it was concluded that the BA-containing mixture had the same physical properties as an asphalt mixture with 100% natural aggregate, although the first required more bitumen to decrease the porosity of the resulting product (Karpinsky et al., 2000). In a European study (Eymael et al., 1994), between 4.5% and 6.6% bitumen and up to 50% of BA was used. Leaching of inorganic contaminants was below the detection limits for all BA mixtures. Physical tests resulted in a maximum practicable load of 25% BA. Other authors (Huang et al., 2006) showed that although the technical performance (mainly in terms of durability) of asphalt mixtures to be used in road construction applications worsened in the presence of BA, their environmental behavior was acceptable in terms of both metal leachability and leachate toxicity. A study by Hassan (2005) reported that optimized asphalt mixtures met criteria for use as a bituminous surface or base road courses and satisfied moisture susceptibility and raveling potential test limits. Toraldo et al. (2013) showed that stabilized BA could be used in asphalt mixtures for road pavements and fulfilled required specifications for both technical properties and leaching behavior.

24.4 CONCLUSIONS AND RECOMMENDATIONS

Although it may appear advisable to pursue the highest level of use of BA, this is not always the most sustainable solution overall. BA processing, treatment, and recycling should be assessed according to a holistic perspective, which in turn should evaluate not only the direct advantages of the intended application, but also the emissions related to the ash treatment itself (and for provision of the energy, resources, materials, etc., used in the treatment processes) as well as potential downstream environmental savings and burdens.

Mechanical separation of metal constituents from BA should be performed in all cases where the environmental value of metal recycling is considerable and the separation process contributes to improving the properties of other ash fractions (for instance, separation of metallic Al which may otherwise be detrimental to the use of the mineral fraction).

Considering that all fractions of the bulk ash materials need to be managed, it may not necessarily be advisable to separate specific size fractions, unless this significantly improves the properties of the remaining material or enhances the recovery of valuable metals. On the other hand, extraction of easily soluble salts may be preferable, because washing with water is a relatively simple process that may be applied directly in the quenching tank and improve the leaching properties of the final BA. Chemical extraction of metals cannot be recommended if the purpose is to stabilize the ashes (because chemical extraction processes often extract only parts of the metal content); however, if the purpose is to recover metals, these processes could be relevant, provided they are selective for the target elements and can be applied at reasonable costs.

As for the stabilization options for the mineral fraction of BA, natural aging may be employed simply by outside storage of the material for several months, which indicates that the benefits of carbonation and weathering can be achieved with relatively little effort. Forced carbonation may facilitate more complete stabilization with a lower final pH (and thereby potentially lower leaching of some heavy metals); however, careful consideration of the effects on leaching of oxyanions is advised.

Chemical binding of problematic heavy metals may be considered an option in some cases; however, this type of process should generally be considered, eg, combined with washing treatments and not necessarily applicable as a separate option.

Thermal treatment offers the benefits of improved mechanical properties, reduced volume and metal leaching, as well as the destruction of organic compounds. However, thermal techniques also involve significant energy consumption and the potential loss of volatile metals (which emphasizes the need for flue gas treatment and the generation of secondary APC residues) and are generally not economically competitive over other BA-processing options unless high-value products can be obtained.

Enhanced mechanical separation of valuable BA constituents (mineral fraction, FeM, and NFeM) appears to be relatively robust with respect to the achieved separation yield and the degree of purity of the recovered metal fractions.

Combinations of various processing techniques (eg, mechanical separation and natural weathering combined with specific treatment of individual grain size fractions) may be recommended to improve the technical properties and environmental behavior of BA components, particularly in view of their use.

REFERENCES

Astrup, T., Dijkstra, J.J., Comans, R.N.J., van der Sloot, H.A., Christensen, T.H., 2006. Geochemical modeling of leaching from MSWI air-pollution-control residues. Environmental Science & Technology 40, 3551–3557.

Arickx, S., Van Gerven, T., Vandecasteele, C., 2006. Accelerated carbonation for treatment of MSWI bottom ash. Journal of Hazardous Materials 137, 235–243.

Aberg, A., Kumpiene, J., Ecke, H., 2006. Evaluation and prediction of emissions from a road built with bottom ash from municipal solid waste incineration (MSWI). Science of the Total Environment 355, 1–12. http://dx.doi.org/10.1016/j.scitotenv.2005.03.007.

Arm, M., 2004. Variation in deformation properties of processed MSWI bottom ash: results from triaxial tests. Waste Management 24, 1035–1042. http://dx.doi.org/10.1016/j.wasman.2004.07.013.

Allegrini, E., Maresca, A., Olsson, M.E., Holtze, M.S., Boldrin, A., Astrup, T.F., 2014. Quantification of the resource recovery potential of municipal solid waste incineration bottom ashes. Waste Management 34, 1627–1636. http://dx.doi.org/10.1016/j.wasman.2014.05.003.

Allegrini, E., Vadenbo, C., Boldrin, A., Astrup, T.F., 2015. Life cycle assessment of resource recovery from municipal solid waste incineration bottom ash. Journal of Environmental Management 151, 132–143. http://dx.doi.org/10.1016/j.jenvman.2014.11.032.

Arickx, S., Van Gerven, T., Knaepkens, T., Hindrix, K., Evens, R., Vandecasteele, C., 2007. Influence of treatment techniques on Cu leaching and different organic fractions in MSWI bottom ash leachate. Waste Management 27, 1422–1427. http://dx.doi.org/10.1016/j.wasman.2007.03.015.

Arickx, S., De Borger, V., Van Gerven, T., Vandecasteele, C., 2010. Effect of carbonation on the leaching of organic carbon and of copper from MSWI bottom ash. Waste Management 30, 1296–1302. http://dx.doi.org/10.1016/j.wasman.2009.10.016.

Astrup, T., 2007. Pretreatment and utilization of waste incineration bottom ashes: Danish experiences. Waste Management 27, 1452–1457. http://dx.doi.org/10.1016/j.wasman.2007.03.017.

Anjum, F., Shahid, M., Bukhari, S., Potgieter, J.H., 2013. Combined ultrasonic and bioleaching treatment of hospital waste incinerator bottom ash with simultaneous extraction of selected metals. Environmental Technology 35, 262–270. http://dx.doi.org/10.1080/09593330.2013.824992.

Alkemade, M.M.C., Eymael, M.M.T., Mulder, E., Wijs, W., 1994. How to prevent expansion of MSWI BA in road constructions? In: Goumans, J.J.J.M., van der Sloot, H.A., Aalbers, T.G. (Eds.), Environmental Aspects of Construction With Waste Materials. Studies in Environmental Science. Elsevier, pp. 863–876. http://dx.doi.org/10.1016/S0166-1116(08)71517-6.

Bethanis, S., 2004. Effect of sintering temperature on the properties and leaching of incinerator bottom ash. Waste Management & Research 22, 255–264. http://dx.doi.org/10.1177/0734242X04045426.

Berkhout, S.P.M., Oudenhoven, B.P.M., Rem, P.C., 2011. Optimizing non-ferrous metal value from MSWI bottom ashes. Journal of Environmental Protection (Irvine., Calif) 2, 564–570. http://dx.doi.org/10.4236/jep.2011.25065.

Bayuseno, A.P., Schmahl, W.W., 2010. Understanding the chemical and mineralogical properties of the inorganic portion of MSWI bottom ash. Waste Management 30, 1509–1520. http://dx.doi.org/10.1016/j.wasman.2010.03.010.

Boesch, M.E., Vadenbo, C., Saner, D., Huter, C., Hellweg, S., 2014. An LCA model for waste incineration enhanced with new technologies for metal recovery and application to the case of Switzerland. Waste Management 34, 378–389. http://dx.doi.org/10.1016/j.wasman.2013.10.019.

Biganzoli, L., Ilyas, A., van Praagh, M., Persson, K.M., Grosso, M., 2013. Aluminium recovery vs. hydrogen production as resource recovery options for fine MSWI bottom ash fraction. Waste Management 33, 1174–1181. http://dx.doi.org/10.1016/j.wasman.2013.01.037.

Baranger, P., Azaroual, M., Freyssinet, P., Lanini, S., Piantone, P., 2002. Weathering of a MSW bottom ash heap: a modelling approach. Waste Management 22, 173–179. http://dx.doi.org/10.1016/S0956-053X(01)00066-6.

Baciocchi, R., Costa, G., Lategano, E., Marini, C., Polettini, A., Pomi, R., et al., 2010. Accelerated carbonation of different size fractions of bottom ash from RDF incineration. Waste Management 30, 1310–1317.

Bodénan, F., Azaroual, M., Piantone, P., 2000. Forecasting the long-term behaviour of municipal solid waste incineration bottom ash: rapid combined tests. In: Woolley, G.R., Goumans, J.J.J.M., Wainwright, P. (Eds.), Waste Materials in Construction. Waste Managem Ser. Elsevier, pp. 475–482. http://dx.doi.org/10.1016/S0713-2743(00)80058-6.

Becquart, F., Bernard, F., Abriak, N.E., Zentar, R., 2009. Monotonic aspects of the mechanical behaviour of bottom ash from municipal solid waste incineration and its potential use for road construction. Waste Management 29, 1320–1329. http://dx.doi.org/10.1016/j.wasman.2008.08.019.

Birgisdóttir, H., Pihl, K.A., Bhander, G., Hauschild, M.Z., Christensen, T.H., 2006. Environmental assessment of roads constructed with and without bottom ash from municipal solid waste incineration. Transportation Research Part D: Transport and Environment 11, 358–368. http://dx.doi.org/10.1016/j.trd.2006.07.001.

Bertolini, L., Carsana, M., Cassago, D., Quadrio Curzio, A., Collepardi, M., 2004. MSWI ashes as mineral additions in concrete. Cement and Concrete Research 34, 1899–1906. http://dx.doi.org/10.1016/j.cemconres.2004.02.001.

Bone, B.D., Knox, K., Picken, A., Robinson, H.D., 2003. The effect of carbonation on leachate quality from landfilled municipal solid waste (MSW) incinerator residues. In: Christensen, T.H., Cossu, R., Stegmann, R. (Eds.), Proc. Sardinia 2003, Ninth Int. Waste Manag. Landfill Symp. CISA, Cagliari, Italy.

Chimenos, J., Fernández, A., Nadal, R., Espiell, F., 2000. Short-term natural weathering of MSWI bottom ash. Journal of Hazardous Materials 79, 287–299. http://dx.doi.org/10.1016/S0304-3894(00)00270-3.

Cheeseman, C.R., Monteiro da Rocha, S., Sollars, C., Bethanis, S., Boccaccini, A.R., 2003. Ceramic processing of incinerator bottom ash. Waste Management 23, 907–916. http://dx.doi.org/10.1016/S0956-053X(03)00039-4.

Crannell, B.S., Eighmy, T.T., Krzanowski, J.E., Eusden, J.D., Shaw, E.L., Francis, C.A., 2000. Heavy metal stabilization in municipal solid waste combustion bottom ash using soluble phosphate. Waste Management 20, 135–148. http://dx.doi.org/10.1016/S0956-053X(99)00312-8.

Cioffi, R., Colangelo, F., Montagnaro, F., Santoro, L., 2011. Manufacture of artificial aggregate using MSWI bottom ash. Waste Management 31, 281–288. http://dx.doi.org/10.1016/j.wasman.2010.05.020.

Chimenos, J., Segarra, M., Fernández, M., Espiell, F., 1999. Characterization of the bottom ash in municipal solid waste incinerator. Journal of Hazardous Materials 64, 211–222. http://dx.doi.org/10.1016/S0304-3894(98)00246-5.

Cornelissen, G., Gustafsson, Ö., Bucheli, T.D., Jonker, M.T.O., Koelmans, A.A., van Noort, P.C.M., 2005. Extensive sorption of organic compounds to black carbon, coal, and kerogen in sediments and soils: mechanisms and consequences for distribution, bioaccumulation, and biodegradation. Environmental Science & Technology 39, 6881–6895. http://dx.doi.org/10.1021/es050191b.

Chandler, A.J., Eighmy, T.T., Hjelmar, O., Kosson, D.S., Sawell, S.E., Vehlow, J., et al., 1997. Municipal Solid Waste Incinerator Residues.

Comans, R.N.J., Meima, J.A., 1994. Modelling CA-solubility in MSWI bottom ash leachates. In: Goumans, J.J.J.M., van der Sloot, H.A., Aalbers, T.G. (Eds.), Environ. Asp. Constr. With Waste Mater. Stud. Environ. Sci. Elsevier Science, Amsterdam, pp. 103–110.

Chimenos, J.M., Fernández, A.I., Miralles, L., Rosell, J.R., Ezquerra, A.N., 2005. Change of mechanical properties during short-term natural weathering of MSWI bottom ash. Environmental Science & Technology 39, 7725–7730. http://dx.doi.org/10.1021/es050420u.

Chimenos, J.M., Fernández, A.I., Miralles, L., Segarra, M., Espiell, F., 2003. Short-term natural weathering of MSWI bottom ash as a function of particle size. Waste Management 23, 887–895. http://dx.doi.org/10.1016/S0956-053X(03)00074-6.

Cornelis, G., Van Gerven, T., Vandecasteele, C., 2012. Antimony leaching from MSWI bottom ash: modelling of the effect of pH and carbonation. Waste Management 32, 278–286. http://dx.doi.org/10.1016/j.wasman.2011.09.018.

Cornelis, G., Van Gerven, T., Vandecasteele, C., 2006. Antimony leaching from uncarbonated and carbonated MSWI bottom ash. Journal of Hazardous Materials 137, 1284–1292. http://dx.doi.org/10.1016/j.jhazmat.2006.04.048.

Comans, R.N., Meima, J.A., Geelhoed, P.A., 2000. Reduction of contaminant leaching from MSWI bottom ash by addition of sorbing components. Waste Management 20, 125–133. http://dx.doi.org/10.1016/S0956-053X(99)00311-6.

Crignon, C., Pecqueur, G., Diaz, E.G., Germaneau, B., Siwak, J.M., 1999. Study of cement-treated MSWI bottom ash expansion. In: Mehu, J., Keck, G., Navarro, A. (Eds.), Proc. STAB ENV 99, Int. Conf. Waste Stab. Environ. 99. Societé Alpine de Publications, Grenoble, FR, pp. 64–68.

Cai, Z., Bager, D.H., Christensen, T.H., 2004. Leaching from solid waste incineration ashes used in cement-treated base layers for pavements. Waste Management 24, 603–612. http://dx.doi.org/10.1016/j.wasman.2004.01.010.

Chung, Y.-S., Moon, J.-H., Kim, S.-H., Kang, S.-H., Kim, Y.-J., 2007. Determination of the elemental composition of the bottom ash of a municipal incinerator by instrumental neutron activation analysis. Journal of Radioanalytical and Nuclear Chemistry 271, 339–344. http://dx.doi.org/10.1007/s10967-007-0214-1.

Cornelis, G., Johnson, C.A., Van Gerven, T., Vandecasteele, C., 2008. Leaching mechanisms of oxyanionic metalloid and metal species in alkaline solid wastes: a review. Applied Geochemistry 23, 955–976. http://dx.doi.org/10.1016/j.apgeochem.2008.02.001.

De Vries, W., Rem, P.C., De Keizer, M., 2012. Value creation out of MSWI bottom ash. In: Arm, M., Vandecasteele, C., Heynen, J., Suer, P., Lind, R. (Eds.), Proc. WASCON 2012. Towar. Eff. Durable Sustain. Prod. Use Altern. Mater. Constr. 8th Int. Conf. Sustain. Manag. Waste Recycl. Mater. Constr., Gotheborg (Sweden).

De Windt, L., Dabo, D., Lidelöw, S., Badreddine, R., Lagerkvist, A., 2011. MSWI bottom ash used as basement at two pilot-scale roads: comparison of leachate chemistry and reactive transport modeling. Waste Management 31, 267–280. http://dx.doi.org/10.1016/j.wasman.2010.06.002.

De Wijs, J.W.M., Cleven, R.M.F.J., 2008. Environmental Quality of Stony Construction Materials in The Netherlands.

Dijkstra, J.J., van der Sloot, H.A., Comans, R.N.J., 2006a. The leaching of major and trace elements from MSWI bottom ash as a function of pH and time. Applied Geochemistry 21, 335–351. http://dx.doi.org/10.1016/j.apgeochem.2005.11.003.

Dijkstra, J.J., Meeussen, J.C.L., Van der Sloot, H.A., Comans, R.N.J., 2008. A consistent geochemical modelling approach for the leaching and reactive transport of major and trace elements in MSWI bottom ash. Applied Geochemistry 23, 1544–1562. http://dx.doi.org/10.1016/j.apgeochem.2007.12.032.

Dabo, D., Badreddine, R., De Windt, L., Drouadaine, I., 2009. Ten-year chemical evolution of leachate and municipal solid waste incineration bottom ash used in a test road site. Journal of Hazardous Materials 172, 904–913. http://dx.doi.org/10.1016/j.jhazmat.2009.07.083.

Dijkstra, J.J., van der Sloot, H.A., Comans, R.N.J., 2002. Process identification and model development of contaminant transport in MSWI bottom ash. Waste Management 22, 531–541. http://dx.doi.org/10.1016/S0956-053X(01)00034-4.

Dijkstra, J.J., Van Zomeren, A., Meeussen, J.C.L., Comans, R.N.J., 2006b. Effect of accelerated aging of MSWI bottom ash on the leaching mechanisms of copper and molybdenum. Environmental Science & Technology 40, 4481–4487. http://dx.doi.org/10.1021/es052214s.

Dyer, T.D., Dhir, R.K., Collier, T.C., 2000. Precast concrete products produced using incinerator BA. In: Dhir, R.K., Dyer, T.D., Paine, K.A. (Eds.), Proc. Int. Symp. Sustain. Constr. Use Inciner. Ash, pp. 341–353.

European Commission, 2006. Integrated Pollution Prevention and Control. Reference Document on the Best Available Techniques for Waste Incineration.

Eusden, J.D., Eighmy, T.T., Hockert, K., Holland, E., Marsella, K., 1999. Petrogenesis of municipal solid waste combustion bottom ash. Applied Geochemistry 14, 1073–1091. http://dx.doi.org/10.1016/S0883-2927(99)00005-0.

Eighmy, T.T., Crimi, D., Whitehead, I.E., Zhang, X., Gress, D.L., 1997. The influence of monolith physical properties and integrity on diffusional leaching behavior of asphaltic pavements constructed with MSW combustion BA. In: Goumans, J.J.J.M., Senden, G.J., van der Sloot, H.A. (Eds.), Waste Mater. Constr. Putt. Theory into Pract.. Studies Environmental Science 71, 125–147.

Eymael, M.M.T., de Wijs, W., Mahadew, D., 1994. The use of MSWI BA in asphalt concrete. In: Goumans, J.J.J.M., van der Sloot, H.A., Aalbers, T.G. (Eds.), Environ. Asp. Constr. With Waste Mater. Stud. Environ. Sci., pp. 851–862.

Fruergaard, T., Astrup, T., 2011. Optimal utilization of waste-to-energy in an LCA perspective. Waste Management 31, 572–582. http://dx.doi.org/10.1016/j.wasman.2010.09.009.

Freyssinet, P., Piantone, P., Azaroual, M., Itard, Y., Clozel-Leloup, B., Guyonnet, D., et al., 2002. Chemical changes and leachate mass balance of municipal solid waste bottom ash submitted to weathering. Waste Management 22, 159–172. http://dx.doi.org/10.1016/S0956-053X(01)00065-4.

Filipponi, P., Polettini, A., Pomi, R., Sirini, P., 2003. Physical and mechanical properties of cement-based products containing incineration bottom ash. Waste Management 23, 145–156. http://dx.doi.org/10.1016/S0956-053X(02)00041-7.

Forteza, R., Far, M., Seguí, C., Cerdá, V., 2004. Characterization of bottom ash in municipal solid waste incinerators for its use in road base. Waste Management 24, 899–909. http://dx.doi.org/10.1016/j.wasman.2004.07.004.

François, D., Pierson, K., 2009. Environmental assessment of a road site built with MSWI residue. Science of the Total Environment 407, 5949–5960. http://dx.doi.org/10.1016/j.scitotenv.2009.08.007.

Funari, V., Bokhari, S.N.H., Vigliotti, L., Meisel, T., Braga, R., 2015. The rare earth elements in municipal solid waste incinerators ash and promising tools for their prospecting. Journal of Hazardous Materials 301, 471–479. http://dx.doi.org/10.1016/j.jhazmat.2015.09.015.

Fernández Bertos, M., Simons, S.J.R., Hills, C.D., Carey, P.J., 2004. A review of accelerated carbonation technology in the treatment of cement-based materials and sequestration of CO_2. Journal of Hazardous Materials 112, 193–205.

Fujimori, E., Minamoto, K., Iwata, S., Chiba, K., Haraguchi, H., 2004. Enrichment of elements in industrial waste incineration bottom ashes obtained from three different types of incinerators, as studied by ICP-AES and ICP-MS. Journal of Material Cycles and Waste Management 6, 73–79. http://dx.doi.org/10.1007/s10163-003-0106-6.

Fujimori, E., Minamoto, K., Haraguchi, H., 2005. Comparative study on the distributions of precious metals (Ru, Rh, Pd, Ir, Pt, and Au) in industrial waste incineration ashes as determined by tellurium coprecipitation and ICP-MS. Bulletin of the Chemical Society of Japan 78, 1963–1969. http://dx.doi.org/10.1246/bcsj.78.1963.

Giampaolo, C., Lo Mastro, S., Polettini, A., Pomi, R., Sirini, P., 2002. Acid neutralisation capacity and hydration behaviour of incineration bottom ash–Portland cement mixtures. Cement and Concrete Research 32, 769–775. http://dx.doi.org/10.1016/S0008-8846(01)00760-8.

Grosso, M., Biganzoli, L., Rigamonti, L., 2011. A quantitative estimate of potential aluminium recovery from incineration bottom ashes. Resources, Conservation and Recycling 55, 1178–1184. http://dx.doi.org/10.1016/j.resconrec.2011.08.001.

Ginés, O., Chimenos, J.M., Vizcarro, A., Formosa, J., Rosell, J.R., 2009. Combined use of MSWI bottom ash and fly ash as aggregate in concrete formulation: environmental and mechanical considerations. Journal of Hazardous Materials 169, 643–650. http://dx.doi.org/10.1016/j.jhazmat.2009.03.141.

Hjelmar, O., Holm, J., Crillesen, K., 2007. Utilisation of MSWI bottom ash as sub-base in road construction: first results from a large-scale test site. Journal of Hazardous Materials 139, 471–480. http://dx.doi.org/10.1016/j.jhazmat.2006.02.059.

Heinrichs, S., Wens, B., Feil, A., Pretz, T., 2012. Recovery of NF-metals from bottom ash's fine fraction – state-of-the-art in Germany. In: Proc. Venice 2012, 4th Int. Symp. Energy From Biomass Waste.

Hu, Y., Bakker, M.C.M., 2015. Recovery of aluminum residue from incineration of cans in municipal solid waste. Journal of Residuals Science & Technology 12.

Holm, O., Bandow, N., Kalbe, U., Simon, F.G., 2015. Innovative treatment trains of bottom ash (BA) from municipal solid waste incineration (MSWI) in Germany. In: Cossu, R., He, P., Kjeldsen, P., Matsufuji, Y., Reinhart, D., Stegmann, R. (Eds.), Proc. Sardinia 2015, Fifteenth Int. Waste Manag. Landfill Symp. CISA.

Hyks, J., Astrup, T., 2009. Influence of operational conditions, waste input and ageing on contaminant leaching from waste incinerator bottom ash: a full-scale study. Chemosphere 76, 1178–1184. http://dx.doi.org/10.1016/j.chemosphere.2009.06.040.

Hyks, J., Astrup, T., Christensen, T.H., 2009. Leaching from MSWI bottom ash: evaluation of non-equilibrium in column percolation experiments. Waste Management 29, 522–529. http://dx.doi.org/10.1016/j.wasman.2008.06.011.

Hjelmar, O., 1996. Disposal strategies for municipal solid waste incineration residues. Journal of Hazardous Materials 47, 345–368. http://dx.doi.org/10.1016/0304-3894(95)00111-5.

Hu, Y., Rem, P., 2009. Aluminium alloys in municipal solid waste incineration bottom ash. Waste Management & Research 27, 251–257. http://dx.doi.org/10.1177/0734242X08095564.

Hu, B., Rem, P., van de Winckel, T., 2009. Fine heavy non ferrous and precious metals recovery in bottom ash treatment. In: Proc. ISWA World Congr. 2009, Lisbon (PT).

Hill, A.R., Dawson, A.R., Mundy, M., 2001. Utilisation of aggregate materials in road construction and bulk fill. Resources, Conservation and Recycling 32, 305–320. http://dx.doi.org/10.1016/S0921-3449(01)00067-2.

Halliday, J.E., Dhir, R.K., 2002. Full scale trials using incinerator bottom ash in cement based products. In: Dhir, R.K., Dyer, T.D., Halliday, J.E. (Eds.), Proc. Int. Conf. Challenges Concr. Constr., pp. 429–438.

Huang, C.-M., Chiu, C.-T., Li, K.-C., Yang, W.-F., 2006. Physical and environmental properties of asphalt mixtures containing incinerator bottom ash. Journal of Hazardous Materials 137, 1742–1749. http://dx.doi.org/10.1016/j.jhazmat.2006.05.016.

Hassan, H.F., 2005. Recycling of municipal solid waste incinerator ash in hot-mix asphalt concrete. Construction and Building Materials 19, 91–98. http://dx.doi.org/10.1016/j.conbuildmat.2004.05.010.

Izquierdo, M., Querol, X., Vazquez, E., 2011. Procedural uncertainties of Proctor compaction tests applied on MSWI bottom ash. Journal of Hazardous Materials 186, 1639–1644. http://dx.doi.org/10.1016/j.jhazmat.2010.12.045.

Izquierdo, M., Querol, X., Josa, A., Vazquez, E., López-Soler, A., 2008. Comparison between laboratory and field leachability of MSWI bottom ash as a road material. Science of the Total Environment 389, 10–19. http://dx.doi.org/10.1016/j.scitotenv.2007.08.020.

Johnson, C.A., Brandenberger, S., Baccini, P., 1995. Acid neutralizing capacity of municipal waste incinerator bottom ash. Environmental Science & Technology 29, 142–147. http://dx.doi.org/10.1021/es00001a018.

Juric, B., Hanzic, L., Ilić, R., Samec, N., 2006. Utilization of municipal solid waste bottom ash and recycled aggregate in concrete. Waste Management 26, 1436–1442. http://dx.doi.org/10.1016/j.wasman.2005.10.016.

Jansegers, E., 1997. The use of MSWI BA in hollow construction materials. In: Goumans, J.J.J.M., Senden, G.J., van der Sloot, H.A. (Eds.), Waste Mater. Constr. Putt. Theory into Pract. Studies in Environmental Science 71, 431–436.

Jung, C.-H., Osako, M., 2007. Thermodynamic behavior of rare metals in the melting process of municipal solid waste (MSW) incineration residues. Chemosphere 69, 279–288. http://dx.doi.org/10.1016/j.chemosphere.2007.03.071.

Kirby, C.S., Rimstidt, J.D., 1993. Mineralogy and surface properties of municipal solid waste ash. Environmental Science & Technology 27, 652–660. http://dx.doi.org/10.1021/es00041a008.

Karpinsky, S., Zhang, X., Janoo, V., Gress, D., Eighmy, T., 2000. A physical evaluation of municipal solid waste grate ash as an aggregate substitute in asphaltic concrete. In: Dhir, R., Dyer, T.D., Paine, K.A. (Eds.), Proc. Int. Symp. Sustain. Constr. Use Inciner. Ash, pp. 216–226.

Keulen, A., van Zomeren, A., Harpe, A., Aarnink, W., Simons, D.J., Brouwers, H.J.H., 2016. High performance of treated and washed MSWI bottom ash granulates as natural aggregate replacement within earth-moist concrete. Waste Management 49, 83–95. http://dx.doi.org/10.1016/j.wasman.2016.01.010.

Kida, A., Noma, Y., Imada, T., 1996. Chemical speciation and leaching properties of elements in municipal incinerator ashes. Waste Management 16, 527–536. http://dx.doi.org/10.1016/S0956-053X(96)00094-3.

Laenen, B., Dreesen, R., van Rossem, D., 2002. Aggregate cement reactions in MWI-bottom ash-based concrete – a petrographical assessment. In: Challenges Concr. Constr.. Sustain. Concr. Constr., vol. 5 Thomas Telford Publishing, pp. 335–344.

Lin, K.L., Chang, W.C., Lin, D.F., 2008. Pozzolanic characteristics of pulverized incinerator bottom ash slag. Construction and Building Materials 22, 324–329. http://dx.doi.org/10.1016/j.conbuildmat.2006.08.012.

Lin, K.L., Lin, D.F., 2006. Hydration characteristics of municipal solid waste incinerator bottom ash slag as a pozzolanic material for use in cement. Cement & Concrete Composites 28, 817–823. http://dx.doi.org/10.1016/j.cemconcomp.2006.03.003.

Meima, J.A., Comans, R.N., 1999. The leaching of trace elements from municipal solid waste incinerator bottom ash at different stages of weathering. Applied Geochemistry 14, 159–171.

Meima, J.A., Comans, R.N.J., 1998. Application of surface complexation/precipitation modeling to contaminant leaching from weathered municipal solid waste incinerator bottom ash. Environmental Science & Technology 32, 688–693. http://dx.doi.org/10.1021/es9701624.

Meima, J.A., van der Weijden, R.D., Eighmy, T.T., Comans, R.N., 2002. Carbonation processes in municipal solid waste incinerator bottom ash and their effect on the leaching of copper and molybdenum. Applied Geochemistry 17, 1503–1513. http://dx.doi.org/10.1016/S0883-2927(02)00015-X.

Meima, J.A., Comans, R.N.J., 1997a. Geochemical modeling of weathering reactions in municipal solid waste incinerator bottom ash. Environmental Science & Technology 31, 1269–1276.

Meima, J., Comans, R.N., 1998. Reducing Sb-leaching from municipal solid waste incinerator bottom ash by addition of sorbent minerals. Journal of Geochemical Exploration 62, 299–304. http://dx.doi.org/10.1016/S0375-6742(97)00044-7.

Meylan, G., Spoerri, A., 2014. Eco-efficiency assessment of options for metal recovery from incineration residues: a conceptual framework. Waste Management 34, 93–100. http://dx.doi.org/10.1016/j.wasman.2013.10.001.

Meima, J.A., Comans, R.N.J., 1997b. Overview of geochemical processes controlling leaching characteristics of MSWI bottom ash. In: Stud. Environ. Sci. Waste Mater. Constr. – Putt. Theory into Pract. Elsevier, pp. 447–457. http://dx.doi.org/10.1016/S0166-1116(97)80228-2.

Muchová, L., Rem, P.C., 2006. Metal content and recovery of MSWI bottom ash in Amsterdam. WIT Transactions on Ecology and the Environment 92.

Muchova, L., Bakker, E., Rem, P., 2008. Precious metals in municipal solid waste incineration bottom ash. Water, Air, & Soil Pollution: Focus 9, 107–116. http://dx.doi.org/10.1007/s11267-008-9191-9.

Morf, L.S., Gloor, R., Haag, O., Haupt, M., Skutan, S., Di Lorenzo, F., et al., 2013. Precious metals and rare earth elements in municipal solid waste–sources and fate in a Swiss incineration plant. Waste Management 33, 634–644. http://dx.doi.org/10.1016/j.wasman.2012.09.010.

Meima, J.A., van Zomeren, A., Comans, R.N.J., 1999. Complexation of Cu with dissolved organic carbon in municipal solid waste incinerator bottom ash leachates. Environmental Science & Technology 33, 1424–1429. http://dx.doi.org/10.1021/es971113u.

Milne, C.J., Kinniburgh, D.G., van Riemsdijk, W.H., Tipping, E., 2003. Generic NICA–Donnan model parameters for metal-ion binding by humic substances. Environmental Science & Technology 37, 958–971. http://dx.doi.org/10.1021/es0258879.

Muchova, L., 2010. Wet Physical Separation of MSWI Bottom Ash. Technical University Ostrawa, Czech Republic.

Müller, U., Rübner, K., 2006. The microstructure of concrete made with municipal waste incinerator bottom ash as an aggregate component. Cement and Concrete Research 36, 1434–1443. http://dx.doi.org/10.1016/j.cemconres.2006.03.023.

Musselman, C.N., Killeen, M.P., Crimi, D., Hasan, S., Zhang, X., Gress, D.L., et al., 1994. The Laconia, New Hampshire BA paving project. In: Environ. Asp. Constr. With Waste Mater. Stud. Environ. Sci, p. 60.

Onori, R., Will, J., Hoppe, A., Polettini, A., Pomi, R., Boccaccini, A.R., 2011a. Bottom ash-based geopolymer materials: mechanical and environmental properties. In: Ceram. Eng. Sci. Proc., pp. 71–82.

Onori, R., Polettini, A., Pomi, R., 2011b. Mechanical properties and leaching modeling of activated incinerator bottom ash in Portland cement blends. Waste Management 31, 298–310.

Olsson, S., Kärrman, E., Gustafsson, J.P., 2006. Environmental systems analysis of the use of bottom ash from incineration of municipal waste for road construction. Resources, Conservation and Recycling 48, 26–40. http://dx.doi.org/10.1016/j.resconrec.2005.11.004.

Olsson, S., van Schaik, J.W.J., Gustafsson, J.P., Kleja, D.B., van Hees, P.A.W., 2007. Copper(II) binding to dissolved organic matter fractions in municipal solid waste incinerator bottom ash leachate. Environmental Science & Technology 41, 4286–4291. http://dx.doi.org/10.1021/es062954g.

Polettini, A., Pomi, R., 2004. The leaching behavior of incinerator bottom ash as affected by accelerated ageing. Journal of Hazardous Materials 113, 209–215.

Piantone, P., Bodénan, F., Chatelet-Snidaro, L., 2004. Mineralogical study of secondary mineral phases from weathered MSWI bottom ash: implications for the modelling and trapping of heavy metals. Applied Geochemistry 19, 1891–1904. http://dx.doi.org/10.1016/j.apgeochem.2004.05.006.

Polettini, A., Pomi, R., Fortuna, E., 2009. Chemical activation in view of MSWI bottom ash recycling in cement-based systems. Journal of Hazardous Materials 162, 1292–1299.

Polettini, A., Pomi, R., Carcani, G., 2005a. The effect of Na and Ca salts on MSWI bottom ash activation for reuse as a pozzolanic admixture. Resources, Conservation and Recycling 43, 403–418.

Pera, J., Coutaz, L., Ambroise, J., Chababbet, M., 1997. Use of incinerator bottom ash in concrete. Cement and Concrete Research 27, 1–5. http://dx.doi.org/10.1016/S0008-8846(96)00193-7.

Pecqueur, G., Crignon, C., Quénée, B., 2001. Behaviour of cement-treated MSWI bottom ash. Waste Management 21, 229–233. http://dx.doi.org/10.1016/S0956-053X(00)00094-5.

Polettini, A., Pomi, R., Ragaglia, M., 2005b. Accelerated ageing as a tool for sustainable disposal of incinerator bottom ash. In: Proc. Sardinia 2005, Tenth Int. Waste Manag. Landfill Symp., CISA (Cagliari, Italy).

Pfrang-Stotz, G., Reichelt, J., Roos, R., 2000. Chemical-mineralogical valuation of the leachate potential of municipal solid waste incineration (MSWI) BA. In: Waste Mater. Constr. Manag. Ser. Elsevier, pp. 975–983. http://dx.doi.org/10.1016/S0713-2743(00)80105-1.

Paine, K.A., Dhir, R.K., Doran, V.P.A., 2000. Unprocessed and processed incinerator BA as a cement bound material. In: Dhir, R.K., Dyer, T.D., Paine, K.A. (Eds.), Proc. Int. Symp. Sustain. Constr. Use Inciner. Ash, pp. 267–278.

Qiao, X.C., Ng, B.R., Tyrer, M., Poon, C.S., Cheeseman, C.R., 2008. Production of lightweight concrete using incinerator bottom ash. Construction and Building Materials 22, 473–480. http://dx.doi.org/10.1016/j.conbuildmat.2006.11.013.

Rem, P.C., De Vries, C., van Kooy, L.A., Bevilacqua, P., Reuter, M.A., 2004. The Amsterdam pilot on bottom ash. Minerals Engineering 17, 363–365. http://dx.doi.org/10.1016/j.mineng.2003.11.009.

Rocca, S., van Zomeren, A., Costa, G., Dijkstra, J.J., Comans, R.N.J., Lombardi, F., 2013. Mechanisms contributing to the thermal analysis of waste incineration bottom ash and quantification of different carbon species. Waste Management 33, 373–381. http://dx.doi.org/10.1016/j.wasman.2012.11.004.

Rendek, E., Ducom, G., Germain, P., 2006a. Influence of organic matter on municipal solid waste incinerator bottom ash carbonation. Chemosphere 64, 1212–1218. http://dx.doi.org/10.1016/j.chemosphere.2005.11.053.

Reijnders, L., 2005. Disposal, uses and treatments of combustion ashes: a review. Resources, Conservation and Recycling 43, 313–336. http://dx.doi.org/10.1016/j.resconrec.2004.06.007.

Rendek, E., Ducom, G., Germain, P., 2006b. Carbon dioxide sequestration in municipal solid waste incinerator (MSWI) bottom ash. Journal of Hazardous Materials 128, 73–79. http://dx.doi.org/10.1016/j.jhazmat.2005.07.033.

van der Sloot, H.A., Kosson, D.S., Hjelmar, O., 2001. Characteristics, treatment and utilization of residues from municipal waste incineration. Waste Management 21, 753–765. http://dx.doi.org/10.1016/S0956-053X(01)00009-5.

Šyc, M., Kameníková, P., Svoboda, K., Krausová, A., Pohořelý, M., Zach, B., et al., 2015. Resource recovery potential of the bottom ash from municipal solid waste incineration in the Czech Republic. In: Proc. Sardinia 2015, Fifteenth Int. Waste Manag. Landfill Symp.

Sormunen, L.A., Rantsi, R., 2015. To fractionate municipal solid waste incineration bottom ash: key for utilisation? Waste Management & Research 33, 995–1004. http://dx.doi.org/10.1177/0734242X15600052.

Sabbas, T., Polettini, A., Pomi, R., Astrup, T., Hjelmar, O., Mostbauer, P., et al., 2003. Management of municipal solid waste incineration residues. Waste Management 23, 61–88. http://dx.doi.org/10.1016/S0956-053X(02)00161-7.

Saveyn, H., Eder, P., Garbarino, E., Muchova, L., Hjelmar, O., van der Sloot, H.A., et al., 2014. Study on Methodological Aspects Regarding Limit Values for Pollutants in Aggregates in the Context of the Possible Development of End-of-Waste Criteria Under the EU Waste Framework Directive. Report EUR 26769 EN.

Saikia, N., Mertens, G., Van Balen, K., Elsen, J., Van Gerven, T., Vandecasteele, C., 2015. Pre-treatment of municipal solid waste incineration (MSWI) bottom ash for utilisation in cement mortar. Construction and Building Materials 96, 76–85. http://dx.doi.org/10.1016/j.conbuildmat.2015.07.185.

Triffault-Bouchet, G., Clément, B., Blake, G., 2005. Ecotoxicological assessment of pollutant flux released from bottom ash reused in road construction. Aquatic Ecosystem Health & Management 8, 405–414. http://dx.doi.org/10.1080/14634980500457724.

Tang, P., Florea, M.V.A., Spiesz, P., Brouwers, H.J.H., 2015. Characteristics and application potential of municipal solid waste incineration (MSWI) bottom ashes from two waste-to-energy plants. Construction and Building Materials 83, 77–94. http://dx.doi.org/10.1016/j.conbuildmat.2015.02.033.

Toraldo, E., Saponaro, S., Careghini, A., Mariani, E., 2013. Use of stabilized bottom ash for bound layers of road pavements. Journal of Environmental Management 121, 117–123. http://dx.doi.org/10.1016/j.jenvman.2013.02.037.

Todorovic, J., Ecke, H., 2006. Demobilisation of critical contaminants in four typical waste-to-energy ashes by carbonation. Waste Management 26, 430–441. http://dx.doi.org/10.1016/j.wasman.2005.11.011.

Van Gerven, T., Cooreman, H., Imbrechts, K., Hindrix, K., Vandecasteele, C., 2007. Extraction of heavy metals from municipal solid waste incinerator (MSWI) bottom ash with organic solutions. Journal of Hazardous Materials 140, 376–381. http://dx.doi.org/10.1016/j.jhazmat.2006.10.037.

Van Gerven, T., Van Keer, E., Arickx, S., Jaspers, M., Wauters, G., Vandecasteele, C., 2005. Carbonation of MSWI-bottom ash to decrease heavy metal leaching, in view of recycling. Waste Management 25, 291–300. http://dx.doi.org/10.1016/j.wasman.2004.07.008.

Vehlow, J., 2015. Air pollution control systems in WtE units: an overview. Waste Management 37, 58–74. http://dx.doi.org/10.1016/j.wasman.2014.05.025.

Vehlow, J., 2012. Reduction of dioxin emissions from thermal waste treatment plants: a brief survey. Reviews in Environmental Science and Bio-Technology 11, 393–405. http://dx.doi.org/10.1007/s11157-012-9296-5.

Vegas, I., Ibañez, J.A., San José, J.T., Urzelai, A., 2008. Construction demolition wastes, Waelz slag and MSWI bottom ash: a comparative technical analysis as material for road construction. Waste Management 28, 565–574. http://dx.doi.org/10.1016/j.wasman.2007.01.016.

Wei, Y., Shimaoka, T., Saffarzadeh, A., Takahashi, F., 2011. Mineralogical characterization of municipal solid waste incineration bottom ash with an emphasis on heavy metal-bearing phases. Journal of Hazardous Materials 187, 534–543. http://dx.doi.org/10.1016/j.jhazmat.2011.01.070.

van der Wegen, G., Hofstra, U., Speerstra, J., 2013. Upgraded MSWI bottom ash as aggregate in concrete. Waste and Biomass Valorization 4, 737–743. http://dx.doi.org/10.1007/s12649-013-9255-6.

Yao, J., Li, W., Xia, F., Wang, J., Fang, C., Shen, D., 2010. Investigation of Cu leaching from municipal solid waste incinerator bottom ash with a comprehensive approach. Frontiers in Energy 5, 340–348. http://dx.doi.org/10.1007/s11708-010-0131-9.

van Zomeren, A., Comans, R.N.J., 2004. Contribution of natural organic matter to copper leaching from municipal solid waste incinerator bottom ash. Environmental Science & Technology 38, 3927–3932. http://dx.doi.org/10.1021/es035266v.

Zevenbergen, C., van Reeuwijk, L.P., Bradley, J.P., Bloemen, P., Comans, R.N.J., 1996. Mechanism and conditions of clay formation during natural weathering of MSWI bottom ash. Clays and Clay Minerals 44, 546–552.

Zevenbergen, C., Comans, R.N.J., 1994. Geochemical factors controlling the mobilization of major elements during weathering of MSWI bottom ash. In: Stud. Environ. Sci. Elsevier, pp. 179–194. http://dx.doi.org/10.1016/S0166-1116(08)71455-9.

Zevenbergen, C., Van Reeuwijk, L., Bradley, J., Comans, R.N., Schuiling, R., 1998. Weathering of MSWI bottom ash with emphasis on the glassy constituents. Journal Geochemical Exploration 62, 293–298. http://dx.doi.org/10.1016/S0375-6742(97)00033-2.

van Zomeren, A., Comans, R.N.J., 2009. Carbon speciation in municipal solid waste incinerator (MSWI) bottom ash in relation to facilitated metal leaching. Waste Management 29, 2059–2064. http://dx.doi.org/10.1016/j.wasman.2009.01.005.

Zwahr, E., 2004. Ash recycling: Just a dream?. In: Proc. 12th Annu. North Am. Waste-to-Energy Conf. (NAWTEC 12), May 17-19 2004, Savannah, Georgia (USA).

Zhang, F.-S., Yamasaki, S., Kimura, K., 2001. Rare earth element content in various waste ashes and the potential risk to Japanese soils. Environment International 27, 393–398. http://dx.doi.org/10.1016/S0160-4120(01)00097-6.

Zhao, L., Zhang, F.-S., Zhang, J., 2008. Chemical properties of rare earth elements in typical medical waste incinerator ashes in China. Journal of Hazardous Materials 158, 465–470. http://dx.doi.org/10.1016/j.jhazmat.2008.01.091.

CHAPTER

COAL FLY ASH UTILIZATION FOR BORON MANAGEMENT IN SOILS, PLANTS, AND WATERS

25

C.D. Tsadilas, E. Evangelou

Hellenic Agricultural Organization, Institute of Soil Mapping and Classification, Larissa, Greece

CHAPTER OUTLINE

25.1 INTRODUCTION

Coal fly ash (CFA) is a by-product of coal combustion produced in large amounts in electric power plants. It has been regarded as a problematic waste because to its high content of potentially toxic trace elements, such as Cd, Cr, Ni, and Pb and organic compounds (eg, polychlorinated biphenyls, polycyclic aromatic hydrocarbons). However, because it contains appreciable amounts of essential plant nutrients such as Ca, Mg, K, P, S, boron (B), Fe, Cu, and Zn, it may be used beneficially in agriculture. Uncontrolled land disposal of CFA is likely to cause pollution with the previously mentioned toxic

Environmental Materials and Waste. http://dx.doi.org/10.1016/B978-0-12-803837-6.00025-1

substances. B is a trace element essential to plant growth but it may be toxic to plants if its concentration exceeds threshold values in soils and irrigation waters. Because B is contained in appreciable amounts in CFAs, their management must take into account B changes caused by the application of CFA in soil from the point of view of correcting B deficiency or avoiding causing B toxicity in plants. Owing to CFA's affinity to B, it may be also used to remove B from water with a high B content, improving its quality and making it suitable for irrigation. In this chapter, issues related to the B content of CFA and its management with respect to this element are reviewed.

25.2 COAL FLY ASH

Because coal continues to be one of the most important sources of energy production in the world, management of its by-products is a serious and challenging issue. Coal combustion results in the production of solid products traditionally considered wastes; more accurately, however, they are classified as coal combustion products (CCPs). Among the several definitions devoted to CCPs, the following one seems to describe these materials in the simplest and clearest way (Heidrich et al., 2013): CCPs are secondary by-products derived from coal combustion, including FA, bottom ash, boiler slag, fluidized-bed combustion ash, and flue gas desulfurization material produced from the combustion of coal or the cleaning of stalk gases. CFA is the finer ash produced in coal-fired power stations, which is collected using electrostatic precipitators. It constitutes about 85% of the whole ash produced in coal combustion. The estimated annual global CFA production is about 750 million tons. India, China, and the United States are the highest producers (with 118, 100, and 75 million tons/year, respectively). In Europe, the amount of CFA produced is smaller; it is distributed among in decreasing order as: Germany (40 million tons/year), the United Kingdom (15 million tons/year), and Denmark, Italy, and The Netherlands (2 million ton/year each) (Shaheen et al., 2014). Properties of CFA (mineralogical, physical, and chemical) vary widely depending on the conditions under which they were produced, the boiler type, and the gas emission control system. These properties are the driving factors in determining the way they are managed and their impact on the environment. There are several reasons in favor of the reuse of CFA. Among them are the minimal cost of its disposal, the preservation of land for other uses, the gain in financial returns directly from the sale of by-products or indirectly by offsetting processing and disposal costs and saving natural resources through replacing some that are scarce or expensive (Ahmaruzzaman, 2010). Based on this philosophy, several uses for CFAs have been tried in many sectors of industrial, agricultural, or environmental activities. Thus, CFAs are reused to produce cement and concrete products, fill, and cover materials; for roadway and pavements construction; for soil amendment as liming materials; to add essential elements to plant growth; as adsorbents of toxic compounds such as sulfur oxide and nitrogen oxide for flue gas desulfurization; to remove their toxic heavy metal retention; and thus to protect their dispersion into the environment. All of these uses have been reviewed more or less extensively.

25.3 COAL FLY ASH CHARACTERISTICS

CFA is a complicated material owing to its origin and the conditions under which it is produced. CFA is produced during the burning of pulverized coal in thermoelectric power stations and is collected by electrostatic precipitators. The type of coal type and technological processes are the main factors that determine FA characteristics. The original coal materials, which are combusted at 1200–1700°C, contain a huge number of mineral groups on which the composition of FA depend. The produced FA is a complex organic–inorganic mixture with a variable composition containing many associated solid, liquid, and gaseous components in varying sizes as major (1%), minor (1.0–0.1%) and traces (<0.1%). There are four types, or ranks, of coal, varying in heating value, chemical composition, ash content, and geological origin. The four types of coal are anthracite, bituminous, subbituminous, and lignite. In addition to being handled in a dry, conditioned, or wet form, FA is also sometimes classified according to the type of coal from which the ash was derived (Ahmaruzzaman, 2010). Fly ash (FA) in general is classified as Class F and Class C. Class F ash is derived from the burning of bituminous coals and anthracites and is pozzolanic (ie, hardening when reacted with $Ca[OH]_2$ and water) and contains more than 70% $SiO_2 + Al_2O_3 + Fe_2O_3$. Class C FA produced from burning lignites and subbituminous coals contains 50–70% of these oxides and has both pozzolanic and cementitious properties (hardening when reacted with water) (Vassilev and Vassileve, 2007).

25.4 ELEMENTAL COMPOSITION OF COAL FLY ASH

To a great degree, the elemental composition of CFA determines its management; from this point of view it is necessary to know it in detail. CFA contains a large number of elements, some of which are essential to plant growth but others of which, especially heavy metals, are toxic and harmful to living organisms. Among those which are essential elements are the macronutrients phosphorus (P), potassium (K), calcium (Ca), and magnesium (Mg) and the metallic microelements zinc (Zn), copper (Cu), manganese (Mn), and iron (Fe). Toxic metals contained in appreciable amounts in CFA are arsenic (As), cadmium (Cd), chromium (Cr), nickel (Ni), and mercury (Hg). Table 25.1 lists the range of basic elements contained in CFA.

From the data in Table 25.1, it is obvious that CFA may be a considerable source of essential macro- and micro-elements but at the same time a serious source of environmental pollution with toxic heavy metals in the case of its reuse for agricultural purposes.

Among elements of special importance in CFA is a trace element that is essential for plant growth but toxic to some plant crops, ie, boron (B). The B content of CFAs is appreciable; special care should be paid to this element when CFA is to be used in agriculture. The B content of CFA is presented in Table 25.2.

Table 25.1 Elemental Composition of Coal Fly Ash

Essential Macronutrients		Essential Micronutrients		Toxic Heavy Metals	
	$g\ kg^{-1}$		$mg\ kg^{-1}$		$mg\ kg^{-1}$
P	0.0024–10.8	Zn	10–3500	As	0.1–6300
S	0.0013–15	Cu	0.002–2800	Cd	0.01–618
K	0.0021–10.8	Mn	0.006–4000	Cr	5–3000
Ca	0.029–500	Fe	0.0032–550	Ni	6.3–4300
Mg	0.017–11.5			Hg	0.02–1.0

Compiled from Shaheen, S.M., Hooda, P.S., Tsadilas, C.D., 2014. Opportunities and challenges in the use of coal fly ash for soil improvements: a review. Journal of Environmental Management 145, 249–267.

Table 25.2 Boron Content of Coal Fly Ash

Origin of Coal Fly Ash	Class of Fly Ash	Boron Content, $mg\ kg^{-1}$	References
Australia	F	12–49	Manoharam et al. (2010)
		29	Tripathi et al. (2004)
Australia	C	56–127	Manoharam et al. (2010)
Australia	F	7.4–25	Yunusa et al. (2006)
Japan	F	1050	Hayashi et al. (2010)
Japan	N (cement and natural pozzolan (class N)	540	Hayashi et al. (2010)
United States	–	100–168	Page et al. (1979)
	–	50	Adriano et al. (1980)
India	–	28.9	Pandey et al. (2009)
France	–	0.4	Lopareve-Pohu et al. (2010)
Laughlin, Nevada, United States	–	725	Ogava et al. (2014)
Japan	–	41–520	Iwasita et al. (2005)
China	F	96–880	Noda and Ito (2007)
Indonesia	–	690–1100	Noda and Ito (2007)
Australia	–	52–690	Noda and Ito (2007)
Japan	–	8.1	Polat et al. (2004)
South Africa	–	240	Kashiwakura et al. (2009)
Japan	–	540	Tagliabue et al. (2014)

25.5 BORON

B is a nonmetallic element with an intermediate behavior between metals and nonmetals, similar in some aspects to aluminum, carbon, and silicon. In nature, B is not found in an elemental state but rather in the form of boric acid [H_3BO_3 or $B(OH)_3$] and borates or borosilicates. Boric acid is water soluble (solubility of 55 $g\ L^{-1}$ at 298 K) and performs as weak Lewis acid accepting electrons according to the reaction: $B(OH)_3 + H_2O \leftrightarrow B(OH)_4 + H^+$ (Wang et al., 2014). Its concentration in the environment is low; in the earth's crust it is on average 10 $mg\ kg^{-1}$; in soils, 30 $mg\ kg^{-1}$; in groundwater, 0.3–100 $mg\ L^{-1}$; and in seawater, about 4.5 $mg\ L^{-1}$ (Scorei, 2012). B's importance lies in the fact that it is an essential element for plants, animals, and humans and causes serious effects in living organisms when it is at abnormal levels, ie, deficiency or toxicity. Concentrations of B which may cause deficiency and toxicity are close. In aqueous solutions B exists in several forms, depending on its concentration and pH. In concentrations of B occurring in nature, main B species are $B(OH)_3$ at low pH and $B(OH)_4$ at high pH. For several reasons, B separation from aqueous solutions is closely associated with the chemical behavior of B species, such as borate formation in the presence of strong electrolytes and the tendency to form stable esters. However, B behavior and its environmental fate still remain unclear and can be explained only roughly through the chemistry of boric acid. Deeper knowledge is needed on B speciation in the environment including soil, waters, and sediments (Wang et al., 2014). The capability of the formation of complexes is important for B separation from aqueous solutions.

25.6 IMPORTANCE OF BORON FOR LIFE ON EARTH

Borates are important prebiotic reagents that probably have contributed to the evolution of prebiotic structures. It is believed that they initially protected and stabilized particular chemical conformers and later became chemical messengers in the establishment of the cell membrane structure and cytoskeleton. Some authors suggest that B contributed to the appearance of life on earth through its involvement in the formation of furanose forms of ribose, a borate sugar, which is considered to be a chemical structure that led to the appearance of the first genetic material on earth. The furanosyl borate ester is considered the most important borate sugar in the building process of the "RNA world" which is the bridge to the "biological world" (Neilsen, 2014). In any case, today it is clearly accepted that B is an essential element that completes the life cycle of organisms in all phylogenetic kingdoms (Matoh, 1997), as discussed in more detail subsequently.

25.7 IMPORTANCE OF BORON FOR PLANTS

B was proved to be an essential element for plant growth in 1923–1926 in experiments carried out by Warington, Somer, and Lipman (Bryson et al., 2014). Since then, much research has shown the special role of B in the physiology of plants. Briefly, this role is related to meristematic growth involved in cell division and elongation. Boron is necessary for the synthesis of uracil, a component of RNA. When B is inadequate, the rate of cell division is reduced. Moreover, B affects the growth of pollen tubes by increasing sugar absorption and respiration. Several other roles are attributed to B, including complexing with enzymes and coenzymes involved in certain metabolic pathways (Pandey and Archana, 2011). Consequently, B is a special concern for agriculture, with the main interest focused on B deficiency problems affecting crop yields especially in humid areas of the world, but also on B toxicity in arid or semiarid areas. Table 25.3 lists threshold values of B deficiency and toxicity in plants.

Table 25.3 Threshold Values of Boron Deficiency and Toxicity in Plants

Plant	Deficiency	Toxicity	Sufficiency Range	Vegetative Stage	References
Catharanthus roseus L.	57 60.4	79 68.9		Vegetative growth Reproductive growth	Sinha et al. (20090
Bittergourd (*Momordica charantia* L.).	5.4	120		Young leaves	Voss (1993)
Corn	<7		7–25	3- to 4-leaf stage, whole plant	Gascho (1993)
Corn	<4		4–25	Ear leaf at silk	Gascho (1993)

Continued

Table 25.3 Threshold Values of Boron Deficiency and Toxicity in Plants—cont'd

Plant	Deficiency	Toxicity	Sufficiency Range	Vegetative Stage	References
Sugarcane	<4	>45	0.4–34	Leaf blade laminas	Ulrich et al. (1993)
Sugar beet	12–40	420		Mature blade	Smith et al. (1993)
Peanuts	<20				Cassman (1993)
Cotton	<10–28	>1000		Seasonal	Wilcox (1993)
Tomato	<15		>35	Expanded leaves in fruit accumulation period	Hall and Schwartz (1993)
Common bean	<15		15–50	Upper, fully developed leaves, before flowering	Ulrich (1993)
Potato	1.2–20		>20	Blade	Hanson (1993)
Apples	<20	>140	20–60	Mature leaves	Wutscher and Smith (1993)
Pears	<15		20–60	Mature leaves	Wutscher and Smith (1993)
Citrus	<25	>200	25–200	4- to 6-month-old leaves on nonfruiting twigs	Johnson (1993)
Peach	<18	>100	20–80	Leaves taken in April or May	Gartel (1993)
Nectarine	<18	>100	20–80	Leaves taken in April or May	Gartel (1993)
Grapes	<15	>60	26–40	Leaf blades	Kabu et al. (2015)

25.8 IMPORTANCE OF BORON FOR ANIMALS

Besides the essentiality of B for plant growth, today is acceptable that B is also essential for the completion of the life cycle for organisms in all phylogenetic kingdoms. It has been shown that B affects many enzymes, energy metabolism, mineralization, and Ca, P, Mg uptake. Boron mineral compounds help in optimizing the performance of an organism, correction of bone structure disorders, and in reduction of cholesterol and triglyceride levels. Boron also affects the activity of vitamin D and some disorders related with its deficiency (Nielsen, 2000). Experiments in animals and humans have shown that B is a bioactive element that affects positively bone growth, function of nervous system, alleviates arthritic symptoms, facilitates hormone action and reduces the risk of some types of cancer. These effects are made through the influence on several biochemical processes such as the formation of bioesters in biomolecules. A dietary recommendation for consumption of foods providing of an intake of B 1.0 mg/day is believed that has health benefits (Gupta, 1993).

25.9 BORON IN THE ENVIRONMENT

25.9.1 BORON SOURCES

The main sources of B in soils are tourmaline, micas, and B bound within the clay structure, adsorbed on clay surfaces and edges on hydroxylated surfaces, and B complexed by soil organic matter (Shorrocks, 1991). Main sources of these B-containing substances are rocks and soils, borate fertilizers, sewage sludge, composts, irrigation water, and CFA and coal mines. The average total B concentration in the earth's crust is about 10 mg kg^{-1}; the total B concentration in soils varies from 2 to 100 mg kg^{-1} with an average of about 30 mg kg^{-1}. Higher values are found in soils derived from volcanic rocks compared with plutonic rocks but show great variability owing to the volatility of H_3BO_3 and the ease of dissolving B and its movement in hot water in later stages of igneous activity (Shorrocks, 1991). Usual chemical B-containing fertilizers employed in agriculture are solubor and foliarel applied as a foliar spray, sodium tetraborate pentahydrate ($Na_2B_4O_7 \cdot 5H_2O$) applied in soil, boron frits, colemanite and ulexite also applied also to soil, and borax (sodium tetraborate decahydrate, $Na_2B_4O_7 \cdot 10H_2O$) and boric acid (H_3BO_3), suitable for both soil and foliar application (Tsadilas et al., 1994).

25.9.2 BORON IN SOILS

From the total B content of soils, only a small percentage (less than 5%) is available to plants; the rest is in unavailable forms. A number of soil factors affect B uptake by plants, including B forms in soil in relation to soil properties. Tsadilas et al. (1994) following the B fractionation procedure suggested by Jin et al. (1987), fractionated soil B in a large number of Greek soils and tested the availability of B fractions to olive trees and barley. The fractions of B determined were in a soil solution extracted with cold water, nonspecifically adsorbed, specifically adsorbed, occluded in Mn oxides, occluded in crystalline Fe-Al oxyhydroxides, occluded in amorphous Fe-Al oxyhydroxides, and a residual fraction calculated as the difference between the sum of all these fractions and total B. Multivariate regression suggested that plant B uptake was best explained by soil solution B, specifically adsorbed B and B occluded in Mn oxides. Hot water extraction, which is considered to be the best available B estimation method, seemed to extract B from the soil solution and specifically adsorbed B, but not from Mn oxides occluded by B. However, this later fraction was correlated with B uptake by olive trees but not by barley, which suggested that the availability of soil B fractions depends on the plant species. Several methods of assessing available soil B have been proposed, among which the hot water extraction method is considered the most effective. Tsadilas et al. (1997) and Goldberg (1997) tested several procedures extracting available soil B from kiwifruit, a plant crop sensitive to B toxicity. Their results showed that soil B corresponding with B toxicity in kiwifruit was 0.51 mg B kg^{-1} for the hot water extraction procedure, 0.80 mg B kg^{-1} for mannitol extractable B, and 0.18 mg B kg^{-1} for dilute HCl extractable B.

It is of important to know that the range between deficient and toxic B concentration in soils is small, much more so than for any other nutrient element. Soil factors affecting availability are pH, texture, moisture, temperature, organic matter content, and clay mineralogy (Brown and Shelp, 1997). B mobility in plants is restricted in some plants but in others it is mobile. This affects the expression of

B nutrition imbalance in plants and its management (Bell, 1997). Soil and plant analysis are effective tools in predicting B deficiency (Bennett, 1993). A general guideline for critical, sufficient, and toxic levels for B in plant nutrients is: critical, <10 mg kg^{-1} d.m. sufficient 10–100 mg kg^{-1}; and toxic, >100 mg kg^{-1} (Tsadilas and Chartoulakis, 1999).

Table 25.4 lists critical B concentrations for the prognosis of B deficiency and toxicity in soils determined by several methods used for various agricultural crops.

Table 25.4 Critical Soil Boron Concentrations for Prognosis of B Deficiency and Toxicity

Plant Species	Soil B Concentration, mg kg^{-1}	Method of Determination	References
Wheat (*Triticum aesticum*)	0.12–0.15 (acid soil)	HwsB[a]	Bennett (1993)
Wheat	0.32–0.38 (alkaline soil)	HwsB[a]	Bennett (1993)
Barley			Tsadilas et al. (1994)
Olive trees	0.33	HwsB[a]	Depnath et al. (2015)
	0.17	HCl-B[b]	
	0.05	Man-B[c]	
	0.41	Hydroxylamine-B[d]	
	0.14	Ammonium oxalate-B[e]	
Apple trees	<1.20 deficiency	Res-B[f]	Depnath et al. (2015)
Variety black	>5.01 toxicity	Hws-B[a]	
Variety golden	>5.19 toxicity	Res-B[f]	
	>4.77 toxicity	Man-B[c]	
	>4.27 toxicity	HCl-B[b]	
	>5.93 toxicity	Sat-B[g]	
	<1.03 deficiency	Res-B[f]	
	>6.19 toxicity	Hws-B[a]	
	>5.93 toxicity	Res-B[f]	
	>6.52 toxicity	Man-B[c]	
	>5.19 toxicity	HCl-B[b]	
Kiwifruit	>0.51 toxicity	HwsB[a]	Goldberg (1997)
	>0.18 toxicity	HCl-B[b]	
	>2.0 toxicity	Res-B[f]	
	>0.80 toxicity	Man-B[c]	
	>2.0 toxicity	Res-B[f]	
Cowpea (*Vigna unguiculat*)	0.48 34.5–plant	Hws-B[a]	Debnath and Ghosh (2014)
Pea (*Pisum sativum*)	0.55 24.5–plant	Hws-B[a]	Ahmed et al. (2013)
Cotton (*Gossypium hirsutum* L.)	0.47	HCl-B[b]	EPA (2008)

[a]*Boiling for 5 min in 0.001 M $CaCl_2$.*
[b]*Extraction for 5 min with 0.005 M HCl.*
[c]*Extraction with mannitol in 0.001 M $CaCl_2$.*
[d]*Hydrochloric-hydroxylamine in 0.01 M HNO_3 extraction for 30 min (Jin et al., 1987).*
[e]*Ammonium oxalate (pH 3.25) extraction for 4 h (Jin et al., 1987).*
[f]*Extraction with resin–amberlite IRA 743.*
[g]*Saturation extract*

25.9.3 BORON IN WATER

In several regions of the world, especially in arid and semiarid ones, because of natural climatic conditions, human pressures, and irrational water management there is high water stress. Often water sources experience salinization to different degrees. One inorganic constituent that may cause serious problems is the B content. A high B content was found in groundwater (up to 8 mg L^{-1} in Italy, Cyprus, and Greece), sweater (4.7 mg L^{-1}), and domestic wastewater (0.5–2 mg L^{-1}) (Kashiwakura et al., 2009). High B in drinking water may be toxic to humans and animals, and may cause various health problems under long-term exposure such as reproductive difficulties in animals (Weinthal et al., 2005). The World Health Organization derived a provisional guideline value of 0.5 mg L^{-1} using a tolerable daily intake value of 0.16 mg B kg^{-1} day^{-1} (Weinthal et al., 2005). The European Union standard for B content in drinking water is 1 mg L^{-1} (Tsadilas, 1997).

The quality of irrigation water seriously affects both crop yield and soil quality. The suitability of waters for irrigation is evaluated using criteria that determine the potential of water to injure plants and decrease crop yields. Irrigation of soils with a high B content is a main factor that may cause B toxicity. Tsadilas et al. (1997) reported that irrigation of corn and cotton with treated municipal wastewater with a high B concentration (>1.0 mg L^{-1}) caused soil B contamination and increased its concentration in soil from less than 1.0 mg kg^{-1} to 2.0 mg kg^{-1} soil, a value toxic for plants sensitive to B, but not for cotton and corn. The B content of irrigation water is of significance, and extended research into plant tolerance to B has been carried out, resulting in establishing threshold concentration values in soil water differentiating plants into classes according to their sensitivity to B content in soil water. A widely accepted classification of crops according to their tolerance to B content in soil solutions at field capacity moisture conditions is the following (Jame et al., 1982): very sensitive, <0.5 mg L^{-1}; sensitive, 0.5–1.0 mg L^{-1}; 1.0–2.0 mg L^{-1}; moderately tolerant, 2.0–4.0 mg L^{-1}; tolerant, 4.0–6.0 mg L^{-1}; and very tolerant, >6.0 mg L^{-1}. These values are not used as B criteria for irrigation water because the in situ levels of B in soil solution differ from B levels in irrigation water; nevertheless, they show that soil solution B at equilibrium is of most concern and may be used to establish B limits for irrigation waters (Nable et al., 1997). Other investigators accept different criteria for the safety of B concentration in waters for plants. Thus, Nable et al. (1997) and Rhoades et al. (1970) consider safe concentrations of B in irrigation waters for sensitive plants to be 0.3, 1, or 2 mg B L^{-1} for semitolerant plants, and 2–4 mg B L^{-1} for tolerant plants. Before soils with a high B content can be used for agriculture, their available B concentration must be reduced to nonphytotoxic levels. Several methods have been tried for detoxifying soils with a high B content (Bingham et al., 1971; Kukier and Sumner, 1996). However, those efforts led to the conclusion that detoxification of high B content soils is practically ineffective. Therefore, serious attention has to be paid to protect soils from increasing in B content. Methods for detoxifying waters from B are discussed subsequently.

25.10 BORON IN COAL FLY ASH

The elemental composition of CFA varies widely. Many of the elements contained are essential for plant growth, such as macronutrients, ie, K, Mg, P, Ca, and S or micronutrients, ie, Zn, Mn, Fe, and Cu. One important microelement related to plant growth contained in appreciable amounts in CFA is B. Table 25.5 lists some of the existing literature related to the B content of CFA.

Table 25.5 Boron Content of Coal Fly Ash

Origin of Coal Fly Ash	Class of Fly Ash	Boron Content, mg kg^{-1}	References
Australia	F	12–49	Manoharam et al. (2010)
		29	Tripathi et al. (2004)
Australia	C	56–127	Manoharam et al. (2010)
Australia	F	7.4–25	Yunusa et al. (2006)
Japan	F	1050	Hayashi et al. (2010)
Japan	N (cement and natural pozzolan (class N)	540	Hayashi et al. (2010)
United States	–	100–168	Page et al. (1979)
	–	50	Adriano et al. (1980)
	–	28.9	Pandey et al. (2009)
	–	0.4	Lopareve-Pohu et al. (2010)
Laughlin, Nevada, United States	–	725	Ogava et al. (2014)
Japan	–	41–520	Iwasita et al. (2005)
China	F	96–880	Noda and Ito (2007)
Indonesia	–	690–1100	Noda and Ito (2007)
Australia	–	52–690	Noda and Ito (2007)
Japan	–	8.1	Polat et al. (2004)
South Africa	–	240	Kashiwakura et al. (2009)
Japan	–	540	Tagliabue et al. (2014)

From the data in Table 25.5, it is obvious that the B content in most CFAs, especially those classified as C, is high. As already mentioned, in addition to B being an essential element for plant growth, it is often cited as the most phytotoxic trace element contained in CFA. Repeated application of CFA in soils may increase B concentration at toxic levels. Kukier and Sumner (1996) and Johnson and Fixen (1990) tested the availability of B to corn and spinach in pot experiments using five different CFAs. These CFAs varied widely in pH (from 5.23 to 12.05) and available B content (extracted with hot water) (from 8 to 142 mg B kg^{-1}). They found that addition of CFA to an acid sandy loam soil at rates up to 200 Mg ha^{-1} increased the soil pH only at high CFA rates (50–100 g kg^{-1}). Available B in soil increased significantly but at different levels. Some ashes increased available B from 0.23 to 28.6–30.8 mg kg^{-1} at the highest rates (100 g kg^{-1}) whereas others had a more benign effect, increasing it to 1.28–3.80 mg B kg^{-1}). Available B (hot water–extractable) below 0.5 mg kg^{-1} is considered low or insufficient for many crops whereas values over 5 mg B kg^{-1} are toxic to plants (Warren, 1992). In all cases, available B in soil correlated negatively with soil pH. The concentration of B in corn dry matter was elevated above 100 mg kg^{-1}, which is considered toxic at the rate of 25 Mg ha^{-1}. At higher rates, B concentration strongly increased, reaching 2633 mg B kg^{-1} d.m., which is associated with a very large decrease in yield. Corn plants with these high B concentrations had typical B toxicity symptoms. However, corn dry matter might be decreased in all cases owing to the increase in electrical conductivity over the threshold values for corn (1.7–4.0 dS m^{-1}) and its concentration in plant tissues. Simultaneous application of $Ca(OH)_2$ in spinach reduced B uptake. In the spinach experiment it was shown that when CFA was applied with $Ca(OH)_2$, B uptake by plants was reduced (the highest rate was around 100 mg B kg^{-1} d.m.) significantly compared with the CFA addition without $Ca(OH)_2$ (the highest rate was more than 250 mg B kg^{-1}). Manoharam et al. (2010) worked with four acid soils (pH 5.02–6.57) with different-textured (sand, loam, sand clay loam, and clay loam) Australian soils and tested the use of two CFAs, one acid and the other alkaline, in canola plants. In two pot experiments, they grew canola plants after incorporating CFAs into the soils at rates up to 625 Mg ha^{-1} in the case of alkaline CFA and 108 Mg ha^{-1} in the case of acid CFA. In the case of alkaline CFA, B concentration in leaves reached a phytotoxic threshold (110–170 mg kg^{-1}) at only the highest rate, but in the case of acid CFA, although available B in soil increased B concentration, plant tissues did not reach the phytotoxic threshold value. They concluded that rates up to 36 Mg ha^{-1} may not cause phytotoxicity and excessive B accumulation in soils.

25.11 BORON FORMS IN COAL FLY ASH

As stated, the addition of CFAs to soils may increase available B at phytotoxic levels, causing detrimental effects to plants and soils. Therefore, B elution from CFAs is important. The toxicity of inorganic pollutants depends on their chemical state rather than on total concentrations because these forms regulate their solubility and therefore their availability to plants. Thus, knowledge of the chemical state of B in CFAs is of particular significance. Hayashi et al. (2010) investigated the chemical state of B in CFA, which controls its reaching to the environment using focused ion-beam time-of-flight secondary ion mass spectrometry and satellite-transition magic angle spinning nuclear magnetic resonance spectrometry. They found that B in CFA with low leaching characteristics appears as trigonal BO_3 coexisting with Ca and Fe at the outer layers of particles and inside CaO-MgO of aluminosilicate particles. B which may be leached more easily appears as CaO or MgO trigonal BO_3 and tetragonal BO_4 distributed only on the outer surface of the aluminosilicate particles without correlation to any particular element.

25.12 USE OF COAL FLY ASH WITH RESPECT TO BORON

25.12.1 AS SOIL AMELIORANT TO IMPROVE SOIL FERTILITY

FA may be acidic or alkaline and can be used to buffer soil pH. In acid soils, lime of FA reacts readily with acidic components in soil and releases nutrients such as S, B, and Mo in available form and in quantities beneficial to crop plants. However, the amount of FA to be applied must be calculated taking into account the risk of increasing B to toxic levels. Warren (1992) and Keren and Bingham (1985) applied FA in a strongly acidic (pH 4.5), finely textured soil (clay content, 41%) at rates 0.5, 1.5, and 2 times of the liming rate (amount of lime to increase soil pH to 6.5 determined by the SMP buffer method), respectively, and grown in amended soil corn, wheat, and soybeans. At the end of the respective experiment they found that soil pH increased gradually to 5.9 at the higher FA rate. Total B increased from 68 mg kg^{-1} in the control to 100 mg B kg^{-1} in the highest rate of FA. Available B ranged from 0.5 mg kg^{-1} in the untreated soil to 12 mg kg^{-1} at the highest FA rate. Concentrations of available B (extracted with hot water) exceeding 3–5 mg kg^{-1} are considered phytotoxic (Tsadilas et al., 2014). These high soil B concentrations confirm that excess B is the major limiting factor to plant growth when we apply FA. Boron in the dry matter of wheat increased for a little while in soybean and corn tissues increased to toxic levels at the highest FA rate. The highest safe FA quantity for increasing the soil pH below the threshold toxic value of B was found to be around 85 ton ha^{-1}. Tsadilas et al. (2014) and Matsi and Keramidas (1999) conducted a field experiment in central Greece with durum wheat grown on an acidic (pH 5.25) clay, loamy Typic Haploxeralfs soil amended with an alkaline FA (pH 12.2) from a lignite-fired electric power plant located in Northern Greece, aged for 3 months before its use for 3 months by being maintained in open air and leaching periodically with deionized water. FA rates were 5.56 up to 12.6 ton ha^{-1} and applied alone or with sewage sludge (SS) and mineral N-P fertilizers. Results showed that FA and SS increased both the yield and biomass of wheat. The soil pH increased to 7.25–7.50 at the highest FA rate alone and with SS, respectively. Electrical conductivity also increased significantly but at unrestricted levels for wheat (from 127 μmhos cm^{-1} in the control to 367 μmhos cm^{-1} in the treated wheat, including the highest FA and SS rate). Available P increased substantially (from 18.2 mg kg^{-1} in the control to 51.7 mg kg^{-1} with the highest FA rate). Potassium was not affected by the treatments applied. Available B (extracted with hot water) increased significantly from 0.77 in the control up to 1.70 mg B kg^{-1} with the treatments including FA and SS. These values are considered phytotoxic, as mentioned earlier.

Matsi and Keramidas (1999) applied two Red Mediterranean acid (pH 4.7 and 5.8) soils (an Ultic Haploxeralf and a Typic Haploxeralf) to two alkaline FAs (pH 8.5 and 8.9) with available B (hot water extractable) 2.8–3.1 mg kg^{-1}) in a pot experiment at rates up to 50 g kg^{-1} soil and grew in them ryegrass. They found that soil pH increased up to 8.0 at the highest FA rate and electrical conductivity up to 2.5 dS m^{-1}. Available B increased but remained below 1.0 mg kg^{-1} whereas available P was not significantly affected. Ryegrass yield significantly increased. They concluded that FA with relatively low B content can be used as a liming material in acid soils at rates not exceeding 40 Mg ha^{-1}. CFA applied to soils also may change soil physical properties. Fail and Wolchok (1977) reported that FA application at a rate of 70 ton ha^{-1} altered the texture of sandy and clay soil to loamy soils. Page et al. (1979) found that the addition of FA decreases the bulk density of soils, improves soil porosity and workability, and enhances water retention capacity. In addition Chang et al. (1977) showed that water-holding capacity increased owing to the addition of FA as well as the hydraulic conductivity of soils, and helped reduce surface encrustation.

In a field experiment with rice cultivars grown in excessively drained, loamy, fine, sandy and silty loamy acid soils (pH 5.2–5.7), Lee et al. (2005) in studied the effects of an alkaline FA (pH 10.1) at rates up to 120 Mg ha^{-1} on rice productivity and soil properties. They found that FA resulted in a significant rice yield, which was maximal at the rate of 90 Mg ha^{-1}. FA increased available S and P. Available B content of soil increased to a maximum of 2.57 mg B kg^{-1} and B content in rice plants increased to a maximum of 53 mg B kg^{-1} at the highest FA rate without showing toxicity symptoms.

25.13 USE OF COAL FLY ASH FOR BORON REMOVAL FROM WATER AND WASTEWATER

Boron occurs naturally everywhere in the environment, including oceans. Depending on the pH in aqueous solutions it contains boric acid and borate ions. B in the forms of boric acid and boron salts is extensively used in industries such as in glass manufacturing, wire drawing, cosmetics, photographic chemicals, fireproofing fabrics, and weatherproofing wood. Boric acid is also used as a disinfectant and food preservative, and borax is used extensively in the welding and brazing of metals. In addition, boron compounds are used for hand cleansing, high-energy fuels, cutting fluids, and catalysts (Hollis et al., 1988). Some investigations were carried out on the possibility of using CFAs for boron adsorption. Hollis et al. (1988) investigated the effect of ash particle size, pH, and $Ca(OH)_2$ on adsorption of boron by FA in aqueous media. It was found that a small amount of born was adsorbed by FA at pH 7 through a ligand exchange mechanism. In an investigation of B removal from aqueous solutions by adsorption, Ozturk and Kavak (2005) found that the adsorbent capacity of FA in batch studies was 20.9 and 46.2 mg B g^{-1} in column studies. They concluded that FA is an effective adsorbent for removing B from aqueous solutions. Polat et al. (2004), in a study aimed at removing B from water using different FAs and coals, found that they are effective because more than 90% of B can be removed from water under suitable conditions (pH $>$ 9.0; liquid/soil ratio, 1:10, and reaction time, $>$6 hours). Polowczyk et al. (2013) used FA from brown coal and a biomass-burning power plant from Poland with batch adsorption experiments and using solutions with H_3BO_3, and concluded that the FA capacity of B adsorption was about 6.9 mg B g^{-1}, which corresponded to 90% of the B content in water; thus, it was an effective medium for B removal from high B-content waters. Chong et al. (2009) investigated the possibility of removing B from wastewater produced by the ceramics industry by using bottom ash, and found that bottom ash is effective in removing B.

25.14 CONCLUSIONS

CFA is a by-product produced in coal-fired power stations in large amounts over the world; it requires proper management. CFA contains several compounds, both beneficial and toxic, among which are essential elements for plant growth, especially B. B is an essential element for completion of the life cycle for organisms of all phylogenetic kingdoms. It may seriously affect animal, human, and plant growth. B is of special concern for agriculture because it affects plant growth when it is at deficient or toxic levels. CFA may be used as a soil ameliorant to improve soil fertility (soil pH or addition of essential elements), especially to correct B deficiency in B-deficient soils, but it also may cause B toxicity when applied at the correct rates. In addition, CFAs may be used effectively to remove B from water and wastewater containing high B concentrations which may pollute soils. Consequently, CFA is a by-product that may be employed beneficially in agriculture and other uses.

REFERENCES

Ahmaruzzaman, M., 2010. A review of the utilization of fly ash. Progress in Energy and Combustion Science 36, 327–363.

Adriano, D.C., Page, A.L., Elsewi, A.A., Chang, A.C., Straughan, I., 1980. Utilization and disposal of fly ash and other residues in terrestrial ecosystems. A review. Journal of Environmental Quality 9, 333–344.

Ahmed, N., Abid, M., Rashid, A.,M., Ammanullah, M., 2013. Boron requirement of irrigated cotton in a typic haplocambid for optimum productivity and seed composition. Communications in Soil Science and Plant Analysis 44 (8), 1293–1309.

Bryson, G.M., Mills, H.A., Sasseville, D.N., Jones Jr., J.B., Baker, A.V., 2014. Plant Analysis Handbook III. A Guide to Sampling, Preparation, Analysis and Interpretation for Agronomic and Horticultural Crops. Micr-Macro Publishing, Inc., Athens, Georgia, p. 571.

Brown, H.P., Shelp, B.J., 1997. Boron mobility in plants. Plant and Soil 193, 85–101.
Bell, R.W., 1997. Diagnosis and prediction of boron deficiency for plant production. Plant and Soil 193, 149–168.
Bennett, W.F., 1993. Plant nutrient utilization and diagnostic plant symptoms. In: Bennet, W.F. (Ed.), Nutrient Deficiencies & Toxicities in Crop Plants. The American Phytopathological Society, St Pault, Minnesota, USA, pp. 1–7.
Bingham, F.T., Page, A.L., Coleman, N.T., Flach, K., 1971. Boron adsorption characteristics of selected soils from Mexico and Hawai. Soil Science Society of America Journal 35, 546.
Cassman, K.G., 1993. Cotton. In: Bennet, W.F. (Ed.), Nutrient Deficiencies & Toxicities in Crop Plants. The American Phytopathological Society, St Pault, Minnesota, USA, pp. 111–119.
Chang, A.C., Lund, L.J., Page, A.L., Warneke, J.E., 1977. Physical properties of fly ash amended soils. Journal of Environmental Quality 6 (3), 267–270.
Chong, M.F., Lee, K.P., Chieng, H.J., Ramli, I.I.S., 2009. Removal of boron from ceramic industry wastewater by adsorption–flocculation mechanism using palm oil mill boiler (POMB) bottom ash and polymer. Water Research 43, 3326–3334.
Depnath, P., Pattanaaik, S.K., Pandey, A.K., 2015. Assessment of critical limit of boron for cowpea in piedmont soils of Arunachal Pradesh. Archives of Agronomy and Soil Science 61 (11), 1649–1658.
Debnath, P., Ghosh, S.K., 2014. Assessment of critical limit of available boron for pea in acidic Alfisols of East Sikkim, India. Legume Research 37 (5), 508–514.
EPA, Environmental Protection Agency, 2008. Drinking Water Health Advisory for Boron, p. 65. Document number 822-R-08-013. Washington, DC.
Fail, Jr, Wolchok, L.J., 1977. Soybean growth on fly ash amended stripmine spoils. 1977. Plant and Soil 48, 473–484.
Gascho, G.J., 1993. Sugarcane. In: Bennet, W.F. (Ed.), Nutrient Deficiencies & Toxicities in Crop Plants. The American Phytopathological Society, St Pault, Minnesota, USA, pp. 37–42.
Gartel, W., 1993. Grapes. In: Bennet, W.F. (Ed.), Nutrient Deficiencies & Toxicities in Crop Plants. The American Phytopathological Society, St Pault, Minnesota, USA, pp. 177–183.
Gupta, U.C., 1993. Sources of Boron. In: Gupta, U.C. (Ed.), Boron and its Role in Crop Production. CRC Press, Inc., Boca Raton, Florida, pp. 45–51.
Goldberg, S., 1997. Reactions of boron in soils. Plant and Soil 193, 35–48.
Heidrich, C., Feuerborn, H.-J., Weir, A., 2013. In: Proceedings of the Word of Coal Ash (WOCA) Conference, April 2013, Lexington, Kentucky. http://www.flyash.info.
Hayashi, S., Takahasi, T., Kanehashi, K., Kubota, N., Mizumo, K., Kashiwakura, S., Sakamoto, T., Nagasaka, T., 2010. Chemical state of boron in coal fly ash investigated by focused-ion-beam time-of-flight secondary ion mass spectrometry (FIB-TOF-SIMS) and satellite-transition magic angle spinning nuclear magnetic resonance (STMS NMR). Chemosphere 80, 881–887.
Hall, R., Schwartz, H.F., 1993. Common bean. In: Bennet, W.F. (Ed.), Nutrient Deficiencies & Toxicities in Crop Plants. The American Phytopathological Society, St Pault, Minnesota, USA, pp. 143–147.
Hanson, E., 1993. Apples and pears. In: Bennet, W.F. (Ed.), Nutrient Deficiencies & Toxicities in Crop Plants. The American Phytopathological Society, St Pault, Minnesota, USA, pp. 159–163.
Hollis, J.F., Keren, R., Gal, M., 1988. Boron release and sorption by fly ash as affected by pH and particle size. Journal of Environmental Quality 17, 181–184.
Iwasita, A., Sakaguchi, Y., Nakajima, Takanashi, H., Ohki, A., Kambara, S., 2005. Leaching characteristics of boron and selenium for various coal fly ashes. Fuel 84, 479–485.
Johnson, R.S., 1993. Stonee fruit: peaches and nectarines. In: Bennet, W.F. (Ed.), Nutrient Deficiencies & Toxicities in Crop Plants. The American Phytopathological Society, St Pault, Minnesota, USA, pp. 171–175.

Jin, J., Martens, D.C., Zelancy, L., 1987. Distribution and plant availability of soil boron fractions. Soil Science Society of America Journal 51, 1228−1231.

Jame, Y.W., Nicholaichuk, W., Leyshon, A.J., Cambell, C.A., 1982. Boron concentration in the soil solution under irrigation: a theoretical analysis. Canadian Journal of Soil Science 62, 461−471.

Johnson, G.V., Fixen, P.E., 1990. Testing soils for sulfur, boron, molybdenum, and chlorine. In: Westerman, R.L. (Ed.), Soil Testing and Plant Analysis, third ed. Soil Science Society of America, Madison, WI, pp. 256−273.

Kashiwakura, S., Kubo, H., Kumagai, y., Kubo, H., Matsubae-Yokoyama, K., Nakajima, K., Nagasaka, T., 2009. Removal of boron from coal fly ash by washing with HCl solution. Fuel 88, 1245−1250.

Kabu, M., Uyarlar, C., Zarczynska, K., Milewsak, W., Sobiech, P., 2015. The role of boron in animal health. Journal of Elementology 20 (2), 535−541. http://dx.doi.org/10.5601/jelem.2014.19.3.706.

Kukier, U., Sumner, M., 1996. Boron availability to plants from coal combustion by-products. Water Air, and Soil Pollution 87, 93−110.

Keren, R., Bingham, F.T., 1985. Boron in water, soils and plants. Advances in Soil Science 1, 229−276.

Lopareve-Pohu, A., Pourrut, B., Waterlot, C., Guillaume, G., Bidar, Pruvot, C., Shirali, Douay, F., 2010. Assessment of fly ash-aided phytostabilisation of highly contaminated soils after an 8-year field trial: Part 1. Influence on soil parameters and metal extractability. Science of the Total Environment 409 (3), 647−654.

Lee, H., Ha, H.S., Lee, C.H., Lee, Y.B., Kim, P.J., 2005. Fly ash effect on improving soil properties and rice productivity in Korean paddy soils. Bioresource Technology 97, 1490−1497.

Manoharam, V., Yunusa, I.A., Loganathan, P., Lawrie, R., Murray, B.R., Skilebeck, C.G., Eamous, D., 2010. Boron contents and solubility in Australian fly ashes and its uptake by canola (*Brassica napus* L.) from the ash-amended soils. Australian Journal of Soil Research 48, 480−487.

Matoh, T., 1997. Boron in plant cell walls. Plant and Soil 193, 59−70.

Matsi, T., Keramidas, V.Z., 1999. Fly ash application on two acid soils and its effect on soil salinity, pH, B, P and on ryegrass growth and composition. Environmental Pollution 104, 107−112.

Noda, N., Ito, S., 2007. The release and behavior of mercury, selenium, and boron in coal combustion. Power Technology 180, 227−231.

Nielsen, F.H., 2000. The emergence of boron as nutritionally important throughout the life cycle. Nutrition 16 (7−8), 512−514.

Neilsen, F.H., 2014. Update on human health effects of boron. Journal of Trace Elements in Medicine and Biology 28, 383−387.

Nable, R.O., Banuelos, G.S., Paul, J.G., 1997. Boron toxicity. Plant and Soil 193 (1−2), 181−198.

Ogava, Y., Sakakibara, K., Wang, L., Suto, K., Inoue, C., 2014. Immobilization of B, F, Cr, and As in alkaline coal fly ash through an aging process with water. Environmental Monitoring and Assessment 186, 6757−6770.

Ozturk, N., Kavak, D., 2005. Adsorption of boron from aqueous solutions by using fly ash: batch and column studies. Journal of Hazardous Materials 127, 81−88.

Page, A.L., Elseewi, A.A., Straughan, I.R., 1979. Physical and chemical properties of fly ash from coal-fired power plants with special reference to environmental impacts. Residue Reviews 71, 83−120.

Pandey, V.C., Abhilash, P.C., Upadhay, R.N., Tewari, D.D., 2009. Application of fly ash on the growth and translocation of toxic heavy metals within *Cajanus cajan* L.: implication for the safe utilization of fly ash for agricultural production. Journal of Hazardous Materials 166, 255−259.

Polat, H., Vengosh, A., Pankratov, I., Polat, M., 2004. A new methodology for removal of boron from water by coal and fly ash. Desalination 164, 173−188.

Pandey, N., Archana, 2011. Physiological and biochemical changes in *Catharanthus roseus* L. in response to boron nutrition. Journal of Plant Nutrition 34 (12), 1797−1810.

Polowczyk, I., Ulatowska, J., Koźlecki, T., Bastrzyk, A., Sawiński, W., 2013. Studies on removal of boron from aqueous solution by fly ash agglomerates. Desalinization 310, 93–101.

Rhoades, J.D., Ingvalson, R.D., Hatcher, J.T., 1970. Adsorption of boron ferromagnesium minerals and magnesium hydroxide. Soil Science Society of America Journal 34, 938–941.

Shaheen, S.M., Hooda, P.S., Tsadilas, C.D., 2014. Opportunities and challenges in the use of coal fly ash for soil improvements: a review. Journal of Environmental Management 145, 249–267.

Scorei, R., 2012. Is Boron a prebiotic element? A mini review of the essentiality of Boron for the appearance of life on Earth. Origins of Life and Evolution of Biospheres 42, 3–17.

Sinha, P., Nautiyal, N., Khurana, N., Gupta, S., 2009. Development and physiological response of bittergourd to boron level. International Journal of Vegetable Science 15, 303–311.

Smith, D.H., Wells, M.A., Porter, D.M., Cox, F.R., 1993. In: Bennet, W.F. (Ed.), Nutrient Deficiencies & Toxicities in Crop Plants. The American Phytopathological Society, St Pault, Minnesota, USA, pp. 105–110.

Shorrocks, V.M., 1991. Boron – A Global Appraisal of the Occurrence, Diagnosis, and Correction of Boron Deficiency. Borax Consolidated Ltd, Micronutrient Bureau, U.K.

Tripathi, R.D., Vajpayee, P., Singh, N., Rai, U.N., Kumar, A., Ali, M.B., Kumar, B., Yunus, M., 2004. Efficacy of various amendments for amelioration of fly ash toxicity: growth performance and metal composition of *Cassia siamea* Lamk. Chemosphere 54, 1581–1588.

Tagliabue, M., Reverbeti, A.P., Bagatin, R., 2014. Boron removal from water: needs, challenges and perspectives. Journal of Cleaner Production 77, 56–64.

Tsadilas, C.D., Yassoglou, N., Kosmas, C.S., Kallianou, C., 1994. The availability of soil boron fractions to olive trees and barley and their relationship to soil properties. Plant Soil 162, 211–217.

Tsadilas, C.D., Dimoyiannis, D., Samaras, V., 1997. Methods of assessing boron availability to kiwifruit plants growing on high boron soils. Communications in Soil Science and Plant Analysis 28 (11&12), 973–987.

Tsadilas, C.D., Chartoulakis, K.S., 1999. Boron deficiency in olive trees in Greece in relation to soil boron concentration. Acta Horticulturae 474, 341–344.

Tsadilas, C.D., 1997. Soil contamination with boron due to irrigation with treated municipal wastewater. In: Bell, K.W., Rerkasen, B. (Eds.), Boron in Soils and Plants. Kluwer Academic Publishers, The Netherlands, pp. 265–270.

Tsadilas, C., Samaras, V., Evangelou, E., Shaheen, S., 2014. Influence of fly ash and sewage sludge application on wheat biomass production, nutrients availability, and soil properties. International Journal of Coal Science & Technology. http://dx.doi.org/10.1007/s40789-014-0027-0.

Ulrich, A., Moraghan, J.T., Whitney, E.D., 1993. In: Bennet, W.F. (Ed.), Nutrient Deficiencies & Toxicities in Crop Plants. The American Phytopathological Society, St Pault, Minnesota, USA, pp. 91–98.

Ulrich, A., 1993. Potato. In: Bennet, W.F. (Ed.), Nutrient Deficiencies & Toxicities in Crop Plants. The American Phytopathological Society, St Pault, Minnesota, USA, pp. 149–156.

Vassilev, S.V., Vassileve, C.G., 2007. A new approach for the classification of coal fly ashes based on their origin, composition, properties, and behavior. Fuel 86, 1490–1512.

Voss, R.D., 1993. Corn. In: Bennet, W.F. (Ed.), Nutrient Deficiencies & Toxicities in Crop Plants. The American Phytopathological Society, St Pault, Minnesota, USA, pp. 11–14.

Wang, B., Guo, X., Bai, P., 2014. Removal technology of boron dissolved in aqueous solutions – a review. Colloids and Surfaces A: Physicochemical and Engineering Aspects 444, 338–344.

Wilcox, G.E., 1993. In: Bennet, W.F. (Ed.), Nutrient Deficiencies & Toxicities in Crop Plants. The American Phytopathological Society, St Pault, Minnesota, USA, pp. 137–141.

Wutscher, H.K., Smith, P.F., 1993. Citrus. In: Bennet, W.F. (Ed.), Nutrient Deficiencies & Toxicities in Crop Plants. The American Phytopathological Society, St Pault, Minnesota, USA, pp. 165–170.

Weinthal, E., Parag, Y., Vengosh, A., Muti, A., Kloppmann, W., 2005. The EU drinking water directive: the Boron standard and scientific uncertainty. European Environment 15, 1–12.
Warren, C.J., 1992. Some limitations of sluiced fly ash as a liming material for acidic soils. Waste Management 10, 317–327.
Yunusa, I.A.M., Eamus, D., DeSilva, D.L., Murray, B.R., Burchett, M.D., Skilbeck, G.C., Heidrich, C., 2006. Fly-ash: an exploitable resource for management of Australian soils. Fuel 85, 2337–2344.

CHAPTER 26

THE CRYSTALLIZATION OF STRUVITE AND ITS ANALOG (K-STRUVITE) FROM WASTE STREAMS FOR NUTRIENT RECYCLING

K. Shih, H. Yan
University of Hong Kong, Hong Kong, Hong Kong SAR, China

CHAPTER OUTLINE

Environmental Materials and Waste. http://dx.doi.org/10.1016/B978-0-12-803837-6.00026-3

26.1 INTRODUCTION

Phosphorus (P) is an essential element for all living organisms and is considered to be the key ingredient of the energy system in organism cells. Most activities in cells are powered by energy derived from the reversible transfer of phosphate bonds between adenosine triphosphate and adenosine diphosphate. However, excess P discharged into bodies of fresh water may cause extraordinary growth in the phytoplankton community, or eutrophication, which has detrimental consequences for the aquatic ecological environment. Schindler et al. (2008) carried out a 37-year whole-ecosystem experiment and concluded that the control of P input is particularly critical to reduce eutrophication in lakes. Therefore, continuous efforts have been made to decrease P discharge by means of source reduction and advanced wastewater treatment technologies. The process of removing P has become one obligatory part in most wastewater treatment plants (WWTPs) under the legislative pressure of stringent P discharge thresholds (Le Corre et al., 2009).

At the same time, the depletion of P resources has received increasing interest because phosphate rock is being consumed at an accelerated pace to produce P fertilizer to support global food production. Phosphate rock is a nonrenewable and nonsubstitutable resource; it is estimated that current global reserves will be exhausted in the next 50–100 years (Cordell et al., 2009, 2011). Furthermore, the extremely uneven geographic distribution of phosphate rock reserves enhances the geopolitical concerns of worldwide P security (Cooper et al., 2011); nearly 77% of global P reserves are concentrated in Morocco. The future availability of P will be increasingly reliant on Morocco unless new sources can be accessed or recycling of P is significantly improved.

Under the double pressure of eutrophication and the limited reserves of phosphate rock, a method for the effective and efficient recovery of P from waste streams is gaining interest among researchers in various fields. A global P scenario analysis suggested that meeting the growing long-term demand for P would likely require consumption management to reduce two-thirds of demands for business as usual, and the remaining third should be supplied by a high recovery rate of P from human excreta, manure, food waste, and mining waste (Cordell et al., 2011). Recycling P from sewage products in WWTPs (including concentrated P solution and sludge ash) is currently the major P recovery route owing to the convenient accessibility of sewage. Common techniques for the removal of P from aqueous solutions include biological methods, chemical precipitation, adsorption–ion exchange, constructed wetland, and filtration on different scales; chemical precipitation is currently the most widely used of these technologies (Loganathan et al., 2014). However, in traditional methods of P precipitation, which use iron, aluminum salts, and/or lime as additives, the resulting P has low reusability, particularly precipitates in the form of iron–P and aluminum–P compounds. Hence, these deposits are normally treated as sludge waste for discard. Researchers have discovered that a type of P precipitate, struvite ($MgNH_4PO_4 \cdot 6H_2O$; MAP), has great potential as a fertilizer and might be a promising P recovery product (De-Bashan and Bashan, 2004; Lind et al., 2000).

Struvite was first identified as a crystalline material by Rawn and Banta (1939) in a study of municipal wastewater digestion problems. It gained public attention in the 1960s as a result of pipe blockages in the Hyperion WWTP in Los Angeles (Borgerding, 1972). Since that time, many similar cases of struvite deposition in pipelines have been reported (Pitman et al., 1991; Williams, 1999) and have even required the replacement of pipe work (Ohlinger et al., 1998). Struvite crystallization is regarded as the major mechanism in severe clogging problems in the sludge piping systems of WWTPs. Turbulence causes a reduction in pressure, resulting in the release of carbon dioxide from wastewater and an associated increase in pH, which facilitates struvite crystallization in aquatic environments with excess P (Borgerding, 1972). In contrast, the formation of struvite is also considered to be a highly promising strategy for the recovery of P owing to its potential as a suitable slow-release fertilizer. The good fertilizer properties of struvite have been widely proved in a series of nutrient-leaching experiments, and struvite fertilizer has been shown to perform especially well in environments that include moderately alkaline and acidic soils (Massey et al., 2007). Magnesium potassium phosphate hexahydrate ($MgKPO_4 \cdot 6H_2O$, also known as K-struvite or MKP) is an isomorphous crystalline compound of struvite with the replacement of NH_4^+ (ion size = 1.48 Å) with a smaller K^+ ion (ion size = 1.33 Å) (Banks et al., 1975). It is a new inorganic phosphate mineral approved in 2003 by the Commission on New Minerals and Mineral Names, International Mineralogical Association (Burke and Ferraris, 2004). Because $MgKPO_4.6H_2O$ has crystallographic properties similar to those of struvite and also contains multiple nutrients (potassium [K], P, and magnesium [Mg]), a number of studies have reported its possible use as an efficient fertilizer like struvite (Lozano et al., 1999; Xu et al., 2011, 2015).

With the extensive exhaustion of phosphate rock resources, the use of struvite-based materials as an eco-friendly fertilizer is becoming more competitive with traditional fertilizers because of its convenient sources of nutrients (nitrogen [N], P, and K) from wastewater, so long as its crystallization process can be well controlled (Rahman et al., 2014). This chapter provides an overview of our current understanding and the typical problems encountered in the recovery of nutrients from waste streams as struvite-type fertilizer.

26.2 CHARACTERISTICS OF STRUVITE-TYPE COMPOUNDS

26.2.1 CHEMICAL AND PHYSICAL PROPERTIES

Struvite is a kind of orthophosphate mineral that is precipitated by Mg, ammonium, and phosphate ions in equal molar ratios via the following reaction (Eq. [26.1]):

$$Mg^{2+} + NH_4^+ + H_nPO_4^{n-3} + 6H_2O \rightarrow MgNH_4PO_4 \cdot 6H_2O + nH^+ \quad (n = 0, 1 \text{ or } 2) \qquad [26.1]$$

Struvite group compounds are generally represented as $A^{2+}B^+PO_4 \cdot nH_2O$, where n = 6–8; A^{2+} corresponds to Mg (Scrimgeour et al., 2007), nickel (Ni), or cobalt (Co) cations (Bassett and Bedwell, 1933); and B^+ corresponds to univalent ions including K^+, Rb^+, Cs^+, Tl^+, and NH_4^+ (Banks et al., 1975). Struvite has low solubility in water, with a solubility constant of $10^{-13.26}$ (Ohlinger et al., 1998). In contrast, it is readily soluble in acidic environments: 0.033 g 100^{-1} mL^{-1} in 0.001 M HCl and 0.178 g 100^{-1} mL^{-1} in 0.01 M HCl at 25°C (Le Corre et al., 2009). Pure struvite is generally a white or transparent fine powder, but the morphology of struvite particles differs under various formation conditions and includes a needle-like shape (Fig. 26.1), a cube-like shape, and irregular shapes (Zhang et al., 2009). According to the summary of Rahman et al. (2014), the length of struvite particles varied from 15 μm (Zhang et al., 2009) to 3.5 mm (Adnan et al., 2003). Abe (1995) indicated that the phosphate concentration in the inflow together with the retention time of particles in reactors had a significant effect on the particle size of the struvite products obtained. For example, with an influent phosphate concentration higher than 200 ppm, the struvite crystals grew at a speed of 0.173 mm d^{-1}, but the growth rate was reduced to 0.061 mm d^{-1} with a lower phosphate concentration in the feeding flow (about 34 parts per million [ppm]).

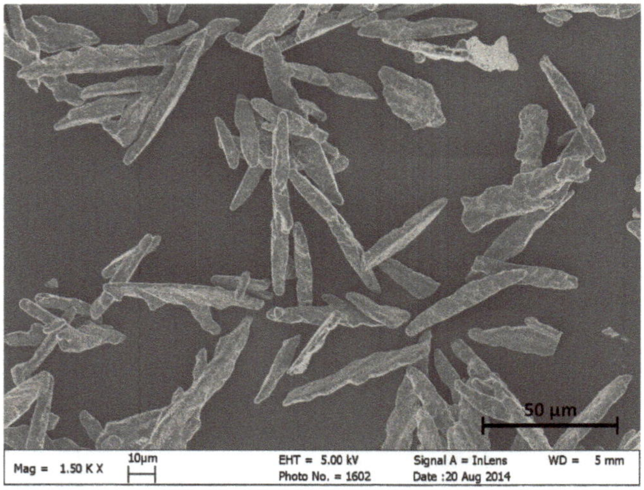

FIGURE 26.1

Scanning electron microscopic image of needle-like struvite precipitated with Mg^{2+}, NH_4^+, and PO_4^{3-} at a molar ratio of 1:1:1, at pH 7.5.

The physical and chemical properties of K-struvite are similar to those of struvite, including the needle-like crystal shape and transparent to whitish appearance (Graeser et al., 2008). K-struvite crystals have a slightly higher density (1.864 g cm^{-3}) (Graeser et al., 2008) than struvite crystals (1.711 g cm^3) (Ohlinger et al., 1998), and this gravity range is advantageous for struvite-type compounds applied as fertilizer in flooded areas (Rahman et al., 2014). The work of Xu et al. (2015) in the recovery of nutrients from urine indicated that struvite has a much greater tendency to crystallize than K-struvite in common urine systems. In contrast to the widely recognized solubility constant of struvite crystal ($10^{-13.26}$), limited information has been provided on the solubility constant of K-struvite; values of $10^{-11.7}$ (Luff and Reed, 1980), $10^{-13.26}$ (Taylor et al., 1963), and $10^{-12.2 \pm 0.253}$ (Xu et al., 2015) have been reported. For considerations of solubility and crystallization kinetics, ammonia should be removed as much as possible to achieve optimal production of K-struvite. Both MAP and MPP crystals can spontaneously form calculi in human and animal kidneys (Coe et al., 2005; Dunlevey and Laing, 2008); this biological phenomenon implies that urine is a good source from which to precipitate magnesium ammonium phosphate (MAP) and/or magnesium potassium phosphate (MPP) for the recovery of nutrients owing to the high concentration of P, N, and K.

26.2.2 CRYSTALLOGRAPHIC PROPERTIES BY X-RAY DIFFRACTION ANALYSIS

X-ray diffraction (XRD) technology is a useful method of analysis, especially for crystallographic studies. The mechanism of XRD analysis is based on Bragg's law (Eq. [26.2]):

$$n\lambda = 2d_{hkl}\sin\theta \qquad [26.2]$$

where d_{hkl} is the distance between two crystal planes; θ is the Bragg angle; n represents an integer; and λ represents the wavelength of the X-ray. The XRD pattern of a pure material, which mainly includes information regarding peak position and intensity, is like a fingerprint of the substance. The peak positions can reveal the crystal structure and symmetry of the contributing phase, whereas the peak intensities reflect the scattering from each plane in the crystal structure and directly imply the crystallinity of the corresponding sample. Thus, the powder diffraction technique is ideally suitable for the identification and characterization of polycrystalline-phase samples. Currently, the International Centre for Diffraction Data (ICDD) is organizing an XRD pattern database for the identification of materials, known as the powder diffraction file database (PDF).

Fig. 26.2 presents the XRD pattern of struvite of PDF#71-2089 from ICDD and the pattern of the struvite sample prepared with the co-precipitation method, applying an Mg^{2+}, NH_4^+, and PO_4^{3-} molar ratio of 1:1:1 at pH 7.5. According to the measurement and calculation performed by Whitaker and Jeffery (1970) with the oscillation photograph method, struvite crystal is indexed on an orthorhombic cell with cell dimensions $a = 6.941 \pm 0.002$ Å, $b = 6.137 \pm 0.002$ Å, and $c = 11.199 \pm 0.004$ Å. The crystal space group is *Pmn*21 (31), and one unit cell contains two molecules. They also conducted least-squares refinement of struvite crystal on three-dimensional data (585 structure factors within the Cu $K\alpha$ sphere). The refinement results revealed that the struvite structure consists of PO_4^{3-} tetrahedra, $[Mg(H_2O)_6]^{2+}$ octahedra, and NH_4^+ groups connected by hydrogen bonds, in which PO_4^{3-} tetrahedra are regular with a mean P—O bond length of 1.5370 ± 0.0011 Å and the complex ion $[Mg(H_2O)_6]^{2+}$ octahedra are distorted with a mean Mg-O (water) bond length of 2.0711 ± 0.0011 Å. NH_4^+ groups were deduced with restricted rotation in the struvite structure, which was confirmed by the observation of splitting of the υ_4 asymmetric bending of NH_4^+ species in the study of Banks et al. (1975). When comparing the XRD pattern of samples prepared by co-precipitation with a struvite standard pattern, strong orientation along the (002), (012), and (004) planes was observed as shown in Fig. 26.2, although all of the samples had been thoroughly ground. The oriented growth of struvite crystals has

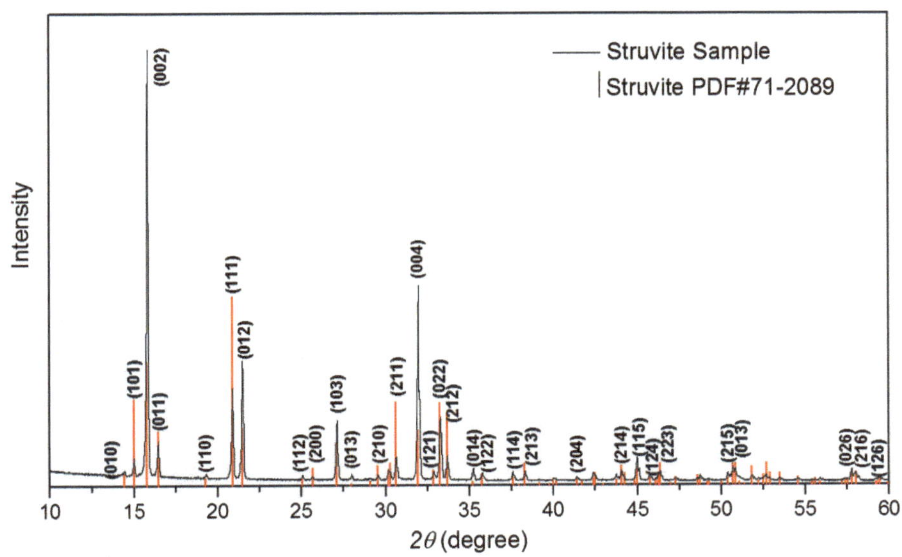

FIGURE 26.2

XRD pattern of struvite sample precipitated with Mg^{2+}, NH_4^+, and PO_4^{3-} molar ratio of 1:1:1 at pH 7.5 and of standard struvite model structure from ICDD PDF#71-2089.

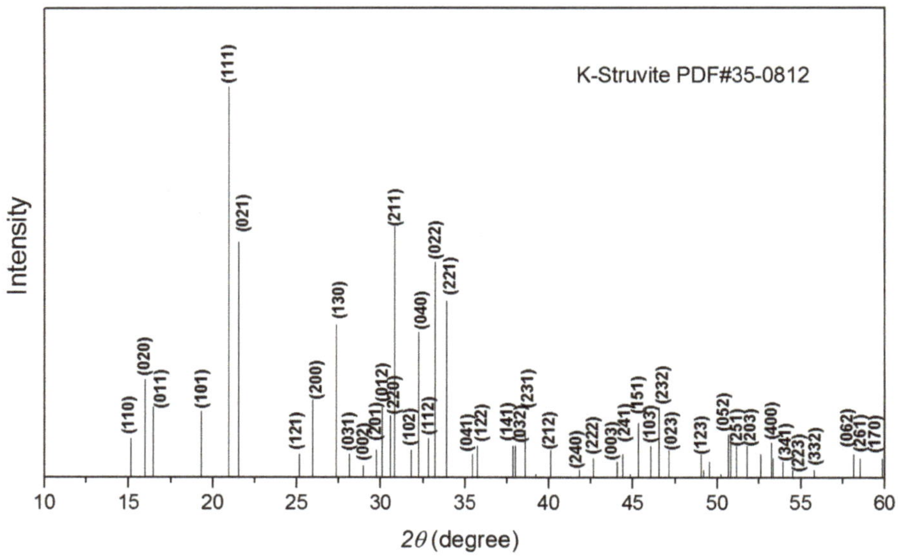

FIGURE 26.3

XRD pattern of K-struvite ($MgKPO_4 \cdot 6H_2O$) from ICDD PDF#35-0812.

been widely reported, especially in studies of urinary calculi (Rouff, 2013; Sutor and Wooley, 1972). This phenomenon is consistent with the commonly observed needle-like or rod-like appearance of struvite precipitates and implies that struvite crystal grows preferentially along one axis.

Fig. 26.3 shows the XRD pattern of $MgKPO_4 \cdot 6H_2O$ (K-struvite) crystal of PDF#35-0812. The 2-theta position of its diffraction peaks is analogous to that in the struvite pattern, which implies that the ion substitution results in minor structural changes. Graeser et al. (2008) measured the cell parameters of K-struvite from different sources: in a dolomitic rock discovered at Lengenbach in Binntal, Switzerland, and in an abandoned galena mine in Rossblei, Austria; the results are listed in Table 26.1 with the cell dimensions of struvite derived from PDF#71-2089.

Table 26.1 Lattice Parameters of K-Struvite From Various Sources and Struvite of PDF#71-2089 (Graeser et al., 2008)

	Space group	a (Å)	b (Å)	c (Å)	V (Å^3)	Z
K-struvite (PDF#35-0812)	$Pmn2_1$	6.879(1)	11.100(1)	6.163(4)	470.63	2
K-struvite (Lengenbach)	$Pmn2_1$	6.892(3)	6.166(2)	11.139(4)	473.4(3)	2
K-struvite-refined (Lengenbach)	$Pmn2_1$	6.903(3)	6.174(2)	11.146(3)	475.0(2)	2
K-struvite (Rossblei)	$Pmn2_1$	6.878(1)	6.161(1)	11.100(1)	470.41(9)	2
Struvite (PDF#71-2089)	$Pmn2_1$	6.941	6.137	11.199	477.04	2

It is clear that the substitution of NH_4^+ by K^+ marginally reduces the cell volume and produces insignificant effects on the structure. Banks et al. (1975) also suggested that the struvite structure ($MgXPO_4 \cdot 6H_2O$) incorporated monovalent ions (X^+) ranging in size from 1.33 Å for K^+ to 1.69 Å for Cs^+. However, a hexagonal structure is more stable than an orthorhombic structure for $MgCsPO_4 \cdot 6H_2O$ because of the large size of cesium ions.

26.2.3 THERMAL ANALYSIS OF STRUVITE-TYPE COMPOUNDS

Thermal stability is an important factor in the application of struvite as a fertilizer, based on the consideration of transportation. Several studies have been conducted to investigate the thermal decomposition of struvite under different conditions. Frost et al. (2004) reported that struvite is decomposed at a lower temperature and has a slower heating rate, which is valuable for the treatment of kidney stones. Using a slow heating rate of 1°C min^{-1}, the mass loss of struvite involves three steps that occur at 39.5°C, 57.8°C, and 82.6°C, respectively. Ammonia is decomposed before the loss of crystallized water. In contrast, when the heating speed is increased to 2°C min^{-1}, decomposition starts from 85°C with simultaneous loss of ammonia and water molecules (Frost et al., 2004). Bhuiyan et al. (2008) used thermogravimetric analysis to confirm the dependence of struvite decomposition on the heating rate. Moreover, they proposed the possible phase transitions of struvite as shown in Fig. 26.4. When struvite is heated in excess water, ammonia is gradually lost and struvite is partly transformed into bobierrite ($Mg_3[PO_4]_2 \cdot 8H_2O$). When struvite is boiled in excess water, it will lose five water molecules of crystallization and transform into its monohydrate product, dittmarite ($MgNH_4PO_4 \cdot H_2O$).

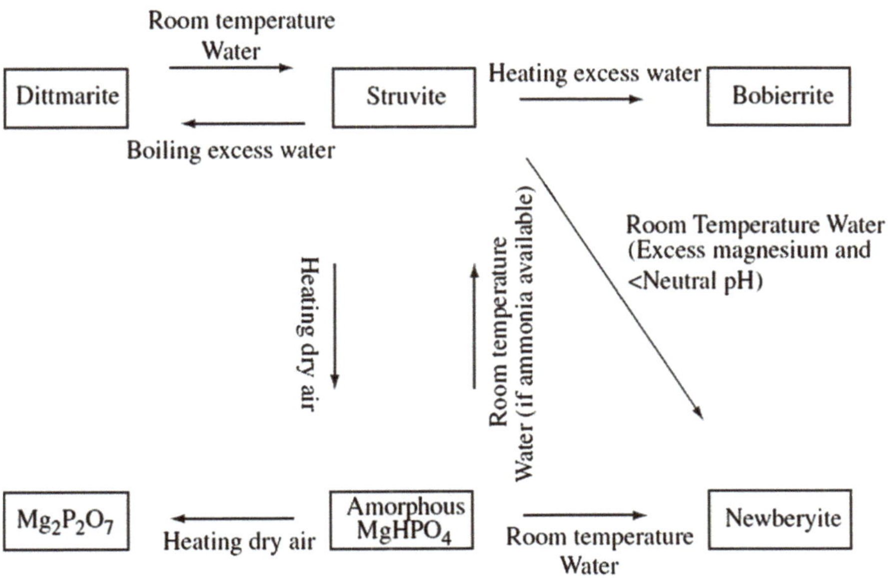

FIGURE 26.4

Diagram of possible transformation mechanisms of struvite-associated phases (Bhuiyan et al., 2008).

Thermal analysis of K-struvite was carried out by Chauhan et al. (2011) by thermogravimetric analysis, differential thermal analysis, and differential scanning calorimetry. Results suggested that K-struvite begins dehydration at approximately 75°C and phase transformation is presumed to occur at 670°C. The dehydration behavior of struvite-K is similar to that of struvite and depends on both temperature and heating rate. If a lower heating rate is used, dehydration begins at a lower temperature. In addition, one-step dehydration of K-struvite occurs in an N_2 atmosphere under nonisothermal conditions, as presented in the following chemical reaction (Zhang et al., 2013) (Eq. [26.3]):

$$MgKPO_4 \cdot 6H_2O \rightarrow MgKPO_4 + 6H_2O\uparrow \qquad [26.3]$$

26.3 KEY FACTORS INFLUENCING STRUVITE CRYSTALLIZATION

Struvite crystallization is a complex process that includes nucleation and crystal growth and is affected by a combination of physicochemical factors such as supersaturation, pH, temperature, mixing strength, and the presence of foreign ions. The manner in which these key factors affect struvite crystallization is briefly discussed in the following sections.

26.3.1 SUPERSATURATION AND MOLAR RATIO

Supersaturation is considered to be the driving force for all crystallization processes (Pastor et al., 2008). It significantly affects the induction time of crystallization and the crystal growth rate (Le Corre et al., 2009). Kofina and Koutsoukos (2005) revealed the role of supersaturation in crystallization in a study of spontaneous precipitation of struvite from synthetic wastewater systems. Their results suggested that the induction time of struvite formation is inversely proportional to the solution supersaturation and corresponded well with the dependence calculated by the classical nucleation theory. Moreover, the high order (>1) of dependence of crystallization over the solution supersaturation indicated a surface diffusion-controlled mechanism for struvite formation. The stoichiometric ratio of (Mg^{2+}:NH_4^+:PO_4^{3-}) also influenced the purity and quality of the obtained products. Yetilmezsoy and Sapci-Zengin (2009) applied various molar ratios of Mg:N:P, including both overdose (1.2:1:1, 1.5:1:1, 1:1:1.2, and 1:1:1.5) and underdose (0.5:1:1, 0.8:1:1, 1:1:0.5, and 1:1:0.8) assuming an N dosage constant at 1, to investigate the effects on chemical oxygen demand and NH_4^+-N removal, which is indicative of struvite production. Their results showed that overdose of Mg increased the removal efficiency of N. Bouropoulos and Koutsoukos (2000) also confirmed that Mg ions influence the microelectrophoretic mobility of struvite distributions with an isoelectric point at $pMg = 1.75$, which suggests that greater Mg supersaturation is beneficial to struvite crystallization. However, a stoichiometric ratio of Mg:N:P at 1:1:1 was still proposed as the optimized design for cost considerations.

26.3.2 pH

Both industry practices and laboratory experiments have demonstrated that pH is one of the most important factors in the struvite crystallization process. The specific pH range of 8.5–9.5 was widely proposed as the optimal pH environment for struvite precipitation together with maximal removal of associated species within the pH range (Booker et al., 1999; Bouropoulos and Koutsoukos, 2000). Within the suitable range, a higher pH condition is more beneficial to struvite formation because of the effects of solubility and supersaturation. However, excessively high pH values have disadvantages based on economic considerations, stripping of ammonia gas, and the competition of other ions such as carbonates. According to Eq. [26.1], the struvite formation reaction occurs with a release of protons, which triggers a reduction in the solution pH. Hence, the pH change can be used as an indicator of struvite nucleation (Bouropoulos and Koutsoukos, 2000; Le Corre et al., 2005).

26.3.3 TEMPERATURE

Temperature directly affects the solubility of crystalline solid and thus affects struvite production and crystalline morphology. Hanhoun et al. (2011) developed a thermodynamic model for struvite precipitation to investigate the phosphate conversion rate and assess struvite solubility between 15°C and 35°C. The values of pK_{sp} for struvite between 15°C and 40°C are listed and compared in Table 26.2. In general, pK_{sp} fluctuates from 12.80 to 14.04 in measured temperatures; both Hanhoun et al. (2011) and Aage et al. (1997) reported that struvite reaches its minimum solubility at 30°C.

Temperature affects crystal growth via different mechanisms. Diffusion-controlled growth is the major mechanism for crystallization at high temperatures, whereas surface integration-controlled growth is dominant for crystallization at low temperatures (Jones, 2002).

26.3.4 COEXISTING FOREIGN IONS

Compared with other well-understood factors, few studies have been conducted to investigate the influences of foreign ions on struvite crystallization (Le Corre et al., 2009). Kabdaszli et al. (2006) evaluated the effects of Na^{+}, SO_4^{2-}, CO_3^{2-}, and Ca^{2+} ions on the induction time of struvite formation by means of absorbance measurements, and monitored the pH and conductivity of solutions. Their results demonstrated that all of the tested ions retarded the induction time of struvite crystallization, but only a minor effect was observed from the carbonate ions. Le Corre et al. (2005) conducted a series of struvite precipitation experiments to estimate the mechanism by which calcium ions affect struvite's

Table 26.2 Comparison of Struvite pK_{sp} Values in Literature References (from Hanhoun et al., 2011)

	pK_{sp} Values			
Temperature (°C)	Hanhoun et al. (2011)	Bhuiyan et al. (2007)	Aage et al. (1997)	Burns and Finlayson (1982)
15	13.29 (±0.02)	14.04 (±0.03)	13.27	
20	13.22 (±0.02)	13.69 (±0.02)		
25	13.17 (±0.01)	13.36 (±0.07)	12.93	13.12 (±0.05)
30	13.00 (±0.04)	13.16 (±0.05)	12.80	
35	13.08 (±0.06)	13.20 (±0.03)		12.97 (±0.03)
40			12.52	

shape, crystal size, and purity. The pH was measured in the entire reaction process and used as an indicator of struvite's crystallization rate. Struvite's purity was determined with the energy-dispersive X-ray method, based on the comparison of elemental distribution. Muryanto and Bayuseno (2014) investigated the kinetics of struvite crystallization and found that both Zn^{2+} and Cu^{2+} inhibited the struvite crystallization rate in trace amounts (0—5 ppm), and Zn^{2+} was thought to be the more effective antiscalant for struvite control.

Of the coexisting ions in wastewater, the effects of calcium ions on struvite crystallization are particularly concerning owing to its high concentration in most types of wastewater. The molar ratio of Ca—Mg ranges from 1.4 to 3.7 in settled sludge liquors (Parsons et al., 2001). Moreover, calcium phosphates are precipitated in an overlapped pH range of struvite formation, and hence show significant effects on struvite production. Within the associated calcium phosphate minerals, hydroxyapatite ($Ca_5[PO_4]_3OH$-HA) is regarded as the most thermodynamically stable form and is precipitated by reaction [26.4]:

$$5Ca^{2+} + 3PO_4^{3-} + OH^- = Ca_5(PO_4)_3OH\ (s) \qquad [26.4]$$

In conditions of homogeneous precipitation, unstable amorphous calcium phosphates are first formed, and two possible transformation approaches might then occur according to the pH conditions. When the pH ranges from 7 to 9, amorphous calcium phosphates are initially transferred into a precursor octacalcium phosphate phase (OCP-$Ca_4H[PO_4]_3 \cdot 3H_2O$) and then into an HA phase. In contrast, when the pH is higher than 9, the intermediate conversion step was skipped and amorphous calcium phosphates were converted directly to HA (House, 1999; Root, 1990). Several studies (Hao et al., 2008; Le Corre et al., 2005; Yan and Shih, 2014) have reported analogous phenomena in which HA and/or amorphous calcium phosphates are co-precipitated with struvite from the solutions simultaneously with calcium and Mg ions. In addition, pH was confirmed to have an important role in determining the effects of calcium ions on struvite crystallization, probably because the precipitation of both Mg and calcium phosphate is sensitive to pH. The high-pH environment ($pH > 10.0$) has adverse effects on both struvite crystallization and HA formation.

26.4 QUANTIFICATION OF STRUVITE IN PRECIPITATES

According to the discussion in Section 26.3, struvite precipitation is a complex process that depends on a series of parameters. Owing to the intricate compositions and variable characters of waste streams, the struvite products recovered from wastewater presumably contain impurities, such as magnesium phosphate compounds, other phosphate salts, and carbonate salts, that significantly reduce the quality of the fertilizer. As a result, a reliable and operational method to quantify the struvite contents in the recovered precipitates must be established for the further development of struvite crystallization technology. Analytical methods that have been applied to analyze the purity of struvite include P K-edge X-ray absorption near edge structure spectra (He and Zhang, 2002), energy-dispersive X-ray elemental analysis (Le Corre et al., 2005), magic angle spinning—nuclear magnetic resonance spectroscopy (Scrimgeour et al., 2007), and ion chromatography and inductively coupled plasma measurements (Hao et al., 2008). The basic mechanism of the techniques mentioned is determination of the element (Mg, N, and P) concentrations in products to indicate the amount of struvite. However, because of the high possibility of co-precipitation of other complicated phosphate phases, elemental analysis has limited ability to identify the struvite phase accurately. To our knowledge, limited work has successfully provided phase composition information and quantified the relative phase abundances for struvite-containing precipitates. Lu (2013) suggested that Rietveld refinement, which is based on fitting a complete experimental XRD pattern with a calculated profile (also called quantitative XRD), is a suitable approach for the quantification of struvite. In addition, quantitative XRD is a convenient method that does not require complex pretreatment such as the digestion of samples. Most importantly, this technique is also applicable for precipitates that contain amorphous phase(s) and can be used to determine the abundance of all included phases and the amorphous content by adding an internal standard in known weight ratios (Hongchao and Changlin, 1997). Fig. 26.5 demonstrates the analysis results of a struvite sample with amorphous content by the Rietveld refinement method and shows the agreements and differences between the observed and calculated diffraction patterns. Four reliability index parameters—the pattern factor (R_p), the weighted pattern factor (R_{wp}), the expected error (R_{exp}), and the goodness of fit (GOF)—are commonly used to evaluate the quality of refinement work and are defined by the following equations (Eqs. [26.5–26.8]):

$$R_p = \frac{\sum |Y_i(\text{obs}) - Y_i(\text{calc})|}{\sum Y_i(\text{obs})} \quad [26.5]$$

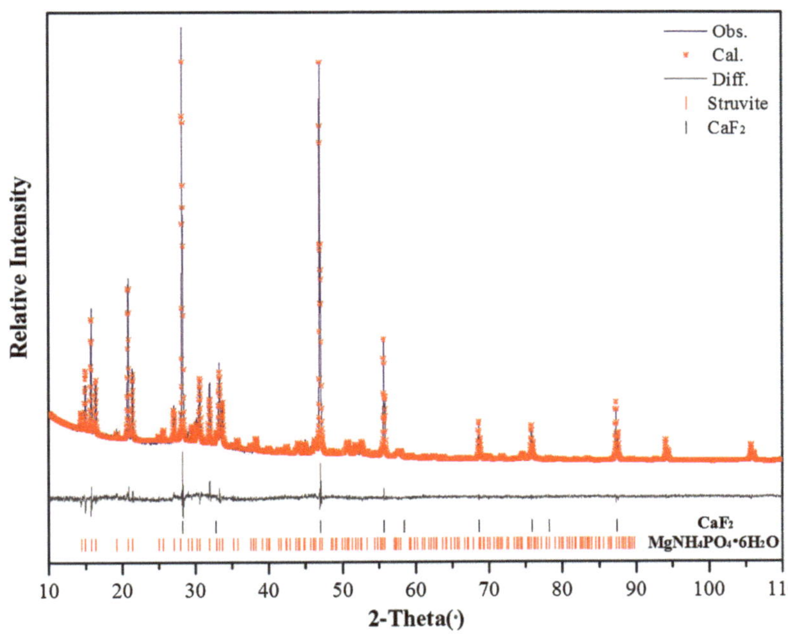

FIGURE 26.5

Observed (*continuous line*), calculated (*red crosses*), and difference (*gray line*) curves of the refined pattern (struvite precipitates containing amorphous phase) with wt. 20% CaF_2 added as internal standard. Vertical bars represent the Bragg positions of the calculated reflections for CaF_2 and struvite.

$$R_{wp} = \left\{ \frac{\sum \omega_i [Y_i(\text{obs}) - Y_i(\text{calc})]^2}{\sum \omega_i [Y_i(\text{obs})]^2} \right\}^{\frac{1}{2}} \quad [26.6]$$

$$R_{exp} = \left\{ \frac{\sum N_d - N_p}{\sum \omega_i [Y_i(\text{obs})]^2} \right\}^{\frac{1}{2}} \quad [26.7]$$

$$\text{GOF} = \text{chi}^2 = \frac{R_{wp}}{R_{exp}} \quad [26.8]$$

where Y_i(obs) and Y_i(calc) are the observed and calculated data at data point x, respectively; ω_i is the weighting given to data point x; N_d is the number of data points; and N_p is the number of parameters. The ideal value for GOF is 1.0, and the quality parameters of refinement work in this study are $R_p = 4.88$, $R_{wp} = 6.15$, $R_{exp} = 3.89$, and GOF = 1.58, which indicate that the calculated data match well with the experimental diffraction pattern and that Rietveld refinement is a reliable method with which to quantify struvite in multiphase precipitates.

To determine the reliability of quantification work of struvite by Rietveld method, binary mixture samples consisting of struvite and amorphous silica with known proportions were prepared and analyzed by QXRD technique; the results are presented in Fig. 26.6. It was obvious that the analysis

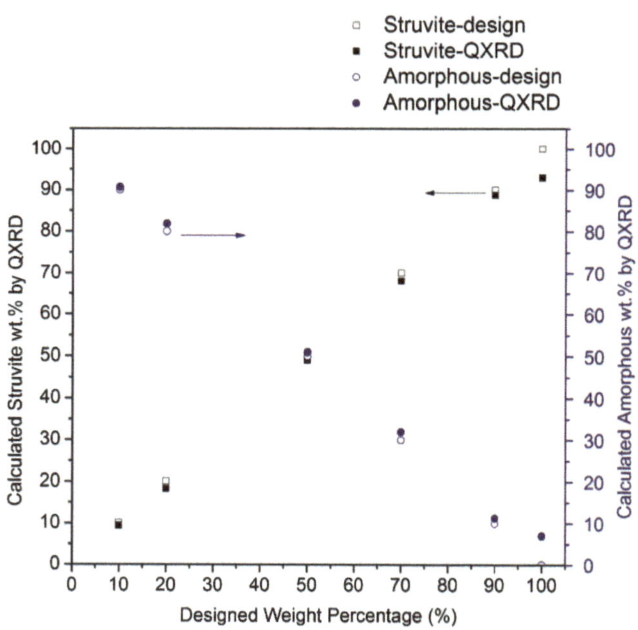

FIGURE 26.6

The analysis results by Rietveld method versus known weight percentage value of prepared samples consisting of struvite and amorphous silica.

results calculated from diffraction data matched well to the actual composition of corresponding samples. Lu's work also reported a good linear correlation ($R^2 = 0.990$) between designed values and calculated values by QXRD, as well as a low measurement error of 3% (Lu, 2013). The high degree of agreement between the calculated weight percentage and the designed weight proportions proves the suitability of the Rietveld method in quantifying struvite in multiphase solids. Yan and Shih (2014) also used the quantitative XRD technique to assess the effects of calcium and iron ions on struvite crystallization. Their results implied that both calcium and ferric ions significantly inhibit the formation of struvite crystals and that the effects vary with the pH conditions. A negative linear correlation was identified between the struvite weight content and the molar ratio of the initial Ca–Mg solution in investigating the influence of calcium. At the same dosage, ferric ions showed an extremely strong impeding influence on the amount of generated struvite and were confirmed as a more effective inhibitor of struvite formation than calcium ions. Moreover, Ca^{2+} and Fe^{3+} modified the needle-like shape of struvite particles into an irregular appearance.

26.5 K-STRUVITE PRODUCTION TECHNIQUES

As another important nutrient element in agriculture, K is mainly produced from potash ore which is consumed at an increasing rate. Similar to the case of P resources, the recovery of K from waste streams is attracting increasing interest owing to the shortage of K minerals in some areas (Xu et al., 2015). K-Struvite contains both K and P and thus is a promising product for nutrient recycling from wastewater.

26.5.1 RECOVER K-STRUVITE FROM URINE

Urine waste is rich in nutrient elements such as N, P, and K, and urine source separation is therefore considered a potential scheme to improve the sustainability of wastewater management (Lienert and Larsen, 2009). Furthermore, the P concentration in urine is much higher than that in supernatant of anaerobically digested sludge, which makes the recovery of P from urine a more feasible and economical route than recycling from other wastewater forms. Xu et al. (2011) conducted bench-scale experiments to investigate the possibility of precipitating K-struvite from synthetic urine. Their results revealed that ammonium should be removed as much as possible, because the struvite crystallization reaction is more likely to take place than that of K-struvite and because struvite is generally precipitated before K-struvite. Moreover, the solution pH, the molar ratio of Mg:K:P, and the coexistence of sodium ions are all influential in K-struvite crystallization. However, under ideal conditions, 77% of P and 98% of K can be removed from the system at pH 10.0 with a molar Mg:K:P ratio of 2:1:2, which suggests that the precipitation of K-struvite efficiently realizes simultaneous removal of P and K. Fig. 26.7 shows the precipitation potential of all possible precipitates, including magnesium potassium phosphate hexahydrate, magnesium sodium phosphate hexahydrate, trimagnesium phosphate octahydrate, and magnesium hydroxide hexahydrate, in a system without ammonia at different pHs and Mg:P molar ratios based on thermodynamic modeling (Xu et al., 2015). The results demonstrate that the suitable pH range for K-struvite precipitation is between 9.0 and 11.0, which is slightly higher than the pH range for struvite formation. However, within the pH range of 9.0–11.0, magnesium sodium phosphate hexahydrate might be co-precipitated and compete with the crystallization of K-struvite. As

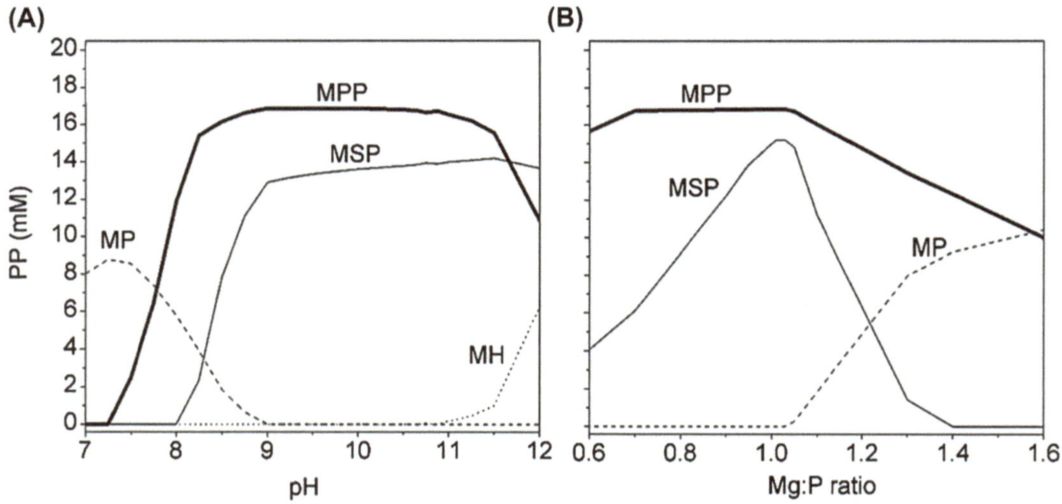

FIGURE 26.7

Precipitation potential of possible precipitates under different pHs and Mg:P molar ratios (A: pH at Mg:P ratio of 1; B: Mg:P ratios at pH around 10.5) based on thermodynamic modeling. *MPP*, K-struvite (magnesium potassium phosphate hexahydrate); *MSP*, magnesium sodium phosphate hexahydrate; *MP*, trimagnesium phosphate octahydrate; *MH*, magnesium hydroxide hexahydrate. (Xu et al., 2015.)

a result, the commonly used pH adjustment agent NaOH should be avoided as much as possible for more efficient production of K-struvite. Wilsenach et al. (2007) established a laboratory-scale struvite–K-struvite precipitator composed of a fluidized-bed reactor and a specially designed liquid–solid separation device. The acceptable performance of the entire setup further confirms the possibility of precipitating struvite and K-struvite from denitrified urine (or a mixture with sludge rejected water) by adding Mg oxides to increase the alkalinity of solutions.

26.5.2 RECOVERY OF K-STRUVITE FROM OTHER SOURCES

With current technology, struvite is precipitated from the supernatant of anaerobically digested sludge, especially that which is produced via enhanced biological P removal. However, K-struvite has low competitive strength in the sludge digestion water as a result of the high concentration of foreign ions, including Ca^{2+}, NH_4^+, and Na^+. Barat et al. (2009) evaluated the precipitation of possible phosphate phases in WWTPs in Murcia, Spain, that used a biological nutrient-removal process. Their thermodynamic calculation indicated that struvite and HA are likely to form in the anaerobic digesting reactor when the pH is increased, whereas K-struvite is not precipitated in the digesters. Fig. 26.8 expresses the saturation index values (SI) for struvite and for K-struvite in the wastewater systems studied by Barat et al. (2009). SI can be obtained from Eq. [26.9]:

$$SI = \log\frac{IAP}{K_{SP}} \qquad [26.9]$$

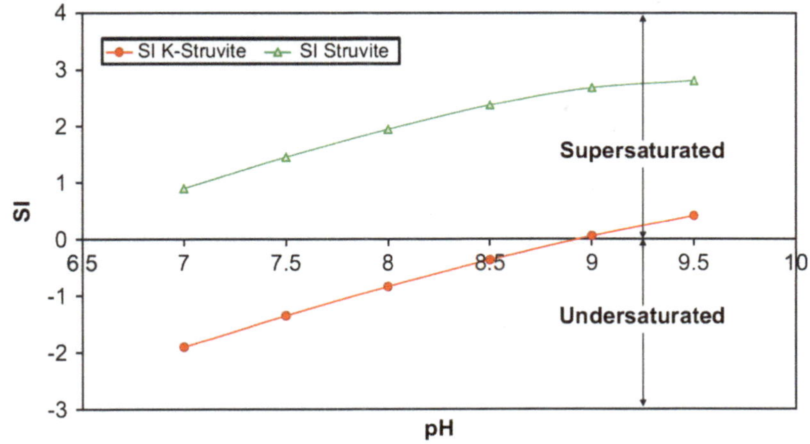

FIGURE 26.8

SI for struvite and K-struvite in the influent to anaerobic digestion under different pH conditions (Barat et al., 2009).

where K_{SP} is the thermodynamic solubility product and IAP is the ion activity product of the associated ions. It is clear that K-struvite has lower SI values than struvite throughout the pH range tested. Moreover, the anaerobic digestion system is undersaturated for K-struvite until a very high pH (pH 9.0). As a result, K-struvite is difficult to precipitate from a sludge digestion system because of the strong influences of struvite crystallization in the reactor.

Except for the synthesis of K-struvite from urine waste, Lozano et al. effectively recovered K-struvite–based fertilizers from seawater bitterns (Lozano et al., 1999; Lozano and Sanvicente, 2002). Seawater bittern is a common by-product of seawater solar halite plants that contains considerable amounts of boron (B), Mg, and K elements. Lozano et al. (1999) added NaH_2PO_4 into seawater bittern with adequate mixing and then added NaOH to increase the pH to 10.0. Precipitates consisting of $MgKPO_4 \cdot 3H_2O$ and $Mg_3(PO_4)_2 \cdot 4H_2O$ were subsequently obtained, which were incorporated with minor amount of $MgNaPO_4 \cdot 3H_2O$ and a marginal B element. The possible chemical reaction may be described in Eq. [26.10]:

$$3NaH_2PO_4 + KCl + 4MgCl_2 + 6NaOH \rightarrow KMgPO_4 \downarrow + Mg_3(PO_4)_2 + 9NaCl + 6H_2O \quad [26.10]$$

Co-precipitation of struvite and K-struvite was also reported by Huang et al. (2011) and by Qureshi et al. (2006) to be caused by the extremely high concentration of K in pig wastewater and dairy manure waste.

26.5.3 MECHANOCHEMICAL ROUTE FOR PREPARING K-STRUVITE

Synthesis of K-struvite with a mechanochemical method is a novel route. Solihin et al. (2010) successfully prepared $KMgPO_4$ by milling $Mg(OH)_2$ and $KHPO_4$ in a planetary ball mill at a rotary speed of 500–600 rpm for 2 hours. The molar ratio of the two reaction substrates was maintained at 1:1. In a similar manner, they synthesized NH_4MgPO_4 by milling together $Mg(OH)_2$ and $NH_4H_2PO_4$.

The reaction formulas for $KMgPO_4$ and NH_4MgPO_4 synthesis are described as Reactions (26.11) and (26.12), respectively:

$$Mg(OH)_2 + KHPO_4 \rightarrow KMgPO_4 + 2H_2O \quad [26.11]$$

$$Mg(OH)_2 + NH_4H_2PO_4 \rightarrow H_4MgPO_4 \cdot 2H_2O \quad [26.12]$$

They also examined the leaching behavior of $KMgPO_4$ and NH_4MgPO_4 in deionized water. About 20–25% of K and phosphate ions was released into aquatic solution by shaking $KMgPO_4$ in deionized water for 700 h, which indicates that magnesium potassium phosphate produced via a mechanochemical route may also serve as a potential slow-release fertilizer. This technique is particularly explicated here because it provides an alternative idea about the synthesis of struvite-type compounds via a solid-phase reaction, such as the production of struvite/K-struvite from nutrient-enriched sludge ash.

26.6 CHALLENGES AND FUTURE PROSPECTS

26.6.1 HEAVY METAL CONTAMINATION PROBLEMS IN STRUVITE CRYSTALLIZATION

In most WWTPs, no specific process targets the removal of heavy metals from wastewater; as a result, heavy metals in waste streams normally accumulate and flow into the sludge digestion reactor, which supplies the influents for struvite crystallization. Heavy metal contamination in the struvite formation process is always a significant concern for the further application of struvite in agricultural fertilization, because once heavy metals are mobilized into the soil ecosystem, they will destroy sustainable development of the land. Stringent legal regulations have been established in many areas to control the heavy metal concentration in fertilizer products. For example, in parts of Austria and Switzerland, strict de jure or de facto bans exist regarding the management of sewage sludge application in agricultural land (Mattenberger et al., 2008). Ronteltap et al. (2007) conducted a series of batch experiments to investigate the behavior of heavy metals and pharmaceuticals during struvite crystallization from a urine-separating source. They spiked the hydrolyzed urine with lead, chromium (Cr), cadmium, Co, copper (Cu), and Ni and examined the solubility of added heavy metals in stored urine before and after struvite precipitation. Thermodynamic modeling showed the low concentration of heavy metals in stored urine, and about 20–63% of the introduced heavy metals were co-precipitated with struvite. In contrast, the concentrations of all determined metals in struvite obtained from normal stored urine without the addition of heavy metals were below the detection limit, which indicated that only a small fraction of metals are attached in the struvite precipitation process. Specifically, Muryanto and Bayuseno (2014) investigated the role of Cu^{2+} and Zn^{2+} in struvite crystallization by monitoring the pH of the solution and observing the morphologic changes in the struvite particles. Struvite production was reduced by 25% with the coexistence of 5 ppm ($mg\ L^{-1}$) Zn^{2+} ions at 25°C. Cu^{2+} and Zn^{2+} do not significantly affect the morphology of struvite crystals by causing some cracks and fractures on the surface. Muryanto and Bayuseno (2014) proposed that the cracks are caused by the incorporation of foreign ions into the struvite structure and/or adsorption onto the crystal surface. Rouff (2011, 2012) combined Fourier transform infrared spectroscopy and X-ray absorption fine-structure spectroscopy to assess the manner in which the Cr concentration and oxidation state affect Cr sorption on struvite. The detected content of Cr(III) in struvite precipitates was much higher than that of Cr(VI). When the initial concentration of Cr(III) was less than 20 μM, the adsorption mechanism dominated the affecting process, whereas co-precipitation of $Cr(OH)_3 \cdot nH_2O$ (amorphous) took place as the Cr(III) concentration exceeded 20 μM. Importantly, the adsorption of Cr on struvite distorts the tetrahedral group of phosphate ions in the mineral structure, and thus presumably influences the stability of struvite crystal and its nutrient-releasing properties.

As a whole, although a trace amount of heavy metals can be found in wastewater and/or sludge, these metals have a considerable effect on struvite crystallization via adsorption, incorporation, and co-precipitation. Current understanding about the behavior of heavy metals during struvite formation hints that crystallization substrates significantly affect the quality of recovered struvite products.

26.6.2 FERTILIZER PROSPECTS

The slow-releasing fertilizer properties of struvite have been widely recognized. It may be used to supply P, N, and Mg to plants as efficiently as monocalcium phosphates (Johnston and Richards, 2003), but without the risk of burning the roots of treated crops (Ries et al., 1969). However, struvite should be officially approved by legislation before direct application as a fertilizer. European Union Regulation 2003/2003 (2003), which includes approved fertilizers with preparation methods and nutrient contents, does not include or recognize either struvite or K-struvite as a fertilizer. This means that struvite and K-struvite cannot be freely transported and delivered as a fertilizer in European Union regions. They are regarded as waste products that might be used as the raw material for fertilizer production or in P industries. To date, struvite has seen commercial use as a fertilizer only in Japan for the cultivation of vegetables and rice (Ueno and Fujii, 2001). Thus, there is still a long way to go for struvite-type compounds to be applied as a commercial fertilizer. More efforts should be made to improve regulation management to reduce production costs and enhance promotion to farmers.

26.6.3 ECONOMIC CONSIDERATIONS

The environmental benefits of precipitating struvite-type materials from waste streams are obvious, including reducing the nutrient concentration in effluents, impeding the clogging of pipes, and reusing the nutrients as fertilizer. However, economic sustainability is a crucial factor to determine the further development of struvite crystallization in the wastewater treatment industry. To increase the profitability of struvite-type fertilizers in the market, the manufacture cost must be reduced. Currently, two major costs in struvite and K-struvite production are chemical consumption (eg, $MgCl_2$, NaOH) and the cost of the energy supplied for mixing (Le Corre et al., 2009). Jaffer et al. (2002) indicated that the addition of NaOH to raise the solution pH is responsible for about 97% of the total chemical expense. With the assumption that more than 90% of P in wastewater can be recovered as struvite on a large scale, an annual production of struvite of 84 tons could be achieved at a yearly profit of about £16,000–20,000. Air stripping is another method to adjust the solution pH, but it carries a greater energy cost in terms of pumping. Other efforts to reduce the chemical cost involve the use of seawater brine as the source of Mg (Shin and Lee, 1998), dosing $Mg(OH)_2$ rather than $MgCl_2$ (Münch and Barr, 2001), and replacing seeded crystallization with the autonucleation technique (Battistoni et al., 2005).

Based on the current market price, P fertilizer produced by phosphate rock minerals (€31–39 ton^{-1}) is still more cost-effective than struvite (€109–359 ton^{-1}) (Le Corre et al., 2009). However, the production of struvite-type fertilizer might bring about some potential savings for WWTPs, such as a decrease in the cost of sludge disposal and expenses for struvite scaling problems. Because of these complex factors, it is difficult to estimate the economic value of struvite/K-struvite in the fertilizer market. Improving the efficiency of production and expanding the potential market will likely assist in implementing the recovery of nutrients from wastewater successfully as struvite-type products.

26.7 SUMMARY

The crystallization of struvite ($MgNH_4PO_4 \cdot 6H_2O$) and K-struvite ($MgKPO_4 \cdot 6H_2O$) from waste streams is an environmentally friendly process to remove and recover nutrients (P, N, K, and Mg) in wastewater. K-Struvite is the isomorphous analogue of struvite with substitution of an NH_4^+ ion into a smaller K^+ ion. The replacement marginally reduces the cell volume of the crystals and yields insignificant effects on the structure; K-struvite thus has physicochemical, crystallographic, and thermal properties similar to those of struvite. Struvite-type materials are widely recognized as a slow-releasing fertilizer based on suitable nutrient leaching behavior, as long as their crystallization process is well controlled. Supersaturation, solution pH, reaction temperature, and the presence of foreign ions all significantly affect the formation course of struvite and K-struvite. In particular, the high content of calcium ions in settled sludge liquors not only hinders the crystallization of struvite but also introduces other phases into the precipitates. The quantitative XRD technique proves its superior performance in quantifying the struvite content in multiphase solid products. Because struvite has a greater tendency to precipitate than does K-struvite in a solution with a molar ratio comparable to that of NH_4^+:K^+, K-struvite is currently obtained mostly from denitrified urine waste. Finally, the challenges and future prospects of struvite-type fertilizers are briefly discussed with consideration of heavy metal contamination problems and economical sustainability as a commercial fertilizer. Crystallization of struvite-based materials from waste streams has shown superiority in nutrient removal and recycling but still requires improvements in production technologies and fertilizer quality to develop an economically applicable product in the agriculture market.

ACKNOWLEDGMENTS

The authors gratefully acknowledge the funding for this research provided by the General Research Fund (17206714 and 17212015) and Collaborative Research Fund (C7044-14G) schemes of the Research Grants Council of Hong Kong, and the HKU strategic research themes on Clear Energy and Earth as a Habitable Planet.

REFERENCES

Aage, H., Andersen, B., Blom, A., Jensen, I., 1997. The solubility of struvite. Journal of Radioanalytical and Nuclear Chemistry 223 (1–2), 213–215.

Abe, S., 1995. Phosphate removal from dewatering filtrate by MAP process at Seibu treatment plant in Fukuoka City. Sewage Works in Japan 43, 59–64.

Adnan, A., Koch, F.A., Mavinic, D.S., 2003. Pilot-scale study of phosphorus recovery through struvite crystallization-II: applying in-reactor supersaturation ratio as a process control parameter. Journal of Environmental Engineering and Science 2 (6), 473–483.

Banks, E., Chianelli, R., Korenstein, R., 1975. Crystal chemistry of struvite analogs of the type $MgMPO4 \cdot 6H_2O$ (M^+= potassium (1^+), rubidium (1^+), cesium (1^+), thallium (1^+), ammonium (1^+). Inorganic Chemistry 14 (7), 1634–1639.

Barat, R., Bouzas, A., Marti, N., Ferrer, J., Seco, A., 2009. Precipitation assessment in wastewater treatment plants operated for biological nutrient removal: a case study in Murcia, Spain. Journal of Environmental Management 90 (2), 850–857.

Bassett, H., Bedwell, W.L., 1933. 210. Studies of phosphates. Part I. Ammonium magnesium phosphate and related compounds. Journal of the Chemical Society (Resumed) 854–871.

Battistoni, P., Boccadoro, R., Fatone, F., Pavan, P., 2005. Auto-nucleation and crystal growth of struvite in a demonstrative fluidized bed reactor (FBR). Environmental Technology 26 (9), 975–982.

Bhuiyan, M., Mavinic, D., Beckie, R., 2007. A solubility and thermodynamic study of struvite. Environmental Technology 28 (9), 1015–1026.

Bhuiyan, M.I.H., Mavinic, D., Koch, F., 2008. Thermal decomposition of struvite and its phase transition. Chemosphere 70 (8), 1347–1356.

Booker, N., Priestley, A., Fraser, I., 1999. Struvite formation in wastewater treatment plants: opportunities for nutrient recovery. Environmental Technology 20 (7), 777–782.

Borgerding, J., 1972. Phosphate deposits in digestion systems. Journal of the Water Pollution Control Federation 44 (5), 813–819.

Bouropoulos, N.C., Koutsoukos, P.G., 2000. Spontaneous precipitation of struvite from aqueous solutions. Journal of Crystal Growth 213 (3), 381–388.

Burke, E.A., Ferraris, G., 2004. New minerals approved in 2003 and nomenclature modifications approved in 2003 by the Commission on New Minerals and Mineral Names, International Mineralogical Association. Mineralogy and Petrology 82 (1), 161–177.

Burns, J., Finlayson, B., 1982. Solubility product of magnesium ammonium phosphate hexahydrate at various temperatures. The Journal of Urology 128 (2), 426–428.

Chauhan, C., Vyas, P., Joshi, M., 2011. Growth and characterization of Struvite-K crystals. Crystal Research and Technology 46 (2), 187–194.

Coe, F.L., Evan, A., Worcester, E., 2005. Kidney stone disease. Journal of Clinical Investigation 115 (10), 2598.

Cooper, J., Lombardi, R., Boardman, D., Carliell-Marquet, C., 2011. The future distribution and production of global phosphate rock reserves. Resources, Conservation and Recycling 57, 78–86.

Cordell, D., Drangert, J.-O., White, S., 2009. The story of phosphorus: global food security and food for thought. Global Environmental Change 19 (2), 292–305.

Cordell, D., Rosemarin, A., Schröder, J., Smit, A., 2011. Towards global phosphorus security: a systems framework for phosphorus recovery and reuse options. Chemosphere 84 (6), 747–758.

De-Bashan, L.E., Bashan, Y., 2004. Recent advances in removing phosphorus from wastewater and its future use as fertilizer (1997–2003). Water Research 38 (19), 4222–4246.

Dunlevey, J., Laing, M., 2008. Struvite infection calculi in dogs: problems with urinary calculus identification, and the value of the results: research in action. South African Journal of Science 104 (11 & 12), 471–472.

EU Regulation 2003/2003. 2003. Retrieved from website of http://eur-lex.europa.eu/LexUriServ/LexUriServ.do?uri=OJ: L:2003:304:0001:0194:EN: PDF.

Frost, R., Weier, M., Erickson, K., 2004. Thermal decomposition of struvite. Journal of Thermal Analysis and Calorimetry 76 (3), 1025–1033.

Graeser, S., Postl, W., Bojar, H.-P., Berlepsch, P., Armbruster, T., Raber, T., Walter, F., 2008. Struvite-(K), $KMgPO4 \cdot 6H_2O$, the potassium equivalent of struvite—a new mineral. European Journal of Mineralogy 20 (4), 629–633.

Hanhoun, M., Montastruc, L., Azzaro-Pantel, C., Biscans, B., Frèche, M., Pibouleau, L., 2011. Temperature impact assessment on struvite solubility product: a thermodynamic modeling approach. Chemical Engineering Journal 167 (1), 50–58.

Hao, X., Wang, C., Lan, L., Van Loosdrecht, M., 2008. Struvite formation, analytical methods and effects of pH and Ca^{2+}. Water Science & Technology 59 (6), 1077–1083.

He, H., Zhang, P., 2002. Determination of trace phosphorus in water by ICP-AES. Chinese Journal of Spectroscopy Laboratory 2, 244–246.

Hongchao, L., Changlin, K., 1997. Quantitative analysis of SiC polytype distributions by the Rietveld method. Journal of Materials Science 32 (10), 2661–2664.

House, W.A., 1999. The physico-chemical conditions for the precipitation of phosphate with calcium. Environmental Technology 20 (7), 727–733.

Huang, H., Xu, C., Zhang, W., 2011. Removal of nutrients from piggery wastewater using struvite precipitation and pyrogenation technology. Bioresource Technology 102 (3), 2523–2528.

Jaffer, Y., Clark, T., Pearce, P., Parsons, S., 2002. Potential phosphorus recovery by struvite formation. Water Research 36 (7), 1834–1842.

Johnston, A., Richards, I., 2003. Effectiveness of different precipitated phosphates as phosphorus sources for plants. Soil Use and Management 19 (1), 45–49.

Jones, A.G., 2002. Crystallization Process Systems. Butterworth-Heinemann.

Kabdaszli, I., Parsons, S.A., Tünaya, O., 2006. Effect of major ions on induction time of struvite precipitation. Croatica Chemica Acta 79 (2), 243–251.

Kofina, A., Koutsoukos, P., 2005. Spontaneous precipitation of struvite from synthetic wastewater solutions. Crystal Growth & Design 5 (2), 489–496.

Le Corre, K.S., Valsami-Jones, E., Hobbs, P., Parsons, S.A., 2005. Impact of calcium on struvite crystal size, shape and purity. Journal of Crystal Growth 283 (3), 514–522.

Le Corre, K.S., Valsami-Jones, E., Hobbs, P., Parsons, S.A., 2009. Phosphorus recovery from wastewater by struvite crystallization: a review. Critical Reviews in Environmental Science and Technology 39 (6), 433–477.

Lienert, J., Larsen, T.A., 2009. High acceptance of urine source separation in seven European countries: a review. Environmental Science & Technology 44 (2), 556–566.

Lind, B.-B., Ban, Z., Bydén, S., 2000. Nutrient recovery from human urine by struvite crystallization with ammonia adsorption on zeolite and wollastonite. Bioresource Technology 73 (2), 169–174.

Loganathan, P., Vigneswaran, S., Kandasamy, J., Bolan, N.S., 2014. Removal and recovery of phosphate from water using sorption. Critical Reviews in Environmental Science and Technology 44 (8), 847–907.

Lozano, J.A.F., Colmenares, A.R., Rosas, D., 1999. A novel process for the production of multinutrient phosphatic base fertilizers from seawater bittern and phosphoric acid. Interciencia 24, 317–320.

Lozano, J.A.F., Sanvicente, L., 2002. Multinutrient phosphate-based fertilizers from seawater bitterns. Interciencia-Caracas 27 (9), 496–499.

Lu, X., 2013. Application of X-ray Diffraction to Identify the Phases Formed during Lead Stabilization and Resource Recovery. HKU Theses Online (HKUTO).

Luff, B.B., Reed, R.B., 1980. Thermodynamic properties of magnesium potassium orthophosphate hexahydrate. Journal of Chemical and Engineering Data 25 (4), 310–312.

Massey, M., Davis, J., Sheffield, R., Ippolito, J., 2007. Struvite production from dairy wastewater and its potential as a fertilizer for organic production in calcareous soils. In: Paper Presented at the International Symposium on Air Quality and Waste Management for Agriculture. CD-Rom Proceedings of the 16–19 September 2007, Conference (Broomfield, Colorado). ASABE Publication, USA.

Mattenberger, H., Fraissler, G., Brunner, T., Herk, P., Hermann, L., Obernberger, I., 2008. Sewage sludge ash to phosphorus fertiliser: Variables influencing heavy metal removal during thermochemical treatment. Waste Management 28 (12), 2709–2722.

Münch, E.V., Barr, K., 2001. Controlled struvite crystallisation for removing phosphorus from anaerobic digester sidestreams. Water Research 35 (1), 151–159.

Muryanto, S., Bayuseno, A., 2014. Influence of Cu^{2+} and Zn^{2+} as additives on crystallization kinetics and morphology of struvite. Powder Technology 253, 602–607.

Ohlinger, K., Young, T., Schroeder, E., 1998. Predicting struvite formation in digestion. Water Research 32 (12), 3607–3614.

Parsons, S., Wall, F., Doyle, J., Oldring, K., Churchley, J., 2001. Assessing the potential for struvite recovery at sewage treatment works. Environmental Technology 22 (11), 1279–1286.

Pastor, L., Mangin, D., Barat, R., Seco, A., 2008. A pilot-scale study of struvite precipitation in a stirred tank reactor: conditions influencing the process. Bioresource Technology 99 (14), 6285–6291.

Pitman, A., Deacon, S., Alexander, W., 1991. The thickening and treatment of sewage sludges to minimize phosphorus release. Water Research 25 (10), 1285–1294.

Qureshi, A., Lo, K.V., Mavinic, D.S., Liao, P.H., Koch, F., Kelly, H., 2006. Dairy manure treatment, digestion and nutrient recovery as a phosphate fertilizer. Journal of Environmental Science and Health Part B: Pesticides, Food Contaminants, and Agricultural Wastes 41 (7), 1221–1235.

Rahman, M.M., Salleh, M.A.M., Rashid, U., Ahsan, A., Hossain, M.M., Ra, C.S., 2014. Production of slow release crystal fertilizer from wastewaters through struvite crystallization—a review. Arabian Journal of Chemistry 7 (1), 139–155.

Rawn, A., Banta, P., Pomeroy, R., 1939. Multiple stage sewage digestion. Transactions of the American Society of Agricultural Engineers 105, 93–132.

Ries, K., Dunseth, R., Salutsky, M., Shapiro, J., 1969. Ultimate Disposal of Phosphates From Wastewater by Recovery as Fertilizer, Phase I. Final Report, FWPCA Report. Contract(14–12), 171.

Ronteltap, M., Maurer, M., Gujer, W., 2007. The behaviour of pharmaceuticals and heavy metals during struvite precipitation in urine. Water Research 41 (9), 1859–1868.

Root, M.J., 1990. Inhibition of the amorphous calcium phosphate phase transformation reaction by polyphosphates and metal ions. Calcified Tissue International 47 (2), 112–116.

Rouff, A.A., 2011. The use of TG/DSC–FT-IR to assess the effect of Cr sorption on struvite stability and composition. Journal of Thermal Analysis and Calorimetry 110 (3), 1217–1223.

Rouff, A.A., 2012. Sorption of chromium with struvite during phosphorus recovery. Environmental Science & Technology 46 (22), 12493–12501.

Rouff, A.A., 2013. Temperature-dependent phosphorus precipitation and chromium removal from struvite-saturated solutions. Journal of Colloid and Interface Science 392, 343–348.

Schindler, D.W., Hecky, R., Findlay, D., Stainton, M., Parker, B., Paterson, M., Kasian, S., 2008. Eutrophication of lakes cannot be controlled by reducing nitrogen input: results of a 37-year whole-ecosystem experiment. Proceedings of the National Academy of Sciences 105 (32), 11254–11258.

Scrimgeour, S.N., Chudek, J.A., Cowper, G.A., Lloyd, C.H., 2007. 31 P solid-state MAS-NMR spectroscopy of the compounds that form in phosphate-bonded dental casting investment materials during setting. Dental Materials 23 (8), 934–943.

Shin, H., Lee, S., 1998. Removal of nutrients in wastewater by using magnesium salts. Environmental Technology 19 (3), 283–290.

Solihin, Q.Z., Tongamp, W., Saito, F., 2010. Mechanochemical route for synthesizing $KMgPO_4$ and NH_4MgPO_4 for application as slow-release fertilizers. Industrial & Engineering Chemistry Research 49 (5), 2213–2216.

Sutor, D., Wooley, S.E., 1972. The structure and formation of urinary calculi: I. Oriented crystal growth. British Journal of Urology 44 (5), 532–536.

Taylor, A., Frazier, A., Gurney, E., 1963. Solubility products of magnesium ammonium and magnesium potassium phosphates. Transactions of the Faraday Society 59, 1580–1584.

Ueno, Y., Fujii, M., 2001. Three years experience of operating and selling recovered struvite from full-scale plant. Environmental Technology 22 (11), 1373–1381.

Whitaker, A., Jeffery, J., 1970. The crystal structure of struvite, $MgNH_4PO_4 \cdot 6H_2O$. Acta Crystallographica Section B: Structural Crystallography and Crystal Chemistry 26 (10), 1429–1440.
Williams, S., 1999. Struvite precipitation in the sludge stream at Slough wastewater treatment plant and opportunities for phosphorus recovery. Environmental Technology 20 (7), 743–747.
Wilsenach, J., Schuurbiers, C., Van Loosdrecht, M., 2007. Phosphate and potassium recovery from source separated urine through struvite precipitation. Water Research 41 (2), 458–466.
Xu, K., Li, J., Zheng, M., Zhang, C., Xie, T., Wang, C., 2015. The precipitation of magnesium potassium phosphate hexahydrate for P and K recovery from synthetic urine. Water Research 80, 71–79.
Xu, K., Wang, C., Liu, H., Qian, Y., 2011. Simultaneous removal of phosphorus and potassium from synthetic urine through the precipitation of magnesium potassium phosphate hexahydrate. Chemosphere 84 (2), 207–212.
Yan, H., Shih, K., 2014. Quantitative X-Ray diffraction (QXRD) reveals the influences of calcium and ferric ions on struvite crystallization. In: 9th IWA International Symposium on Waste Management Problems in Agro-industries, 1.
Yetilmezsoy, K., Sapci-Zengin, Z., 2009. Recovery of ammonium nitrogen from the effluent of UASB treating poultry manure wastewater by MAP precipitation as a slow release fertilizer. Journal of Hazardous Materials 166 (1), 260–269.
Zhang, S., Shi, H.-S., Huang, S.-W., Zhang, P., 2013. Dehydration characteristics of struvite-K pertaining to magnesium potassium phosphate cement system in non-isothermal condition. Journal of Thermal Analysis and Calorimetry 111 (1), 35–40.
Zhang, T., Ding, L., Ren, H., 2009. Pretreatment of ammonium removal from landfill leachate by chemical precipitation. Journal of Hazardous Materials 166 (2), 911–915.

CHAPTER

PHOSPHORUS RECOVERY FROM WASTES#

27

R. Karunanithi[1], A. Szogi[2], N.S. Bolan[1], R. Naidu[1], Y.S. Ok[3], S. Krishnamurthy[4], B. Seshadri[1]

University of Newcastle, Callaghan, NSW, Australia[1]; United States Department of Agriculture, Florence, SC, United States[2]; Kangwon National University, Chuncheon, Korea[3]; ICAR-Indian Institute of Soil and Water Conservation, Chandigarh, India[4]

CHAPTER OUTLINE

#This chapter is a condensed version of the following review paper: Karunanithi, R., Szogi, A.A., Bolan, N., Naidu, R., Loganathan, P., Hunt, P.G., Vanotti, M.B., Saint, C.P., Ok, Y.S., Krishnamurthy, S., 2015. Phosphorus recovery and reuse from waste stream. Advances in Agronomy 131, 173–250.

Environmental Materials and Waste. http://dx.doi.org/10.1016/B978-0-12-803837-6.00027-5

27.1 INTRODUCTION

Increasing demand for phosphorus (P) fertilizers to achieve food security, coupled with finite phosphate rock (PR) reserves, is one of the emerging problems for many countries. The ever-increasing production of large quantities of solid and liquid waste from animal manures and effluents and their disposal pose major problems for environmental sustainability (Cordell et al., 2009). Future food security depends on a long-term sustainable P supply. Therefore, it is essential to look for alternative and quality P sources and also solve the eutrophication of lakes and reservoirs by using the P present in these waste resources (Vaccari, 2011).

The Food and Agriculture Organization predicts that the world human population in 2050 will be 9 billion. Accordingly, it will be a tough task for the agricultural and fertilizer industries to ensure food availability, including sufficient production of animal protein. Only 15% of the total P extracted from mines and used for food production is eventually consumed by humans; the remaining P is lost to the environment (Roy et al., 2006). More than 66% of the total P extracted is lost during crop cultivation and livestock, meat, and dairy production. The other losses, which include household food waste, mining waste, and fertilizer manufacturing waste, constitute the remaining 19%. Once P enters the terrestrial ecosystem, it is used and reused in local cycles about 46 times, and then it reaches the ocean, where it is reused by marine organisms over 800 times and finally ends up in ocean sediments (Vaccari, 2009).

Traditional approaches to using waste streams including manures, farm effluents, and biosolids because nutrient sources are not currently economically viable owing to the high volumes required to obtain adequate nutrient levels. Optimum use of these wastes as a source of nutrients and water requires knowledge of their composition. To cope with these emerging problems, it is essential to develop advanced technologies which can be used to recover and reuse P from these wastes. This may serve the dual purpose of sustainable agricultural production and environmental protection (Rittmann et al., 2011).

This chapter provides an overview of the importance of P in living organisms, and the sources of P and its negative impact on the environment. It also discusses possibilities of using various wastes as a P source and technologies involved in recovering P from these wastes.

27.2 IMPORTANCE OF PHOSPHORUS

P is one of the important components of biochemical processes in living systems, because it occurs in enzymes which catalyze chemical reactions of biosynthesis and energy transfer in plants. Moreover, P is a component of nucleic acid, phytin, and phospholipids of plant cells. Therefore, an adequate amount of P is essential for the growth and development of higher plants (Tisdale et al., 1985).

In addition, P is a component of DNA and RNA, which form the backbone of nucleic acids and impart structure to molecules. ATP is a nucleotide involved in intracellular energy transfer by storing energy generated during photosynthesis and respiration; the energy is used in other biological processes such as cell division and biosynthesis. The P can be assimilated in ATP by phosphorylation, photophosphorylation, and oxidative phosphorylation, and via the adenylate kinase reaction. In plants, P containing phospholipids is important in the formation of biological membranes and in some signal transduction pathways. Phytic acid, another important P-containing organic compound in seeds, serves as a P source during seed germination (Mackey and Paytan, 2009).

Along with calcium, P is an important component of bones and teeth in animals because it makes up more than 75% of total mineral materials in animals' bodies. It is also involved in diverse important roles in animal metabolism as a component of the following compounds: high-energy phosphate bonds (ATP, adenosine diphosphate, and creatine phosphate), low-energy phosphate bonds (glucose-6-phosphate and glycerol phosphate), phospholipids (which are important constituents of membranes); nicotinamide adenine dinucleotide, flavin adenine dinucleotide, and other metabolic enzymes. Because it is a component of nucleic acid, P is involved in protein synthesis through DNA transcription and RNA translation. In addition, P is a major intercellular buffer in the body and is involved in acid–base balance (Cromwell et al., 2005).

27.3 SOURCE OF PHOSPHORUS IN THE ENVIRONMENT

Phosphorus is the 11th most abundant element in the earth's crust (0.09%; 5–10 $\times$ 10^8 million tons (Mt)); owing to its high reactivity, P is always found in combination with other elements, mainly as apatites and other common P-bearing minerals. Phosphate minerals occur both in sedimentary rocks (about 85%) and igneous rocks (about 15%) (Johnston, 2000). The largest sedimentary deposits as marine phosphorites are found in northern Africa (Jordan, Morocco, and Western Sahara), China, the Middle East, and the United States. Significant igneous rock occurrences are found in Brazil, Canada, Finland, Russia, and South Africa. Moreover, large PR resources have been identified near the continental shelves and on seamounts in the Atlantic and Pacific Oceans (Jasinski, 2012).

Farm operation by-products such as manures and effluents contain significant amount of P along with other nutrients. Adequate P is essential for the healthy and economic performance of farm animals, and the P requirement of animals is supplied by agronomic crops, feed concentrates, or readily available P supplements that are added to animal feed to meet dietary requirements (Lanyon et al., 2005). Ruminants such as cattle and sheep excrete more than 70% of consumed P in urine and excreta, whereas monogastric animals such as pigs and poultry excrete more than 90% of consumed P owing to the rich phytic acid content in cereal grain and oilseed meals which cannot be easily digested (Barnett, 1994). Other important sources of P include farm and industrial effluents which are produced as by-products of farming and agro-industrial activities. Table 27.1 gives the P content in some manures and effluents.

27.4 NEGATIVE IMPACT OF PHOSPHORUS ON THE ENVIRONMENT

P is essential for both terrestrial and aquatic organisms including plants, animals, and microorganisms living in fresh and marine waters. Nevertheless, excess P can cause environmental pollution in aquatic ecosystems. The increasing concentration of P in natural waters results from overuse of P fertilizers, and anthropogenic sources such as human sewage and detergents (Oelkers and Valsami-Jones, 2008). Autotrophs in aquatic ecosystems such as algae require P for growth and development similar to terrestrial organisms. In general, nitrogen (N) and P control the productivity of a water ecosystem, and N is usually readily available as nitrate runoff from arable land. Compared with other nutrients, P has a

Table 27.1 Total Phosphorus Content in Various Wastes

Liquid/Solid Wastes	Total Phosphorus	References
Dairy manure (mg kg^{-1})	4100–18,300	He et al. (2004)
Sheep manure (mg kg^{-1})	4164	McDowell and Stewart (2005)
Beef manure (mg kg^{-1})	6400–7200	Barnett (1994)
Hog manure (mg kg^{-1})	45,000	Ajiboye et al. (2004)
Poultry manure–layer (mg kg^{-1})	23,900	Dou et al. (2000)
Poultry manure–broiler (mg kg^{-1})	6700	Bolan et al. (2010)
Poultry litter (mg kg^{-1})	16,320	Sharpley and Moyer (2000)
Pig manure (mg kg^{-1})	6500	Bolan et al. (2010)
Piggery effluent (mg L^{-1})	61.2	Szogi and Vanotti (2009)
Dairy effluent (mg L^{-1})	28	Bolan et al. (2004)
Biosolids (mg kg^{-1})	13,350	Arulrajah et al. (2011)
Deer manure (mg kg^{-1})	4005	McDowell and Stewart (2005)

unique and important role in the eutrophication of receiving water bodies. Carbon (C) and N can also be made available from the atmosphere whereas P is mainly transported by surface waters (Correll, 1998). Because of the cost of transporting bulk manure to distant sites, animal farms tend to dispose of waste on nearby fields, which often leads to soil P saturation beyond its carrying capacity, resulting in eutrophication of water bodies. According to Toth et al. (2006), application of animal manure for 4 years at N-based rate increased the degree of P saturation (a measure of the extent to which P sorption sites have been filled) from 8% to 16% in alfalfa-based farming. With intensification of crop and livestock farming, excess P in soil and livestock waste are becoming important sources of P in surface runoff which eventually reach water resources and cause nutrient enrichment, excess algal growth, and eutrophication (Sharpley et al., 2000; McDowell, 2012).

27.5 WASTES AS A PHOSPHORUS SOURCE

Annually, large quantities of various agricultural wastes such as animal manures, agricultural and industrial effluents, municipal wastewater, and biosolids that are enriched with C and nutrients such as N and P are being produced. Potential quantities of waste production in selected countries are given in Table 27.2.

27.5.1 ANIMAL MANURES

The world's livestock population is 65 billion and produces a huge quantity of manure annually that contains P equal to 10 times the global agricultural demand (Naidu et al., 2012). For instance, India has the largest animal population in the world and produces around 173 Mt cattle manure, 12 Mt sheep manure, and 10 Mt poultry manure, which can supply (apart from other nutrients) 346, 62, and 274 kilotons (kt) of P, respectively. A major part of feed (>70%) that is consumed by animals is excreted as urine and feces which contain organic matter, N, P, K, and other micronutrients (Barnett, 1994). Diets of animals vary significantly among animal species and stages of animal

Table 27.2 Potential Quantities of Waste Produced in Selected Countries

Country	Manures/ Effluents	Potential Quantity Produced ($\times 10^3$) (mg or kL year^{-1})	Country	Manures/ Effluents	Potential Quantity Produced ($\times 10^3$) (mg or kL year^{-1})
Australia	Dairy effluent	520,238	India	Dairy effluent	3,847,538
	Piggery effluent	29,194		Piggery effluent	121,363
	Winery effluent	5669		Winery effluent	0
	Municipal waste waster	10,270		Municipal waste waster	139,600
	Poultry manure	976		Poultry manure	10,136
	Sheep manure	12,117		Sheep manure	12,349
	Cattle manure	23,409		Cattle manure	173,124
	Biosolids	407		Biosolids	22,086
United States	Dairy effluent	1,691,454	China	Dairy effluent	182,881
	Piggery effluent	847,762		Piggery effluent	4175
	Winery effluent	11,057		Winery effluent	949
	Municipal waste waster	767,500		Municipal waste waster	0
	Poultry manure	22,596		Poultry manure	148
	Sheep manure	908		Sheep manure	5160
	Cattle manure	76,109		Cattle manure	8229
	Biosolids	5645		Biosolids	24,691
United Kingdom	Dairy effluent	181,277	Japan	Dairy effluent	77,198
	Piggery effluent	56,734		Piggery effluent	124,786
	Winery effluent	7		Winery effluent	395
	Municipal waste waster	40,900		Municipal waste waster	140,000
	Poultry manure	2		Poultry manure	1893
	Sheep manure	5224		Sheep manure	2
	Cattle manure	8157		Cattle manure	3474
	Biosolids	1120		Biosolids	2319

Based on the average values of waste generated and their P content.
Dairy effluent: 50 L animal^{-1} day^{-1}, 1 36 mg L^{-1}; piggery effluent: 35 L animal^{-1} day^{-1}, 1145 mg L^{-1}; winery effluent: 5000 L ton^{-1} grapes crushed, 10.5 mg L^{-1}; municipal wastewater: 109 m^3 year^{-1} country^{-1}, 14 mg L^{-1}; poultry manure: 10.8 kg animal^{-1} year^{-1}, 2.7%; sheep manure: 166 kg animal^{-1} year, 10.5%; cattle manure: 821.18 kg animal^{-1} year^{-1}, 0.2%; biosolids: 18.3 kg person^{-1} year^{-1}, 2.23%.
FAO Statistical Yearbook 2013. http://faostat.fao.org (viewed on 20.11.13.).

growth, resulting in different complex compositions of manure (Dao and Schwartz, 2010). Factors that influence the chemical compositions of manures include animal species, rations of feed, types of housing, types of bedding materials, amounts of water used by animals, waste management systems, and climatic conditions (Mullins et al., 2005). For example, the digestible P requirement of pigs with a body weight of 20, 30, and 50 kg was calculated as 3.2, 2.7, and 2.3 g kg^{-1} of feed, respectively (Partanen et al., 2010).

Manure from ruminants is lower in P content than that in monogastric swine and poultry because ruminants can use P with the help of phytase enzymes (Lanyon et al., 2005). For example, total P content in dairy, poultry, and pig manure is 9.3, 18, and 39 g kg^{-1}, respectively (Barnett, 1994; Shen and Shen, 2001). The manure contains P mostly as inorganic P (60–90%) but organic P occurs and is a poorly studied subject (Barnett, 1994).

However, increasing costs of transporting bulk manure to distance fields and animal farms tends to lead farmers to dispose the wastes to nearby fields, thereby increasing the P-carrying capacity, resulting in nutrient transport and eventually the eutrophication of water bodies. One possible solution is to remove the N and P compounds by co-crystallizing with magnesium, producing magnesium ammonium phosphate, commonly called struvite (Ali et al., 2003). Another issue of using unprocessed animal manure is the presence high levels of *Escherichia coli*, cadmium (Cd), and lead (Pb) which are above the admissible limits (Tulayakul et al., 2011; Wu et al., 2012). Therefore, it is important to process animal manure at the source.

27.5.2 AGRICULTURAL EFFLUENTS

In a dairy or piggery farm, liquid wastewater from activates such milking, washing of animals, spilled feed, or urine with a high organic load, nutrients, and salts is commonly referred to as effluent (Ali et al., 2003). Table 27.2 gives the quantity of effluent produced annually for selected countries. For example, Australia produces 520,238 × 10^3 kL of dairy effluent and 29,194 × 10^3 kL of piggery effluent which contains (apart from other nutrients) 18,729 and 4233 P $year^{-1}$, respectively. Bolan et al. (2004) studied farm effluents in New Zealand dairy farms and found them to contain 120 g m^{-3} N, 28 g m^{-3} P, 320 g m^{-3} total suspended solids, 950 g m^{-3} chemical oxygen demand (COD), and 210 g m^{-3} biological oxygen demand (BOD).

A number of risks are associated with the use of effluents in agriculture, such as contamination from human and animal pathogens, heavy metal pollution, pharmaceuticals and endocrine disruptors, and water quality issues such as excess nutrient and sodium concentrations, as well as public attitudes toward the use of effluents (Toze, 2006). The use of effluents in agriculture provides point source P, the most important risk for river eutrophication rather than other diffused agricultural P sources (Jarvie et al., 2006). Current environmental concerns related to the direct discharge of these effluents into surface waters have prompted the development of technologies to reduce the nutrient content to an acceptable level in the effluents (Bolan et al., 2009).

27.5.3 INDUSTRIAL EFFLUENTS

Industries producing paper, fermentation products, wine, and baker's yeast produce large volumes of wastewater rich in organic load and other nutrients. Characteristic features of industrial effluents are high BOD and COD and a strong odor as a result of the presence of organic loads and volatile compounds which make them difficult to use directly (Molinuevo et al., 2009). Table 27.2 lists the quantity of industrial effluents produced annually for selected countries. For example, the United States produces 11,057 × 10^3 kL winery effluents annually which contain 116 kt P.

Environmental laws in most countries necessitate the use of appropriate technology to reduce the P content in effluents to <1 mg L^{-1} and prevent unnecessary precipitation, which affects the transportation of effluents owing to clogging of pipes (Parsons and Smith, 2008). One method used to prevent clogging of pipes is to recover P present in the effluent by struvite crystal precipitation (Altinbas et al., 2002).

27.5.4 MUNICIPAL WASTEWATER

Annually, each county produces a large quantity of municipal wastewater which contains high amount of nutrients and organic materials (Guo et al., 2010). For example, Rosenberger et al. (2002) studied the composition of municipal wastewater and found it to contain 31.8–202.4 mg L^{-1} N, 4.9–106.2 mg L^{-1} P, and 299–4294 COD mg L^{-1}. P coming from both domestic and industrial sources enters municipal wastewater treatment plants. Loads of P from various sources such as human waste (0.60 kg P $capita^{-1}$ $year^{-1}$); laundry detergent (0.30 kg P $capita^{-1}$ $year^{-1}$), and other household cleaning products (0.10 kg P $capita^{-1}$ $year^{-1}$) have been estimated to calculate the total load in the wastewater treatment plants and subsequent treated water and biosolids used for land application. If not properly treated, approximately 75% of P load may reach water streams and cause eutrophication (Parsons and Smith, 2008). Table 27.2 gives the quantity of municipal wastewater produced annually for selected countries. For example, the United Kingdom produces 40,900 × 10^3 kL of municipal wastewater annually which contains 572 kt P. P present in wastewater is commonly recovered by chemical coagulation or biological P removal.

27.5.5 BIOSOLIDS

Urban wastewater treatment plants receive a mixture of storm water, industrial wastewater, domestic household waste, and resulting residues called biosolids which are sufficiently treated to allow these waste streams to be safely applied to land (Mullins et al., 2005). The biosolids are rich in organic matter and plant nutrients and can be used to supply organic matter, prevent erosion, and improve soil structure; they can be used as soil amendments and to supply plant nutrients (Haynes et al., 2009). Arulrajah et al. (2011) reported that biosolids contain on an average 174,000 mg kg^{-1} organic C, 17,700 mg kg^{-1} of total N, and 13,350 mg kg^{-1} of total P. Biosolids application has been shown to provide around 100, 53, and 125 kg N, P, and K per hectare, respectively, at the highest rate of application (50 Mg ha^{-1}) and increased the biomass by 2- to 2.5-fold (Bolan et al., 2013). Estimation of potential quantities of biosolids generated in selected countries is given in Table 27.2. For example, China, the most populated country in the world, produces 24,691 kt of biosolids, which can supply 556 kt of P.

However, a number of problems are associated with biosolids application, including the presence of environmental pollutants such as heavy metals (arsenic, chromium, cadmium, and mercury), toxic residues (pesticides: DDT and organochlorine derivatives), pathogens (fecal coliforms and *Salmonella* spp.), and greenhouse gas emissions (CO_2, methane, and N_2O) (Haynes et al., 2009; Arulrajah et al., 2011; Belyaeva et al., 2012). Another important problem related to the application of biosolids is that agricultural fields may receive P in excess of the needs of crops and may result in P accumulation in the soil, thereby increasing the risk of P loss to surface runoff (Shober and Sims, 2003).

27.6 METHODS OF RECOVERING PHOSPHOROUS FROM WASTES

The P present in waste streams can be captured by various processes: (1) enhanced biological phosphorus removal: in this method, P present in wastewater is used by P-accumulating organisms (PAO) under anaerobic–aerobic cycles and stored as a microbial biomass rich in polyphosphates; (2) precipitation as sparingly soluble P compounds: here, cations such as calcium and magnesium are used to precipitate P as calcium phosphate or magnesium phosphate; (3) adsorption of dissolved P using suitable high-affinity adsorbents: in this technology, various selective and nonselective adsorbents are used to capture P from waste streams; and (4) thermochemical combustion followed by wet extraction, in which thermal energy is used to separate P from solid waste streams and the P is extracted from ash using mineral acids (Rittmann et al., 2011). The main objective of these technologies is to reduce P concentration in waste streams to meet environmental regulations and avoid eutrophication rather than P recovery and reuse. However, some of these technologies generate concentrated P materials from the recovered P, which can be used as a valuable fertilizer for agriculture.

27.6.1 ENHANCED BIOPROCESS PHOSPHORUS REMOVAL

Over the past 30 years, several biological suspended growth process configurations have been implemented to achieve biological P removal in municipal wastewater treatment systems. Biological P removal, also known as enhanced bioprocess P removal (EBPR), involves accumulating P from influent wastewater into microbial biomass as polyphosphate (polyP) and subsequent biomass removal (De-Bashan and Bashan, 2004). To provide a competitive advantage to PAOs, favorable environmental conditions in wastewater treatment reactors are provided by alternating anaerobic–aerobic cycles, in which PAOs use energy previously stored as polyP to grow biomass and release P during the anaerobic phase (Fig. 27.1). In a wastewater treatment system, influent wastewater enters the anaerobic phase containing rapidly biodegradable C. This biodegradable C undergoes fermentation by bacteria to produce volatile fatty acids (VFAs). VFAs such as acetate are assimilated by PAOs using stored polyP as an energy source. VFAs are converted into polyhydroxyalkanoates as storage products. For instance, acetate (the most studied VFA in EBPR) is converted into polyhydroxybutyrate (Zach-Maor et al., 2011). This conversion depletes polyP stored in PAO cells, releasing soluble phosphate anion and magnesium, calcium, and potassium cations. In experimental systems, 7–10 mg acetate is removed per milligram of P (Wentzel et al., 1991).

Once the system is aerated, PAOs take up soluble P and accumulate it as polyP using the stored C as an energy source. The phosphate in EBPR is removed in the waste-activated sludge, with a P content as high as 5% (Strom, 2006). The EBPR process is well designed for municipal treatment systems in which influent P concentrations are relatively low ($<25\ mg\ L^{-1}$) compared with agricultural or

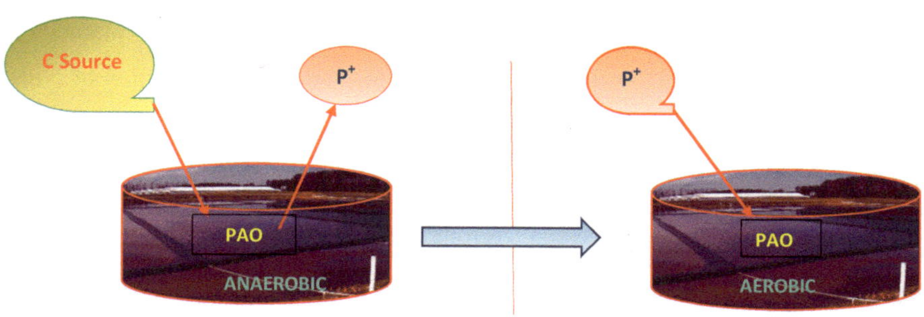

FIGURE 27.1

Enhanced bioprocess P removal. Under alternating anaerobic–aerobic conditions, PAOs take up organic carbon and release P and cations in the anaerobic environment. In turn, PAOs take up soluble P and cations under aerobic conditions.

industrial waste streams. Some modifications to EBPR to further enhance P removal in wastewater treatment plants include (1) an anaerobic–aerobic sequence combined with biological N removal; (2) adding volatile fatty acids to the anaerobic zone; (3) reducing sludge retention time; and (4) adding alum or iron salts for additional P removal.

The option in which EBPR is combined with biological N removal from wastewater has the advantage of effectively using organic C substrates and low sludge production along with N and P removal (Kuba et al., 1997). Activated sludge containing P removed through EBPR is usually further treated (thermal or anaerobic digestion) to reduce pathogen content and is dewatered. P in biosolids obtained after sludge dewatering is recycled via land application. However, land application of biosolids is becoming impractical because of the limited availability of farmlands close to municipal treatment plants and concerns about heavy metals and pathogen content in biosolids (Valsami-Jones, 2001). As an alternative to land application, municipal biosolids are disposed in landfills, resulting in the loss of recyclable P. This problem could be circumvented by technologies to recover P from biosolids (Vaccari, 2011). Fig. 27.1 presents a schematic diagram of the EBPR process.

27.6.2 PRECIPITATION

Chemical precipitation of P from waste streams is a well-studied method of P removal. P can be removed as a precipitate by adding metal salts of Al, Fe, Mg, or Ca. However, there is limited value of Al or Fe phosphates for industrial and agricultural applications. The most common precipitation technologies to recover P for use as a fertilizer use Mg- or Ca-based precipitation processes (Rittmann et al., 2011).

27.6.3 RECOVERY OF PHOSPHORUS AS MAGNESIUM AMMONIUM PHOSPHATE (STRUVITE)

Precipitation of P with magnesium (Mg) as magnesium ammonium phosphate (struvite) ($MgNH_4(PO_4)\cdot 6H_2O$) is the most common method of P recovery because it allows the simultaneous recovery of two common nutrients in the waste streams (N and P). Struvite is formed when the ion activity of magnesium, ammonia, and phosphate exceeds the solubility product of struvite, as described in Eq. [27.1]:

$$Mg^{2+} + NH_4{}^{+} + PO_4{}^{3+} + 6H_2O \rightarrow MgNH_4PO_4\cdot 6H_2O(\text{struvite}) \qquad [27.1]$$

Blockage of pipes resulting from struvite buildup is a common problem in wastewater treatment plants, but removal and recovery of P can reduce maintenance costs and increase plant operation efficiency. Major factors that control struvite precipitation are the pH and $Mg:NH_4:PO_4$ molar ratio. Hence, external application of Mg is necessary if there is no Mg or if Mg is present in concentrations lower than the $Mg:NH_4:PO_4$ molar ratio required for struvite precipitation. The reaction pH affects the formation of struvite because its precipitation occurs at a slightly alkaline pH. At pH 7.0 or less, little struvite will form. However, struvite formation will increase as the pH in the waste stream increases. For example, Çelen and Türker (2001) found that the optimal formation of struvite can be obtained at pH 8.5–9.0. The theoretical $Mg:NH_4:PO_4$ molar ratio required for struvite formation is 1:1:1, but in practice the molar ratios are slightly higher with respect to Mg. For instance, Gadekar and Pullammanappallil (2010) found that at pH 9.8 and an $Mg:NH_4:PO_4$ ratio of 1.7:1.0:3.4, P recovery in struvite formed from wastewater was as high as 98%. Techniques for P removal by precipitating from wastewater include crystallizers and fluid-bed reactors requiring seeding material, and several full-scale struvite recovery systems have been in operation in wastewater treatment plants around the world. Fig. 27.2 shows the overall process of struvite precipitation.

27.6.4 STRUVITE RECOVERY FROM EFFLUENT

Bowers and Westerman (2005) designed and tested a struvite crystallizer consisting of a cone-shaped fluidized bed to reduce the P content of swine lagoon liquid. A field-scale apparatus was tested operating at flow rates of 341 and 568 L h^{-1}. They demonstrated an increase in the amount of struvite formation by raising the pH from 7.71 to 8.21 and 8.71 by adding ammonia to the lagoon water before pumping the liquid into the cone-shaped crystallizer. Additional increases in the amount of struvite formation were obtained by adding $MgCl_2$. On average, soluble P removal increased from 13% to 82%. Because the retention time was longer with the lower flow rate (341 L h^{-1}), it provided greater P removal (80%) at pH 8.7 with Mg supplementation at the rate of 30 mg L^{-1}. As a matter of fact, Zhang et al. (2010) found that most P in anaerobically digested dairy effluent was tied up in finely suspended calcium-phosphate solids. They demonstrated that the P removal efficiency improved up to 82% by removing Ca interference with an acid pretreatment of dairy effluents using either hydrochloric acid or ethylenediamine tetraacetic acid. This acid pretreatment dissolved the calcium phosphate solid phase at pH 4.5, allowing subsequent precipitation of struvite at pH 7.6 by adding NaOH and $MgCl_2$. Recovery of P as struvite is a suitable method for the final treatment of swine wastewater, but precipitation of struvite from dairy manure effluents is a challenge. It may require additional acid pretreatment to improve the initial phosphate solubility at a significant expense in chemicals used to adjust pH (acid and ammonia) and Mg supplementation.

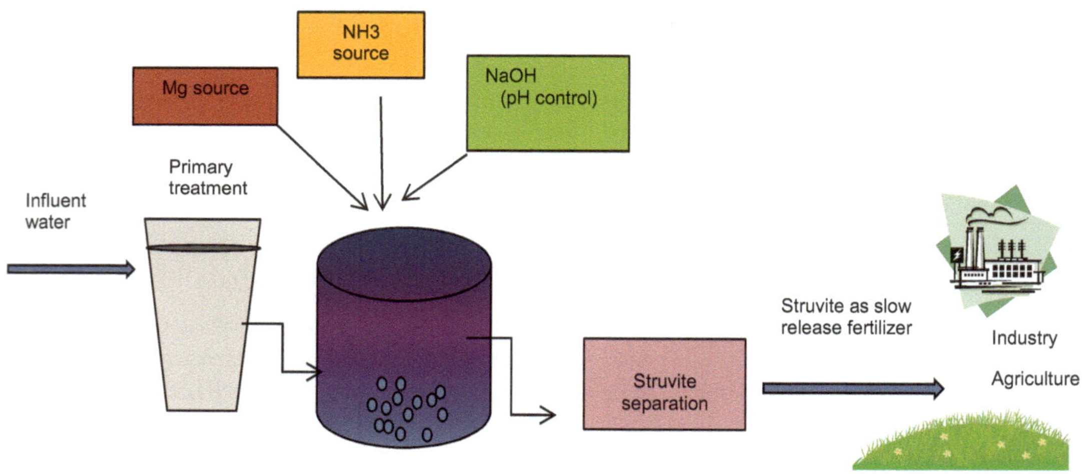

Mg and NH_4^+ (Struvite) precipitation

$$NH_4^+ + Mg^{+2} + PO_4^{-3} \rightarrow NH_4MgPO_4$$

FIGURE 27.2

P precipitation and crystallization as struvite.

27.6.5 RECOVERY OF PHOSPHATE AS CALCIUM PHOSPHATE

Removal of P using Ca compounds may initially precipitate a number of calcium phosphate minerals such as brushite, octacalcium phosphate, hydroxyl apatite, whitlockite, monetite, or amorphous calcium phosphate. The most common of these minerals is hydroxylapatite [$Ca_5(PO_4)_3OH$] (Valsami-Jones, 2001); its formation is described in Eq. [27.2]:

$$10Ca^{2+} + 6PO_4{}^{3+} + 2OH^- \rightarrow 2Ca_5(PO_4)_3OH \qquad [27.2]$$

A prerequisite condition for calcium phosphate precipitation is the presence of enough P to produce thermodynamic supersaturation (De-Bashan and Bashan, 2004). The precipitation process starts with the nucleation of P, the first step in the crystallization process. Induced nucleation can be initiated by adding artificial nucleic materials such as sand, calcite, and tobermorite crystals (Morse et al., 1998). Factors which affect P precipitation are pH, P concentration, and the presence of other ions such as carbonate. The formation of calcium carbonate reduces free Ca concentration, thereby reducing P precipitation. The effect of carbonate concentration on calcium phosphate precipitation was studied by Song et al. (2002), who found that at pH 8 with a carbonate concentration of up to 5 mmol L^{-1}, the rate and percentage of precipitation decreased, whereas at pH > 9 the effect was small. Generally, the efficiency of P removal by this method is 75–80%. The major advantage of the recovered P product is that it contains less contaminant compared with natural PR (Moriyama et al., 2001). Both crystallizers and fluid-bed reactors require seeding material to promote aggregation of fine calcium phosphate particles for subsequent recovery by filtration devices (Seckler et al., 1996). Fig. 27.3 explains calcium phosphate precipitation from waste.

27.6.6 CALCIUM PHOSPHATE RECOVERY FROM EFFLUENTS

Effluent is a mixture of urine, water, and feces. Livestock urine usually contains more than 55% excreted N, of which more than 70% is in the form of urea and hydrolysis of urea by the enzyme urease produces ammonium $(NH_4{}^+)$ and carbonate according to the following equation (Eq. [27.3]):

$$CO(NH_2)_2 + 2H_2O \rightarrow 2NH_4{}^+ + CO_3{}^{2-} \qquad [27.3]$$

Thus, a large part of the inorganic carbon in liquid manure is released during the decomposition of organic compounds. Although remaining organic compounds contribute to the buffering capacity in liquid manure, carbonate and $NH_4{}^+$ alkalinity are the most important chemical components. Both contribute to the buffering capacity in the alkaline pH range. Alkaline pH is necessary to form a P precipitate with calcium and magnesium compounds. When calcium or magnesium hydroxide is added to liquid manure, the hydroxide reacts with the existing bicarbonate to form carbonate, with ammonium $(NH_4{}^+)$ to form ammonia (NH_3), and with P to form P precipitate compounds (Szogi and Vanotti, 2009). For instance, using calcium hydroxide as an example, the following equations define the reactions (Eqs. [27.4] and [27.5]):

$$Ca(OH)_2 + Ca(HCO_3)_2 \rightarrow 2CaCO_3\downarrow + 2H_2O \qquad [27.4]$$

$$5Ca^{2+} + 4OH^- + 3HPO_4{}^{=} \rightarrow Ca_5OH(PO_4)_3\downarrow + 3H_2O \qquad [27.5]$$

The reaction in Eq. [27.4] is complete at pH > 9.5, whereas that in Eq. [27.5] starts at pH > 7.0, but the reaction is slow at pH < 9.0. As the pH value of the wastewater increases beyond 9.0, excess Ca ions react with the P to precipitate as calcium phosphate (Eq. [27.5]). Not expressed in Eq. [27.4] is the

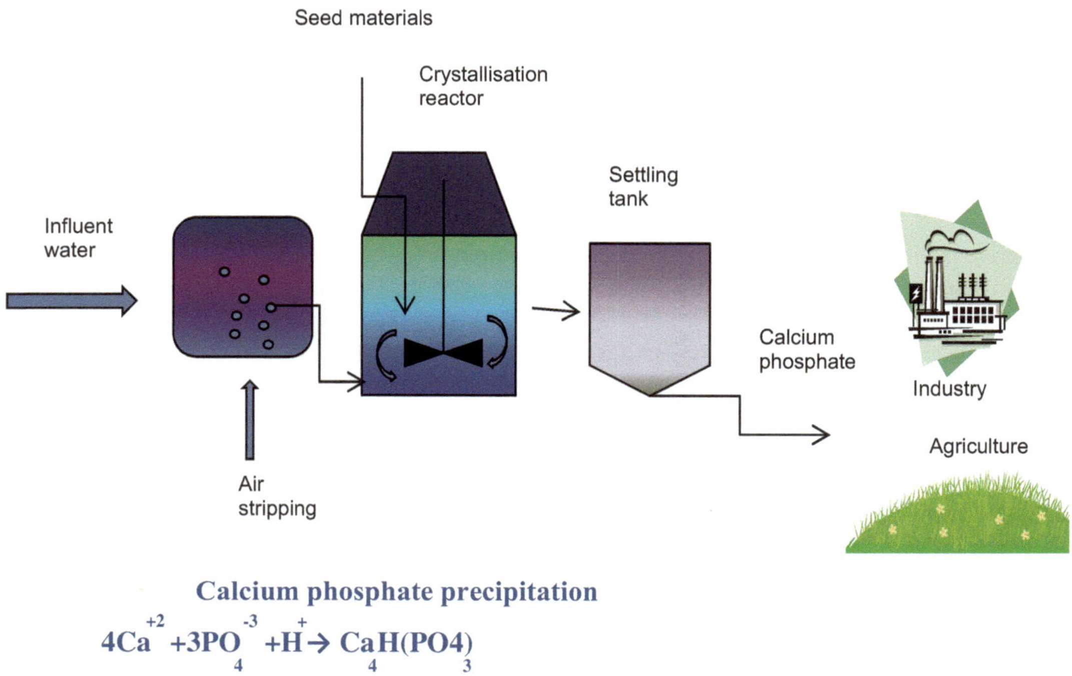

FIGURE 27.3

P precipitation and crystallization as calcium phosphate.

fact that in wastewater containing a high NH_4^+ concentration, large amounts of lime are required to elevate the pH to required values because the NH_4^+ reaction tends to neutralize the hydroxyl ions according to (Eq. [27.6]):

$$Ca(OH)_2 + 2NH_4^+ \rightarrow 2NH_3\uparrow + Ca^{2+} + 2H_2O \qquad [27.6]$$

Consequently, precipitation of P in effluents using an alkaline compound such as lime is difficult owing to the inherent high buffering capacity of liquid manure. This buffering effect prevents rapid changes in pH. However, this problem is solved using a prenitrification step that reduces the concentration of NH_4^+ (Eq. [27.7]) and bicarbonate alkalinity (Eq. [27.8]) (Vanotti et al., 2003):

$$NH_4^+ + 2O_2 \rightarrow NO_3^- + 2H^+ + H_2O \qquad [27.7]$$

$$HCO_3^- + H^+ \rightarrow CO_2\uparrow + H_2O \qquad [27.8]$$

The buffering effect of NH_4^+ (Eq. [27.6]) is reduced by biological nitrification of the NH_4^+ (Eq. [27.7]). Simultaneously, the buffering effect of bicarbonate (Eq. [27.4]) is greatly reduced with the acid produced during nitrification (Eq. [27.8]). These two simultaneous reactions provide a less buffered liquid in optimum pH conditions for effective P removal with the addition of small amounts of lime (Eq. [27.3]). With this process, the use of a crystallizer or fluidized-bed reactor with seeding to induce P nucleation is not required. In practice, soluble P was removed from prenitrified liquid swine manure in a settling tank by adding hydrated lime to form a fine precipitate at pH 10.5. Thereafter, the fine precipitate was aggregated by adding a polymer and dewatered using filter bags, recovering 99% of total P with respect to unfiltered material (Szögi et al., 2006).

In systems without a lagoon, raw liquid manure is first treated through an enhanced solid–liquid separation process with polymers to remove most of the carbonaceous material from the wastewater. The separated water is then treated with the nitrification and soluble P removal sequence (Vanotti et al., 2007). The recovered P precipitate solid had a concentration of 10.73% ± 1.98% P and was >99% plant available based on a standard citrate P analysis used by the fertilizer industry (Bauer et al., 2007). Fig. 27.3 gives the overall process of calcium phosphate precipitation.

27.6.7 PHOSPHORUS RECOVERY FROM SOLIDS

27.6.7.1 Thermal Pretreatment

In many countries, an increasing percentage of sewage sludge is thermally treated as a result of concerns about land spreading and land filling of organic and inorganic contaminants. Thermochemical processes (incineration, pyrolysis, gasification, and liquefaction) employ high temperatures to convert organic matter into gases, hydrocarbon fuels, and charcoal or ash residues (Goyal et al., 2008). Although thermochemical conversion has a role in energy recovery from organic waste, it cannot be considered a complete disposal process, because its by-products may still require safe disposal because of a high heavy metal content.

Incineration refers to the direct combustion of organic feedstock in the presence of air, and the incineration of sewage sludge produces ashes with P contents ranging from 3.96% to 11.44% as P. However, direct use of sewage sludge ash as fertilizer or building materials is questionable because the heavy metals content, such as Cu, Cr, Ni, Pb, and Zn, may exceed regulatory limits (Franz, 2008). Nevertheless, P can be recovered from sewage sludge ash after acid wet extraction and as a P precipitate at alkaline pH. For instance, Takahashi et al. (2001) reported 89–93% dissolution of P from incinerated sewage sludge at pH lower than 2.0 using a 0.5 M sulfuric acid solution (10:1 liquid–solid ratio (L/S)). Yet, a major problem with acid washing of sludge ash is the co-dissolution of P along with cations such as Fe, Al, and Ca, and heavy metals. Some adjustments made during the recovery process such as pH adjustment, sulfide precipitation, and cation exchange, can separate soluble P from both metals and heavy metals before precipitation. Kaikake et al. (2009) extracted P from chicken manure by incinerating the ash with 1.0 M hydrochloric acid solution (10: 1 L S^{-1}). In the first step, heavy metal impurities at pH 3.1 were precipitated using sodium hydroxide and the precipitate was removed by filtration. In the second step, the pH of the filtrate was raised to 8.15 to form a precipitate containing 92% P in a brushite form ($CaHPO_4 \cdot 2H_2O$). Animal manure incinerated ash seems a viable source of recovered P because of its lower content in heavy metal than sewage sludge incinerated ash (Vaccari, 2011).

Pyrolysis, a thermal method for converting biomass in the absence of air, has been tested as a method to dispose of animal wastes. The pyrolysis process produces a combination of gas, liquid fuel, and charcoal also known as biochar. Production of biochar from manures by pyrolysis is a viable option for nutrient recycling of P by extracting it from biochar using mineral acids. According to Azuara et al. (2013), about 92–97% of the P present in the fresh pig manure ended up in the biochar fraction. Of the biochar P, 60–75% was extracted as orthophosphate, and 90% as total P using a 0.2 M sulfuric acid (50:1 L S^{-1}) solution.

The application of thermochemical conversion technologies such as incineration or pyrolysis requires a relatively dry feedstock. Dry wastes such as poultry litter and feedlot manures can be processed directly by incineration or pyrolysis. However, solids in the liquid waste streams from dairy and swine operations must be separated and dried before conversion (Cantrell et al., 2007). As an alternative to incineration or pyrolysis, wet pyrolysis and supercritical water gasification processes allow the direct use of wet feedstocks such as raw animal manure and sewage sludge and open the possibility of P recovery from their char products by precipitation after acid dissolution (Yanagida et al., 2009).

27.6.8 PHOSPHORUS RECOVERY FROM RAW ANIMAL MANURE SOLIDS

Direct wet extraction is an alternative method to recover P from raw solid waste without thermal pretreatment (Szogi and Vanotti, 2009). A treatment process called "quick wash" was developed to extract and recover P from poultry litter and animal manure solids that produced two materials: a P-depleted washed residue and a concentrated P material with fertilizer value (Szögi et al., 2008). The quick-wash process consists of three consecutive steps: (1) P extraction, (2) P recovery, and (3) P recovery enhancement. In step 1, a large fraction (60–90%) of the initial total P in raw animal waste is selectively extracted by hydrolysis reactions using mineral or organic acids when the mixture of animal waste and extracting solution is in the pH range 3.0–5.0. The washed residue is further settled and dewatered to prevent unnecessary C and N oxidation and digestion. This first step produces a liquid extract containing low suspended solids and extracted soluble P. In step 2, P is precipitated from the liquid extract by adding lime to increase the pH to 9.0–11.0, forming a Ca-containing P product. In step 3, an organic flocculent is added to enhance precipitation and P concentration of the precipitated product. The precipitated P-rich solid is dewatered while the liquid is recycled back into the quick-wash system or land applied. The advantage of this process with respect to thermal pretreatment is the low content of heavy metals in the recovered P product and conservation of C and N in the acid-washed residue. The washed solid residue has a higher N:P ratio than the initial raw waste, which makes the washed residue better balanced with respect to its N:P ratio, thus improving crop use efficiencies of the two nutrients and avoiding excess application of P.

27.6.9 ADSORPTION AND ION EXCHANGE

P adsorption is widely used in to remove P from wastewater in treatment plants using different adsorbents. In this method, it is possible to use low-cost materials such as industrial by-products and naturally occurring materials (Loganathan et al., 2013). High operation costs and removal efficiency limit its wide application. P adsorption depends on various factors such as the type of adsorbent used, its surface area, pH, and contact time. Ion exchange provides a more specific separation of solutes from waste streams. Ion exchange is a reversible process and helps recover P in the form of precipitation by the subsequent addition of Ca, Fe, and Al solutes (Kuzawa et al., 2006). The advantage of ion exchange over adsorption is that the adsorbent can be regenerated and reused. Fig. 27.4 presents the overall process of adsorption and ion exchange.

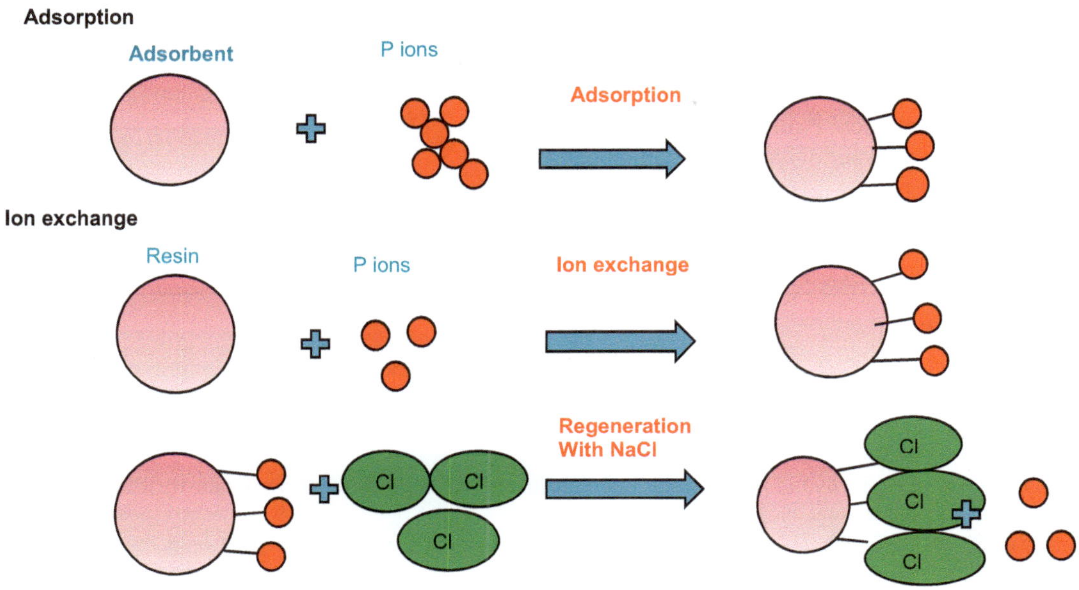

FIGURE 27.4

Mechanisms of P adsorption and ion exchange.

27.6.10 SOLAR EVAPORATION

Most P recovery methods are chemically, energetically, and operationally intensive and practically impossible to adopt. As an alternative, solar radiation can be used as an energy source for evaporating and drying waste streams such as urine or wastewater in a solar still; the resultant solid residue product can be used as an N and P fertilizer (Antonini et al., 2012). According to the World Health Organization, urine as a sterile liquid rich in N and P can supply half of the P requirement of cereal crops (Cordell et al., 2009). The potential P availability from urine and feces produced in urban settlements is currently (2010) about 0.88 million metric tons, which may rise with population growth to over 1.5 million metric tons by 2050. If harnessed, P availability from urine and feces could contribute 22% of the total global P demand (Mihelcic et al., 2011).

27.7 SUMMARY AND CONCLUSIONS

Farming and industry mainly depend on expensive P fertilizer for their P needs. Owing to the nonrenewable nature of P and inherently low use efficiency of P by living organisms, it is essential to use each kilogram of P produced carefully to avoid waste and for a sustainable environment. On the other hand, enormous quantities of manures and effluents are produced in each country which are all rich in P content. These types of waste can be used to recover P along with other nutrients which can replace one-third of the P requirements.

Various technologies are available for the economic recovery of P from both liquid and solid wastes. The P present in waste streams can be removed and eventually recovered in concentrated form by various processes, including:

- enhanced biological P removal
- precipitation as sparingly soluble P compounds
- using adsorption of dissolved P with high-affinity adsorbents
- thermochemical combustion followed by wet extraction.

REFERENCES

Ajiboye, B., Akinremi, O.O., Racz, G.J., 2004. Laboratory characterization of phosphorus in fresh and oven-dried organic amendments. Journal of Environmental Quality 33, 1062–1069.

Ali, M., Schneider, P., Hudson, N., 2003. Assessing nutrient recovery from piggery effluents. In: International Congress on Modelling and Simulation (MODSIM03) (New Zealand).

Altinbas, M., Ozturk, I., Aydin, A., 2002. Ammonia recovery from high strength agro industry effluents. Water Science and Technology 45, 189–196.

Antonini, S., Phong Thanh, N., Arnold, U., Eichert, T., Clemens, J., 2012. Solar thermal evaporation of human urine for nitrogen and phosphorus recovery in Vietnam. Science of the Total Environment 414, 592–599.

Arulrajah, A., Disfani, M.M., Suthagaran, V., Imteaz, M., 2011. Select chemical and engineering properties of wastewater biosolids. Waste Management 31, 2522–2526.

Azuara, M., Kersten, S.R.A., Kootstra, A.M.J., 2013. Recycling phosphorus by fast pyrolysis of pig manure: concentration and extraction of phosphorus combined with formation of value-added pyrolysis products. Biomass and Bioenergy 49, 171–180.

Barnett, G.M., 1994. Phosphorus forms in animal manure. Bioresource Technology 49, 139–147.

Bauer, P.J., Szogi, A.A., Vanotti, M.B., 2007. Agronomic effectiveness of calcium phosphate recovered from liquid swine manure. Agronomy Journal 99, 1352–1356.

Belyaeva, O.N., Haynes, R.J., Sturm, E.C., 2012. Chemical, physical and microbial properties and microbial diversity in manufactured soils produced from co-composting green waste and biosolids. Waste Management 32, 2248–2257.

Bolan, N.S., Wong, L., Adriano, D.C., 2004. Nutrient removal from farm effluents. Bioresource Technology 94, 251–260.

Bolan, N.S., Laurenson, S., Luo, J., Sukias, J., 2009. Integrated treatment of farm effluents in New Zealand's dairy operations. Bioresource Technology 100, 5490–5497.

Bolan, N., Szogi, A., Chuasavathi, T., Seshadri, B., Rothrock Jr., M., Panneerselvam, P., 2010. Uses and management of poultry litter. World Poultry Science Journal 66, 673–698.

Bolan, N.S., Kunhikrishnan, A., Naidu, R., 2013. Carbon storage in a heavy clay soil landfill site after biosolid application. Science of the Total Environment 465, 216–225.

Bowers, K., Westerman, P., 2005. Performance of cone-shaped fluidized bed struvite crystallizers in removing phosphorus from wastewater. Transactions of the ASAE 48, 1227–1234.

Cantrell, K., Ro, K., Mahajan, D., Anjom, M., Hunt, P.G., 2007. Role of thermochemical conversion in livestock waste-to-energy treatments: obstacles and opportunities. Industrial & Engineering Chemistry Research 46, 8918–8927.

Çelen, I., Türker, M., 2001. Recovery of ammonia as struvite from anaerobic digester effluents. Environmental Technology 22, 1263–1272.

Cordell, D., Drangert, J.O., White, S., 2009. The story of phosphorus: global food security and food for thought. Global Environmental Change 19, 292–305.

Correll, D.L., 1998. The role of phosphorus in the eutrophication of receiving waters: a review. Journal of Environmental Quality 27, 261–266.

Cromwell, G.L., Sims, J., Sharpley, A., 2005. Phosphorus and swine nutrition. In: Sims, J.T., Sharpley, A. (Eds.), Phosphorus: Agriculture and the Environment, pp. 607–634 (Madison, USA).

Dao, T.H., Schwartz, R.C., 2010. Mineralizable phosphorus, nitrogen, and carbon relationships in dairy manure at various carbon-to-phosphorus ratios. Bioresource Technology 101, 3567–3574.

Dou, Z., Toth, J., Galligan, D., Ramberg, C., Ferguson, J., 2000. Laboratory procedures for characterizing manure phosphorus. Journal of Environment Quality 29, 508–514.

De-Bashan, L.E., Bashan, Y., 2004. Recent advances in removing phosphorus from wastewater and its future use as fertilizer (1997–2003). Water Research 38, 4222–4246.

Faostat, 2013. FAO Statistical Yearbook 2013 (viewed on 20.11.13.). http://faostat.fao.org.

Franz, M., 2008. Phosphate fertilizer from sewage sludge ash (SSA). Waste Management 28, 1809–1818.

Gadekar, S., Pullammanappallil, P., 2010. Validation and applications of a chemical equilibrium model for struvite precipitation. Environmental Modeling & Assessment 15, 201–209.

Goyal, H., Seal, D., Saxena, R., 2008. Bio-fuels from thermochemical conversion of renewable resources: a review. Renewable & Sustainable Energy Reviews 12, 504–517.

Guo, C.H., Stabnikov, V., Ivanov, V., 2010. The removal of nitrogen and phosphorus from reject water of municipal wastewater treatment plant using ferric and nitrate bioreductions. Bioresource Technology 101, 3992–3999.

Haynes, R.J., Murtaza, G., Naidu, R., 2009. Inorganic and organic constituents and contaminants of biosolids: implications for land application. In: Donald, L.S. (Ed.), Advances in Agronomy, vol. 104. Academic Press, pp. 165–267.

He, Z., Griffin, T.S., Honeycutt, C.W., 2004. Phosphorus distribution in dairy manures. Journal of Environmental Quality 33, 1528–1534.

Jarvie, H.P., Neal, C., Withers, P.J.A., 2006. Sewage-effluent phosphorus: a greater risk to river eutrophication than agricultural phosphorus? Science of the Total Environment 360, 246–253.

Jasinski, S.M., 2012. Phosphate Rock, Mineral Commodity Summaries. U.S. Geological Survey. minerals.usgs.gov/minerals/pubs/commodity/phosphate_rock/.

Johnston, A.E., 2000. Soil and Plant Phosphate. International Fertilizer Industry Association (IFA), Paris.

Kaikake, K., Sekito, T., Dote, Y., 2009. Phosphate recovery from phosphorus-rich solution obtained from chicken manure incineration ash. Waste Management 29, 1084–1088.

Kuba, T., Van Loosdrecht, M., Brandse, F., Heijnen, J., 1997. Occurrence of denitrifying phosphorus removing bacteria in modified UCT-type wastewater treatment plants. Water Research 31, 777–786.

Kuzawa, K., Jung, Y.J., Kiso, Y., Yamada, T., Nagai, M., Lee, T.G., 2006. Phosphate removal and recovery with a synthetic hydrotalcite as an adsorbent. Chemosphere 62, 45–52.

Lanyon, L.E., Sims, J., Sharpley, A., 2005. Phosphorus, animal nutrition, and feeding: overview. In: Sims, J.T., Sharpley, A. (Eds.), Phosphorus: Agriculture and the Environment, pp. 561–586 (Madison, USA).

Loganathan, P., Vigneswaran, S., Kandasamy, J., Bolan, N.S., 2013. Removal and recovery of phosphate from water using sorption. Critical Reviews in Environmental Science and Technology 44, 847–907.

Mackey, K.R.M., Paytan, A., 2009. Phosphorus cycle. In: Moselio, S. (Ed.), Encyclopedia of Microbiology. Academic Press, Oxford, pp. 322–334.

McDowell, R., Stewart, I., 2005. Phosphorus in fresh and dry dung of grazing dairy cattle, deer, and sheep. Journal of Environmental Quality 34, 598–607.

McDowell, R.W., 2012. Minimising phosphorus losses from the soil matrix. Current Opinion in Biotechnology 23, 860–865.

Mihelcic, J.R., Fry, L.M., Shaw, R., 2011. Global potential of phosphorus recovery from human urine and feces. Chemosphere 84, 832–839.

Molinuevo, B., García, M.C., Karakashev, D., Angelidaki, I., 2009. Anammox for ammonia removal from pig manure effluents: effect of organic matter content on process performance. Bioresource Technology 100, 2171–2175.

Moriyama, K., Kojima, T., Minawa, Y., Matsumoto, S., Nakamachi, K., 2001. Development of artificial seed crystal for crystallization of calcium phosphate. Environmental Technology 22, 1245–1252.

Morse, G., Brett, S., Guy, J., Lester, J., 1998. Review: phosphorus removal and recovery technologies. Science of the Total Environment 212, 69–81.

Mullins, G., Joern, B., Moore, P., 2005. By-product phosphorus: sources, characteristics, and management. In: Sims, J.T., Sharpley, A. (Eds.), Phosphorus: Agriculture and the Environment, pp. 829–879 (Madison, USA).

Naidu, R., Lamb, D., Bolan, N., Gawandar, J., 2012. Recovery and reuse of phosphorus from wastewater sources. In: Currie, L.D., Christensen, C.L. (Eds.), Advanced Nutrient Management: Gains from the Past - Goals for the Future". Fertilizer and Lime Research Centre, Massey University, Palmerston North, New Zealand. Occasional Report No. 25.

Oelkers, E.H., Valsami-Jones, E., 2008. Phosphate mineral reactivity and global sustainability. Elements 4, 83–87.

Parsons, S.A., Smith, J.A., 2008. Phosphorus removal and recovery from municipal wastewaters. Elements 4, 109–112.

Partanen, K., Siljander-Rasi, H., Karhapää, M., Ylivainio, K., Tupasela, T., 2010. Responses of growing pigs to different levels of dietary phosphorus-performance, bone characteristics, and solubility of faecal phosphorus. Livestock Science 134, 109–112.

Rittmann, B.E., Mayer, B., Westerhoff, P., Edwards, M., 2011. Capturing the lost phosphorus. Chemosphere 84, 846–853.

Rosenberger, S., Krüger, U., Witzig, R., Manz, W., Szewzyk, U., Kraume, M., 2002. Performance of a bioreactor with submerged membranes for aerobic treatment of municipal waste water. Water Research 36, 413–420.

Roy, R.N., Finck, A., Blair, G., Tandon, H., 2006. Plant Nutrition for Food Security -A Guide for Integrated Nutrient Management. Food and Agriculture Organization of the United Nations. Publishing Management Service, Information Division, FAO, Italy.

Seckler, M., Bruinsma, O., Van Rosmalen, G., 1996. Calcium phosphate precipitation in a fluidized bed in relation to process conditions: a black box approach. Water Research 30, 1677–1685.

Sharpley, A., Foy, B., Withers, P., 2000. Practical and innovative measures for the control of agricultural phosphorus losses to water: an overview. Journal of Environmental Quality 29, 1–9.

Sharpley, A., Moyer, B., 2000. Phosphorus forms in manure and compost and their release during simulated rainfall. Journal of Environmental Quality 29, 1462–1469.

Shen, Q.R., Shen, Z.G., 2001. Effects of pig manure and wheat straw on growth of mung bean seedlings grown in aluminium toxicity soil. Bioresource Technology 76, 235–240.

Shober, A.L., Sims, J.T., 2003. Phosphorus restrictions for land application of biosolids. Journal of Environmental Quality 32, 1955–1964.

Song, Y., Hahn, H.H., Hoffmann, E., 2002. The effect of carbonate on the precipitation of calcium phosphate. Environmental Technology 23, 207–215.

Strom, P.F., 2006. Technologies to Remove Phosphorus from Wastewater. Rutgers University. Accessed at: http://water.rutgers.edu/Projects/trading/p-trt-lit-rev-2a.pdf.

Szögi, A.A., Vanotti, M.B., Hunt, P.G., 2006. Dewatering of phosphorus extracted from liquid swine waste. Bioresource Technology 97, 183–190.

Szogi, A.A., Vanotti, M.B., 2009. Removal of phosphorus from livestock effluents. Journal of Environmental Quality 38, 576–586.

Szögi, A., Vanotti, M., Hunt, P., 2008. Phosphorus recovery from poultry litter. Transactions of the ASABE 51, 1727–1734.

Takahashi, M., Kato, S., Shima, H., Sarai, E., Ichioka, T., Hatyakawa, S., Miyajiri, H., 2001. Technology for recovering phosphorus from incinerated wastewater treatment sludge. Chemosphere 44, 23–29.

Tisdale, S.L., Nelson, W.L., Beaton, J.D., 1985. Soil Fertility and Fertilizers. Collier Macmillan Publishers.

Toth, J.D., Dou, Z., Ferguson, J.D., Galligan, D.T., Ramberg Jr., C.F., 2006. Nitrogen- vs. phosphorus-based dairy manure applications to field crops: nitrate and phosphorus leaching and soil phosphorus accumulation. Journal of Environmental Quality 35, 2302–2312.

Toze, S., 2006. Reuse of effluent water—benefits and risks. Agricultural Water Management 80, 147–159.

Tulayakul, P., Boonsoongnern, A., Kasemsuwan, S., Wiriyarampa, S., Pankumnoed, J., Tippayaluck, S., Hananantachai, H., Mingkhwan, R., Netvichian, R., Khaodhiar, S., 2011. Comparative study of heavy metal and pathogenic bacterial contamination in sludge and manure in biogas and non-biogas swine farms. Journal of Environmental Sciences 23, 991–997.

Vaccari, D., 2009. Phosphorus: a looming crisis. Scientific American 300, 54–59.

Vaccari, D.A., 2011. Sustainability and the phosphorus cycle: inputs, outputs, material flow, and engineering. Environmental Engineering 12, 29–38.

Valsami-Jones, E., 2001. Mineralogical controls on phosphorus recovery from wastewaters. Mineralogical Magazine 65, 611–620.

Vanotti, M.B., Szogi, A.A., Hunt, P.G., 2003. Extraction of soluble phosphorus from swine wastewater. Transactions of the ASAE 46, 1665–1674.

Vanotti, M.B., Szogi, A.A., Hunt, P.G., Millner, P.D., Humenik, F.J., 2007. Development of environmentally superior treatment system to replace anaerobic swine lagoons in the USA. Bioresource Technology 98, 3184–3194.

Wentzel, M., Lötter, L., Ekama, G., Loewenthal, R., 1991. Evaluation of biochemical models for biological excess phosphorus removal. Water Science and Technology 23, 567–576.

Wu, L., Tan, C., Liu, L., Zhu, P., Peng, C., Luo, Y., Christie, P., 2012. Cadmium bioavailability in surface soils receiving long-term applications of inorganic fertilizers and pig manure. Geoderma 173–174, 224–230.

Yanagida, T., Minowa, T., Shimizu, Y., Matsumura, Y., Noda, Y., 2009. Recovery of activated carbon catalyst, calcium, nitrogen and phosphate from effluent following supercritical water gasification of poultry manure. Bioresource Technology 100, 4884–4886.

Zach-Maor, A., Semiat, R., Shemer, H., 2011. Synthesis, performance, and modeling of immobilized nano-sized magnetite layer for phosphate removal. Journal of Colloid and Interface Science 357, 440–446.

Zhang, T., Bowers, K.E., Harrison, J.H., Chen, S., 2010. Releasing phosphorus from calcium for struvite fertilizer production from anaerobically digested dairy effluent. Water Environment Research 82, 34–42.

Index

'*Note:* Page numbers followed by "f" indicate figures, "t" indicate tables.'

B

G

H

N

O

P

Q

R

S

T

W

X

Kolibri 360 GmbH
Ferdinand-Jühlke-Straße 7
99095 Erfurt, Deutschland
produktsicherheit@kolibri360.de